SOLIDWORKS 2022
Advanced Techniques

Advanced Level Tutorials
Mastering Parts, Surfaces, Sheet Metal, SimulationXpress,
Top-Down Assemblies, Core - Cavity Molds & Repair Errors

Paul Tran, CSWE, CSWI

SDC
PUBLICATIONS

SDC Publications
P.O. Box 1334
Mission, KS 66222
913-262-2664
www.SDCpublications.com
Publisher: Stephen Schroff

ISBN-13: 978-1-63057-467-3
ISBN-10: 1-63057-467-8

Printed and bound in the United States of America.

Acknowledgments

Thanks as always to my wife Vivian and my daughter Lani for always being there and providing support and honest feedback on all the chapters in the textbook.

Additionally, thanks to Peter Douglas for writing the forewords.

I also have to thank SDC Publications and the staff for its continuing encouragement and support for this edition of **SOLIDWORKS 2022 Advanced Techniques**. Thanks also to Tyler Bryant for putting together such a beautiful cover design.

Finally, I would like to thank you, our readers, for your continued support. It is with your consistent feedback that we were able to create the lessons and exercises in this book with more detailed and useful information.

Foreword

I first met Paul Tran when I was busy creating another challenge in my life. I needed to take a vision from one man's mind, understand what the vision looked like, how it was going to work and comprehend the scale of his idea. My challenge was I was missing one very important ingredient, a tool that would create a design with all the moving parts.

Research led me to discover a great tool, SOLIDWORKS. It claimed to allow one to make 3D components, in picture quality, on a computer, add in all moving parts, assemble it and make it run, all before money was spent on bending steel and buying parts that may not fit together. I needed to design and build a product with thousands of parts, make them all fit and work in harmony with millimeters tolerance. The possible cost implications of failed experimentation were daunting.

To my good fortune, one company's marketing strategy of selling a product without an instruction manual and requiring one to attend an instructional class to get it, led me to meet a communicator who made it all seem so simple.

Paul Tran has worked with and taught SOLIDWORKS as his profession for 35 years. Paul knows the SOLIDWORKS product and manipulates it like a fine musical instrument. I watched Paul explain the unexplainable to baffled students with great skill and clarity. He taught me how to navigate the intricacies of the product so that I could use it as a communication tool with skilled engineers. ***He teaches the teachers***.

I hired Paul as a design engineering consultant to create the machinery equipment with thousands of parts for my company's product. Paul Tran's knowledge and teaching skill has added immeasurable value to my company. When I read through the pages of these manuals, I now have an "instant replay" of his communication skill with the clarity of having him looking over my shoulder - ***continuously***. We can now design, prove and build our product and know it will always work and not fail. Most important of all, Paul Tran helped me turn a blind man's vision into reality and a monument to his dream.

Thanks Paul.

These books will make dreams come true and help visionaries change the world.

Peter J. Douglas

CEO, Cake Energy, LLC

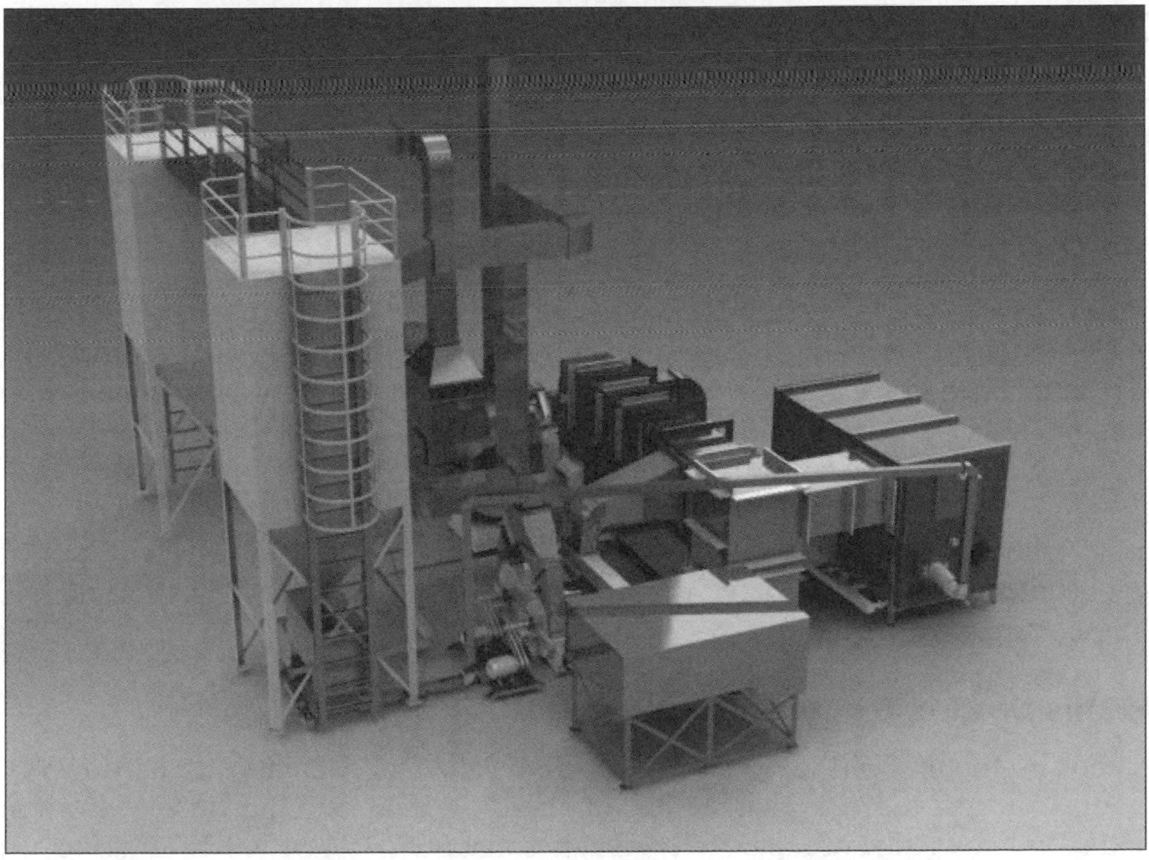

Images courtesy of C.A.K.E. Energy Corp., designed by Paul Tran

Author's Note

SOLIDWORKS 2022 Basic Tools, Intermediate Skills, and Advanced Techniques are comprised of lessons and exercises based on the author's extensive knowledge on this software. Paul has 35 years of experience in the fields of mechanical and manufacturing engineering. He spent 2/3 of those years teaching and supporting the SOLIDWORKS software and its add-ins.

As an active Sr. SOLIDWORKS instructor and design engineer, Paul has worked and consulted with hundreds of reputable companies including IBM, Intel, NASA, US- Navy, Boeing, Disneyland, Medtronic, Edwards Lifesciences, Terumo, Kingston and many more. Today, he has trained more than 13,000 engineering professionals and given guidance to nearly ½ of the number of Certified SOLIDWORKS Professionals and Certified SOLIDWORKS Expert (CSWP & CSWE) in the state of California.

Every lesson and exercise in this book was created based on real world projects. Each of these projects have been broken down and developed into easy and comprehendible steps for the reader. Learn the fundamentals of SOLIDWORKS at your own pace, as you progress from simple to more complex design challenges. Furthermore, at the end of every chapter, there are self-test questionnaires to ensure that the reader has gained sufficient knowledge from each section before moving on to more advanced lessons.

Paul believes that the most effective way to learn the "world's most sophisticated software" is to learn it inside and out, create everything from the beginning, and take it step by step. This is what the **SOLIDWORKS 2022 Basic Tools, Intermediate Skills & Advanced Techniques** manuals are all about.

About the Training Files

The files for this textbook are available for download on the publisher's website at www.SDCpublications.com/downloads/978-1-63057-467-3. They are organized by the chapter numbers and the file names that are normally mentioned at the beginning of each chapter or exercise. In the **Built Parts** folder, you will also find copies of the parts, assemblies and drawings that were created for cross referencing or reviewing purposes.

It would be best to make a copy of the content to your local hard drive and work from these documents; you can always go back to the original training files location at any time in the future, if needed.

Who this book is for

This book is for the mid-level user, who is already familiar with the SOLIDWORKS program. It is also a great resource for the more CAD literate individuals who want to expand their knowledge of the different features that SOLIDWORKS 2022 has to offer.

The organization of the book

The chapters in this book are organized in the logical order in which you would learn the SOLIDWORKS 2022 program. Each chapter will guide you through some different tasks, from navigating through the user interface, to exploring the toolbars, from some simple 3D modeling and moving on to more complex tasks that are common to all SOLIDWORKS releases. There is also a self-test questionnaire at the end of each chapter to ensure that you have gained sufficient knowledge before moving on to the next chapter.

The conventions in this book

This book uses the following conventions to describe the actions you perform when using the keyboard and mouse to work in SOLIDWORKS 2022:

Click: means to press and release the mouse button. A click of a mouse button is used to select a command or an item on the screen.

Double-Click: means to quickly press and release the left mouse button twice. A double mouse click is used to open a program or show the dimensions of a feature.

Right-Click: means to press and release the right mouse button. A right mouse click is used to display a list of commands, a list of shortcuts that is related to the selected item.

Click and Drag: means to position the mouse cursor over an item on the screen and then press and hold down the left mouse button; still holding down the left button, move the mouse to the new destination and release the mouse button. Drag and drop makes it easy to move things around within a SOLIDWORKS document.

Bolded words: indicated the action items that you need to perform.

Italic words: Side notes and tips that give you additional information, or to explain special conditions that may occur during the course of the task.

Numbered Steps: indicates that you should follow these steps in order to successfully perform the task.

Icons: indicates the buttons or commands that you need to press.

SOLIDWORKS 2022

SOLIDWORKS 2022 is a program suite, or a collection of engineering programs, that can help you design better products faster. SOLIDWORKS 2022 contains different combinations of programs; some of the programs used in this book may not be available in your suites.

Start and exit SOLIDWORKS

SOLIDWORKS allows you to start its program in several ways. You can either double click on its shortcut icon on the desktop or go to the Start menu and select the following: All Programs / SOLIDWORKS 2022 / SOLIDWORKS or drag a SOLIDWORKS document and drop it on the SOLIDWORKS shortcut icon.

Before exiting SOLIDWORKS, be sure to save any open documents, and then click File/Exit; you can also click the X button on the top right of your screen to exit the program.

Using the Toolbars

You can use toolbars to select commands in SOLIDWORKS rather than using the drop-down menus. Using the toolbars is normally faster. The toolbars come with commonly used commands in SOLIDWORKS, but they can be customized to help you work more efficiently.

To access the toolbars, either right click in an empty spot on the top right of your screen or select View / Toolbars.

To customize the toolbars, select Tools / Customize. When the dialog pops up, click on the Commands tab, select a Category, and then drag an icon out of the dialog box and drop it on a toolbar that you want to customize. To remove an icon from a toolbar, drag an icon out of the toolbar and drop it into the dialog box.

Using the task pane

The task pane is normally kept on the right side of your screen. It displays various options like SOLIDWORKS resources, Design library, File explorer, Search, View palette, Appearances and Scenes, Custom properties, Built-in libraries, Technical alerts and news, etc.

The task pane provides quick access to any of the mentioned items by offering the drag and drop function to all of its contents. You can see a large preview of a SOLIDWORKS document before opening it. New documents can be saved in the task pane at any time, and existing documents can also be edited and re-saved. The task pane can be resized, closed or moved to a different location on your screen if needed.

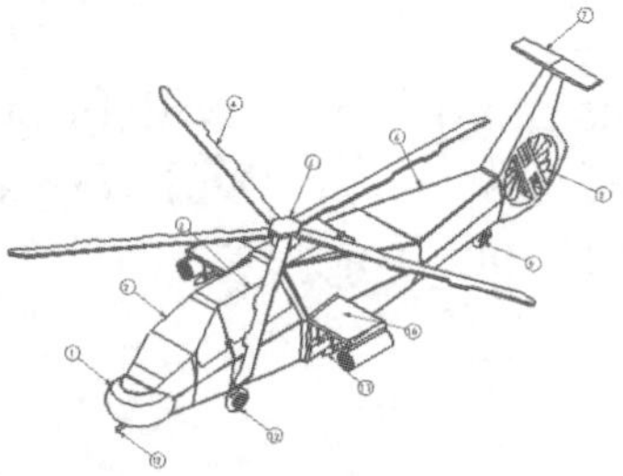

Become a CSWP – Certified SOLIDWORKS Professional

Demonstrate and validate your competency and knowledge in SOLIDWORKS software by achieving SOLIDWORKS certifications.

A Certified SOLIDWORKS Professional is an individual who has successfully passed our advanced skills examination. Each CSWP has proven their ability to design and analyze parametric parts and moveable assemblies using a variety of complex features in SOLIDWORKS software.

Standing out amongst SOLIDWORKS users from around the world can be challenging. Earning a SOLIDWORKS Certification can help you get a job, keep a job, or possibly move up in your current job. Be a part of our growing community of certified users. The certification center is your place to validate your certificate. This manual includes the CSWP-Exam preparation materials to help get you started.

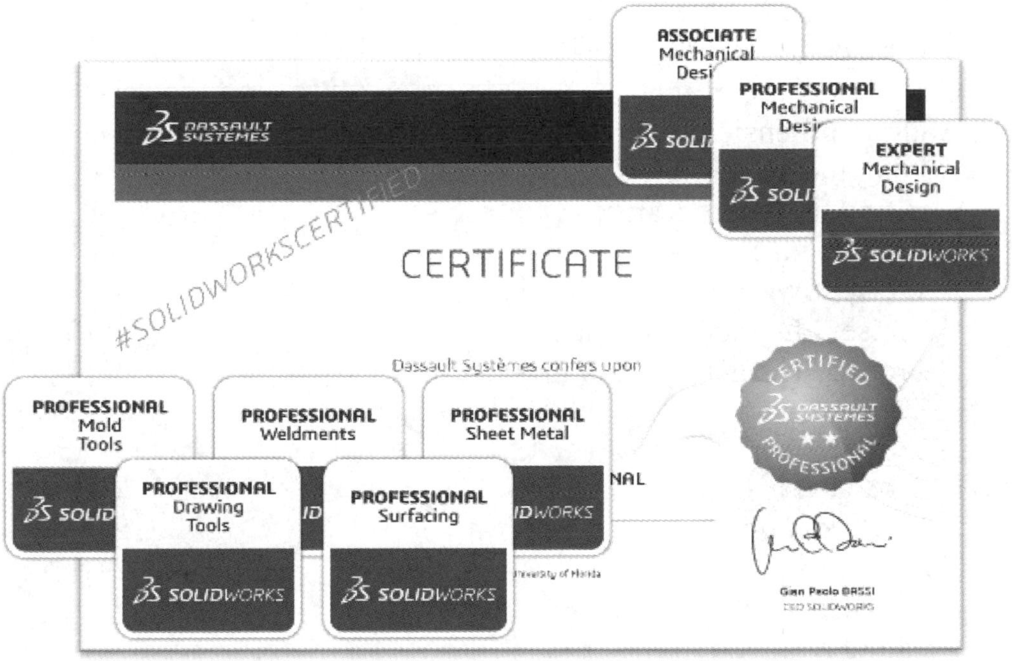

For more information, log on to: https://solidworks.virtualtester.com/#home_button

Table of Contents

Advanced Modeling Topics

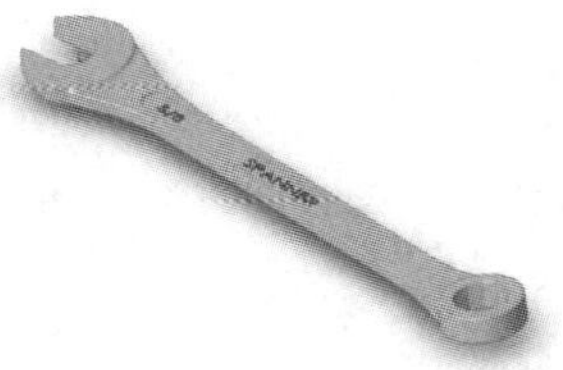

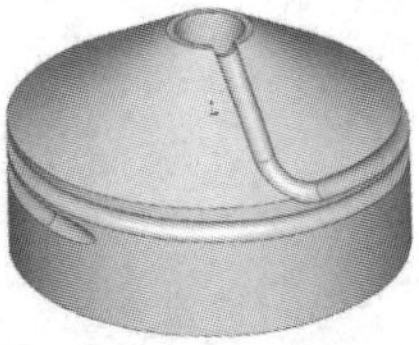

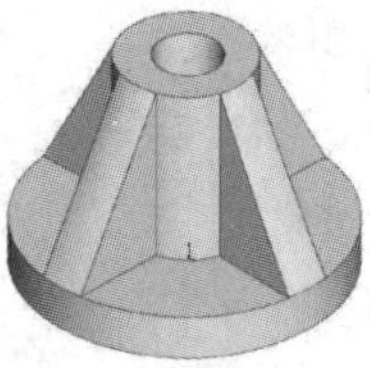

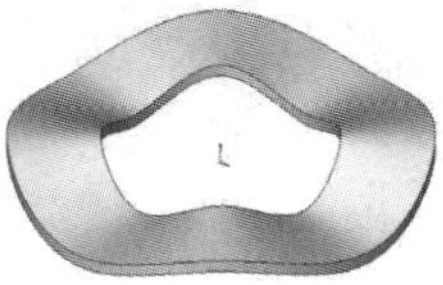
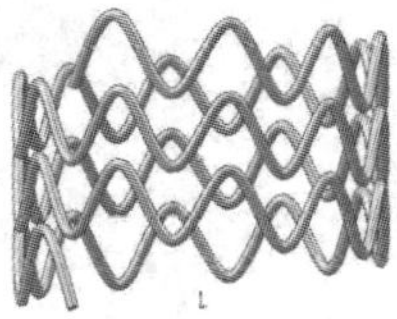

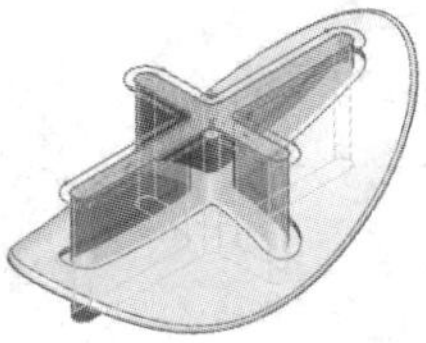

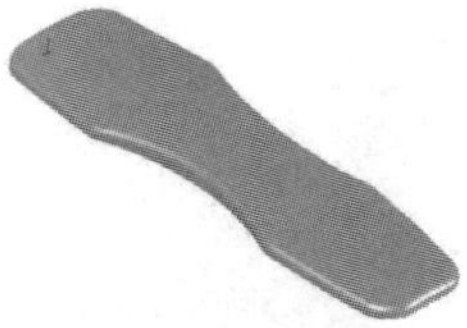

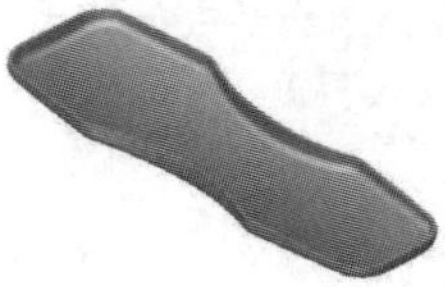

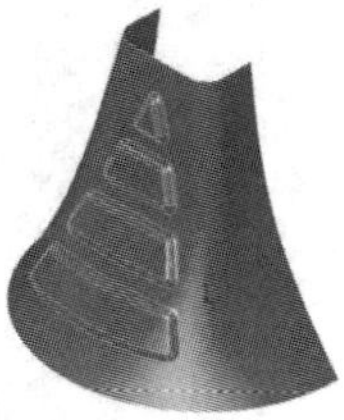

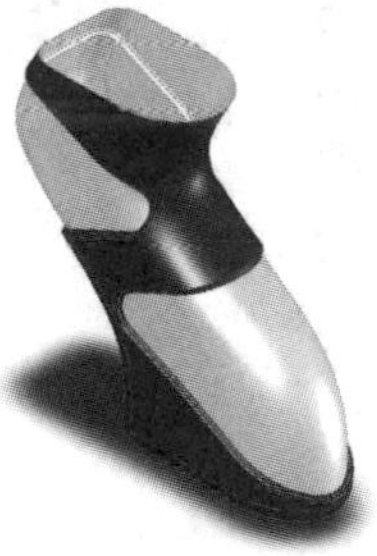

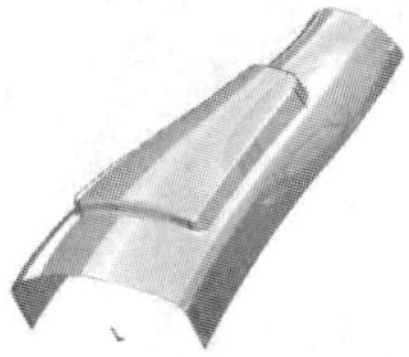

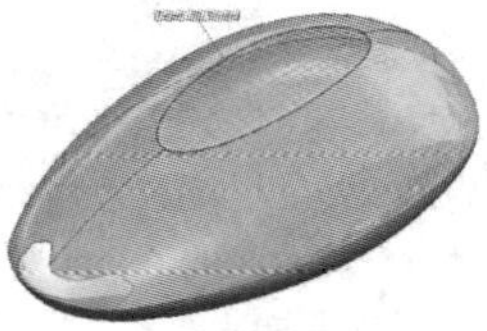

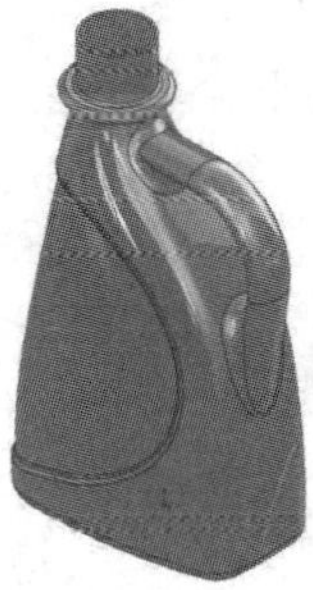

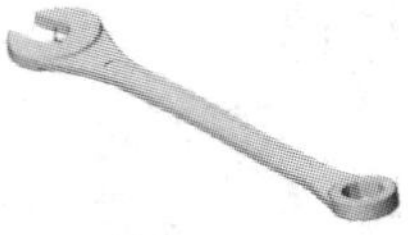

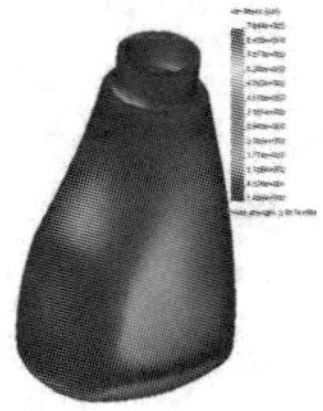

Sheet Metal Topics

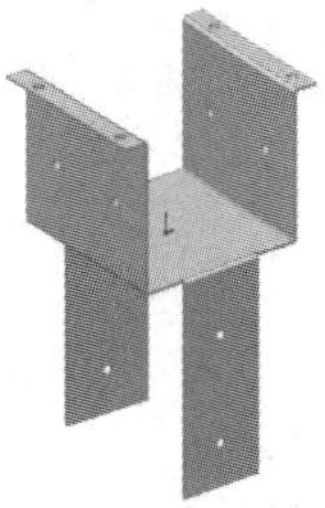

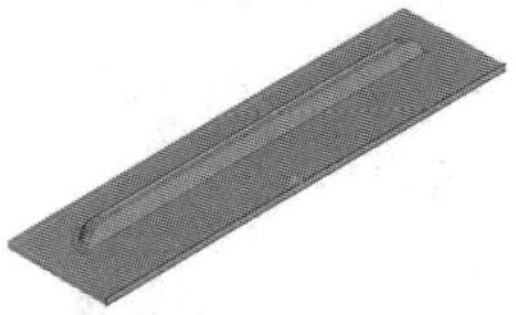

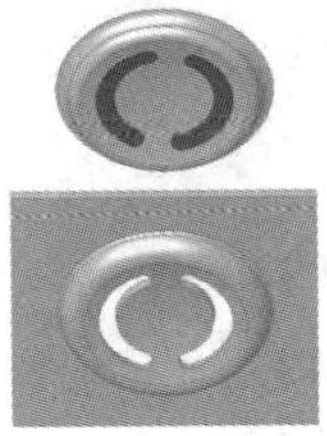

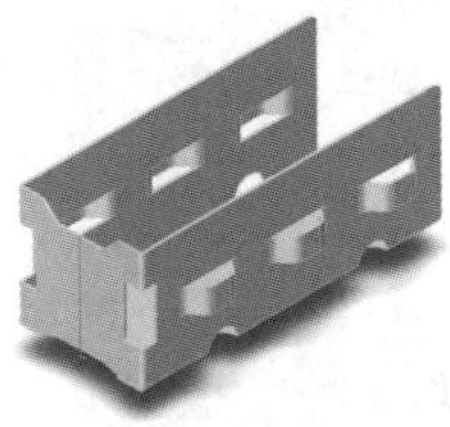

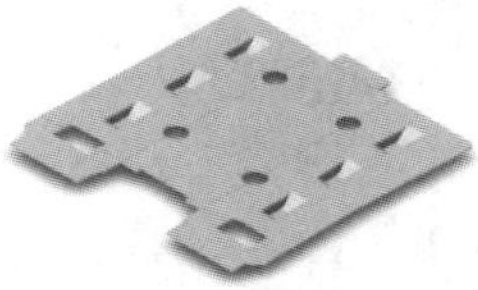

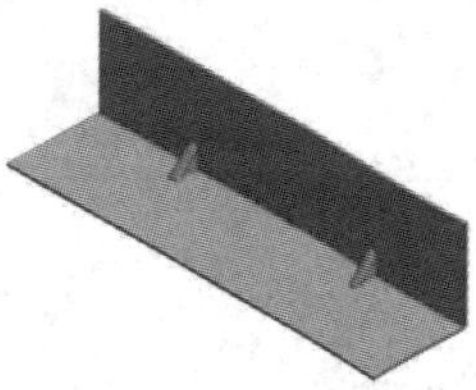

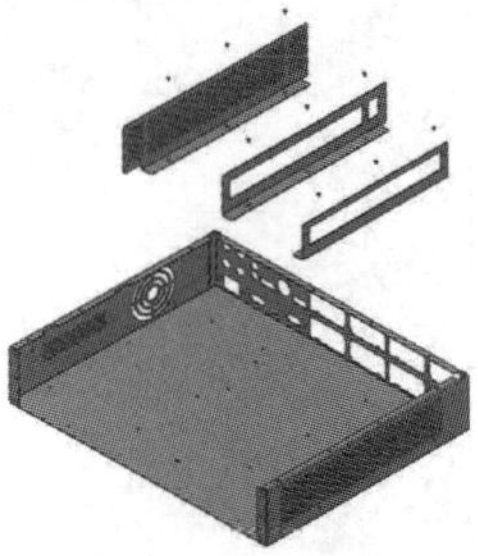

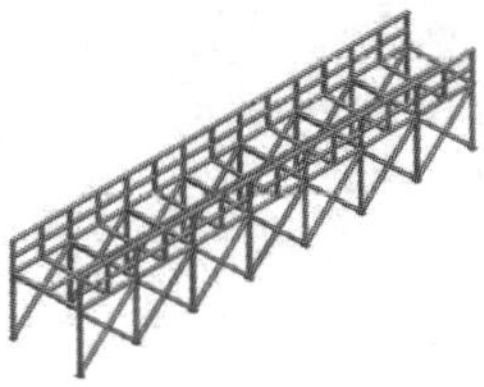

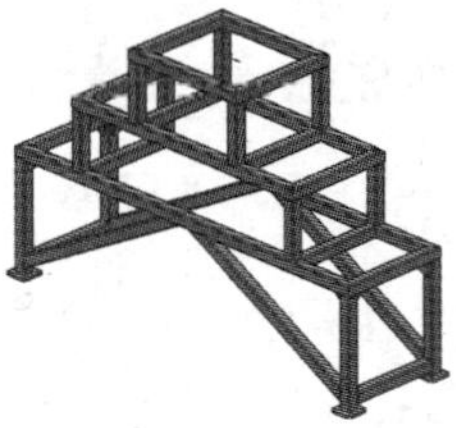

Mold Tools Design Topics

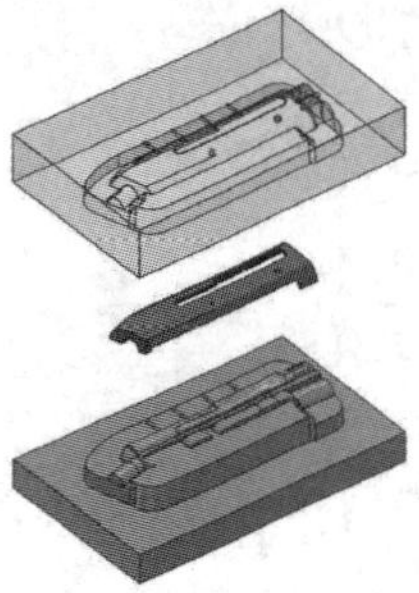

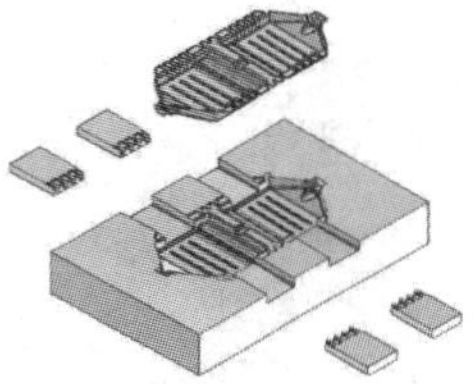

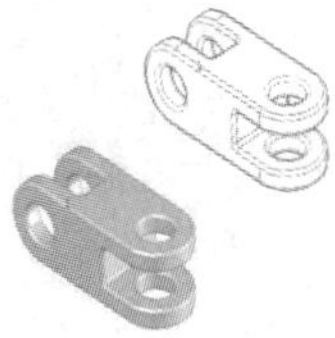

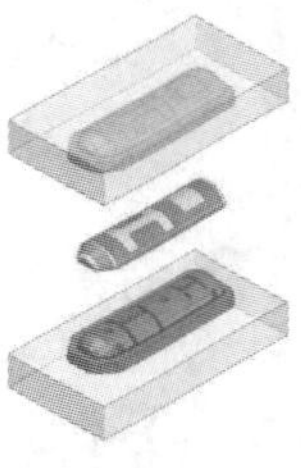

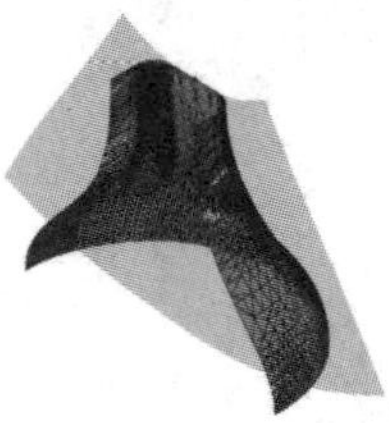

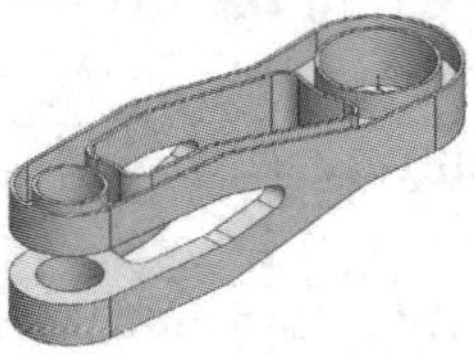

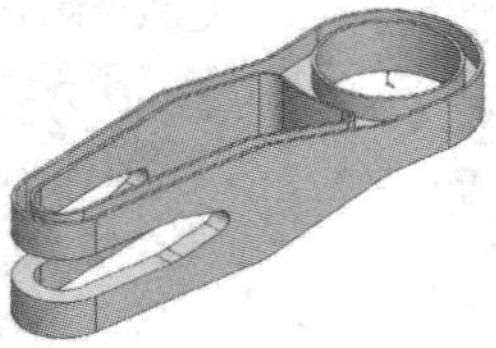

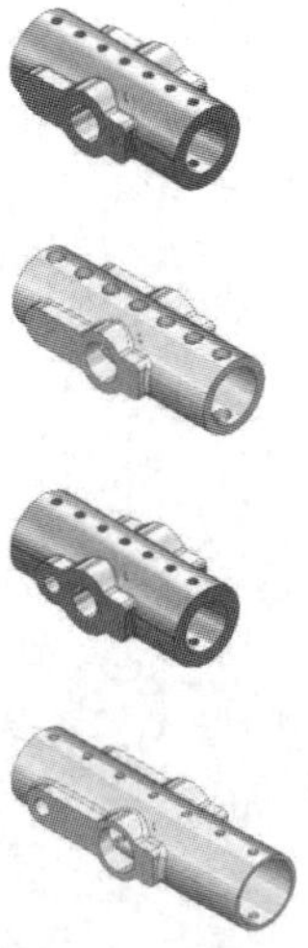

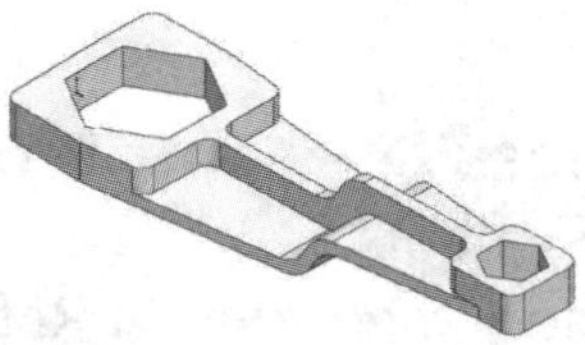

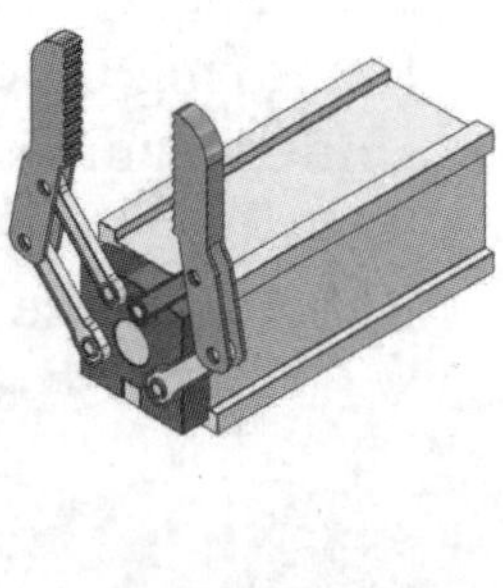

INTRODUCTION

SOLIDWORKS User Interface

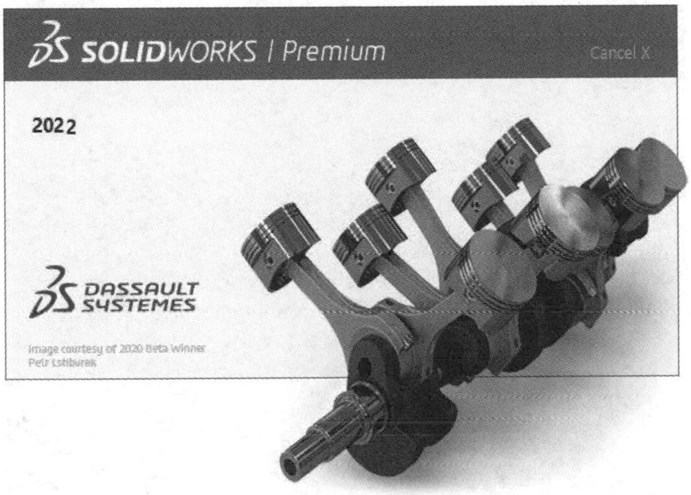

The SOLIDWORKS 2022 User Interface

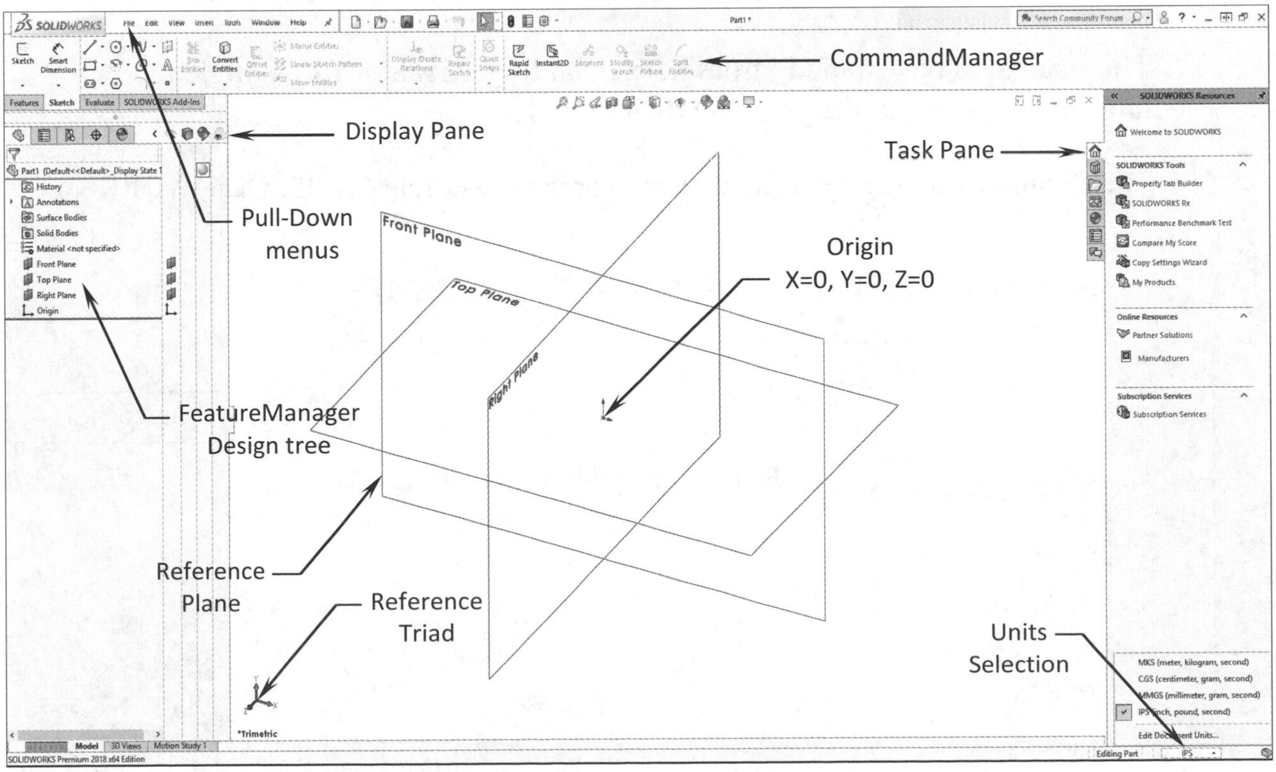

The 3 reference planes:

The Front, Top and the Right plane are 90° apart.
They share the same center point called the Origin.

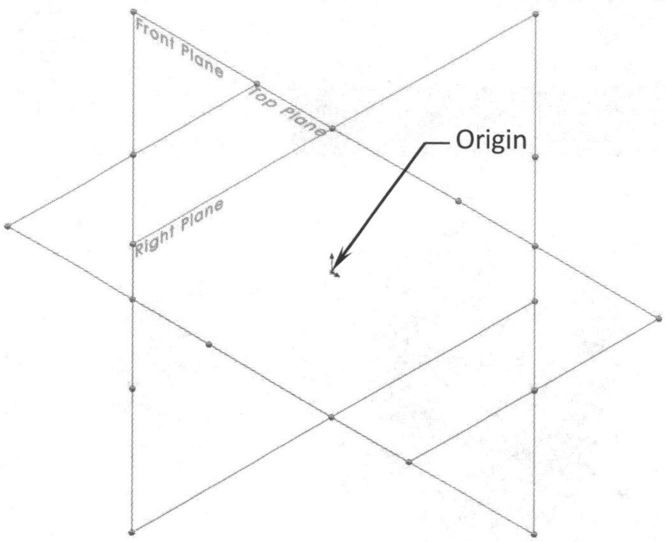

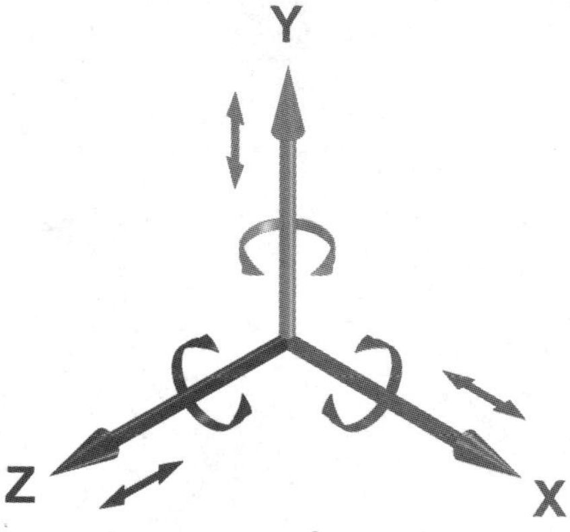

6 Degrees of Freedom

The Toolbars:

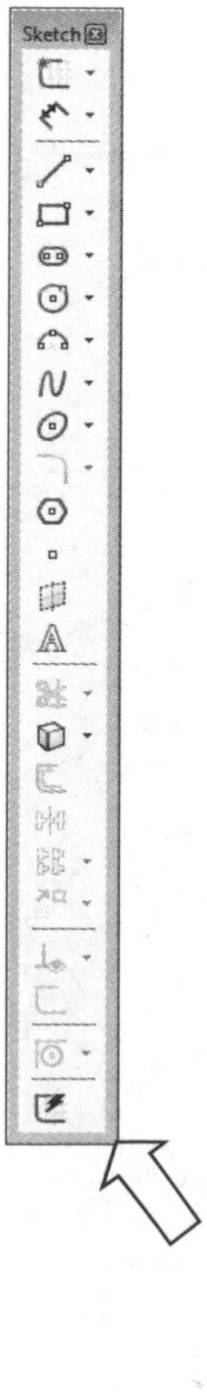

Toolbars can be moved, docked, or left floating in the graphics area.

They can also be "shaped" from horizontal to vertical, or from single to multiple rows when dragging on their corners.

The CommandManager is recommended for the newer releases of SOLIDWORKS.

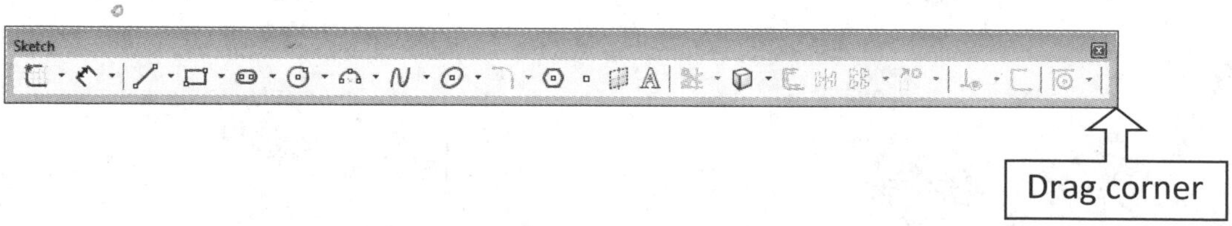

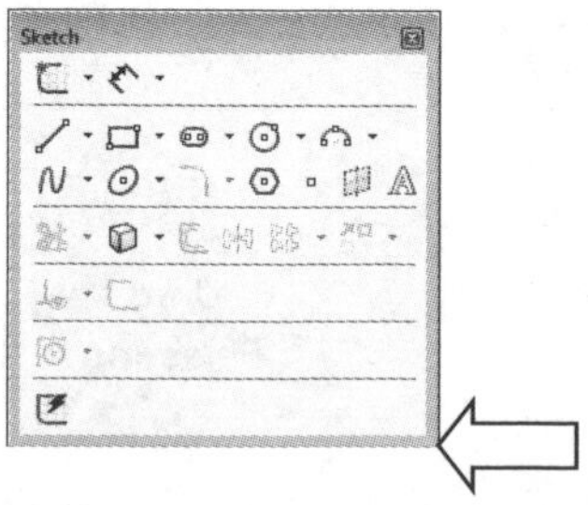

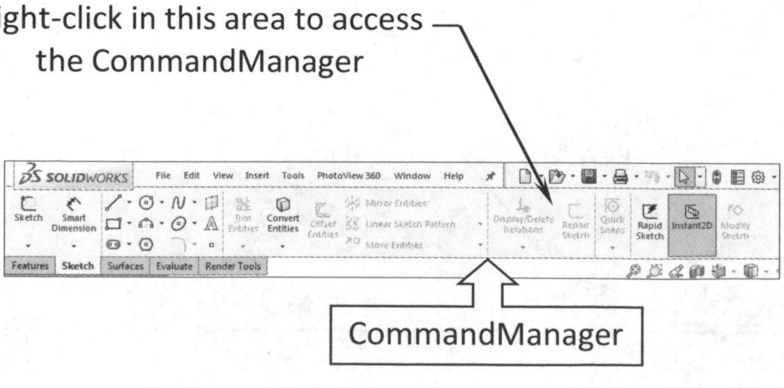

If the CommandManager is not used, toolbars can be docked or left floating.

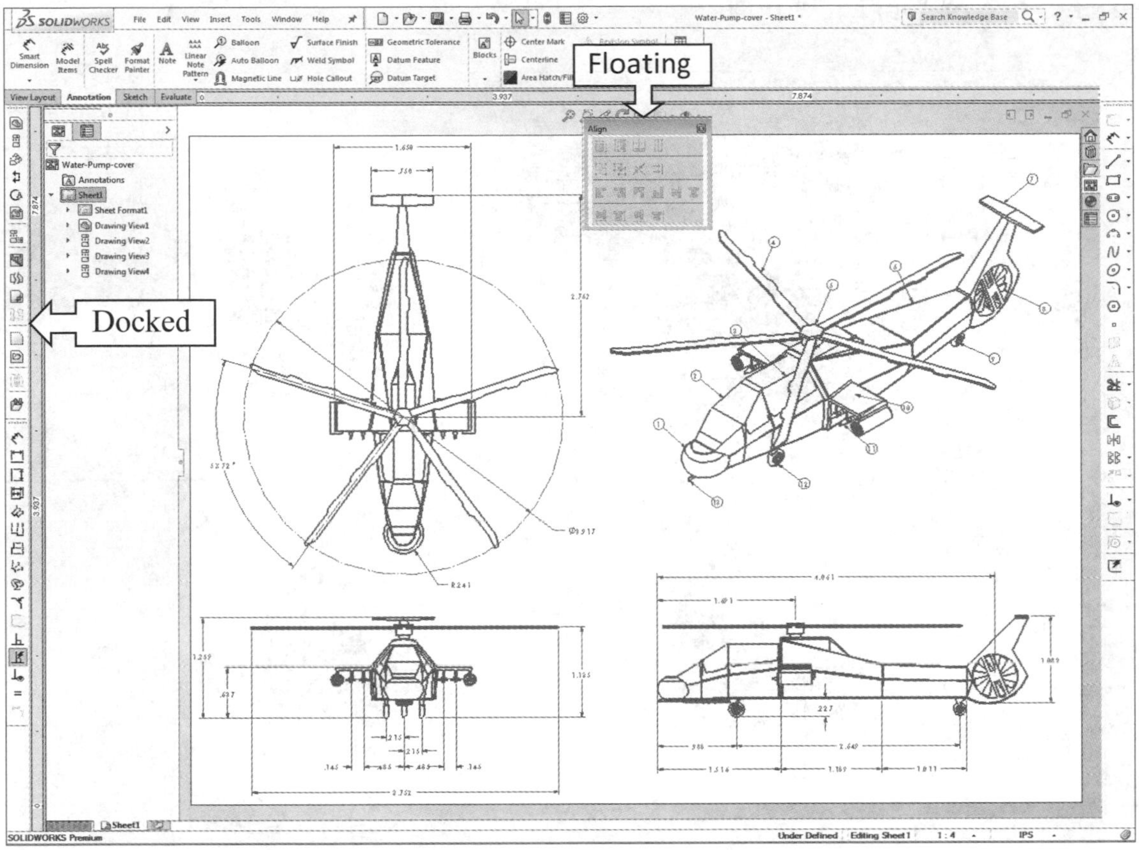

Toolbars can be toggled on or off by activating or de-activating their check boxes:

Select **Tools / Customize / Toolbars** tab.

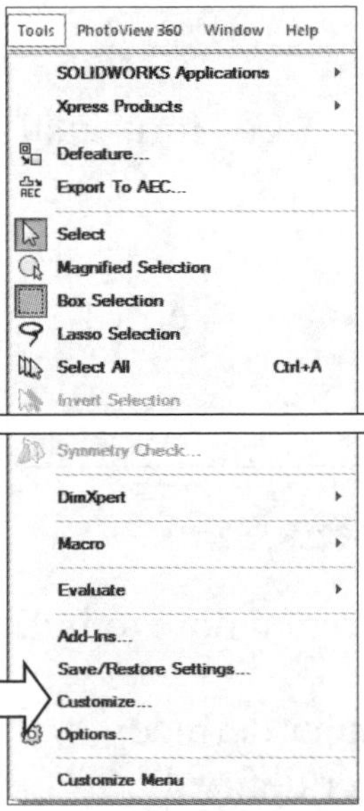

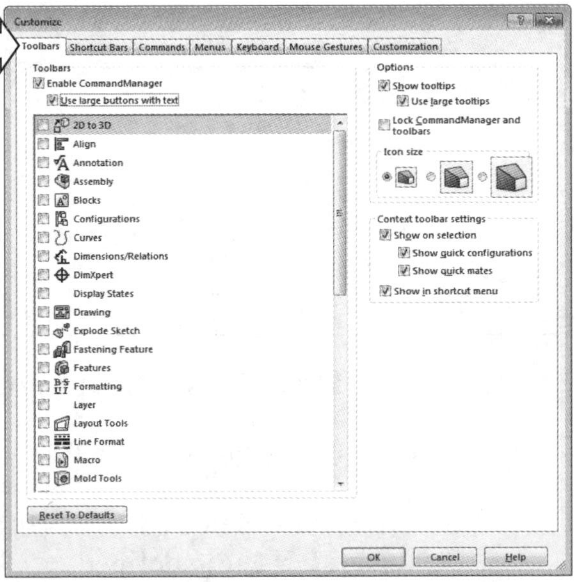

The icons in the toolbars can be enlarged when its check box is selected 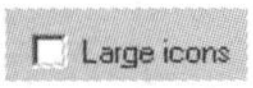

The View ports: You can view or work with SOLIDWORKS model or an assembly using one, two or four view ports.

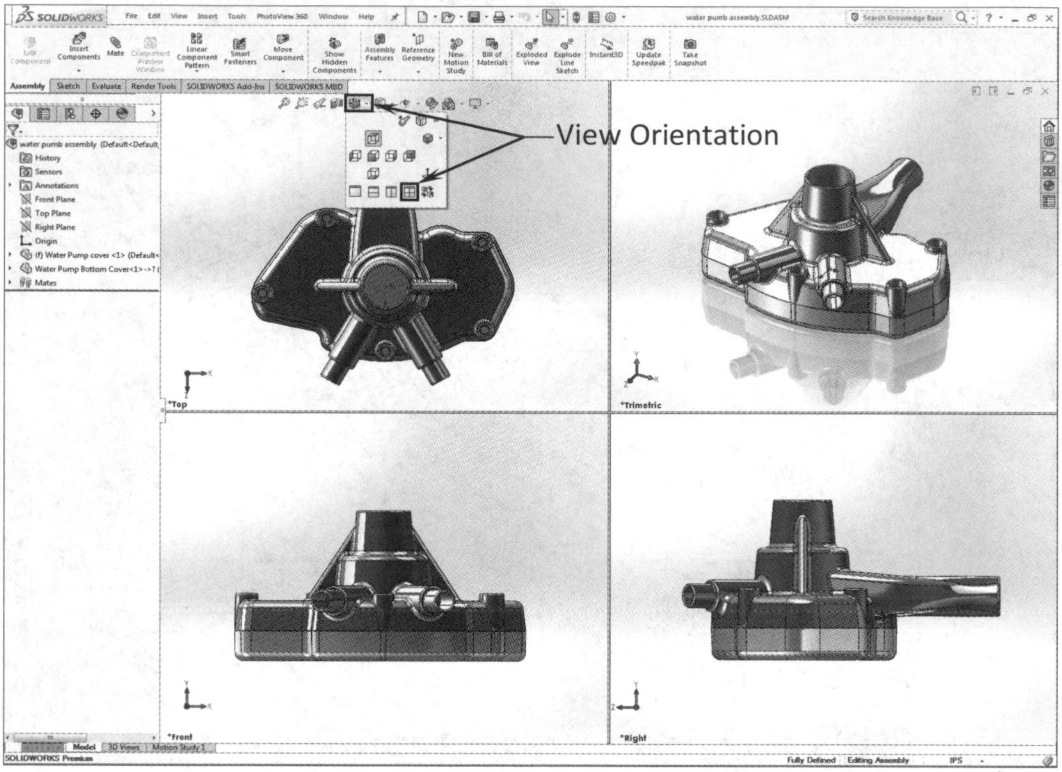

Some of the **System Feedback symbols** (Inference pointers):

 Snap to Vertex (endpoint)

 Snap to Edge (curve)

 Snap to Mid-point

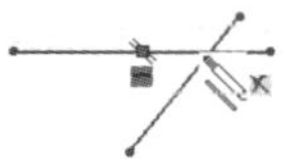 Snap to Intersection

 Horizontal Line

 Vertical Line

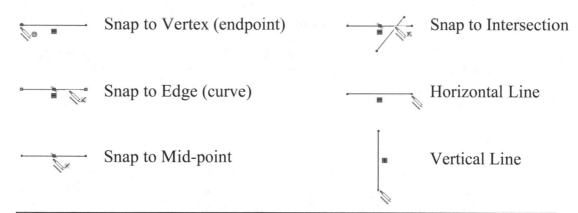

The Status Bar: (View / Status Bar)

Displays the status of the sketch entity using different colors to indicate:

Green = Selected
Black = Fully defined
Blue = Under defined
Red = Over defined

2D Sketch examples:

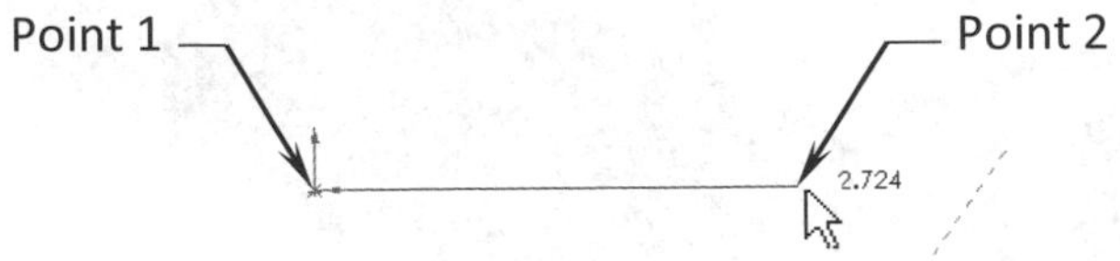

Click-Drag-Release: Single entity.

'ck Point 1, hold the mouse button, drag oint 2 and release.)

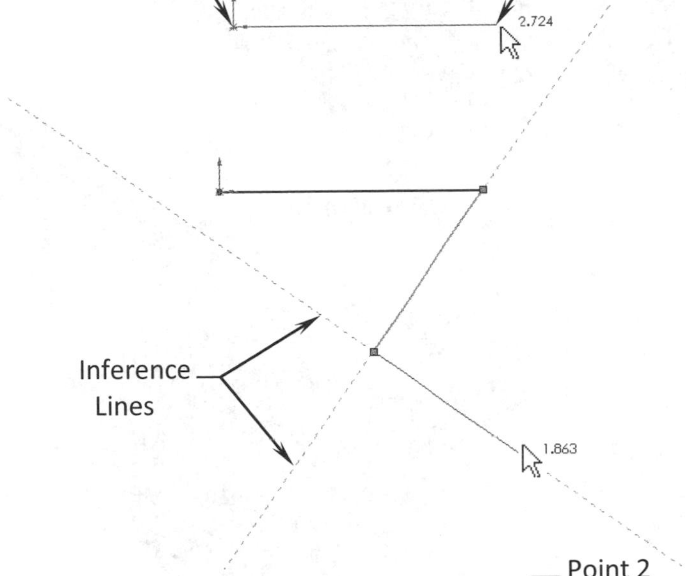

ck-Release: Continuous multiple entities.

e Inference Lines appear when the sketch ties are Parallel, Perpendicular, or Tangent ı each other.)

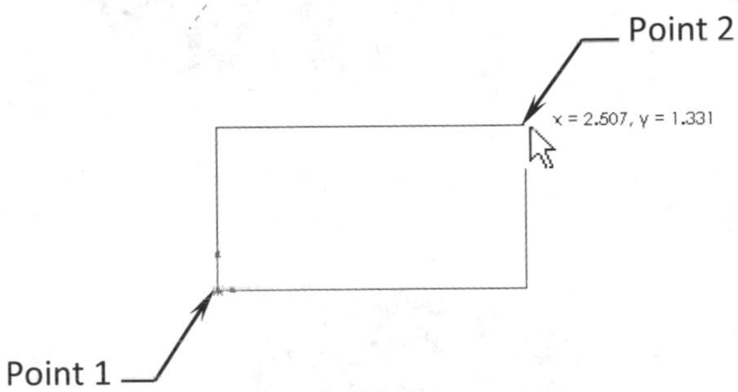

Click-Drag-Release: Single Rectangle

(Click point 1, hold the mouse button, drag to Point 2 and release.)

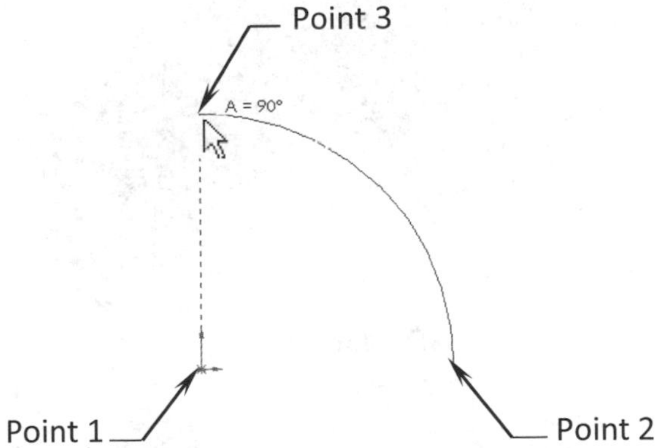

Click-Drag-Release: Single Centerpoint Arc

(Click point 1, hold the mouse button and drag to Point 2, release; then drag to Point 3 and release.)

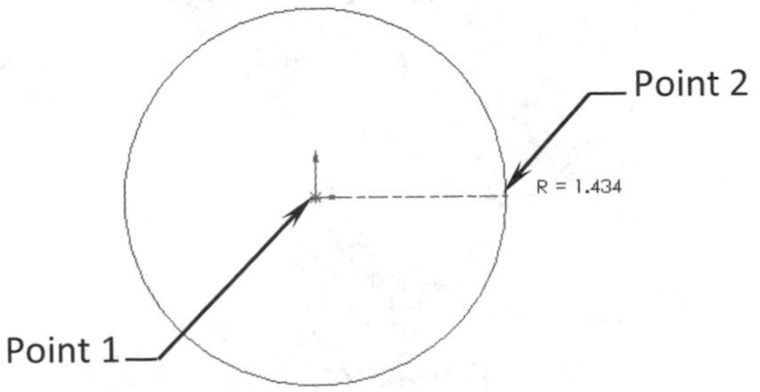

Click-Drag-Release: Single Circle

(Click point 1 [center of circle], hold the mouse button, drag to Point 2 [Radius] and release.)

3D Feature examples:

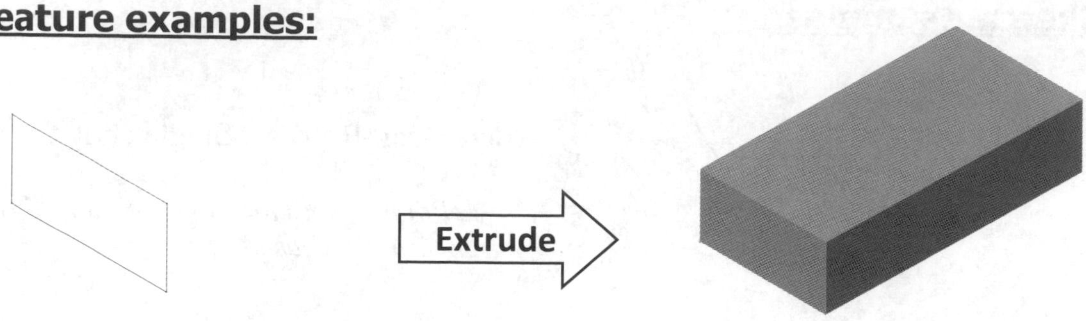

2D sketch 3D feature

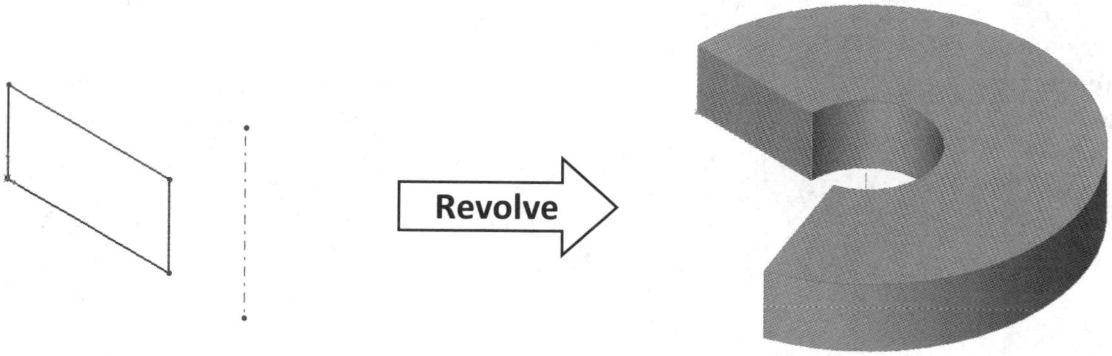

2D sketch 3D feature

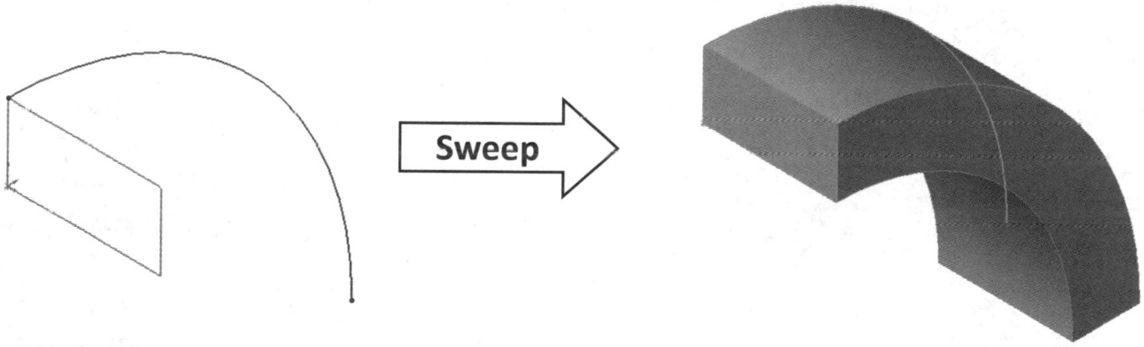

2D sketch 3D feature

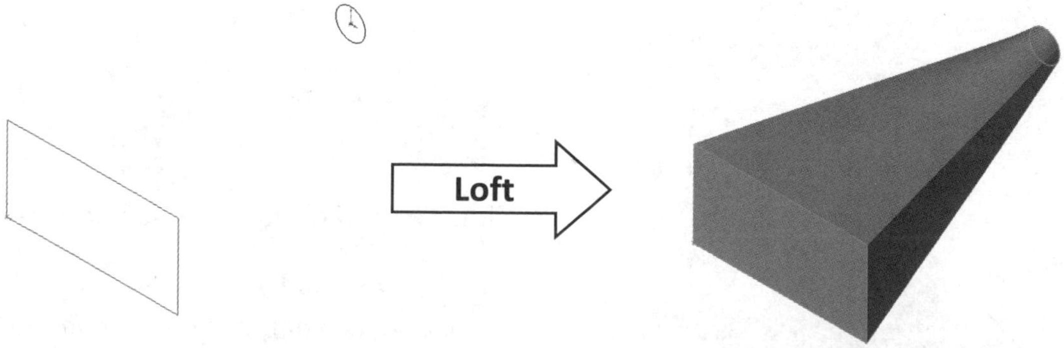

2D sketch 3D feature

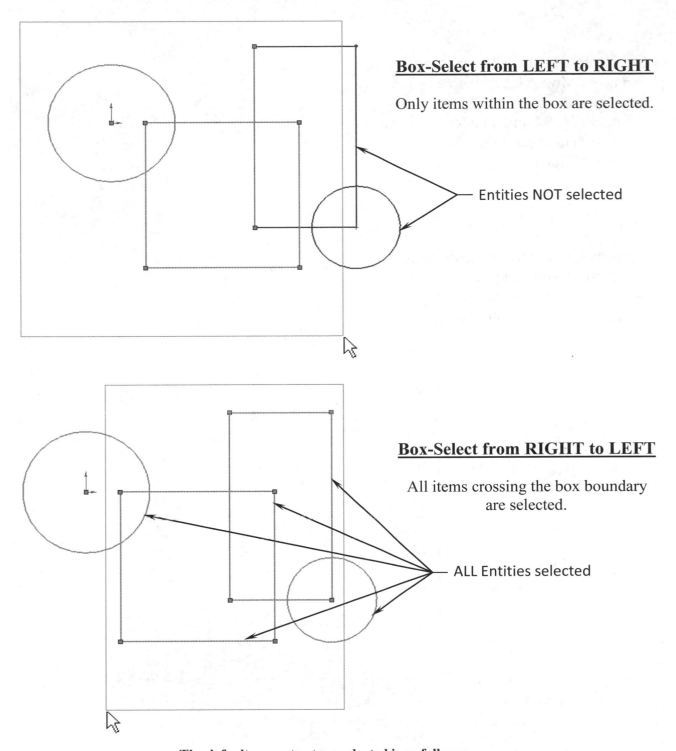

Box-Select: Use the Select Pointer to drag a selection box around items.

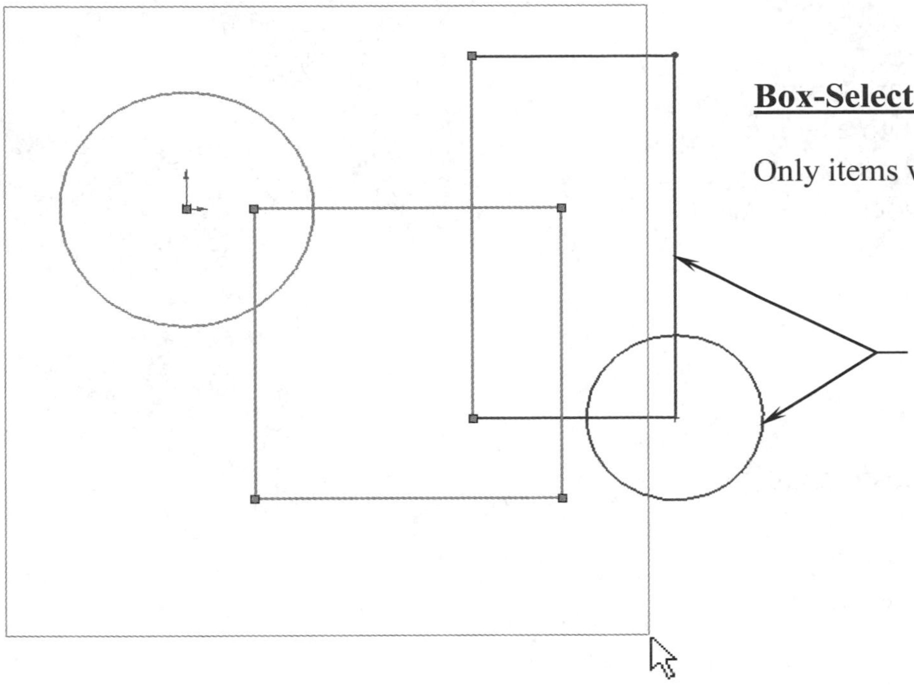

Box-Select from LEFT to RIGHT

Only items within the box are selected.

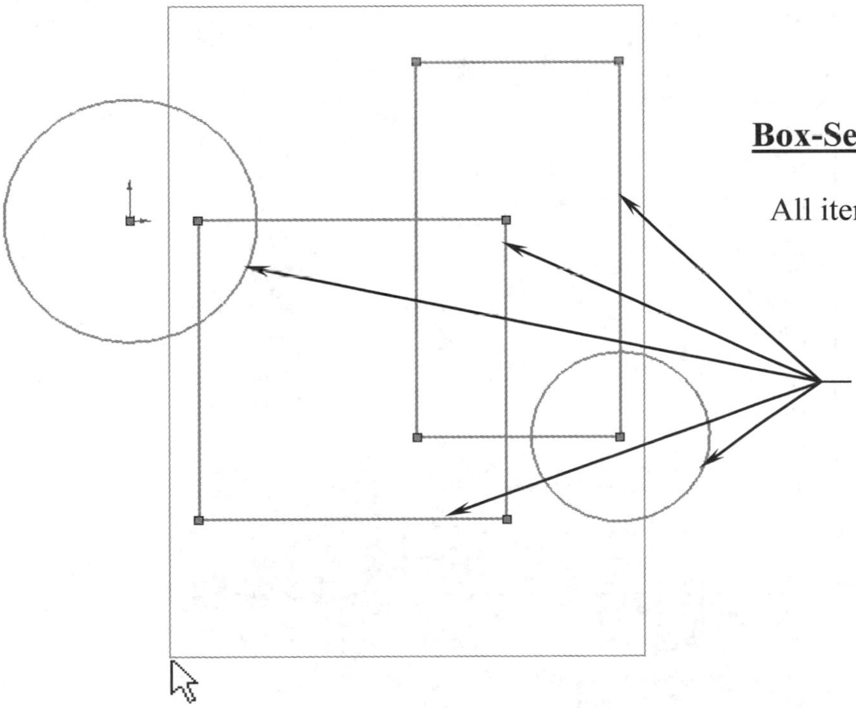

Box-Select from RIGHT to LEFT

All items crossing the box boundary are selected.

The default geometry type selected is as follows:

* Part documents – edges * Assembly documents – components * Drawing documents - sketch entities, dims & annotations. * To select multiple entities, hold down **Ctrl** while selecting after the first selection.

The Mouse Gestures for Parts, Sketches, Assemblies and Drawings

Similar to a keyboard shortcut, you can use a Mouse Gesture to execute a command. A total of 12 keyboard shortcuts can be independently mapped and stored in the Mouse Gesture Guides.

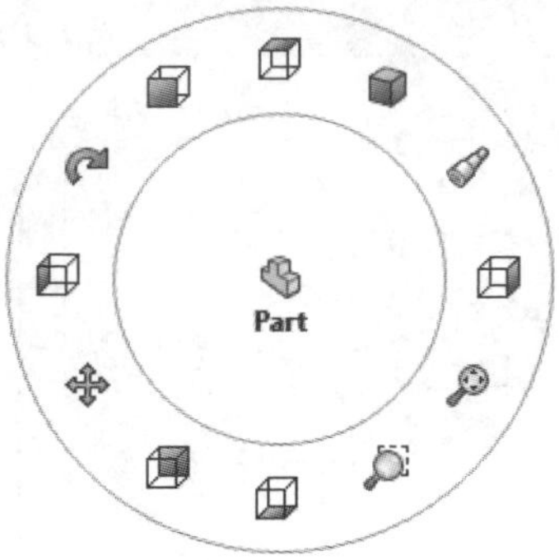

To activate the Mouse Gesture Guide, **right-click-and-drag** to see the current 12 gestures, then simply select the command that you want to use.

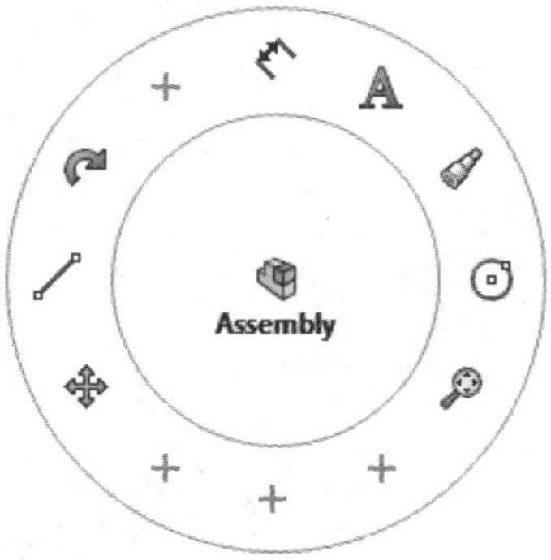

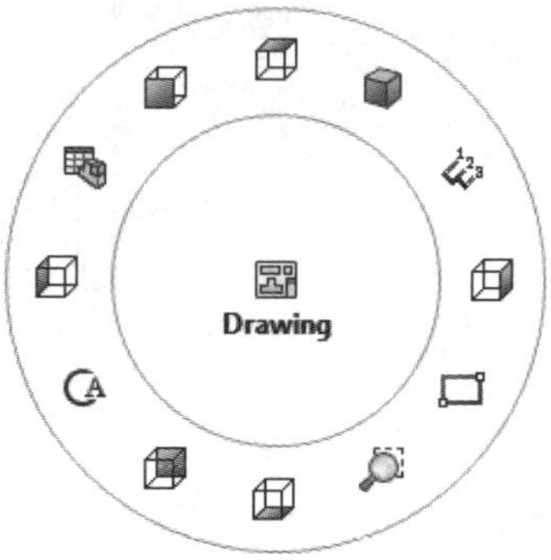

To customize the Mouse Gestures and include your favorite shortcuts, go to **Tools / Customize**.

From the **Mouse Gestures** tab select **All Commands**.

Click the **Enable Mouse Gestures** checkbox.

Select the **12 Gestures** option (arrow).

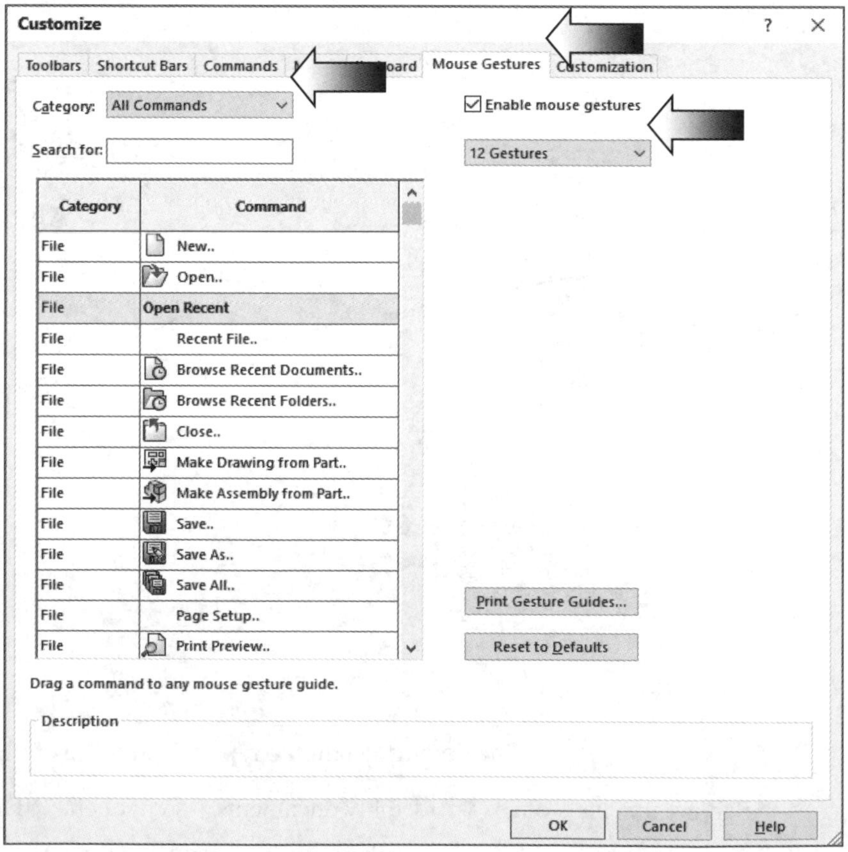

Customizing Mouse Gestures

To reassign a mouse gesture:

1.With a document open, click **Tools > Customize** and select the **Mouse Gestures** tab. The tab displays a list of tools and macros. If a mouse gesture is currently assigned to a tool, the icon for the gesture appears in the appropriate column for the command.

For example, by default, the right mouse gesture is assigned to the Right tool for parts and assemblies, so the icon for that gesture () appears in the Part and Assembly columns for that tool.

To filter the list of tools and macros, use the options at the top of the tab. By default, four mouse gesture directions are visible in the Mouse Gestures tab and available in the mouse gesture guide. Select 8 gestures to view and reassign commands for eight gesture directions.

2.Find the row for the tool or macro you want to assign to a mouse gesture and click in the cell where that row intersects the appropriate column.

For example, to assign Make Drawing from Part to the Part column, click in the cell where the Make-Drawing-from-Part row and the Part column intersect.

A list of either 4 or 12 gesture directions appears as shown, depending on whether you have the 4 gestures, or 12 gestures option selected.

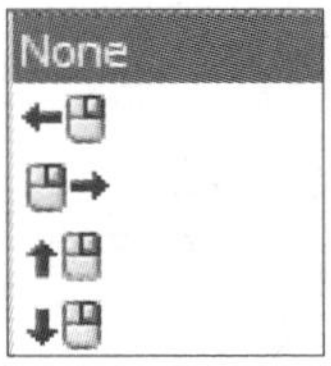

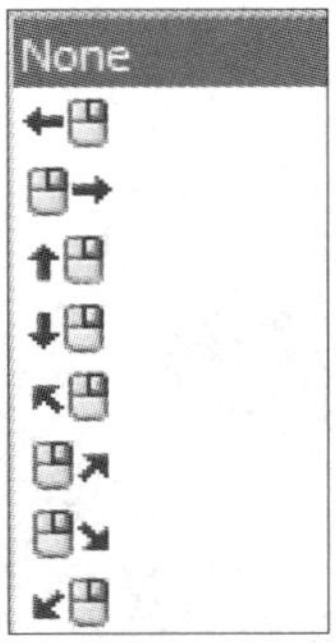

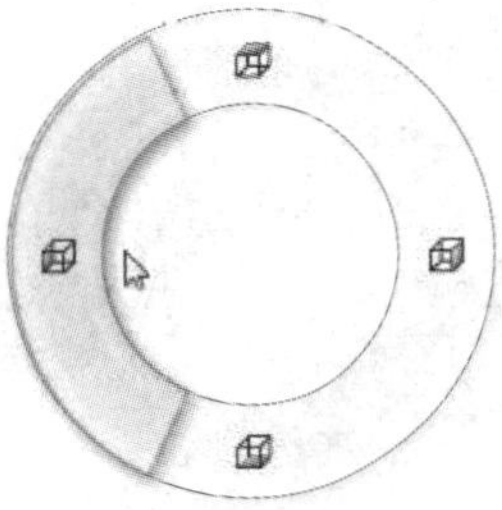

4 gestures

12 gestures

Some tools are not applicable to all columns, so the cell is unavailable, and you cannot assign a mouse gesture. For example, you cannot assign a mouse gesture for Make Drawing from Part in the Assembly or Drawing columns.

3.Select the mouse gesture direction you want to assign from the list. The mouse gesture direction is reassigned to that tool and its icon appears in the cell.

4.Click OK.

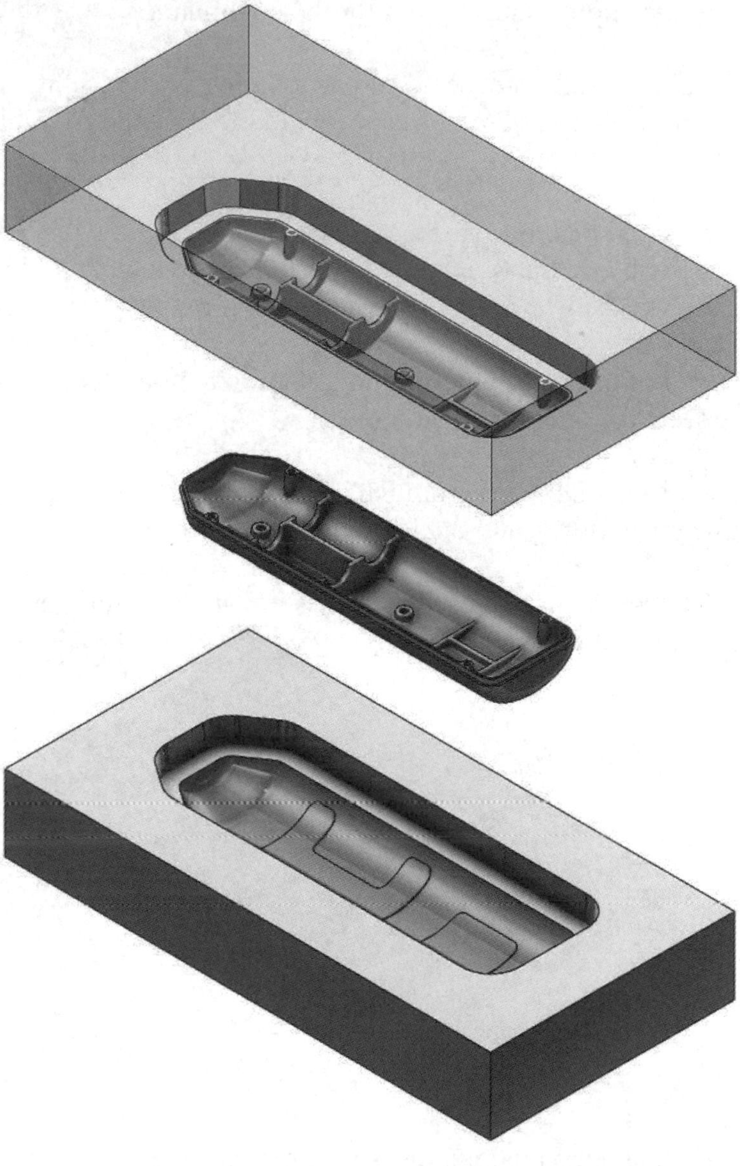

Designed with SOLIDWORKS 2022, SP0

CHAPTER 1

Introduction to 3D Sketch

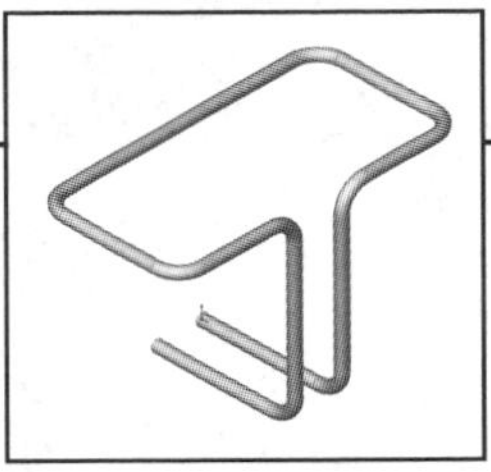

Introduction to 3D Sketch

SOLIDWORKS has 3D sketch capabilities. A 3D sketch consists of lines and arcs in series and splines. You can use a 3D sketch as a sweep path, as a guide curve for a loft or sweep, a centerline for a loft, or as one of the key entities in a piping system. Geometric relations can also be added to 3D Sketches.

Parameters

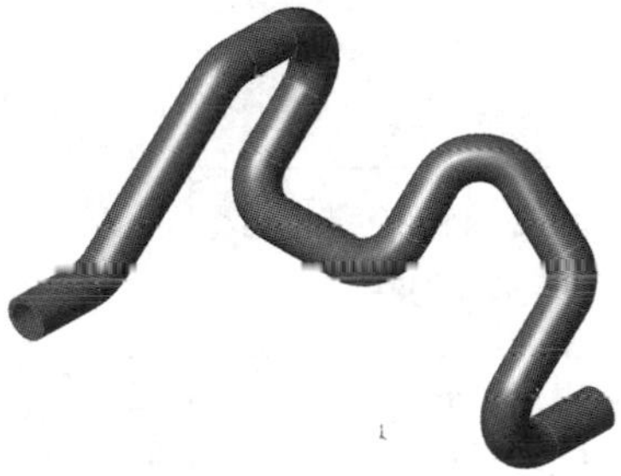

- **X Coordinate**
- **Y Coordinate**
- **Z Coordinate**
- **Curvature** (Spline curvature at the frame point)
- **Tangency** (In the **XY** plane)
- **Tangency** (In the **XZ** plane)
- **Tangency** (In the **YZ** plane)

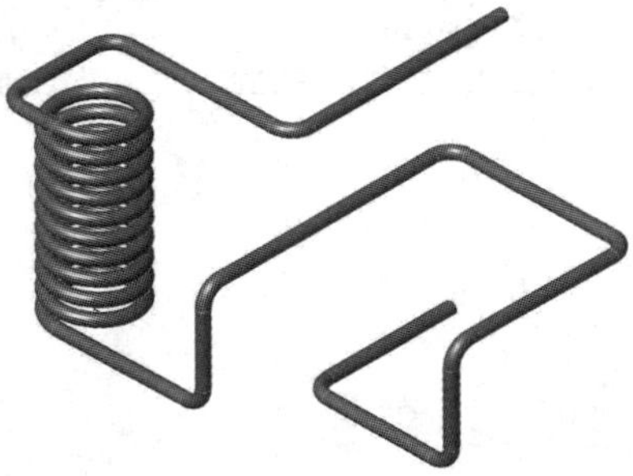

Space Handle

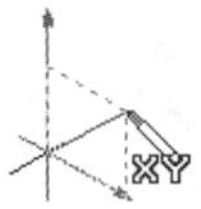

When working in a 3D sketch, a graphical assistant is provided to help you maintain your orientation while you sketch on several planes. This assistant is called a ***space handle***. The space handle appears when the first point of a line or spline is defined on a selected plane. Using the space handle you can select the axis along which you want to sketch.

Introduction to 3D Sketch

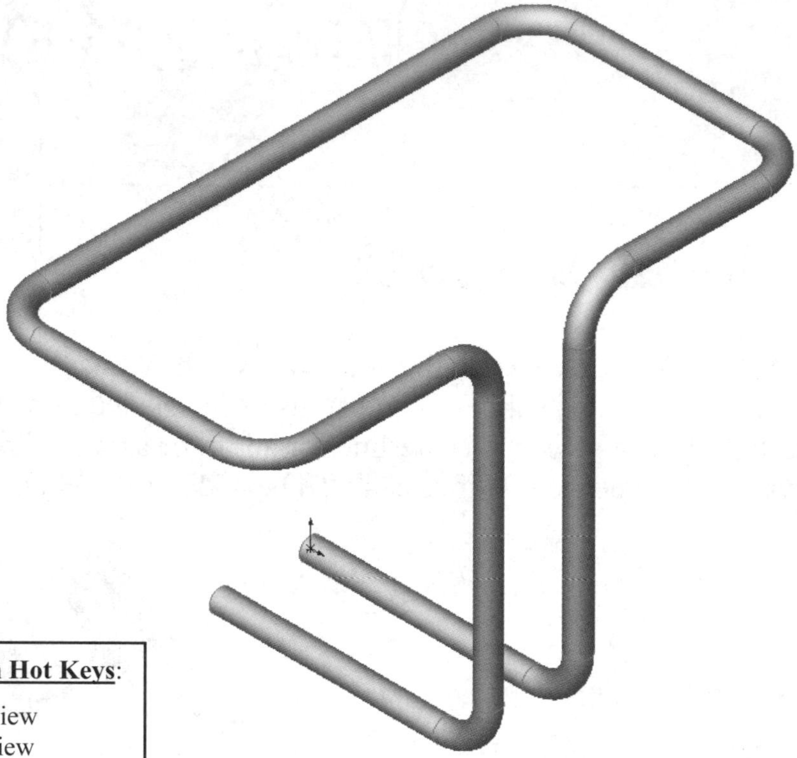

View Orientation Hot Keys:

Ctrl + 1 = Front View
Ctrl + 2 = Back View
Ctrl + 3 = Left View
Ctrl + 4 = Right View
Ctrl + 5 = Top View
Ctrl + 6 = Bottom View
Ctrl + 7 = Isometric View
Ctrl + 8 = Normal To Selection

Dimensioning Standards: **ANSI**
Units: **INCHES** – 3 Decimals

Tools Needed:

3D Sketch	2D Sketch	Sketch Line
Circle	Dimension	Add Geometric Relations
Sketch Fillet	Tab Key	Base/ Boss Sweep

1. Starting a new part file:

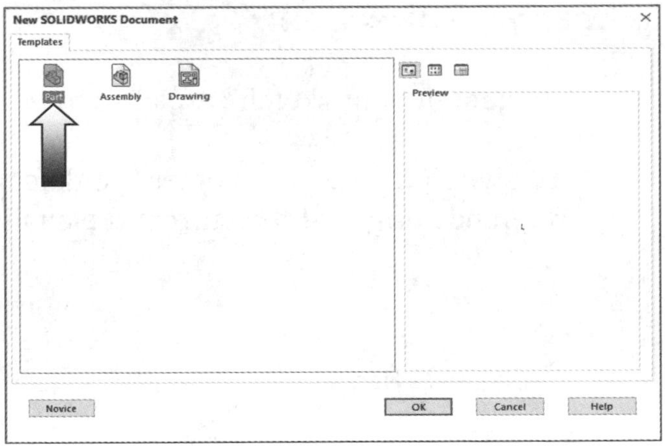

Click **File / New.**

Select the **Part** template and click **OK**.

Set the Units to **IPS**, 3 decimals.

2. Creating a 3D Sketch:

Click  or select **Insert / 3D Sketch** and change to **Isometric view** (Control+7).

Select the **Line** tool and sketch the first line along the **X** direction. A yellow symbol appears next to the mouse cursor when the line is drawn along the X axis; this indicates an **Along X** relation (horizontal) is being added to the line.

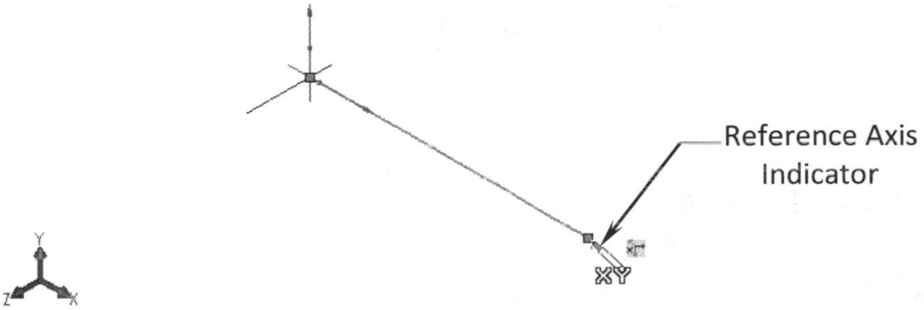

Sketch the second line along the **Y** axis as shown.

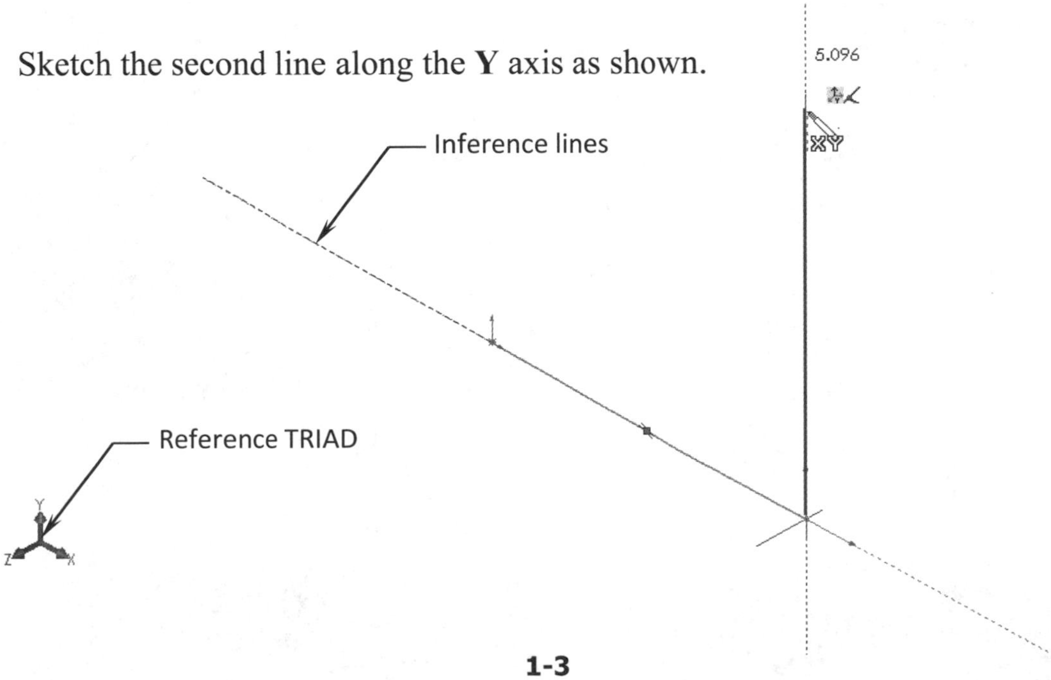

3. Changing direction:

By default your sketch is relative to the default coordinate system in the model.

To switch to one of the other two default planes, press the **TAB** key and the reference origin of the current sketch plane is displayed on that plane.

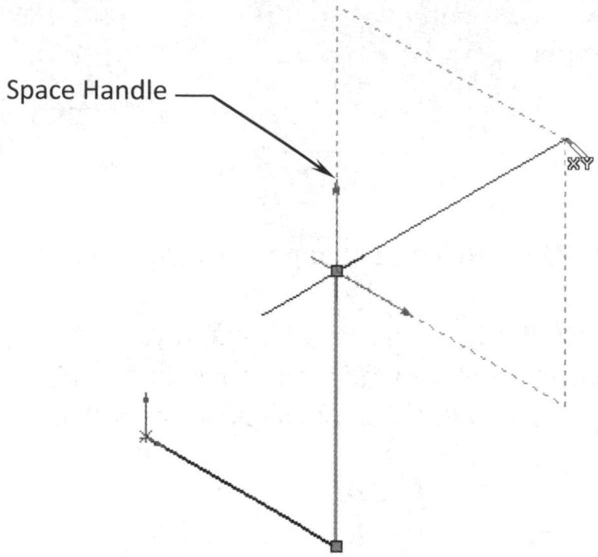

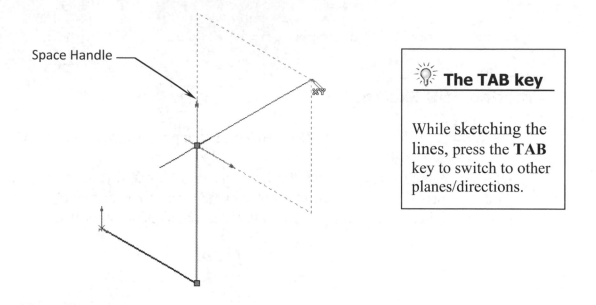

> **The TAB key**
>
> While sketching the lines, press the **TAB** key to switch to other planes/directions.

4. Completing the profile:

Sketch the other lines and follow the directions as labeled; press the **TAB** key if needed to change the direction.

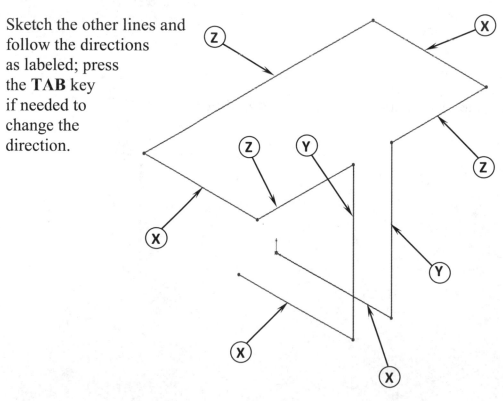

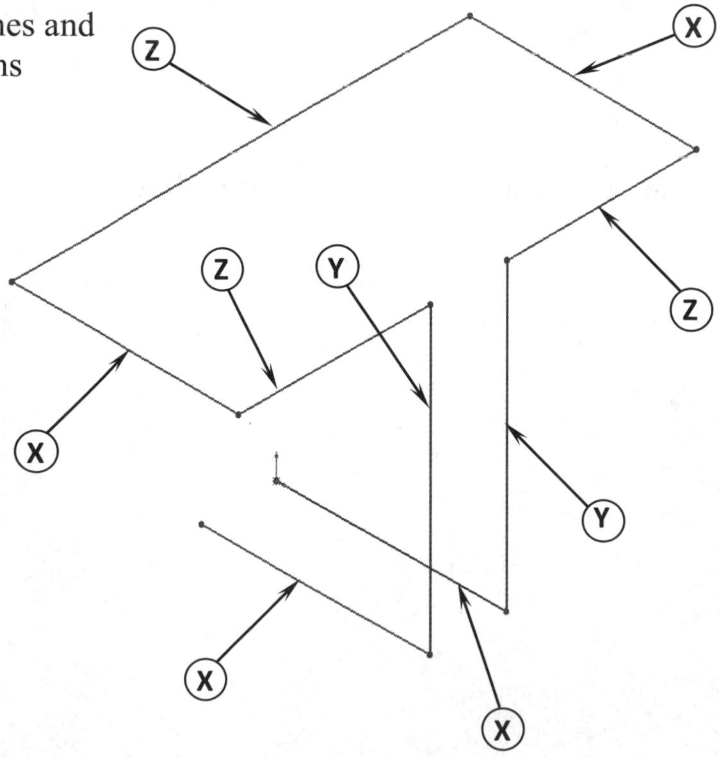

5. Adding dimensions:

Click **Smart Dimension** or select **Tools / Dimensions / Smart Dimension**.

Click the first line and enter a dimension of **3.00in**.

There is not a general sequence to follow when adding dimensions, so for this lesson, add the dimensions in the same order you sketched the lines.

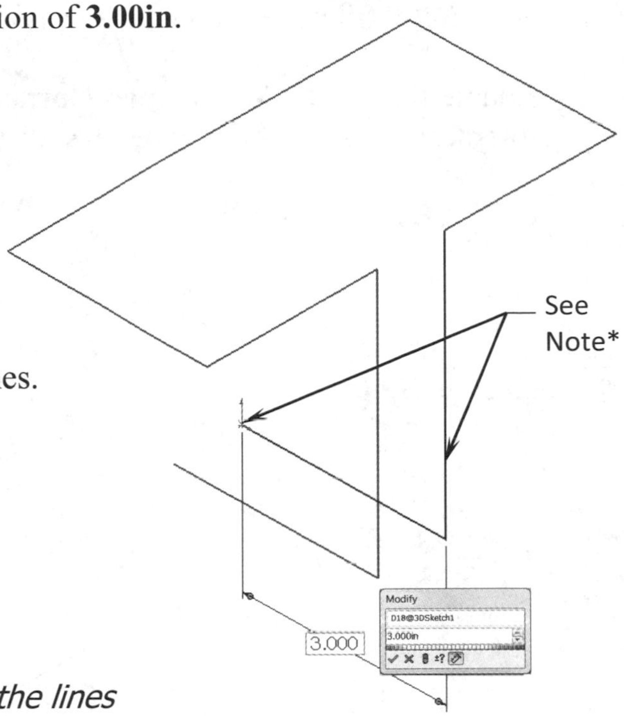

Note: *To make the dimensions parallel to the lines as shown, select the line and an endpoint instead of selecting just the endpoints.*

Continue adding the dimensions to fully define the 3D sketch as shown.

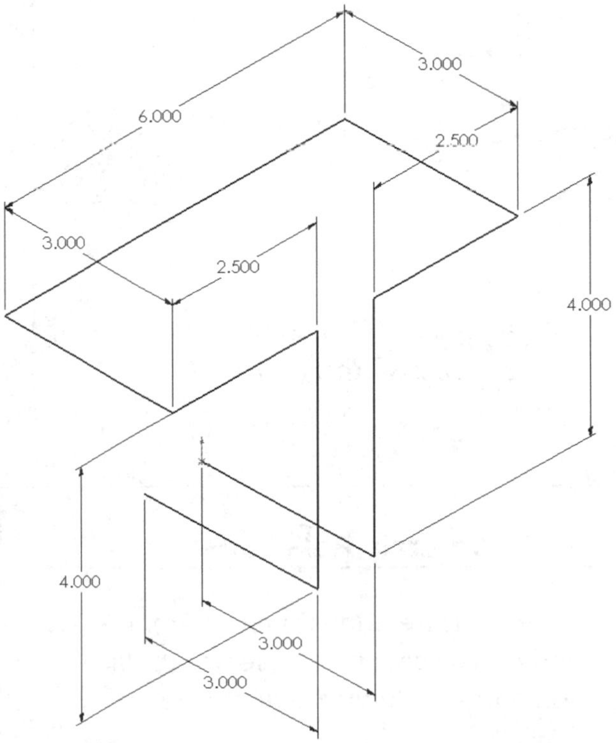

Rearrange the dimensions so they are easy to read, which will make editing a little easier later on.

6. Adding the Sketch Fillets:

Click **Fillet** on the Features toolbar or select **Tools / Sketch Tools / Fillet**.

Add **.500"** fillets to <u>all</u> the intersections as indicated.

Enable the **Keep Constrained Corner** check box (to maintain the virtual intersection point if the vertex has dimensions or relations).

Click **OK** when finished.

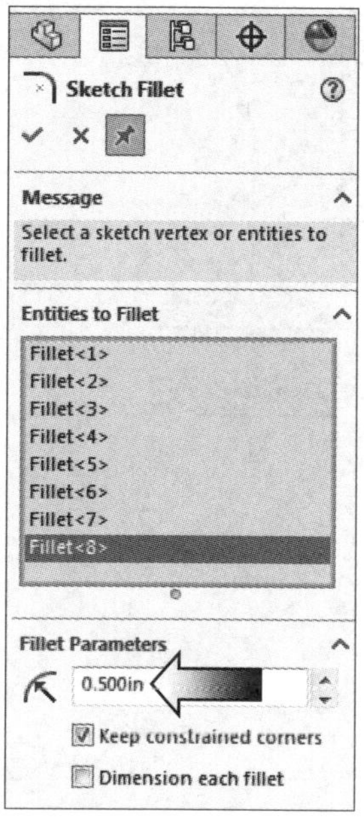

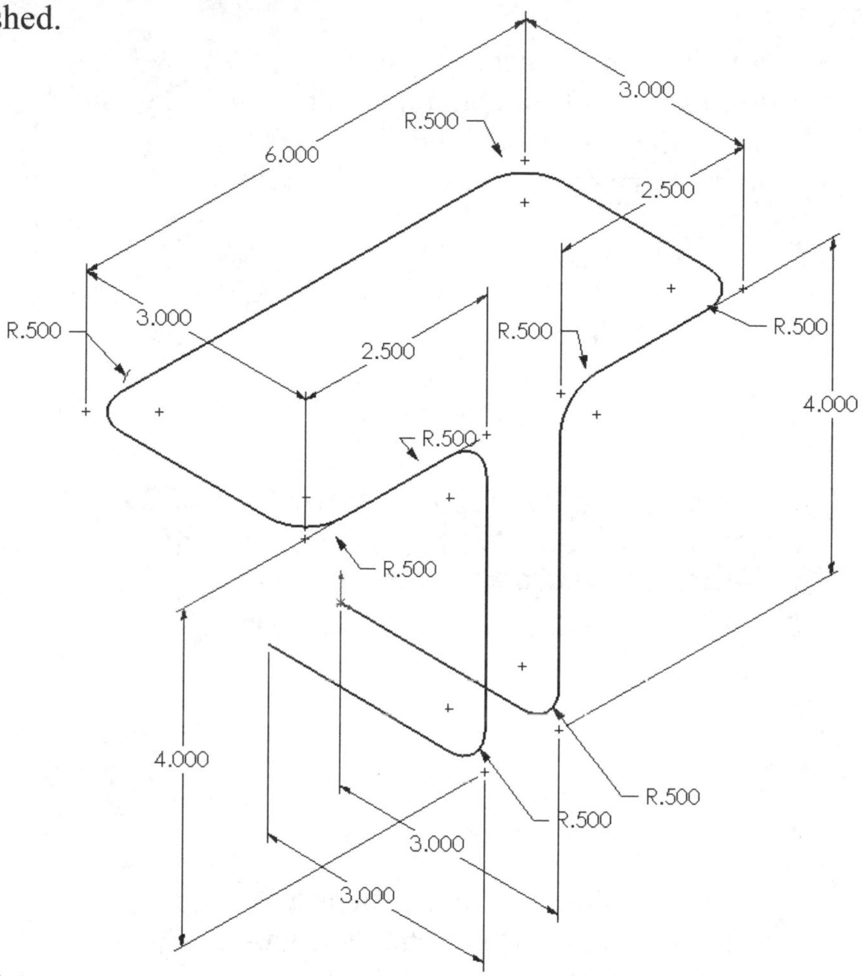

<u>Exit</u> the 3D Sketch
or press **Control + Q**.

Geometric Relations

Geometric Relations such as Along X, Y, Z and Equal can also be used to replace some of the duplicate dimensions.

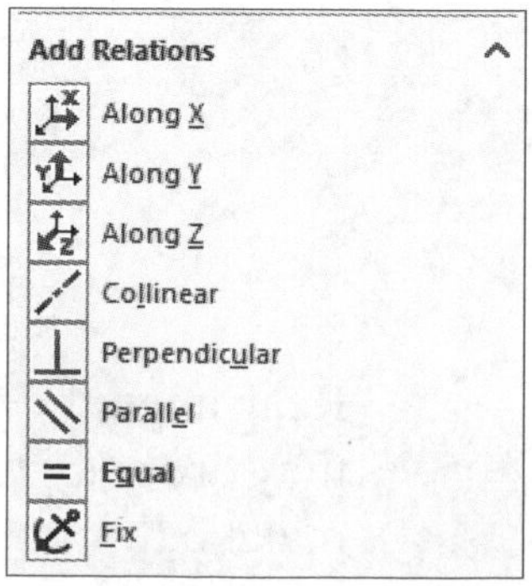

7. Creating the Swept feature:

The Circular Profile option allows you to create a solid rod or hollow tube along a path, edge, or curve directly on a model without having to sketch the circular profile. This enhancement is available for Swept Boss/Base, Swept Cut, and Swept Surface features.

Click [icon] or select **Insert / Boss-Base / Sweep**.

Select the **Circle Profile** option and enter **.250in** for the diameter of the profile [icon].

Select the **3D Sketch** for Sweep Path (3Dsketch1).

Click **OK**.

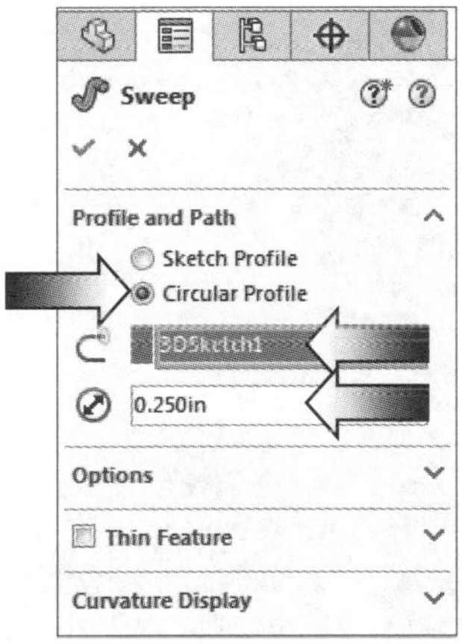

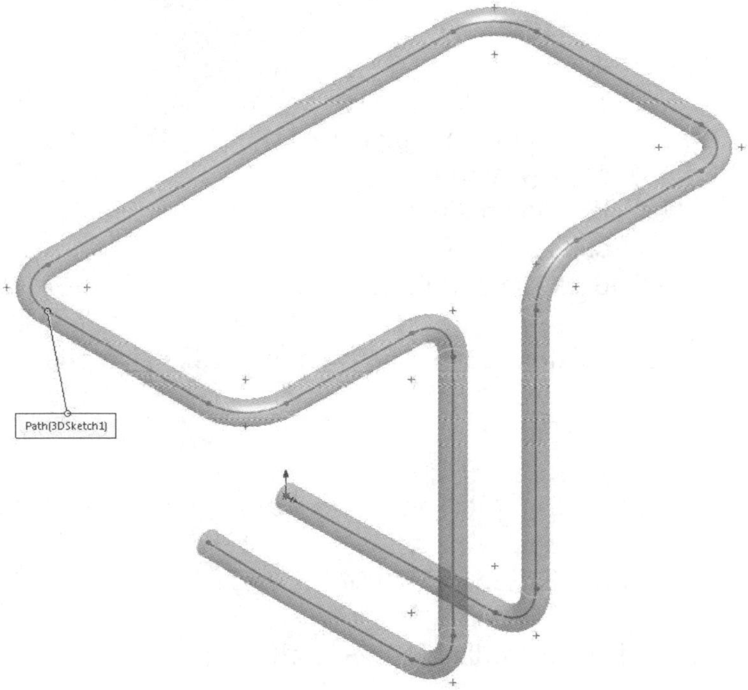

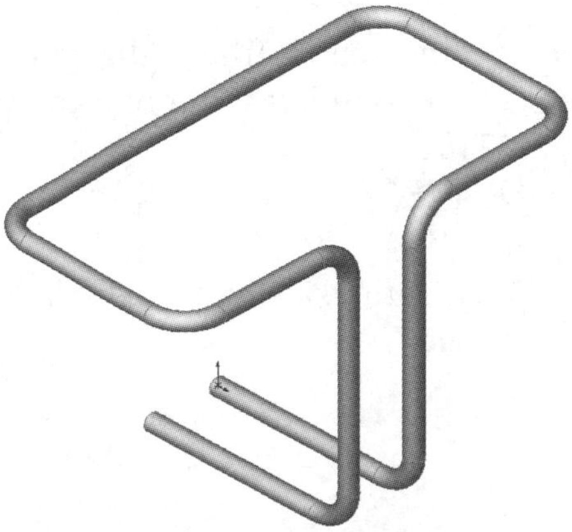

8. Saving your work:

Select **File / Save As**.

Enter **3D Sketch** for the file name.

Click **Save**.

Questions for Review

Introduction to 3D Sketch

1. When using 3D Sketch you do not have to pre-select a plane as you would in 2D Sketch.
    a. True
    b. False

2. The space handle appears only after the first point of a line is started.
    a. True
    b. False

3. To switch to other planes (or direction) in 3D Sketch mode, press:
    a. Up Arrow
    b. Down Arrow
    c. TAB key
    d. CONTROL key

4. Dimensions cannot be used in 3D Sketch mode.
    a. True
    b. False

5. Geometric Relations cannot be used in 3D Sketch mode.
    a. True
    b. False

6. All sketch tools in 2D Sketch are also available in 3D Sketch.
    a. True
    b. False

7. When adding sketch fillets, the option Keep Constrained Corner will create a virtual intersection point but will not create a radius dimension.
    a. True
    b. False

8. 3D Sketch entities can be used as a path in a swept feature.
    a. True
    b. False

1. TRUE 2. TRUE
3. C 4. FALSE
5. FALSE 6. FALSE
7. FALSE 8. TRUE

Exercise: Sweep with 3D Sketch

1. Create the part shown using 3D Sketch.

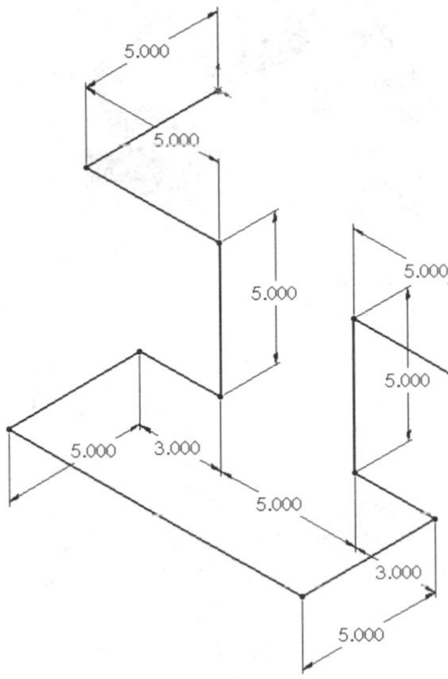

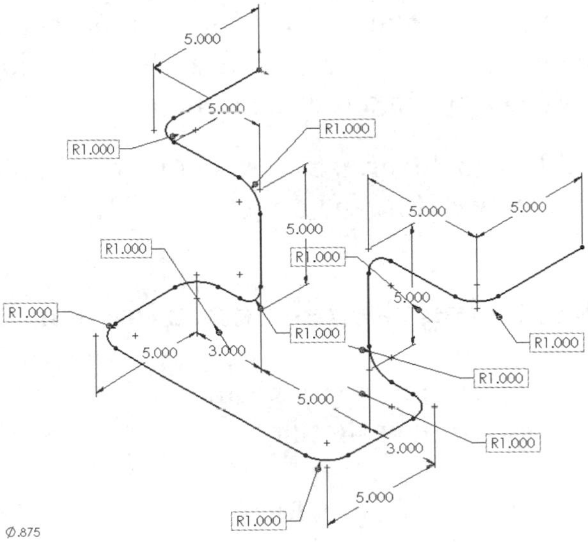

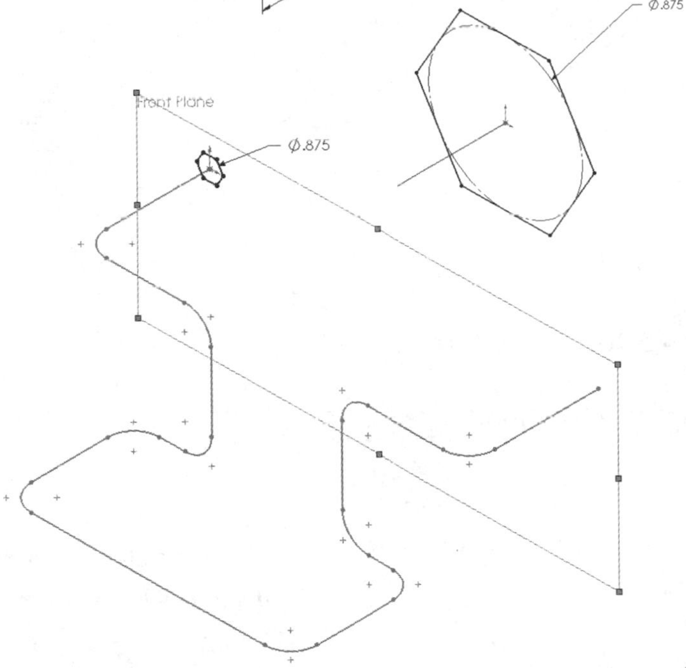

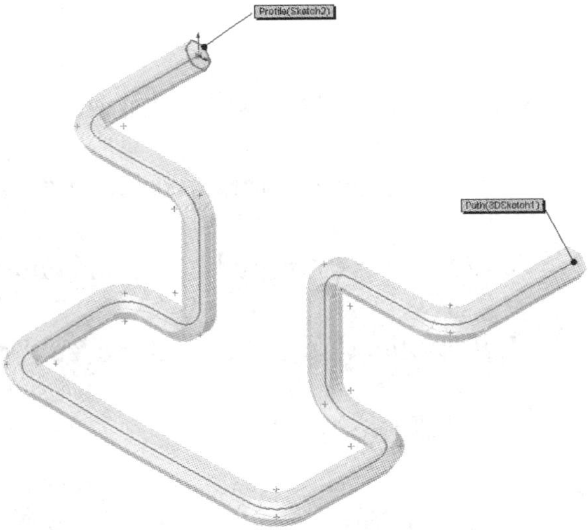

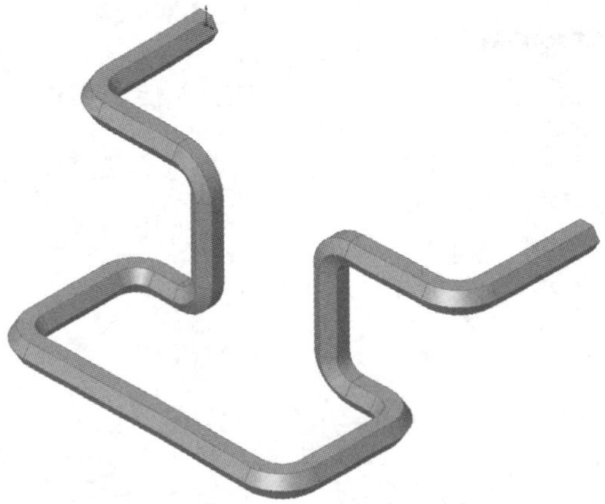

2. Save your work as:
Sweep_3D Sketch_Exe.

Exercise: 3D Sketch & Planes

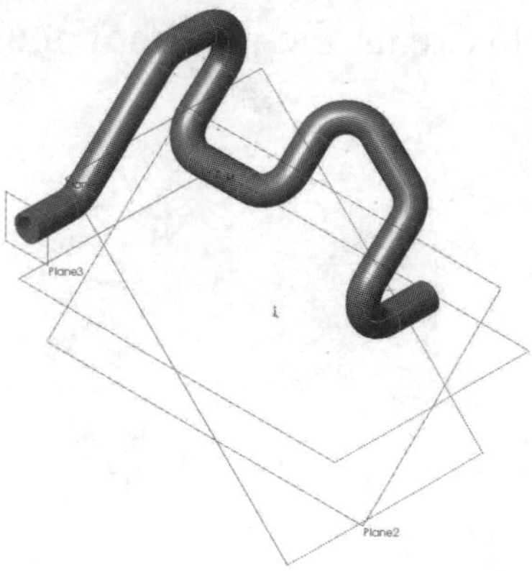

A 3D sketch normally consists of lines and arcs in series, and splines. You can use a 3D sketch as a sweep path, as a guide curve for a loft or sweep, a centerline for a loft, or as one of the key entities in a routing system.

The following exercise shows how several planes can be used to help define the directions of 3D Sketch Entities.

1. Sketching the reference Pivot lines:

Select the Top plane and open a **new sketch**.

Sketch **2 Centerlines** and add dimensions as shown.

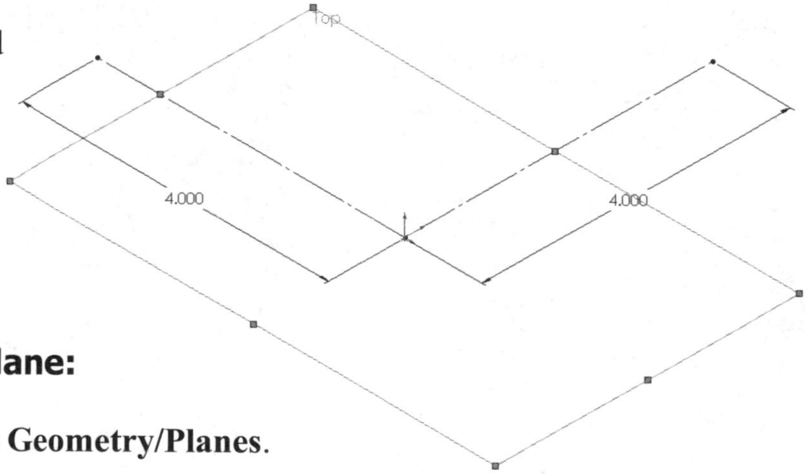

2. Creating the 1st 45º Plane:

Select **Insert/Reference Geometry/Planes**.

Click the **At Angle** button and enter **45** for Angle (arrow).

Select the **Top** plane and the **Vertical line** as noted.

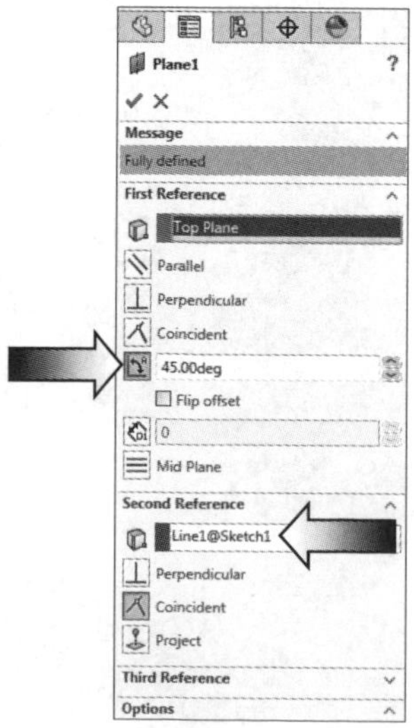

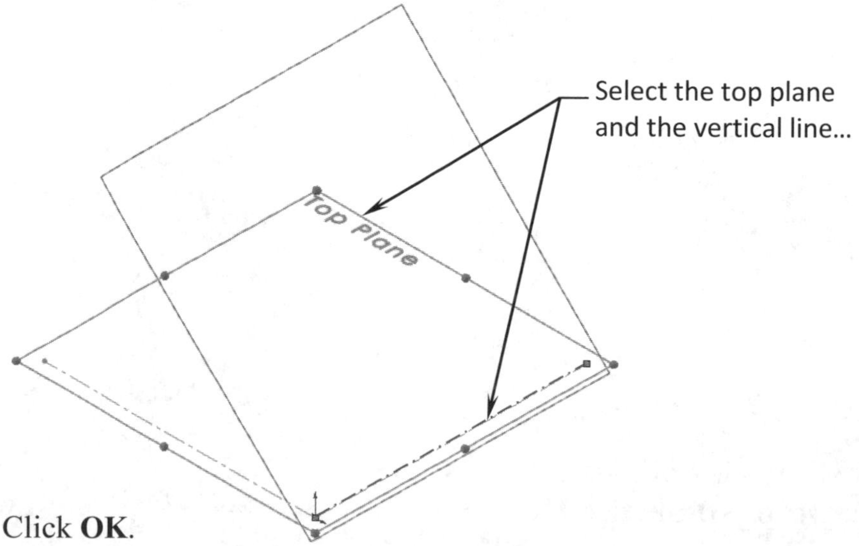

Click **OK**.

3. Creating the 2nd 45° Plane:

Click the **Plane** command or select **Insert/Reference Geometry/Planes** .

Click the **At Angle** option and enter **45** for Angle (arrow).

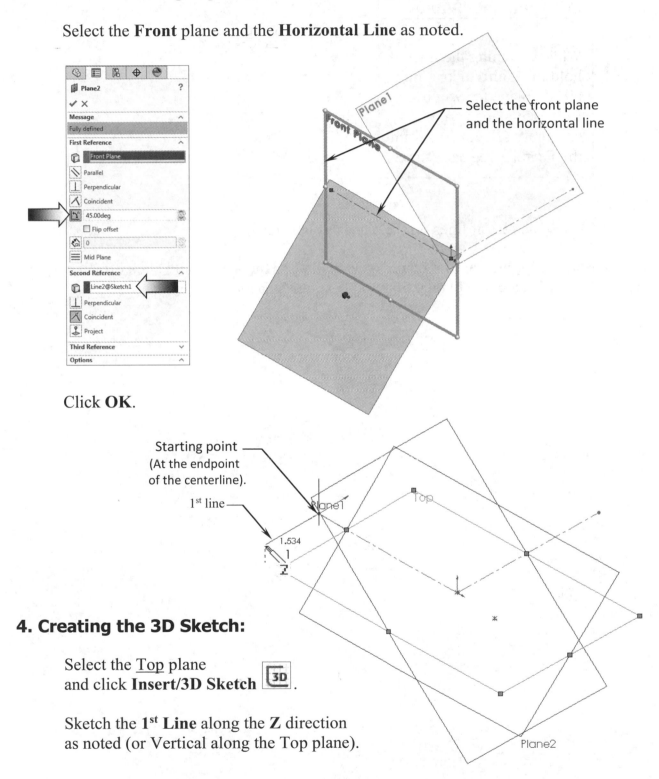

Select the **Front** plane and the **Horizontal Line** as noted.

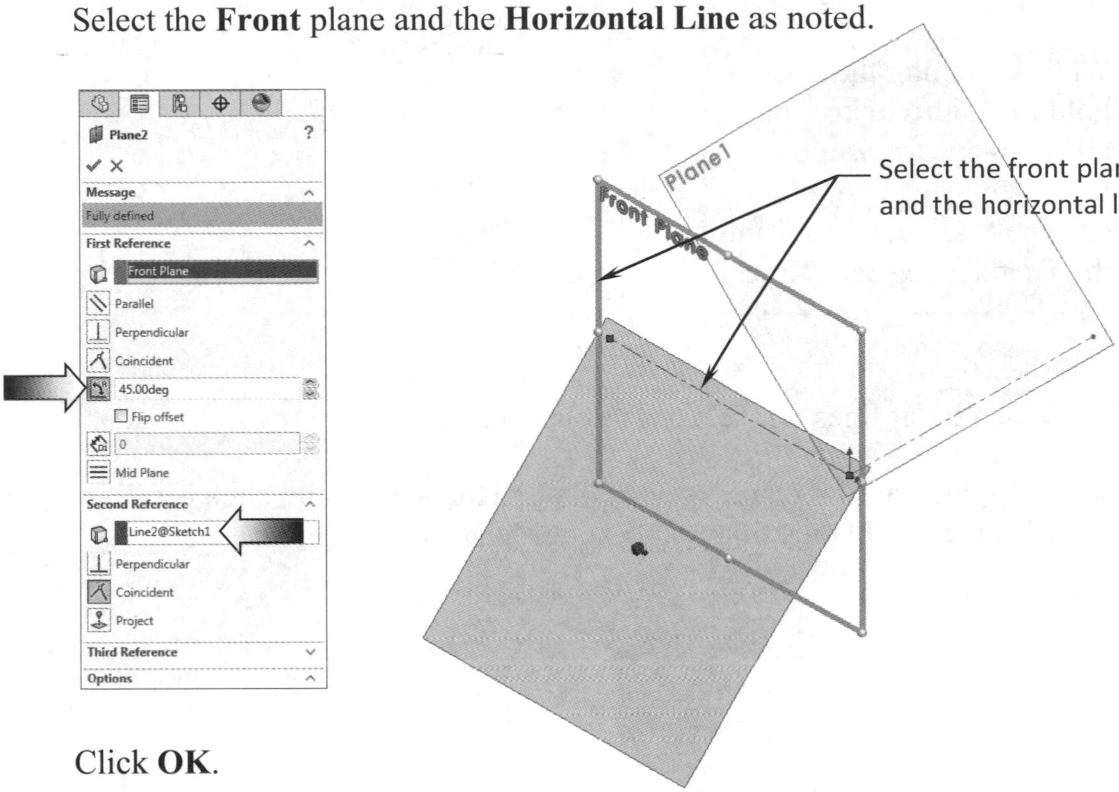

Click **OK**.

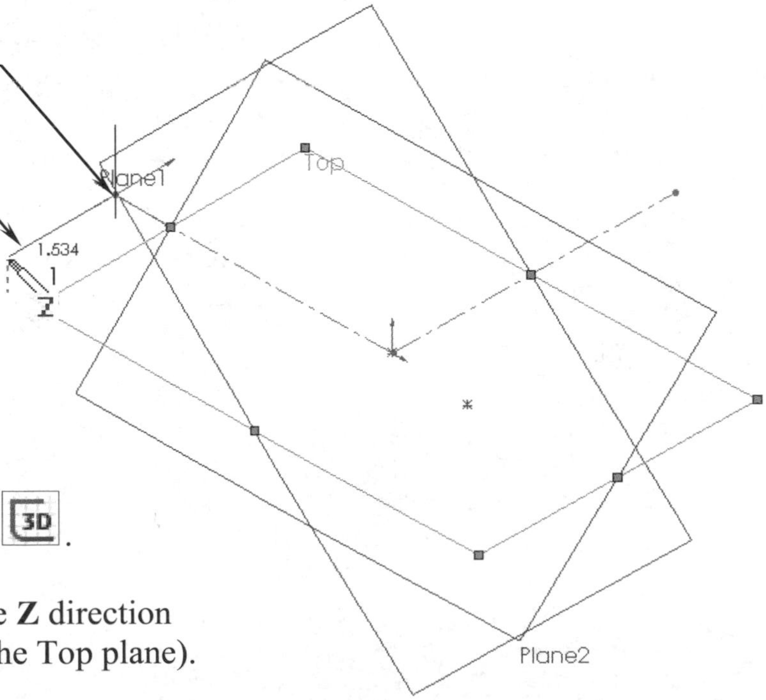

4. Creating the 3D Sketch:

Select the Top plane
and click **Insert/3D Sketch** .

Sketch the **1st Line** along the **Z** direction
as noted (or Vertical along the Top plane).

Select the **Plane2** (45 deg.) from the Feature Manager tree and Sketch the **2nd Line** along the **Y** direction (watch the cursor feedback symbol).

Switching Planes

While sketching the lines, hold the **Control** key and click a plane to switch from one plane to another, or simply select them from the Feature tree each time.

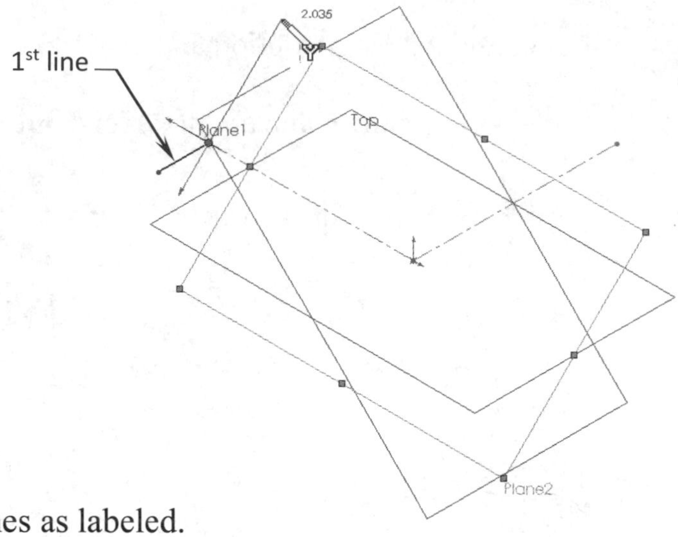

Sketch the rest of lines on the planes as labeled.

For clarity, hide all the planes (select **View / Hide-Show** and click off **Planes**). We will select the planes from the FeatureManager tree when needed.

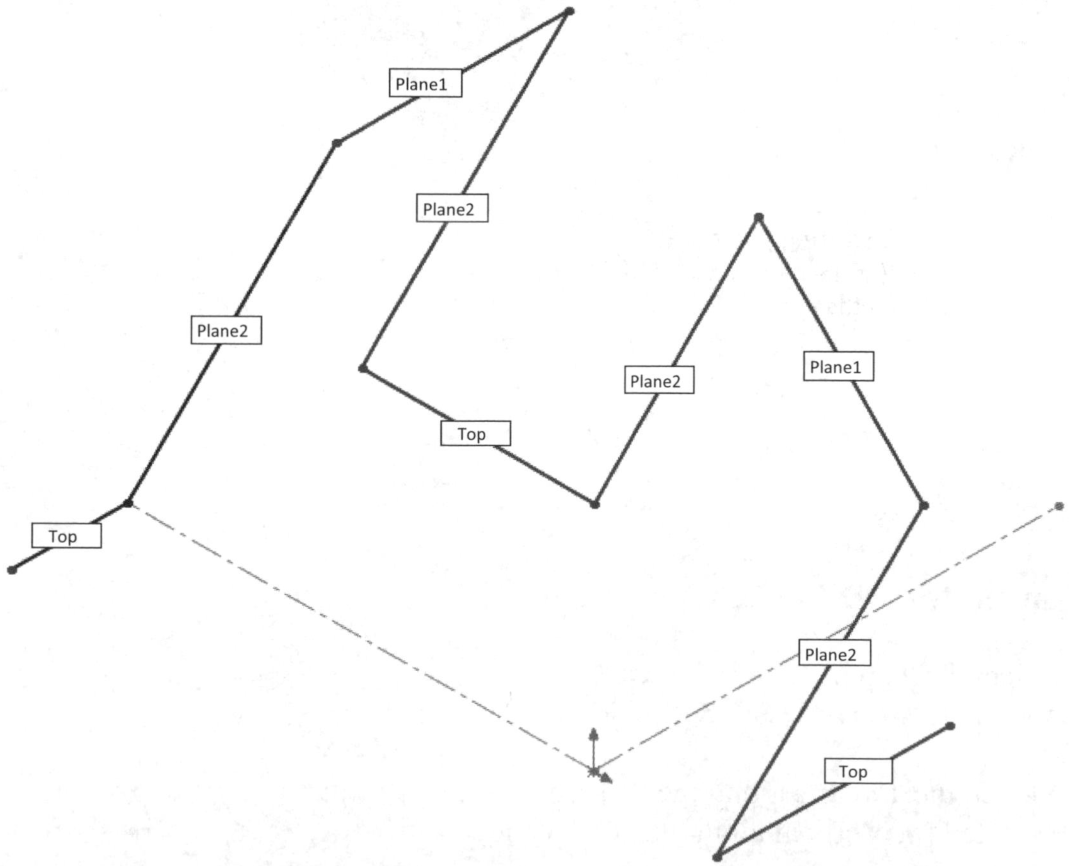

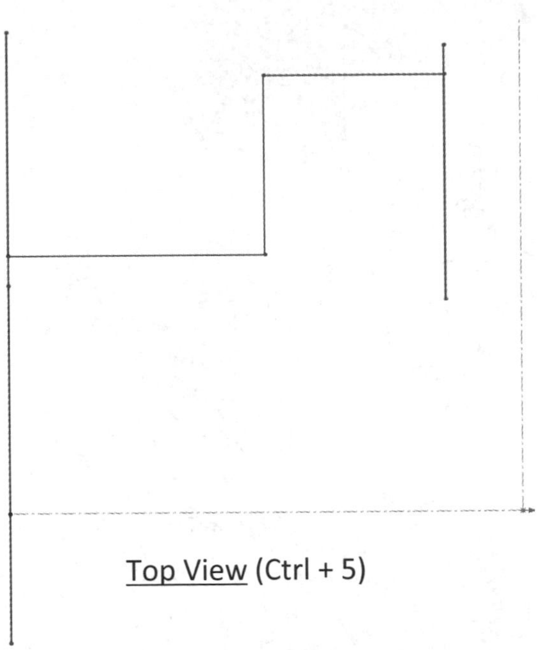

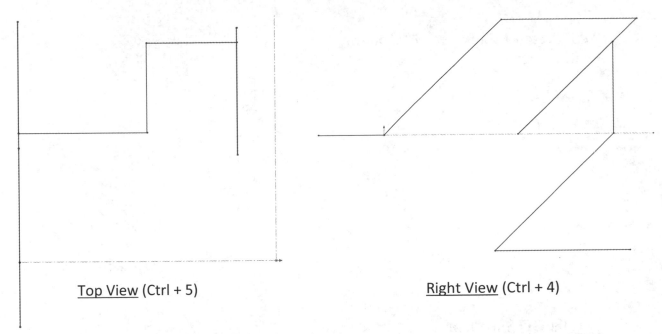

Top View (Ctrl + 5)

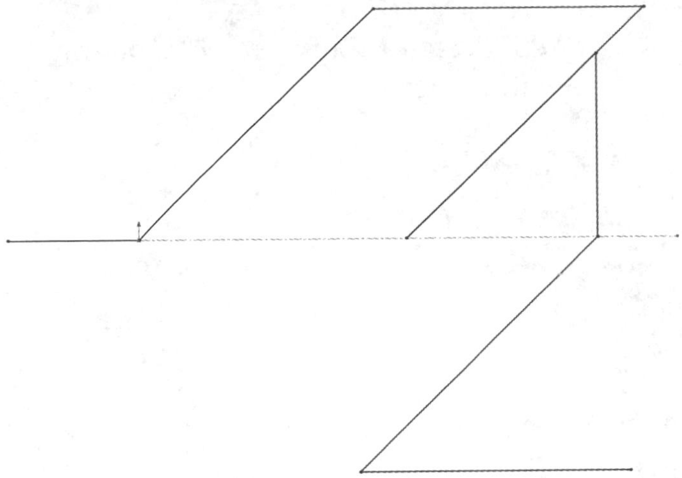

Right View (Ctrl + 4)

Add the dimensions  below to fully define the sketch.

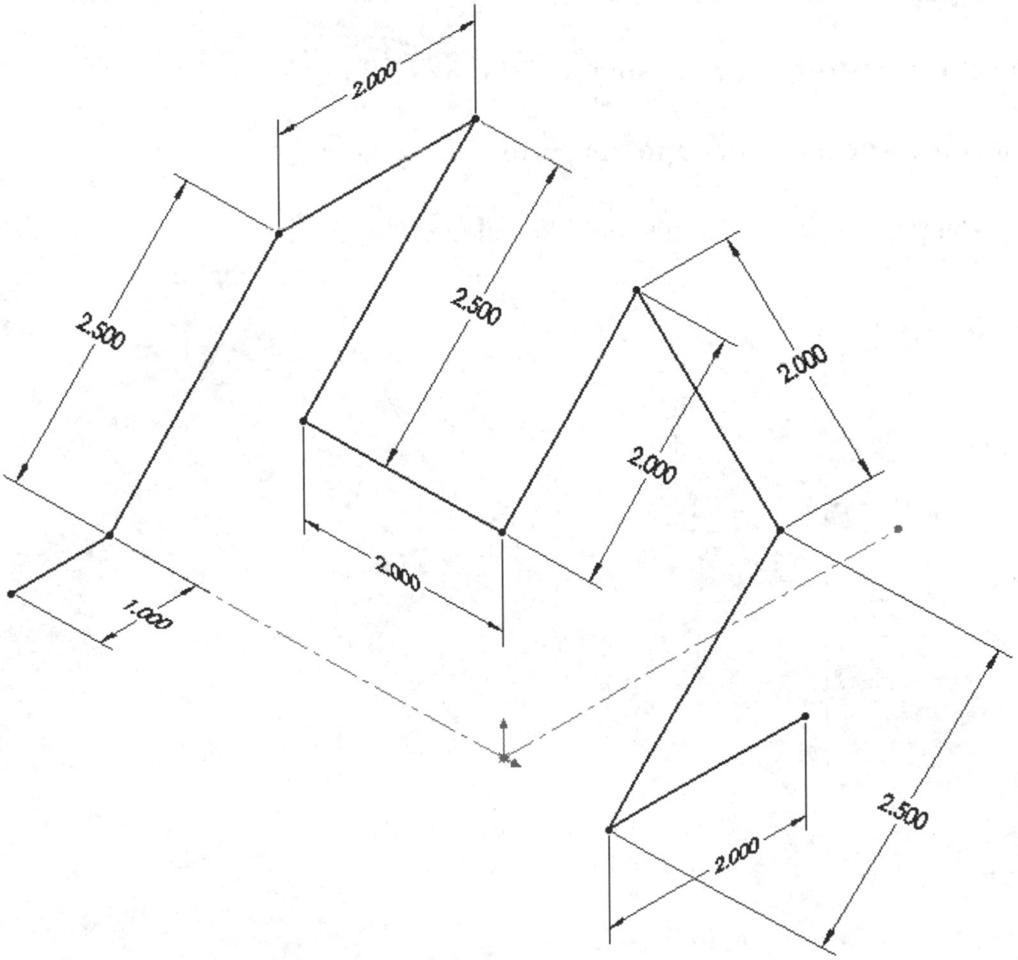

Add **Sketch Fillets** of **.500 in.** to all corners.

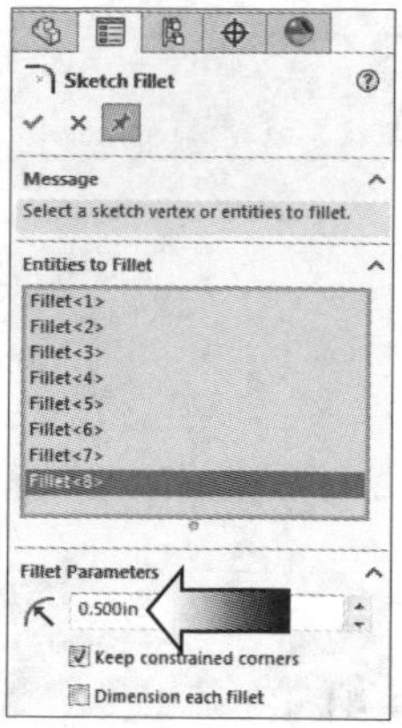

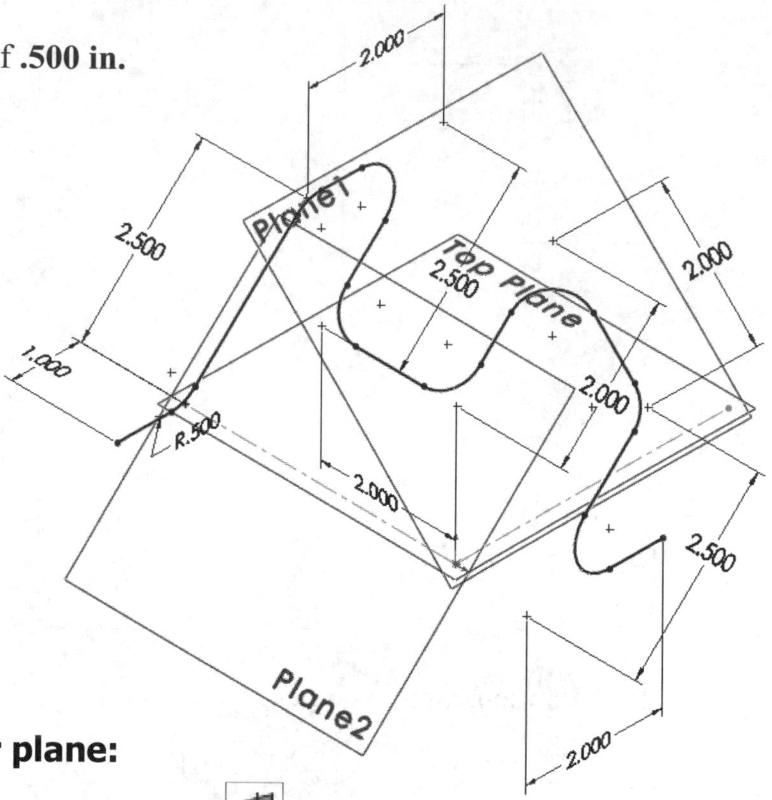

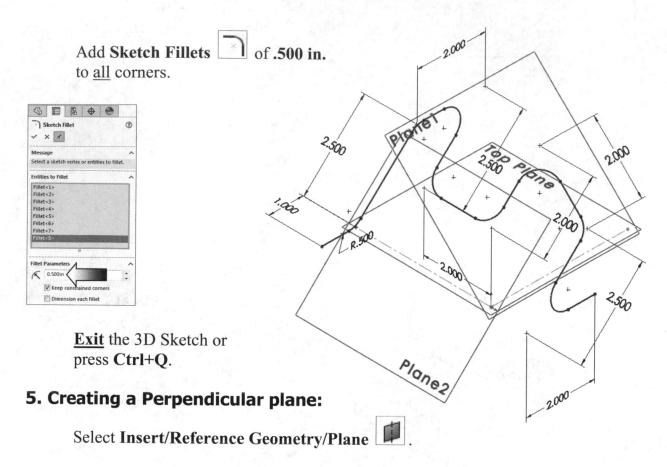

Exit the 3D Sketch or press **Ctrl+Q**.

5. Creating a Perpendicular plane:

Select **Insert/Reference Geometry/Plane**.

Select the **line** and its **endpoint** approximately as shown.

The **Perpendicular** option should be selected by default.

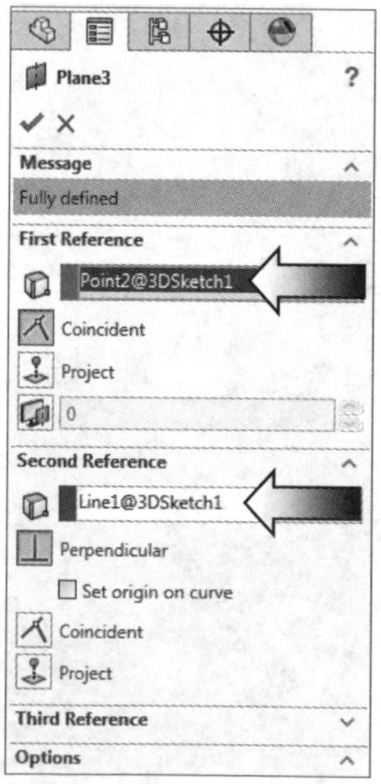

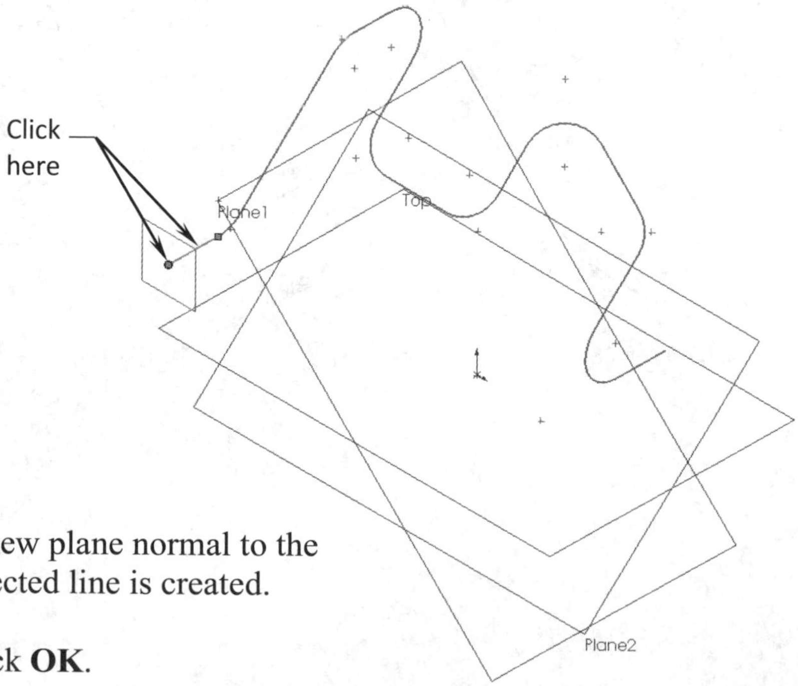

A new plane normal to the selected line is created.

Click **OK**.

6. Sketching the Sweep Profile:

Select the new plane (Plane3) and open a **new sketch** .

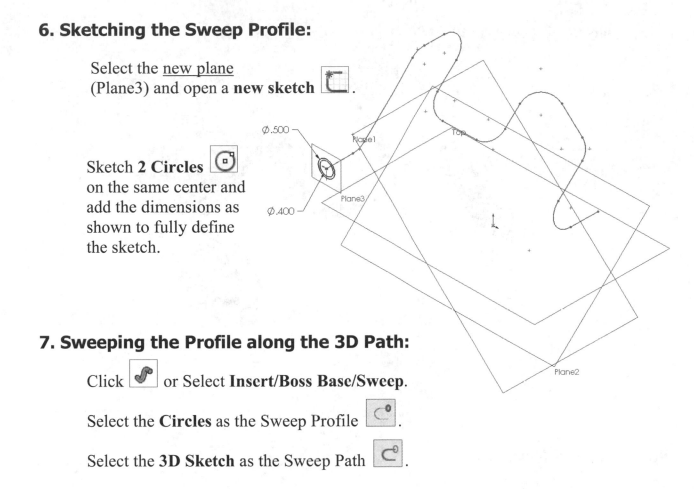

Sketch **2 Circles** on the same center and add the dimensions as shown to fully define the sketch.

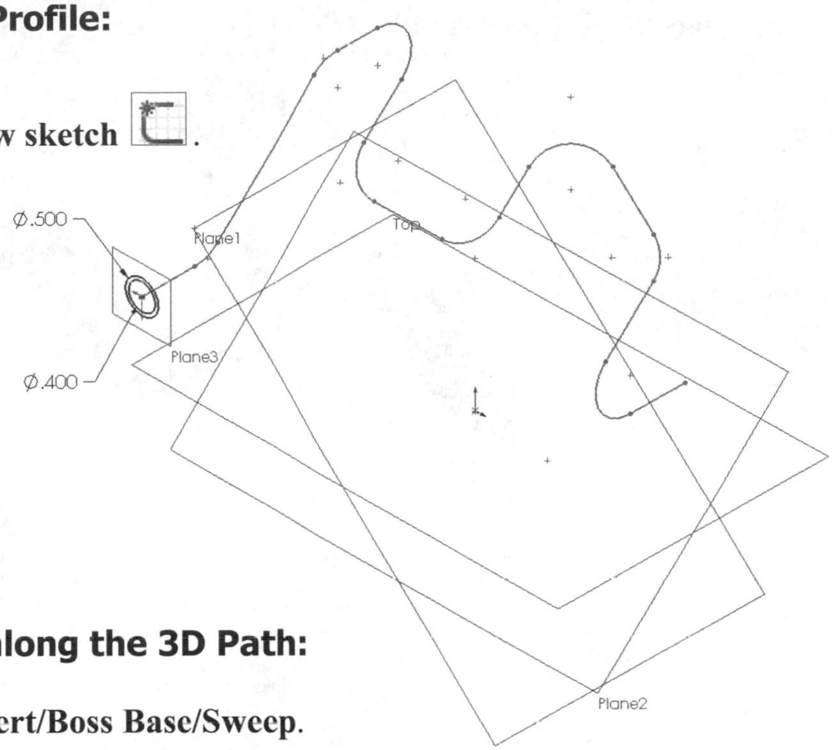

7. Sweeping the Profile along the 3D Path:

Click or Select **Insert/Boss Base/Sweep**.

Select the **Circles** as the Sweep Profile .

Select the **3D Sketch** as the Sweep Path .

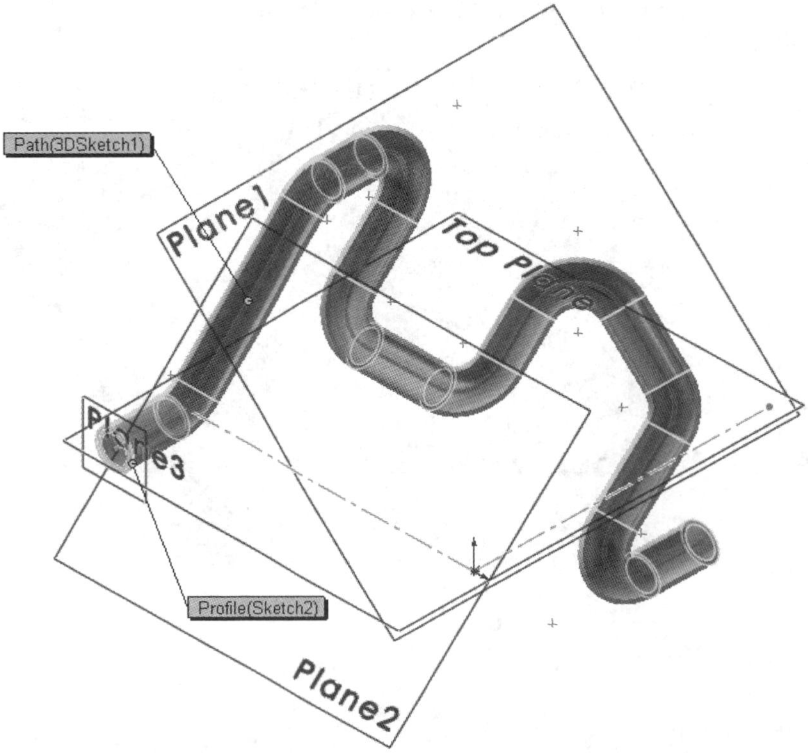

Click **OK**.

The resulting Swept feature.

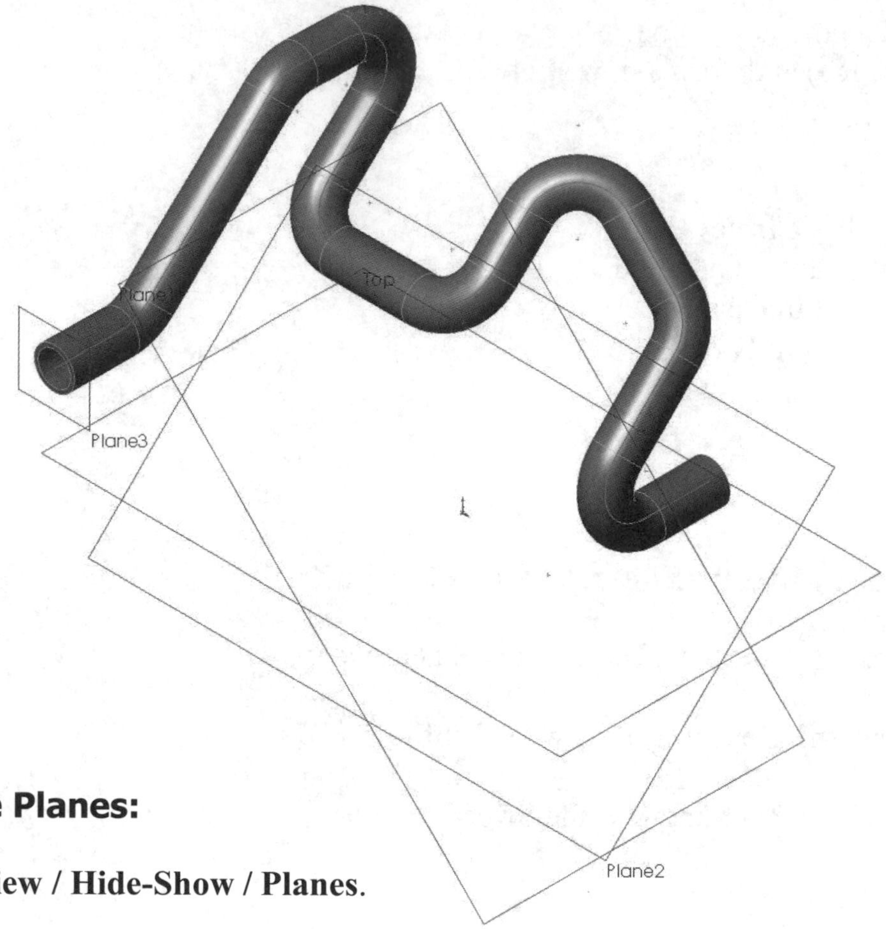

8. Hiding the Planes:

Select **View / Hide-Show / Planes**.

The planes are temporarily put away from the scene.

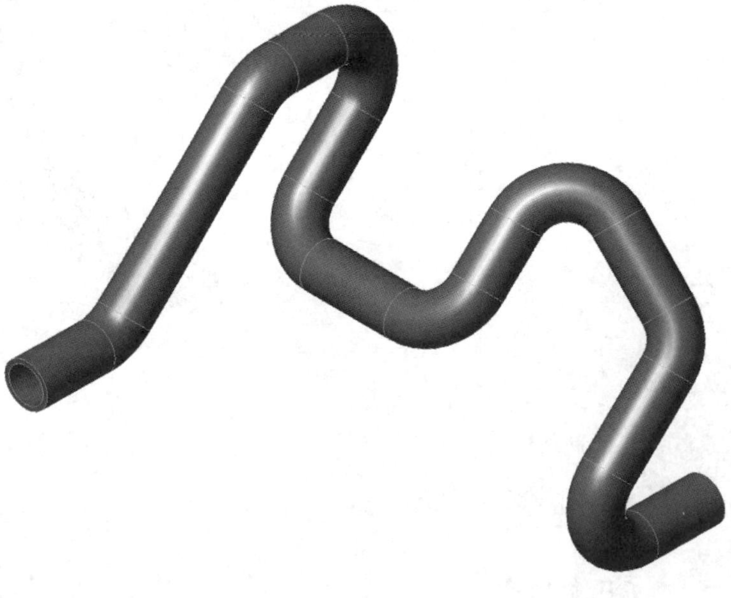

9. Saving your work:

Select **File / Save As**.

Enter **3D Sketch_Planes** for the name of the file.

Click **Save**.

Exercise: 3D Sketch & Composite Curve

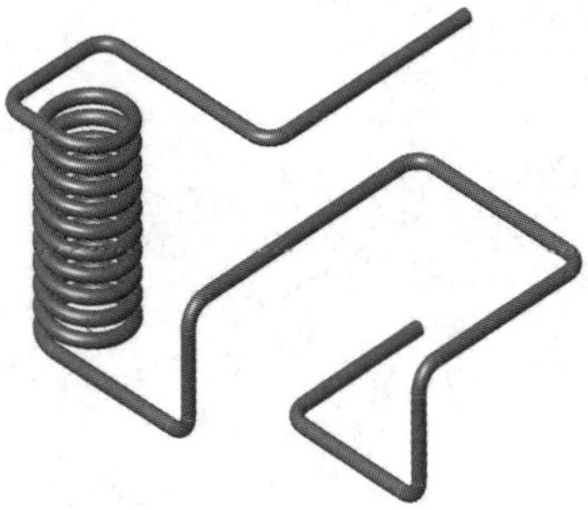

A 3D sketch normally consists of lines and arcs in series and Splines. You can use a 3D sketch as a sweep path, as a guide curve for a loft or sweep, a centerline for a loft, or as one of the key entities in a routing system.

The following exercise demonstrates how several 3D Sketches can be created, combined into 1 continuous Composite Curve, and used as a Sweep Path.

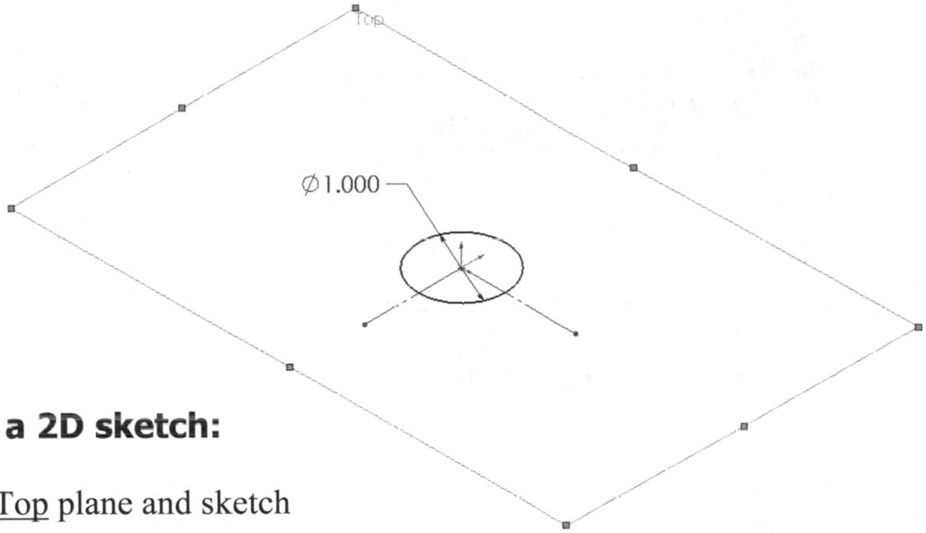

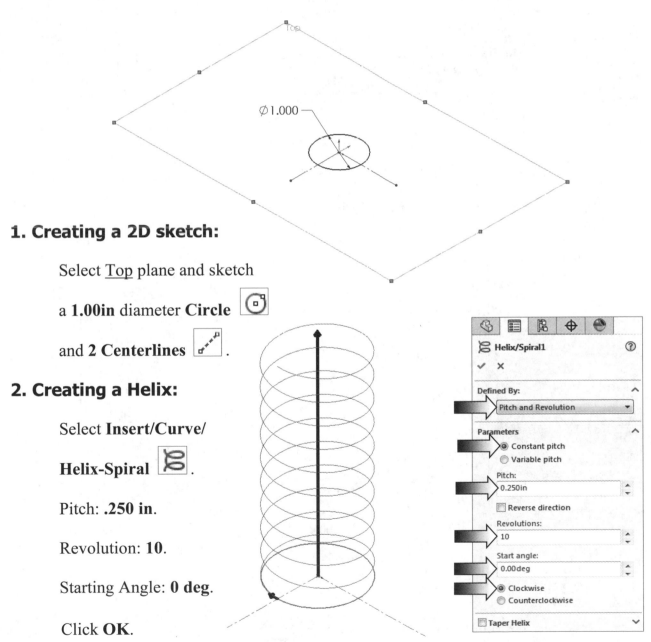

1. Creating a 2D sketch:

Select Top plane and sketch

a **1.00in** diameter **Circle**

and **2 Centerlines**.

2. Creating a Helix:

Select **Insert/Curve/**

Helix-Spiral.

Pitch: **.250 in**.

Revolution: **10**.

Starting Angle: **0 deg**.

Click **OK**.

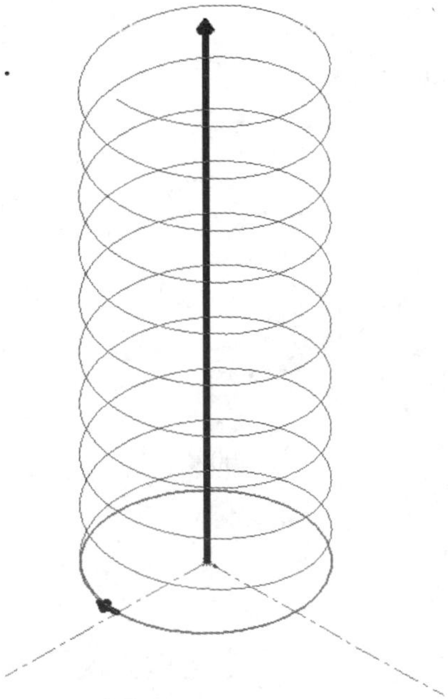

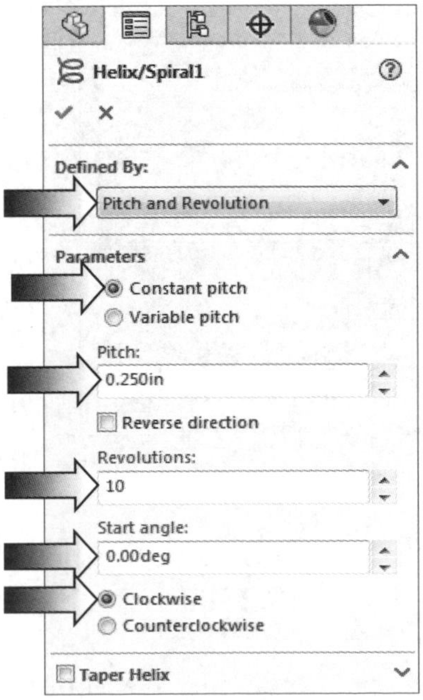

3. Creating the 1st 3D sketch:

Select **Insert/3D Sketch**.

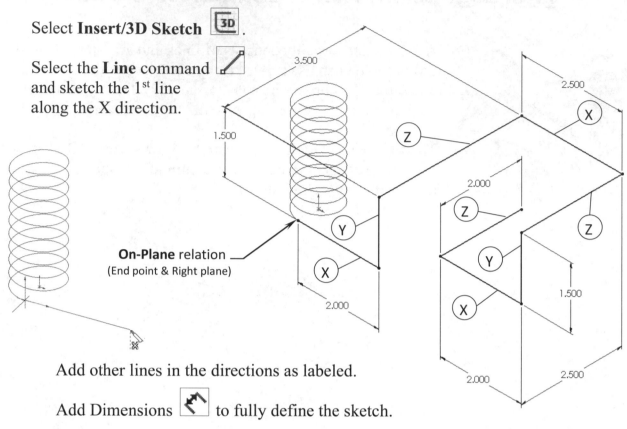

Select the **Line** command and sketch the 1st line along the X direction.

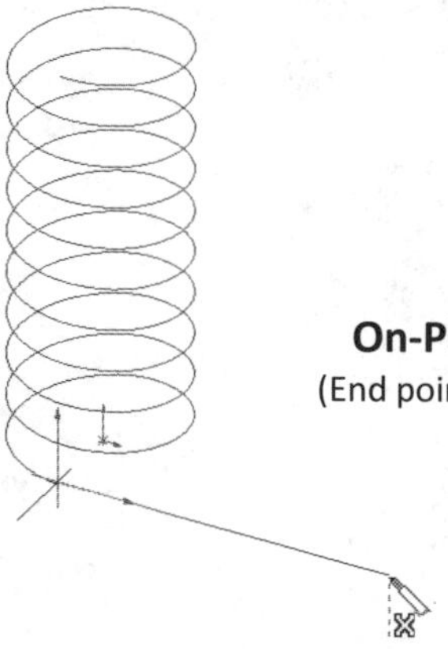

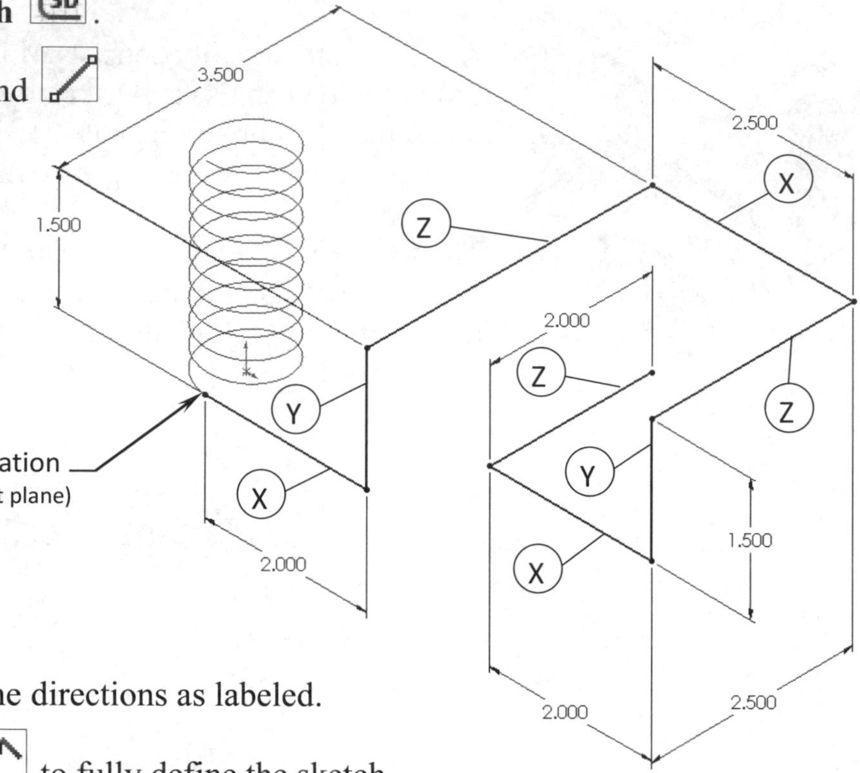

Add other lines in the directions as labeled.

Add Dimensions to fully define the sketch.

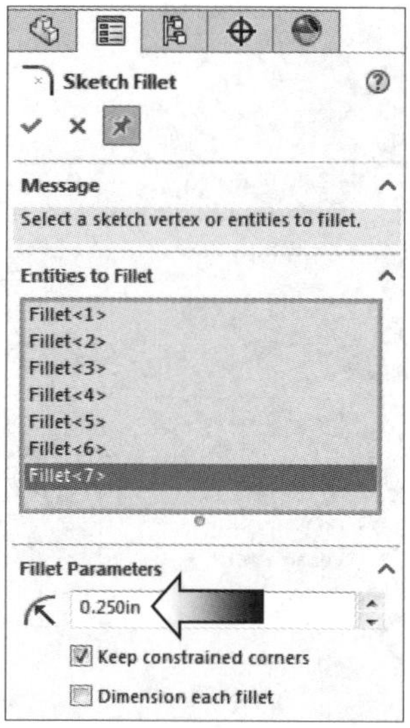

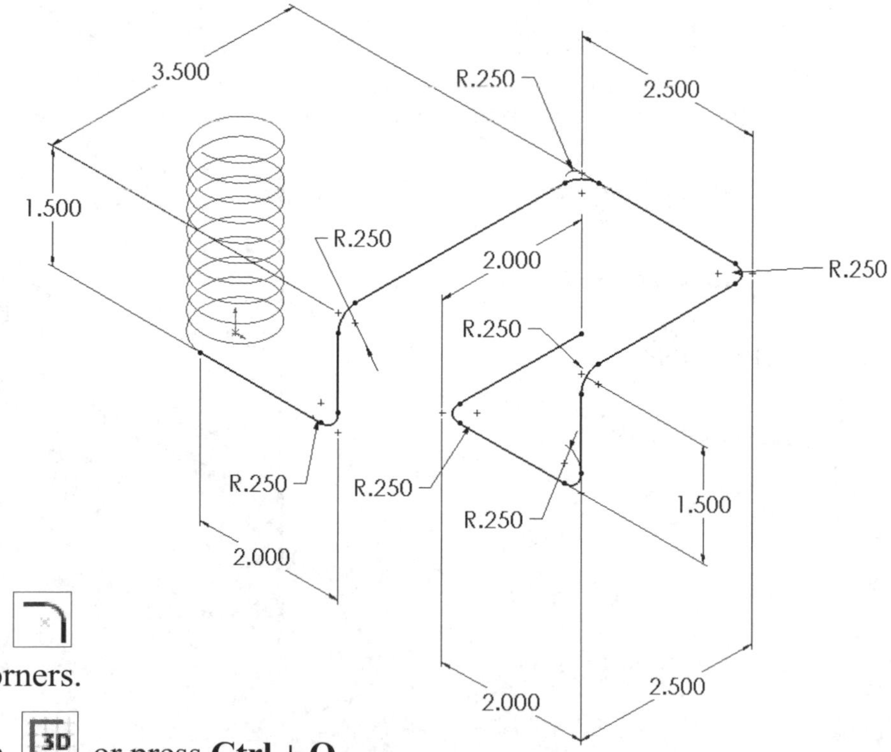

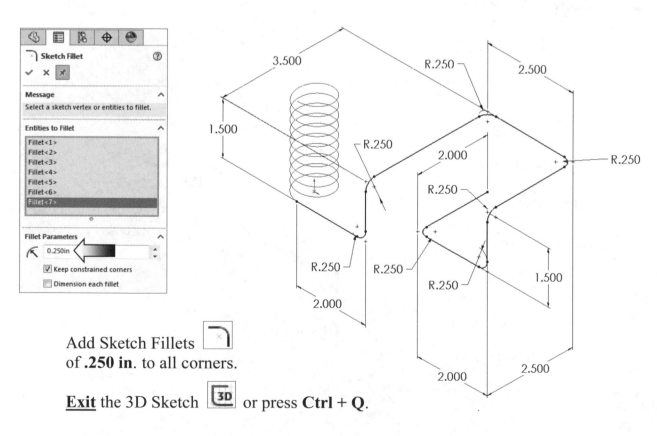

Add Sketch Fillets of **.250 in**. to all corners.

Exit the 3D Sketch or press **Ctrl + Q**.

4. Creating the 2nd 3D sketch:

Select **Insert/3D Sketch** .

Select the **Line** command and sketch the 1st line along the X direction.

Sketch the rest of the lines following their direction shown below.

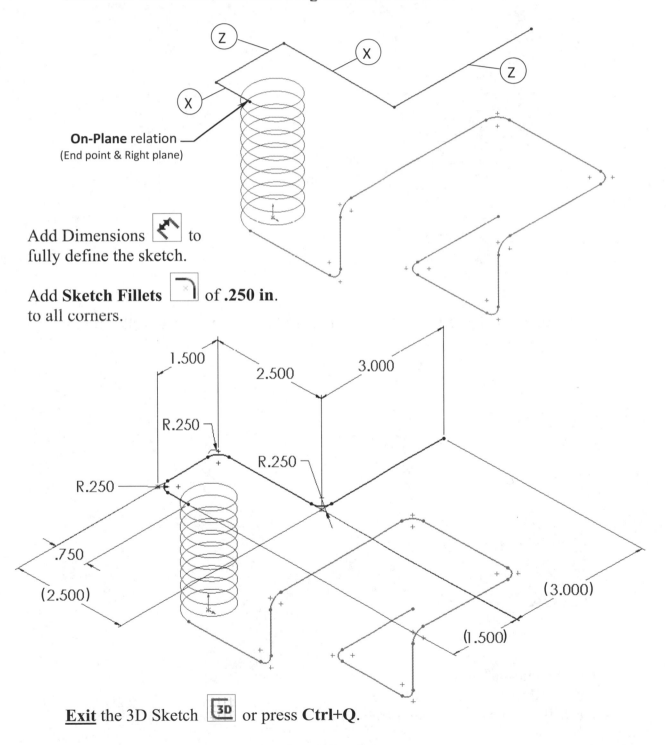

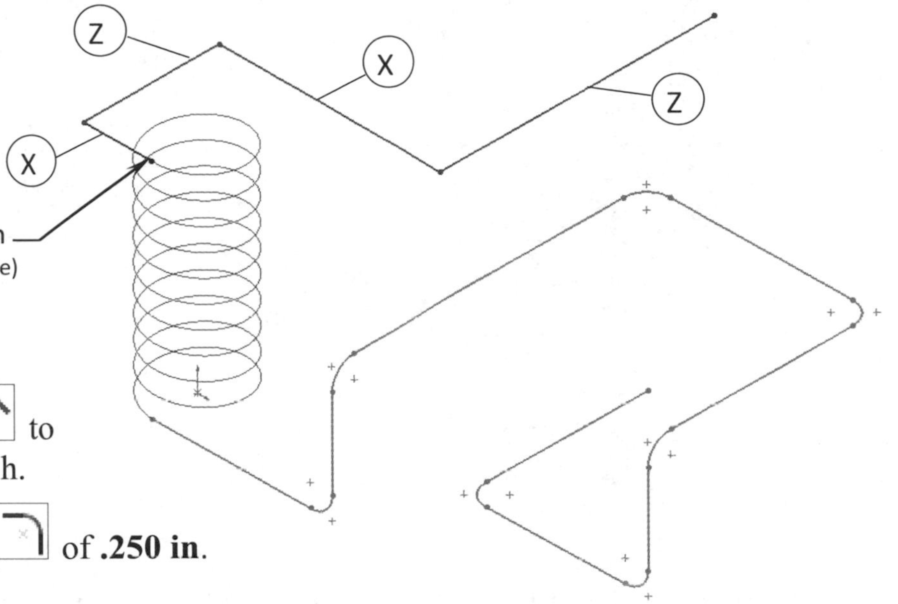

Add Dimensions to fully define the sketch.

Add **Sketch Fillets** of **.250 in.** to all corners.

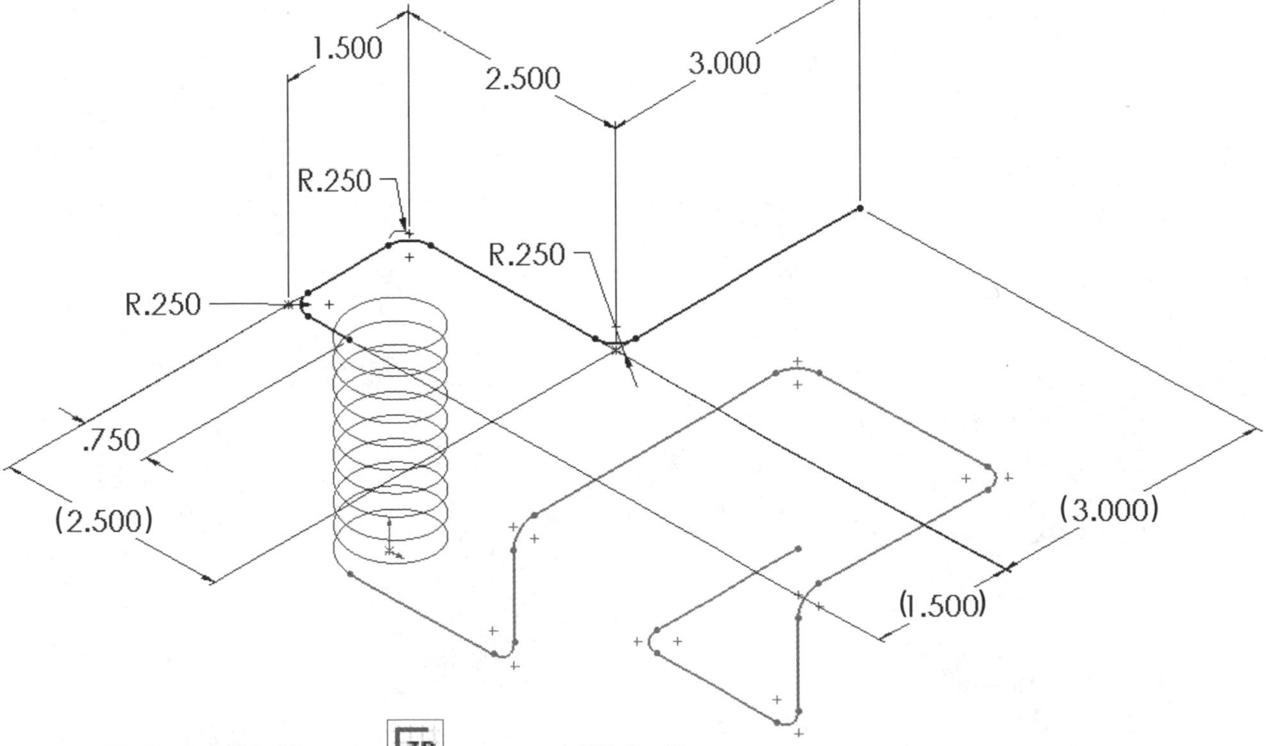

<u>Exit</u> the 3D Sketch or press **Ctrl+Q**.

5. Combining the curves:

Select the **Composite Curve** command below the Curves button, or select: **Insert / Curve / Composite**.

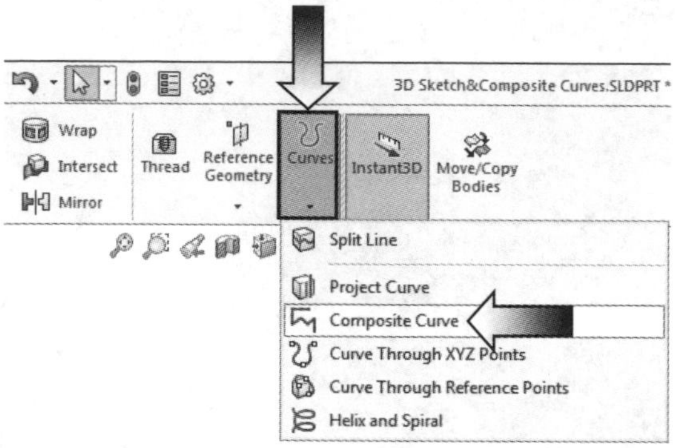

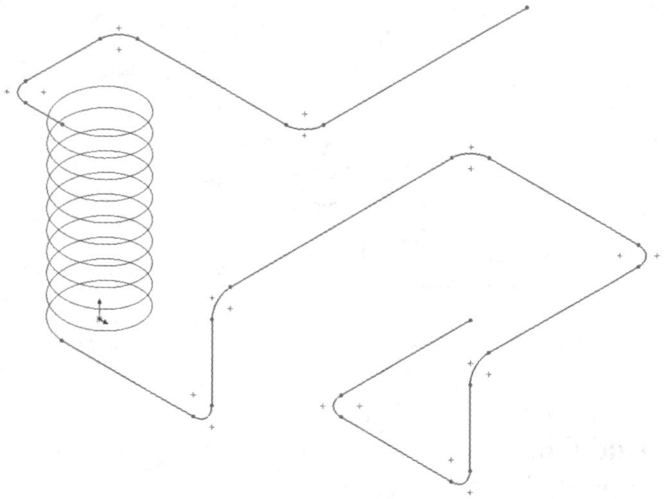

Select the 3 Sketches either from the FeatureManager tree or directly from the graphics area.

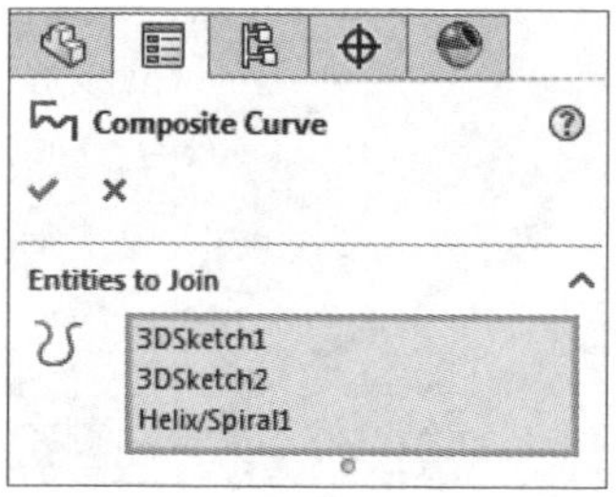

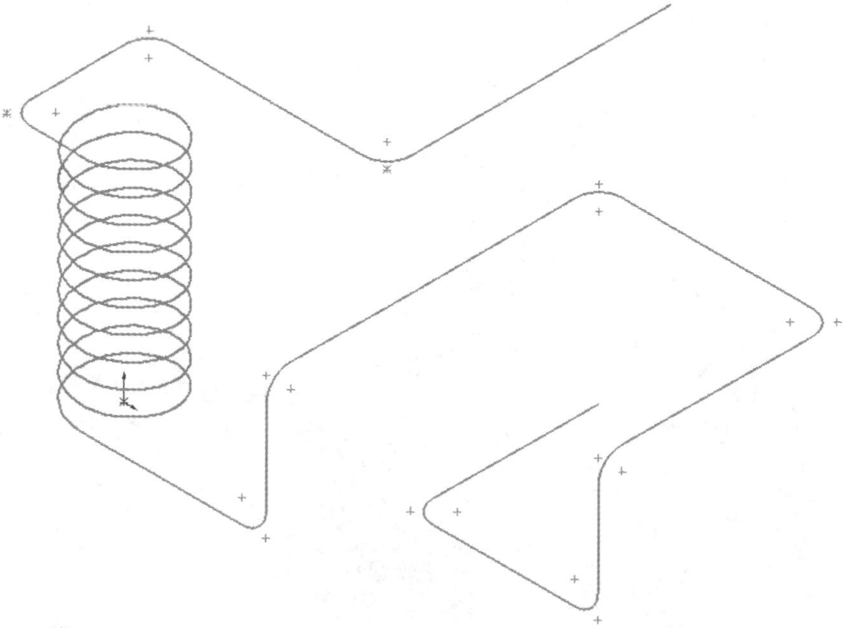

Click **OK**.

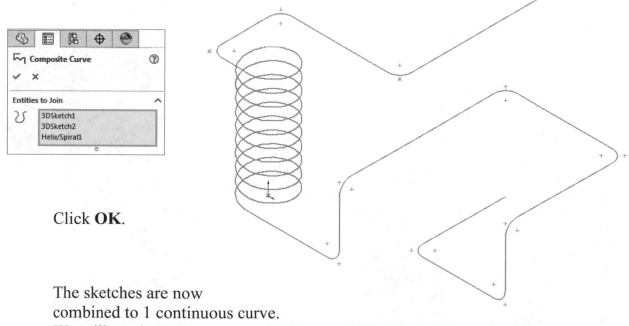

The sketches are now combined to 1 continuous curve.
We will use it as the sweep path in the next few steps.

6. Creating a Sweep using Circular Profile:

Select **Insert/Boss Base/ Sweep** .

Select the **Circle Profile** option (arrow).

Enter **.165 in** for the diameter of the sweep profile .

Select the **Composite Curve** as the Sweep Path .

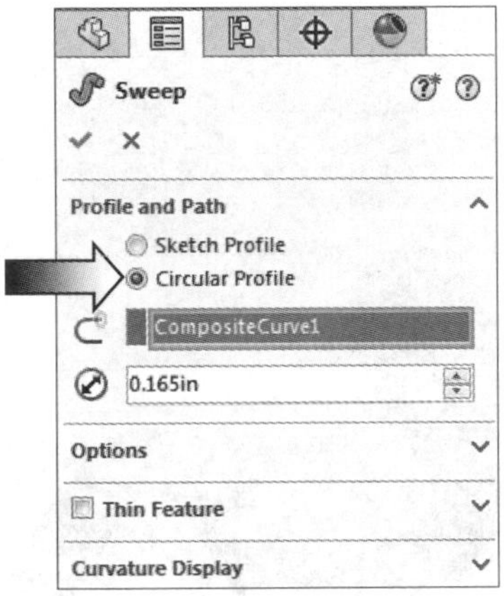

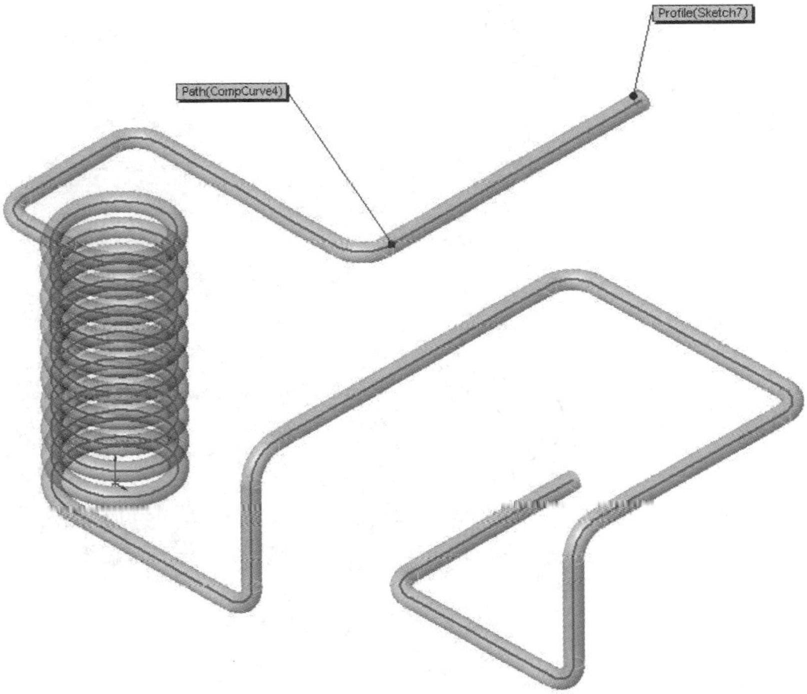

Click **OK**.

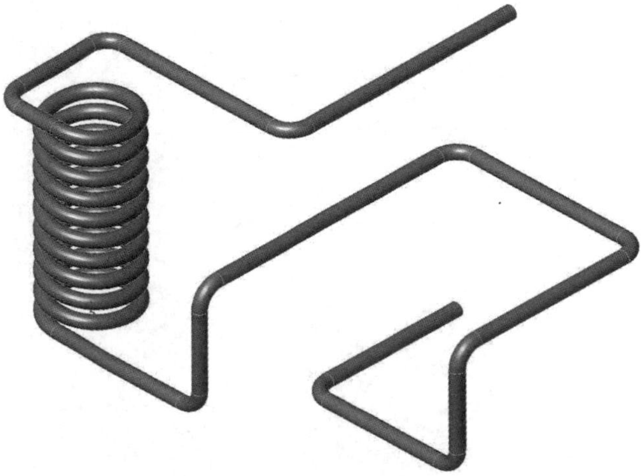

7. Saving your work:

Click **File/Save As**.

Enter **3D Sketch_ Composite Curve** for the name of the file.

Click **Save**.

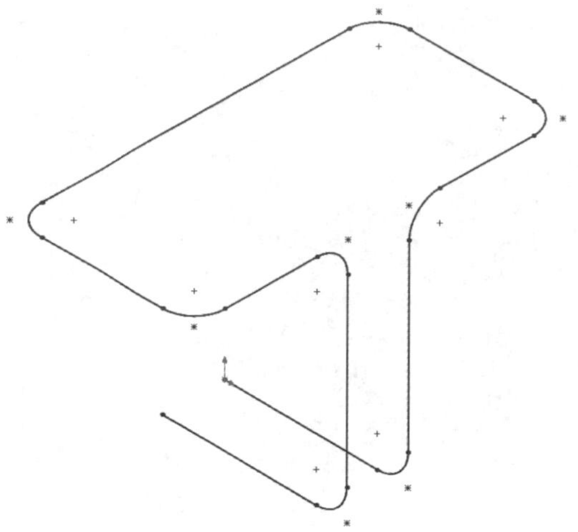

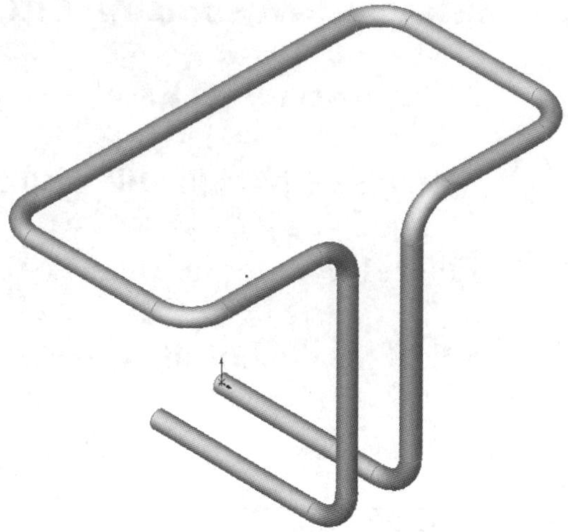

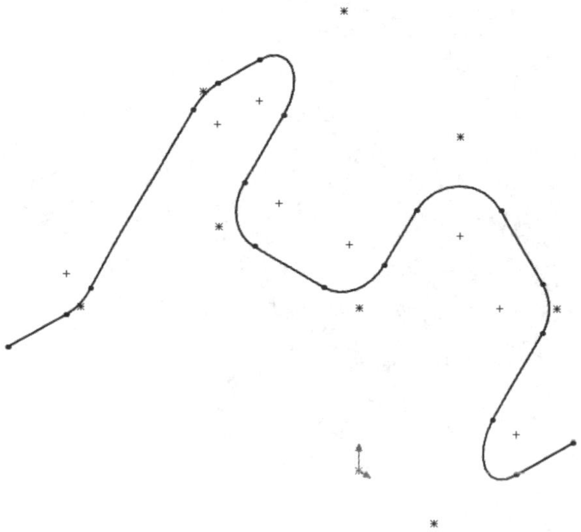

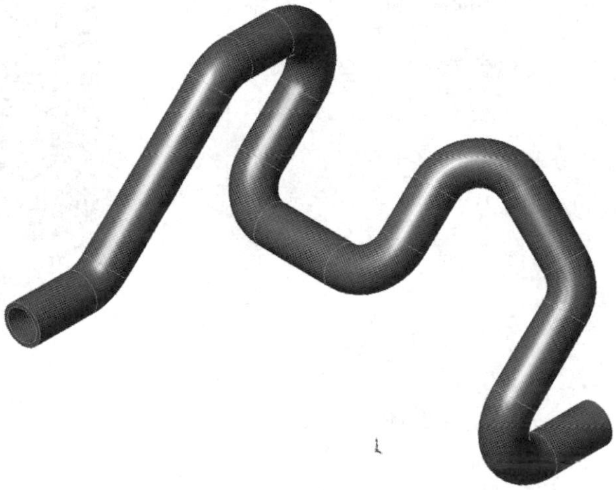

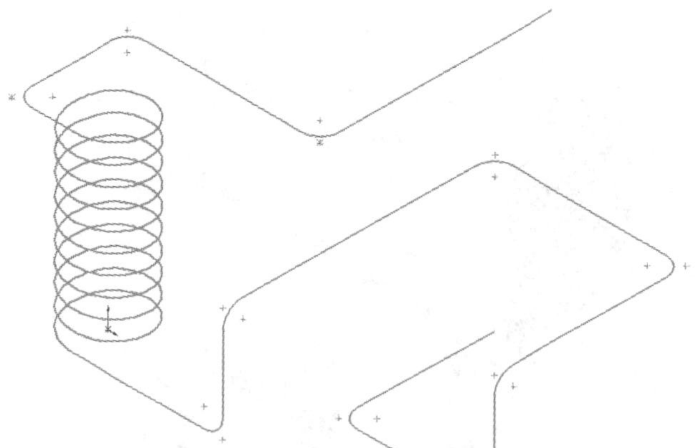

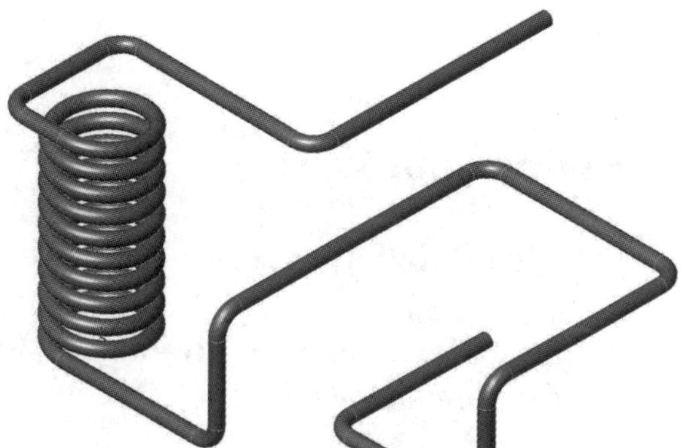

CHAPTER 2

Plane Creation

Planes

Advanced Topics

In SOLIDWORKS, planes are not only used to sketch geometry, but also used to create section views of a model or an assembly.

Planes are also used as end conditions for feature extrusion and as neutral planes to define the draft angles, etc.

There are several options to create planes:

- Parallel Plane.
- At Angle Plane.
- Perpendicular Plane.
- Offset Distance Plane.
- Coincident Plane.
- Mid Plane.
- Project Plane.

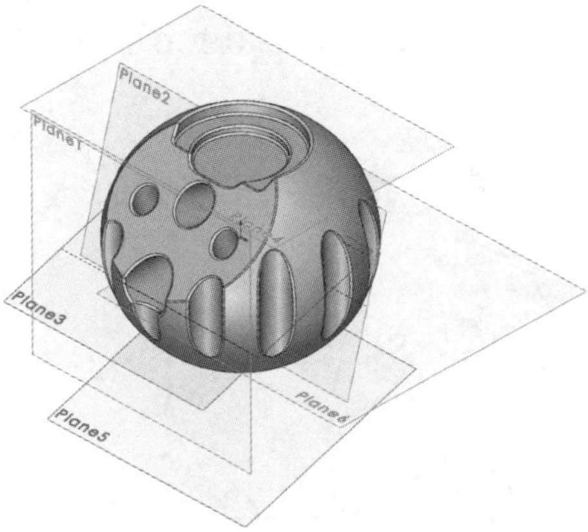

Each plane requires slightly different types of references; some of them may require only one and some others may require two or three.

This chapter discusses how planes are created using the sketch geometry and other features that are available in the model as references.

Advanced Topics
Plane Creation

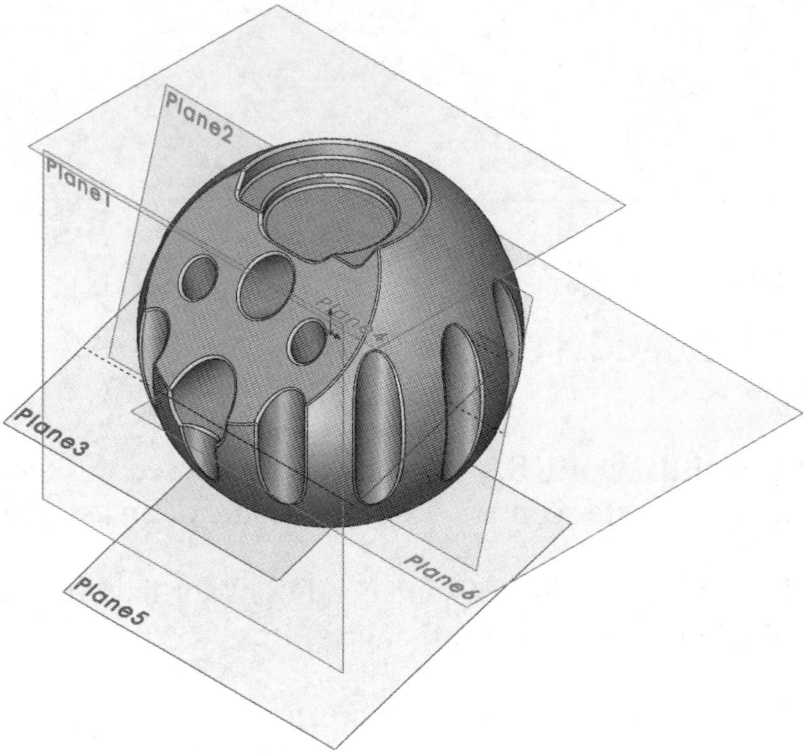

View Orientation Hot Keys:

Ctrl + 1 = Front View
Ctrl + 2 = Back View
Ctrl + 3 = Left View
Ctrl + 4 = Right View
Ctrl + 5 = Top View
Ctrl + 6 = Bottom View
Ctrl + 7 = Isometric View
Ctrl + 8 = Normal To Selection

Dimensioning Standards: **ANSI**

Units: **INCHES** – 3 Dccimals

Tools Needed:

Insert Sketch	Rectangle	Circle
Planes	Add Geometric Relations	Dimension
Sketch Mirror	Offset Entities	Boss/Base Revolve
Circular Pattern	Extruded Cut	Fillet/Round

1. Starting with a new Part document:

Select **File / New / Part** and click **OK**.

Select the Front plane from the FeatureManager tree.

Click  or select **Insert / Sketch**.

Sketch the profile below and add dimensions as shown.
(It may be easier to sketch a circle, instead of a center-point arc, add the 2 centerlines, and then trim away the bottom half of the circle.)

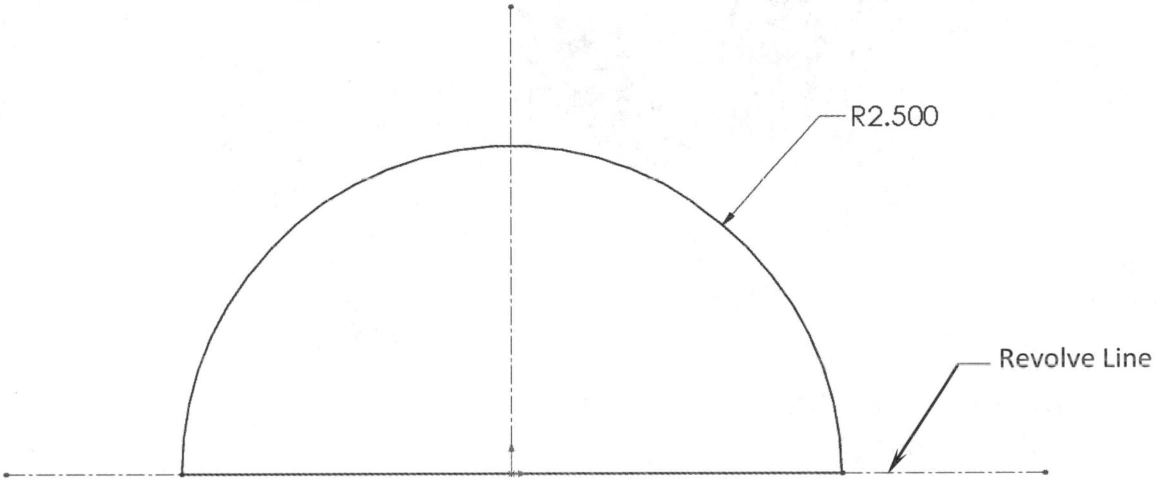

2. Revolving the Base:

Click  or select **Insert / Boss Base / Revolve**.

Set Revolve Type to: **Blind**.

Set Revolve Angle to **360 deg**.

Click **OK**.

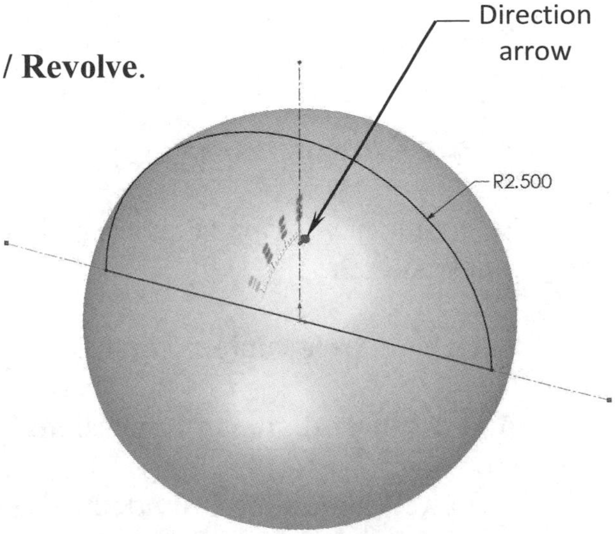

Note: Drag the Direction arrow to see the preview of the rotate angle.

3. Creating a Tangent plane: (Requires a cylindrical face and a parallel plane).

Click 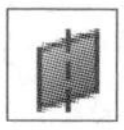 or select: **Insert / Reference Geometry / Plane.**

Click the **face** of the sphere.

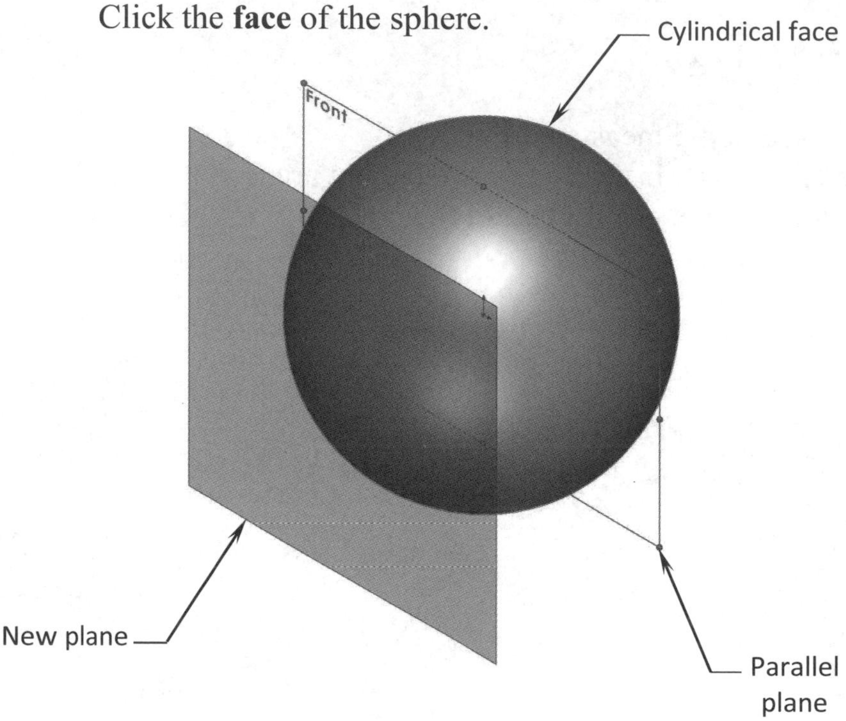

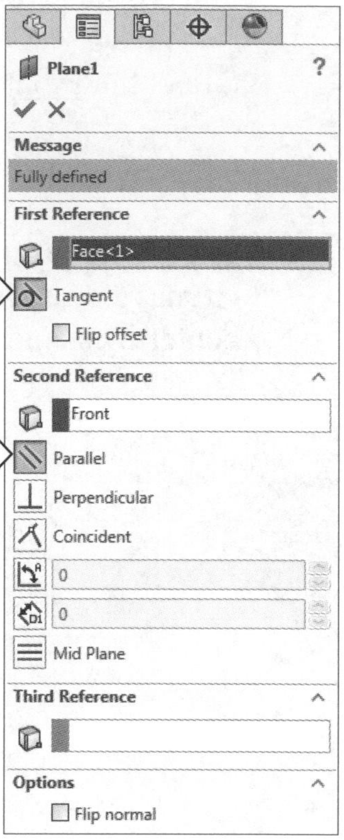

Expand the FeatureManager tree and select the **Front** plane.

The **Tangent** option is selected automatically.

Click the **Parallel** option in the Second Reference section.

Click **OK**.

4. Adding a Center hole:

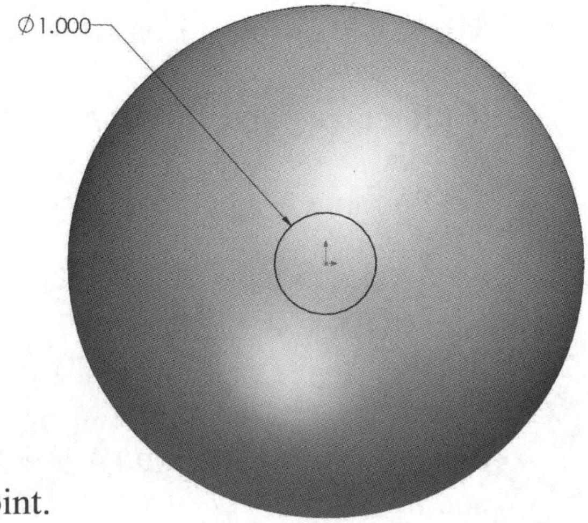

Select the <u>new plane</u> (Plane1) and open a **new sketch**.

Sketch a **Circle** centered on the origin.

Add a **1.000"** diameter dimension.

The circle should be fully defined at this point.

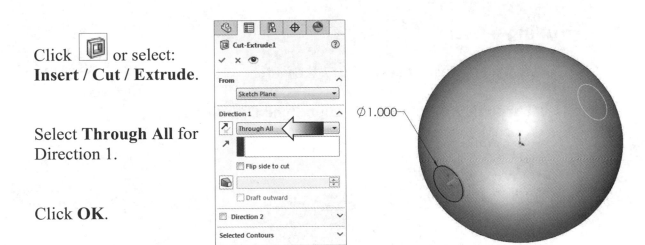

Click  or select:
Insert / Cut / Extrude.

Select **Through All** for Direction 1.

Click **OK**.

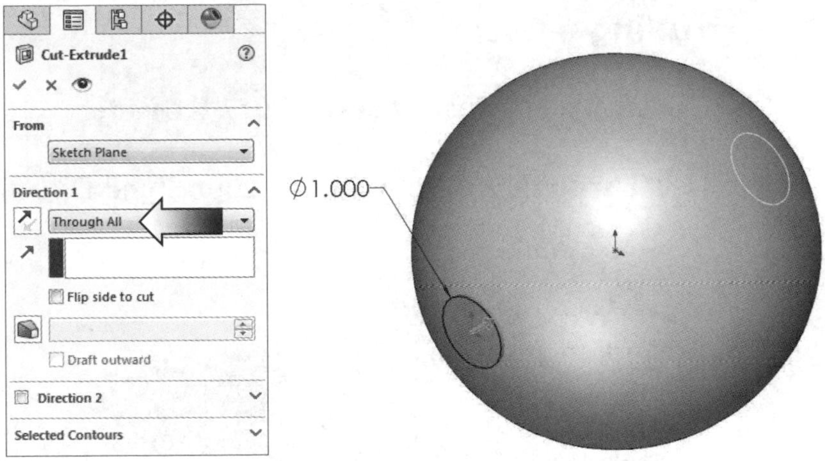

5. Creating a flat surface:

This step will demonstrate the use of geometric relations to fully define the sketch without using dimensions.

Select the Right plane from the FeatureManager tree.

Click  or select **Insert / Sketch** and change to the right view (Ctrl+4).

Sketch a **Line** and add the Relations  as shown.

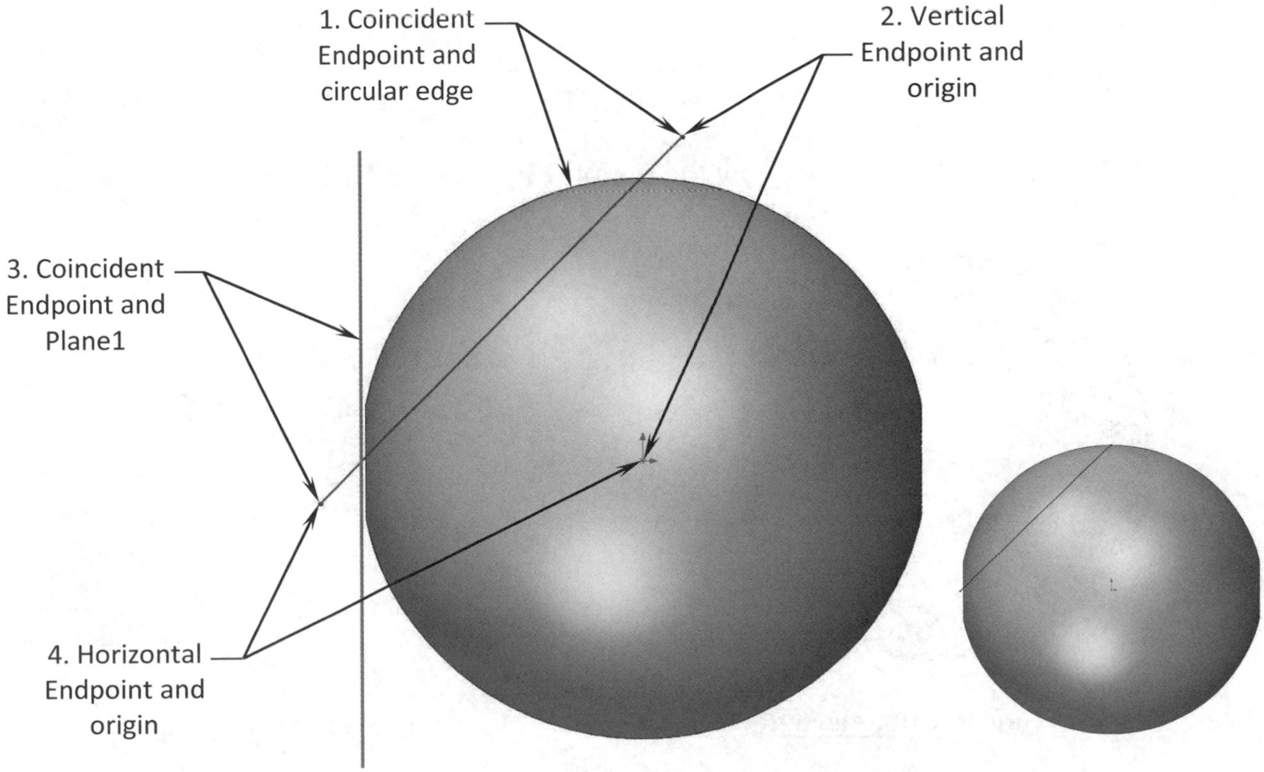

6. Extruding a Cut:

Click or select: **Insert / Cut / Extrude**.

Use **Through All** Both for Direction 1 and Direction 2.

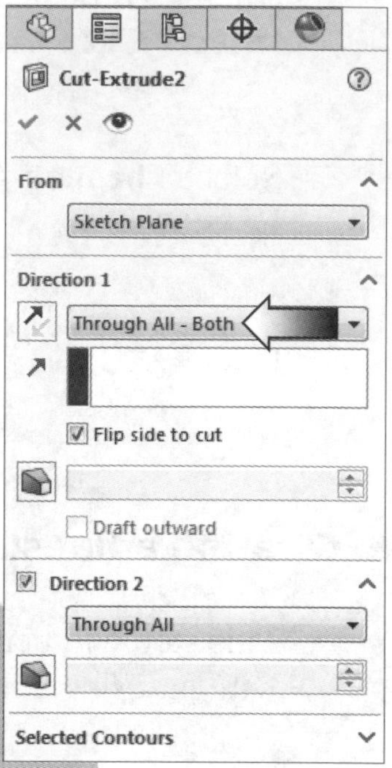

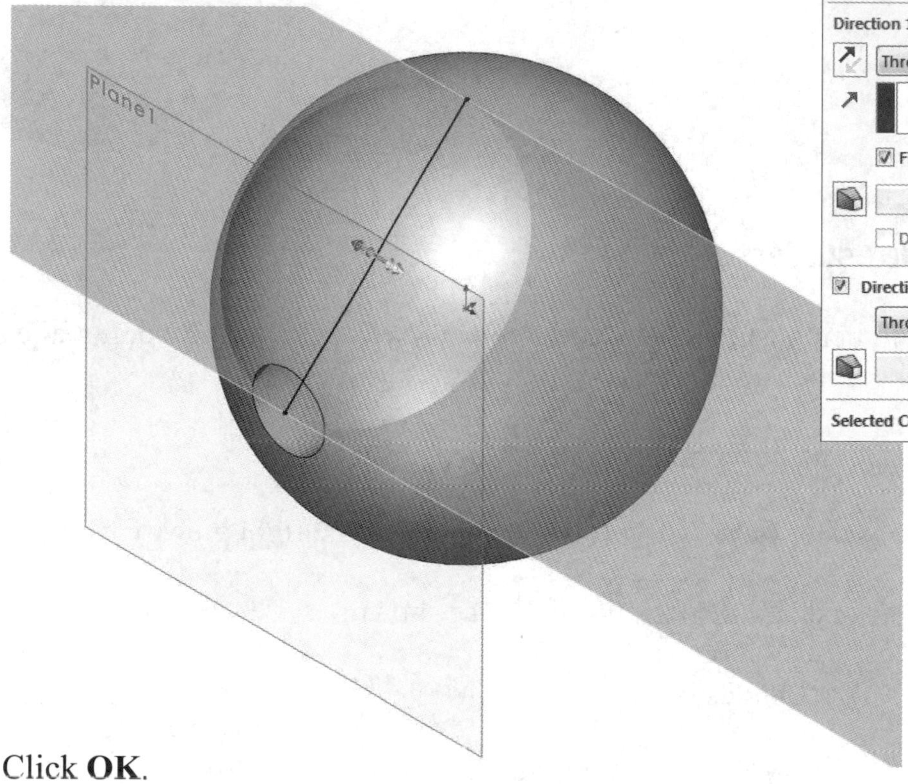

Click **OK**.

The examples below show how the option **Flip Side to Cut** works.

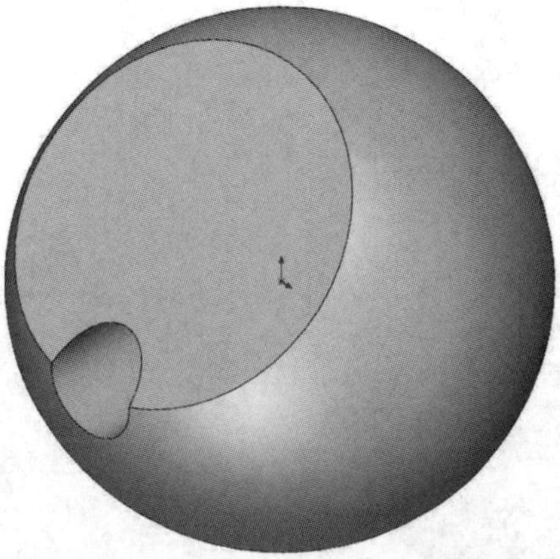

Flip Side to Cut **Selected**

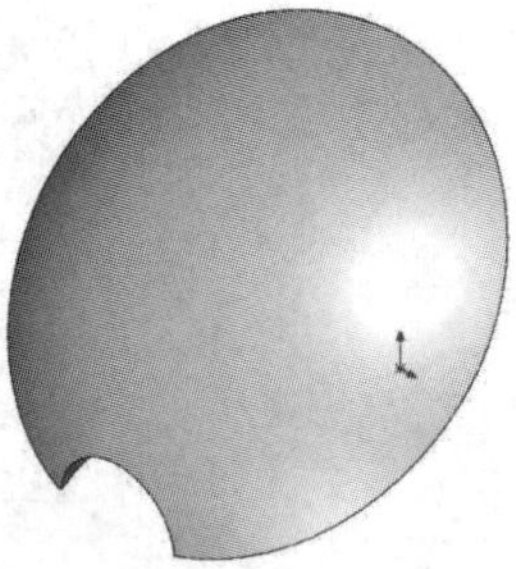

Flip Side to Cut **Cleared**

7. Creating an At-Angle plane: (Requires a Reference Plane, a Reference Axis, and an Angular Dimension).

Select the **Front** plane from FeatureManager tree.

Click  or select **Insert / Reference Geometry / Plane**.

Select the **horizontal centerline** as reference axis.

Select the **At Angle** option.

Enter **15 deg**. in the dialog box and click **Flip Offset**.

Click **OK**.

> **Show Sketch**
>
> Right-click on **Sketch1** (on the FeatureManager tree, below the **Revolve1**) and select **Show**.

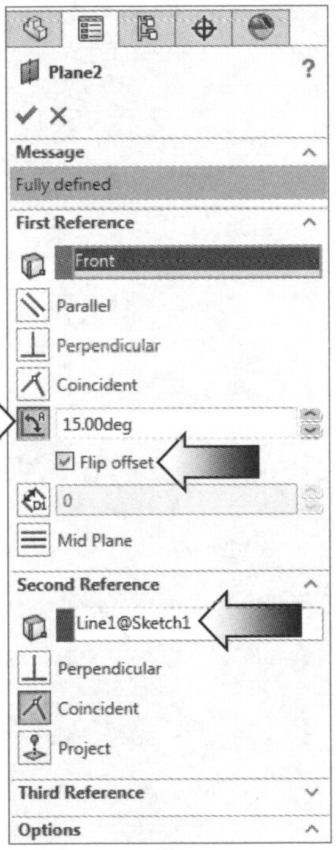

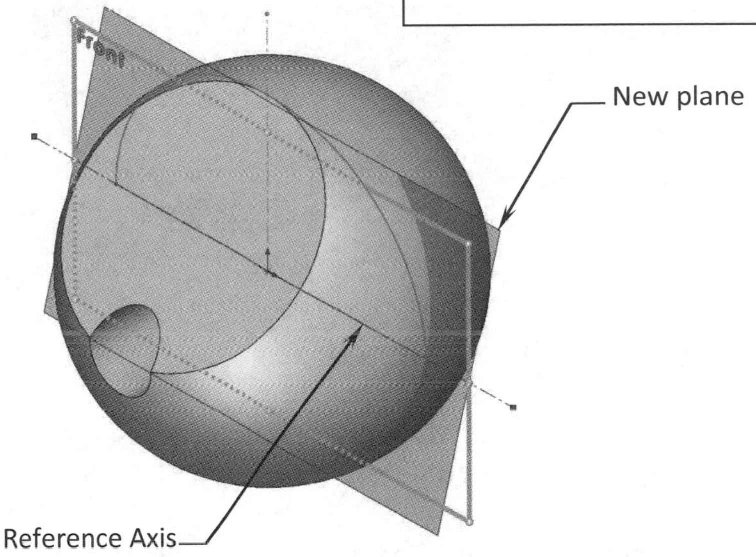

8. Creating a Ø.750 hole:

Select the new plane (Plane2).

Click 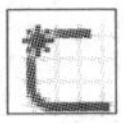 or select **Insert / Sketch**.

Sketch a **Circle**  and add the dimensions and a vertical relation as shown.

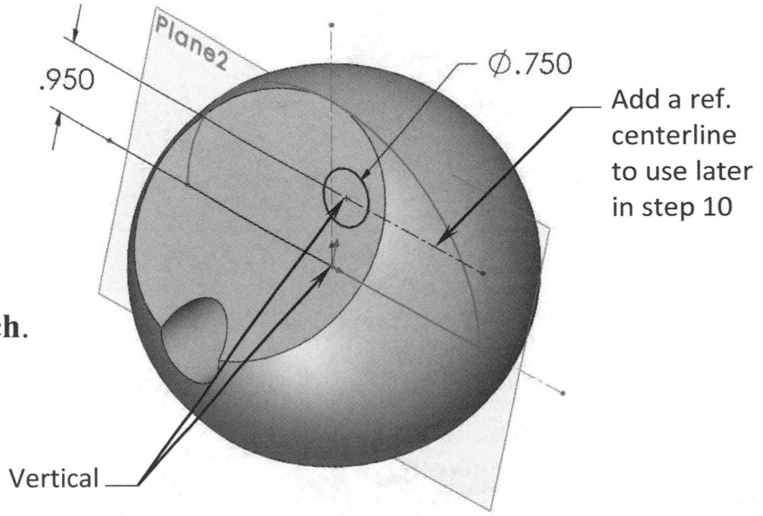

Click  or select **Insert / Cut / Extrude**.

Direction 1: **Through All Both**.

Click **OK**.

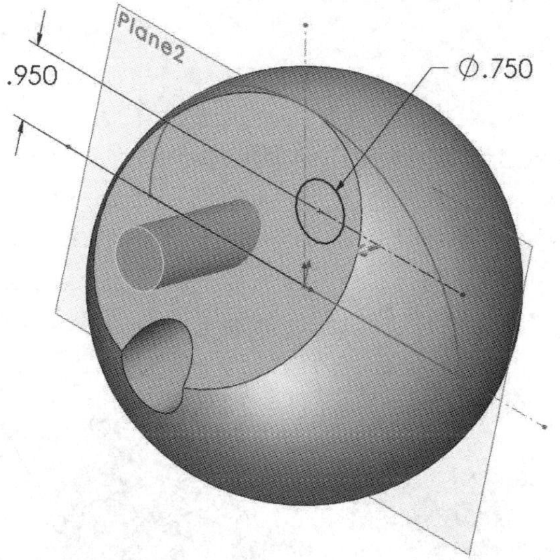

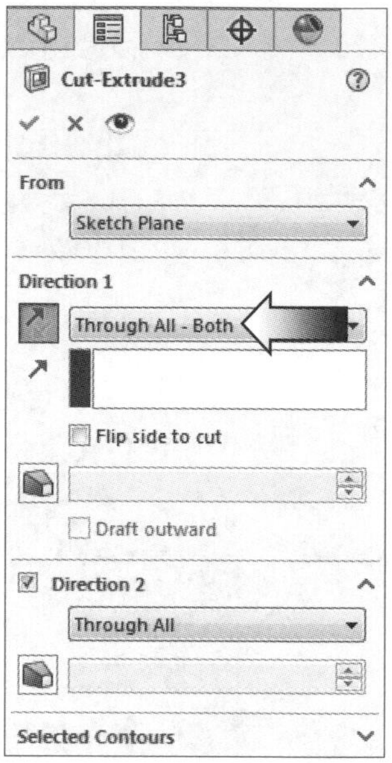

9. Showing the Sketches:

On the FeatureManager tree expand the **Cut-Extrude1** (click the + symbol), right-click **Sketch2** and select **Show**.

Expand the **Cut-Extrude2** (click the + symbol), right-click **Sketch3**, and select **Show** ; also **Hide** the **Sketch1**.

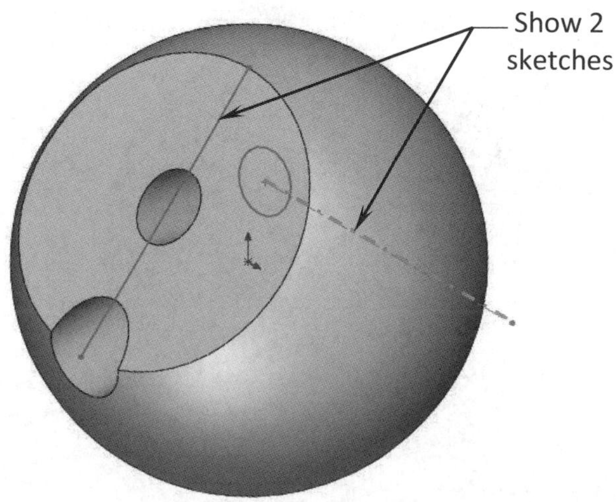

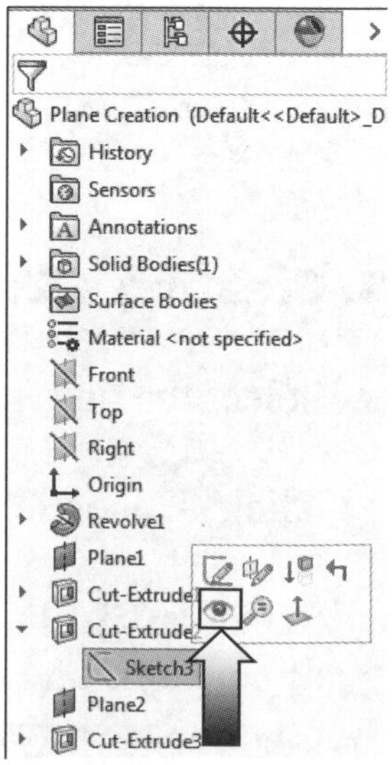

10. Creating a Coincident plane: (Requires a Reference Line and a Sketch Point or a Vertex).

Click or select **Insert / Reference Geometry / Plane**.

Select the **Centerline** and the **Endpoint** as indicated.

The **Coincident** option should be selected automatically.

Click **OK**.

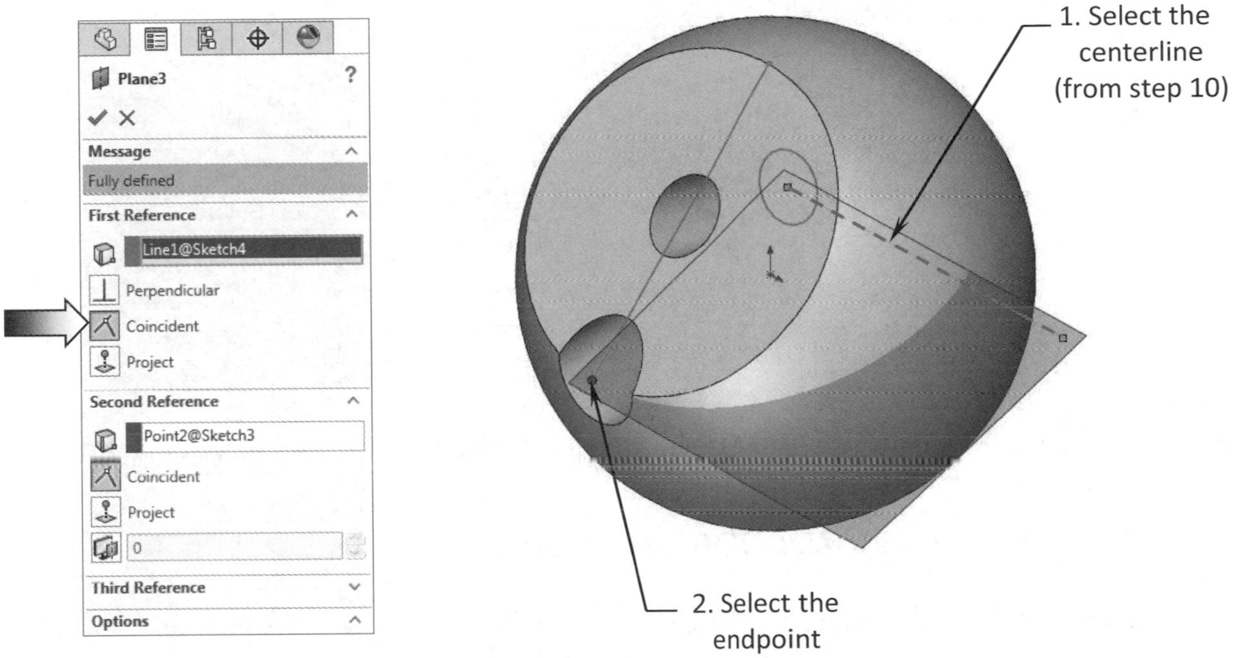

11. Creating the Ø.500 holes:

Select the new plane (Plane3).

Click or select **Insert / Sketch**.

Sketch a **Circle** and Mirror it (Use either Dynamic Mirror or Mirror Entity to mirror the circles.)

Add Dimensions as shown to fully define the sketch.

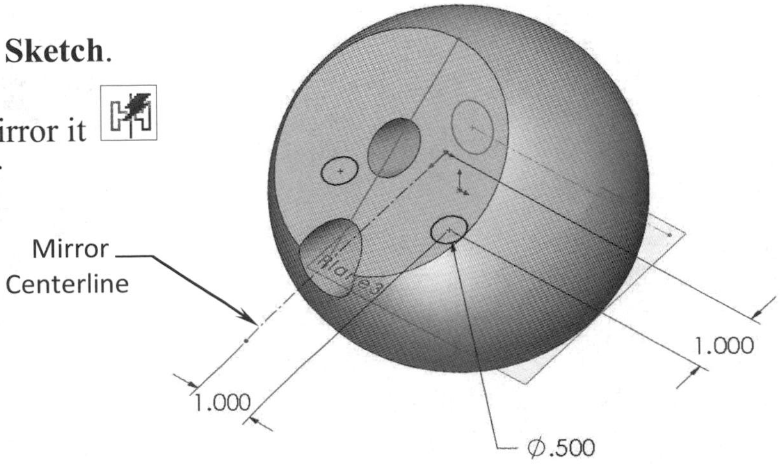

Click or select **Insert / Cut / Extrude**.

Direction 1: **Through All Both**.

Click **OK**.

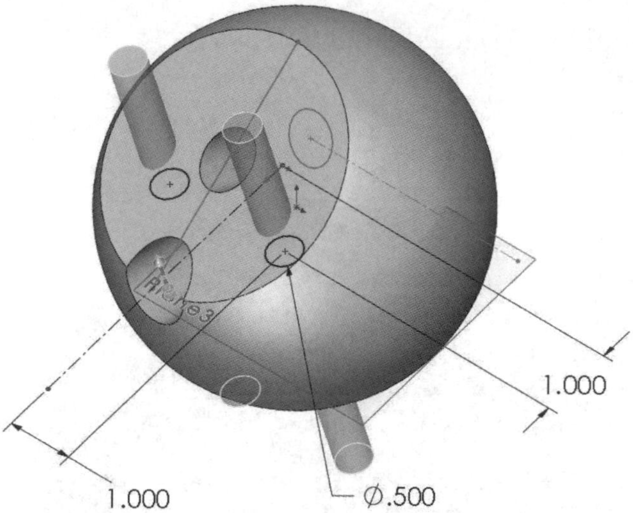

12. Creating a Parallel plane: (Requires a Reference Plane and Reference Point).

Click or select **Insert / Reference Geometry / Plane**.

Select the **Top** plane and the **Endpoint** as indicated.

Based on the selection, the system selects the **Parallel** and **Coincident** options.

Click **OK**.

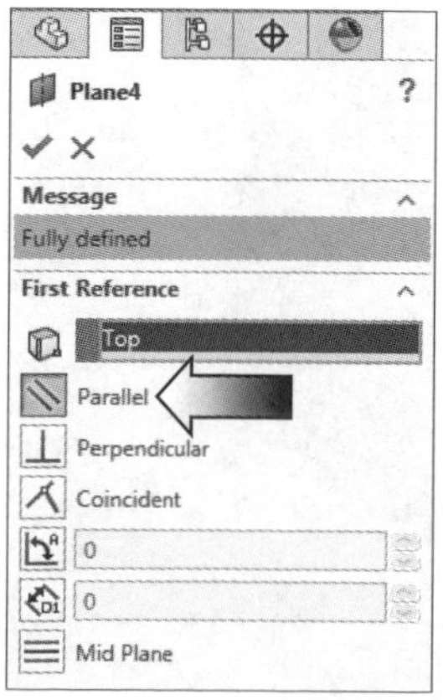

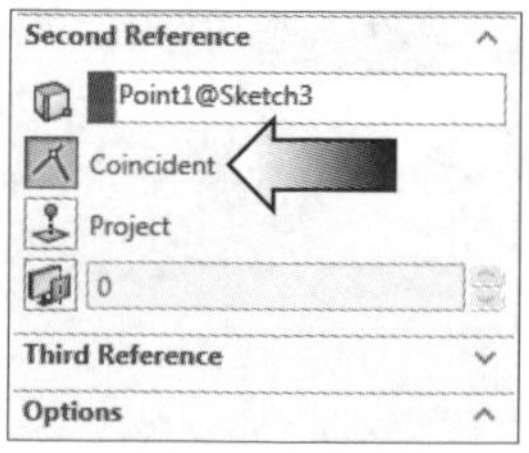

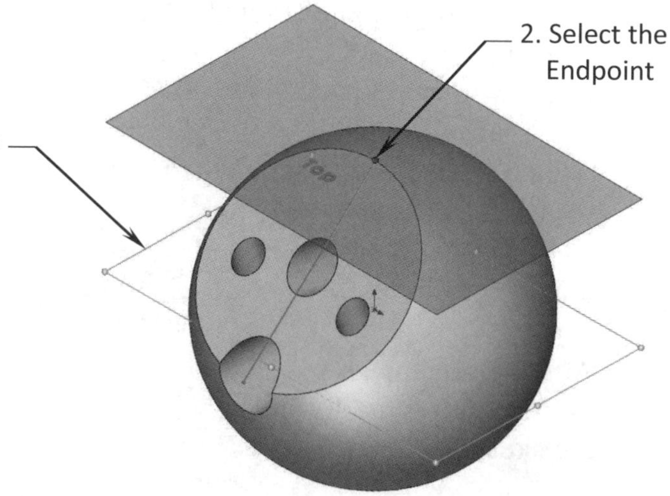

13. Creating the Ø2.500 Recess:

Select the new Plane (Plane4) and insert a **new sketch** 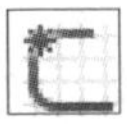.

Sketch a **Circle**  and add the diameter dimension to fully define the sketch.

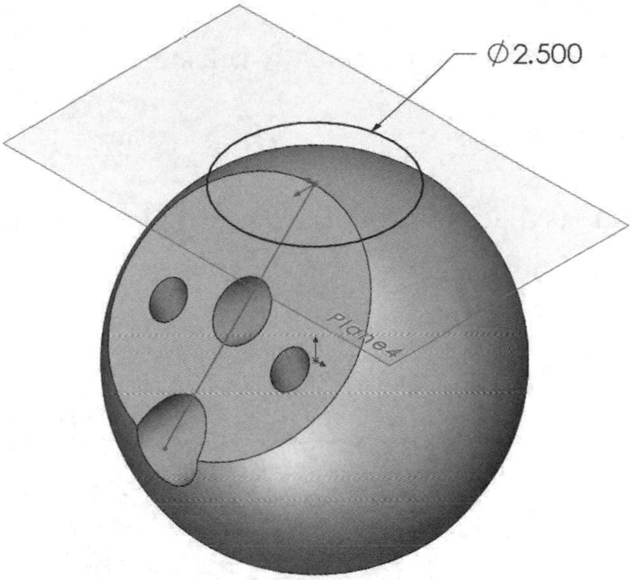

Click  or select **Insert / Cut / Extrude**.

End Condition: **Blind**.

Extrude Depth: **.625** in.

Click **OK**.

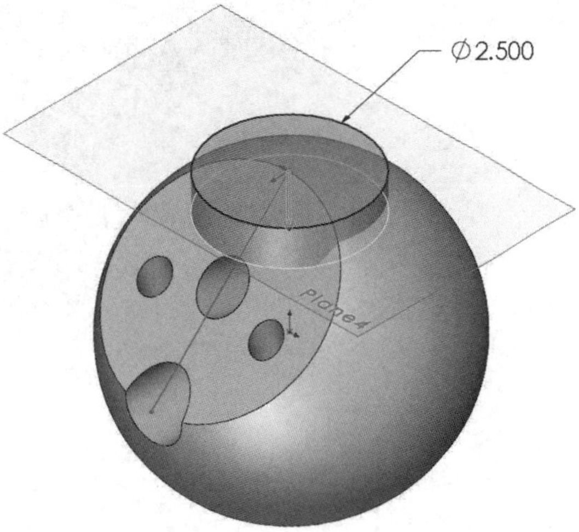

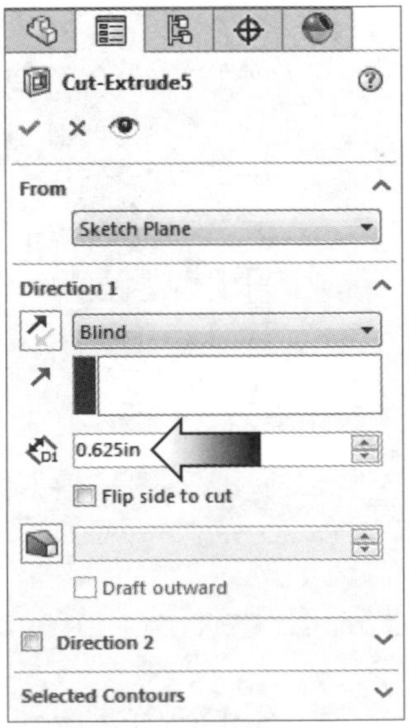

Hide  the Sketch2, Sketch3 and all planes.

14. Creating an Offset-Distance plane: (Requires a Reference Plane and a Distance dimension.)

Click or select **Insert / Reference Geometry / Plane**.

Select **Plane3** (from the FeatureManager tree) to offset from.

The **Offset Distance** option is automatically selected.

Enter **3.375** for offset value.

Ensure the new plane is placed below the Plane3 (click Flip Offset if needed).

Click **OK**.

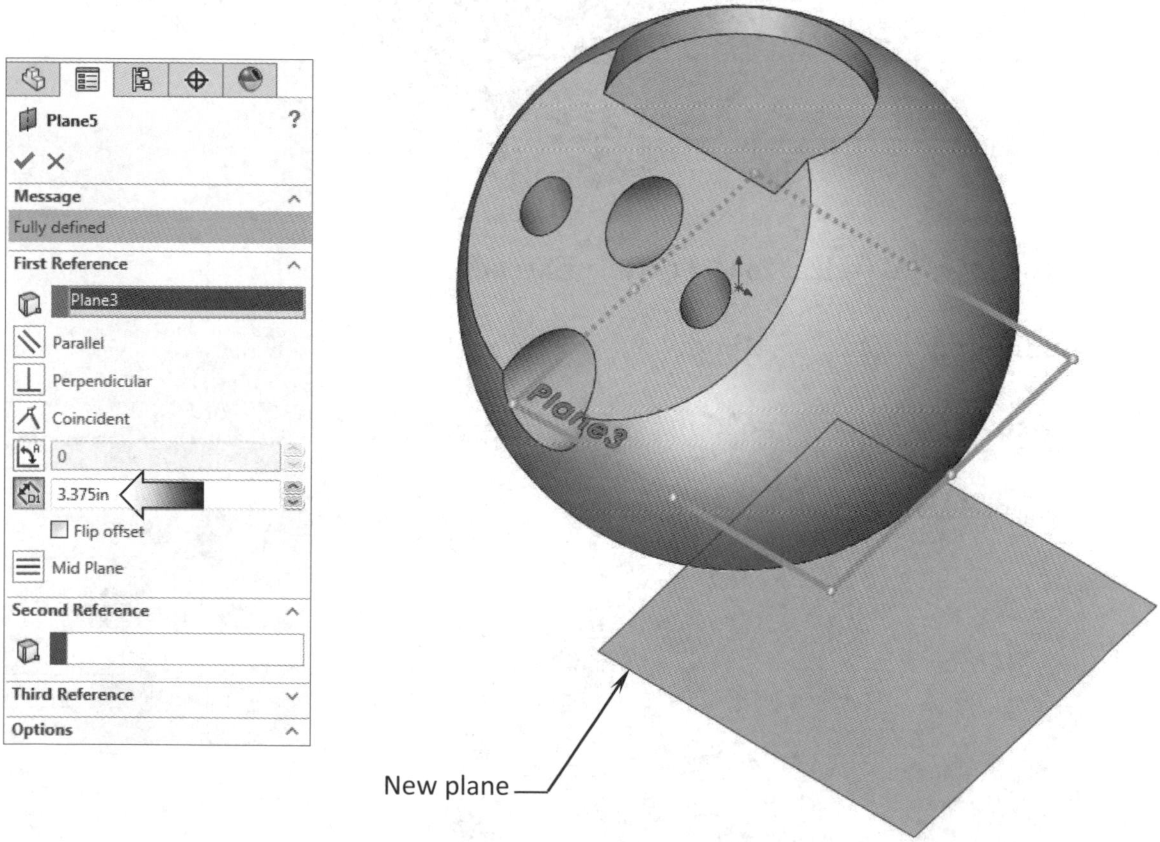

15. Creating the Bore holes:

Select the <u>new plane</u> (Plane5) and insert a **new sketch** .

Select the **circular edge** of the hole and press **Offset-Entities** .

Enter **.100 in**. for Offset Distance. (Only one offset can be done at a time, since the 2 circles are not connected to each other.)

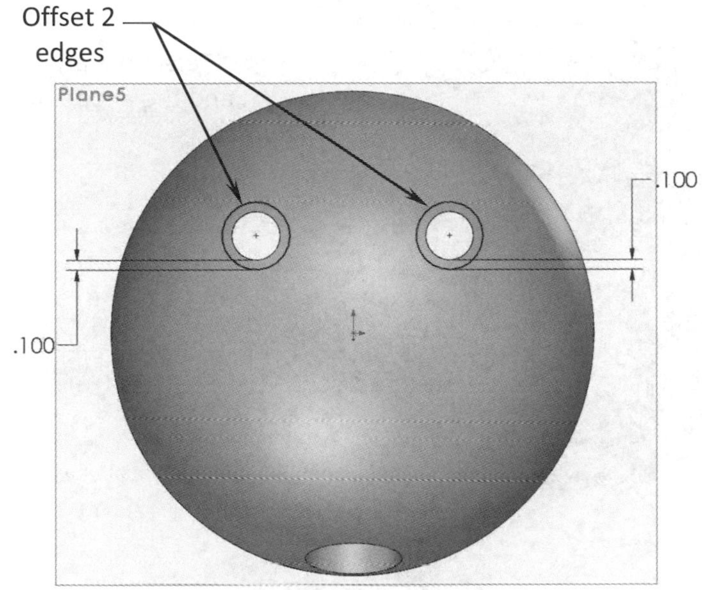

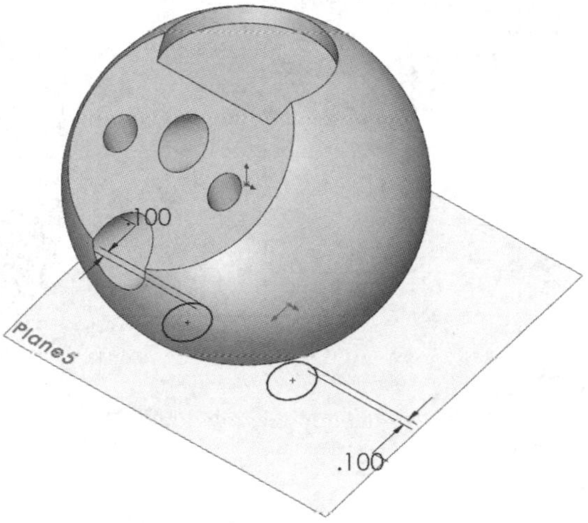

Click or select **Insert / Cut Extrude**.

End Condition: **Blind**.

Extrude Depth: **1.500** in.

Click **OK**.

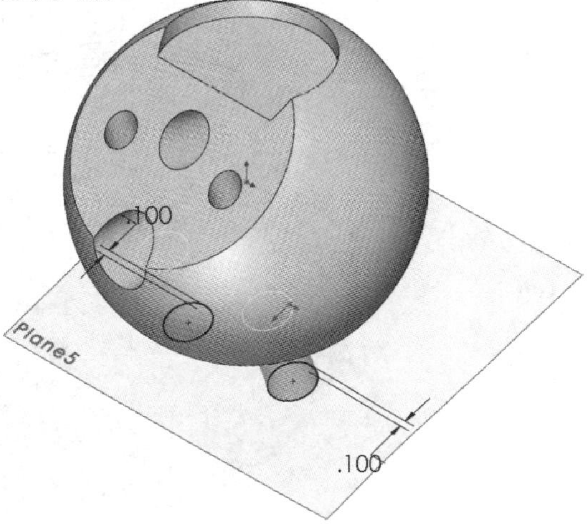

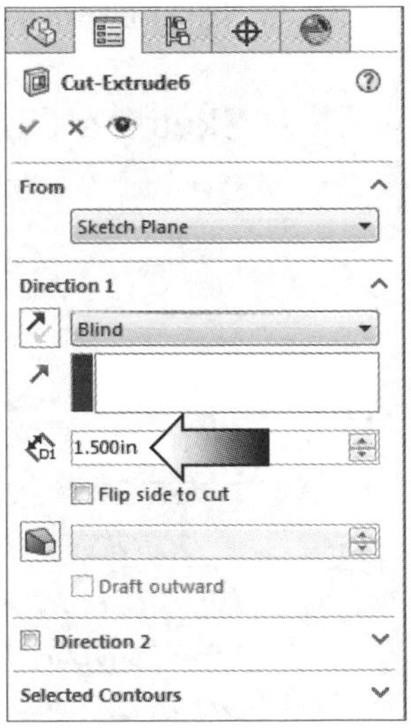

Hide **Plane5**.

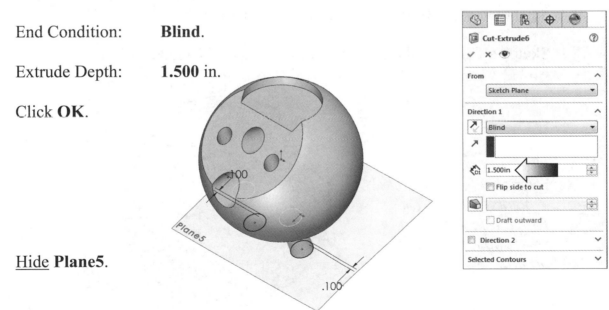

16. Creating a Perpendicular plane: (Requires a Reference Line or Curve & a Point).

Click or select **Insert / Reference Geometry / Plane**.

Show the **Sketch1** and select the **Arc** and the **Endpoint** as noted.

The **Perpendicular** and **Coincident** options should be selected automatically.

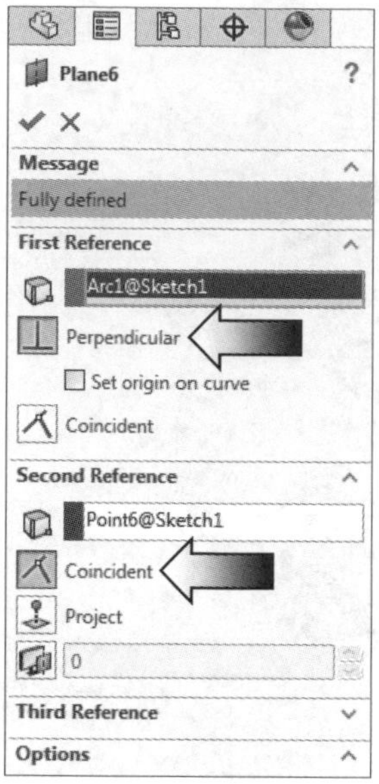

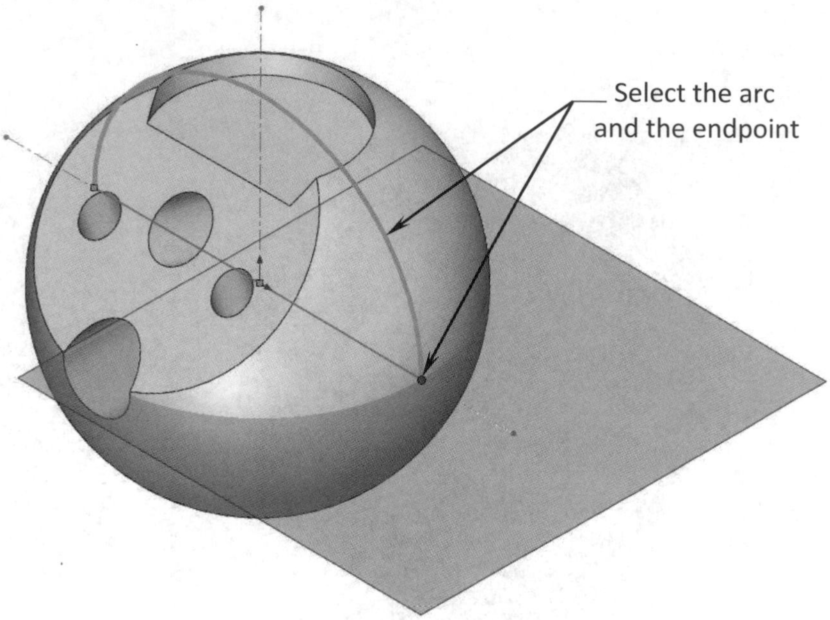

Click **OK**.

17. Creating the side-grips:

Select the new plane (Plane6) and insert a **new sketch**.

Sketch a **Circle** at the endpoint of the arc and add a **Ø.625** dimension.

Note: Use the Coincident relation when selecting 2 points, but use the Pierce relation when selecting a point and an arc.

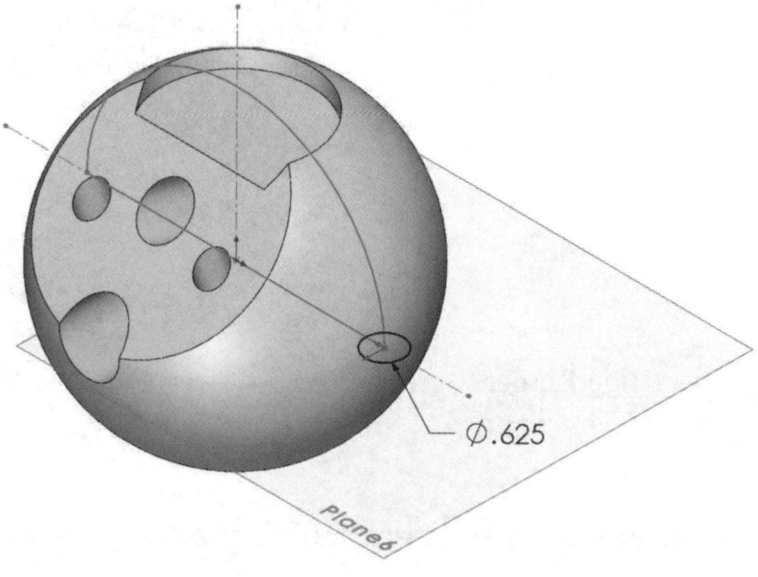

Click or select **Insert / Cut Extrude**.

Direction 1: **Through All Both**.

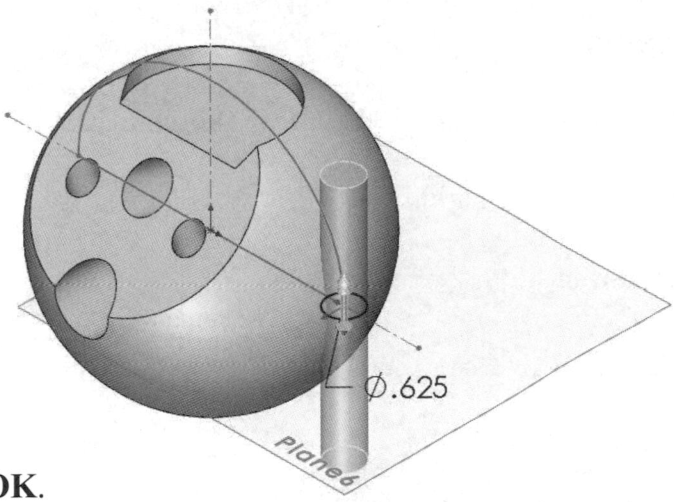

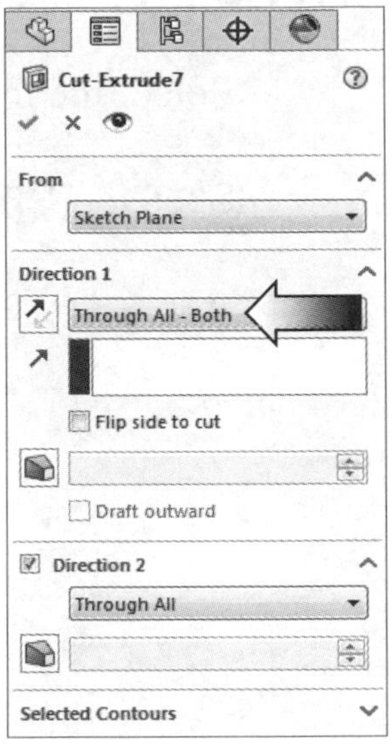

Click **OK**.

Hide **Sketch1** and **Plane6**.

18. Creating a Circular Pattern of the Grips:

Click  or select **Insert / Pattern Mirror / Circular Pattern**.

Click **View / Temporary Axis** and select the center **axis** as indicated.

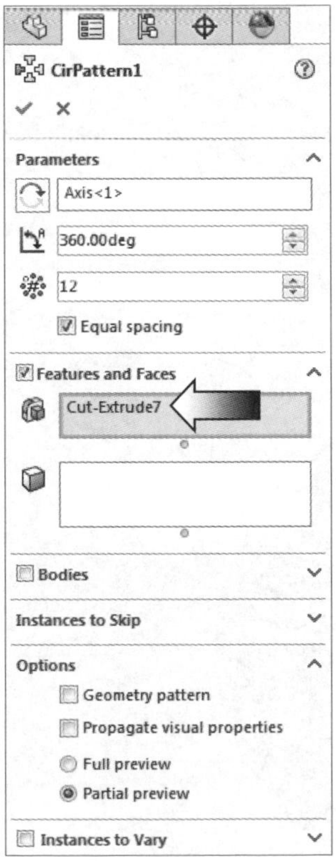

Equal Spacing: **Enabled**.

Total Angle: **360 deg**.

Number of instances: **12**.

Select the Cut-Extrude6 as Feature To Pattern.

Click **OK**.

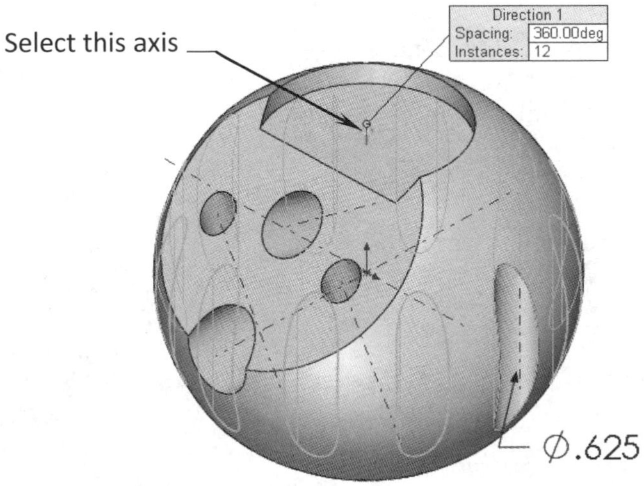

19. Adding another recess:

Select the upper face of the recess and insert a **new sketch** .

Sketch a **Circle** centered on the Origin.

Add a **1.750 diameter** dimension.

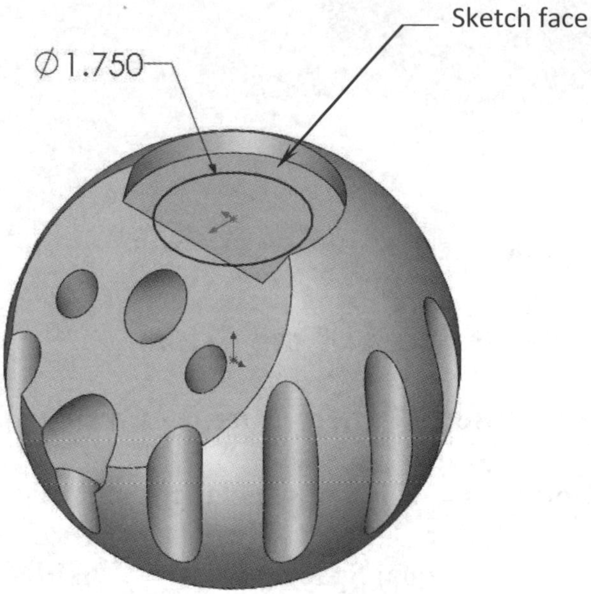

Click  or select **Insert / Cut Extrude**.

End Condition: **Blind**.

Extrude Depth: **.175** in.

Click **OK**.

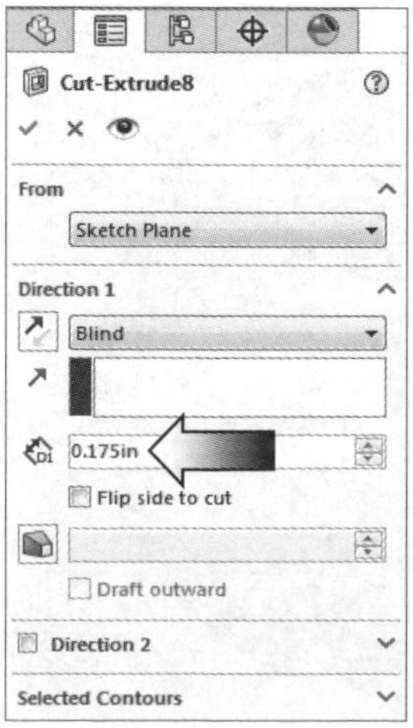

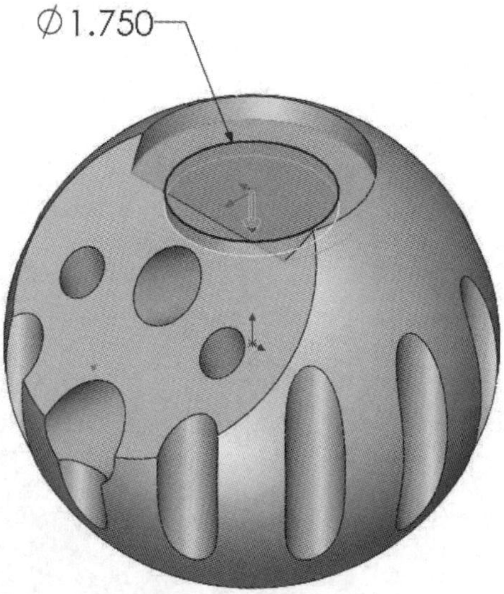

20. Creating a Mid-Plane: (Requires 2 parallel planes or 2 planar faces).

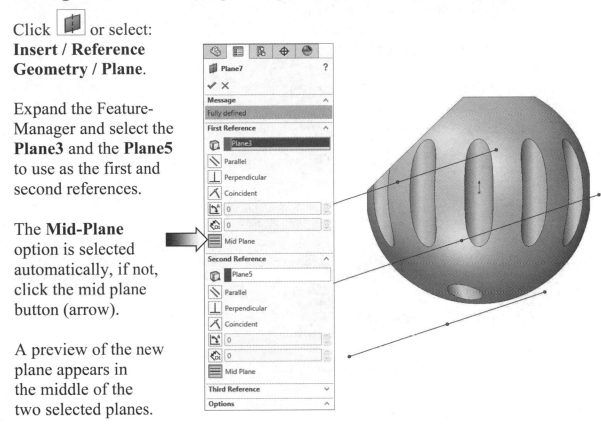

Click or select: **Insert / Reference Geometry / Plane**.

Expand the Feature-Manager and select the **Plane3** and the **Plane5** to use as the first and second references.

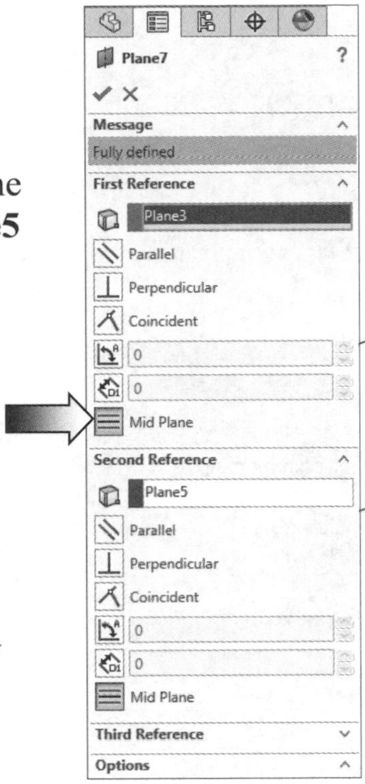

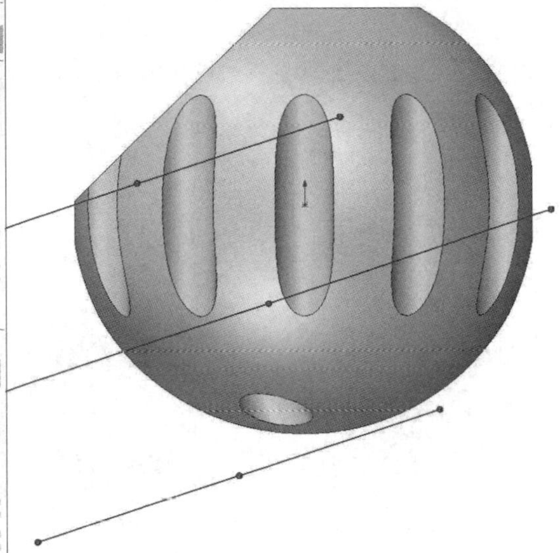

The **Mid-Plane** option is selected automatically, if not, click the mid plane button (arrow).

A preview of the new plane appears in the middle of the two selected planes.

Click **OK**.

21. Creating a rectangular pocket:

Select the new plane (Plane7) and insert a **new sketch**.

Sketch a **Center Rectangle** that is centered on the origin.

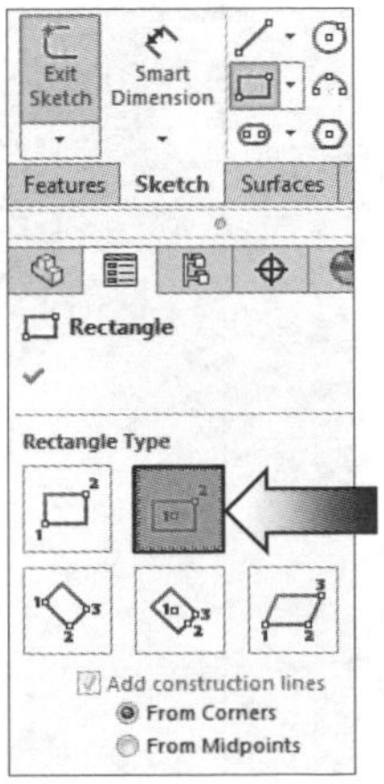

Add the width and height dimensions to fully define the sketch.

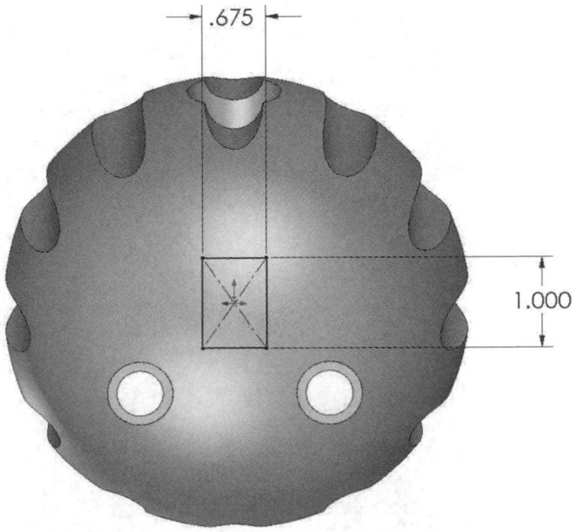

Click  or select **Insert / Cut Extrude**.

End Condition: **Through All**.

Reverse Direction: **Enabled** (arrow).

Draft: **1deg. Outward** (arrow).

Click **OK**.

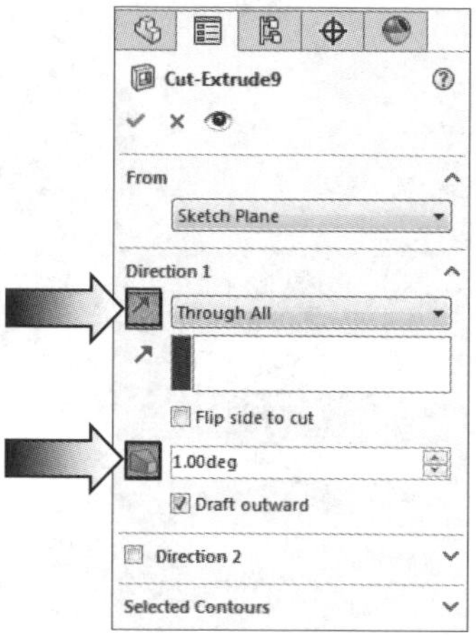

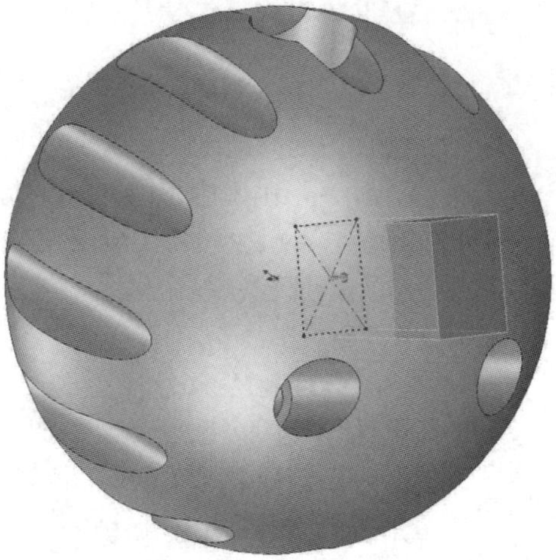

22. Adding fillets to the pocket:

Click  or select **Insert Features / Fillet-Round**.

Use the default **Constant Radius** option.

Enter **.093in.** for radius.

Select the **4 edges** of the pocket as noted.

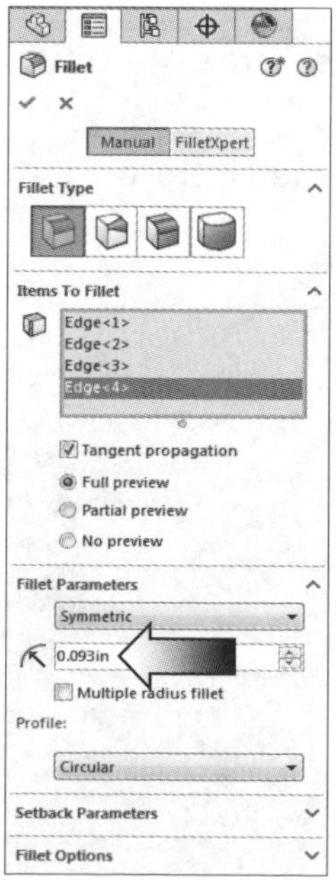

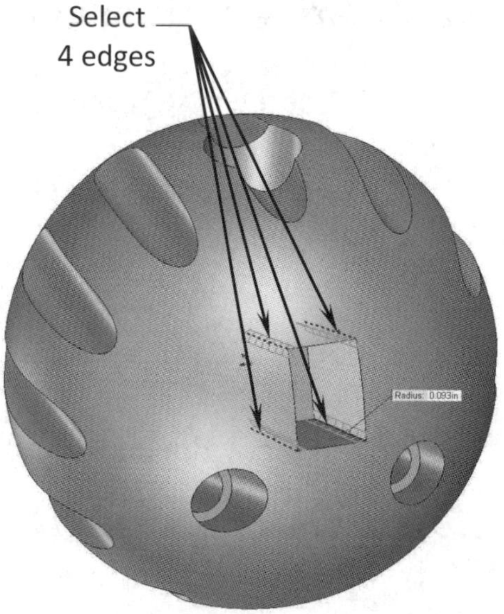

Enable the Full Preview checkbox.

Click **OK**.

23. Adding Fillets to all edges:

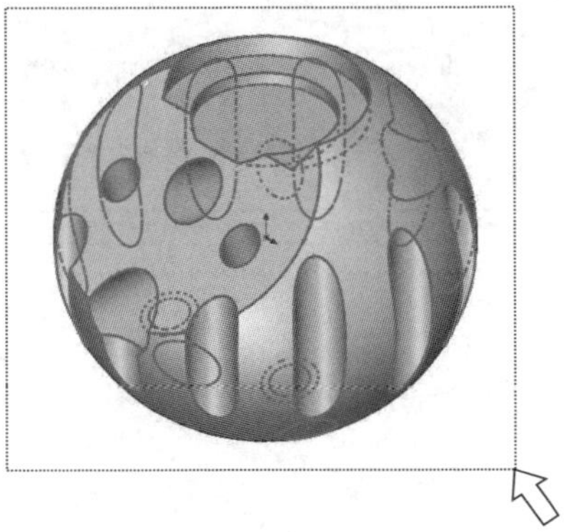

Lasso or **Box-Select** around the entire part, (or press Ctrl+A), to select all of its edges.

Click  or select **Insert Features / Fillet-Round**.

Enter **.040** for Radius.

Tangent Propagation: **Enabled**.

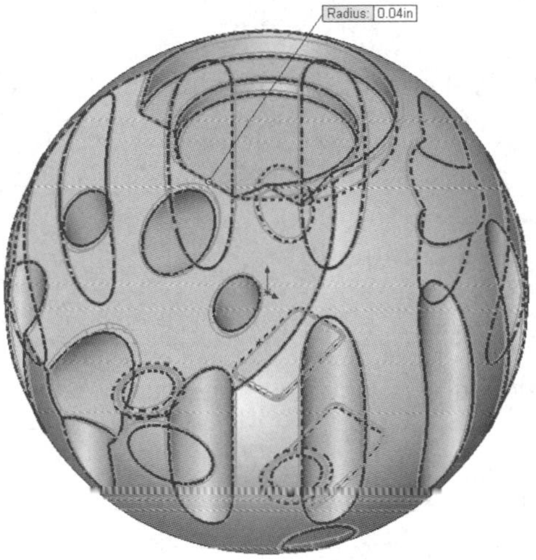

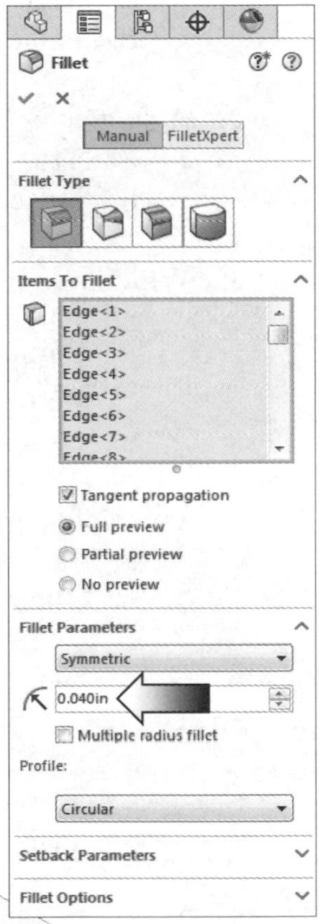

Click **OK**.

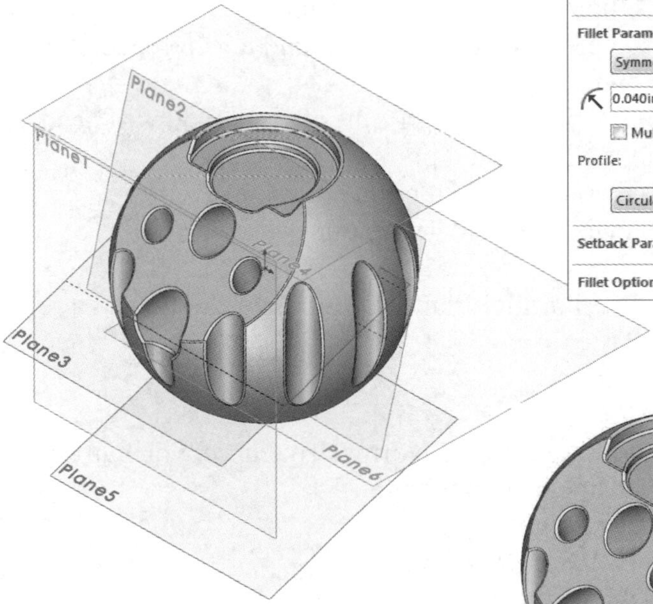

Hide the planes before saving the part. **(View / Hide-Show / Planes)**.

24. Saving your work:

Click **File / Save As / Planes Creation / Save**.

Questions for Review

Plane Creation

1. Planes can be used to section a part or an assembly.
    a. True
    b. False

2. A sketch can be extruded to a plane as the end condition by using the Up-To-Surface option.
    a. True
    b. False

3. Which one of the options below is not a valid command?
    a. Parallel plane at Point.
    b. Offset plane at Distance.
    c. Perpendicular to another plane at Angle.
    d. Normal to Curve.

4. To create a plane at Angle, you will need:
    a. The Angle and a Reference plane.
    b. The Angle and a pivot Line.
    c. The Angle, a pivot Line, and a Reference plane.

5. To create a plane through Lines/Points, you will need at least:
    a. One line and a point
    b. Two lines and a point
    c. Two lines and Two points

6. To create a Parallel Plane At Point, you will need a reference plane and a point.
    a. True
    b. False

7. When creating a Plane Normal To Curve, you can select:
    a. A linear model edge
    b. A straight line.
    c. A 2D or 3D curve
    d. All of the above

1.TRUE 2.TRUE
3.C 4.C
5.A 6.TRUE
7.D

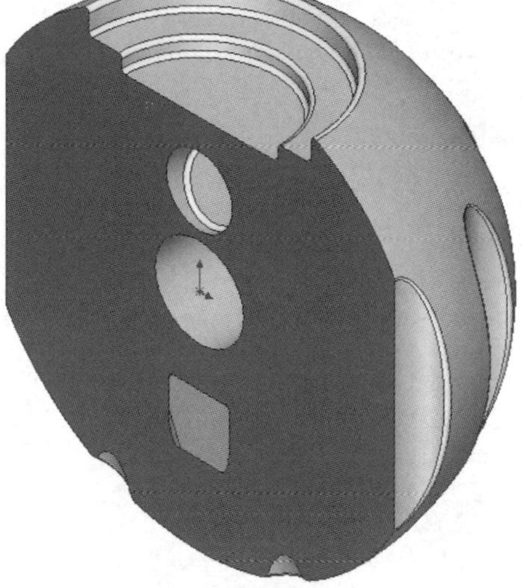

Section with **Front** plane

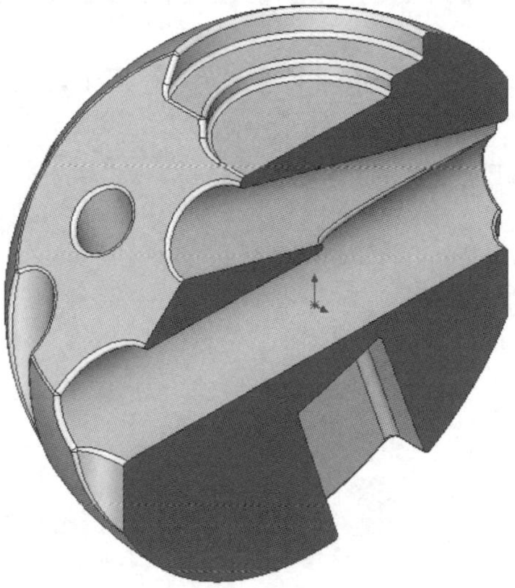

Section with **Right** plane

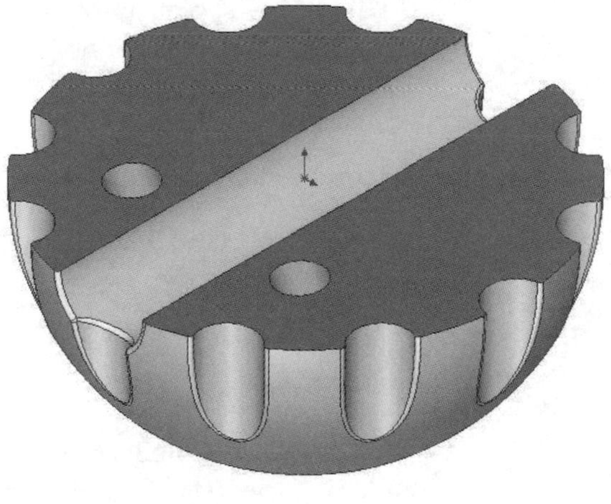

Section with **Top** plane

Isometric View

Exercise: Creating New Planes

1. Create a reference sketch:

Create the sketch below on the **Front** plane.

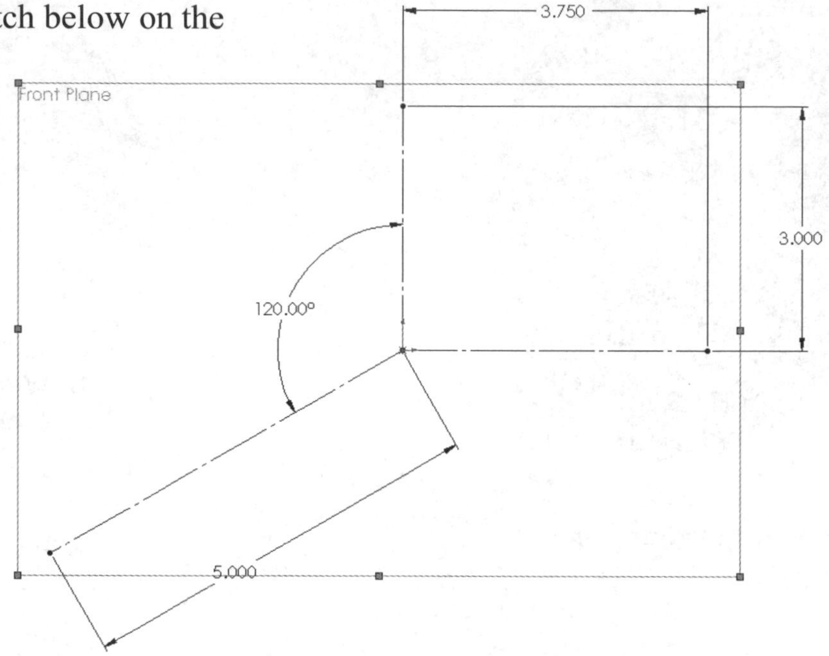

2. Create 3 new planes:

Use the references as indicated.

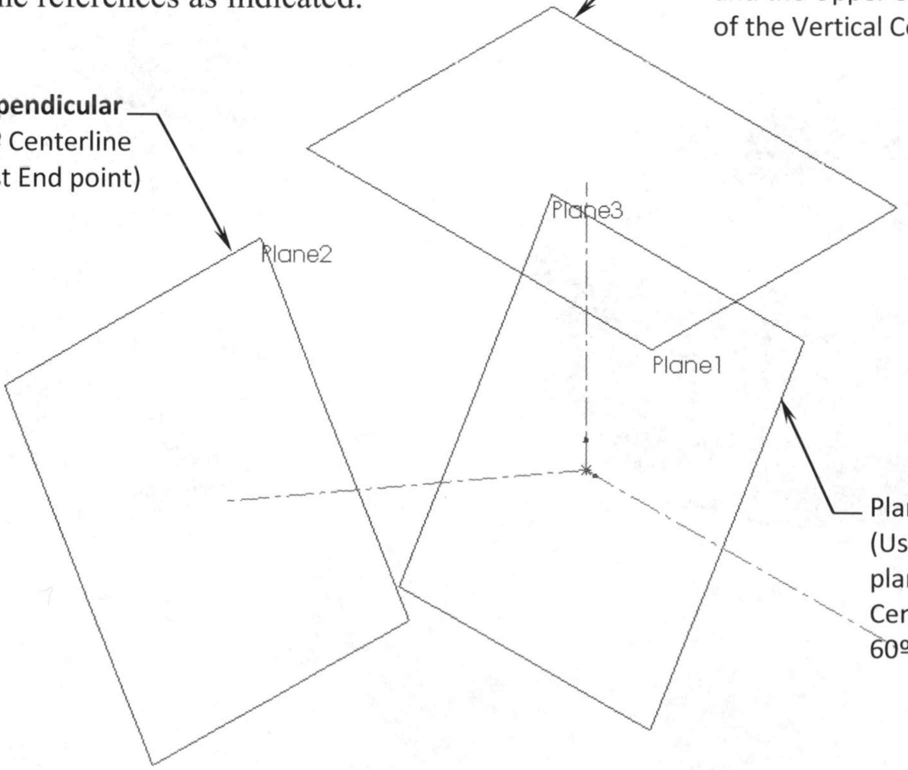

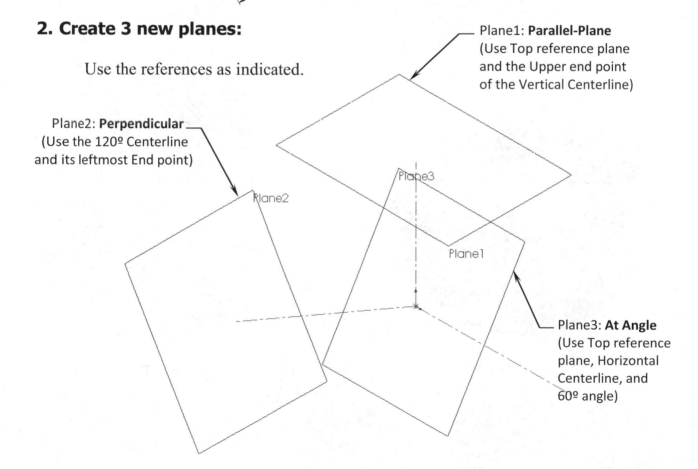

3. Open a new sketch on Plane1:

Sketch a **Circle** centered on the end point of the vertical centerline.

Add the **2.250in** diameter dimension.

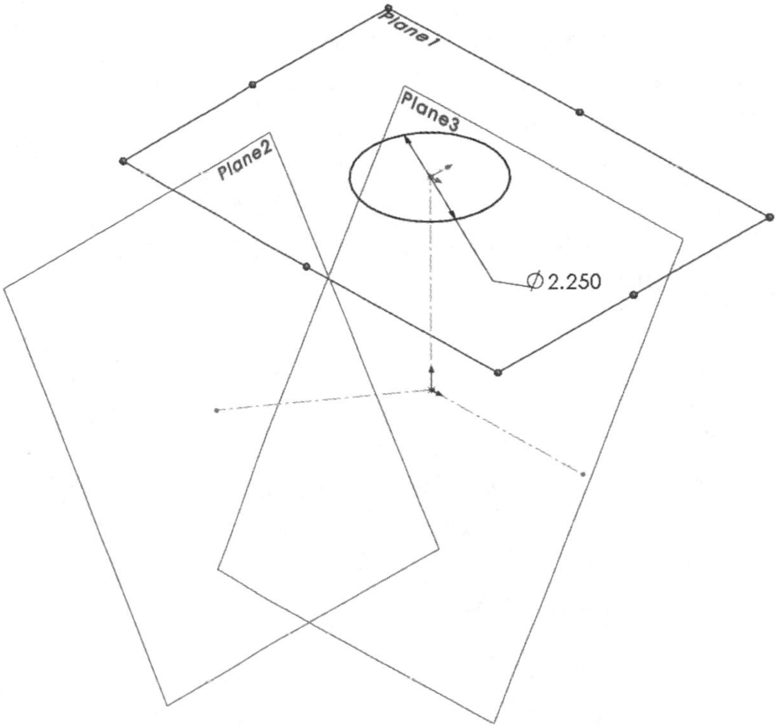

4. Extrude a boss:

Extrude Type: **Blind**.

Extrude Depth: **8.000in**

Reverse direction enabled.

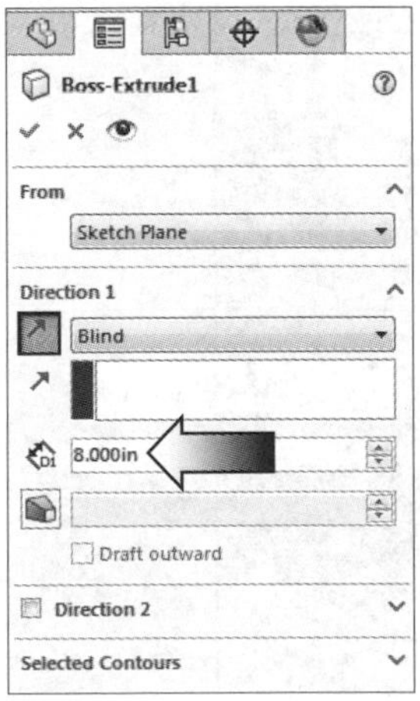

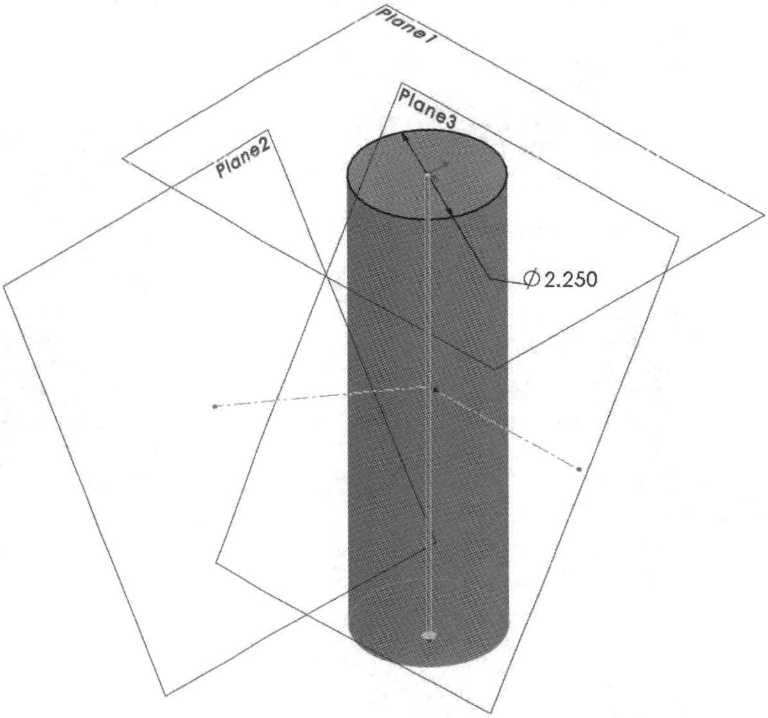

Click **OK**.

5. Open a new sketch on the Plane2:

Sketch a **Circle** centered on the left end of the 120deg. centerline.

Add the **1.750in** diameter dimension.

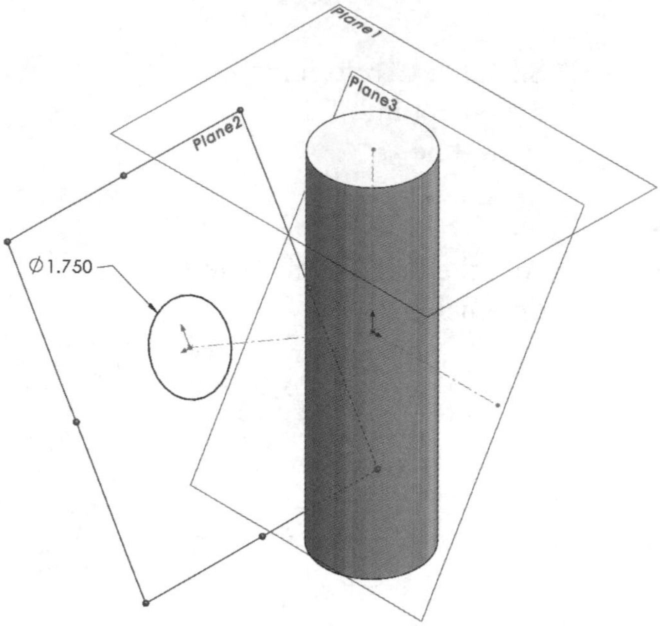

6. Extrude a boss:

Extrude Type: **Up To Surface**.

Select the **outer face** of the 1st cylinder.

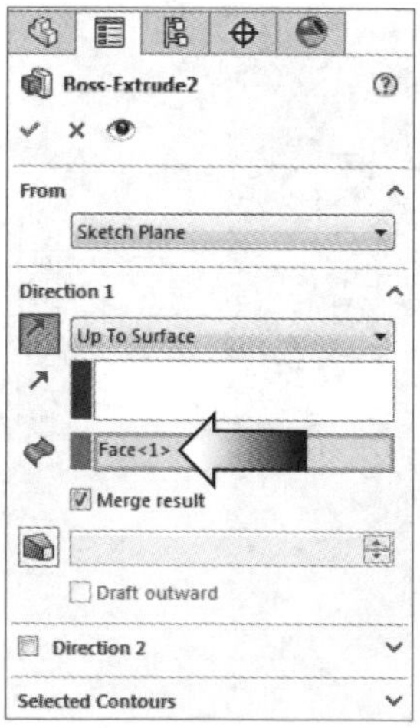

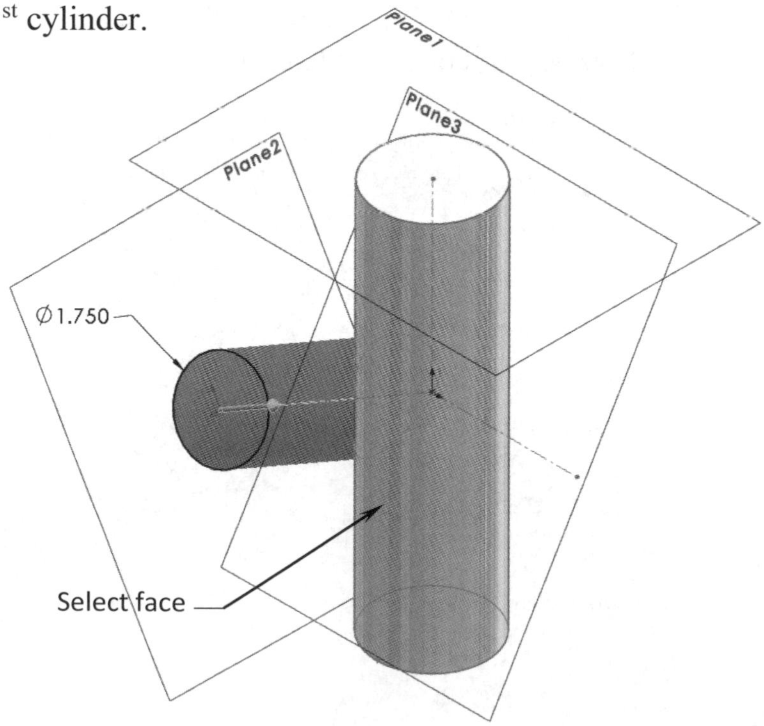

Click **OK**.

7. Open a new sketch on the plane3:

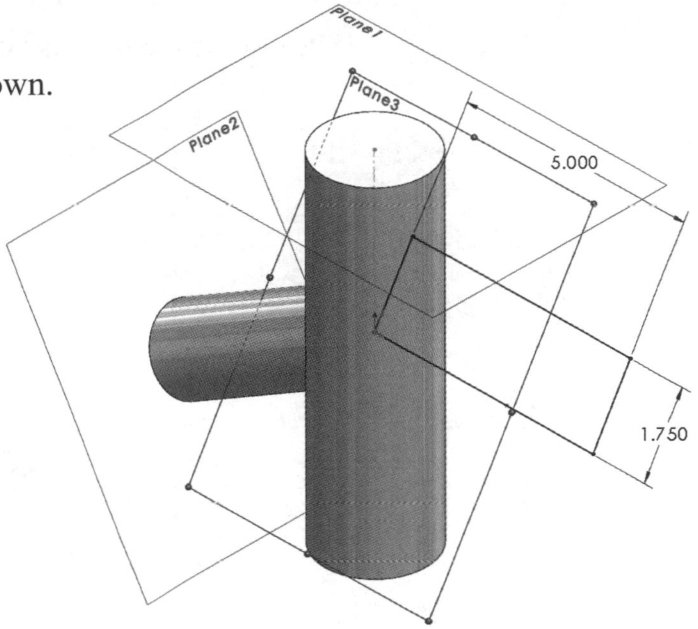

Sketch a **Corner Rectangle** as shown.

Add thc hcight and thc width dimensions.

8. Extrude a boss:

Extrude Type: **Blind**.

Extrude Depth: **1.000in**.

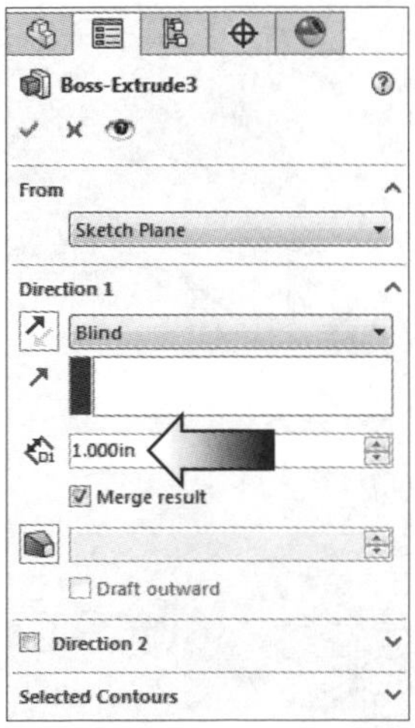

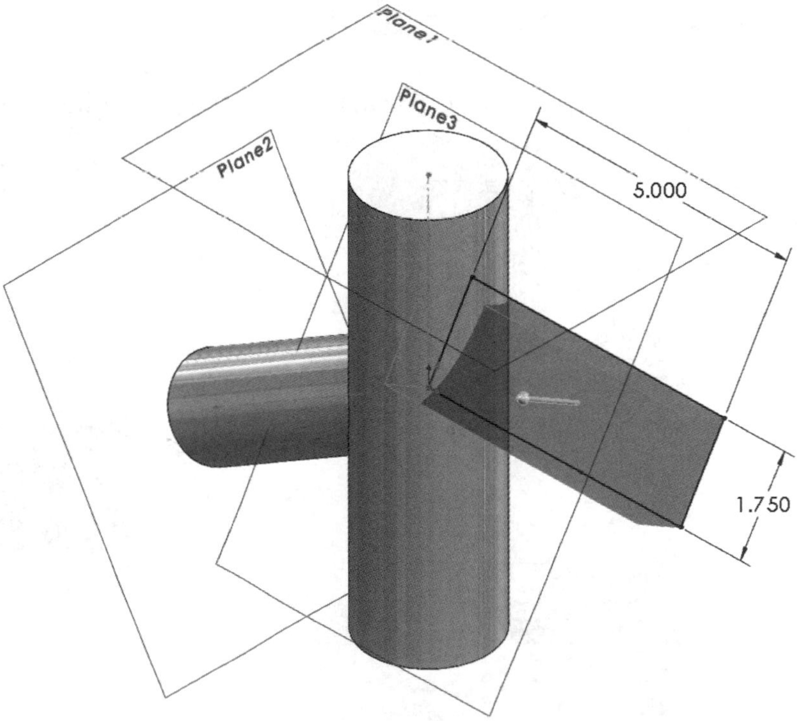

Click **OK**.

9. Add the .250in fillets:

Add a fillet of **.250in** to the **4 edges** as noted.

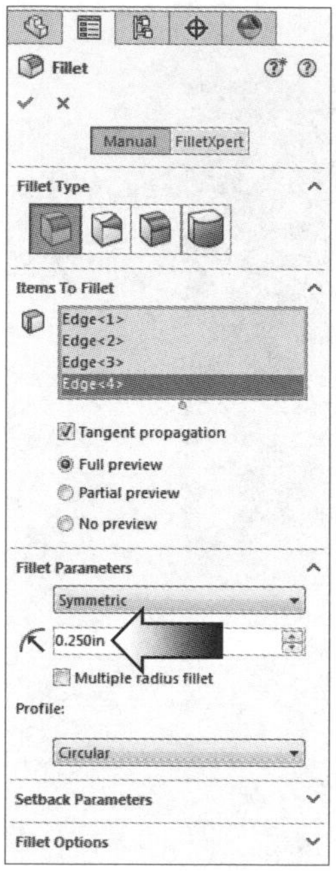

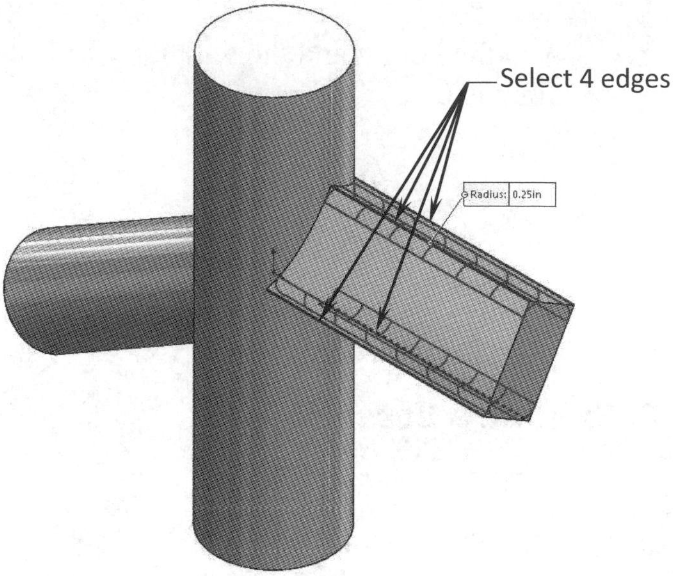

Click **OK**.

10. Add the .125in fillets:

Add a fillet of **.125in** to the **2 edges** as indicated below.

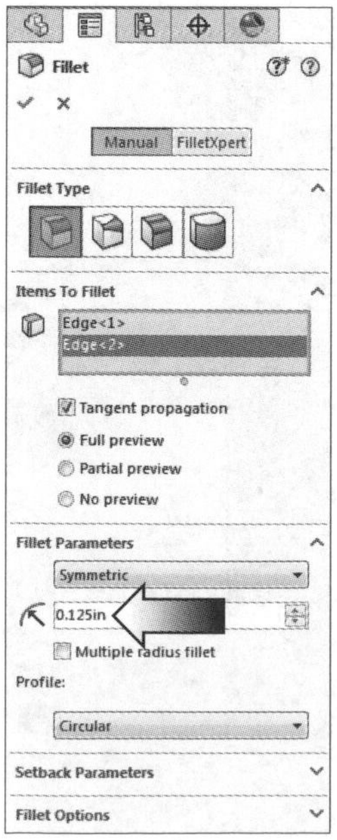

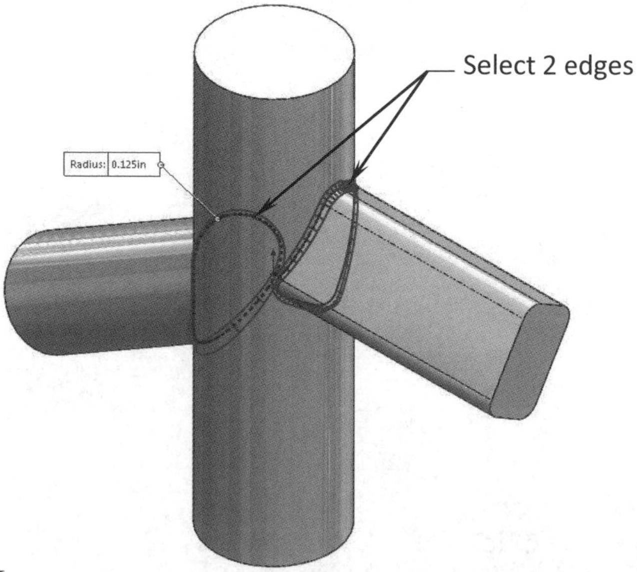

Click **OK**.

11. Shell the model:

Shell the model using a thickness of **.080in**.

Select the **4 faces** to remove as indicated.

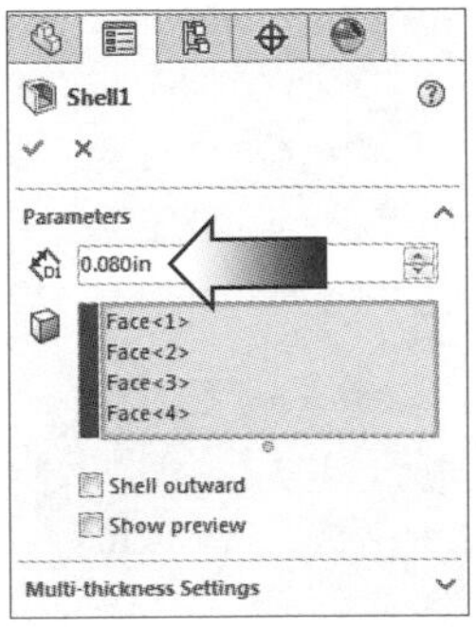

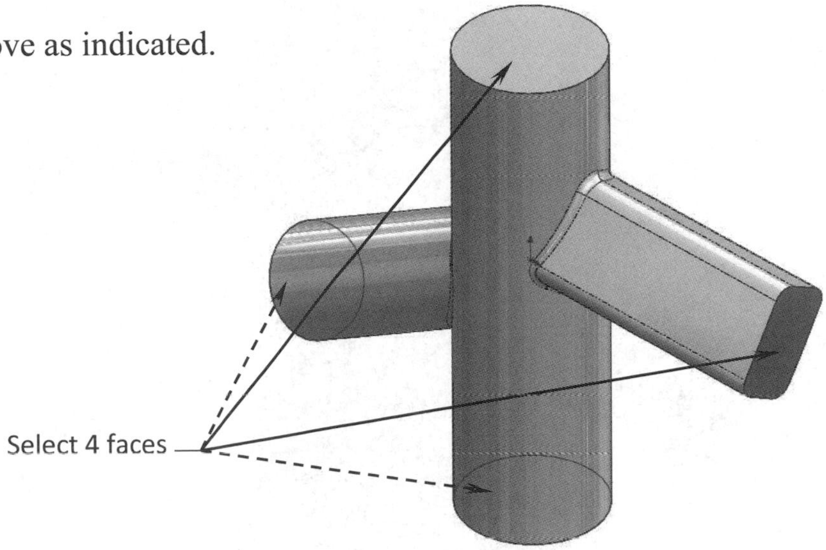

Click **OK**.

12. Save your work:

Click **File / Save As**.

Enter **Plane_Exe.sldprt** for the file name.

Click **Save**.

Close all documents.

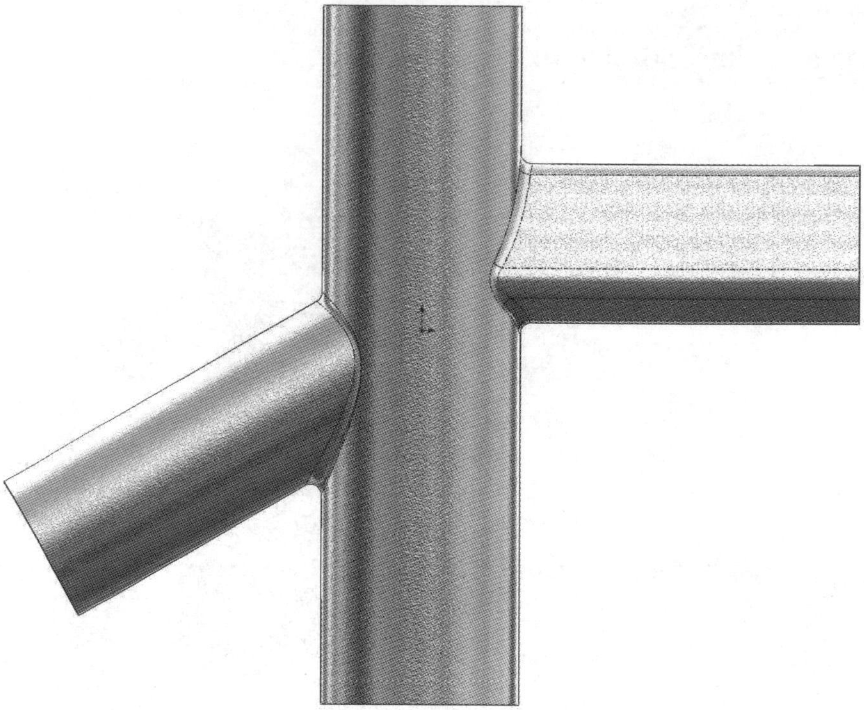

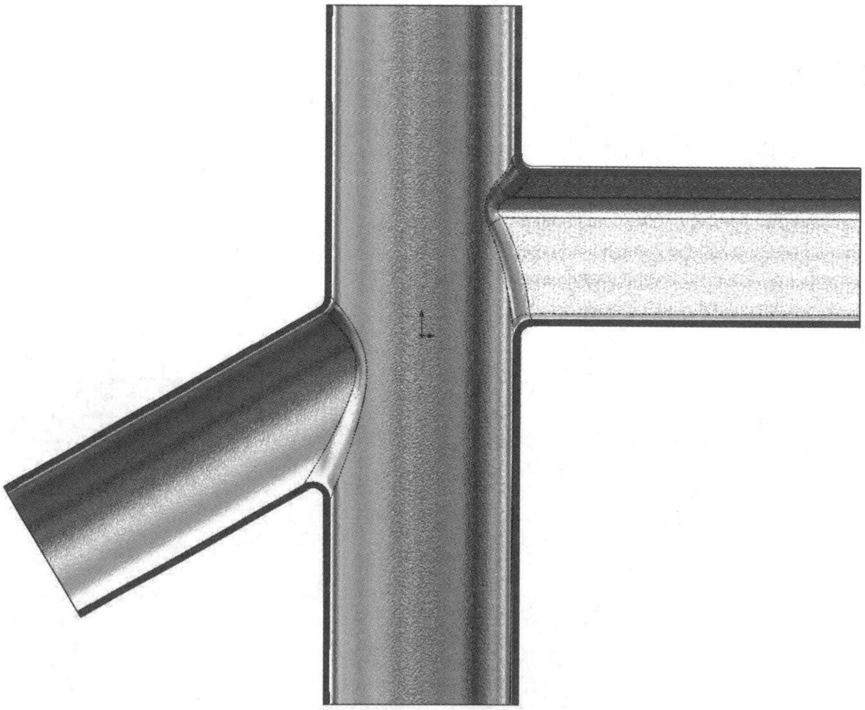

CHAPTER 3

Advanced Modeling

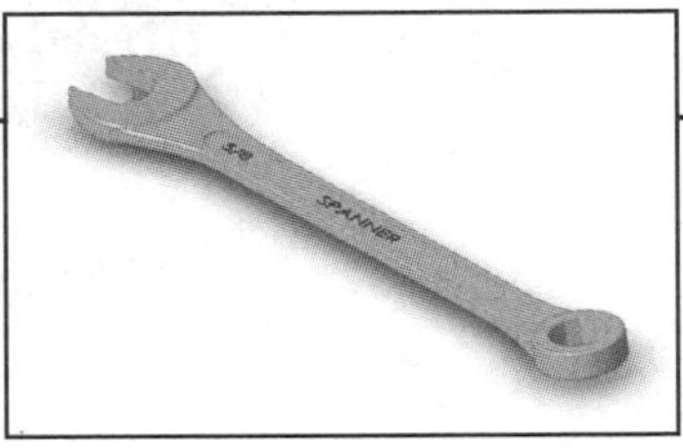

Advanced Modeling
5/8″ Spanner

The draft option is omitted in this lesson to help focus in other areas.

The arc conditions Min / Max are options that help in placing the dimensions on the tangents of arcs or circles. Once a dimension is created, the arc conditions can be changed by right clicking on the dimension and selecting the Leaders tab. Only two conditions can be specified at a time:
either Center/Center, Min/Max, Max/Max,
or Min/Min, etc.

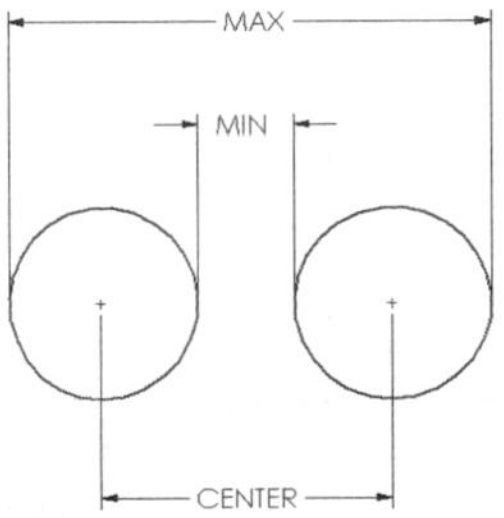

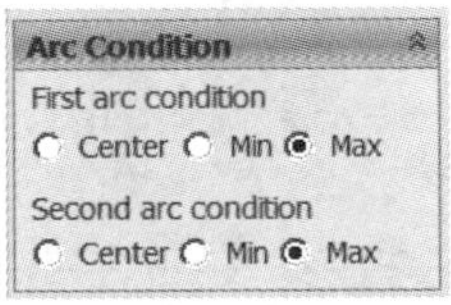

Adding text on the model is another unique feature in SOLIDWORKS. This option allows the letters in the sketch to be extruded as an emboss or a cut, similar to other extruded features.

All letters in the same sketch are treated as one entity; they will be extruded at the same time and will have the same extrude depth. However, the option Dissolve-Sketch-Text is used to convert the sketch-text into individual sketch entities so that the shape and size of each letter can be modified.

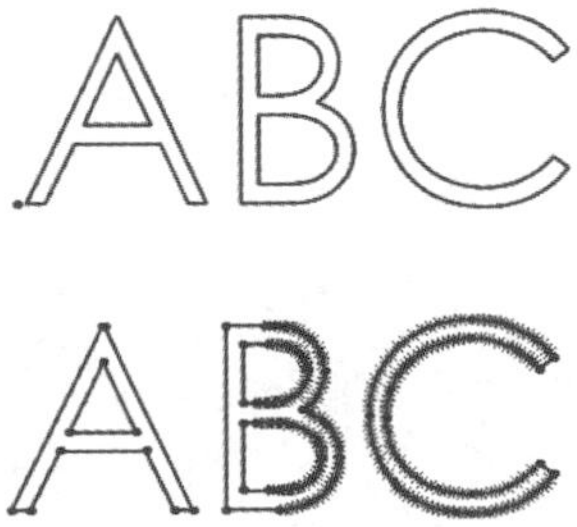

To use this option simply right-click the text and select Dissolve Sketch Text.

This chapter and its exercise will guide you through some of the advanced modeling techniques as well as learn to use the Text tool to create the straight or curved extruded letters.

5/8" Spanner
Advanced Modeling

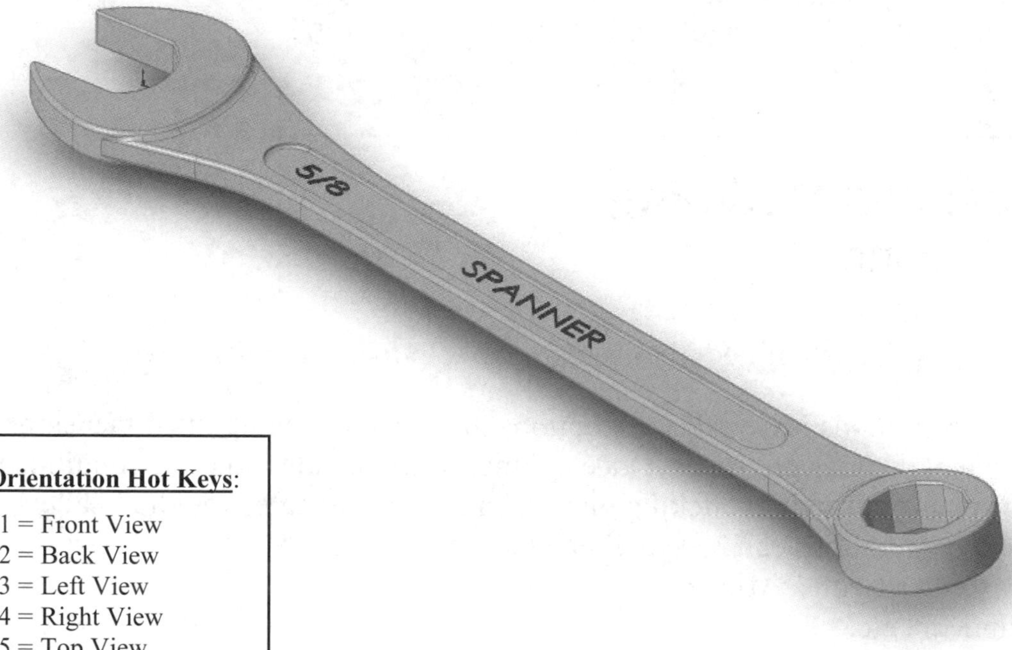

View Orientation Hot Keys:

Ctrl + 1 = Front View
Ctrl + 2 = Back View
Ctrl + 3 = Left View
Ctrl + 4 = Right View
Ctrl + 5 = Top View
Ctrl + 6 = Bottom View
Ctrl + 7 = Isometric View
Ctrl + 8 = Normal To Selection

Dimensioning Standards: **ANSI**

Units: **INCHES** – 3 Decimals

Tools Needed:

Insert Sketch	Line	3 Point Arc
Text	Add Geometric Relations	Dimension
Sketch Fillet	Polygon	Plane
Base/Boss Extrude	Extruded Cut	Fillet/Round

1. Opening the Spanner sketch document:

From the Training Files folder, open a document named: **Spanner Sketch**.

Edit the **Sketch1**. This is the open end of the spanner and it is fully defined.

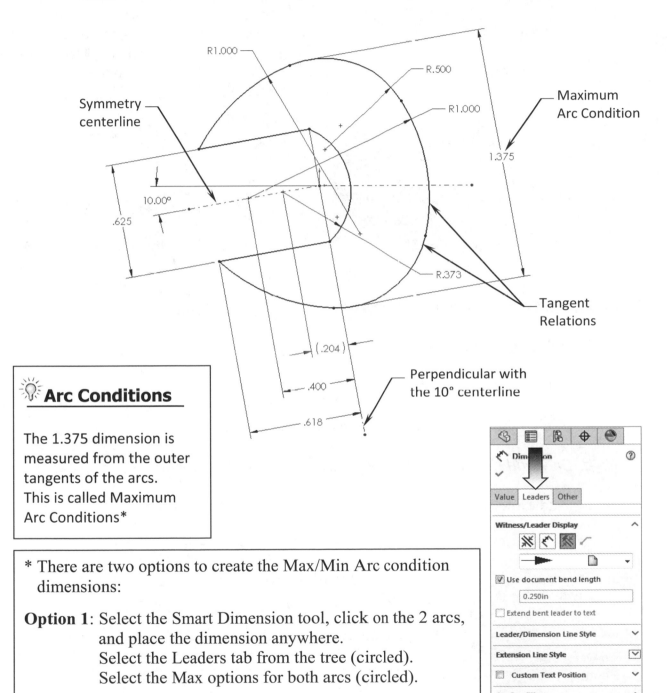

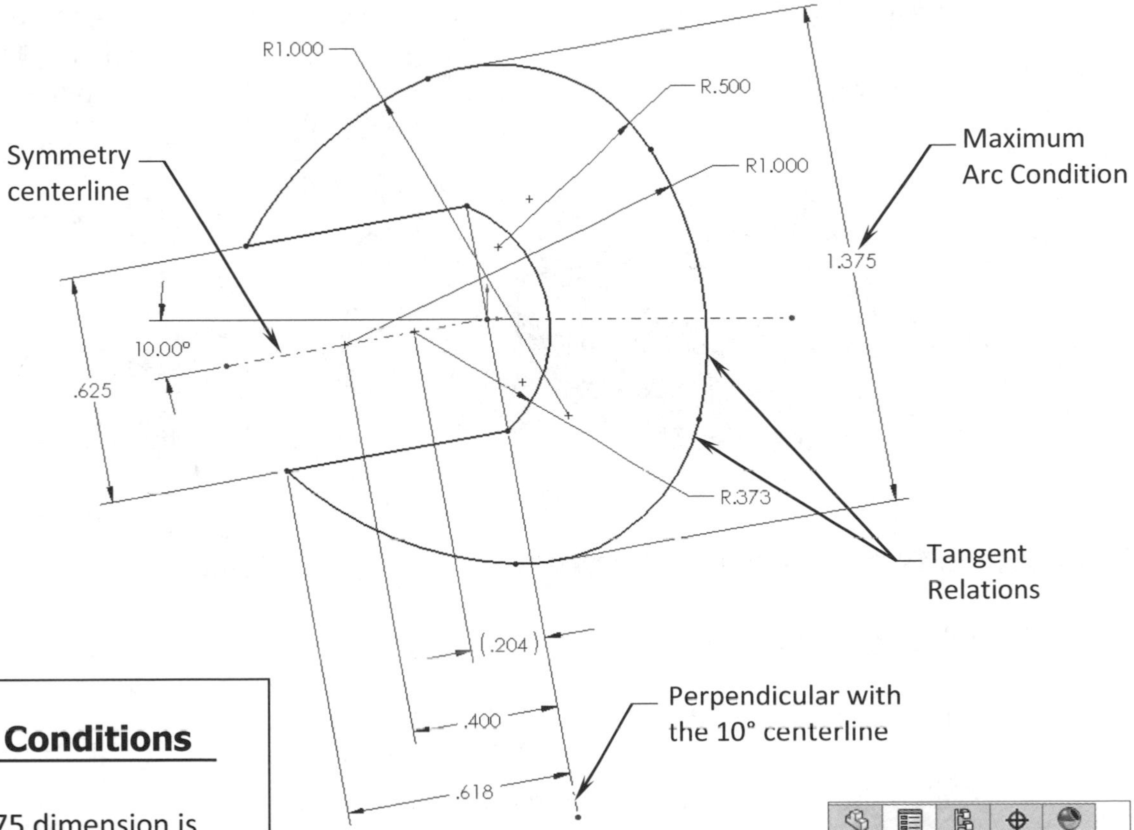

Arc Conditions

The 1.375 dimension is measured from the outer tangents of the arcs.
This is called Maximum Arc Conditions*

* There are two options to create the Max/Min Arc condition dimensions:

Option 1: Select the Smart Dimension tool, click on the 2 arcs, and place the dimension anywhere.
Select the Leaders tab from the tree (circled).
Select the Max options for both arcs (circled).

Option 2: Hold down the SHIFT key and click on the 2 arcs; the dimension's leader lines will snap to the arc tangents automatically.

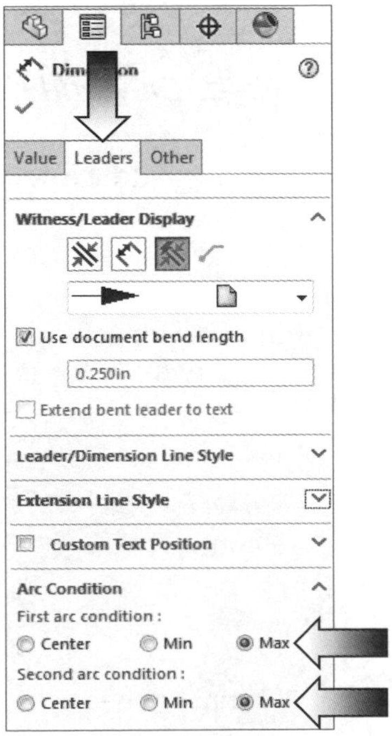

2. Extruding the base feature:

Click  or select **Insert / Boss-Base / Extrude**.

End Condition: **Mid Plane**.

Extrude Depth: **.250 in**.

Click **OK**.

Renaming Features

Slow double click on each feature's name (or press F2) and rename them to something more descriptive like Open-End, Transition-Body, Closed-End, etc...

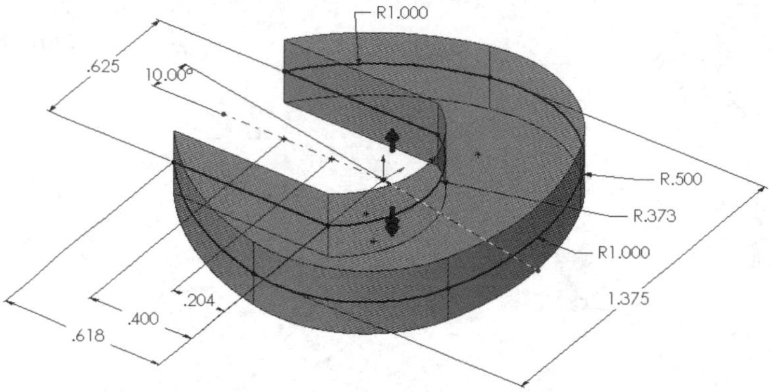

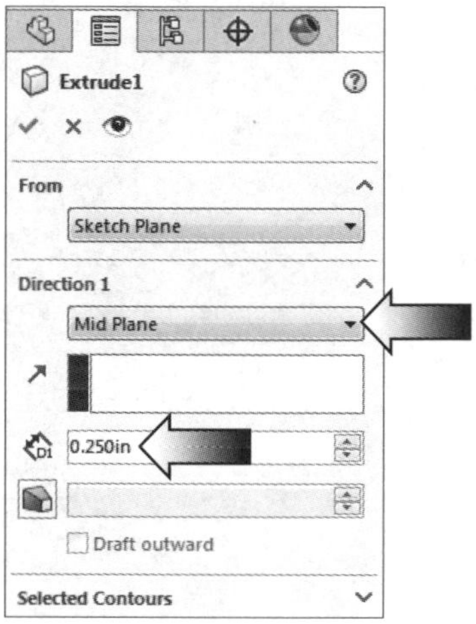

3. Creating the transition sketch:

Select <u>Top</u> plane from the FeatureManager tree.

Click  or select **Insert / Sketch**.

Sketch the profile below using the **Line** command.

<u>Note:</u> Only add the Sketch Fillets after the sketch is fully defined.

Add dimensions or Relations needed to fully define the sketch.

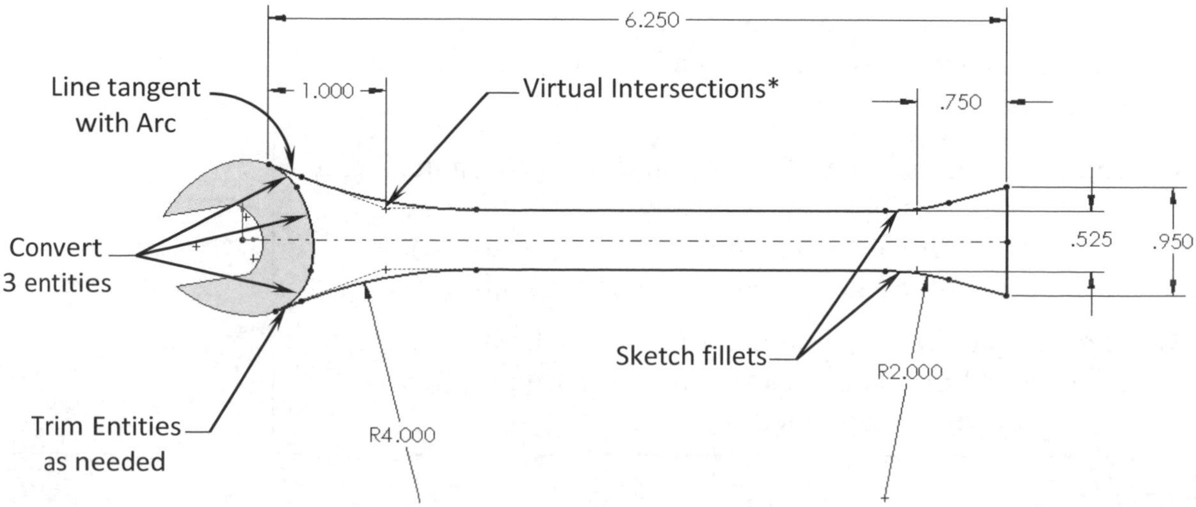

4. Extruding the Transition feature:

Click  or select **Insert / Boss-Base / Extrude**.

End Condition: **Mid Plane**

Extrude Depth: **.175 in**.

Click **OK**.

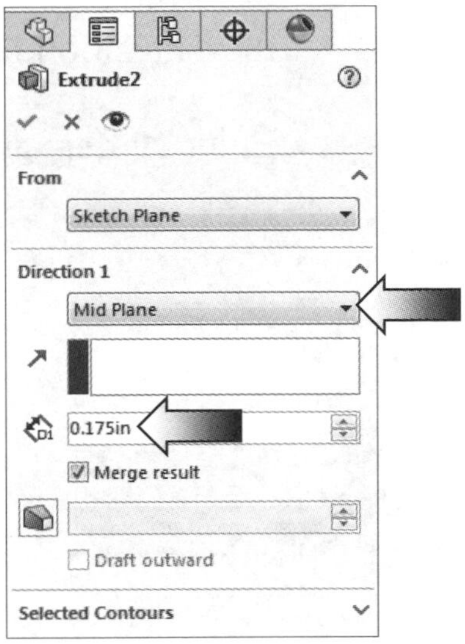

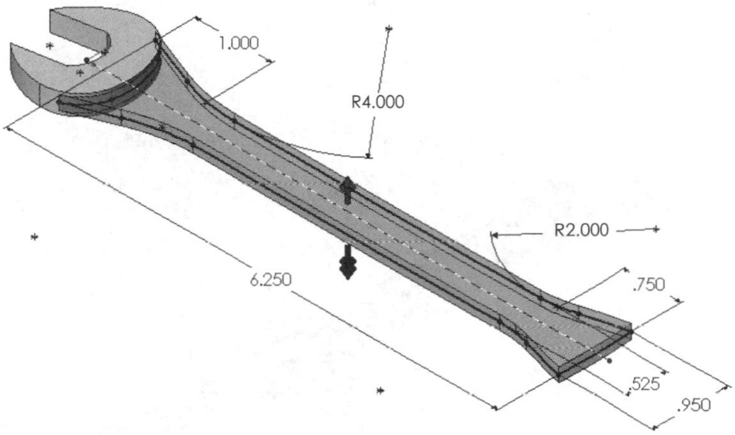

5. Adding the reference geometry:

Select the face as indicated.

Click  or select **Insert / Sketch**.

Sketch a **Centerline** at the mid-point of the two vertical edges.
This line will be used to create a new plane in the next step.

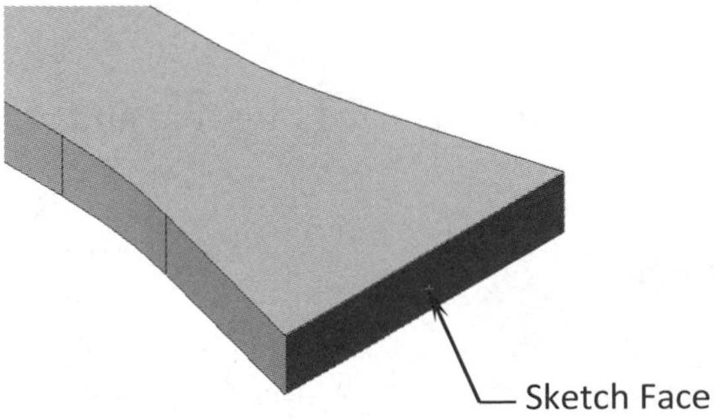

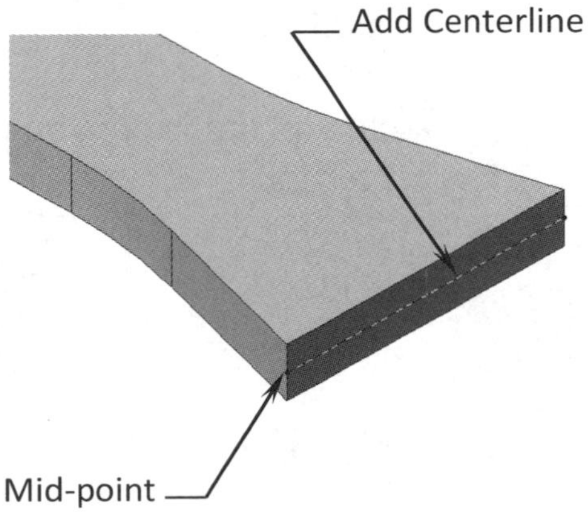

Exit the Sketch  or select **Insert / Sketch**.

6. Creating a new work plane: Plane at Angle

Click  or select **Insert / Reference Geometry / Plane**.

For Reference Entities, select the **Sketch4** (centerline) and the **upper face** of the Transition feature.

Enter **10 deg**. and click **Flip Offset** to place the new plane on the bottom.

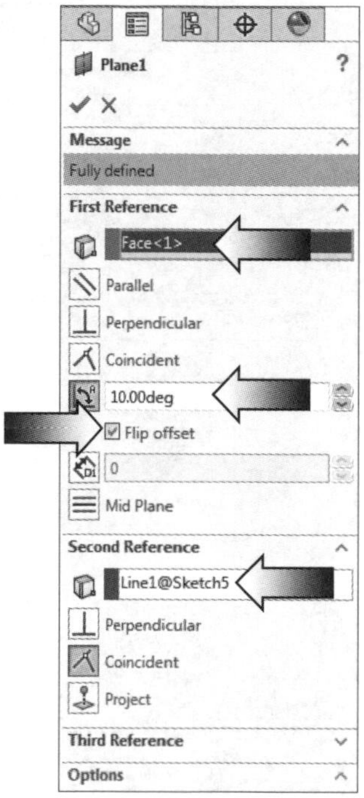

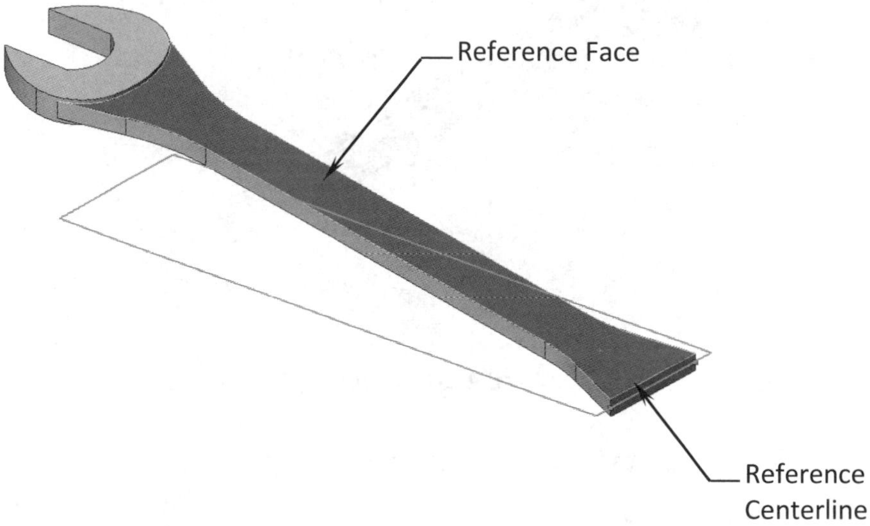

Click **OK**.

A new plane is created using the upper reference face and pivoting about the centerline.

The preview of the new 10 degrees plane.

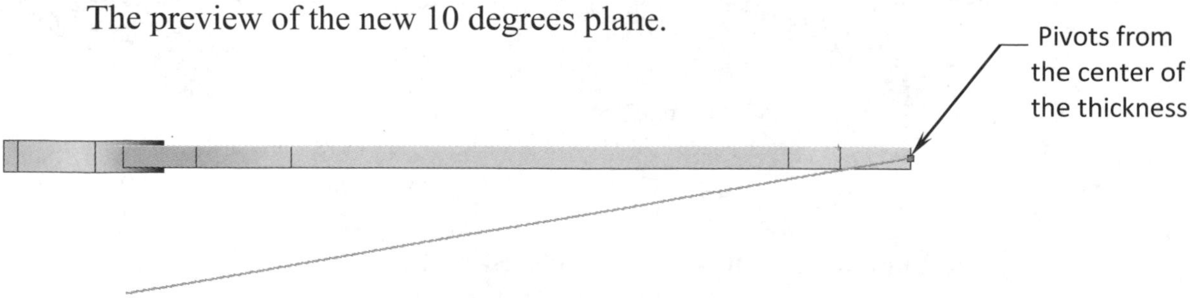

7. Creating the Closed-End sketch:

Select the new **10° plane** from the FeatureManager tree.

Click 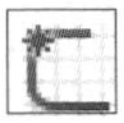 or select **Insert / Sketch.**

Sketch a **circle** and add the dimensions and relations shown below.

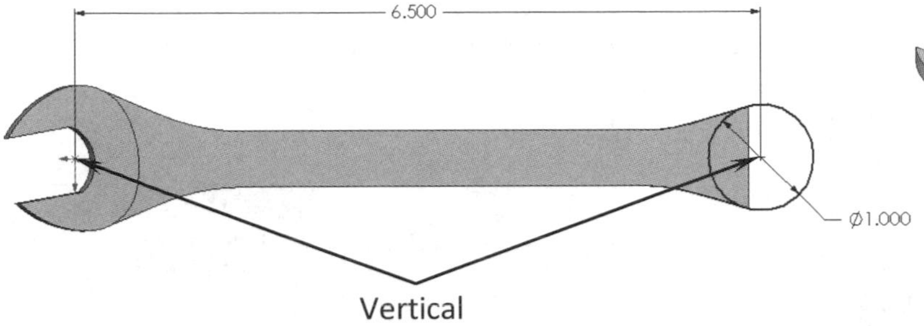

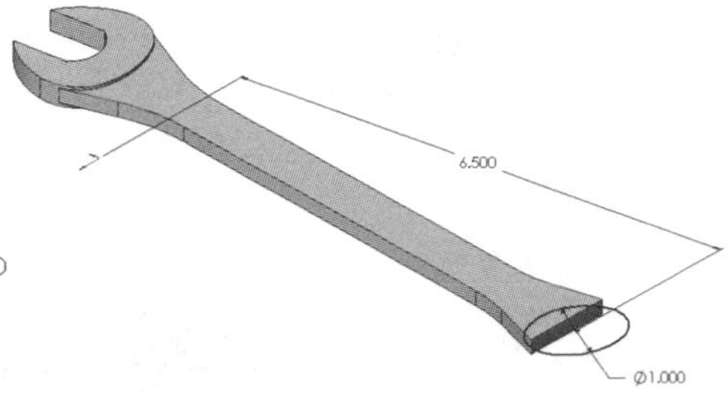

8. Extruding the Closed-end feature:

Click  or select **Insert / Boss-Base / Extrude.**

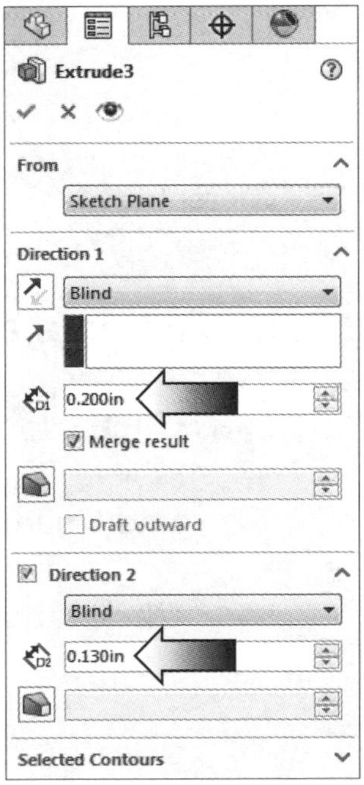

Direction 1: **Blind**.

Extrude Depth: **.200 in**.

Direction 2: **Blind**.

Extrude Depth: **.130 in**.

Click **OK**.

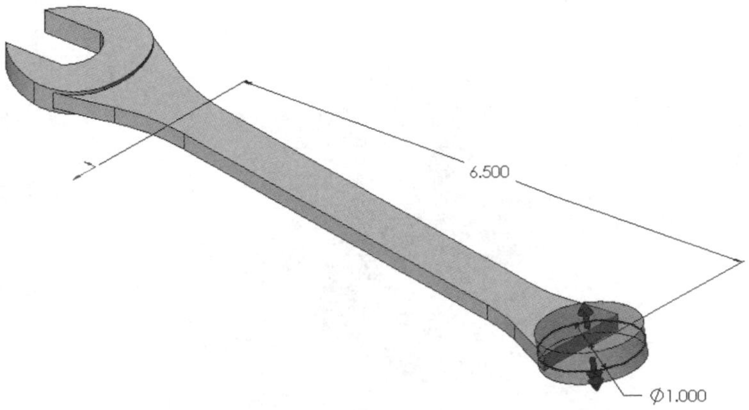

9. Adding a 12-Sided polygonal hole:

Select the face indicated as sketch plane.

Click or select **Insert / Sketch**.

Sketch a **Polygon** with **12 sides** (arrow).

Add a **.625 Dia**. dimension to the inside construction circle.

Add a **Concentric** relation between the construction circle and the circular edge.

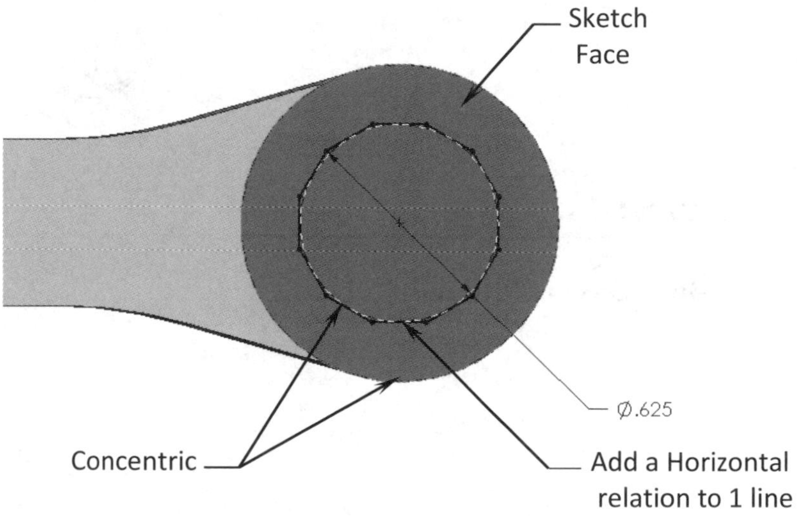

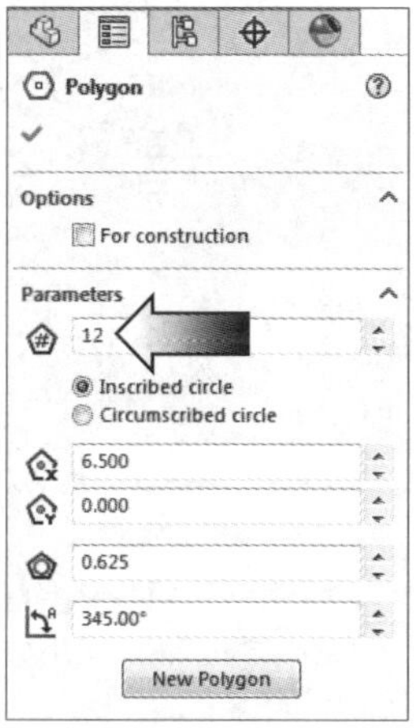

10. Extruding a cut:

Click or select **Insert / Cut / Extrude**.

End Condition: **Through All**.

Click **OK**.

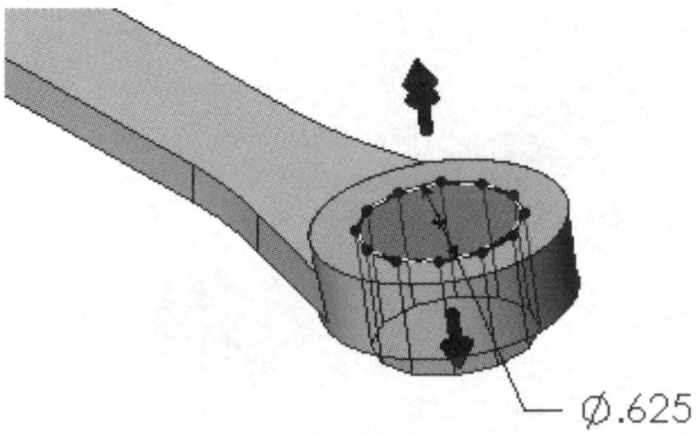

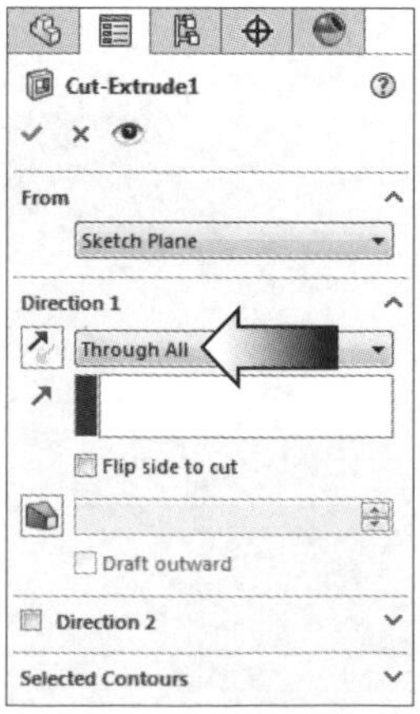

11. Creating the Recess profile:

Select the <u>face</u> indicated as sketch plane.

Click or select **Insert / Sketch**.

Sketch the profile shown below using the **Straight-Slot** command .

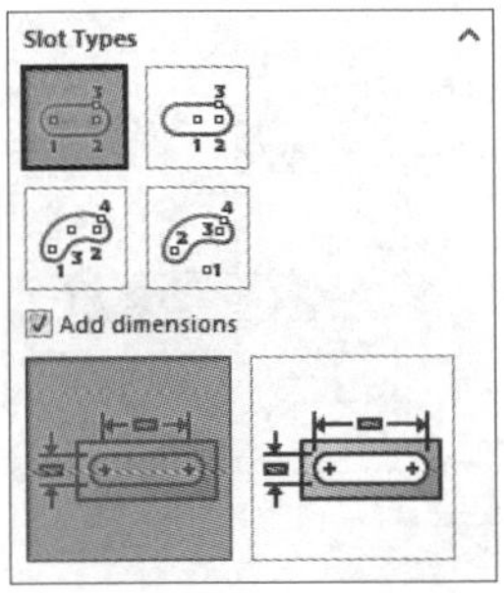

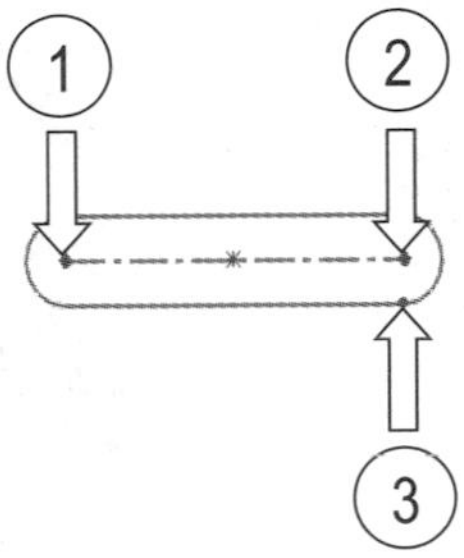

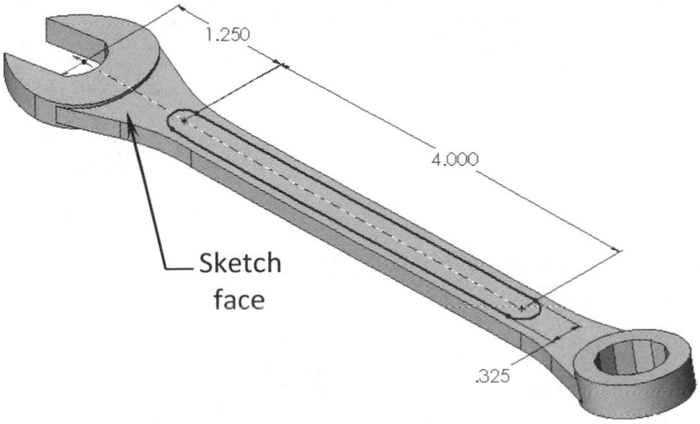

Add dimensions or Relations needed to fully define the sketch.

12. Extruding the Recessed feature:

Click or select **Insert / Cut / Extrude**.

End Condition: **Blind**.

Extrude Depth: **.030 in**.

Click **OK**.

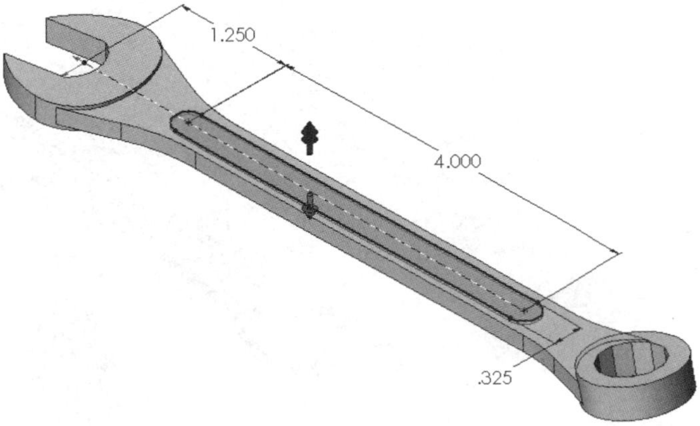

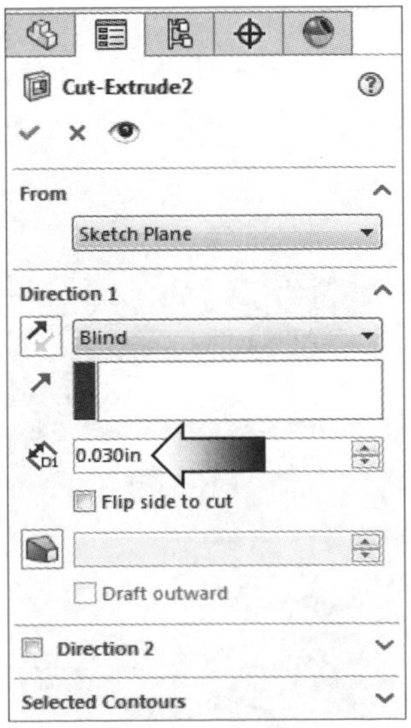

13. Mirroring the Recessed feature:

Hold the **Control** key, select the Top reference plane and the Recessed feature from the FeatureManager tree.

Click 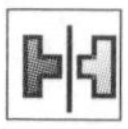 or select **Insert / Pattern Mirror** menu, then select **Mirror**.

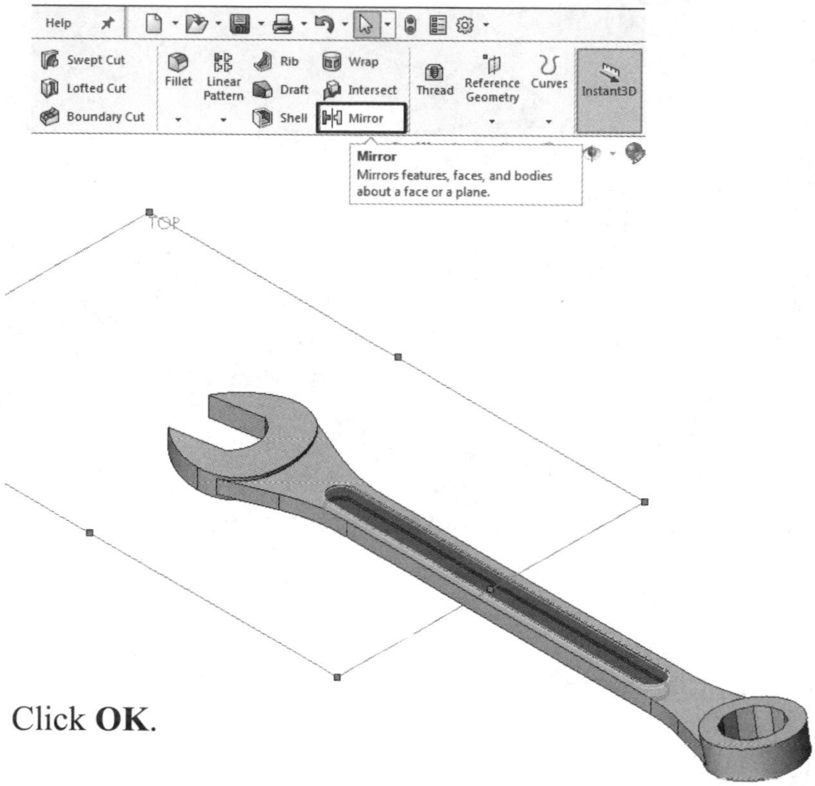

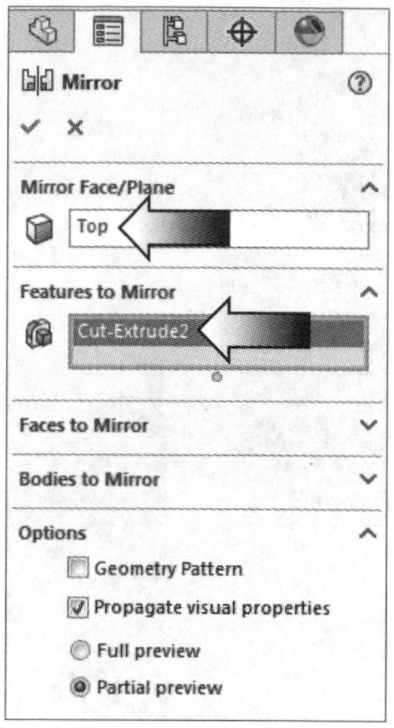

Click **OK**.

Rotate  the model to verify the mirrored recessed feature.

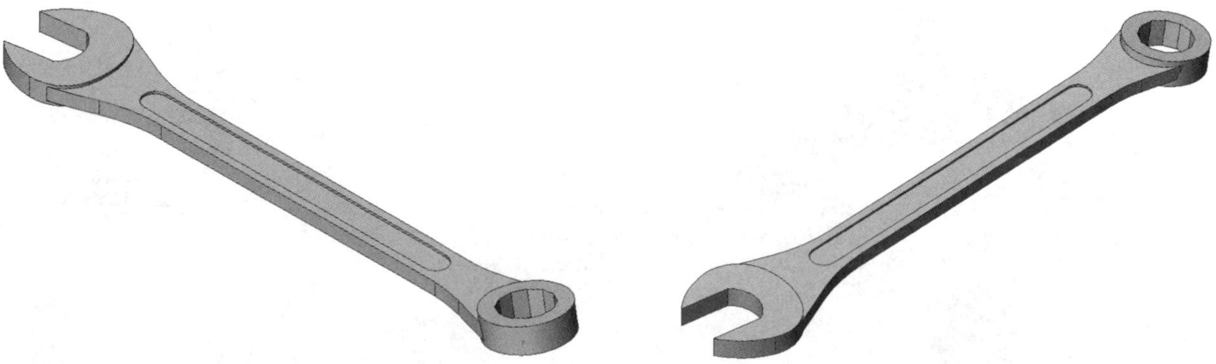

14. Adding the .030" fillets:

Click  or select **Insert / Features / Fillet/Round**.

For Radius, enter **.030 in**.

For Edges to Fillet, select the edges as indicated.

Tangent Propagation: **Enabled**.

Click **OK**.

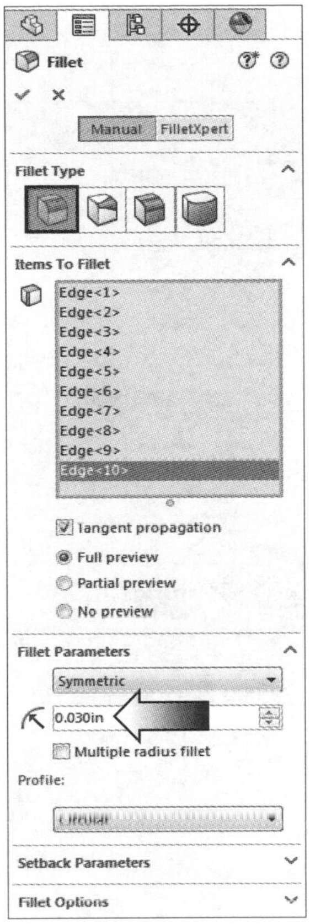

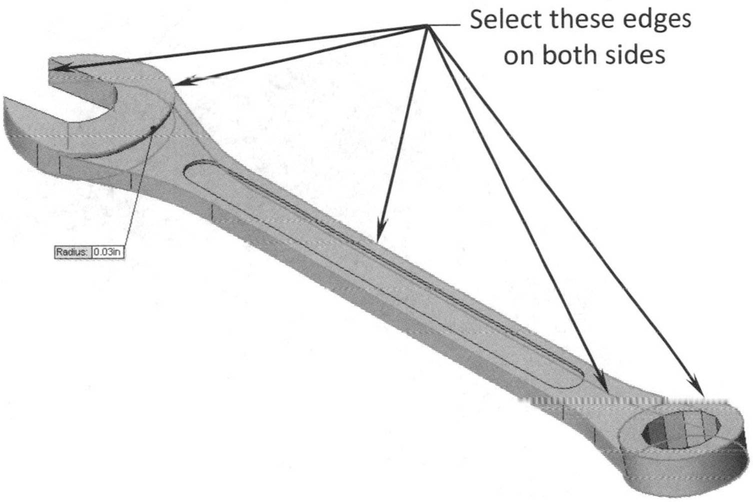

15. Adding the .050" fillets:

Repeat step 14 and add a **.050"** fillet to the **4 edges** shown below.

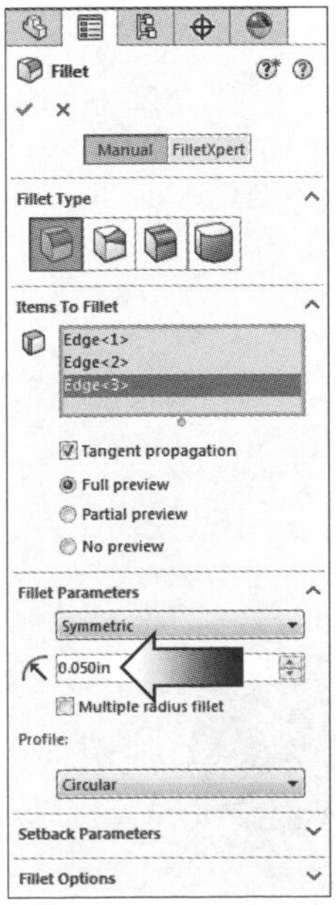

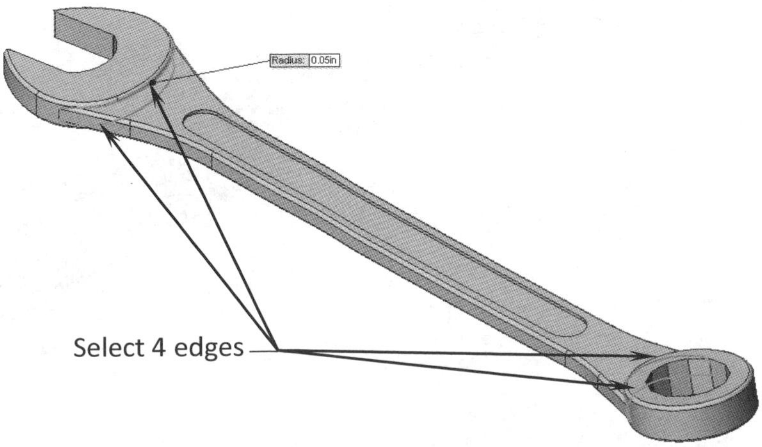

16. Adding the .015" fillets:

Click  and add a **.015"** fillet to the **edges** and **faces** shown below.

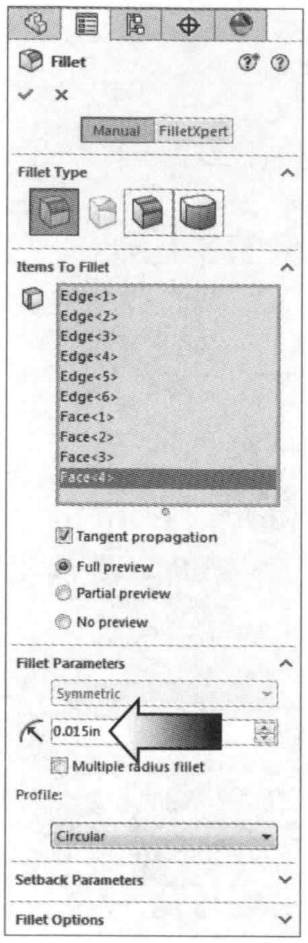

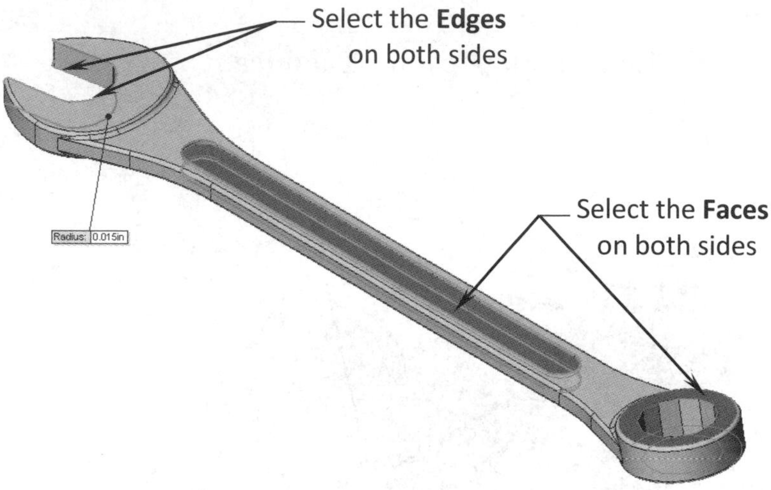

Click **OK**.

Verify your fillets with the model shown below.

17. Adding text:

Select the face indicated as sketch plane.

Click  or select **Insert / Sketch**.

Click  and type **5/8** in the text dialog box.

Click **OK**.

Add dimensions to position the text.

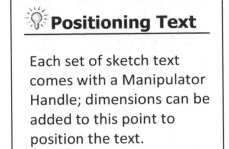

> **Positioning Text**
>
> Each set of sketch text comes with a Manipulator Handle; dimensions can be added to this point to position the text.

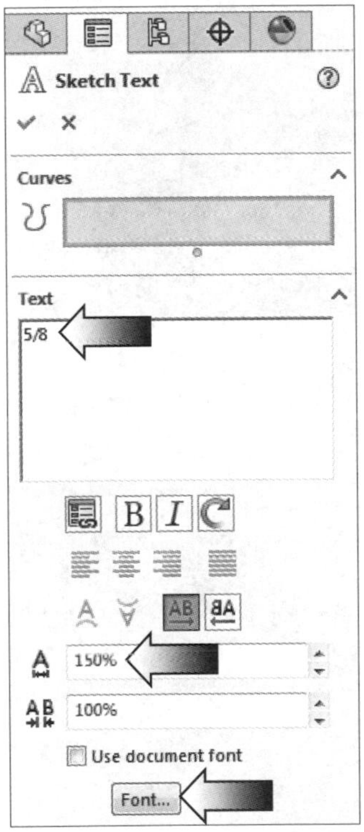

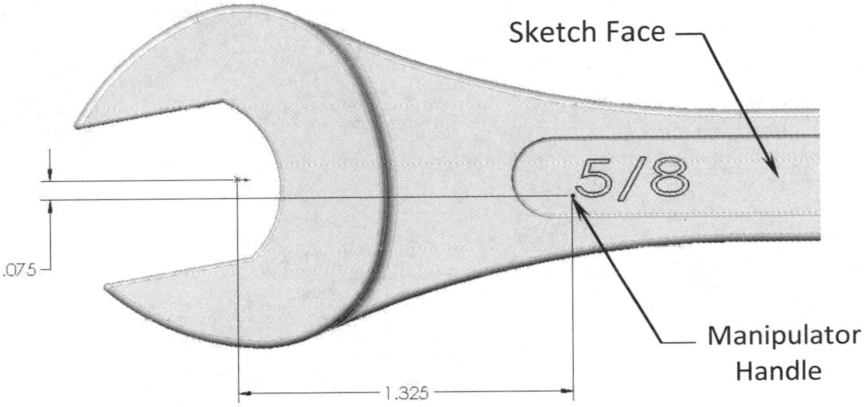

Clear Use document's font check box .

Change Width factor to **150%** .

Leave Spacing at **100%** .

Font: **Century Gothic** .

Style: **Regular** Points size: **14 pt.**

NOTE:
Use the Curves option when you want your sketch letters to wrap along a curve.

It will work better if the curve is created in the same sketch, as construction geometry.

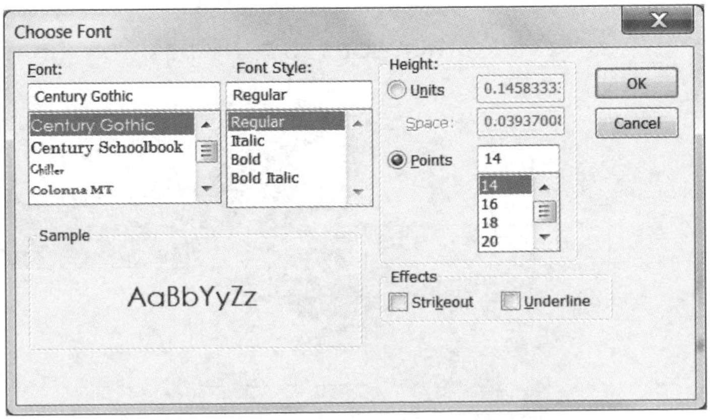

18. Extruding the text:

Click  or select **Insert / Boss-Base / Extrude.**

End Condition: **Blind.**

Extrude Depth: **.015 in.**

Click **OK.**

> **Extruding Text**
>
> Text or letters can be used as a normal sketch and extruded with drafts.
>
> Text can also be extruded as a boss or a cut feature.

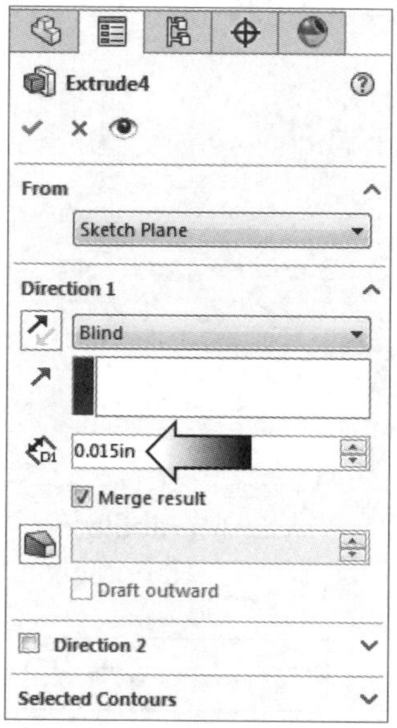

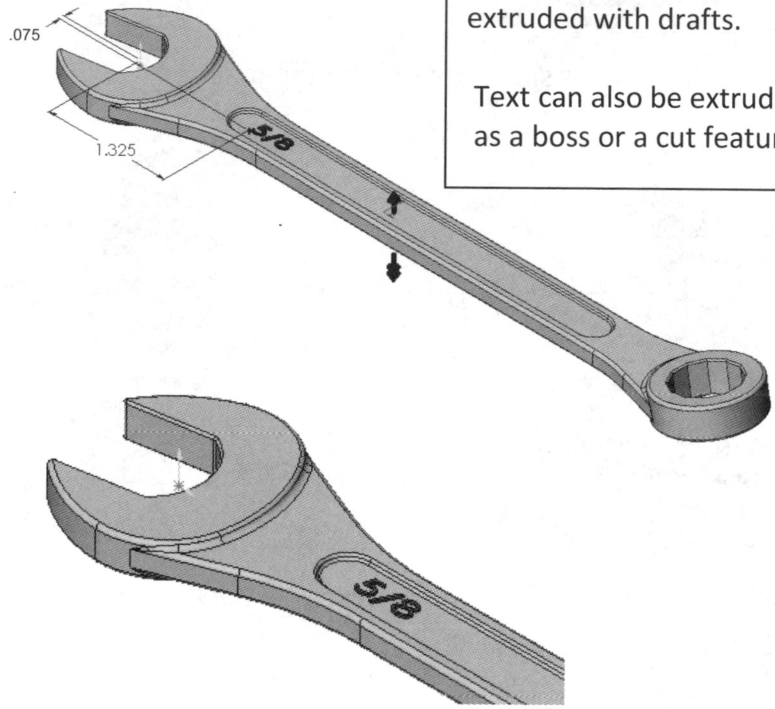

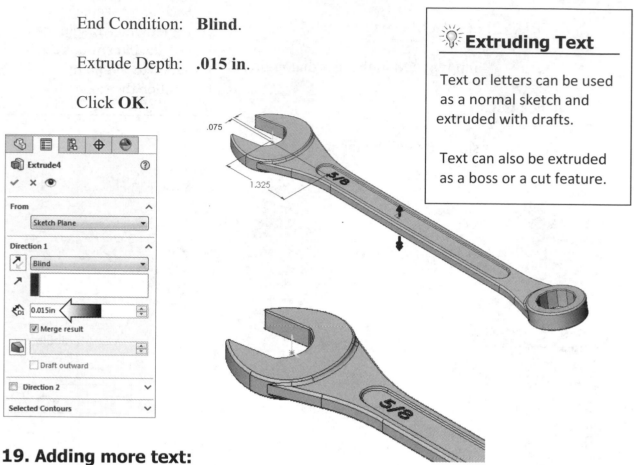

19. Adding more text:

Select the indicated face as sketch plane.

Click  or select **Insert / Sketch.**

Click  and type SPANNER in the Text dialog box.

Add dimensions to fully position the text.

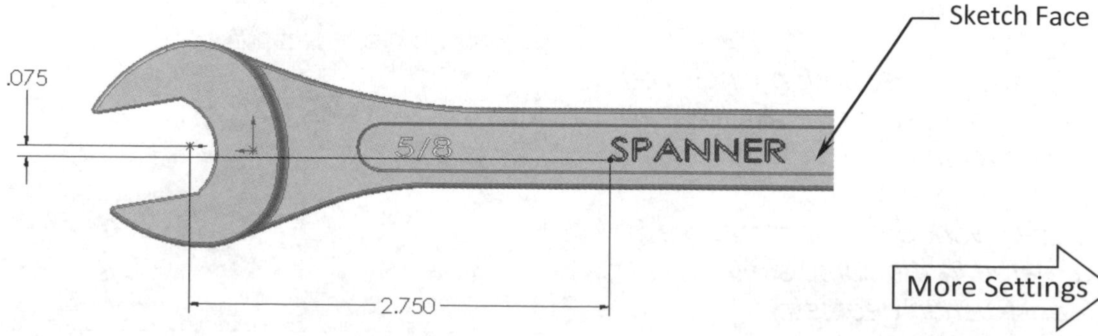

More Settings

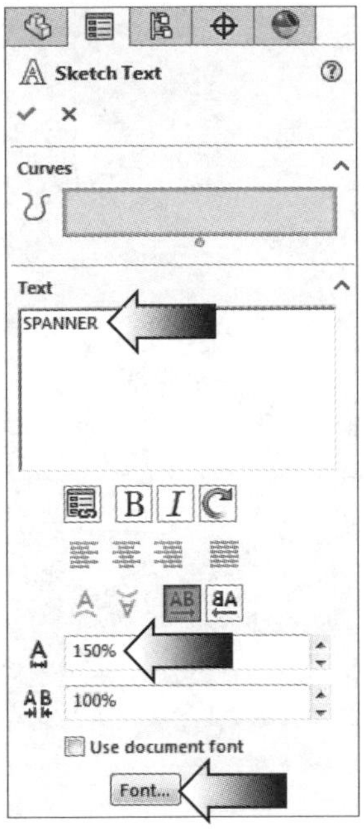

Clear Use document's font check box .

Change Width factor to **150%** .

Keep Spacing at **100%** .

Font: **Century Gothic** .

Style: **Regular**.

Points size: **14 pt**.

Click **OK**.

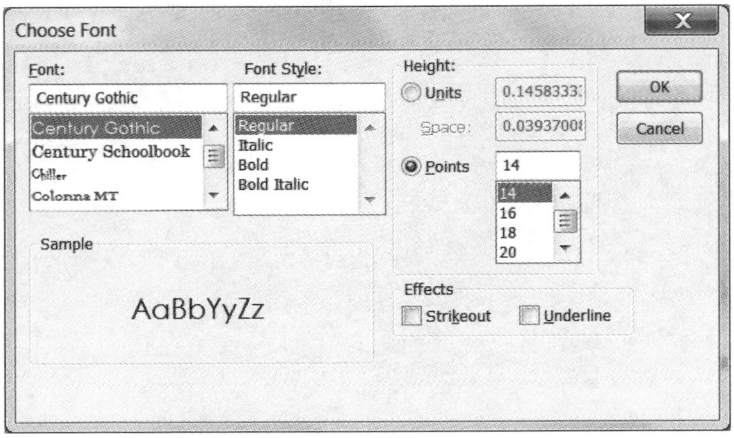

20. Extruding the text:

Click  or select **Insert / Boss-base / Extrude**.

End Condition: **Blind**.

Extrude Depth: **.015 in**.

Click **OK**.

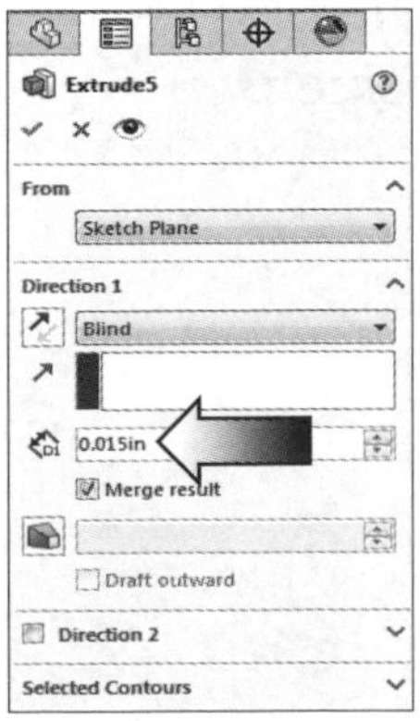

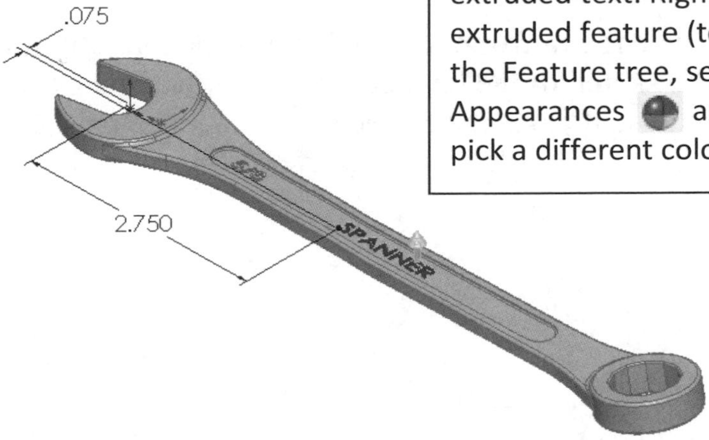

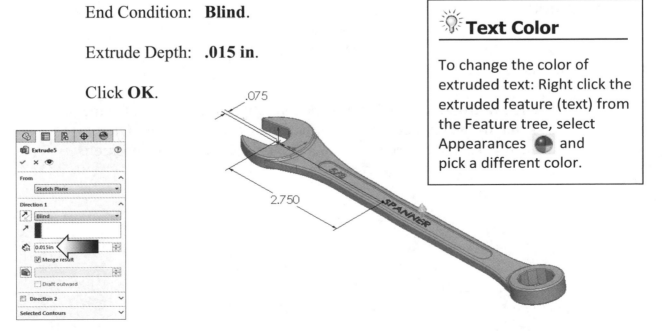

> **Text Color**
>
> To change the color of extruded text: Right click the extruded feature (text) from the Feature tree, select Appearances and pick a different color.

21. Saving your work:

Select **File / Save as / Spanner / Save**.

22. Optional:

To add the same text on the opposite side of the part, repeat from step 17 through step 20.

Since the mirror option will not work correctly for text, you can either copy the sketch of the text, edit it, reposition, and extrude it again - OR - copy and paste the extruded text and then edit its sketch to reposition.

Questions for Review

Advanced Modeling

1. The Min / Max conditions can be selected from the dimensions properties, under the Leaders tab.
    a. True
    b. False

2. The Mid-Plane extrude type protrudes the sketch profile to both directions equally.
    a. True
    b. False

3. It is sufficient to create a plane at an angle with a surface and an angular dimension.
    a. True
    b. False

4. When sketching a polygon, the number of sides can be changed on the Properties tree.
    a. True
    b. False

5. A 3D solid feature can be mirrored using a centerline as the center of mirror.
    a. True
    b. False

6. Text cannot be used to extrude as a boss or a cut feature.
    a. True
    b. False

7. Extruded text can be mirrored just like any other 3D features.
    a. True
    b. False

8. Text in a sketch can be extruded with drafts, inward or outward.
    a. True
    b. False

1. TRUE 2. TRUE
3. FALSE 4. TRUE
5. FALSE 6. FALSE
7. TRUE 8. TRUE

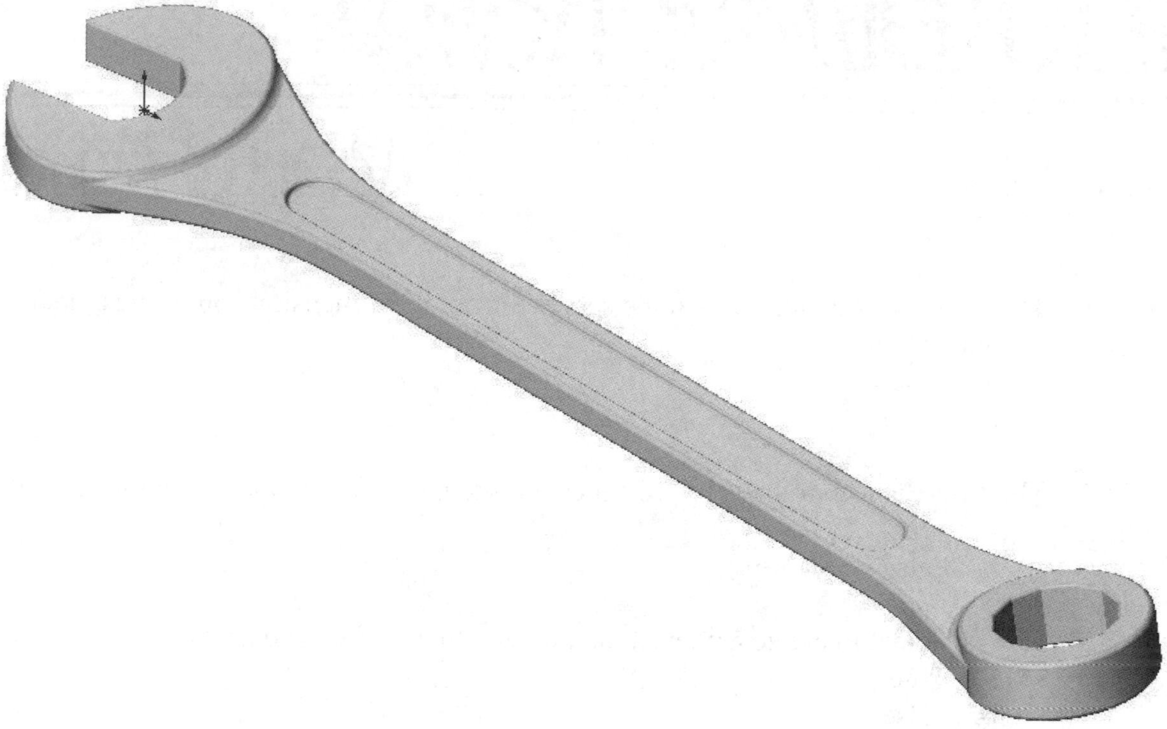

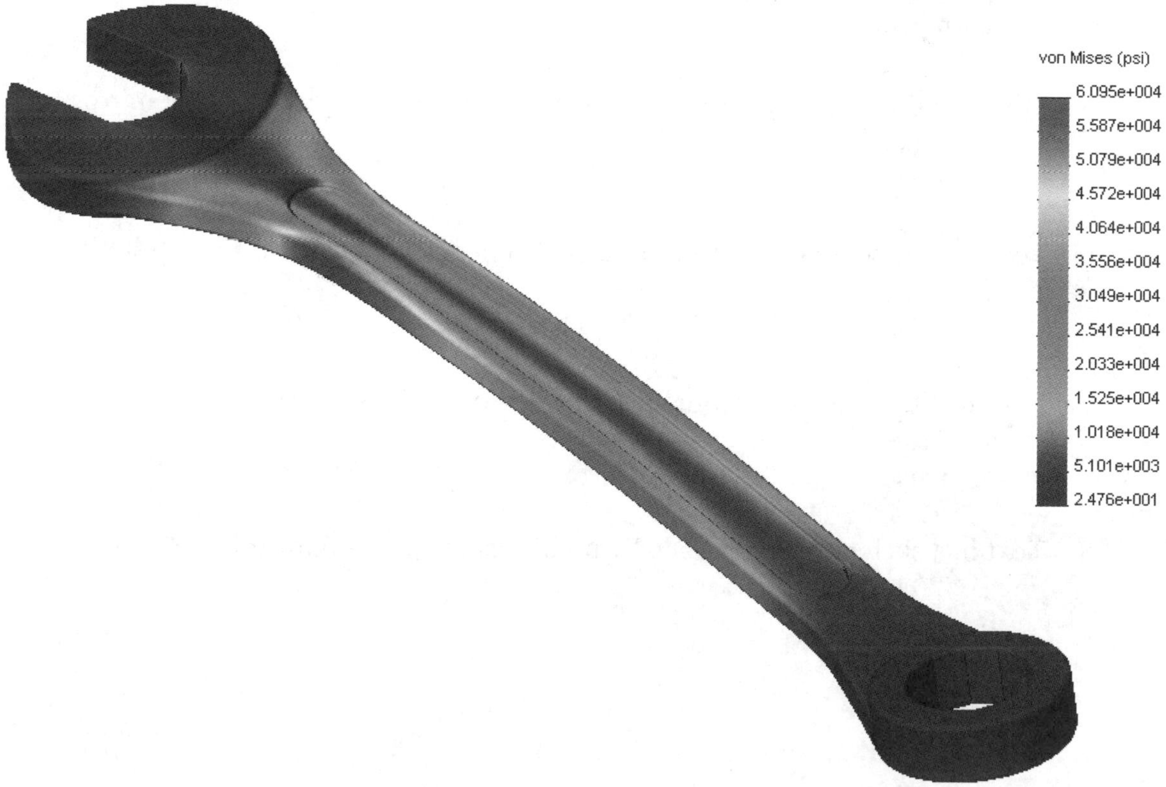
von Mises (psi)
6.095e+004
5.587e+004
5.079e+004
4.572e+004
4.064e+004
3.556e+004
3.049e+004
2.541e+004
2.033e+004
1.525e+004
1.018e+004
5.101e+003
2.476e+001

<u>Exercise:</u> Circular Text Wraps

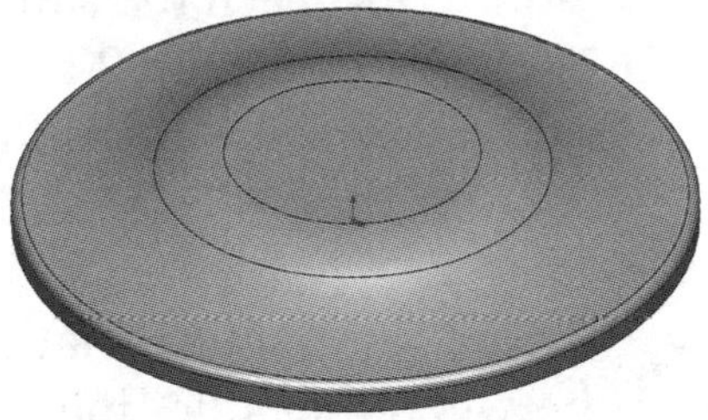

1. Opening a part file:

From the Training Files folder, open an existing part named: **Text Wrap**.

2. Adding Text:

Select the <u>Top</u> plane and open a new sketch.

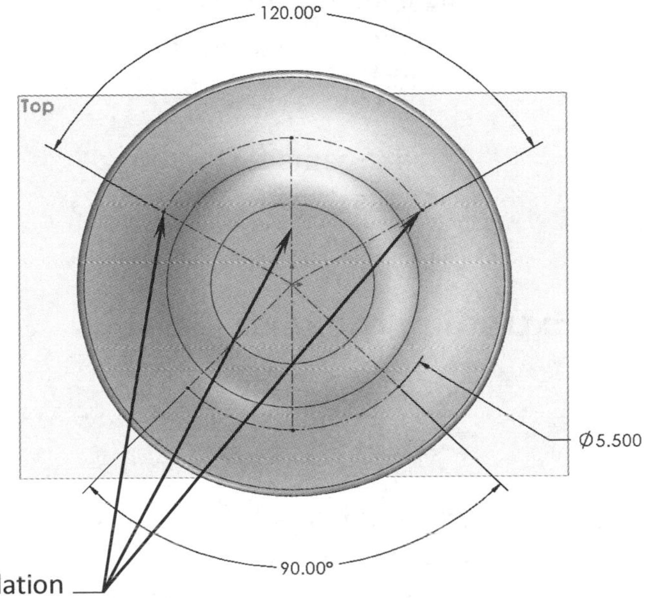

Sketch a circle at **Ø5.500**" and convert it into construction geometry.

Sketch the other centerlines, trim, then add the dimensions and relations as indicated.

Click the **Text** command.

Enter the text: **SolidWorks Text**.

Symmetric relation (both top and bottom) between the endpoints and the centerlines

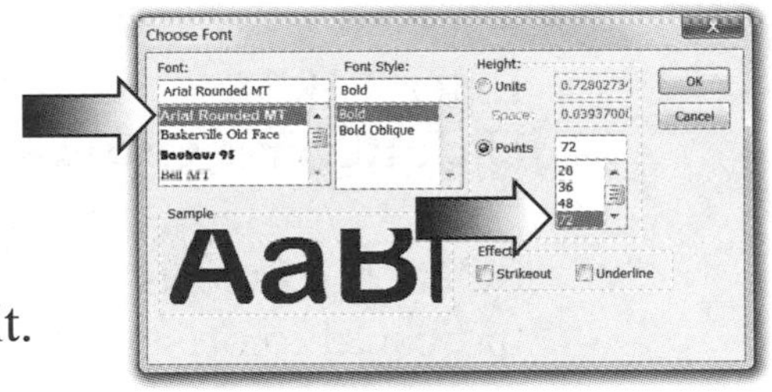

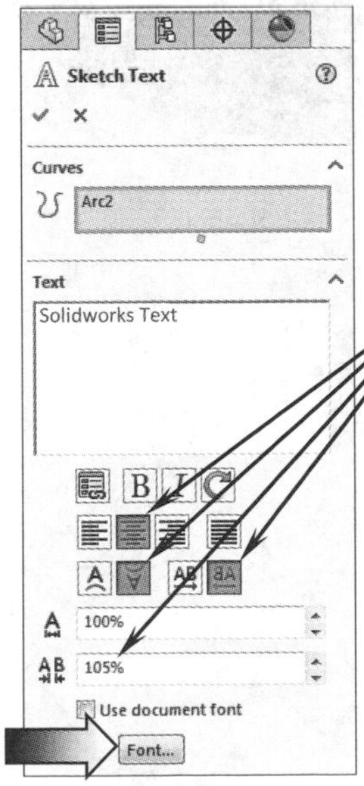

Select the upper construction curve to bend the text around it.

Select these options

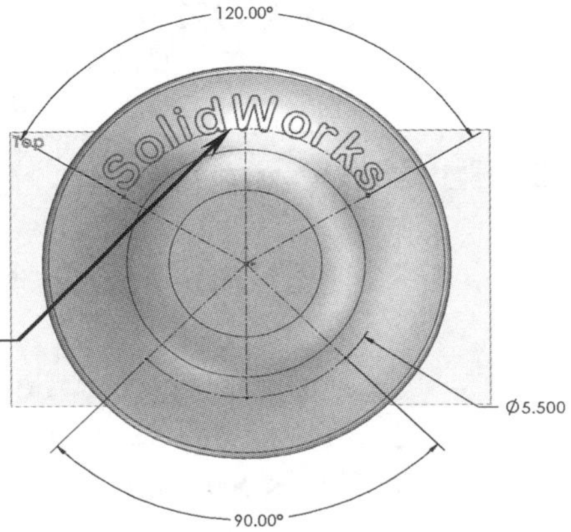

Click the **Font** button and set the size to **72 points.**

Select this curve

Select all other options as indicated to align the text.

3. Repeating:

Still working in the same sketch, repeat step 2 and add the word: **Text** at the bottom.

Add a Symmetric relation between the endpoints of the construction curve and the vertical centerline.

Use the same text settings as the last text.

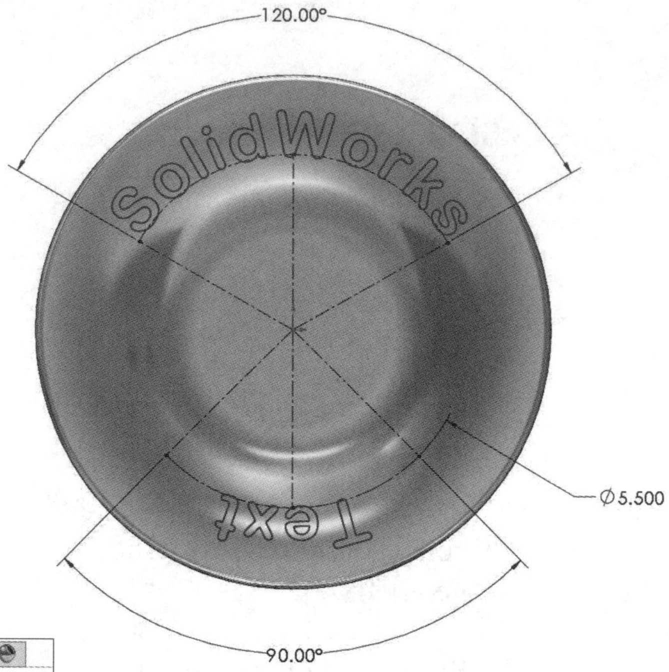

4. Extruding the text:

Click **Extruded Boss-Base**.

Change Direction 1 to **Offset From Surface**.

Enter **.030"** and click **Reverse Offset**.

Select the **face** as indicated to offset from.

Click **OK**.

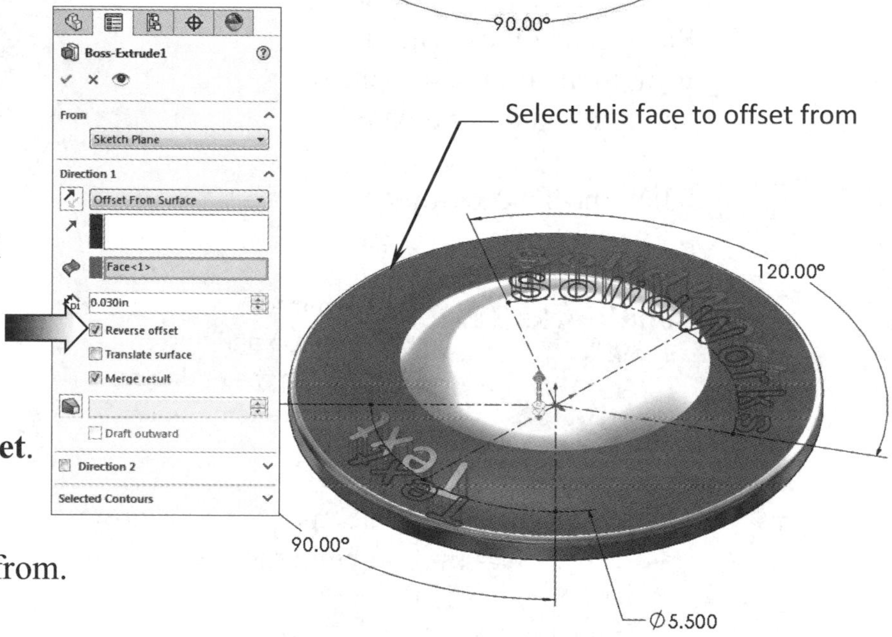

5. Saving your work:

Click **File / Save As**.

Enter **Circular Text Wrap**.

Press **Save**.

CHAPTER 4

Sweep With Composite Curves

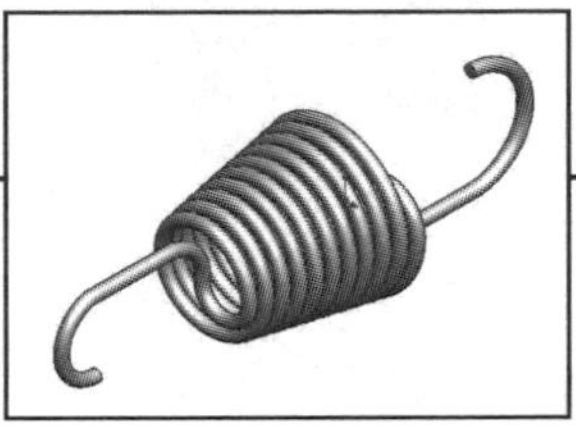

Sweep with Composite Curves
Helical Extension Spring

Unlike extruded or revolved shapes, the sweep option offers a more advanced way to creating complex geometry, where a single profile can be swept along 2D guide path or 3D curves to define the feature.

To create a sweep feature the Sweep Path gets created first, then a single closed sketch Profile.

The Profile will be related to the Sweep Path with a Pierce or a coincident relation.

When the Profile is swept along the Sweep Path, the Guide Curves can help control the feature's accuracy and its behaviors like twisting, tangencies, etc.

The Composite Curve option allows multiple sketches or model edges to be jointed into one continuous path to use in sweep features. (The sketches must be connecting with one another in order for the composite curve to work.)

This lesson will guide you through the creation of a helical extension spring where several 2D sketches and a 3D helix are combined into a single curve. This curve is called: Composite Curve.

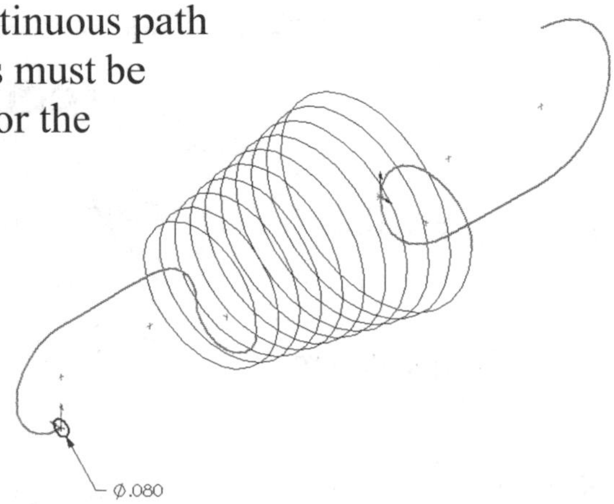

Helical Extension Spring
Sweep with Composite Curves

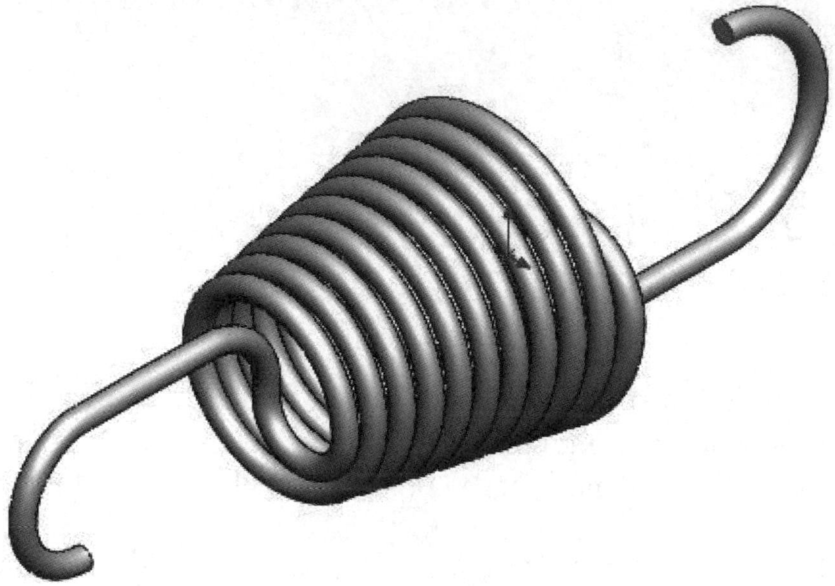

View Orientation Hot Keys:

Ctrl + 1 = Front View
Ctrl + 2 = Back View
Ctrl + 3 = Left View
Ctrl + 4 = Right View
Ctrl + 5 = Top View
Ctrl + 6 = Bottom View
Ctrl + 7 = Isometric View
Ctrl + 8 = Normal To Selection

Dimensioning Standards: **ANSI**

Units: **INCHES** – 3 Decimals

Tools Needed:

Insert Sketch	Line	Circle
Tangent Arc	3 Point Arc	Add Geometric Relations
Dimension	Composite Curve	Sweep

1. Sketching the first profile:

Select the Front plane from the FeatureManager Tree.

Click  or **Insert / Sketch**.

Sketch a **Circle** and **two centerlines**, then add the diameter dimension shown below.

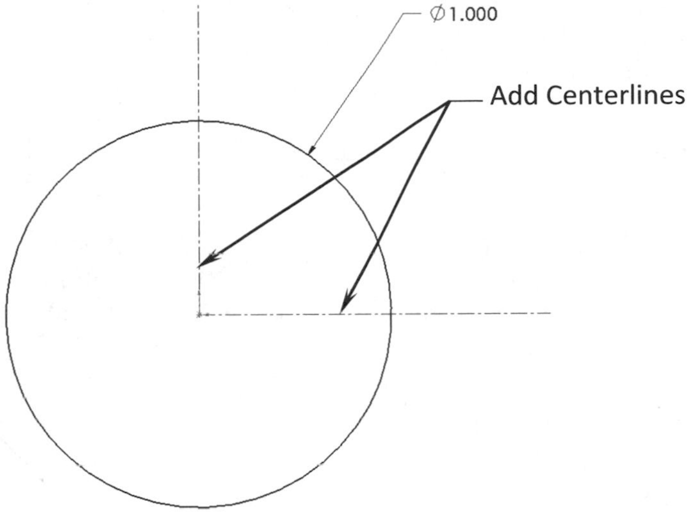

2. Converting to a Helix:

Select **Insert / Curve / Helix / Spiral**.

Defined by:	**Pitch and Revolution**
Pitch:	**.100**
Revolution:	**10**
Starting angle:	**0°**
Taper helix:	**Enabled**
Taper angle:	**10°**

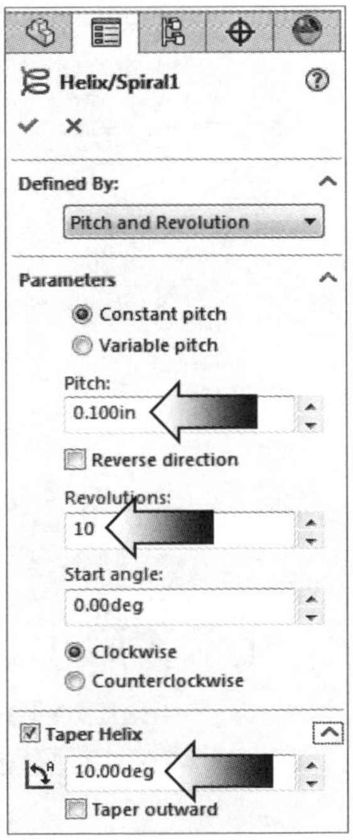

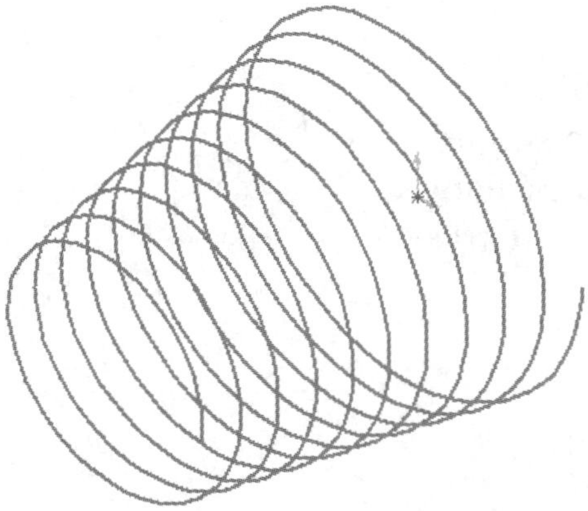

Click **OK**.

3. Creating a 2-degree plane:

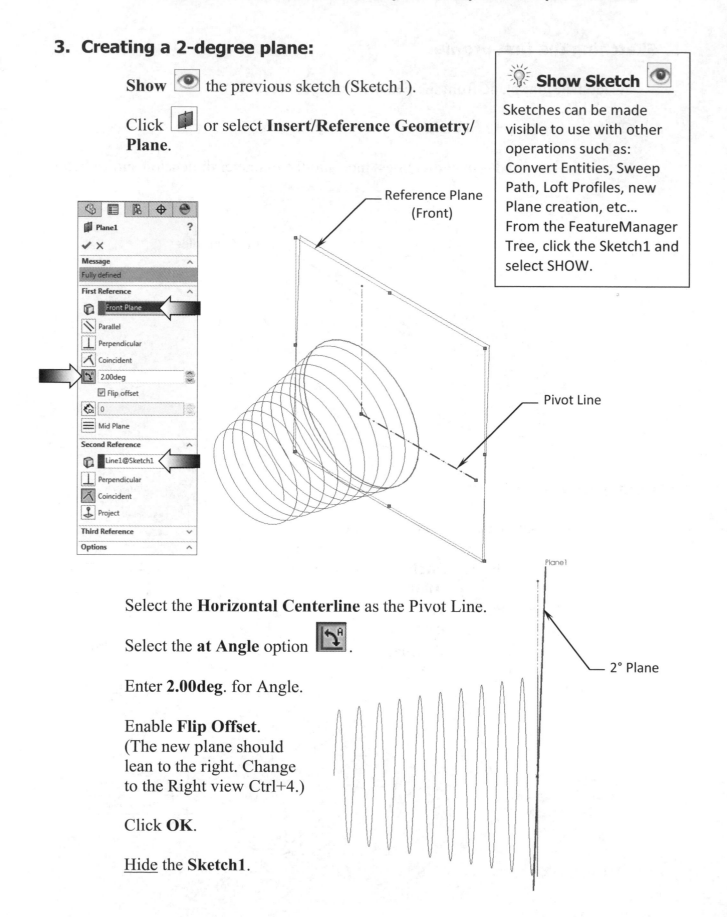

Show the previous sketch (Sketch1).

Click or select **Insert/Reference Geometry/ Plane**.

Show Sketch

Sketches can be made visible to use with other operations such as: Convert Entities, Sweep Path, Loft Profiles, new Plane creation, etc... From the FeatureManager Tree, click the Sketch1 and select SHOW.

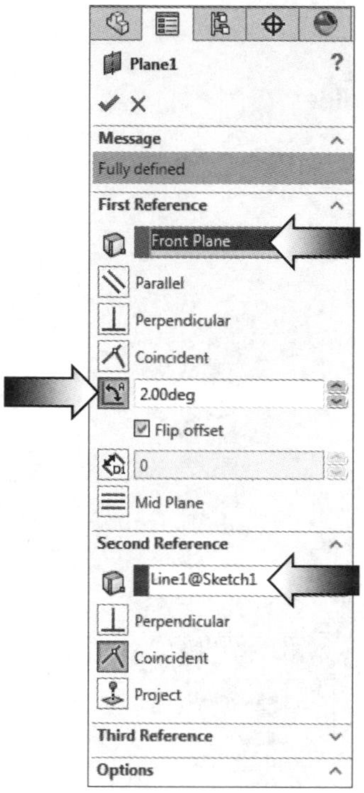

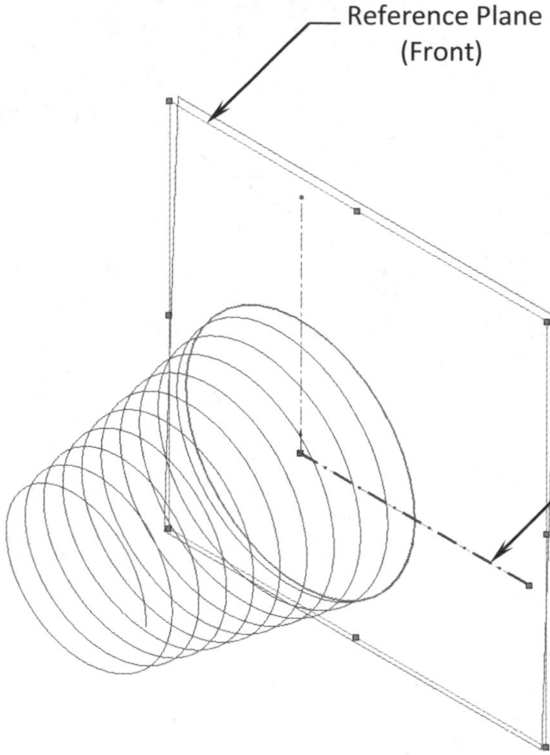

Select the **Horizontal Centerline** as the Pivot Line.

Select the **at Angle** option.

Enter **2.00deg**. for Angle.

Enable **Flip Offset**.
(The new plane should lean to the right. Change to the Right view Ctrl+4.)

Click **OK**.

Hide the **Sketch1**.

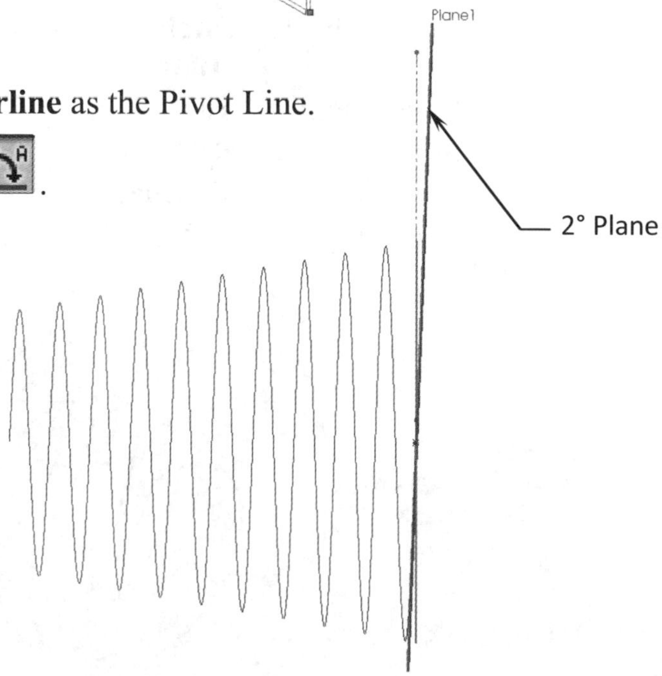

4. Sketching the large loop:

Select the Plane1 from the FeatureManager Tree.

Click or select **Insert / Sketch**.

Sketch a **3-point Arc** and add dimension shown:

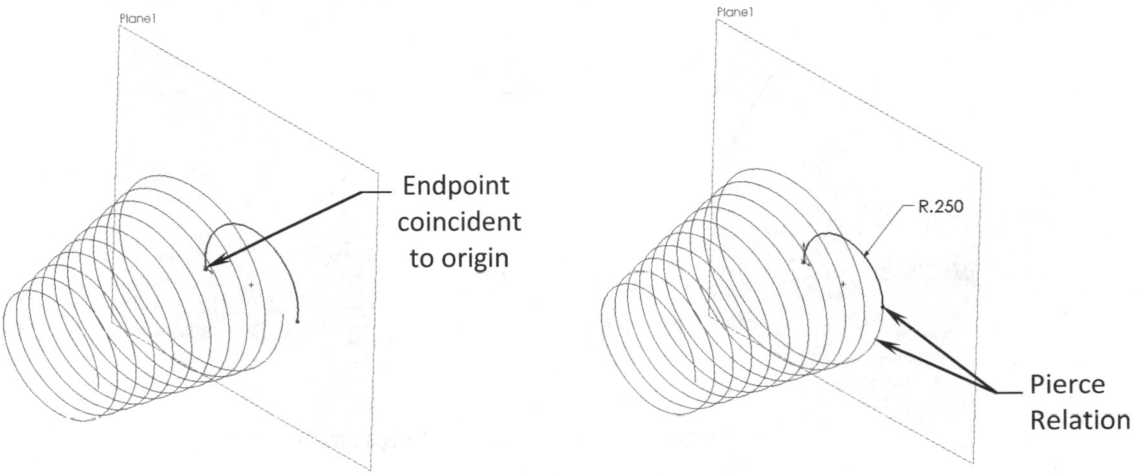

Add a **Pierce** relation between the end point of the Arc and the Helix.

Exit the sketch or select **Insert / Sketch**.

5. Sketching the large hook:

Select the Right plane from the FeatureManager Tree.

Click or select **Insert / Sketch**.

Sketch the profile and add dimension and relations as shown below:

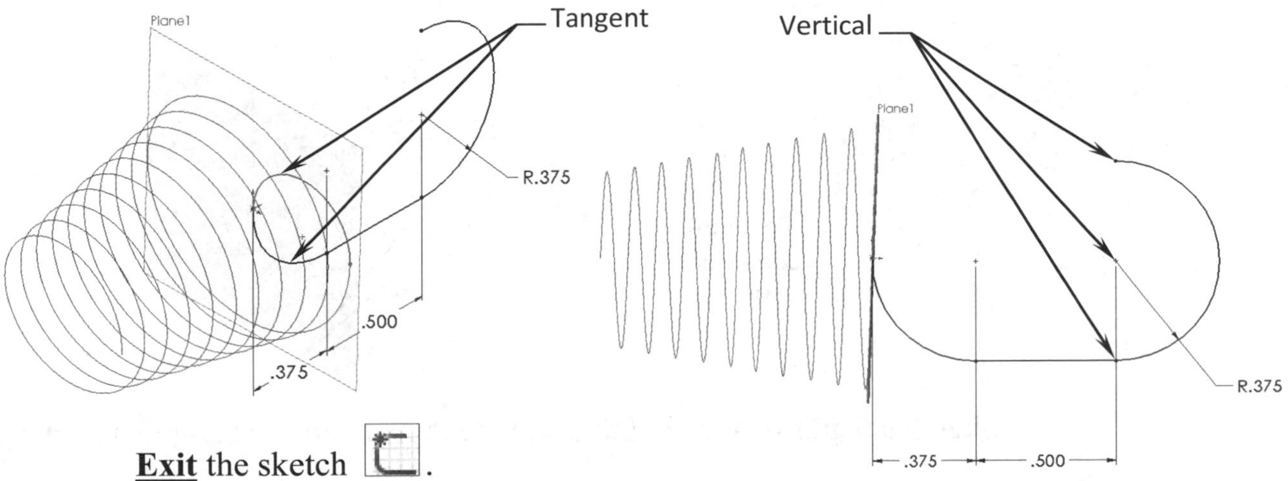

Exit the sketch .

6. Creating a Parallel plane:

Select the Plane1 from the FeatureManager Tree.

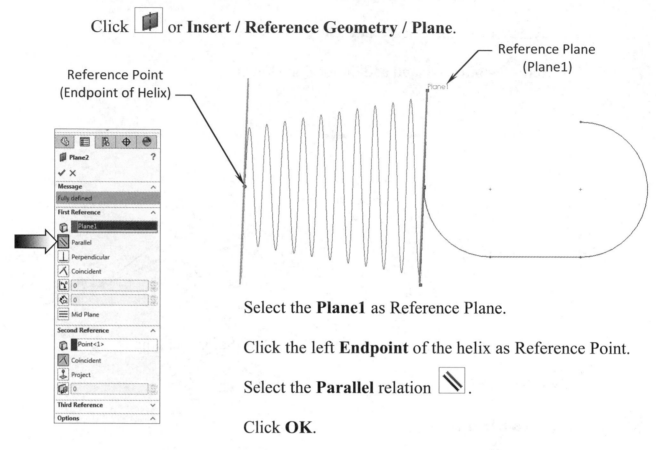

Click  or **Insert / Reference Geometry / Plane.**

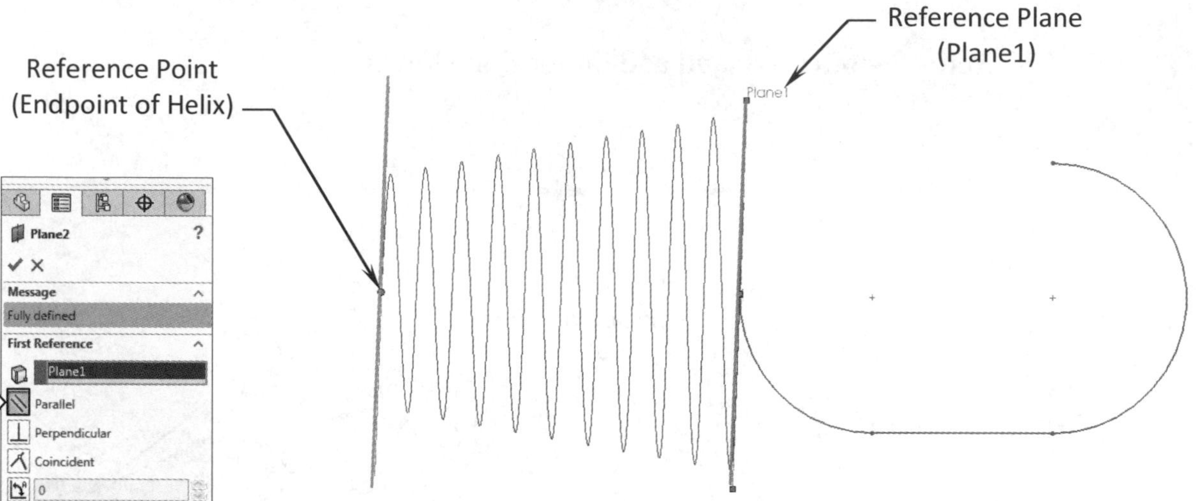

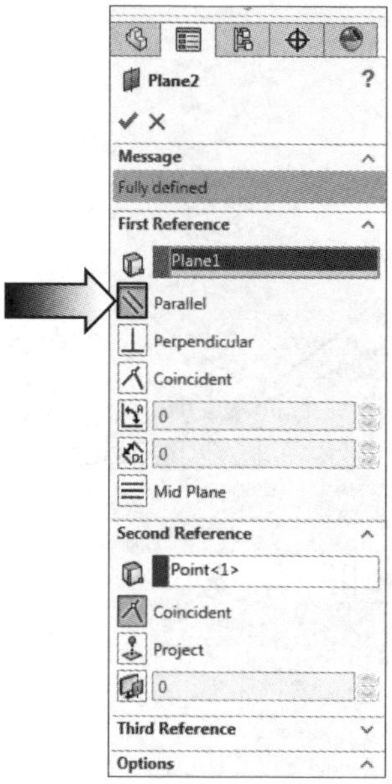

Select the **Plane1** as Reference Plane.

Click the left **Endpoint** of the helix as Reference Point.

Select the **Parallel** relation .

Click **OK**.

7. Adding the small loop:

Select the new plane (Plane2) from the FeatureManager Tree.

Click  or **Insert / Sketch.**

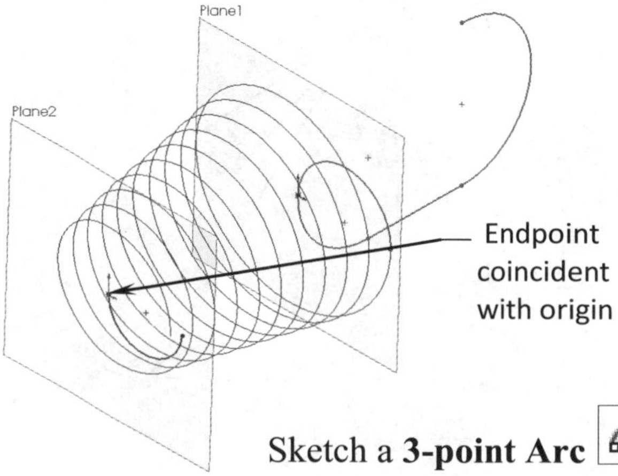

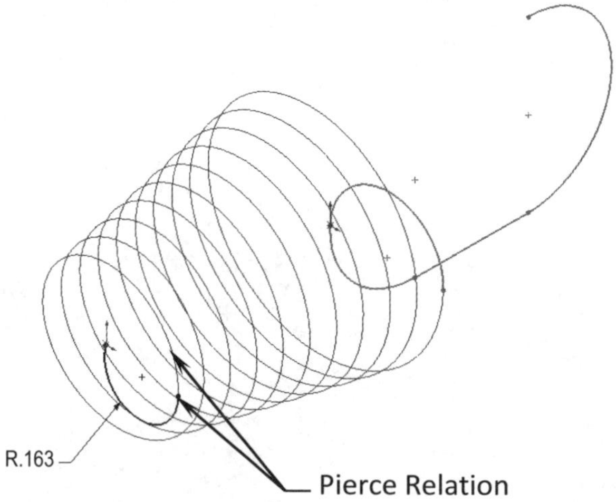

Sketch a **3-point Arc**  and add the radius dimension shown above.

Add a **Pierce** relation between the endpoint of the Arc and the Helix.

Exit the sketch 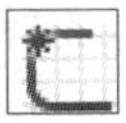 or **Insert / Sketch**.

8. Creating a small hook:

Select the Right plane from the FeatureManager Tree.

Click  or **Insert / Sketch**.

Sketch the profile and add the dimensions shown.

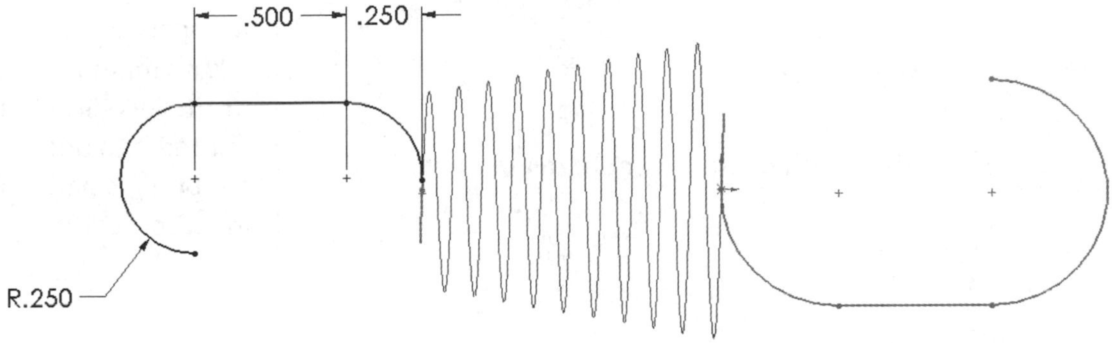

Add the relations Vertical and Tangent to thc indicated entities.

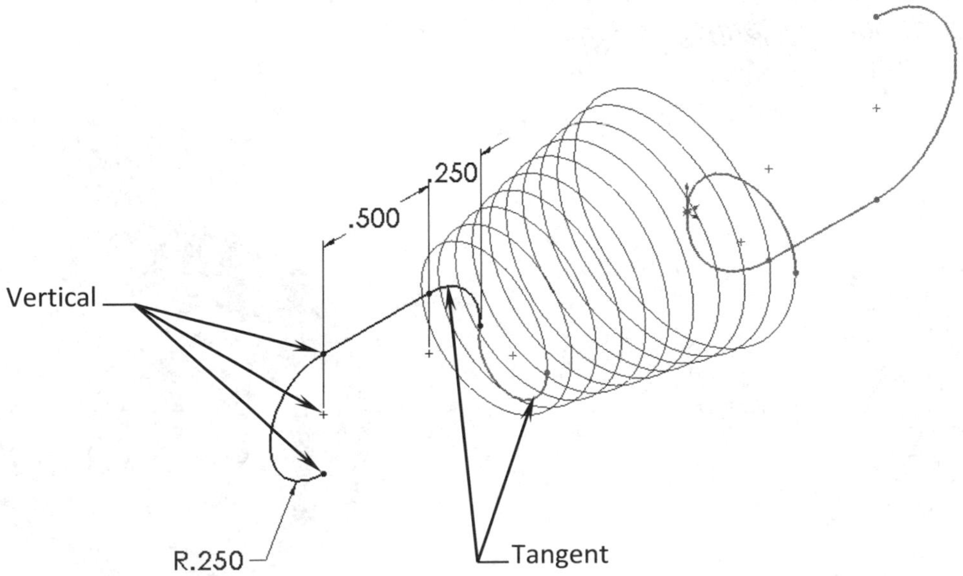

Exit the sketch 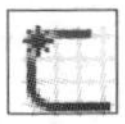 or click **Insert / Sketch**.

9. Creating a Composite Curve:

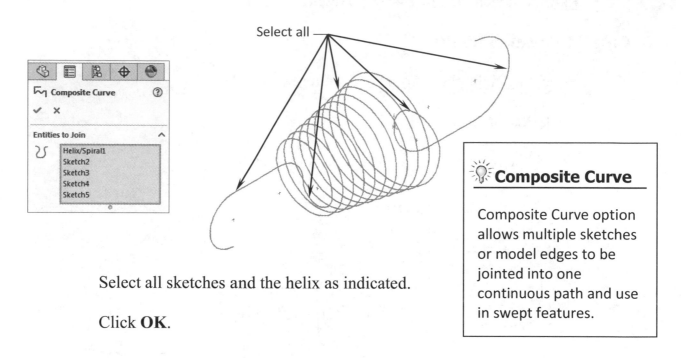

Click under the Curves drop-down or select: **Insert / Curve / Composite**.

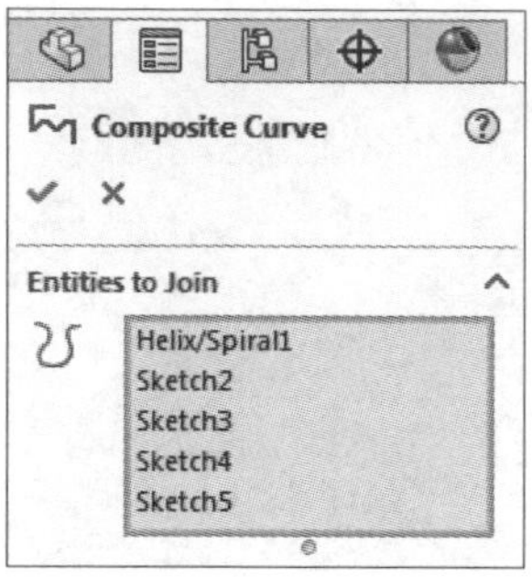

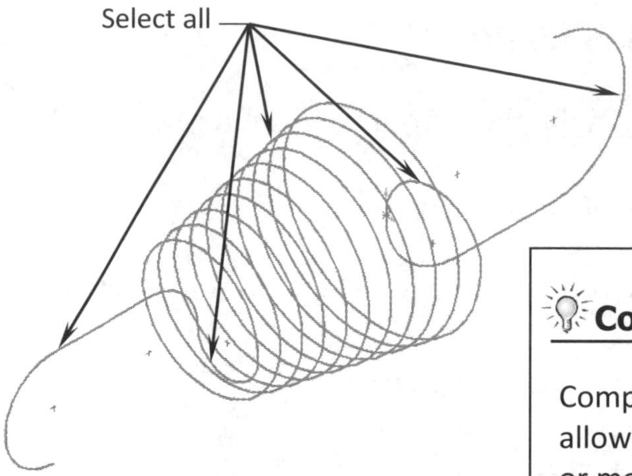

Select all sketches and the helix as indicated.

Click **OK**.

> **Composite Curve**
>
> Composite Curve option allows multiple sketches or model edges to be jointed into one continuous path and use in swept features.

NOTE:

*The transitions sketch between the helix and the sketch of the hook is not perfectly tangent. Since we cannot add a fillet between the helix (3D) and the hook (2D) we will have to try another approach called **Fit Spline**, to smooth out any tangency issues in the model.*

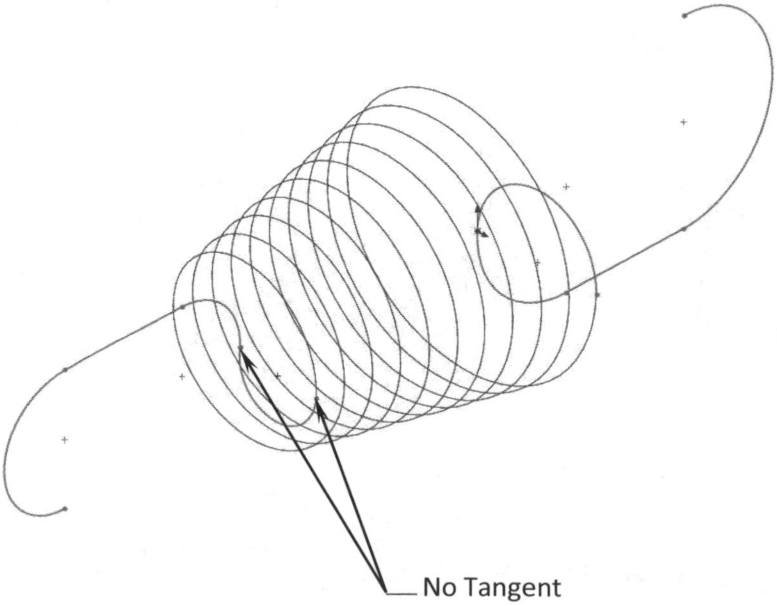

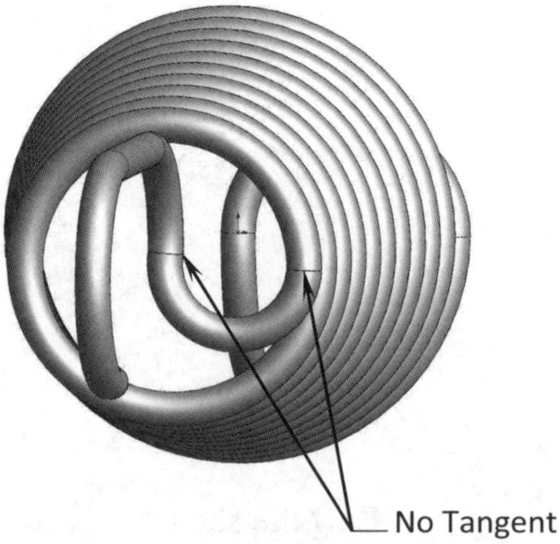

10. Converting to 3D sketch:

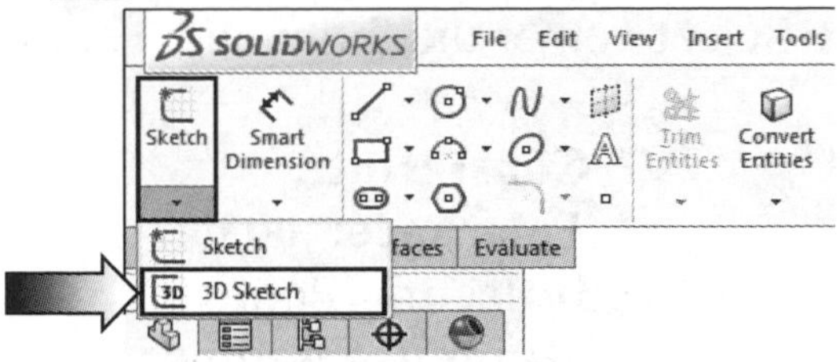

The Fit Spline command can only be used with sketch entities, not 3D curves. So the next step is to convert the composite curve to a 3D sketch.

Click **3D Sketch** under the Sketch drop-down menu (arrow).

Select the Composite Curve and click **Convert Entities** .

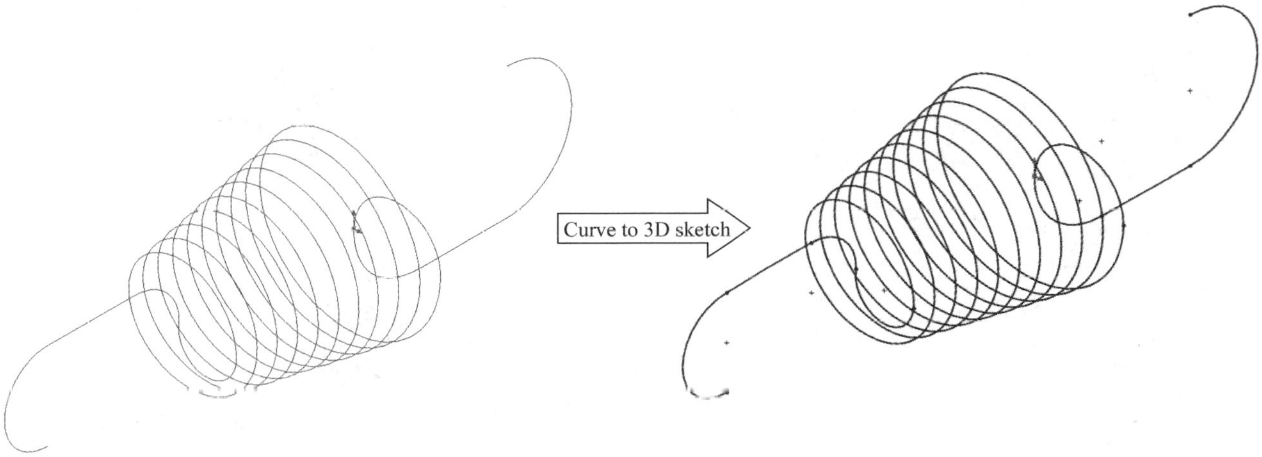

The composite curve is converted to a 3D sketch.

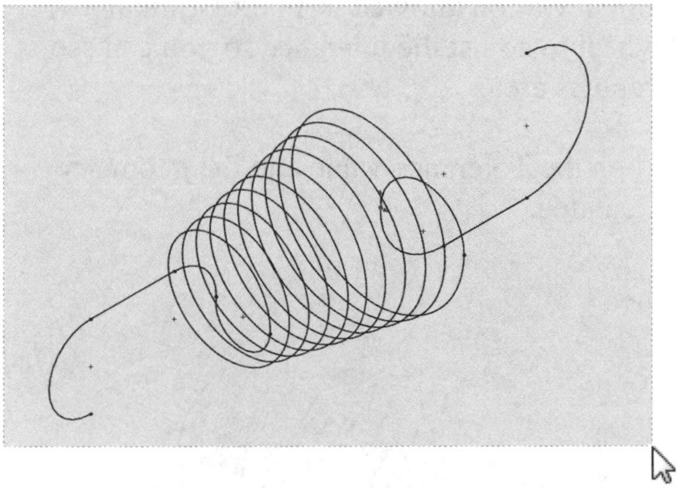

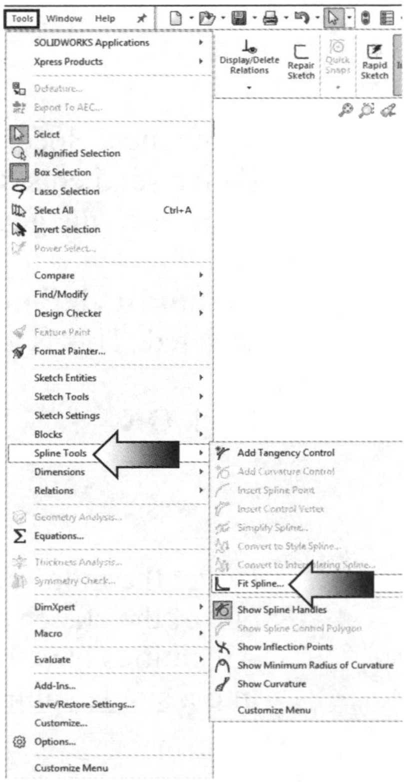

Box-select the entire 3D sketch (or press Control + A) and click:

Tools / Spline Tools / Fit Spline .

Select / enter the following:

> **Fit Spline tool**
>
> The Fit Spline tool to fit sketch segments to a spline. Fit splines are parametrically linked to underlying geometry so that changes to the geometry update the spline.

- *** Constraint.**
- *** Tolerance: .003in.**
- *** Inflection Points.**
- *** Minimum Radius.**
- *** Curvature Comb.**

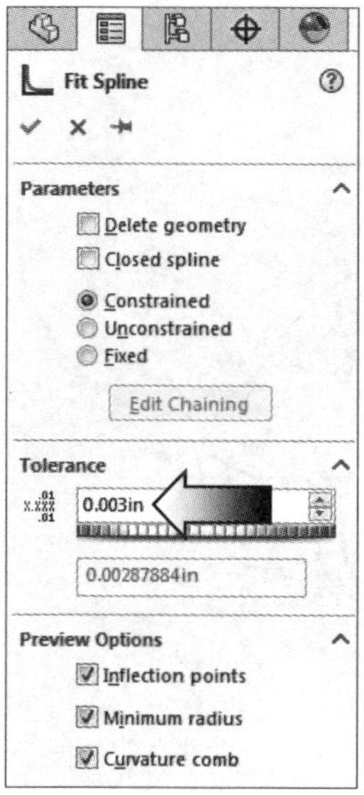

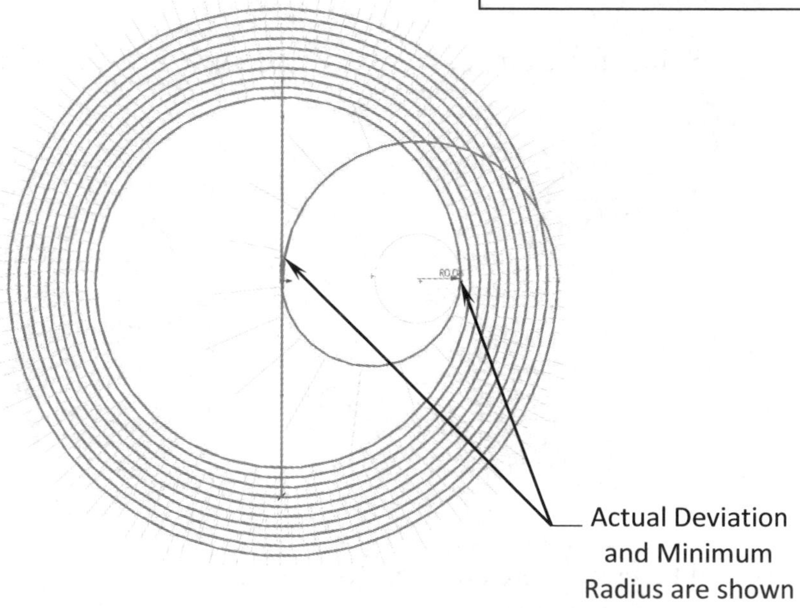

Tolerance: Specifies the maximum deviation allowed from the original sketch segments. Use the thumbwheel to adjust the tolerance so you can see changes to the geometry in the graphics area.

Actual Deviation: Updates based on the Tolerance value and the geometry selected. This is automatically calculated.

Click **OK**.

Hover the mouse cursor over one of the sketch segments. All entities have been fitted into a single spline.

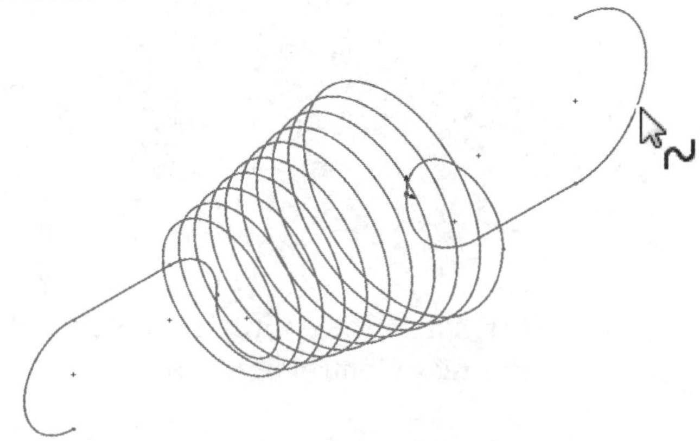

11. Sweeping the profile along the path:

Click  on the Features toolbar or select: **Insert / Boss-Base / Sweep**.

Click the **Circular Profile** button and enter **.080in**. for Profile Diameter.

For Sweep Path, select the **Composite Curve** in the graphics area.

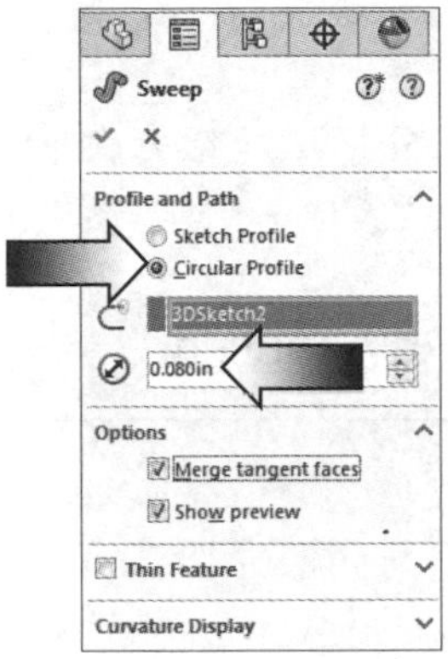

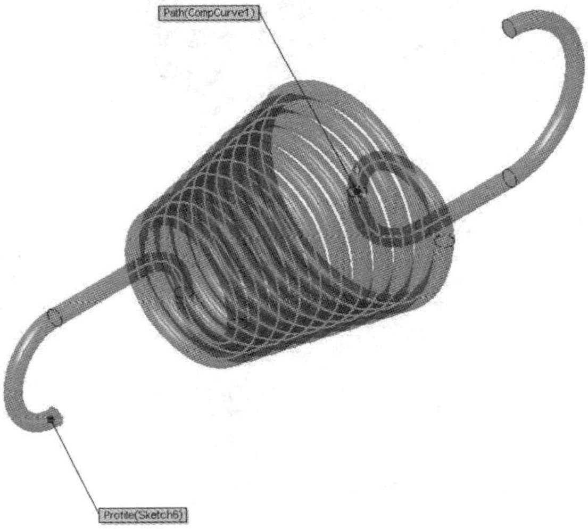

Click **OK**.

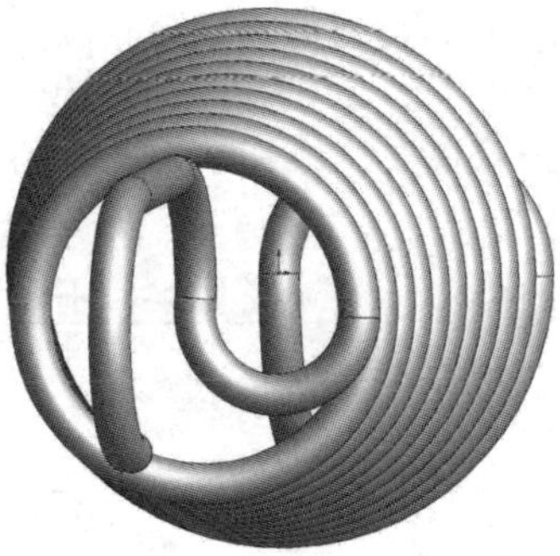

Before Fit Spline

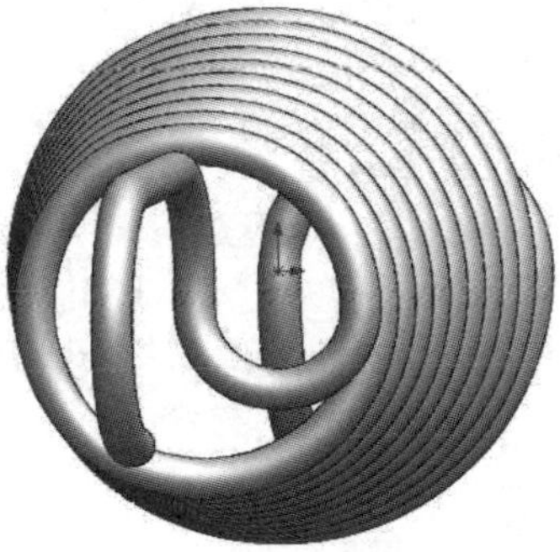

After Fit Spline

12. Saving your work:

Click **File / Save As /**

Enter **Helical Extension Spring** for the file name and click **Save**.

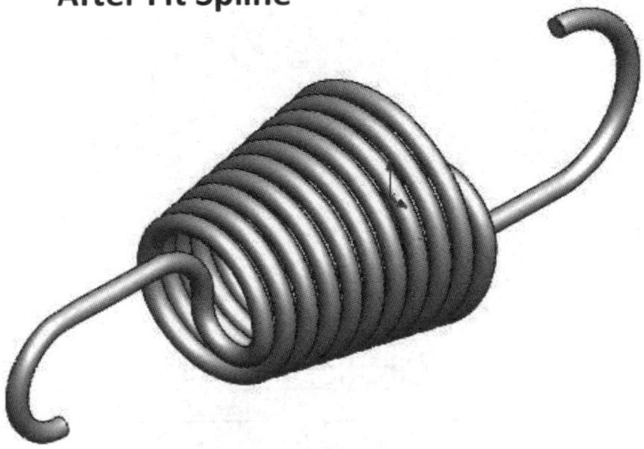

Other Examples:

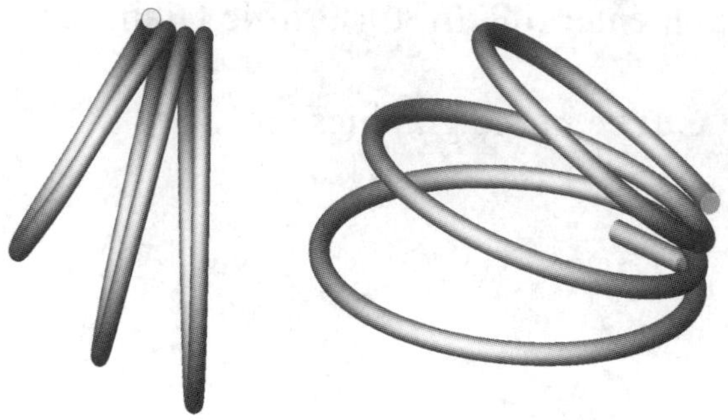

Conical Spring

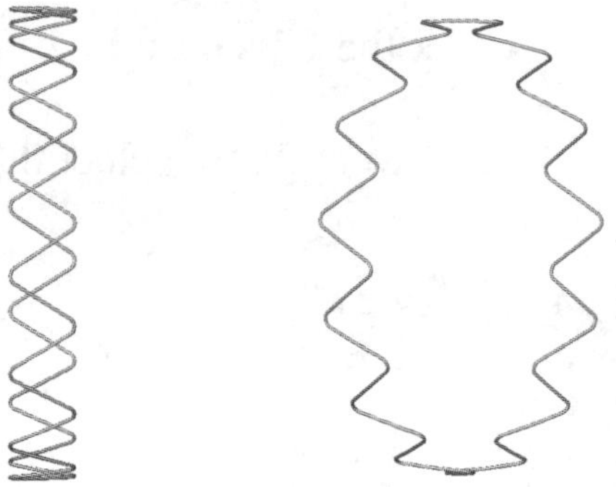

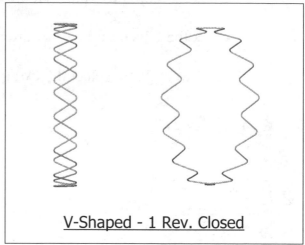

V-Shaped - 1 Rev. Closed

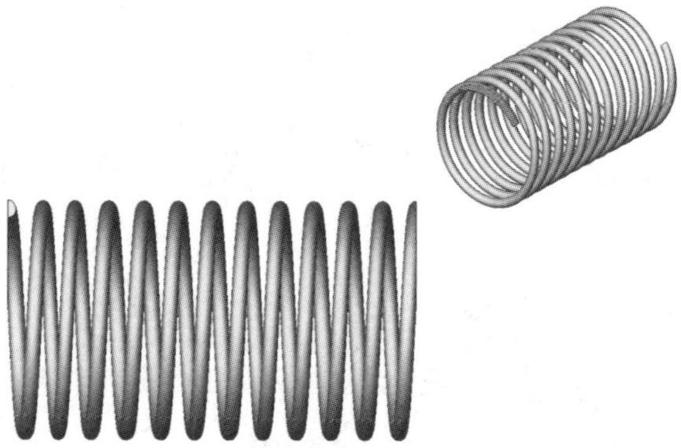

Ground-End

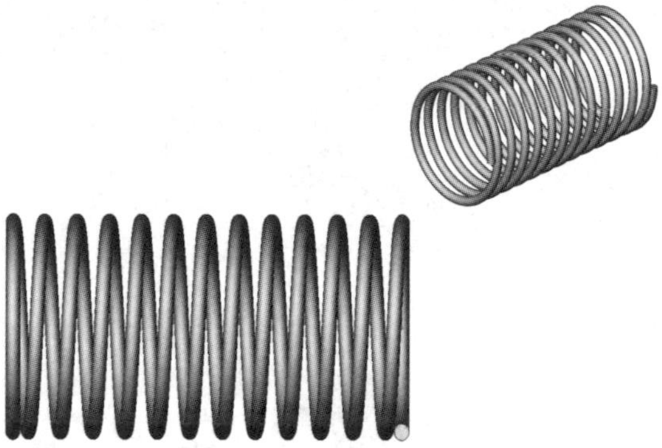

Closed-End

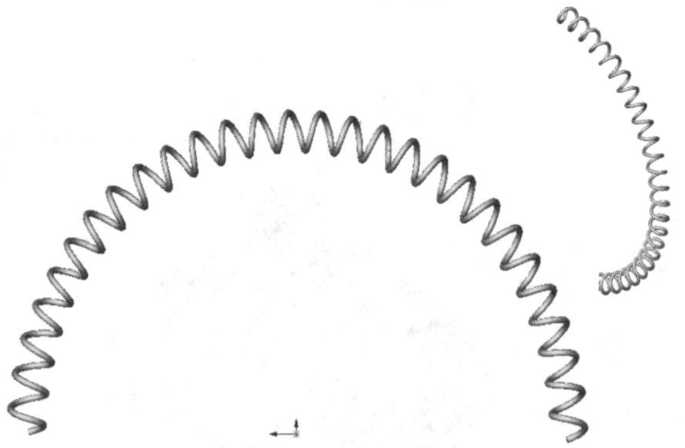

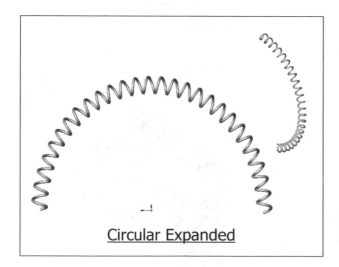

Circular Expanded

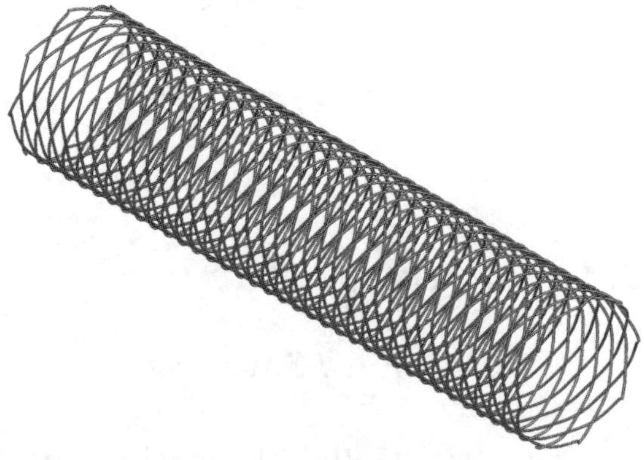

16 Wires - Braided

Questions for Review

Sweep with Composite Curve

1. Besides the Pitch and Revolution option, a helix can be defined with Pitch and Height.
   a. True
   b. False

2. It is sufficient to create an Offset Distance plane using a reference plane and a distance.
   a. True
   b. False

3. The sweep profile should have a Pierce relation with the sweep path.
   a. True
   b. False

4. Several sketches or model edges can be combined to make a Composite curve.
   a. True
   b. False

5. A Composite curve cannot be used as a sweep path.
   a. True
   b. False

6. The composite curve combines all sketches and model edges into one continuous curve, even if they are not connected.
   a. True
   b. False

7. In a sweep feature, SOLIDWORKS allows only one sweep path, but multiple guide curves can be used.
   a. True
   b. False

8. Several sketch profiles can be used to sweep along a path.
   a. True
   b. False

1. TRUE 2. TRUE
3. TRUE 4. TRUE
5. FALSE 6. FALSE
7. TRUE 8. FALSE

Exercise: Circular Spring – 180deg.

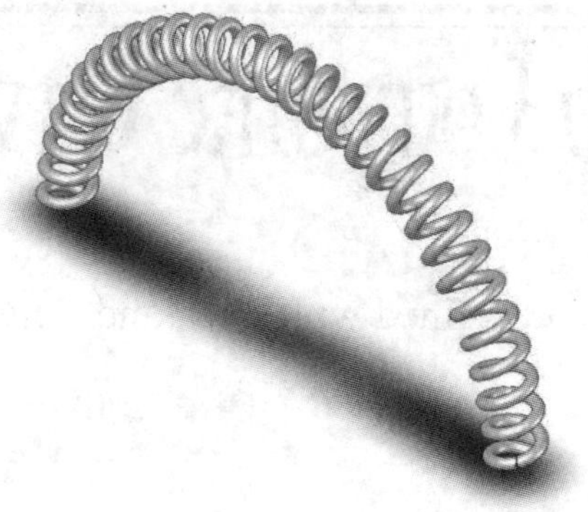

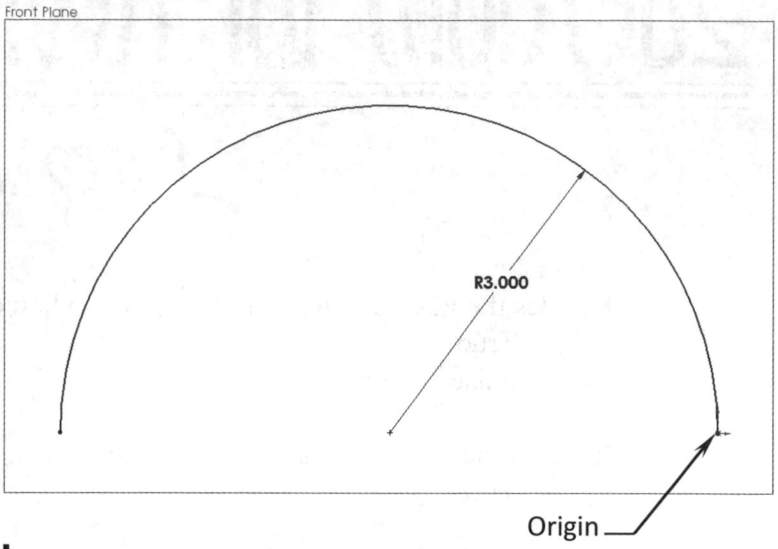

1. Sketching the Sweep Path:

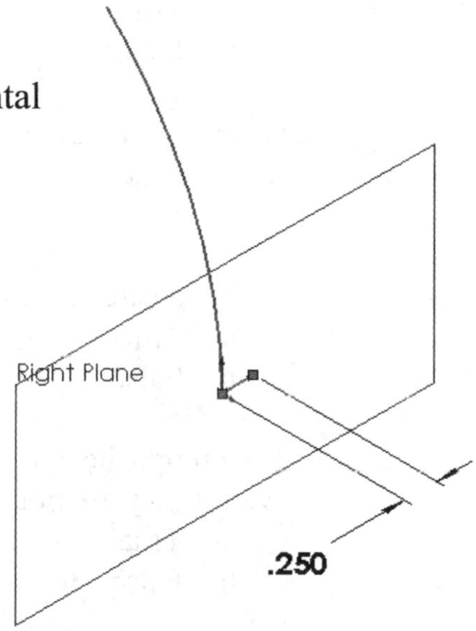

Select the Front plane and open a **new sketch**.

Sketch a 3-Point **Arc** as shown and add a Horizontal relation between the left and the right endpoints, and then add a radius dimension.

Exit the sketch.

2. Sketching the Sweep Profile:

Select the Right plane and open a **new sketch**.

Sketch a horizontal **Line** towards the right.

Add a **.250 in**. dimension.

Exit the sketch.

3. Creating a Swept Surface:

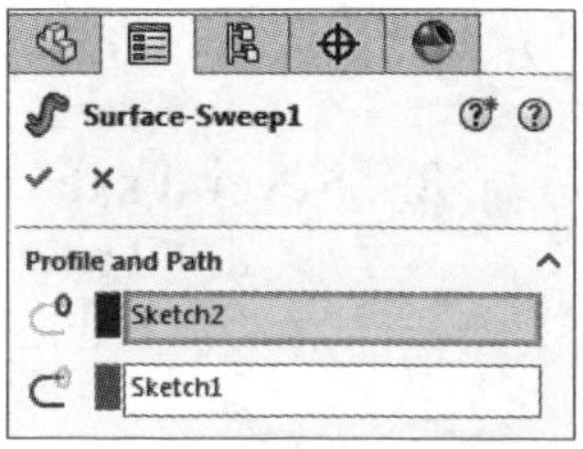

Click  or select **Insert / Surface / Sweep**.

Select the horizontal **Line** to use as the Sweep Profile.

Select the **Arc** as the Sweep Path.

More...

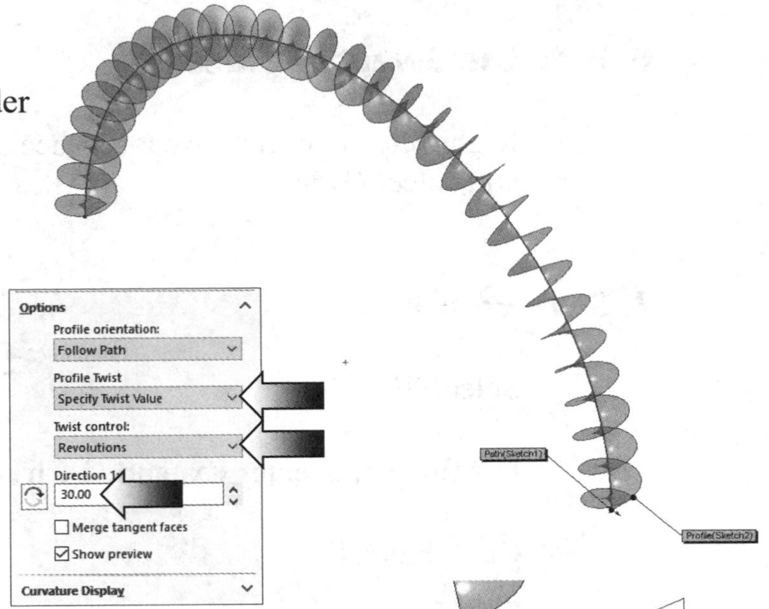

Expand the **Options** dialog box.

Select **Specify Twist Value**, under Profile Twist.

For Twist Control: Select **Revolutions**.

For Direction 1: Enter **30**.

Click **OK**.

The line is swept and twisted 30 revolutions along the path.

4. Sketching the Wire-Diameter:

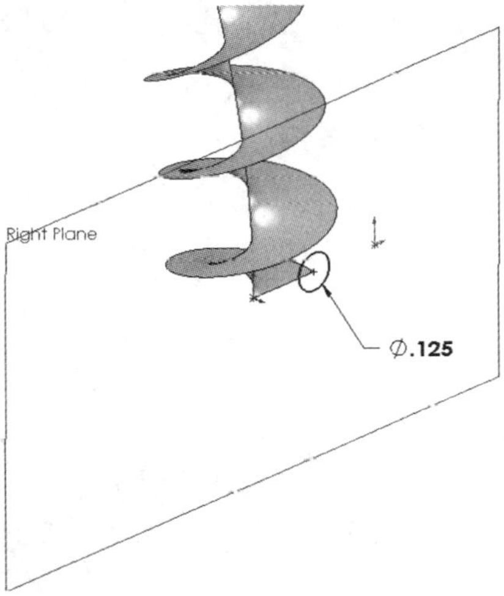

Select the Right plane and open a **new sketch**.

Sketch a **Circle** at the right end of the swept surface.

Add a **Ø.125 in**. diameter dimension.

Exit the sketch.

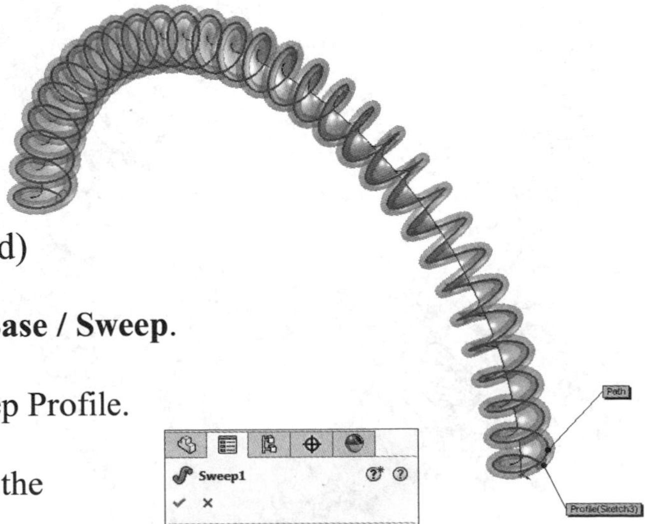

5. Creating a Swept Boss-Base: (Solid)

Click [icon] or select **Insert / Bose-Base / Sweep**.

Select the **Circle** to use as the Sweep Profile.

For Sweep Path, select the **Edge** of the Swept-Surface.

Click **OK**.

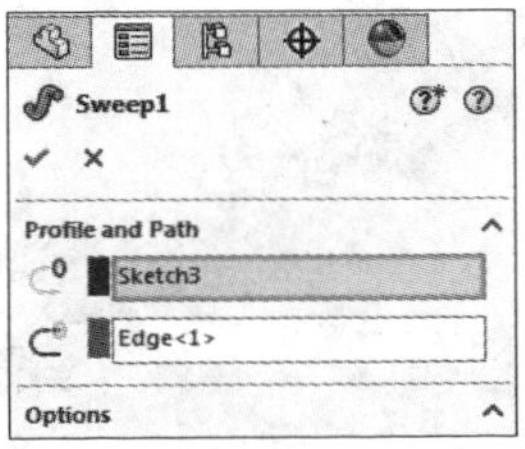

6. Hide the Swept-Surface:

Right-click over the Swept-Surface and select **Hide**.

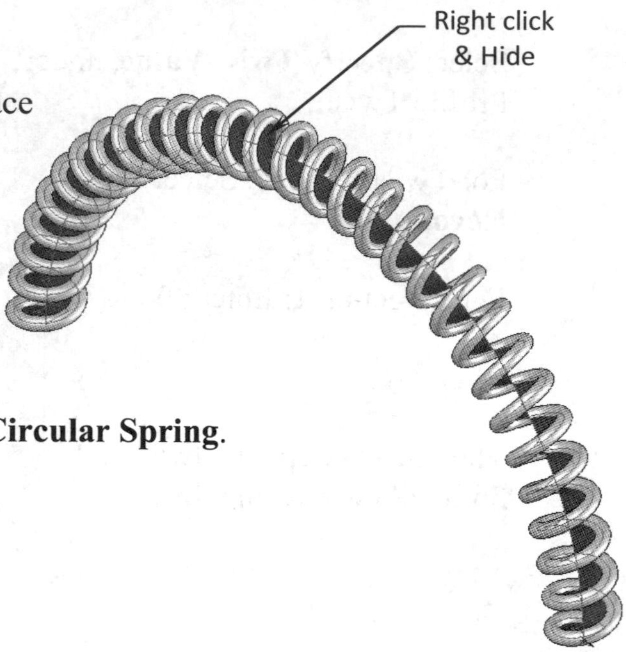

7. Save your work:

Select **File / Save As.**

For file name, enter **Expanded Circular Spring**.

Click **Save**.

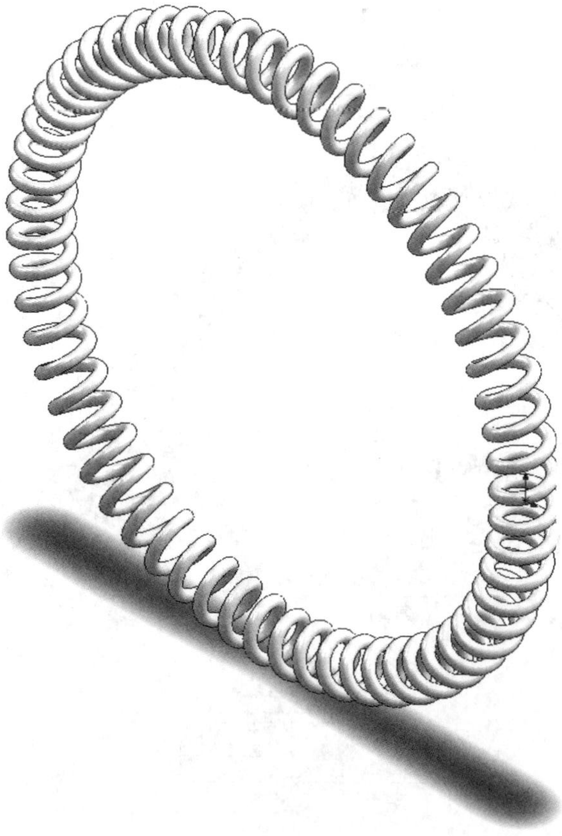

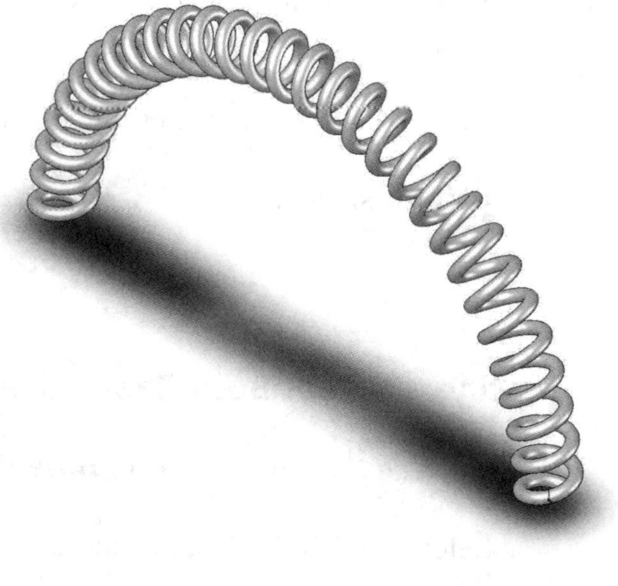

Using Variable Pitch

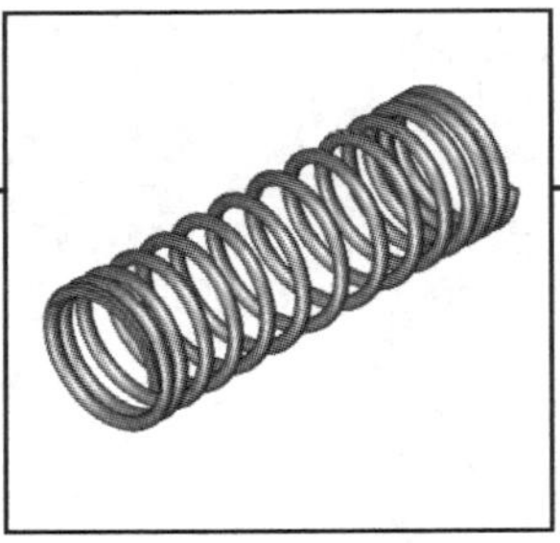

Sweep
with Variable Pitch Helix

In a Sweep feature, there is only one Sweep Profile, one Sweep Path, and one or more Guide Curves.

The Sweep Profile describes the feature's cross-section; the Sweep Path helps control how the Sweep Profile moves along the path.

The Guide Curve(s) keeps the profile from twisting while it moves along the sweep path.

The Sweep path can either be a 2D or a 3D sketch, the edges of the part, or a Composite Curve.

Beside the Pitch and Revolution option, the Helix / Spiral command offers other options to create more advanced curves such as:

* Height and Revolutions.
* Height and Pitch.
* Spiral.
* Constant Pitch.
* Variable Pitch.

Region parameters:

	P	Rev	H	Dia
1	0.115in	0	0in	1in
2	0.115in	1.5	0.1725	1in
3	0.375in	6.5	1.3975	1in
4	0.25in	11.5	2.96in	1in
5	0.115in	12.5	3.1425	1in
6	0.115in	14	3.315i	1in
7				

We will take a look at the Variable Pitch option in this lesson and learn how a helix with variable pitch is created using a table to help control the changes of the dimensions.

To create the flat ground ends, an extruded cut feature is added at the end of the process.

Multi-Pitch Spring with Closed Ends
Using Variable Pitch

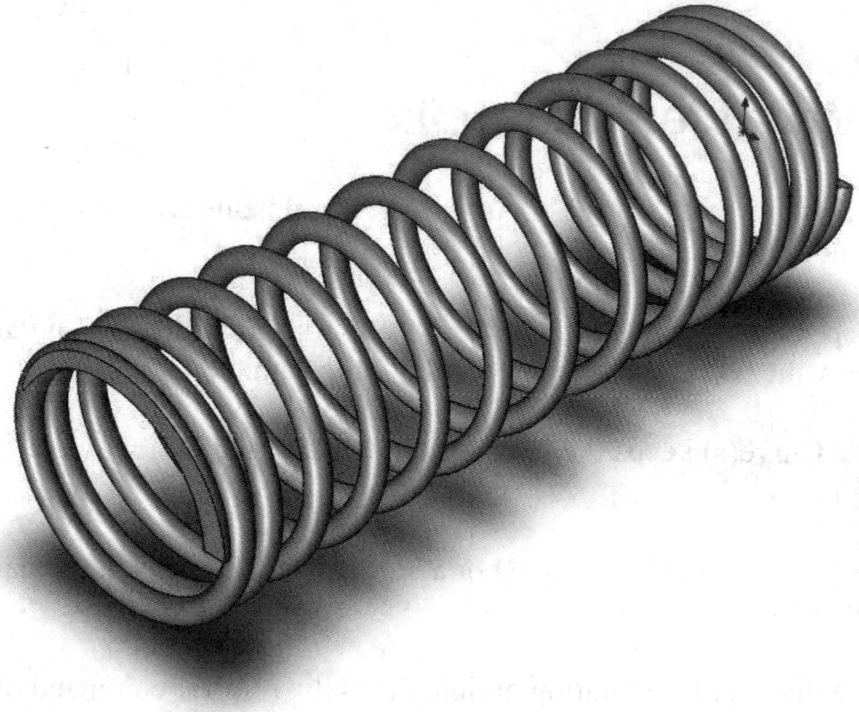

Dimensioning Standards: **ANSI**

Units: **INCHES** – 3 Decimals

Tools Needed:

Insert Sketch	Helix / Spiral	Add Geometric Relations
Dimension	Base/Boss Sweep	Extruded Cut

1. Creating the base sketch:

Select the Front plane from the FeatureManager tree.

Click [icon] or select: **Insert / Sketch**.

Sketch a **Circle** centered on the origin.

Add a diameter dimension of **1.000"**.

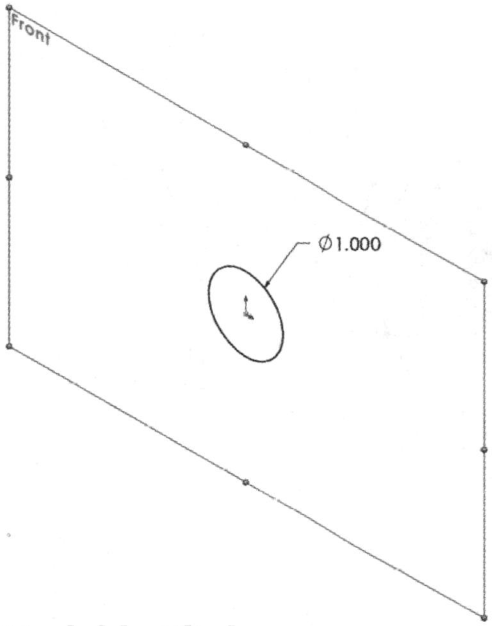

2. Creating a helix with Variable Pitch:

Click [icon] or select **Insert / Curve / Helix-Spiral**.

Under Define By, select:
Pitch and Revolution.

Under Parameter, select:
Variable Pitch.

For Region Parameters,
enter the values of the **Pitches**, **Revolutions**, and **Diameters**.
(Ignore the Height column; SOLIDWORKS will fill in the values automatically.)

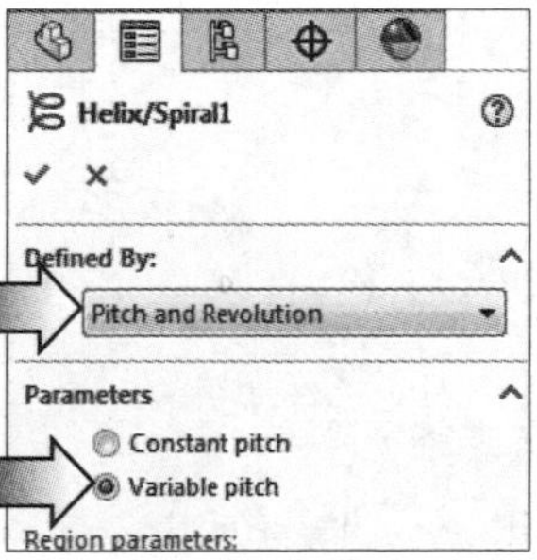

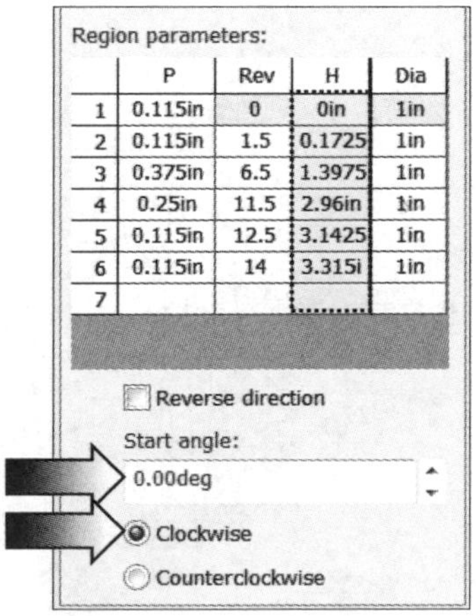

Region parameters:
P Rev H Dia
1 0.115in 0 0in 1in
2 0.115in 1.5 0.1725 1in
3 0.375in 6.5 1.3975 1in
4 0.25in 11.5 2.96in 1in
5 0.115in 12.5 3.1425 1in
6 0.115in 14 3.315i 1in
7
Reverse direction
Start angle:
0.00deg
Clockwise
Counterclockwise

Set Angle to **0deg** and **Clockwise** direction.

Your Variable Pitch helix should look like the image below.

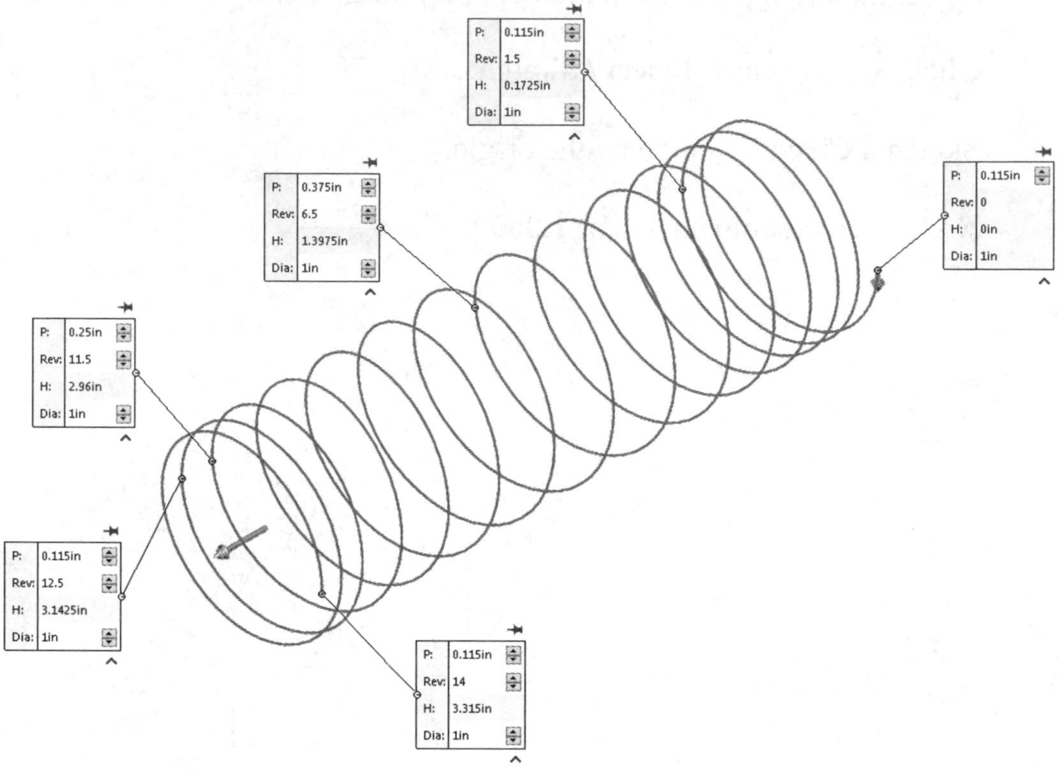

Click **OK**.

Press **Control** + **4** to view the Variable-Pitch Helix from the right side.

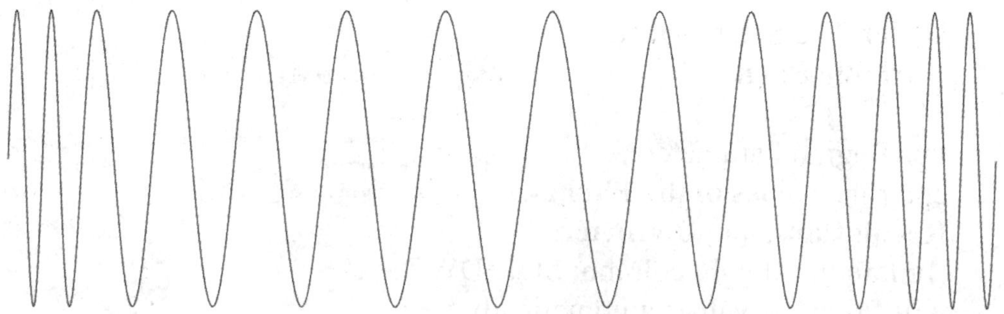

3. Sweeping the profile along the path:

Click  or select: **Insert / Boss-Base / Sweep**.

Select the **Circular Profile** option.

For profile diameter, enter **.100in**.

For sweep path, select the **Helix** either from the feature tree or from the graphics area.

Click **OK**.

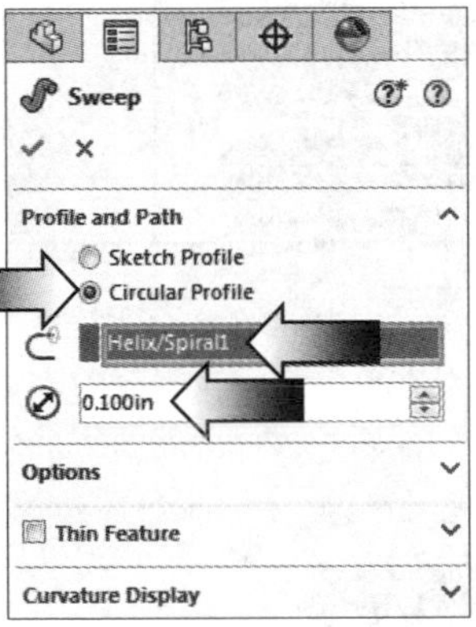

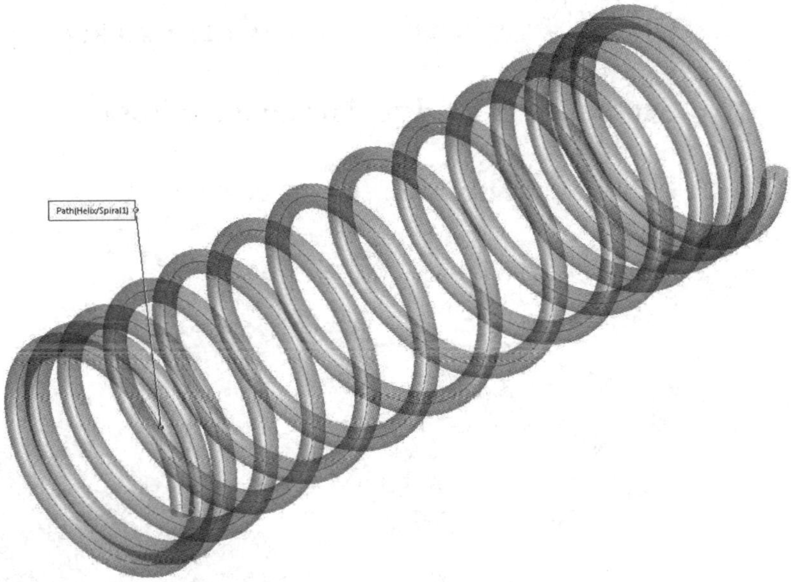

Change to different orientations to inspect the Swept feature.

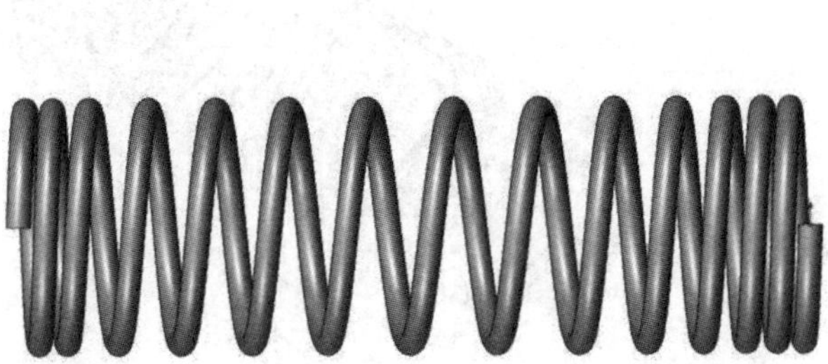

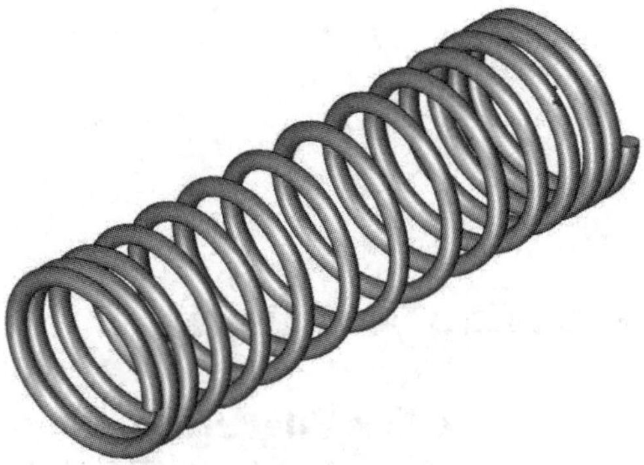

4. Creating a trimmed sketch:

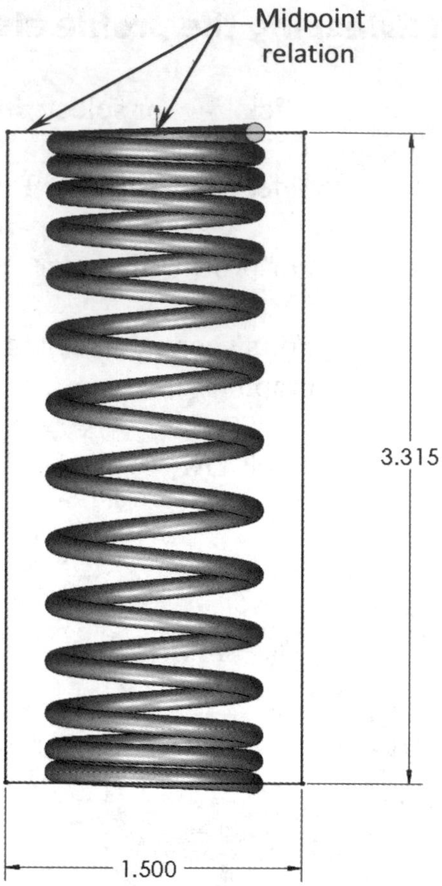

Select the Top plane from the Feature-Manager tree and open a **new sketch**.

Sketch a **Rectangle** and add a **Midpoint** relation between the line on top and the origin.

Add a width and a height dimension. The sketch should now be fully defined.

5. Extruding a cut:

Click or select: **Insert / Cut / Extrude**.

Set the Direction 1 to **Through All Both**.

Enable the **Flip Side to Cut** checkbox.

Click **OK**.

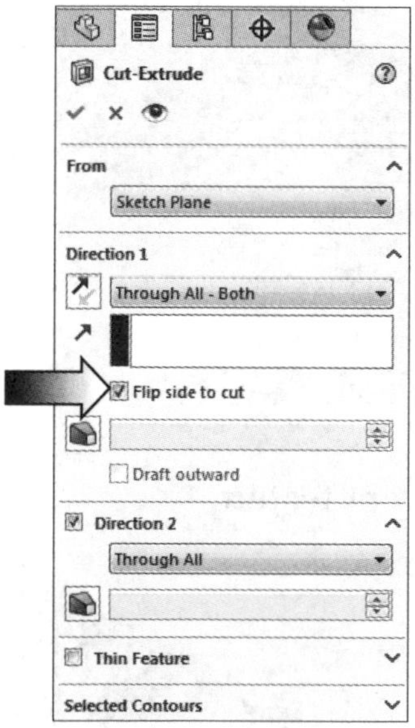

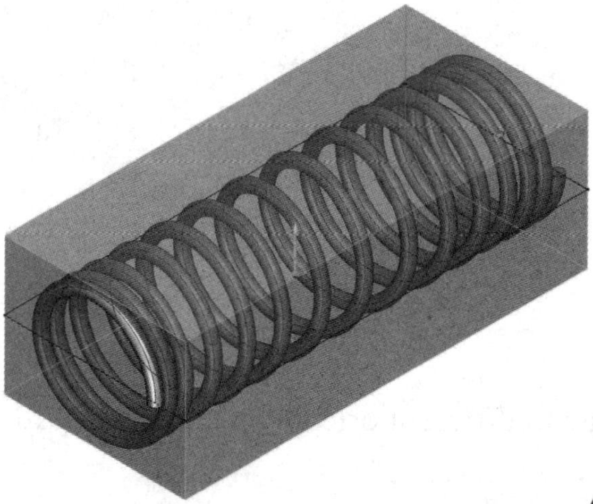

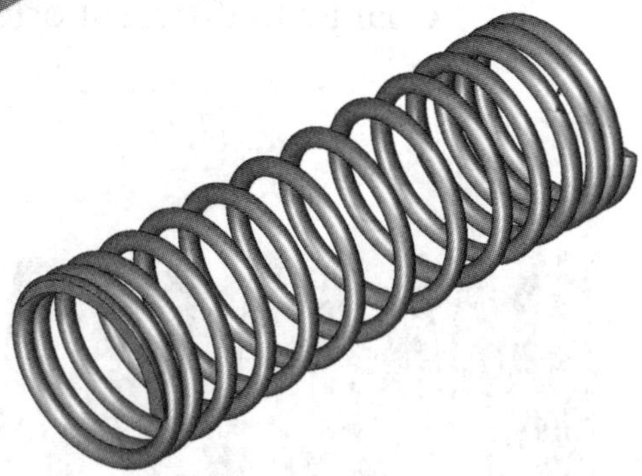

6. Saving your work:

Click **File / Save As**.
Enter **Variable Pitch Spring** and press **Save**.

Questions for Review

Sweep with Composite Curve

1. Multiple Sweep Profiles can be used in a sweep feature.
   a. True
   b. False

2. Multiple Sweep Paths can be used in a sweep feature.
   a. True
   b. False

3. Only one Sweep Profile and one Sweep Path can be used in a sweep.
   a. True
   b. False

4. A Helix can be defined by:
   a. Pitch and Revolution
   b. Height and Revolution
   c. Height and Pitch
   d. Spiral
   e. All of the above

5. Several connected Helixes can be combined into one single Composite Curve.
   a. True
   b. False

6. The Sweep Profile sketch should be related to the Sweep Path using the relation:
   a. Perpendicular
   b. Parallel
   c. Coincident
   d. Pierce

7. The Sweep Path controls the twisting and how the Sweep Profile moves along.
   a. True
   b. False

8. The Edges of the part can also be used as the Sweep Path.
   a. True
   b. False

1. FLASE 2. FALSE
3. TRUE 4. E
5. TRUE 6. D
7. TRUE 8. TRUE

Exercise: Using Equation Driven Curve

The Equation Driven Curve tool in SOLIDWORKS allows us to create a curve by defining the equation for the curve. The values in the equation driven curves must be in radians.

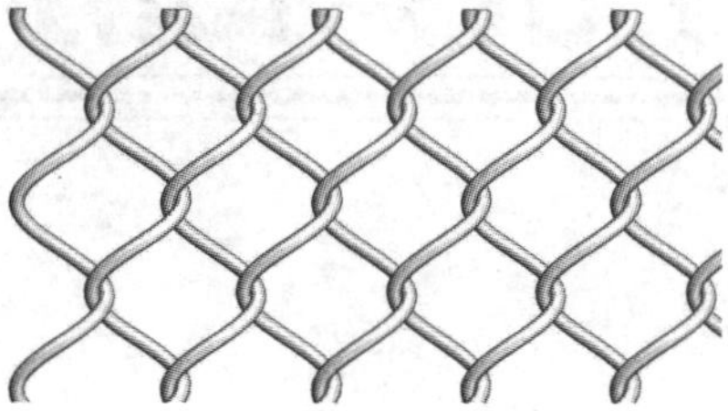

This exercise will teach us how to create an equation that can be used to model the chain link fence.

1. Creating an Equation Driven Curve:

Start with a new **Part Template**, set the units to **Millimeter** and open a new <u>3D Sketch</u>.

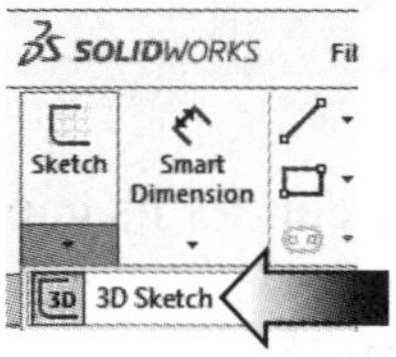

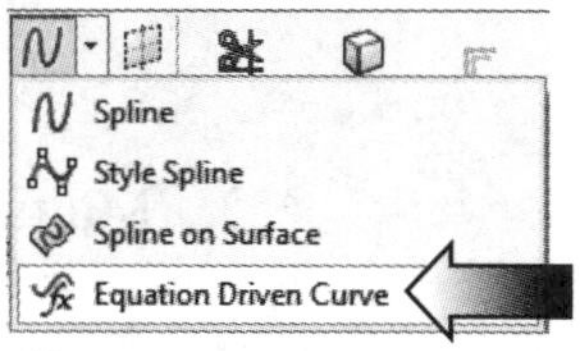

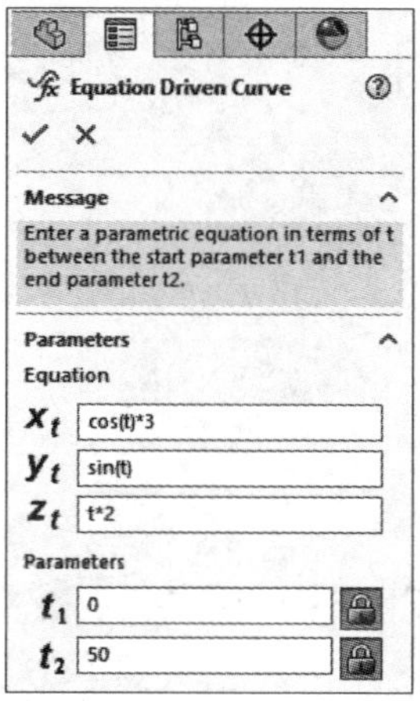

Select **Equation Driven Curve** under the Spline drop-down list.

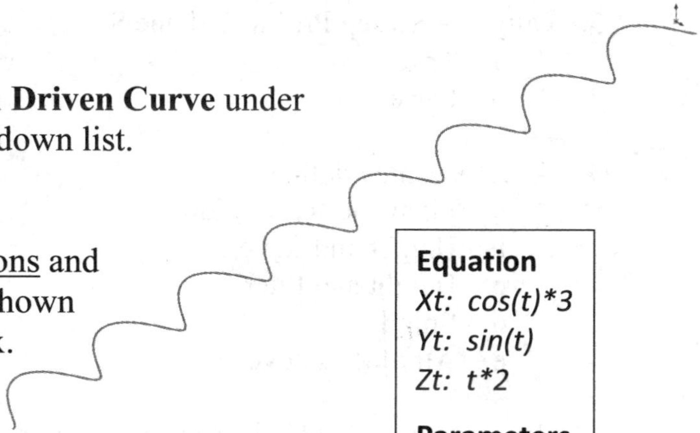

Enter the <u>equations</u> and the <u>parameters</u> shown in the dialog box.

Equation
*Xt: cos(t)*3*
Yt: sin(t)
*Zt: t*2*

Parameters
t1: 0
t2: 50

Click **OK**.

2. Adding a Fix relation:

To prevent the 3D Spline from shifting we will add a fix relation to lock it.

Click the Spline and select the **Fix** relation button (arrow).

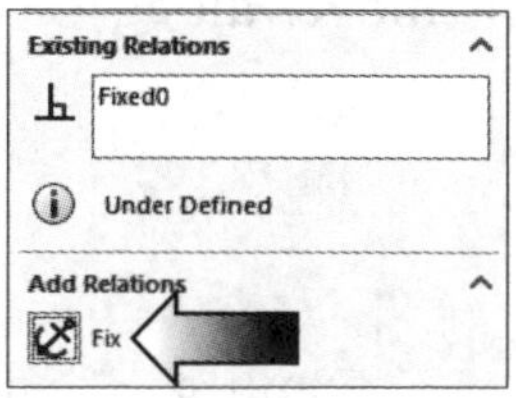

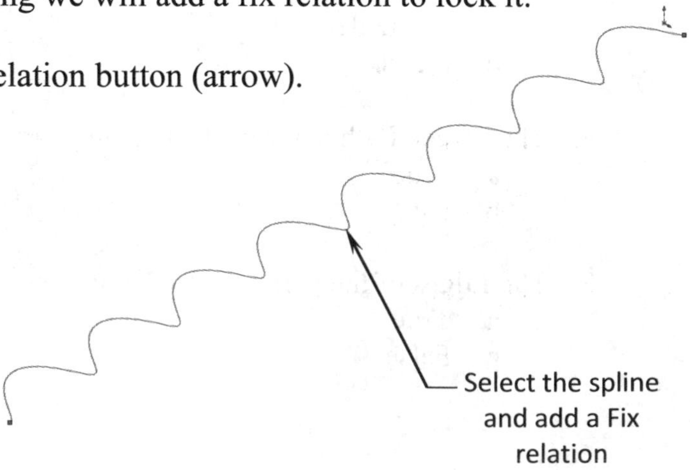

<u>Exit</u> the **3D Sketch**.

3. Making a swept feature:

Switch to the **Features** tab.

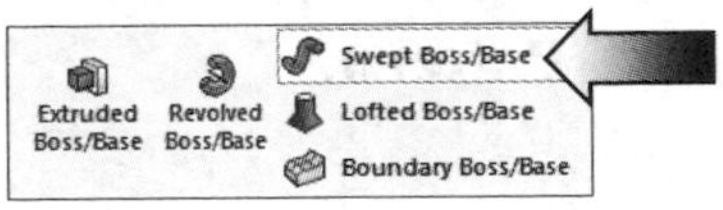

Click **Swept Boss/Base.**

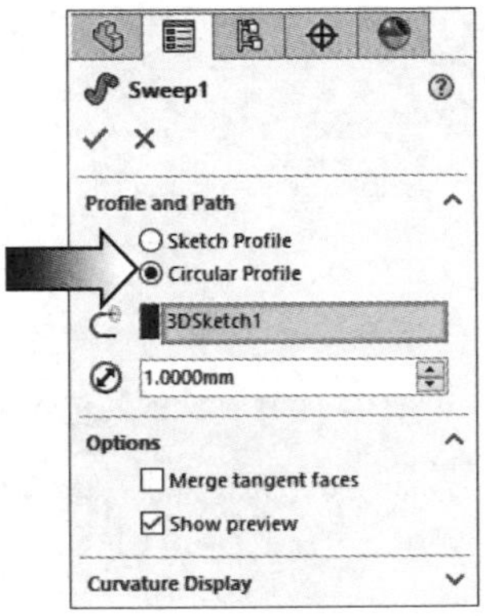

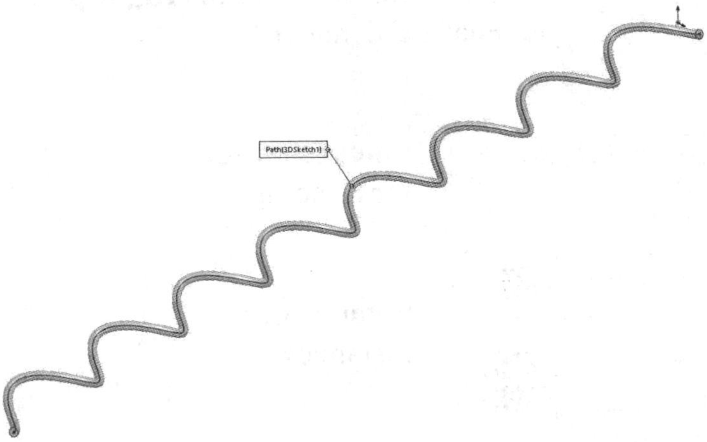

For Profile and Path, select the **Circular Profile** option.

For Path, select the **3D Spline** from the graphics area.

For Profile Diameter, enter **1.00mm**.

Click **OK**.

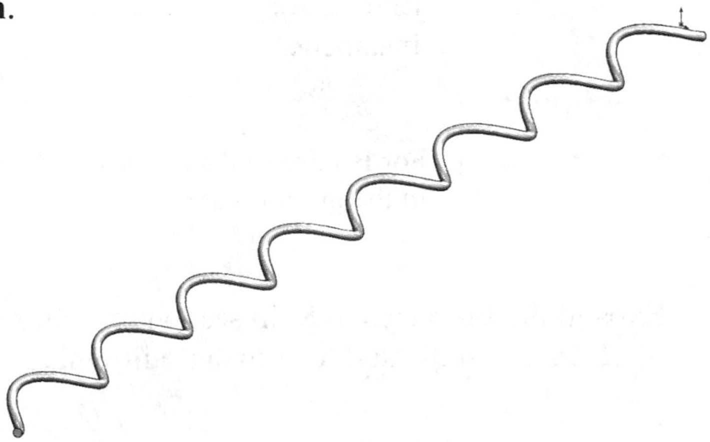

A 1mm circular profile is created and swept along the 3D Spline.

The Swept feature will be patterned a couple times in the next step to create the chain link fence model.

4. Creating the 1st Linear Pattern:

Click **Linear Pattern**.

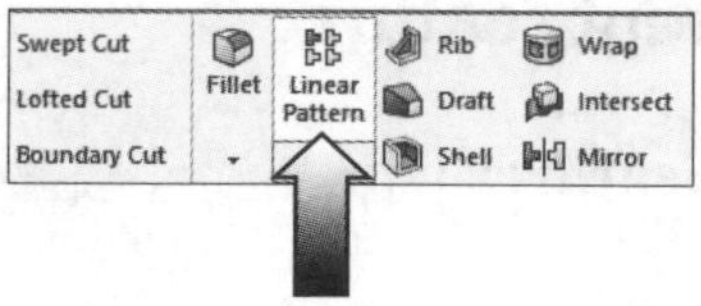

For Direction 1, select the **Right plane** from the FeatureManager tree.

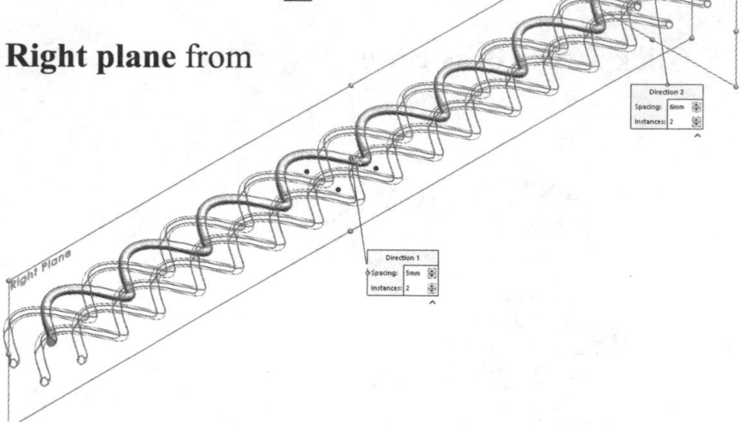

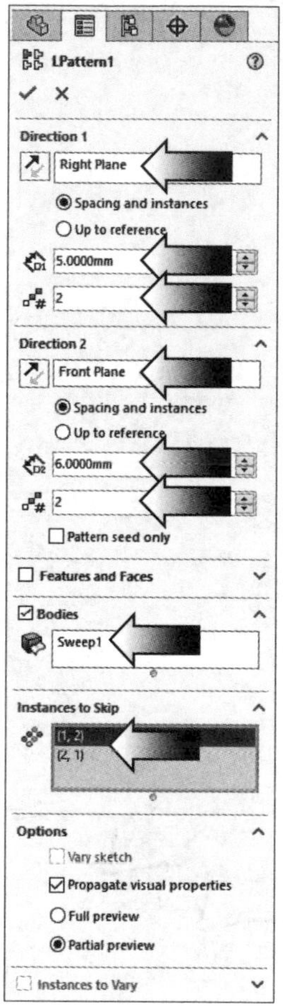

Enter **5.00mm** for Spacing.

Enter **2** for Instances.

For Direction 2, select the **Front plane** from the Feature Manager tree.

Enter **6.00mm** for Spacing.

Enter **2** for Instances.

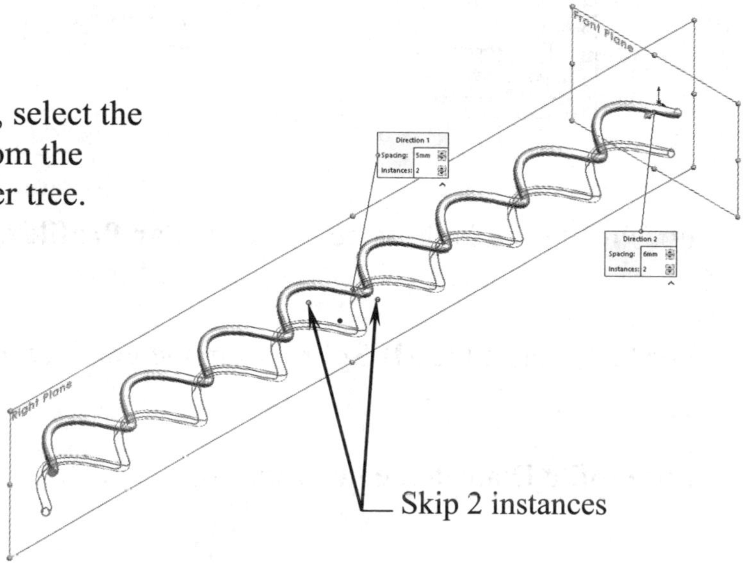

For Bodies to Pattern, select the **Swept feature** in the graphics area.

Expand the Instances to Skip section and click the **2 dots** as indicated to skip those instances.

Click **OK**.

Compare your model against the one shown on the right.

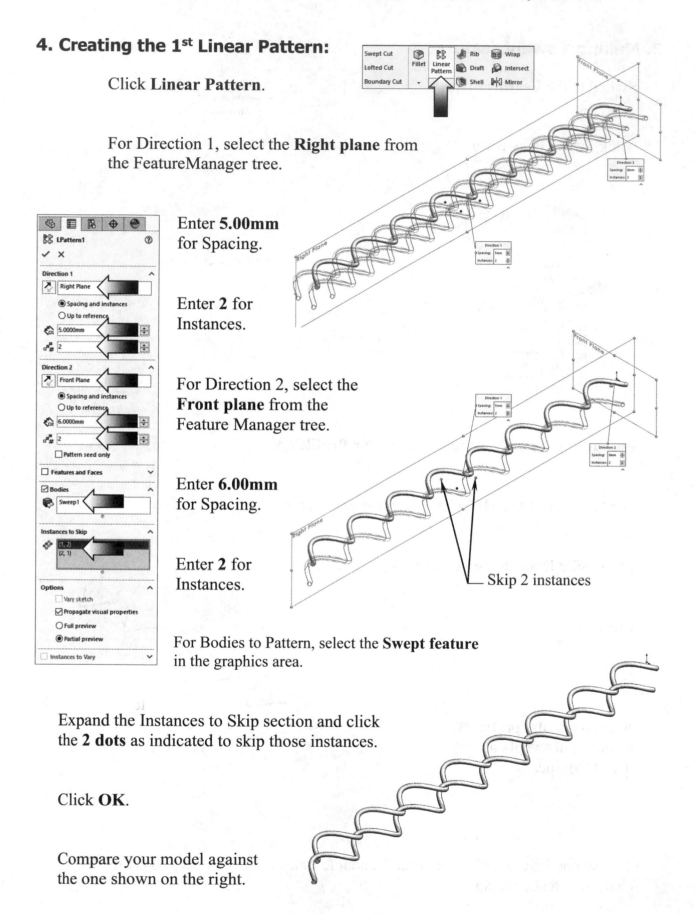

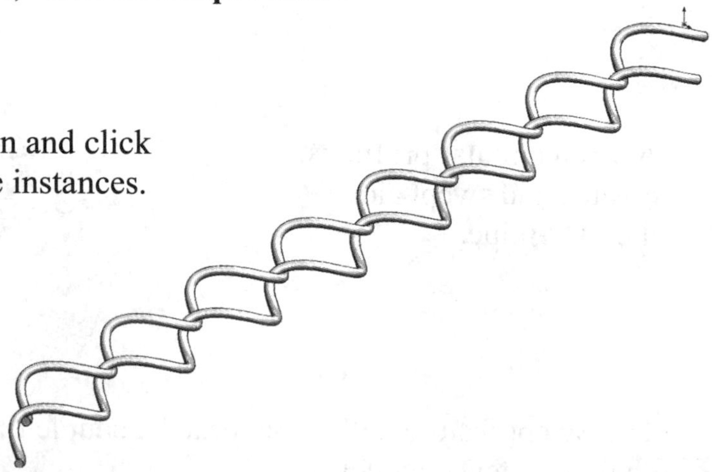

5. Creating the 2nd Linear Pattern:

Click **Linear Pattern** once again.

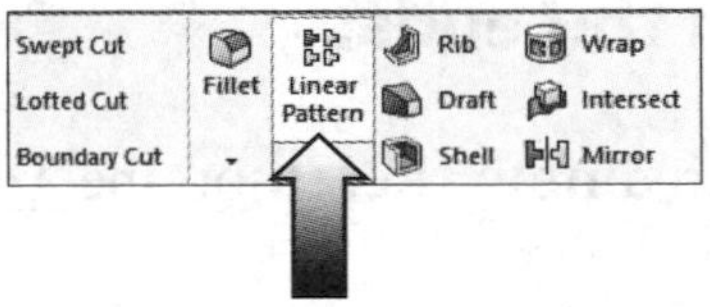

For Direction 1, select the **Right plane** from the FeatureManager tree.

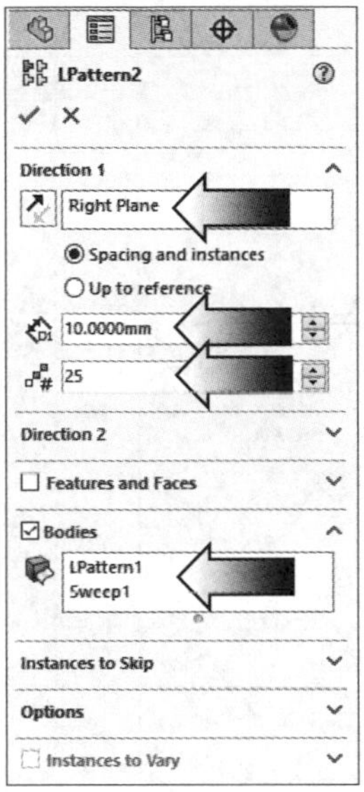

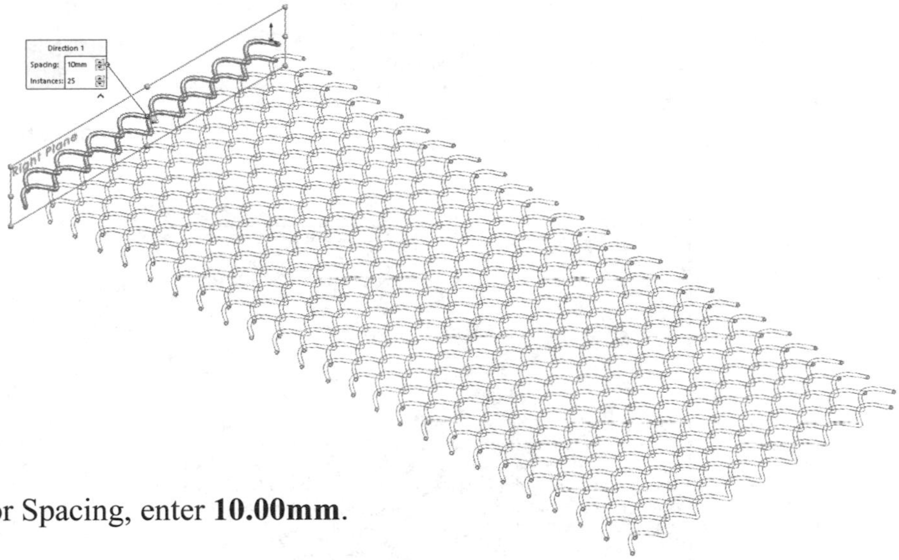

For Spacing, enter **10.00mm**.

For Number of Instances, enter **25**.

For Bodies to Pattern, select the **Sweep1** and the **Pattern1** bodies either from the graphics area or from the FeatureManager tree.

Click **OK**.

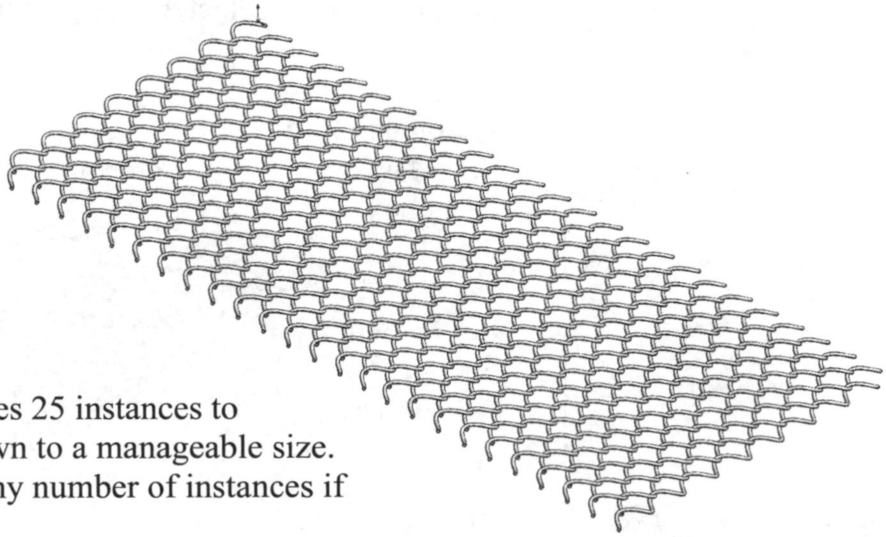

This exercise uses 25 instances to keep the file down to a manageable size. You can enter any number of instances if desired.

6. Trimming the ends:

Open a **new sketch** on the Top plane.

Sketch a **Corner Rectangle** and add the dimensions shown below.

Add the dimensions and relation to fully define the sketch.

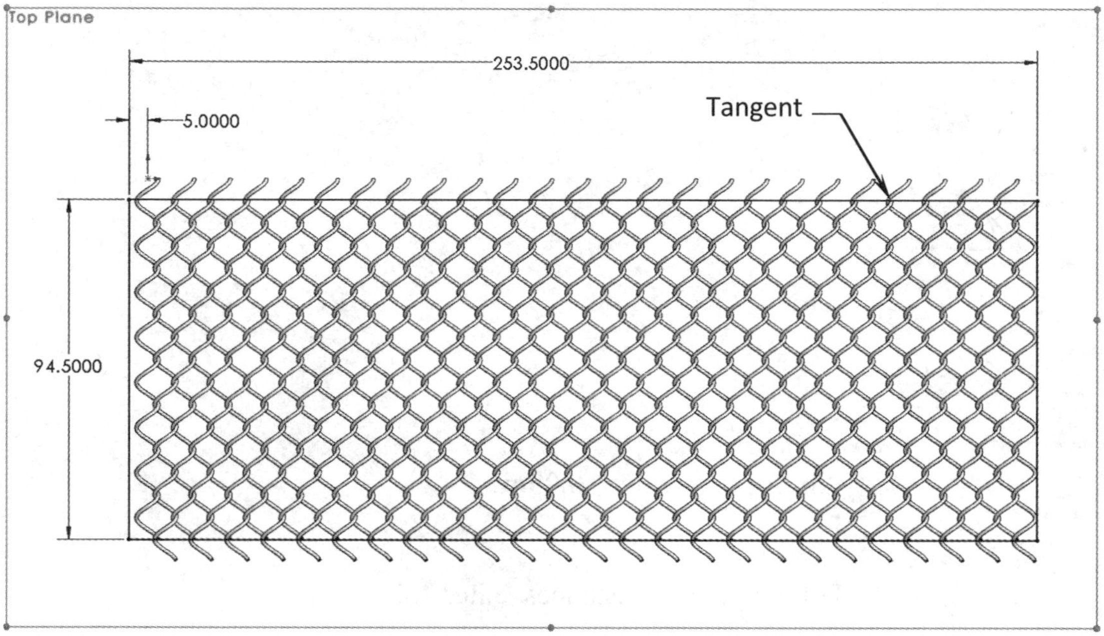

Switch to the **Features** tab and click **Extruded Cut**.

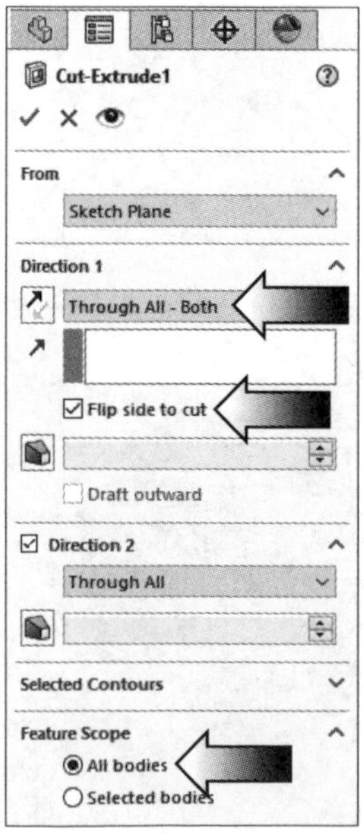

For Direction 1, select **Through All – Both**.

Enable the **Flip-Side To Cut** checkbox.

For Feature Scope, select **All Bodies**.

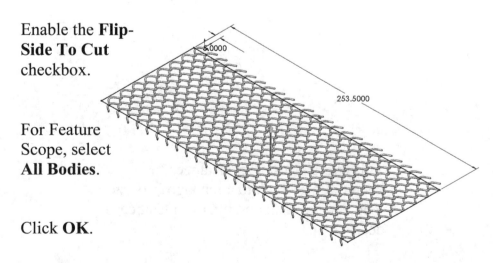

Click **OK**.

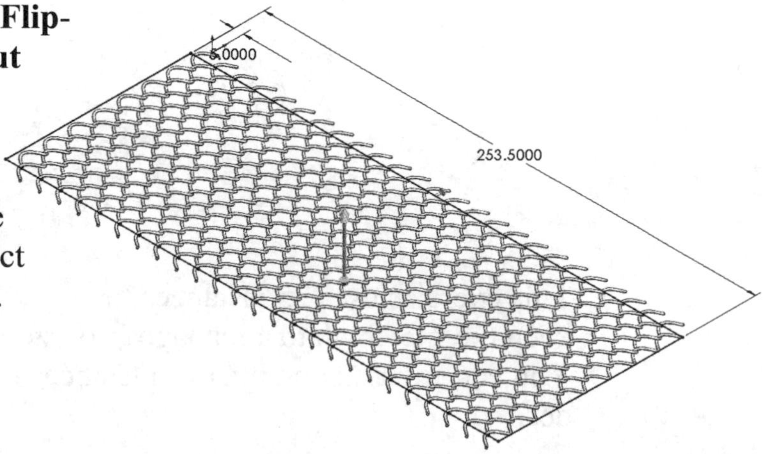

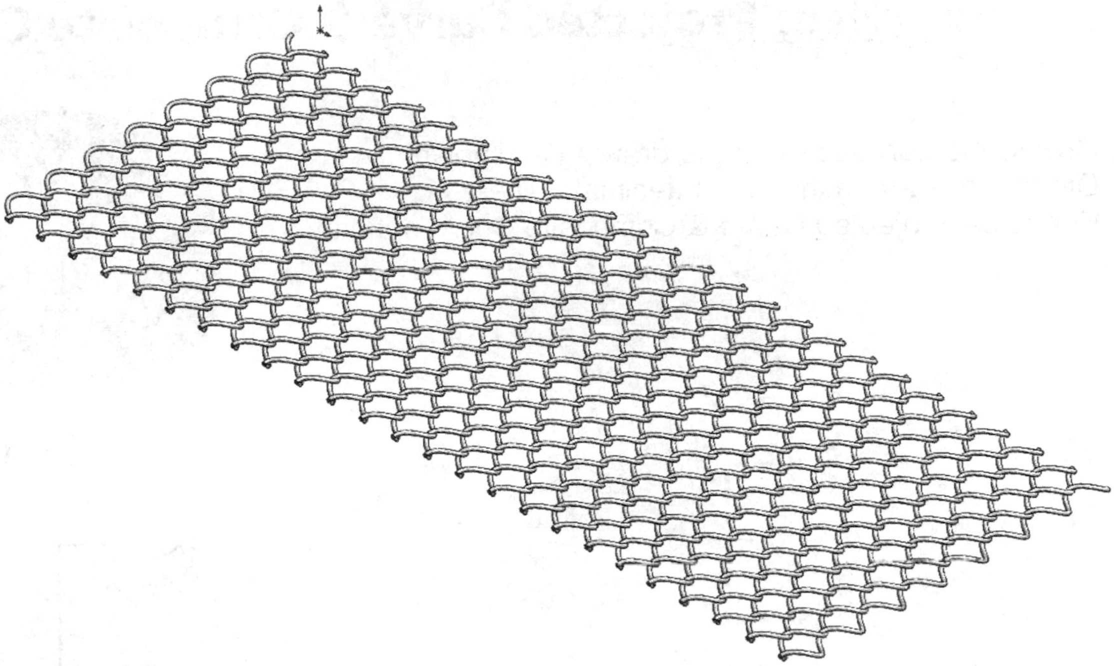

7. Saving your work:

Select **File, Save As**.

Enter **Equation Driven Curve_Chain Link Fence** for the file name.

Click **Save**.

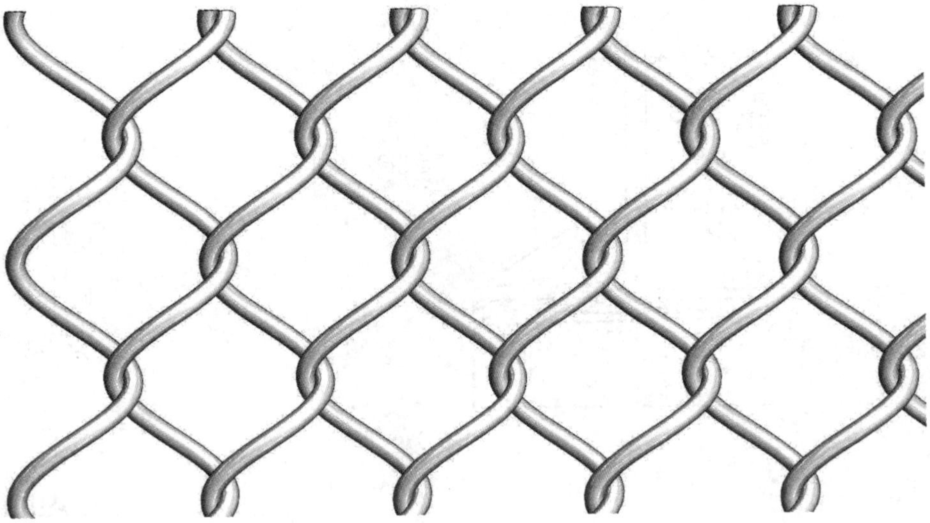

<u>Exercise:</u> Projected Curve & Composite Curve

1. Create the part based on the drawing as shown.
2. Dimensions are in inches, 3 decimal places.
3. Focus on Projected Curve & Composite Curve options.

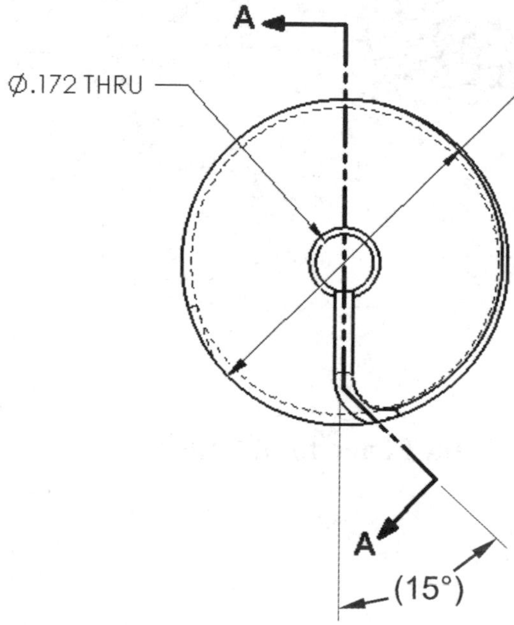

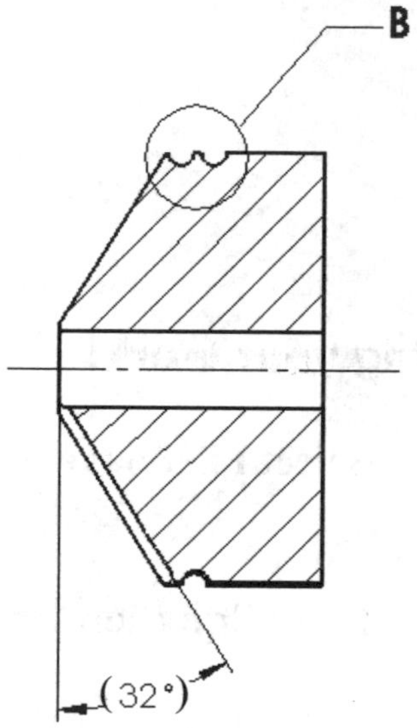

SECTION A-A

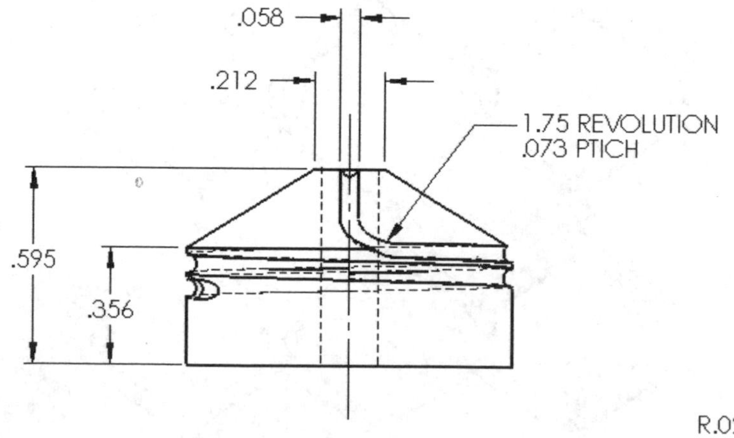

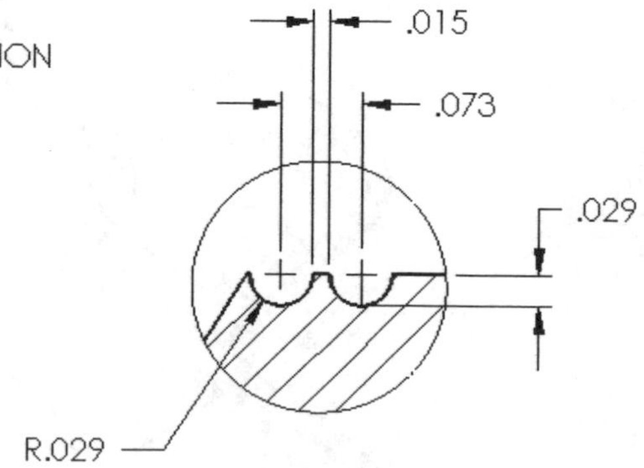

DETAIL B

4. Follow the instructions on the following pages, if needed.

1. Opening a part document:

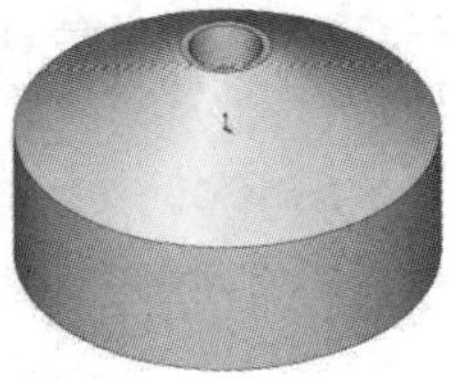

From the Training Files folder, locate and open the part document named:
Project and Composite Curves.

This is an actual die to form the shape of a catheter tip. We will create the groove that wraps around this die, using the Project and Composite Curve options.

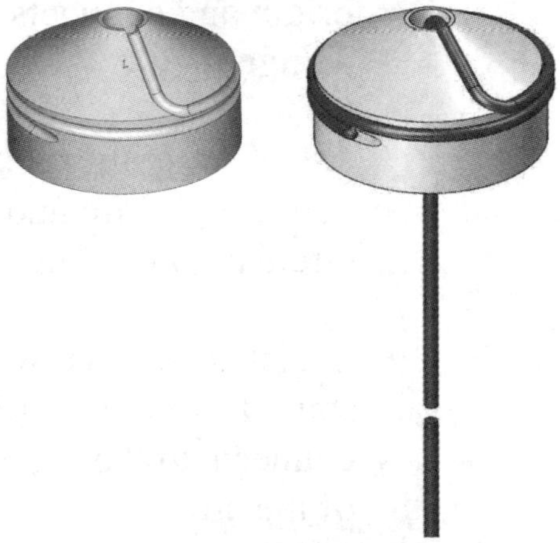

2. Creating a helix:

Select the Top plane and open a **new sketch**.

Select the **bottom edge** (or top edge) and press **Convert- Entity**.

Click  or select:

Insert / Curve / Helix-Spiral.

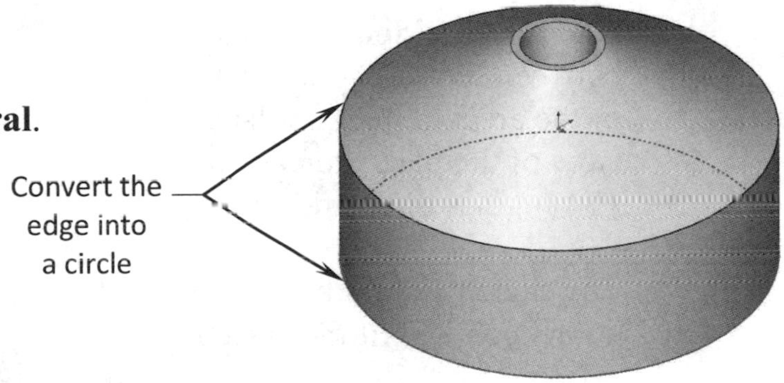

Enter the following:

* **Constant Pitch**.

* Pitch: **.073"**

* **Reverse Direction**

* Revolutions: **1.75**

* Start Angle: **15deg**.

* **Counterclockwise**

Click **OK**.

Notice the helix starts at a 15° angle. This way the end of the helix will line up with the bend radius in the next step.

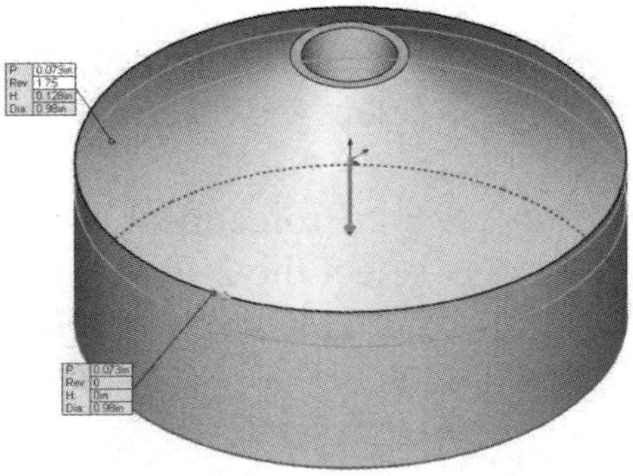

3. Sketching the upper transition:

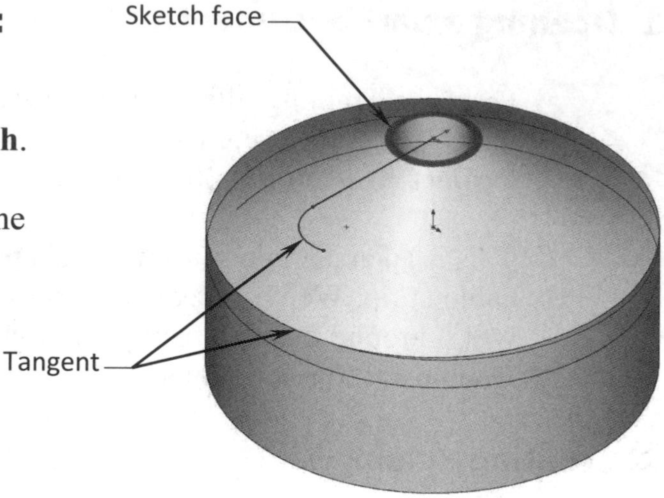

Select the small upper face as indicated and open a **new sketch**.

Sketch a **Line** that starts from the origin and connects with a **Tangent Arc**.

Add a **Tangent** relation between the arc and the circular edge of the part.

Sketch a **centerline** that starts from the origin and connects to the right endpoint of the arc.

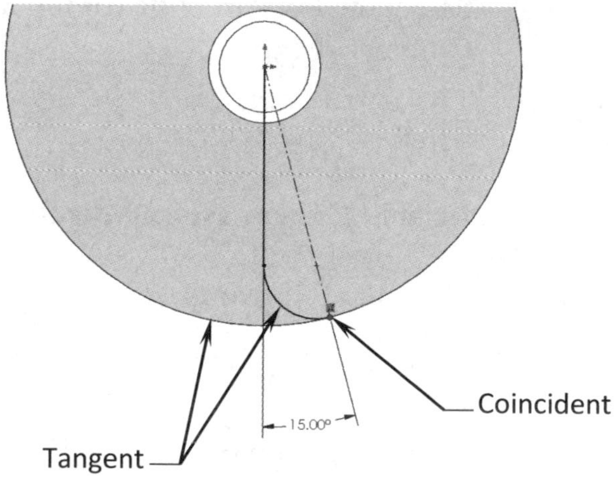

Add a **Coincident** relation between the right endpoint of the arc and the circular edge of the part, then add a **15°** angular dimension.

The sketch should be fully defined at this point. Exit the sketch.

4. Creating a projected Curve:

Click  or select: **Insert / Curve / Projected**.

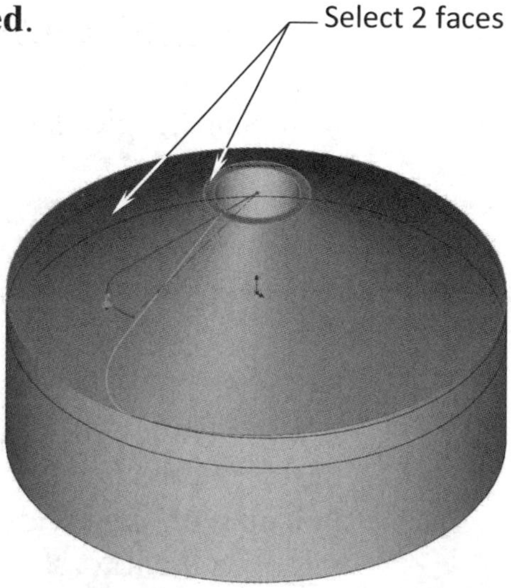

Select the **Sketch on Faces** option.

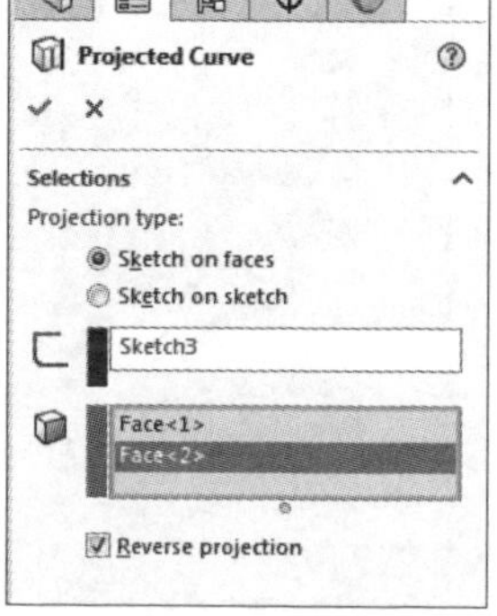

For Sketch to Project, select the **Sketch3**.

For Projection Faces, select the **2 faces** as noted. Click Reverse.

Click **OK**.

5. Adding a sketch line:

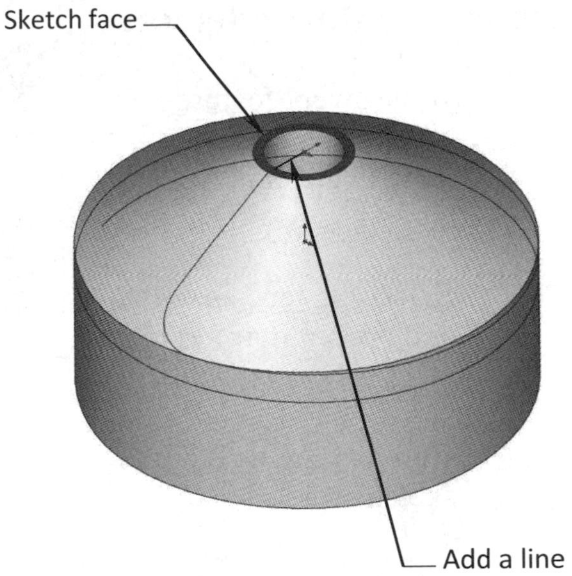

We want the cut to start from the origin; a sketch line is needed to connect the projected curve to the origin.

Sketch a line that starts from the origin to the endpoint of the projected curve.

Exit the sketch.

6. Creating a Composite Curve:

Click [icon] or select: **Insert / Curve / Composite**.

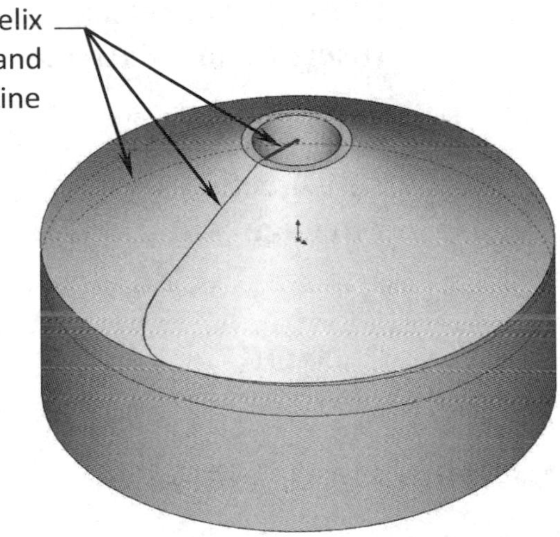

Select either from the Feature tree or from the graphics: the **Helix**, the **Curve1** and the **Sketch4** (the line).

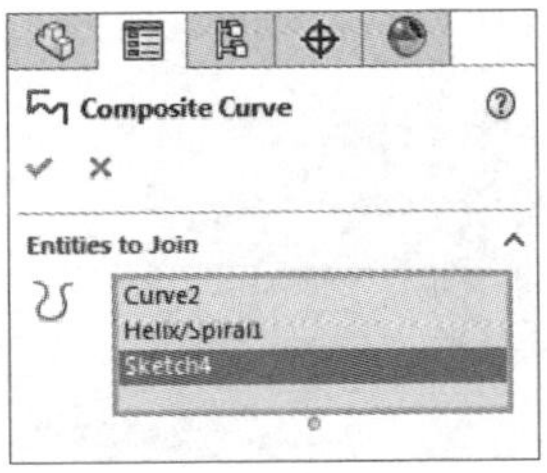

Click **OK**.

7. Creating a swept cut:

Click [icon] or select: **Insert / Cut / Sweep**.

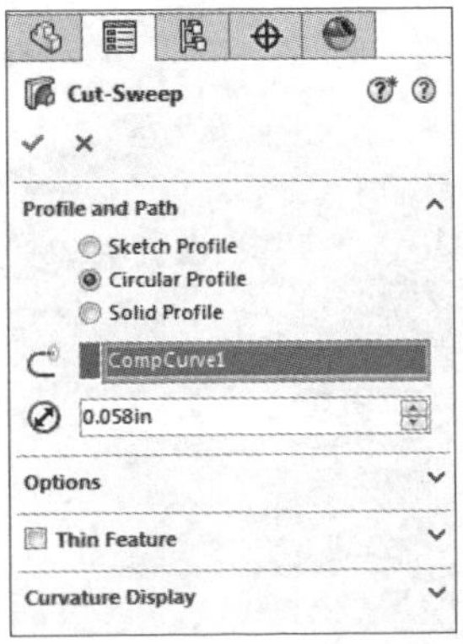

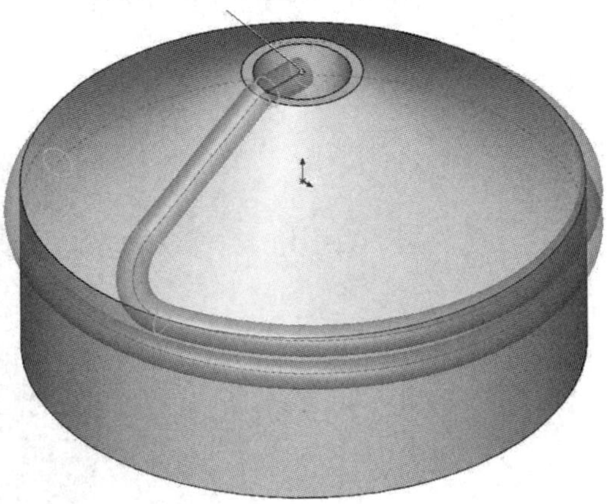

Select the **Circular Profile** option.

For Profile Diameter enter **.058in**.

For Sweep Path, select the **Composite Curve**.

Click **OK**.

8. Removing the undercut:

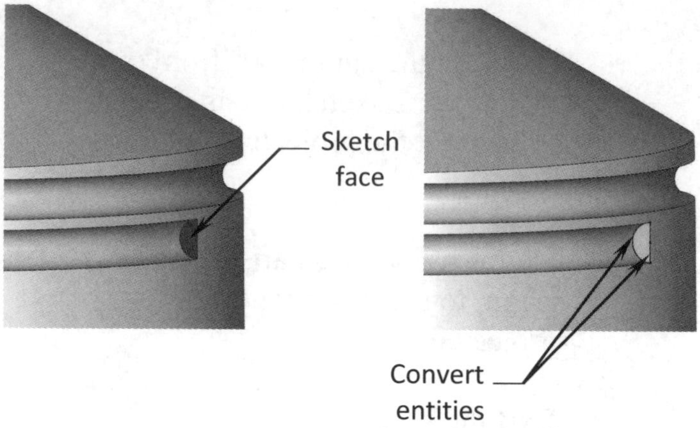

If the swept feature stops short, we will need to clean it up.

Select the face as noted and open a **new sketch**.

Convert the face into a sketch.

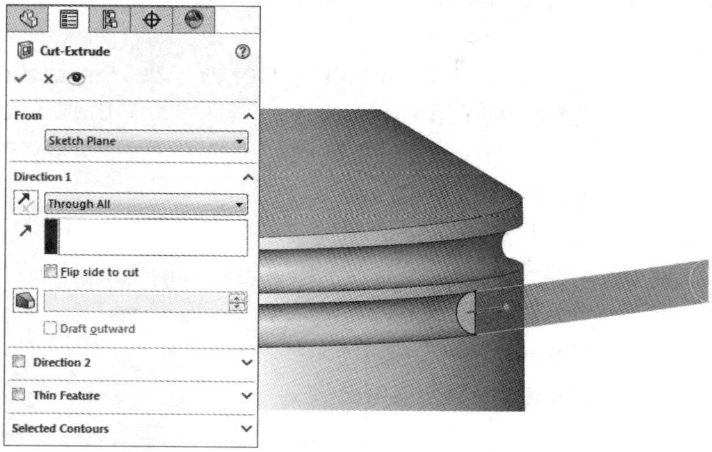

Click [icon] or select: **Insert / Cut / Extrude**.

Set Direction 1 to: **Through All**.

Click **OK**.

9. Saving your work:

Click **File / Save As**.

Enter **Project and Composite Curves** for the name of the file.

Press **Save**.

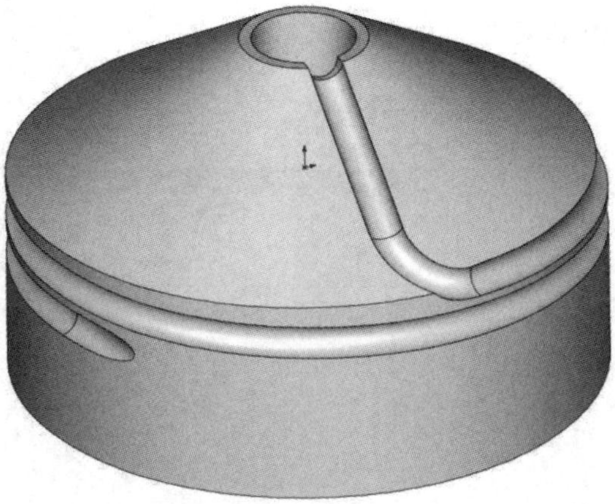

CHAPTER 5

Advanced Modeling with Sweep & Loft

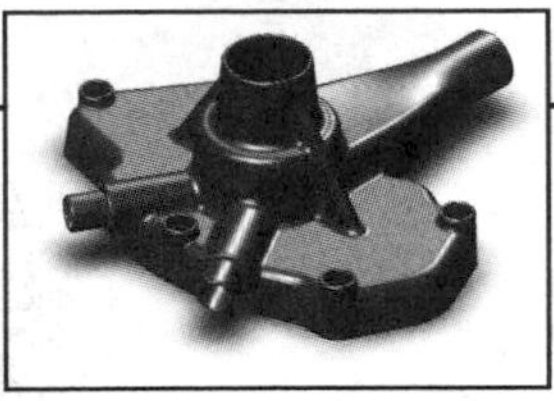

Advanced Modeling with Sweep & Loft
Water Pump Housing

The Sweep option creates a solid, thin, or surface feature by moving a single sketch profile along a path and guiding with one or more guide curves.

In order to create a sweep feature properly, a set of rules should be taken into consideration:

- The Sweep option uses only one sketch profile and it must be a closed non-intersecting contour for a **solid** feature.
- The sketch profile can be either closed or open for a **surface** feature.
- Only one path is used in a sweep and it can be open or closed.
- One or more guide-curves can be used to guide the sketch profile.
- The sketch profile must be drawn on a new plane starting at the end point of the path.

The Loft option creates a solid, thin, or surface feature by making a transition between the sketch profiles.

Keep in mind the following requirements when creating a loft feature:

- The Loft option uses multiple sketch profiles that must be closed, and non-intersecting for a **solid** feature.
- The sketch profiles can be either closed or open, for a **surface** feature.
- Use the Centerline Parameter option to guide the profiles from the inside.
- Use Guide Curves option to guide the sketch profiles from the outside.
- The Guide Curves can be either 2D or a 3D sketch.

Water Pump Housing
Advanced Modeling with Sweep & Loft

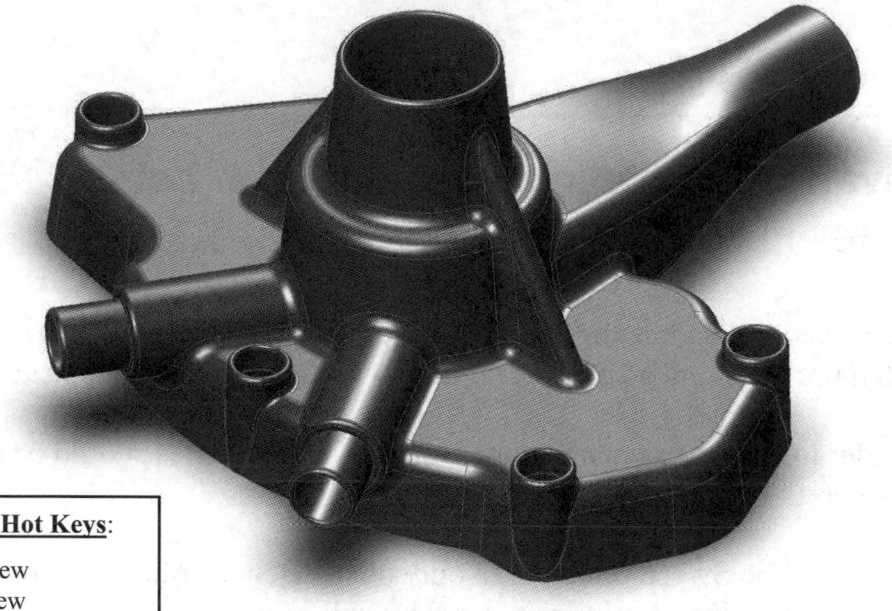

View Orientation Hot Keys:

Ctrl + 1 = Front View
Ctrl + 2 = Back View
Ctrl + 3 = Left View
Ctrl + 4 = Right View
Ctrl + 5 = Top View
Ctrl + 6 = Bottom View
Ctrl + 7 = Isometric View
Ctrl + 8 – Normal To Selection

Dimensioning Standards: **ANSI**

Units: **INCHES** – 3 Decimals

Tools Needed:

Insert Sketch	Split Entities	Add Geometric Relations
Rib	Plane	Revolved Boss/Base
Extruded Boss/Base	Swept Boss/Base	Lofted Boss/Base

Understanding the Draft Options

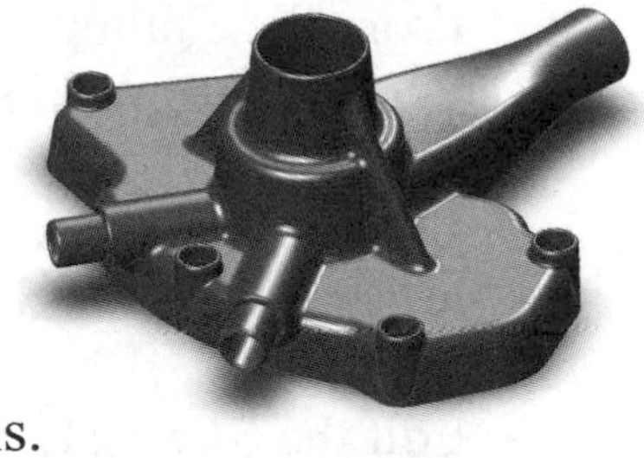

Drafts are usually required in most plastic injection molded parts to ensure proper part removal from the mold halves.
The Draft option in SOLIDWORKS adds tapers to the faces using the angles specified by the user.
Drafts can be inserted in an existing part or added to a feature while being extruded.
Drafts can be applied to solid parts as well as the surface models.
There are several types of draft available:

* Neutral Plane
* Parting Line
* Step Draft

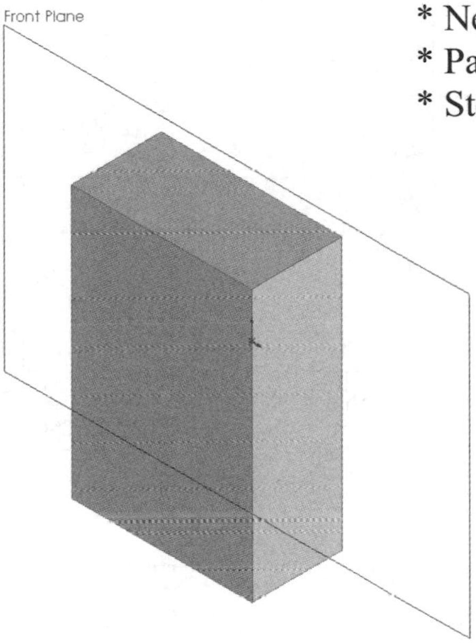

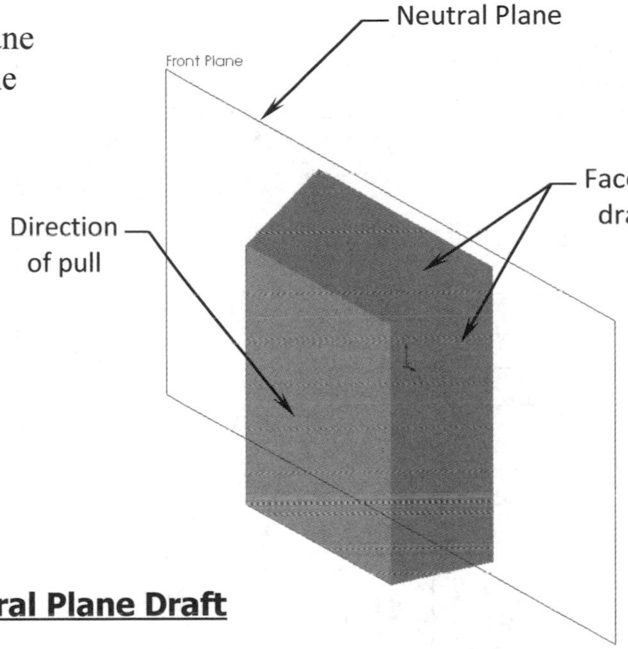

Neutral Plane Draft

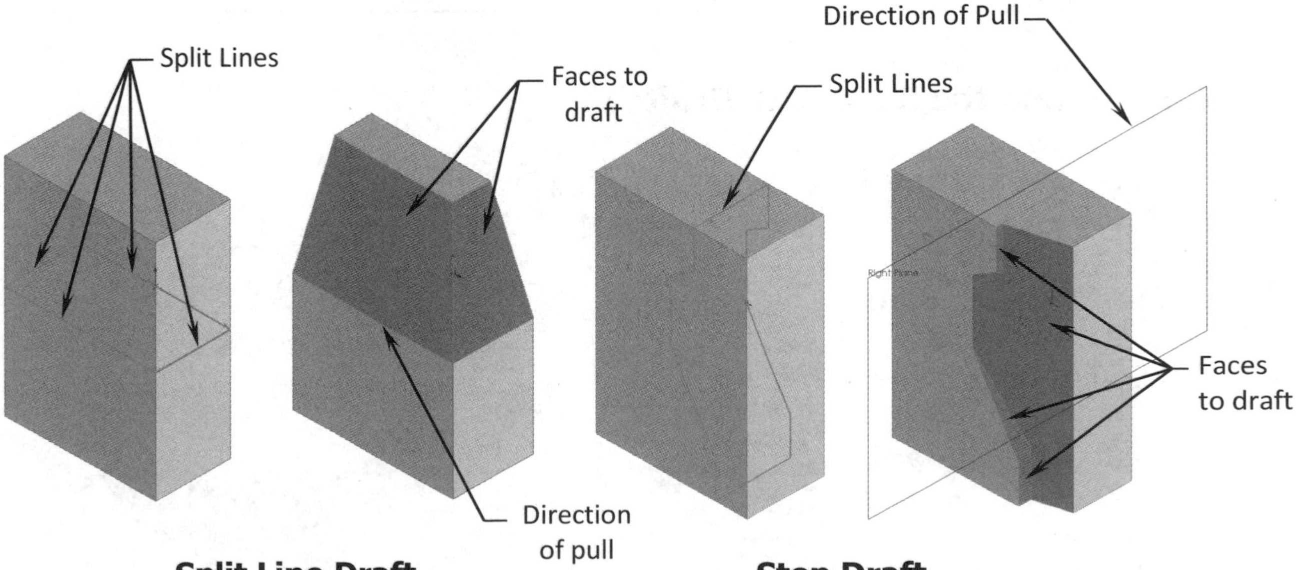

Split Line Draft

Step Draft

1. Opening a part document:

From the Training Files folder, locate and open the part document named: **Water Pump Sketch**.

The sketch was created ahead of time to help focus on the key features of this lesson: Sweep and Loft.

Edit the **Sketch1**.

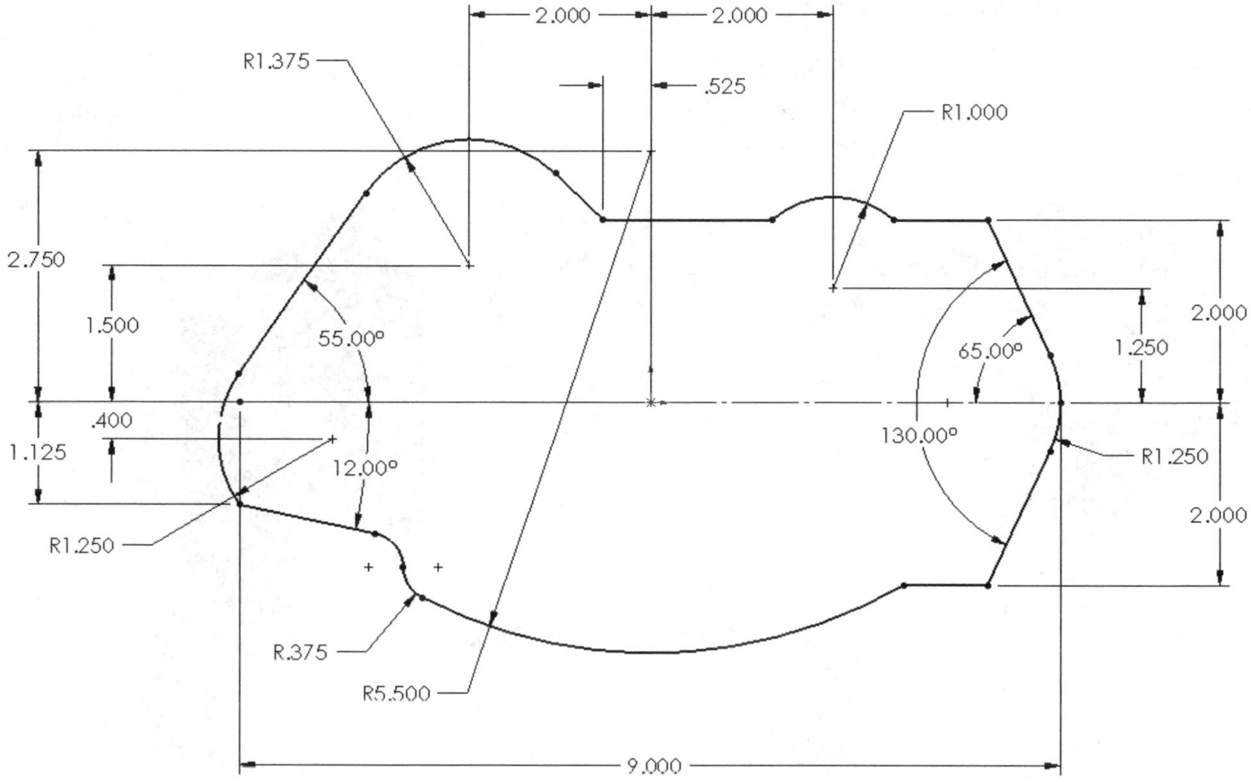

2. Extruding the Base with Draft:

Click **Extruded Boss/Base.**

End Condition: **Blind**.

Depth: **1.00 in**.

Draft: **7 deg**. **inward**.

Click **OK**.

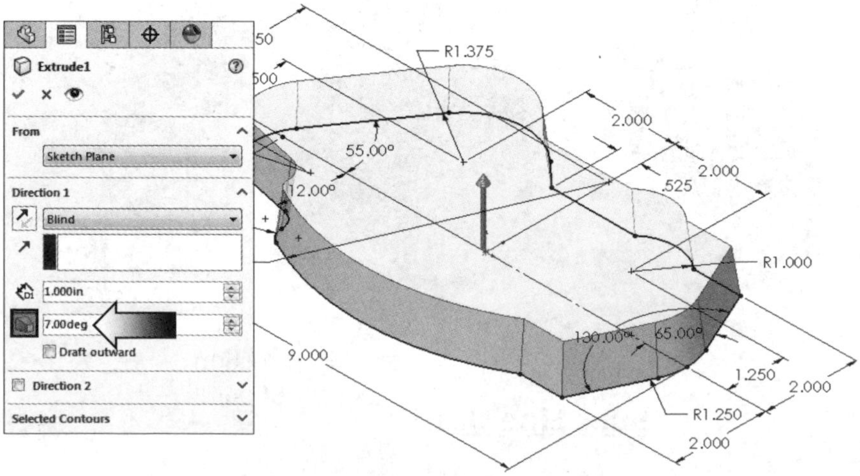

3. Sketching the upper Inlet Port:

Select the Front plane and open a **new-sketch**.

Sketch the profile shown.

Add the dimensions and relations as indicated.

Add a **Vertical Centerline** from the Origin and use it as the center of the revolve in the next step.

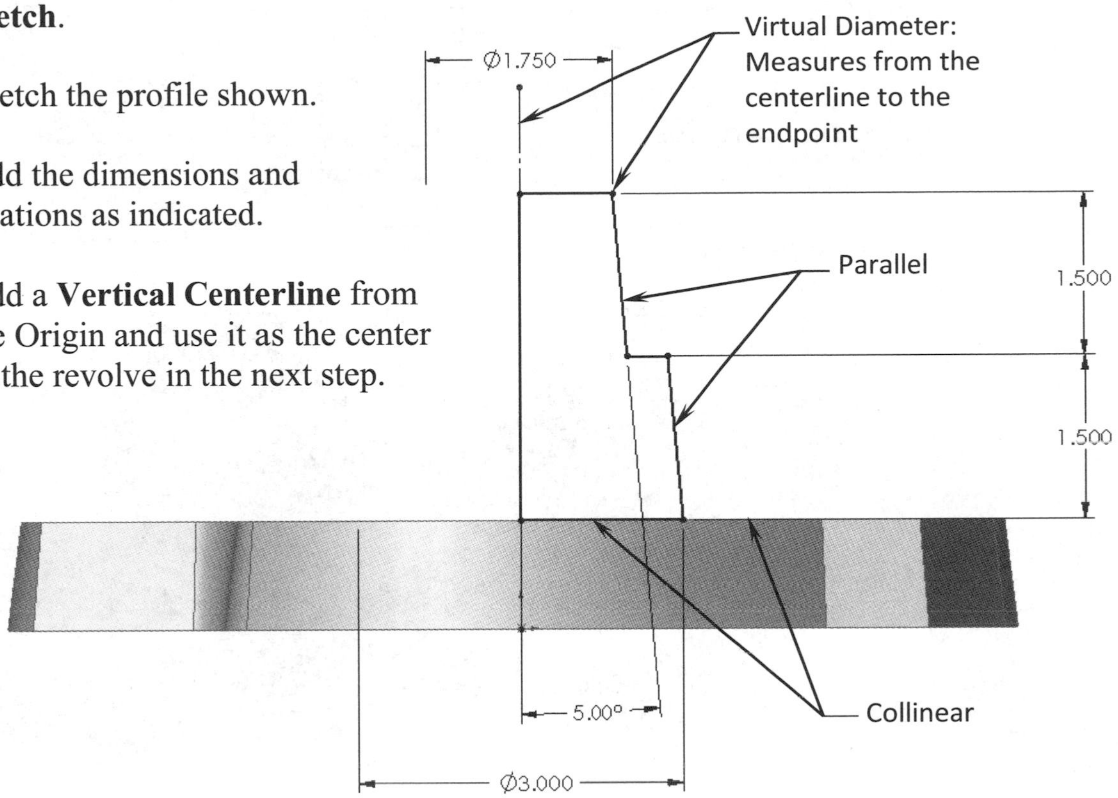

4. Revolving the upper Inlet Port:

Click **Revolve Boss/Base**  or Select: **Insert / Features / Boss-Base / Revolve** from the drop-down menus.

The vertical centerline should be selected automatically to use as the Axis of Revolution.

Revolve Direction: **Blind**.

Revolve Angle: **360 deg**.

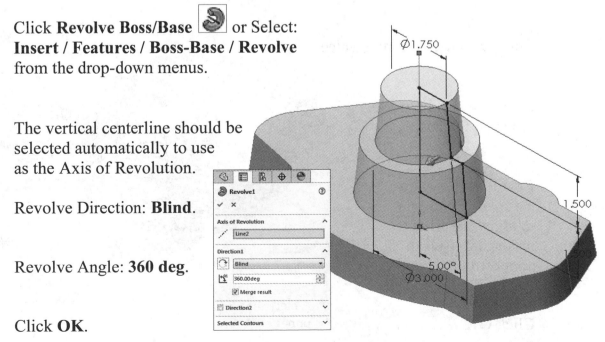

Click **OK**.

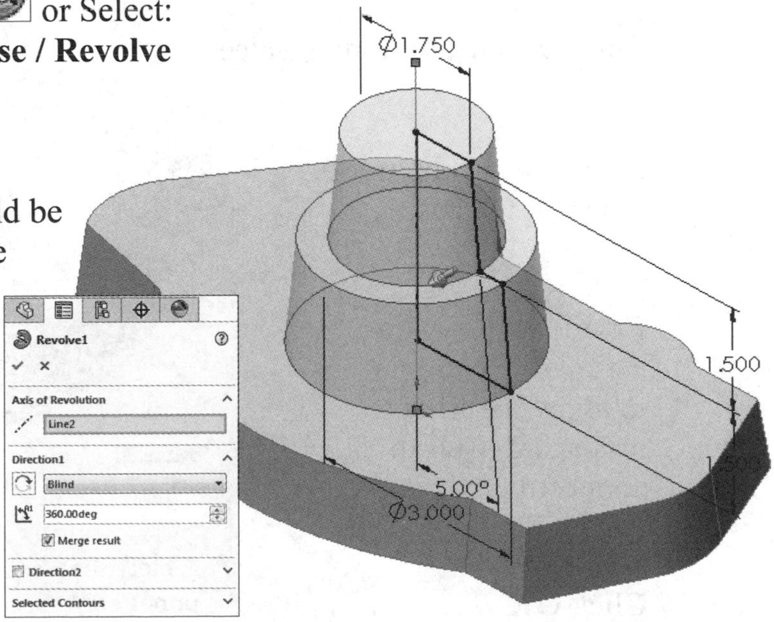

5. Adding the .500" Fillets:

Click **Fillet**  or select **Insert / Features / Fillet-Round.**

Enter **.500 in.** for radius.

Select the **7 edges** as indicated.

Click **OK**.

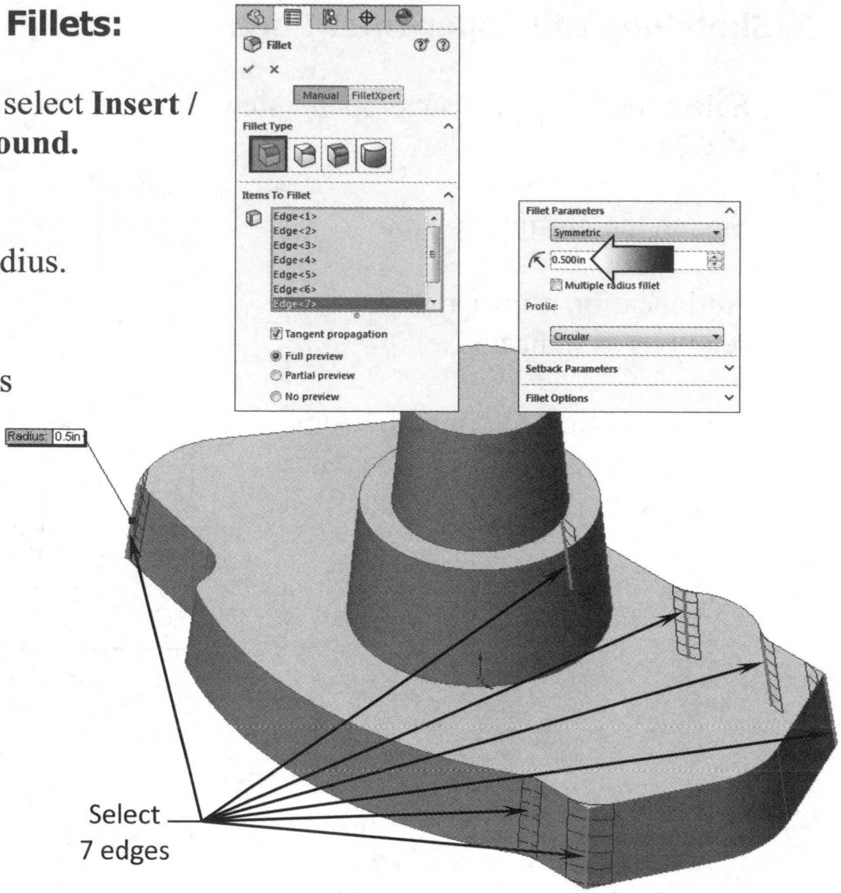

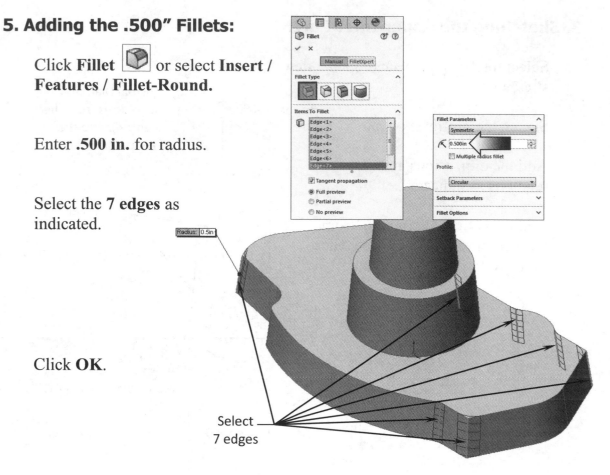

6. Adding the .275" Fillets:

Click **Fillet**  or select **Insert / Features / Fillet-Round.**

Enter **.275 in.** for radius value.

Select one of the **upper edges** of the base.

Enabled the option **Tangent Propagation** to allow the fillet to propagate itself to all connecting edges.

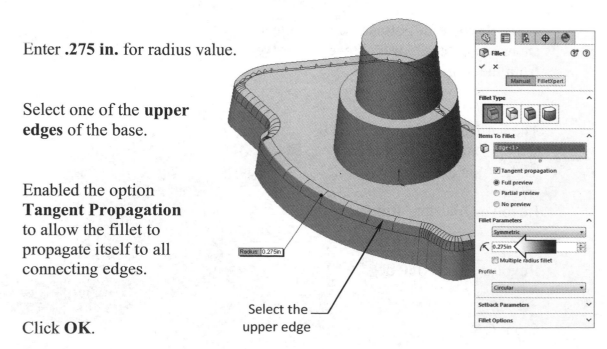

Click **OK**.

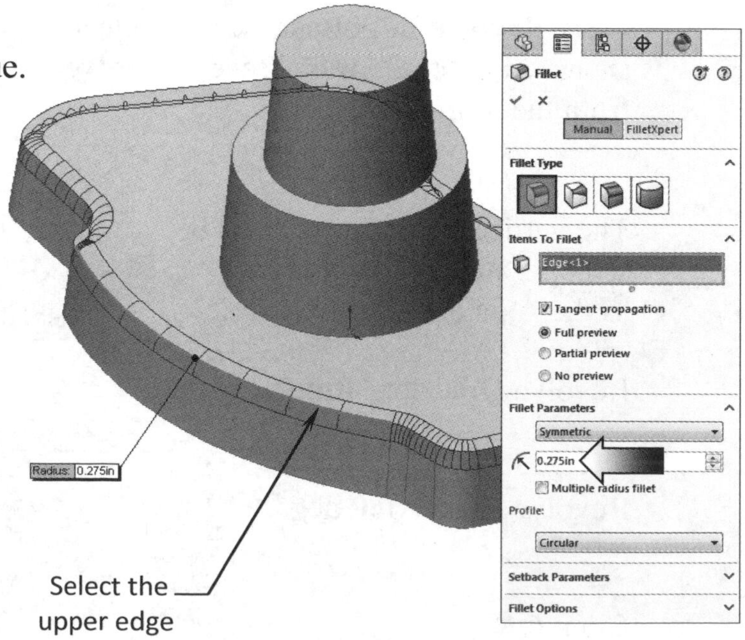

7. Creating the 1st Offset-Distance Plane:

Click **Plane** or select **Insert / Reference Geometry / Plane.**

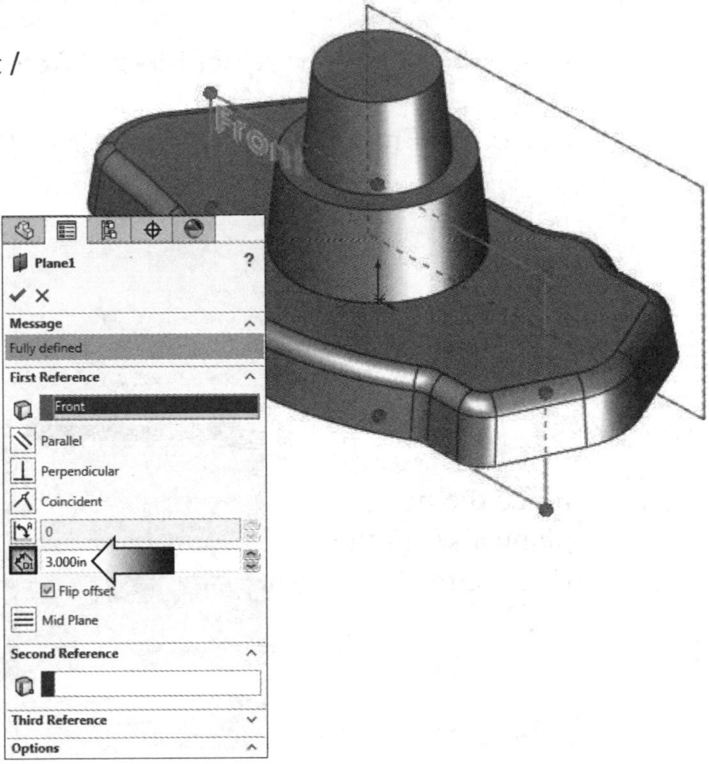

From the FeatureManager tree, select the **Front** plane to offset from.

Enter **3.000 in**. for distance.

Place the new plane on the **right side** of the front plane.

Click **OK**.

8. Creating the 2nd Offset-Distance Plane:

Click **Plane** or select **Insert / Reference Geometry / Plane.**

Select the Front plane again from the FeatureManager tree to offset from.

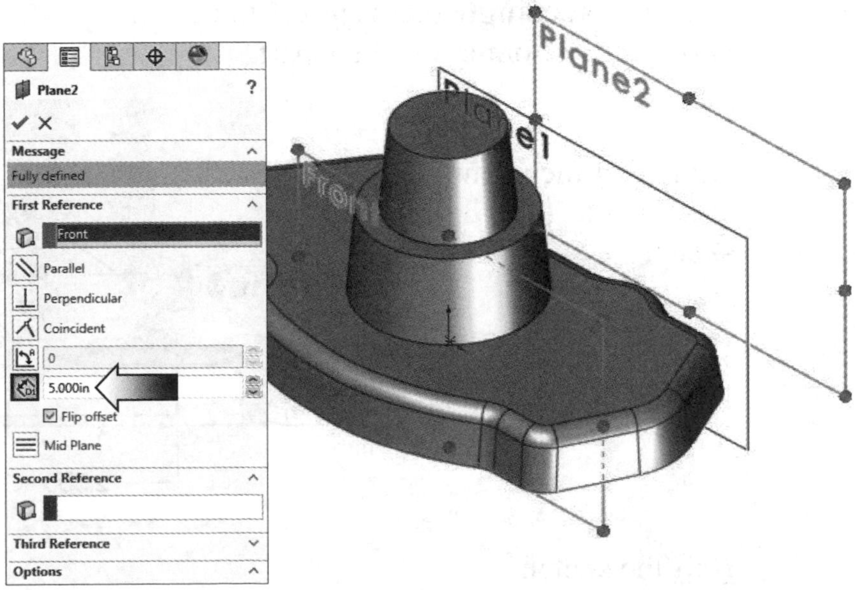

Enter **5.000 in**. for offset distance.

Place the new plane also on the **right side** of the front plane.

Click **OK**.

9. Creating the 3rd Offset-Distance Plane:

Click **Plane** or select **Insert / Reference Geometry / Plane.**

Select the Front plane once again from the FeatureManager tree to offset from.

Enter **6.000 in**. for offset distance.

Place the new plane also on the **right side**.

Click **OK**.

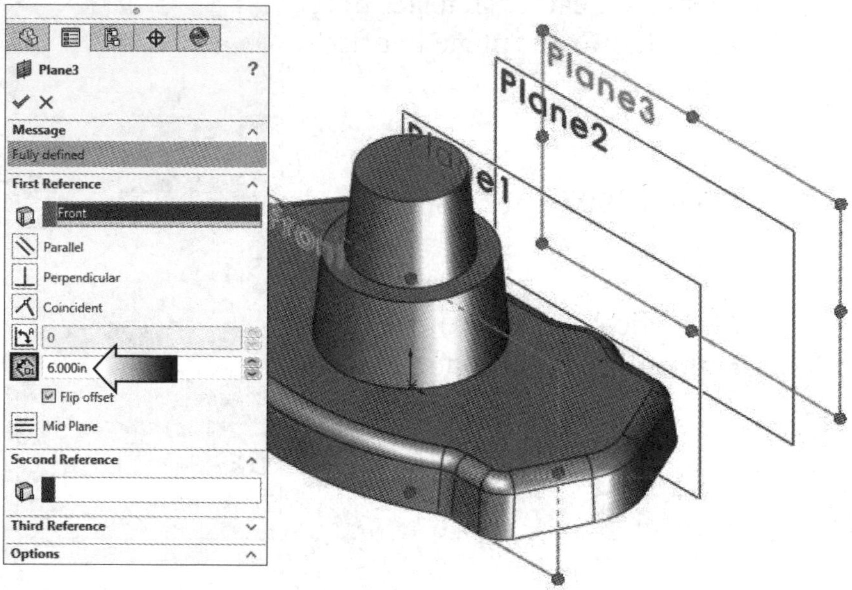

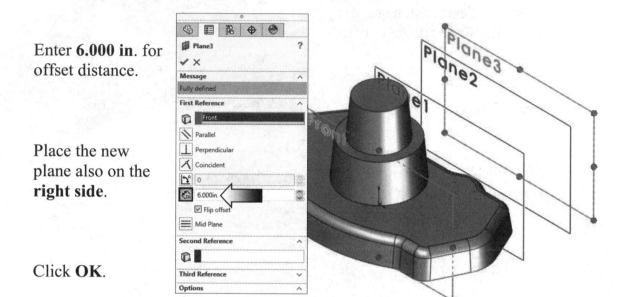

10. Sketching the 1st loft profile:

Select the Front plane and open a **new sketch** .

Sketch a **Rectangle** that is just **.125 in**. above the bottom edge of the part.

Add the dimensions and relations as shown to fully define the sketch.

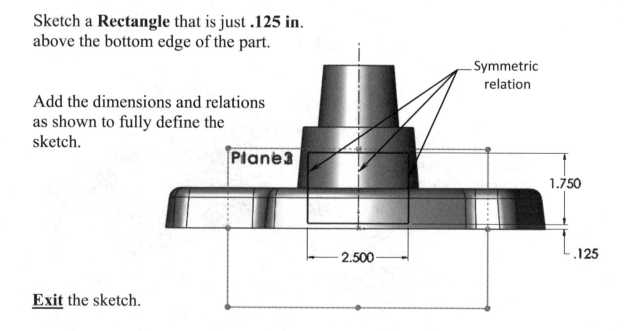

Exit the sketch.

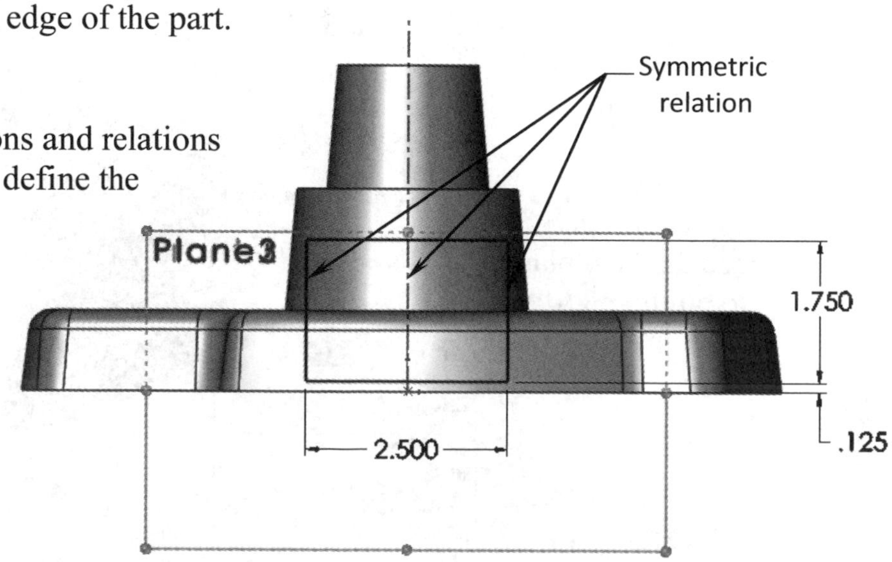

11. Sketching the 2nd loft profile:

Select the Plane1 and open a **new sketch**.

Sketch another **Rectangle** as shown.

Add the dimensions and relations needed to fully define the sketch.

Exit the sketch.

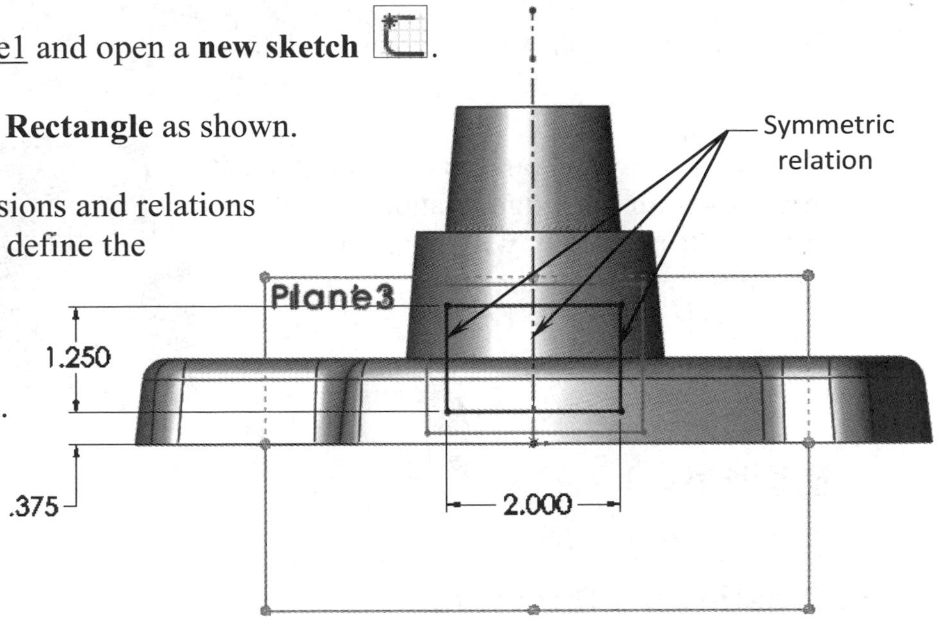

12. Sketching the 3rd loft profile:

Select the Plane2 and open a **new sketch**.

Sketch a **Circle** just above the Origin as shown below.

Use the **Split-Entities** command and split the circle into **4 segments**.

Add Vertical and Horizontal relations between the split points.

The Split Entities command is used to split the entities in each sketch to an even number of connecting points to help control the twisting in a loft feature.

Add the dimensions and relations needed to fully define the sketch.

Exit the sketch.

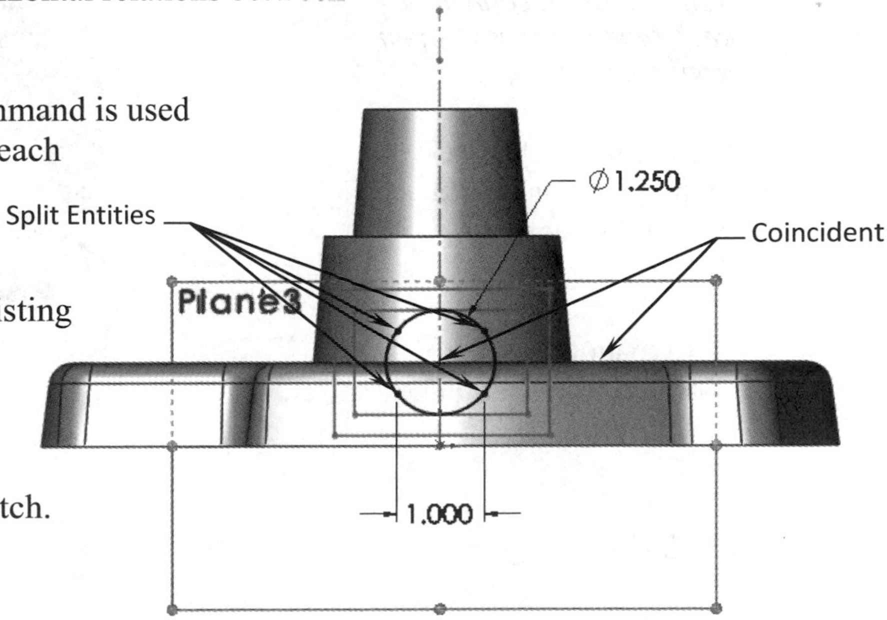

13. Sketching the 4th loft profile:

Select the Plane3 and open a **new sketch** .

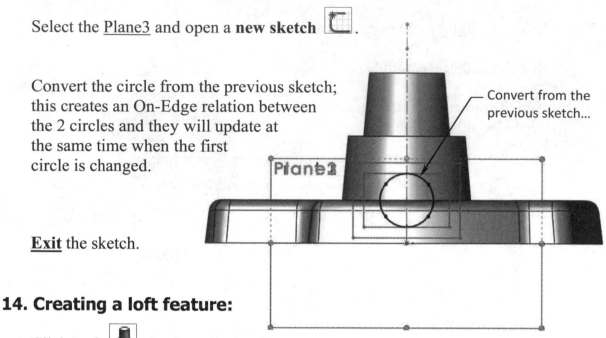

Convert the circle from the previous sketch; this creates an On-Edge relation between the 2 circles and they will update at the same time when the first circle is changed.

Exit the sketch.

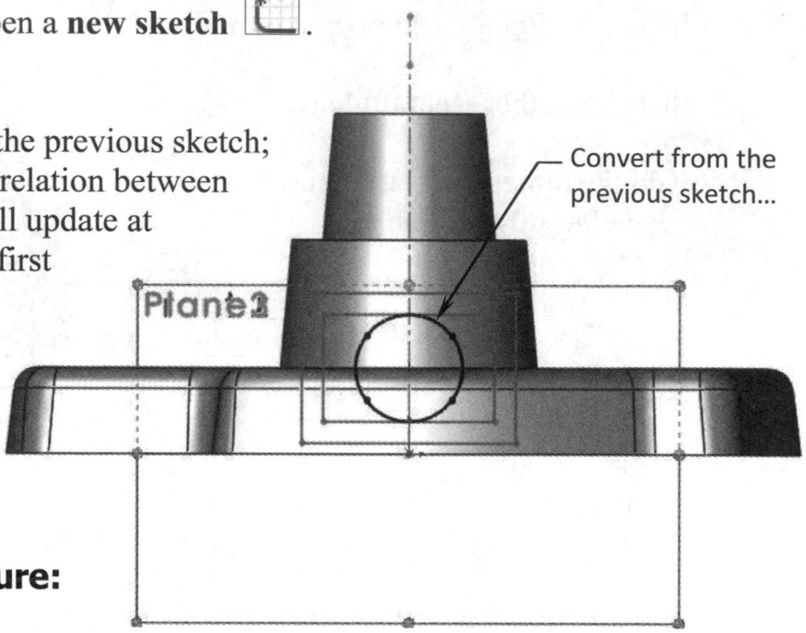

14. Creating a loft feature:

Click Loft or select: **Insert / Boss-Base / Loft** from the drop-down menus.

Select the **4 sketch profiles** in the graphics area.

(Since there are no guide curves to help control the loft, the profiles should be selected from the same connector each time to prevent them from twisting.)

For clarity, right-click in the yellow shaded area and select the following options:

* Opaque Preview
* Clear Meshed Faces

Click **OK**.

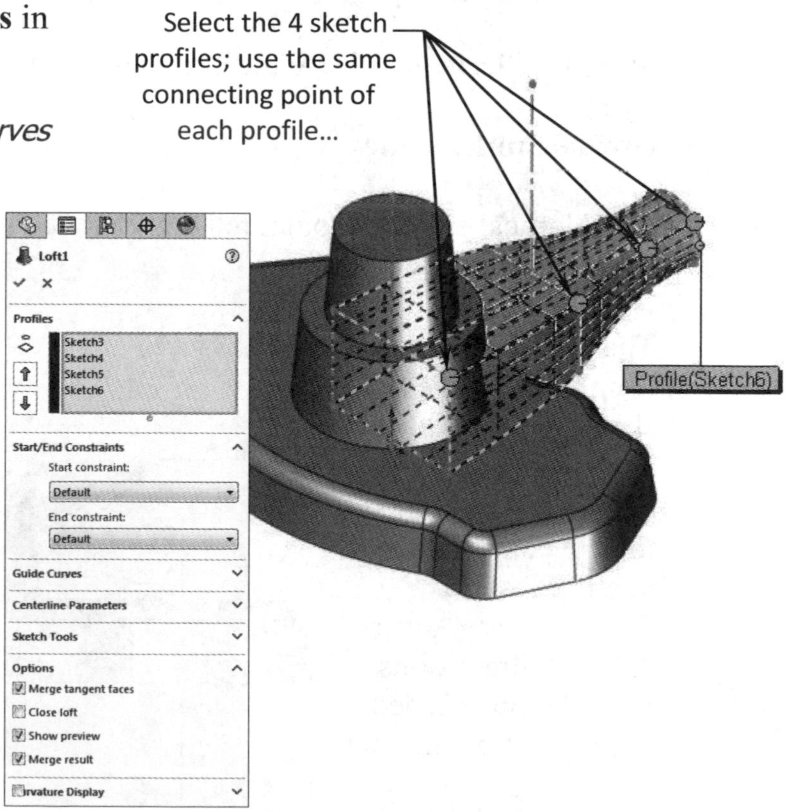

15. Creating the mounting bosses:

Select the bottom face and open a **new sketch**.

Sketch **5 Circles** as shown*.

* Avoid the hidden entities.

(When sketching, your circles may snap to some of the hidden edges of the model causing an over defined error when adding the dimensions.
To overcome this, hold the control key every time a hidden edge highlights; it will cancel the Auto-Relation snapping.)

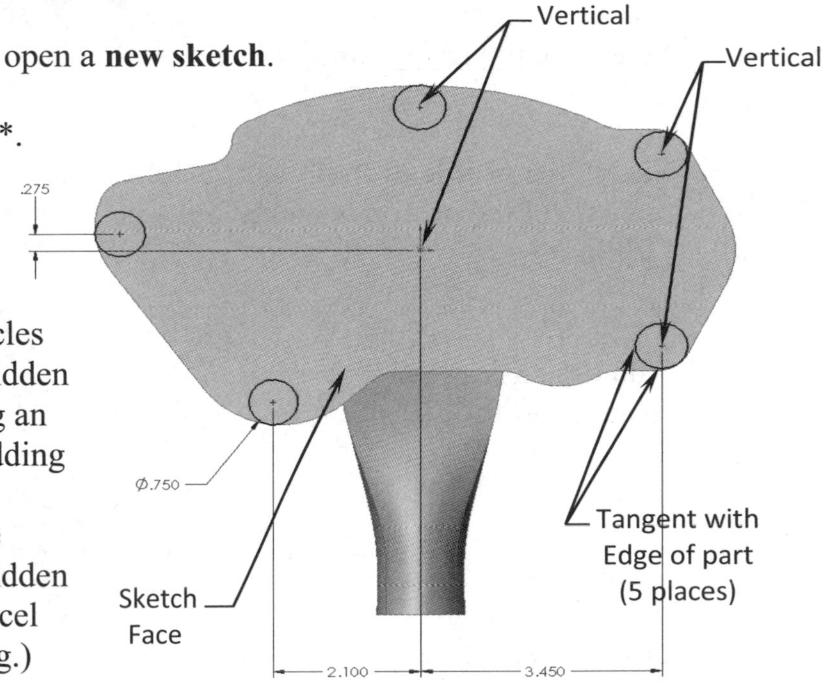

Add a **Tangent** relation for each circle to the outer edge of the part.

Add an **Equal** relation to all 5 circles.

Add dimensions to fully position the 5 circles.

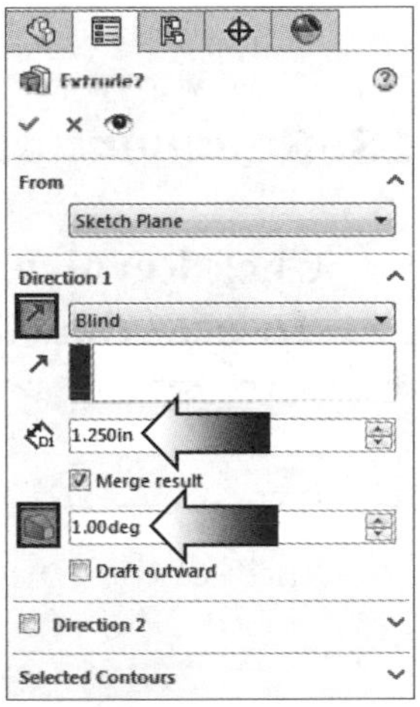

16. Extruding the 5 mounting bosses:

Click **Extrude Boss-Base** or select: **Insert / Extrude / Boss-Base**, from the drop-down menus.

Set the following:

End Condition: **Blind**

Depth: **1.250 in**.

Draft: **1 deg. Inward.**

Click **OK**.

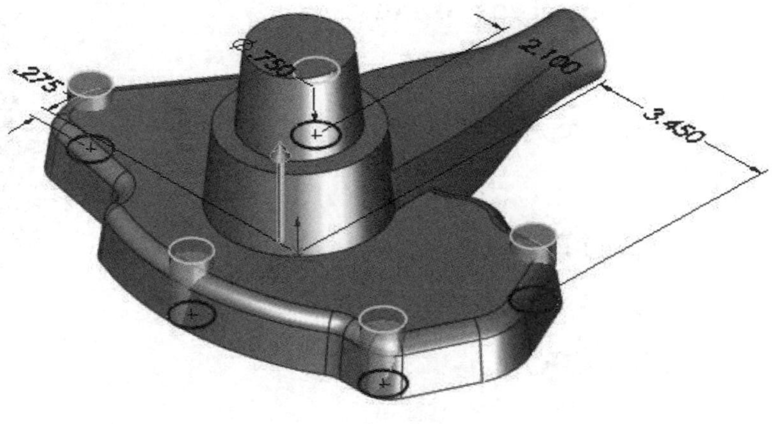

17. Sketching the rear Inlet Port:

Select the <u>upper face</u> and open a **new sketch**.

Sketch the profile of the Inlet at a **30º angle**; one end of the profile is locked onto the Origin.

Add dimensions and other relations to fully position the sketch.

There is more than one centerline in this sketch; select the 30º centerline before clicking the revolve command.

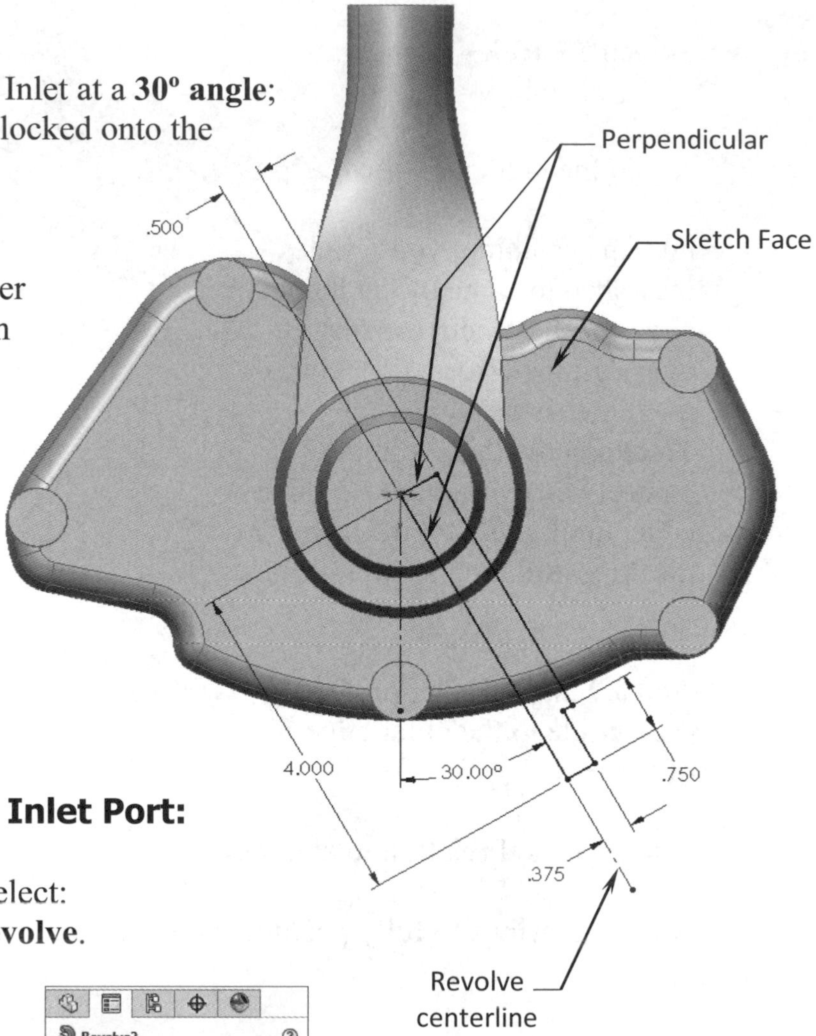

18. Revolving the Rear Inlet Port:

Click **Revolve**  or select:
Insert / Boss-Base / Revolve.

Select the 30° centerline and set the following:

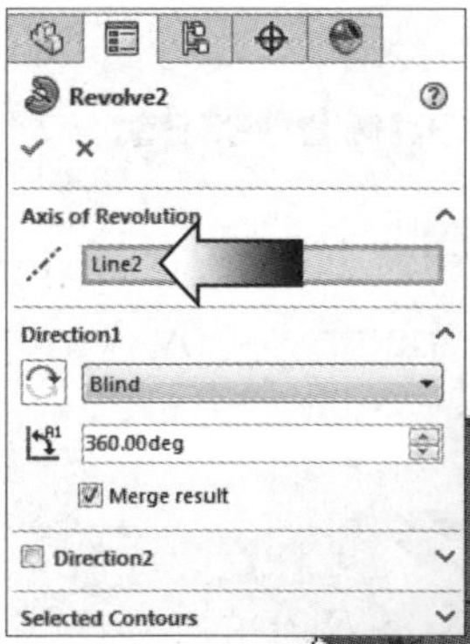

Revolve Type: **Blind**.

Revolve Angle: **360 deg**.

Merge Result: **Enabled**

Click **OK**.

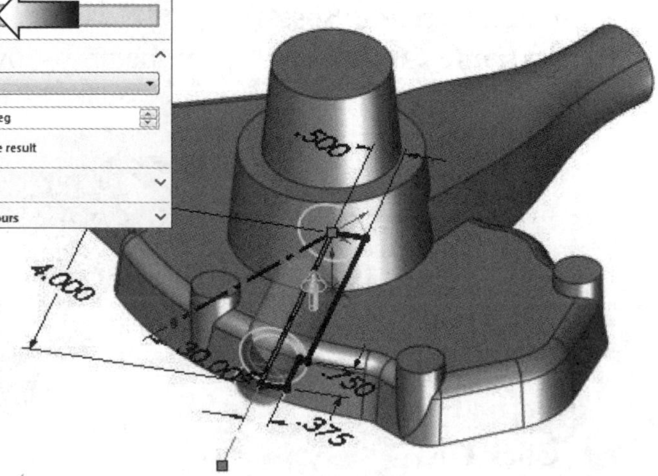

19. Adding the 1st Face Fillet:

> **Face Fillet**
>
> A Face Fillet is used to create a blend between non-adjacent, non-continuous faces.

Click **Fillet** or select: **Insert / Features / Fillet-Round**.

Select the **Face Fillet** option.

Enter **.250 in.** for radius.

For **Face Set 1**, select the **upper face** of the lofted feature.

For **Face Set 2**, select the **side face** of the lofted feature.

Click **OK**.

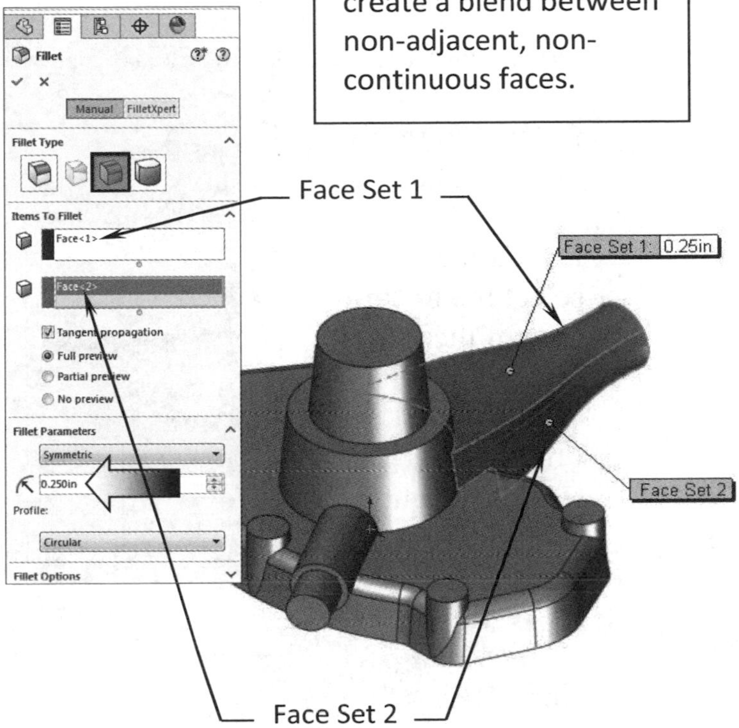

20. Adding the 2nd Face Fillet:

Click **Fillet** or select: **Insert / Features / Fillet-Round.**

Select the **Face Fillet** button again.

Enter **.250 in.** for radius value.

For **Face Set 1,** select the **upper face** of the lofted feature.

For **Face Set 2,** select the **side face** of the lofted feature.

Click **OK**.

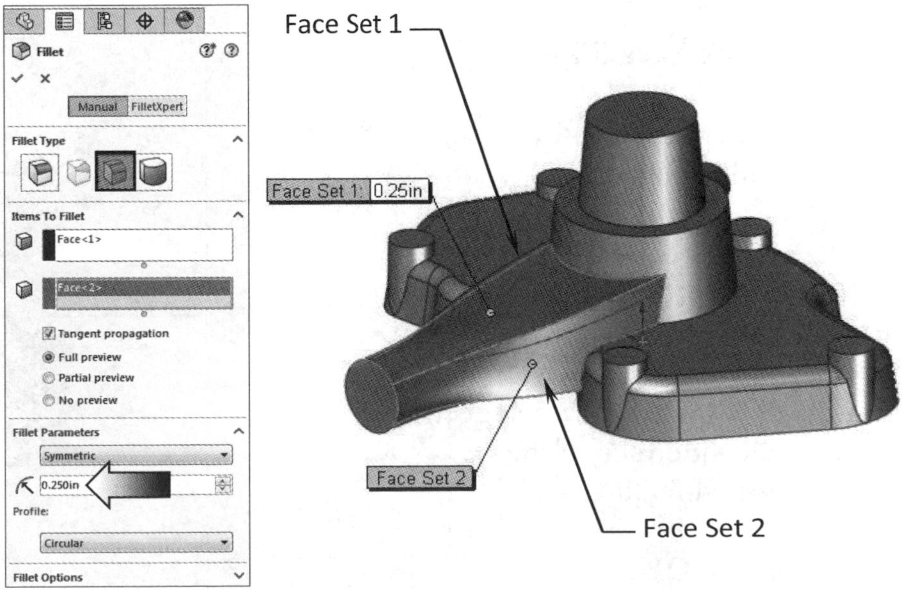

21. Adding the 3rd Face Fillet:

Click **Fillet** or select:
Insert / Features / Fillet-Round.

Click **Face Fillet.**

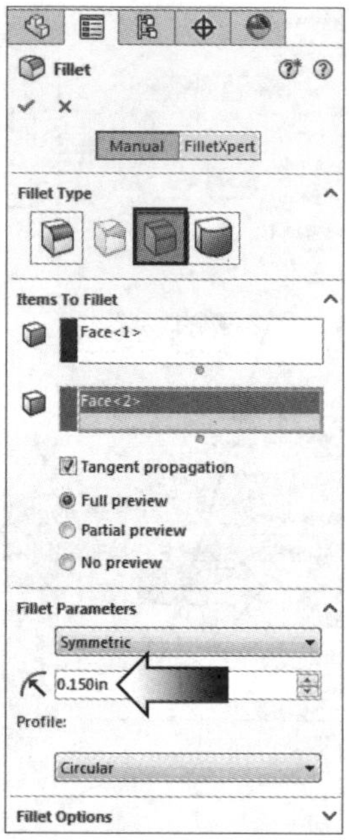

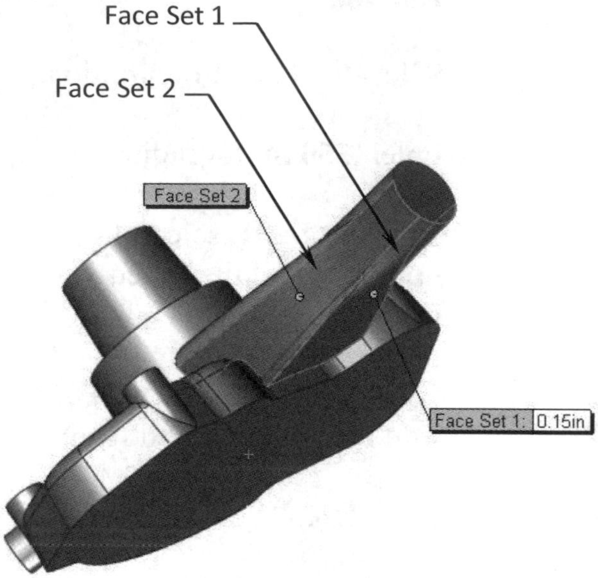

Enter **.150 in.** for radius value.

For **Face Set 1**, select the **bottom face** of the lofted feature.

For **Face Set 2**, select the **side face** of the lofted feature.

Click **OK.**

22. Adding the 4th Face Fillet:

Click **Fillet** or select:
Insert / Features / Fillet-Round.

Click **Face Fillet.**

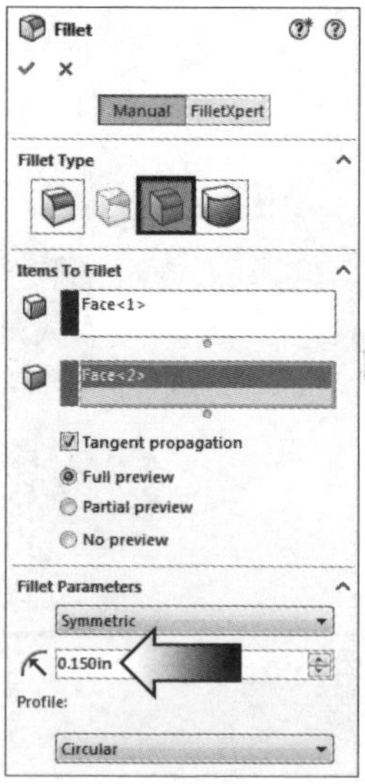

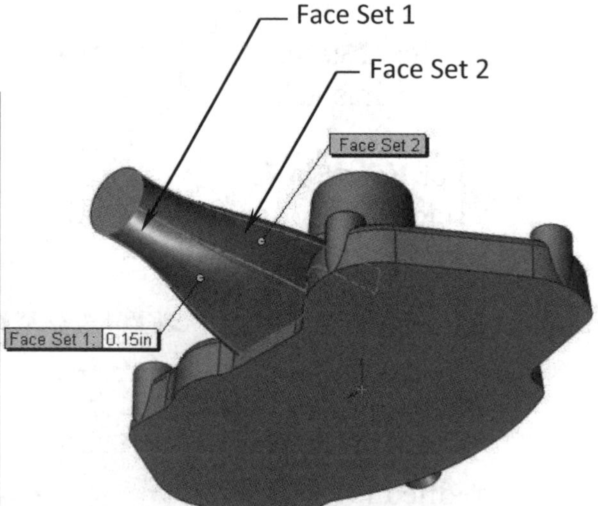

Enter **.150 in.** for radius value.

For **Face Set 1**, select the **bottom face** of the lofted feature.

For **Face Set 2**, select the **side face** of the lofted feature.

Click **OK.**

23. Mirroring the rear Inlet Port:

Click **Mirror** or select: **Insert / Pattern Mirror / Mirror.**

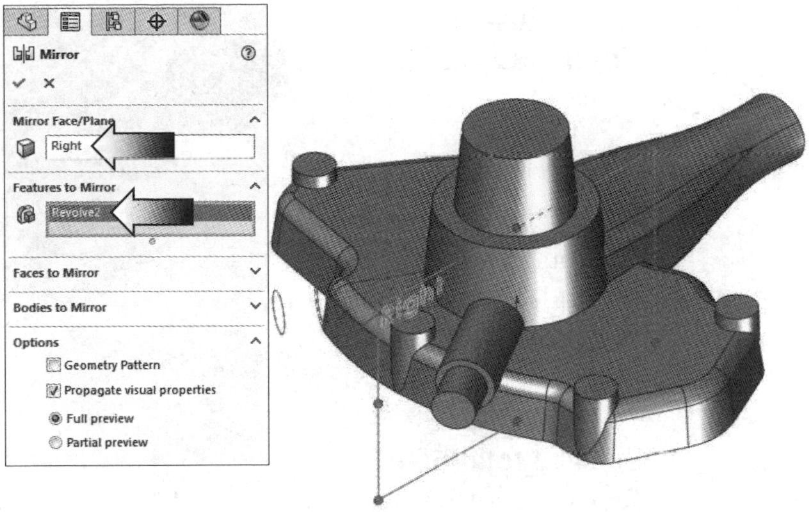

For Mirror Face/Plane, select the **Right** plane from the FeatureManager tree.

For Features to Mirror, select the **Rear Inlet Port** either from the graphics area or from the feature tree.

Click **OK**.

24. Adding the .175" Fillets:

Click **Fillet** or select: **Insert / Features / Fillet-Round.**

Select the **Constant Size** fillet option.

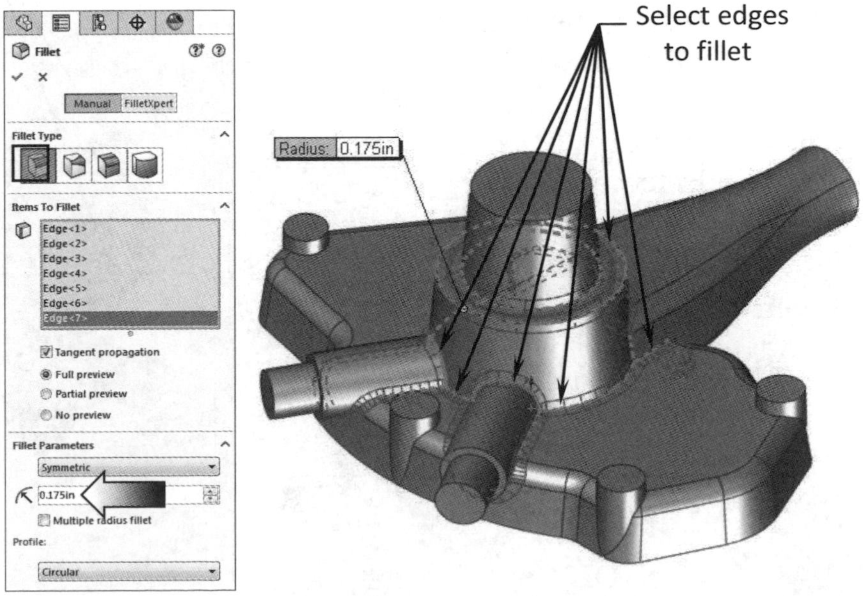

Enter **.175 in.** for radius size.

Select the **edges** as noted to add the fillets.

The option Tangent-Propagation should be enabled by default.

Click **OK**.

25. Shelling the part:

Click **Shell** or select: **Insert / Features / Shell.**

Under the Parameters section, enter **.080 in**. for wall thickness.

Select a total of **10 faces** to remove.

Click **OK**.

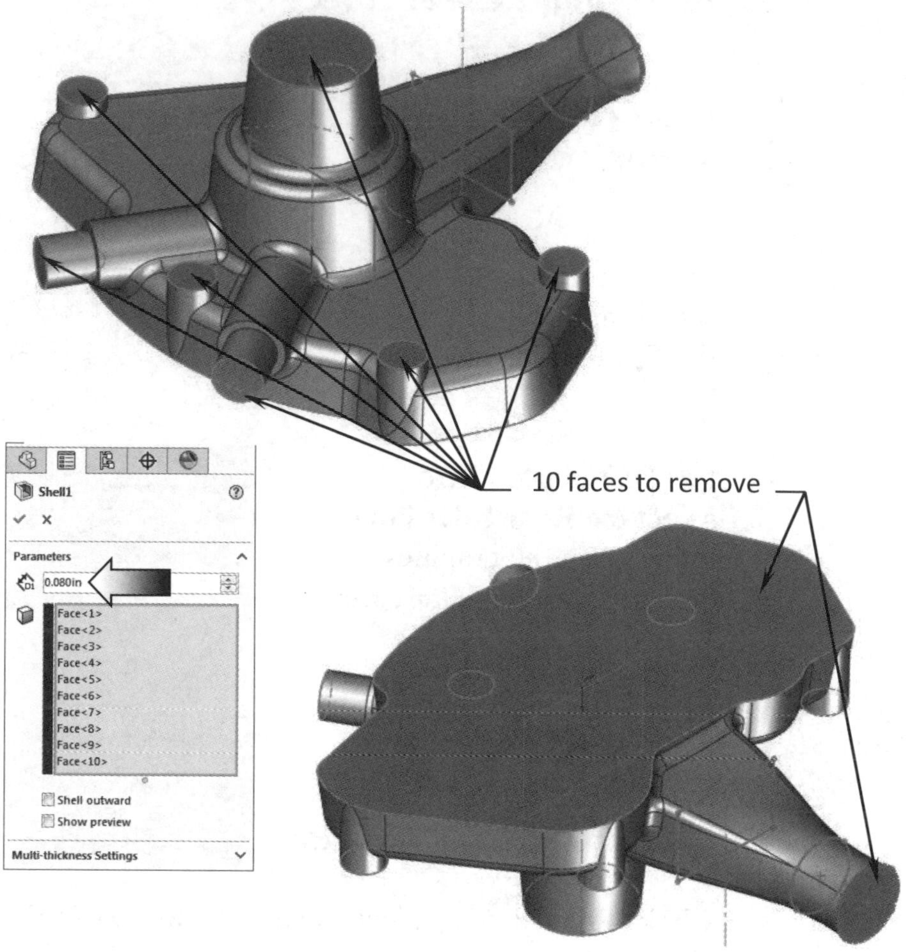

26. Adding the .0625" Fillets:

Click **Fillet** or select: **Insert / Features / Fillet-Round.**

Enter **.0625 in.** for radius size.

Select the **edges** as noted to add the fillets.

Click **OK**.

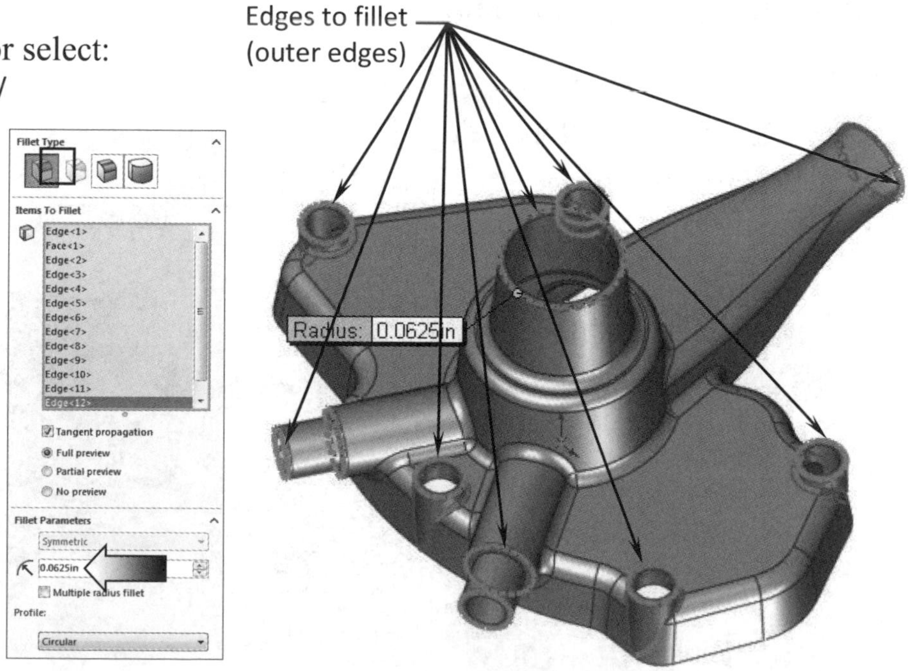

27. Adding a Rib:

Select the Front plane from the FeatureManager tree and open a **new sketch**.

Sketch a **Line** as shown.

Add the coincident relations between the endpoints of the line and the edges of the part.

Add the location dimensions shown.

Rib Features

A rib is an extruded feature which adds material of a specified thickness in a specified direction. Drafts can also be added to the faces of the rib.

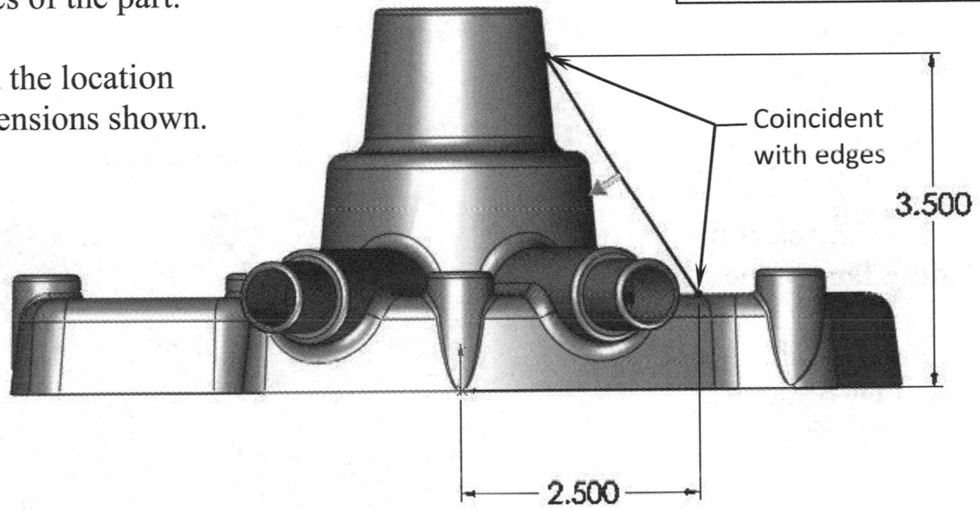

28. Extruding the Rib:

Click **Rib** or select: **Insert / Features / Rib.**

Select **Both Directions** under the Thickness section.

Enter **.275 in**. for the thickness of the rib.

Enable the **Draft** option and enter **1.00 deg**.

Enable the **Draft Outward** check box.

Click **OK**.

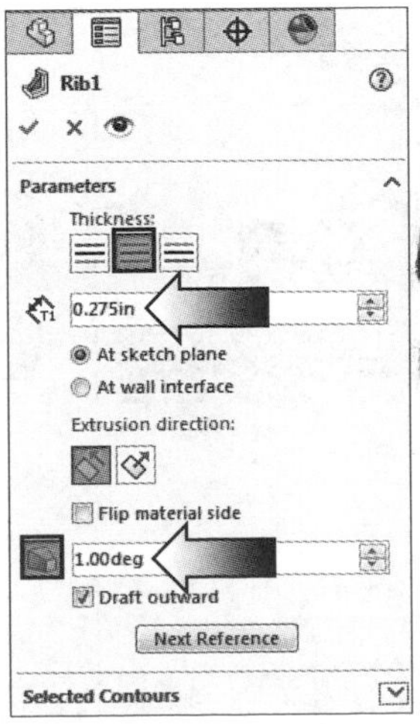

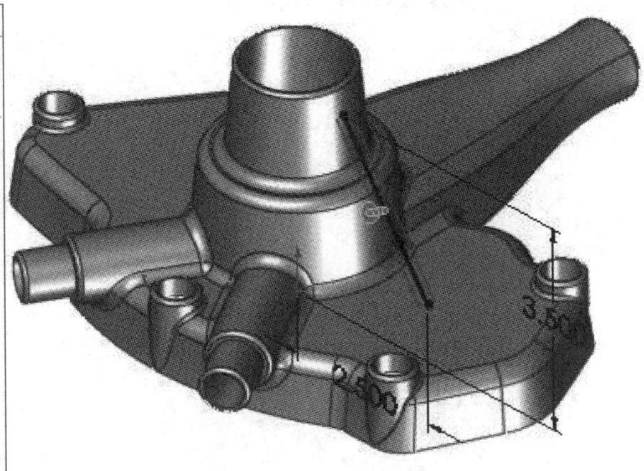

29. Creating a Full-Round fillet:

A Full-Round creates fillets that are tangent to three adjacent face sets.

Click **Fillet** or select: **Insert / Features / Fillet-Round.**

Select the **Full Round** fillet option.

For Face 1, select the **left face** of the Rib.

For Face 2, select the **middle face** of the Rib.

For Face 3, select the **right face** of the Rib.

Click **OK**.

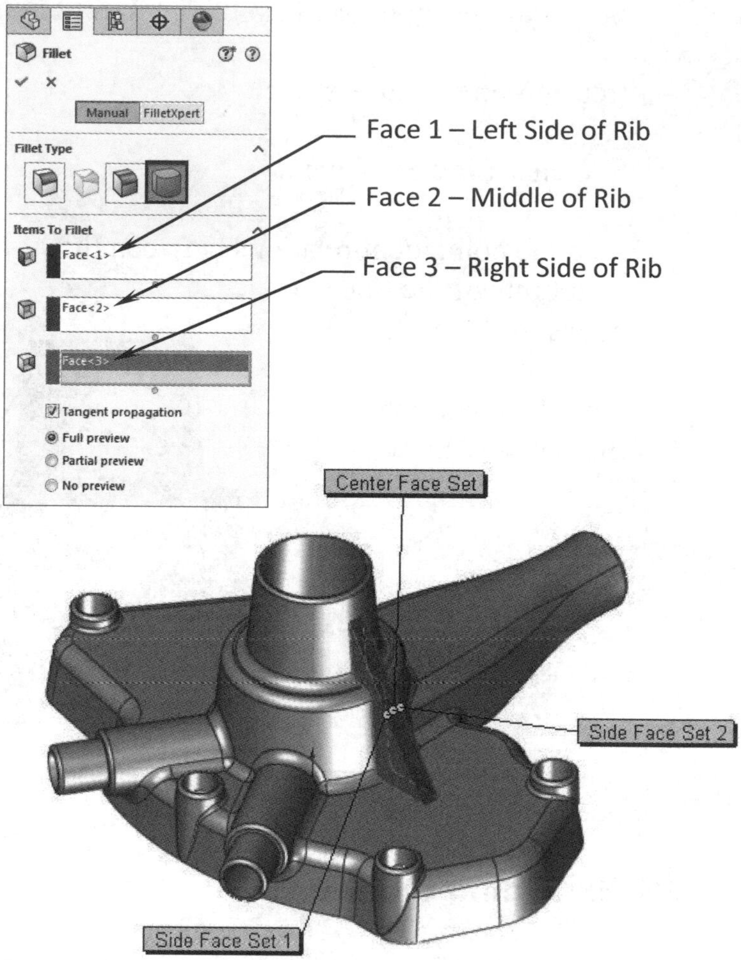

30. Mirroring the Rib:

Click **Mirror** or select: **Insert / Pattern Mirror / Mirror.**

For Mirror Face/Plane, select the **Right** plane.

For Features to Mirror, select the **Rib** and its **fillet**.

Click **OK**.

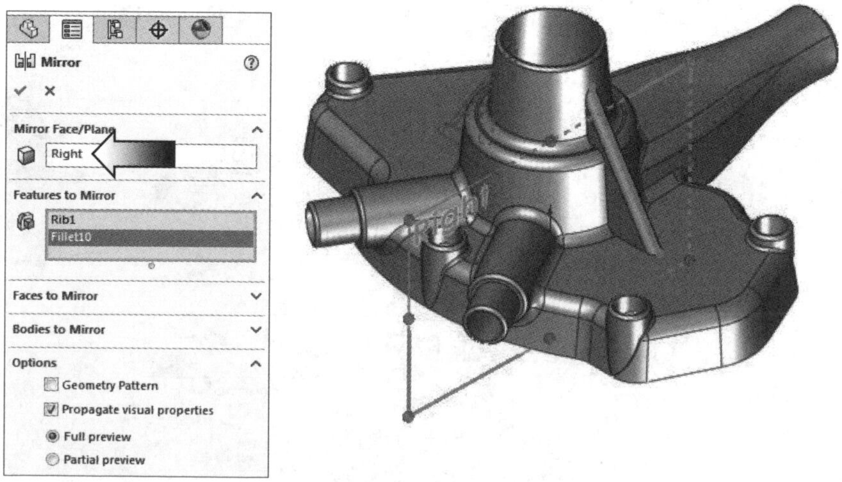

31. Adding the .025" fillets:

Click **Fillet** or select: **Insert / Features / Fillet-Round.**

Select the **Constant Radius** fillet option.

Enter **.025 in**. for radius value.

Select the **edges** as indicated to add the fillets.

Enable the Tangent-Propagation checkbox.

Click **OK**.

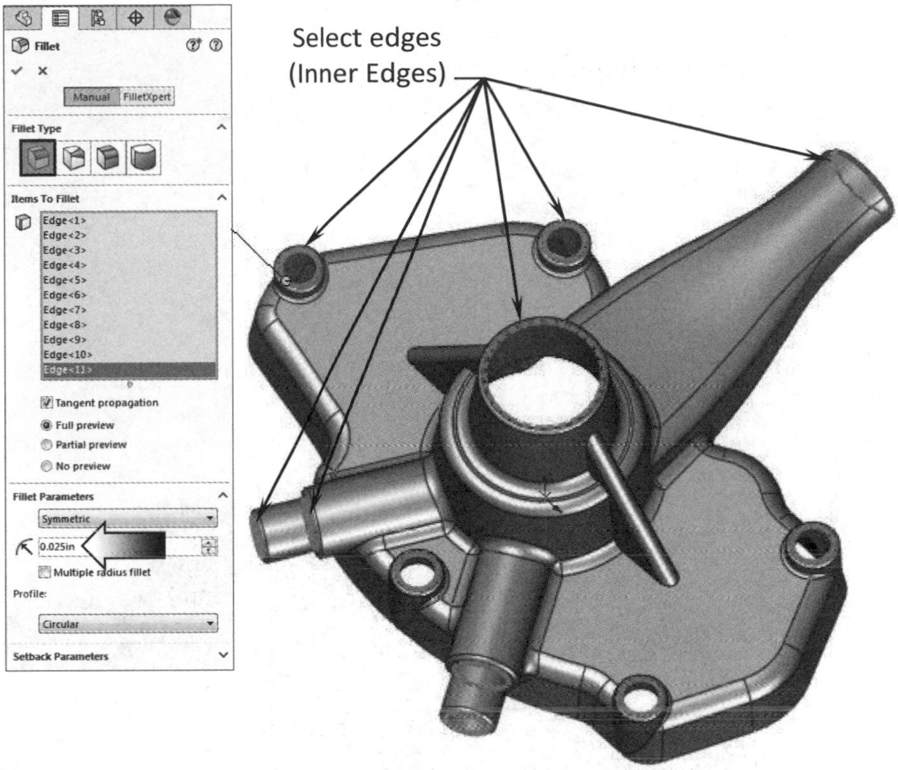

32. Adding the .175" fillets:

Click **Fillet** or select: **Insert / Features / Fillet-Round.**

Click **Constant Radius.**

Enter **.175 in.** for radius value.

Select the **edges** of the 2 ribs as indicated.

Click **OK**.

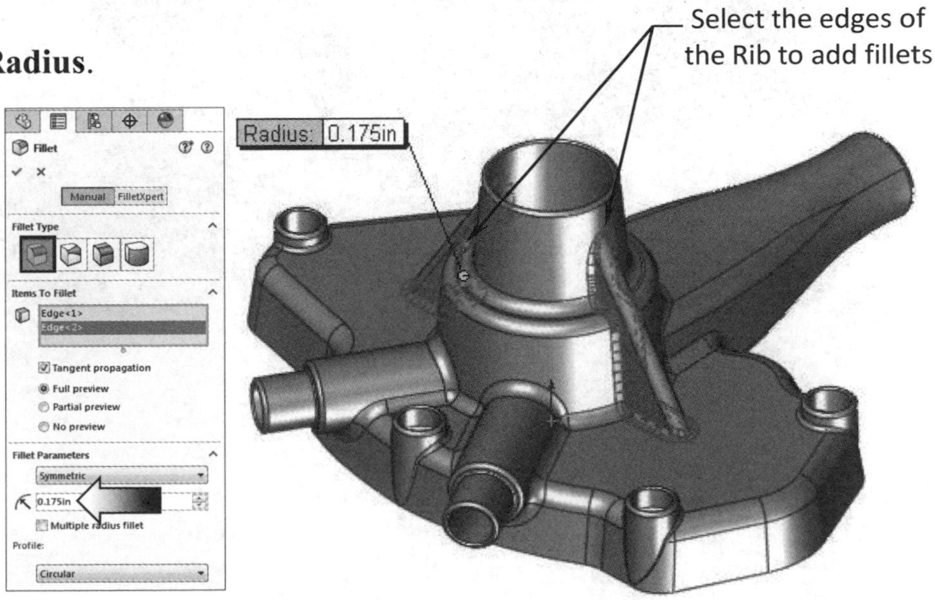

33. Saving your work:

Click **File / Save As**.

Enter **Water Pump Cover** as the name of the file.

Click **Save.**

CHAPTER 6

Loft Vs. Sweep

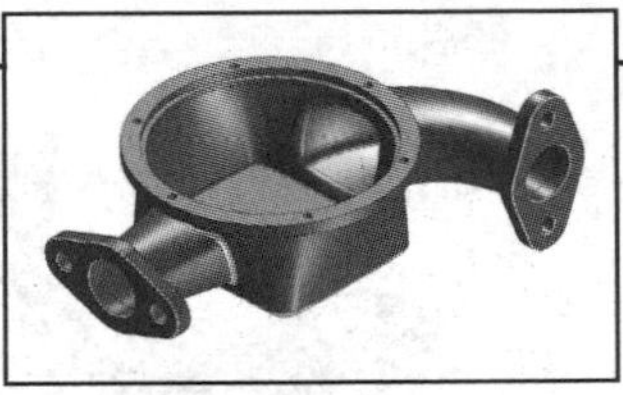

Loft vs. Sweep
Water Meter Housing

The Loft and the Sweep commands are usually used to create advanced, complex shapes. The differences between the two are:

- Sweep uses a single sketched profile to sweep along a path and is controlled by one or more guide curves.

Sweep uses one profile, one path and multiple guide curves

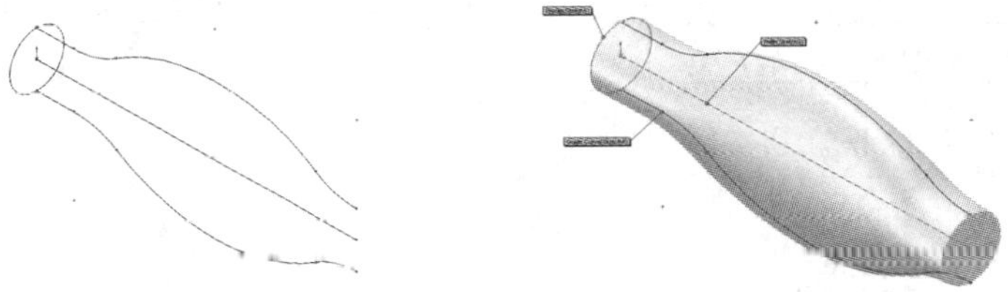

- Loft uses multiple sketched profiles to loft between the sections and is controlled by 1 or more guide curves or 1 Centerline Parameter.

Loft uses multiple profiles, one centerline parameter, and/or multiple guide curves

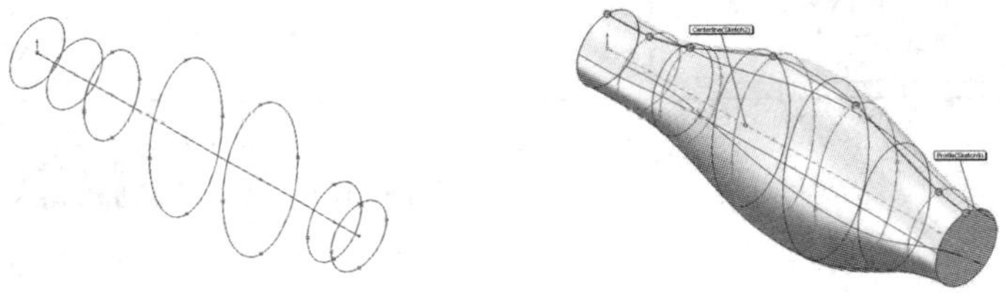

In order to create a solid feature, each sketch profile must be a single, closed, and non-intersecting shape.

The guide curves can be either a 2D sketch or a 3D curve.

The sweep path and guide curves must be related to the sketch profiles with either a Coincident or Pierce relation.

The loft profiles should have the same number of entities or segments. The Split-Entities commands can be used to split the sketch entities and add the necessary connectors to help control the loft more accurately.

Water Meter Housing
Loft vs. Sweep

View Orientation Hot Keys:

Ctrl + 1 = Front View
Ctrl + 2 = Back View
Ctrl + 3 = Left View
Ctrl + 4 = Right View
Ctrl + 5 = Top View
Ctrl + 6 = Bottom View
Ctrl + 7 = Isometric View
Ctrl + 8 = Normal To Selection

Dimensioning Standards: **ANSI**

Units: **INCHES** – 3 Decimals

Tools Needed:

Insert Sketch	Line	Split Entities
Sketch Fillet	Mirror	Add Geometric Relations
Dimension	Base/Boss Extrude	Extruded Cut
Plane	Base/Boss Sweep	Base/Boss Loft

1. Sketching the 1st Loft Profile:

Select the Top plane and open a **new sketch** .

Sketch a **Rectangle** centered on the Origin and add the dimensions shown.

(Only add the corner fillets after the sketch is fully defined.)

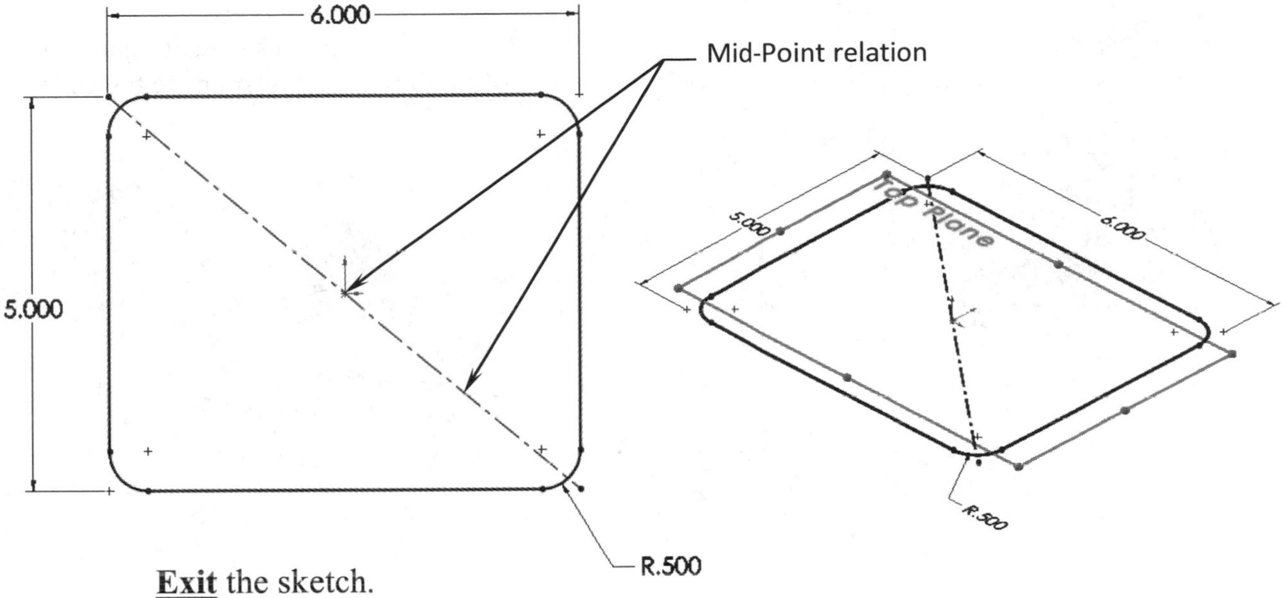

Exit the sketch.

2. Creating a new plane:

Click  or select **Insert / Reference Geometry / Plane**.

Select the Top reference plane from the FeatureManager tree.

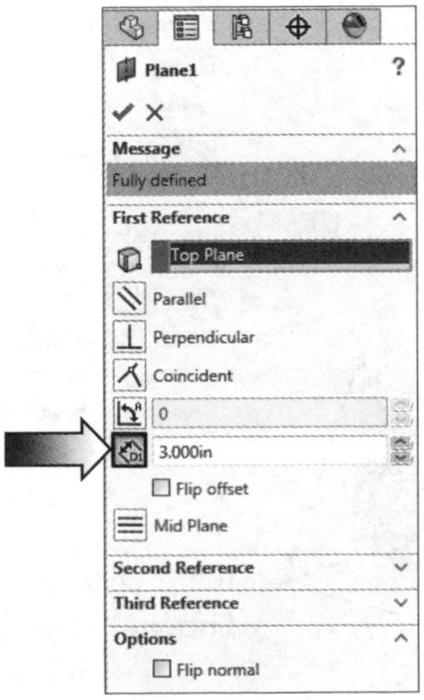

Select the **Offset Distance** option and enter **3.000**in.

Click **OK**.

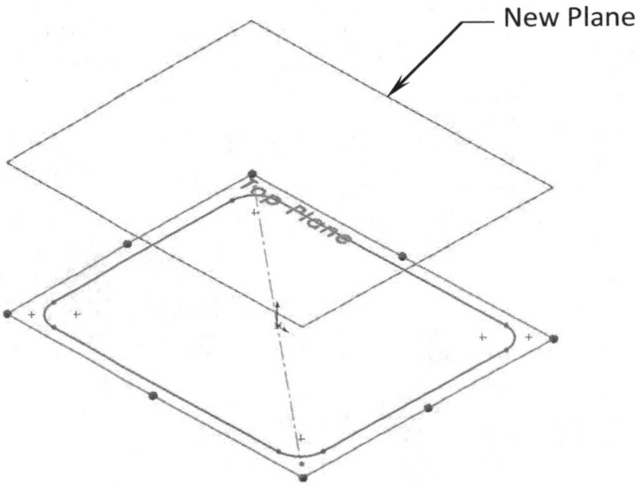

3. Sketching the 2nd Loft Profile:

Select the new plane **(Plane1)** and open a **new sketch** .

Sketch a **Circle** and split it into 8 segments (Tools / Sketch Tools / Split Entities).

Add dimensions and the vertical / horizontal relations between the split points.

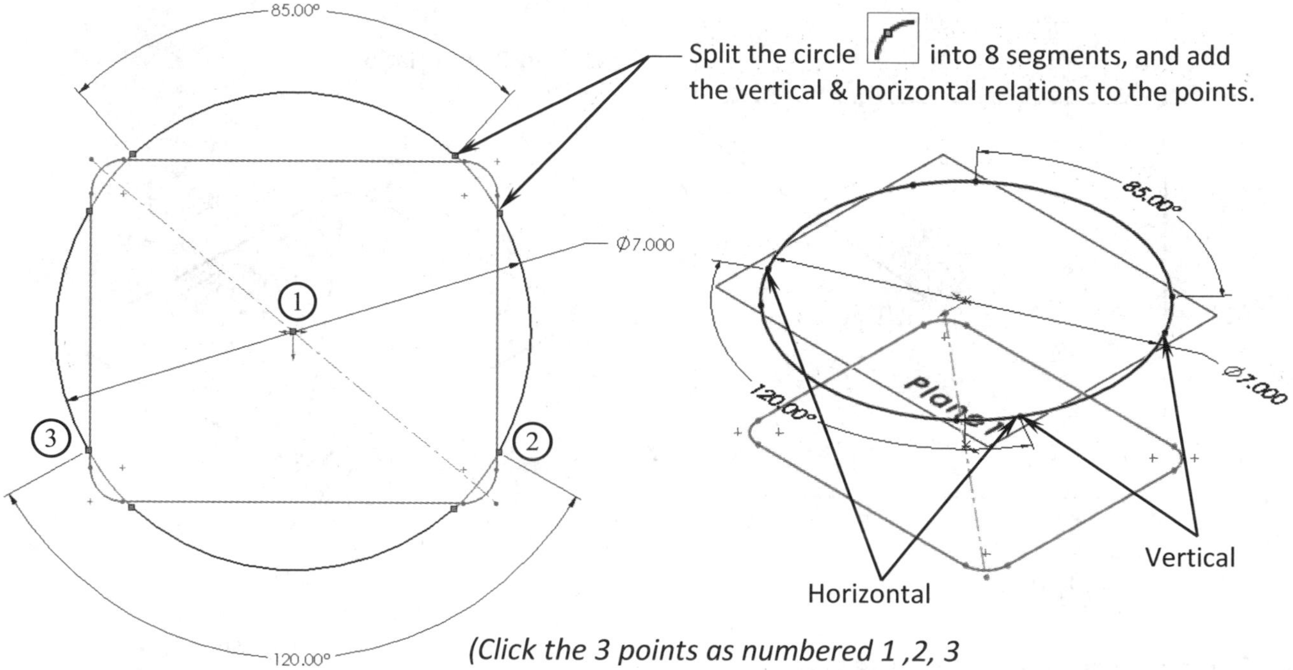

(Click the 3 points as numbered 1 ,2, 3 to create the angular dimensions)

Exit the Sketch .

4. Creating the Lofted Base feature:

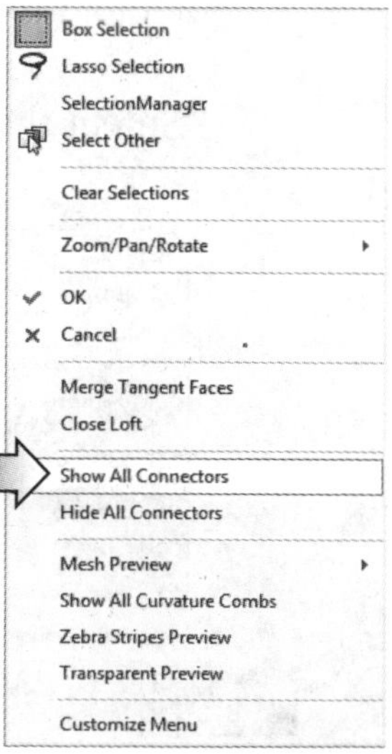

Click or select **Insert / Boss-Base / Loft.**

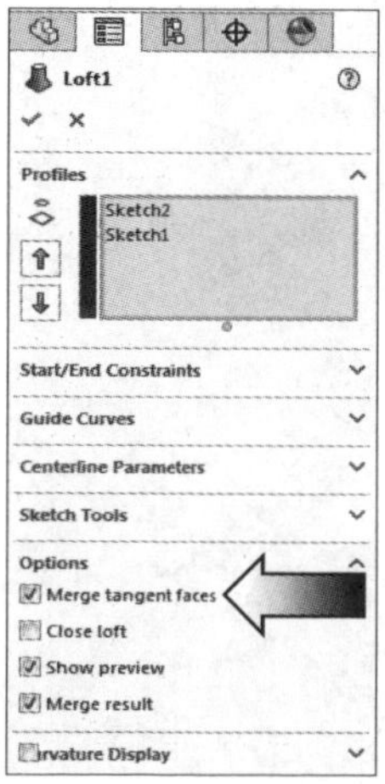

For Loft Profiles: Select the **2 sketches** in the graphics area.

Show all connectors as noted.

Enable **Merge Tangent Faces**.

Click **OK**.

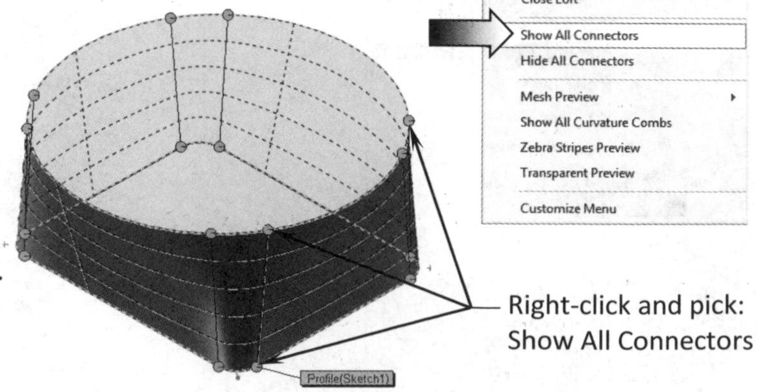

5. Constructing the Inlet's 1st Loft Profile:

Select the Front plane and open a **new sketch** .

Sketch a **Rectangle**, add the dimensions shown below. Only add the sketch fillets after the sketch is fully defined.

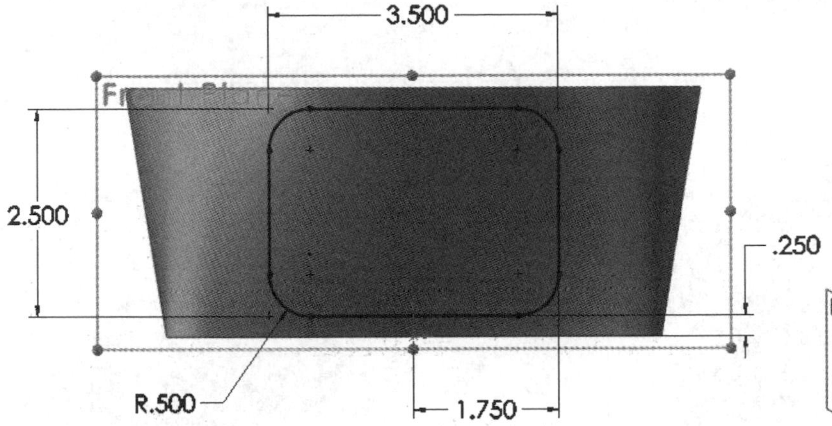

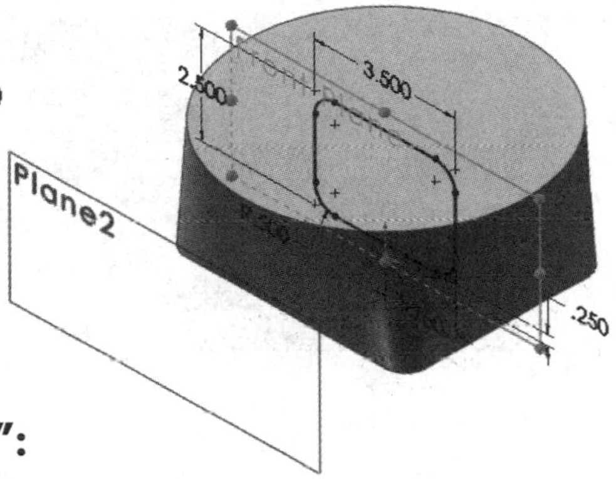

6. Creating an Offset Distance plane at 5.500":

Click  or select **Insert / Reference Geometry / Plane.**

Select the **Front** reference plane from the FeatureManager tree.

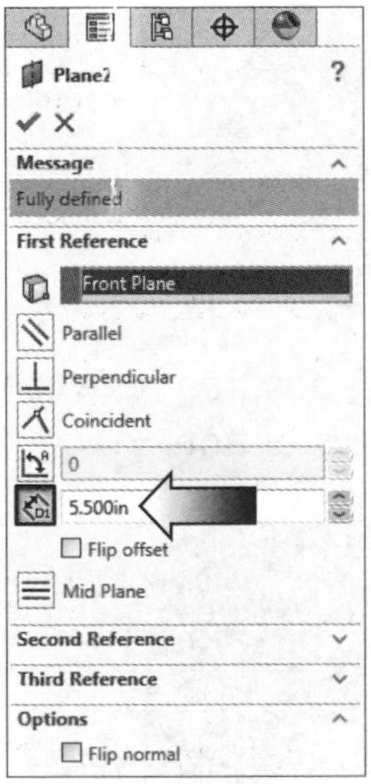

Select **Offset Distance** option and enter **5.500 in**. for distance.

Click **OK**.

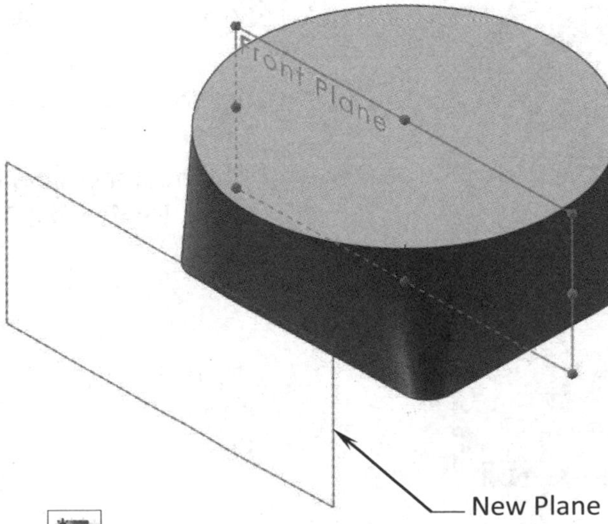

Exit the Sketch .

7. Constructing the Inlet's 2nd Loft Profile:

Select the Plane2 and open a **new sketch**.

Sketch a **Circle** and add dimensions to fully define it.

Click **Split Entities** and split the circle into 8 segments (Tools, Sketch Tools, Split Entities).

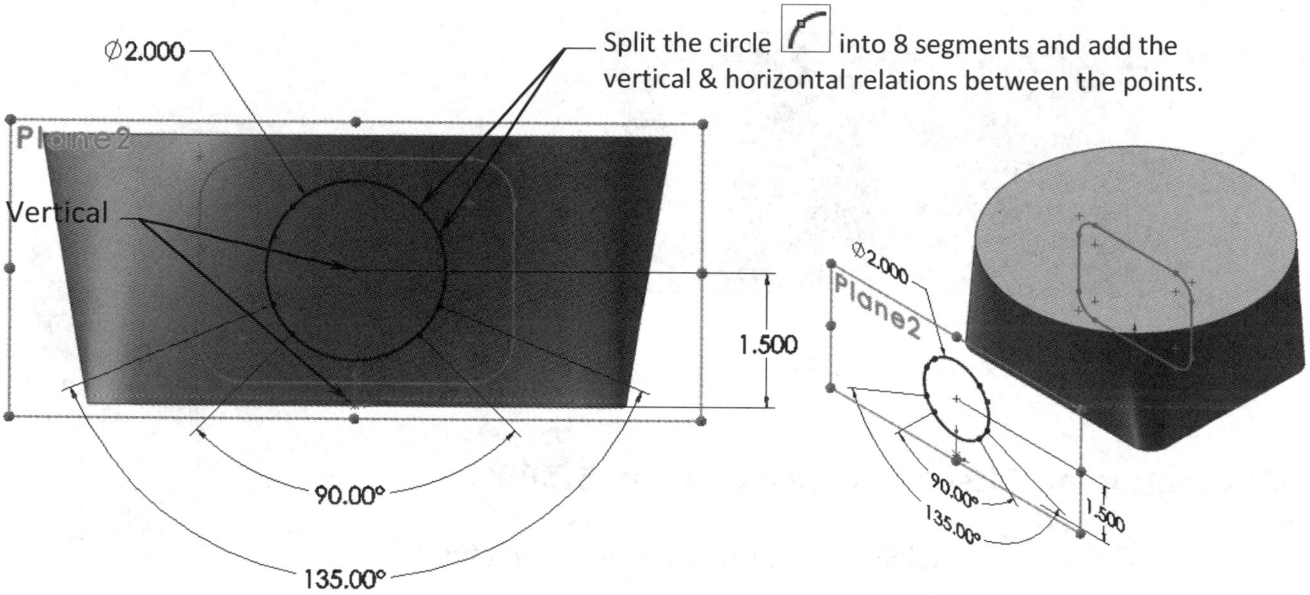

Exit the Sketch.

8. Creating the Inlet feature:

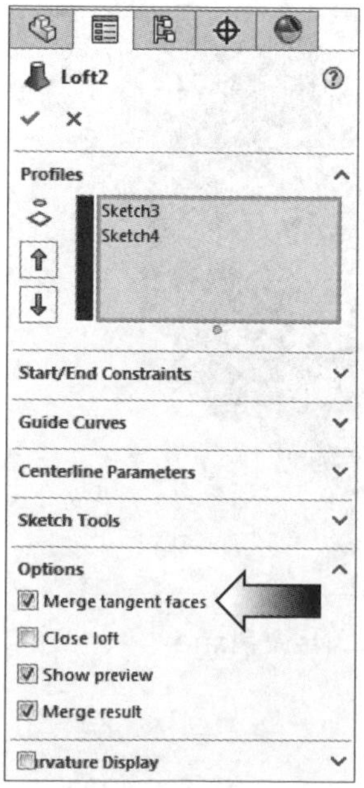

Click or select **Insert / Boss-Base / Loft**.

For profiles: Select the **2 sketches** from the graphics area.

Show all connector points to ensure a straight transition between the 2 profiles. Enable the **Merge Tangent Faces** option.

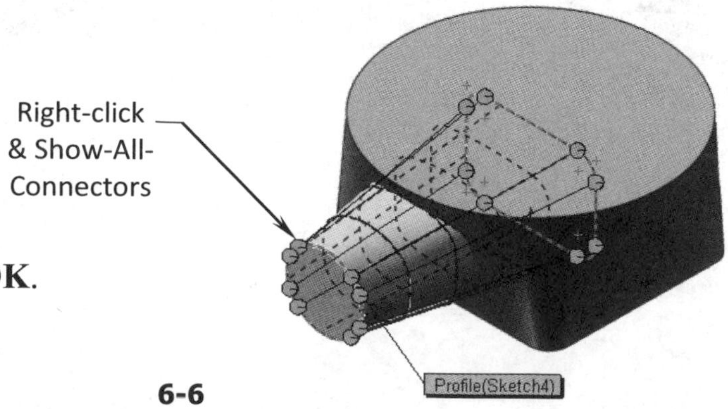

Click **OK**.

9. Creating an Offset Distance plane at 4.00":

Click or select **Insert / Reference Geometry / Plane.**

Select the **Right** reference plane from the FeatureManager tree.

Select **Offset Distance** and enter **4.000 in**. for distance; place it on the right.

Click **OK**.

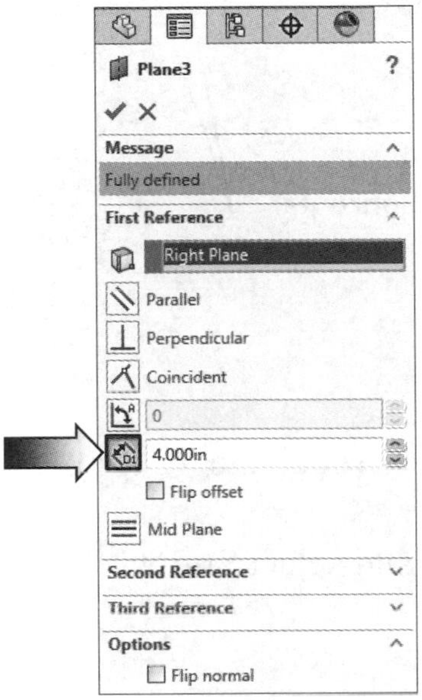

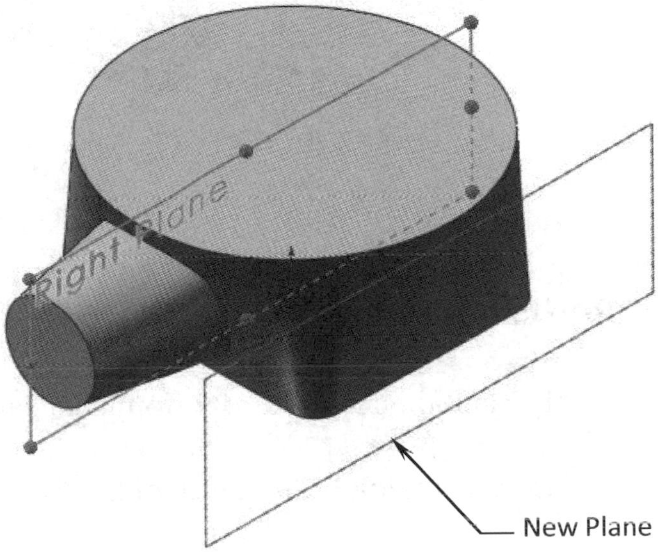

10. Constructing the Outlet's 1st Loft profile:

Select the <u>new plane</u> (Plane3) and open a **new sketch** .

Sketch a **Circle** and split it into 8 segments. Add the dimensions as noted.

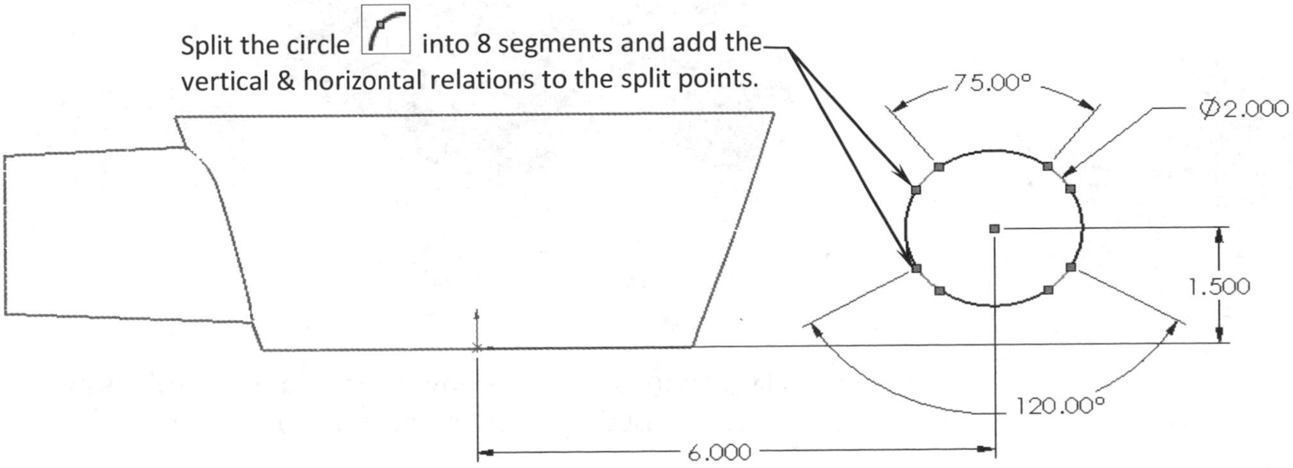

Add the Vertical and Horizontal relations to the endpoints of the split circle to fully define them.

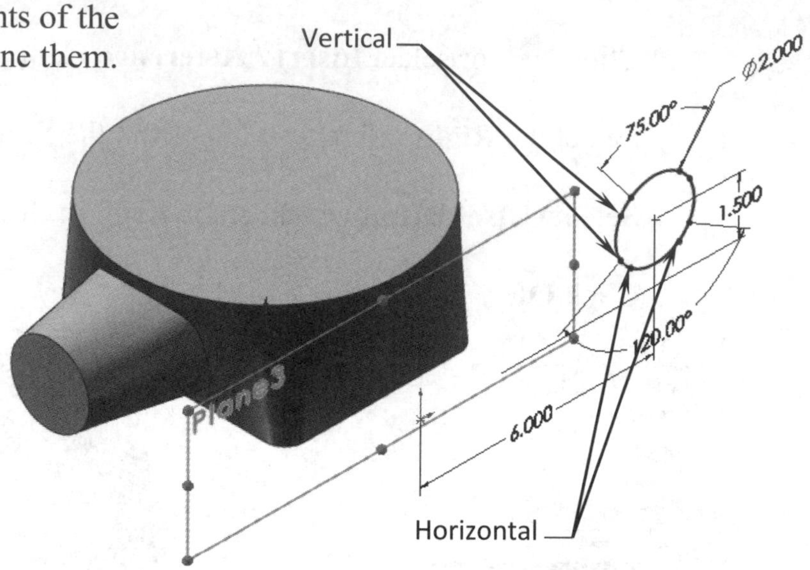

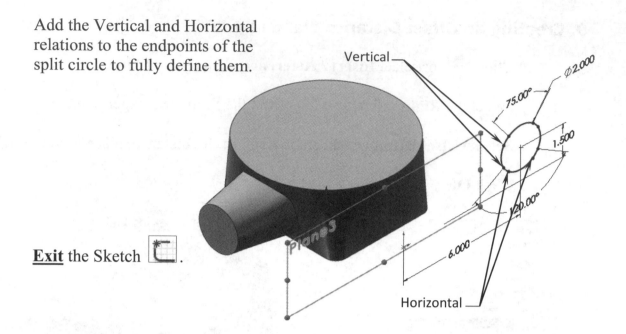

Exit the Sketch.

11. "Re-Using" the previous sketch*:

Expand the feature **Loft2** from the FeatureManager tree (click the + sign).

Locate **sketch3** (the sketch of the Rectangle), click it and select **Show**.

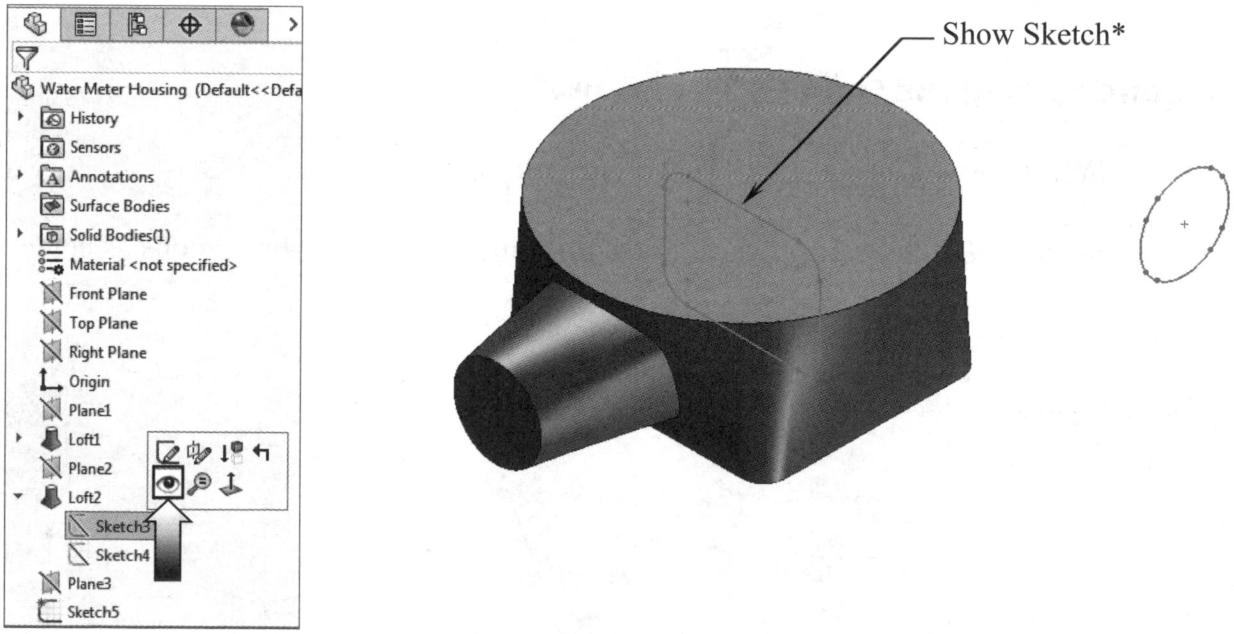

* The previous sketch is now visible in the graphics area. It will be used again in the next loft operation.

12. Creating a Parallel plane:

Click [icon] or select **Insert / Reference Geometry / Plane**.

For 1st Reference, select the **Top** reference plane from the FeatureManager tree.

Click the **Parallel** option (arrow).

For 2nd Reference, click the **Center Point** of the circle as noted.

Click **OK**.

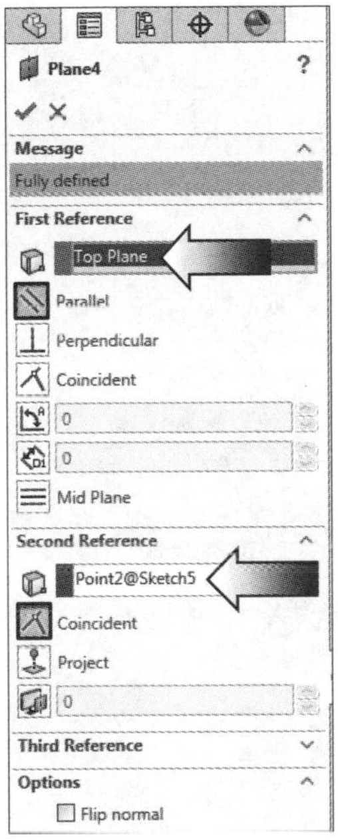

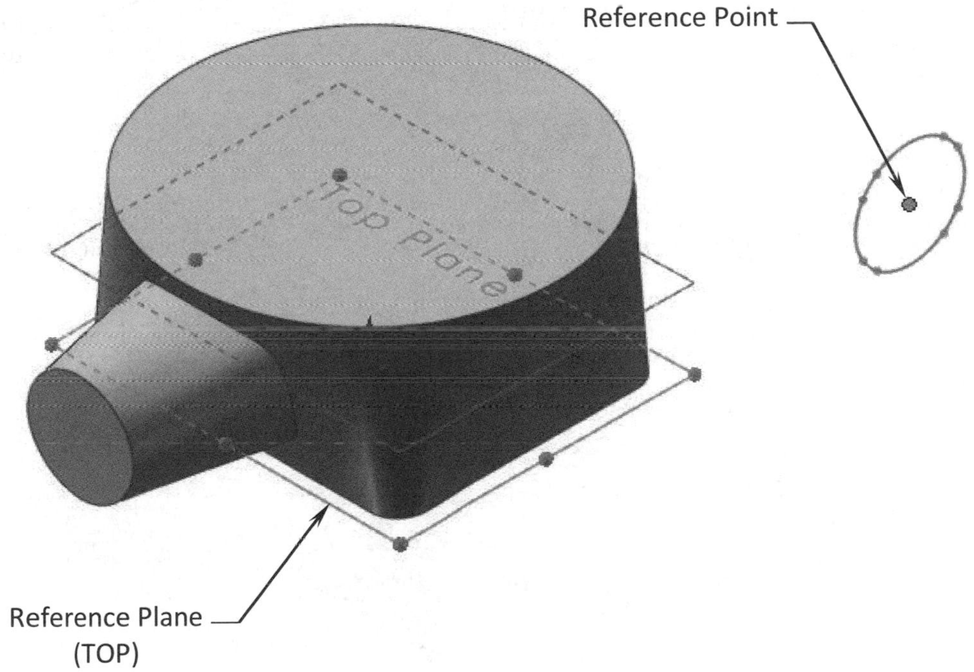

The new plane is created (Plane4).

13. Constructing the Centerline Parameter:

Select the new plane (Plane4) and open a **new sketch** .

Switch to the Top view orientation (Ctrl+5).

Sketch **2 Lines** and add a **Sketch Fillet** as shown below.

The end points of the lines must be coincident or Pierced to the centers of the other sketches.

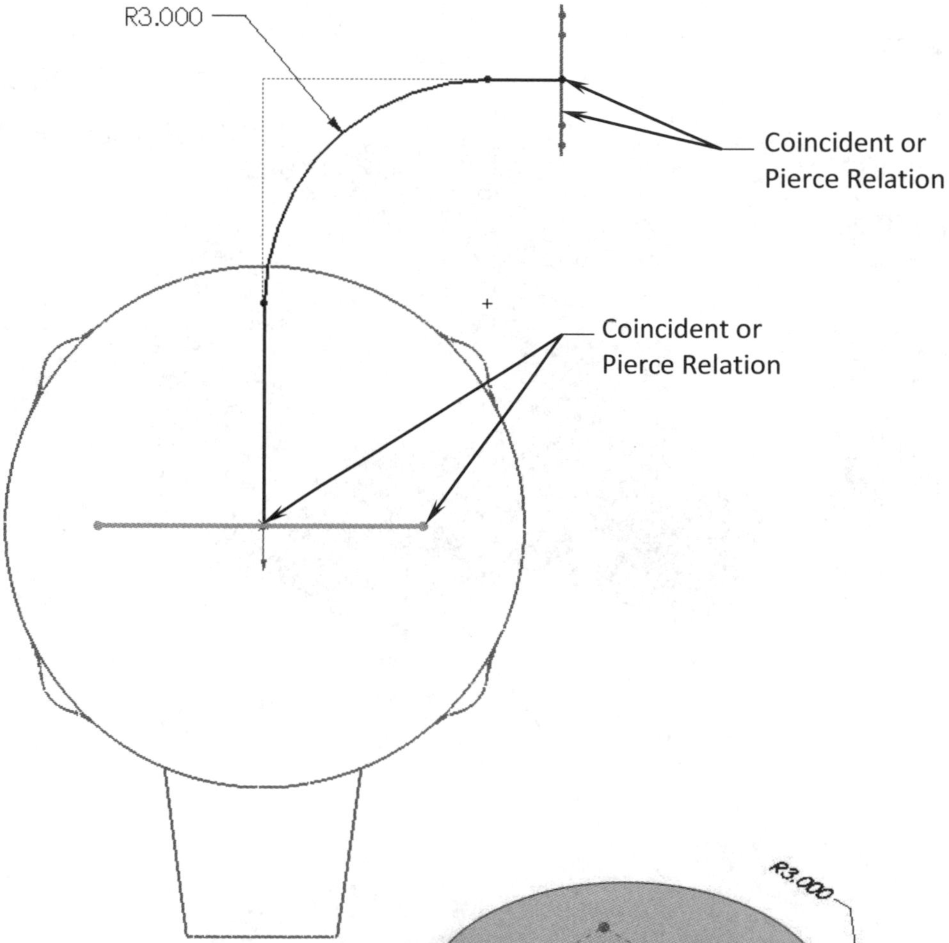

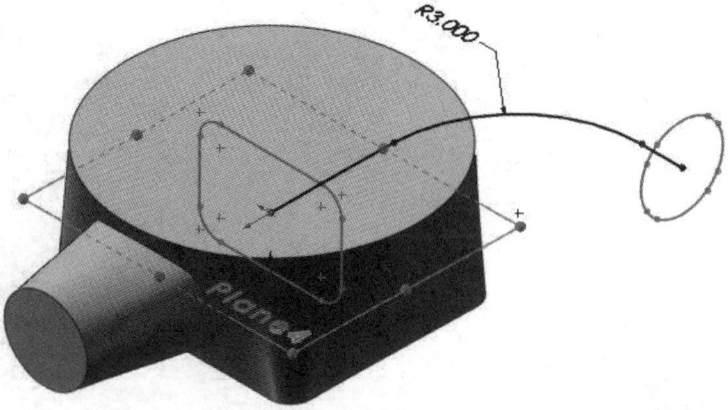

It is good practice to make the sketch fully defined before adding the fillets.

Exit the Sketch .

14. Creating the Outlet loft feature:

Click  or select **Insert / Boss / Loft**.

Select the **2 Sketch Profiles** (Rectangle and Circle) from the graphics area.

Expand the **Centerline Parameters** option and select **Sketch6** from either the graphics area or from the FeatureManager tree.

When the Centerline Parameter is used to guide the loft, the sketch planes of all the intermediate sections will be rotated normal to the centerline.

After the preview appears, check the connectors to ensure a proper loft transition.

> **Centerline Parameters**
>
> If the number of entities in each sketch is the same, a "Centerline Parameter Sketch" can be used instead of the guide curves.

Click **OK**.

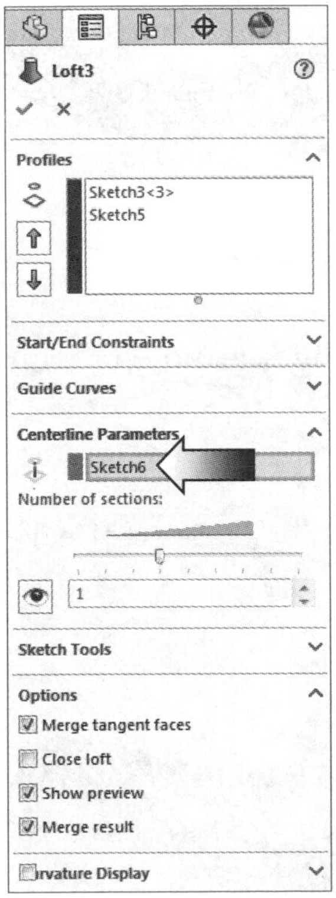

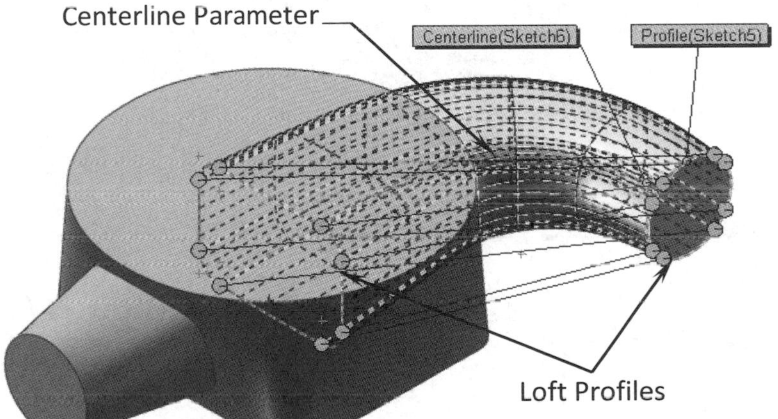

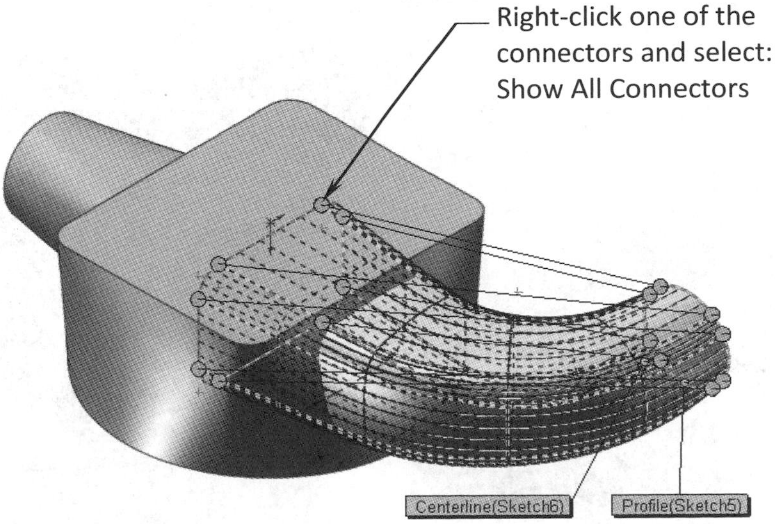

15. Adding .250" fillet to the Bottom:

Click or select **Insert / Features / Fillet-Round**.

Enter **.250 in**. for radius value and select the **bottom edges** as indicated.

Click **OK**.

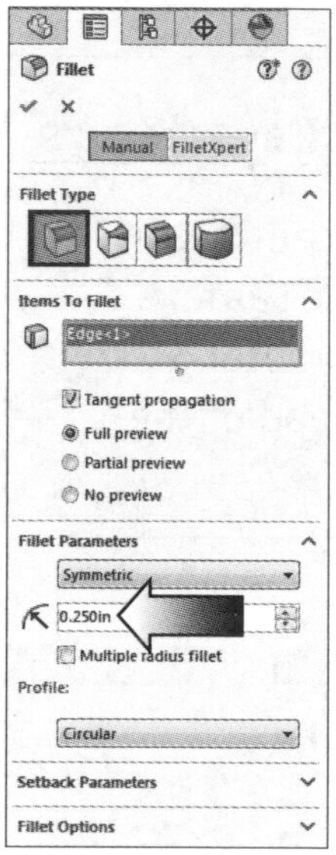

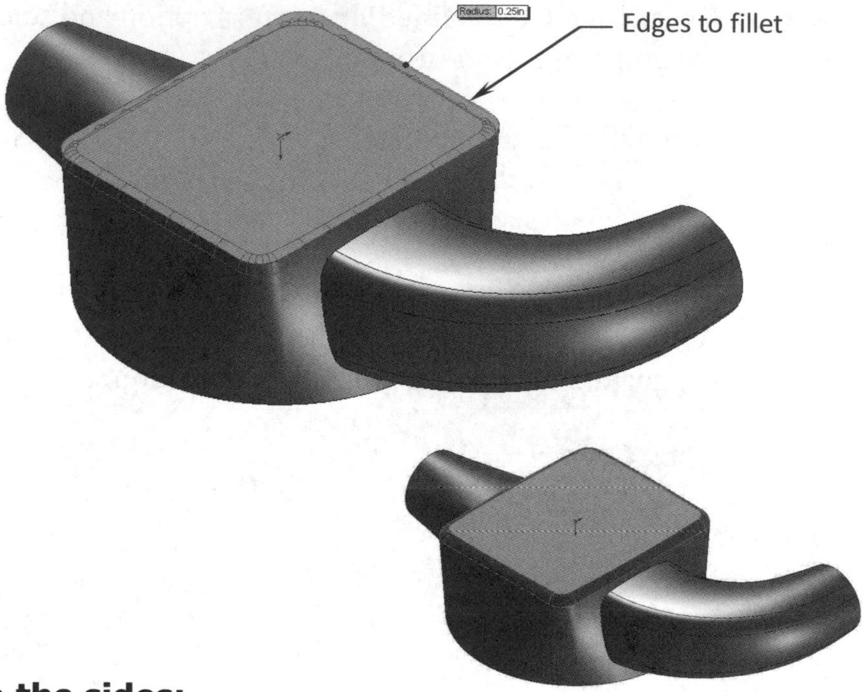

16. Add .175" fillets to the sides:

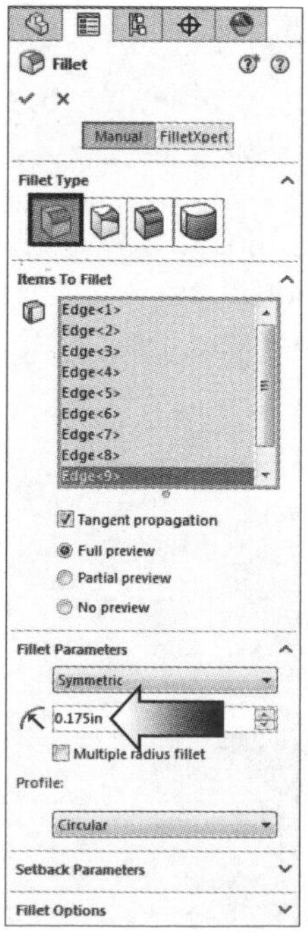

Click or select **Insert / Features / Fillet-Round**.

Enter **.175 in**. for Radius and select all side-edges as noted.

Click **OK**.

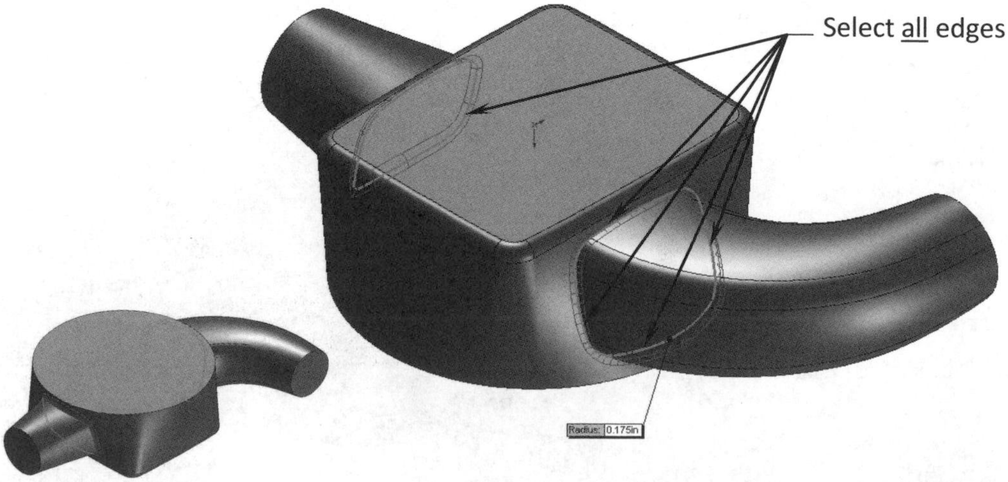

17. Shelling the part:

Click [icon] or select **Insert / Features / Shell.**

Enter **.175 in.** for wall thickness and select the **3 faces** as noted.

Click **OK.**

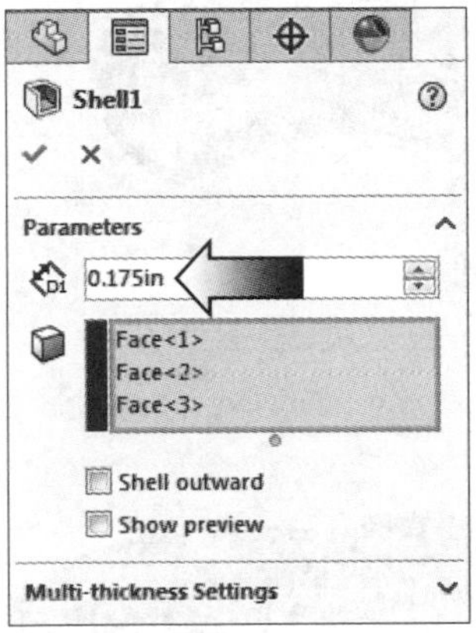

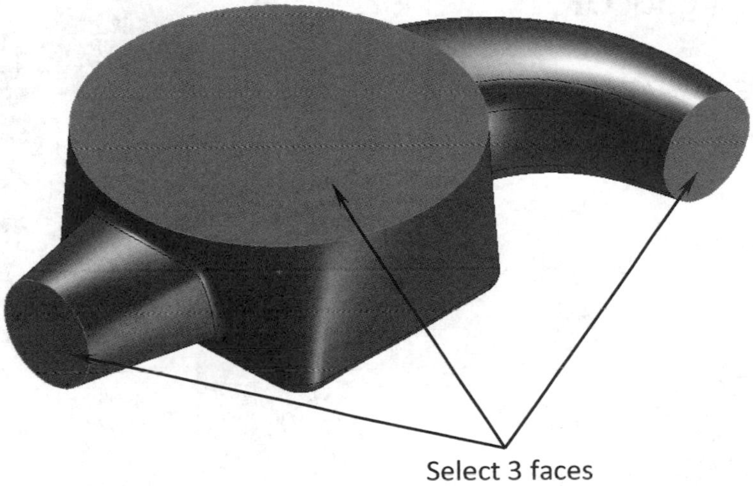

18. Creating the 1st Mounting bracket:

Select the <u>Face</u> as noted and open a **new sketch** [icon].

Sketch the profile shown below, use **Convert Entities** [icon] where applicable.

Add dimensions and relations needed to fully define the sketch.

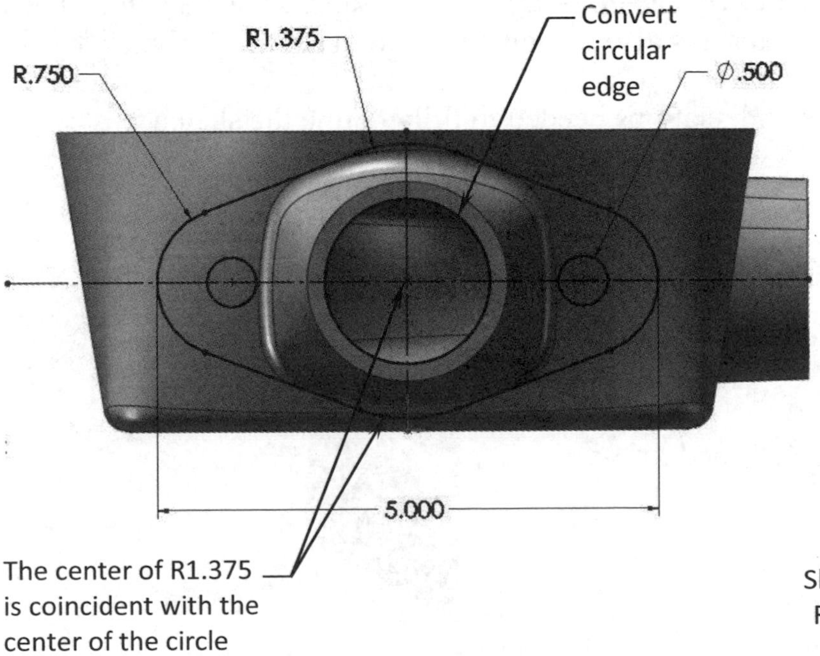

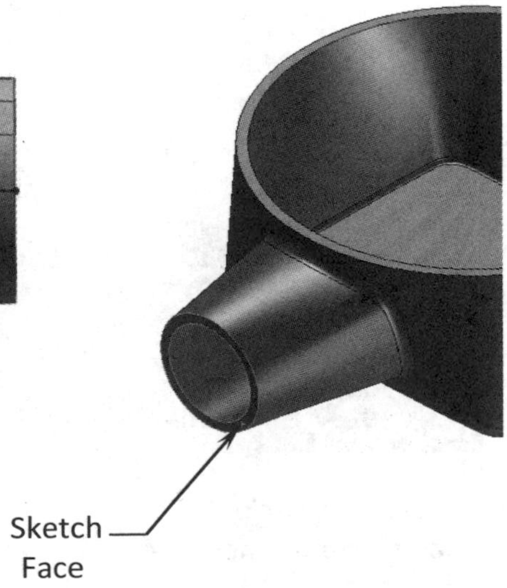

19. Extruding the Left bracket:

Click or select **Insert / Boss-Base / Extrude**.

End Condition: **Blind** and Depth: **.500 in**.

Click **OK**.

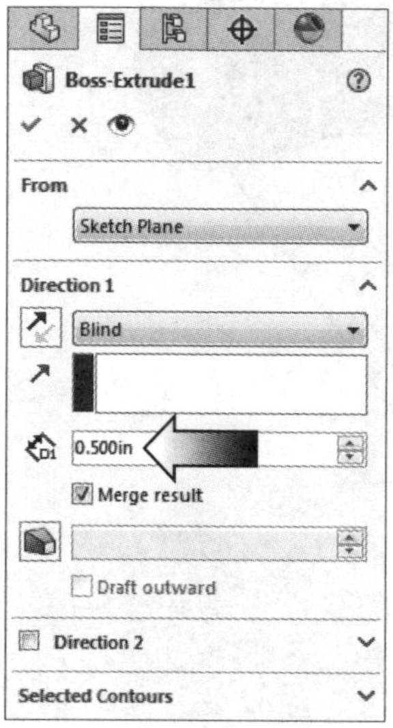

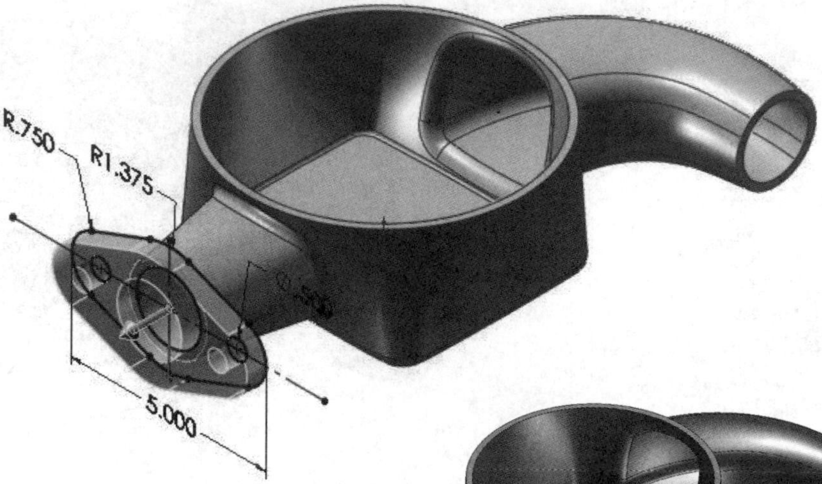

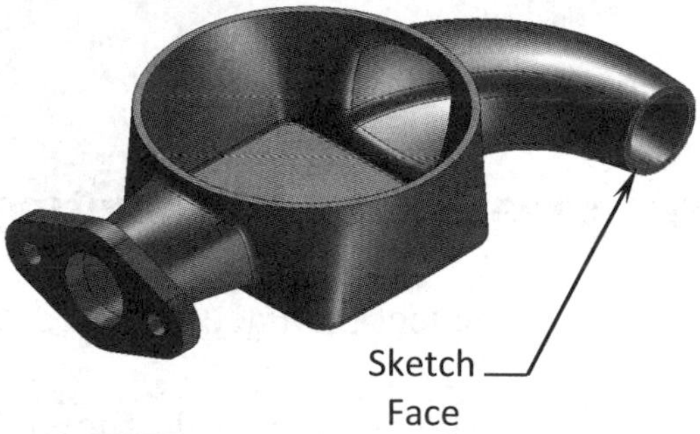

20. Creating the 2nd Mounting bracket:

Select the Face as noted and open a **new sketch** .

Either copy the previous sketch – or – recreate the same sketch again on the right side. (Derived-Sketch is also a good option to make the 2nd bracket.)

Add the dimensions and relations needed to fully define the sketch.

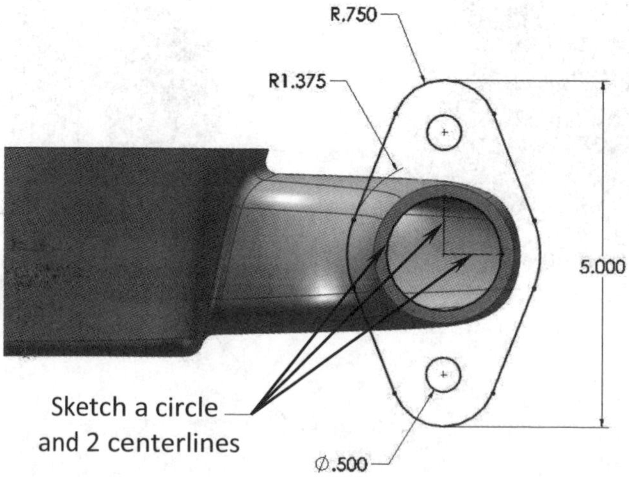

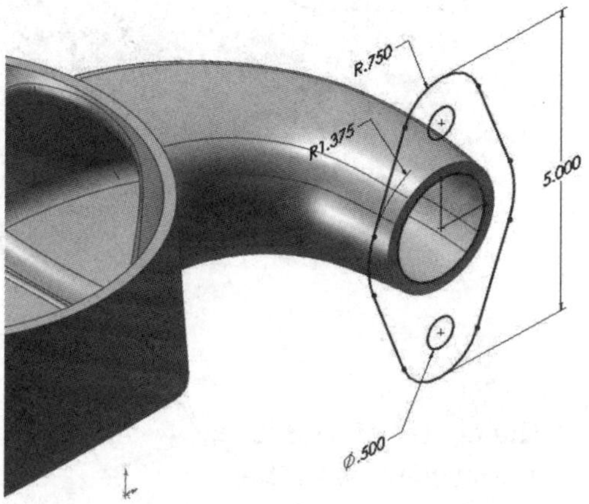

21. Extruding the Right bracket:

Click  or select **Insert / Boss-Base / Extrude**.

End Condition: **Blind**. Depth: **.500 in**.

Click **OK**.

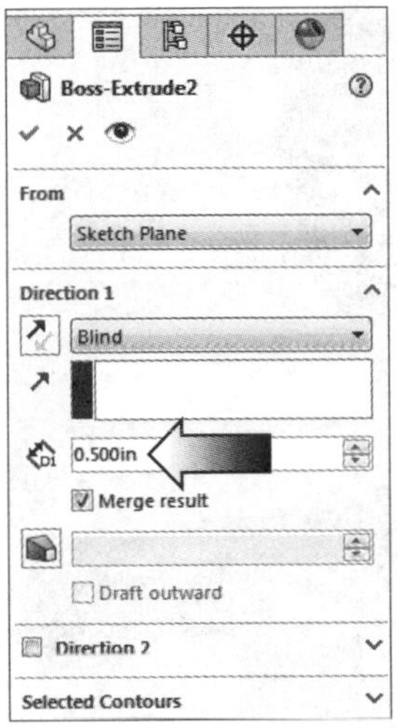

22. Constructing the Upper Ring:

Select the <u>Face</u> as indicated and open a **new sketch** .

Sketch the profile below; use **Convert Entities**  where needed.

The centers of the Ø.250 circles are coincident with the Ø7.375 bolt-circle.

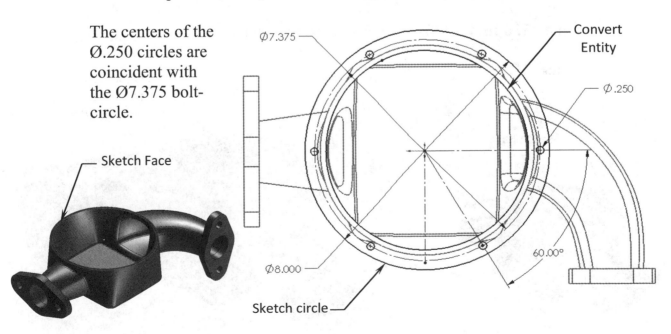

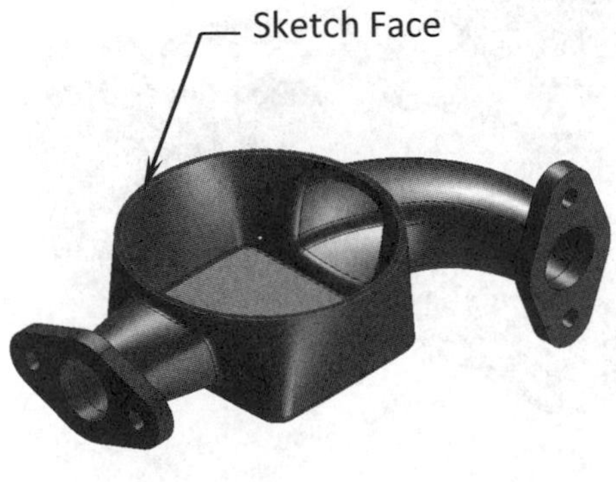

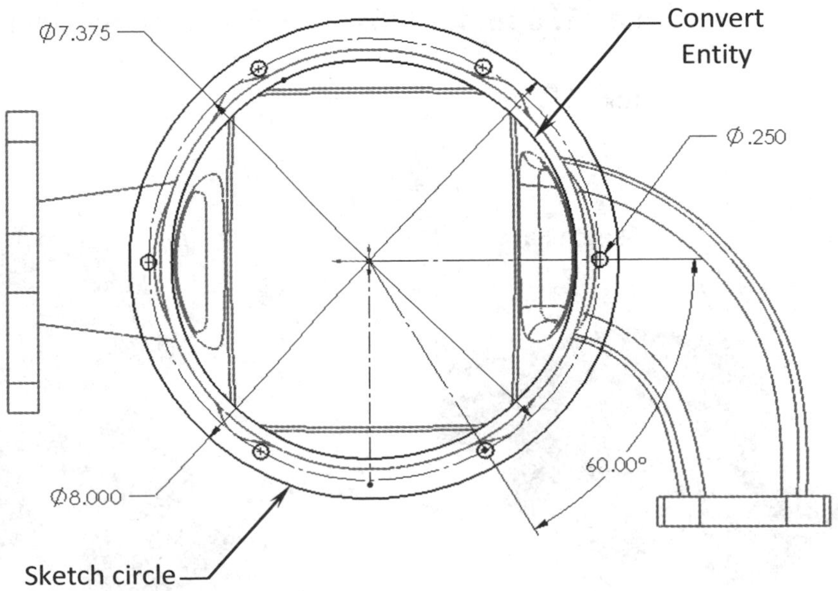

23. Extruding the Upper Ring:

Click or select **Insert / Boss-Base / Extrude**.

End Condition: **Blind**. Depth: **.400 in**. (upward)

Click **OK**.

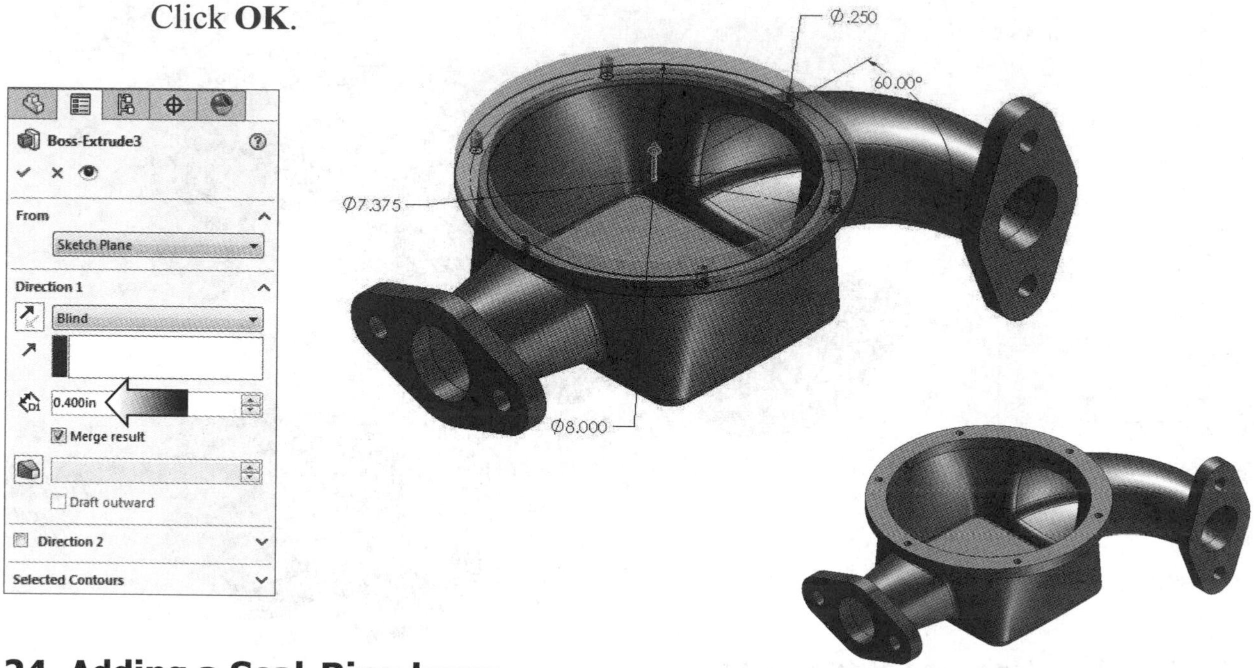

24. Adding a Seal-Ring bore:

Select the <u>Face</u> as noted and open a **new sketch**.

Select the <u>inside circular edge</u> and click **Offset Entities**.

Enter **.150 in**. for Offset Value (larger diameter).

Click **OK**.

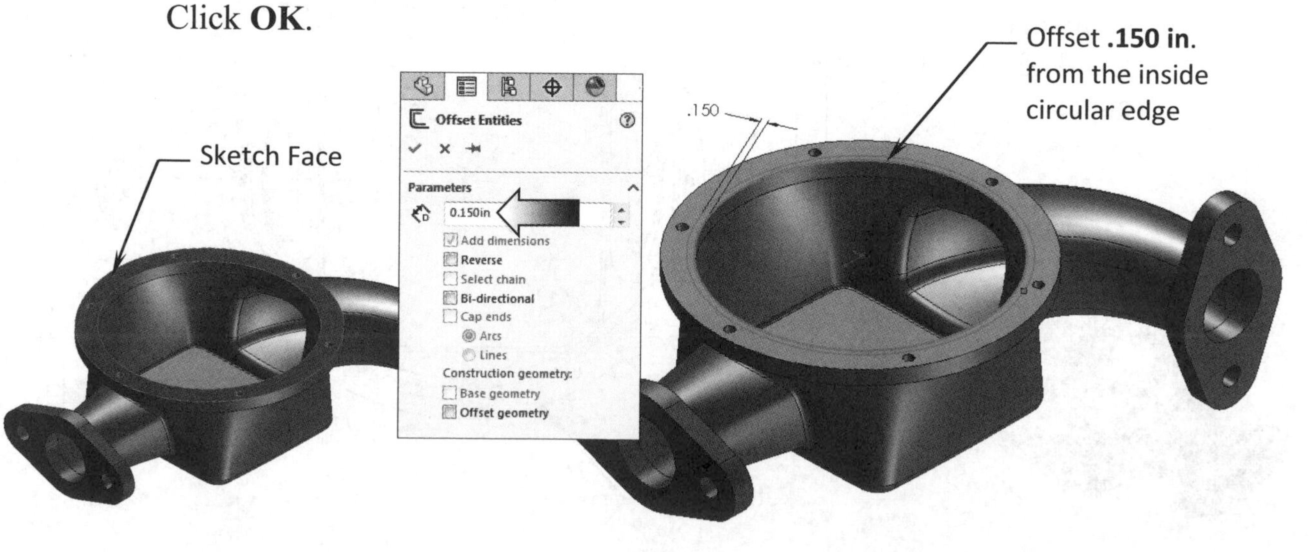

25. Extruding Cut the Seal Ring bore:

Click  or select **Insert / Cut / Extrude**.

End Condition: **Blind**.

Depth: **.250 in**.

Click **OK**.

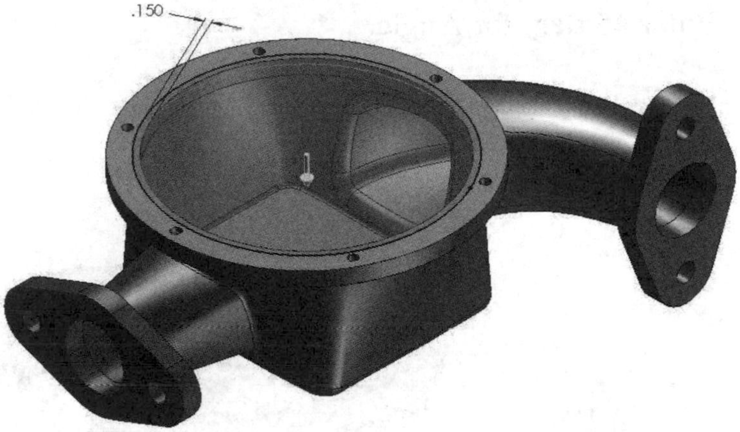

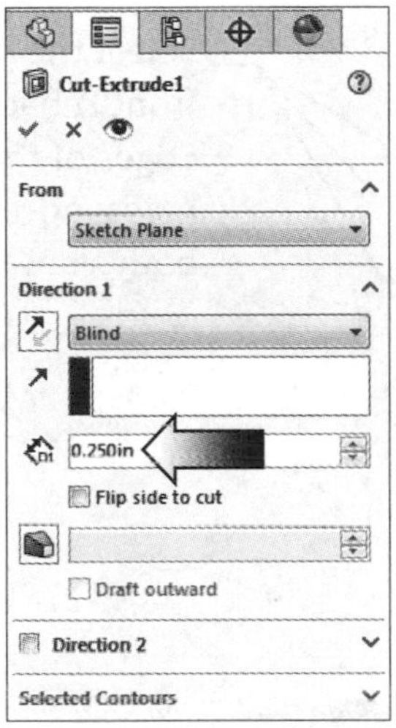

26. Adding fillets:

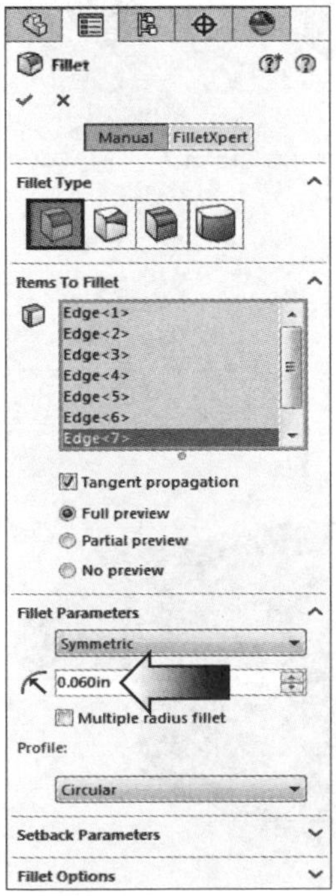

Click  or select **Insert / Features / Fillet-Round**.

Enter **.060 in**. for radius value.

Select the **7 edges** as noted.

Click **OK**.

27. Adding Chamfers:

Click  or select **Insert / Features / Chamfer**.

Enter **.060 in**. for Depth.

Enter **45 deg**. for Angle.

Select the **edges** of the 6 holes.

Click **OK**.

28. Saving your work:

Select **File / Save As / Water Meter Housing / Save**.

Questions for Review

Loft Vs. Sweep

1. In a new part mode, when the Sketch Pencil is selected first, SOLIDWORKS will prompt you to select a sketch plane.
   a. True
   b. False

2. It is sufficient to create a Parallel-Plane-At-Point with a Reference Plane and a Reference Point.
   a. True
   b. False

3. A loft feature uses multiple sketch profiles to define its shape.
   a. True
   b. False

4. Only one guide curve can be used in each loft feature.
   a. True
   b. False

5. Multiple guide curves can be used to connect and control the loft feature.
   a. True
   b. False

6. The guide curves can be either a 2D sketch or a 3D curve.
   a. True
   b. False

7. The loft profiles and the guide curves should be related with Coincident or Pierce relations.
   a. True
   b. False

8. The loft profiles should be created before the guide curves.
   a. True
   b. False

1. TRUE 2. TRUE
3. TRUE 4. FALSE
5. TRUE 6. TRUE
7. TRUE 8. TRUE

Exercise: Loft

There are several different ways to model this part, but this exercise will focus on the Loft technique instead.

1. Create the part below, using the Loft and the Circular Pattern features.

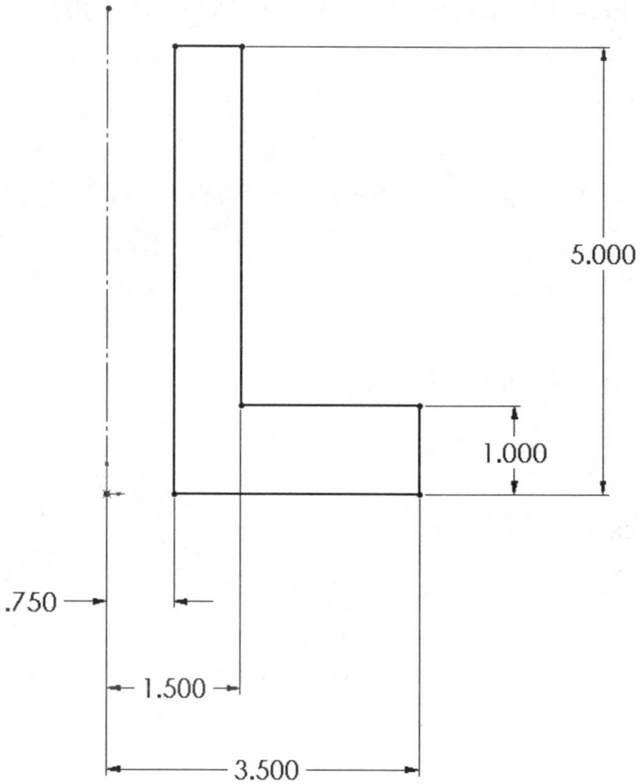

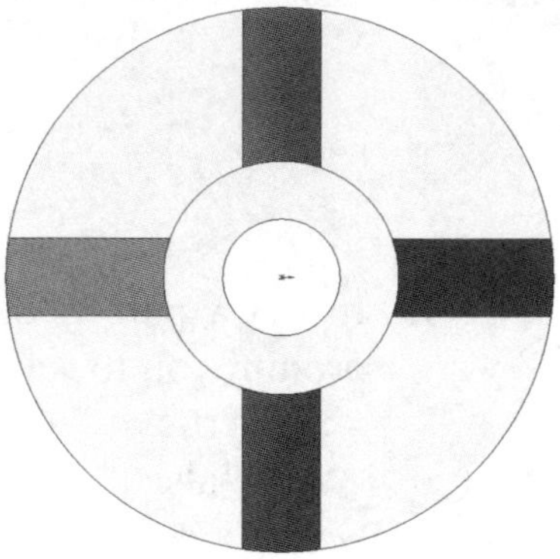

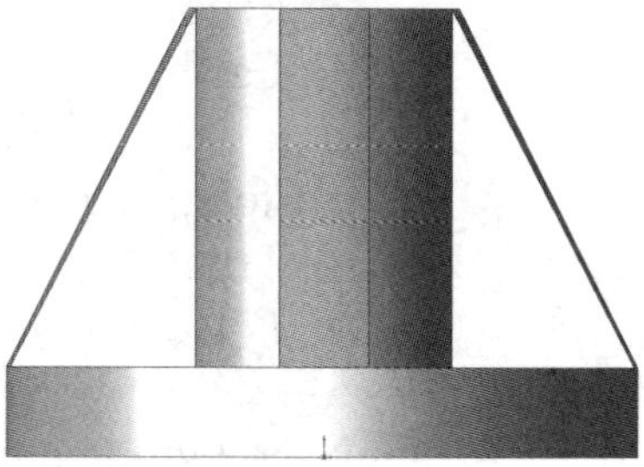

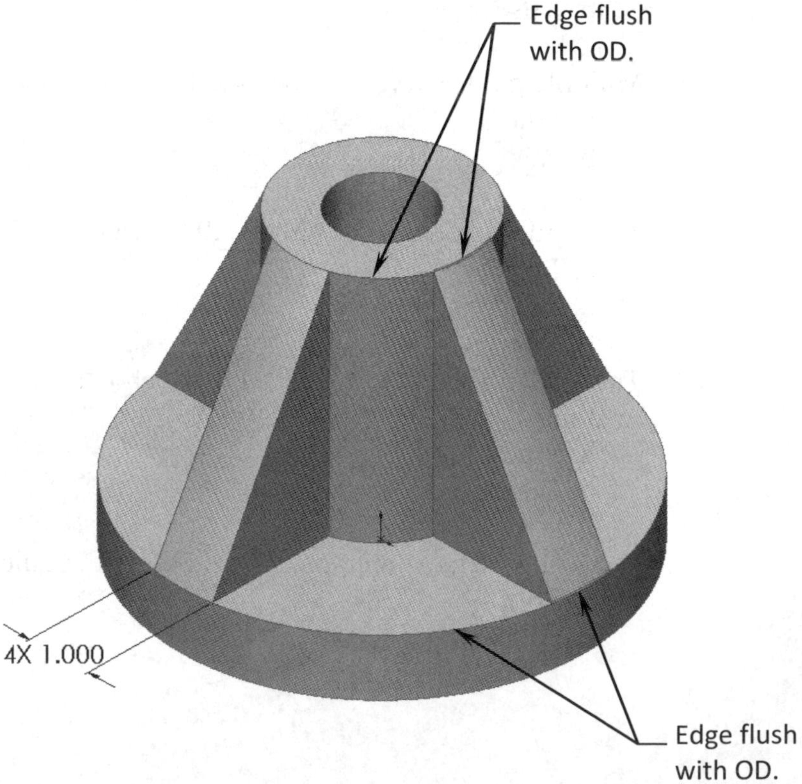

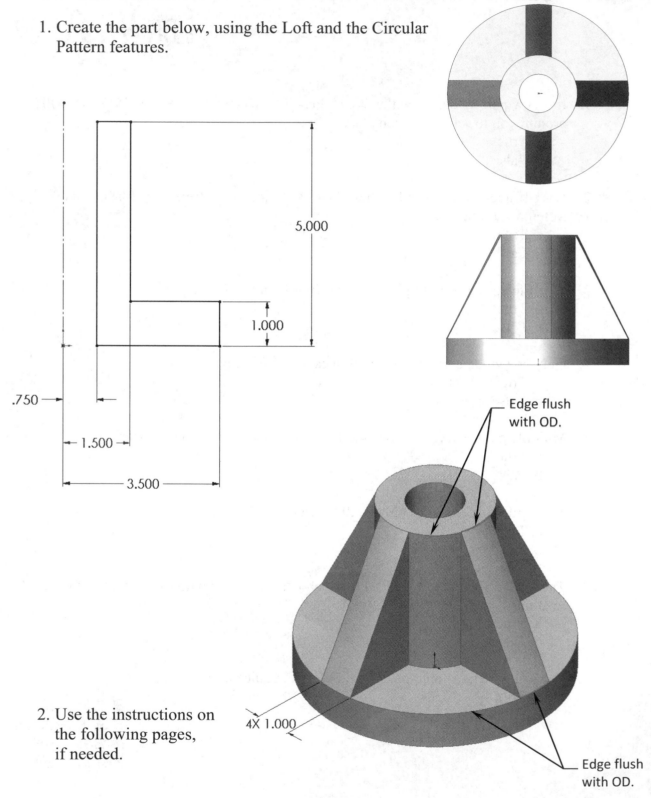

2. Use the instructions on the following pages, if needed.

1. Starting with the base sketch:

Select the Front plane and open a **new sketch**.

Sketch the profile shown on the right.

Add the dimensions to fully define the sketch.

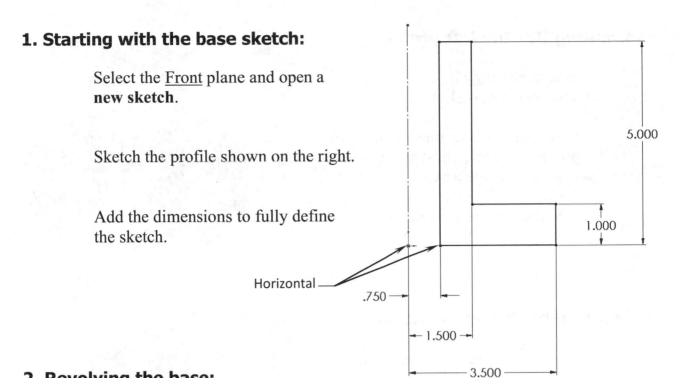

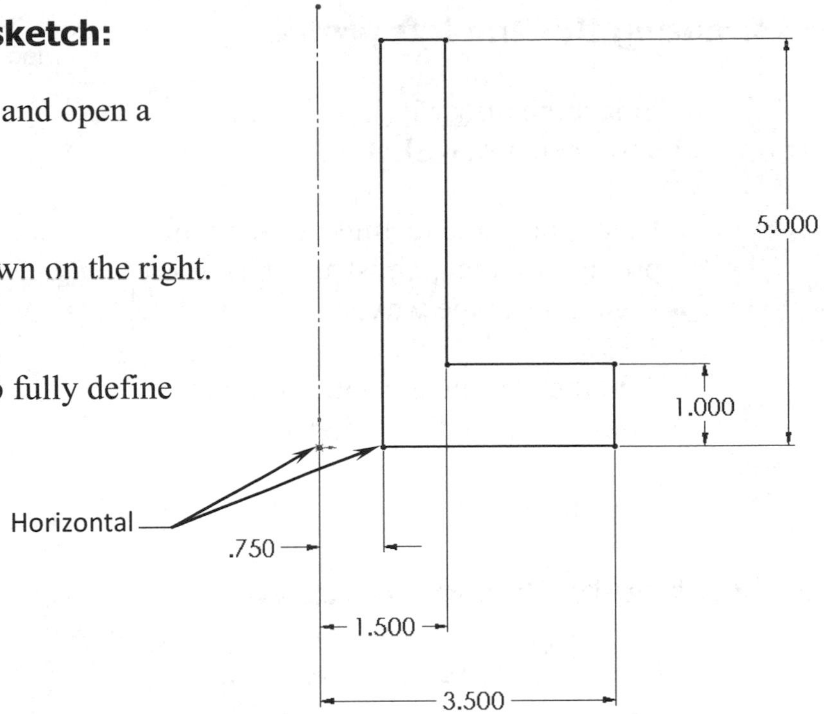

2. Revolving the base:

Click  or select **Insert / Boss-Base / Revolve**.

Direction 1: **Blind**.

Revolve Angle: **360deg**.

Click **OK**.

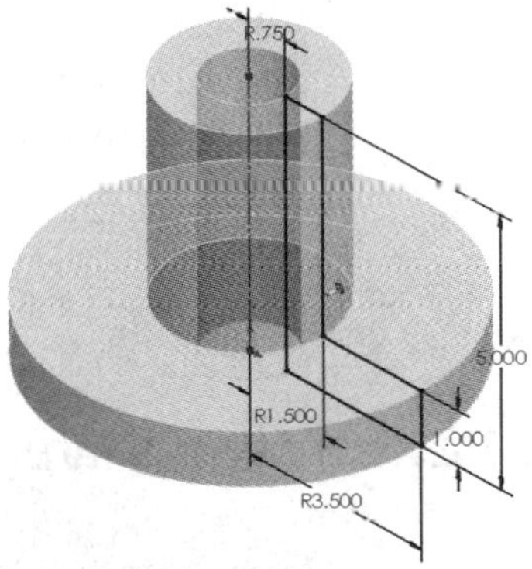

3. Creating the 1st loft profile:

Select the face indicated and open a **new sketch**.

Create the sketch profile by converting the 2 circular edges then add 2 vertical lines.

Trim the circles to form one continuous profile.

Exit the sketch.

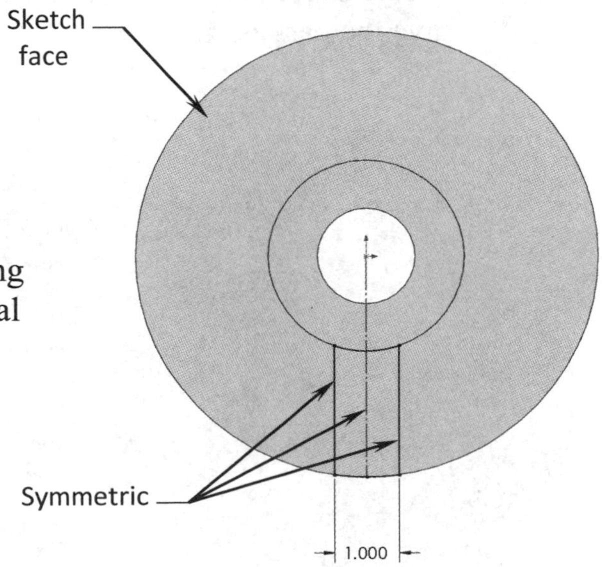

4. Creating the 2nd loft profile:

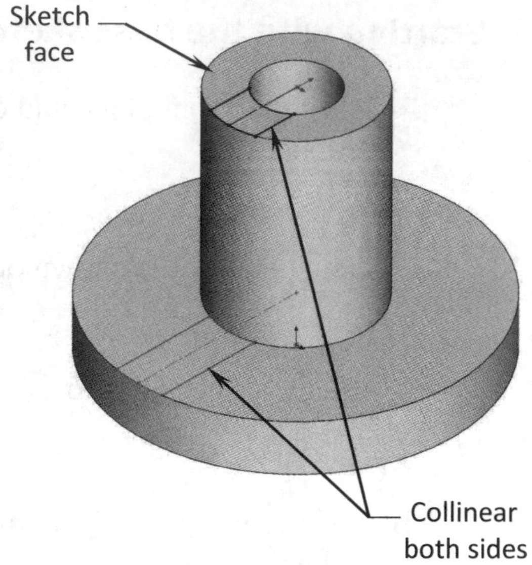

Select the <u>upper face</u> as indicated and open a **new sketch**.

Using the same techniques as in the previous sketch, construct this new sketch the same way.

Add the Collinear relations to both sides, between the 2 sketches.

<u>Exit</u> the sketch.

5. Creating the 1st loft feature:

Click  or select: **Insert Boss-Base Loft**.

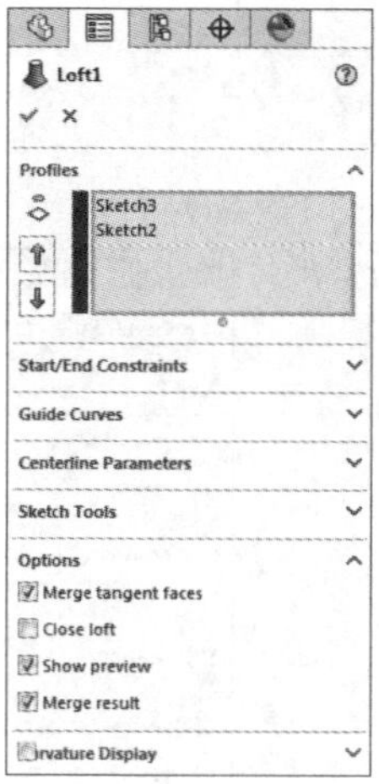

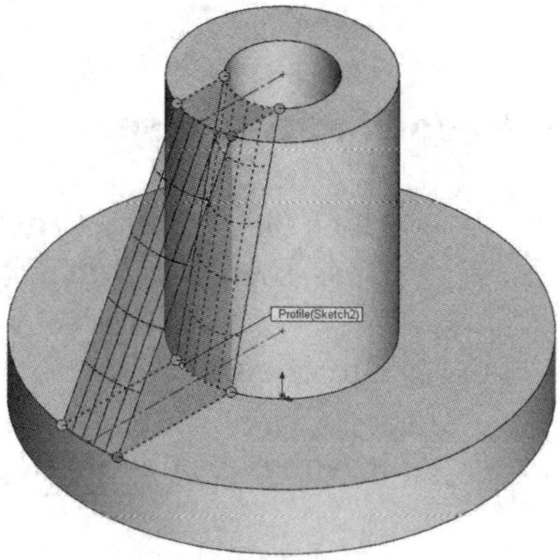

Select the outer corner of each sketch profile to prevent the loft from twisting.

Click **OK**.

6. Pattern and Save:

Create a **circular pattern** of the lofted feature with a total of **4 instances**, then <u>save</u> the part as: **Loft_Exe**.

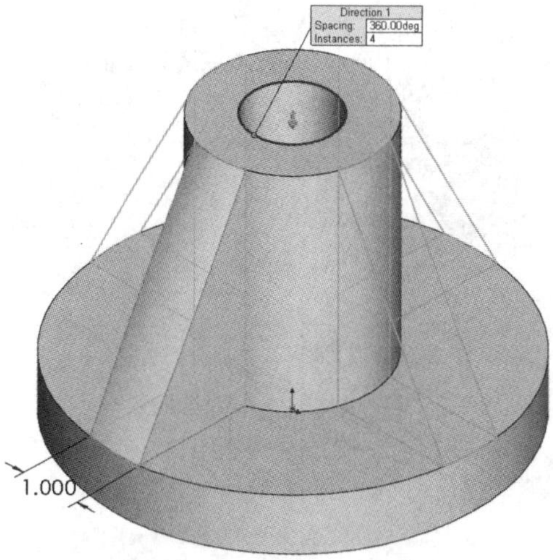

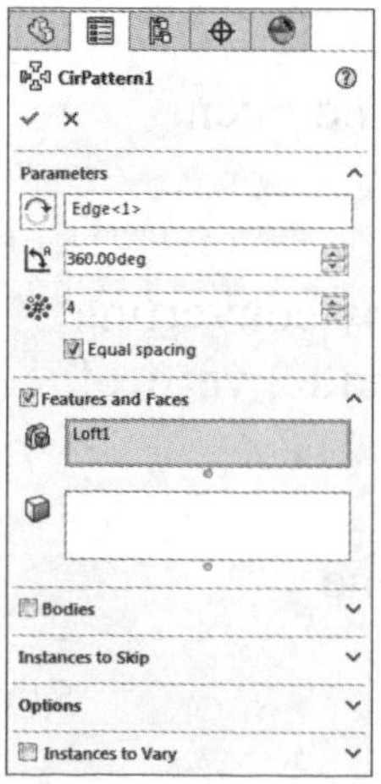

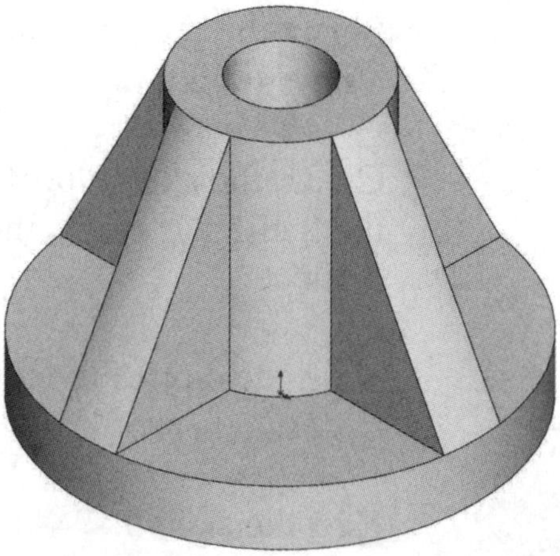

CHAPTER 7

Loft With Guide Curves

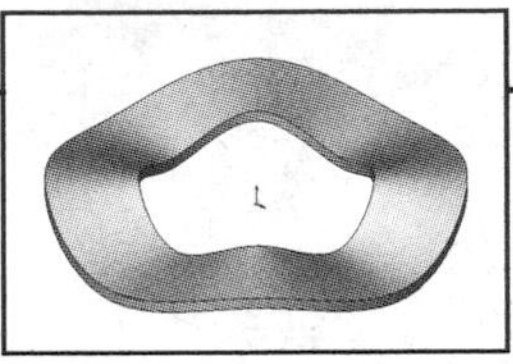

Loft with Guide Curves
Waved Washer

This lesson demonstrates the creation of a waved washer using the loft technique, where 4 sketch profiles and a single 3D guide curve are used.

A loft feature usually contains several sections, one centerline parameter and one or more guide curves.

In this case study, four identical sketches will be used as the loft profiles and a single guide curve will be used to connect the sections.

Since the loft profiles are identical, the derived-sketch option will be used to show how the sketches can be derived or copied. A derived sketch is driven by the original sketch. It can only be positioned with relations or dimensions, but its sketched entities cannot be changed.

When the Original sketch is changed, the derived-sketch will be updated automatically; however, the derived sketches can be Un-derived to break their associations with the Original sketch.

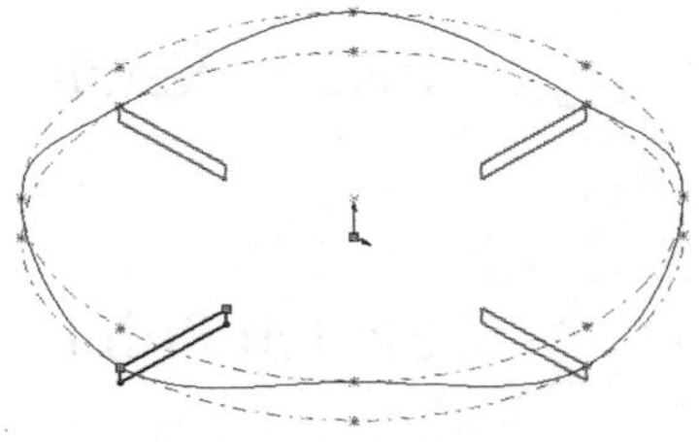

The loft profiles are connected with a 3D curve. The 3D curve will be generated using the command called Curve-Through-Reference-Points. The loft profiles are also Pierced to these reference points.

The Curve-Through-Reference-Points will be used to guide and control the transition between the loft profiles.

Waved Washer
Loft with Guide Curves

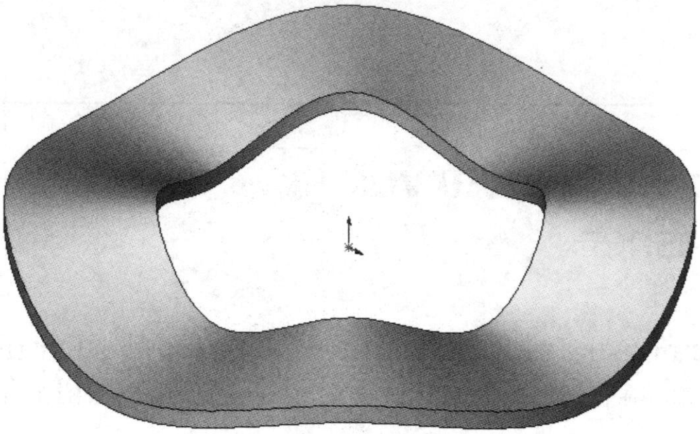

View Orientation Hot Keys:

Ctrl + 1 = Front View
Ctrl + 2 = Back View
Ctrl + 3 = Left View
Ctrl + 4 = Right View
Ctrl + 5 = Top View
Ctrl + 6 = Bottom View
Ctrl + 7 = Isometric View
Ctrl + 8 = Normal To Selection

Dimensioning Standards: **ANSI**

Units: **INCHES** – 3 Decimals

Tools Needed:

Insert Sketch	Line	Circle
Sketch Point	Add Geometric Relations	Dimension
Derived Sketch	Curve Through Reference Points	Base/Boss Loft

1. Creating the 1st Construction profile:

Select the Top plane from the FeatureManager tree.

Click  or select **Insert / Sketch**.

Sketch a **Circle** and convert it to a construction circle 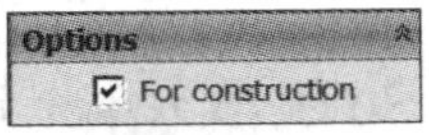

.

Add **8 sketch points** 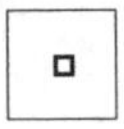 approximately as shown.

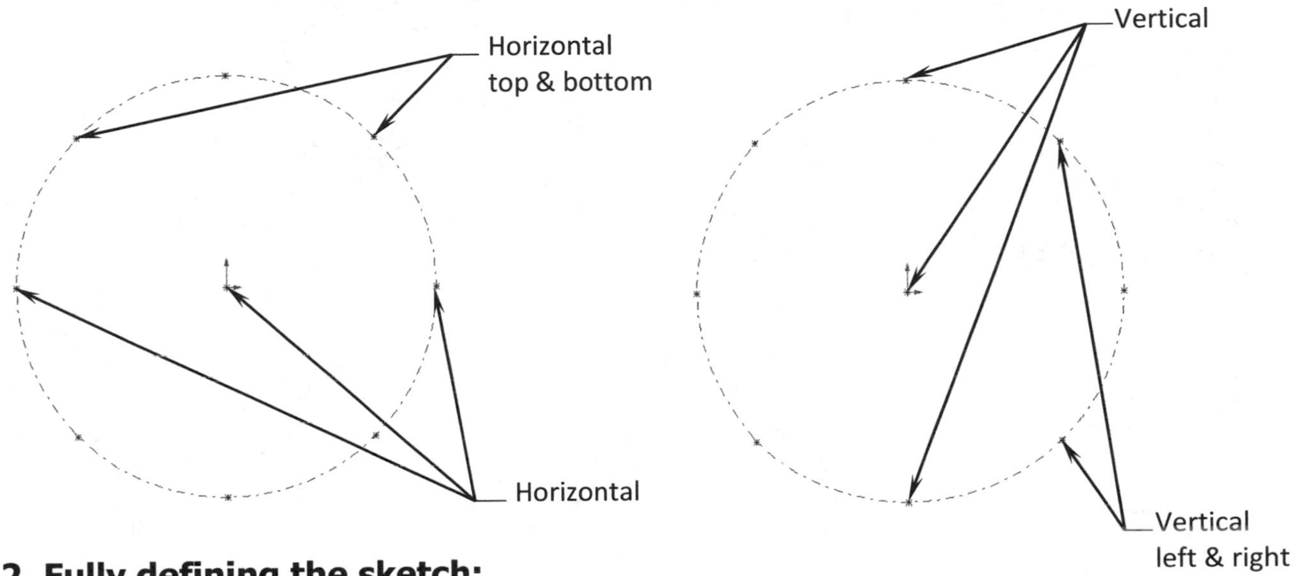

2. Fully defining the sketch:

Add an angular dimension between the sketch points.

Add a vertical and a horizontal relation as indicated.

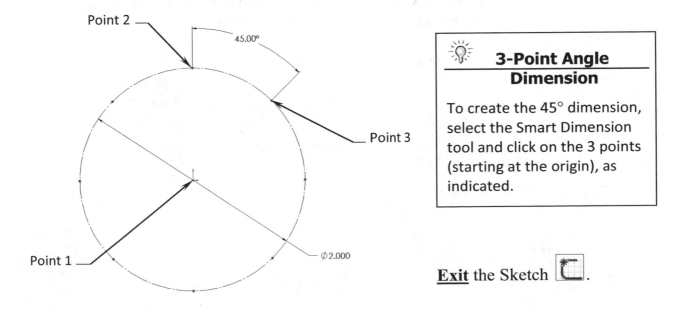

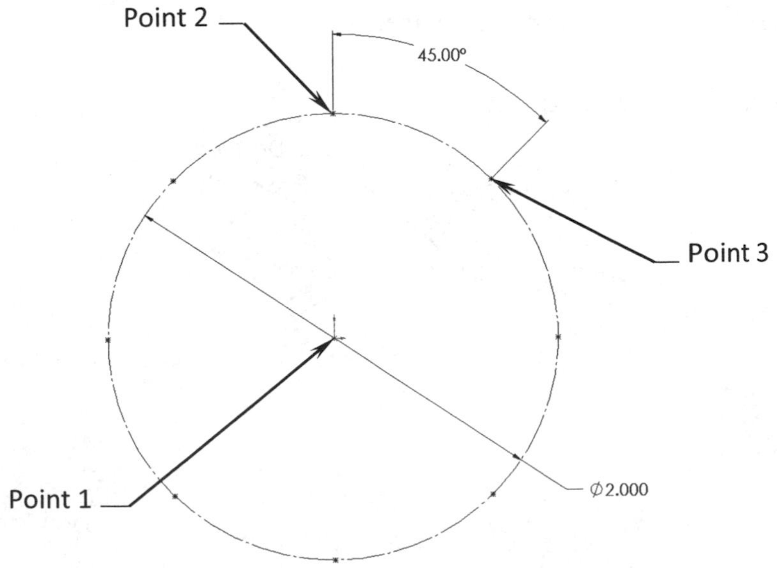

 **3-Point Angle Dimension**

To create the 45° dimension, select the Smart Dimension tool and click on the 3 points (starting at the origin), as indicated.

Exit the Sketch .

3. Creating an Offset Distance plane:

Click  or select **Insert / Reference Geometry / Plane**.

Select the **Top** plane as First Reference.

Select **Offset Distance** option and enter **.150 in**.

Click **OK**.

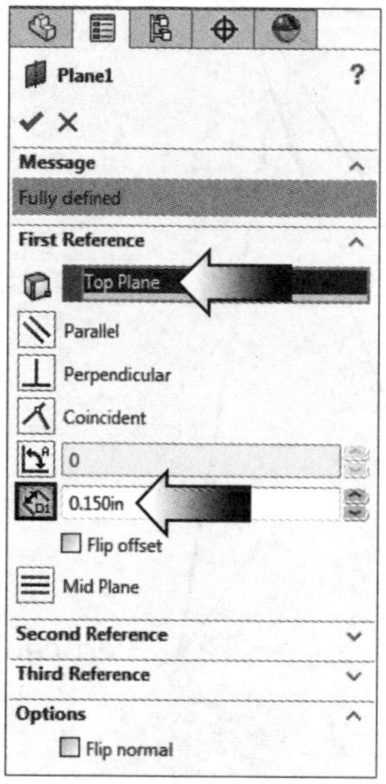

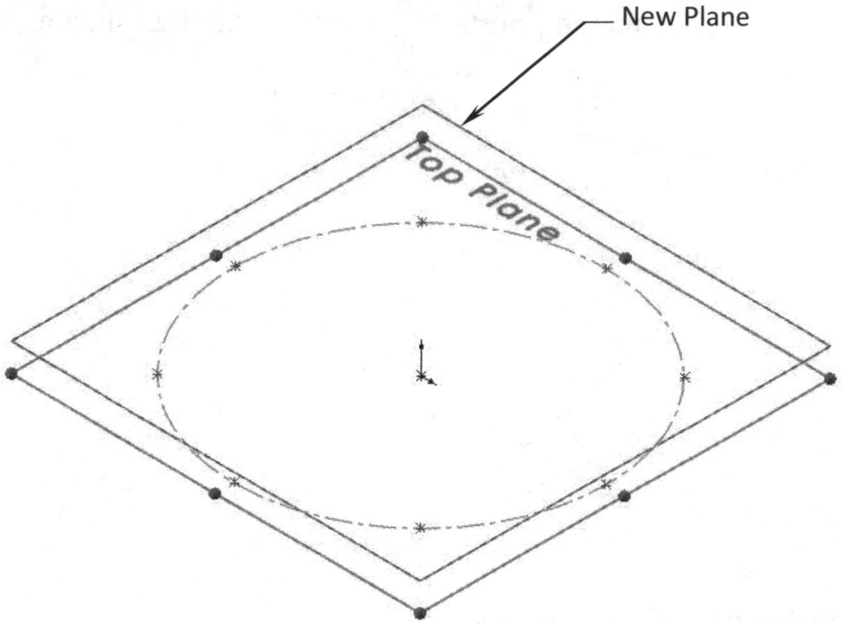

4. Creating the 2nd construction profile using Derived Sketch:

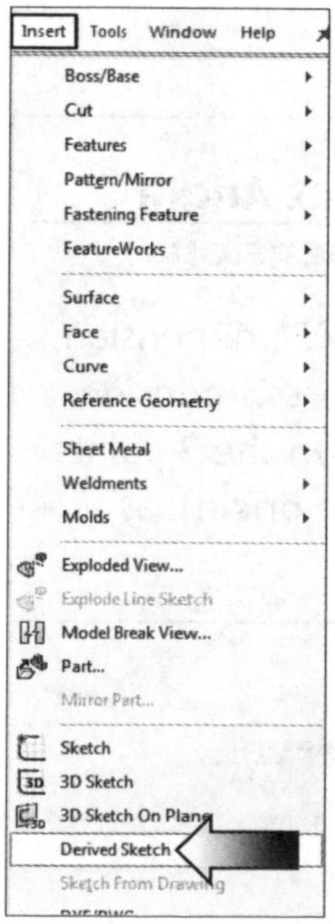

Hold the **Control** key, select the **new Plane** (plane1) and the **Sketch1** from FeatureManager tree.

Click **Insert / Derived Sketch**.

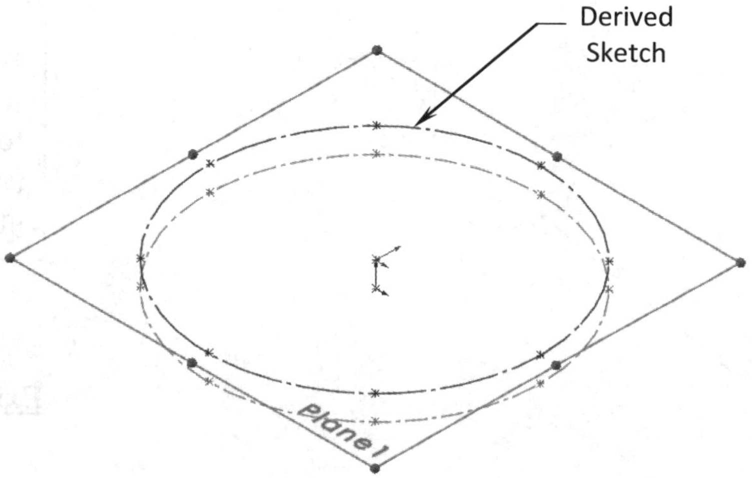

5. Positioning the Derived Sketch:

Add the Vertical & Coincident relations between the indicated points.

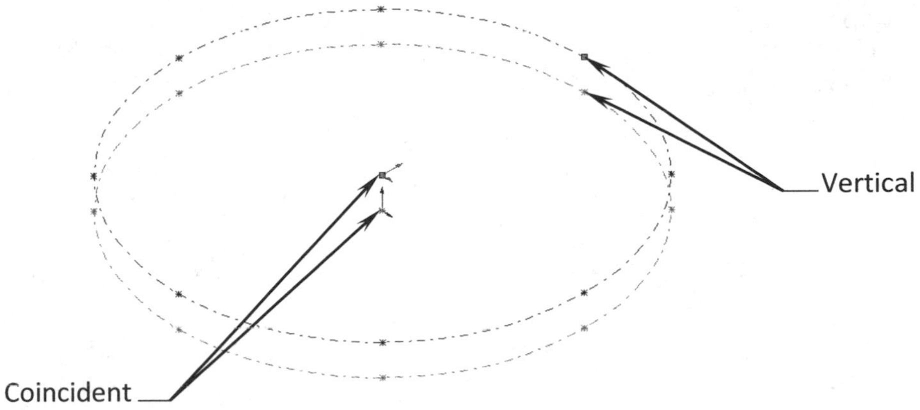

Exit the Sketch  or select **Insert / Sketch.**

6. Creating a Curve through Reference Points:

Click  or select **Insert / Curve / Curve-Through-Reference-Points.**

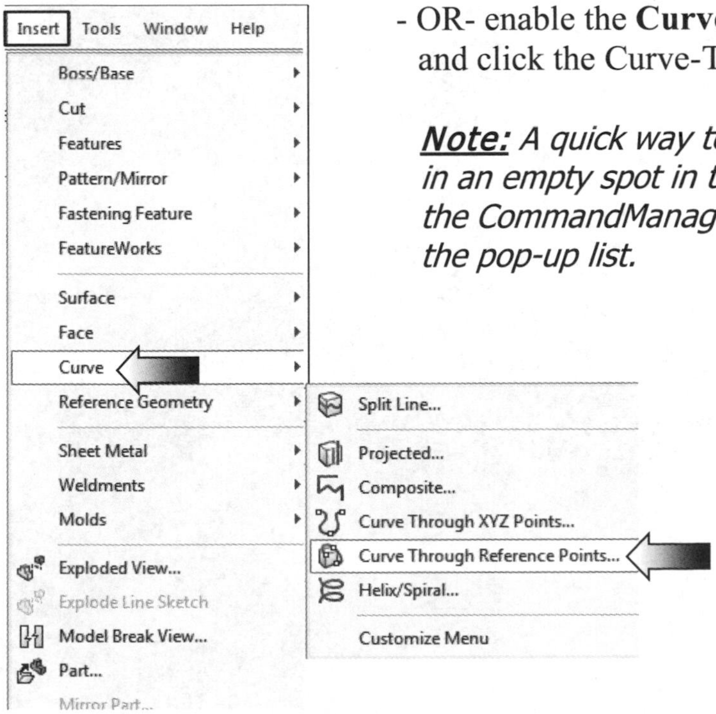

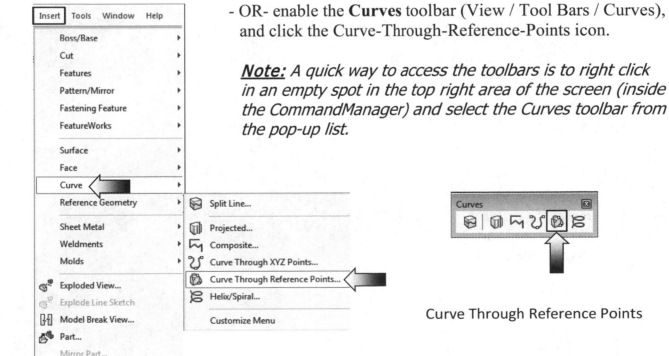

- OR- enable the **Curves** toolbar (View / Tool Bars / Curves), and click the Curve-Through-Reference-Points icon.

Note: *A quick way to access the toolbars is to right click in an empty spot in the top right area of the screen (inside the CommandManager) and select the Curves toolbar from the pop-up list.*

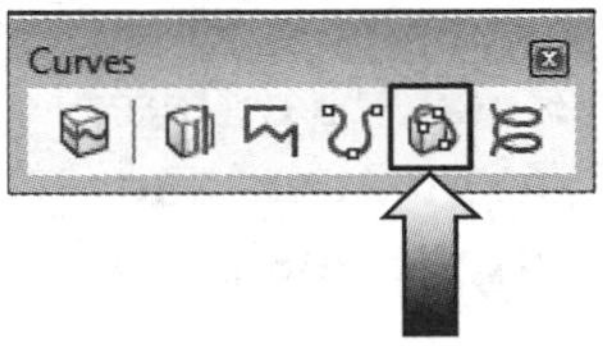

Curve Through Reference Points

Select the sketch points in the order as noted (required for this lesson only). (Point 1 is on top, point 2 is on the bottom, 3 on top, 4 on bottom, and so on.)

Click 

to close the curve.

The Closed Loft option connects the 1st and the last points to form a closed loop.

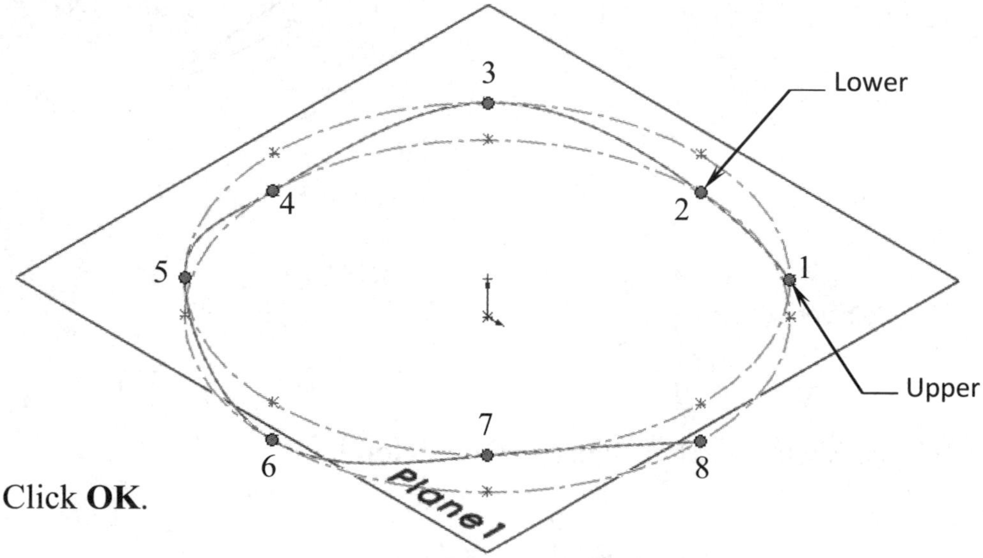

Click **OK**.

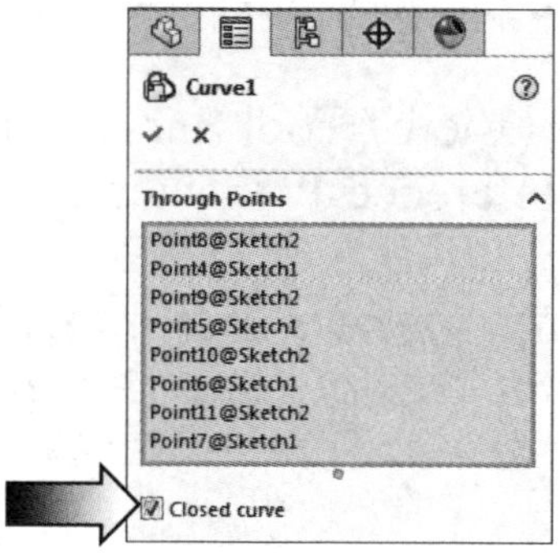

Rotate to different orientations and inspect the curve.

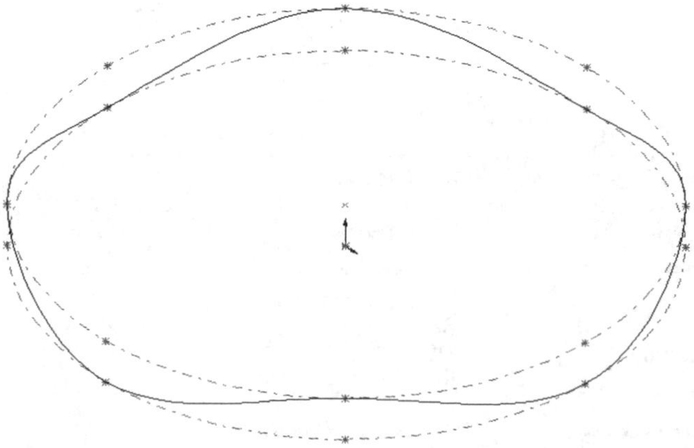

7. Sketching the 1st loft section:

Select the Front plane from the FeatureManager tree and open a **new sketch**.

Sketch a **Rectangle** and add the dimensions / relations as shown.

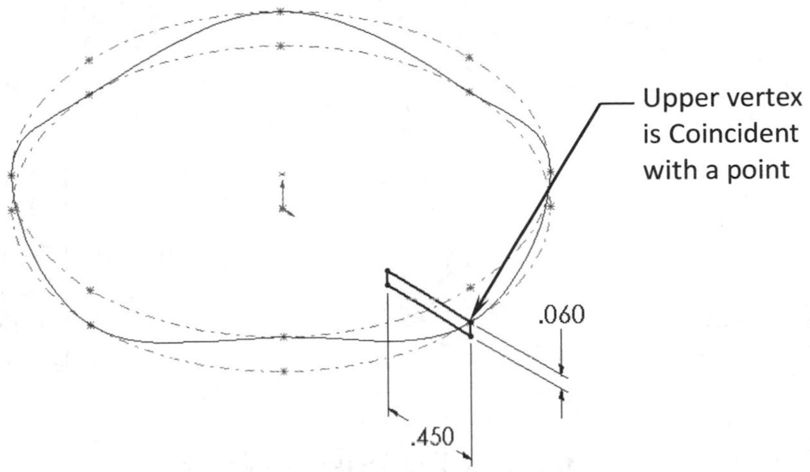

Exit the sketch  or select **Insert / Sketch**.

8. Creating the 2nd loft section using Derived-Sketch:

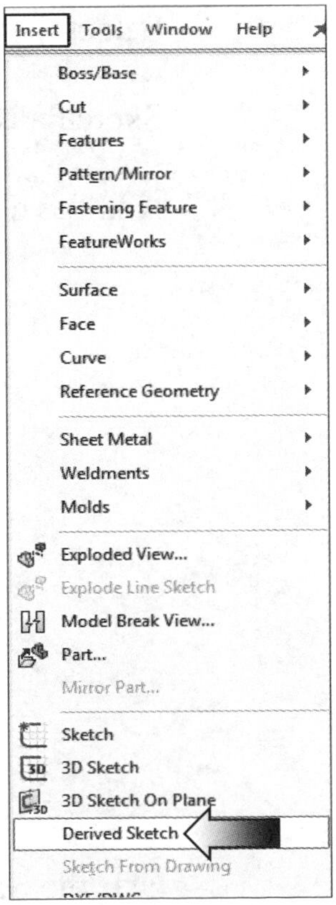

Hold the **Control** key, select the **Front** plane and the sketch of the **rectangle** from the Feature tree.

Select **Insert / Derived Sketch**.

A derived sketch is created and placed on top of the original sketch; drag it to the left side and then add a coincident relation to the point on the bottom as noted.

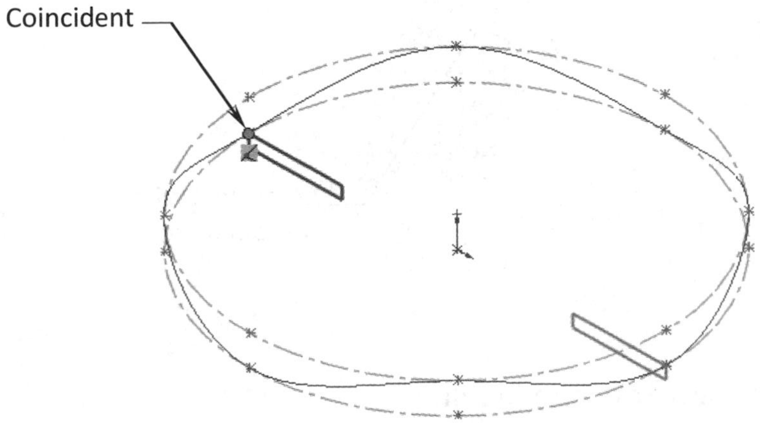

9. Fully defining the Derived Sketch:

Add a **Collinear** relation between the 2 lines as indicated.

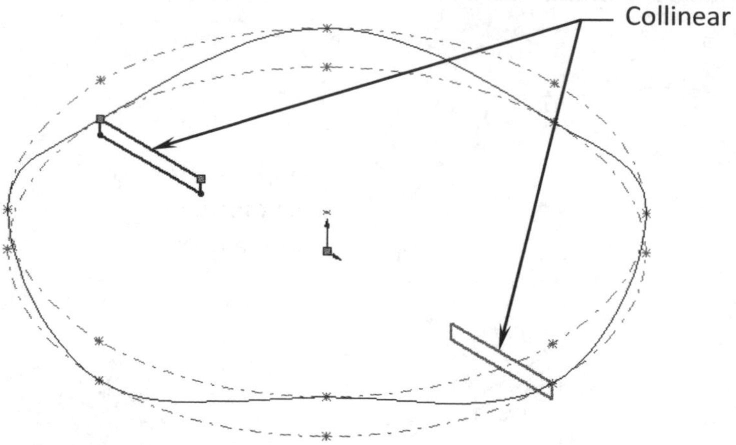

Derived Sketch

A derived sketch is a dependent copy of the original sketch. It can only be moved or positioned on the same or different plane with respect to the same model.

Exit the Sketch  or select **Insert / Sketch.**

10. Sketching the 3rd loft section: (or use the Derived Sketch option)

Select the Right plane from the FeatureManager tree and open a **new sketch**.

Sketch a **Rectangle** (or copy and paste the previous sketch).

Add the dimensions and relations needed to fully define the sketch.

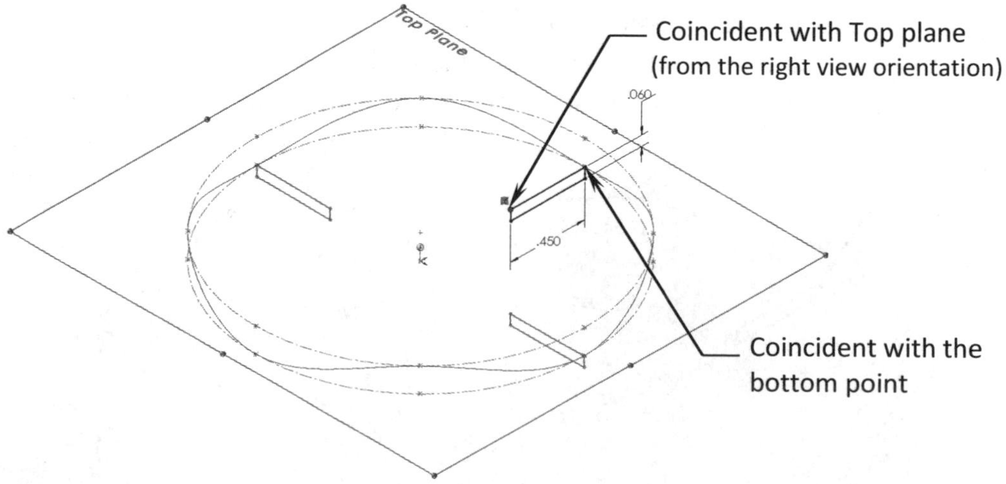

Exit the Sketch  or select **Insert / Sketch.**

11. Creating the 4th loft section using Derived-Sketch:

Hold the **Control** key; select the **Right** plane and the sketch of the **3rd rectangle** from the FeatureManager tree.

Select **Insert / Derived Sketch**.

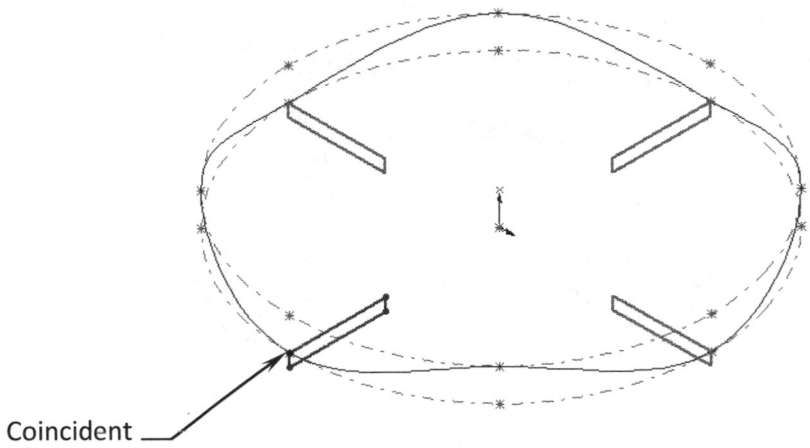

12. Constraining the Derived sketch:

Add a **Collinear** relation between the 2 lines as shown.

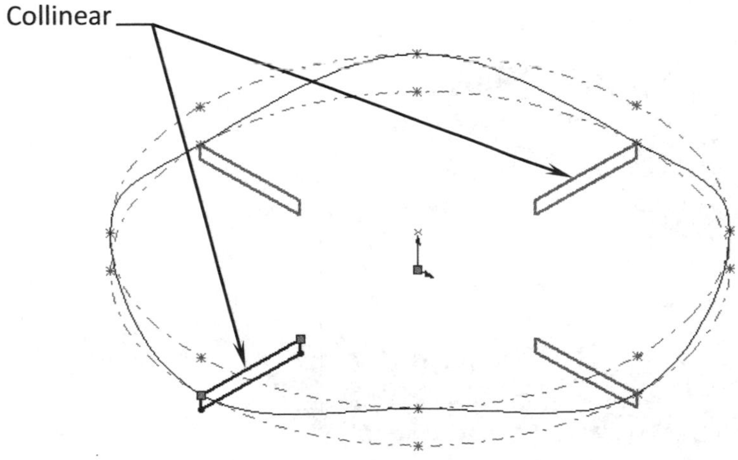

Exit the Sketch  or select **Insert / Sketch**.

13. Creating a Loft with Guide curve:

Click  or select **Insert / Boss-Base / Loft**.

For Loft Profiles: Select the **four** rectangular sketches.

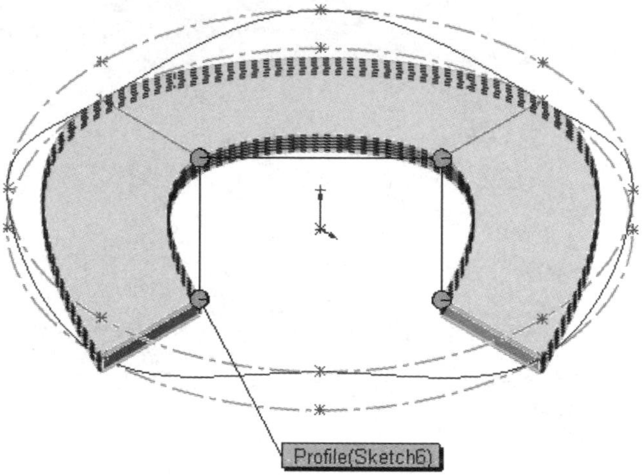

For Guide Curve, select the **3D Curve**.

Click **Close Loft** 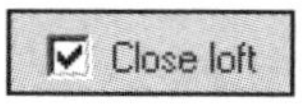

under Options (arrow).

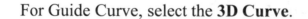

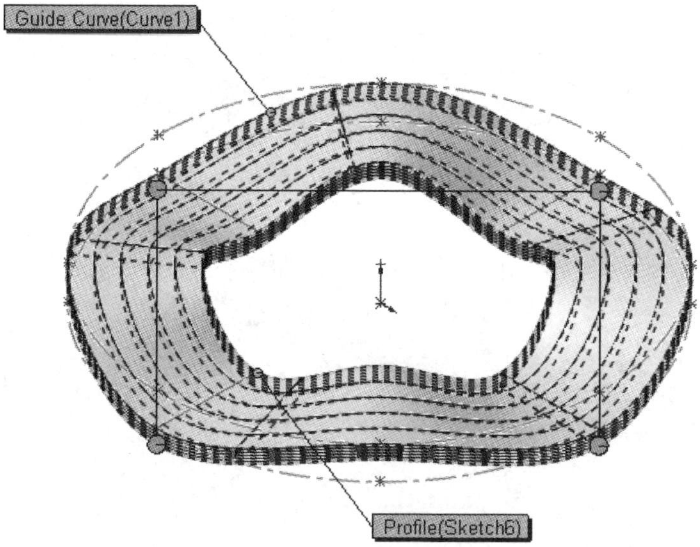

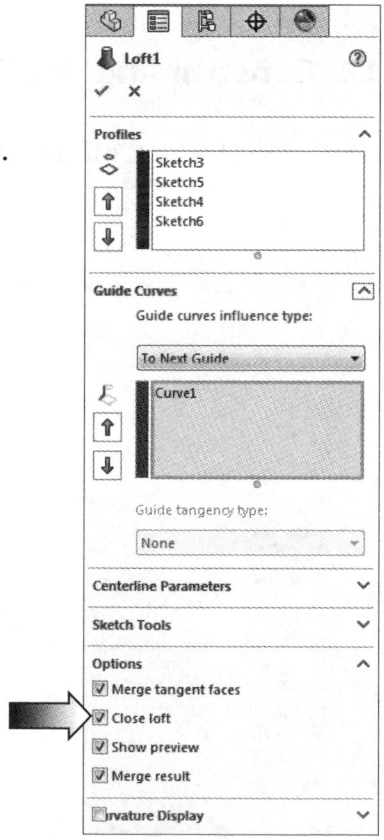

Click **OK**.

The completed Waved Washer (with the construction sketches still visible).

(Use 8 sketch points to create 4 waves, 16 sketch points will make 8 waves and so on.)

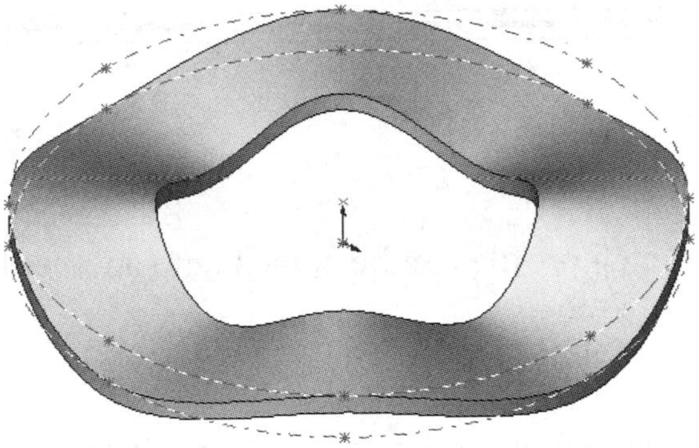

14. Hiding the construction sketches:

Click the construction sketches and select **Hide** .

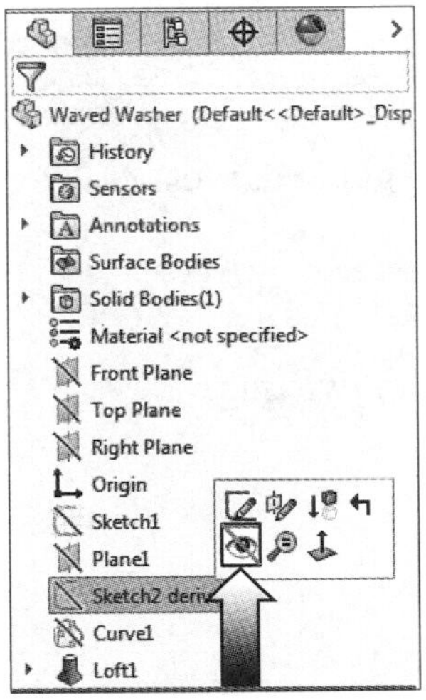

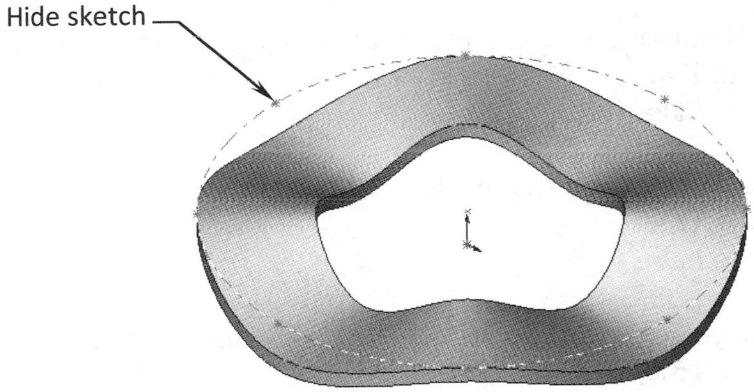

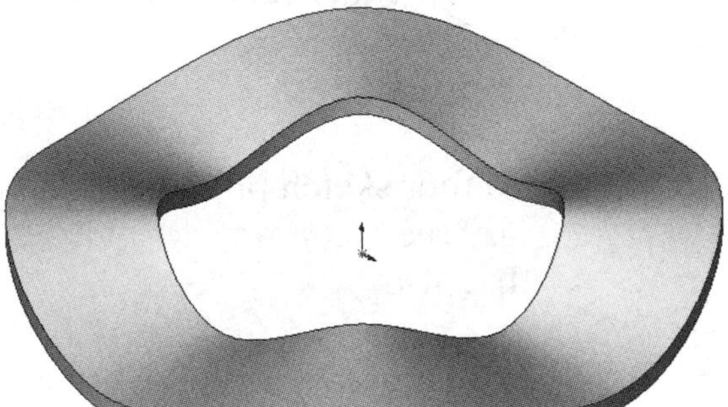

15. Saving your work:

Select **File / Save As**.

Enter Waved Washer for the file name.

Click **Save.**

Questions for Review

Loft with Guide Curves

1. A sketch profile can be copied onto another plane or a planar surface.
   a. True
   b. False

2. Sketch points can be added in any sketch to help define the sketch geometry or locations.
   a. True
   b. False

3. If a derived sketch is driven by the original sketch, its entities cannot be changed.
   a. True
   b. False

4. A 3D curve can be created using the reference points in the sketches or model's vertices.
   a. True
   b. False

5. The loft sections should either be Pierced or Coincident with the guide curves.
   a. True
   b. False

6. The guide curves can also be used to control the loft sections from twisting.
   a. True
   b. False

7. Only two guide curves can be used in each loft feature.
   a. True
   b. False

8. Up to four sketch profiles can be used in a loft feature.
   a. True
   b. False

9. The construction sketches can be toggled (Show/Hide) at any time.
   a. True
   b. False

1. TRUE 2. TRUE
3. TRUE 4. TRUE
5. TRUE 6. TRUE
7. FALSE 8. FALSE
9. TRUE

Exercise: Using Curve Driven Pattern

1. Creating the Base sketch:

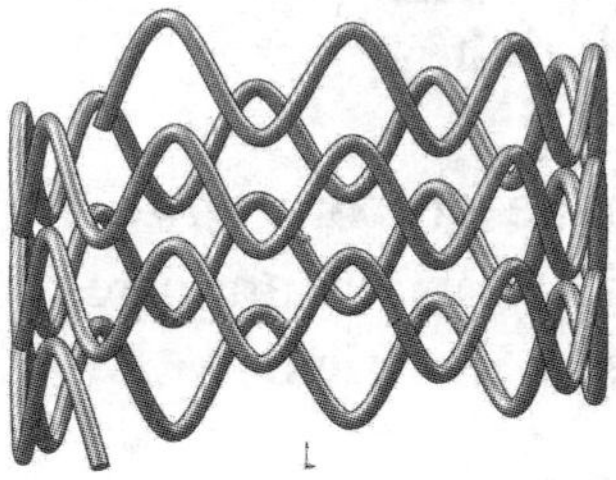

From the Top plane sketch a **circle**, centered on the origin.

Add a diameter dimension of **.100"**.

2. Activating the Surfaces toolbar:

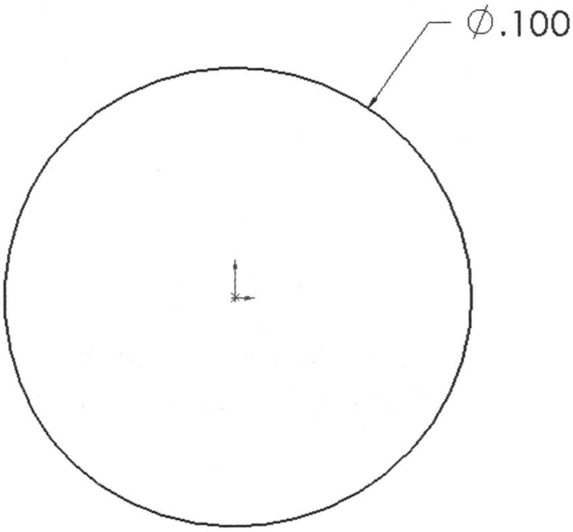

Right-click the **Sketch Tab** and select **Tabs**, **Surfaces** to enable the toolbar.

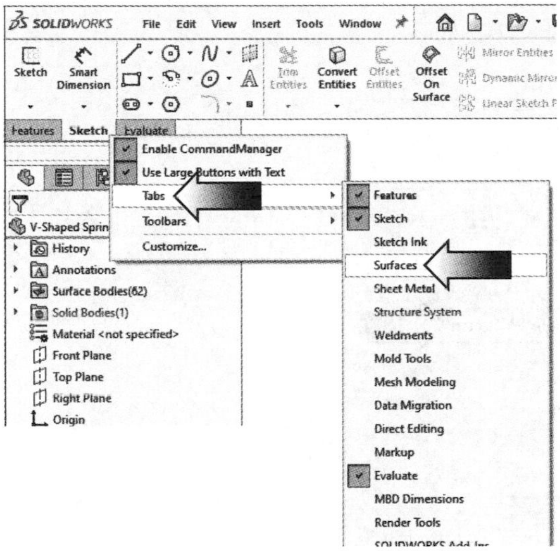

3. Extruding a <u>surface</u>:

From the Surfaces toolbar click: **Extruded Surface**.

Use the **Blind** type and enter **.060"** for extrude depth.

Click **OK**.

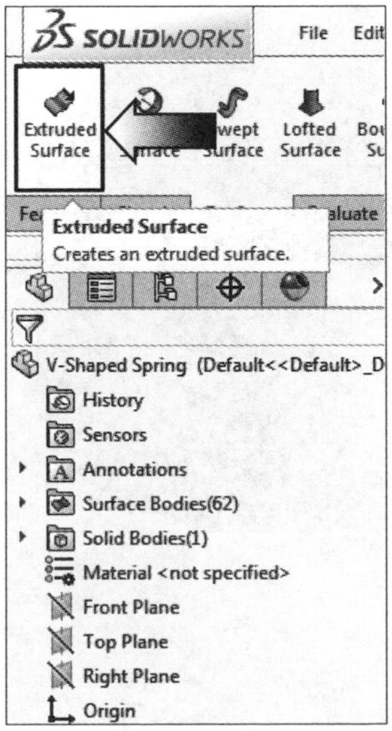

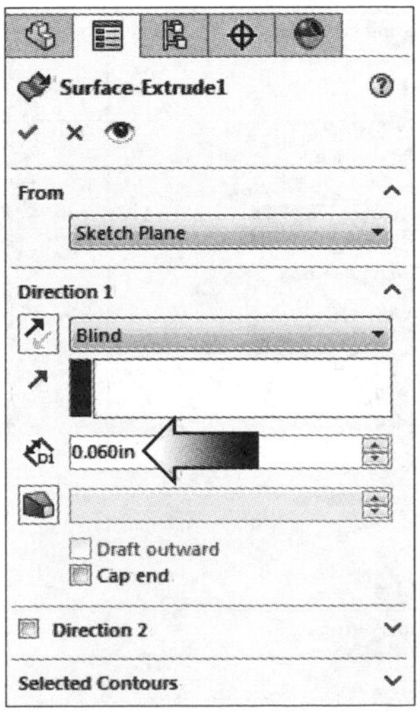

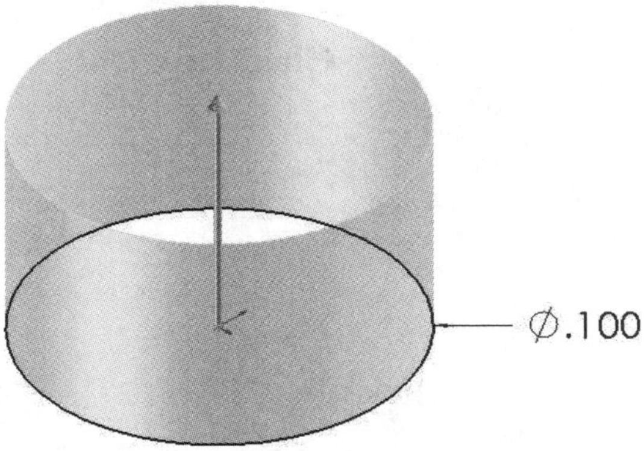

4. Creating a helix:

Select the <u>Top</u> plane and open a **new sketch**.

Select the **bottom edge** of the extruded surface and press **Convert Entities**.

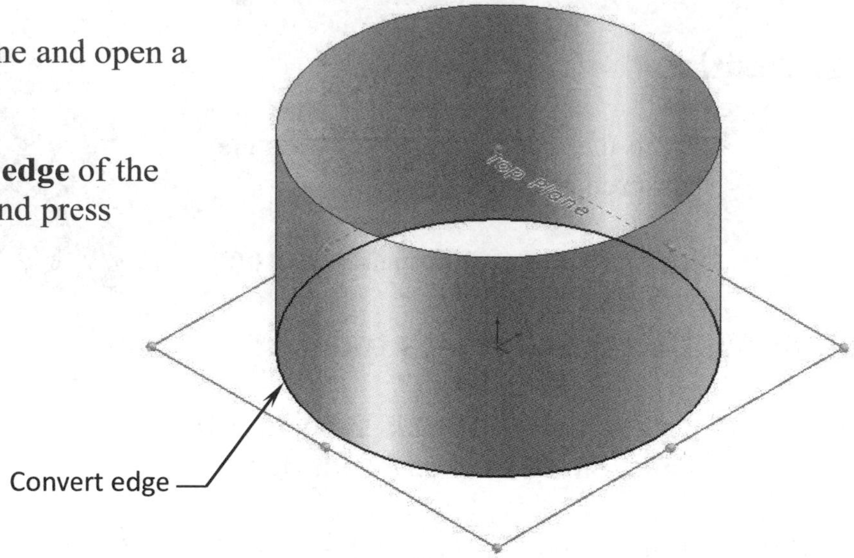

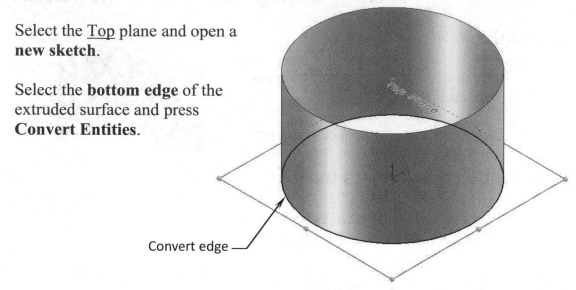

From the Features toolbar click: **Curves / Helix and Spiral**.

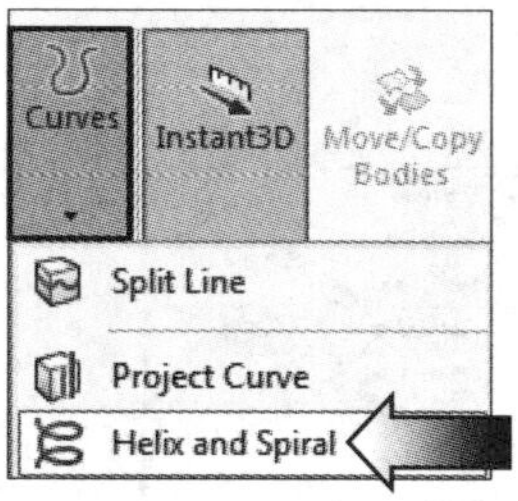

Enter the following:

* Defined by: **Pitch and Revolution**

* **Constant Pitch**

* Pitch: **0.020"**

* Revolutions: **3**

* Start Angle: **0**

* **Clockwise**

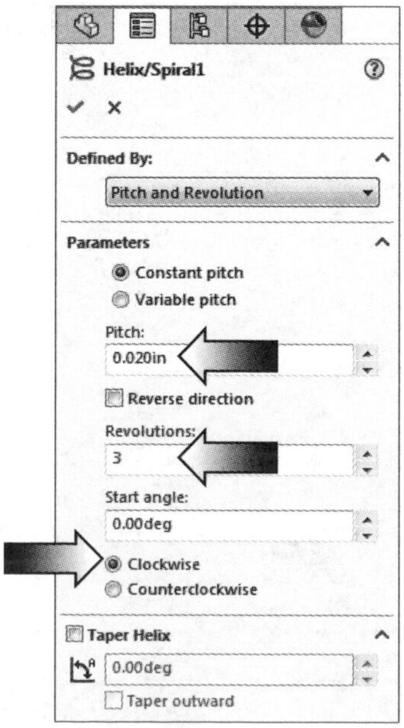

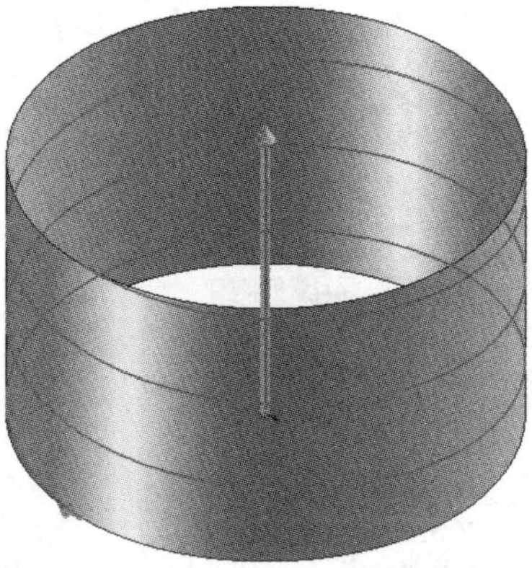

Click **OK**.

5. Creating a surface profile:

Select the Top plane and open a **new sketch**.

Sketch a **vertical line** and add a **.010"** linear dimension.

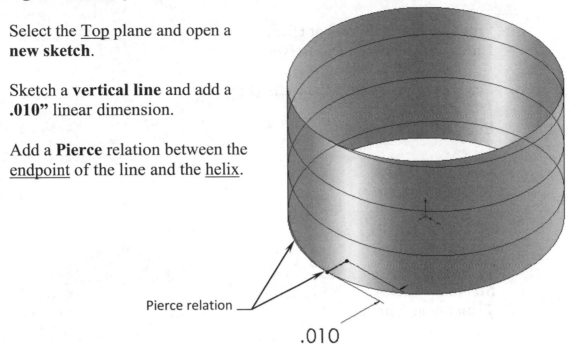

Add a **Pierce** relation between the endpoint of the line and the helix.

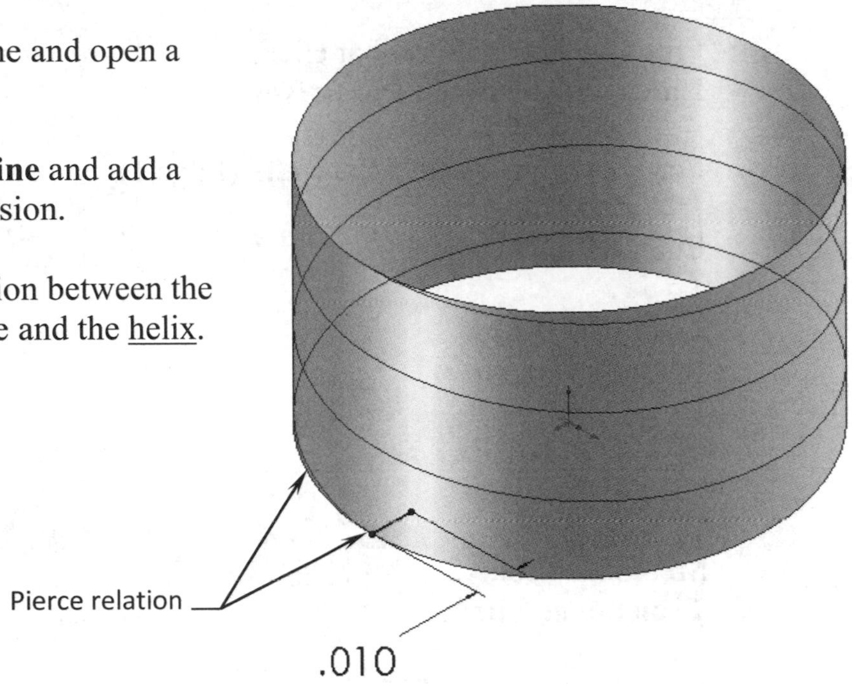

Switch to the **Surfaces** toolbar and click **Extruded Surface**.

For extrude type, use the default **Blind** type.

Enter **.020"** for extrude depth.

Click **OK**.

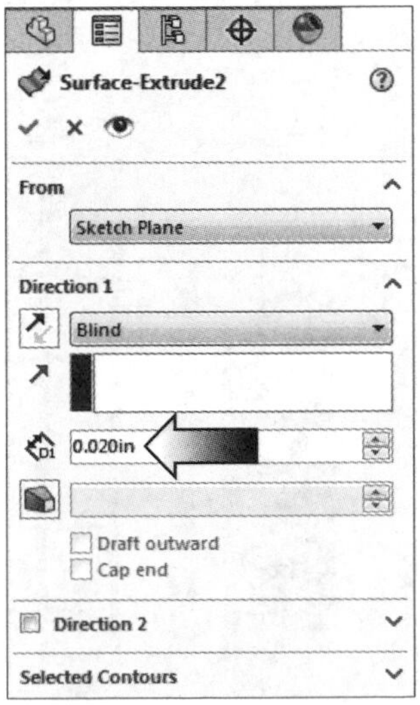

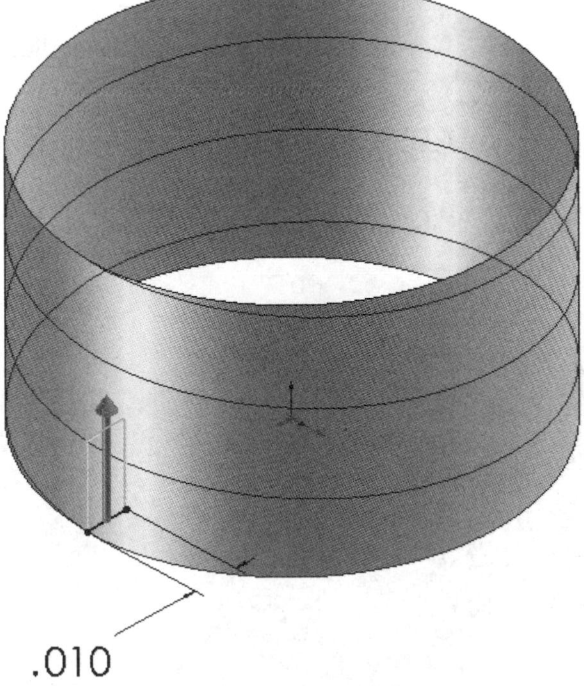

6. Creating a Curve Driven Pattern:

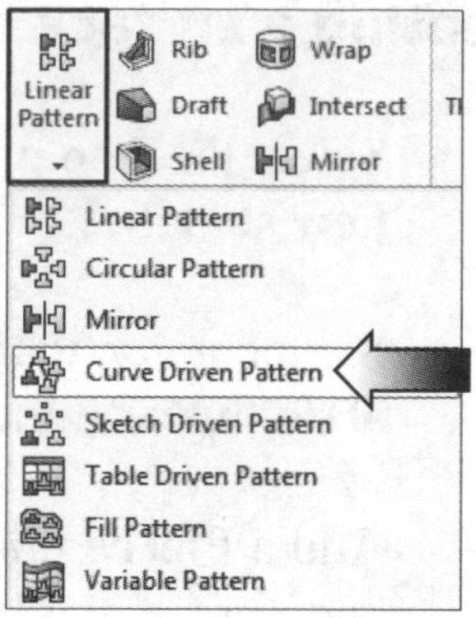

From the Features toolbar click:
Linear Pattern / Curve Driven Pattern.

For Direction 1, select the **Helix** (Edge 1).

Enter **61** for Instances.

Enable the **Equal Spacing** checkbox.

Under Curve Method, select: **Transform Curve**.

For Alignment Method, select: **Tangent to Curve**.

Under Face Normal select the **Cylindrical surface** of the cylinder.

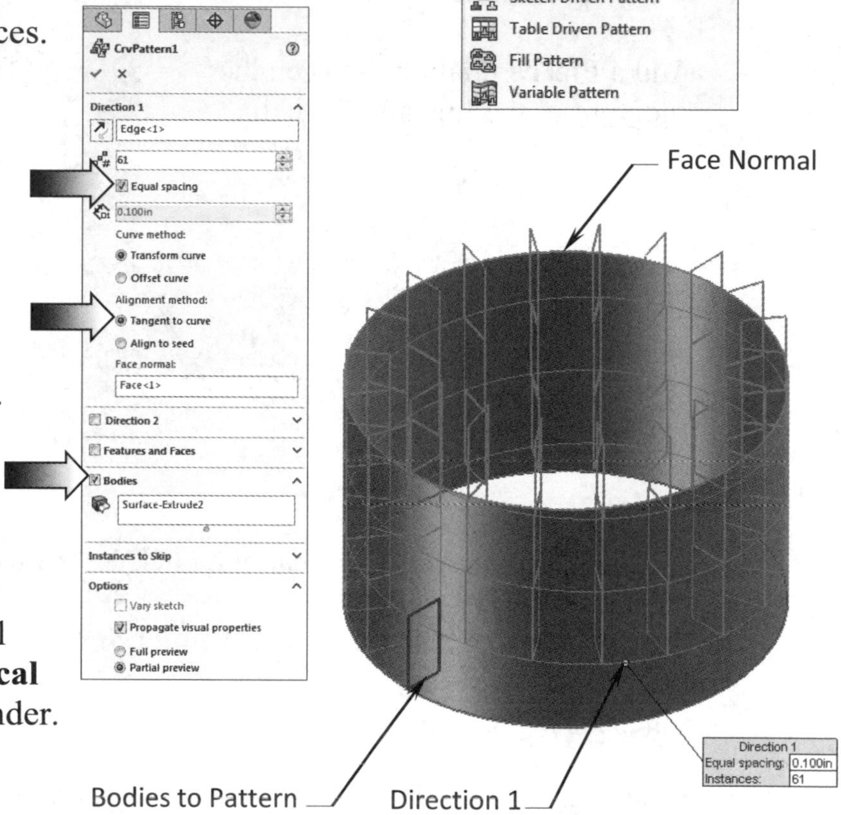

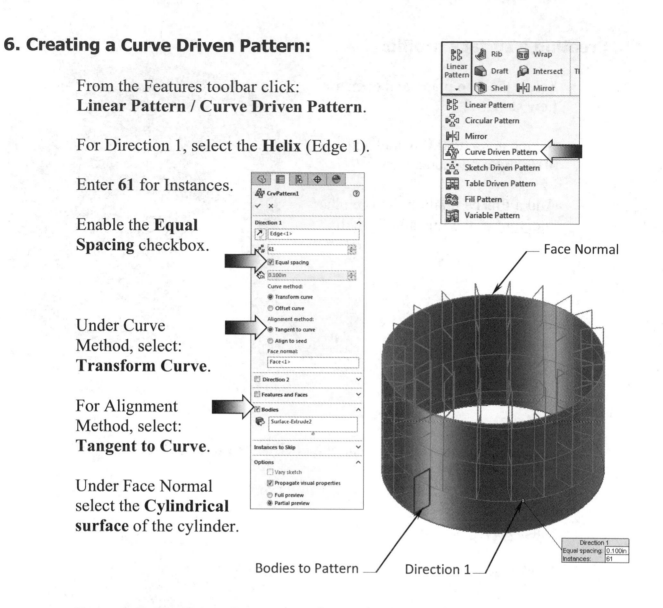

Expand the **Bodies to Pattern** section and select the **Extruded Surface** in step 5.

Click **OK**.

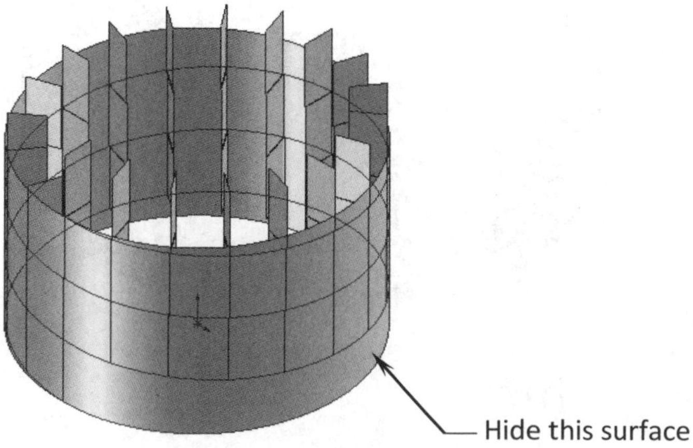

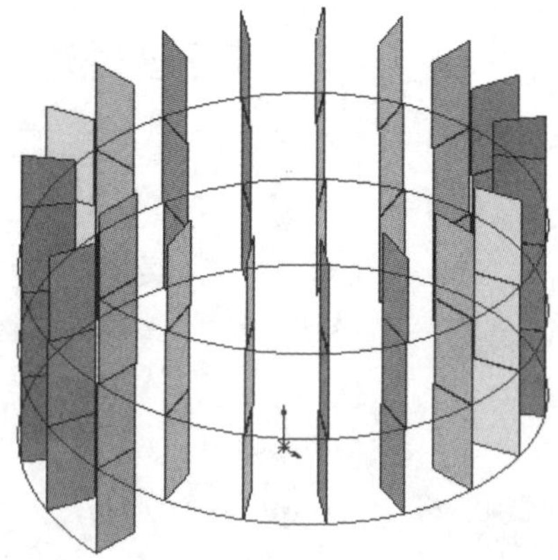

7. Creating a Curve Through Reference Points:

From the **Features** toolbar select:
Curves / Curve Through Reference points.

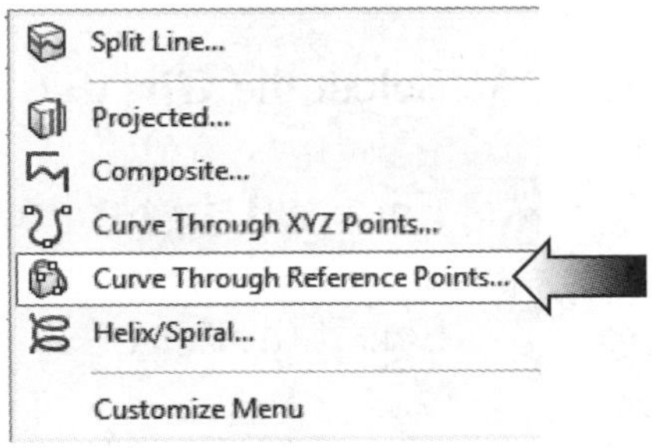

Click the starting point (point 1) at the end of the helix and go clockwise to point 2, then point 3 as indicated.

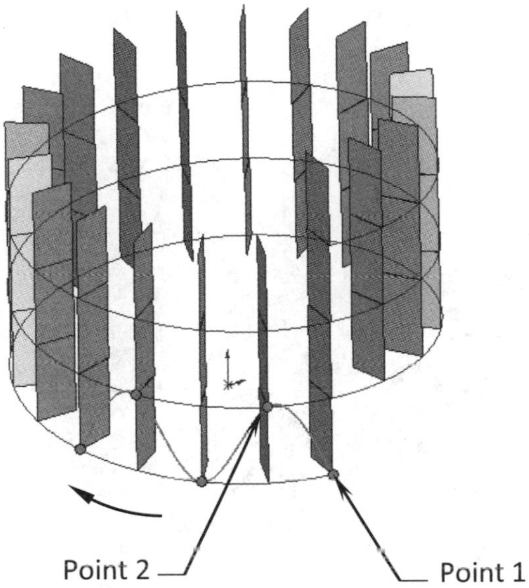

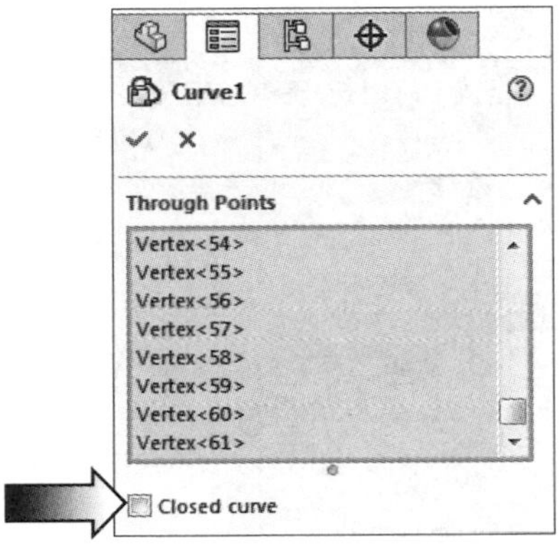

Continue going around and select all connecting points. (If you make a mistake simply delete the point in the Through Points dialog box and reselect it again.)

Be sure to <u>uncheck</u> the **Close Curve** checkbox since the two ends of this curve are not supposed to connect.

Click **OK**.

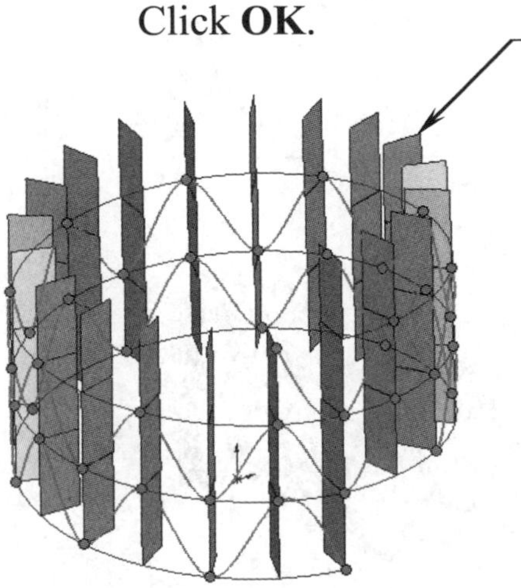

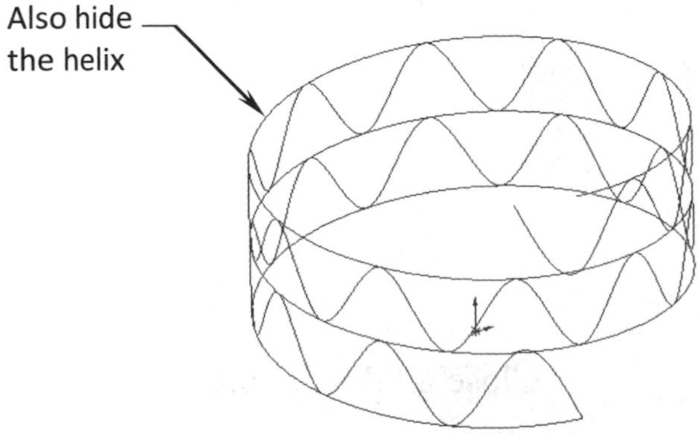

8. Creating the final sweep:

Switch to the **Features** toolbar and click: **Swept Boss Base**.

Select the **Circular Profile** option (arrow).

Enter **.002in** for Profile Diameter (arrow).

Select the **Curve1** as the sweep path.

Click **OK**.

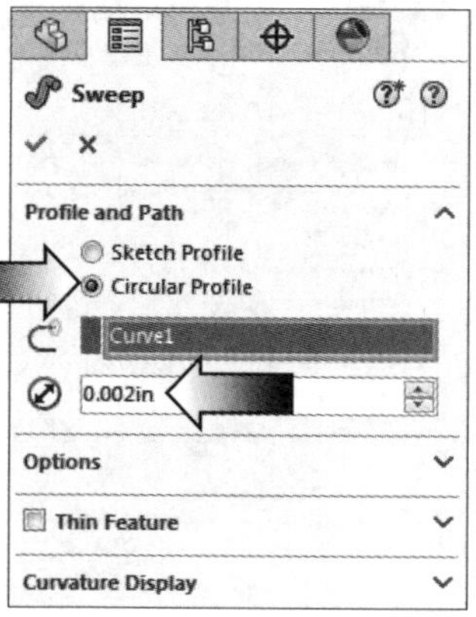

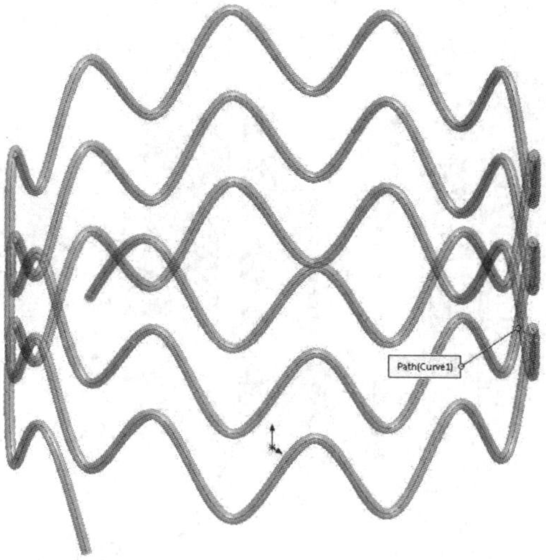

9. Saving your work:

Click **File / Save As**.

Enter **V-Shape Spring** for the name of the file.

Click **Save**.

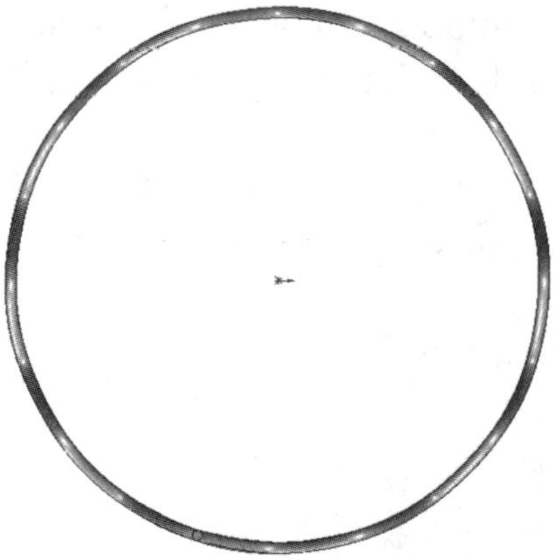

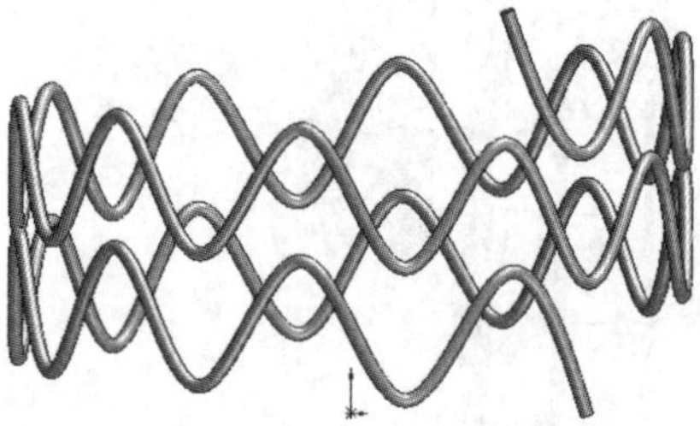

Close all documents.

Advanced Sweep

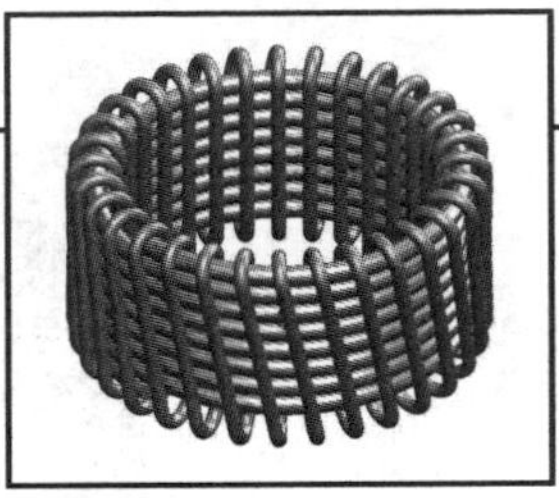

Advanced Sweep
Using 3D-Sketch Sweep Path

This second half of the chapter discusses one of the advanced techniques on creating a continuous wrapped wire around a coil.

There will be two separate sweep features created in this design. The first sweep feature is the coil, shaped like a spring; the sweep profile is made tangent to the sweep path. To achieve this, a couple of construction lines are used to help locating one of its quadrant points on the path.

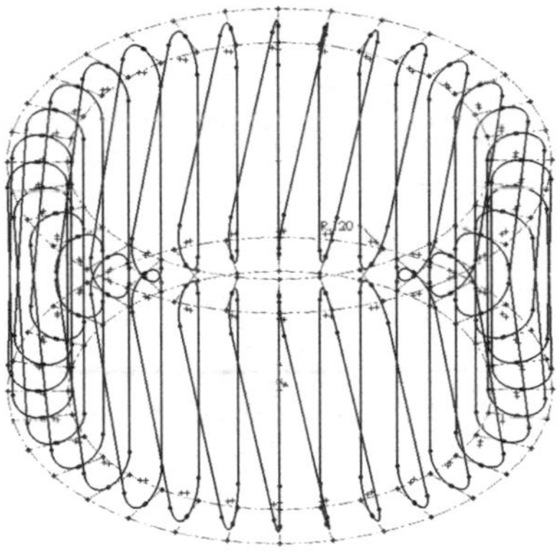

The second sweep feature, due to its complex shape, is done by using a 3D sketch as a sweep path. It is going to wrap around the first sweep feature in a unique way that makes 3D sketch one of the few options to create it with.

The method in this lesson demonstrates the use of locating points created in two different sketches to guide a 3D sketch. The lines in the 3D sketch are connected to the pre-defined endpoints of the lines, and the shape of the wire, or the sweep path, is formed.

To ensure the point locations are identical between the two sketches, the option Derived-Sketch is used. That way if one of the endpoints in the original sketch is moved, the same point in the derived sketch will also move.

The entities in the derived sketch are driven by the original sketch. Changes done to the original sketch are reflected in the derived sketch. To break the link between the derived sketch and its parent sketch right click the derived sketch from the FeatureManager tree and select **Underived**. After the link is broken, the derived sketch will no longer update when the original sketch is changed.

3D-Sketch Sweep Path
Advanced Sweep

Dimensioning Standards: **ANSI**
Units: **INCHES** – 3 Decimals

Tools Needed:

Insert Sketch	3D Sketch	Line
Add Geometric Relations	Sketch Fillet	Circle
Dimension	Centerline	Fillet/Round
Sketch Circular Pattern	Helix/Spiral	Sweep Boss/Base

1. Starting a new part document:

Go to **File / New / Part**.

Set the Drafting Standard to **ANSI** and the Units to **Inches**, **3 decimals**.

Select the <u>Top</u> plane and open a **new sketch**.

Sketch a **Circle** centered on the origin as shown.

Add a **Ø2.000"** diameter dimension to fully define the sketch.

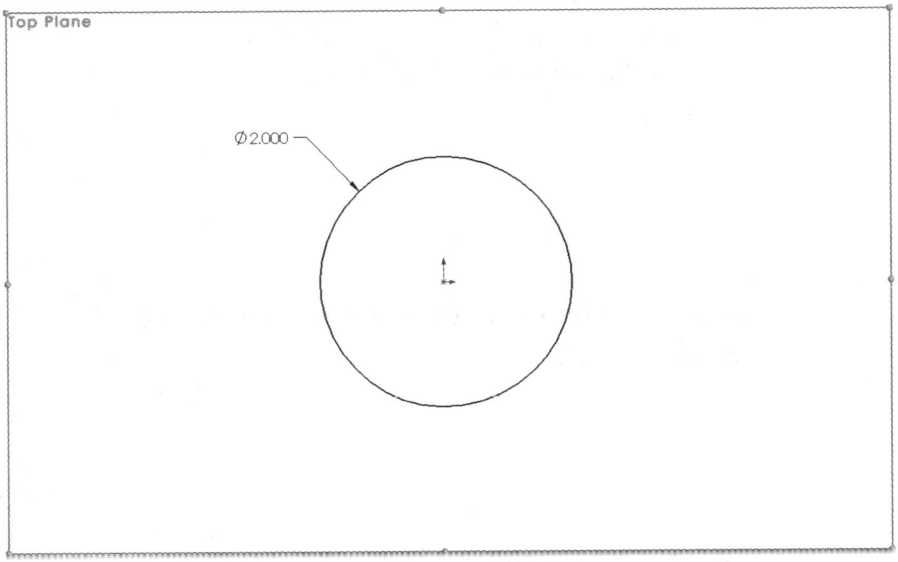

2. Creating the sweep path:

Click the **Helix / Spiral** command on the Curves drop-down menu, or select: **Insert / Curve Helix-Spiral**.

Set the following:

* **Constant Pitch**
* Pitch = **.125"**
* Revolutions: **8**
* Start Angle = **0deg**
* **Clockwise**

Click **OK**.

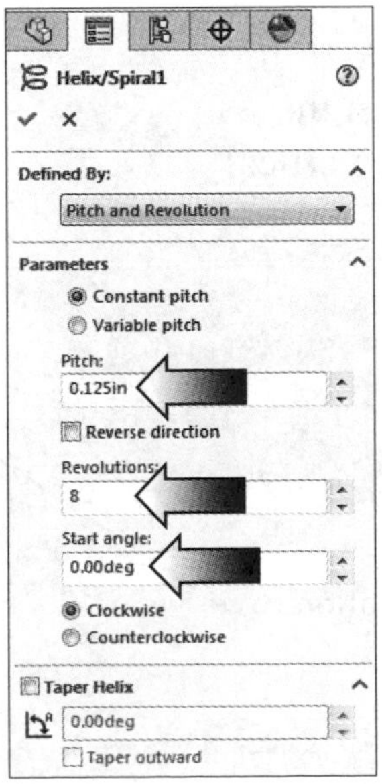

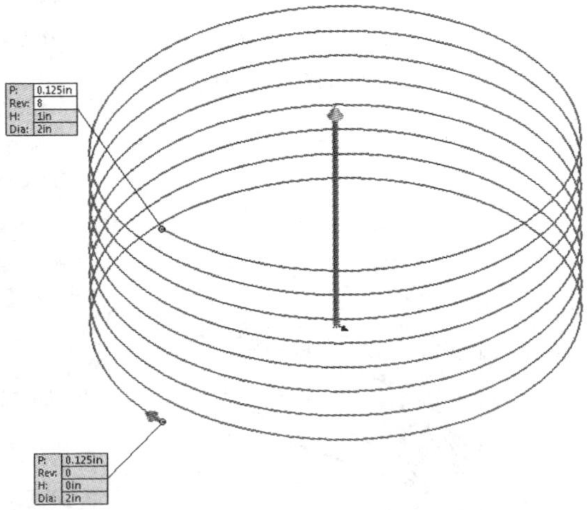

3. Creating the sweep profile:

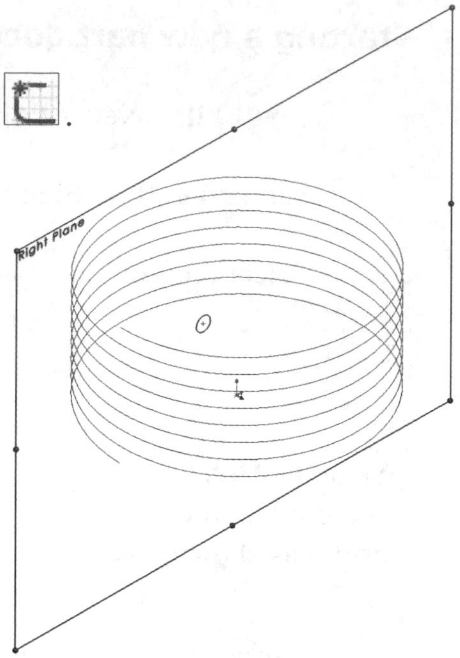

Select the <u>Right</u> plane and open a **new sketch**.

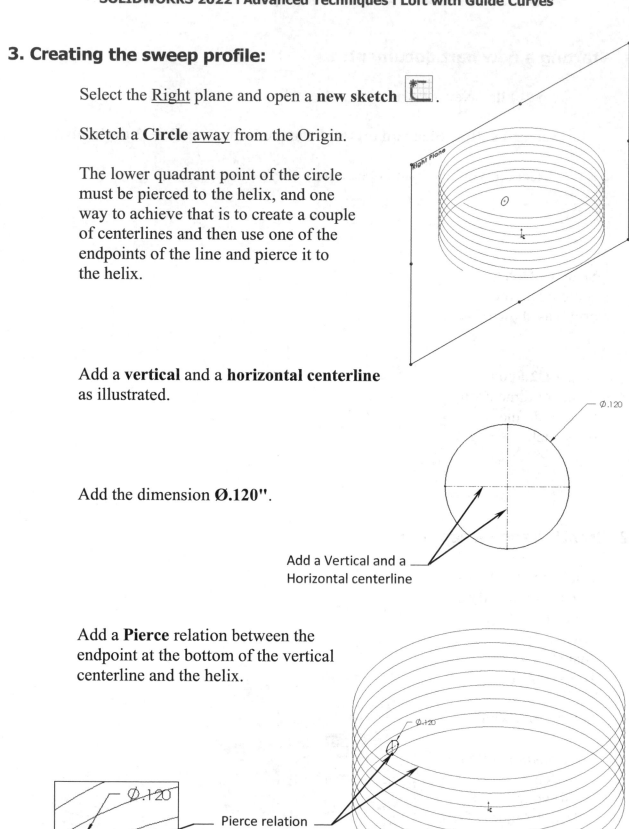

Sketch a **Circle** <u>away</u> from the Origin.

The lower quadrant point of the circle must be pierced to the helix, and one way to achieve that is to create a couple of centerlines and then use one of the endpoints of the line and pierce it to the helix.

Add a **vertical** and a **horizontal centerline** as illustrated.

Add the dimension **Ø.120"**.

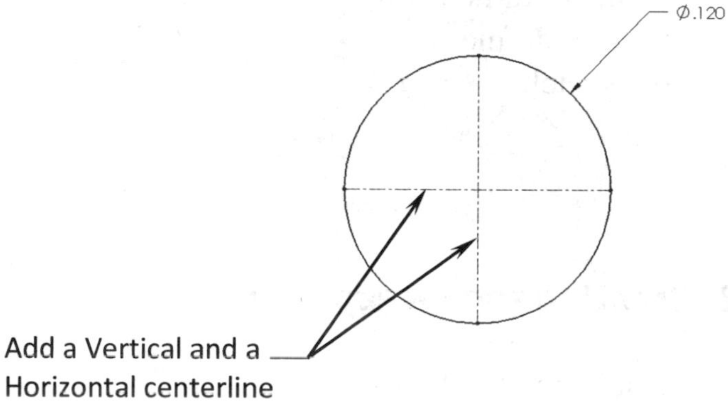

Add a **Pierce** relation between the endpoint at the bottom of the vertical centerline and the helix.

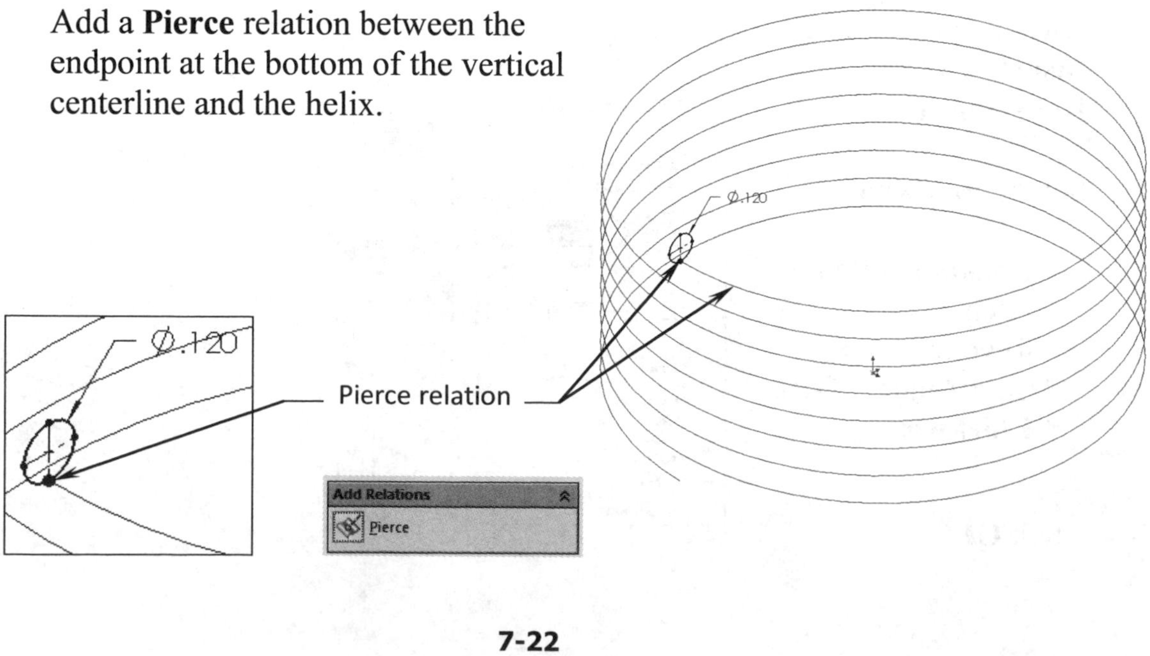

4. Creating the sweep feature:

Switch to the **Features** tool tab.

Click the  **Swept Boss /Base** command.

For Sweep Path, select the **Helix**.

For Sweep Profile, select the small **Circle**.

Keep all other parameters at their default settings.

Click **OK**.

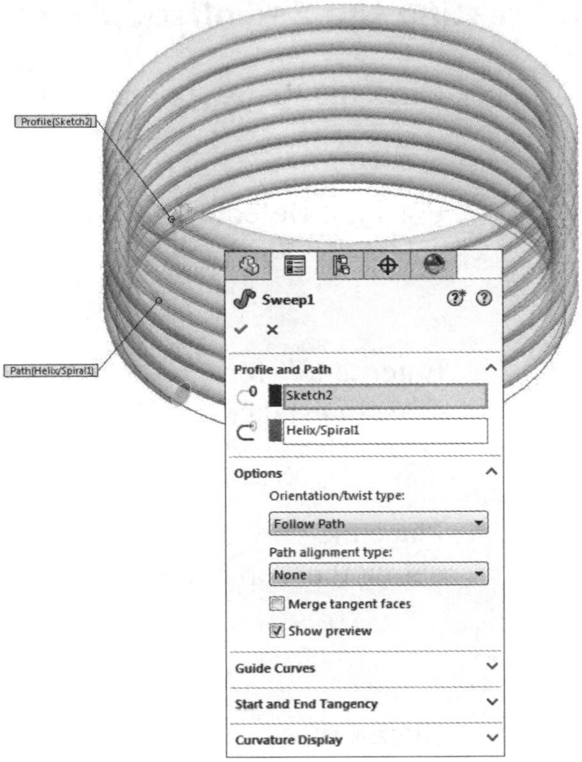

5. Creating the 1st offset plane:

From the Features tool tab, click **Reference Geometry, Plane** or select: **Insert, Reference Geometry, Plane**.

For the First-Reference select the **Top** plane from the Feature-Manager tree.

For Offset-Distance enter: **1.150"**.

Place the new plane above the Top plane.

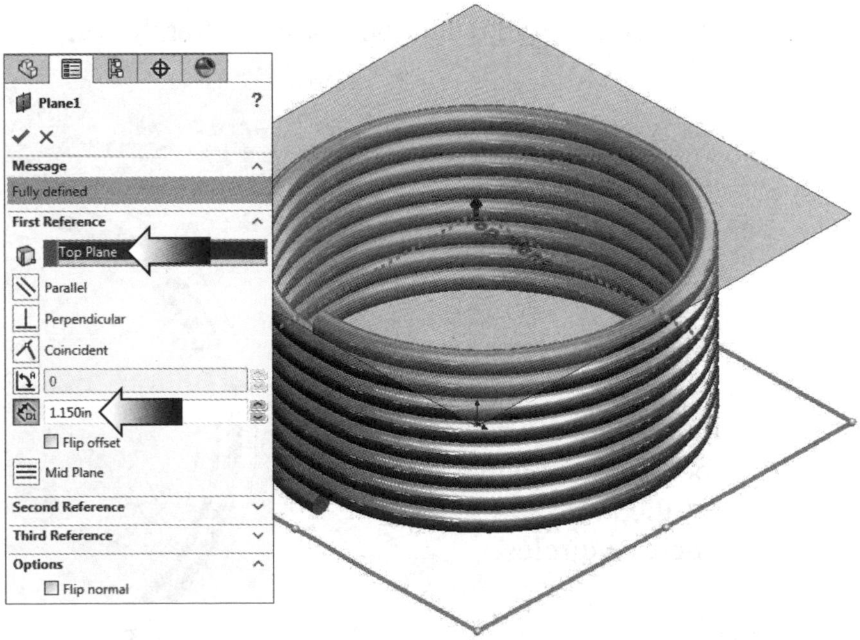

Click **OK**.

6. Creating the 2nd offset plane:

Click the **Plane** command  once again.

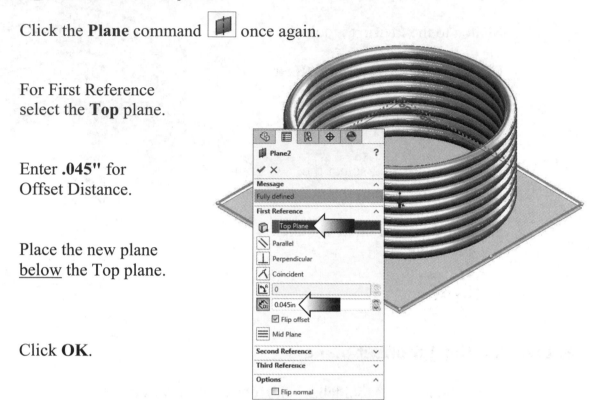

For First Reference select the **Top** plane.

Enter **.045"** for Offset Distance.

Place the new plane below the Top plane.

Click **OK**.

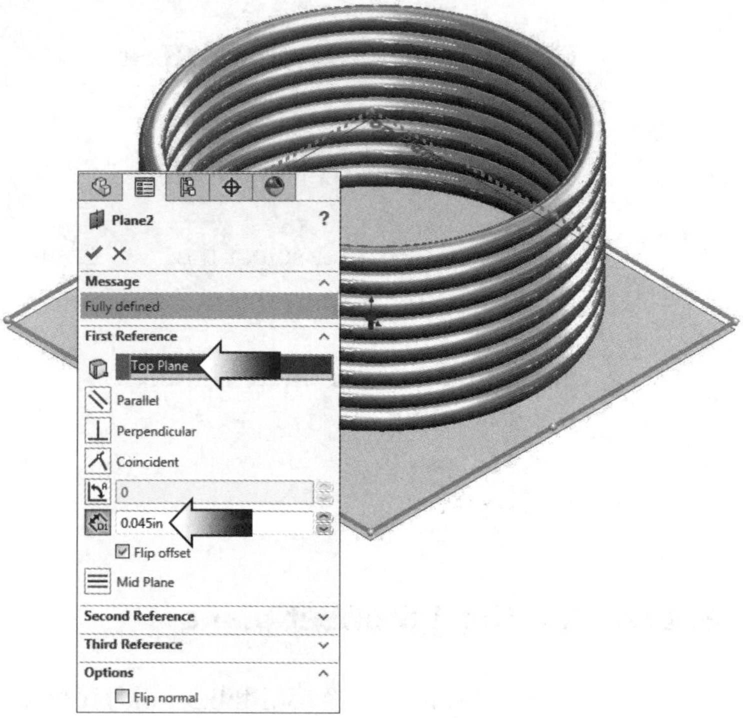

7. Sketching on the 1st offset plane:

Select the Plane1 and open a **new sketch** .

Sketch **2 Circles** that are centered on the origin.

Sketch a vertical **Line** and add the Coincident relations between the ends of the lines and the two circles.

Add the two diameter dimensions shown.

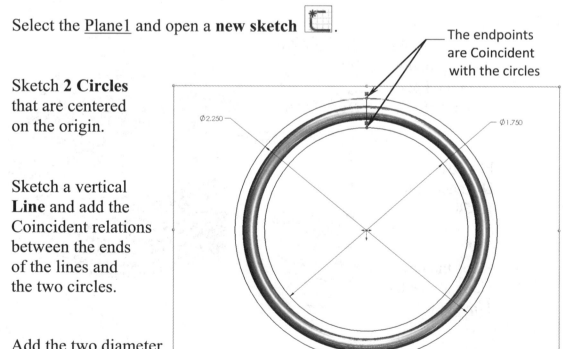

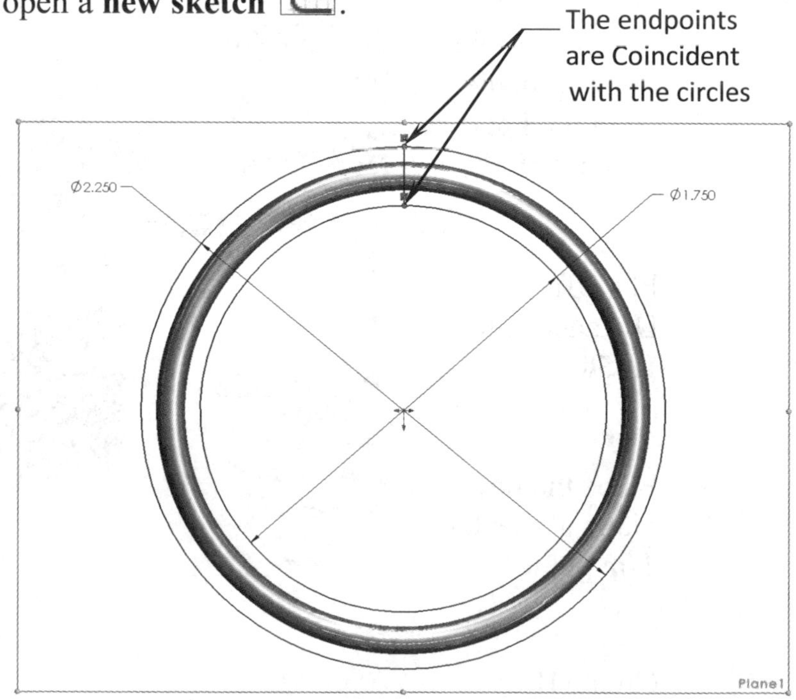

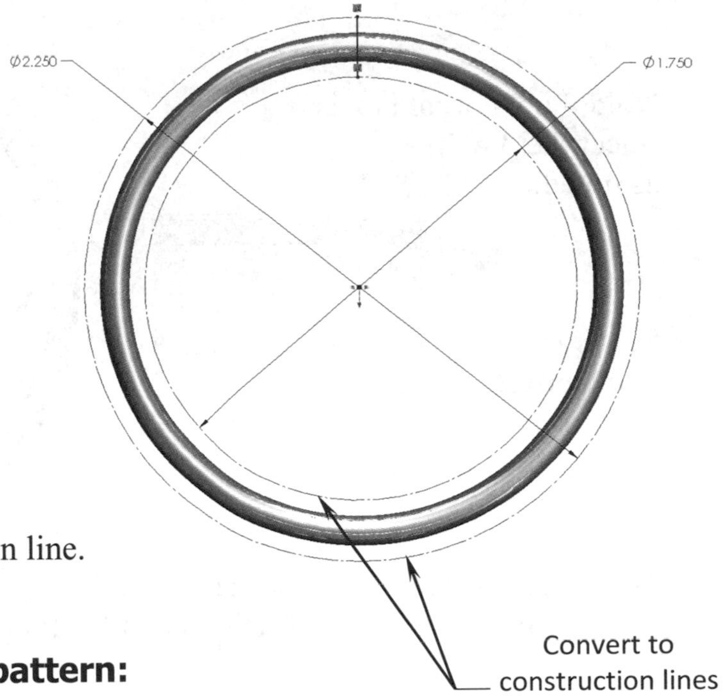

Hold the **Control** key and select both circles.

Release the control key and click the **For Construction** checkbox on the Property tree.

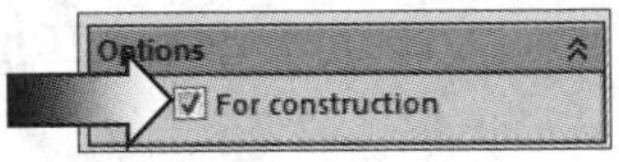

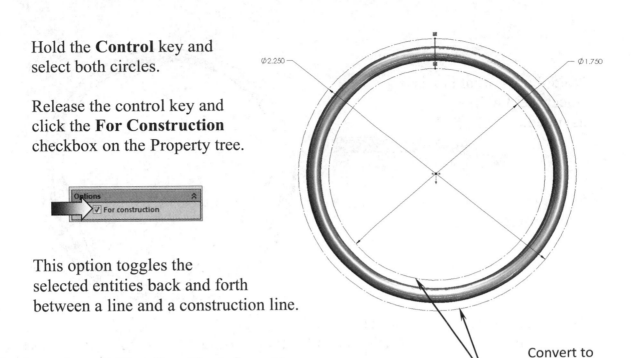

This option toggles the selected entities back and forth between a line and a construction line.

8. Creating a Circular Sketch pattern:

Select the vertical line (Entity to Pattern) and click the **Circular Sketch Pattern** command.

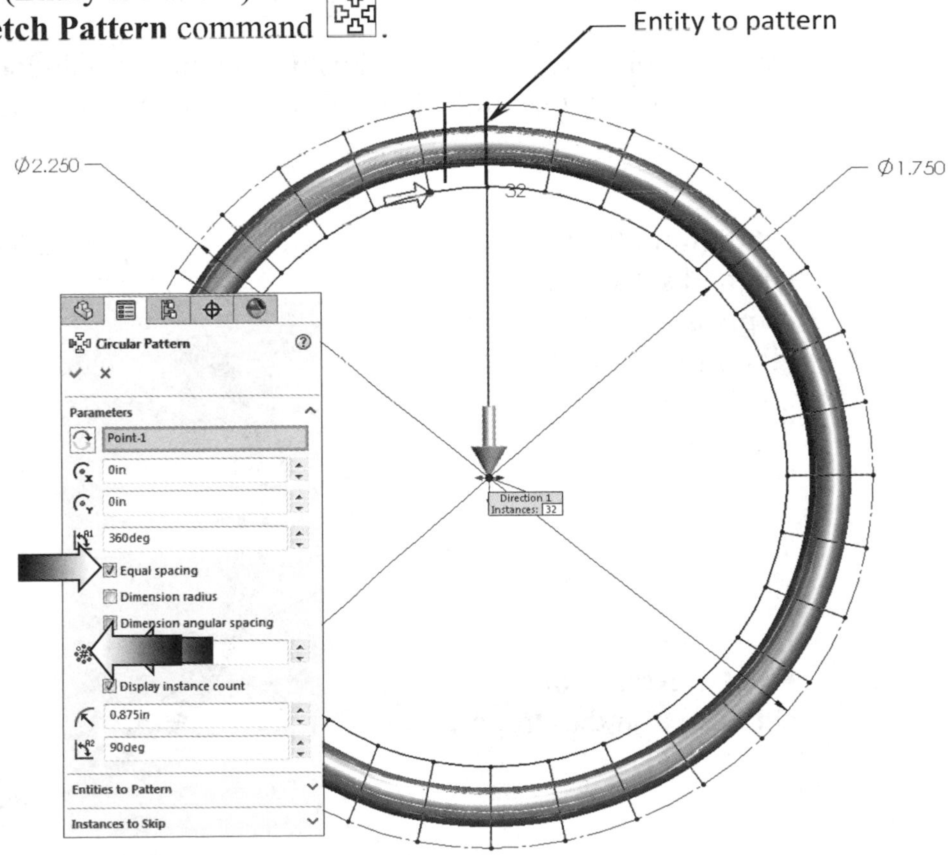

The origin is automatically selected as the center of the pattern.

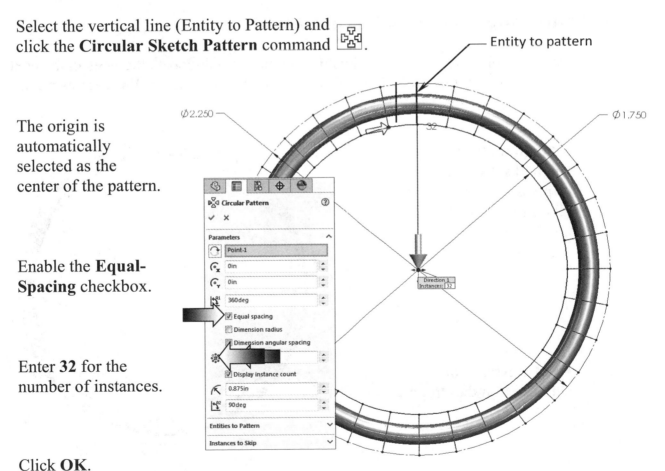

Enable the **Equal-Spacing** checkbox.

Enter **32** for the number of instances.

Click **OK**.

The pattern is still under defined at this point.

Hold the **Control** key and select the 3 points as noted.

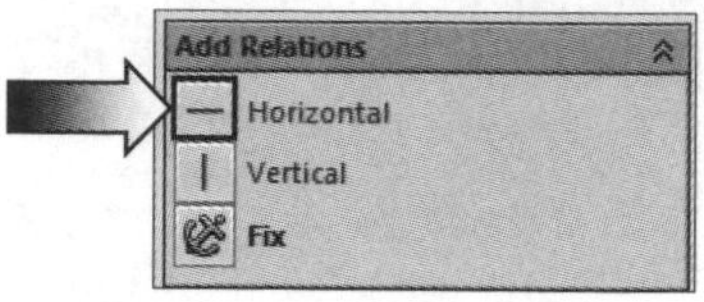

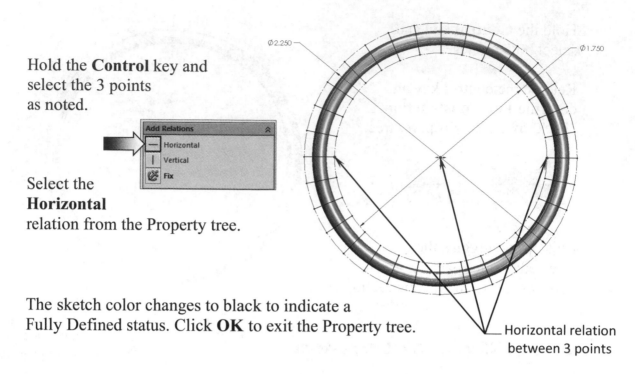

Select the **Horizontal** relation from the Property tree.

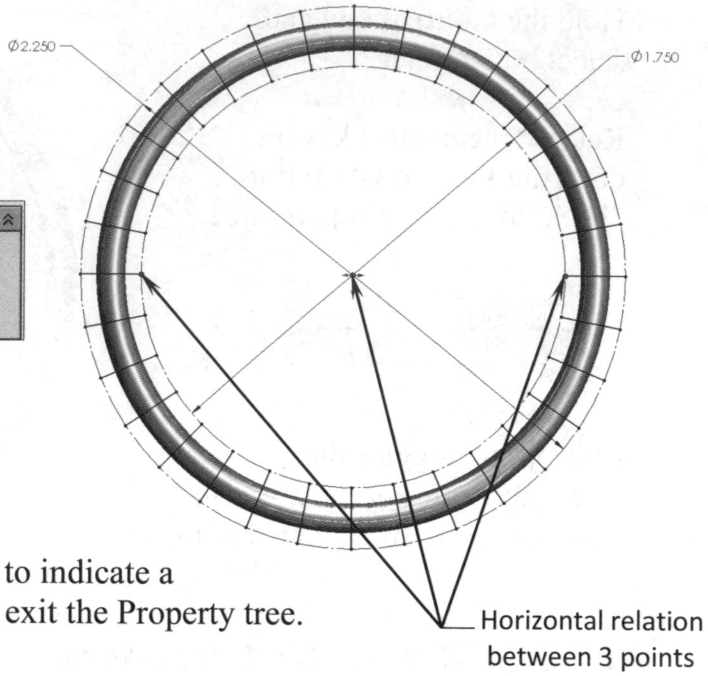

The sketch color changes to black to indicate a Fully Defined status. Click **OK** to exit the Property tree.

9. Converting to construction geometry:

Since we only need the endpoints of each line to help locate the lines in the next sketch, we will need to convert all the lines into construction lines at this point.

Either hold the Control key and select **all the lines**, or press Control+A to select all entities, then hold the control key and deselect the 2 circles.

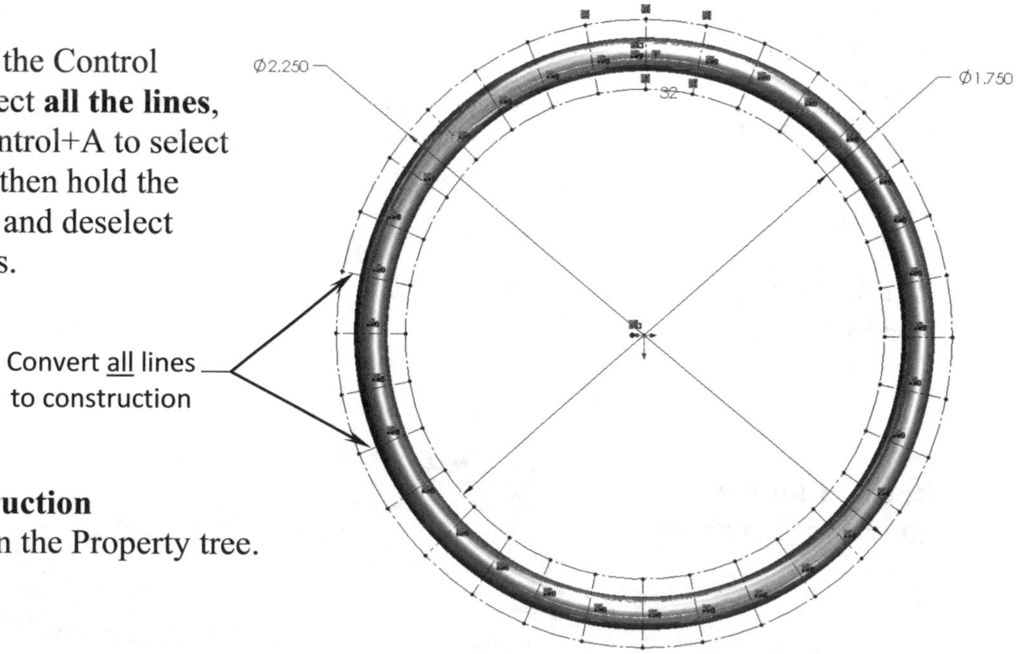

Enable the **For Construction** checkbox on the Property tree.

Exit the sketch.

10. Creating a derived sketch:

Hold the **Control** key and select the **Plane2** and the **Sketch3** from the feature tree.

Click **Insert / Derived Sketch** (arrow).

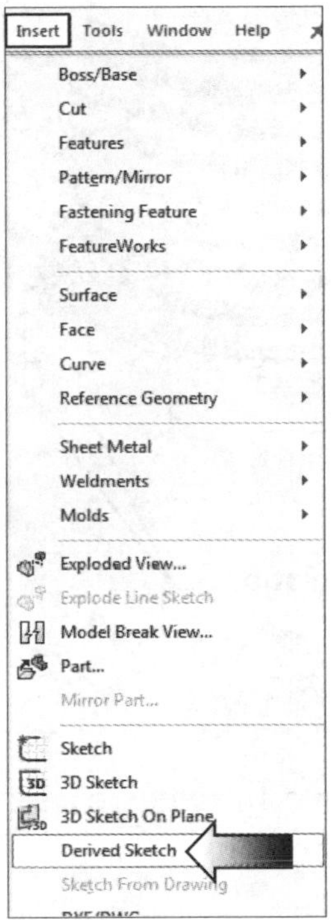

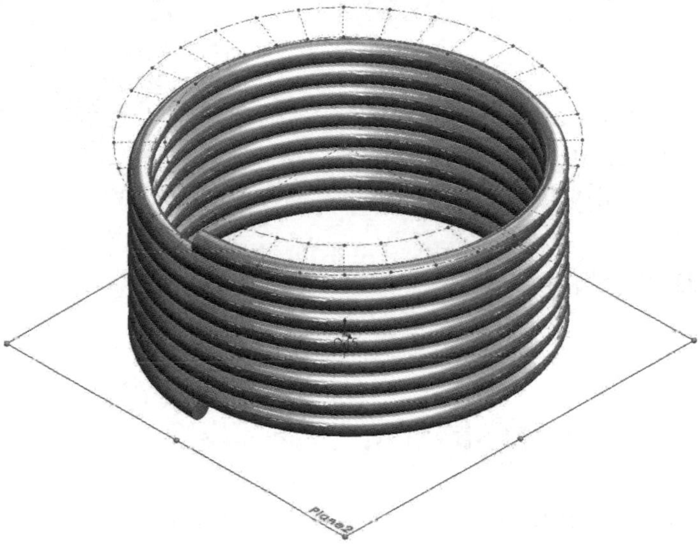

A copy of the sketch is created on the selected plane.

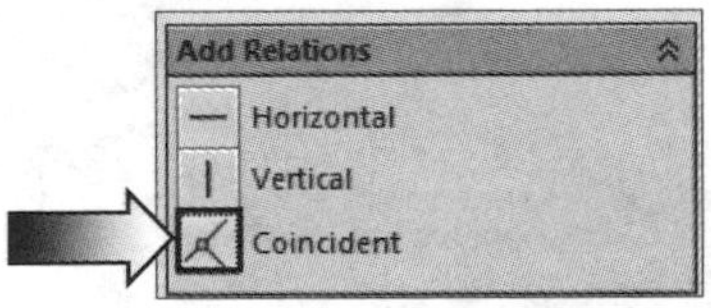

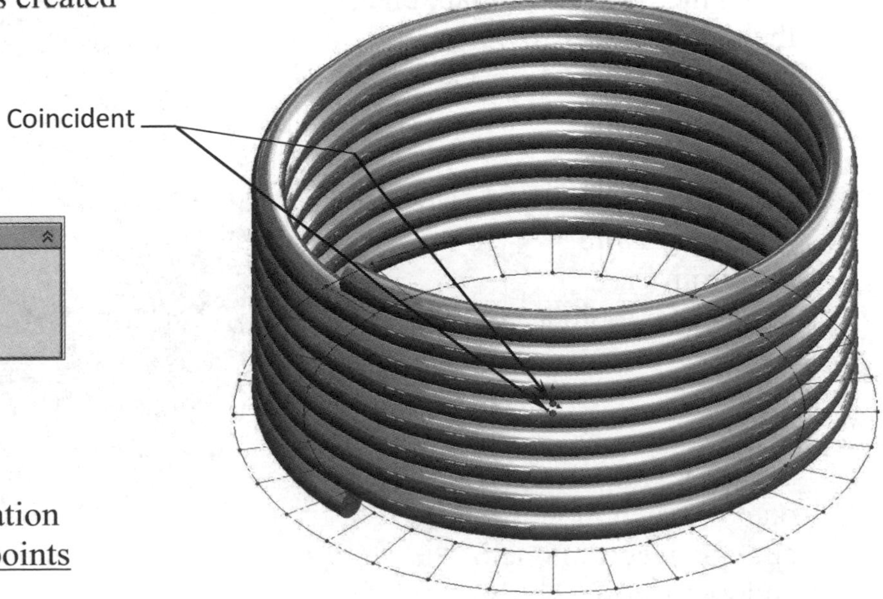

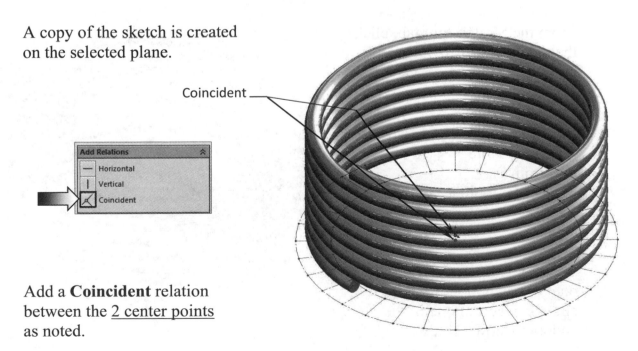

Add a **Coincident** relation between the 2 center points as noted.

Add a **Collinear** relation between the horizontal line on the right and the Front plane as indicated.

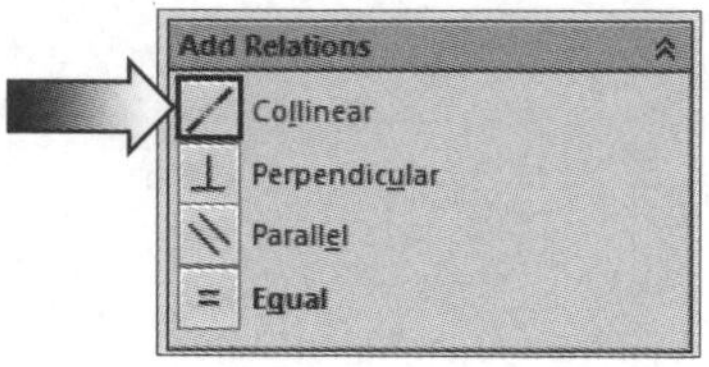

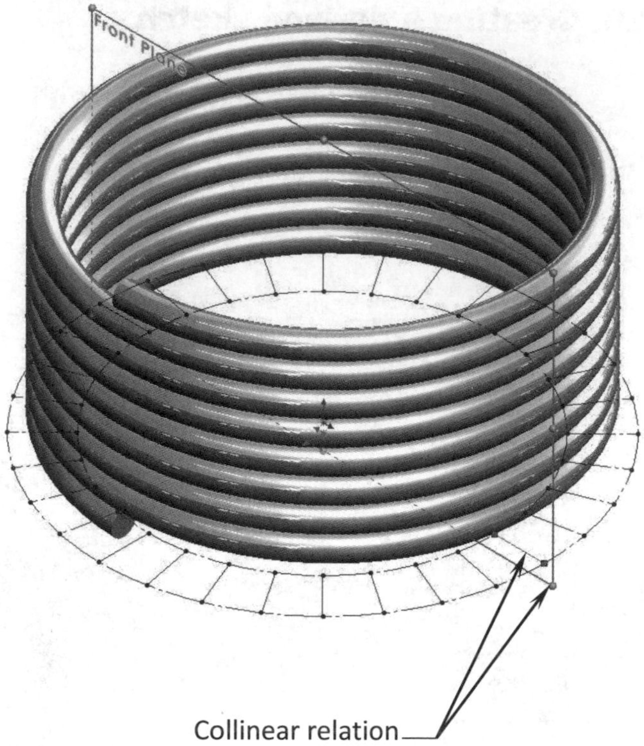

All entities in the sketch should change to the black color at this point (Fully Defined).

Exit the sketch.

11. Creating a 3D sketch:

The endpoints of the lines will help us to create the 3D sketch a little easier and also more accurately. Ensure both of the sketches are visible for this step.

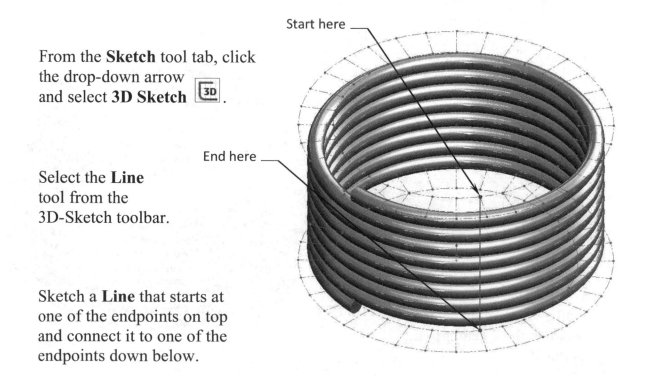

From the **Sketch** tool tab, click the drop-down arrow and select **3D Sketch** 3D.

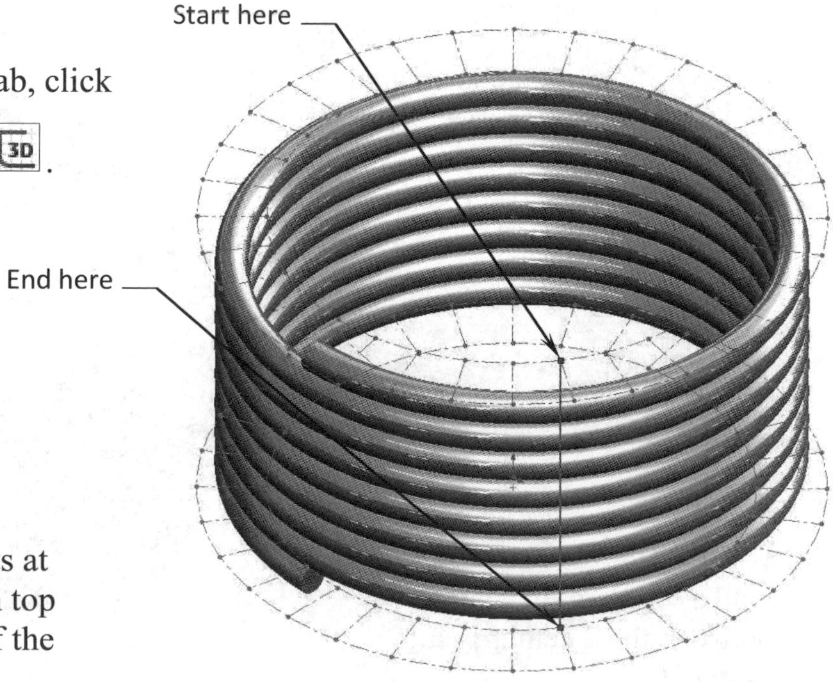

Select the **Line** tool from the 3D-Sketch toolbar.

Sketch a **Line** that starts at one of the endpoints on top and connect it to one of the endpoints down below.

The feature Swept1 is hidden for clarity.

(Use the Click + Release technique to create multiple lines, while the Click + Hold + Drag creates only 1 line each time.)

Add 3 more lines as shown. Be sure to snap to each existing endpoint when sketching the lines.

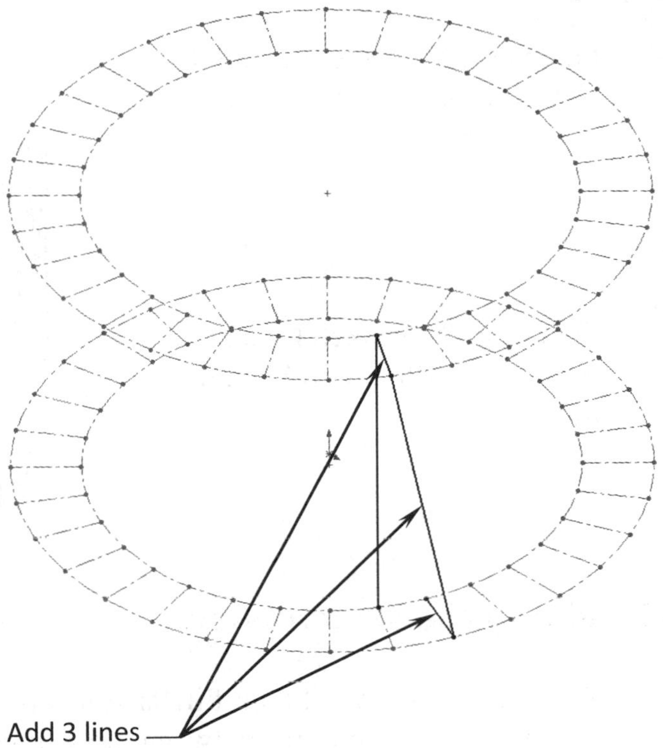

Continue to add the other lines as shown below. If you run into an error, simply press Escape and start sketching the lines again.

It is maybe a little easier if you could change the orientation of the view after adding a few lines.

While in the middle of the line mode you can press and hold the scroll-wheel to rotate slightly to a different angle, then continue with your sketch.

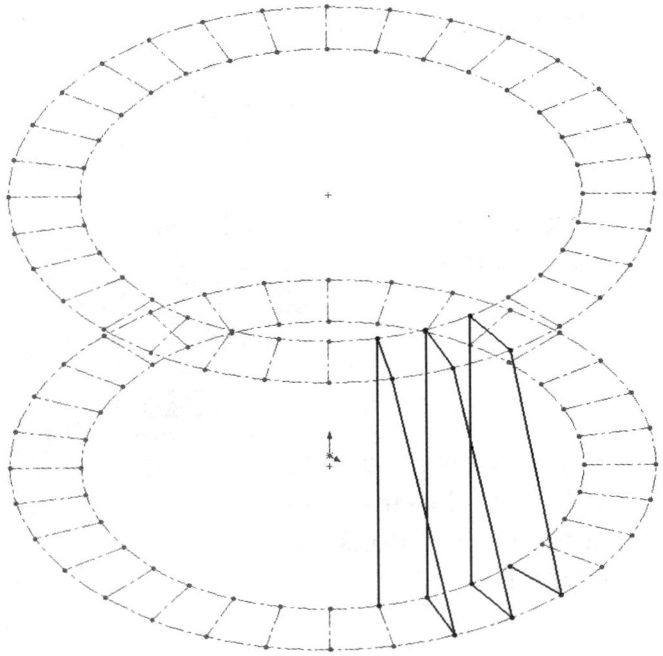

Once completed, your sketch should look like the image shown here.

Do not exit the sketch just yet; the sketch fillets will be added next.

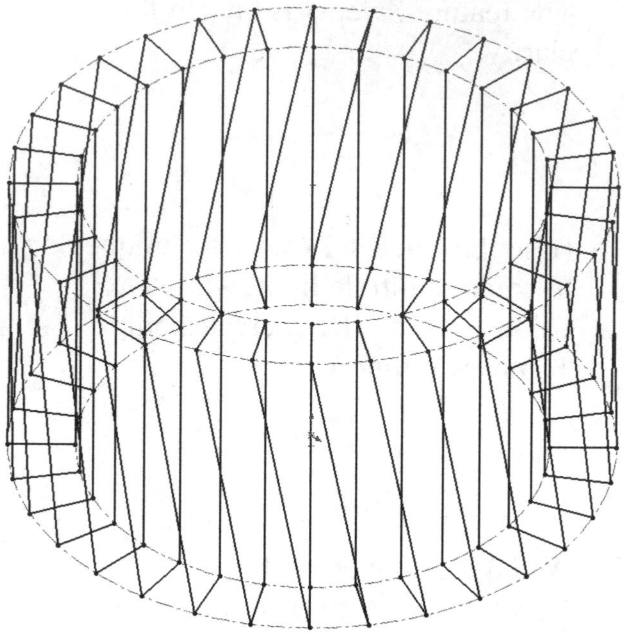

12. Adding the sketch fillets:

When creating the Sketch Fillets you can either select the 2 lines, or click directly at the intersection point between the 2 lines to add the fillets. If a fillet gets deleted the 2 lines will be extended and form a sharp corner once again.

Click the **Sketch Fillet** command from the Sketch Tools tool tab.

Enter **.120"** for radius value.

Begin adding the fillets by clicking at the intersections of the lines.

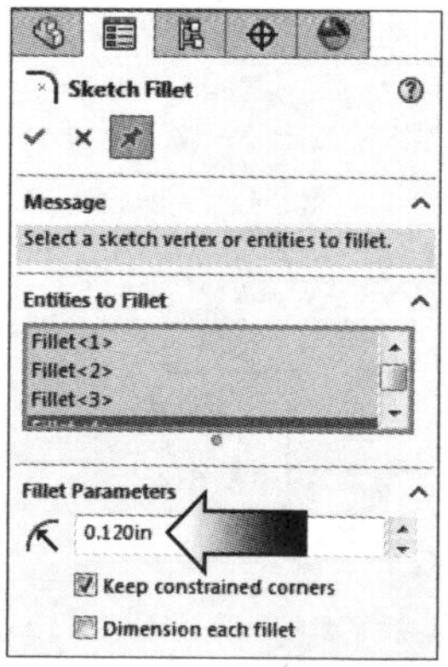

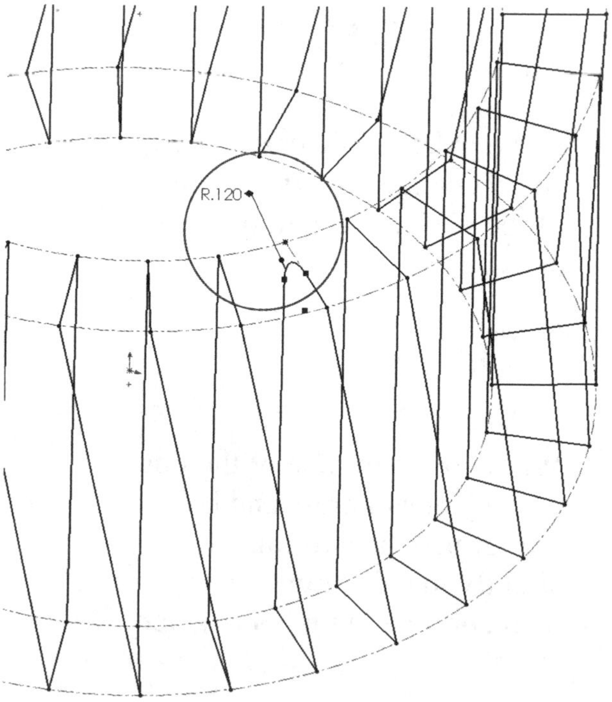

Continue with adding the same fillet to the intersection of the other lines.

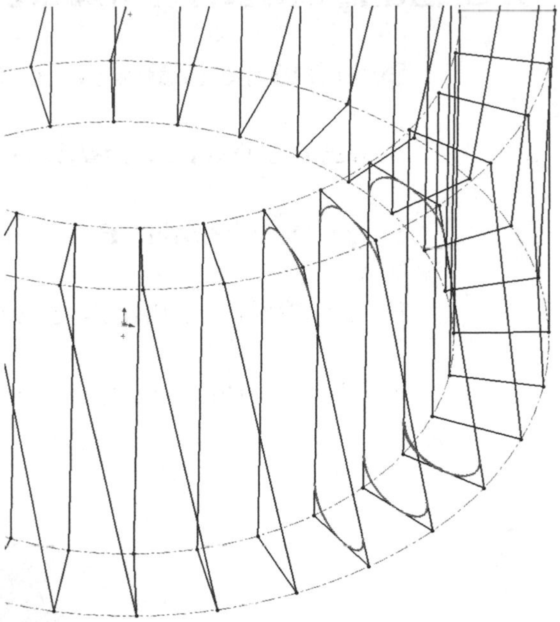

It does not really matter which direction you want to go when adding the fillets (clockwise or counterclockwise), but it will be a little easier if you go around 1 direction and not to jump back and forth as you may accidently skip one or more vertices.

After completed, your sketch should look like the image shown below. This 3D-sketch will be used as the sweep path in the next few steps.

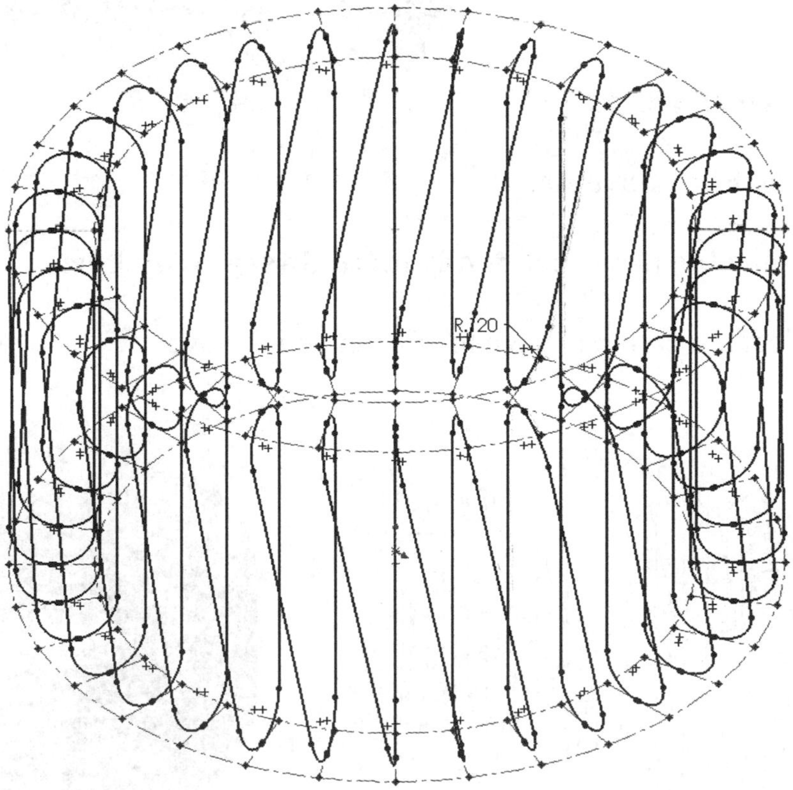

Exit the 3D sketch.

13. Creating the sweep feature:

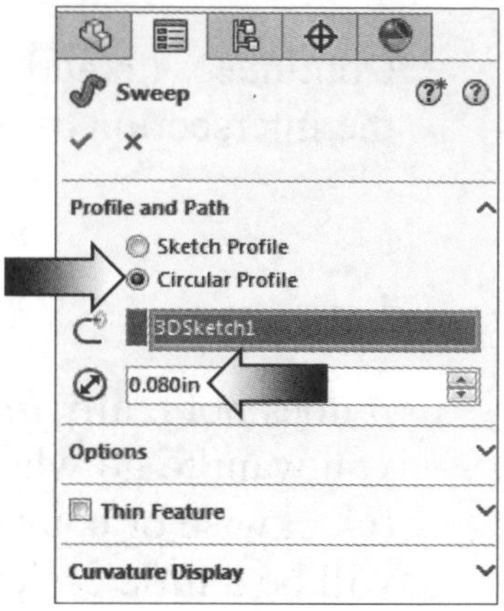

Switch to the **Features** tool tab.

Select the **Swept Boss/Base** command.

Select the **Circular Profile** option (arrow).

Enter **.080in** for Profile Diameter (arrow).

For Sweep Path, select the **3D Sketch**.

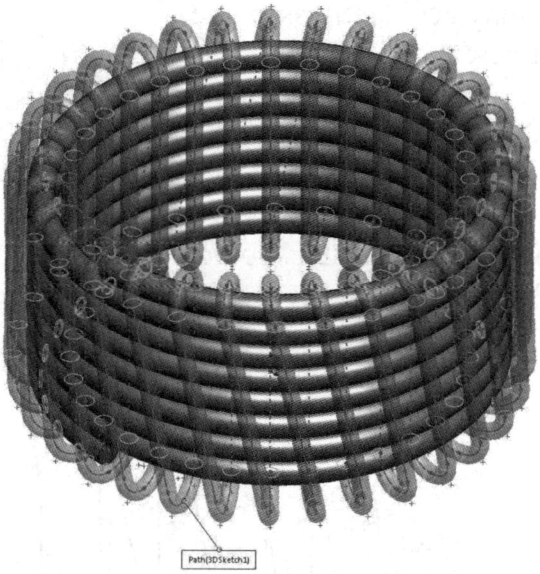

Click **OK**.

14. Saving your work:

Click **File / Save As**.

For the file name, enter: **Advanced Sweep_Wire Form**.

Select a location to save the file and click **Save**.

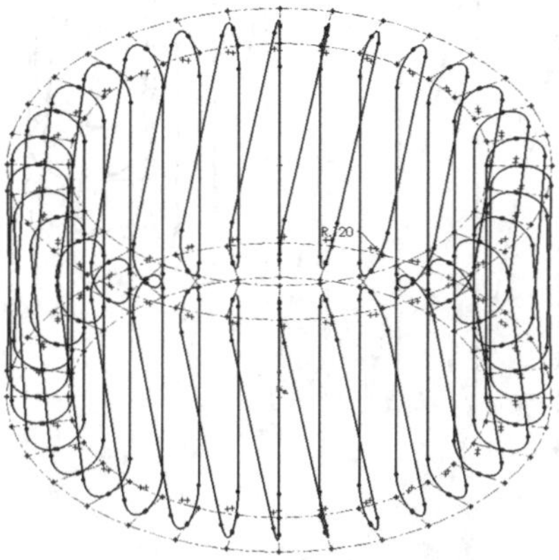

Exercise: Using Curve Through Reference Points

1. Creating the Base sketch:

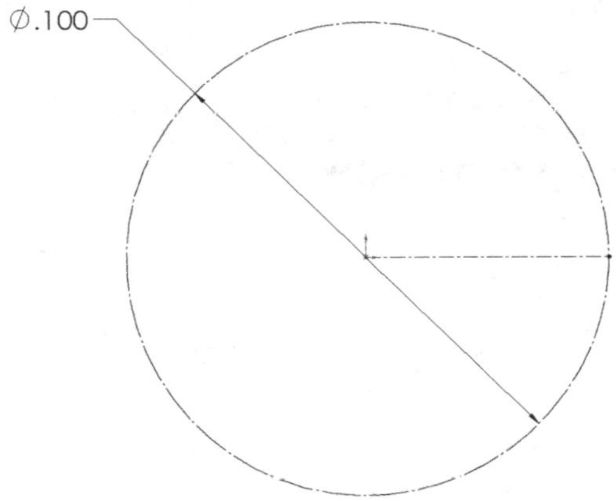

From the Top plane sketch a **Circle** and a horizontal **Centerline**, then convert both entities to Construction lines.

Add a diameter dimension of **.100"**.

Select the horizontal Centerline and click **Circular Sketch Pattern**.

Enter **16** instances, **Equal Spacing**.

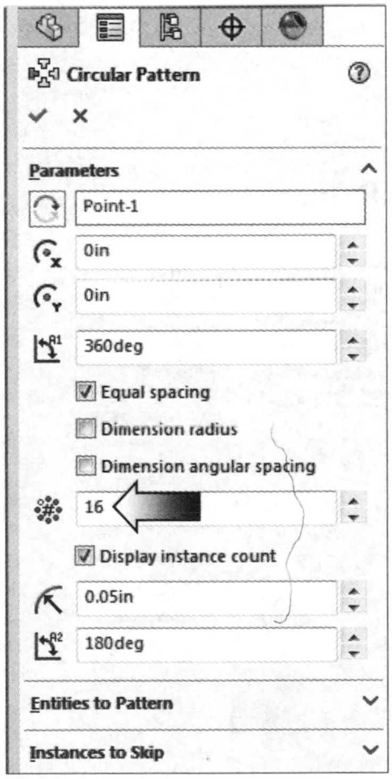

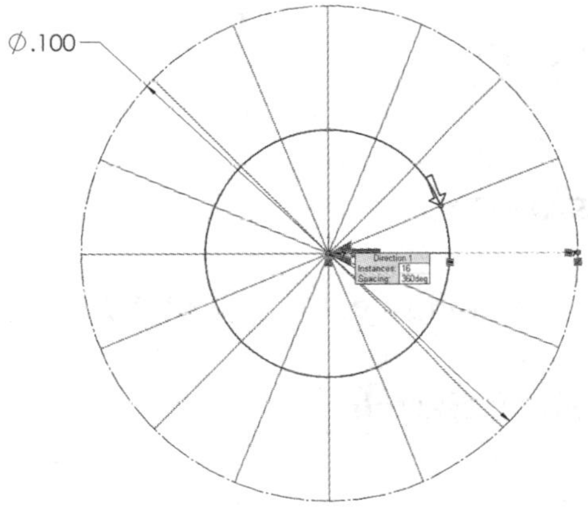

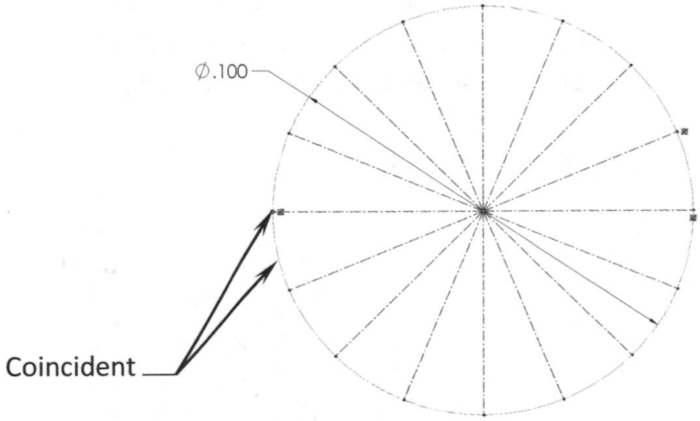

Add a couple of **Coincident** relations between the endpoints of any of the instances and the circle, to fully define the sketch.

Exit the sketch (or press **Control +Q**).

2. Creating an offset plane:

Create a new plane that is **.020"** above the **Top** plane.

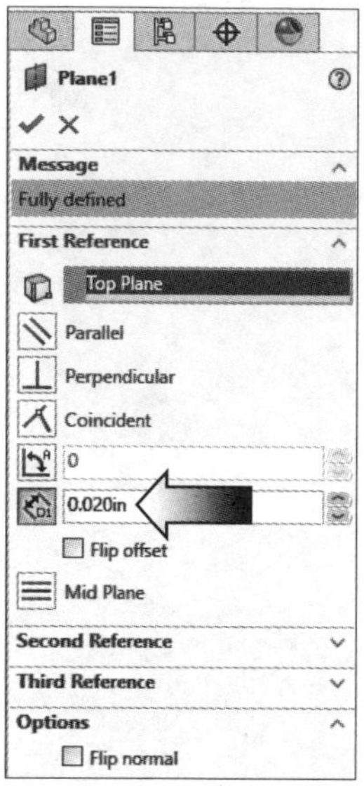

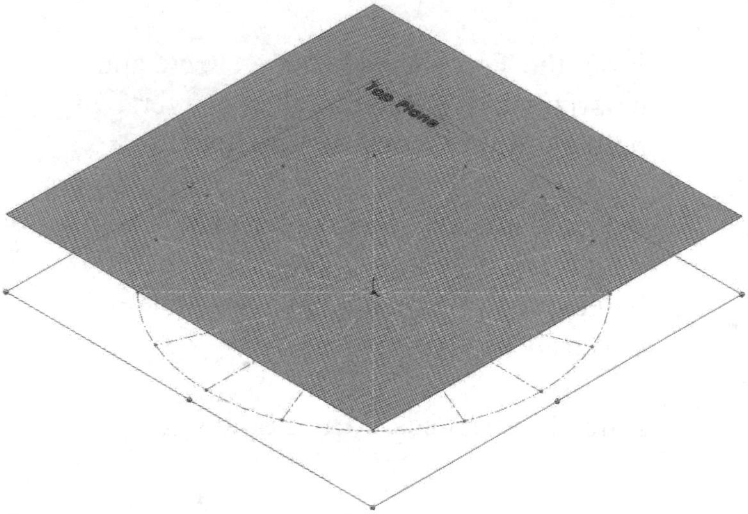

Click **OK**.

3. Creating a Derived sketch:

Hold the **Control** key and select the **Sketch1** and the **Plane1**.

Click **Insert / Derived Sketch**.

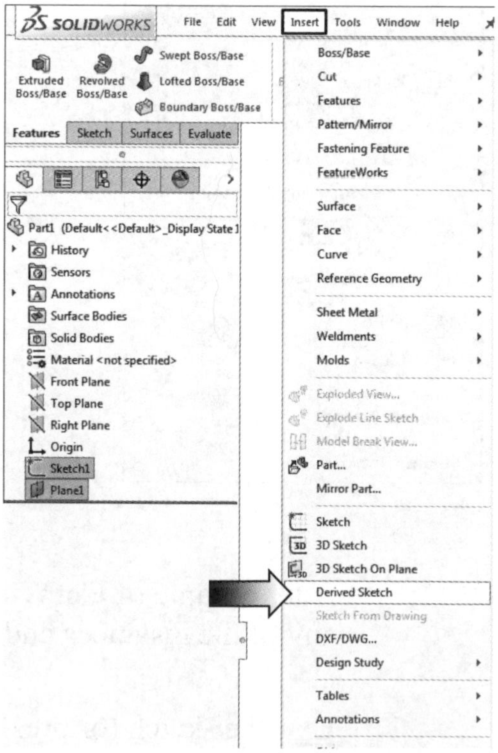

Add a coincident relation between the centers of the 2 circles and a vertical relation between any 2 endpoints.

Exit the sketch.

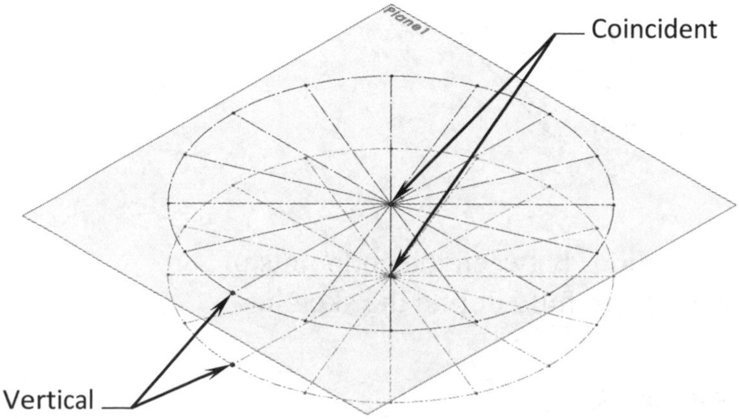

4. Creating a Curve Through Reference Points:

From the **Features** toolbar, click:
Curves / Curve Through Reference Points.

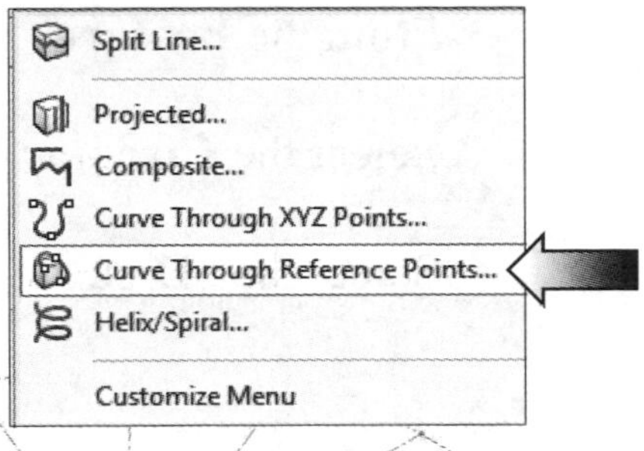

Click the starting point (point 1), move clockwise and click point 2, then point 3 so on.

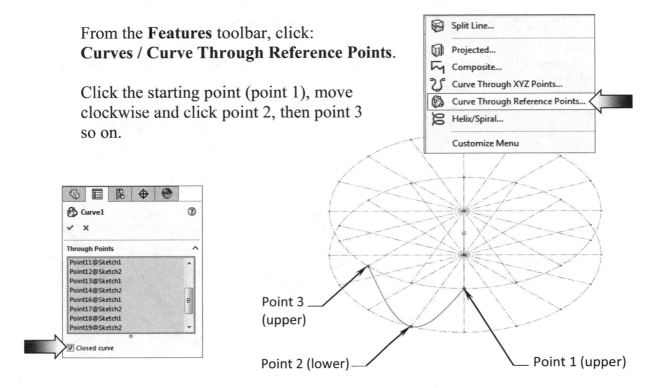

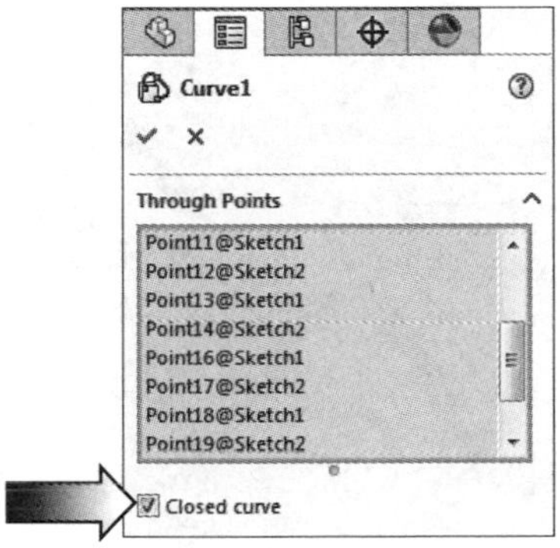

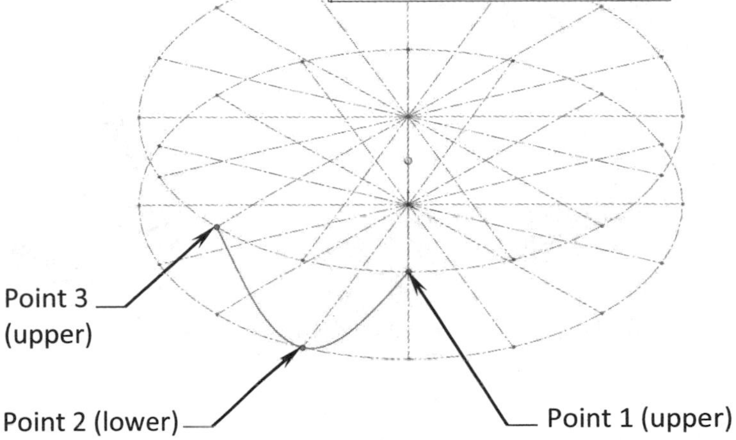

Continue going around a full revolution and select all the connecting points in the 2 sketches.

At the end of the path, do not click point 1 again as the **Closed Curve** option will join the 2 ends together automatically.

Enable the **Closed Curve** checkbox.

Click **OK**.

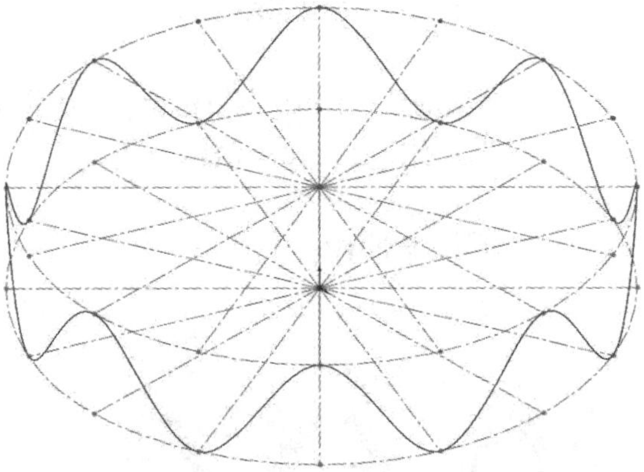

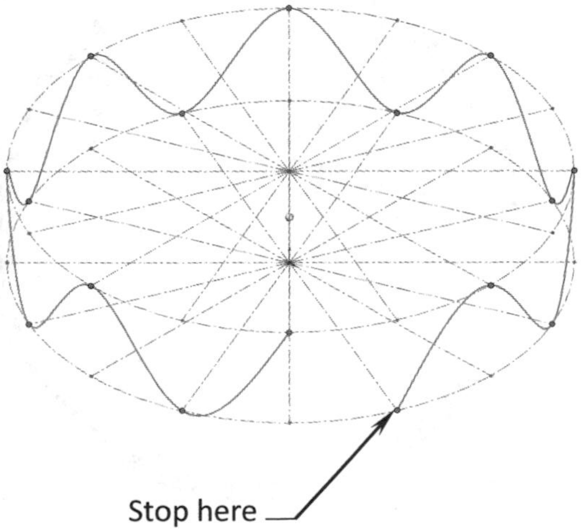

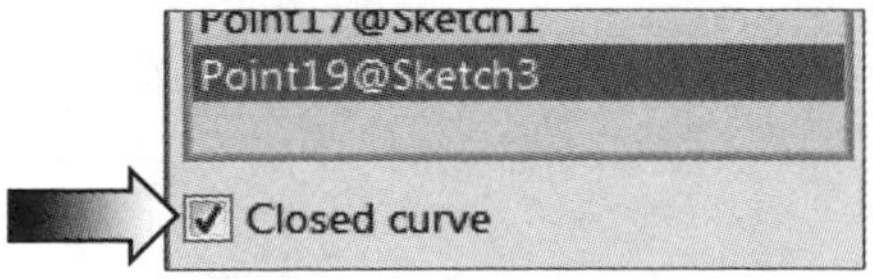

5. Creating a swept feature:

From the **Features** toolbar, click **Swept Boss Base**.

Select the **Circular Profile** option and enter **.003in** for diameter (arrows).

Select the **3D curve** as the Sweep path.

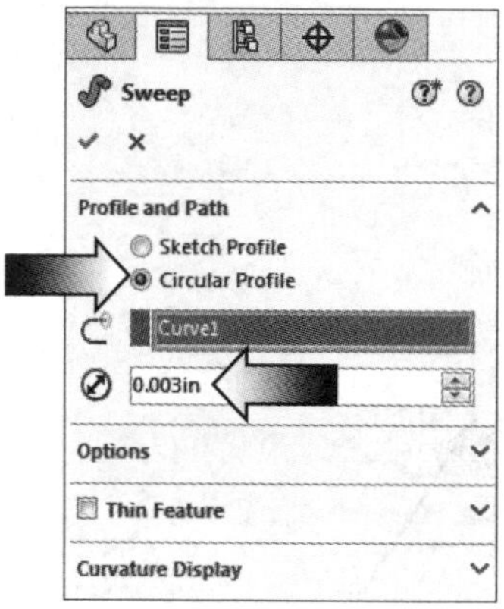

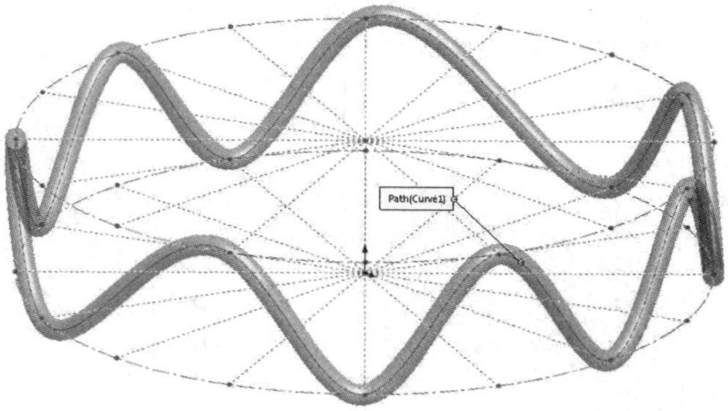

Click **OK**.

<u>Hide</u> the construction sketches and the 3D curve.

6. Saving your work:

Click **File / Save as**:

Enter: **Curve Through Reference Points** for the name of the file.

Click **Save**.

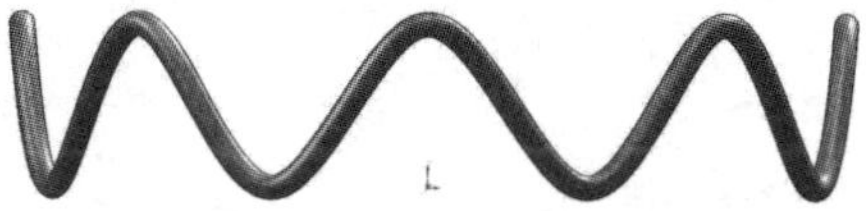

CHAPTER 8

Using Surfaces

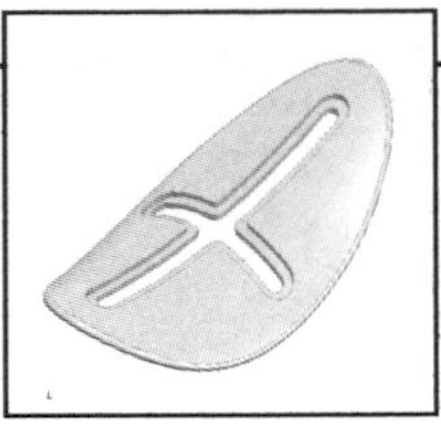

Advanced Modeling - Using Surfaces

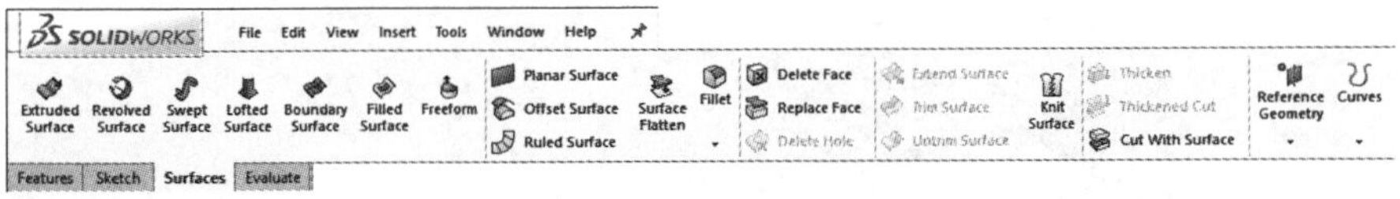

Surfaces are a type of geometry that can be used to create solid features.

The surface options are used to form complex free-form shapes and to manipulate files imported from other Cad formats.

Unlike solid models, surfaces can be opened, overlapped, and have no thickness. Each surface can be constructed individually and then knitted together. A solid feature is created by thickening the surfaces that have been knitted into a closed volume.

Surfaces can be modeled in any shape and their sketches can either be extruded, revolved, swept, or lofted into a surface. These surfaces can also be replaced and filled with other surfaces.

Edges of the surfaces can be extended and trimmed. Surfaces can be moved, rotated, and copied.

The angle between the faces of a surface can be calculated using the Draft-Analysis tool: Positive Drafts, Negative Drafts, and Required Drafts are reported on screen.

This 1st half of the chapter discusses the use of some surfacing tools in the newer releases of SOLIDWORKS.

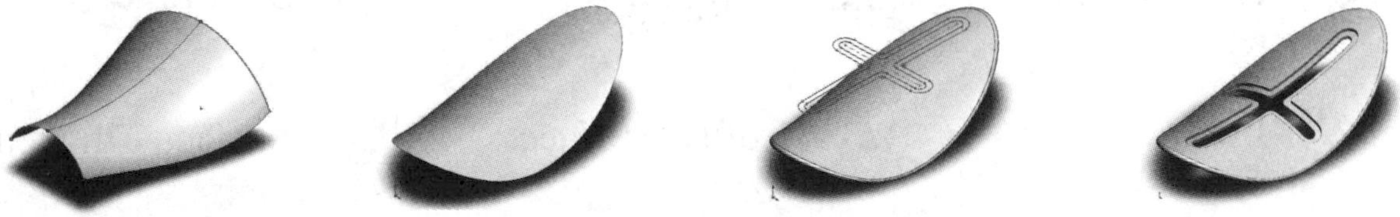

Advanced Modeling
Using Surfaces

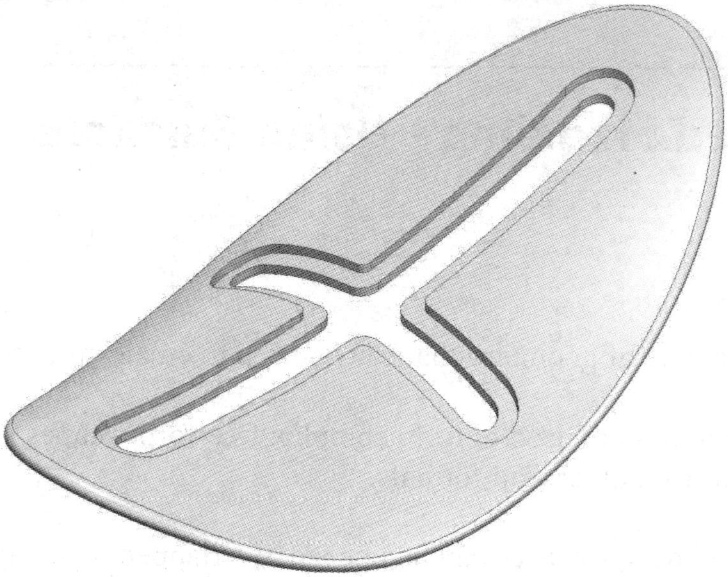

View Orientation Hot Keys:

Ctrl + 1 = Front View
Ctrl + 2 = Back View
Ctrl + 3 = Left View
Ctrl + 4 = Right View
Ctrl + 5 = Top View
Ctrl + 6 = Bottom View
Ctrl + 7 = Isometric View
Ctrl + 8 = Normal To Selection

Dimensioning Standards: **ANSI**

Units: **INCHES** – 3 Decimals

Tools Needed:

Insert Sketch	3 Point Arc	Ellipse
Dimension	Add Geometric Relations	Split Line
Plane	Lofted Surface	Surface Thicken

1. Constructing a new work plane:

Select the Front plane from the FeatureManager tree.

Click or select **Insert / Reference Geometry / Plane**.

Click the **Offset Distance** button and enter **3.000 in**.

Enable the **Flip Offset** check box to reverse the direction.

Click **OK**.

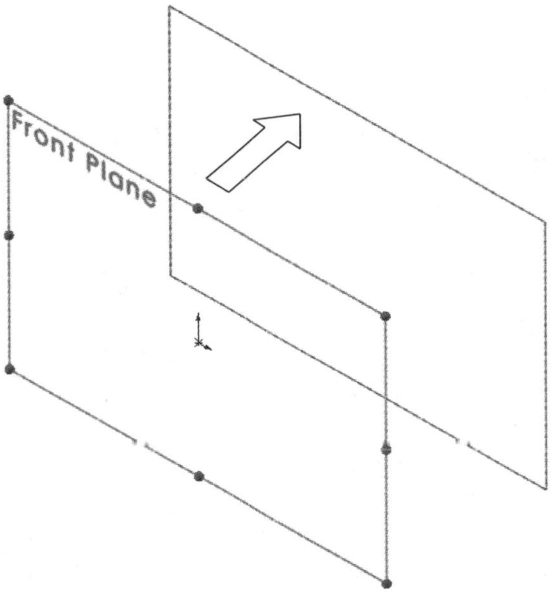

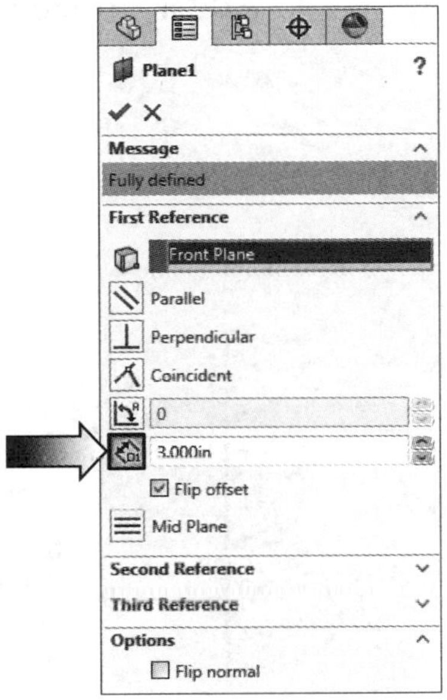

2. Sketching the 1st profile:

Select the new plane (Plane1).

Click or select **Insert / Sketch**.

Sketch a **3-Point-Arc** and add the dimensions shown.

Add a **Vertical** and **Horizontal** relation as noted.

NOTE: The center of the Arc is .625" below the origin.

Exit the Sketch .

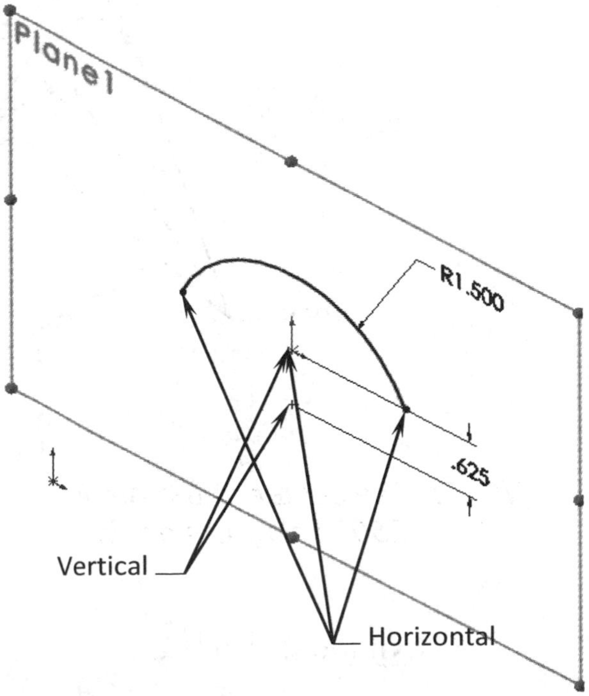

3. Sketching the 2nd profile:

Select the <u>Front</u> plane from the FeatureManager tree.

Click or select **Insert / Sketch**.

Sketch a **3-Point-Arc** as shown.

Add a **.750** radius dimension to the arc.

Add a Horizontal and a Vertical relation as indicated.

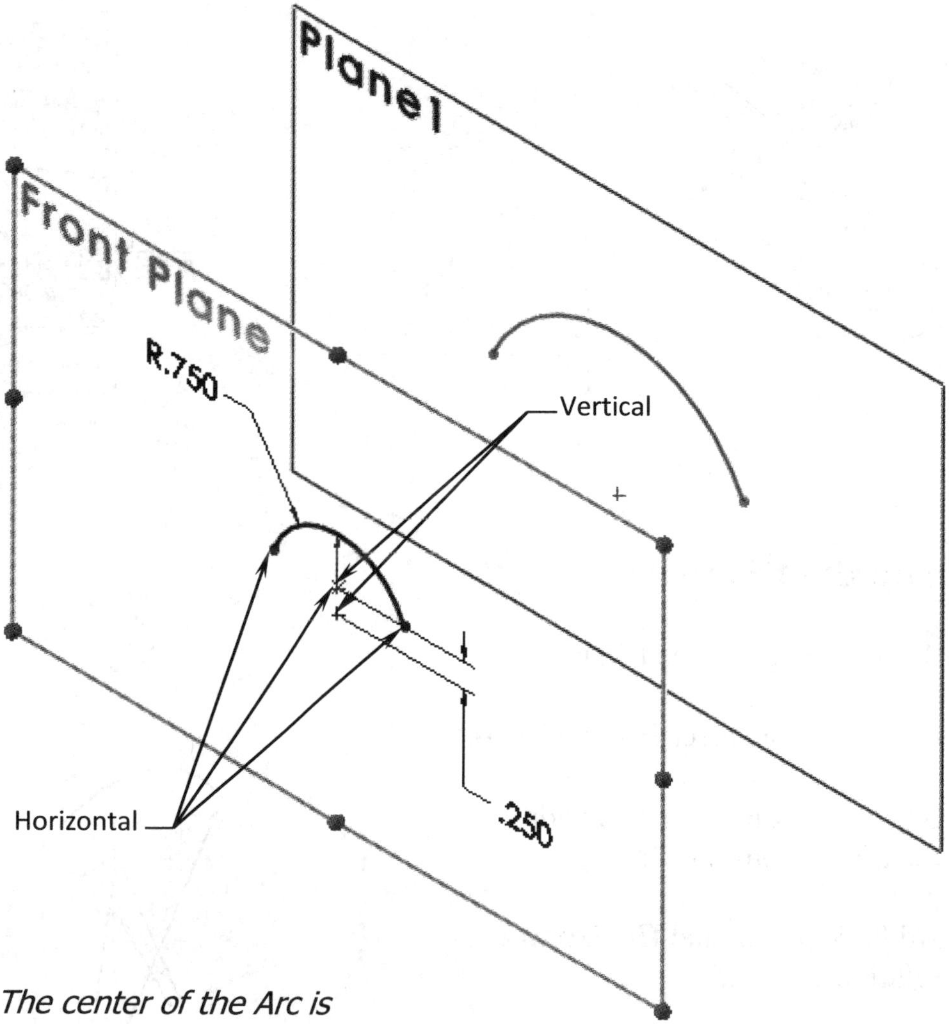

<u>NOTE:</u> The center of the Arc is .250" below the origin.

<u>**Exit**</u> the sketch .

4. Sketching the Guide Curve:

Select the Right plane from the FeatureManager tree and open a **new sketch**.

Sketch a **3-Point-Arc** and add a **4.00**in. dimension.

Add the **Pierce** relations as noted.

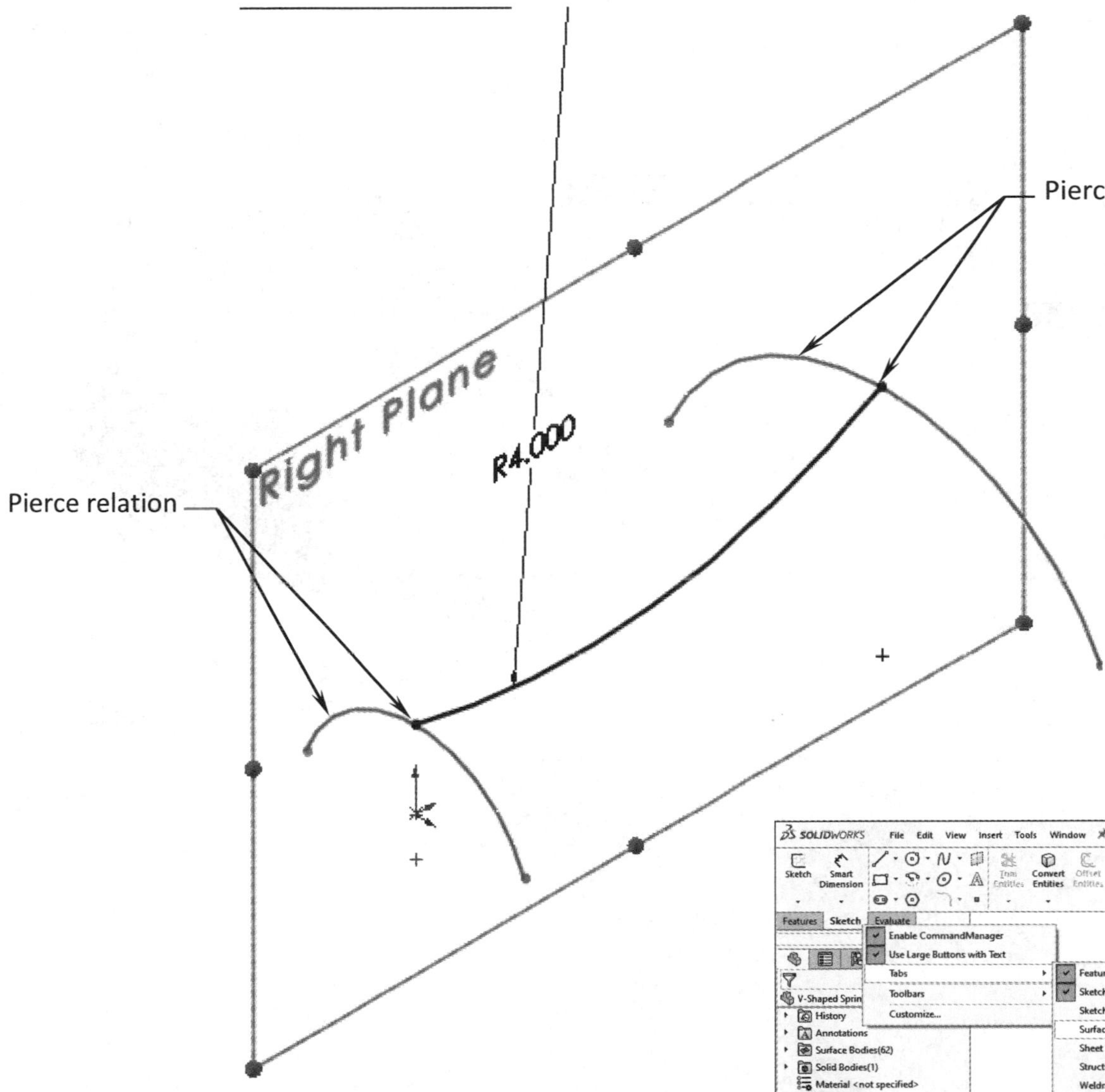

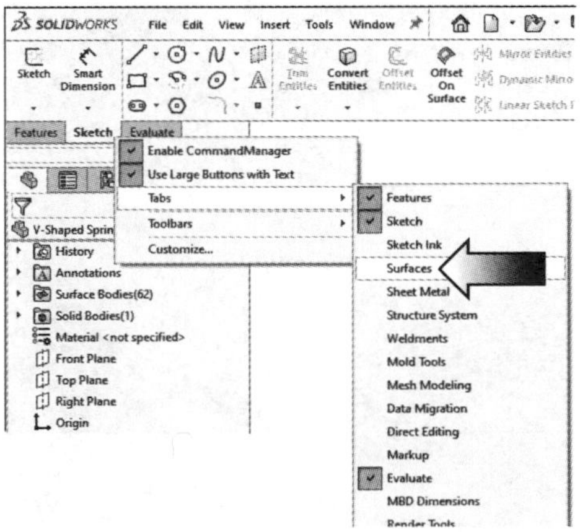

Exit the sketch.

Enabling the Surfaces toolbar: *Right-click on one of the tabs (Features, Sketch, etc.) and enable the Surfaces option. Switch to the Surfaces tab.*

5. Creating a Surface-Loft:

Click  on the **Surfaces** tool tab or select: **Insert / Surface / Loft.**

Select the **2 Sketched Profiles** by clicking on their ***right-most endpoints***.

Expand the Guide-Curve section and select the **Sketched Arc** as noted.

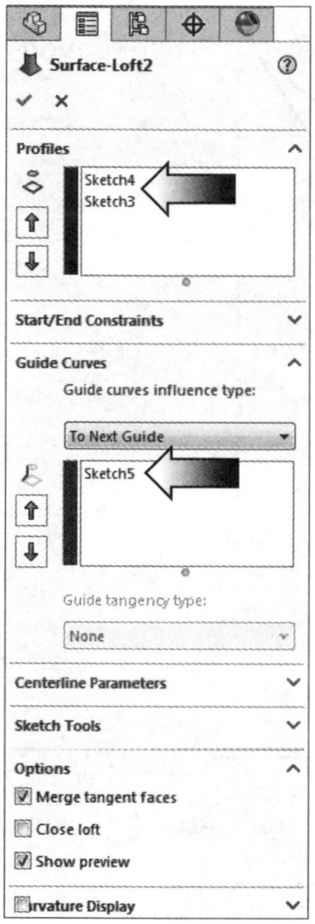

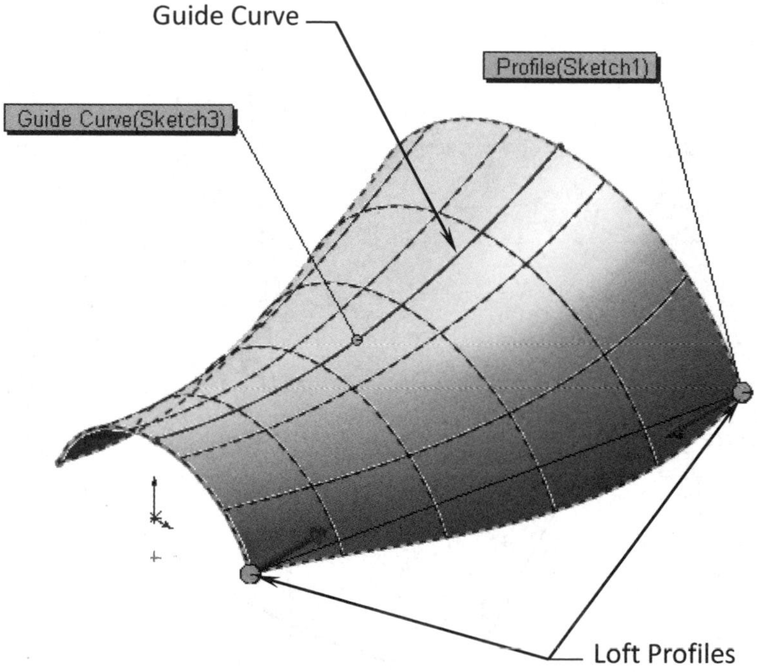

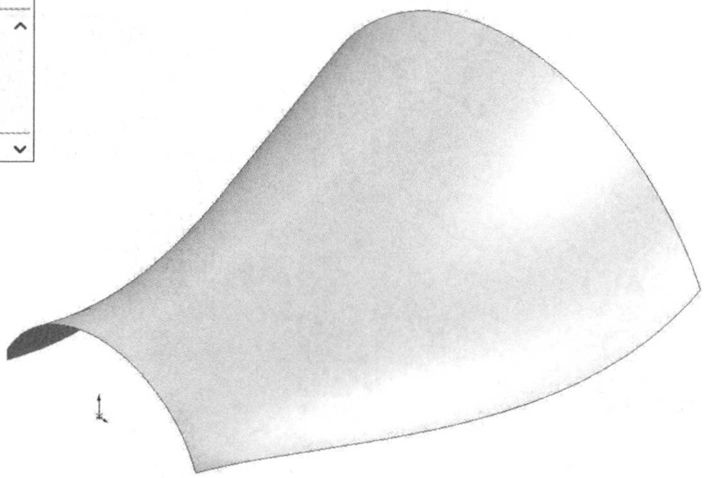

Click **OK**.

6. Sketching the Split profile:

Select the Top plane from the FeatureManager tree.

Click  or **Insert / Sketch**.

Change to the Top view orientation (Control+5).

Sketch an **Ellipse**  and add Dimensions/Relations as shown below.

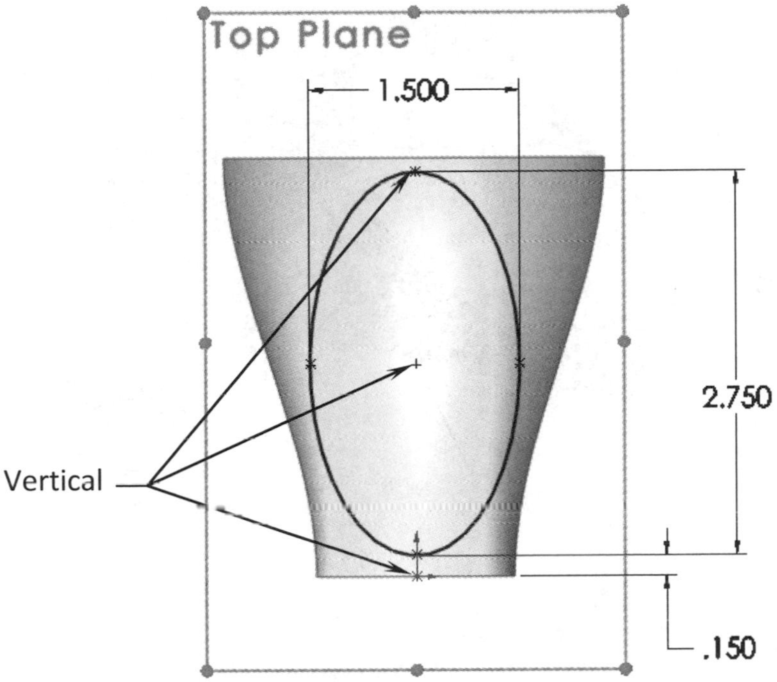

> **Split Lines**
>
> The Split Lines command projects a sketch entity onto a face (or a group of faces) and divides the selected face(s) into multiple separate faces, enabling you to select and work with each face individually.

7. Splitting the surface:

Click  on the Curves toolbar OR select **Insert / Curve / Split Line.**

For Sketch-to-Project, select the **Ellipse**.

For Faces-to-Split, select the **Surface-Loft1**.

Click **OK**.

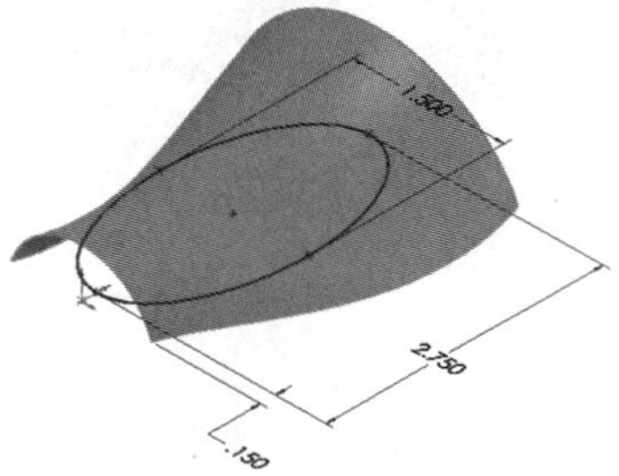

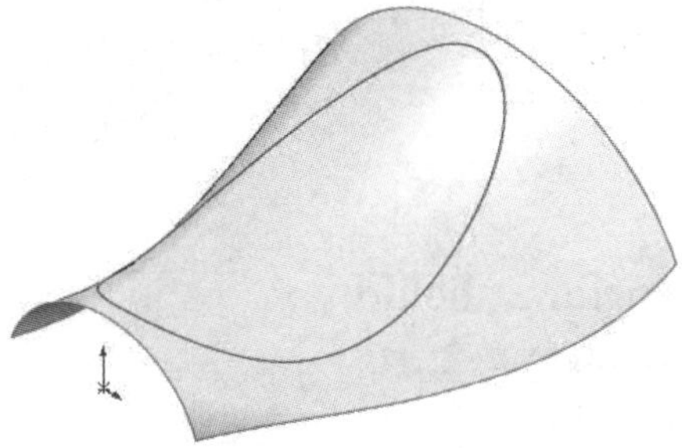

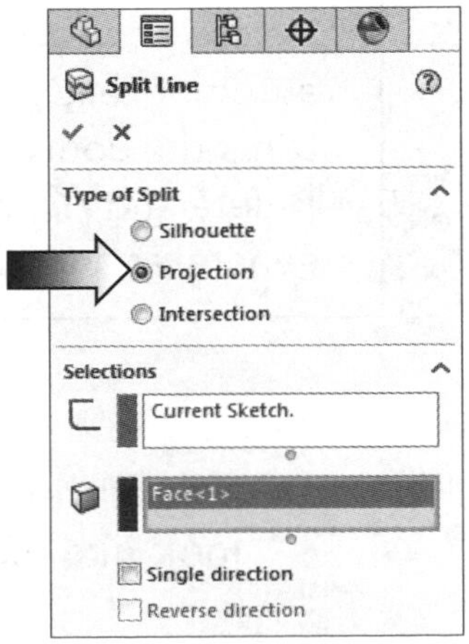

8. Deleting Surfaces:

Right-click on the **outer portion** of the surface.

Select **Face** / **Delete** from the menu.

Click **Delete** under Options (circled).

Click **OK**.

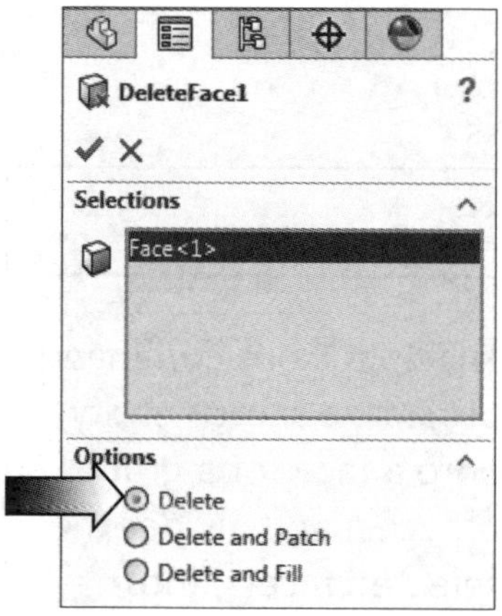

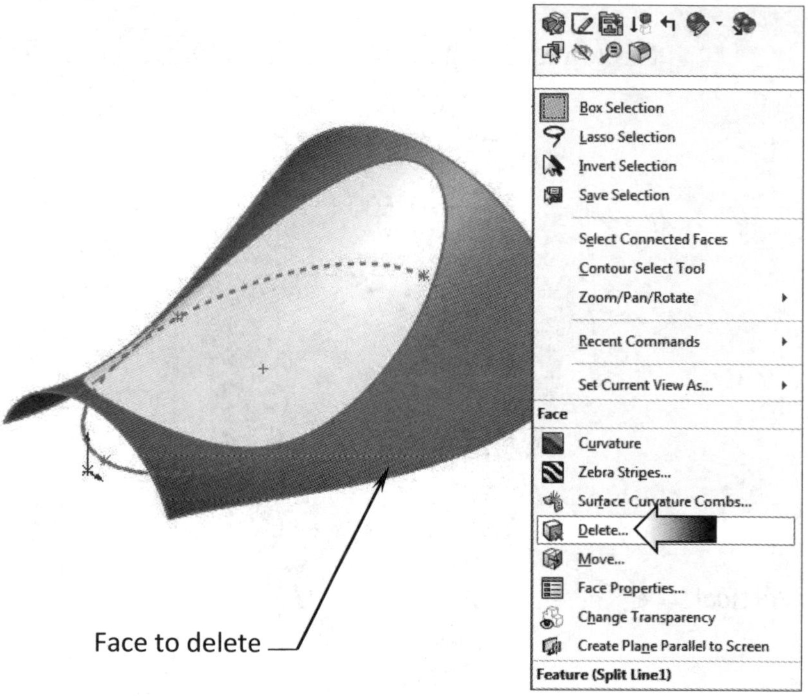

Delete Face

This command deletes one or more faces from a surface or solid body.

Other options are:
* Delete and Patch, which automatically patches and trims the body.
* Delete and Fill, which generates and fills any gap.

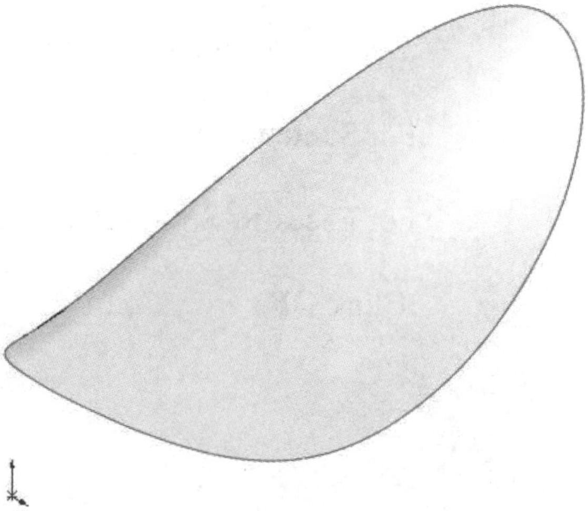

Your trimmed surface should look like the one shown here.

9. Thickening the surface:

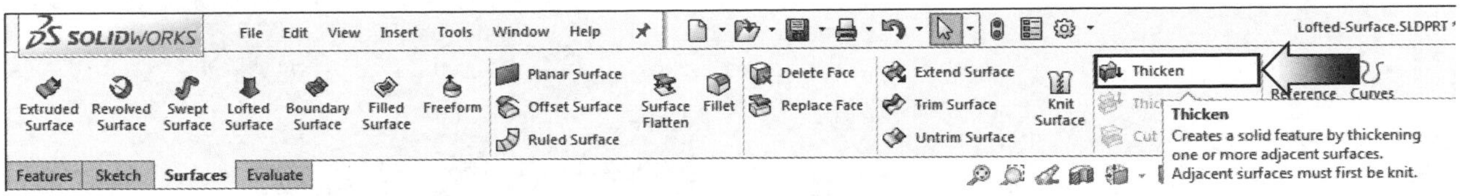

Click  or select **Insert / Boss-Base / Thicken**.

For Surface-To-Thicken, click on the **Trimmed Surface**.

Select the **Thicken Both Sides** option .

For Thickness, enter **.030** in. (.060 total thickness).

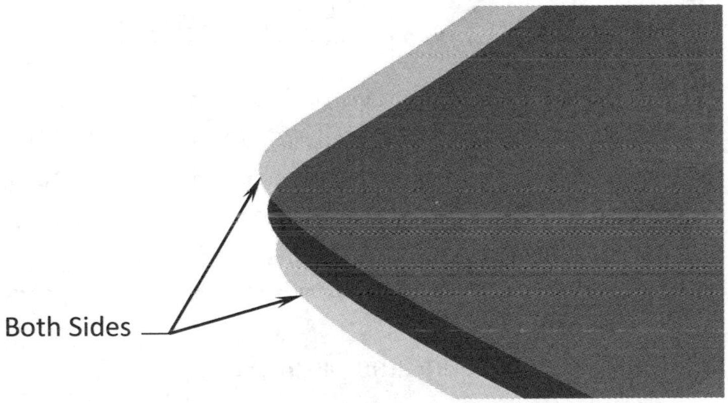

Thicken Surfaces

In order to create a solid volume, all surfaces have to form a closed shape.

If the shape is open, a wall thickness can be added to the surface to close.

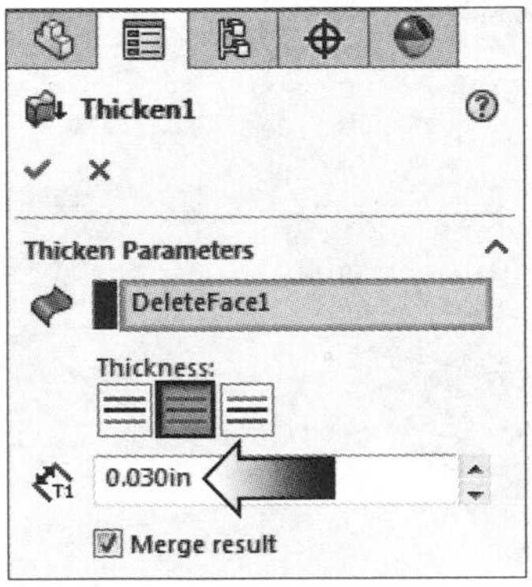

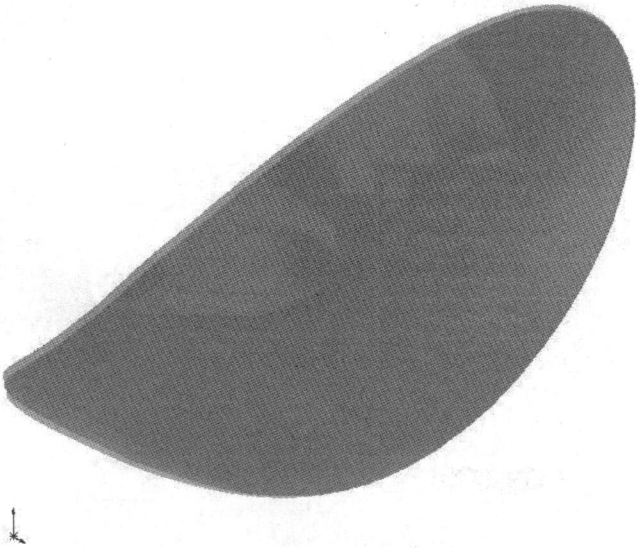

Click **OK**.

The surface model is thickened into a solid model.

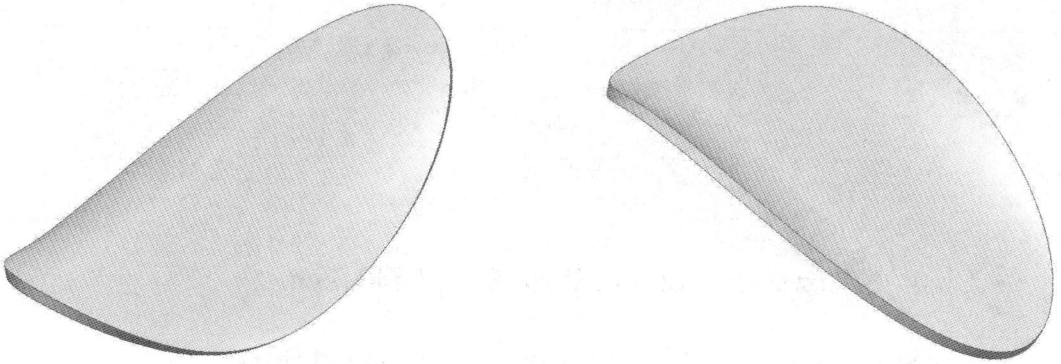

10. Calculating the angles between the faces:

Change to the **Evaluate** tool tab and click the **Draft Analysis** command.

For Direction of Pull, select the **Top** plane.

Enter **1.00** deg. for Draft Angle.

Enable the **Face Classification** checkbox and click **Calculate**.

The **light blue color** indicates the surfaces that have both positive and negative drafts on them. This can be eliminated by creating a split line in the middle, or by adding a full round fillet around the parameter.

> **Draft Analysis**
>
> Using the settings in draft analysis, you can verify the draft angles on model faces or you can examine angle changes within a face.

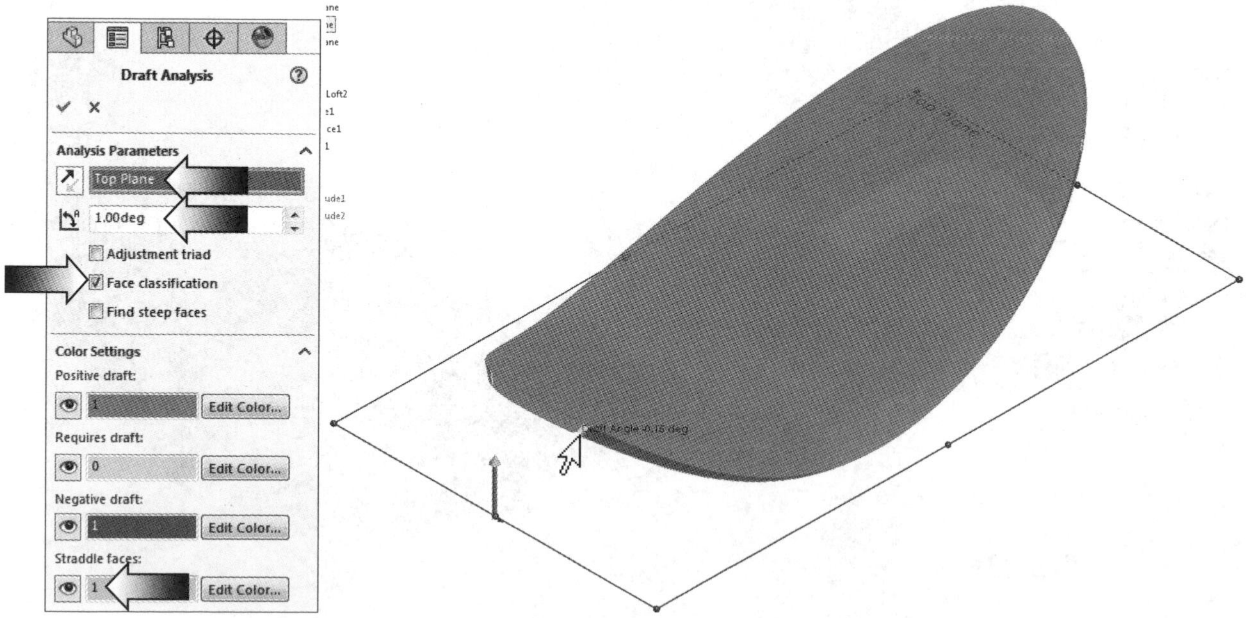

Draft Analysis

The **Draft Analysis** is a tool to check the correct application of draft to the faces of each part. With draft analysis, you can verify draft angles, examine angle changes within a face, as well as locate parting lines, injection, and ejection surfaces in parts.

Draft analysis results listed under Color Settings are grouped into four categories, when you specify Face classification:

Positive draft: Displays any faces with a positive draft based on the reference draft angle you specified. A positive draft means the angle of the face, with respect to the direction of pull, is more than the reference angle.

Negative draft: Displays any faces with a negative draft based on the reference draft angle you specified. A negative draft means the angle of the face, with respect to the direction of the pull, is less than the negative reference angle.

Draft required: Displays any faces that require correction. These are faces with an angle greater than the negative reference angle and less than the positive reference angle.

Straddle faces: Displays any faces that contain both positive and negative types of draft. Typically, these are faces that require you to create a split line.

*Note: When analyzing the draft for surfaces, an additional **Face classification** criterion is added: **Surface faces with draft**. Since a surface includes an inside and an outside face, surface faces are not added to the numerical part of the classification (**Positive draft** and **Negative draft**). **Surface faces with draft** lists all positive and negative surfaces that include draft.*

* Hover the pointer over the upper surface to see the read out draft angle for that particular area.

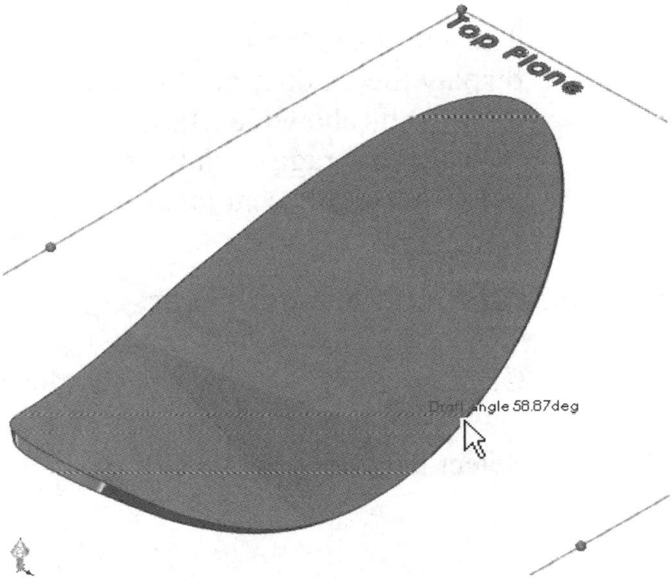

* Position the mouse cursor over the yellow areas (required drafts) and check the draft angles to see if they meet your draft requirements.

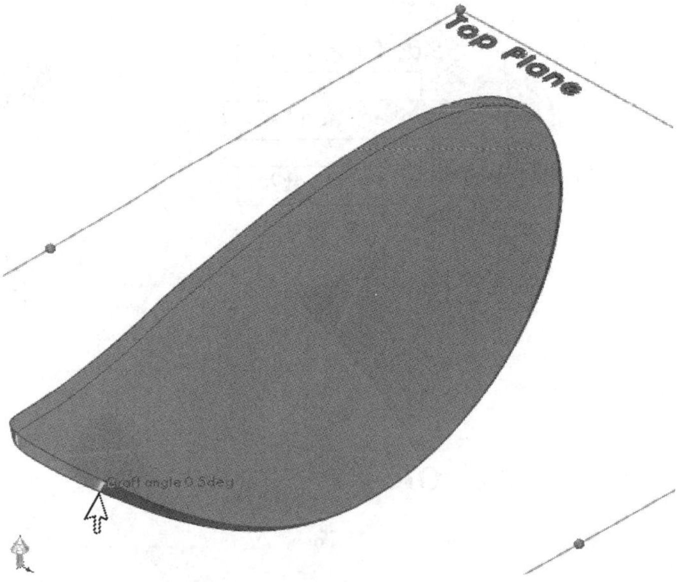

* Click **OK**.

The option Deviation-Analysis can be used to diagnose and calculate the angle between faces.

The colored arrows display the amount of deviation. The results show the Max, Min and Average deviations between the adjacent faces.

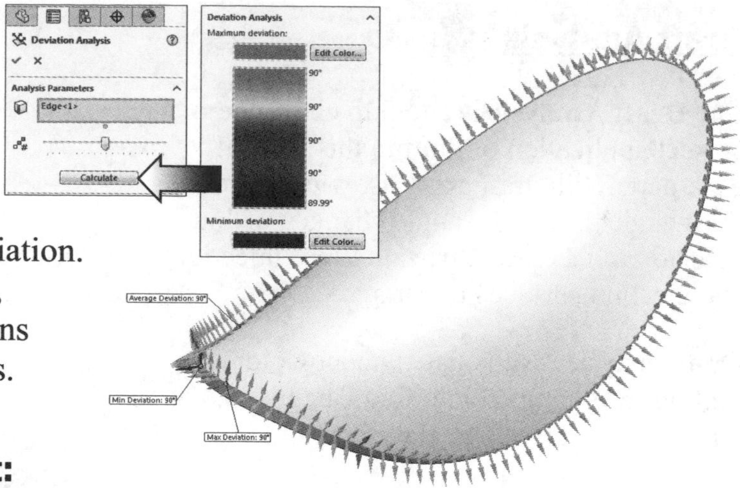

11. Adding a Full Round Fillet:

Click or select **Insert / Features / Fillet-Round**.

Select the Side-Face-Set1, Center-Face-Set, and Side-Face-Set2 as noted.

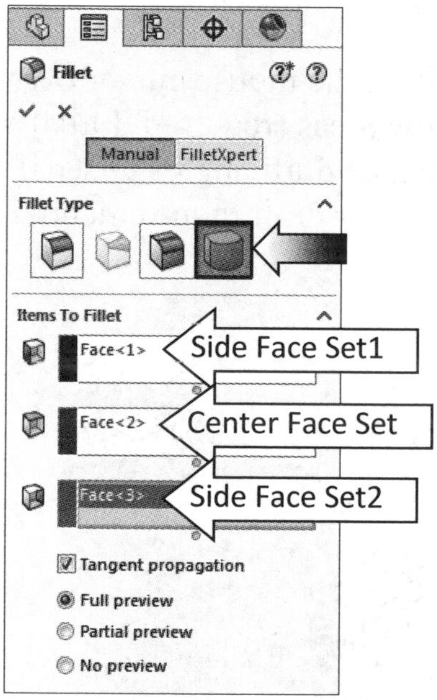

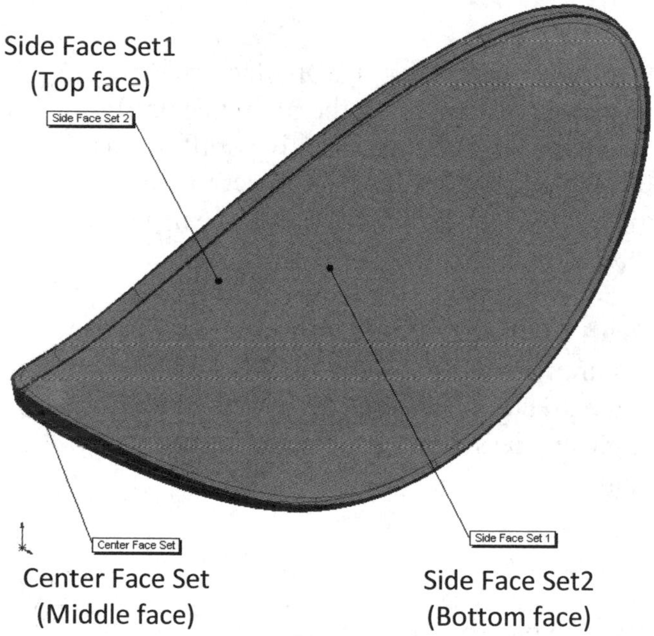

Click **OK**.

Rotate the model to verify the full round fillet from different orientations.

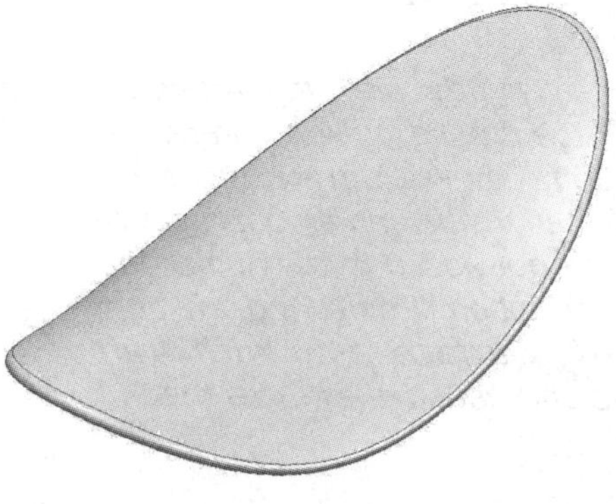

12. Creating an Offset Distance plane:

Select the **Top** plane from the FeatureManager tree and click , or select **Insert / Reference Geometry / Plane**.

Enter **.850 in**. for Distance and place the new plane above the Top plane.

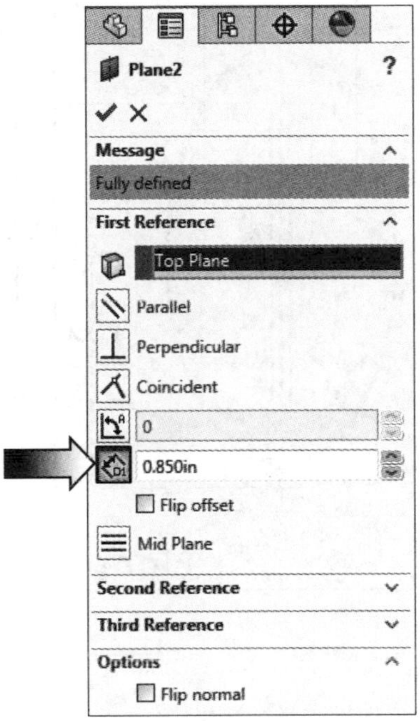

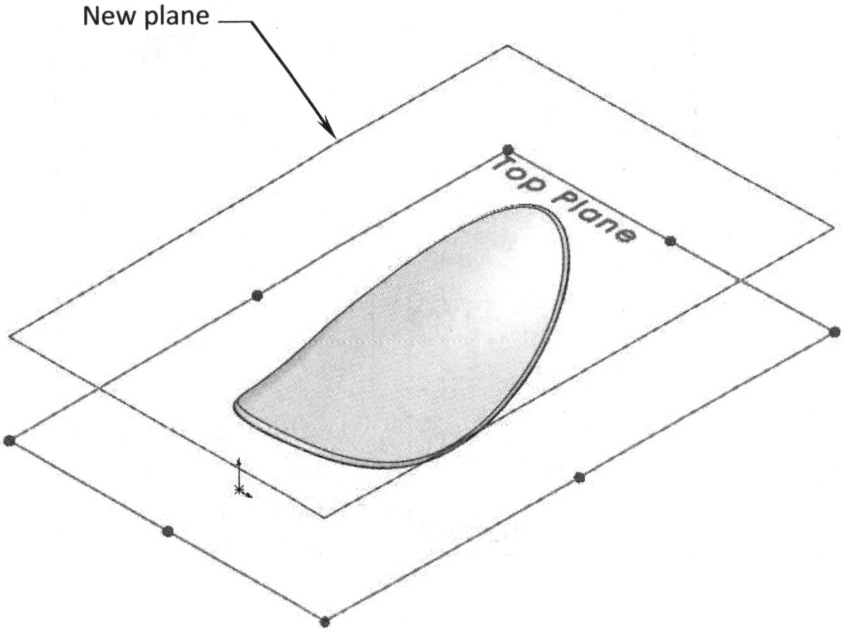

Click **OK**.

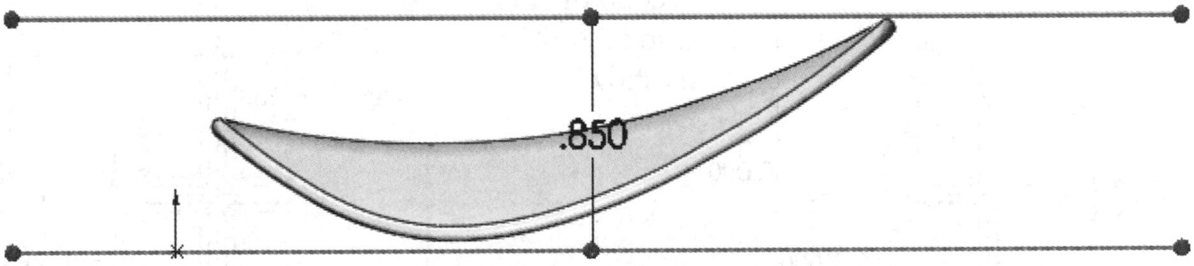

13. Sketching the Slot Contours:

Select the new plane **(Plane2)** and open a **new sketch** 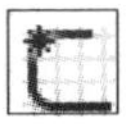.

Create the sketch using either **Mirror** or **Offset** options.

Sketch the profile shown below and add dimensions to fully position the sketch.

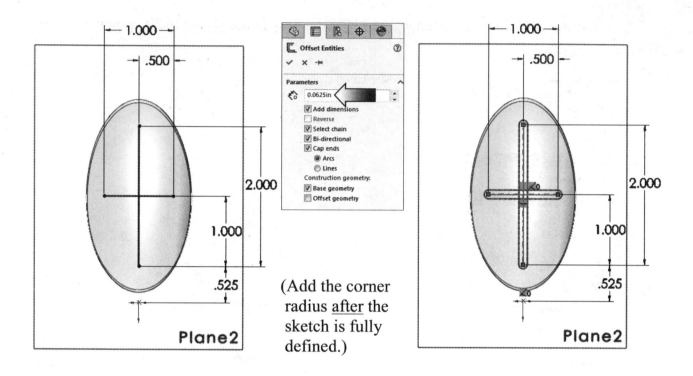

Create an offset of **.0625 in.** from the 2 sketch lines, using the settings below.

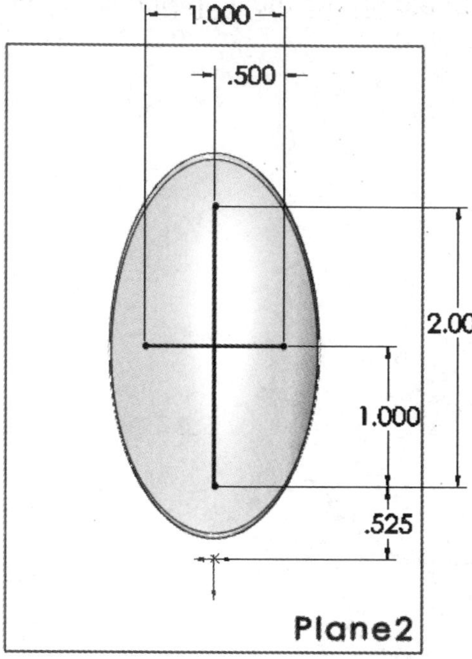

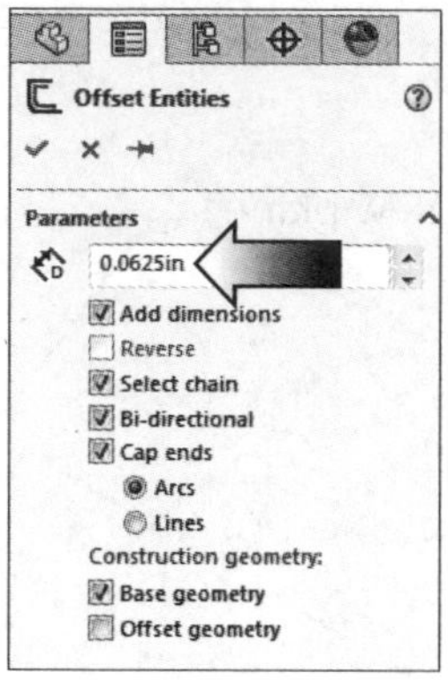

(Add the corner radius after the sketch is fully defined.)

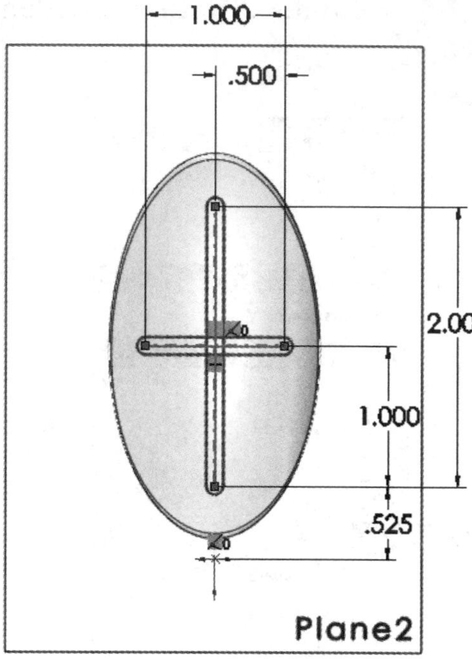

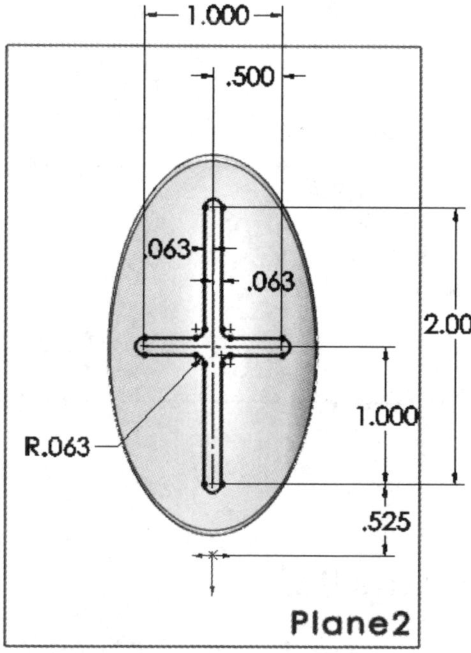

Trim the inner intersections and create a second offset also at .0625in as shown.

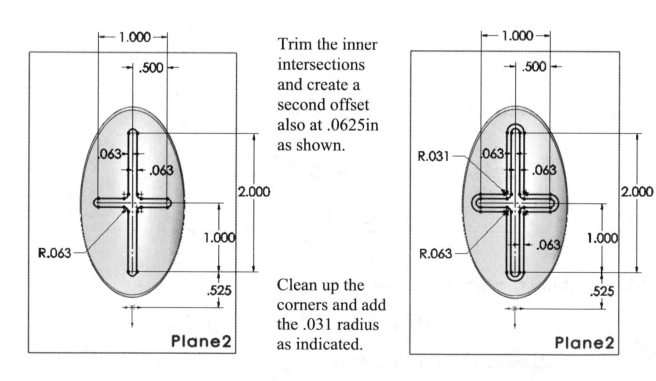

Clean up the corners and add the .031 radius as indicated.

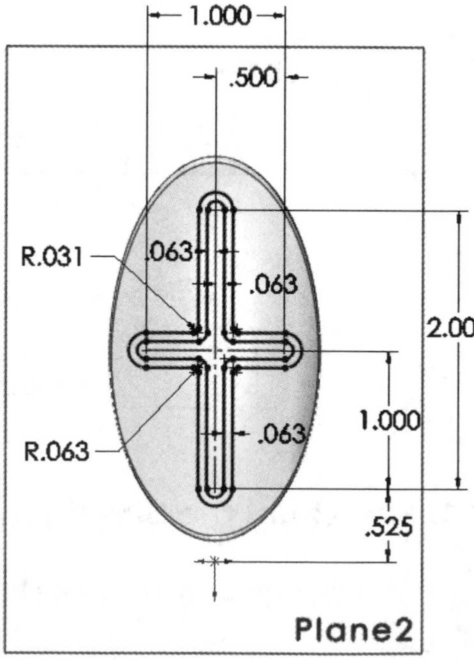

14. Extruding Cut the 1st Contour:

Click  or select **Insert / Cut / Extrude**.

Expand the Selected Contour section and select one of the **Outer Lines** (Outer Contour).

Use **Offset From Surface** end condition.

Enter **.030 in**. for Depth.

Click the **Top face** of the model.

Enable **Reverse Offset** check box.

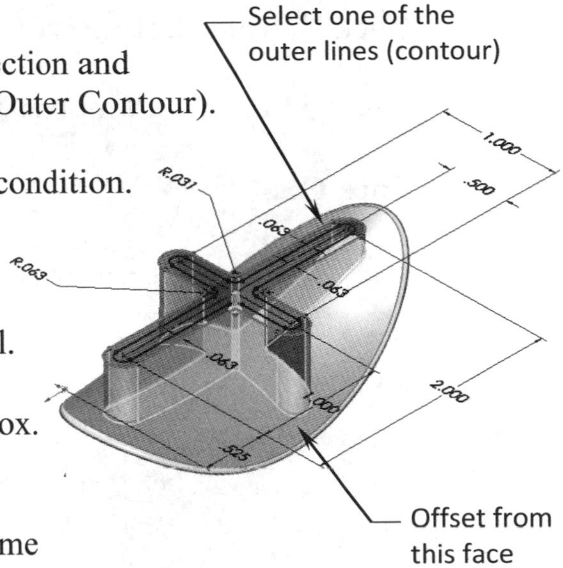

The slot is cut, following the same contours of the upper surface.

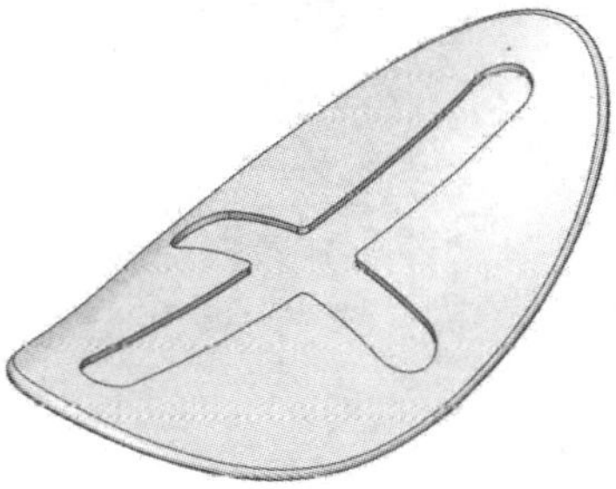

Click **OK**.

15. Extruding Cut the 2nd Contour:

Expand the Cut-Extrude1 from the FeatureManager tree, right-click **Sketch5** and select **Show** .

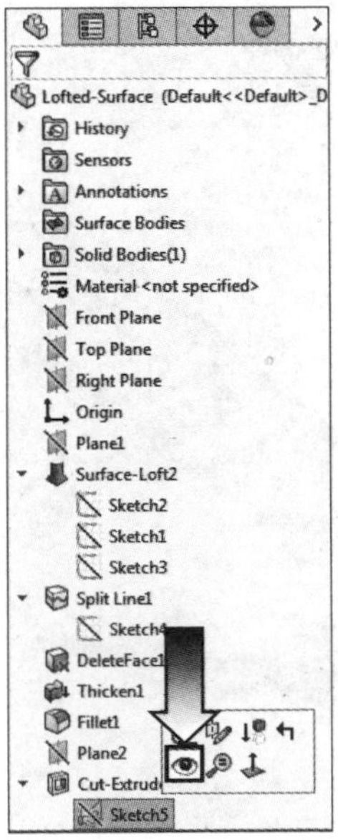

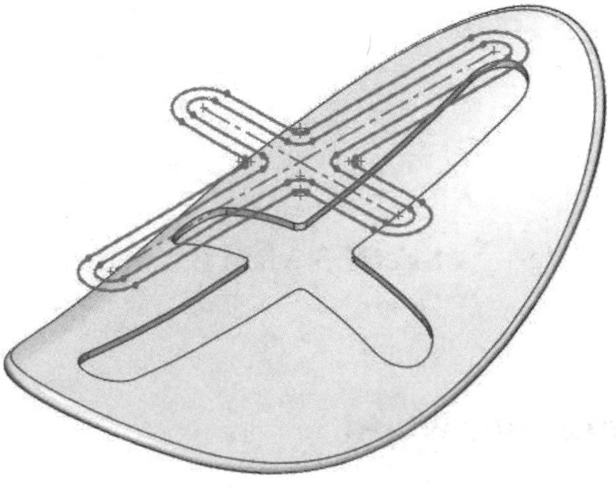

Hover over one of the inner lines and select the entire contour when it highlights.

Click  or select **Insert / Cut / Extrude**.

Select **Through All** for end condition.

Click **OK**.

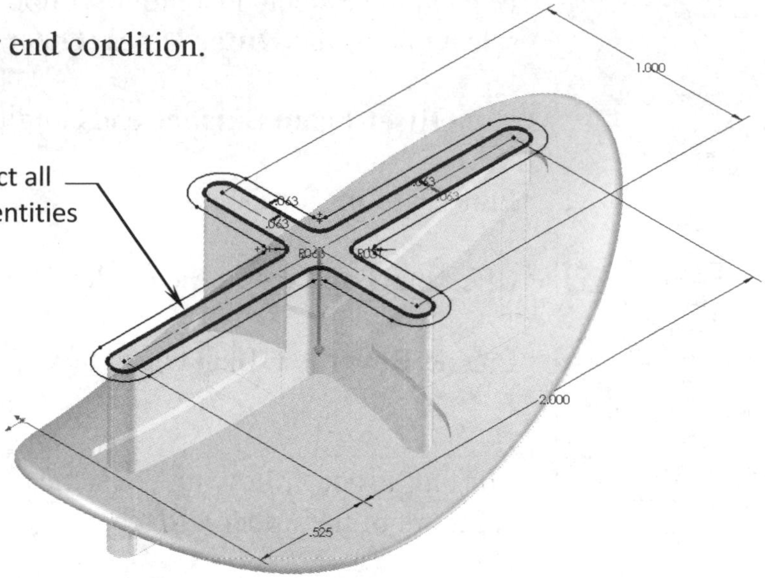

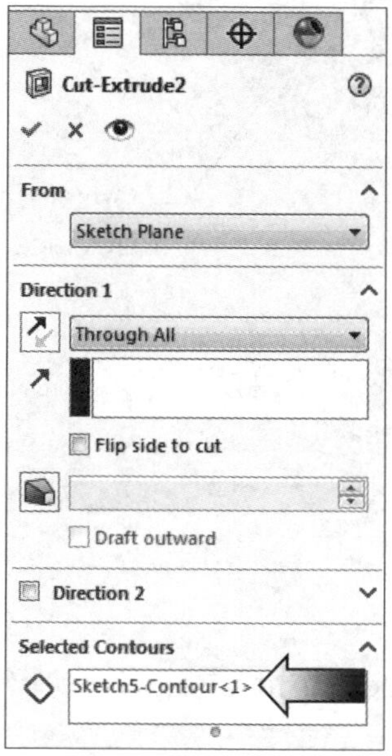

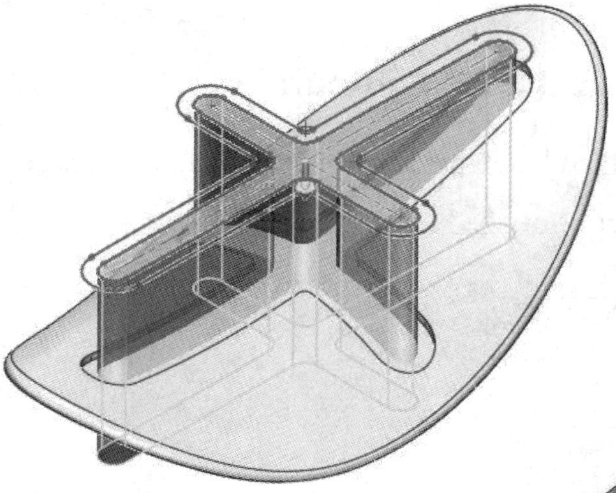

Hide  **Sketch5** when finished.

16. Saving your work:

Select **File / Save As / Lofted-Surface / Save**.

Questions for Review

Advanced Modeling – Using Surfaces

1. Surfaces are a type of geometry that can be used to create complex shapes.
   a. True
   b. False

2. Surfaces can be opened, overlapped, and have no thickness.
   a. True
   b. False

3. Surfaces can be modeled into any shape and can be extruded, revolved, swept, or lofted.
   a. True
   b. False

4. The split line option can be used to "divide" a surface into two or more surfaces.
   a. True
   b. False

5. Several surfaces can be lofted together to form a solid feature.
   a. True
   b. False

6. Surfaces cannot be moved or copied in a part document.
   a. True
   b. False

7. Each surface can be created individually and then knitted together as one surface.
   a. True
   b. False

8. The same Sketched profile can be re-used to create different extruded contours.
   a. True
   b. False

9. Offset From Surface (extrude option) only works with surfaces, not solid features.
   a. True
   b. False

1. TRUE
2. TRUE
3. TRUE
4. TRUE
5. TRUE
6. FALSE
7. TRUE
8. TRUE
9. FALSE

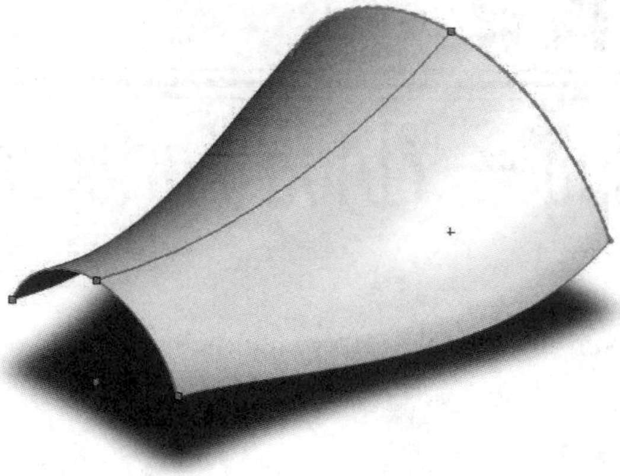

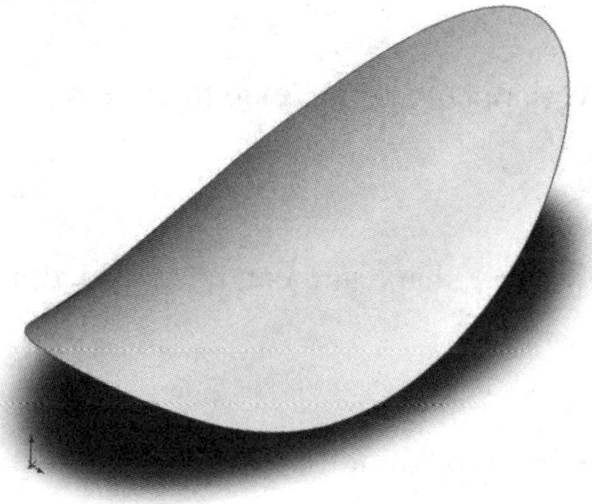

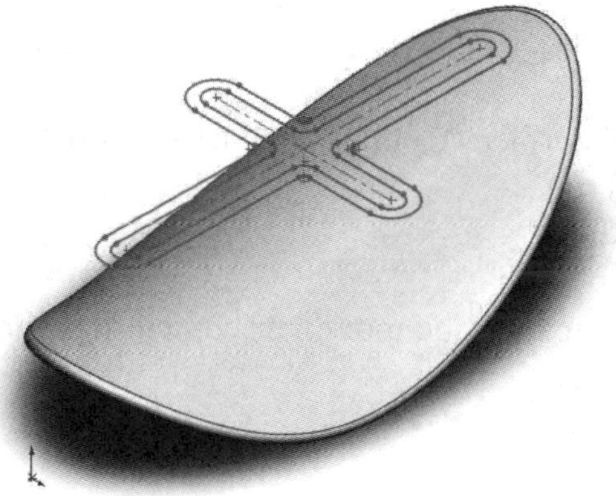

Lofted Surface

Let us take a look at a couple of techniques when modeling advanced shapes.

We will start out by creating some surfaces using various surfacing tools. These surfaces will get knitted into one surface; this surface will then get thickened into a solid part. Finally the part is split into two halves and gets assembled in an assembly document.

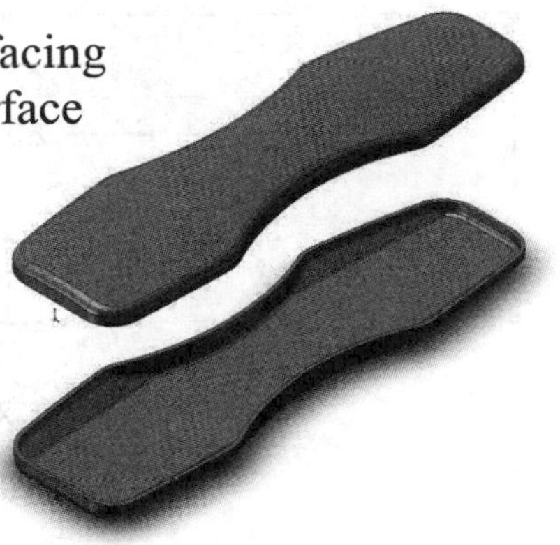

1. Creating new offset planes:

Select the <u>Front</u> plane from the FeatureManager tree.

Click or select **Insert / Reference Geometry / Plane**.

Select the **Offset Distance** option and enter **4.00in**.

Click **Flip Offset** if needed to place the new plane on the ***back side***.

Enter **2** for number of instances.

Click **OK**.

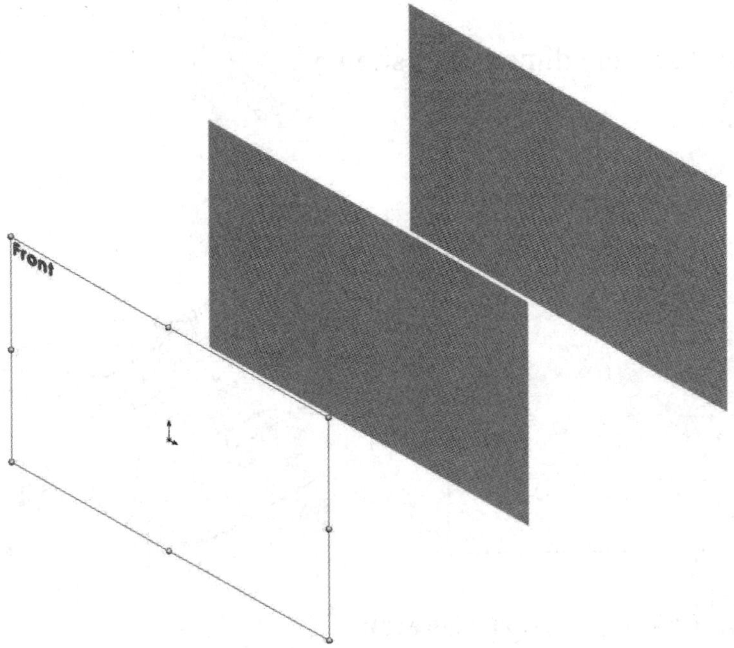

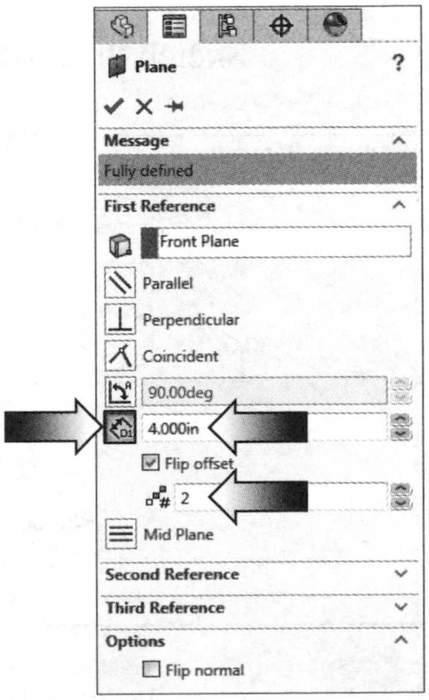

2. Sketch the first profile: (the front section)

Select the Front plane and open a **new sketch**.

Sketch the profile and add the dimensions shown below.

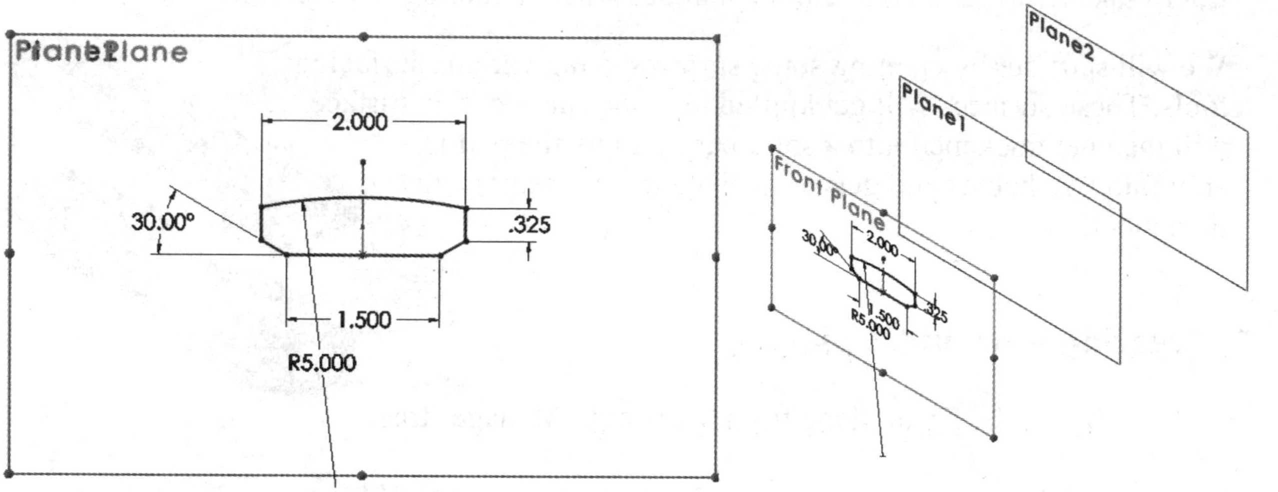

Exit the Sketch or select **Insert / Sketch**.

3. Sketching the second profile: (the middle section)

Select the Plane1 from FeatureManager tree.

Click or select **Insert / Sketch**.

Sketch the profile and add the dimensions shown.

> **Copy & Paste**
>
> The 1st profile can be copied and pasted to make the next 2 sketches. The dimensions will then be adjusted to size.

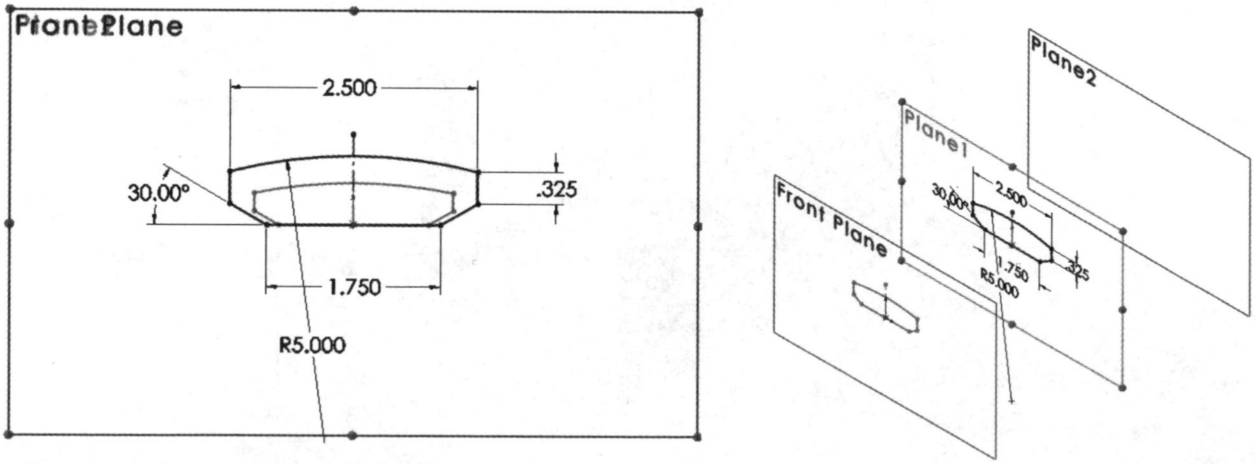

Exit the Sketch or select **Insert / Sketch**.

4. Sketching the third profile: (the end section)

Select the Plane2 from the FeatureManager tree.

Click  or select **Insert / Sketch**.

Sketch the profile and add dimensions as shown.

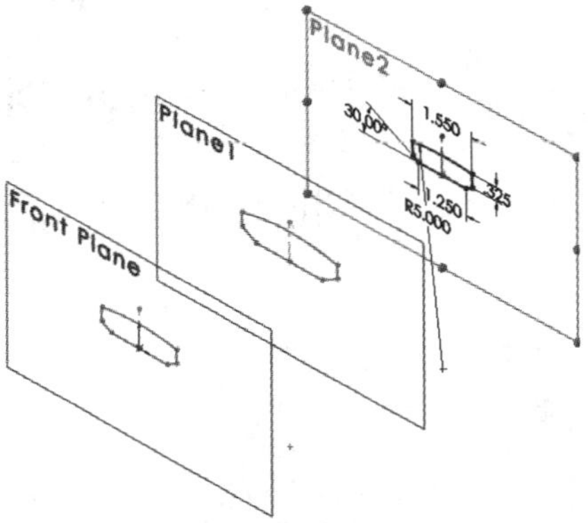

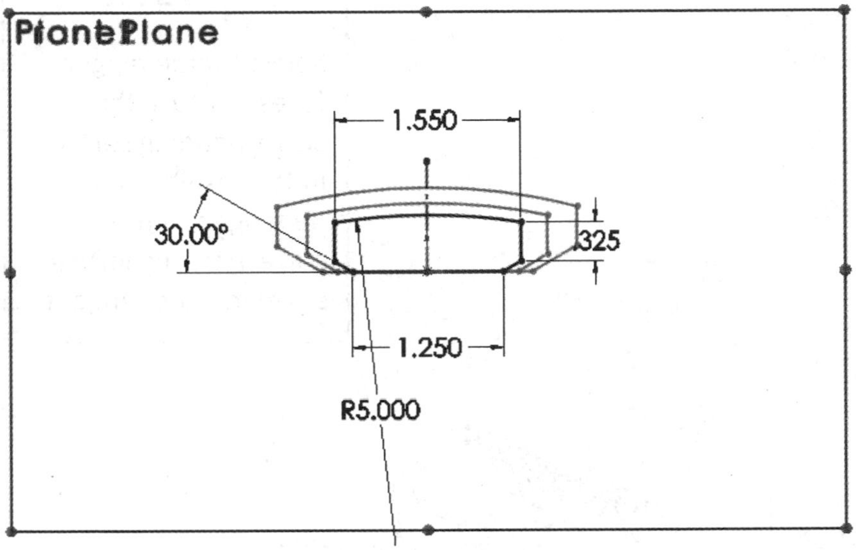

Lofted Surface

Lofted Surface creates a surface by making transitions between the sketch profiles.

Two or more profiles are needed to create a loft.

Exit the Sketch  or select **Insert / Sketch**.

5. Selecting the loft profiles:

To prevent the loft feature from being twisted, try and select the same vertex in each profile.

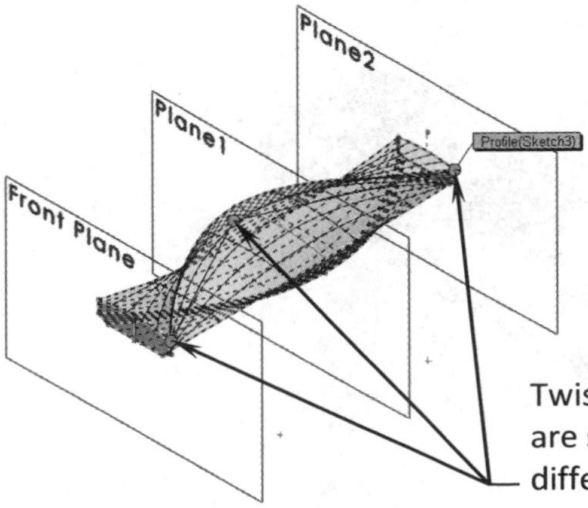

Twisted – Profiles are selected from different sides...

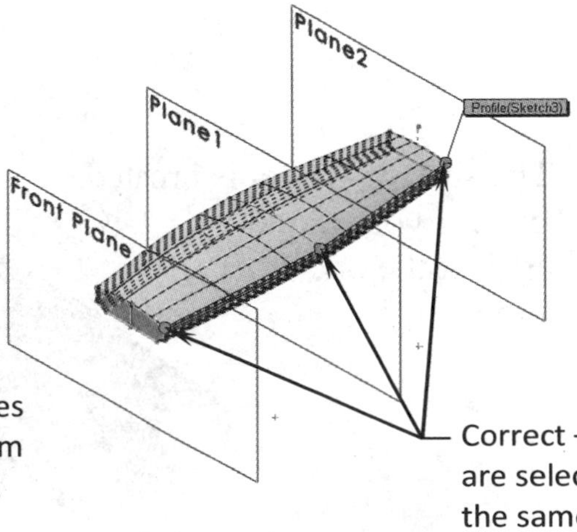

Correct – Profiles are selected from the same sides...

6. Lofting between the profiles:

Click  or select **Insert / Surface / Loft.**

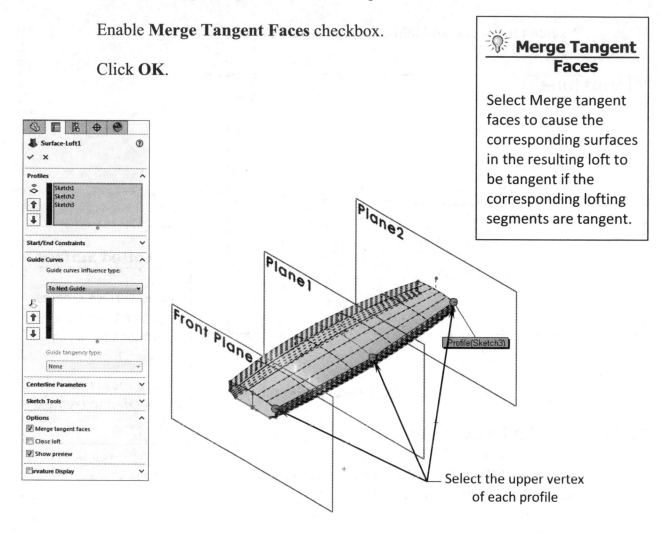

Select the **Upper-right vertex** of each profile.

Enable **Merge Tangent Faces** checkbox.

Click **OK.**

Merge Tangent Faces

Select Merge tangent faces to cause the corresponding surfaces in the resulting loft to be tangent if the corresponding lofting segments are tangent.

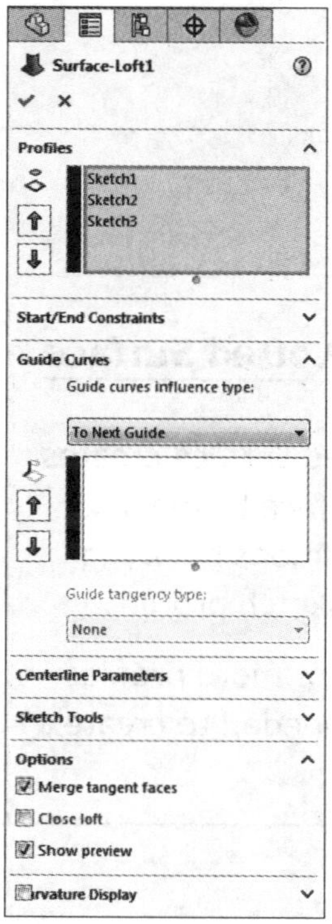

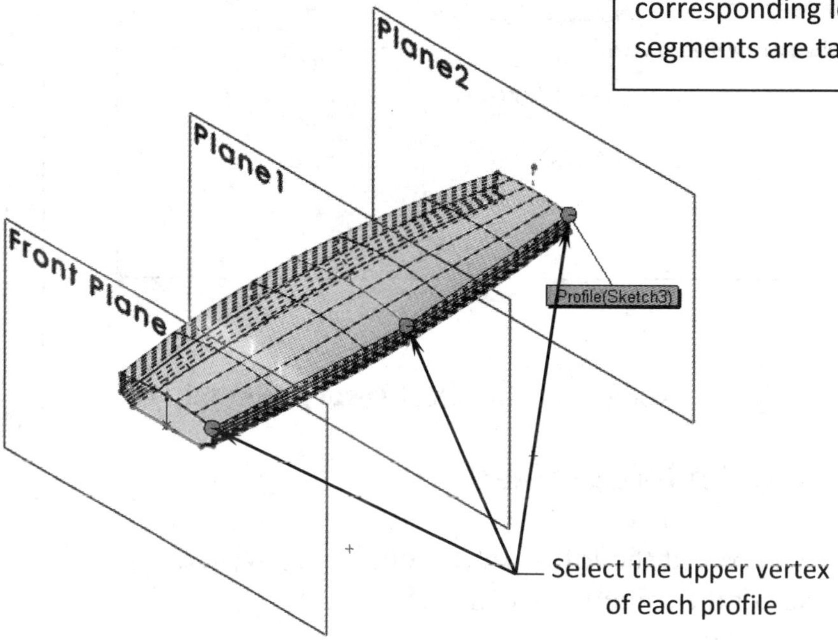

The surface model is created. Inspect your model against the one shown here.

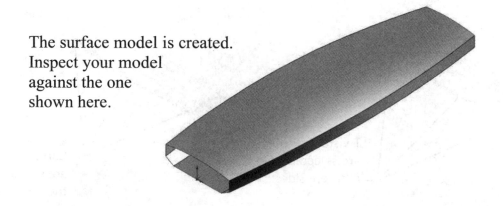

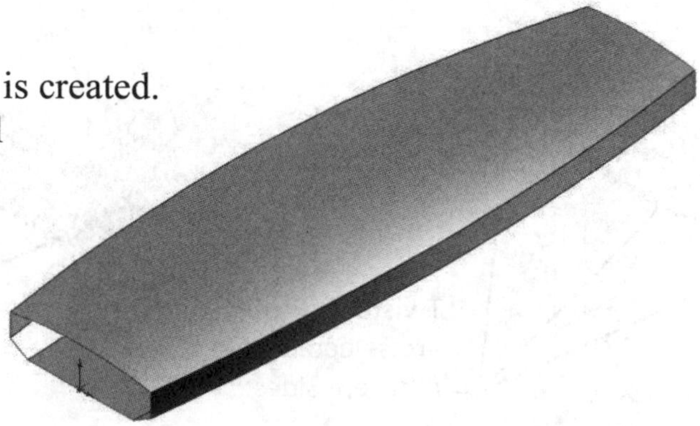

7. Creating a Revolved sketch:

Select the Top plane from the FeatureManager tree.

Click  or select **Insert / Sketch**.

Sketch the revolve profile and add the dimensions shown.

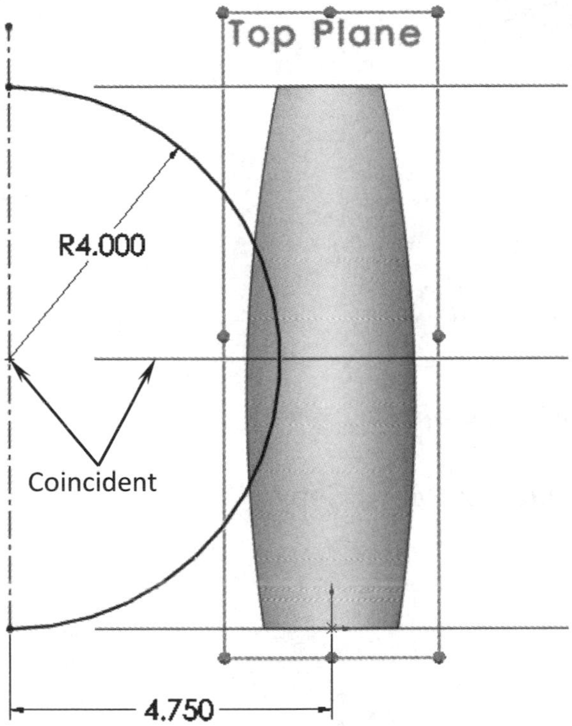

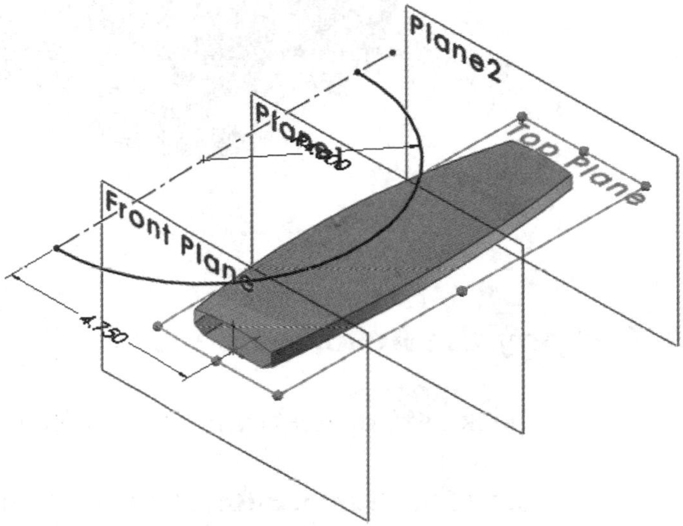

Revolved Surface

Revolved Surface creates a surface by rotating a sketch profile around a centerline (or the Axis of Revolution).

8. Revolving the Spherical surface:

Click  or select **Insert / Surface / Revolve**.

Revolve Type: **Blind**.

Revolve Angle: **360 deg**.

Click **OK**.

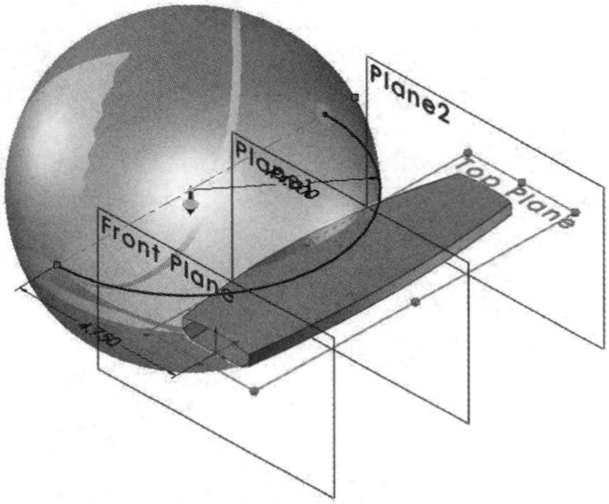

The Revolved Surface.

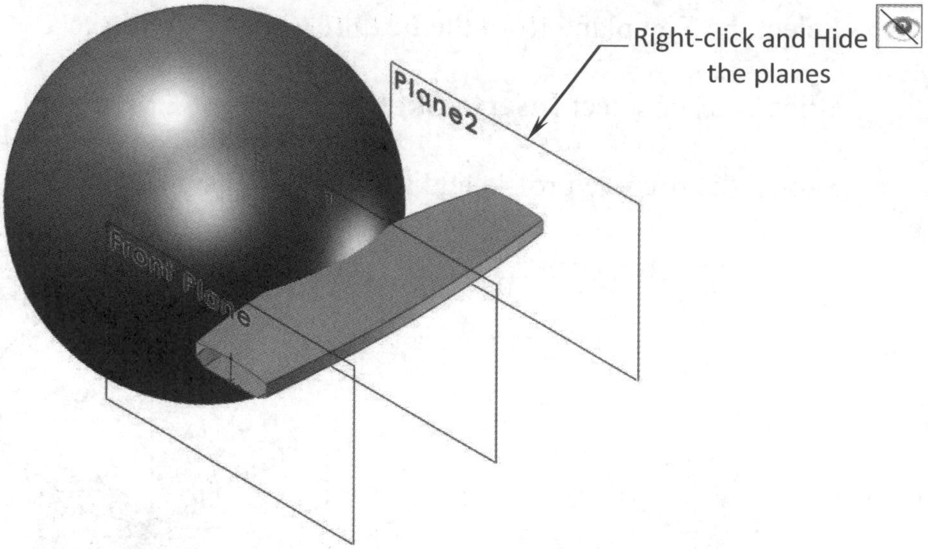

9. Copying the Revolved Surface:

Click  or select **Insert / Surface / Move/Copy**.

Select the **Surface-Revolve1** under Surfaces to Move/Copy.

Enable the **Copy** check box.

Enter **1** for Number of Copies.

Enter **9.500** in the **Delta X** distance box.

Click **OK**.

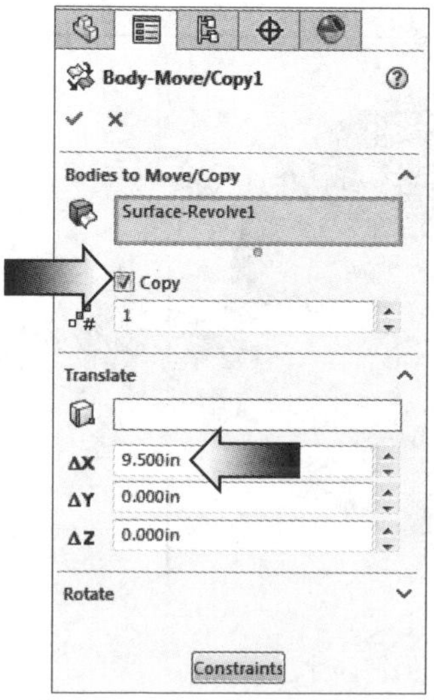

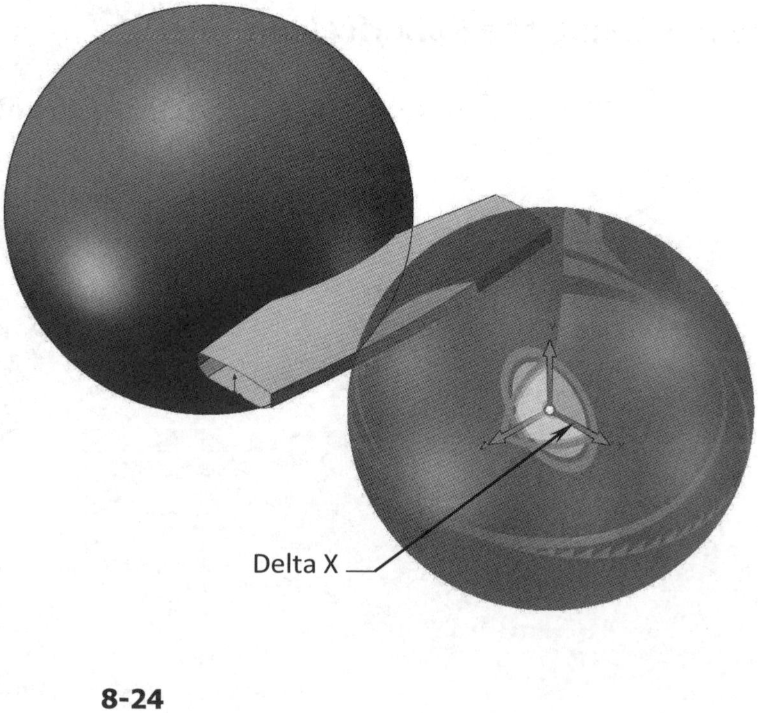

10. Trimming the Base part:

Click  or select **Insert / Surface / Trim**.

For Trim-Type, click **Mutual Trim** .

For **Trimming-Surfaces**, select all **3 surfaces**.

For **Keep Selection**, select the Surface-Loft1.

Click **OK**.

> **Trim Surface**
>
> A surface or a sketch can be used as a trim tool to trim the intersecting surfaces.

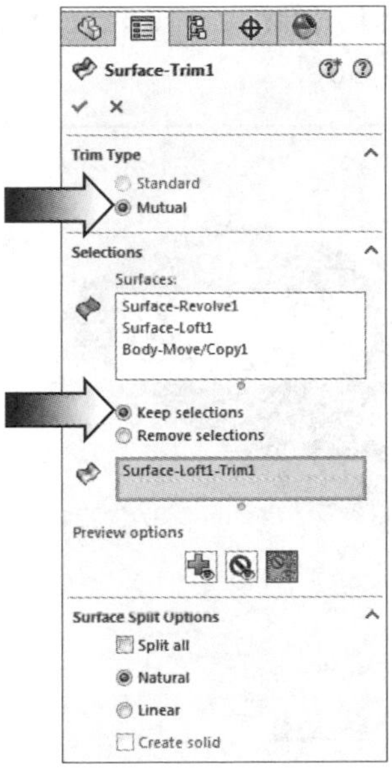

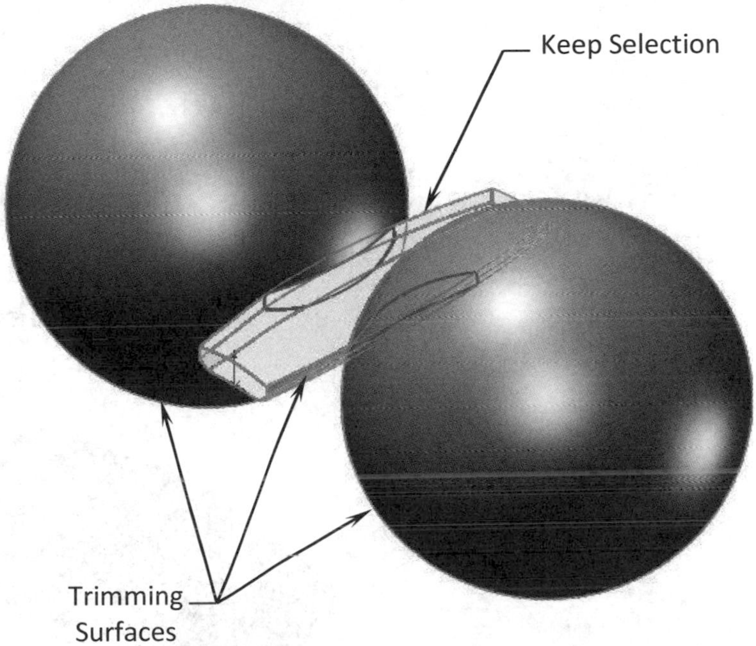

11. Hiding the surfaces:

Right-click on the two revolved surfaces and select **Hide** .

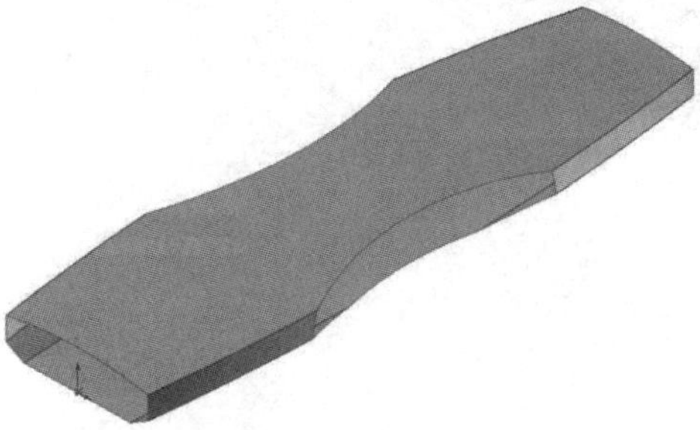

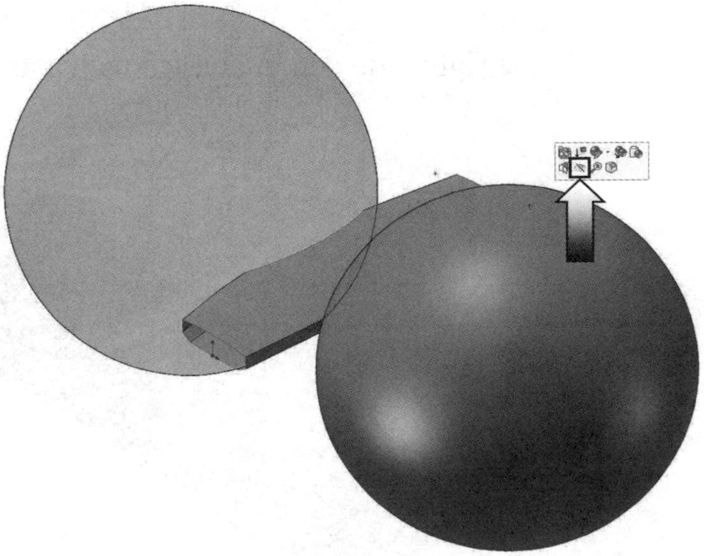

12. Patching the right side opening with Filled Surface:

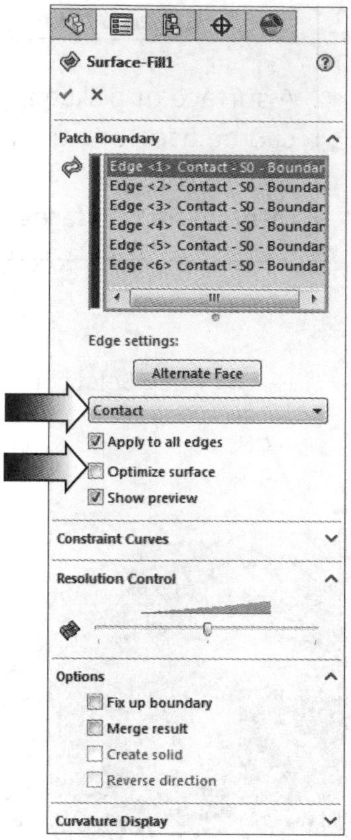

Click **Filled Surface** .

For Patch Boundary: select the **6 edges** as shown.

For Curvature Control use: **Contact**.

Enable **Apply to all edges**.

Clear the **Optimize Surface** option.

Click **OK**.

> **Filled Surface**
>
> Filled surface constructs a surface patch to fill a non-planar opening in a model.

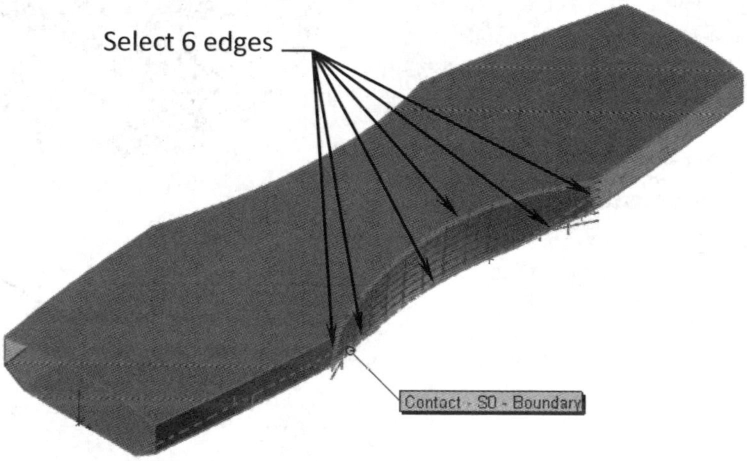

The right side cutout is filled with a new surface (Surface-Fill1).

13. Filling the left side opening:

Rotate the view  to the other side - or - Hold the Shift key and press the Up-Arrow key twice (this hotkey rotates 90º each time).

Repeat step 13 to fill the left side cutout with a new surface.

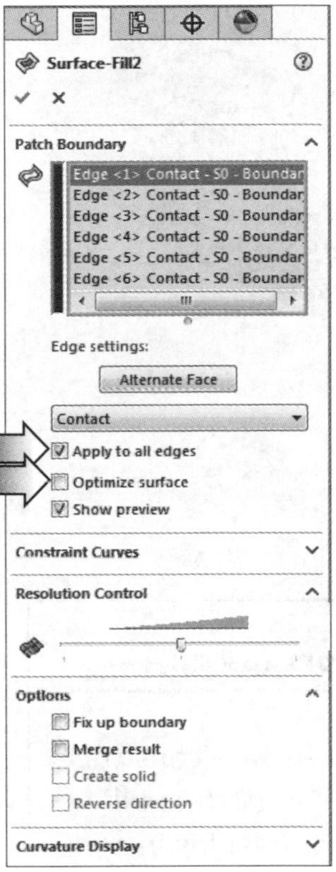

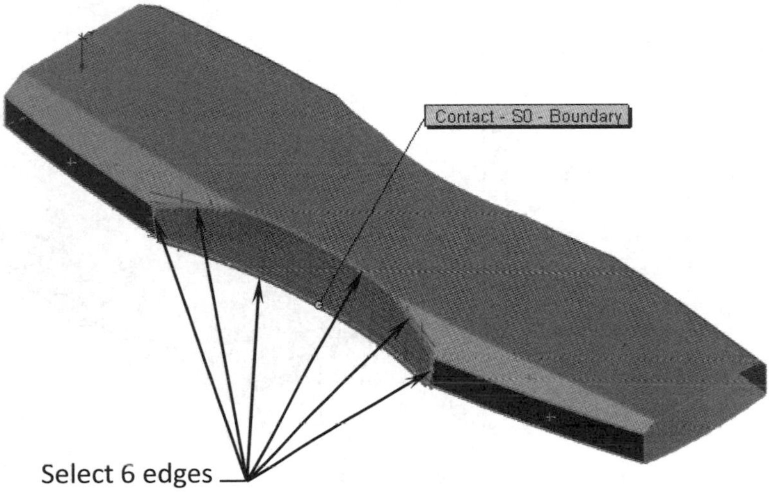

The left side cutout is filled with a new surface (Surface-Fill2).

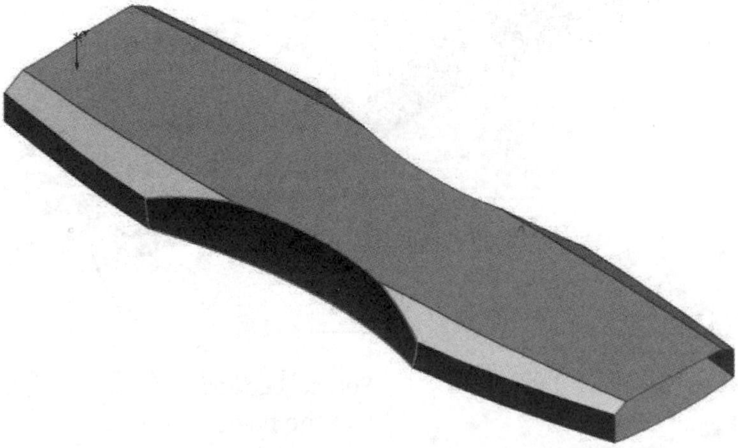

14. Filling the front and back openings with Planar Surface:

Click  or select **Insert / Surface / Planar**.

For Boundary Entities: select all **12 edges** in the front and back openings.

Click **OK**.

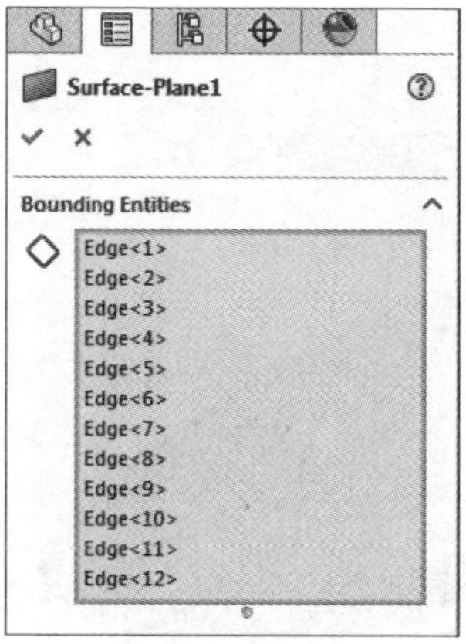

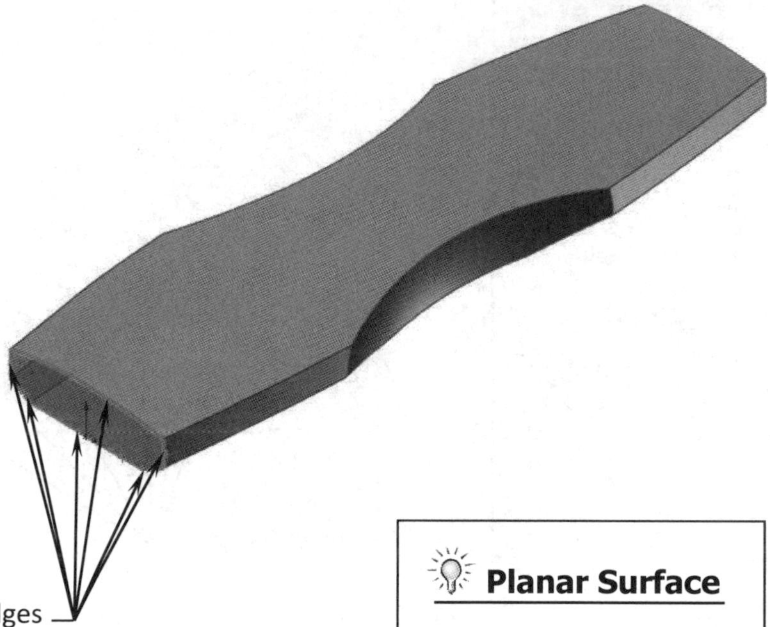

Planar Surface

Planar surface constructs a surface patch to fill a planar opening in a model.

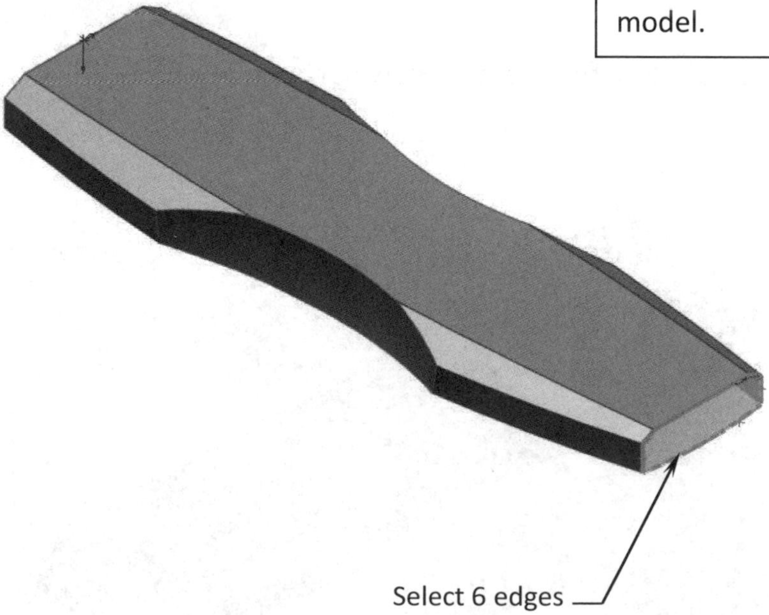

The front and back openings are filled with planar surfaces (Surface-Plane2).

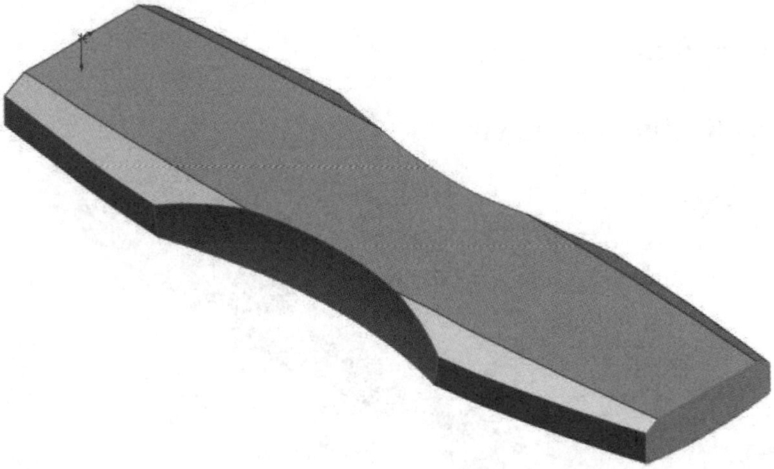

15. Creating a Surface-Knit:

Click  or select **Insert / Surface / Knit**.

For Surfaces/Faces-To-Knit, select all **5 surfaces**.

Click **OK**.

Knit Surface

Combines two or more faces and surfaces into one.

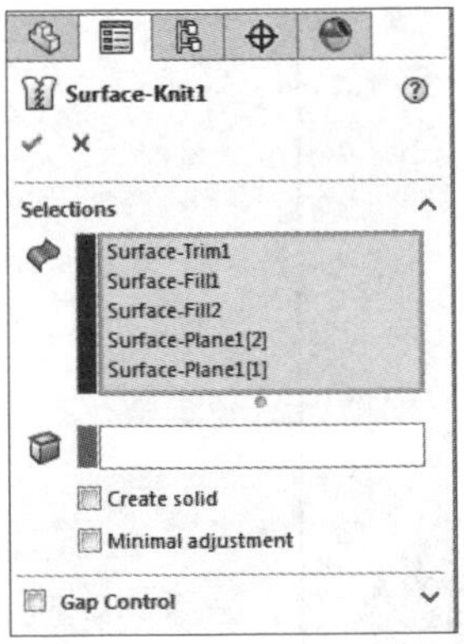

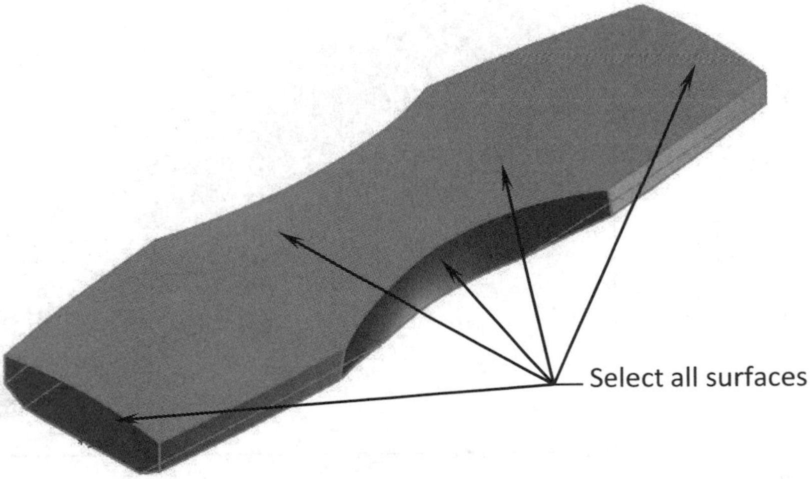

All 5 surfaces are knitted and combined into a single surface (Surface-Knit1).

16. Adding .500" fillets:

Click  or select **Insert / Features / Fillet/Round**.

Type **.500 in**. for Radius.

Select **8 edges** as shown.

Click **OK**.

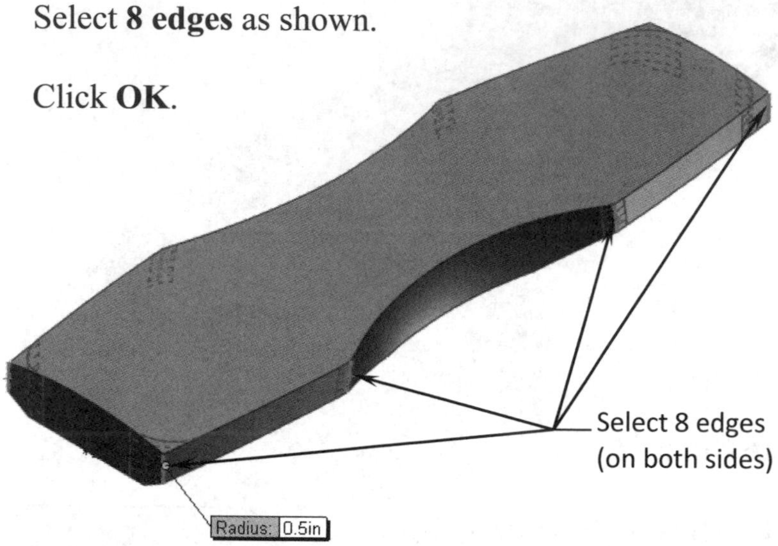

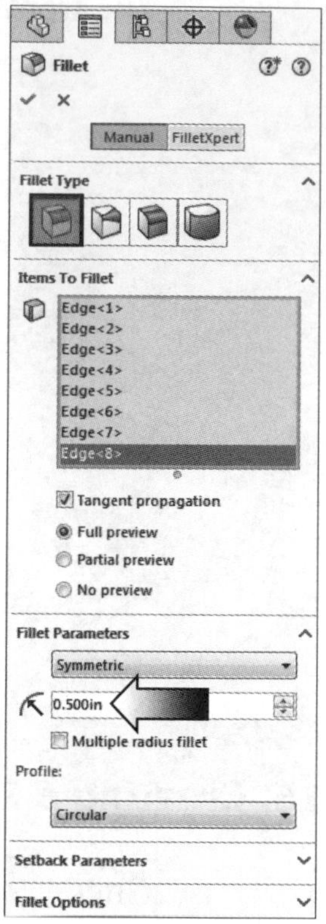

17. Adding .125" fillets:

Click  or select **Insert / Features / Fillet/ Round**.

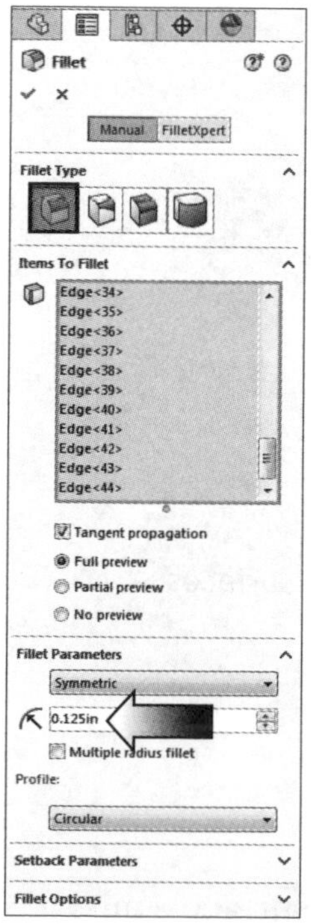

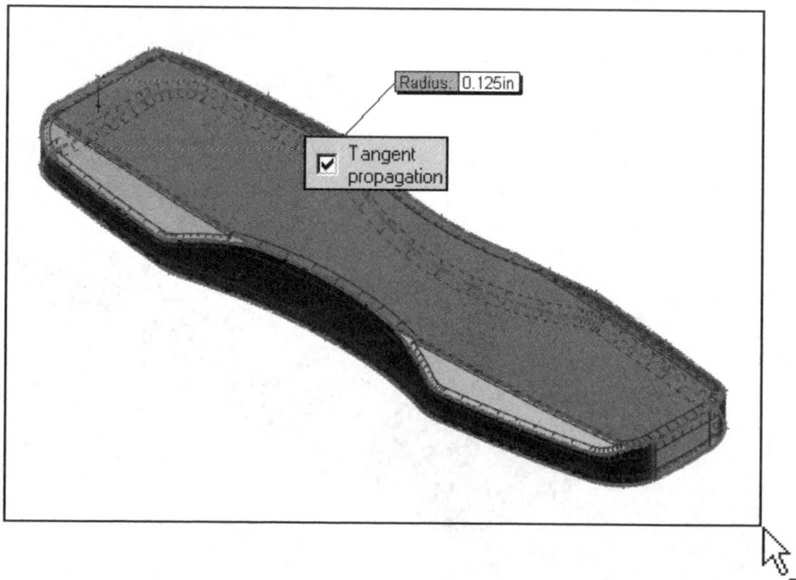

Box-Select to group all edges (on both sides)

Enter **.125 in**. for Radius and select **all edges** shown.

Click **OK**.

Inspect your model against this image. Rotate the model to make sure that all edges are filleted.

18. Creating a solid from the surface model:

Click [Thicken icon] on the Surfaces tool tab or select **Insert / Boss-Base Thicken**.

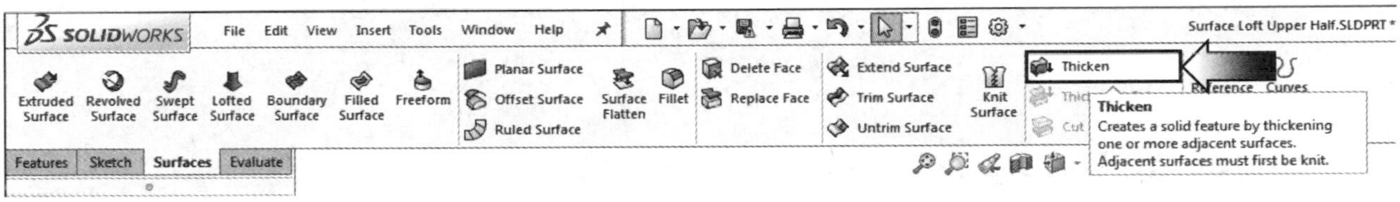

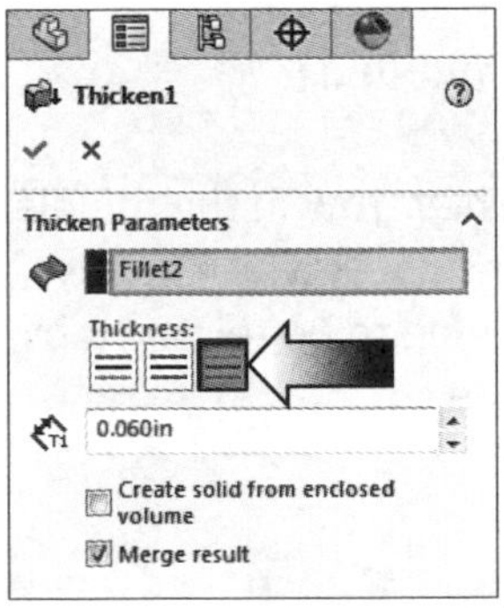

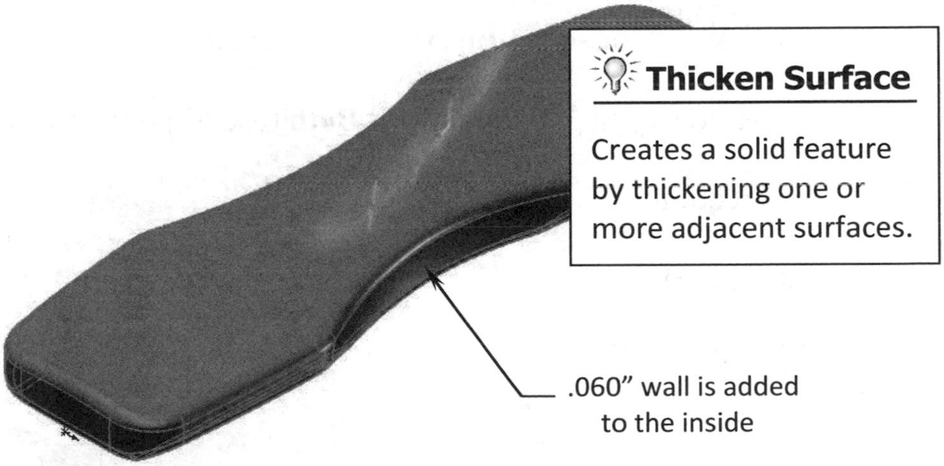
Thicken Surface
Creates a solid feature by thickening one or more adjacent surfaces.
.060" wall is added to the inside

For **Surface-To-Thicken**: Select the model from the graphics area.

For thickness Direction: Select **Thicken Side 2** [icon] (Inside).

For Thickness: Enter **.060 in**.

Click **OK**.

19. Sketching the split profile: (to split the part into 2 halves)

Select the Right plane from the FeatureManager tree.

Click  or select **Insert / Sketch**.

Sketch a **3-Point-Arc** and add the dimension shown.

Add a **Mid-point** relation between the end points of the arc and the outermost edges.

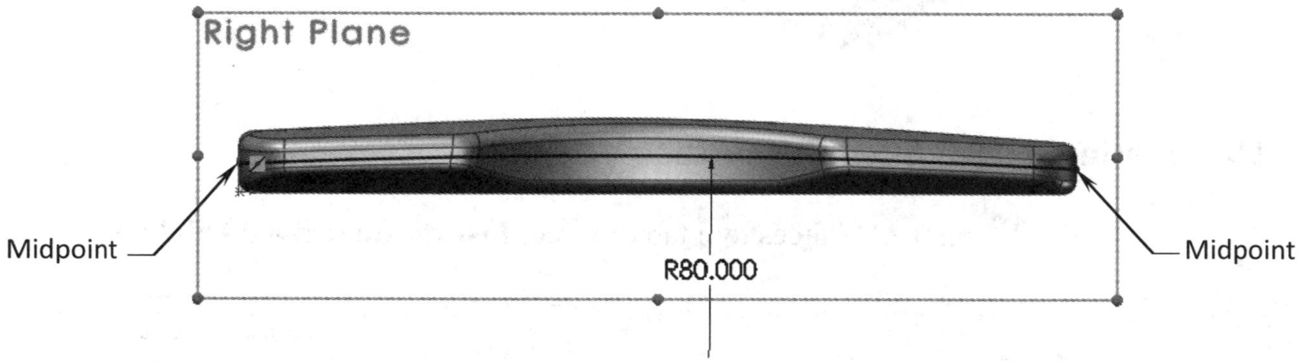

20. Removing the Upper Half:

Click  or select **Insert / Cut / Extrude**.

Extrude Type: **Through All Both** (use Flip-side-to-Cut if needed).

Direction 2: **Through All** (default).

Click **OK**.

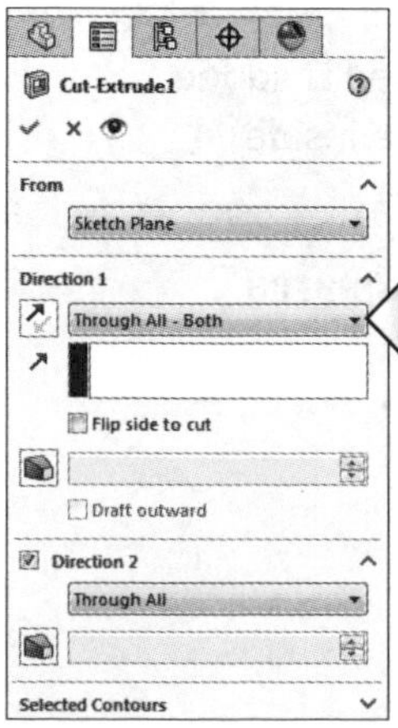

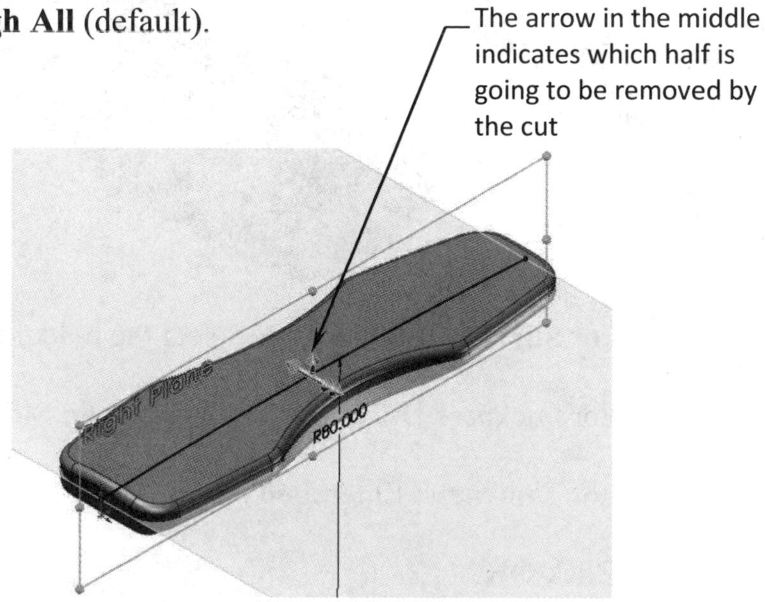

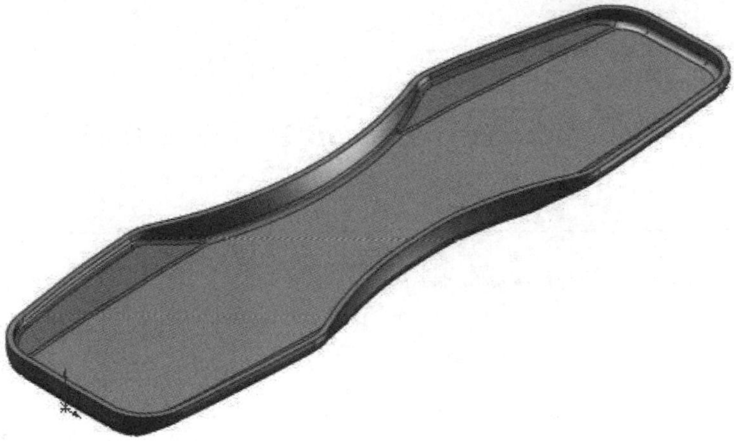

The resulted cut, the lower half of the model is kept.

21. Saving the lower half of the part: Original

Select **File / Save As / Surface Loft Lower-Half / Save**.

---

22. Saving the lower half of the part: Copy

Select **File / Save As** and enter: **Surface Loft Lower Half** for the file name.

Enable the **Save As Copy and Continue*** check box and click **Save**.

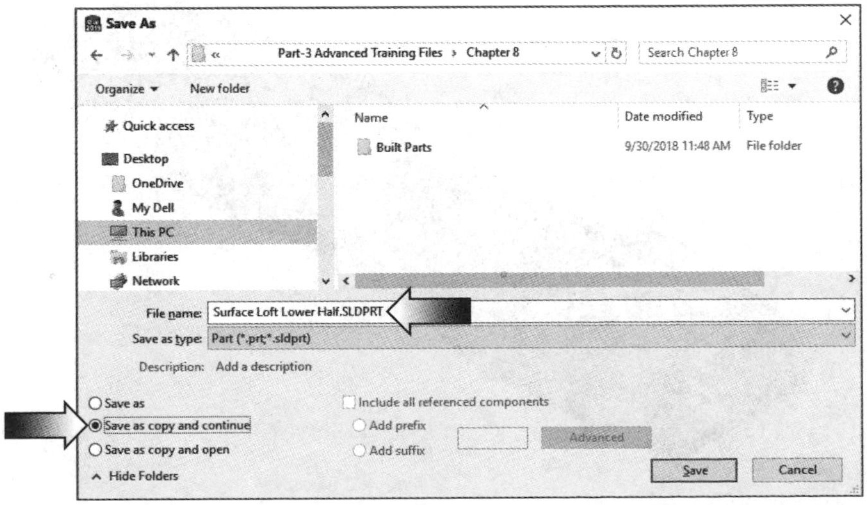

* This saves an exact copy of the same part but with a different name, so that we still have a full feature tree to create the second half of the part.

23. Modifying the copied file:

Select **File / Open**.

Select the document **Surface Loft Lower Half** and click **Open**.

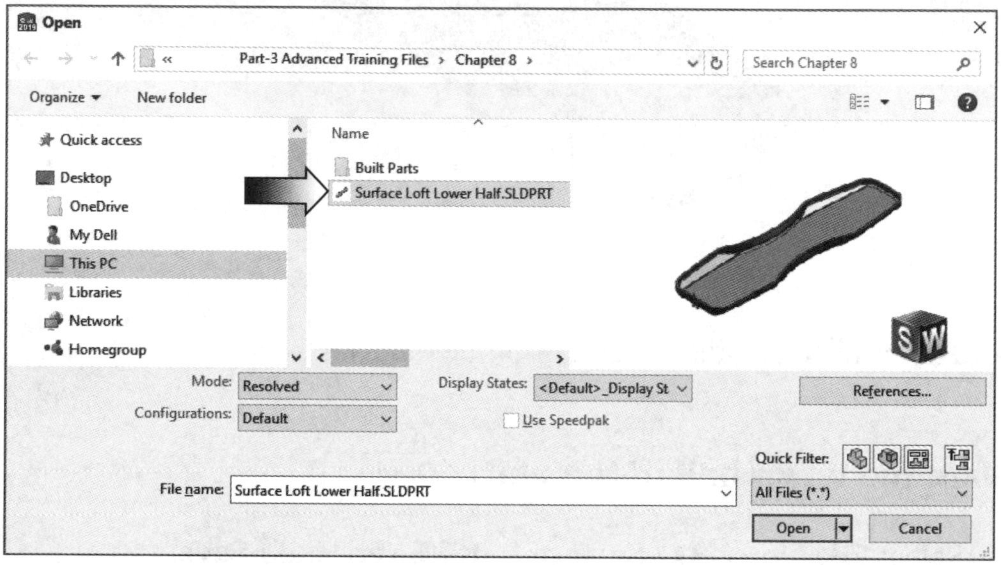

24. Changing the direction of the cut:

Right-click the **Cut-Extrude1** (the last feature on the tree) and select:

Edit Feature .

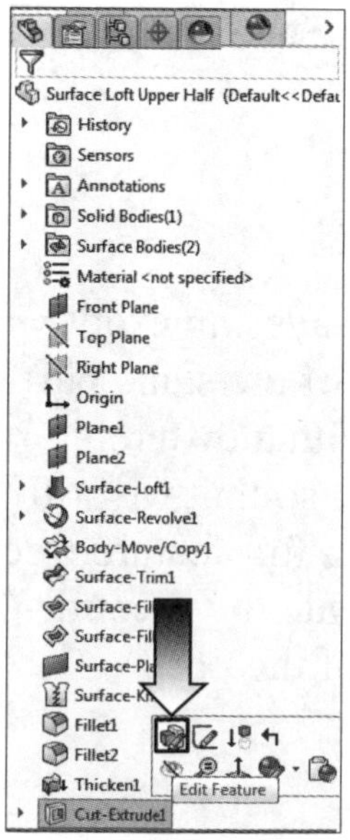

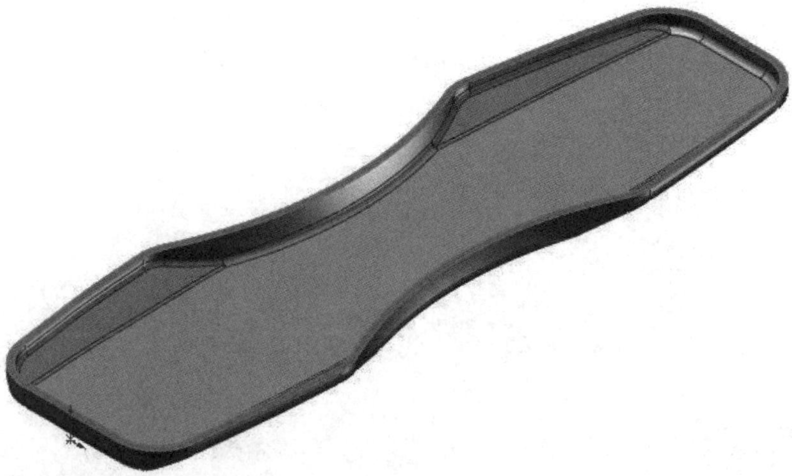

Select **Flip Side To Cut** option 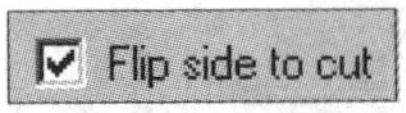.

Click **OK**.

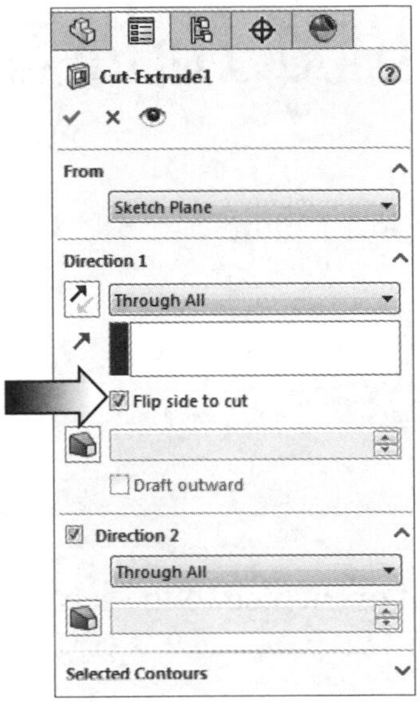

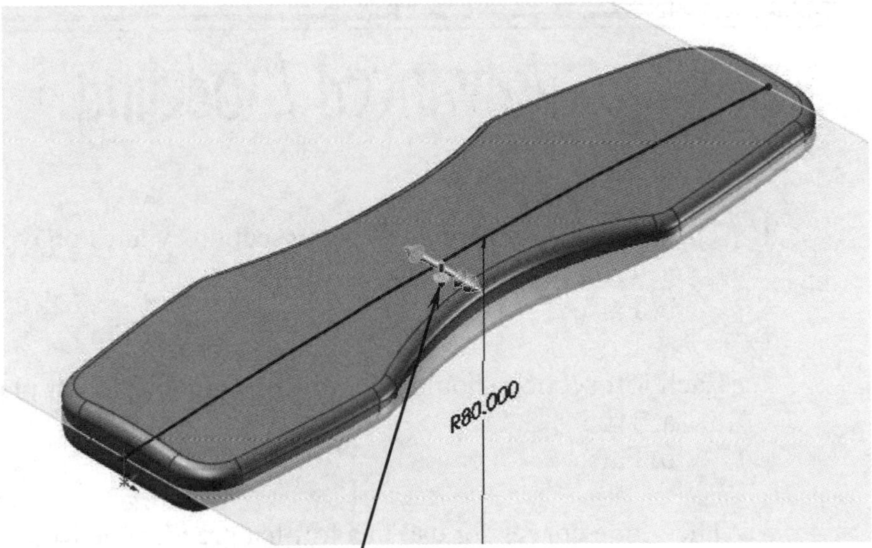

The middle arrow pointing downwards indicates the lower half is being removed

The Lower Half of the part is removed, leaving the Upper-Half as the result of the Flip-Side cut.

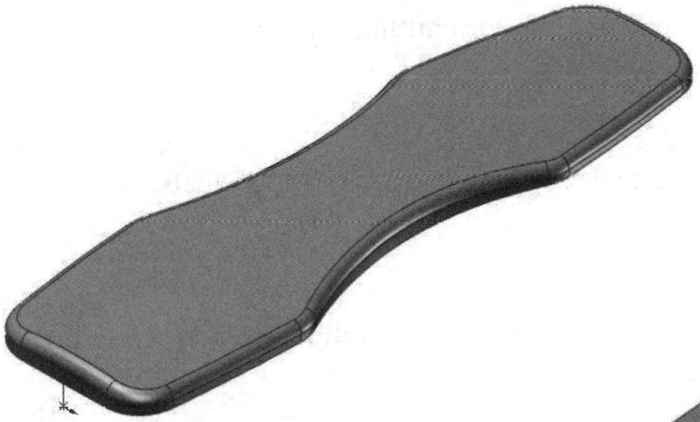

25. Saving the upper half:

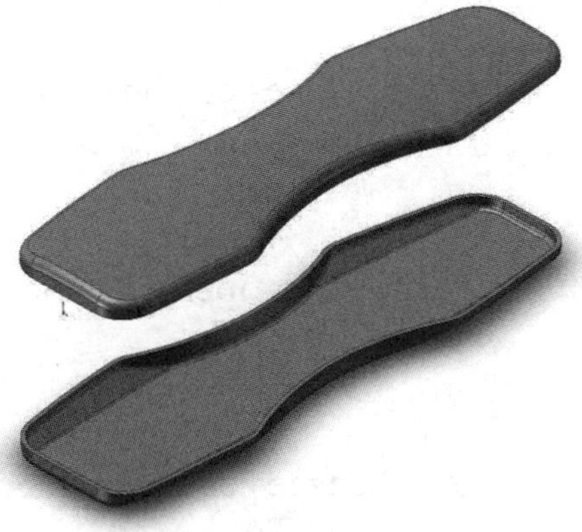

Select **File / Save As**. Enter: **Surface Loft Upper Half** for the file name and click **Save.**

OPTIONAL:

Insert the 2 halves into an assembly document and assemble them as shown.

Questions for Review

Advanced Modeling – Using Surfaces

1. There are no limits on how many sections you can have in a loft feature.
   a. True
   b. False

2. Each loft section should be modeled onto a different plane.
   a. True
   b. False

3. The guide curves for use in a loft feature must be Coincident or Pierced to the sections.
   a. True
   b. False

4. Surfaces cannot be mirrored as solid features.
   a. True
   b. False

5. Only two surfaces can be used for knitting at a time.
   a. True
   b. False

6. Fillets cannot be used with surfaces, only in solid models.
   a. True
   b. False

7. Surfaces can be thickened after they are knitted together.
   a. True
   b. False

8. Mass properties options such as volume, surface area, etc., are available for all surfaces.
   a. True
   b. False

9. Surfaces can be knitted into a closed volume and then thickened into a solid.
   a. True
   b. False

1. TRUE 2. TRUE
3. TRUE 4. FALSE
5. FALSE 6. FALSE
7. TRUE 8. FALSE
9. TRUE

<u>Exercise</u>: Loft & Delete Face

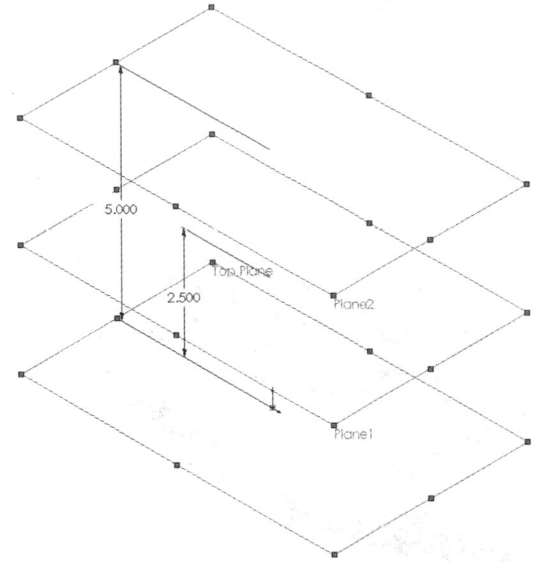

1. Create 2 new Planes, offset from Top plane.

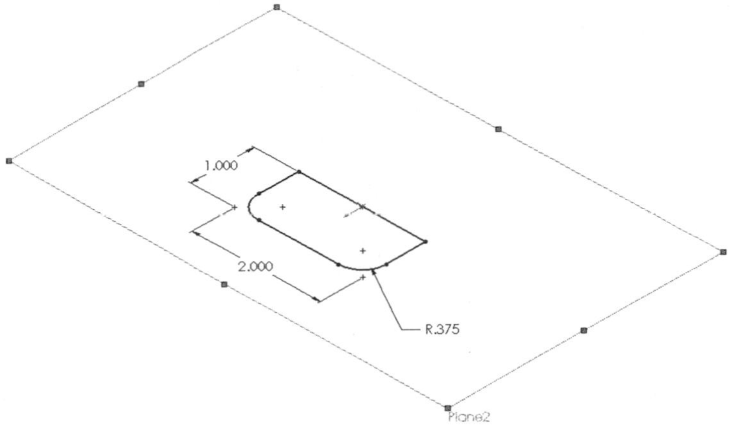

2. Sketch the 1st Profile on Plane2.

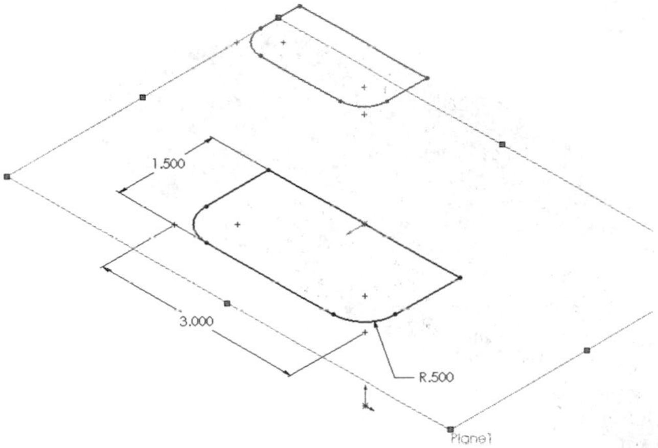

3. Sketch the 2nd Profile on Plane1.

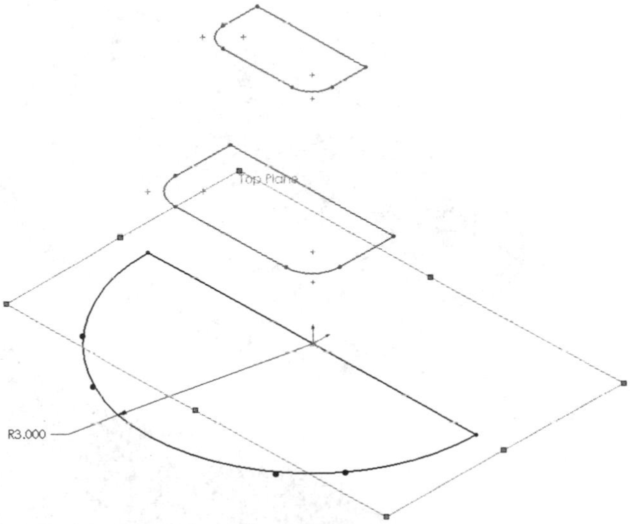

4. Sketch the 3rd Profile on Top plane & add the connector points.

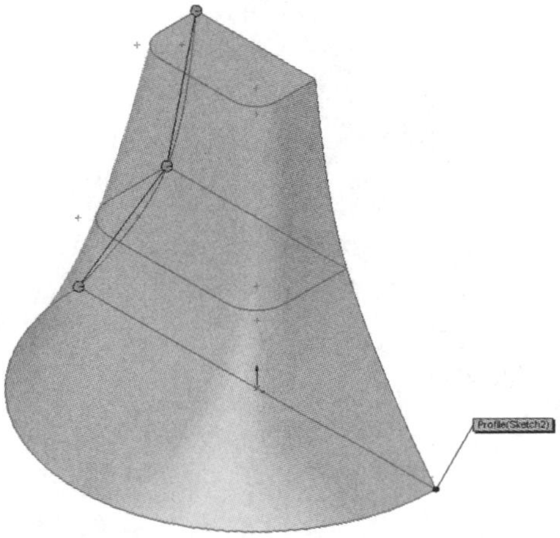

5. **Solid-Loft** the Profiles.

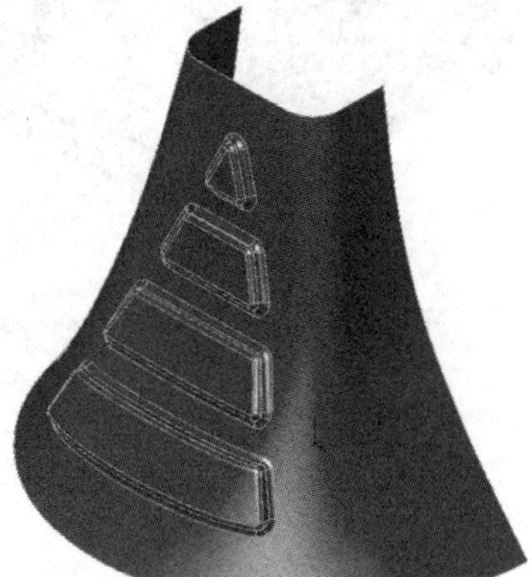

6. Add the Raised features (any size), use Delete Face command and remove 3 Faces (top, bottom & back).

CHAPTER 9

Offset Surface & Ruled Surface

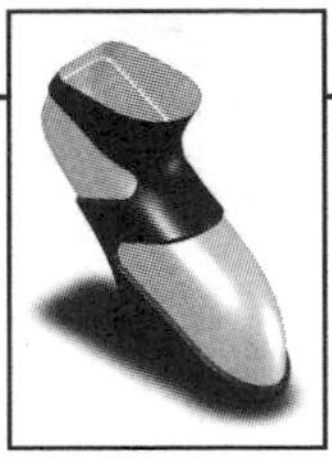

Offset Surface & Ruled Surface

The Offset Surface command creates a new surface from a **single face** or a **group of faces**, with a distance of zero or greater.
The Offset Surfacc can be created inward or outward.

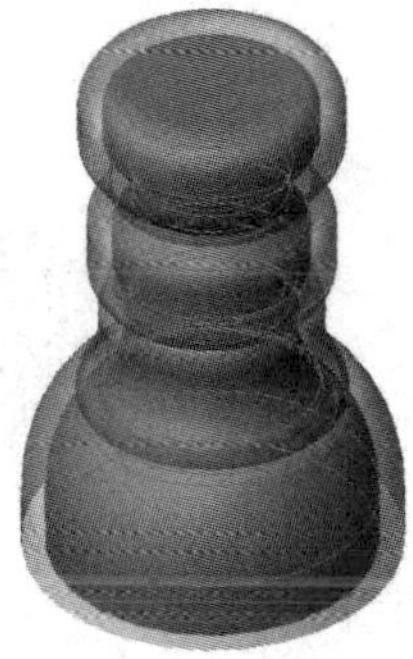

Offset Surface

Offset from a single face *Offset from a group of faces*

The Ruled Surface command creates a new surface from a single edge or a group of edges. The Ruled Surface can either be perpendicular or tapered from the selected edges.

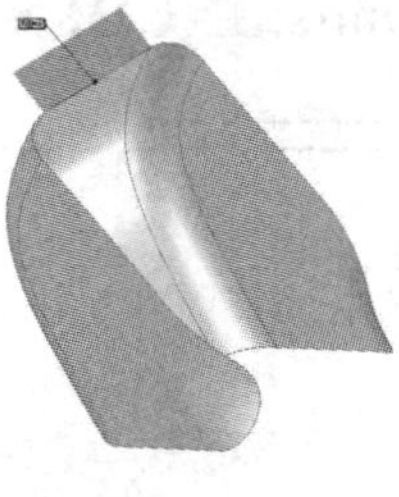

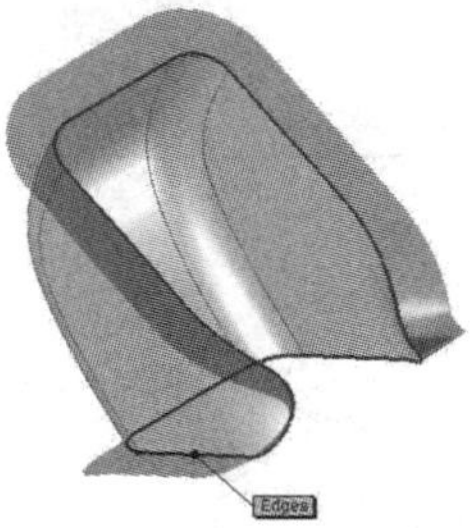

Ruled Surface

Ruled surface from a single edge *Ruled surface from a group of edges*

The Offset Surface and the Ruled Surface are used to create reference surfaces that help define the solid features in a part. In most cases, these surfaces should be knitted together before the next operation such as extruded cuts, fillets, etc., can be performed.

Advanced Surfaces Using
Offset Surface & Ruled Surface

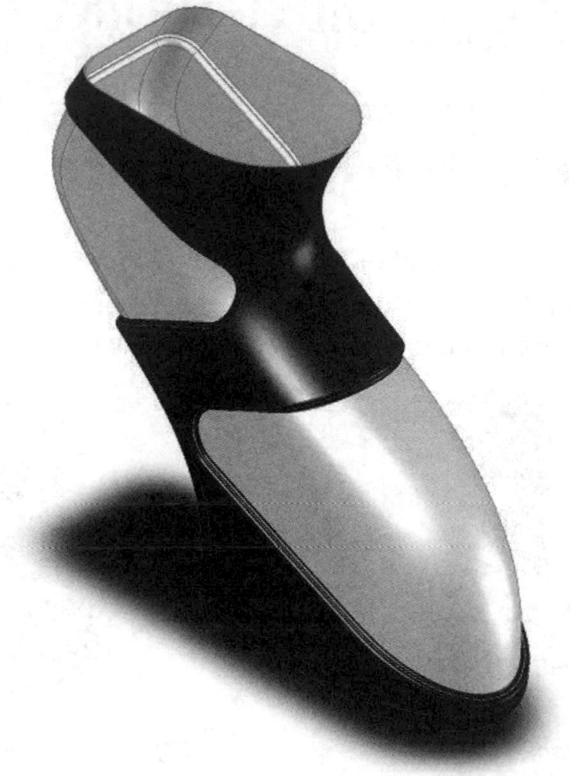

View Orientation Hot Keys:

Ctrl + 1 = Front View
Ctrl + 2 = Back View
Ctrl + 3 = Left View
Ctrl + 4 = Right View
Ctrl + 5 = Top View
Ctrl + 6 = Bottom View
Ctrl + 7 = Isometric View
Ctrl + 8 = Normal To Selection

Dimensioning Standards: **ANSI**
Units: **INCHES** – 3 Decimals

Tools Needed:

Lofted Boss/Base	Split Line	Offset Surface
Ruled Surface	Knit Surface	Shell

Advanced Surface Modeling
Using Offset & Ruled Surface options

1. Opening the existing file:

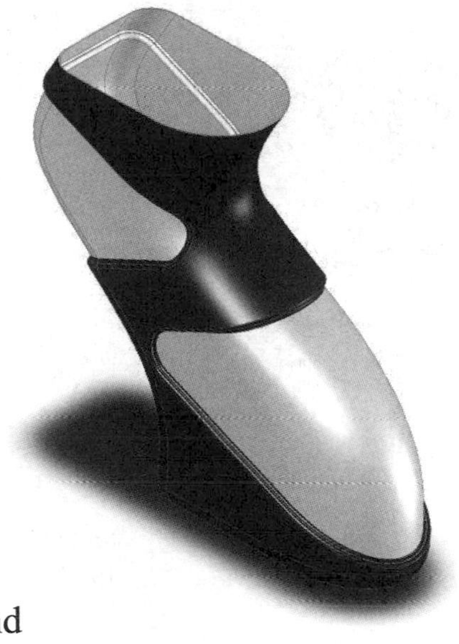

Browse to the Training Files folder and open a part document named:
Surface_Offset_Ruled.sldprt

This part document contains several sketches to be used as the Loft Profiles, and 2 other sketches used as the Guide Curves to help control the transition between each profile.

This case study focuses on the use of the Offset and Ruled Surface commands.

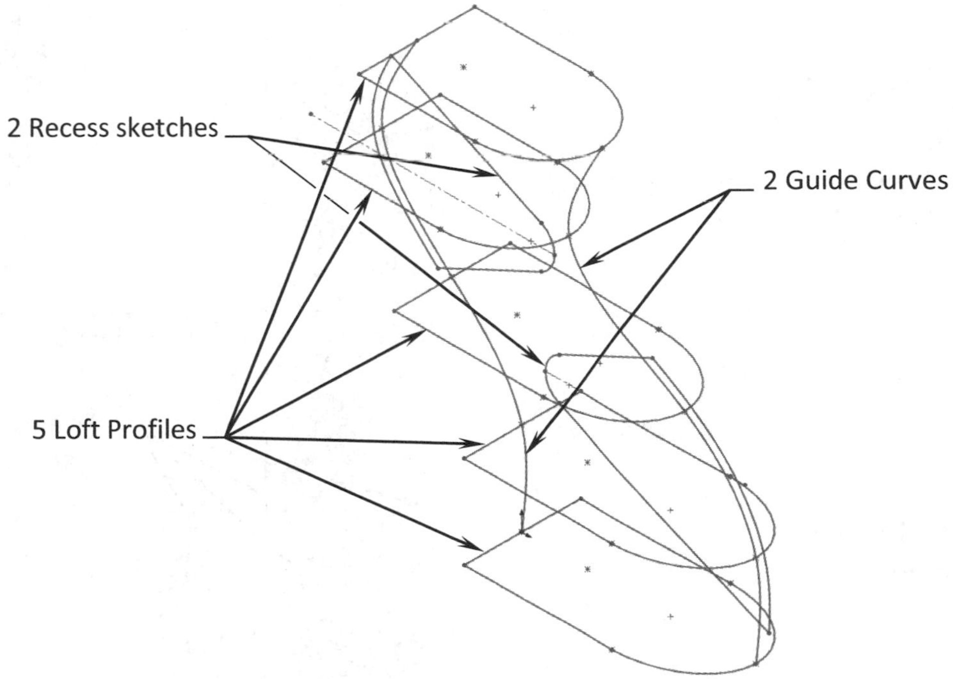

2. Creating the Base Loft:

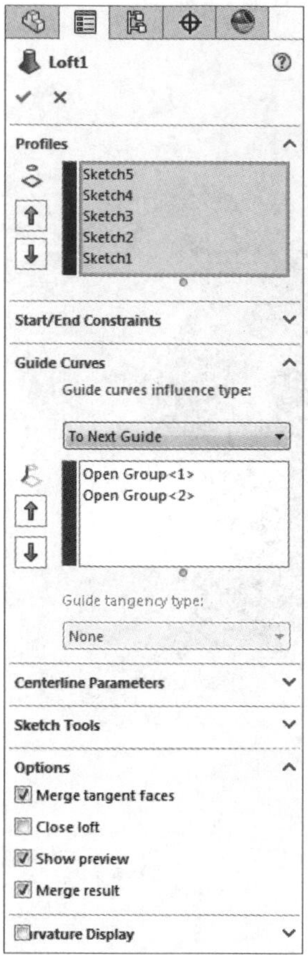

Click [icon] or select **Insert / Boss-Base / Loft**.

Select the **5 Loft Profiles** and the **2 Guide Curves** as noted. (The SelectionManager appears when disjointed or overlapped entities are found in the sketch.)

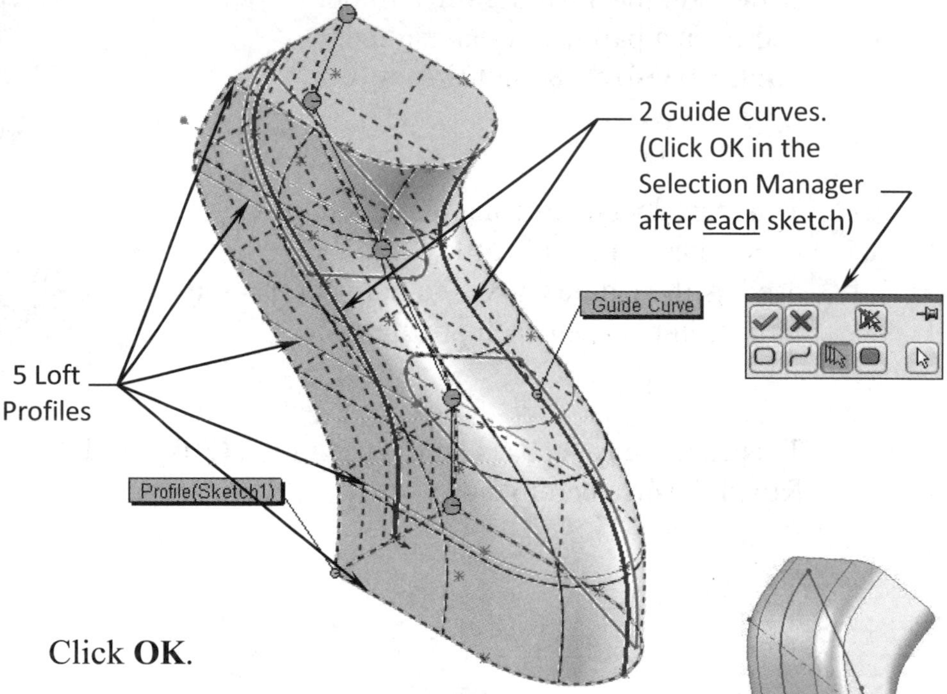

Click **OK**.

3. Adding .250" fillets:

Click [icon] or select **Insert / Features / Fillet-Round**.

Enter **.250in**. for radius value.

Select the **4 edges** shown to add the fillets.

Click **OK**.

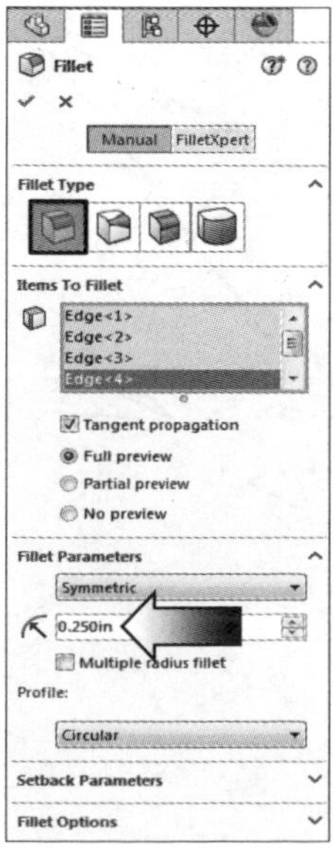

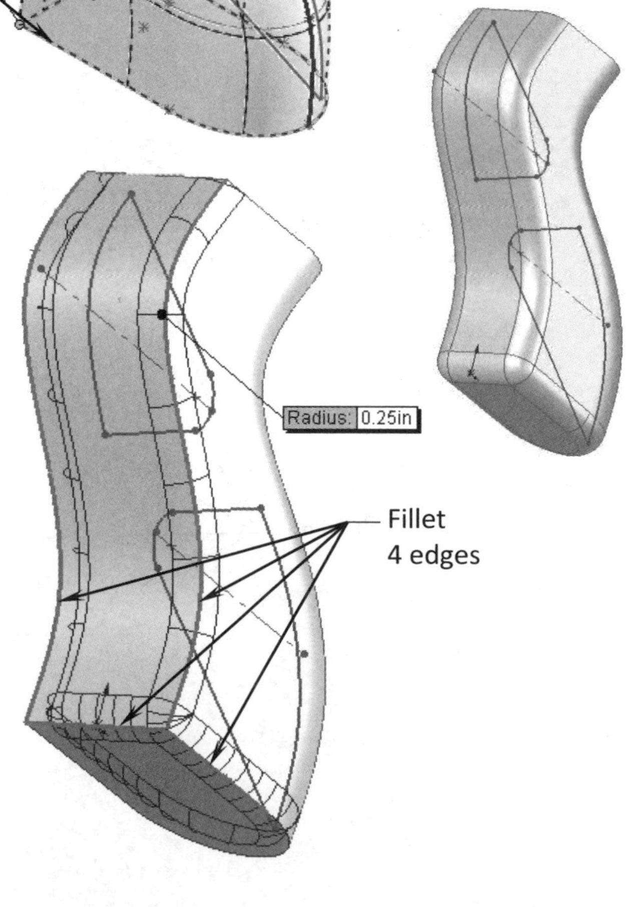

The Split Line command projects an entity to the surface(s) and divides the surface into multiple faces.

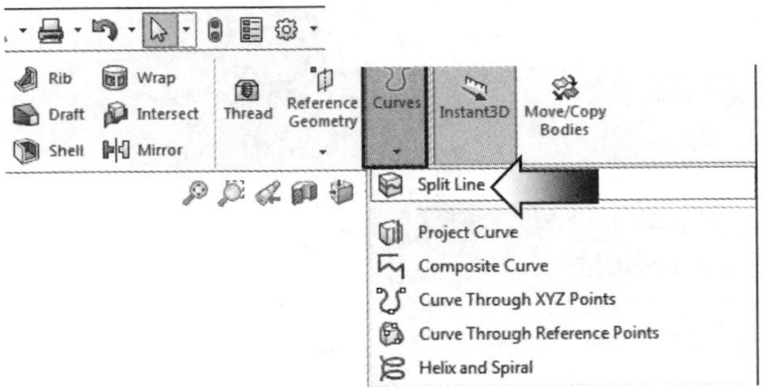

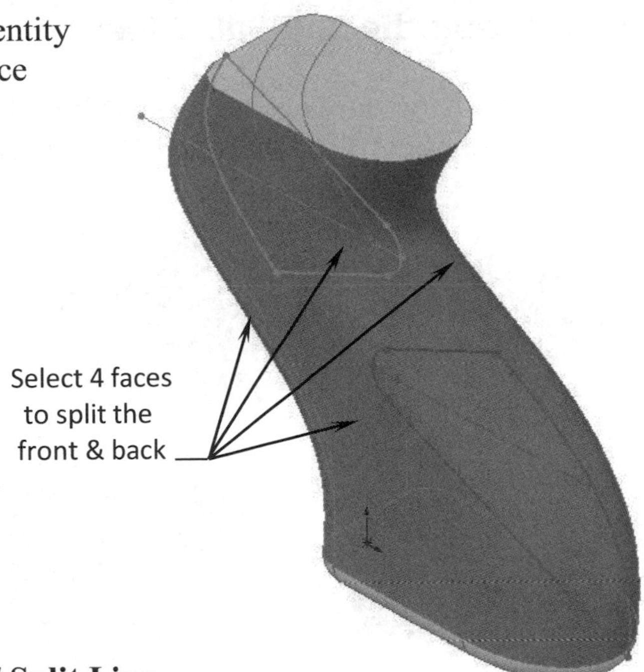

4. Creating the 1st Split Line:

Click or select **Insert / Curves / Split Line**.

For Type-of-Split, select the **Projection** option (Default).

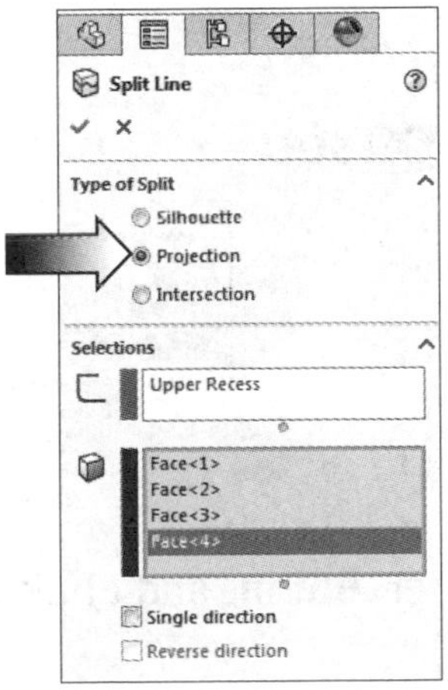

For Sketch-To-Project, select the sketch **Upper Recess** from the Feature-Manager tree.

For Faces-To-Split, select the **4 faces** as indicated.

Click **OK**.

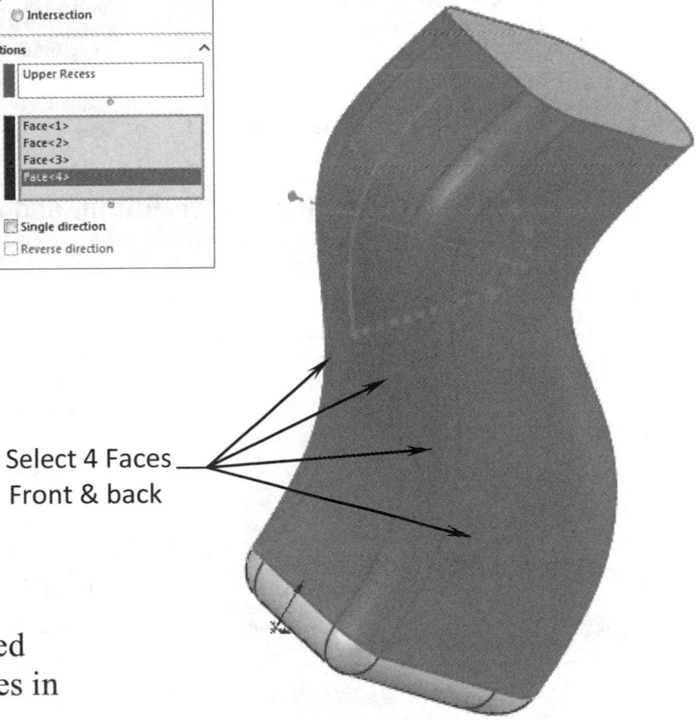

The handle body now has a new group of faces, which will be used to create the two recessed features in the next few steps.

5. Creating the 2nd Split Line:

Using the sketch **Lower Recess** from the FeatureManager tree, repeat step number 3 to create the 2nd Split Line.

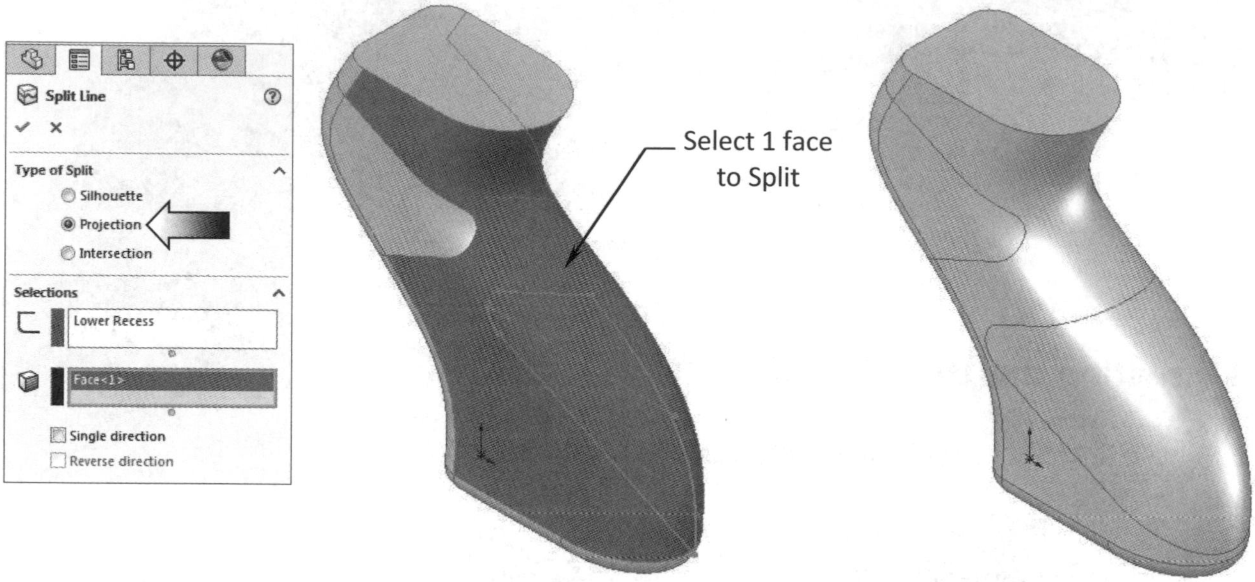

6. Creating the 1st Surface-Offset:

Click  or select **Insert / Surface / Offset**.

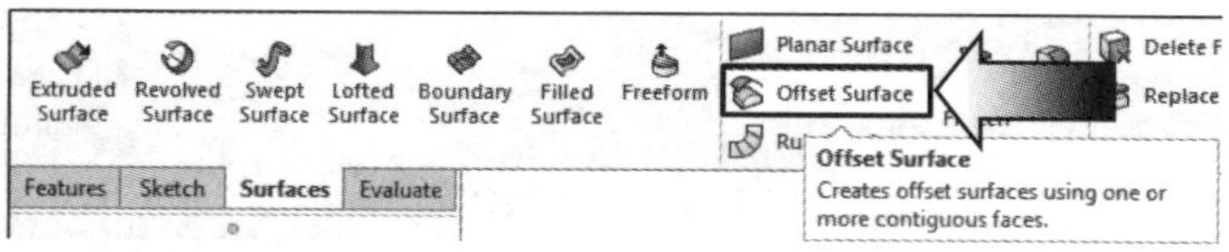

Select the **5 Split-Faces** to offset.

For Offset Distance, enter **.050in**. and click the **Reverse** button (Inside).

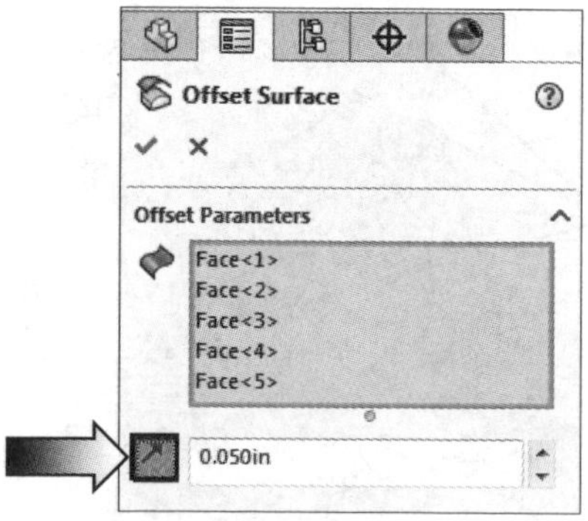

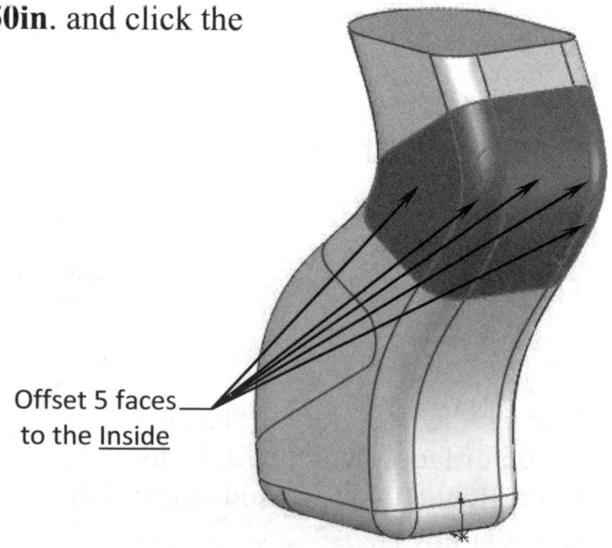

Click **OK**.

7. Hiding the Solid Body:

Right-click on the **upper surface** of the solid body and select **Hide** .

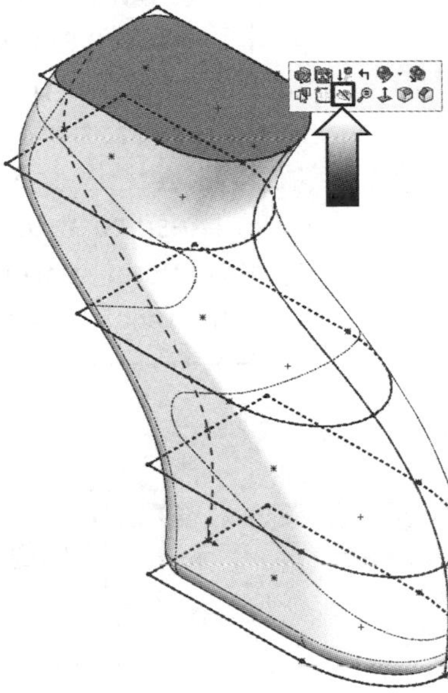

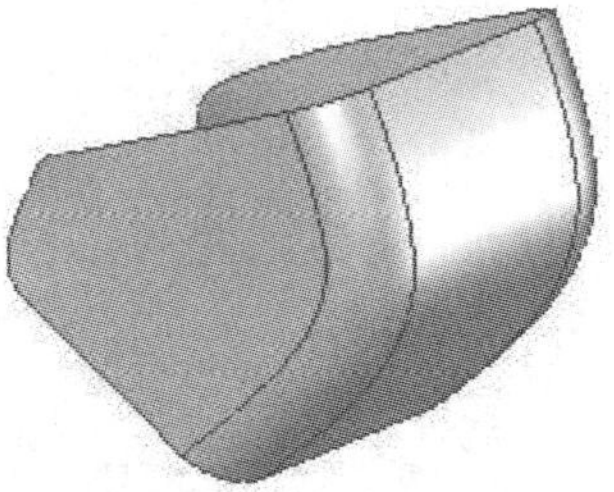

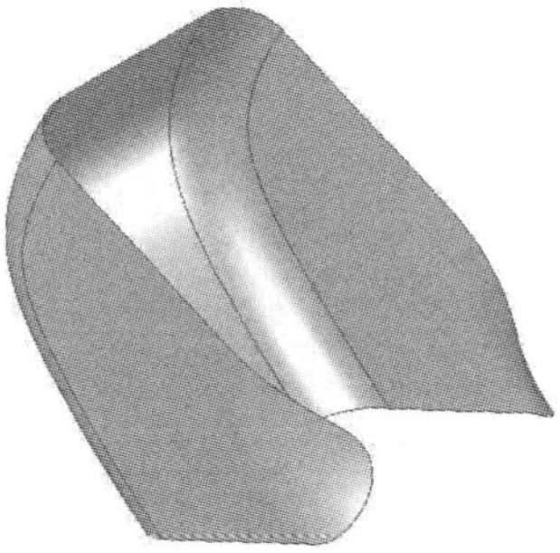

Inspect your surface offset against the one shown here.

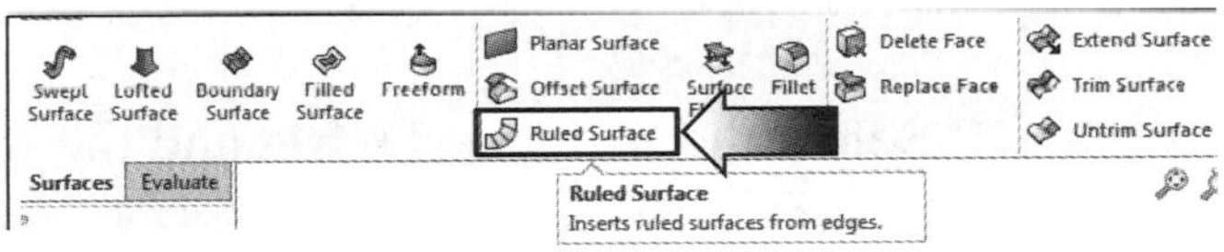

8. Creating the 1st Ruled Surface:

Click  on the Surfaces tool tab or select: **Insert / Surface / Ruled Surface**.

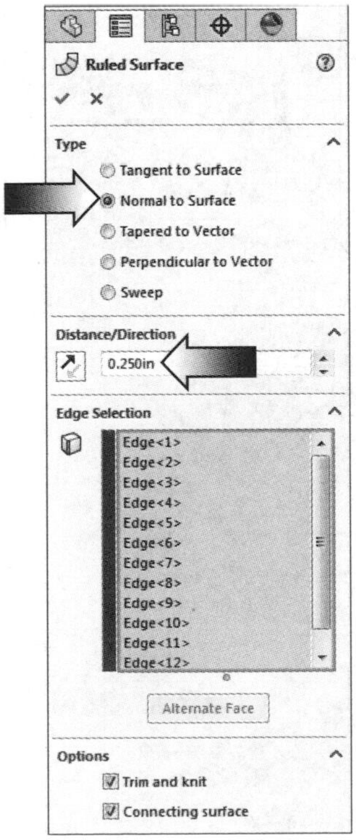

Select the **Normal-To-Surface** option.

Enter **.250in**. for Offset Distance.

Right-click on one of the **outer edges** and select: **Select-Open-Loop**.

Click **OK**.

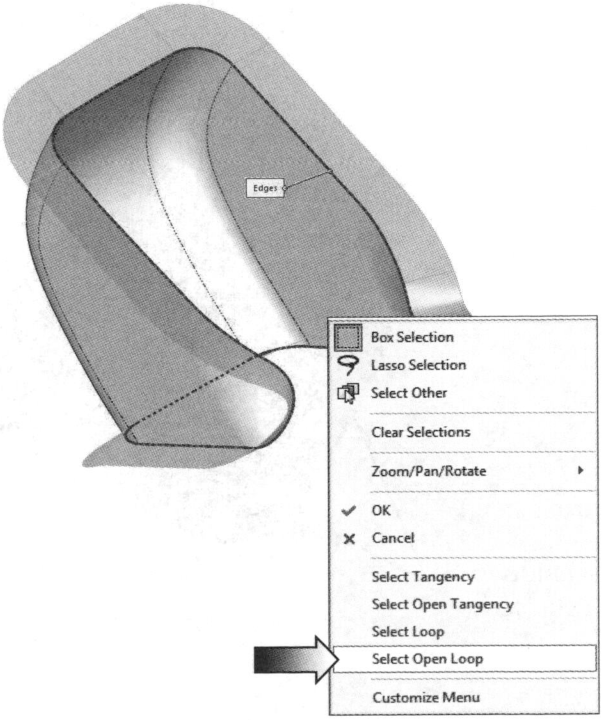

Rotate to different orientations and inspect the ruled surface.

Ruled Surfaces

The Ruled Surfaces creates a set of surfaces that either perpendicular or taper from the selected edges.
These surfaces can also be used as the Interlock Surfaces in molded parts.

9. Knitting the 2 surfaces:

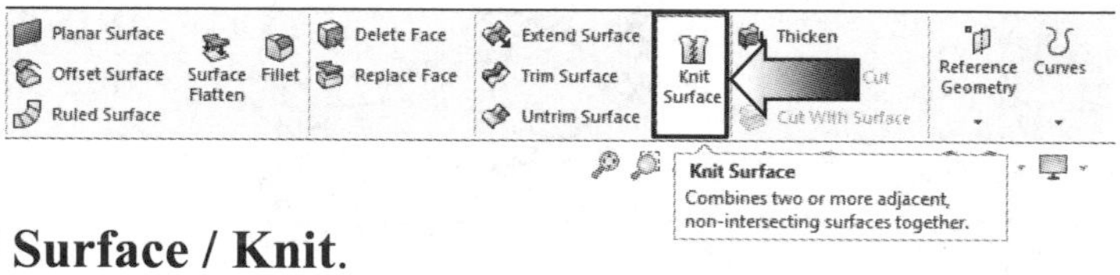

Click [icon] or select **Insert / Surface / Knit.**

Select the **Surface-Offset** and the **Ruled-Surface** to knit.

Clear the **Gap Control** checkbox (arrow).

Click **OK**.

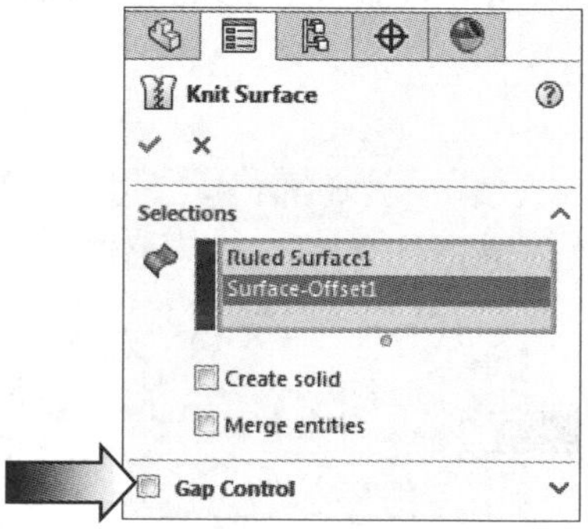

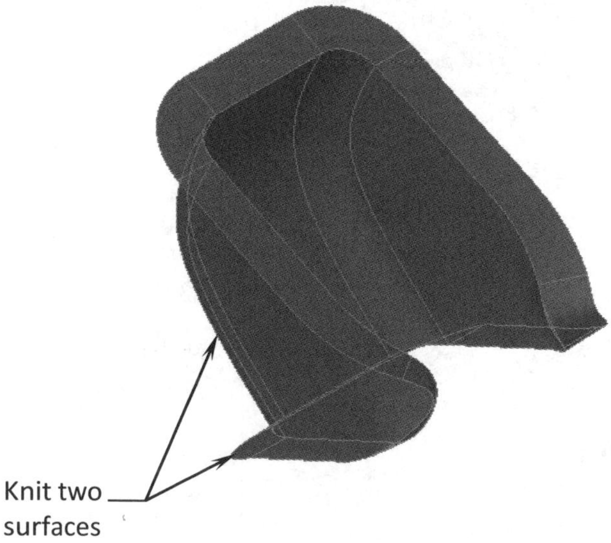

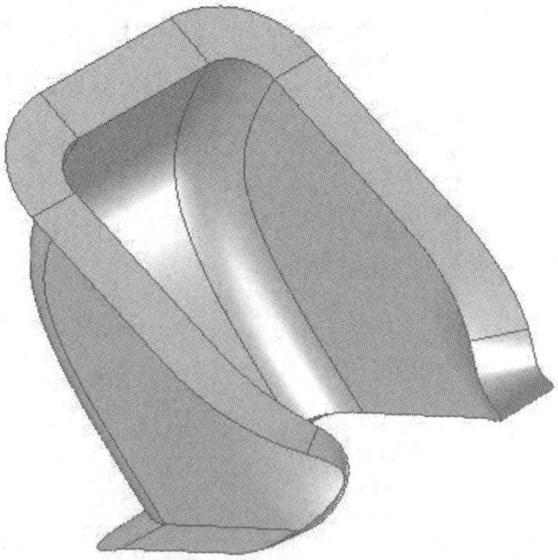

10. Adding .020" Fillets:

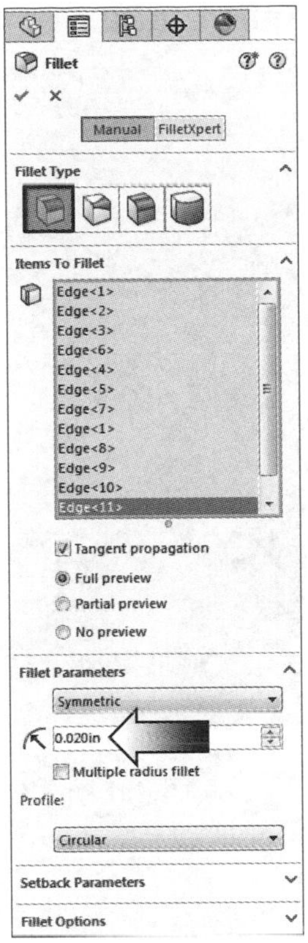

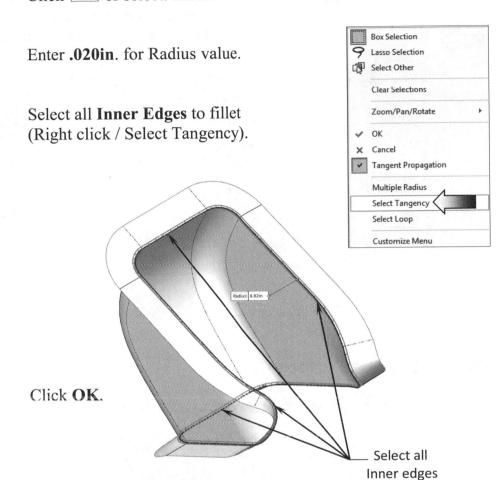

Click  or select: **Insert / Features / Fillet-Round**.

Enter **.020in**. for Radius value.

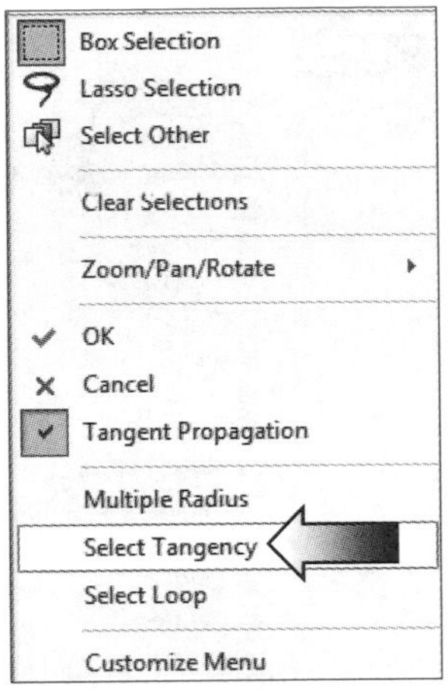

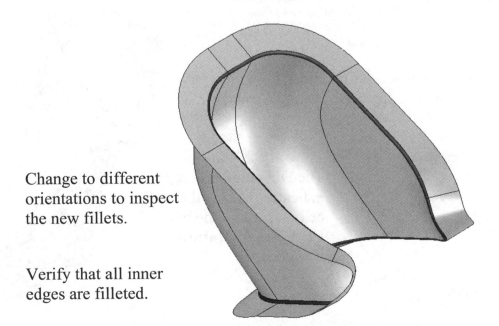

Select all **Inner Edges** to fillet (Right click / Select Tangency).

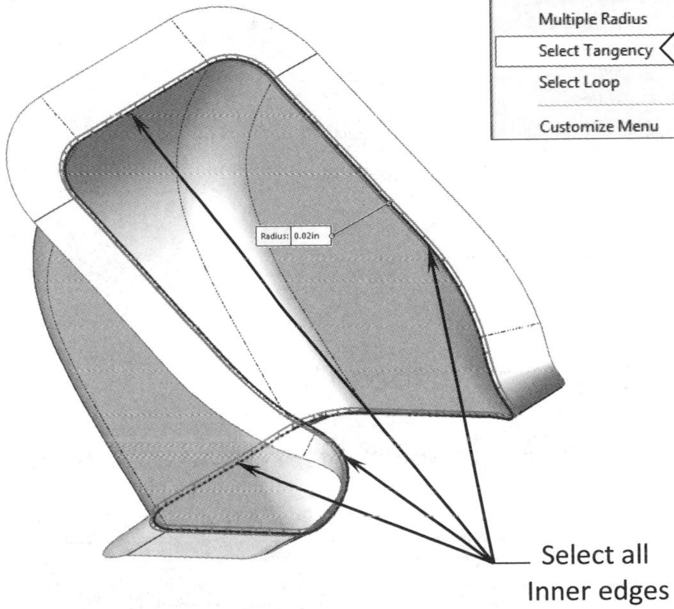

Click **OK**.

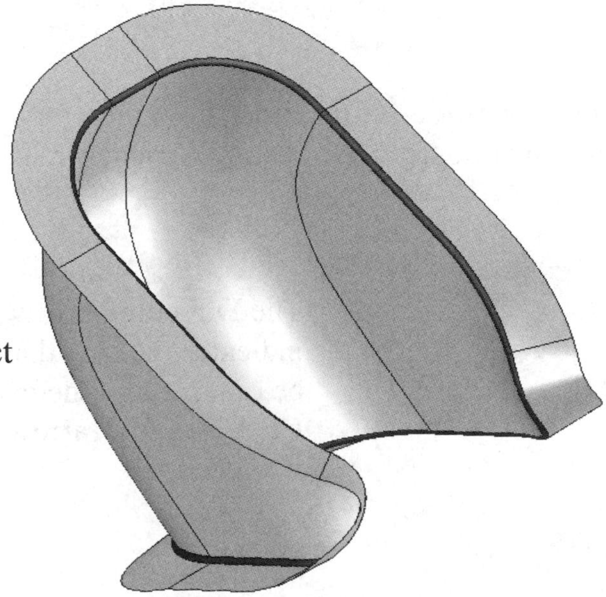

Change to different orientations to inspect the new fillets.

Verify that all inner edges are filleted.

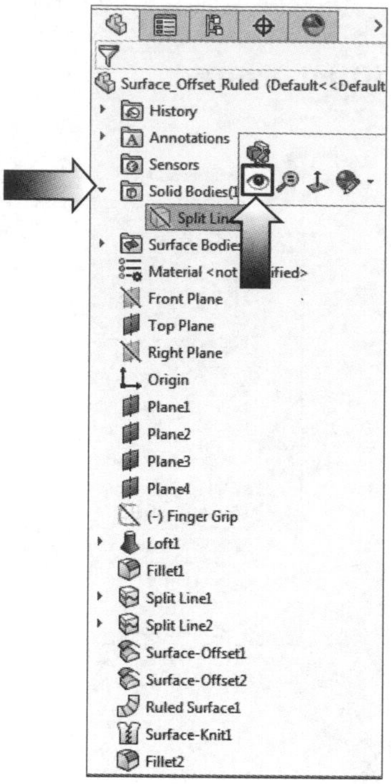

11. Showing the Solid Body:

From the FeatureManager tree, expand the Solid Bodies folder, right-click on the **Split-Line2** body and select: **Show Solid Body**.

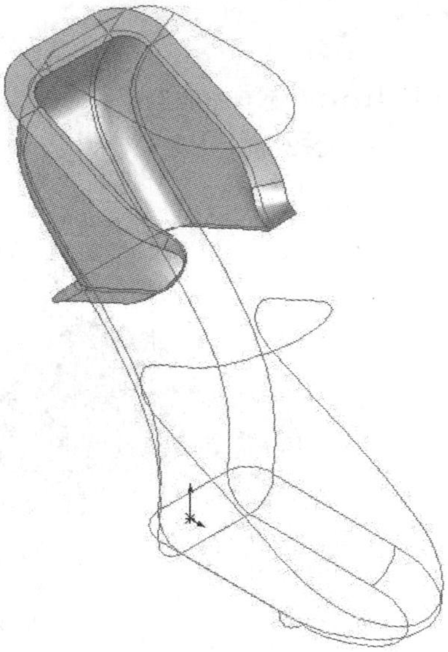

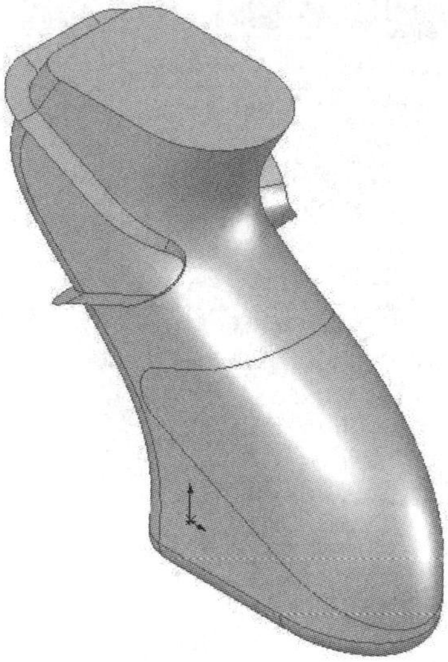

12. Creating the Surface Cut:

Click  or select **Insert / Cut / With Surface**.

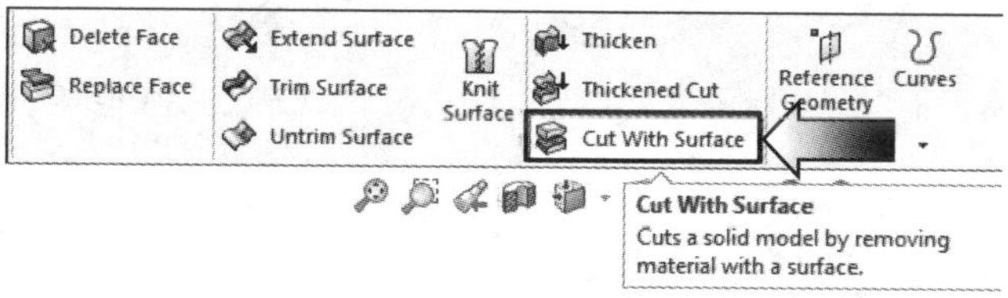

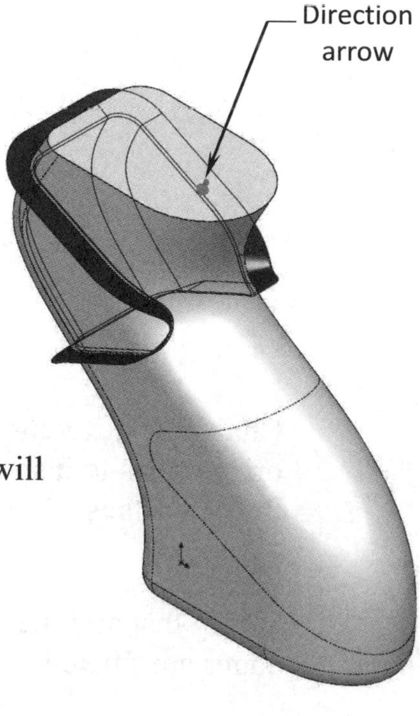

Select the **Surface-Knit1** either from the graphics area or from the FeatureManager tree.

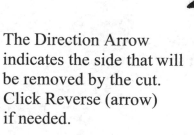

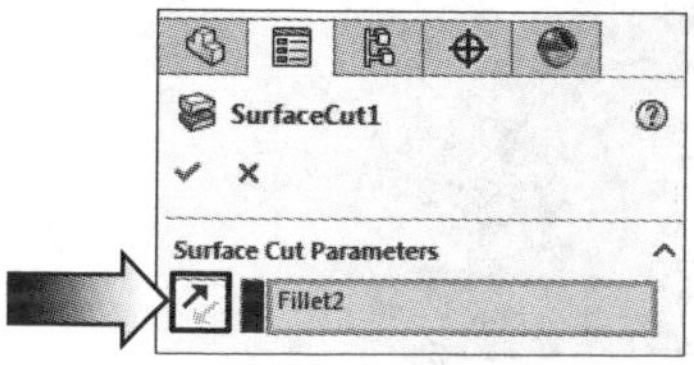

The Direction Arrow indicates the side that will be removed by the cut. Click Reverse (arrow) if needed.

Click **OK**.

13. Hiding the Knit Surface:

Right-click on the Knit Surface and select **Hide**  .

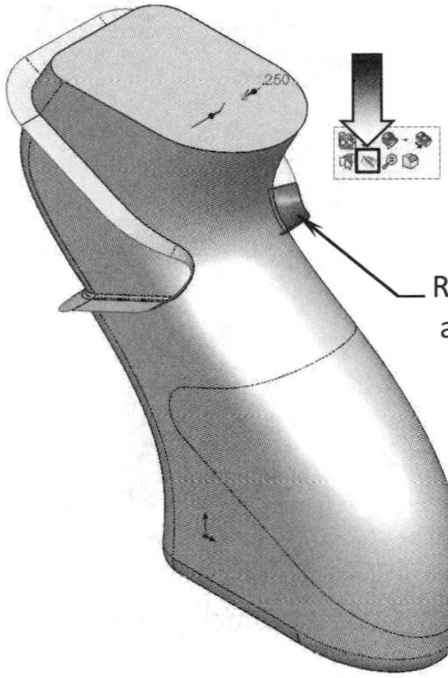

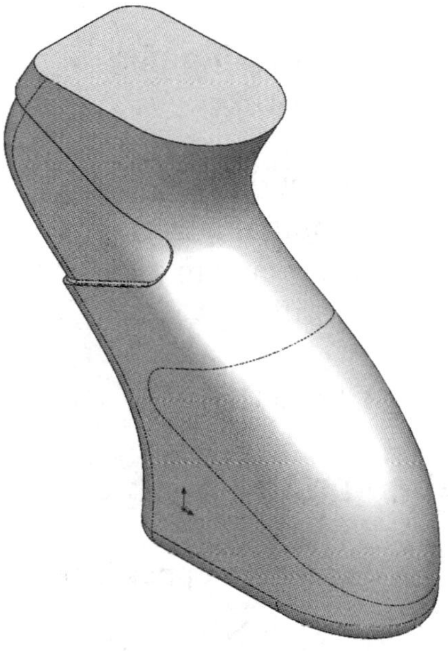

14. Creating the 2nd Ruled Surface:

Repeat from step number 8 to create the 2nd Ruled Surface.

Create the **Surface Cut** as indicated in step 12.

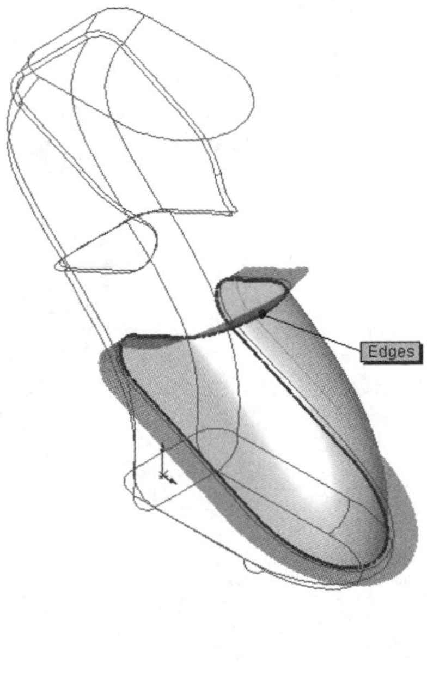

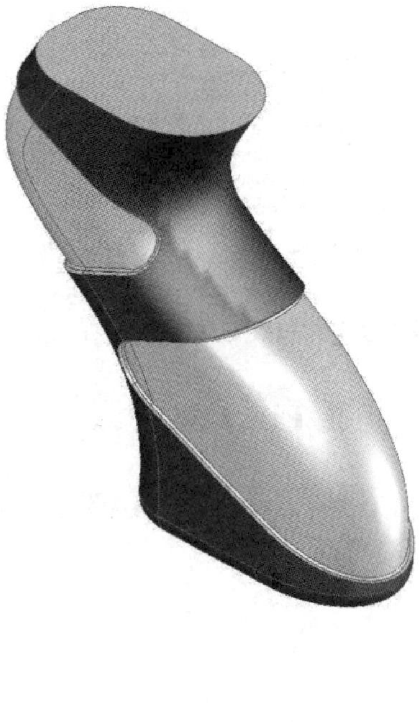

15. Adding more Fillets:

Click  or select: **Insert / Features / Fillet-Round**.

Enter **.020in**. for radius size.

Select the <u>outer edges</u> of both upper and lower cuts to add the fillet.

Click **OK**.

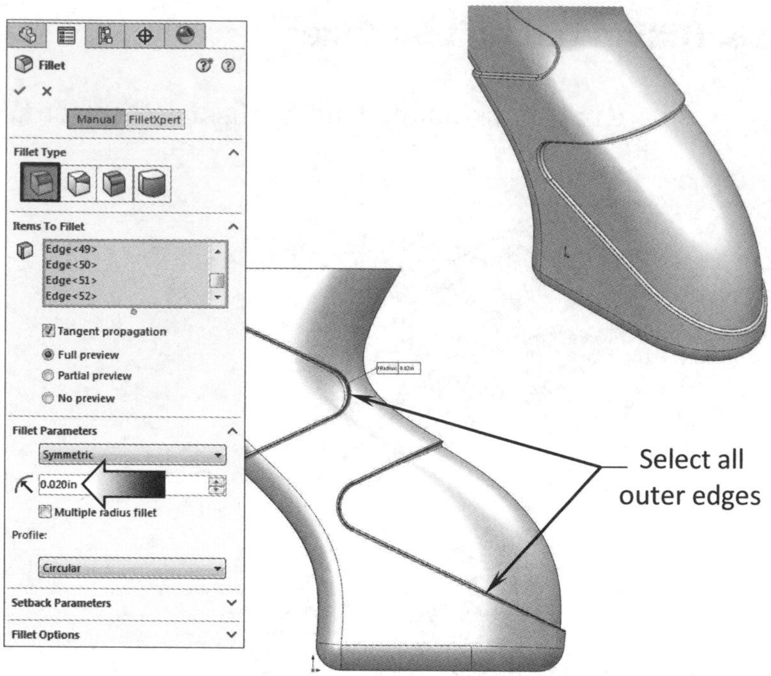

16. Shelling the part:

Select the upper <u>face</u> as noted.

Click  or select **Insert / Features / Shell**.

Enter **.020in**. for thickness.

Click **OK**.

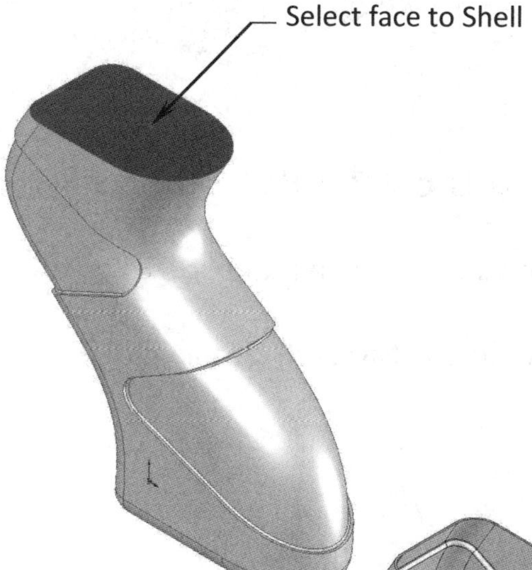

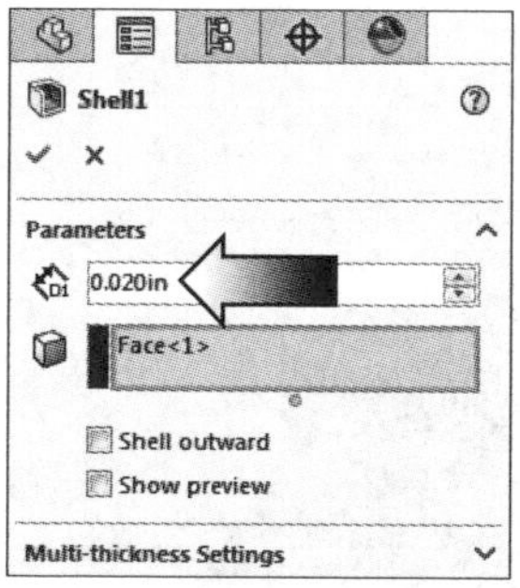

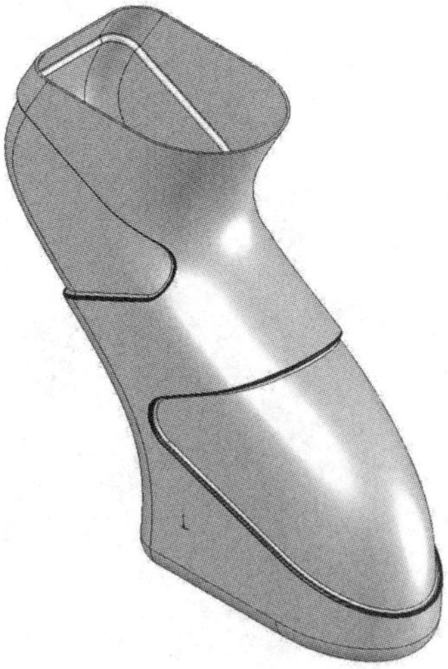

17. Saving your work:

Click **File / Save As**.

Enter: **Surface_Offset_Ruled** for file name.

Click **Save**.

Exercise: Using Loft

1. Opening an Existing file:

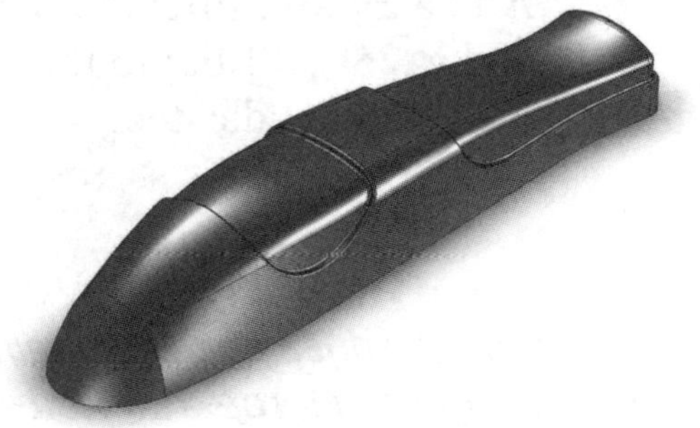

Browse to the Training Files folder.

Open the document named:
Advanced Surfaces Exercise.

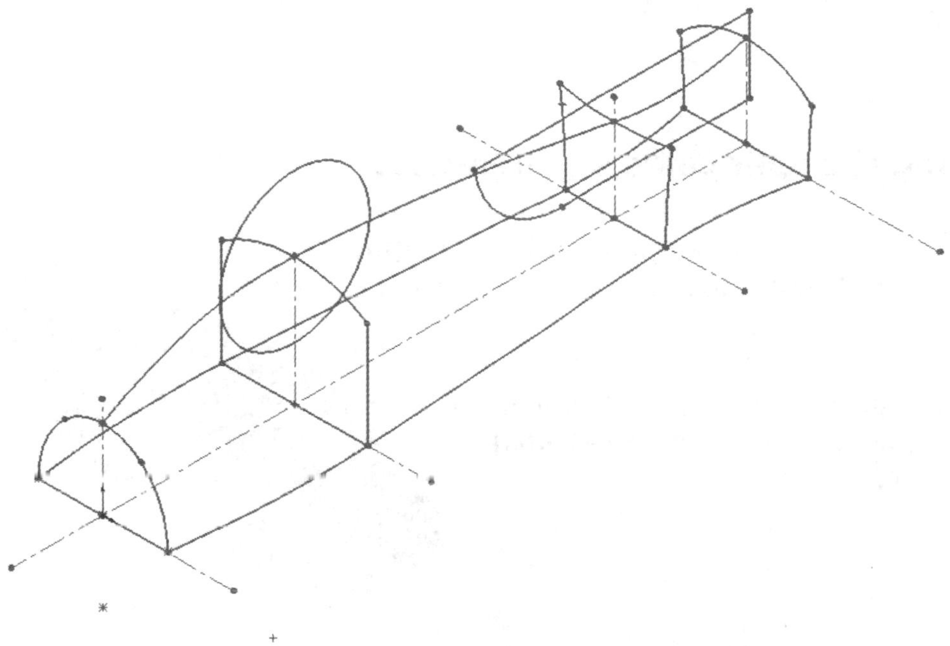

2. Creating the Loft body:

Create a **Solid Loft** from the **4 profiles** as indicated.

Use the **2 bottom Guide Curves** to control the sides.

Use the top **Guide Curve** to control the upper curvatures.

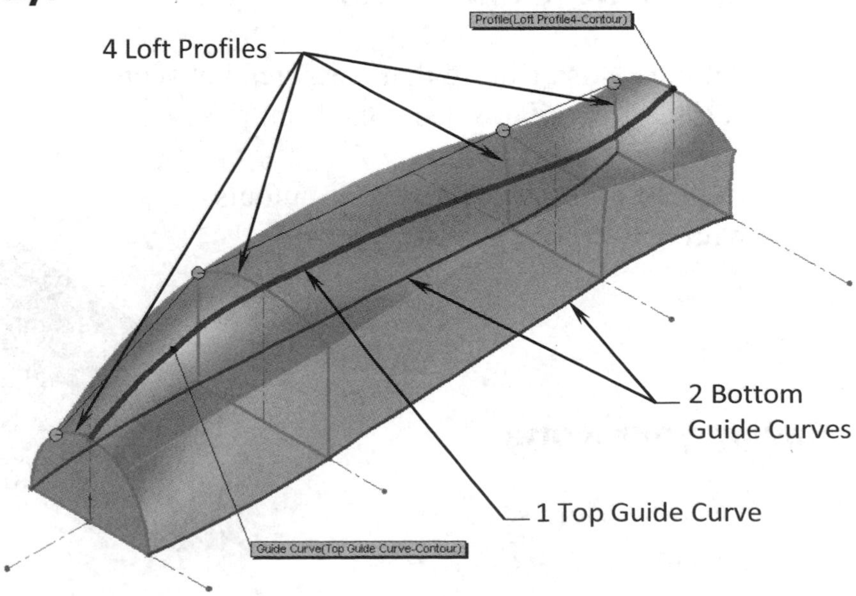

3. Adding Fillets:

Add a **.500in**. fillet to the upper edges, and a **.250in.** fillet to the edges on the end, as indicated.

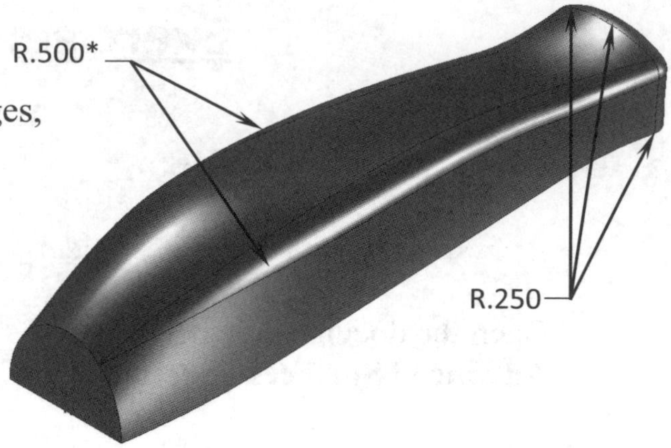

* *The tangent edges can be eliminated by enabling the Merge-Tangent-Faces in the loft options.*

4. Creating the Split Lines & Lofted-Cuts:

Use the 2 sketches named: **Circular-Split** and **Side-Split** to create 2 split surfaces.

Create 2 lofted-cuts at **.090**in. deep, using either the **Offset** or **Ruled** surface options.

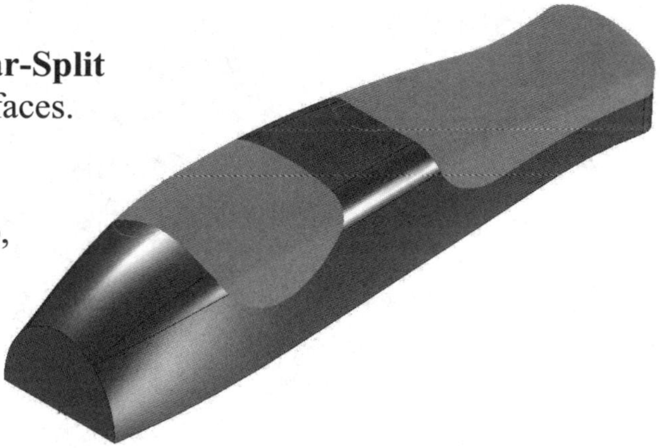

5. Adding the Nose and Fillets:

Add the Nose feature that measured between **1.250in**. to **1.500in**. from the front face.

Replace all sharp edges of the cutouts with **.040in**. fillets.

6. Saving your work:

Save the exercise as:
Advanced_Surfaces_Exe.

Exercise: Advanced Surfacing Techniques

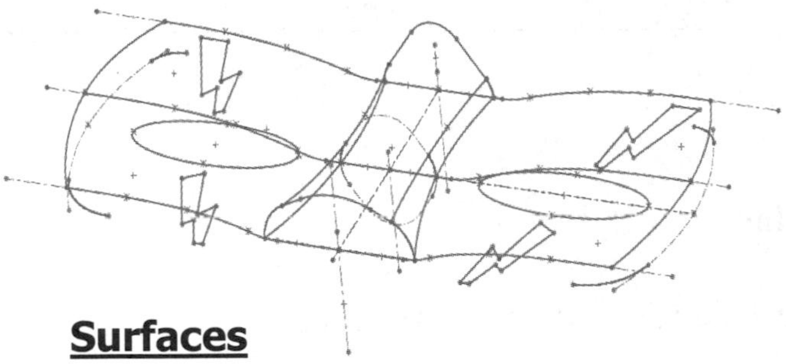

Surfaces

Surfaces are a type of geometry that can be used to create solid features. Surfaces can be created in a variety of different ways; from a sketch or multiple sketches, a surface can be made by extruding, revolving, sweeping, and lofting.

Surfaces are normally created individually and knitted together so that an enclosed volume or a solid feature can be generated afterwards.

This exercise discusses some advanced techniques on surfacing such as Lofted Surface, Boundary Surface, Trimmed Surface, Offset Surface, Extrude From, and variable fillets.

1. Opening an existing document named:

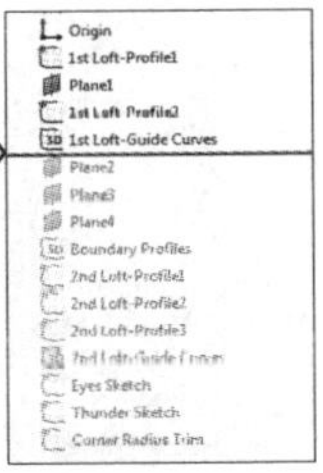

Advanced Surfacing Techniques from the Training Files folder. Rollback below the sketch: 1st Loft Guide-Curves.

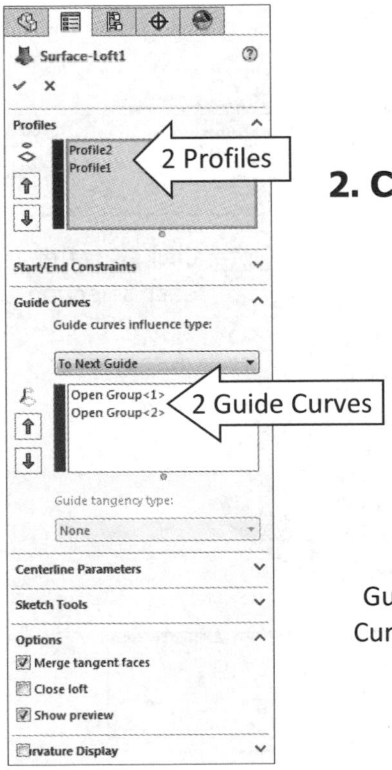

2. Creating the 1st Lofted Surface:

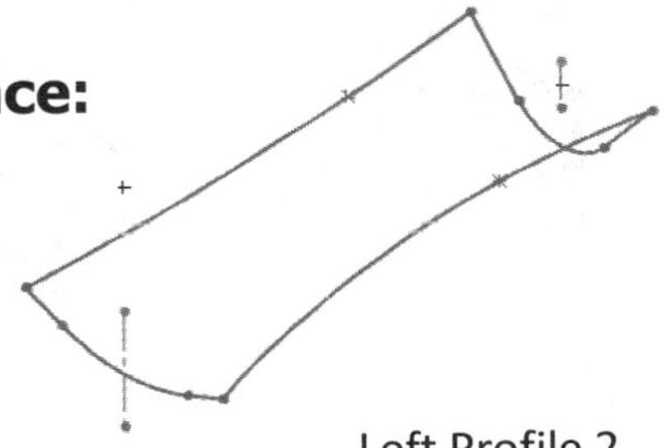

Click or select: **Insert / Surface / Loft.**

Select the **2 Loft Profiles** and the **2 Guide Curves** as indicated.

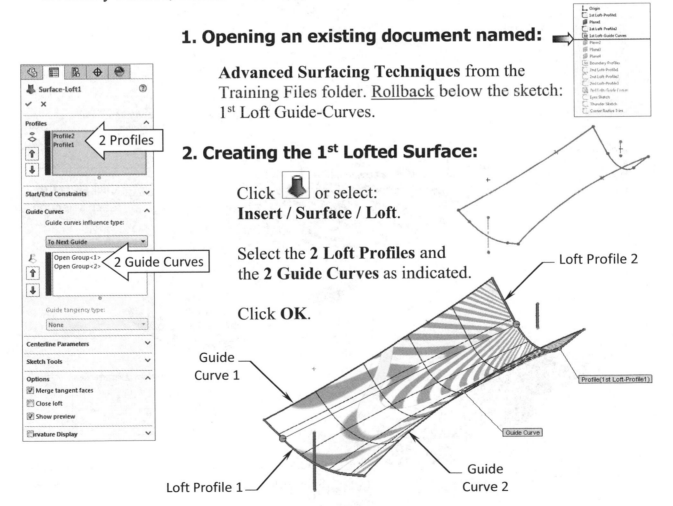

Click **OK**.

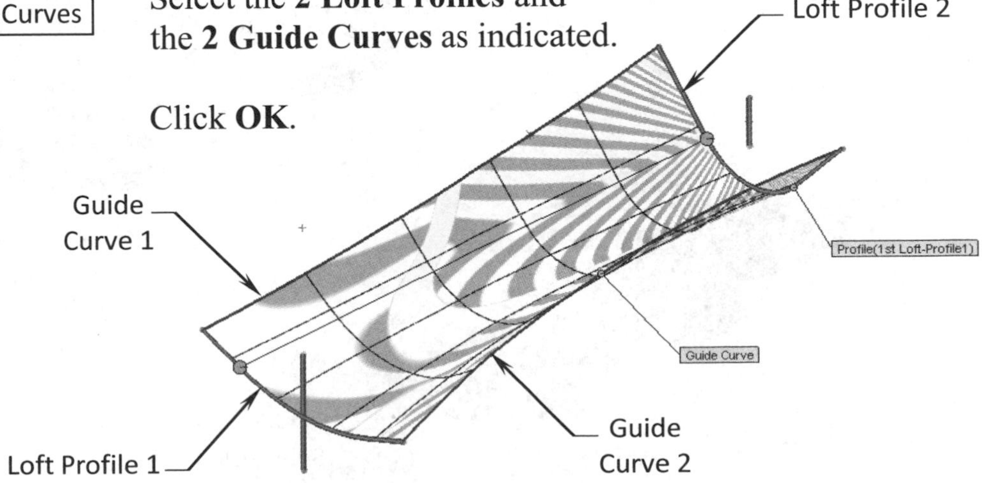

Lofted-Surface creates a feature by making transitions between two or more profiles. A loft can either be a surface or solid and one or more Guide Curves can be used to guide the transitions between the profiles.

3. Creating the 2nd Lofted Surface:

Click or select: **Insert / Surface / Loft.**

Select the **3 Loft Profiles** and the **4 Guide Curves** as indicated.

Click OK after each selection of profile and guide curves.

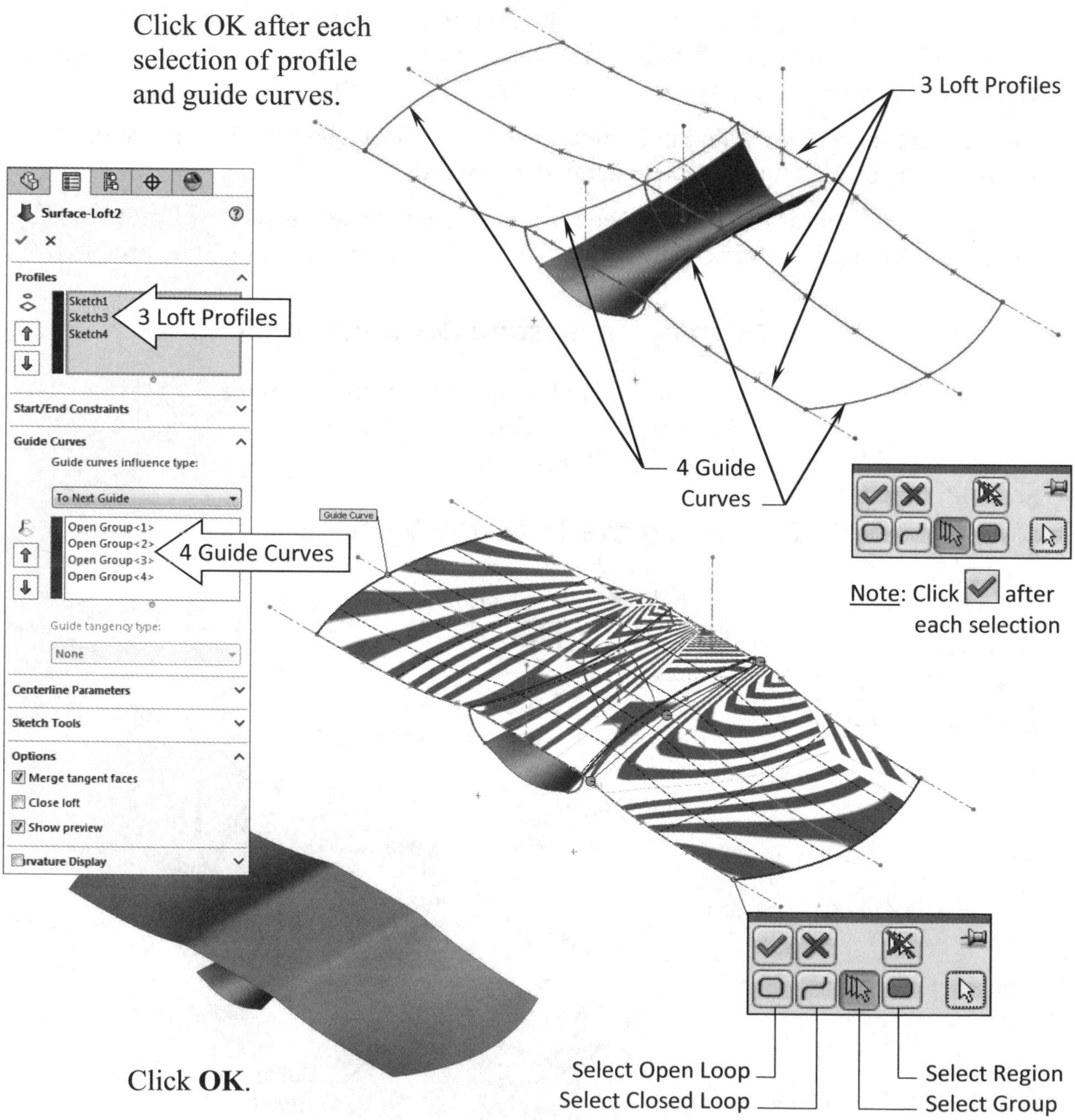

Click **OK**.

4. Creating the Boundary-Surfaces:

Click  or select **Insert / Surface / Boundary Surface.**

For **Direction 1**, select the **2 edges** as shown (Blue tags).

For **Direction 2**, select the **3 Arcs** in the Boundary Sketch as shown (Purple tags).

Boundary-Surface

Creates a new surface from a set of 2D or 3D sketch entities.
The Boundary Surface can be tangent or curvature-continuous in both directions (all sides of the surface).
This option offers a higher quality result than the Loft.

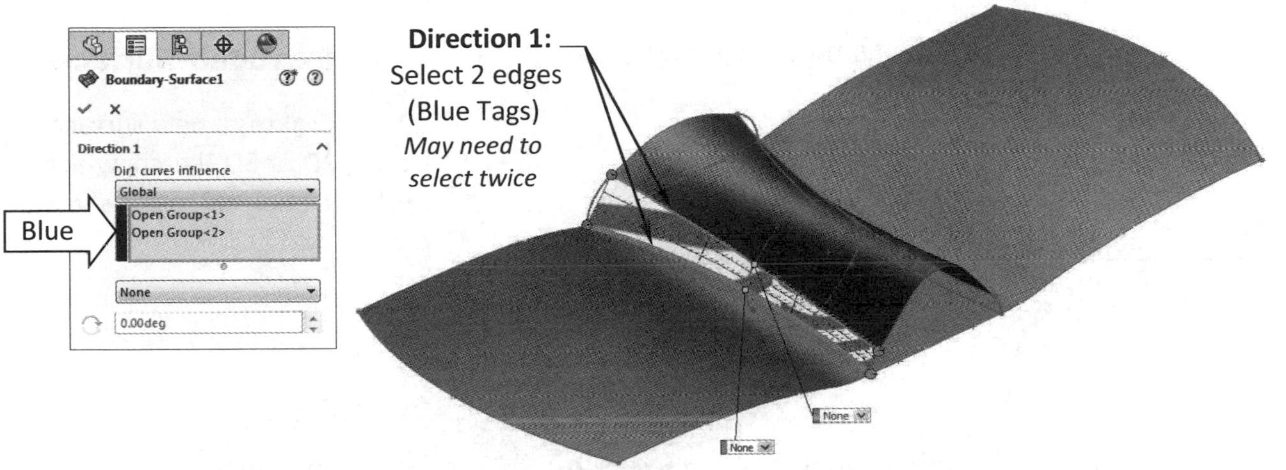

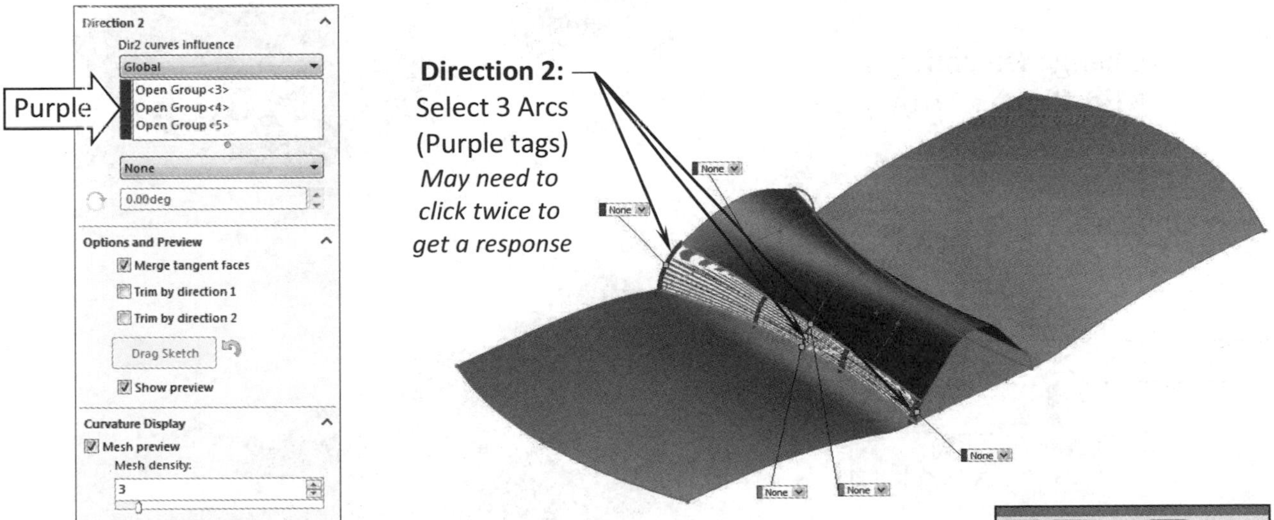

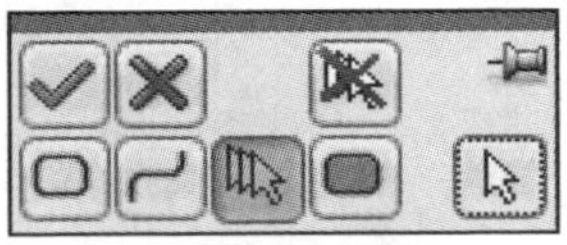

Note: Click ✓ after each selection

Click **OK**.

5. Repeating:

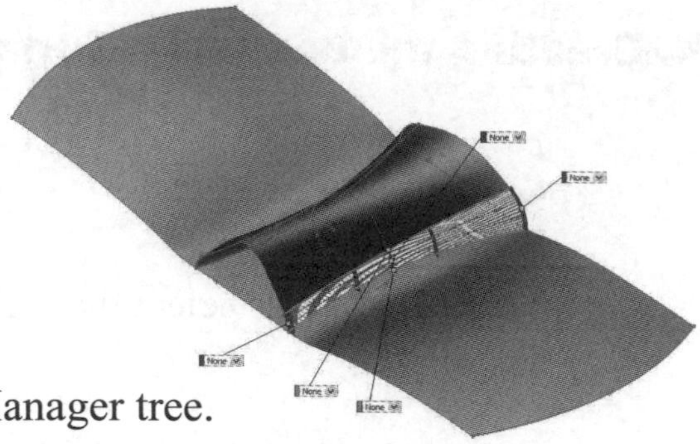

Repeat step number 4 and create another Boundary Surface on the opposite side.

6. Creating an Extruded-Surface:

Select the **Eyes Sketch** from the FeatureManager tree.

Click  or select: **Insert / Surface / Extrude**.

Change the option **Extrude From** to **Surface/Face/Plane** (Arrow).

Select the surface as indicated to extrude.

Extruded-Surface

Creates a new surface from a 2D or 3D sketch, which protrudes normal to the sketch plane.

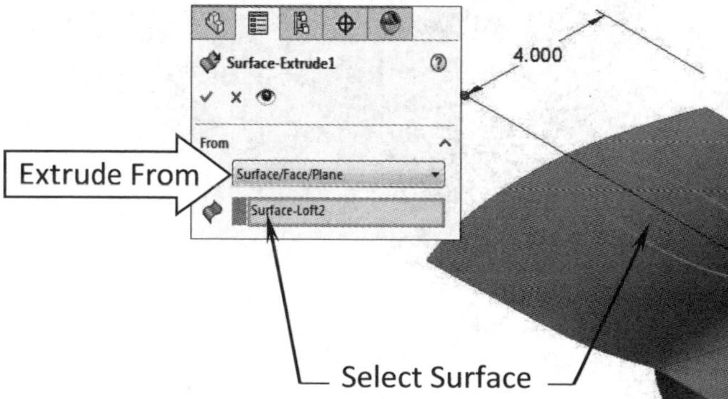

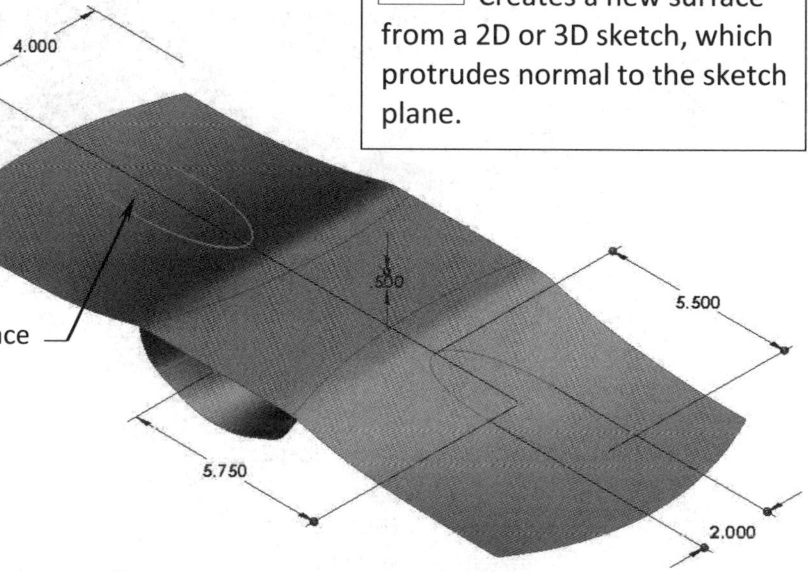

Change **Direction 1** to **Mid Plane** (Arrow).

Enter **.410in.** for extrude depth.

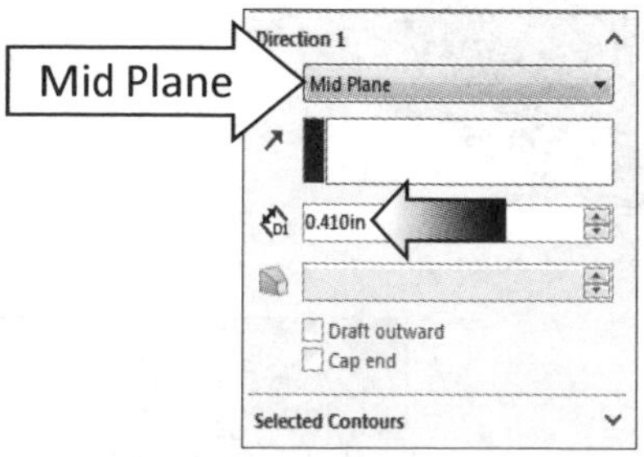

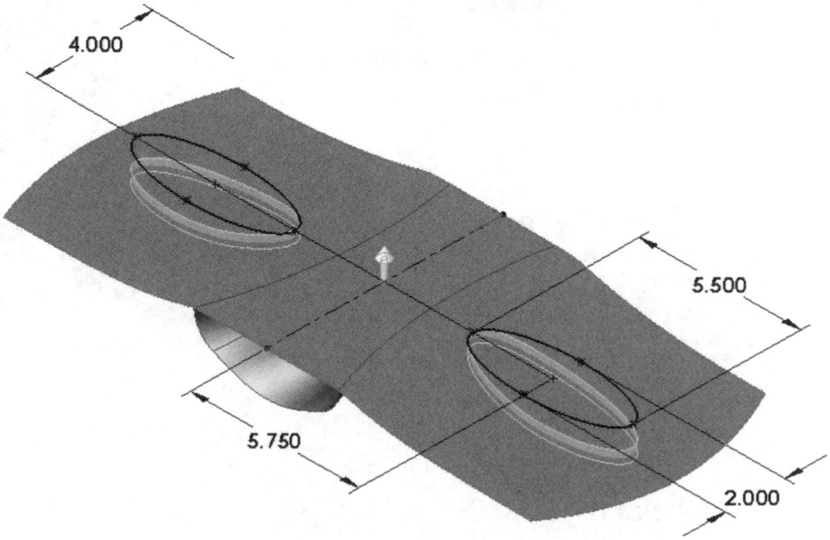

Click **OK**.

7. Creating a Trimmed Surface:

Trimmed-Surface

Uses a plane, a surface, or a sketch as a trim tool to trim the intersecting surfaces.

Click  or select **Insert / Surface / Trim.**

Select **Mutual** under Trim Type (Arrow).

For **Trimming Surfaces**, select **all surfaces** of the model.

For **Remove Selections**, select the **5 Faces** as shown (Arrow).

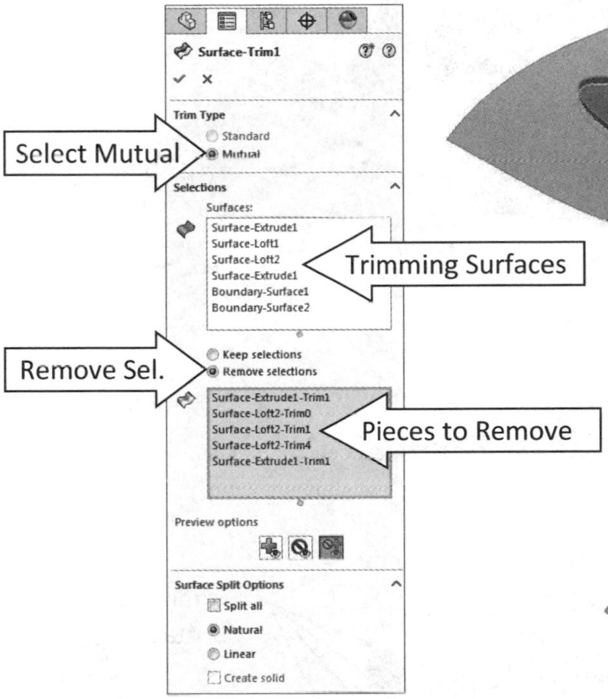

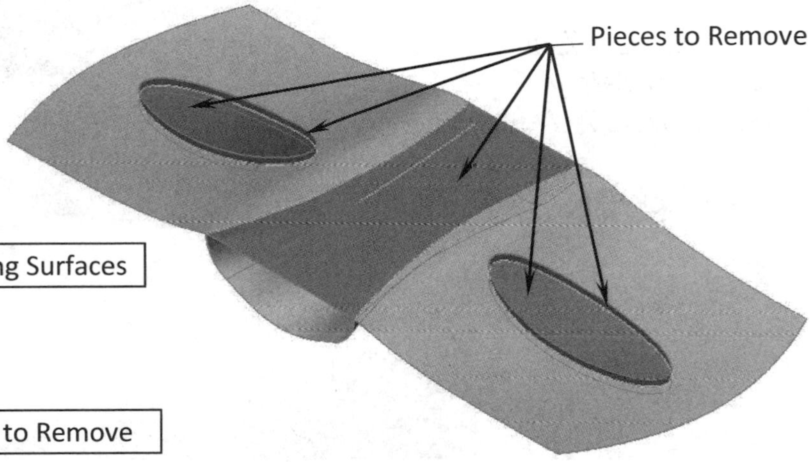

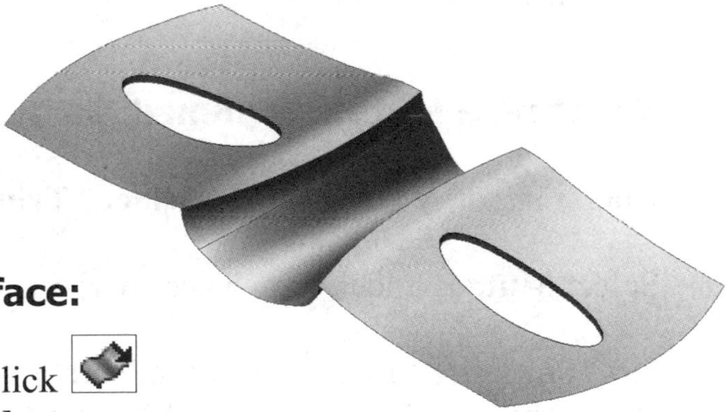

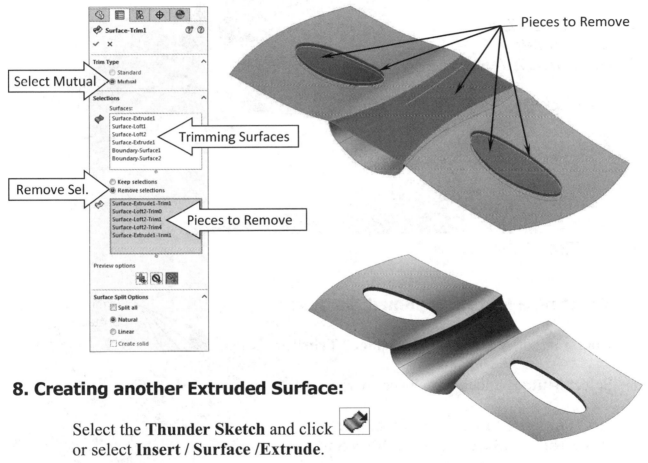

8. Creating another Extruded Surface:

Select the **Thunder Sketch** and click or select **Insert / Surface /Extrude.**

Set the **Direction 1** to **Blind** and click **Reverse Direction.**

Set **Extrude Depth** to: **1.250in.**

Click **OK.**

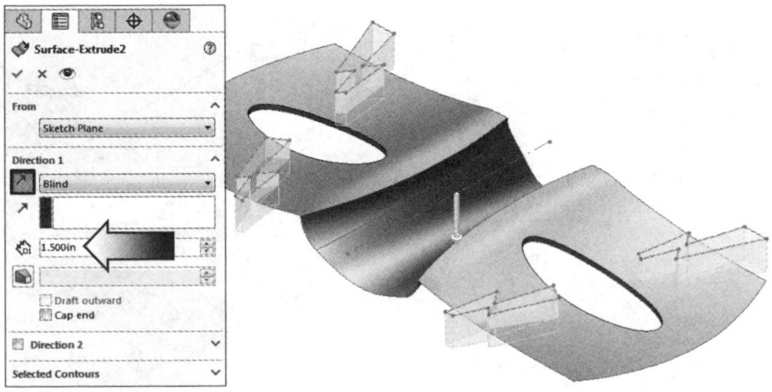

9. Creating an Offset-Surface:

Offset-Surface

Creates a copy of a surface in either direction and parallel to the selected surface. The offset distance can be zero or any other value.

Click or select **Insert / Surface / Offset**.

Select the **2 surfaces** as shown (arrow).

Enter **.100in.** for **Offset Distance**.

Place the copy on the **bottom** of the original.

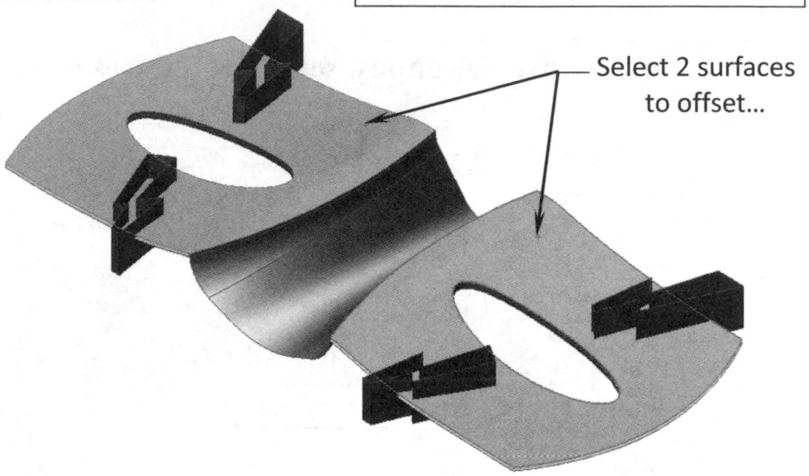

<u>Note:</u> Create 2 offset surfaces separately if the next trim failed.

Click **OK**.

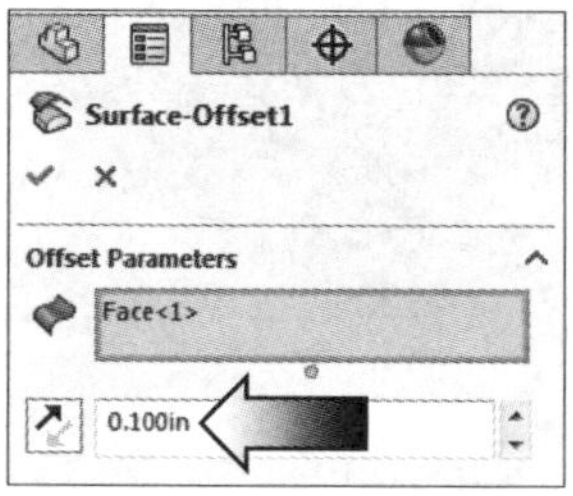

10. Creating a Mutual Trimmed Surface:

Click or select **Insert / Surface / Trim***.

Select **Mutual** under Trim Type (Arrow).

For **Trimming Surfaces**, select **all surfaces** of the Thunder Sketch and the Offset Surfaces.

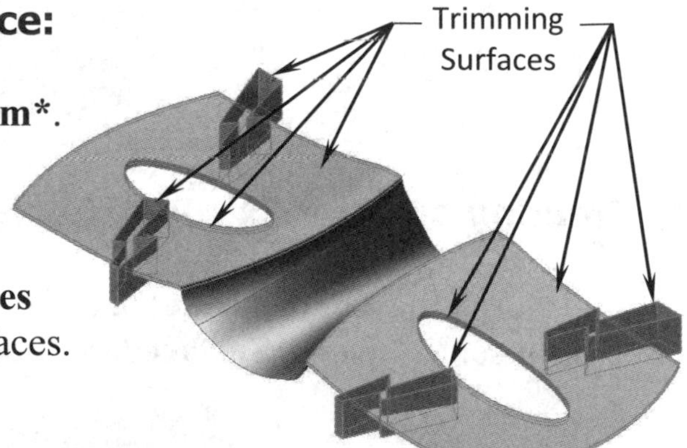

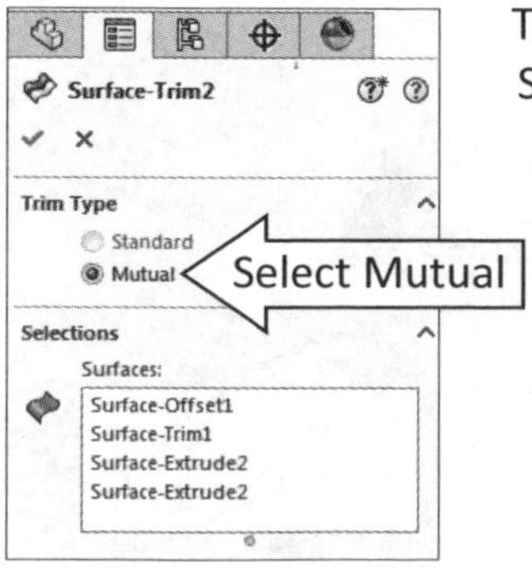

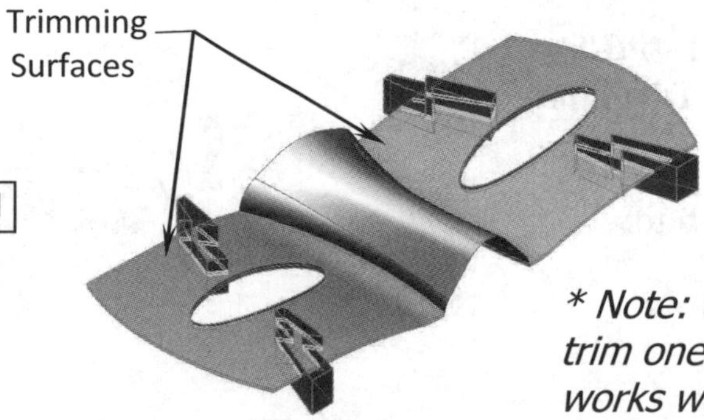

** Note: Using Standard-Trim to trim one feature each time also works well.*

For **Remove Selection**, select the following:

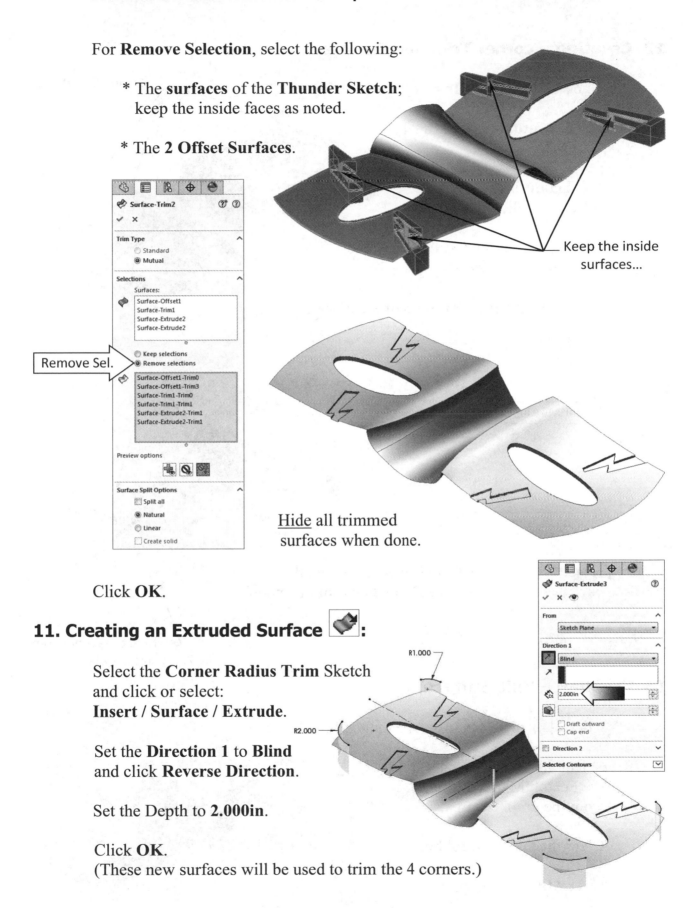

* The **surfaces** of the **Thunder Sketch**; keep the inside faces as noted.

* The **2 Offset Surfaces**.

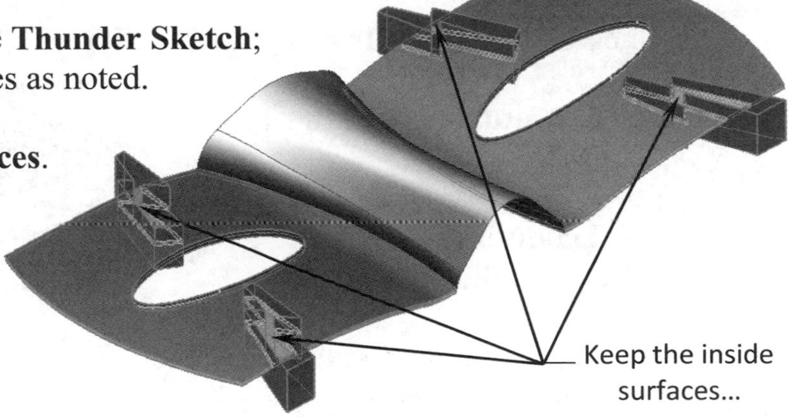

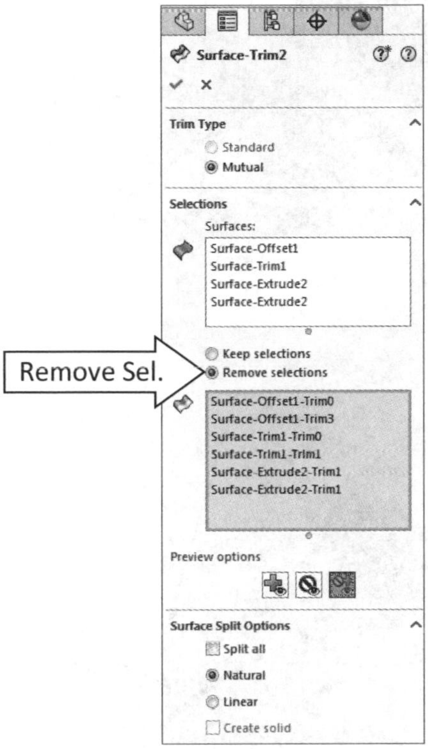

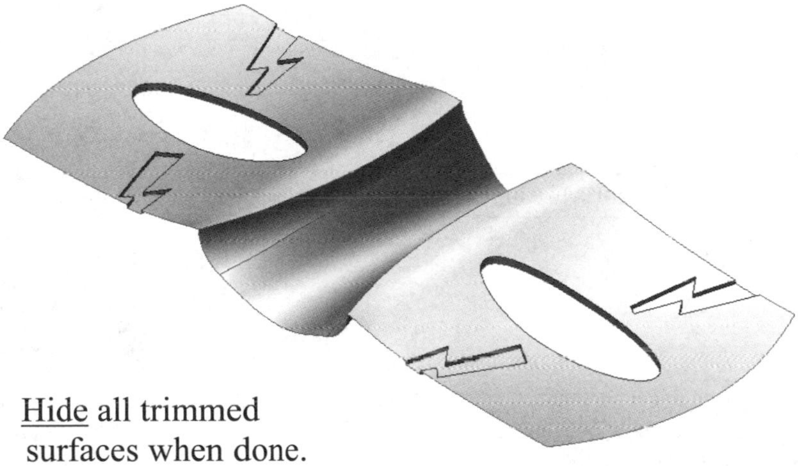

Hide all trimmed surfaces when done.

Click **OK**.

11. Creating an Extruded Surface:

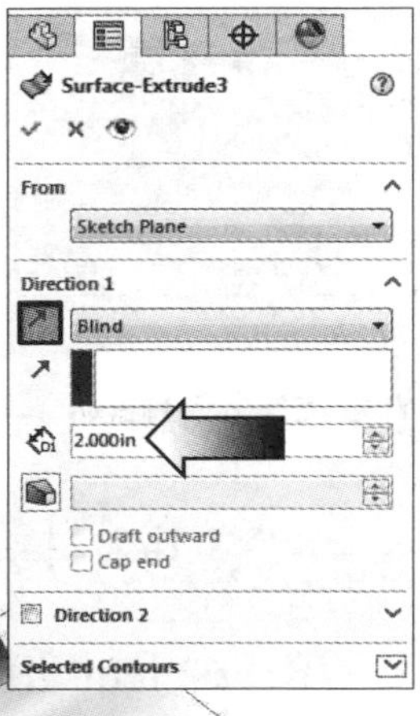

Select the **Corner Radius Trim** Sketch and click or select: **Insert / Surface / Extrude**.

Set the **Direction 1** to **Blind** and click **Reverse Direction**.

Set the Depth to **2.000in**.

Click **OK**.
(These new surfaces will be used to trim the 4 corners.)

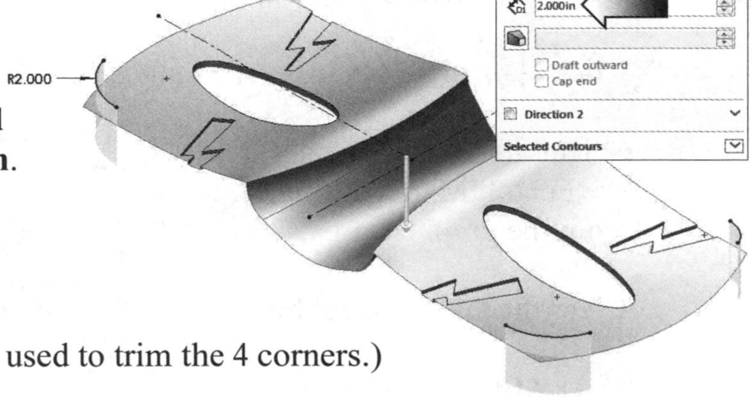

12. Creating a corner Trimmed Surface:

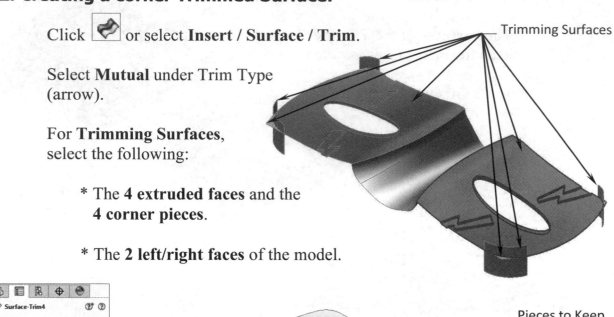

Click [icon] or select **Insert / Surface / Trim**.

Select **Mutual** under Trim Type (arrow).

For **Trimming Surfaces**, select the following:

* The **4 extruded faces** and the **4 corner pieces**.

* The **2 left/right faces** of the model.

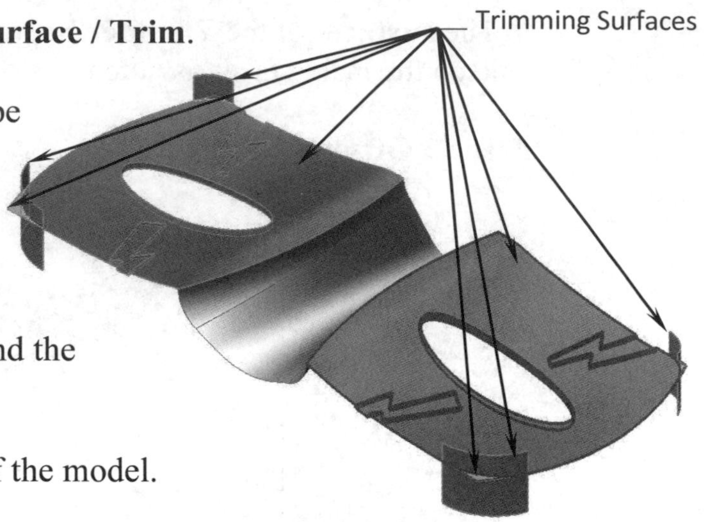

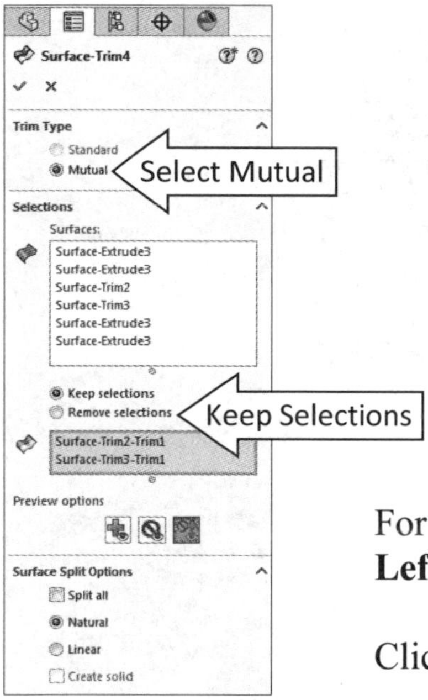

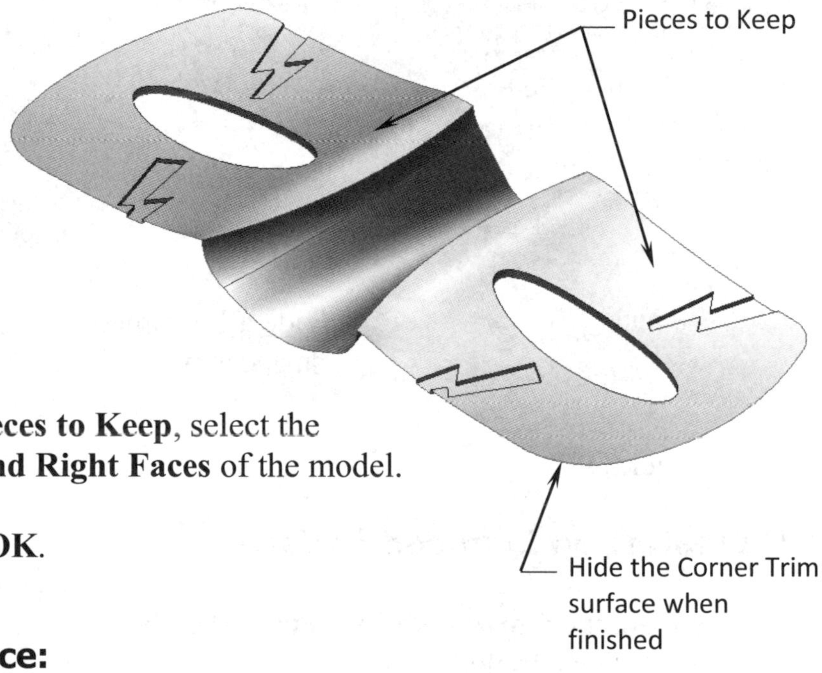

For **Pieces to Keep**, select the **Left and Right Faces** of the model.

Click **OK**.

13. Creating a Knit-Surface:

Click [icon] or select **Insert / Surface / Knit**.

Select **all surfaces** either from the FeatureManager tree or from the graphics area.

Clear the Gap Control box.

Click **OK**.

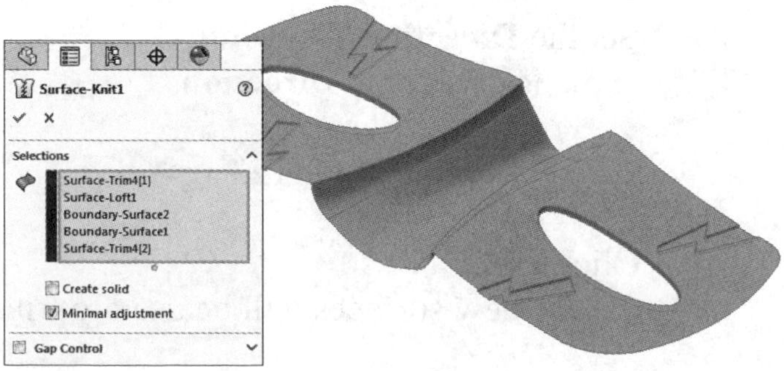

14. Adding a Variable Fillet:

Click  or select **Insert / Features / Fillet-Round**.

Select the **2 edges** shown.

Use the **Call-out tags** and enter the radius values as noted.

Click **OK**.

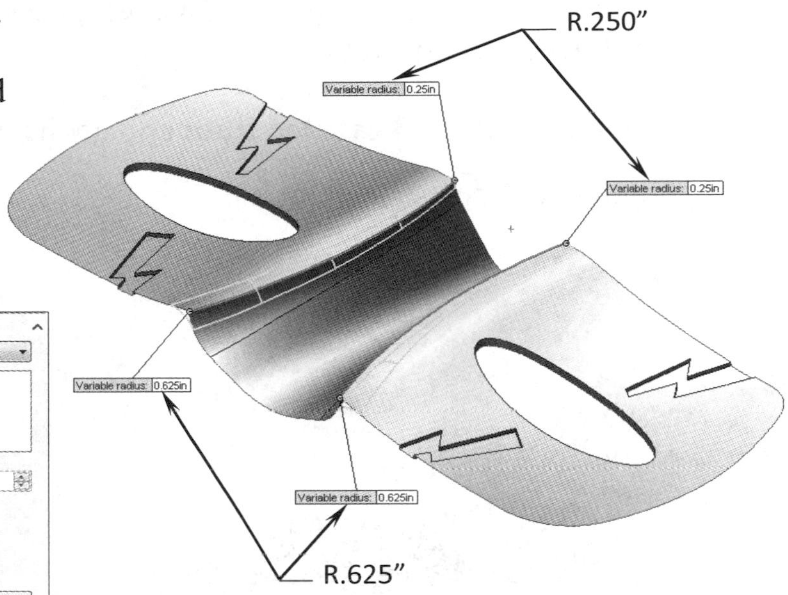

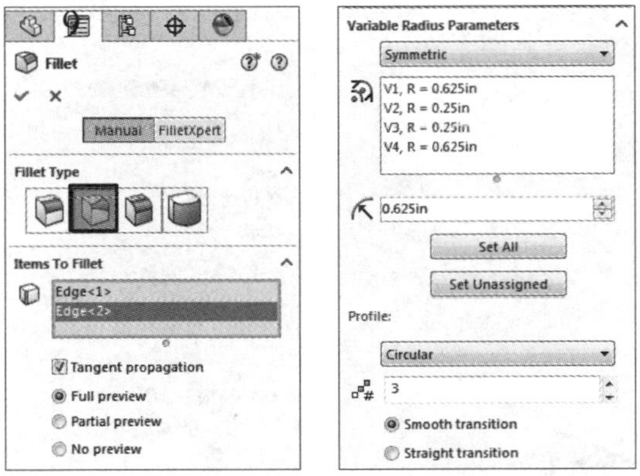

15. Adding a Constant Fillet:

Click  or select **Insert / Features / Fillet-Round**.

Select the **2 edges** as shown.

Enter **.200in.** for Radius values.

Click **OK**.

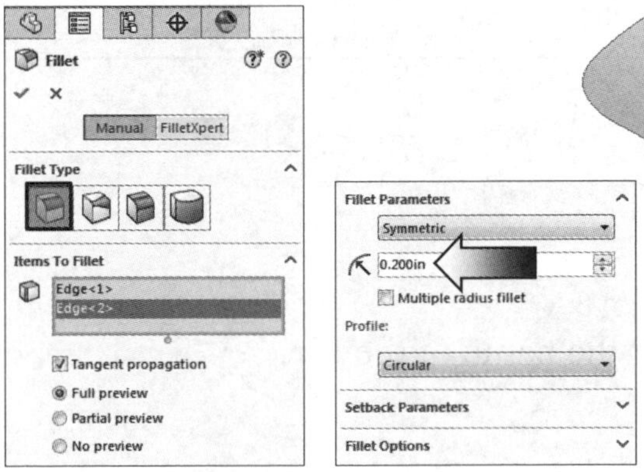

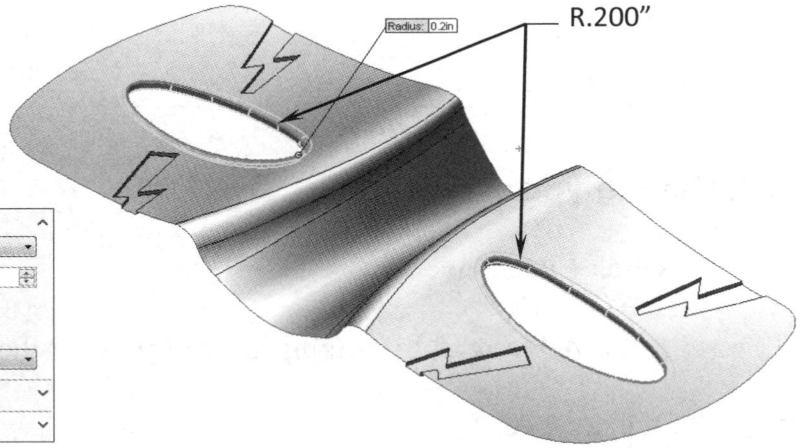

16. <u>Optional</u>: Adding texture

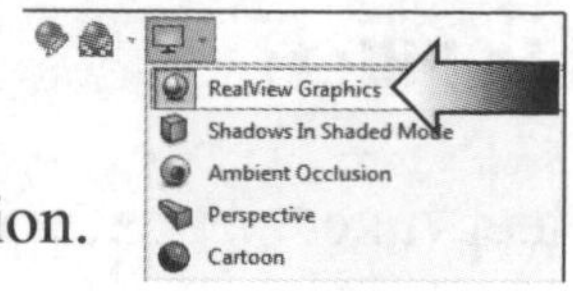

Enable the **RealView Graphics** option.

From the Task Pane expand the **Appearances**  folder and locate the **Rubber / Texture** folder.

Drag & Drop the **Textured Rubber** onto the part and select the <u>Apply to Part</u> option  .

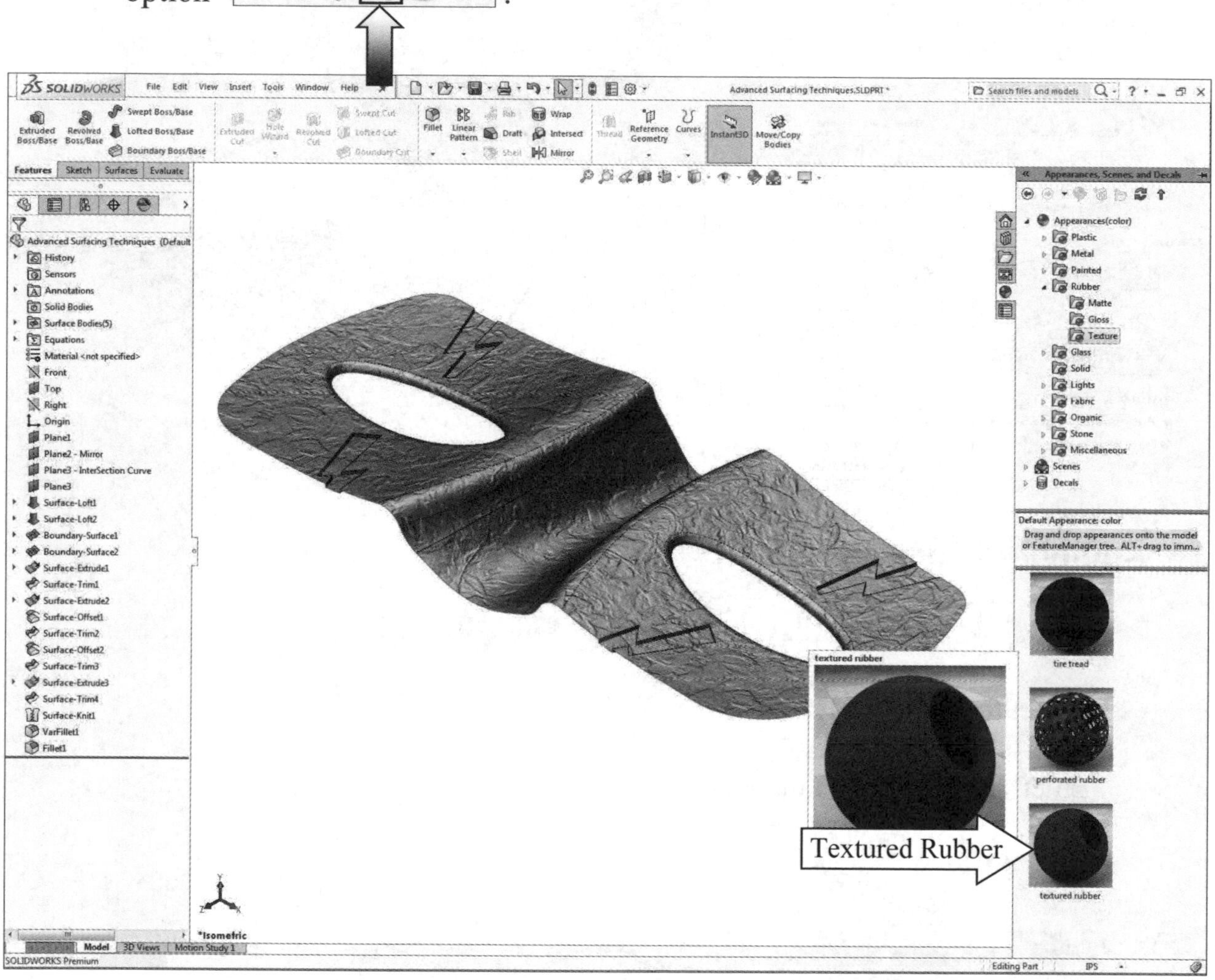

17. Saving your work:

Click **File / Save As**.

Enter **Advanced Surfacing Techniques** for the name of the file.

Click **Save**.

Exercise: Using Split

1. Opening a part document:

Browse to the Training Files folder and open a part document named: **Using Split.sldprt**.

This model contains a single surface body and a 2D sketch.

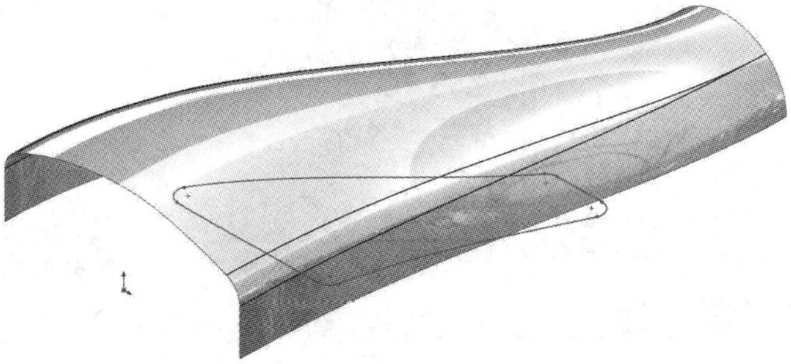

The 2D sketch will be used to create the split line in the next step.

2. Creating a split line:

Switch to the **Surfaces** tab.

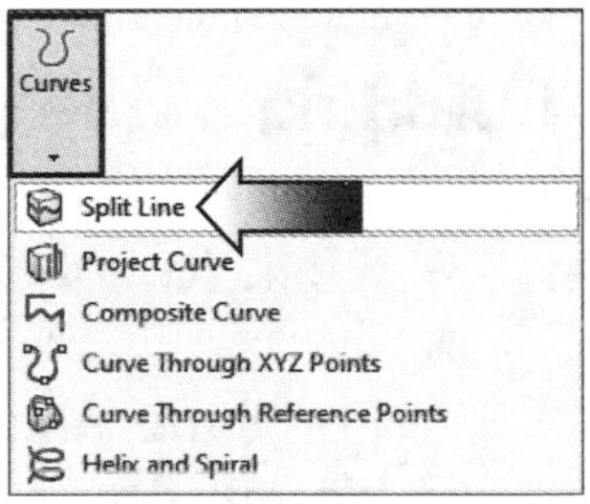

Expand the **Curves** drop-down and select **Split Line**.

For Type of Split, use the default **Projection** option.

For Selections select **Sketch3**.

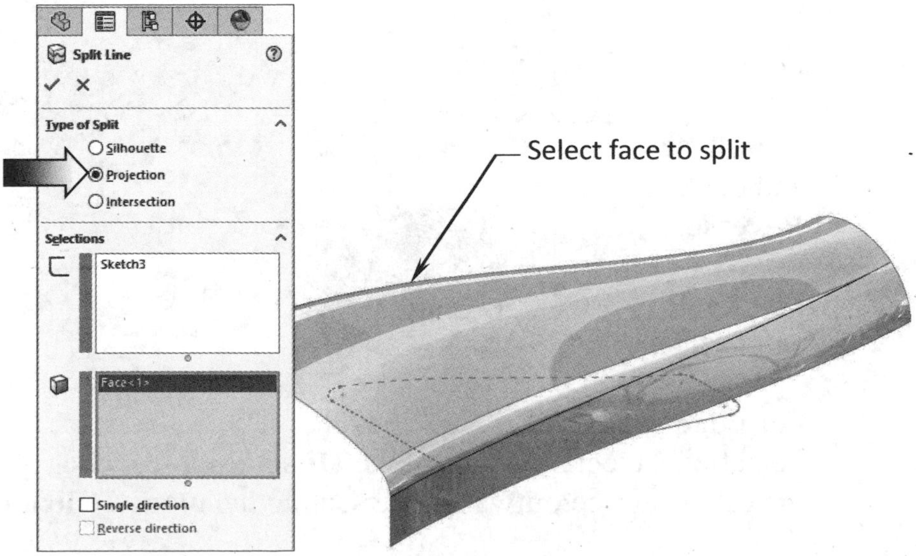

For Faces to Split, select the **upper face** of the model as indicated.

Click **OK**.

3. Making an offset surface:

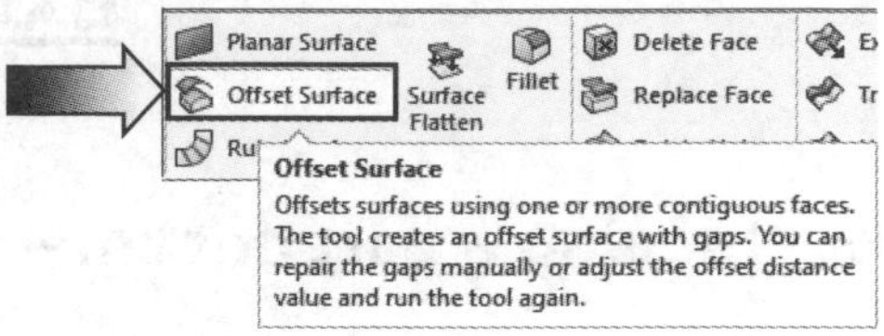

Click **Offset Surface** on the **Surfaces** tab.

For Offset Parameters, select the **inside** of the split surface.

For Distance, enter **.250"** and click **Reverse**.

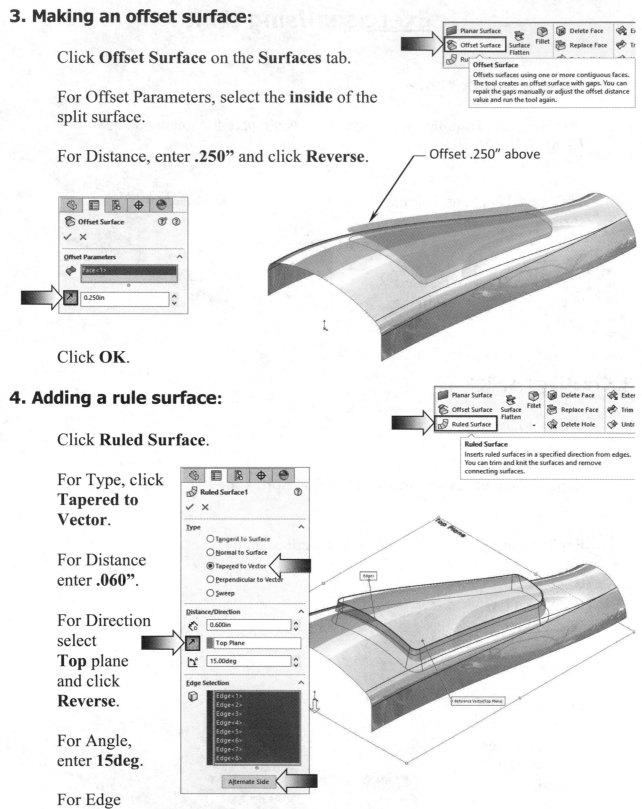

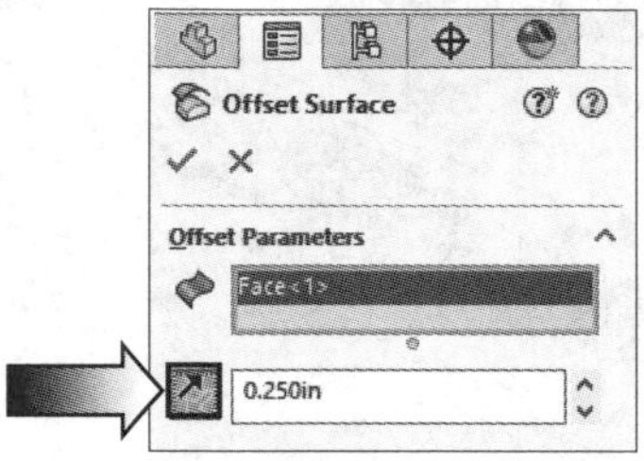

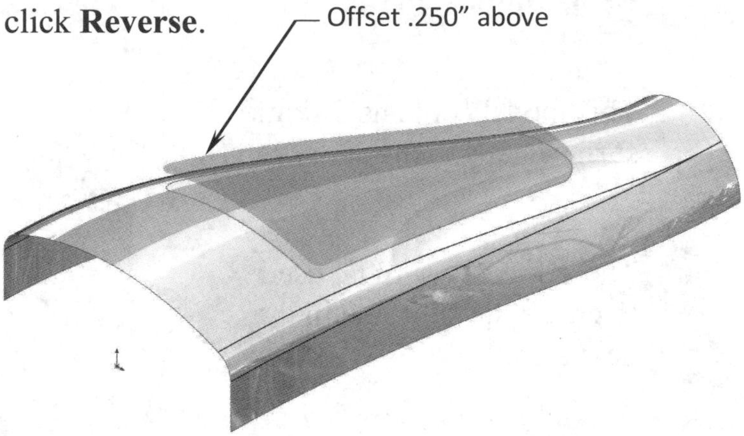

Click **OK**.

4. Adding a rule surface:

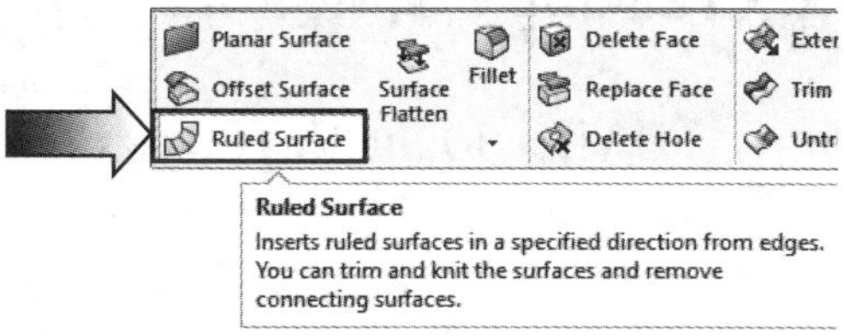

Click **Ruled Surface**.

For Type, click **Tapered to Vector**.

For Distance enter **.060"**.

For Direction select **Top** plane and click **Reverse**.

For Angle, enter **15deg**.

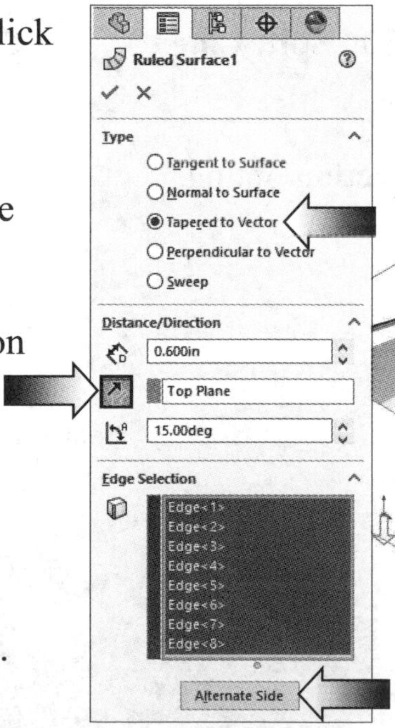

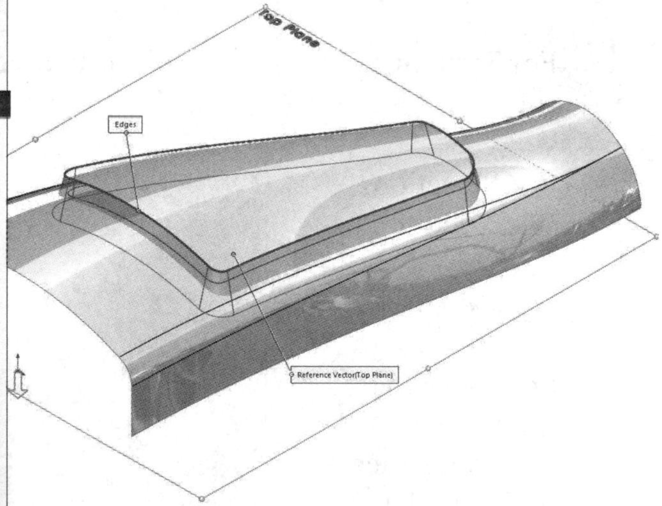

For Edge Selection, select all edges of the **Offset Surface**, click Alternate Side to protrude the ruled surface outward, intersecting the model. **Click OK**.

5. Deleting a surface:

The surface that was used to create the offset surface is no longer needed.

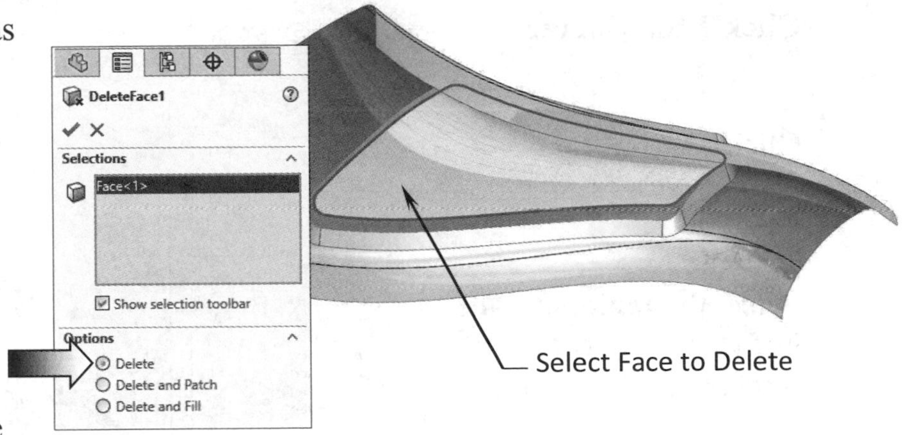

Click **Delect Face**.

For Options, select **Delete**.

For Faces to Delete, select the face in the center of the model as noted.

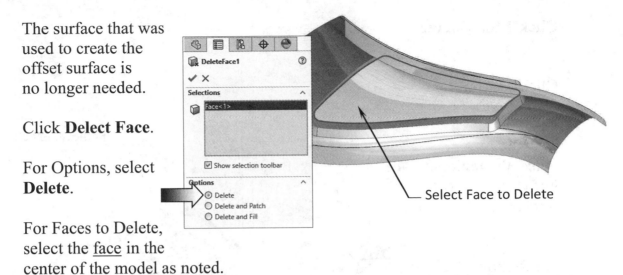

Click **OK**.

6. Creating the 1st surface trim:

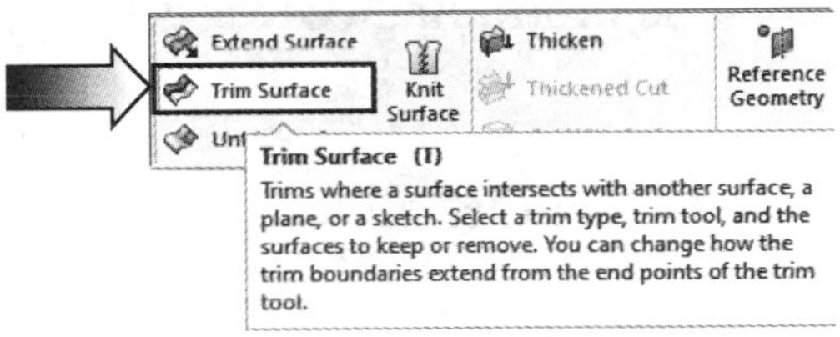

Click **Trim Surface**.

Select **Mutual** for Trim Type.

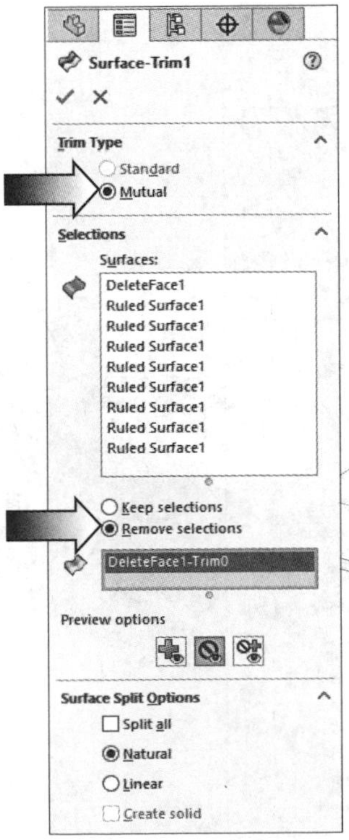

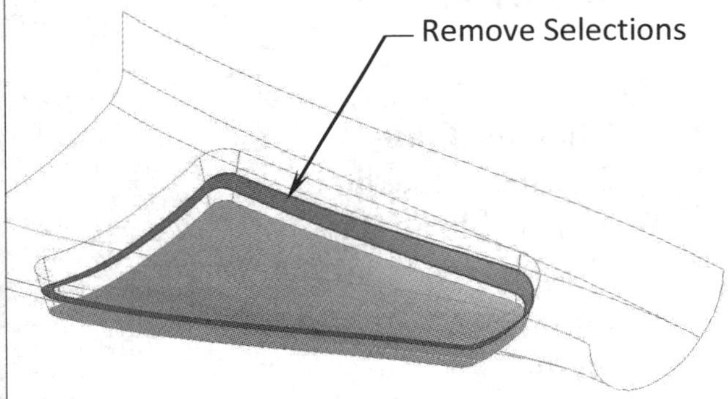

For Selections, select the **main surface body** and all surfaces or the **ruled surface**.

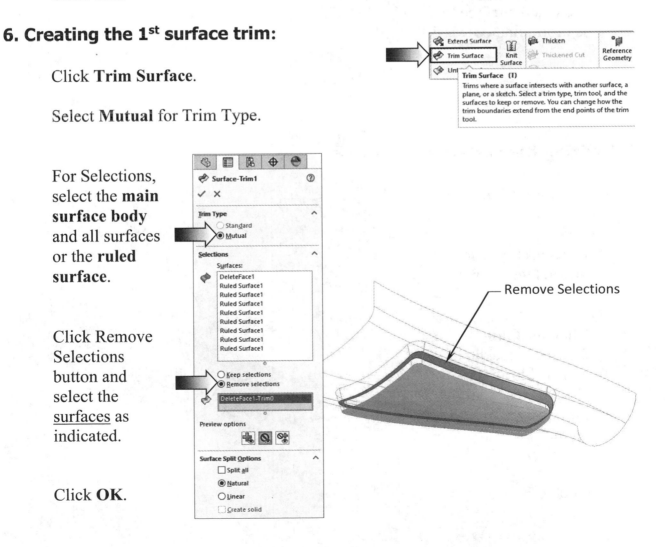

Click Remove Selections button and select the surfaces as indicated.

Click **OK**.

7. Creating the 2nd surface trim:

Click **Trim Surface**.

Click **Mutual** for Trim Type.

Select the **main surface body** and the **Rules Surface** for Selections.

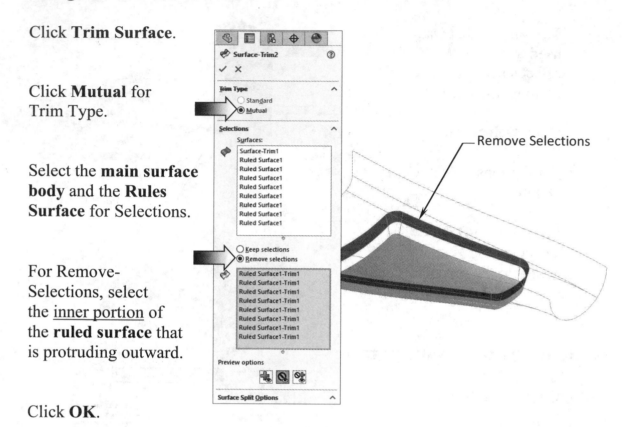

For Remove-Selections, select the inner portion of the **ruled surface** that is protruding outward.

Click **OK**.

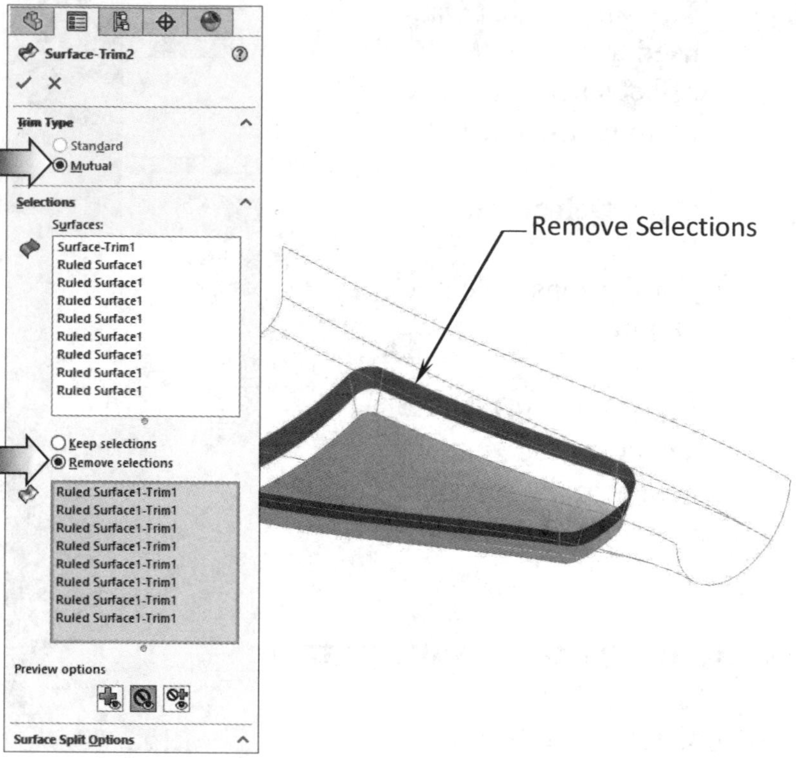

8. Knitting the surfaces:

Click **Knit Surface**.

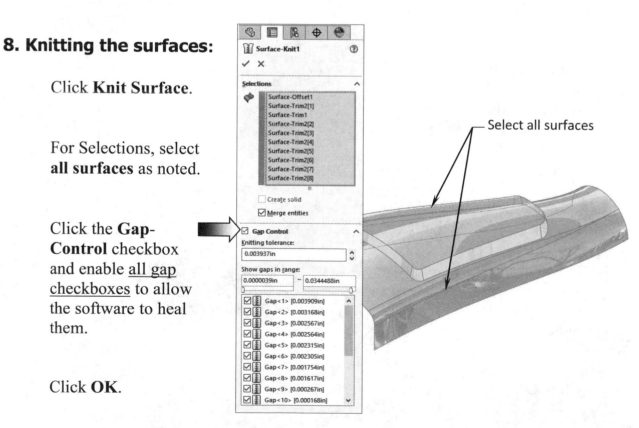

For Selections, select **all surfaces** as noted.

Click the **Gap-Control** checkbox and enable all gap checkboxes to allow the software to heal them.

Click **OK**.

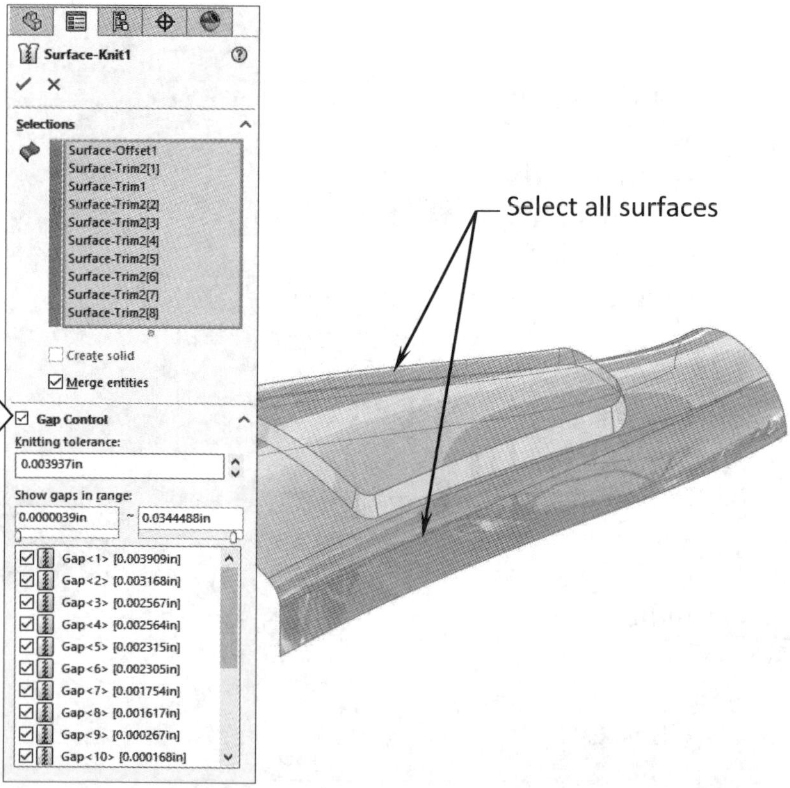

9. Adding fillets:

Click **Fillet**.

Use the default **Constant Size** option.

For Items to Fillet, select all edges of the ruled surface, both upper and lower edges.

For radius size, enter **.100in**.

Click **OK**.

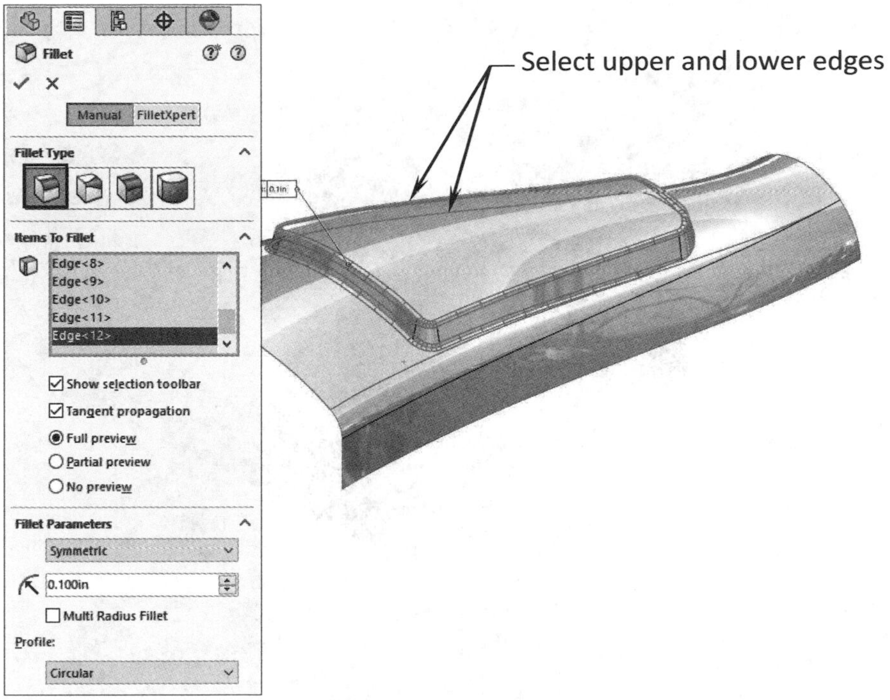

10. Saving your work:

Select File, Save As.

Enter **Using Split_Completed** for the file name.

Click **Save**.

Close all documents.

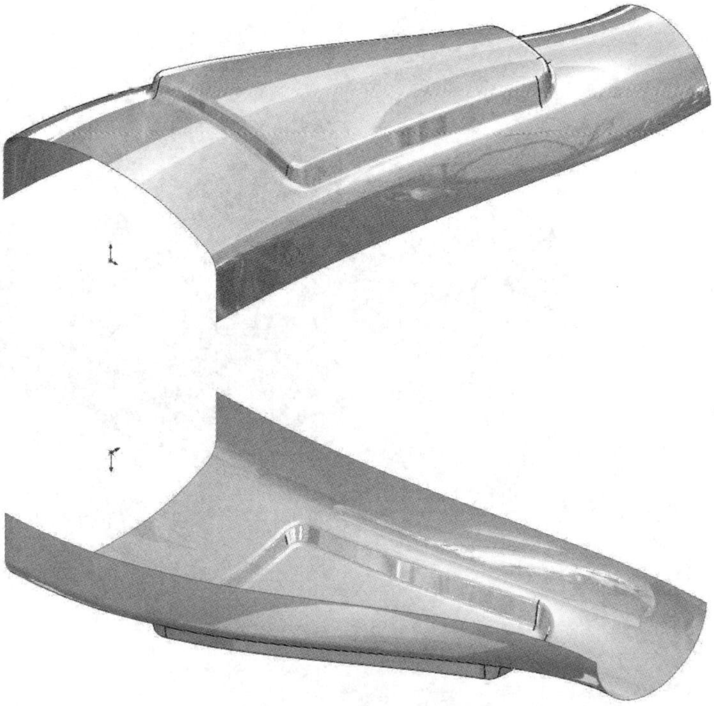

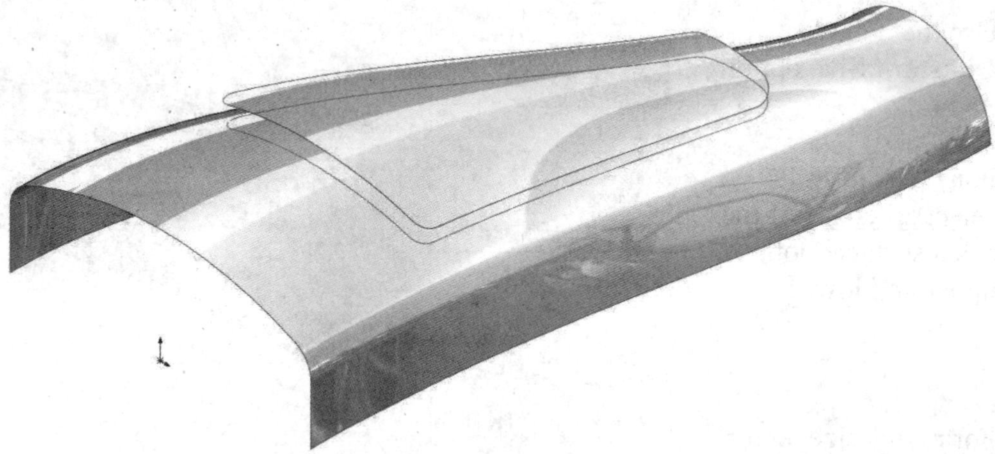

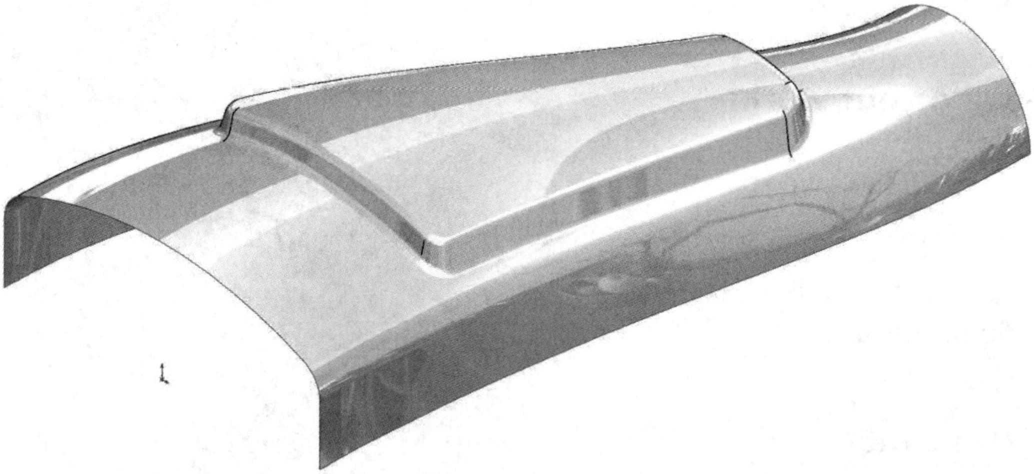

CHAPTER 10

Advanced Surfaces

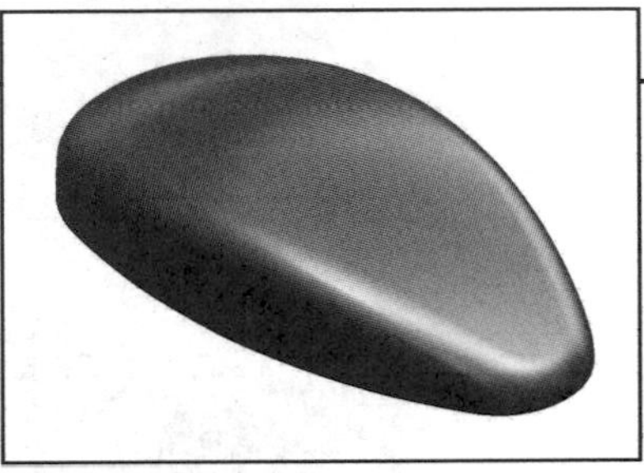

Advanced Surfaces

Computer Mouse

Surfaces are a type of geometry that can be used to create solid features. Surface tools are available on the Surfaces toolbar.

You can use surfaces in the following ways:

* Select surface edges and vertices to use as a sweep guide curve and path.
* Create a solid or cut feature by thickening a surface.
* Extrude a solid or cut feature with the end condition Up to Surface or Offset from Surface.
* Create a solid feature by thickening surfaces that have been knit into a closed volume.
* Replace a face with a surface.

Surface body is a general term that describes connected zero-thickness geometries such as single surfaces, knit surfaces, trimmed, and filleted surfaces, and so on. You can have multiple surface bodies in a single part.

Keep in mind that when modeling with surfaces, every feature should be easily edited. One way to achieve that is to try and create one surface at a time, and towards the end, knit all surfaces into single surface, then thicken to a solid model.

This lesson will show us one of the more convenient methods where multibody surfaces are used to create the computer mouse body. Each surface will be created individually and then trimmed and knitted into a single surface body.

After the surface knit, a wall thickness can be added to the surface body to convert it into a solid body so that other solid features can be added to it from that point on.

Advanced Surfaces
Computer Mouse

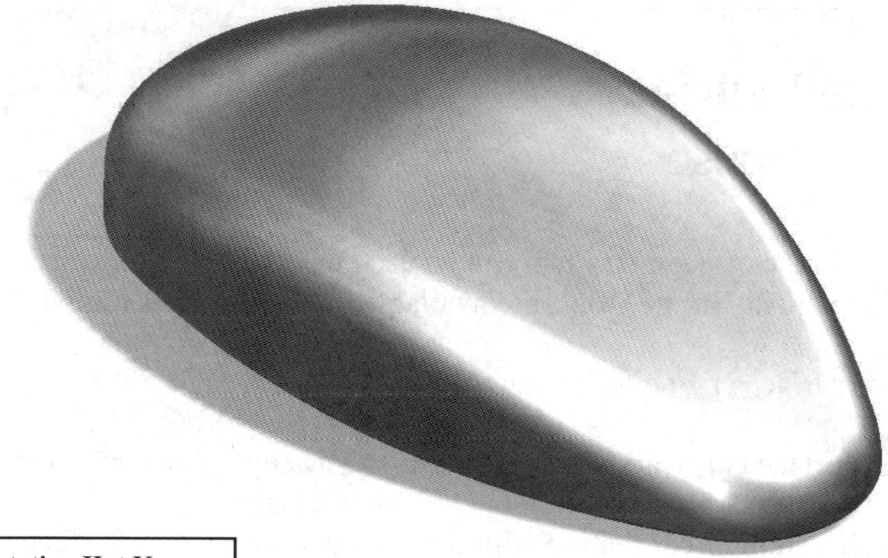

View Orientation Hot Keys:

Ctrl + 1 = Front View
Ctrl + 2 = Back View
Ctrl + 3 = Left View
Ctrl + 4 = Right View
Ctrl + 5 = Top View
Ctrl + 6 = Bottom View
Ctrl + 7 = Isometric View
Ctrl + 8 = Normal To Selection

Dimensioning Standards: **ANSI**
Units: **INCHES** – 3 Decimals

Tools Needed:

- Extruded Surface
- Trim Surface
- Mirror Surface
- Filled Surface
- Planar Surface
- Knit Surface

1. Starting a new part document:

Click **File / New**.

Select the **Part** template and click **OK**.

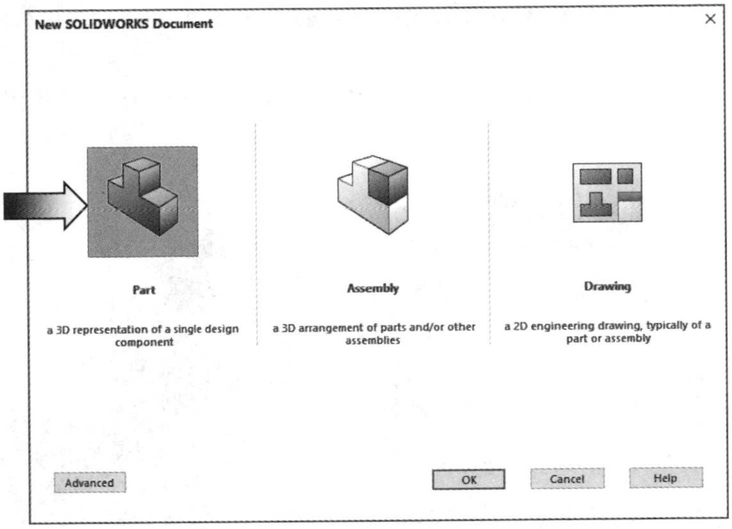

2. Sketching the 1st profile:

Select the Top plane and open a **new sketch**.

Sketch a **Centerline** and a **3-Point Arc** shown below.

Add the dimensions and relations as shown below to fully define the sketch.

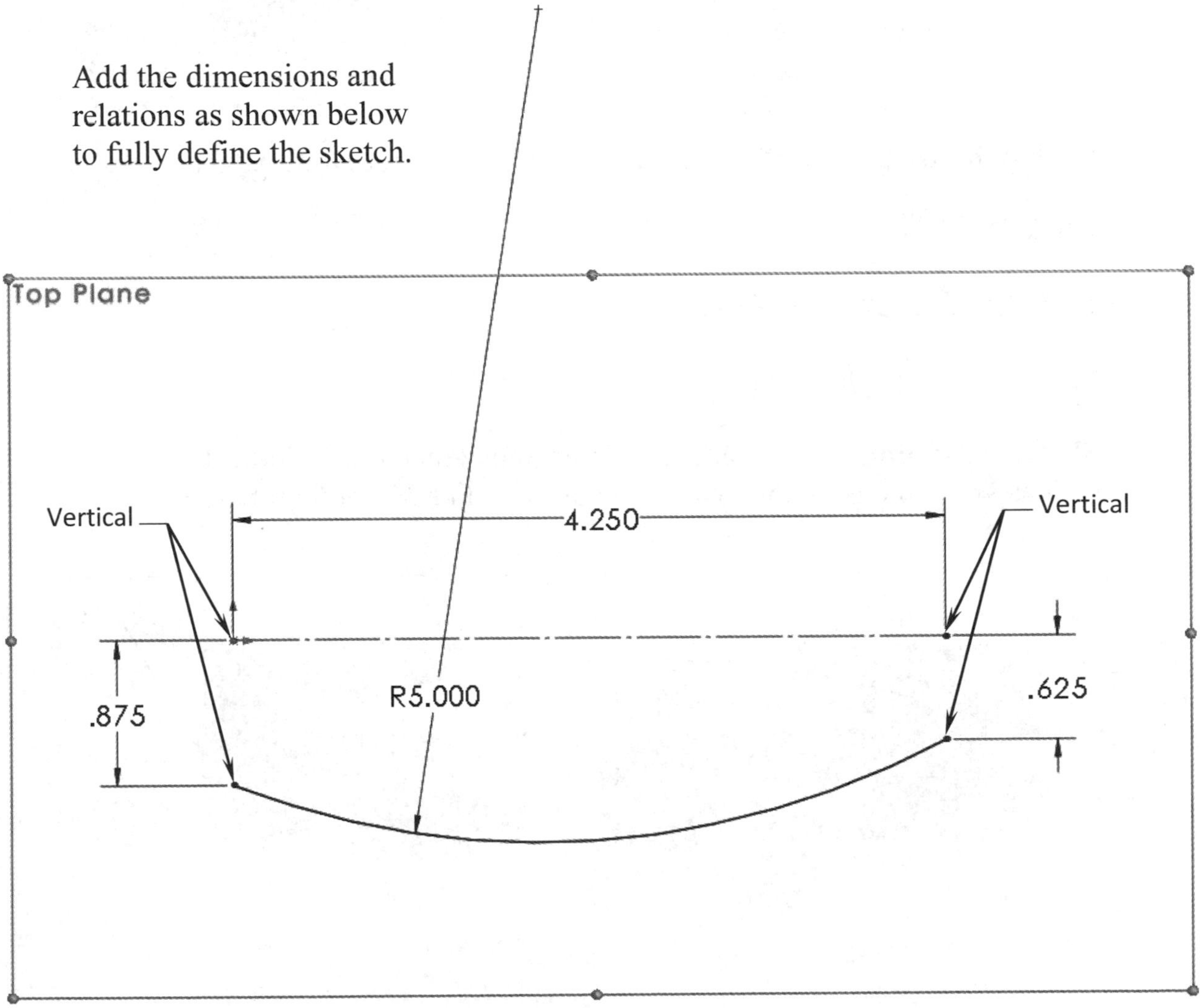

3. Extruding the 1st surface:

Switch to the **Surfaces** toolbar and click **Extruded Surface** .

Use the default **Blind** type and enter a Depth of **1.00in**.

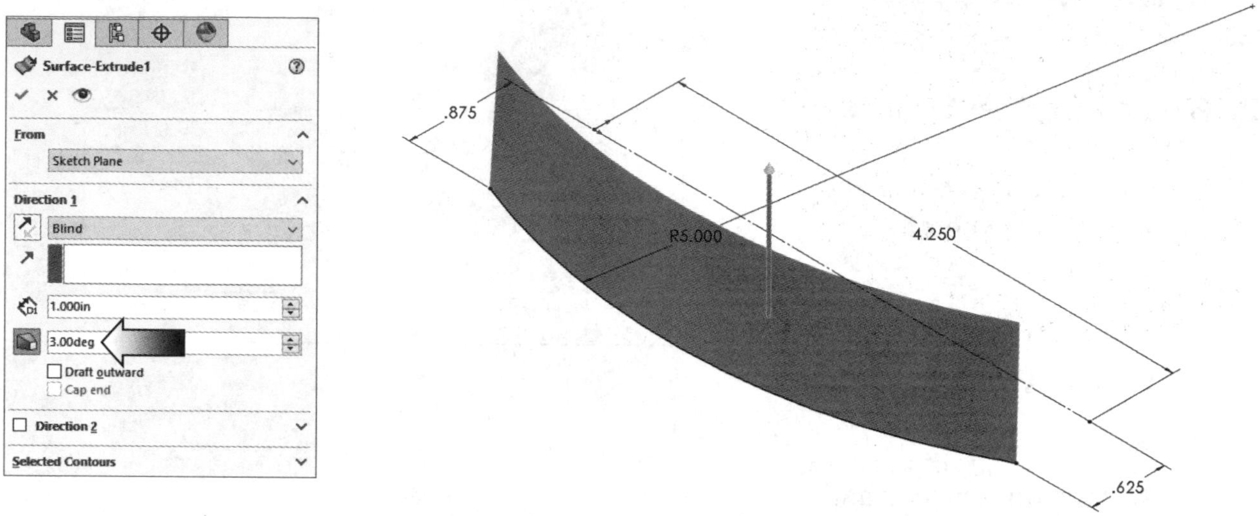

Enable the **Draft On/Off** button and enter **3.00deg** for Draft Angle.

Click **OK**.

4. Sketching the 2nd profile:

Select the Front plane and open a **new sketch**.

Sketch a **3-Point-Arc** and add the dimensions / relations as indicated.
(Note: The dimension .500 can be replaced with a Midpoint relation.)

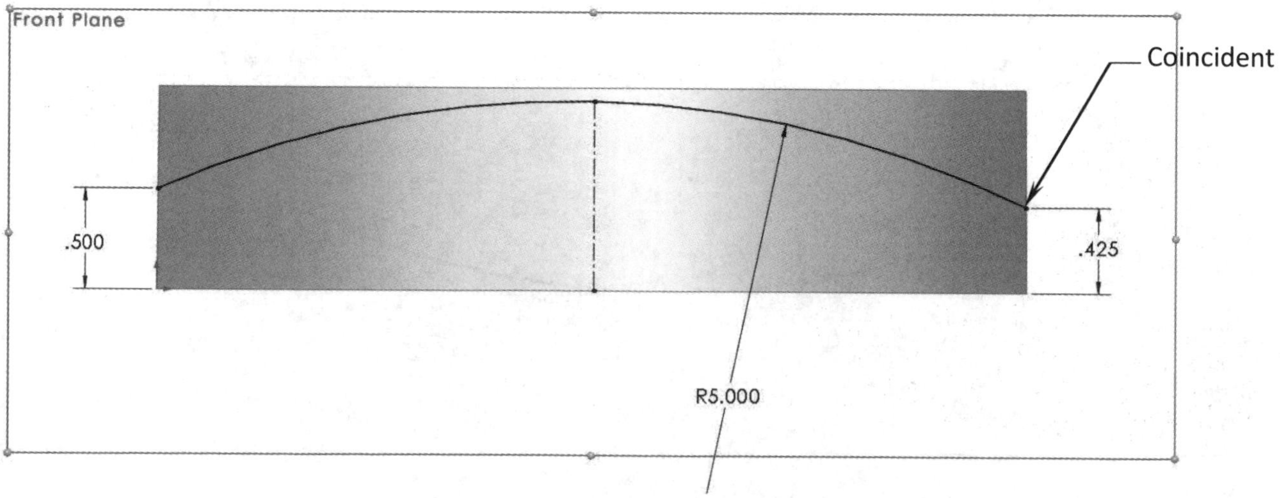

5. Extruding the 2nd surface:

Switch to the **Surfaces** toolbar and click **Extruded Surface** 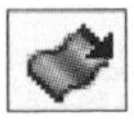.

Use the default **Blind** type and enter a Depth of **1.50in**.

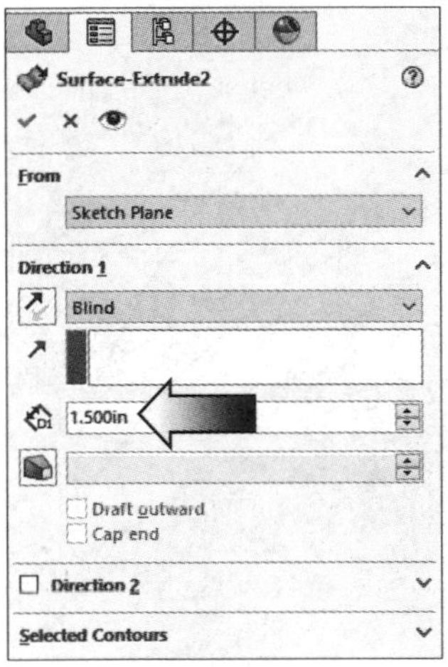

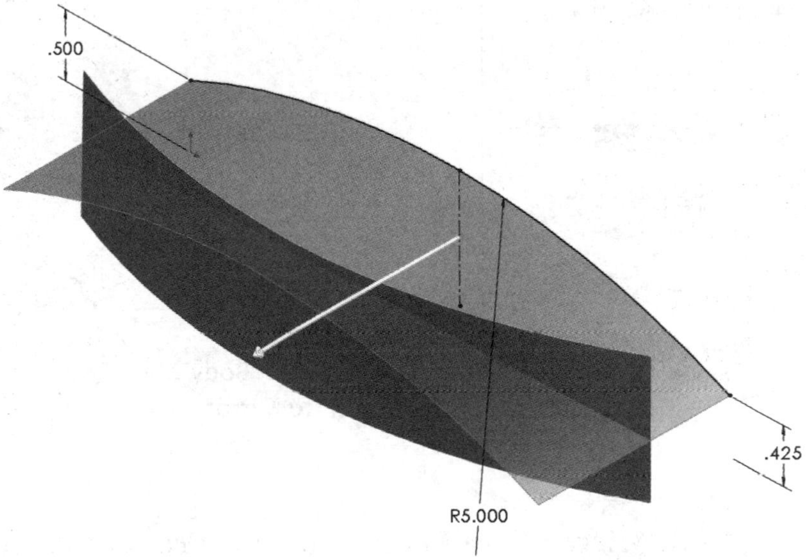

Click **OK**.

6. Trimming the surfaces:

Select the **Trim Surface** command .

Use the default **Standard Trim** option. For Trim Tool, click **Surface-Extrude2**.

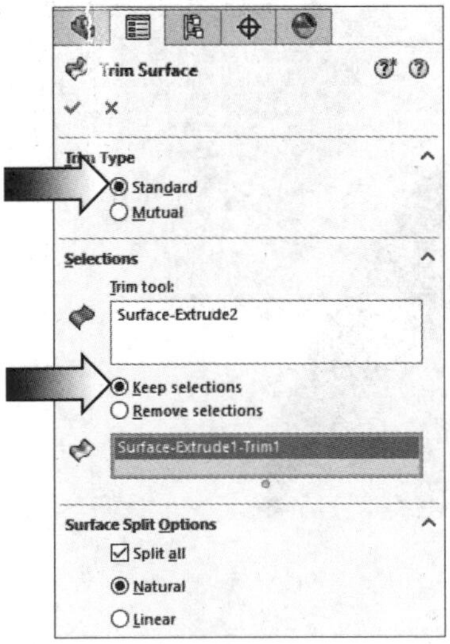

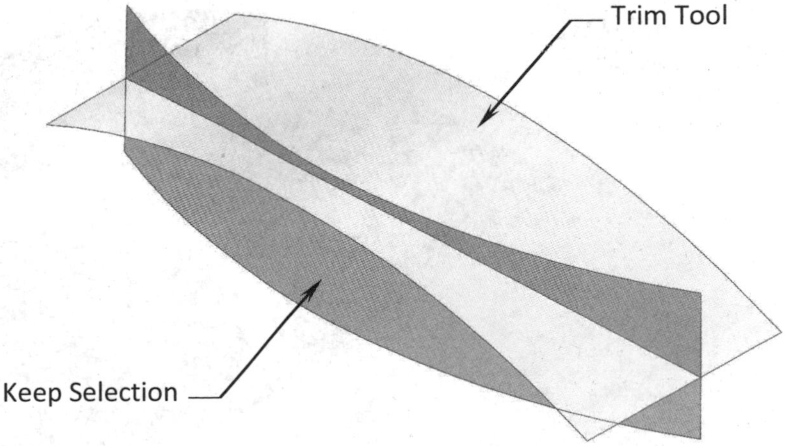

For Keep Selections, select the lower portion of the **Surface Extrude1** as indicated.

Click **OK**.

7. Mirroring the surfaces:

Switch to the **Features** tool tab.

Click the **Mirror** command .

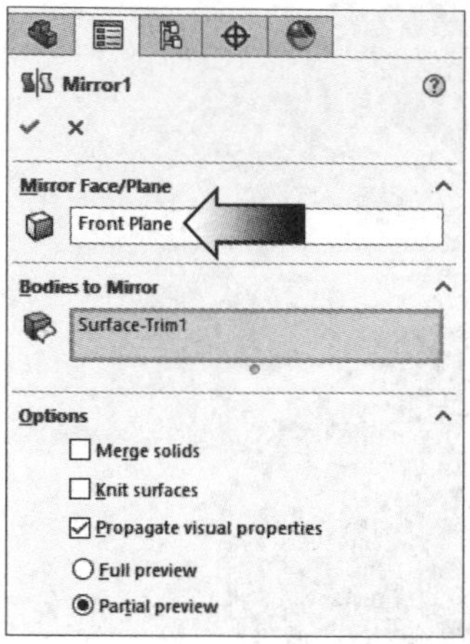

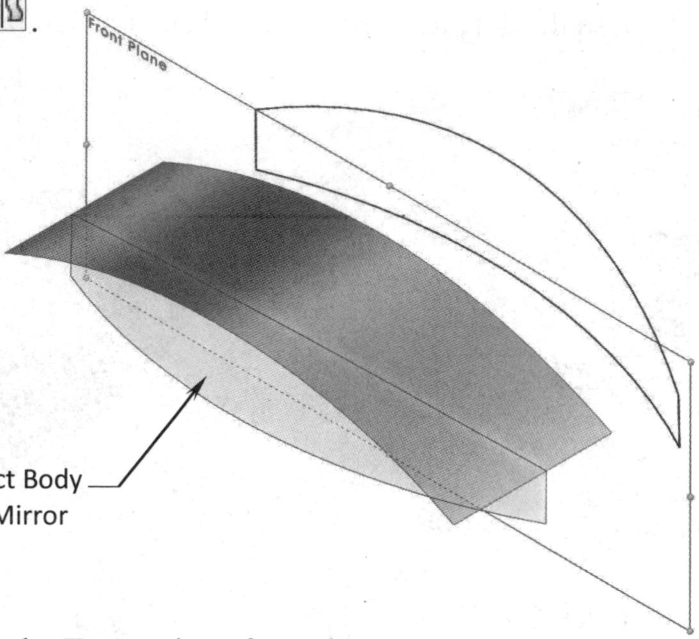

For **Mirror Face/Plane**, select the **Front** plane from the FeatureManager tree.

Expand the **Bodies to Mirror** section and select the lower portion of the **Surface-Trim1** as noted.

Click **OK.**

Inspect your model against the image above.

8. Creating the 1st lower sketches:

Hide the **Surface-Extruded2**.

Open a **new Sketch** on the Top plane.

Sketch a **3-Point Arc** as shown below and add the **Tangent** relations as noted.

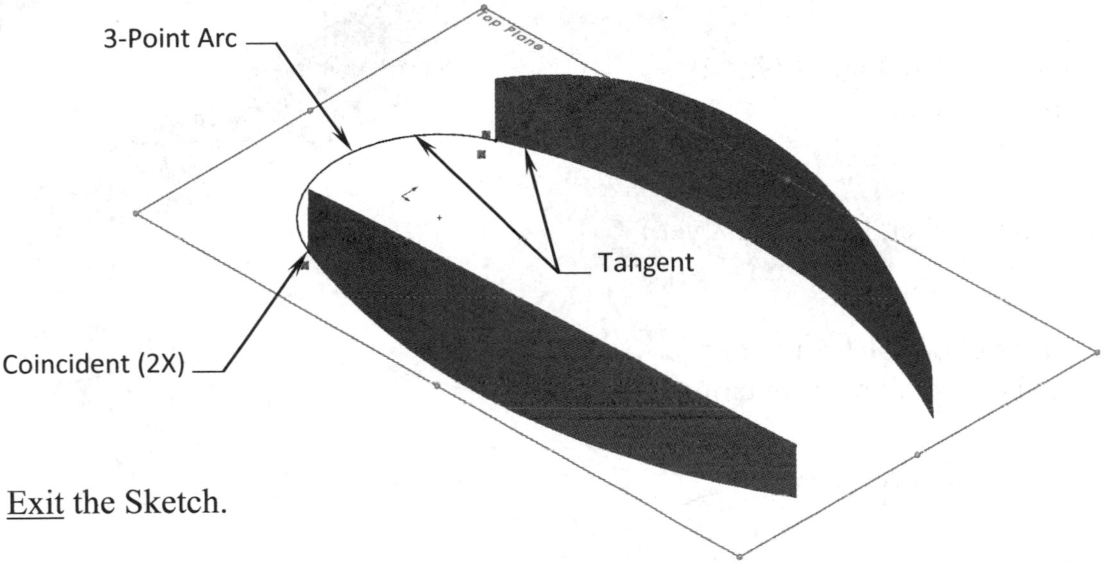

Exit the Sketch.

9. Creating the 2nd lower sketches:

Open a **new sketch** on the Top plane again.

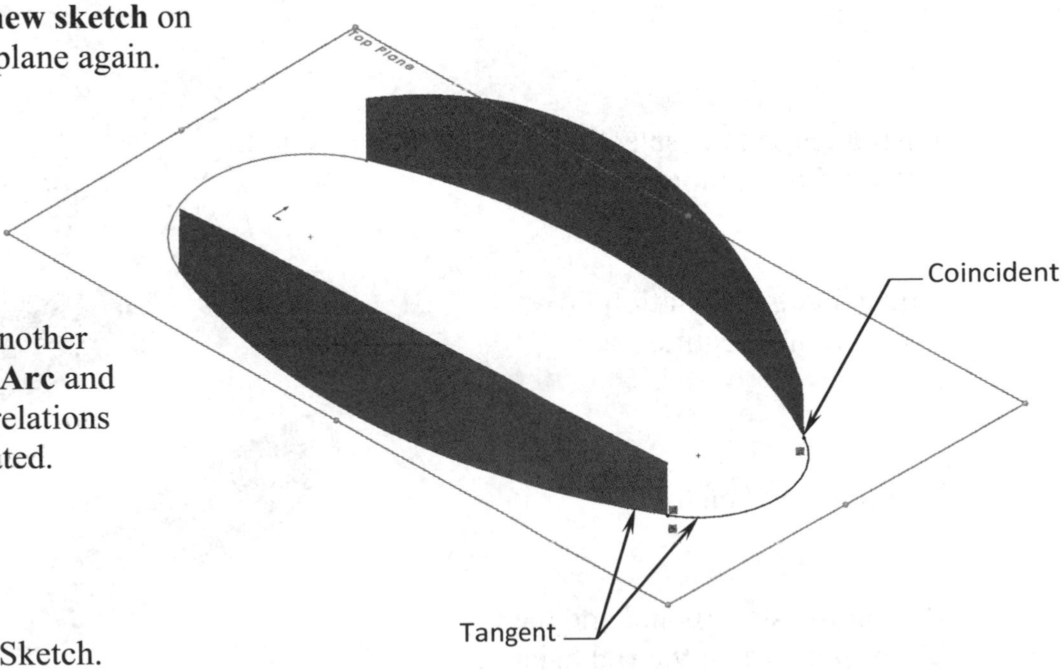

Sketch another **3-Point-Arc** and add the relations as indicated.

Exit the Sketch.

10. Creating the 1st upper sketches:

Open a new <u>3D Sketch</u>.

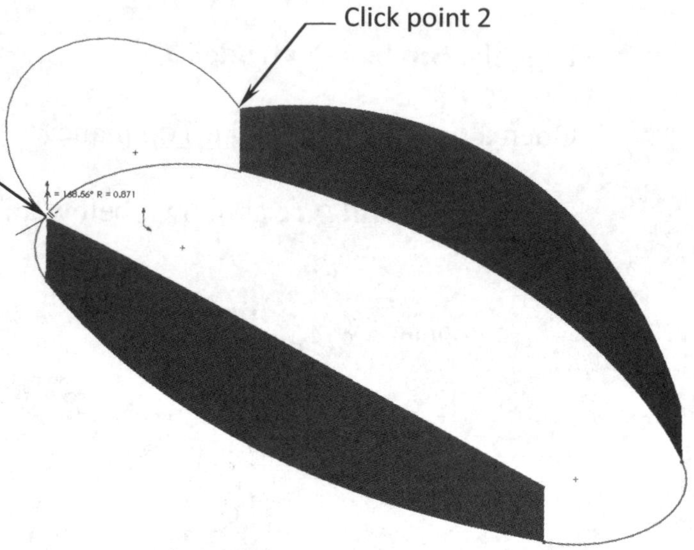

Select the
3-Point Arc command.

Click **Point 1** and **Point 2** as indicated.

(Do not click Point 3 just yet.)

Press **Control + 5** to change to the Top View orientation.

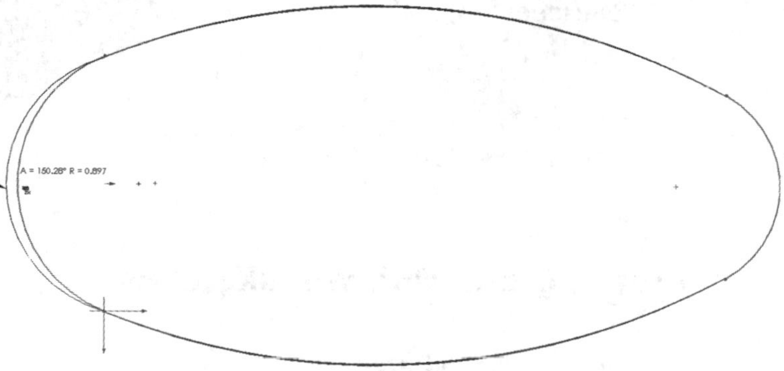

Move the mouse cursor outward approximately as shown and click **Point 3**.

Push **Escape** to deselect the 3 Point Arc command.

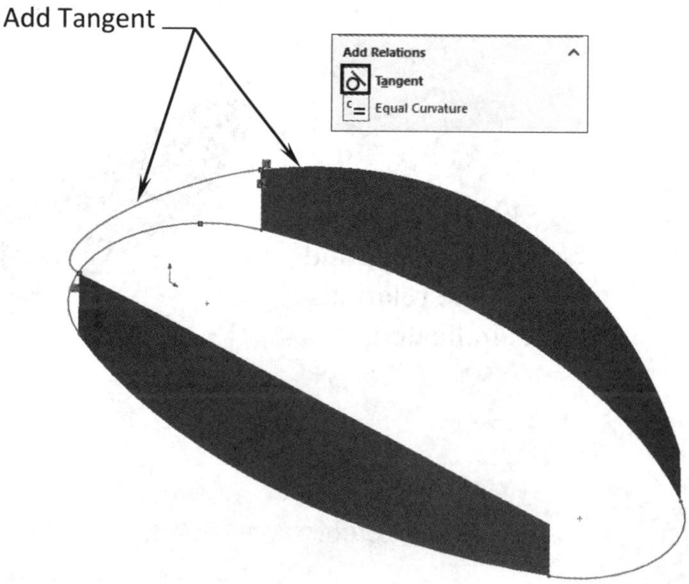

Add a **Tangent** relation between the 2 entities as noted.

<u>Exit</u> the 3D Sketch.

<u>Repeat</u> the step 10 and add the **2nd upper arc** on the right side.

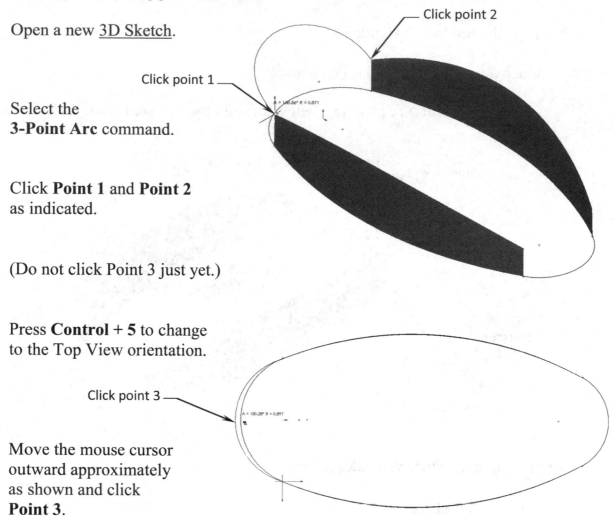

11. Creating a new plane:

Select the **Right** plane from the Feature tree.

Hold the **Control** key and drag the **Right** plane to the right side to make a copy.

Enter **2.125** for Distance.

Click **OK**.

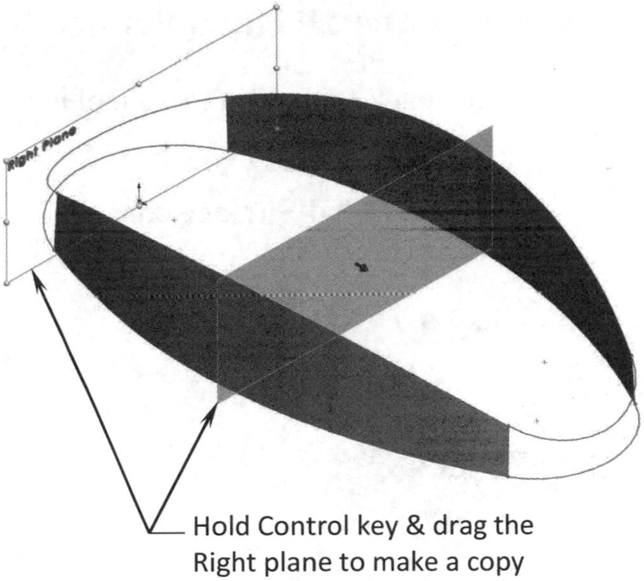

12. Making a 3-Point Arc:

Open a new sketch on the **new plane** (Plane1).

Sketch a **3-Point-Arc**. Add a dimension and a Pierce relation between the end point of the arc and the upper edge of the Surface-Trim1.

Exit the sketch.

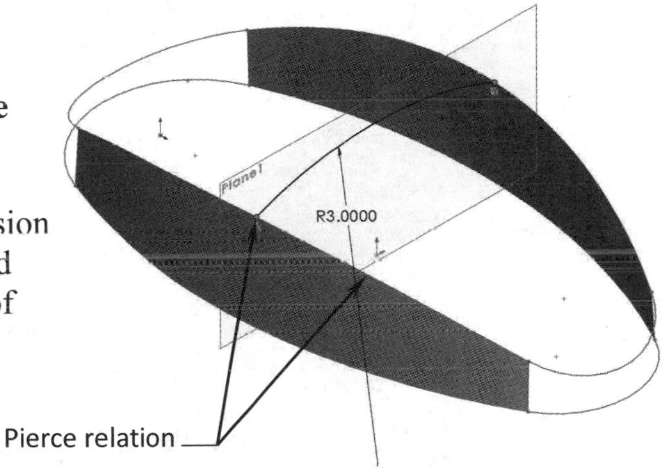

13. Creating the 1st Filled Surface:

Click **Filled Surface**.

For Patch Boundary, select the top **4 edges** as noted.

Expand the **Constraint-Curves** section and select the **Sketch4** as indicated. This curve helps control the curvature of the patch.

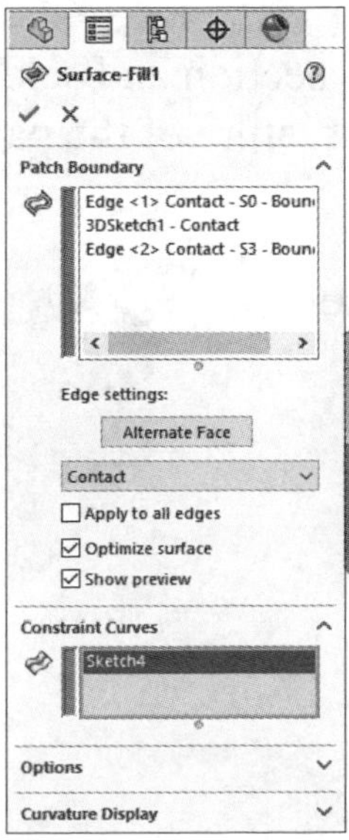

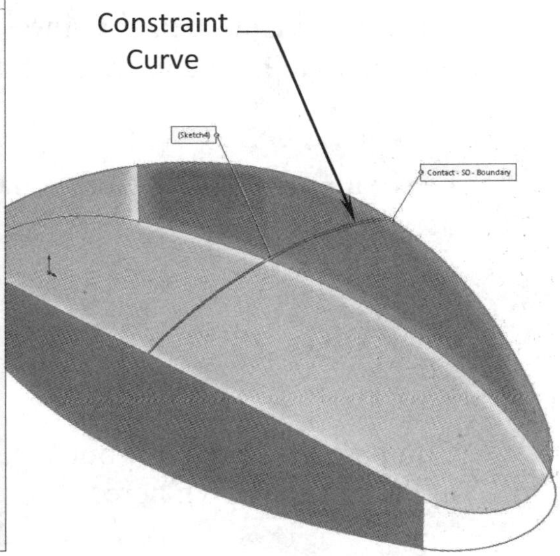

14. Creating the 1st Filled Surface:

> **Filled Surface**
>
> The Filled Surface feature constructs a surface patch with any number of sides, within a boundary defined by existing model edges, sketches, or curves, including composite curves.

Switch back to the **Surfaces** tool bar.

Select the **Filled Surface** command.

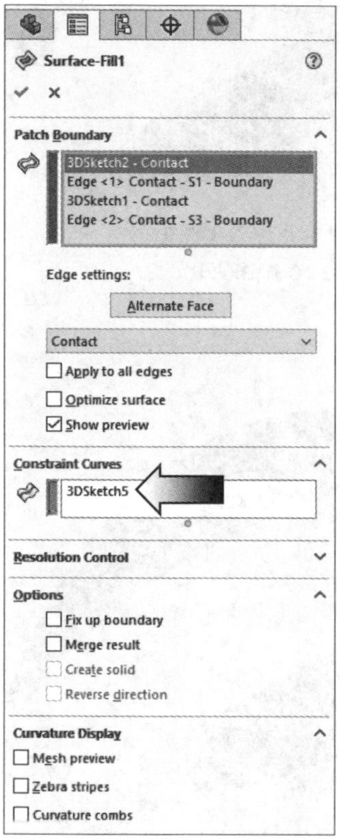

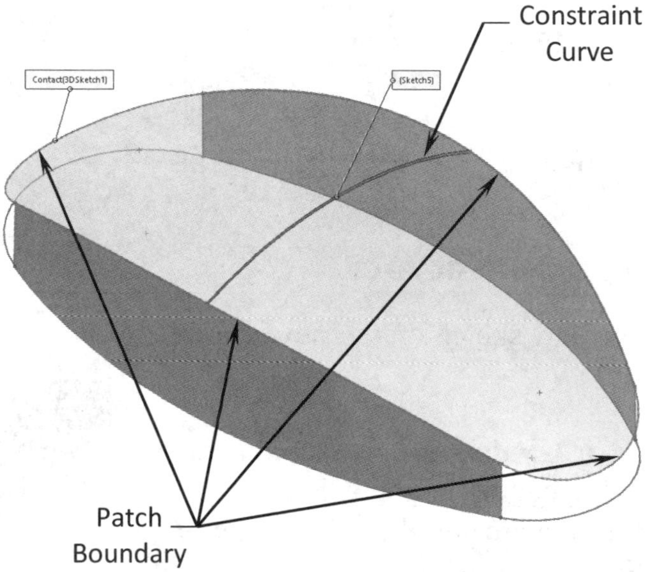

For Patch Boundary, select the top **4 edges** as noted.

Expand the **Constraint Curves** section and select the **3D Sketch Curve** as indicated, to help control the curvature of the patch.

Leave all other parameters at their default settings.

Click **OK**.

Compare the result of your Filled Surface to the image shown here.

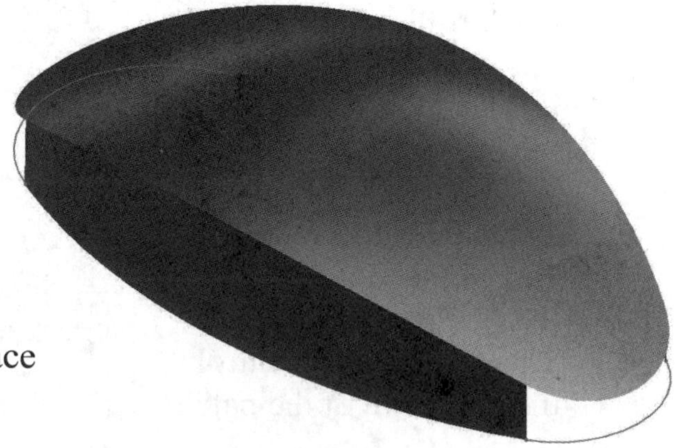

15. Creating the 2nd Filled Surface:

Rotate the right end of the model to a similar position shown below.

Click the **Filled-Surface** command again.

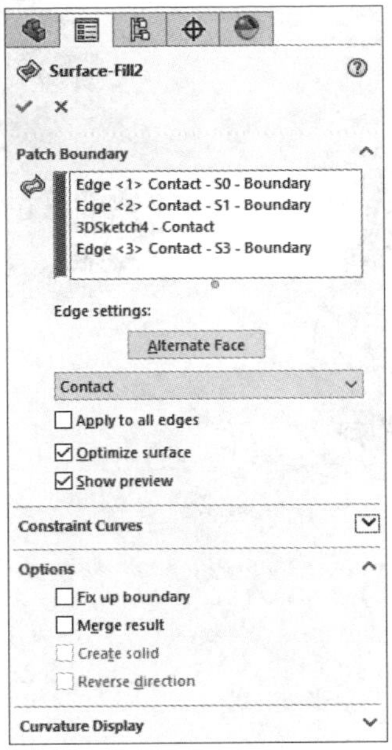

For Patch Boundary, select the **4 edges** on the right end as indicated.

Click **OK**.

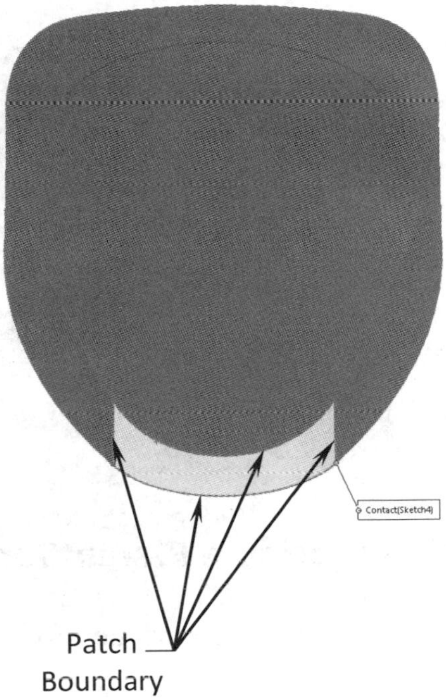

16. Creating the 3rd Filled Surface:

Select the **Filled Surface** command once again.

Rotate the left end of the model approximately as shown below.

For Patch Boundary, select the **4 edges** as noted in the image.

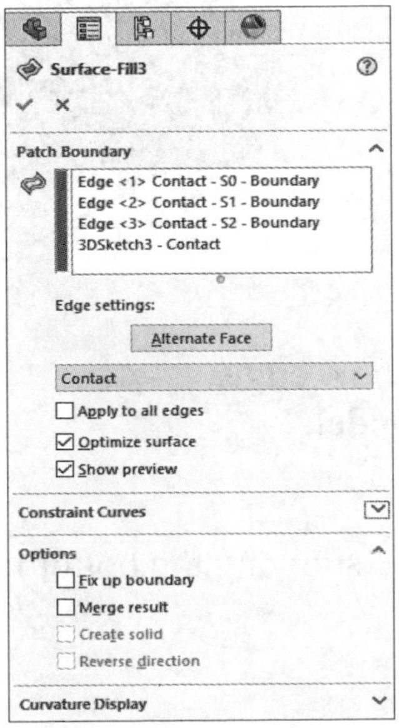

Keep all other parameters at their default settings.

Click **OK**.

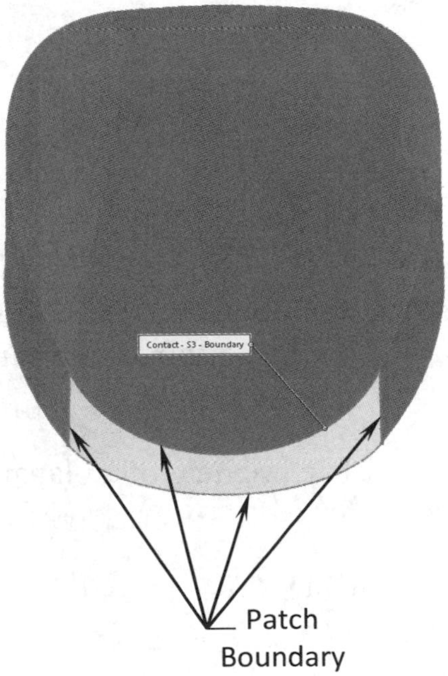

Check your model against the ones shown below.

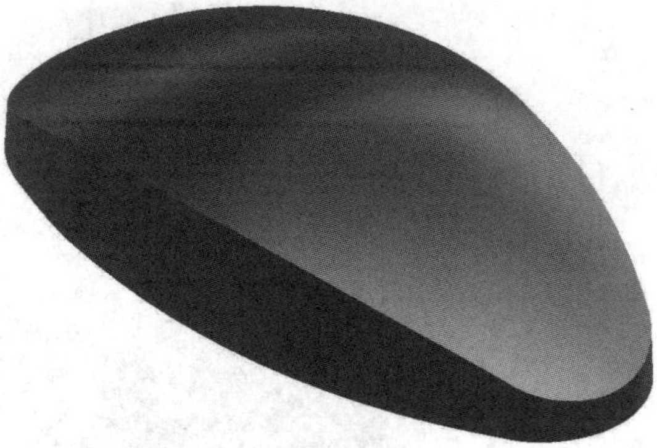

17. Creating a Planar Surface:

Rotate the surface model so that the bottom opening is visible.

Select the **Planar Surface** command .

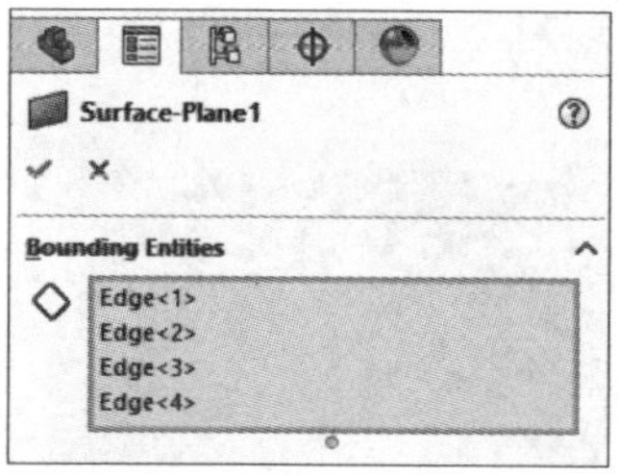

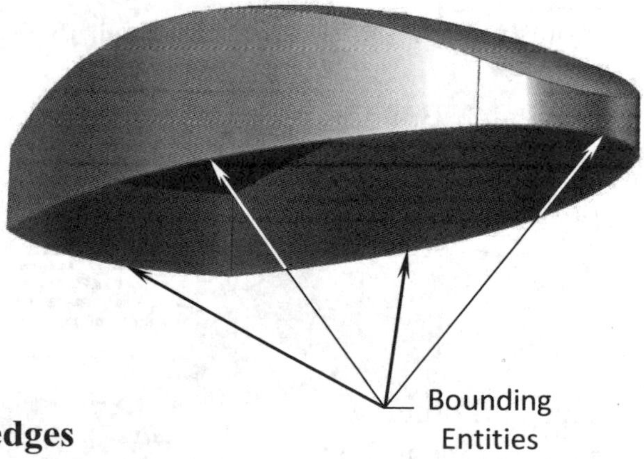

For Bounding Entities, select the **4 edges** on the bottom of the surface model.

The preview of a planar surface showing the bottom opening is being closed off.

Click **OK**.

18. Creating a Knit Surface:

Click the **Knit Surface** command.

> **Knit Surface**
>
> The Knit Surface feature combines two or more surfaces into one surface.

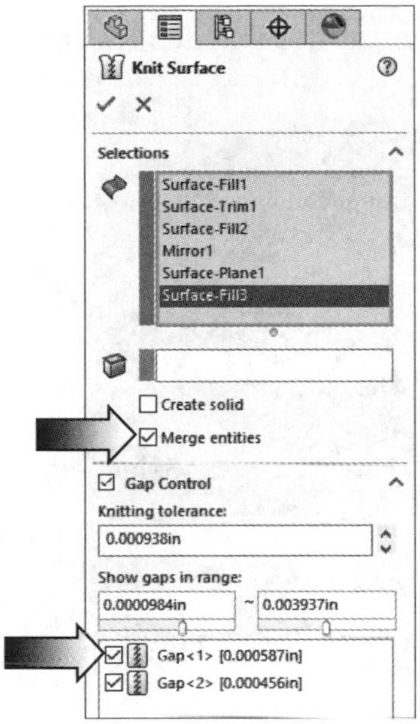

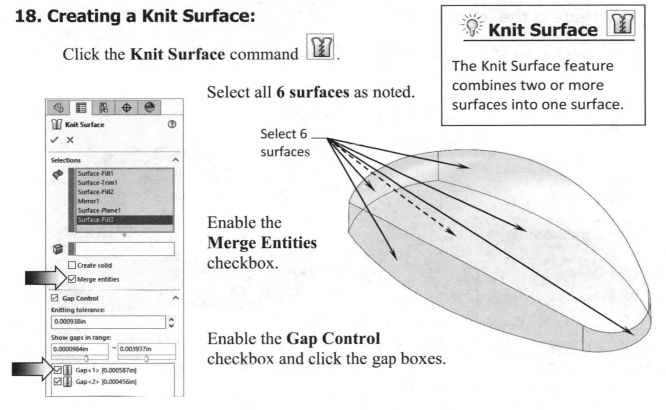

Select all **6 surfaces** as noted.

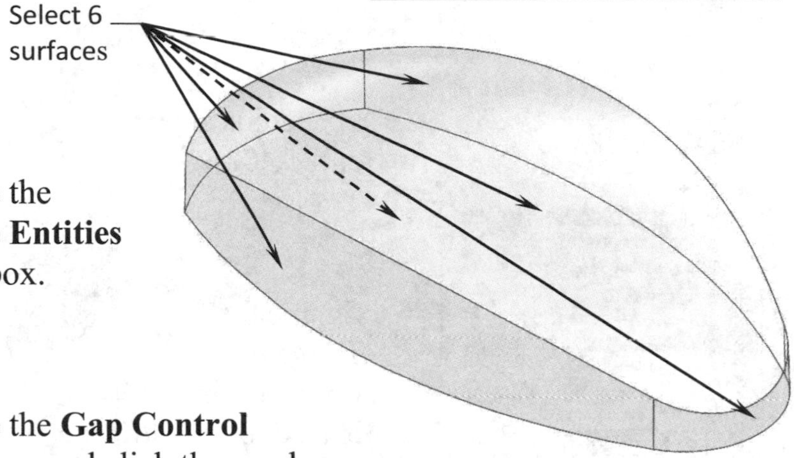

Enable the **Merge Entities** checkbox.

Enable the **Gap Control** checkbox and click the gap boxes.

Click **OK**.

19. Adding the .040" fillet:

Select the **Fillet** command again.

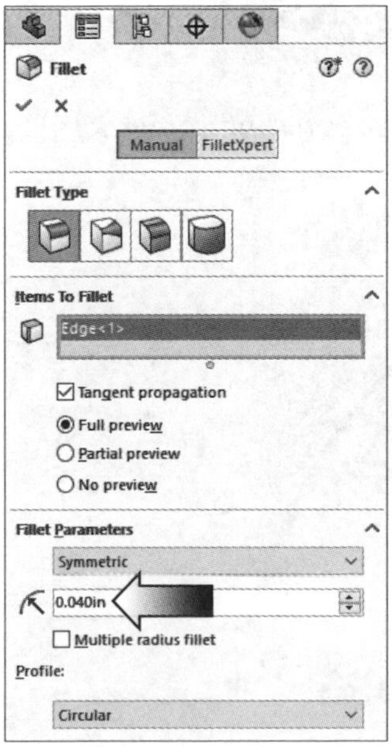

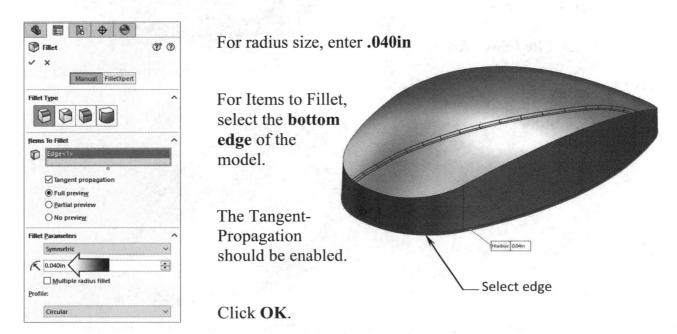

For radius size, enter **.040in**

For Items to Fillet, select the **bottom edge** of the model.

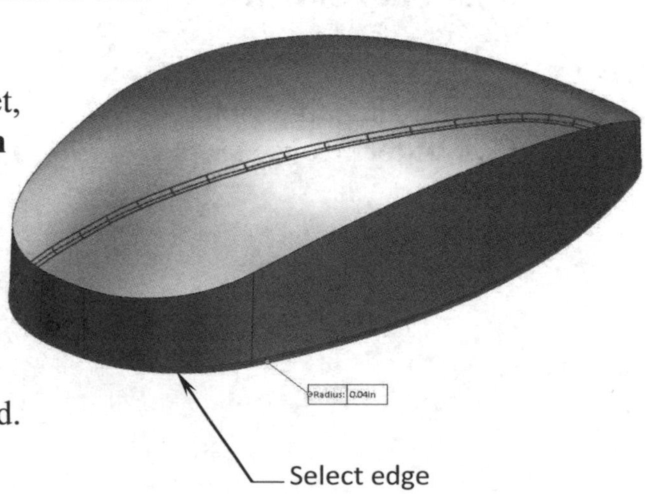

The Tangent-Propagation should be enabled.

Click **OK**.

20. Creating the .400" fillet:

Select the **Fillet** command  once again.

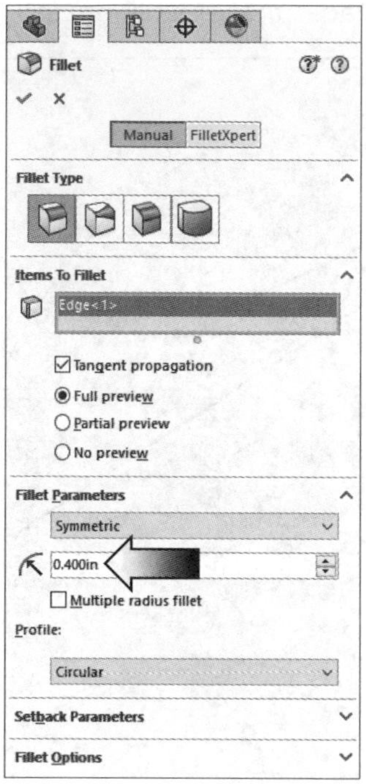

For Items to Fillet, select the **top edge** of the surface model.

For radius size, enter **.400in**

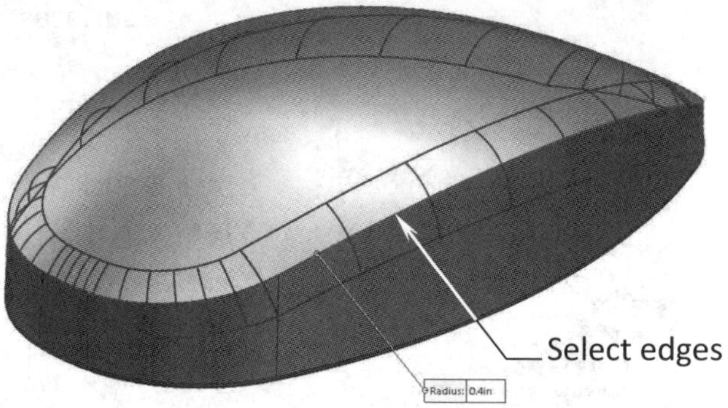

Enable the Tangent-Propagation checkbox to allow the fillet to propagate all around the bottom of the surface model.

Click **OK**.

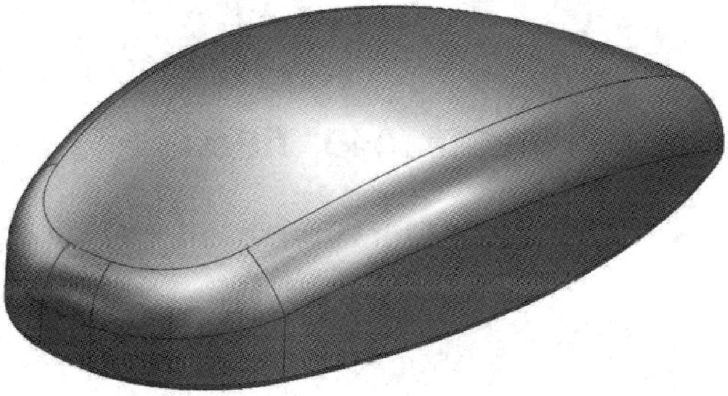

Shaded with Edges (Front Isometric)

21. Saving your work:

Click **File / Save As**.

For the file name, enter: **Computer Mouse.sldprt**

Click **Save**.

Shaded with Ambient Occlusion (Back Isometric)

Using Filled Surfaces

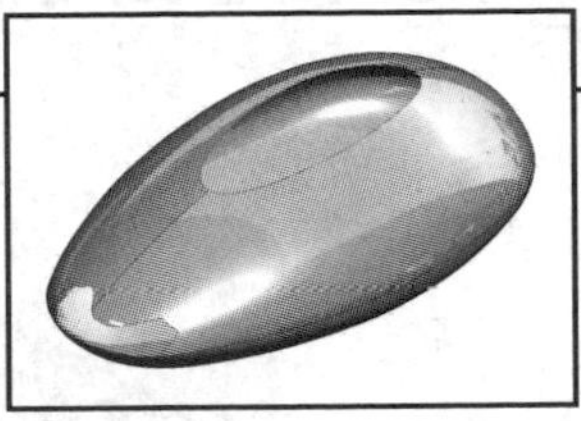

Using Filled Surfaces
Curvature Controls

Use Filled Surface and Planar Surface commands to fill or patch a surface boundary with any number of sides.

The boundary can be a set of existing model edges, sketches, or curves, including composite curves. The boundary should be closed for the patch to work properly.

There are several options to help you control the curvatures when patching a surface boundary such as Contact, Tangent, and Curvature. These options are explained later in the lesson.

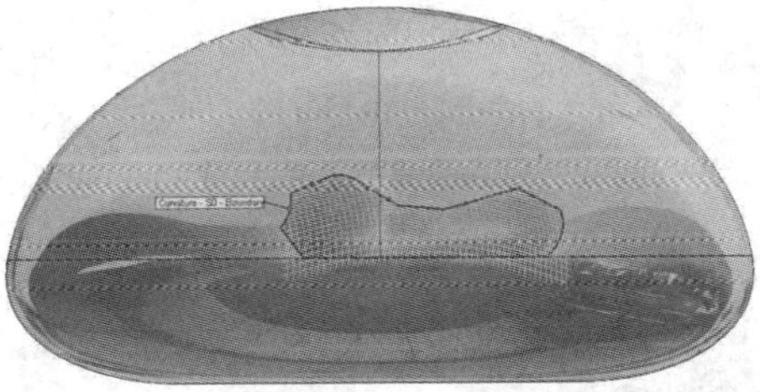

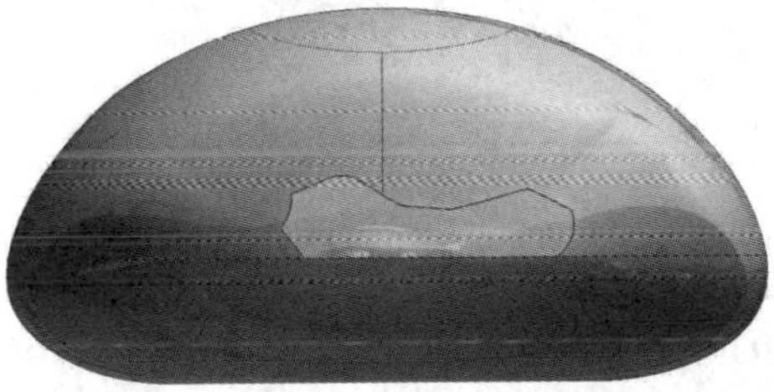

Other than the Curvature Control options the Apply-to-All-Edges check box enables you to apply the same curvature control to all edges. If you select the function after applying both **Contact** and **Tangent** to different edges, it applies the current selection to all edges.

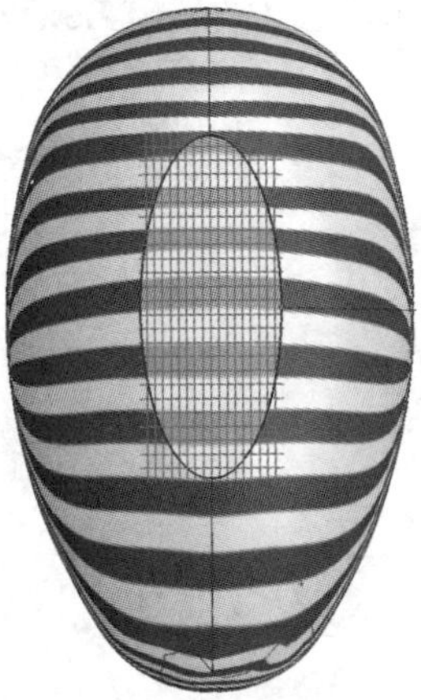

If your surface model has two or four-sided surfaces, try using the **Optimize surface** option. The Optimize surface option applies a simplified surface patch that is similar to a lofted surface. Potential advantages of the optimized surface patch include faster build times and increased stability when used in conjunction with other features in the model.

This lesson will teach us the use of the Filled Surface and Planar Surface commands.

Using Filled Surfaces
Patch with Curvature Controls

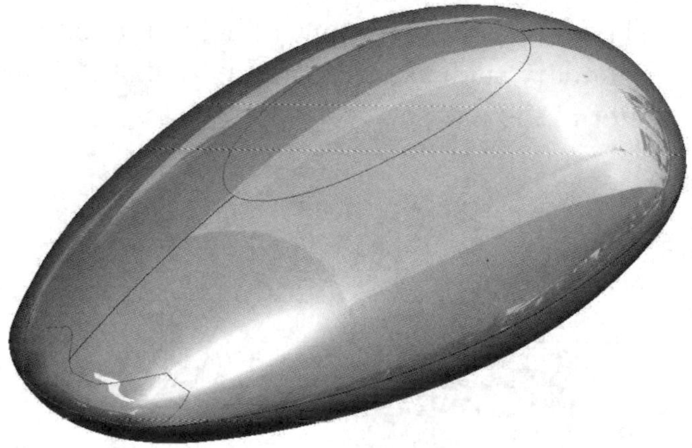

View Orientation Hot Keys:

Ctrl + 1 = Front View
Ctrl + 2 = Back View
Ctrl + 3 = Left View
Ctrl + 4 = Right View
Ctrl + 5 = Top View
Ctrl + 6 = Bottom View
Ctrl + 7 = Isometric View
Ctrl + 8 = Normal To Selection

Dimensioning Standards: **ANSI**

Units: **INCHES** – 3 Decimals

Tools Needed:

Planar Surface

Filled Surface

Knit Surface

1. Opening a part document:

Click **File / Open**.

Browse to the Training Files folder, open a part document named: **Filled Surfaces.sldprt**

This part document was previously saved as a different file format. There is no feature history available on the FeatureManager tree.

There are three openings in the part that we will have to fill using different options available in the Filled-Surface command.

If the Feature Recognition dialog pops up, click **NO** to close it.

2. Enabling the Surfaces toolbar:

Right-click one of the tool tabs and select **Tabs, Surfaces**.

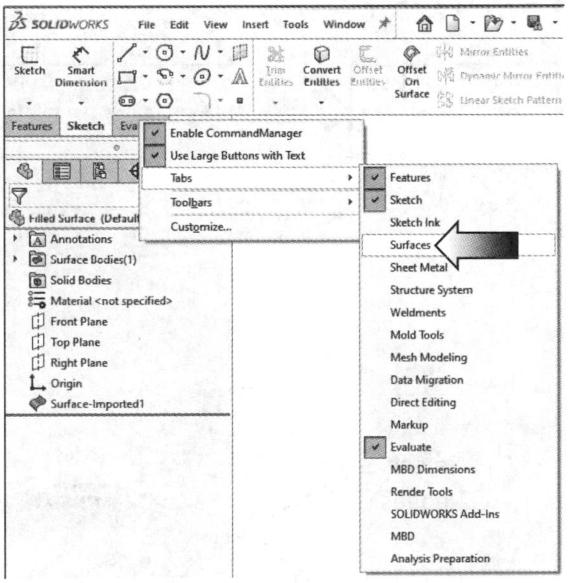

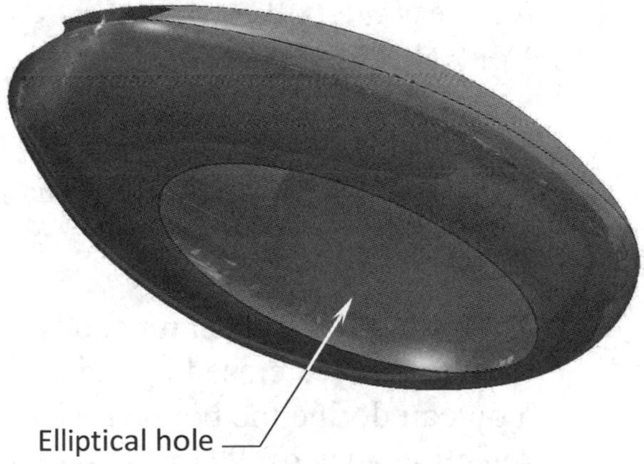

Rotate the model and locate the elliptical hole at the bottom of the model.

The opening should form a flat surface. We can use the Planar-Surface command to patch it up.

3. Creating a Planar surface:

Switch to the **Surfaces** tool tab.

Select the **Planar-Surface** command.

Select the <u>edge</u> of of the elliptical hole.

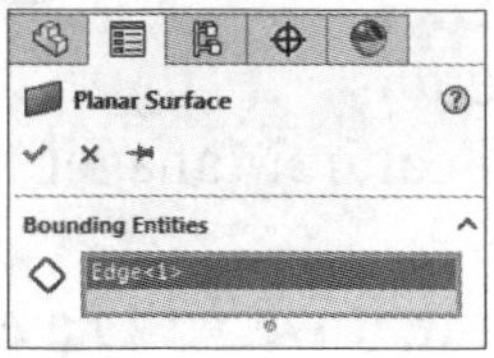

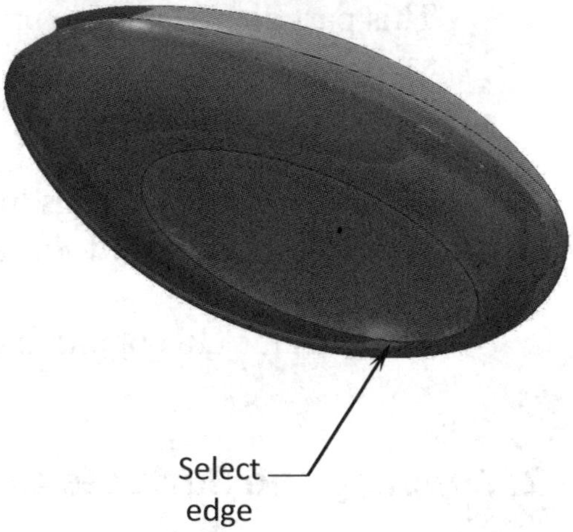

The preview of a planar surface appears.

Click **OK**.

A planar surface can be created from a sketch, a set of closed edges, or a pair of planar entities such as curves or edges.

4. Creating a Surface Fill with Tangent Control:

Switch back to the **Isometric view** (or press the hotkey Control + 7).

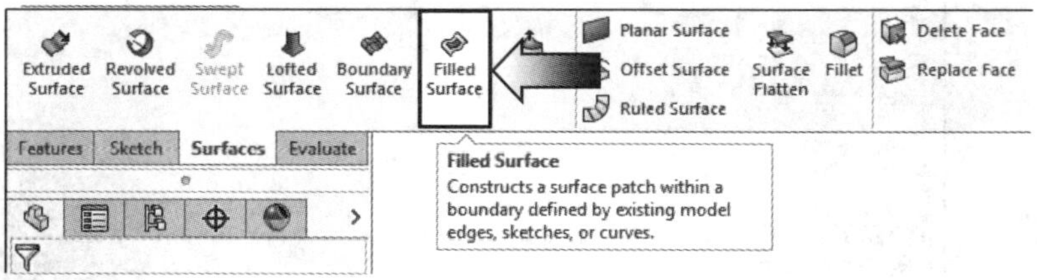

Click the **Filled Surface** command.

The Filled Surface command is used to patch a closed boundary. You can define the boundary by selecting a set of 2D or 3D sketch entities, model edges or composite curves.

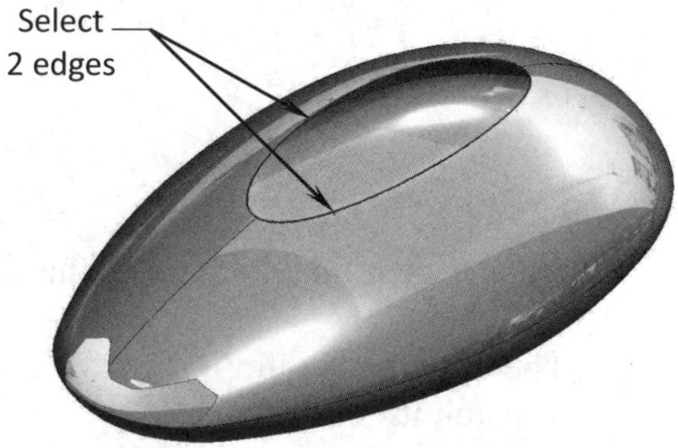

Select the <u>two edges</u> of the elliptical hole on the top of the model, as noted.

A preview mesh is displayed to help you visualize the curvature of the new surface.

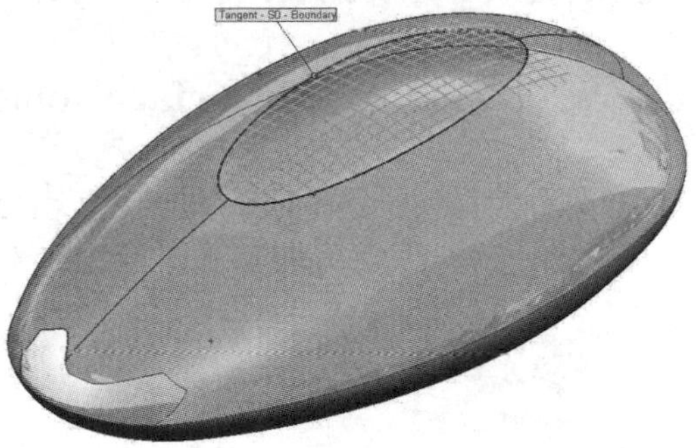

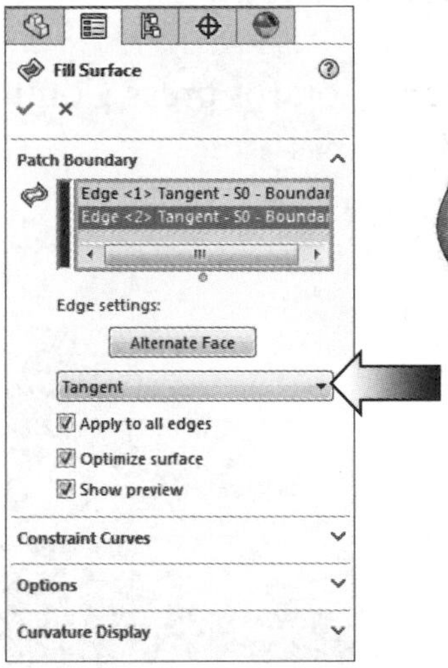

Change the Curvature Control* to **Tangent** (arrow).

Enable the check-boxes as shown in the dialog box.

Click **OK**.

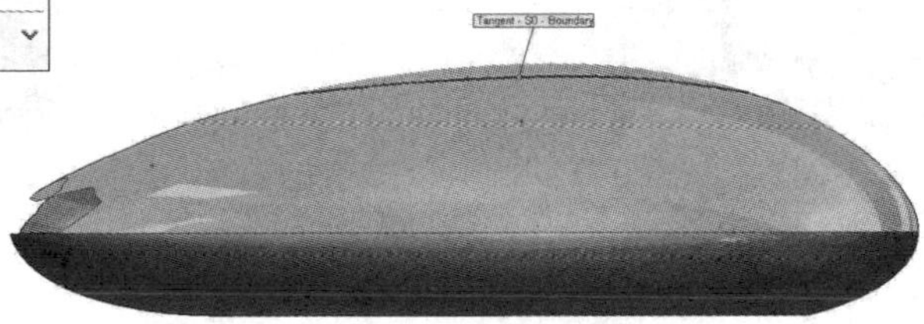

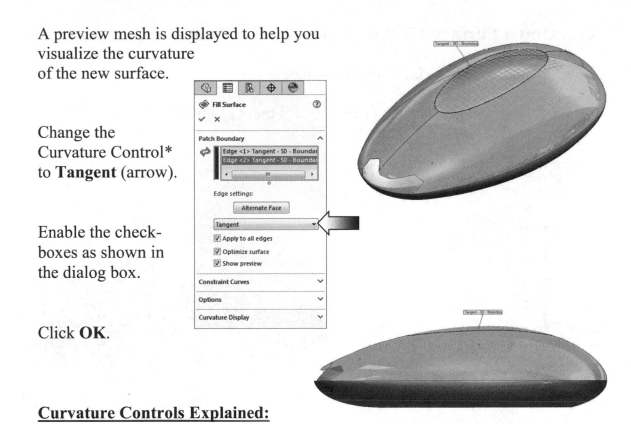

__**Curvature Controls Explained:**__

The **Curvature Control** defines the type of control you want to exert on the patch you create. The types of **Curvature Control** include:

* **Contact**: Creates a surface within the selected boundary.

* **Tangent**: Creates a surface within the selected boundary, but maintains the tangency of the patch edges.

* **Curvature**: Creates a surface that matches the curvature of the selected surface across the boundary edge with the adjacent surface.

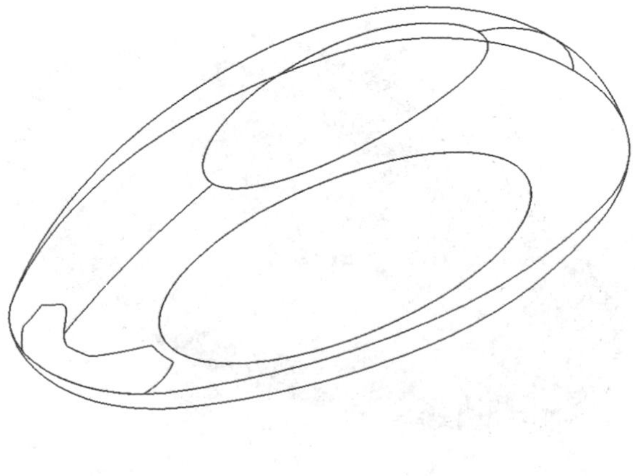

5. Creating a Surface Fill with Curvature Control:

Click the **Filled Surface** command once again.

Change to the **Front** view orientation (or press Control + 1).

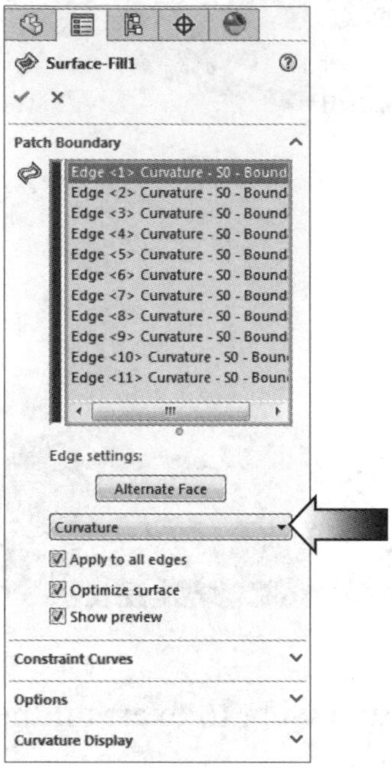

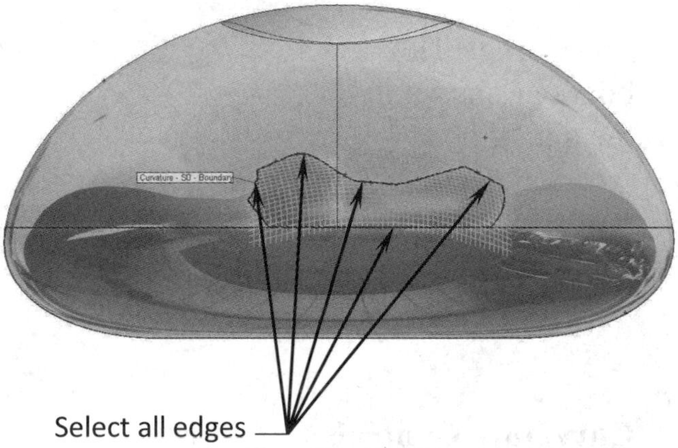

Select all edges of the opening in the front as noted.

The preview mesh appears indicating a closed boundary is found. (Enable the Preview Mesh checkbox, under Curvature Display, if the preview is not visible.)

Under the Curvature Control change the Contact option to **Curvature** (arrow).

Enable the other checkbox as shown in the dialog box.

Click **OK**.

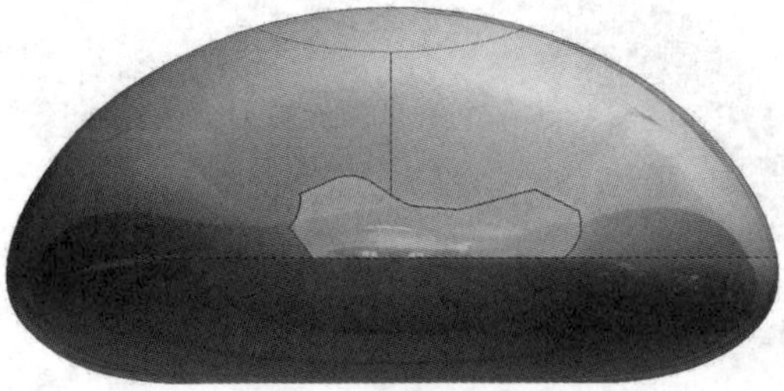

6. Knitting all surfaces:

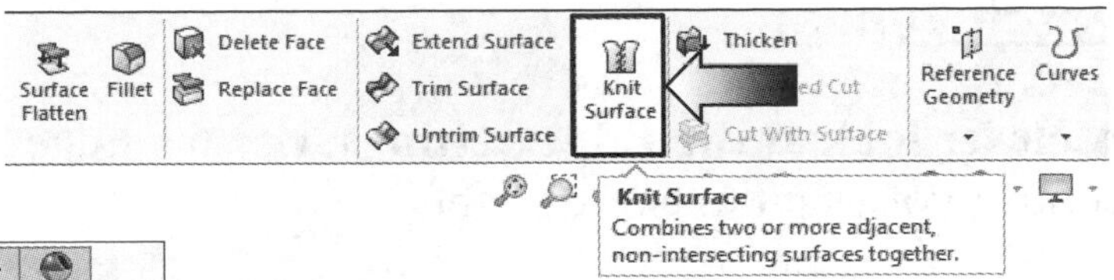

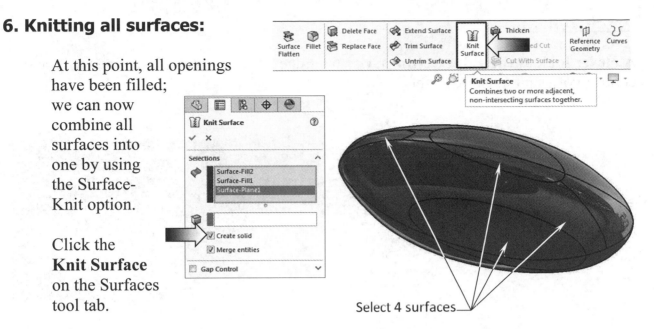

At this point, all openings have been filled; we can now combine all surfaces into one by using the Surface-Knit option.

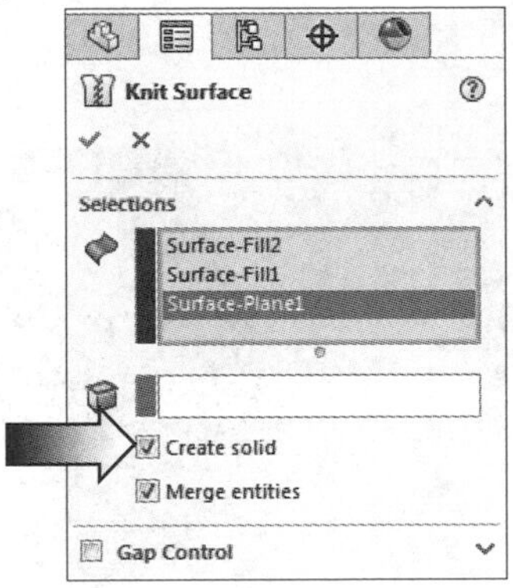

Click the **Knit Surface** on the Surfaces tool tab.

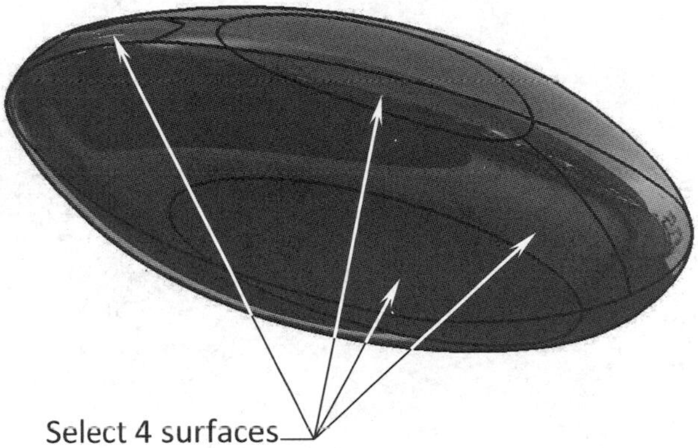

Select all **4 surfaces** in the graphics area as indicated.

Click the **Try To Form Solid** checkbox (arrow) to convert the part to a solid model.

Click **OK**.

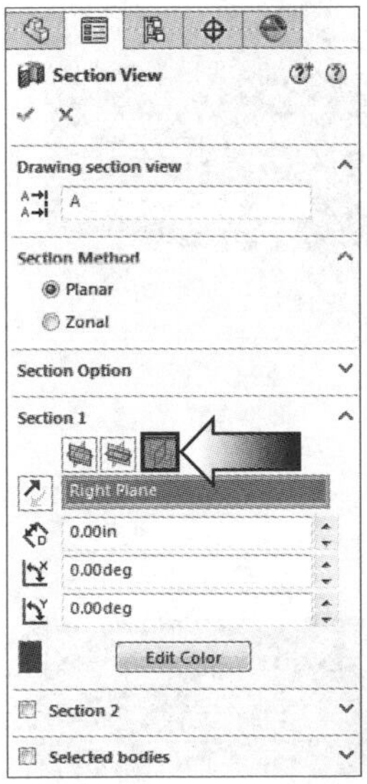

To verify the interior of the part create a section view using the **Right** plane as the cutting plane. Click Cancel when done.

7. Saving your work:

Save your work as **Using Filled Surface**.

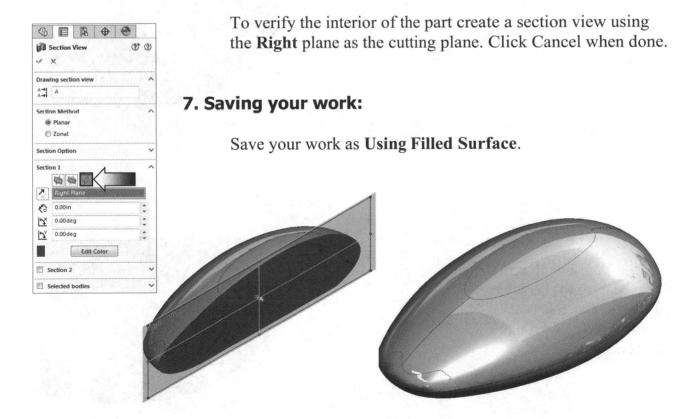

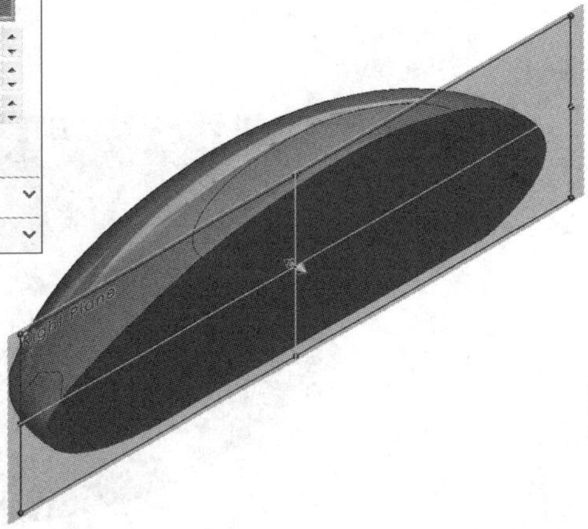

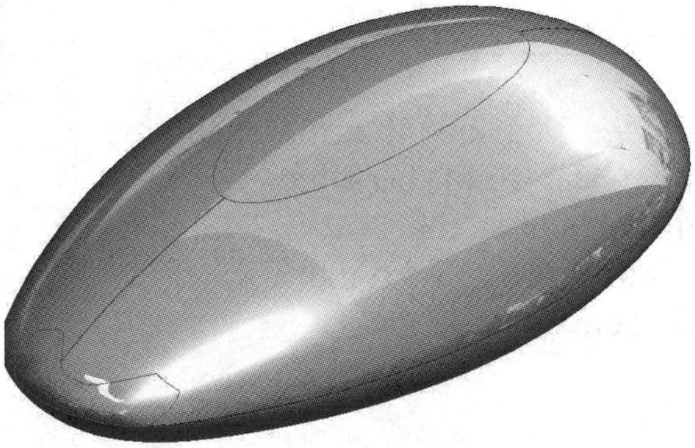

Patch Types:

A closer look at the Edge Settings: Use the Edge Settings options to define the type of control you want to exert on the patch you create.

Contact Patch:

Creates a surface within the selected boundary.

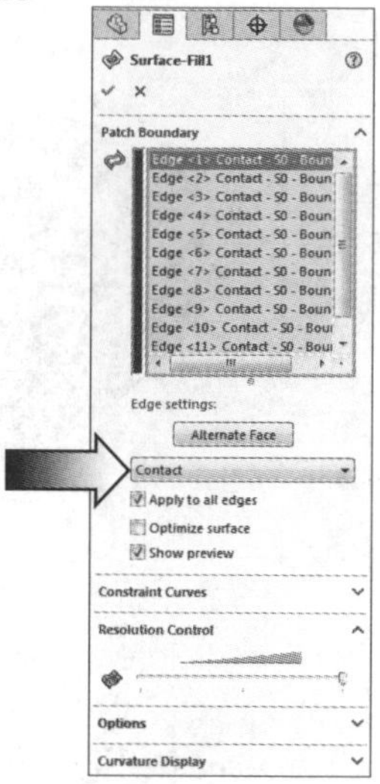

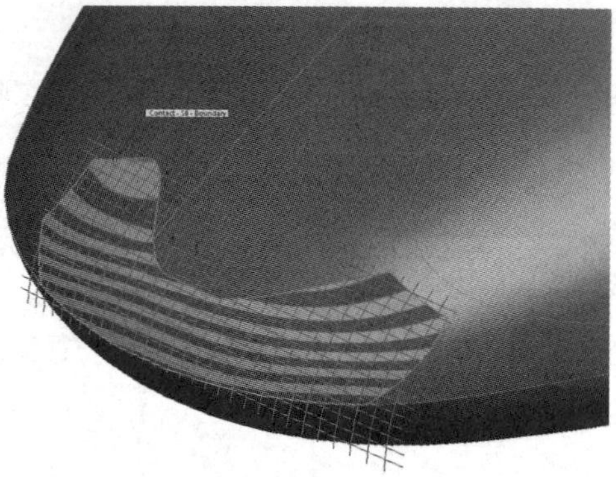

Tangent Patch:

Creates a surface within the selected boundary but maintains the tangency of the patch edges.

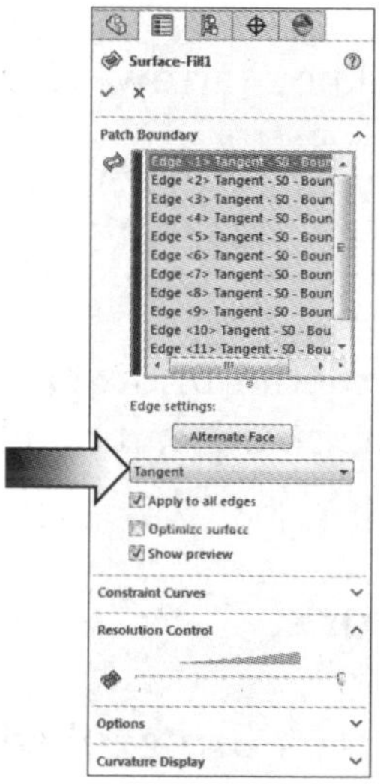

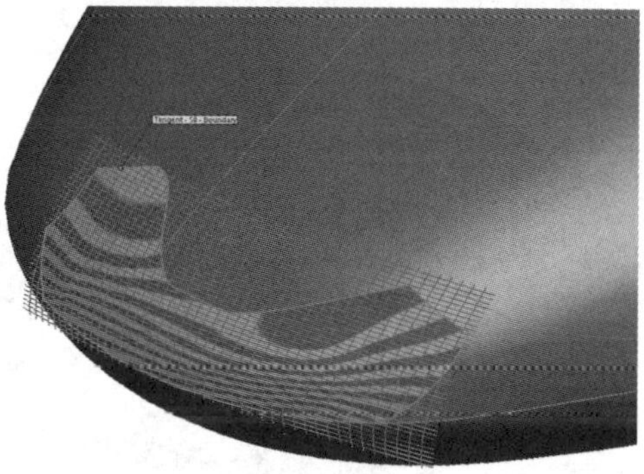

Curvature Patch:

Creates a surface that matches the curvature of the selected surface across the boundary edge with the adjacent surface.

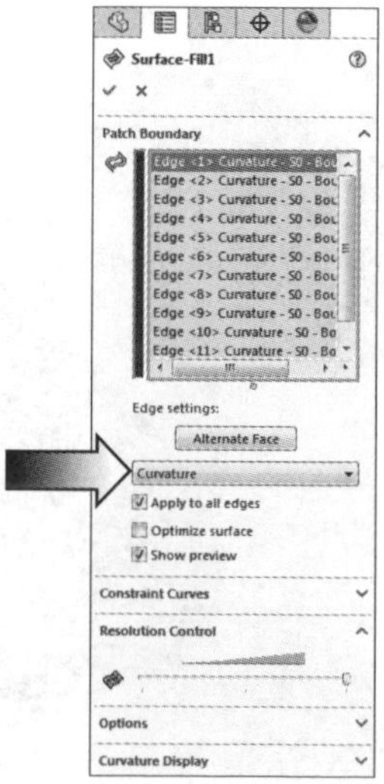

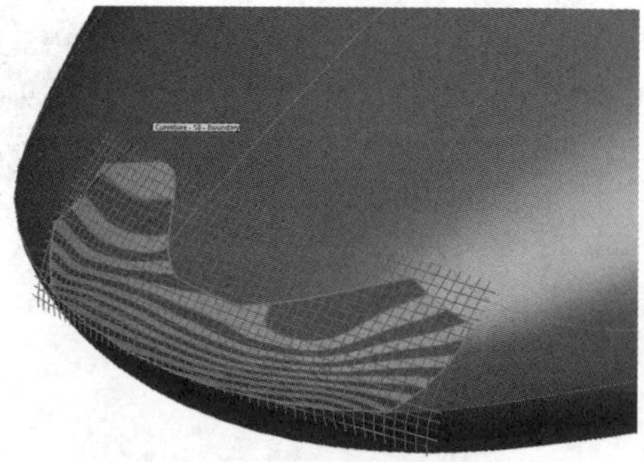

Optional:

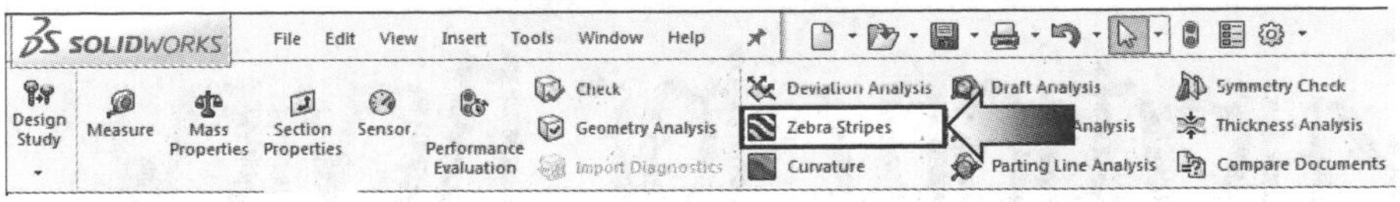

A closer look at the Zebra Stripes:

Use zebra stripes to visually determine which type of boundary to use between surfaces.

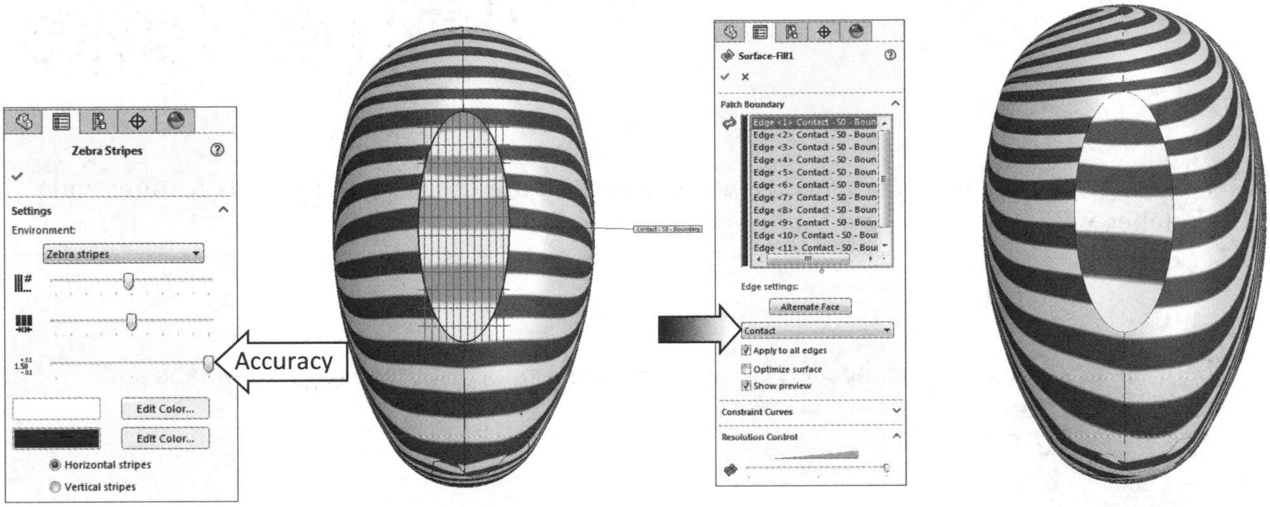

Zebra Stripes examples:

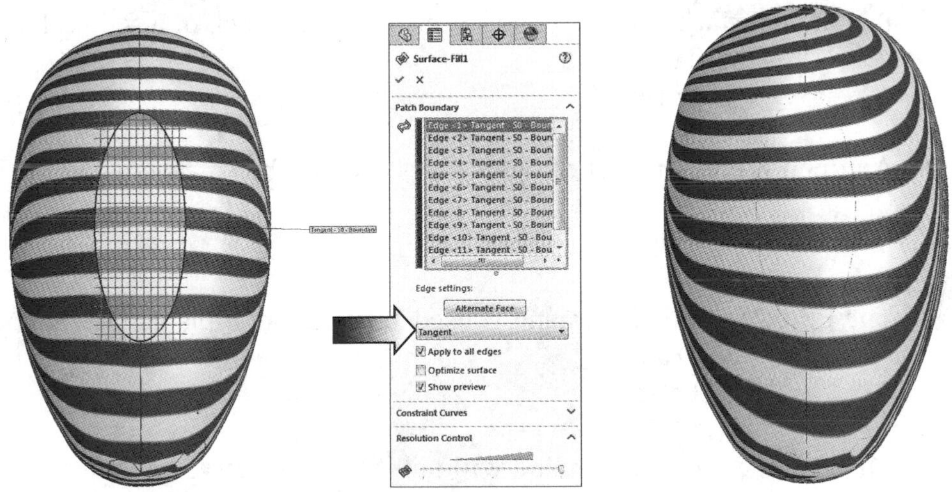

You can see small changes in a surface that may be hard to see with a standard display. Zebra stripes simulate the reflection of long strips of light on a very shiny surface.

With zebra stripes, you can easily see wrinkles or defects in a surface, and you can verify that two adjacent faces are in contact, are tangent, or have continuous curvature.

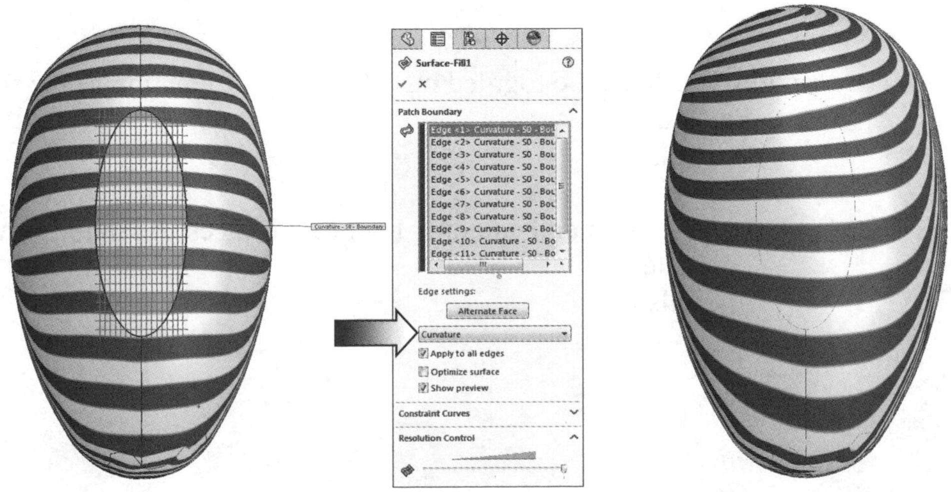

Questions for Review

Using Filled Surfaces

1. When opening a part document created from another CAD software, all of its features and sketches will appear on the FeatureManager tree.
   a. True
   b. False

2. Tool tabs can be added as needed by right clicking on one of the existing tool tabs and select them from the list.
   a. True
   b. False

3. A planar surface can be created from a closed sketch or a set of closed edges.
   a. True
   b. False

4. An open sketch or a set of open edges can also be patched using the Planar surface command.
   a. True
   b. False

5. The Filled Surface command is used to patch a non-planar closed boundary.
   a. True
   b. False

6. The Filled Surface command will fail if the boundary is not closed.
   a. True
   b. False

7. The Preview Mesh can be toggled on/off during the creation of the filled surface.
   a. True
   b. False

8. The Knit Surface command can only Knit the surfaces; it cannot form a solid from the surfaces.
   a. True
   b. False

1. FALSE 2. TRUE
3. TRUE 4. FALSE
5. TRUE 6. TRUE
7. TRUE 8. FALSE

Using Ruled Surface

1. Opening a part document:

Select **File, Open**.

From the Training Files folder, open the part document named: **Using Ruled Surface.sldprt**.

The main body of the Detergent Container has already been created to help focus on the creation of a ruled surface.

2. Creating a Split Line feature:

Edit the **Sketch8** from the FeatureManager tree.

This sketch was created on the Front plane and has already been fully defined.

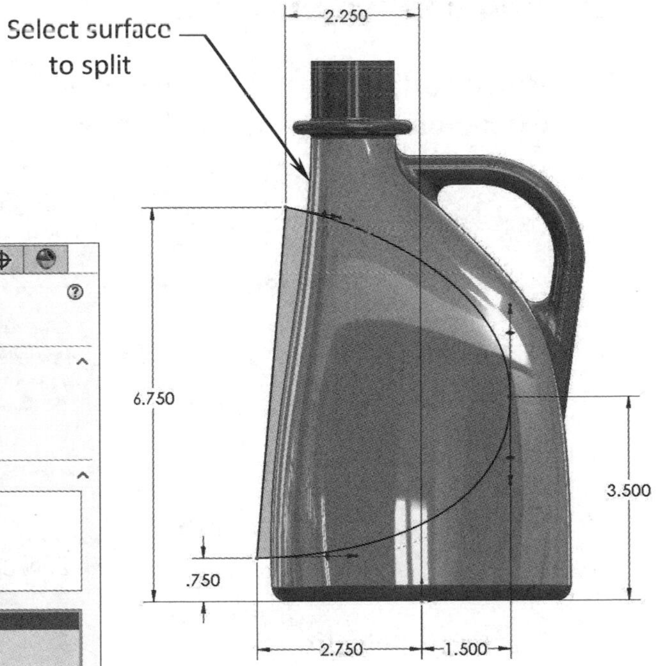

Switch to the **Surfaces** tool tab and select: **Curves, Split Line**.

For Type of Split, select **Projection**.

For Split Sketch, select **Sketch8**.

For Faces to Split, select the **face** of the container as noted.

Click **OK**.

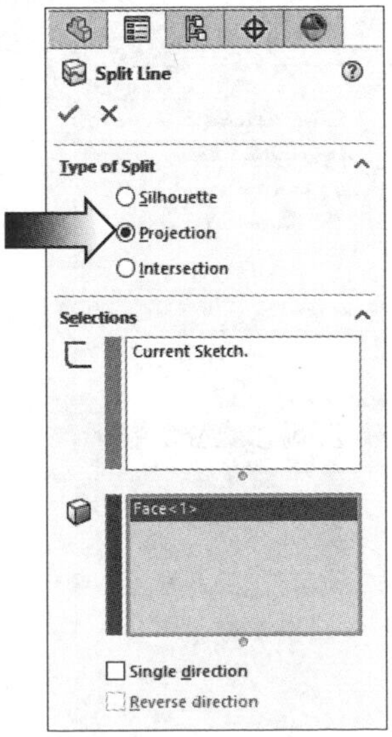

3. Creating an offset surface:

Click **Offset Surface.**

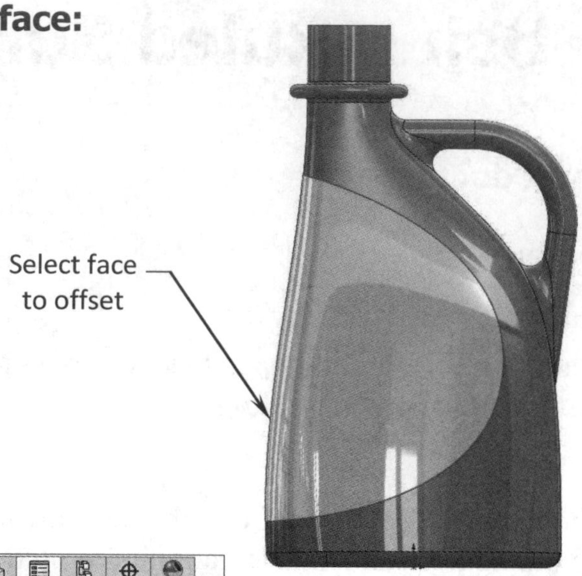

For Faces to Offset, select the **face** as indicated.

For Offset Distance, enter **.075in**.

Click **Reverse** to place the copy on the <u>inside</u>.

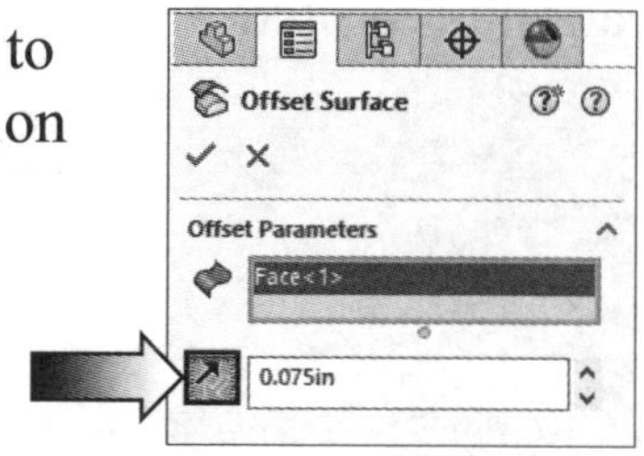

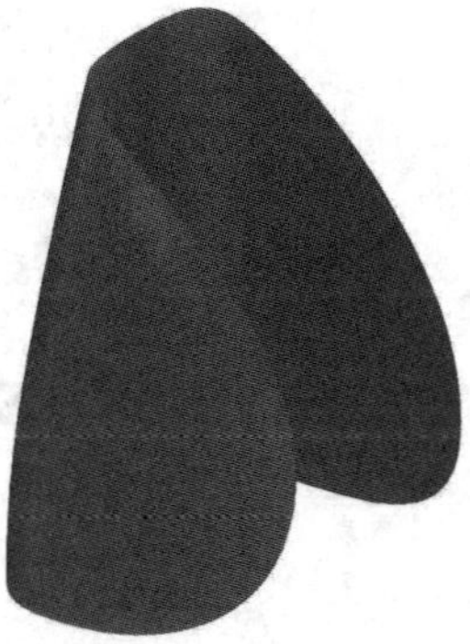

Click **OK.**

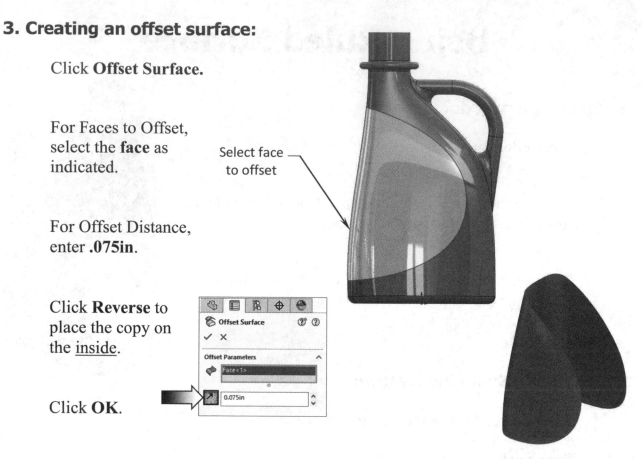

4. Adding a Ruled Surface:

Ruled surface creates surfaces that extend out in a specified direction from selected edges.

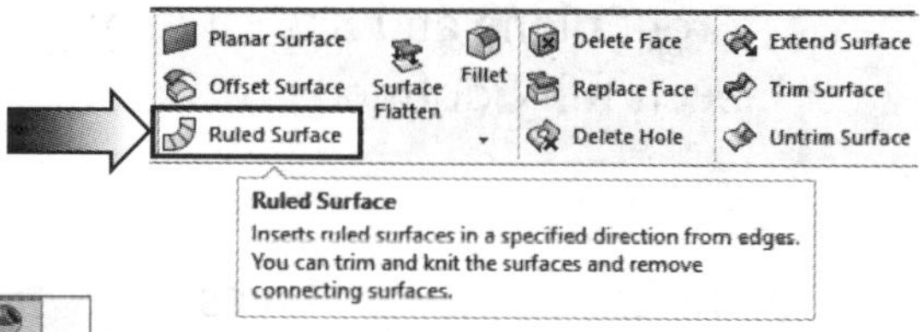

Click **Ruled Surface.**

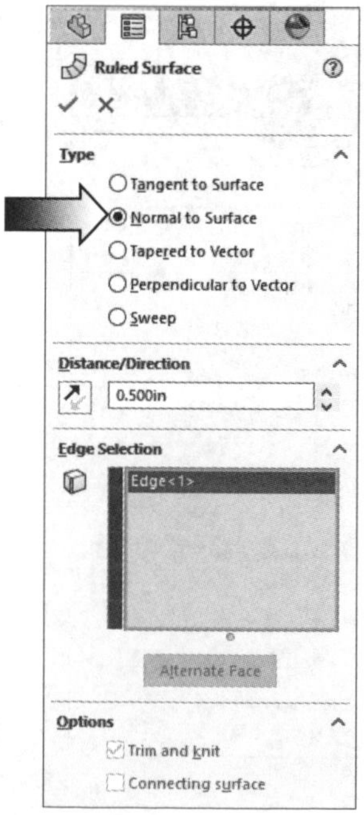

Click **Normal to Surface**.

For Distance, enter **.500in**.

For Edge Selection, select all **edges** as noted.

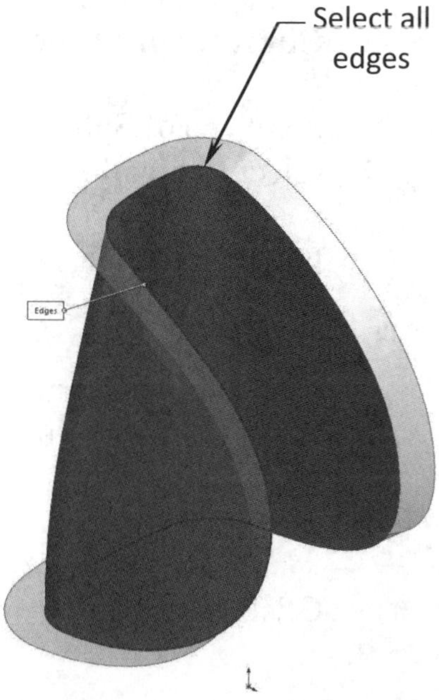

Click **OK.**

5. Knitting the surfaces:

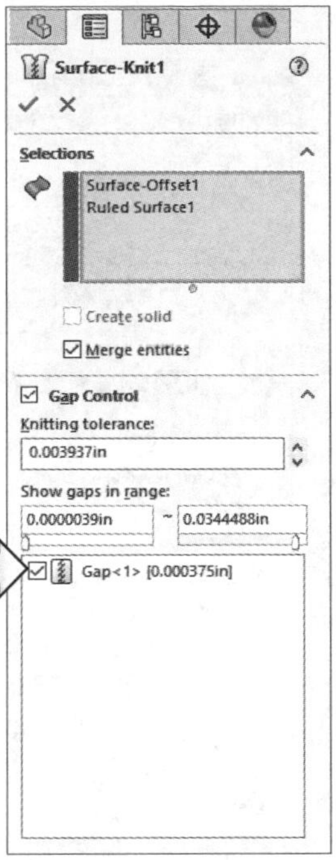

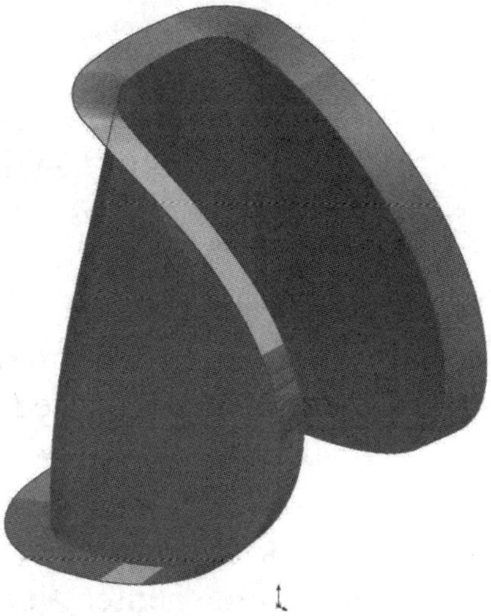

The Offset Surface and the Ruled Surface must be knitted into a single surface in order to use as the trim tool.

Click **Knit Surface**.

Select both surfaces, the **Offset** and the **Ruled** surfaces.

Click **Merge Entities**.

Enable the **Gap Control** checkbox and check the gap-box to allow the software to close any gaps automatically.

Click **OK**.

6. Creating a Surface Cut:

The Knit Surface can now be used as the trim tool to cut into the container.

Click **Cut with Surface**.

For Surface Cut Parameters, select the **Knit Surface**.

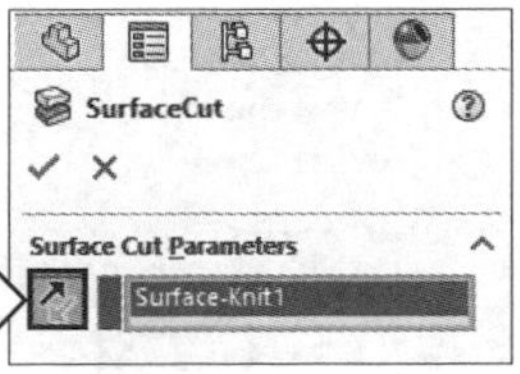

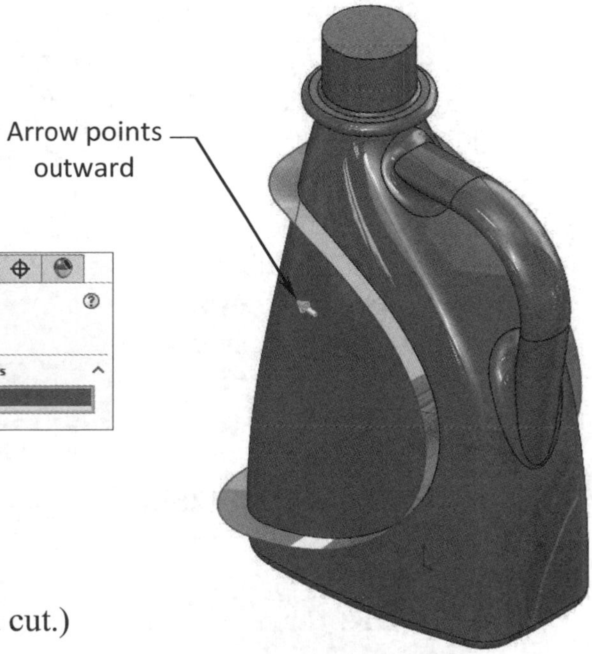

Click **Reverse**.
The direction arrow should be pointing <u>outward</u>.

Click **OK**.
(Hide the Knit Surface to see the recessed cut.)

7. Adding the .060" fillet:

Click **Fillet**.

For Fillet Type, use the default **Constant Size**.

For Items to Fillet, select one of the **inner edges** of the recessed feature.

For Radius, enter **.060in**.

Click **OK**.

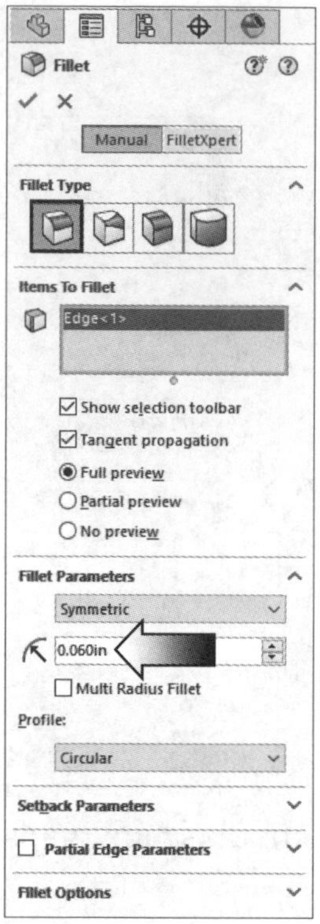

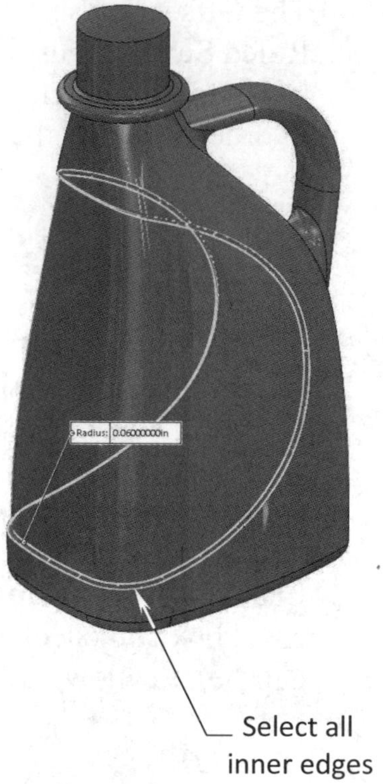

8. Adding the .125" fillet:

Click **Fillet** again.

For Fillet Type, use the default **Constant Size**.

For Items to Fillet, select one of the **outer edges** of the recessed feature.

For Radius, enter **.125in**.

Click **OK**.

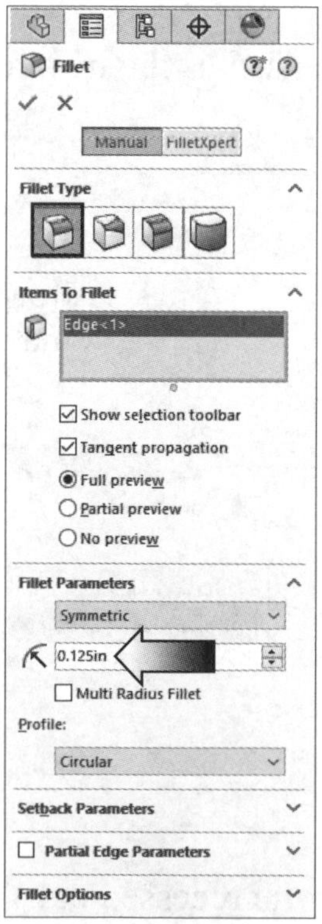

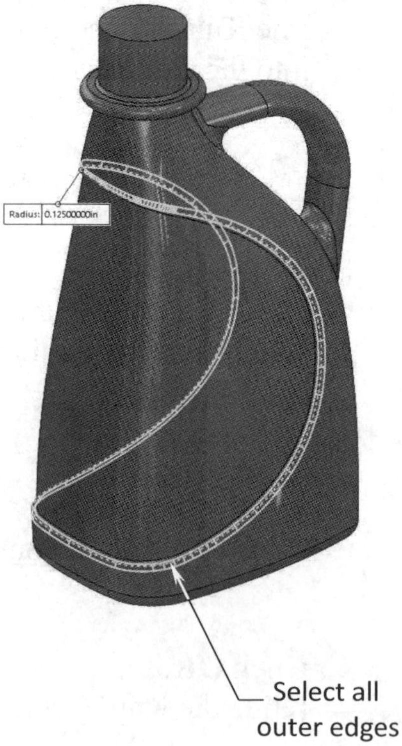

9. Shelling the model:

Switch to the **Features** tool tab.

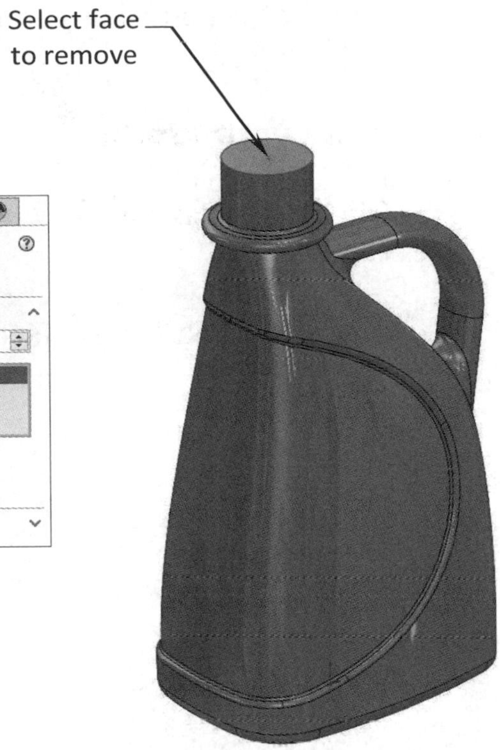

For Wall Thickness, enter **.060in**.

For Faces to Remove, select the **top surface** of the model.

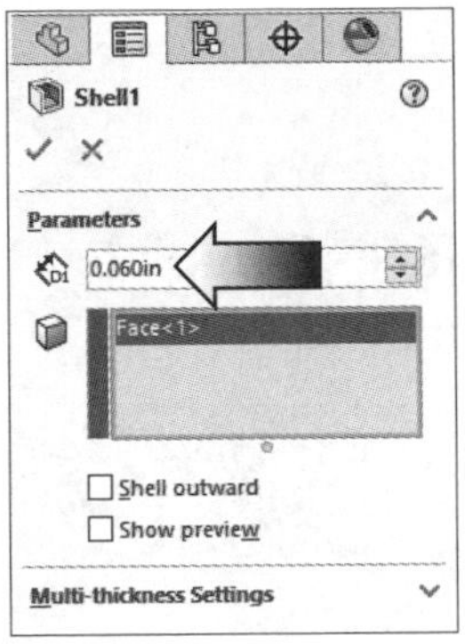

Click **OK**.

10. Verifying the wall thickness:

Section View is one of the quick ways to inspect the wall thickness and to view the interior features.

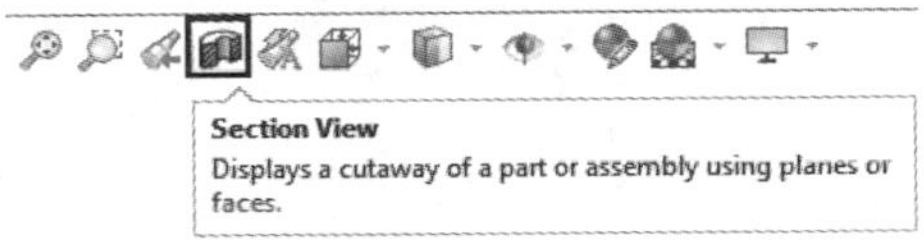

Click **Section View** on the View-Heads Up tool bar.

For Section 1, use the default **Front** plane.

For Distance, use the default **0.00** dimension.

Click **OK**.

Zoom in closer to inspect the wall thickness, especially around the recessed area.

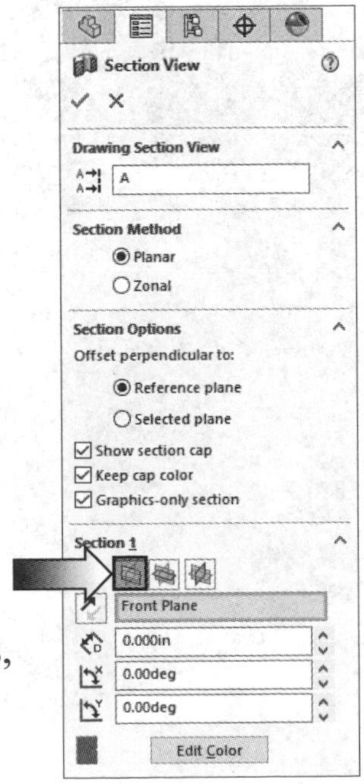

Exit the Section View command.

11. Saving your work:

Click **File / Save As**.

Enter **Using Ruled Surface_Completed** for the name of the file.

Click **Save**.

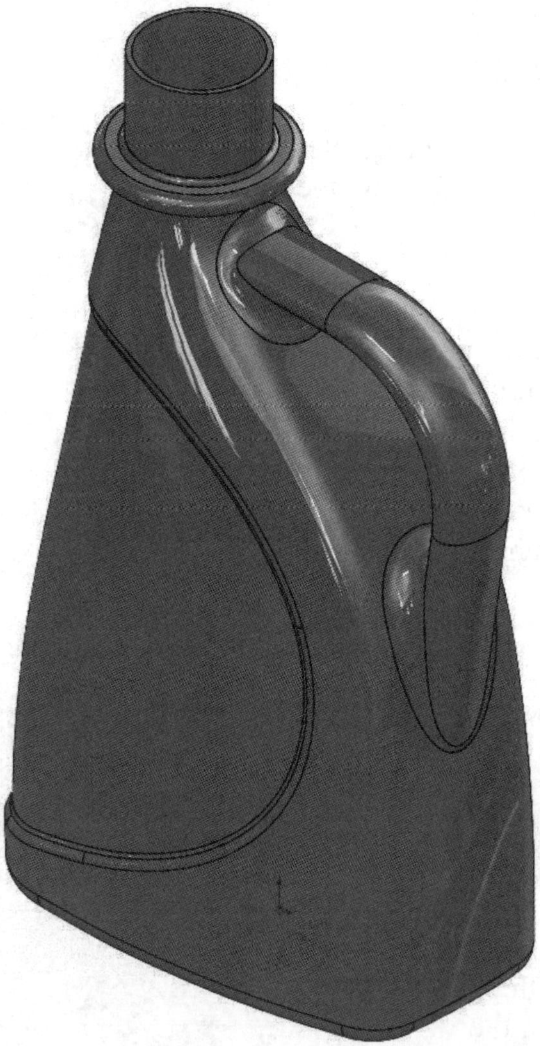

CHAPTER 11

Surfaces Vs. Solid Modeling

Surfaces vs. Solid Modeling

Surfaces are a type of geometry that can be used to create solid features. Surface tools are available on the Surfaces toolbar.

You can create surfaces using these methods:

- Insert a planar surface from a sketch or from a set of closed edges that lie on a plane
- Extrude, revolve, sweep, or loft from sketches
- Offset from existing faces or surfaccs
- Import a file
- Create mid-surfaces
- Radiate surfaces

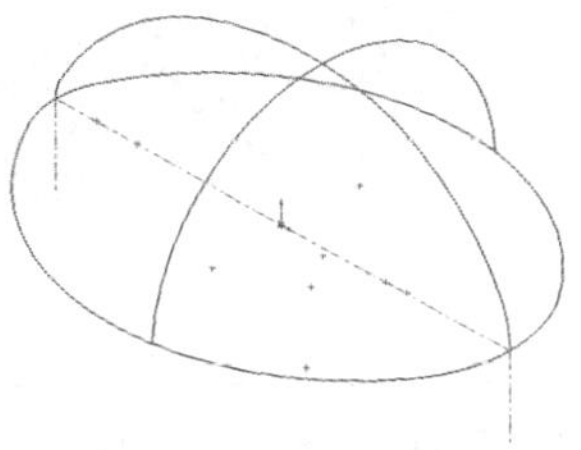

You can modify surfaces in the following ways:

- Extend
- Trim existing surfaces
- Un-trim surfaces
- Fillet surfaces
- Repair surfaces using Filled Surface
- Move/Copy surfaces
- Delete and patch a face
- Knit surfaces

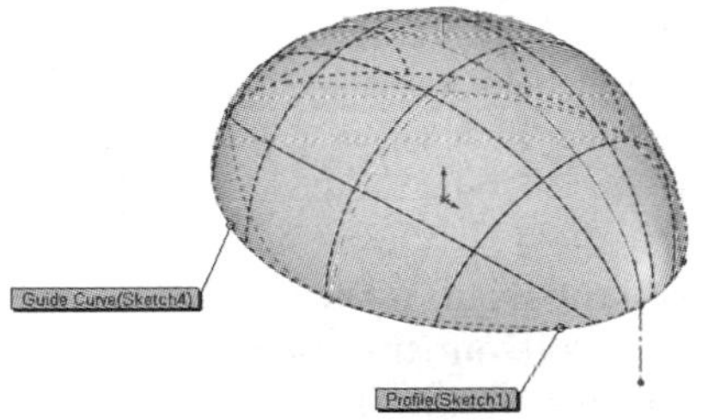

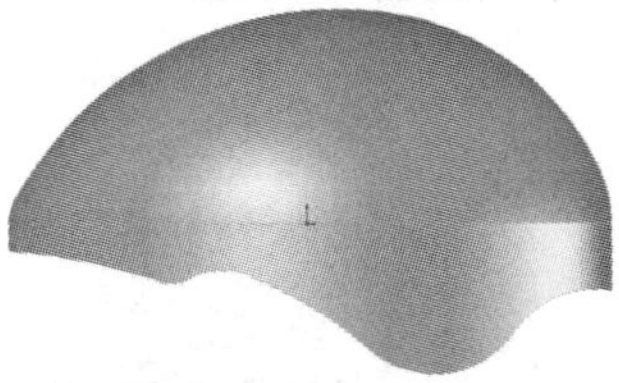

You can use surfaces in the following ways:

- Select surface edges and vertices to use as a sweep guide curve and path
- Create a solid or cut feature by thickening a surface
- Extrude a solid or cut feature with the end condition Up to Surface or Offset from Surface
- Create a solid feature by thickening surfaces that have been knit into a closed volume
- Replace a face with a surface

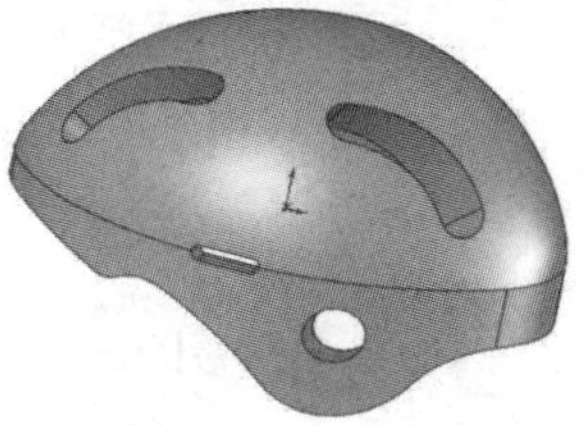

Safety Helmet
Surfaces vs. Solid Modeling

View Orientation Hot Keys:

Ctrl + 1 = Front View
Ctrl + 2 = Back View
Ctrl + 3 = Left View
Ctrl + 4 = Right View
Ctrl + 5 – Top View
Ctrl + 6 = Bottom View
Ctrl + 7 = Isometric View
Ctrl + 8 = Normal To Selection

Dimensioning Standards: **ANSI**

Units: **INCHES** – 3 Decimals

Tools Needed:

Insert Sketch	Plane	Lofted Surface
Swept Surface	Planar Surface	Knit Surface
Revolve Cut	Sweep Cut	Surface Thicken

1. Opening the Existing file:

Browse the Training Files Folder and open a part document named: **Helmet**.

This part file has 3 sketch profiles and 3 guide curves previously created.

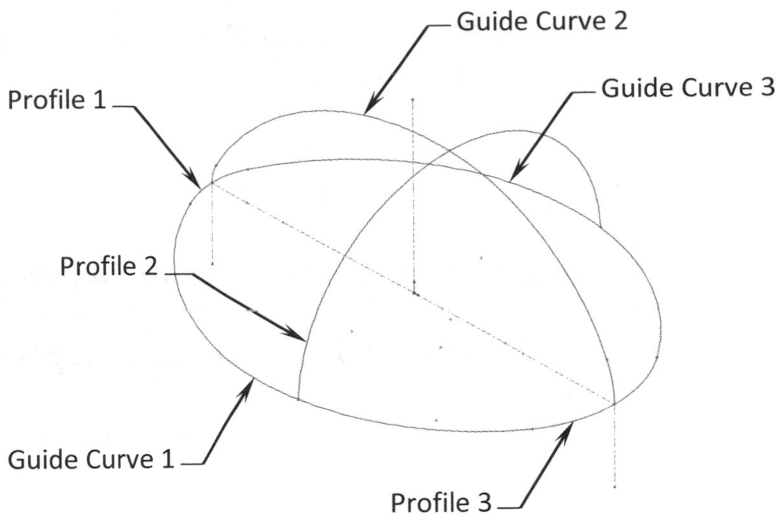

2. Constructing the Body of the Helmet:

Click **Lofted Surface**  or select **Insert, Surface, Loft**.

Select the **3 Sketch Profiles**.

Select the **3 Guide Curves** as noted.

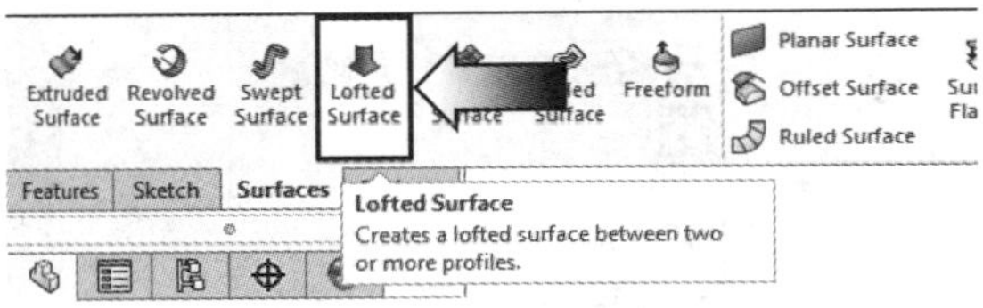

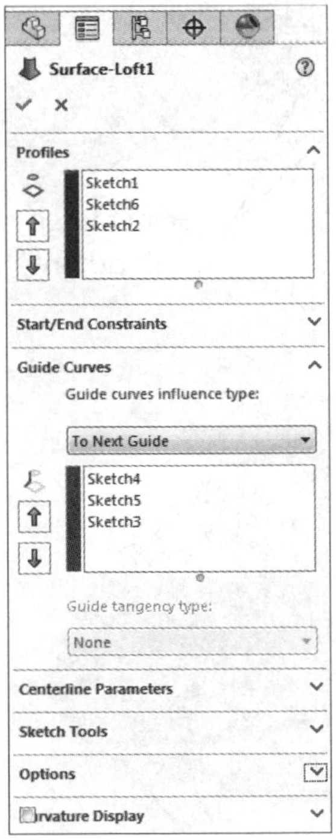

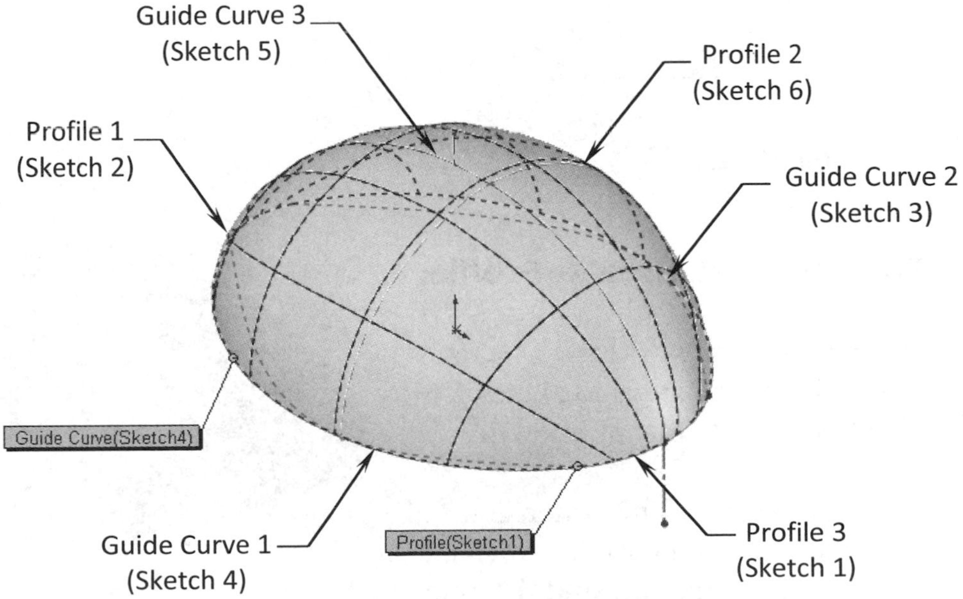

Click **OK**.

Rotate the surface model to verify the details of the lofted surface.

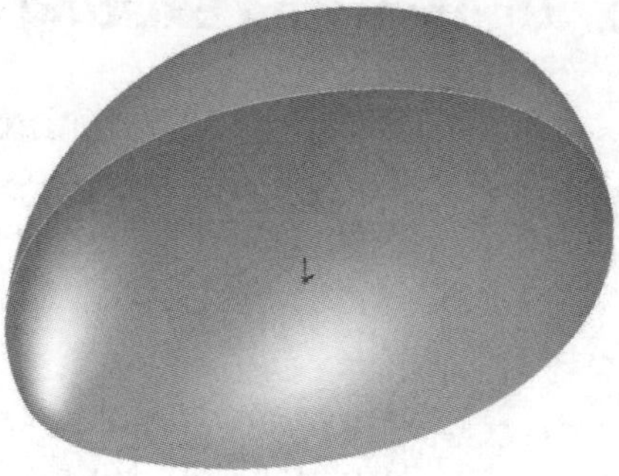

3. Creating a new work plane:

Create a plane **Perpendicular** as illustrated.

Click **Insert, Reference Geometry, Plane** .

Select the **circular edge** and its **endpoint** as noted.

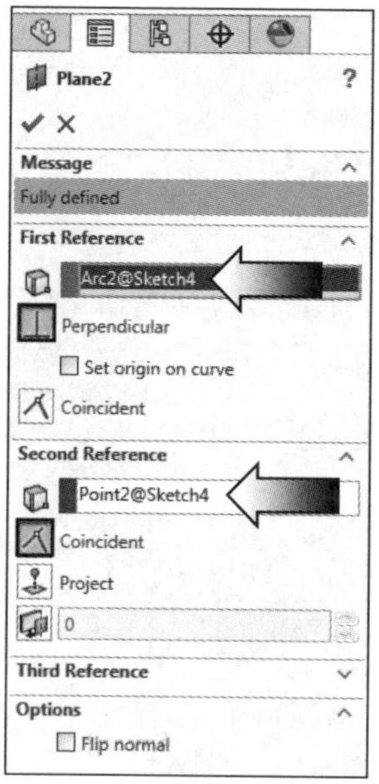

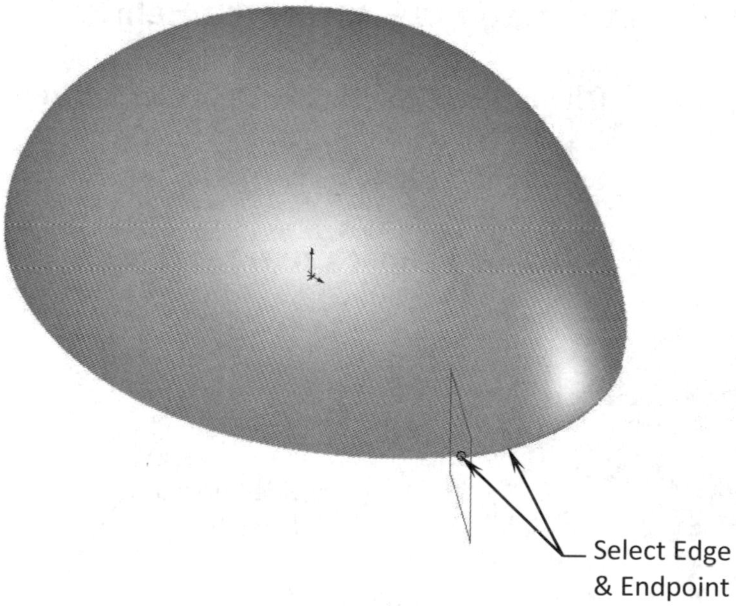

Click **OK**.

4. Sketching the Sweep-Profile:

Open a **new sketch** on the new plane and sketch a **Vertical Line** as shown.

Add a **3.000in.** dimension and a **Pierce** relation between the end point of the line and the bottom edge.

Exit the Sketch .

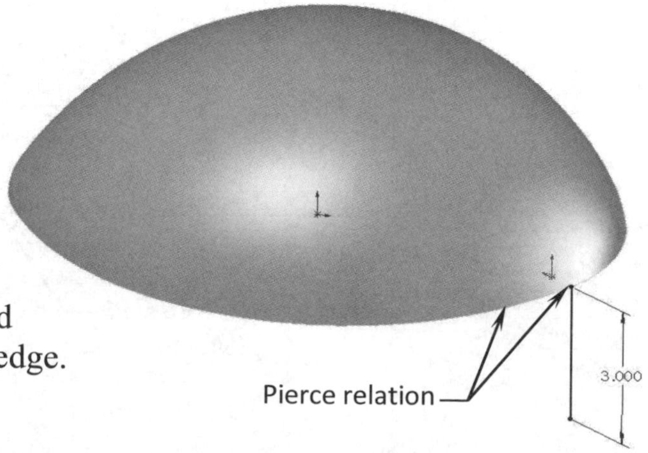

5. Creating the Sweep Path (Composite Curve):

Click **Composite** on the Curves toolbar, or select:

Insert / Curve /Composite .

Select **all edges** as noted.

Click **OK**.

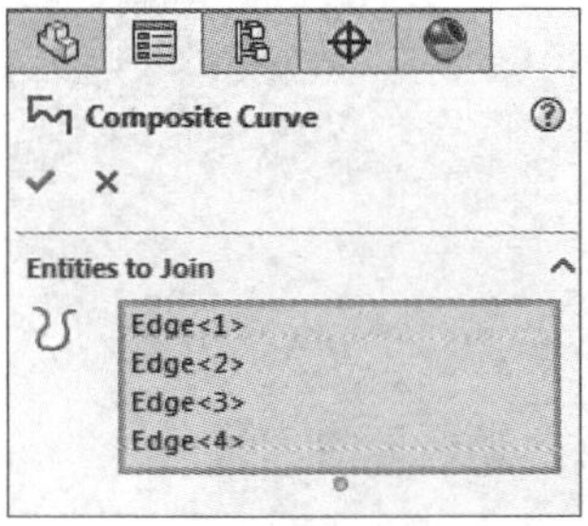

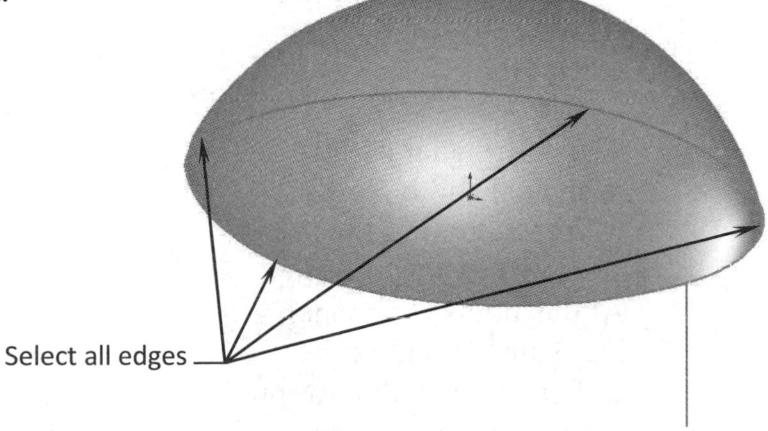

6. Creating a Swept-Surface:

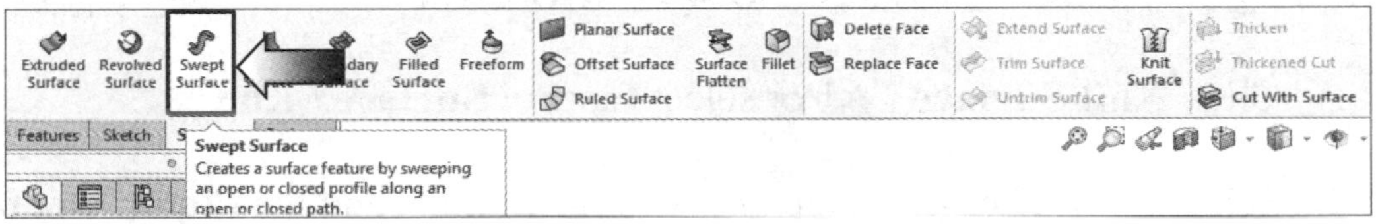

Click **Swept Surface** or select: **Insert, Surface, Sweep**.

Select the **Vertical Line** as the Sketch Profile.

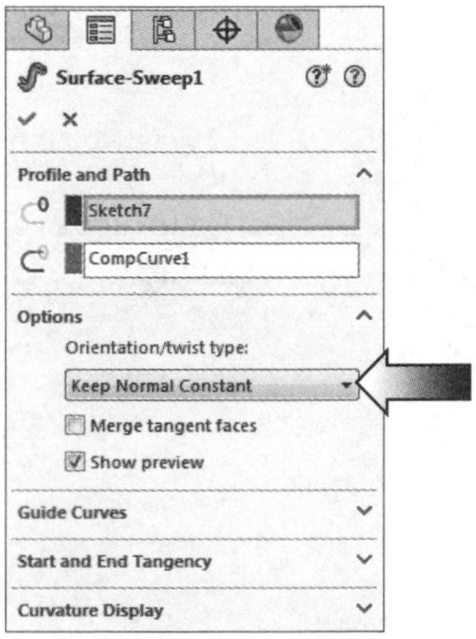

Select the **Composite-Curve** as the Sweep Path.

Select **Keep Normal Constant** under Options. (This option keeps the line perpendicular to the sweep path all around.)

Click **OK**.

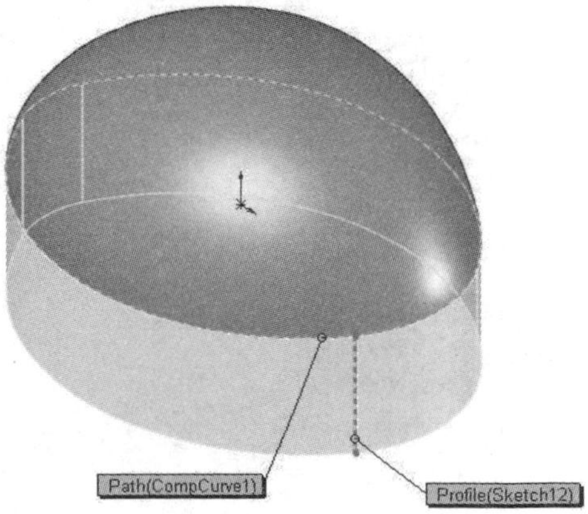

7. Adding a Planar Surface:

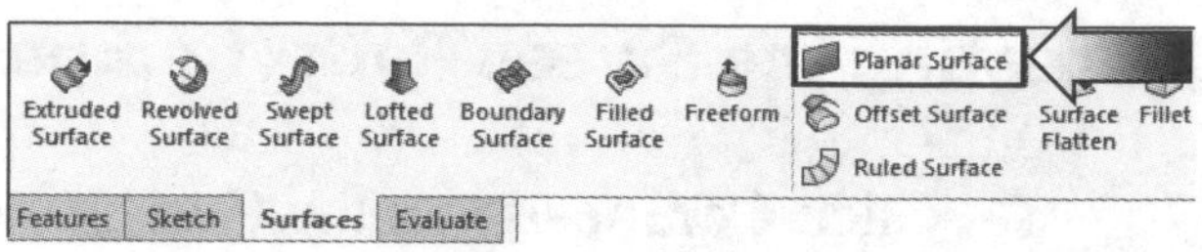

Click **Planar Surface** or select: **Insert / Surface / Planar**.

Select **all edges** at the bottom as the Bounding Entities.

Click **OK**.

The new planar surface is created and it closes off the bottom of the part.

At this point, the model has 3 surfaces in it: the Lofted surface, the Swept surface, and the Planar surface.

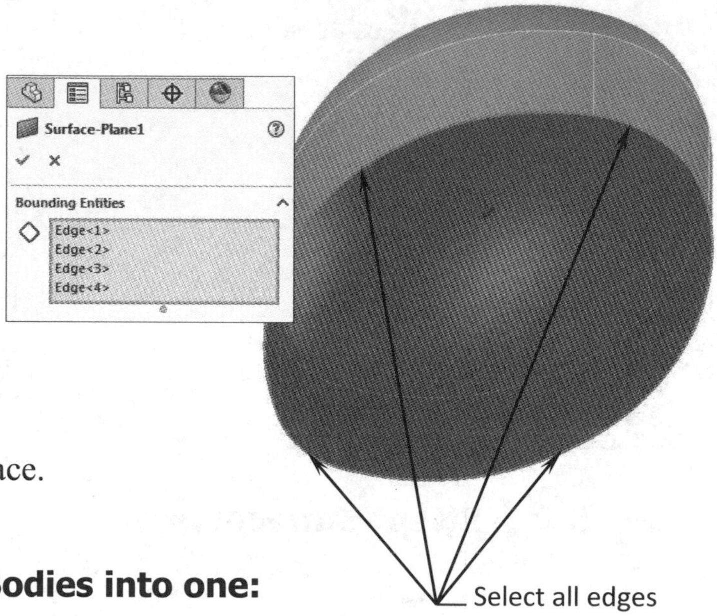

8. Knitting the three Surface-Bodies into one:

Click **Knit Surface** or select **Insert / Surface / Knit**.

Select the **Lofted-Surface**, the **Swept-Surface**, and the **Planar-Surface** either from the FeatureManager tree or from the graphics area.

Clear the **Gap Control** checkbox.

Enable the **Create Solid** option to convert the surface into a solid body.

Click **OK**.

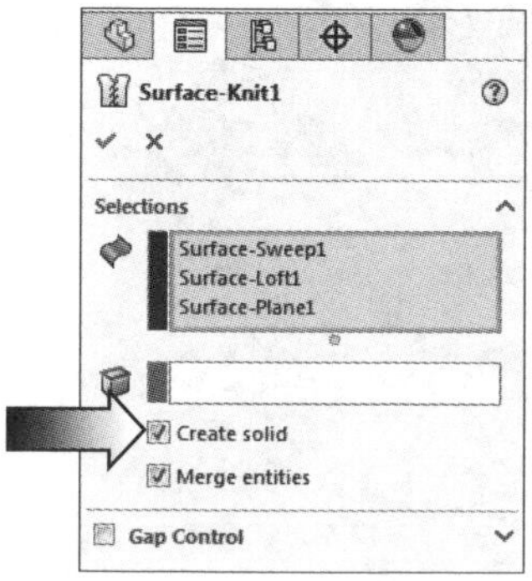

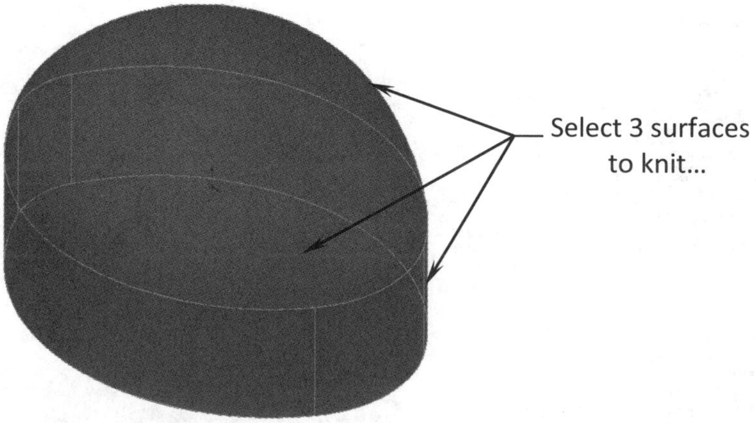

9. Creating a section view:

Select the Front plane from the FeatureManager and click **Section View** (arrow) to verify the solid material. Click-off the section view command when finished viewing.

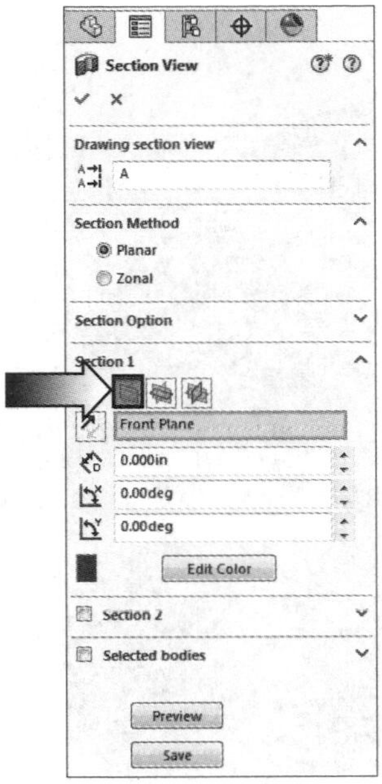

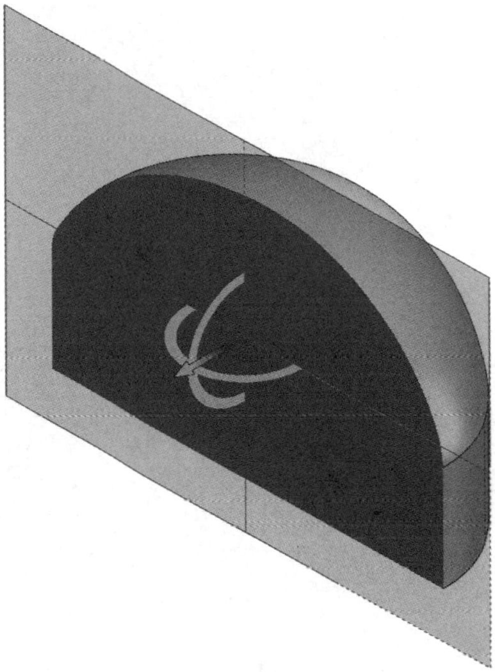

10. Adding an Extruded Cut feature:

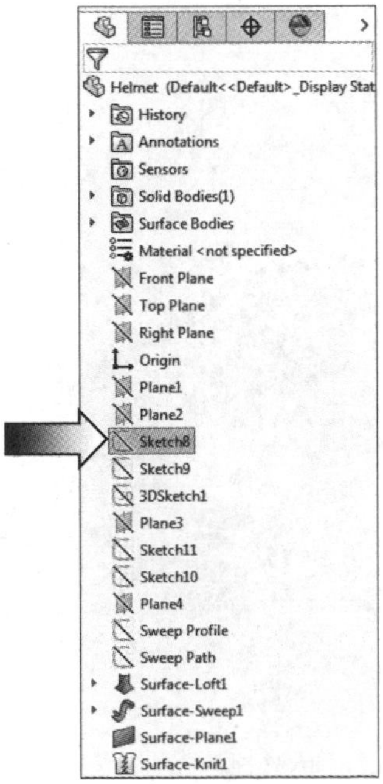

Select the **Sketch8** from the FeatureManager Tree.

This sketch will be used to remove the lower portion of the Helmet.

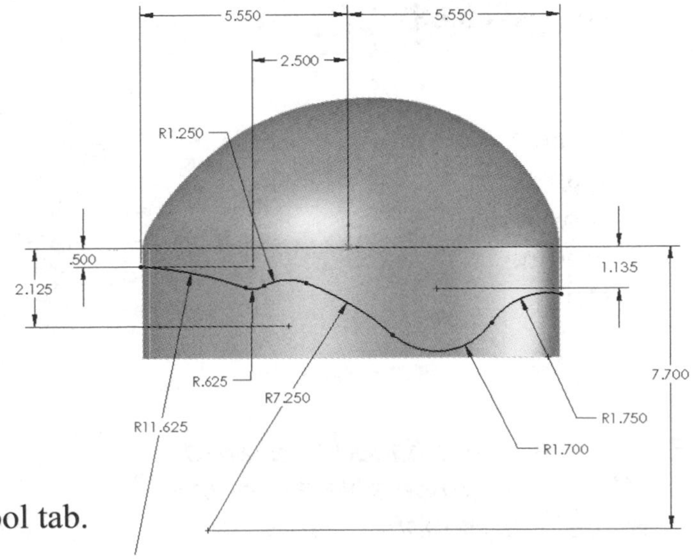

Switch to the **Features** tool tab.

Click **Extruded-Cut**  or select: **Insert / Cut / Extrude**.

For Direction, select **Through-All Both** from the drop-down list.

Enable **Flip-Side-To-Cut** if needed to remove the lower portion of the part.

Click **OK**.

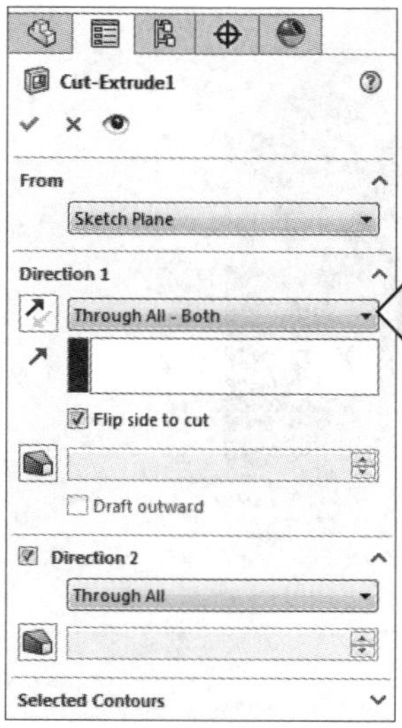

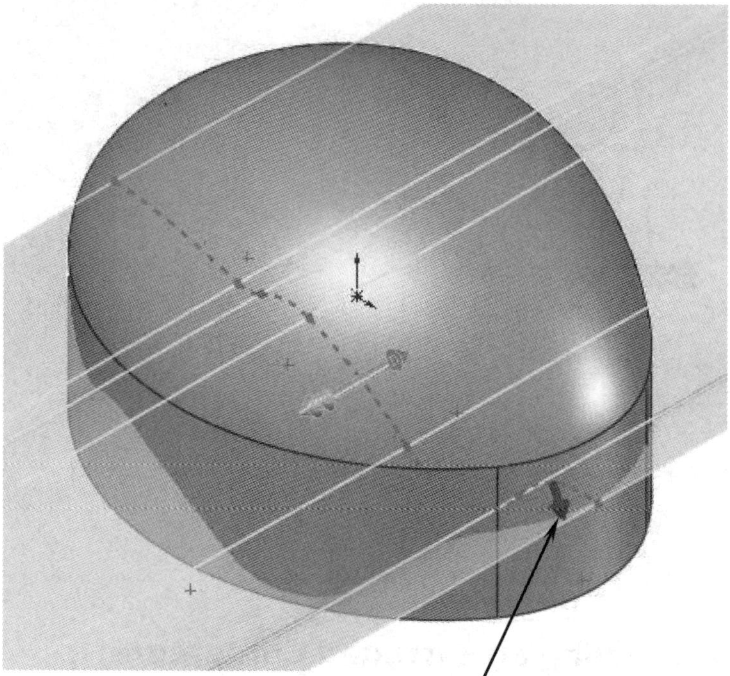

Arrow direction indicates the side to remove

<u>Hide</u> **Sketch8**.

Inspect your model against the image shown here before moving to the next step.

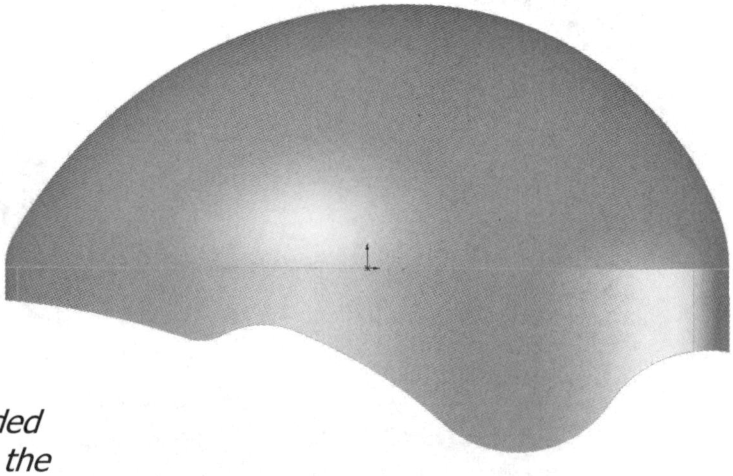

The resulted cut.

<u>NOTE:</u> Change from Shaded to Shaded With-Edges to be able to see the model edges more clearly.

11. Adding a Revolve-Cut feature:

Select the **Sketch10** from the FeatureManager Tree.

This sketch will be used to remove the material on the inside of the Helmet.

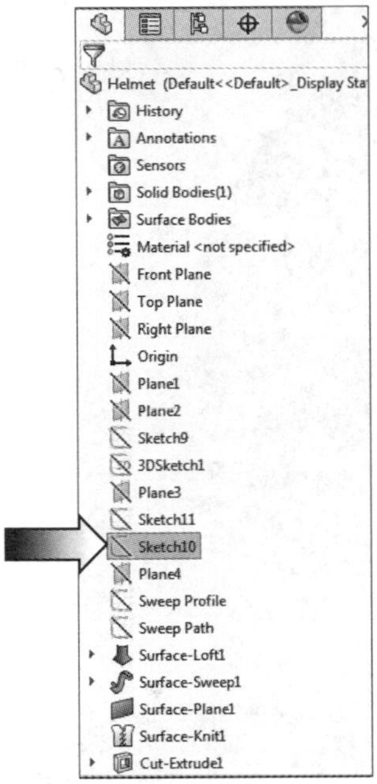

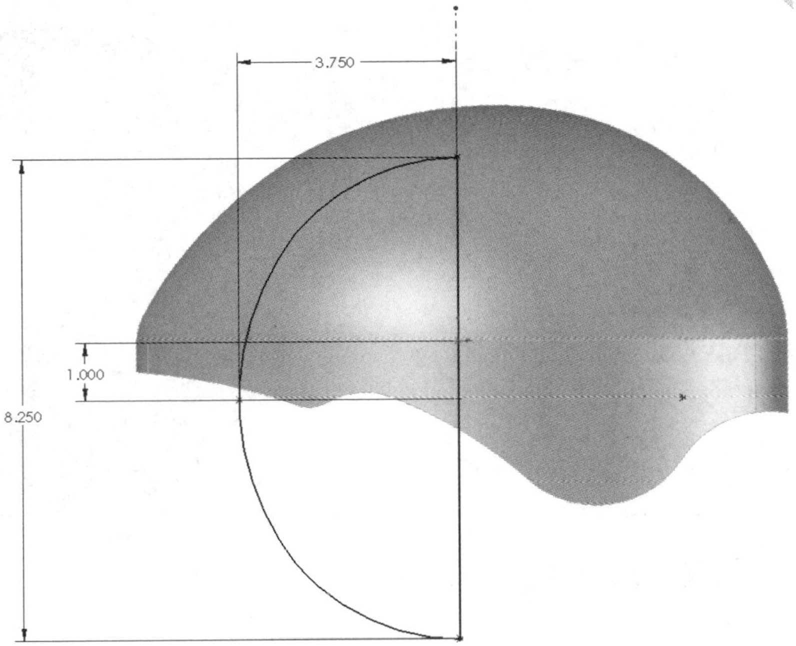

Switch to the **Features** tool tab.

Click **Revolve-Cut**  or select **Insert / Cut / Revolve**.

Select the **vertical centerline** as Revolve-Direction.

Revolve Angle: **360.00deg**.

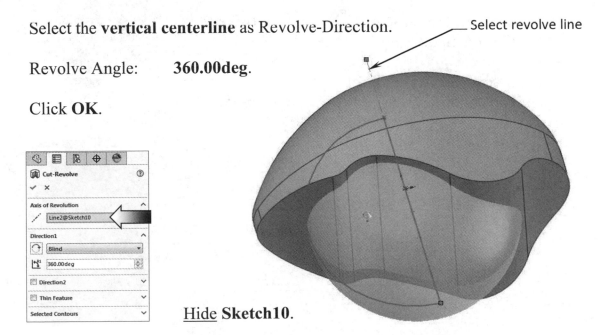

Click **OK**.

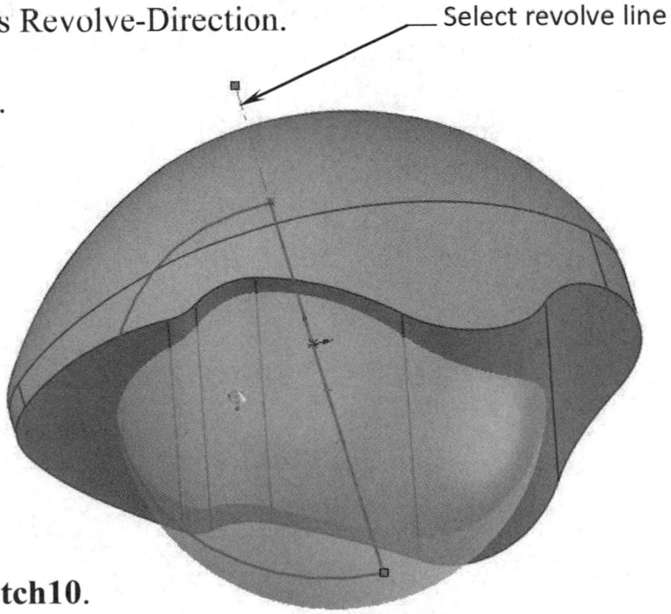

Hide **Sketch10**.

12. Adding the side cut features:

Select the **Sketch9** from the FeatureManager Tree.

This sketch is used to cut from the <u>inside</u> with a **1.00deg**. draft angle.

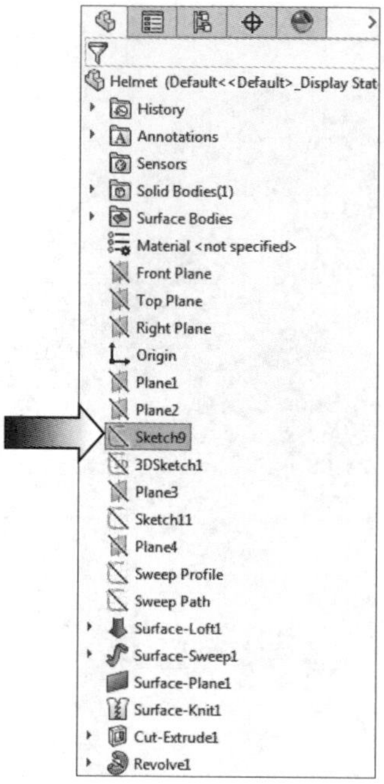

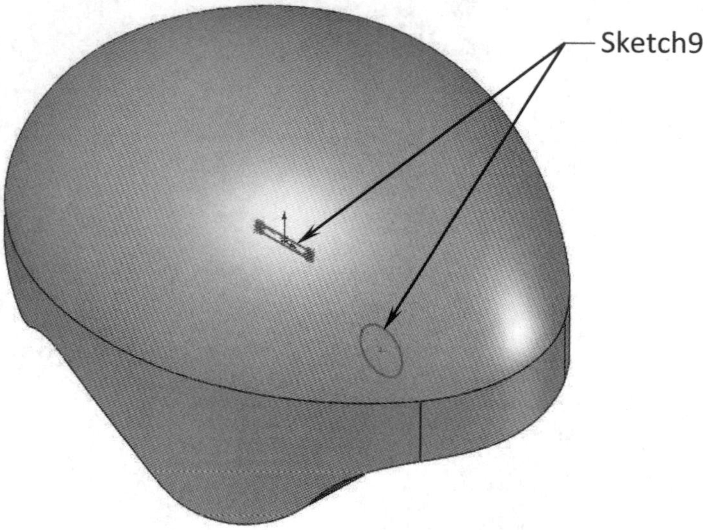

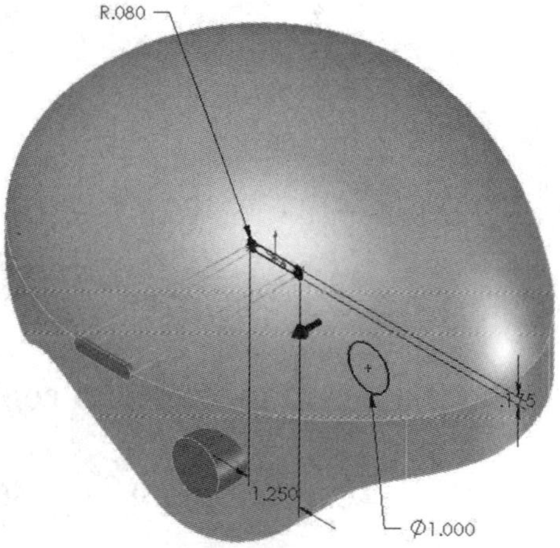

Switch to the **Features** tool tab.

Click **Extruded-Cut** or select: **Insert / Cut / Extrude**.

Select **Through-All** for end condition.

Enable Draft Angle and enter: **1.00deg**.

Enable **Draft Outward** option.

Click **OK**.

<u>Hide</u> the **Sketch9**.

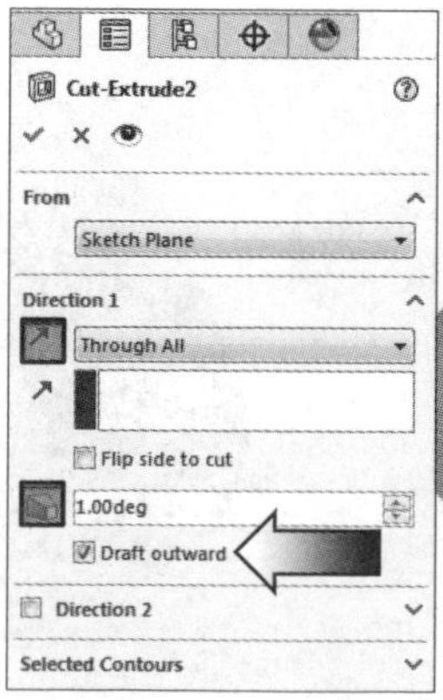

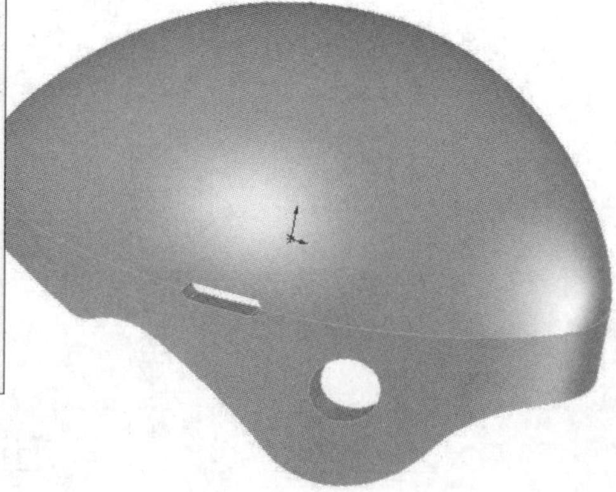

13. Creating the cutout slot:

Select the **Sketch11** from the FeatureManager Tree.

This sketch will be used to cut a slot from the <u>outside</u> of the model and with a **10.00deg**. draft angle.

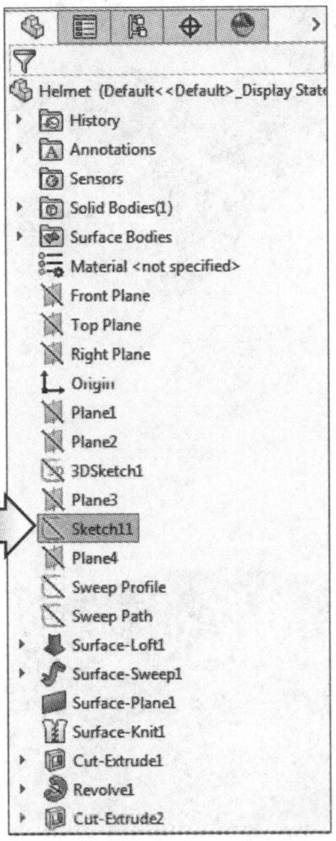

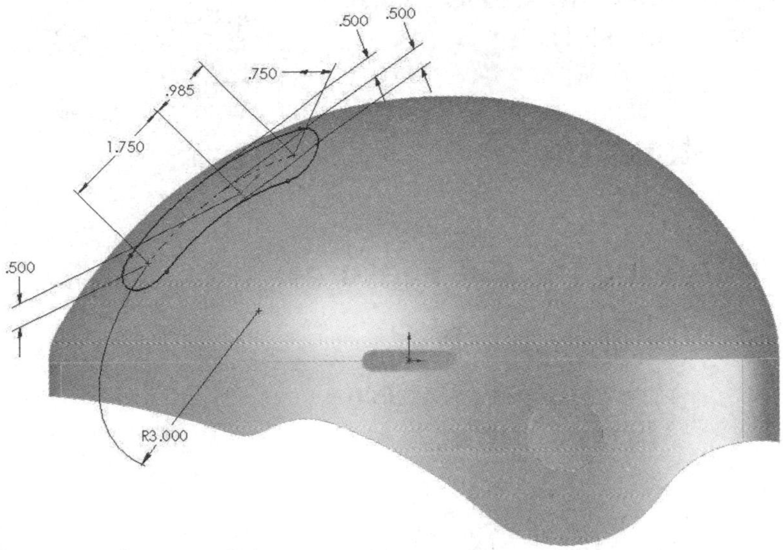

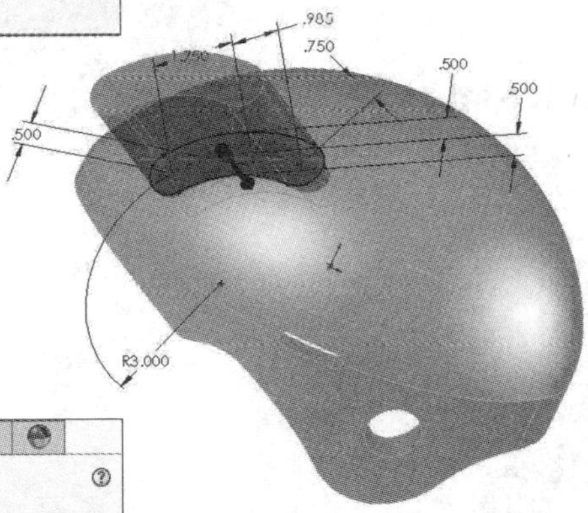

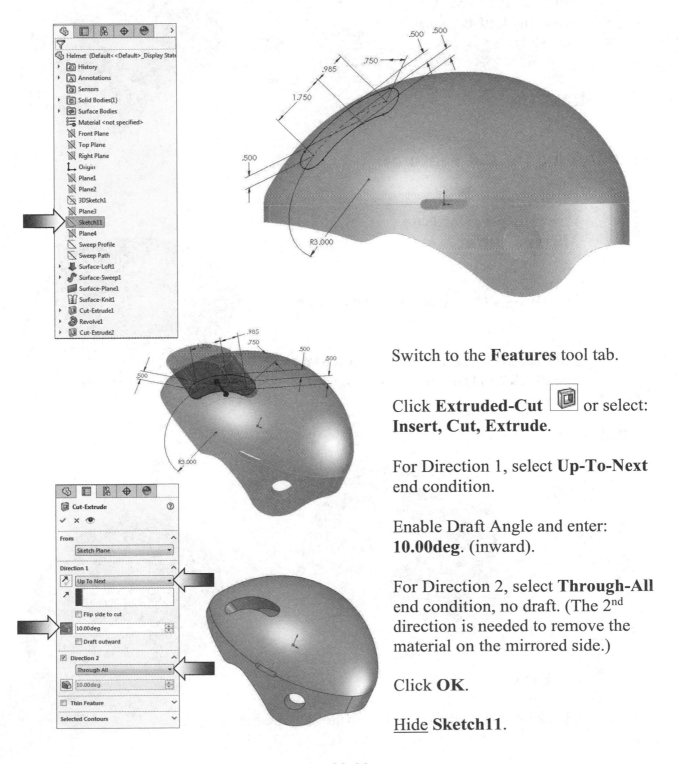

Switch to the **Features** tool tab.

Click **Extruded-Cut**  or select: **Insert, Cut, Extrude**.

For Direction 1, select **Up-To-Next** end condition.

Enable Draft Angle and enter: **10.00deg**. (inward).

For Direction 2, select **Through-All** end condition, no draft. (The 2nd direction is needed to remove the material on the mirrored side.)

Click **OK**.

<u>Hide</u> **Sketch11**.

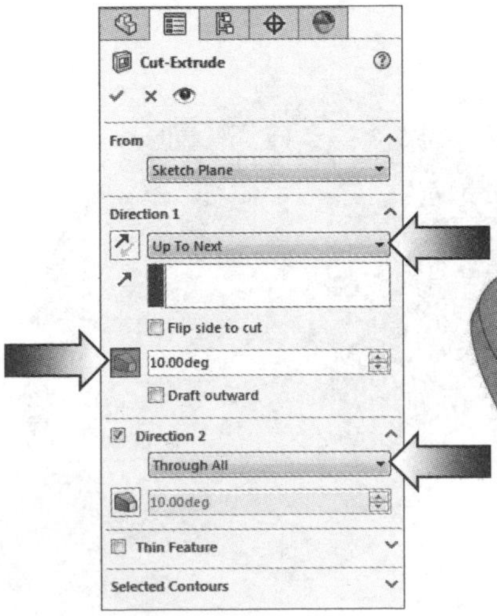

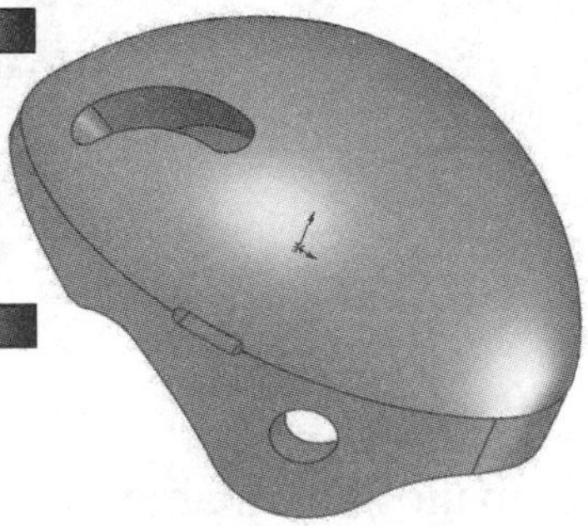

14. Mirroring the slot:

Click **Mirror** – OR – select **Insert, Pattern-Mirror, Mirror** 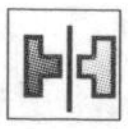.
For Mirror Plane, select the **Right** plane.

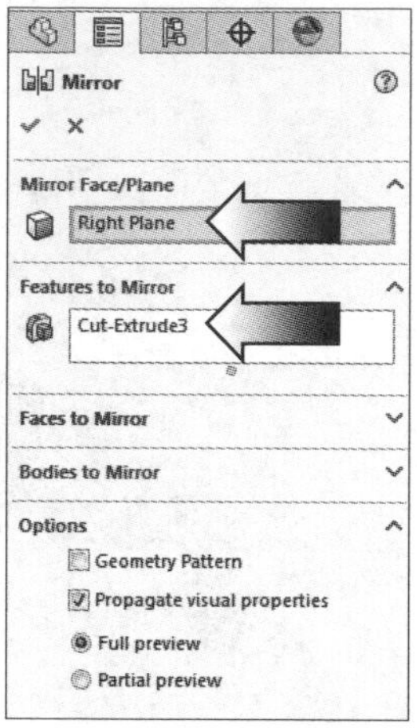

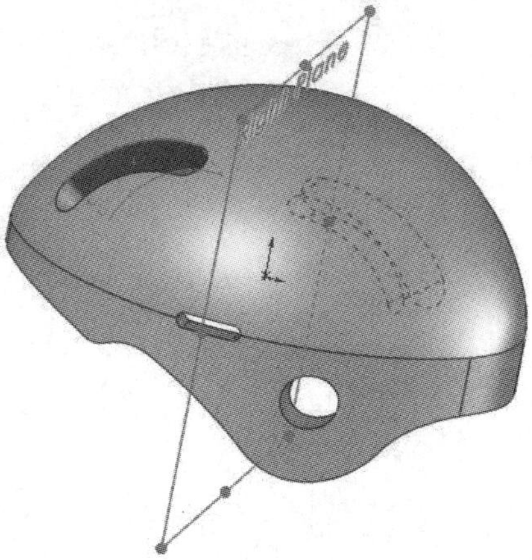

For Features to Mirror, select the **Cut Extrude3** (the slot).

Click **OK**.

15. Mirroring the Cut Features:

Click **Mirror** 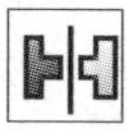 or select: **Insert / Pattern-Mirror / Mirror**.

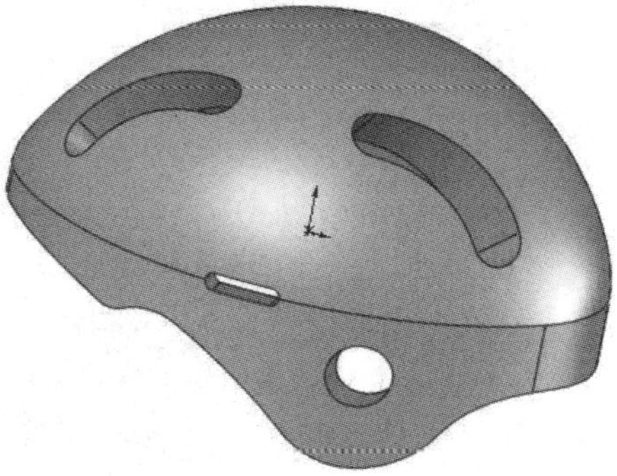

For Mirror Plane, select the **Front** plane.

For Features to Mirror, select **both Slots** and the **Side-Holes** as Features to Mirror.

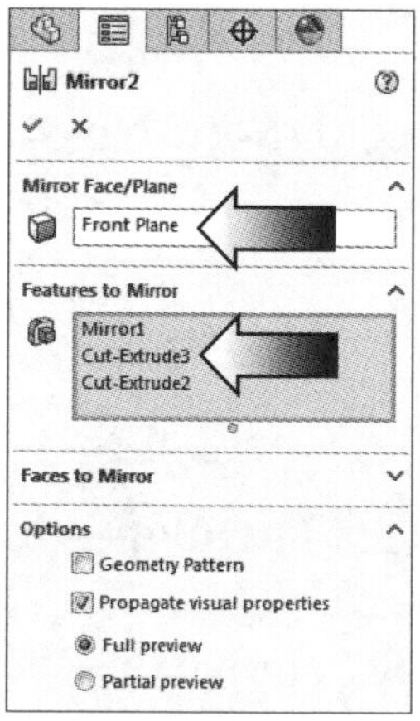

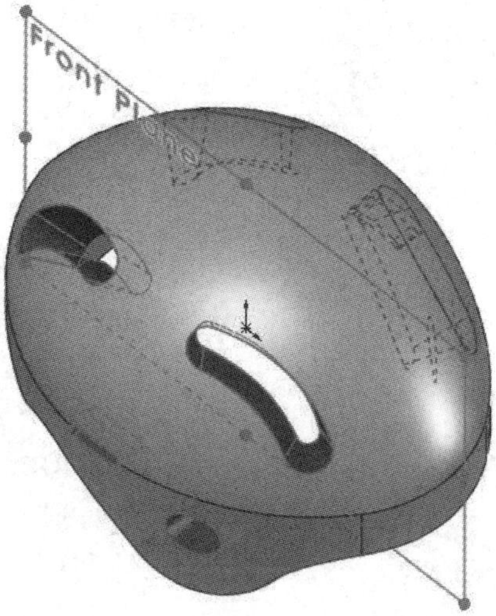

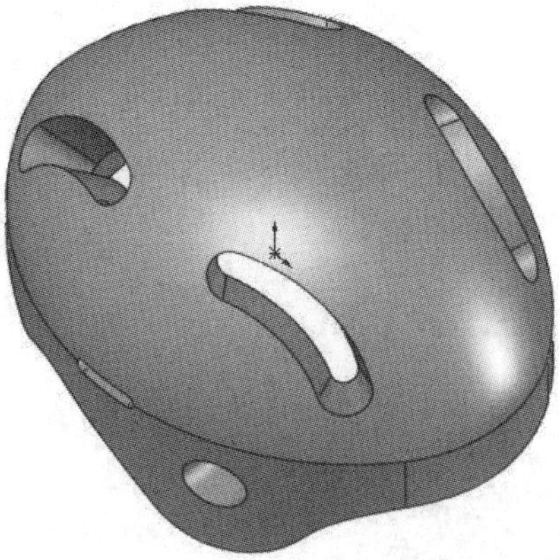

Click **OK**.

Check your model against the image shown above.

16. Creating the Sweep-Cut:

Click  or select: **Insert, Cut, Sweep**.

Select the sketch **Sweep-Profile** and the sketch **Sweep Path** from the FeatureManager tree.

The preview graphics shows the proper transition of the sweep feature.

Click **OK**.

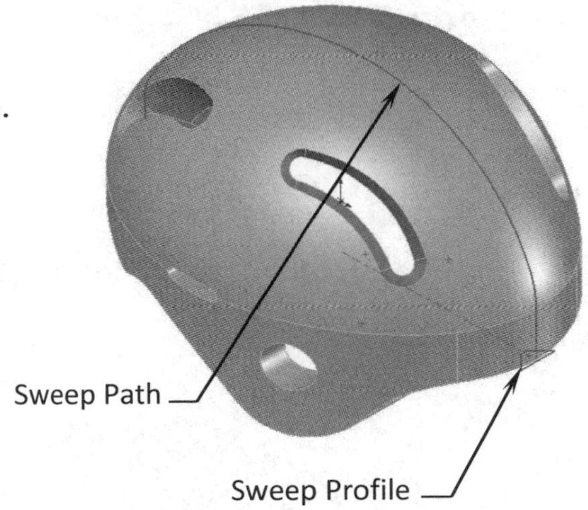

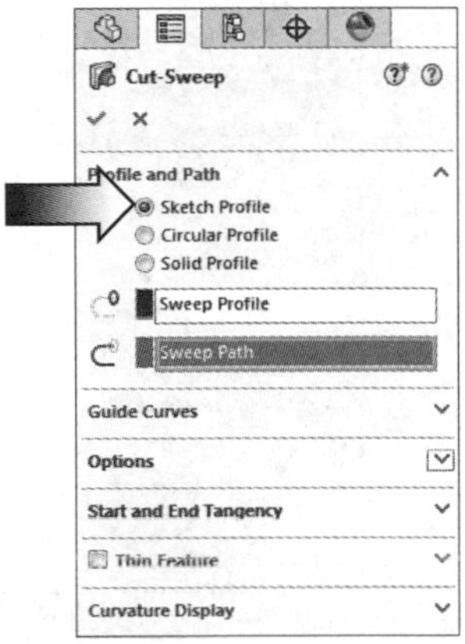

17. Adding the .500" fillets:

Click **Fillet**  or select **Insert, Features, Fillet-Round**.

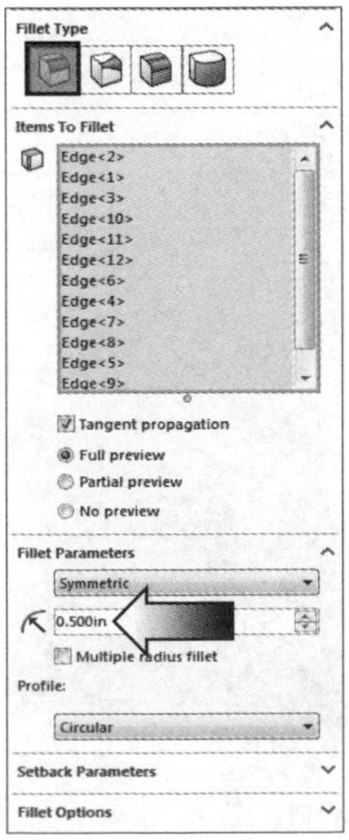

Enter **.500in** for radius size.

Select the **12 edges** of the Swept feature.

Click **OK**.

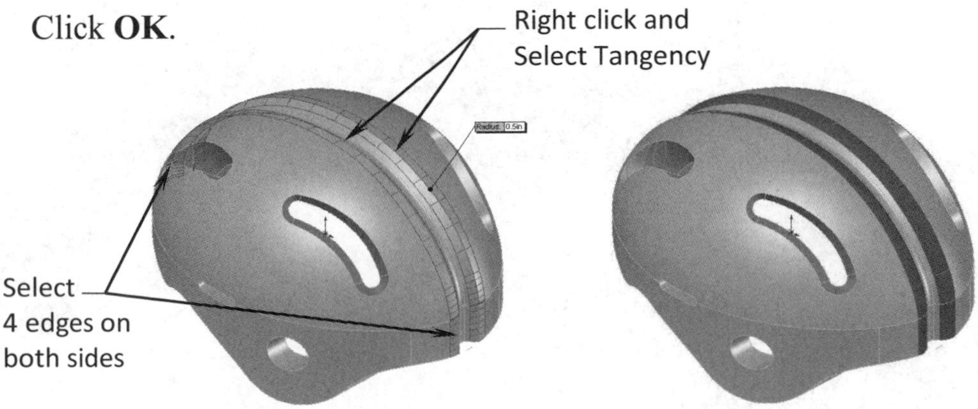

18. Adding the .250" fillets:

Click **Fillet**  or select: **Insert, Features, Fillet-Round**.

Enter **.250**" for radius size.

Select **all edges** of the part (Control+A).

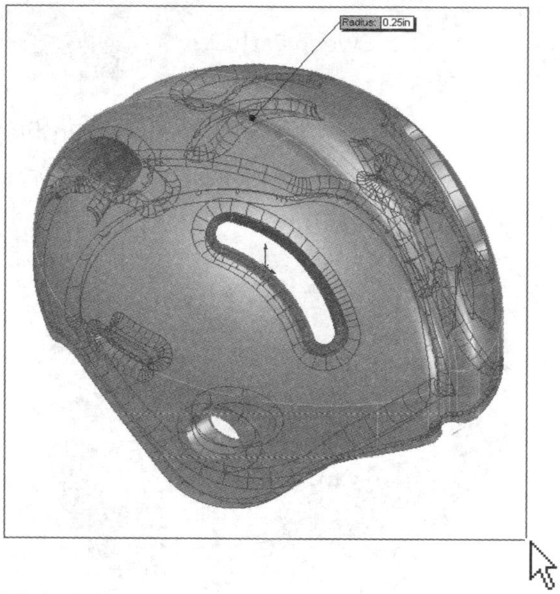

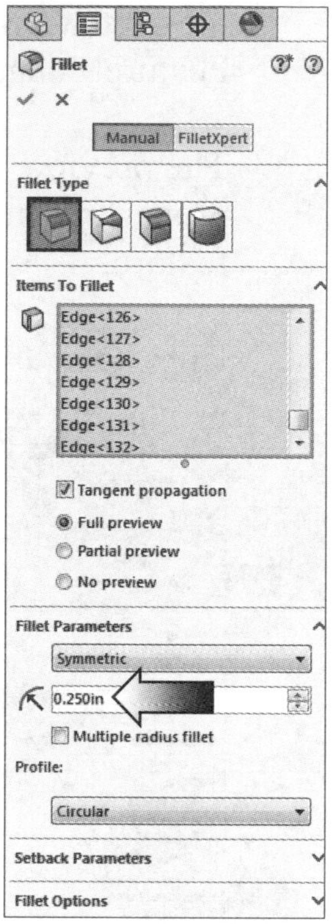

Click **OK**.

19. Saving your work:

Click **File / Save As / Helmet / Save**. (Save on the Desktop.)

Front Isometric

Back Isometric

Exercise: Advanced Loft – Turbine Blades

1. Open a part document:

Turbine.sldprt from the Training Files folder.

This part file has 6 sketch profiles and 4 guide curves previously created.

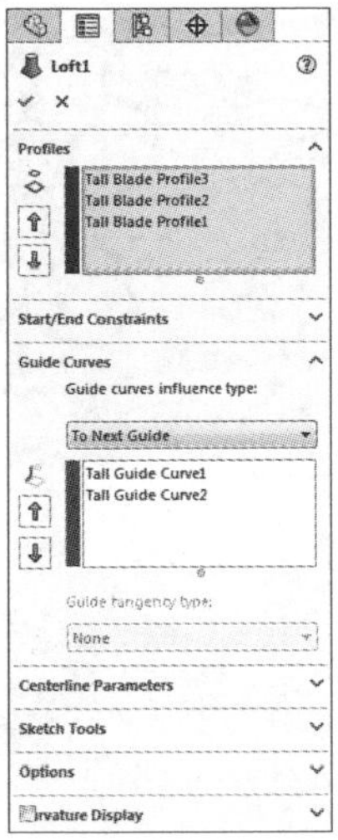

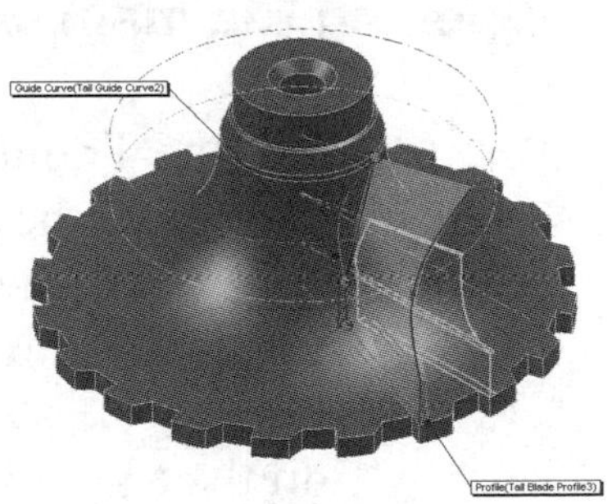

2. Create the 1st loft:

Select the 3 Tall Blade sketches for Profiles.
Select the 2 Tall Guide Curve sketches for Guide Curves.

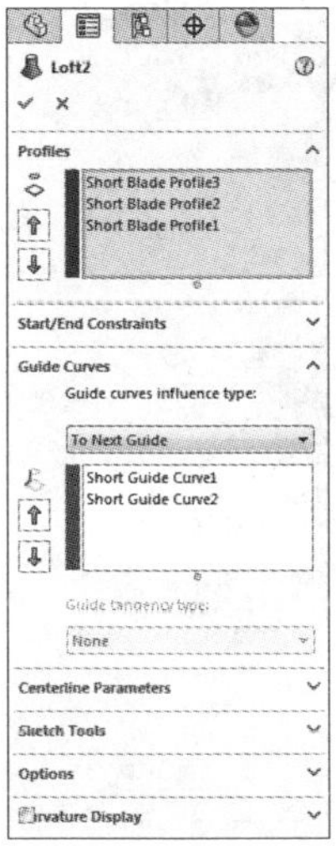

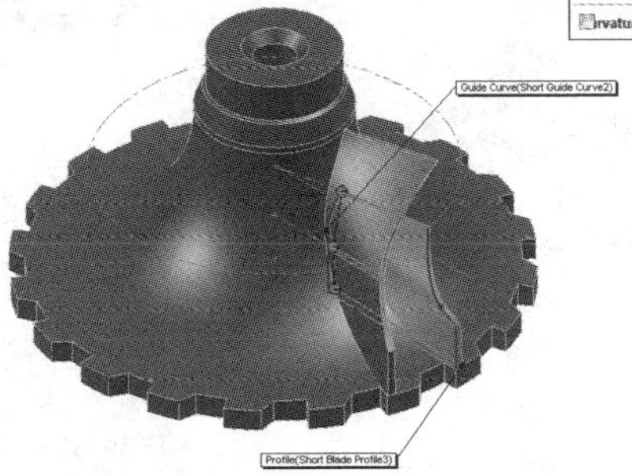

The resulted Lofts.

3. Create the 2nd loft:

Select the 3 Short Blade sketches for Profiles.
Select the 2 Short Guide Curve sketches for Guide Curves.

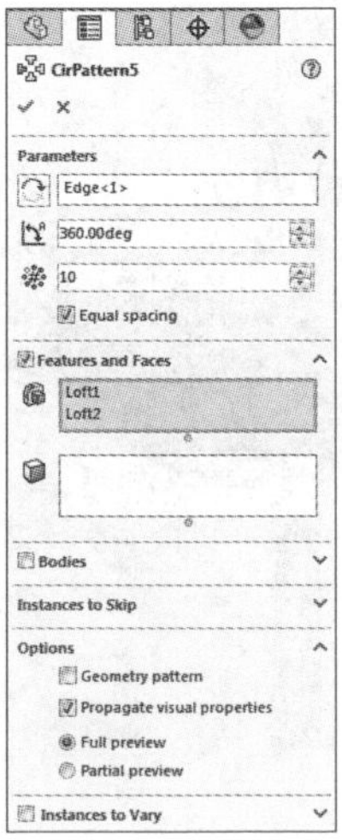

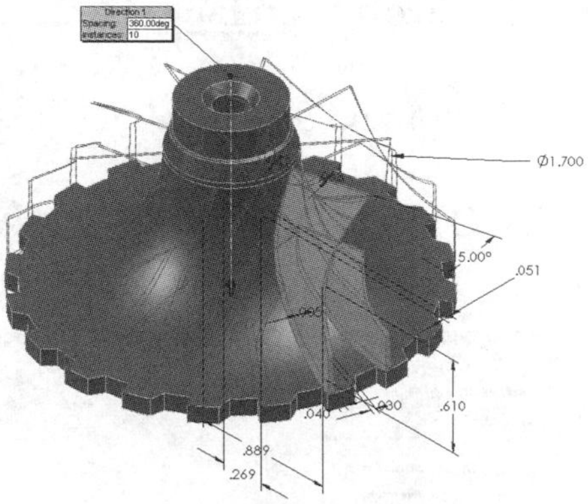

4. Circular pattern the Blades:

Create a circular pattern of the Lofted Blades.
Enter **10** for number of instances.

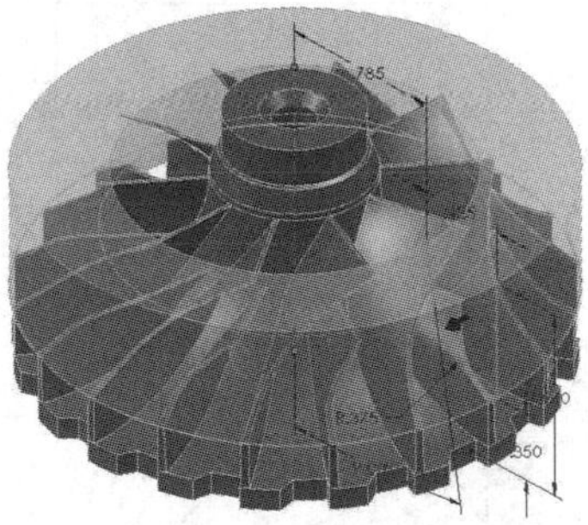

5. Create a Revolve Cut:

Use the Blade Trim Sketch and create a Revolve-Cut.

6. Save your work as:

Turbine Blades_Exe.

<u>Exercise</u>: Advanced Sweep – Candle Holder

1. Opening the main sketch:

From the Training Files folder open the part document named: **Candle Holder Sketch**.

<u>Edit</u> the **Sketch1** and verify that the sketch is fully defined.

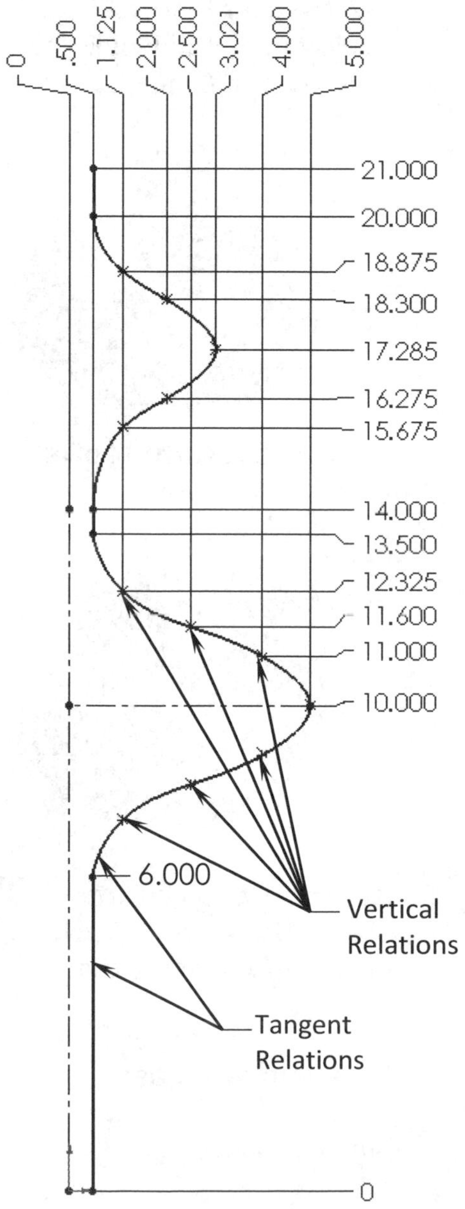

2. Revolving a surface:

Click **Revolve <u>Surface</u>** .

Use **Blind** and **360 deg**.

Click **OK**.

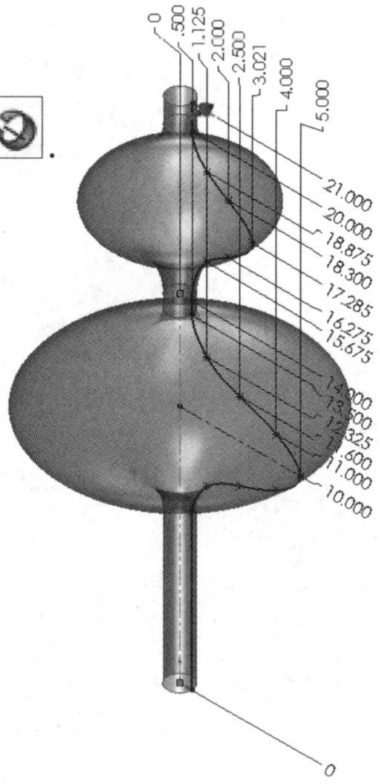

3. Creating a Helix:

Select the Top plane and open a **new sketch**.

Sketch a **10.00in circle** centered on the origin.

Convert the circle into a Helix using the settings shown in the dialog box.

Exit the Sketch.

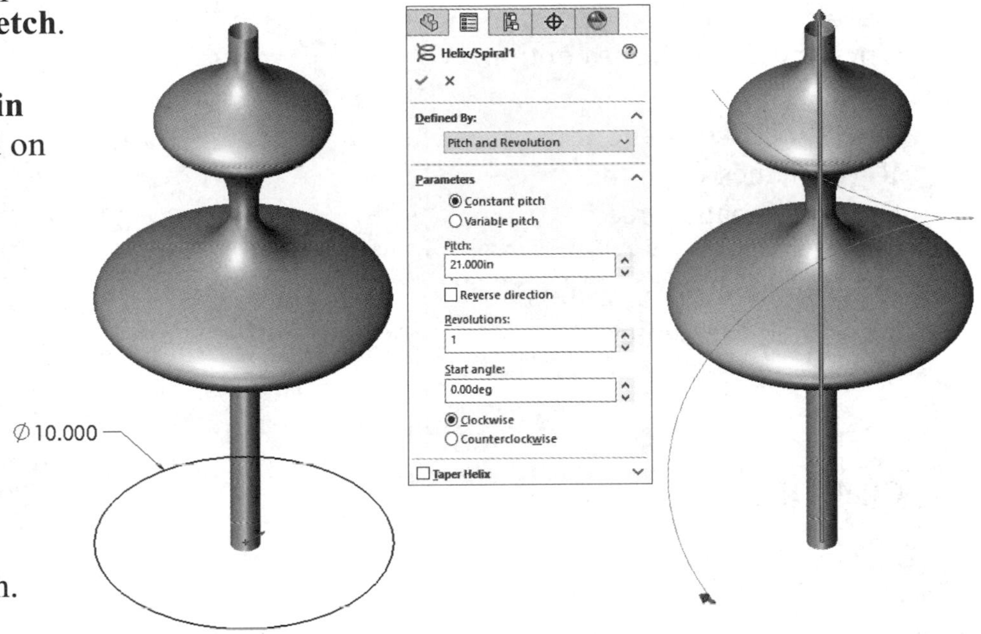

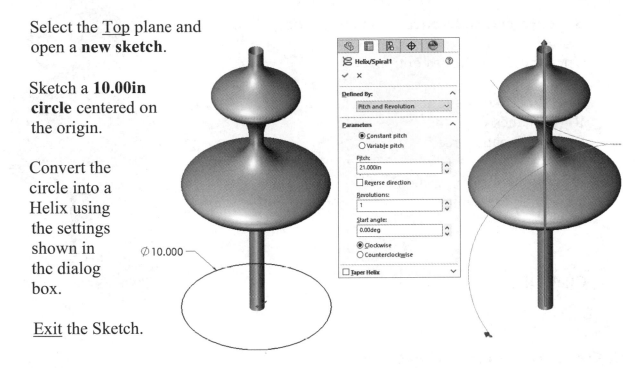

4. Creating a swept surface:

Select the Top plane once again and open a **new sketch**.

Sketch a **Line** from the Origin and **Pierce** the other end of the line to the Helix.

Click and sweep the Line along the Helix using the **Swept-Surface** option.

Click **OK**.

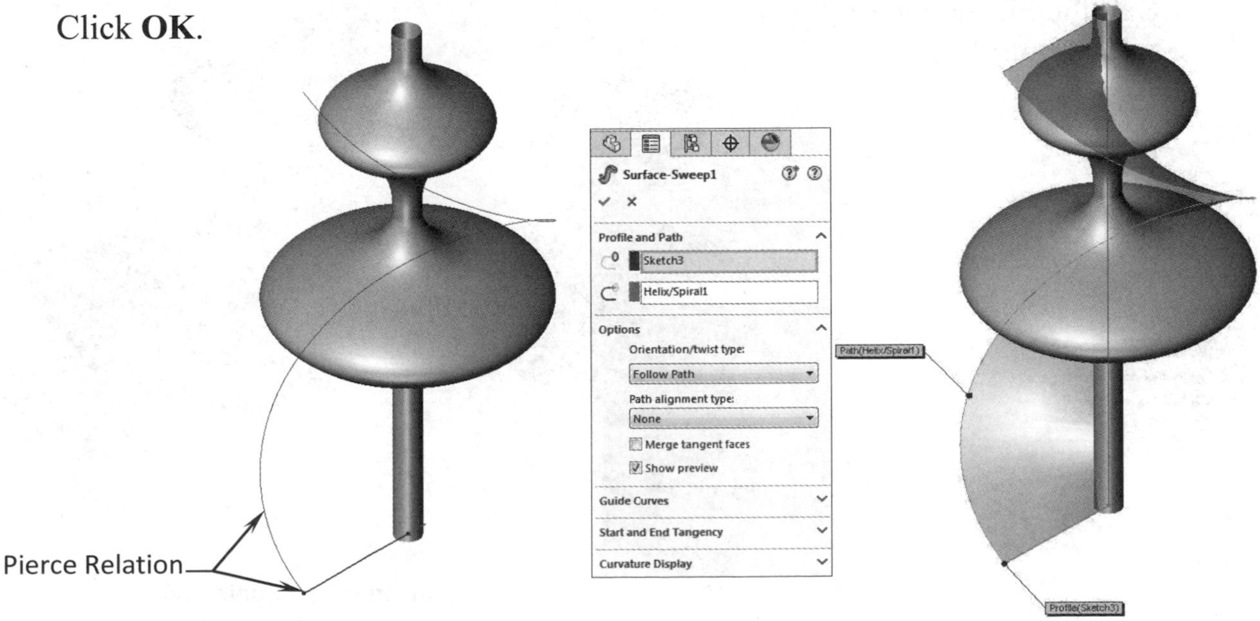

5. Create a new Axis: (to be used in step 9)

Select **Insert, Reference Geometry, Axis** .

Click the **Two-Planes** option.

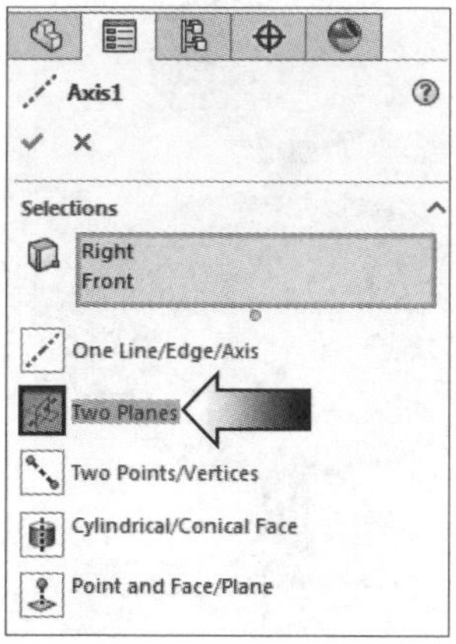

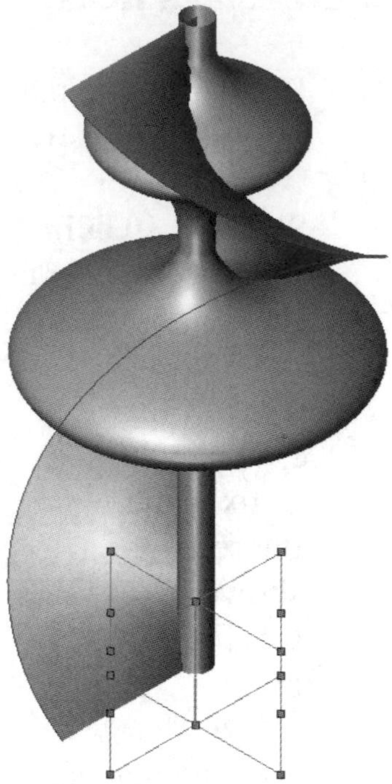

Select the **Front** and the **Right** planes from the FeatureManager tree.

A preview of the new axis appears in the center of the part.

Click **OK**.

6. Create the Intersection Curve:

Hold the **Control** key and select the **Surface-Revolve1** and the **Surface Sweep1** from the FeatureManager tree.

Click  or select **Tools / Sketch Tools / Intersection Curve**.

A 3D-Sketch is created from the intersection of the two surfaces.

<u>**Exit**</u> the 3D Sketch or press Control+Q.

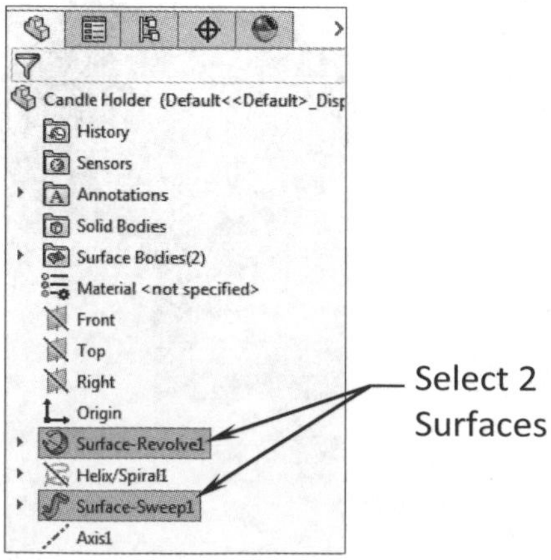

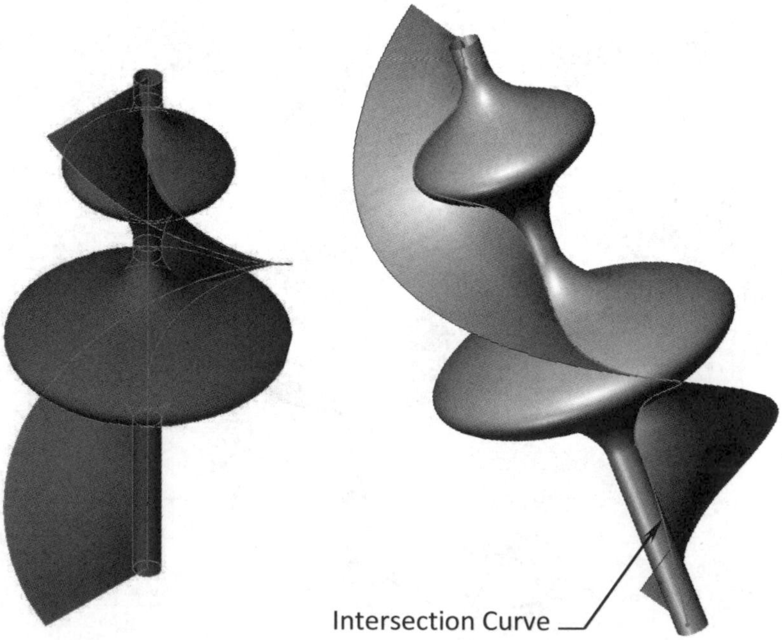

7. Hide the 2 Surface Bodies:

From the FeatureManager tree, right click on each Surface and select **Hide**.

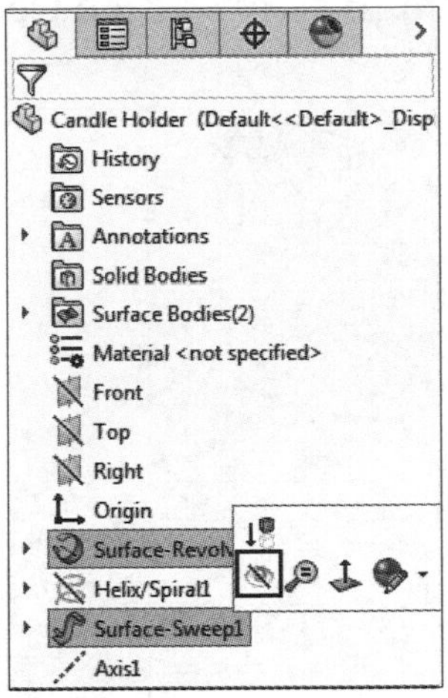

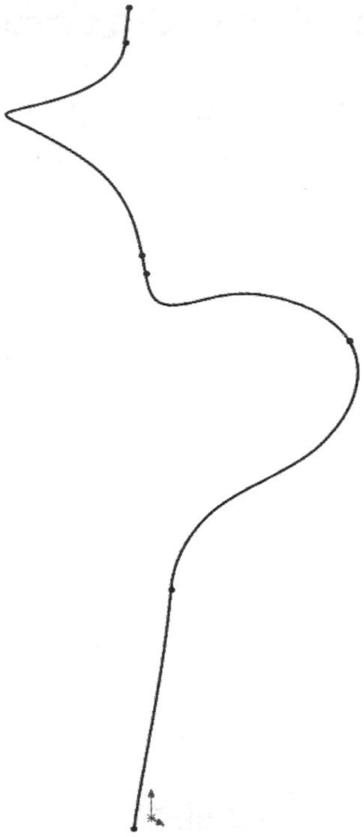

This 3D sketch will be used as the sweep path in the next step.

Click **OK**.

8. Create a <u>solid</u> swept feature:

Switch to the **Features** tool tab and click  or select **Insert / Boss-Base/ Sweep**.

Select the **Circular Profile** option and enter **.260in** for Profile Diameter (arrow).

Select the **3D-Sketch** for Sweep Path.

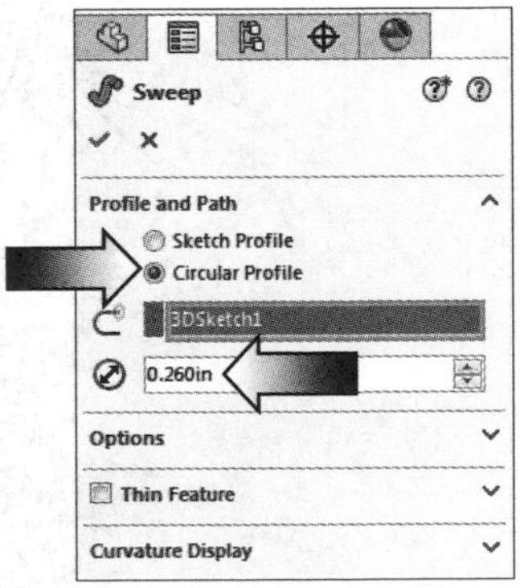

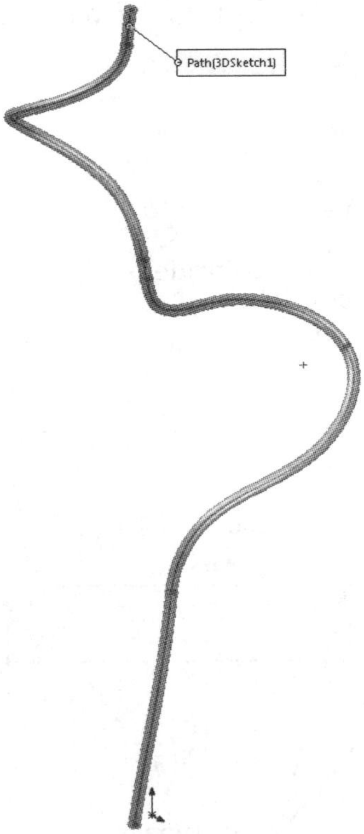

Click **OK**.

9. Create the Circular Pattern :

Using the **Axis** created in step 5 as the Center of the Pattern, repeat the Swept Feature **12** times.

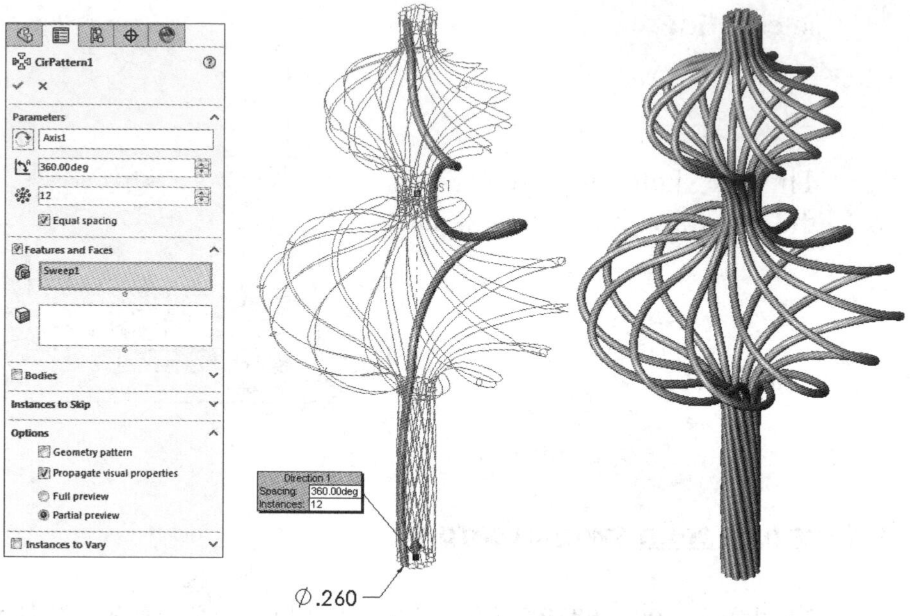

Click **OK**.

10. __OPTIONAL__:

Create the **Base** and the **Candle Holder** solid features as shown below.

Modify or design your own shapes if needed.

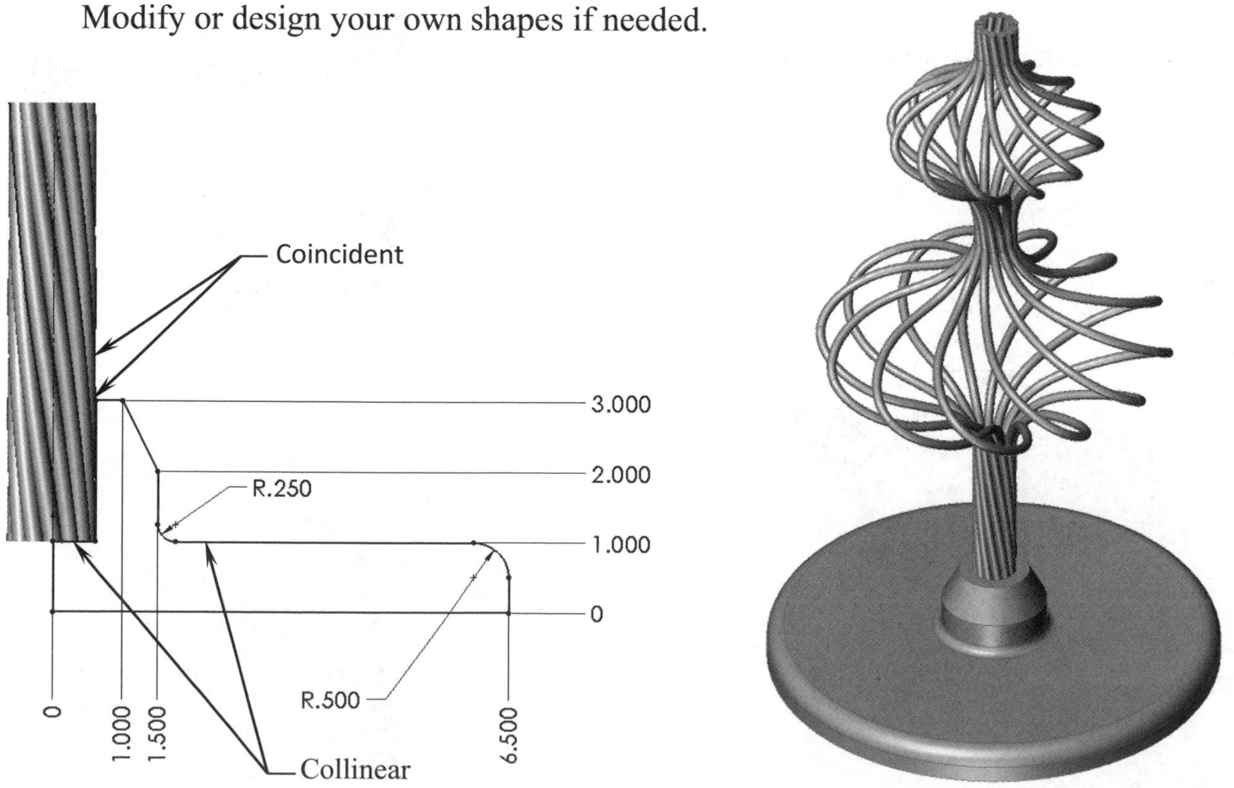

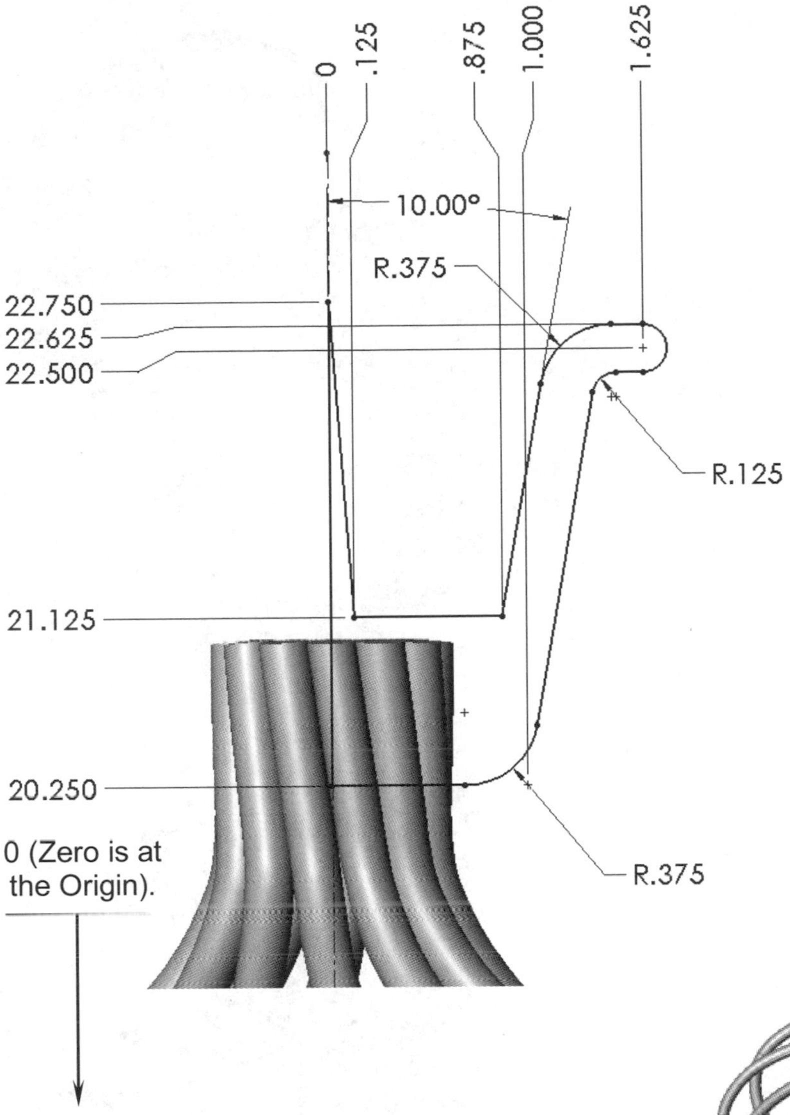

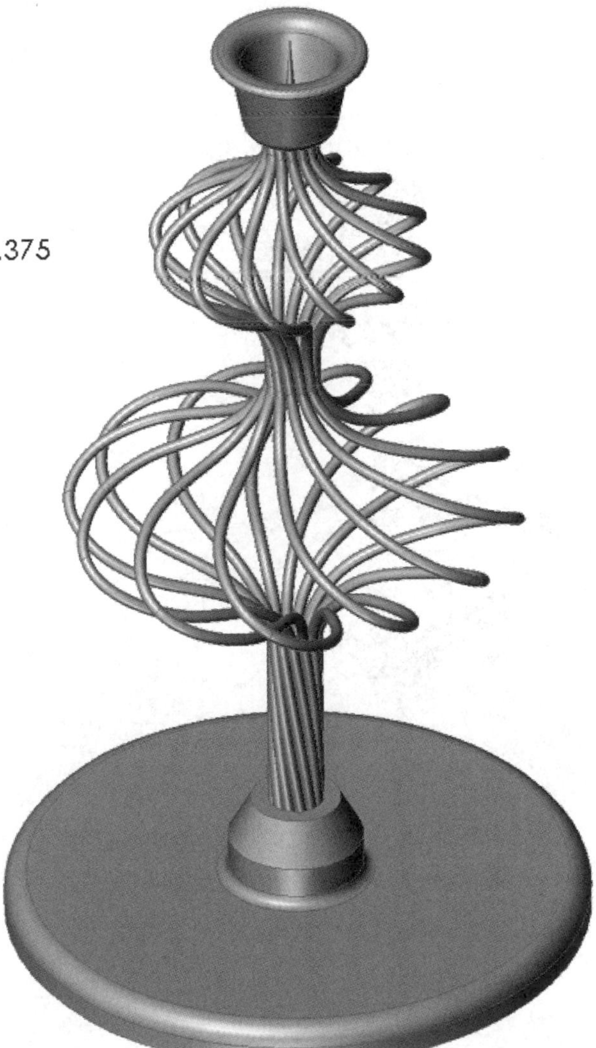

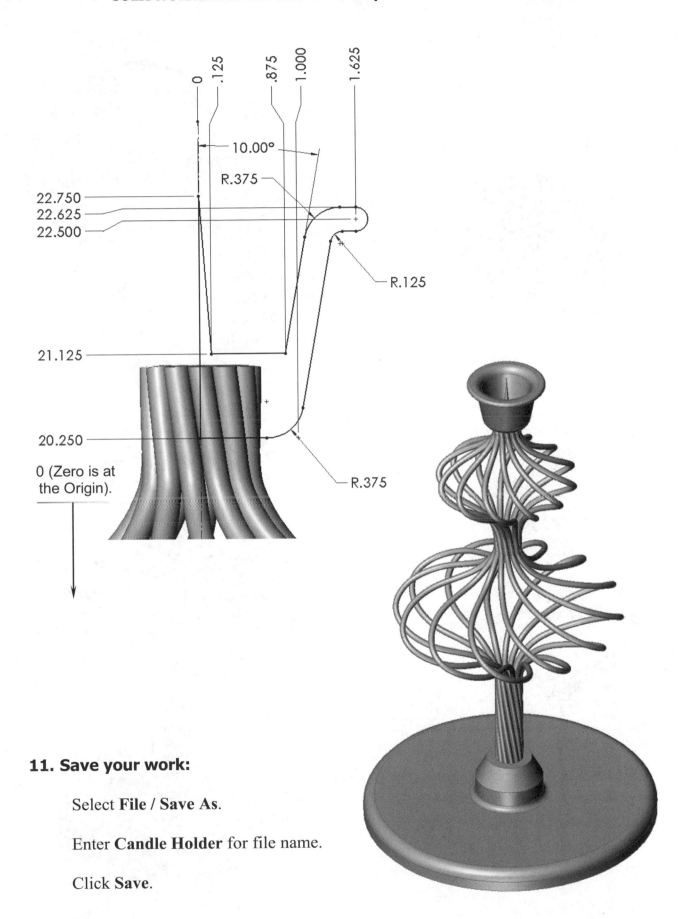

11. Save your work:

Select **File / Save As**.

Enter **Candle Holder** for file name.

Click **Save**.

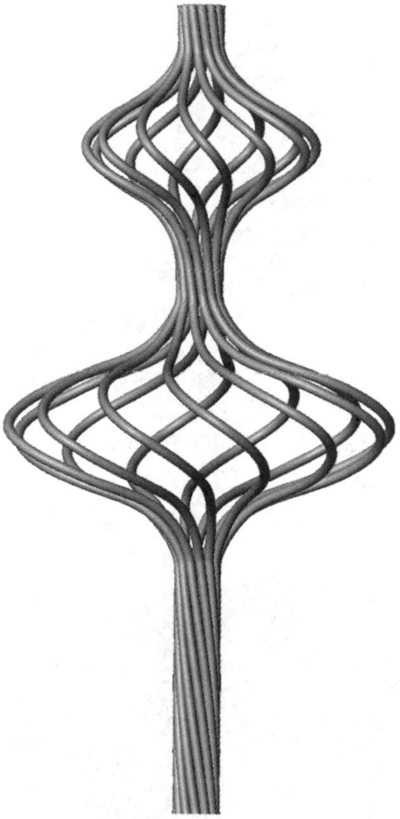

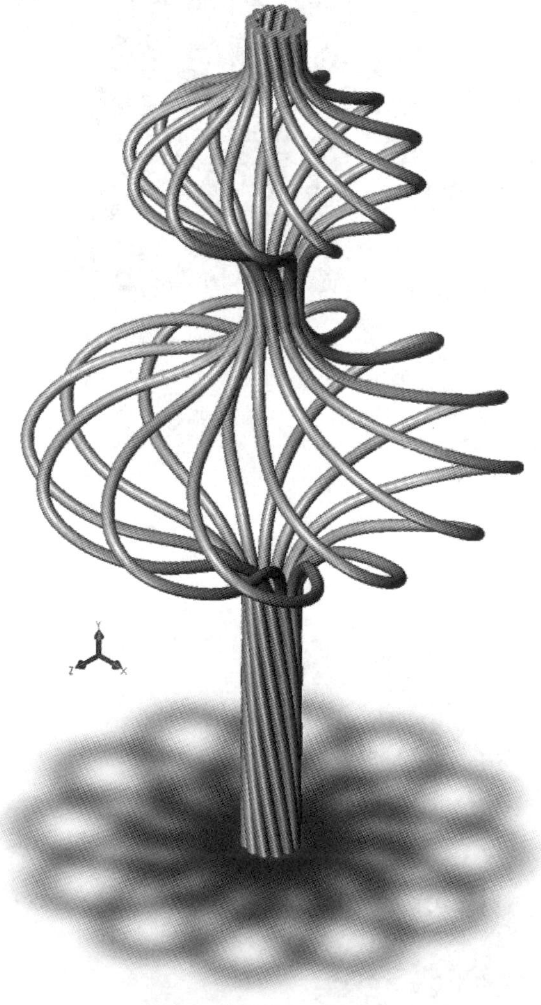

Final Exam

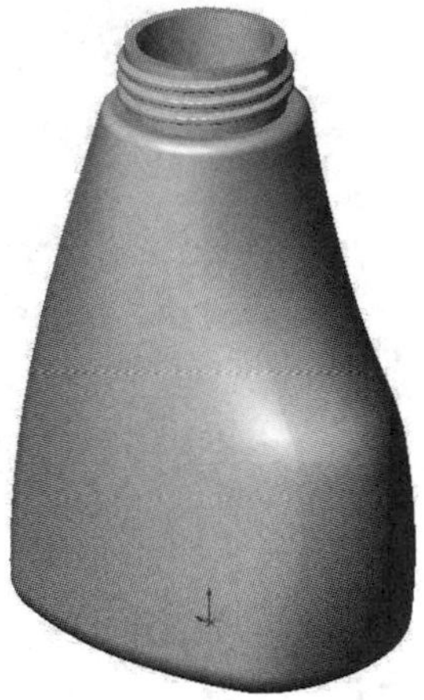

Create the part **Bottle** using the LOFT and SWEEP options where noted.

All sketch profiles must be fully defined.

The part must have no errors when finished.

Apply what you have learned from the previous lessons to create this model. The instructions in the following pages are for references only.

1. Creating an Offset plane:

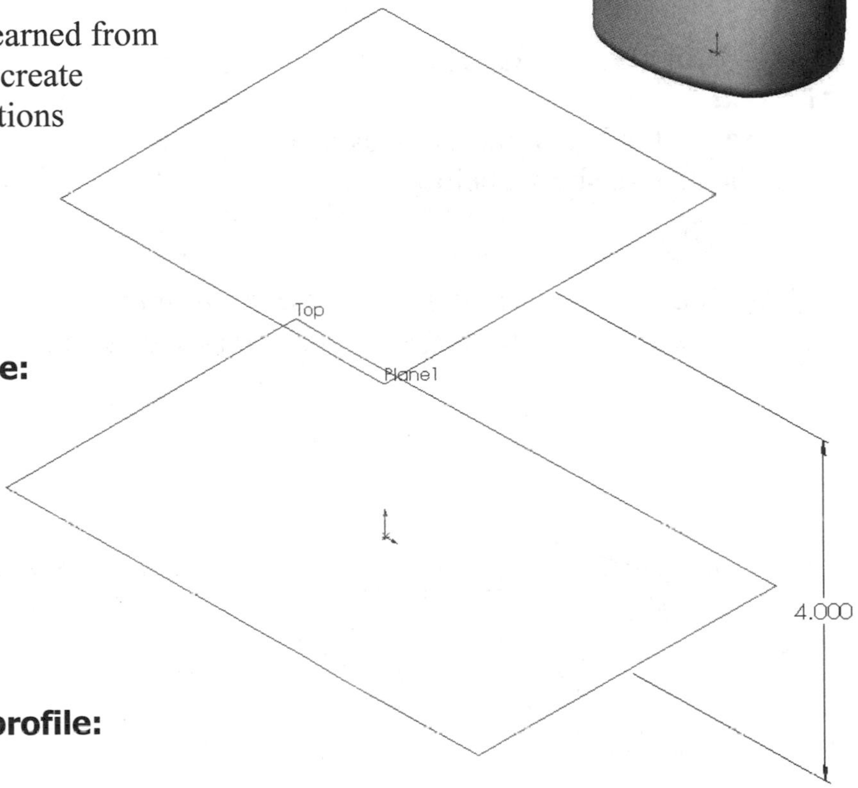

Use the Top reference Plane to create a new plane at **4.000 in**. offset distance.

2. Sketching the bottom profile:

Select the Top plane and sketch the profile as shown.

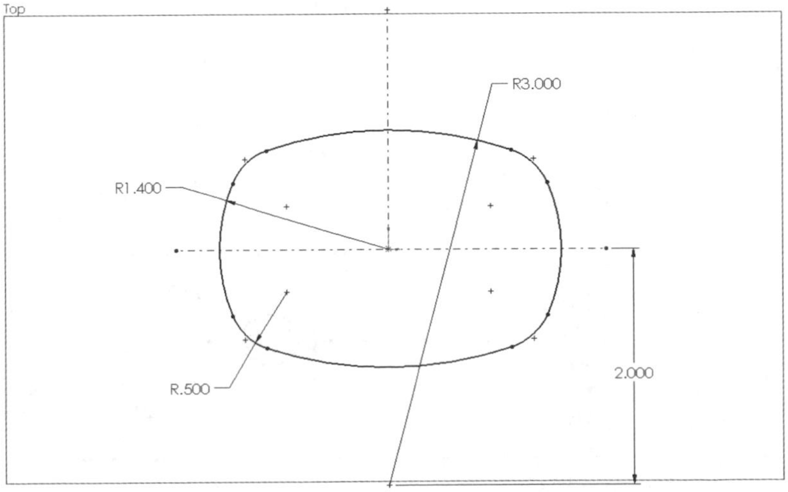

Use the Mirror option to create the **Symmetric** relations between entities.

Exit the Sketch.

3. Creating the top profile:

Use the Plane1 and sketch a **Circle** centered on the Origin.

Add a diameter dimension to fully define the sketch.

Exit the sketch.

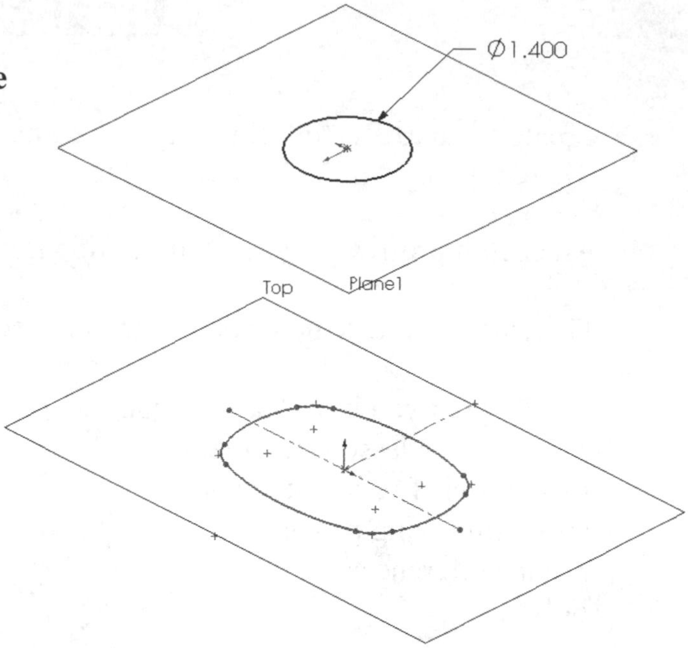

4. Creating the 1st Guide Curve:

Select the Front plane and sketch the profile shown below.

** Note: The construction lines will be used later to create the Derived-Sketch and the Guide Curves.*

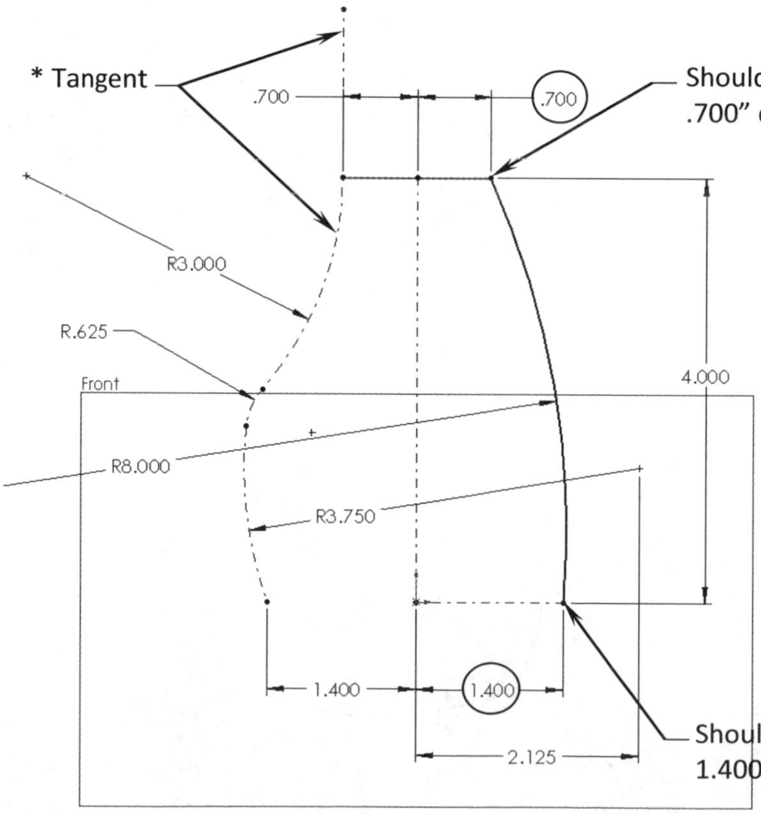

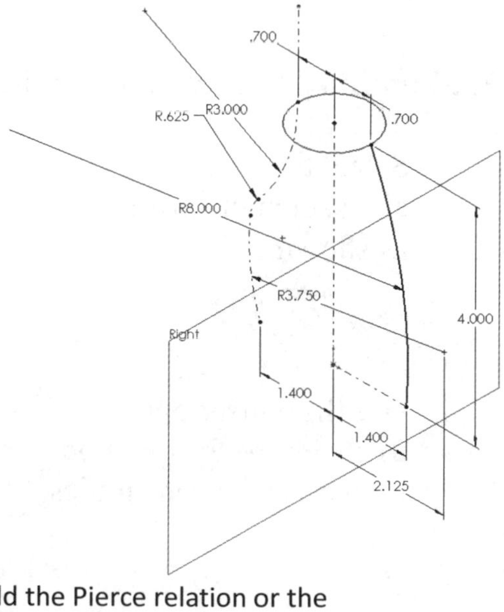

5. Creating the 2nd Guide Curve:

Use the <u>Right</u> plane; sketch a **3-Point-Arc** and add dimensions.

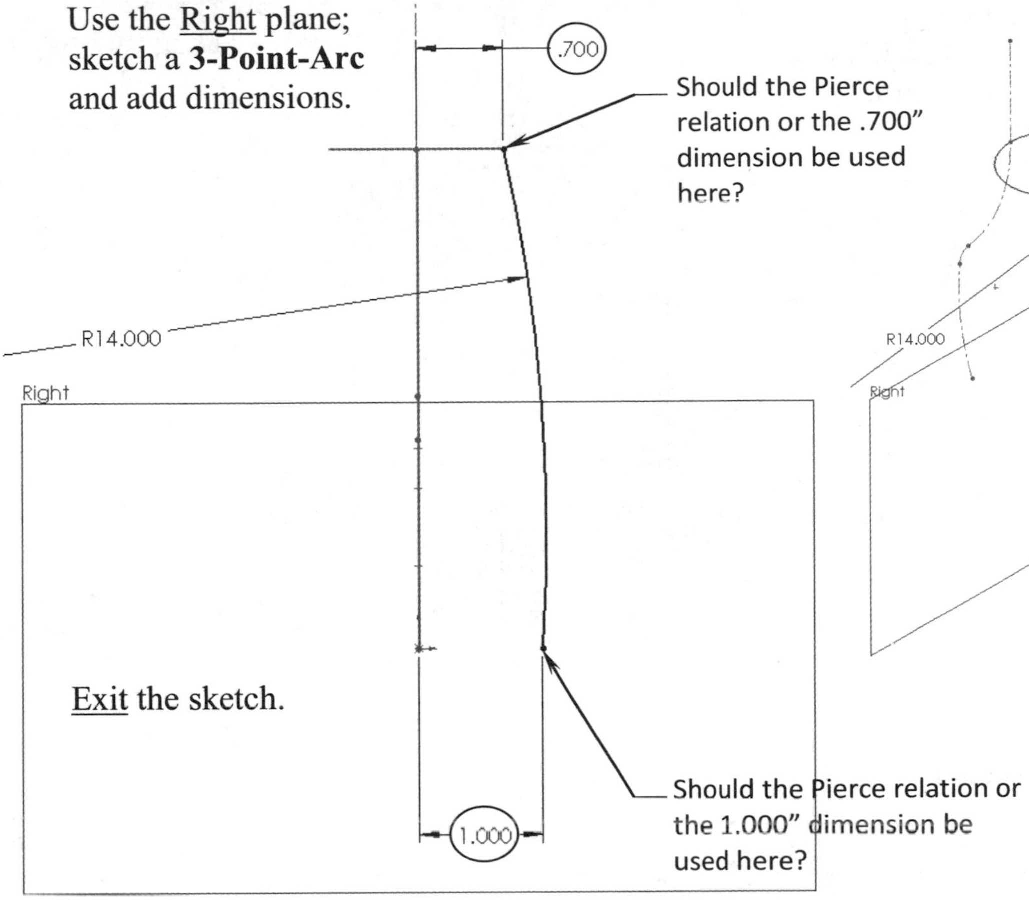

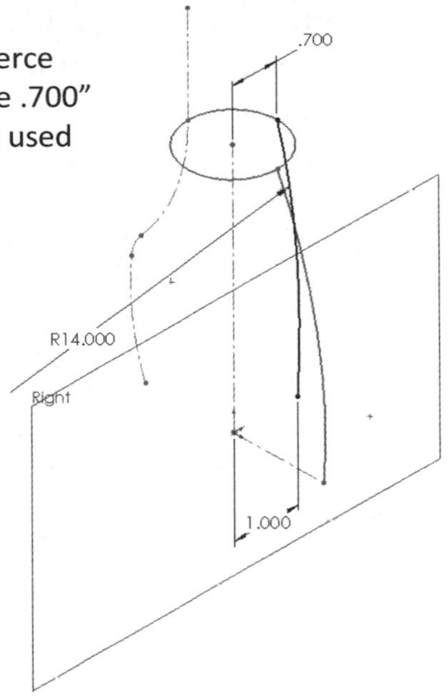

<u>Exit</u> the sketch.

6. Creating the 3rd Guide Curve:

Use the <u>Right</u> plane and sketch a **3-Point-Arc** and add dimensions.

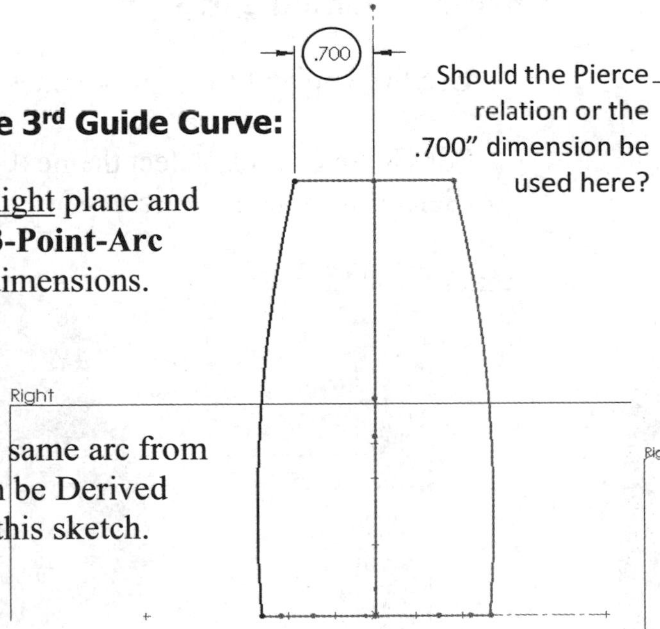

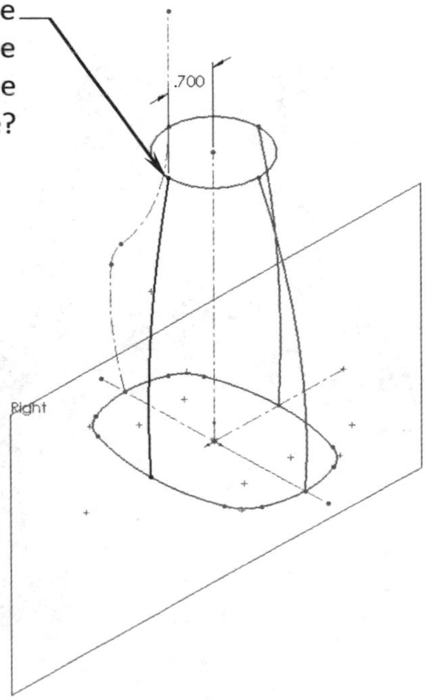

<u>Hint:</u> The same arc from step 5 can be Derived to create this sketch.

<u>Exit</u> the sketch.

7. Creating the 4th Guide Curve:

Select the <u>Front</u> plane; open a **new sketch** and convert the 3 entities as noted.

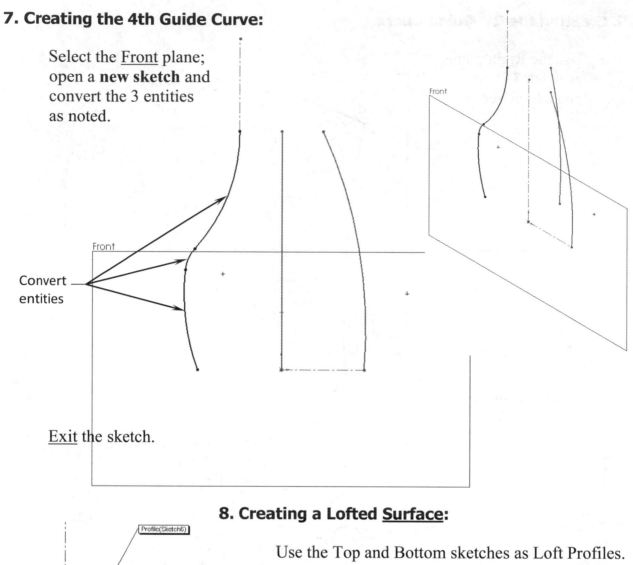

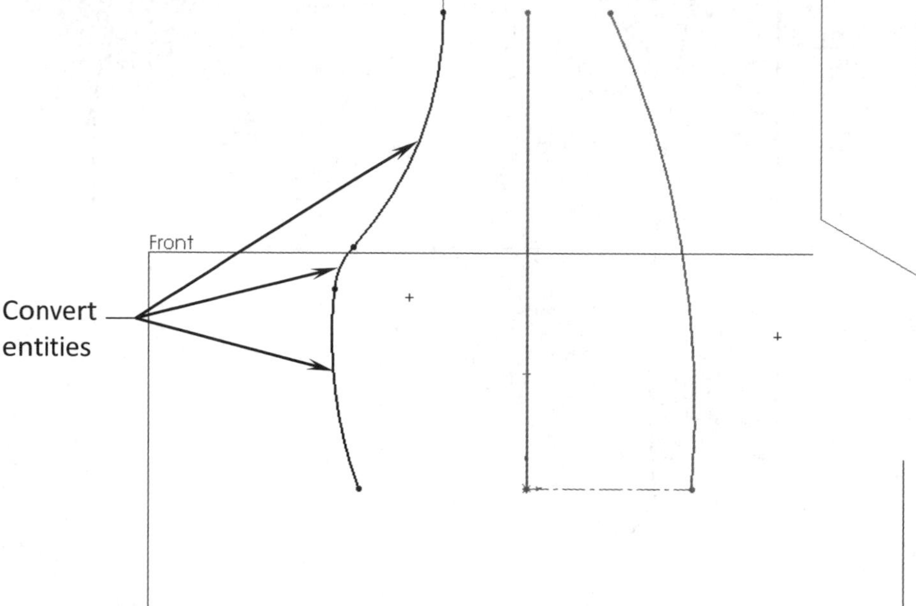

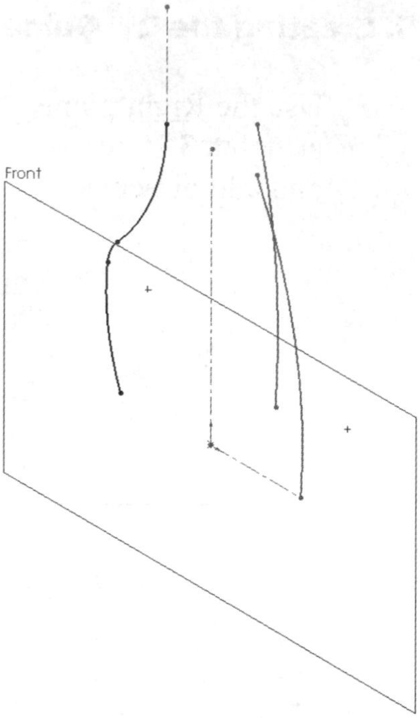

<u>Exit</u> the sketch.

8. Creating a Lofted <u>Surface</u>:

Use the Top and Bottom sketches as Loft Profiles.

For Guide Curves, select the next 4 sketches. (Select from the Feature tree.)

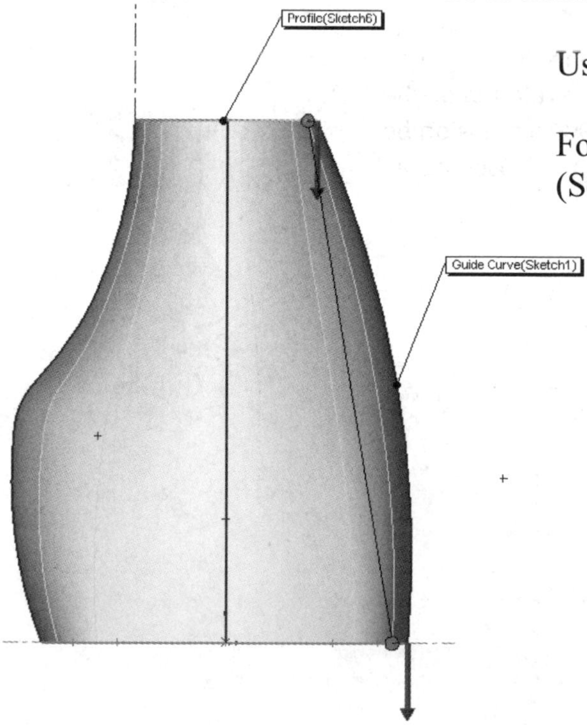

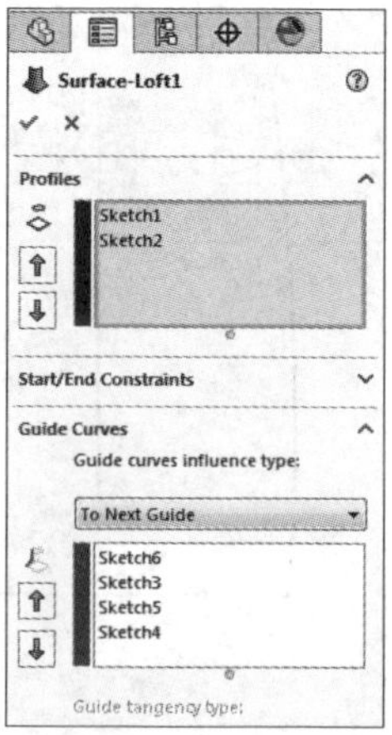

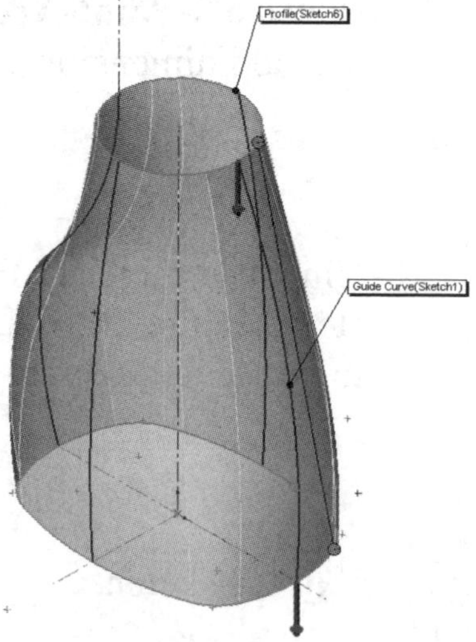

9. Filling the bottom surface:

Select **Insert / Surface / Planar**.

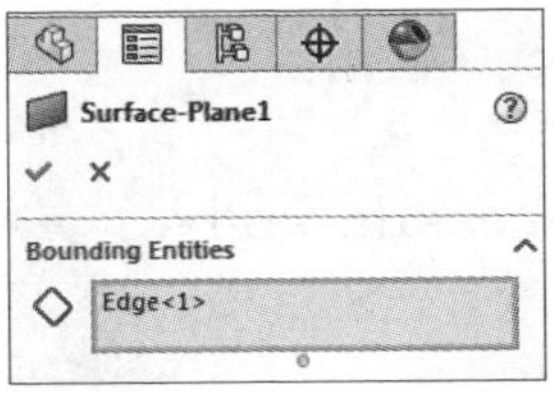

Select **all edges** at the bottom for this operation.

When finished, the bottom surface should be completely covered.

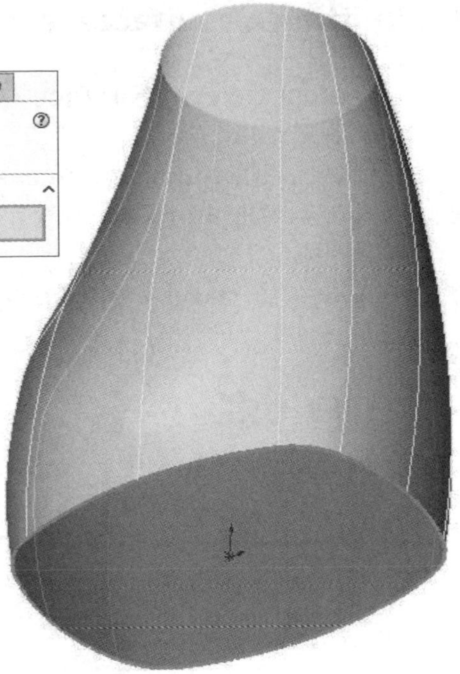

10. Sketching the Neck profile:

Select the <u>Front</u> plane and sketch the profile below (2 lines).

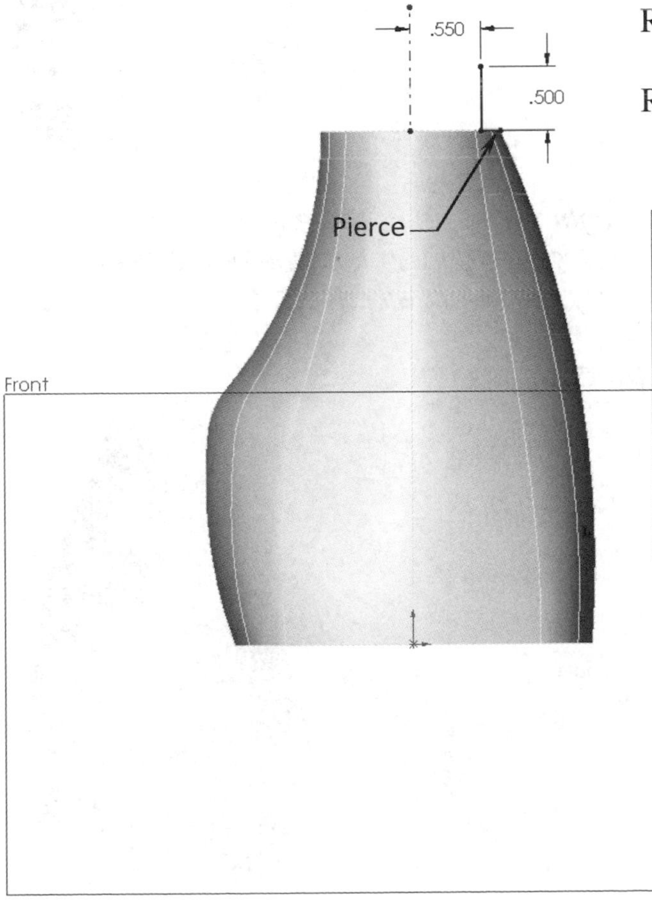

Revolve the sketch profile as a **<u>Surface</u>**.

Revolve **One Direction**.

Revolve a complete **360º**.

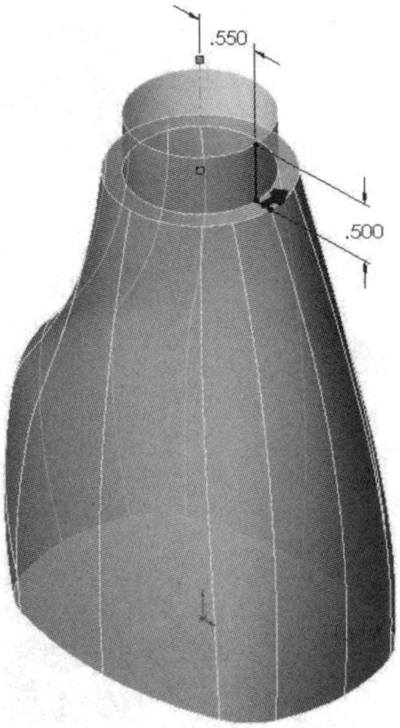

11. Knitting all surfaces into one:

Select **Insert / Surface / Knit**.

Select all **three surfaces**: the body, the Neck, and the Bottom surface.

<u>Clear</u> the **Gap Control** option.

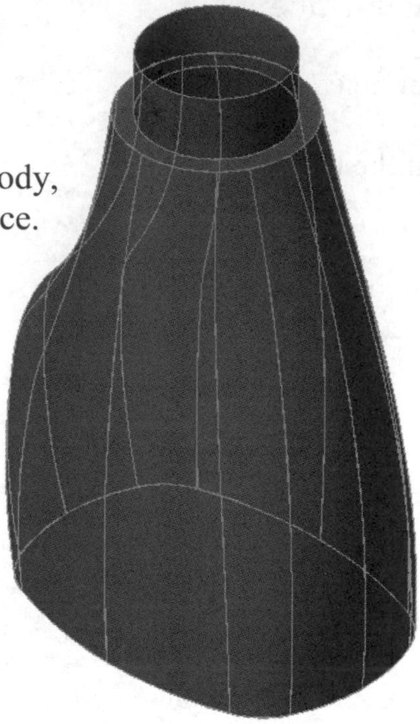

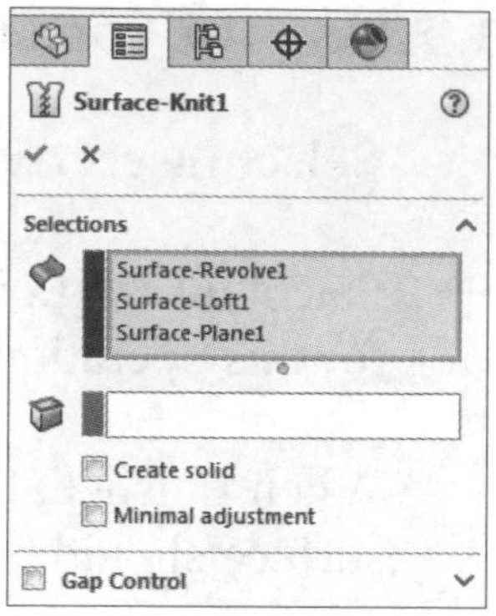

Click **OK**.

When finished, all 3 surfaces should become a single surface body.

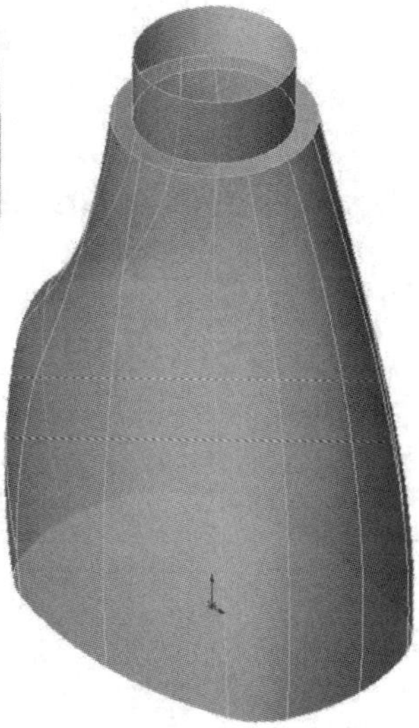

12. Adding fillet to the bottom edges:

Add a **.250 in**. fillet to the bottom edges as shown.

<u>Tips:</u>

Right click one of the edges and pick: Select Tangency; it is one of the faster ways to select all edges at the same time.

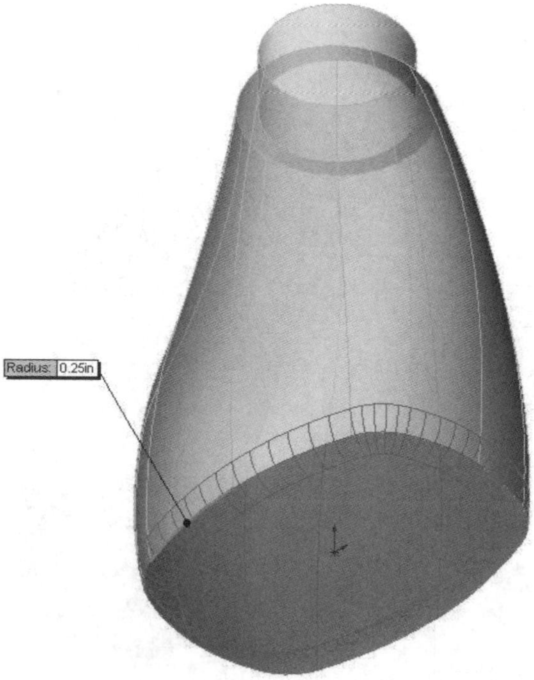

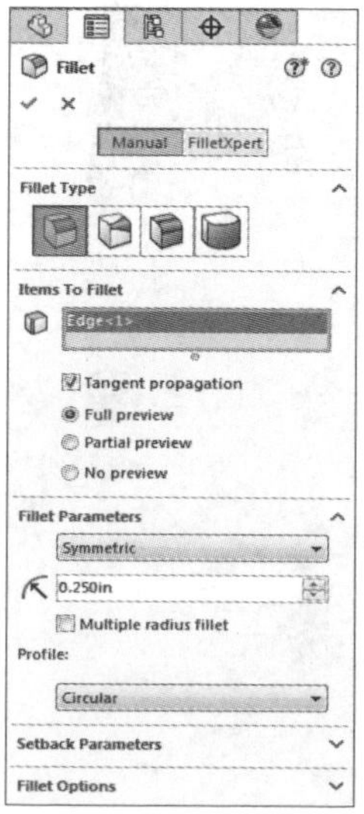

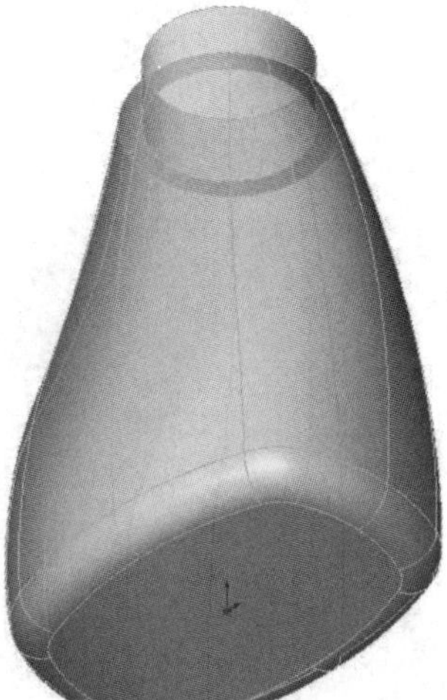

13. Adding fillet to the upper area:

Add a **.093 in**. fillet to the upper edge as indicated.

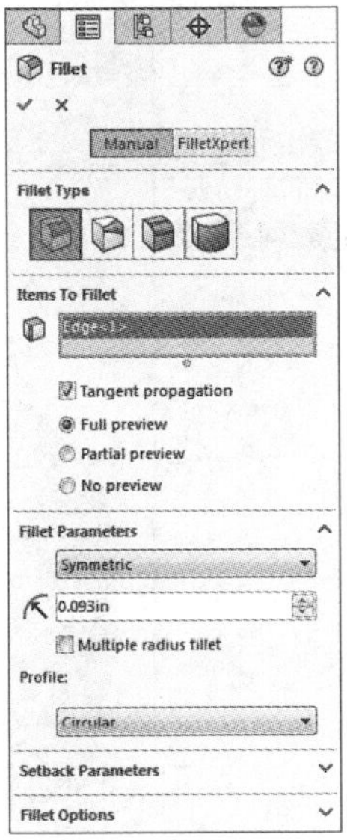

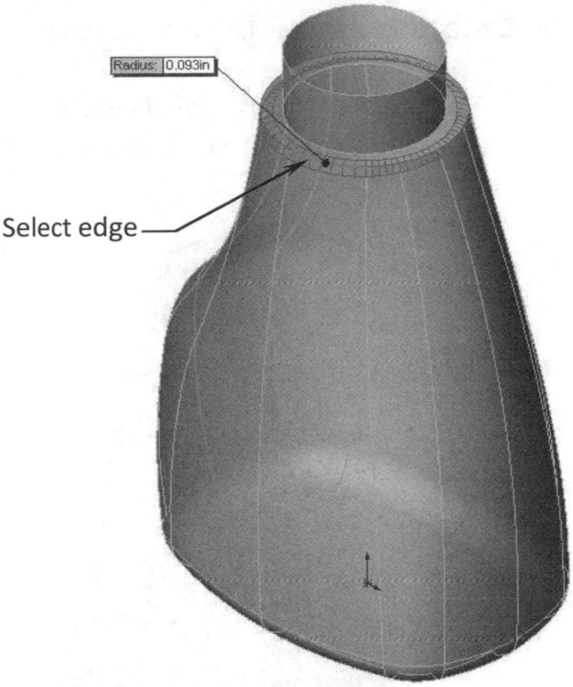

14. Thickening the surface body:

Select **Insert / Boss-Base / Thicken**.

Enter a wall thickness of **.080 in**. to the **<u>INSIDE</u>** of the bottle.

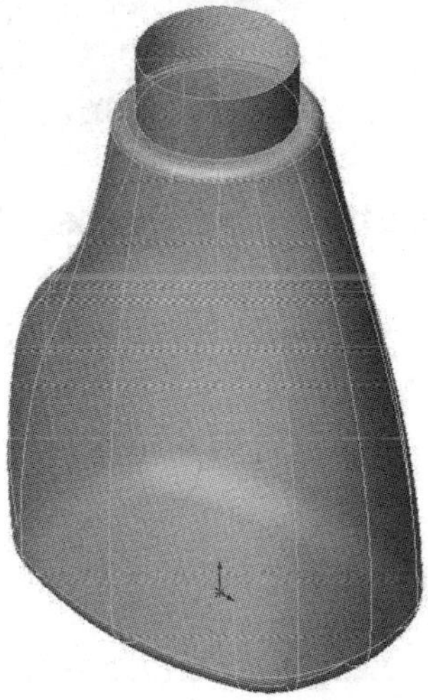

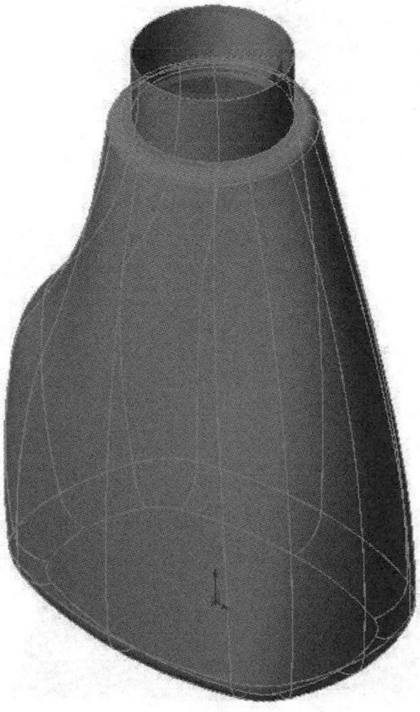

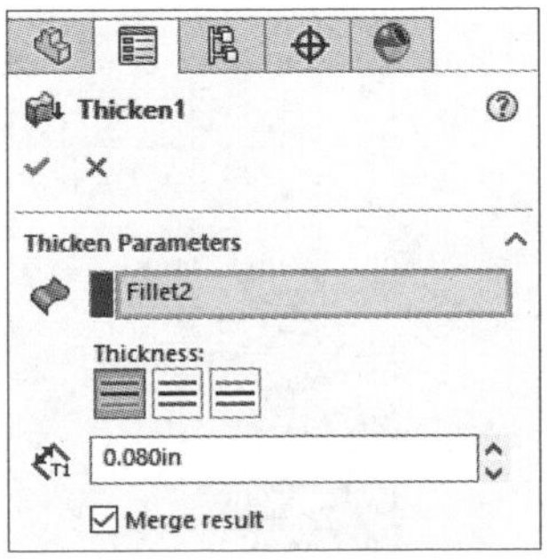

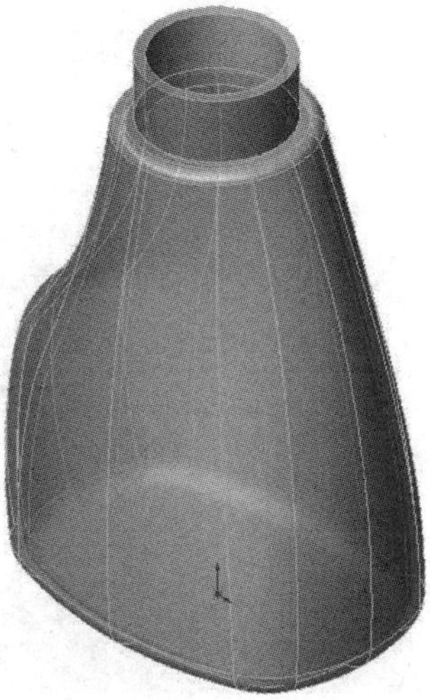

15. Creating a new Offset plane:

Select the top face of the neck and create an Offset Distance plane at **.100 in**. below it.

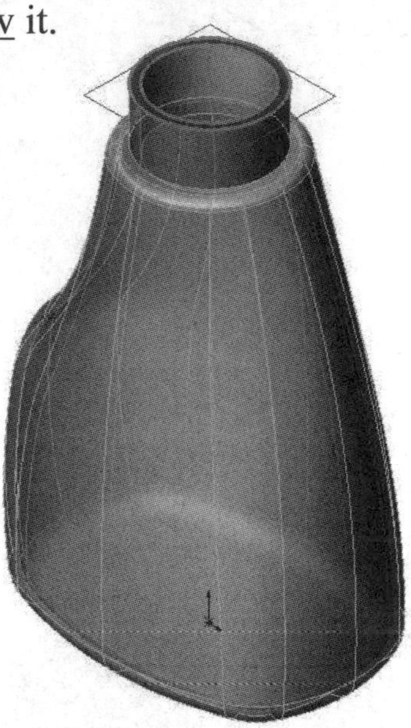

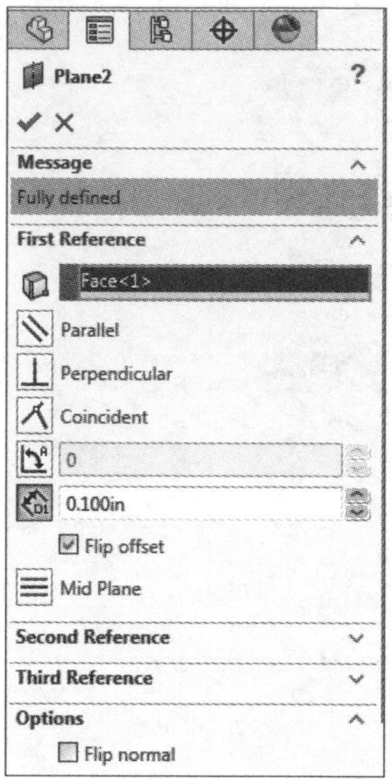

16. Creating the Sweep Path of the Thread (the Helix):

Convert the upper **circular edge** and then click **Insert / Curves / Helix-Spiral**.

Pitch = **.125 in**.

Revolution = **2.5**.

Starting Angle = **0 deg**.

Click **OK**.

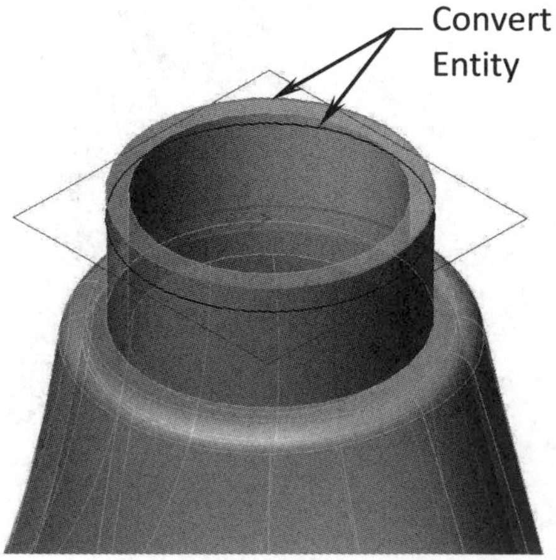

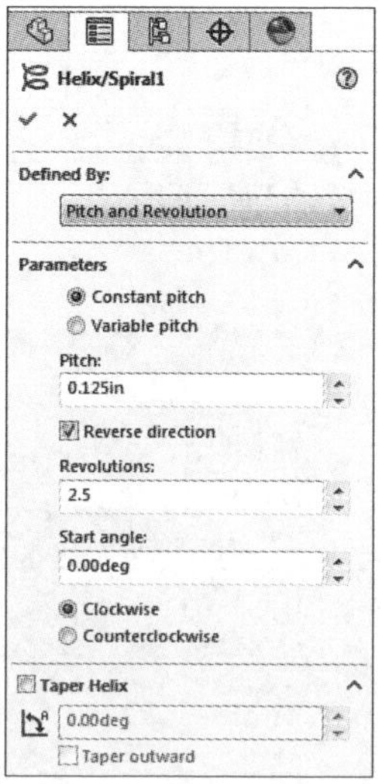

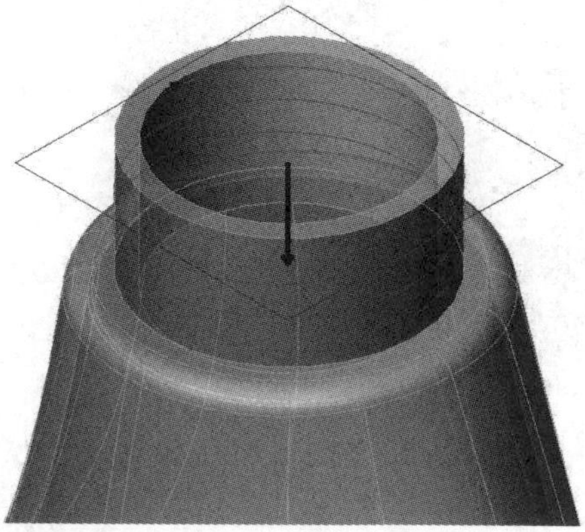

17. Creating the Sweep Profile of the Thread:

Select the Right plane and sketch the thread profile as shown below.

Add Dimensions and Relations needed to fully define the sketch.

Add the **R.015"** sketch fillets after the sketch is fully defined.

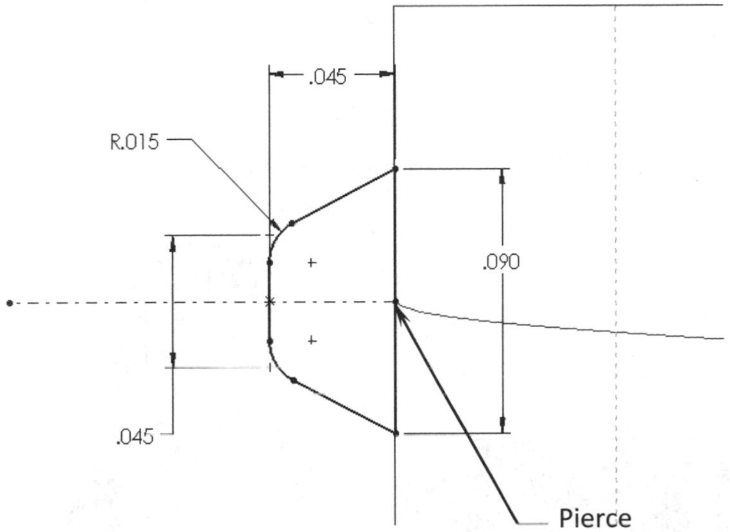

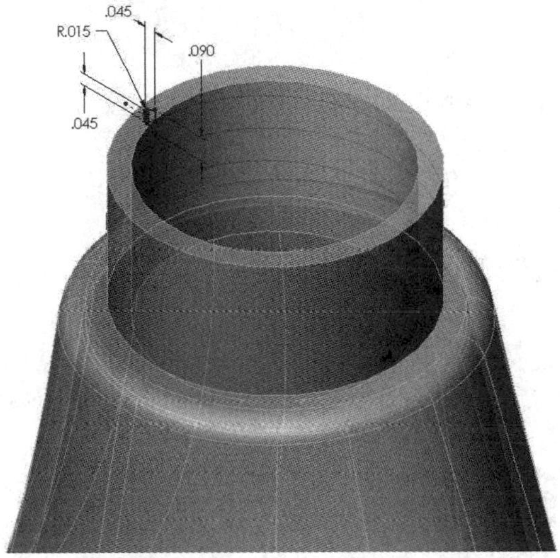

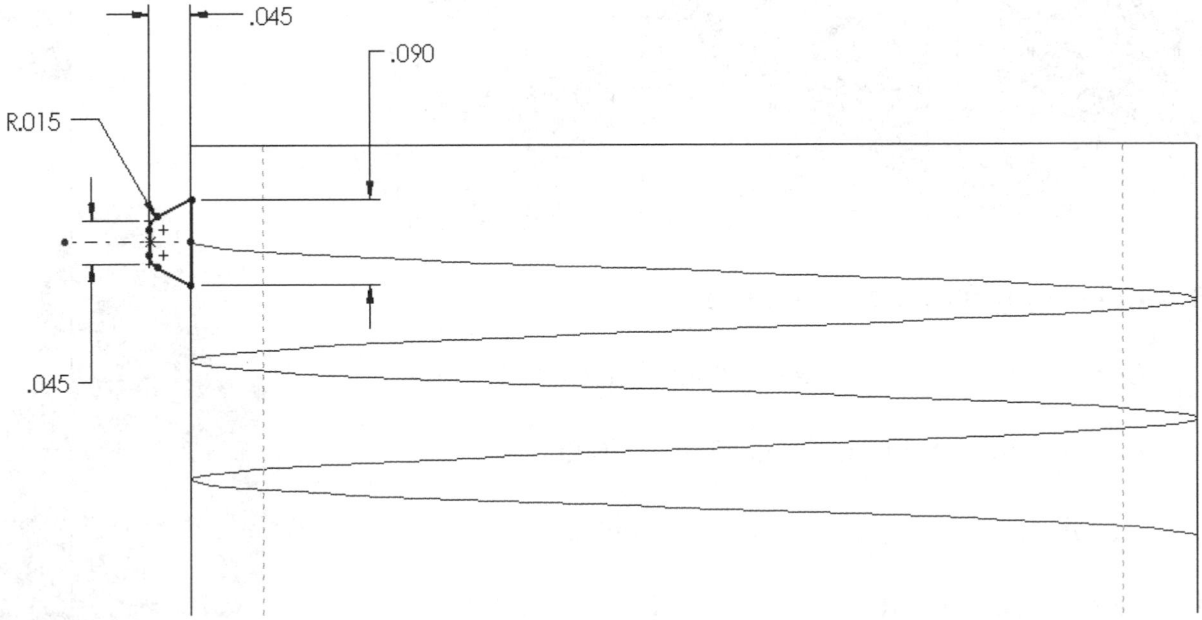

Exit the sketch.

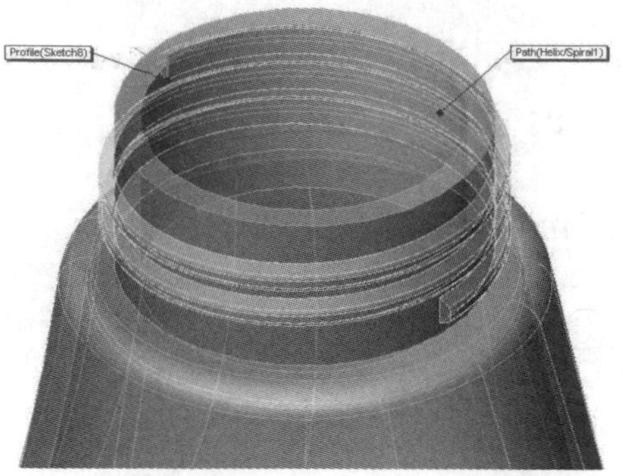

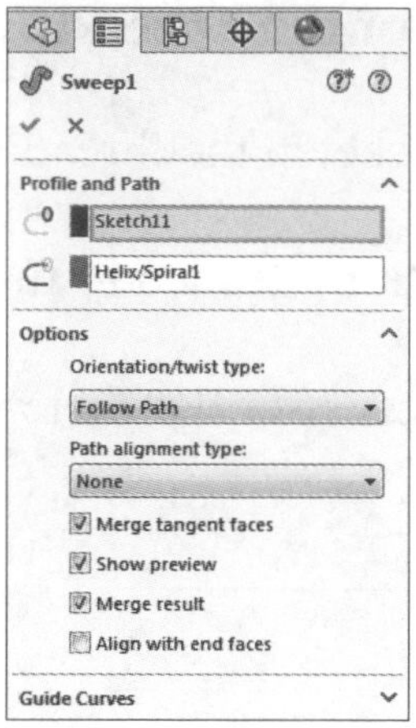

18. Sweeping:

Sweep the thread profile along the helix to create the external threads.

19. Revolving:

Convert the faces at the end of the thread and revolve them about the vertical centerlines to round off the ends.

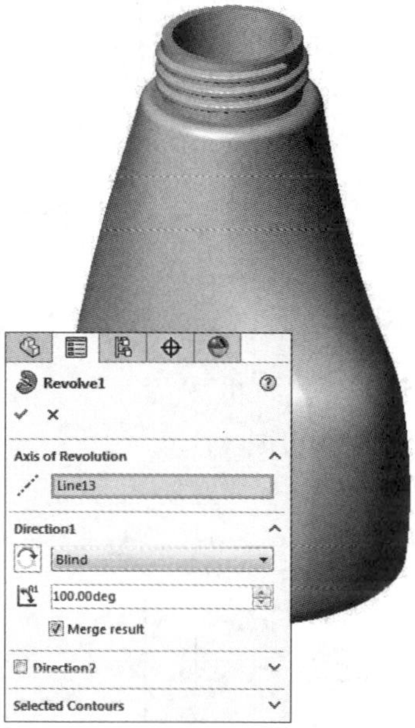

20. Applying dimension changes:

Change the dimension **R1.400** in Sketch1 to **R1.500**.

Change the **Ø1.400** in Sketch2 to **Ø1.500**.

Repair any errors caused by the changes.

21. Saving your work:

Save your work as: **Level 3 – Final Exam**.

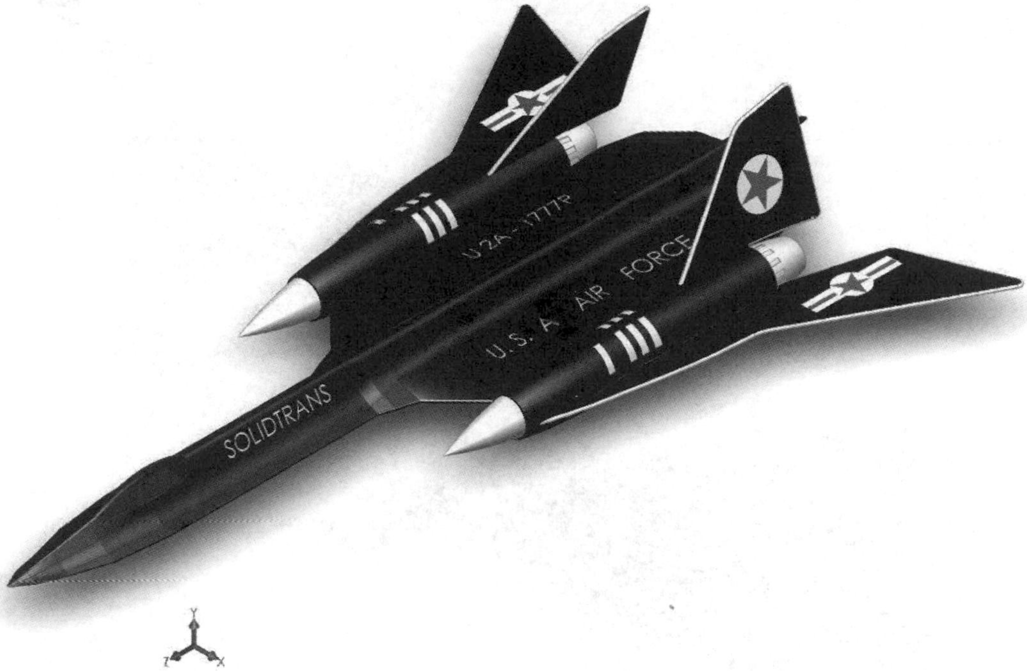

Designed by a CSWP student

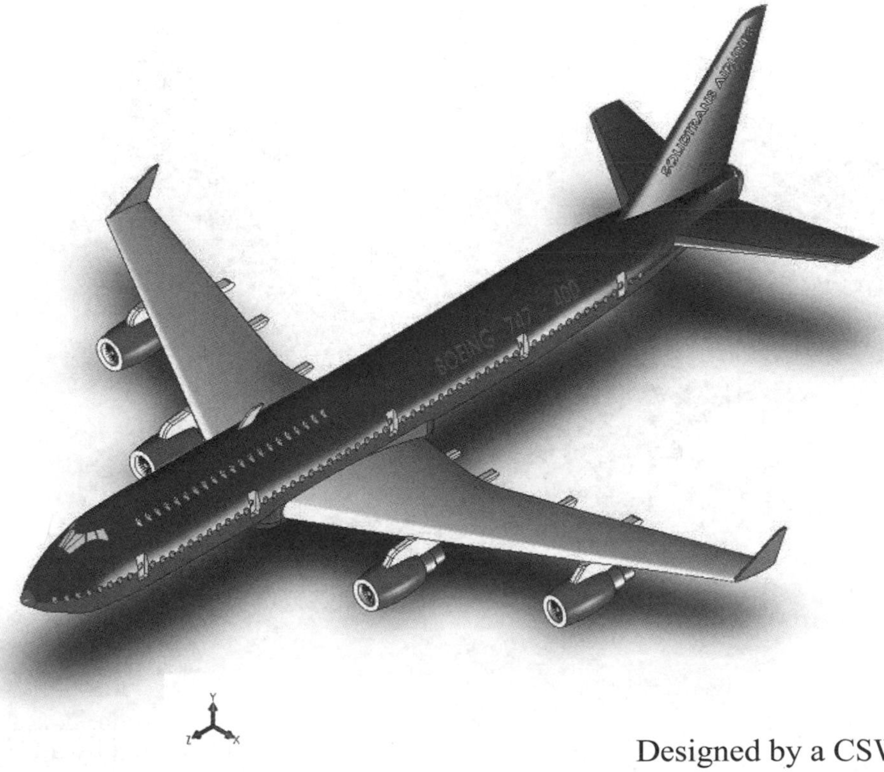

Designed by a CSWP student

Designed by a CSWP student

Designed by a CSWP student

CHAPTER 12

SimulationXpress

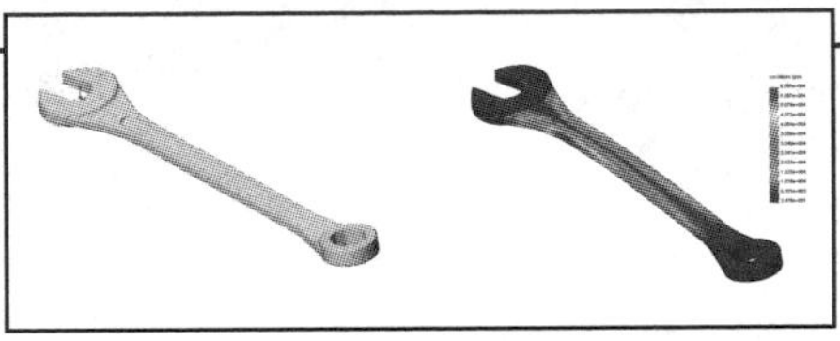

SimulationXpress
Using the Analysis Wizard

SimulationXpress is a design analysis technology that allows SOLIDWORKS users to perform first-pass stress analysis. SimulationXpress can help you reduce cost and time-to-market by testing your 3D designs within the SOLIDWORKS program, instead of expensive and time-consuming field tests.

There are five basic steps to complete the analysis using SimulationXpress:

1. Apply restraints (Fixture)

Users can define restraints. Each restraint can contain multiple faces. The restrained faces are constrained in all directions due to rigid body motion; you must at least restrain one face of the part to avoid analysis failure.

2. Apply loads

User inputs force and pressure loads to the faces of the model.

3. Define material of the part

* EX (Modulus of clasticity).
* NUXY (Poisson's ratio). If users do not define NUXY, SimulationXpress assumes a value of 0.
* SIGYLD (Yield Strength). Used only to calculate the factors of safety (FOS).
* DENS (Mass density). Used only to include mass properties of the part in the report file.

4. Analyze the part

SimulationXpress prepares the model for analysis, then calculates displacements, strains, and stresses.

5. View the results

After completing the analysis, users can view results. A check mark on the Results tab indicates that results exist and are available to view for the current geometry, material, restraints, and loads. A report can also be created in MS-Word format.

SimulationXpress
Using the Analysis Wizard

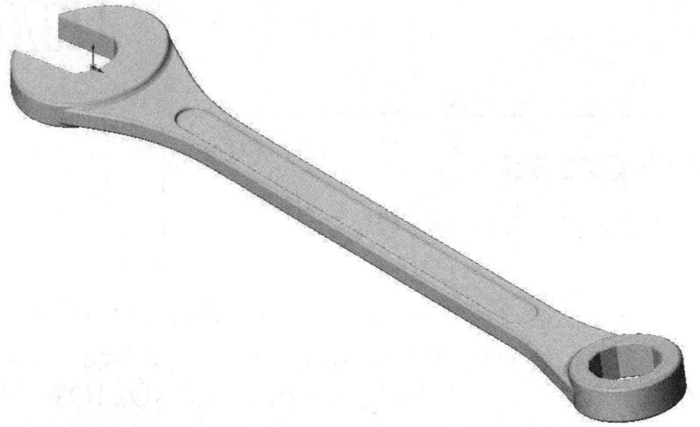

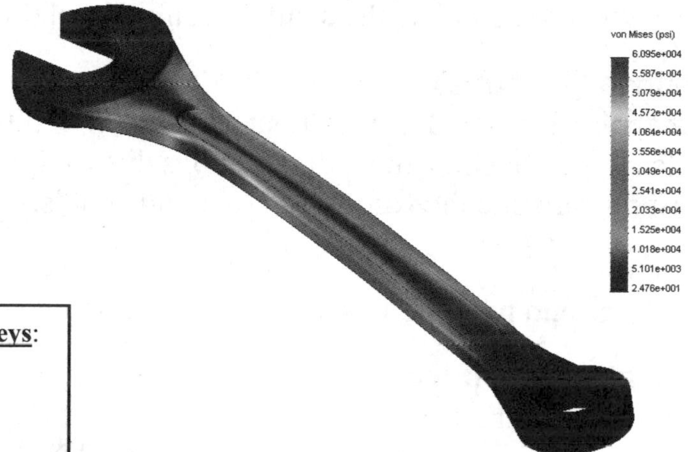

View Orientation Hot Keys:

Ctrl + 1 = Front View
Ctrl + 2 = Back View
Ctrl + 3 = Left View
Ctrl + 4 = Right View
Ctrl + 5 = Top View
Ctrl + 6 = Bottom View
Ctrl + 7 = Isometric View
Ctrl + 8 = Normal To Selection

Dimensioning Standards: **ANSI**

Units: **INCHES** – 3 Decimals

Tools Needed:

SimulationXpress is part of SOLIDWORKS Basic, SOLIDWORKS Office Professional, and SOLIDWORKS Premium.

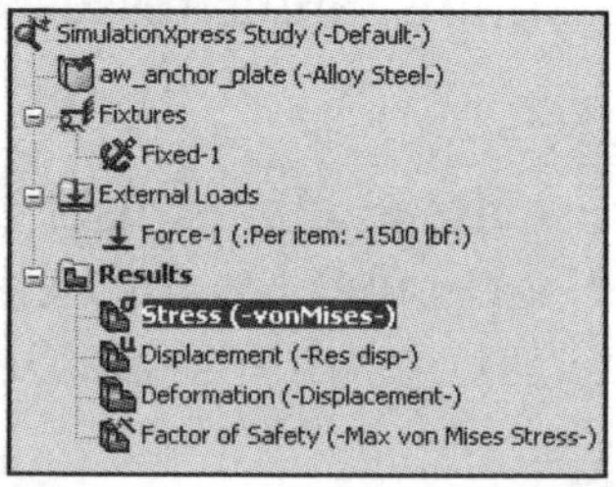

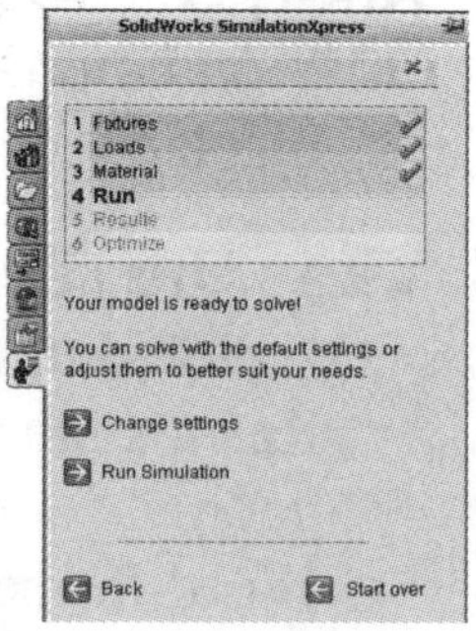

1. Starting SimulationXpress:

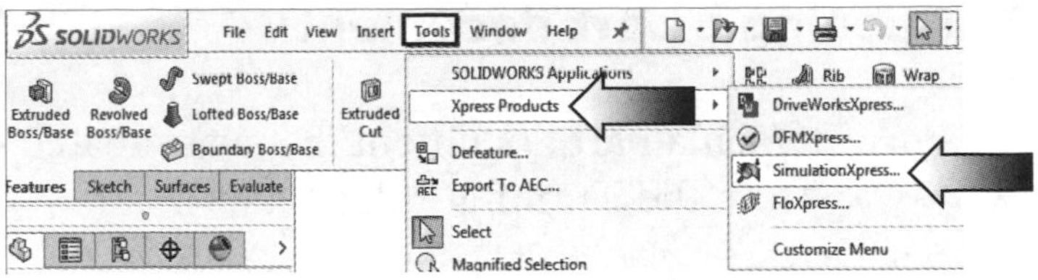

SimulationXpress is an introductory version of fully featured simulation.

SimulationXpress is a design analysis application that is fully integrated with SOLIDWORKS. It is used by designers, analysts, engineers, students, and others worldwide to design safe, efficient, and economical products.

To enable SimulationXpress, visit: www.my.solidworks.com/xpress, log in or create a user account and enter your SOLIDWORKS Serial Number to generate the **Xpress Product Codes**. You can find your serial number from the **Help** menu > **About SOLIDWORKS**.

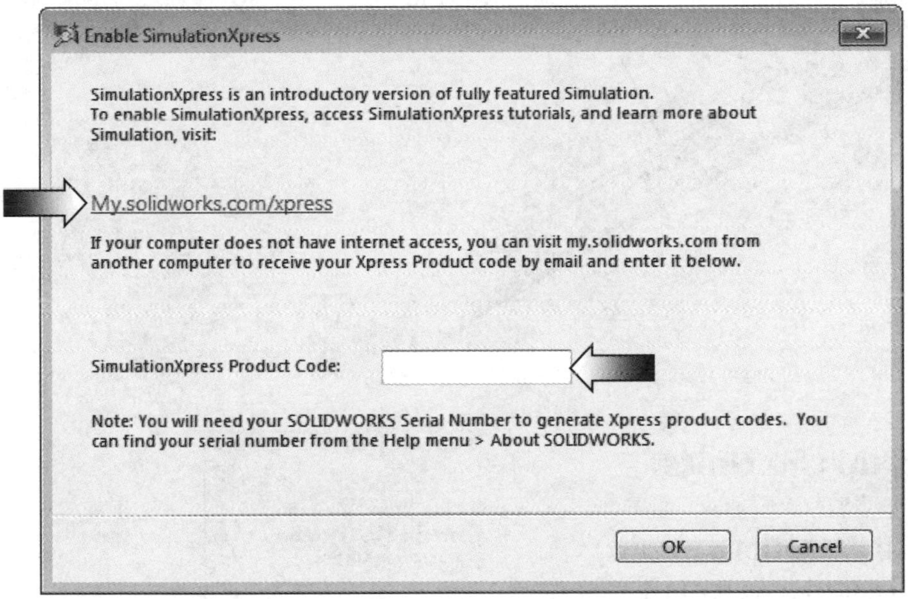

Select: **Tools / Xpress Products / SimulationXpress**. After entering the Xpress Product Code, click **OK** to launch the SimulationXpress application.

The SimulationXpress application is launched and appeared on the right side of the screen.

2. Opening a part document:

Open a part document that was created earlier: **Spanner** (or open a copy from the Training Files folder).

Suppress the extruded text on both sides (the **5/8"** and the **Spanner** text).

From the **Tools** drop-down menu, select **SimulationXpress** (Arrow).

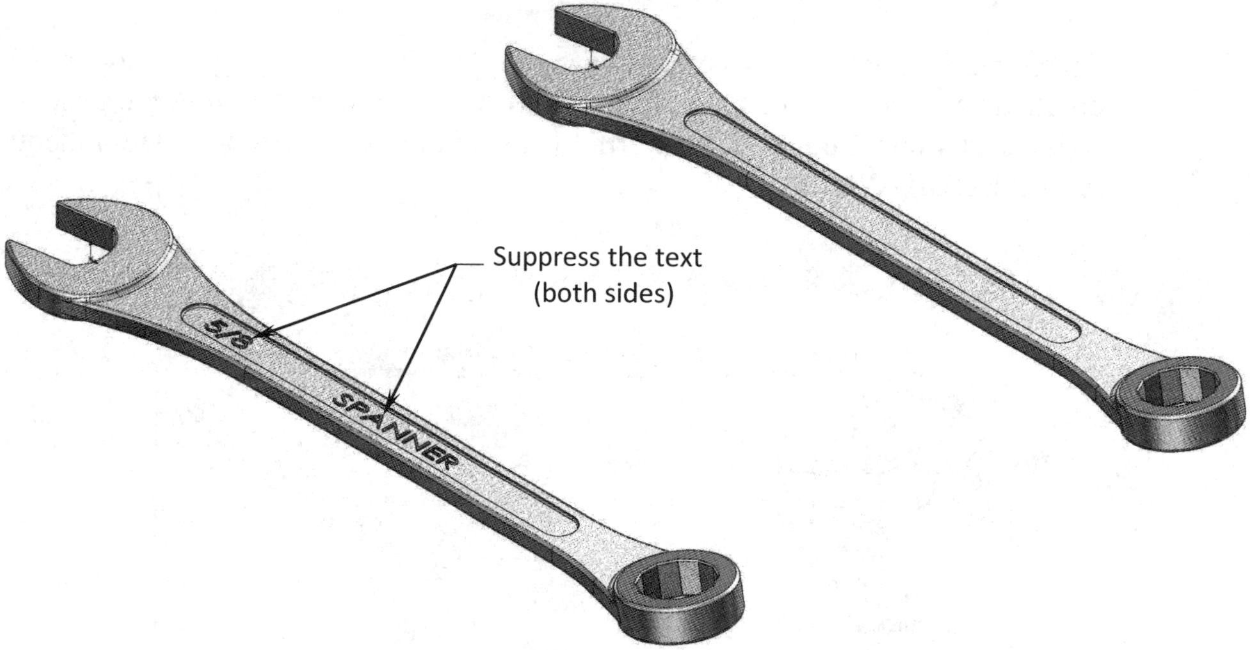

3. Setting up the Units:

Click **Options** (arrow) to set the system of units for the analysis.

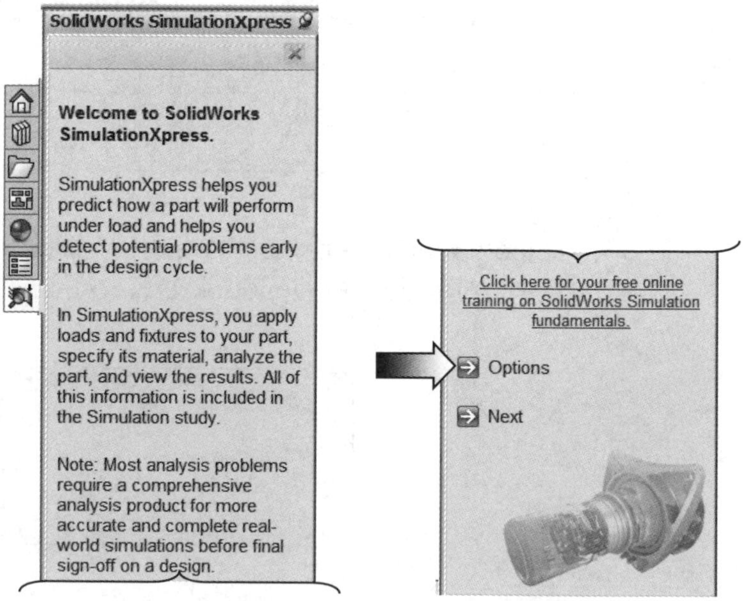

Select **English (IPS)** for System Of Units (Inch, Pound, Second).

Select the folder and the location to save the analysis results.

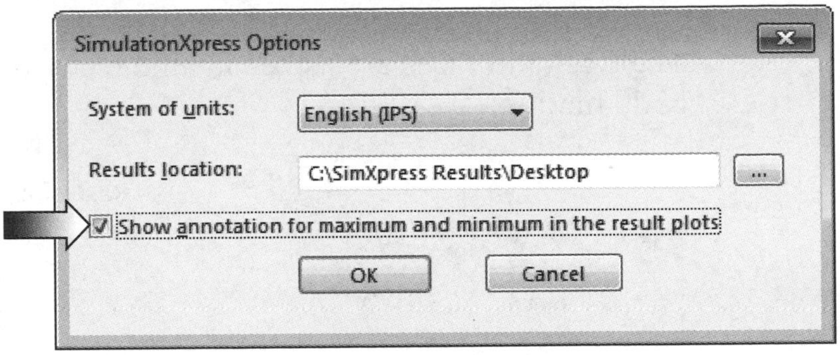

Enable the option:

Show Annotation for Maximum and Minimum in the Result Plot.

Creating a new folder for each study is recommended.

Click **OK**.

Click **Next**.

4. Adding a Fixture (restraint):

The next step is to create the restraint area(s).

Each restraint can contain one or multiple faces. The restrained faces are constrained in all directions. There must be at least one fixed face of the part to avoid analysis failure due to rigid body motion.

> **Restraints**
>
> Restraint is used to anchor certain areas of the model so that they will not move or shift during the analysis. At least one face should be restrained prior to running the analysis.

Click **Add a Fixture**.

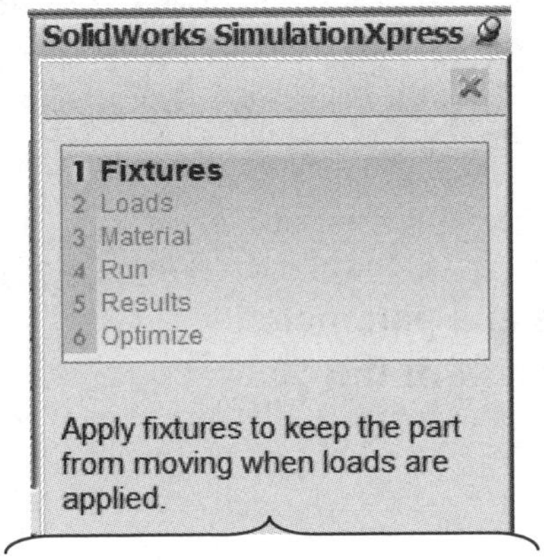

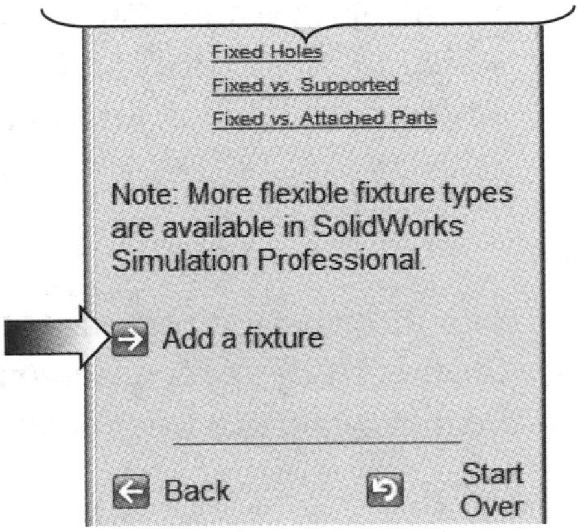

Select the 3 faces as indicated to use as restraint faces.

The Restraint faces are locked in all directions to avoid failure due to rigid body motion.

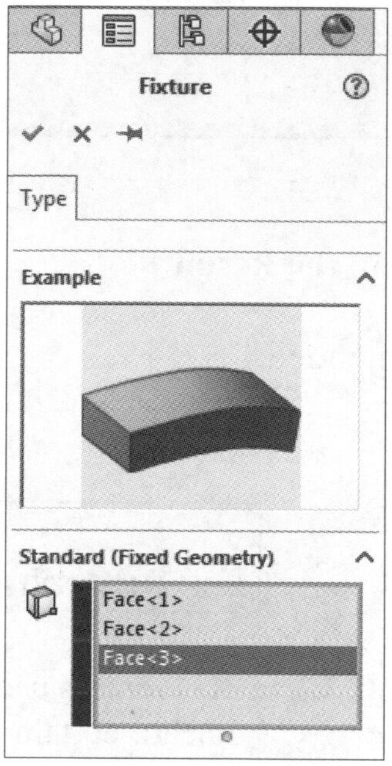

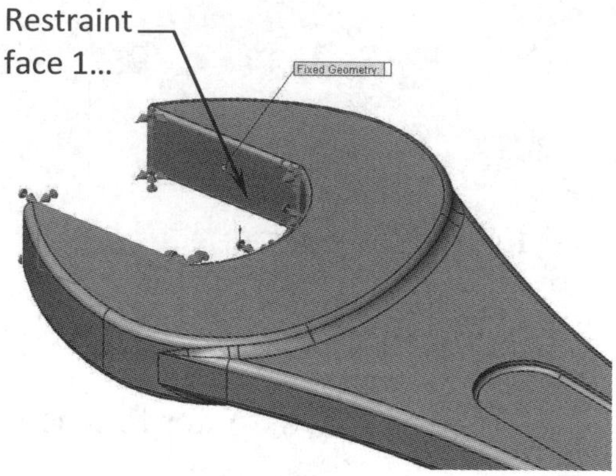

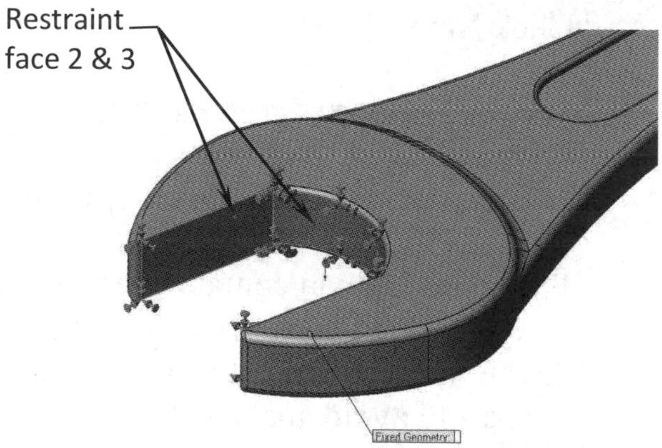

Click **OK**.

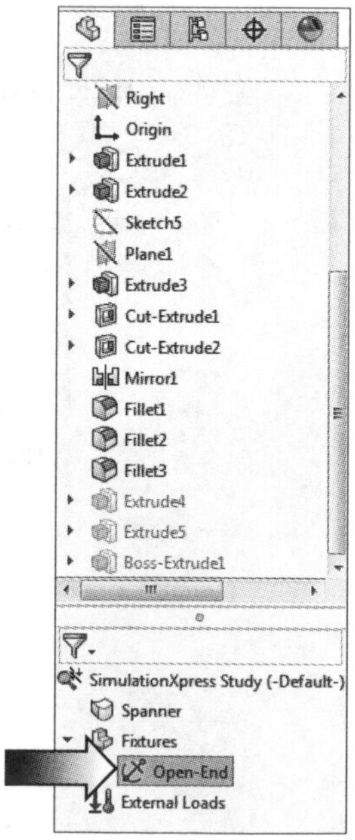

When the restraint faces are selected, more faces can be added to create different restraint sets. They can also be edited or deleted at anytime.

The information regarding the settings and parameters for this study is recorded on the lower half of the FeatureManager tree (arrows).

Rename the fixture Fixed-1 to **Open-End**.

5. Applying a Force:

Click **Add a Force** .

The **Forces** and **Pressures** options allow SOLIDWORKS users to apply force or pressure loads to faces of the model. Multiple forces can be applied to a single face or to multiple faces.

Select the 2 faces from the back side of the closed end as shown below.

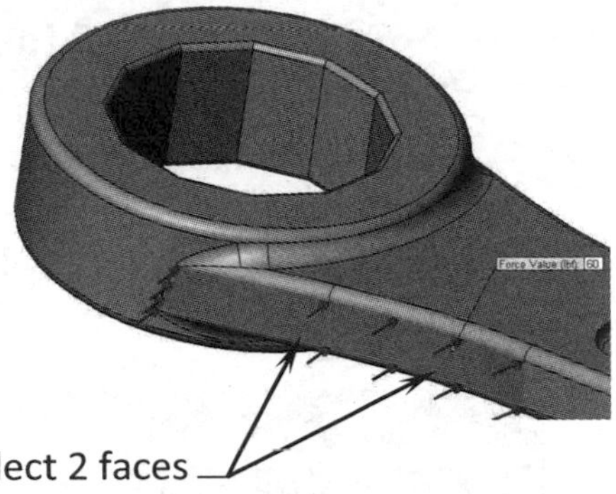

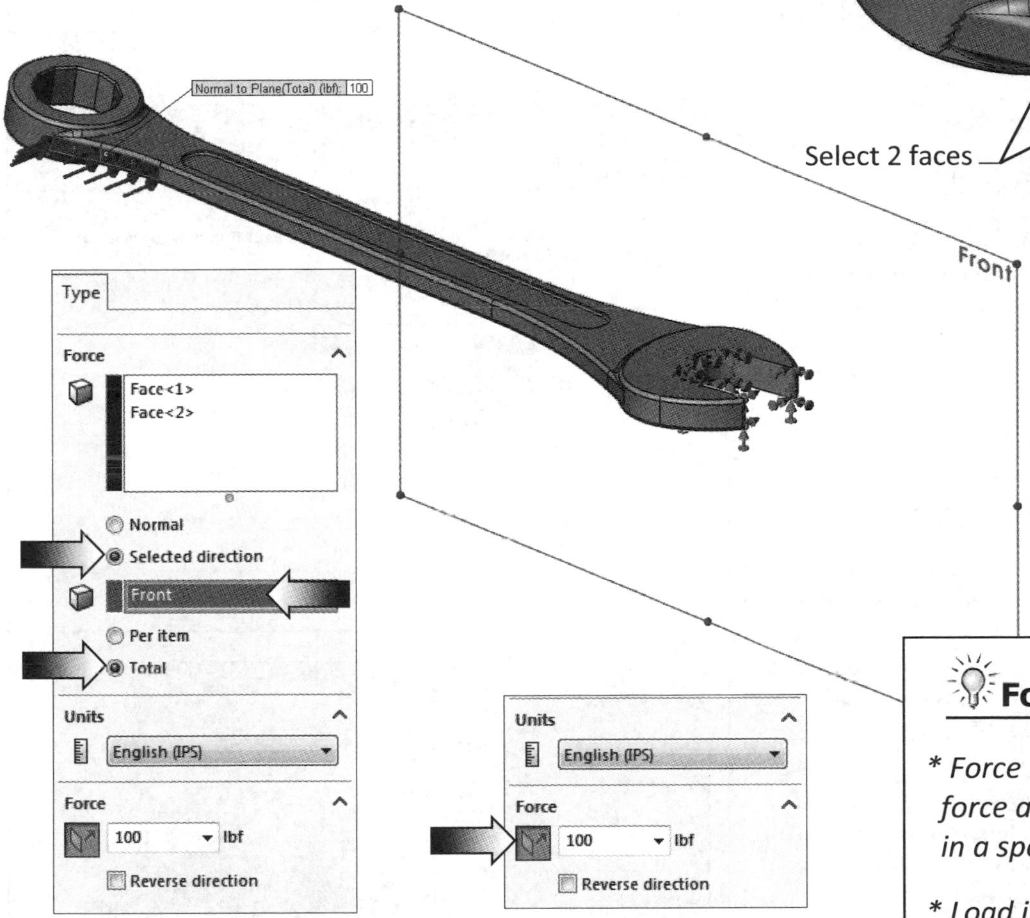

Force & Load

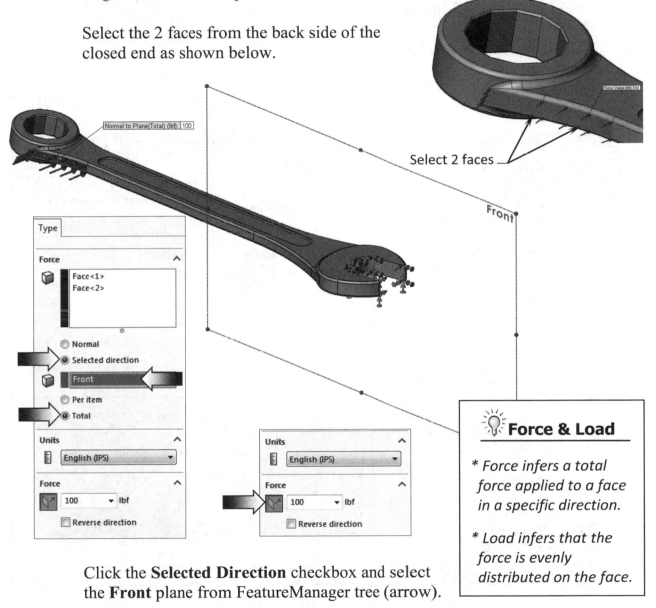

** Force infers a total force applied to a face in a specific direction.*

** Load infers that the force is evenly distributed on the face.*

Click the **Selected Direction** checkbox and select the **Front** plane from FeatureManager tree (arrow).

Click the **Total** option (arrow). Enter **100** lbs. for Total Force value (arrow).

Click Reverse Direction if needed.

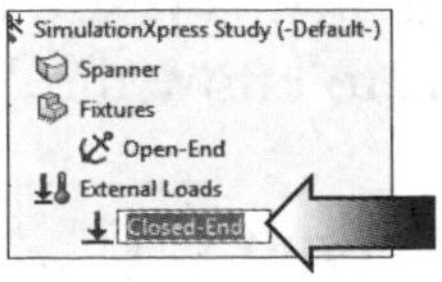

Click **OK**. Rename Force-1 to **Closed-End**

We will need to specify a material at this point so that SimulationXpress can predict how it will respond to the loads.

6. Selecting the material:

Click **Next**.

Select **Choose Material**.

Expand the **Steel** folder.

Choose **Alloy Steel** from the list (arrow).

> **Material Editor**
>
> Material can be assigned to the part using the **Material Editor** PropertyManager. The material will then appear in SimulationXpress.

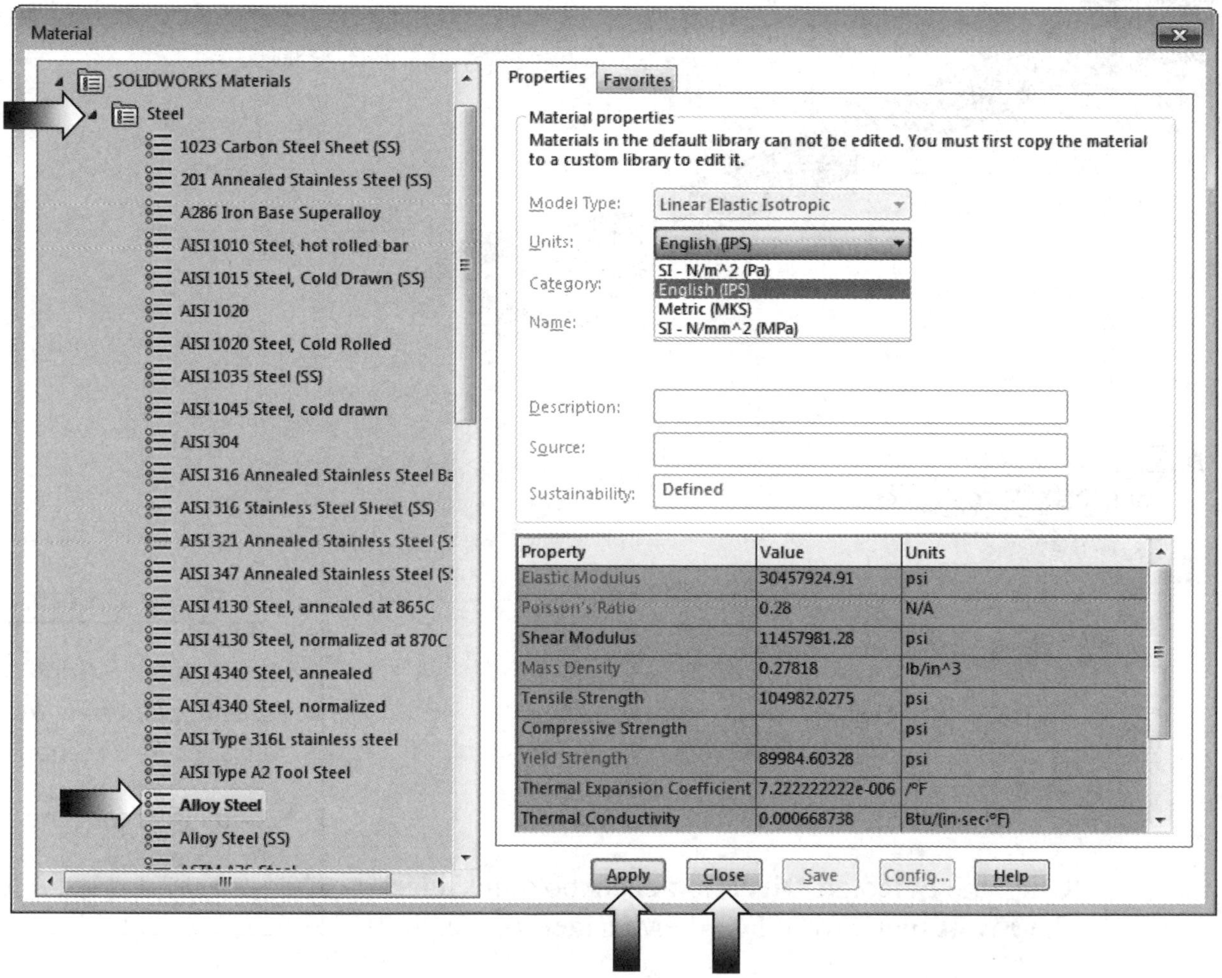

The analysis results are dependent upon the material selection. SimulationXpress needs to know the Visual and the Physical properties to run the analysis.

Click **Apply** and **Close**.

The material **Alloy Steel** is now assigned to the part. SimulationXpress assumes that the material deforms in a linear fashion with increased load. Non-Linear materials (such as many plastic materials) require the use of Simulation Premium.

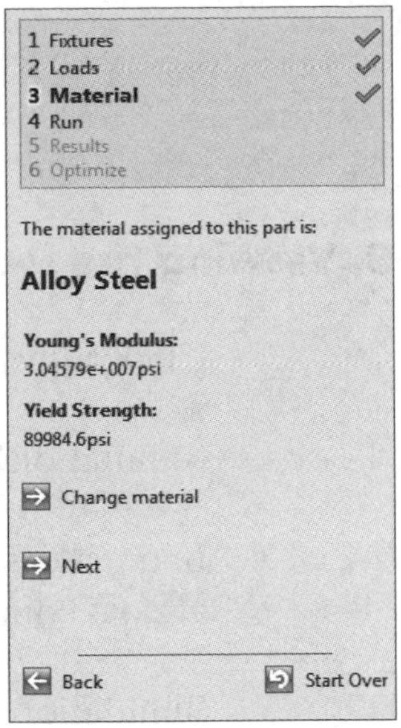

The Modulus of Elasticity and the Yield Strength for the selected material are reported on the right side pane.

Click **Next**.

SimulationXpress is ready to analyze the model based on the information provided. Displacements, Strains, and Stresses will then be calculated.

7. Analyzing the model:

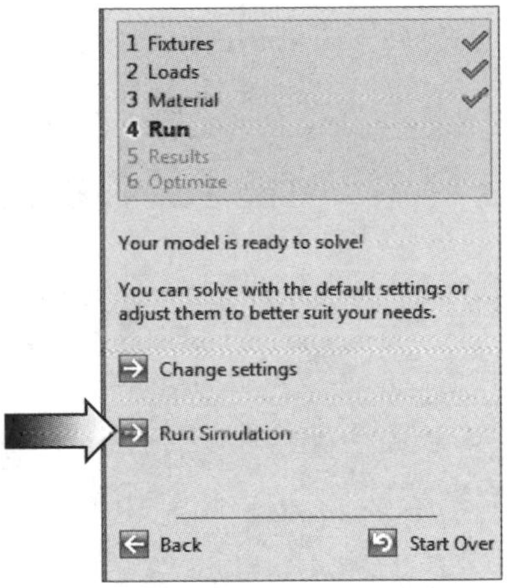

Click **Run Simulation**.

SimulationXpress automatically tries to mesh the model using the default element size. The smaller the element size the more accurate the results, but more time is needed to analyze the model.

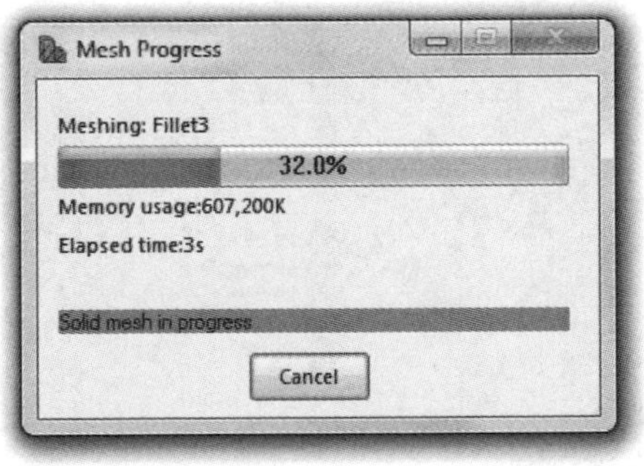

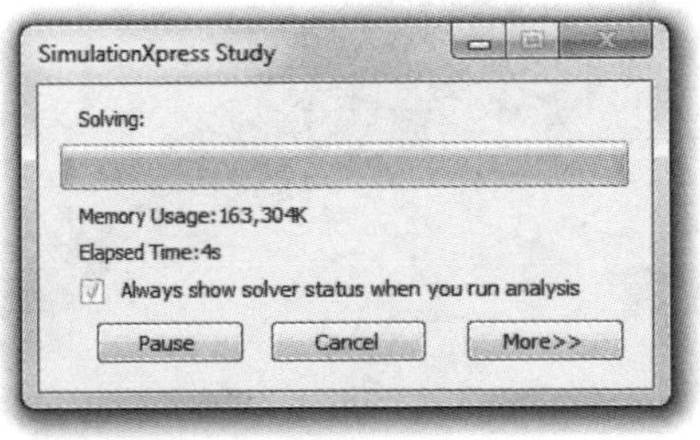

To change mesh settings: First click C**hange settings** then click **Change mesh density**.

* Drag the slider to the right for a finer mesh (more accurate but takes longer).

* Drag the slider to the left for a coarser mesh (quicker).

Select **Yes, continue** .

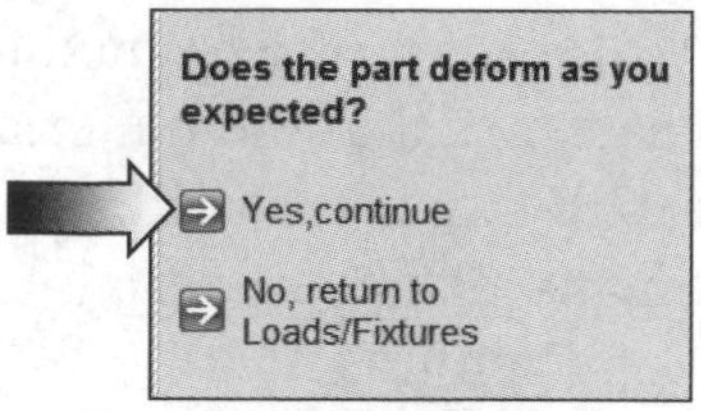

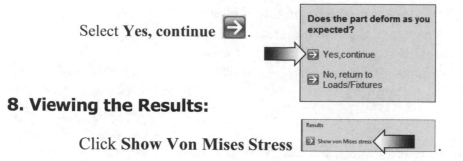

8. Viewing the Results:

Click **Show Von Mises Stress** 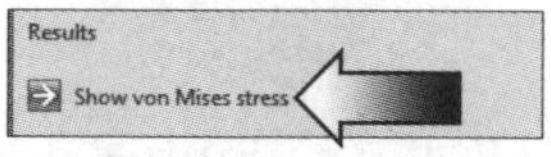.

SimulationXpress plots stresses on the deformed shape of the part.

In most cases, the actual deformation is so small that the deformed shape almost coincides with the un-deformed shape, if plotted to scale.

SimulationXpress exaggerates the deformation to demonstrate it more clearly.

The Deformation Scale shown on the stress and deformed shape plots is the scale used to rescale the maximum deformation to **4.34794%** of the bounding box of the part.

The **Stress Distribution Plot** is displayed below.

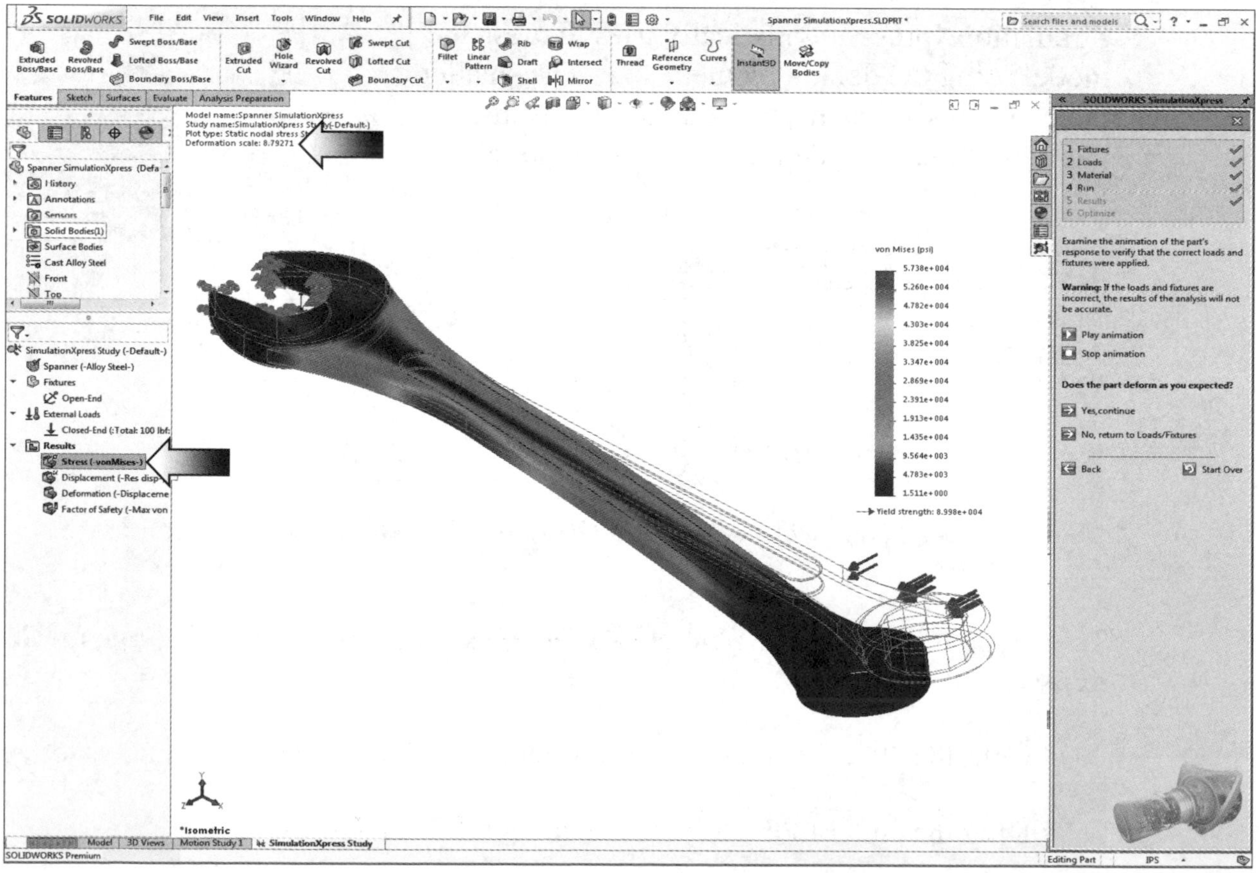

To see the resultant displacement plot click **Show Displacement**.

Click **Play Animation** or **Stop Animation** when finished viewing.

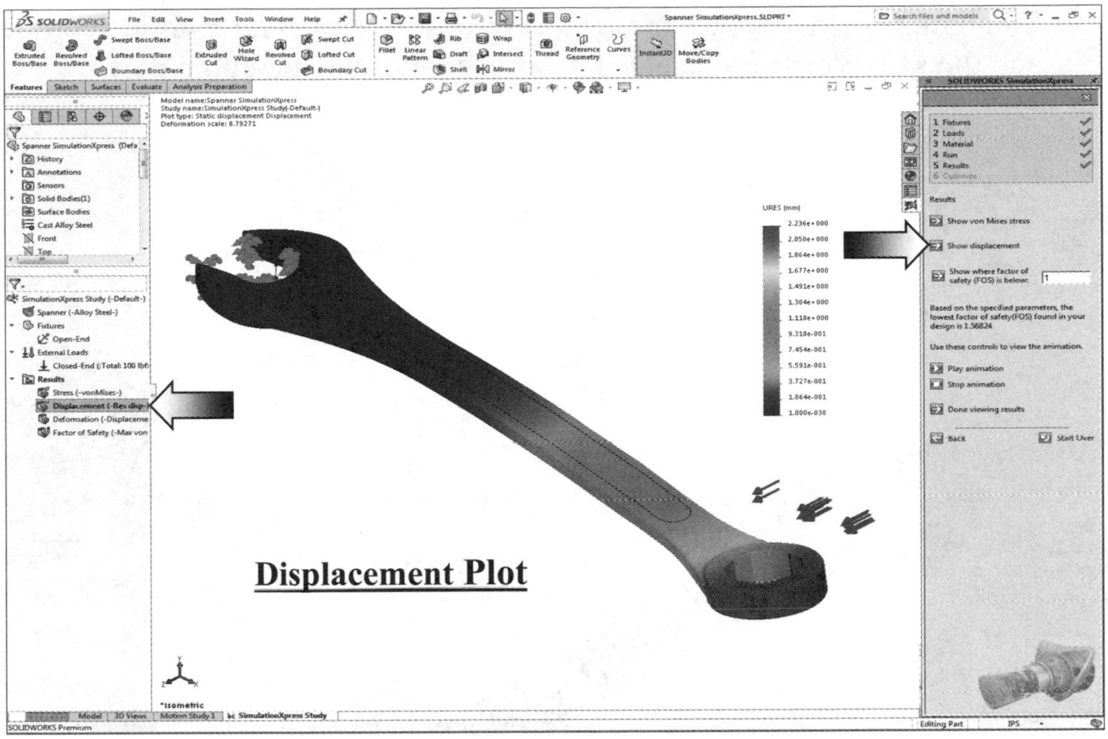

To view regions of the model with a factor of safety less than a given value (1), click: **Show Where Factor Of Safety (FOS) Is Below: 1** (or enter any value).

SimulationXpress displays regions of the model with factors of safety less than the specified value in red (unsafe regions) and regions with higher factors of safety in blue (safe regions).

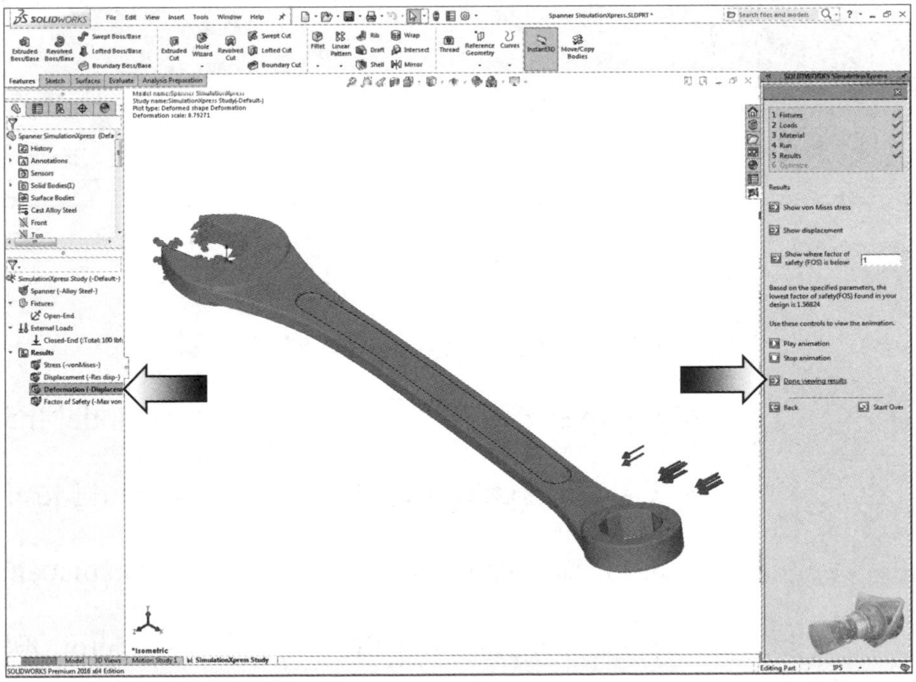

When finished viewing click: **Done Viewing Results**.

9. Creating the report:

Click **Generate Report** .

SimulationXpress cycles through the results, generates a report in Word format, and the MS-Word application is launched to display the full report.

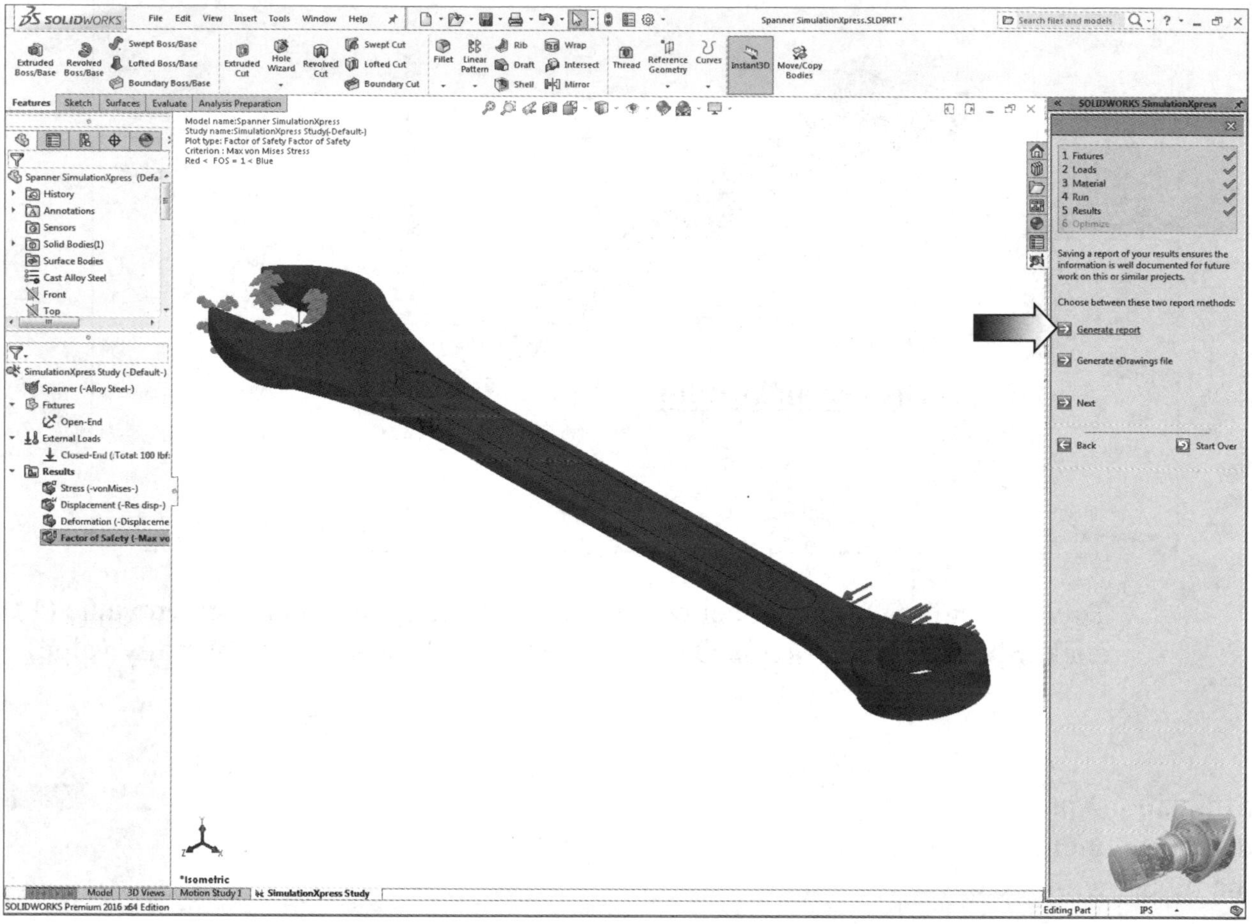

The report includes:

1. Cover Page
2. Model Information
2. Load/Fixture Details
4. Slid Mesh Information
5. Stress Results
6. Displacement Results
7. Deformation Results
8. Factor of Safety Results

In the Report Settings dialog box, enable the Description checkbox and enter the following:

* **Spanner - SimXpress Study**.

* **Your Name**.

* **Your Company Name** and any information that you wish to include in the report.

Select the location to save the report.

Click **Generate**.

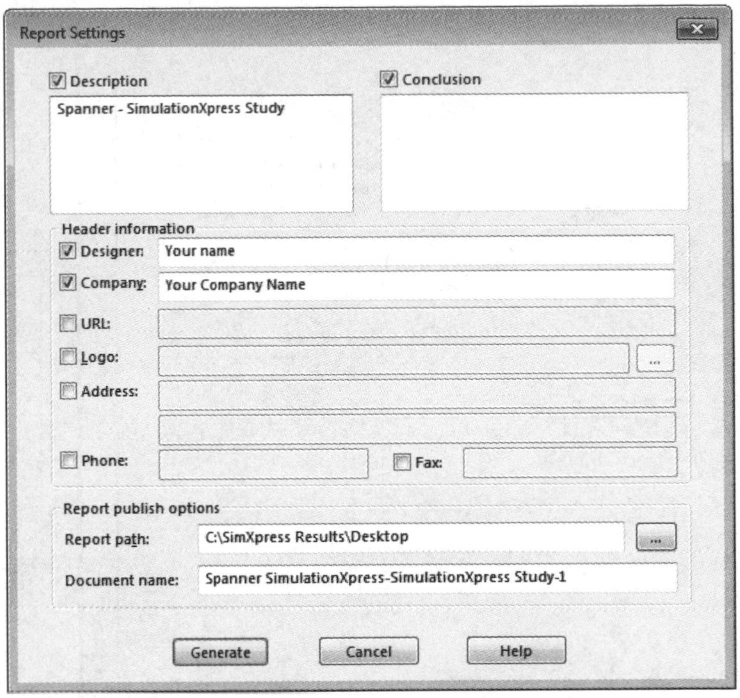

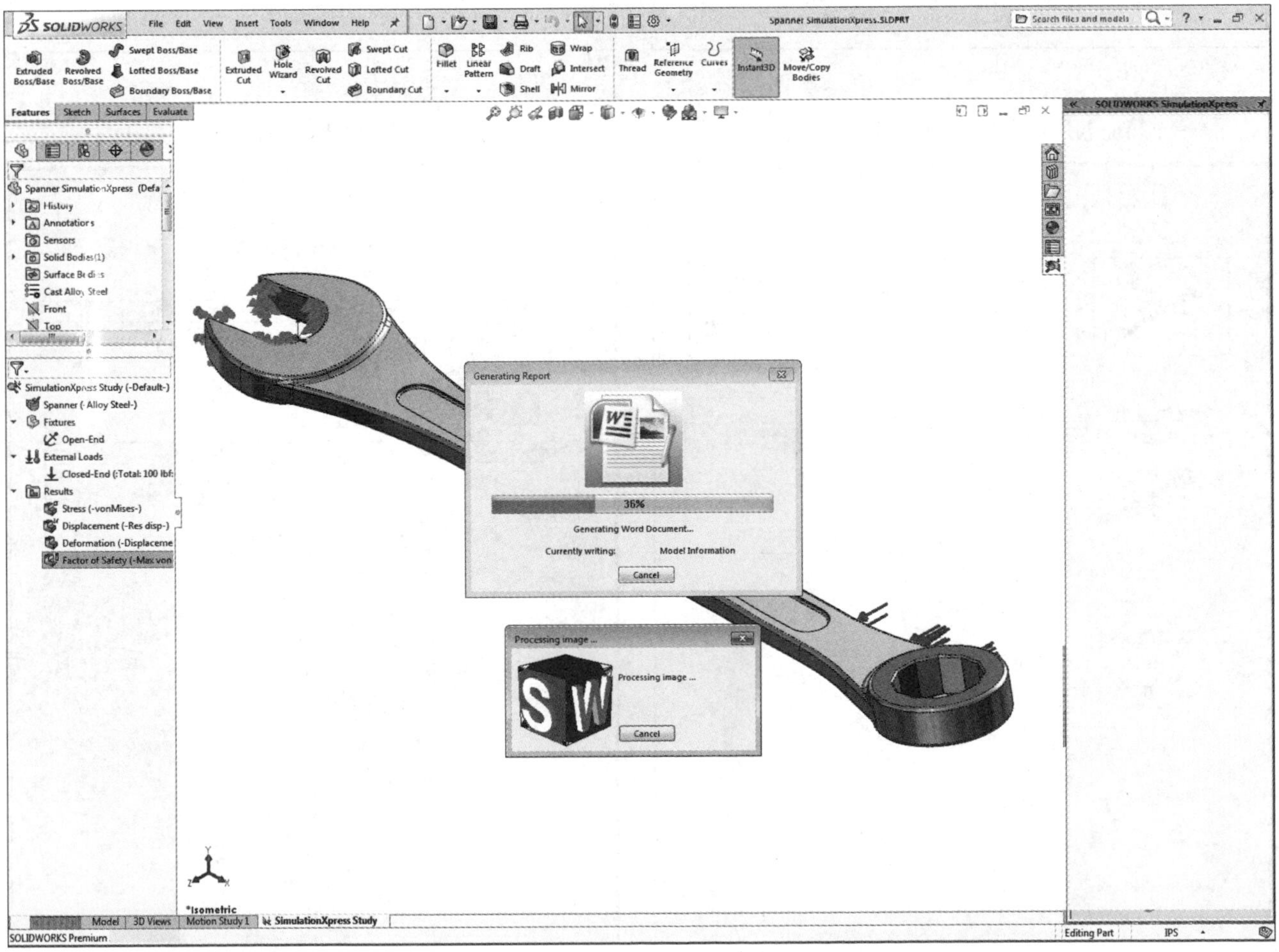

The Report Cover Page

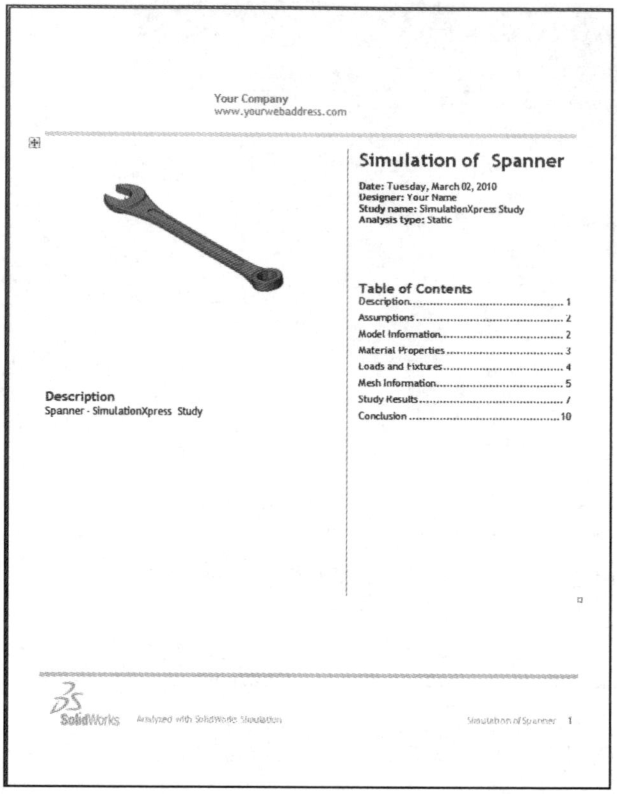

Your Company
www.yourwebaddress.com

Simulation of Spanner

Date: Tuesday, March 02, 2010
Designer: Your Name
Study name: SimulationXpress Study
Analysis type: Static

Table of Contents

Description
Spanner - SimulationXpress Study

SolidWorks Analyzed with SolidWorks Simulation — Simulation of Spanner 1

The Model Information

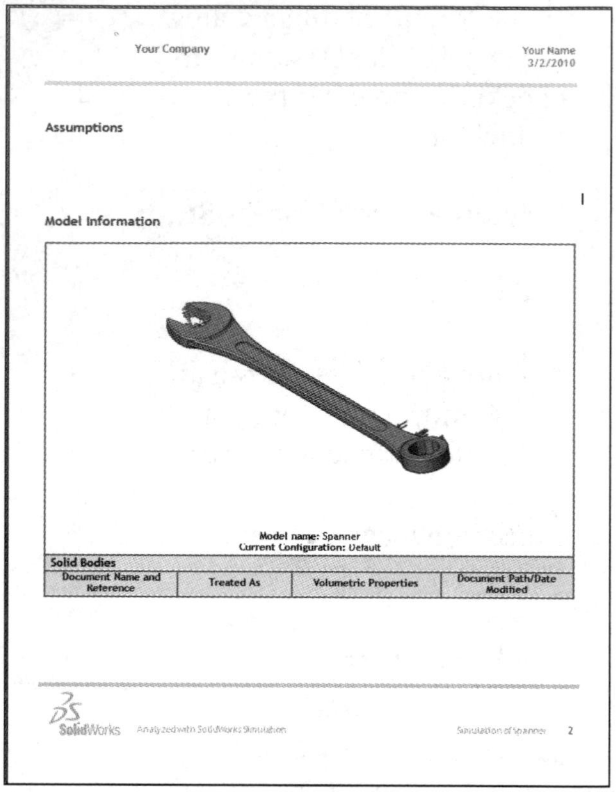

Your Company — Your Name 3/2/2010

Assumptions

Model Information

Model name: Spanner
Current Configuration: Default

Solid Bodies			
Document Name and Reference	Treated As	Volumetric Properties	Document Path/Date Modified

SolidWorks Analyzed with SolidWorks Simulation — Simulation of Spanner 2

The Model Information cont.

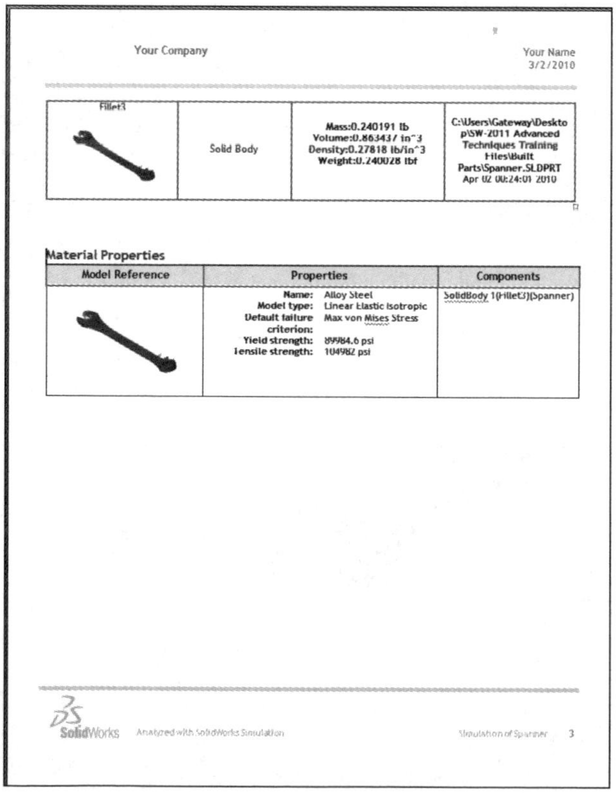

Your Company — Your Name 3/2/2010

Fillet3	Solid Body	Mass:0.240191 lb Volume:0.863437 in^3 Density:0.27818 lb/in^3 Weight:0.240028 lbf	C:\Users\Gateway\Desktop\SW-2011 Advanced Techniques Training Files\Built Parts\Spanner.SLDPRT Apr 02 00:24:01 2010

Material Properties

Model Reference	Properties	Components
	Name: Alloy Steel Model type: Linear Elastic Isotropic Default failure criterion: Max von Mises Stress Yield strength: 89984.6 psi Tensile strength: 104982 psi	SolidBody 1(Fillet3)(Spanner)

SolidWorks Analyzed with SolidWorks Simulation — Simulation of Spanner 3

The Loads and Fixtures Details

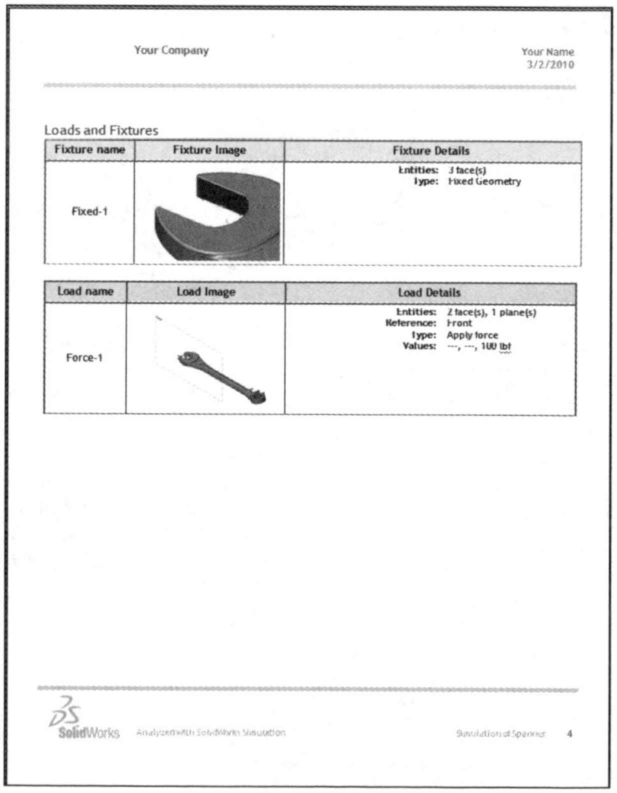

Your Company — Your Name 3/2/2010

Loads and Fixtures

Fixture name	Fixture Image	Fixture Details
Fixed-1		Entities: 3 face(s) Type: Fixed Geometry

Load name	Load Image	Load Details
Force-1		Entities: 2 face(s), 1 plane(s) Reference: Front Type: Apply force Values: ---, ---, 100 lbf

SolidWorks Analyzed with SolidWorks Simulation — Simulation of Spanner 4

The Mesh Information

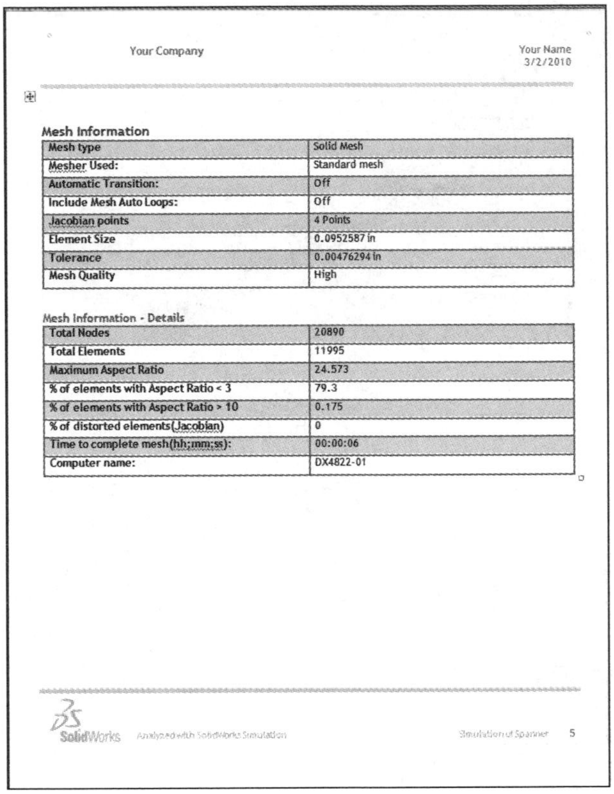

Your Company

Your Name
3/2/2010

Mesh Information

Mesh type	Solid Mesh
Mesher Used:	Standard mesh
Automatic Transition:	Off
Include Mesh Auto Loops:	Off
Jacobian points	4 Points
Element Size	0.0952587 in
Tolerance	0.00476294 in
Mesh Quality	High

Mesh Information - Details

Total Nodes	20890
Total Elements	11995
Maximum Aspect Ratio	24.573
% of elements with Aspect Ratio < 3	79.3
% of elements with Aspect Ratio > 10	0.175
% of distorted elements(Jacobian)	0
Time to complete mesh(hh;mm;ss):	00:00:06
Computer name:	DX4822-01

SolidWorks Analyzed with SolidWorks Simulation Simulation of Spanner 5

The Solid Mesh Plot

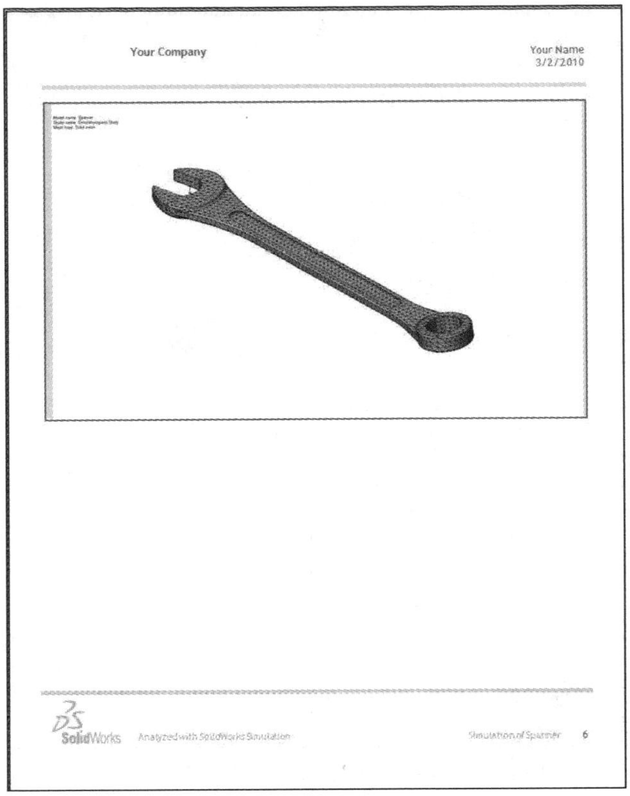

Your Company

Your Name
3/2/2010

SolidWorks Analyzed with SolidWorks Simulation Simulation of Spanner 6

The Von Mises Stress Plot

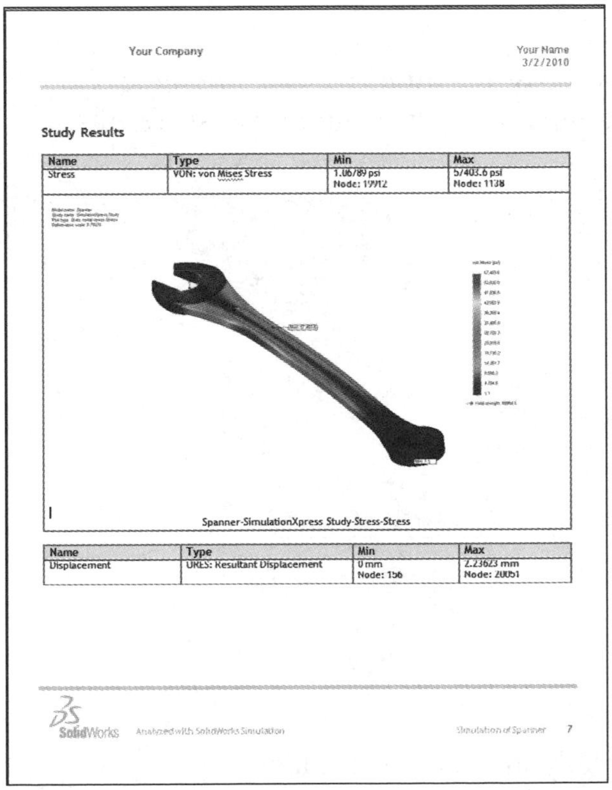

Your Company

Your Name
3/2/2010

Study Results

Name	Type	Min	Max
Stress	VON: von Mises Stress	1.06789 psi Node: 19912	57403.6 psi Node: 1138

Spanner-SimulationXpress Study-Stress-Stress

Name	Type	Min	Max
Displacement	URES: Resultant Displacement	0 mm Node: 156	2.23623 mm Node: 20051

SolidWorks Analyzed with SolidWorks Simulation Simulation of Spanner 7

The Displacement Plot

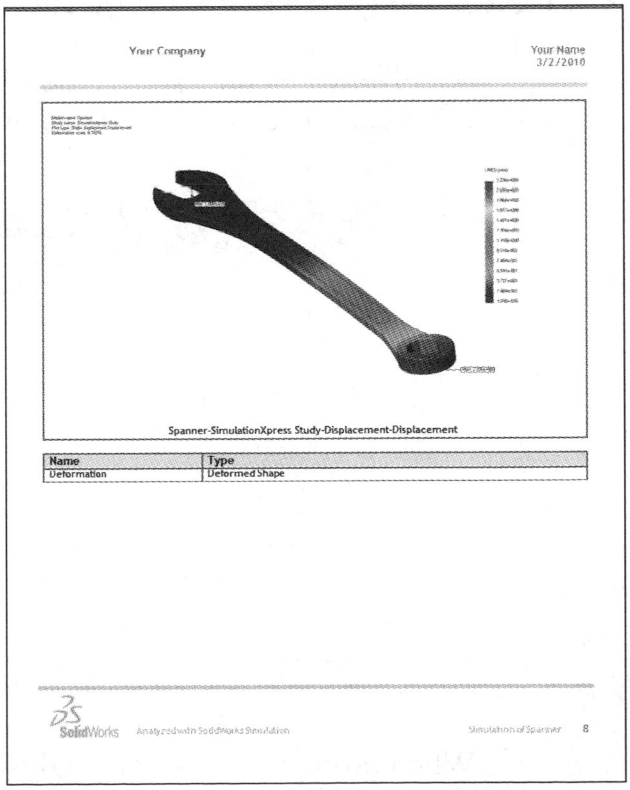

Your Company

Your Name
3/2/2010

Spanner-SimulationXpress Study-Displacement-Displacement

Name	Type
Deformation	Deformed Shape

SolidWorks Analyzed with SolidWorks Simulation Simulation of Spanner 8

The Deformation Plot

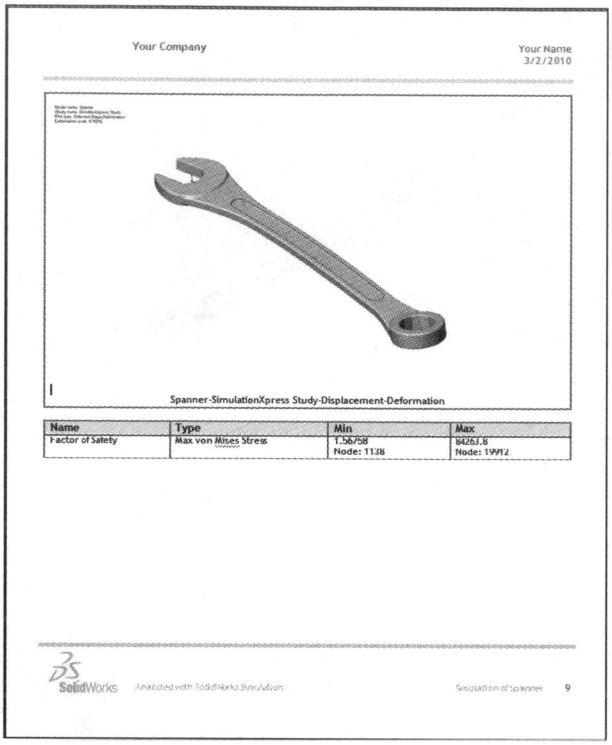

The Factor of Safety Plot

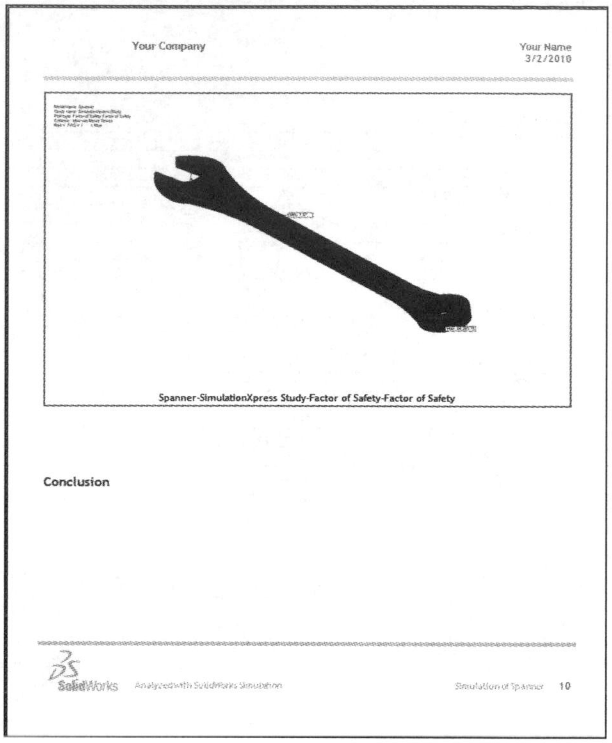

When finished with viewing the Report, click: **Generate eDrawings File**.

10. Generating the eDrawings file:

An eDrawings file can be created for the SimulationXpress result plots. The eDrawings file allows you to view, animate and print your analysis results.

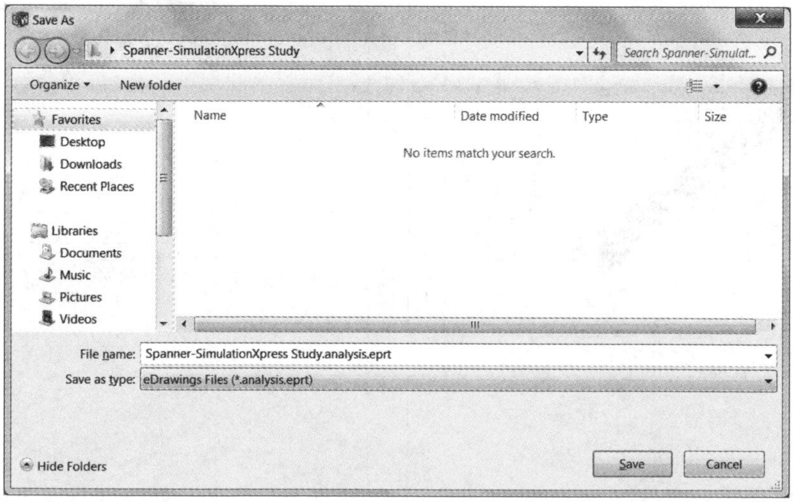

When prompted, **Save** the analysis study in the default folder.

SimulationXpress creates an eDrawing file with the **.eprt** extension. The file contains von Mises stress, displacement, deformation, and Factor of Safety plots. By default, the von Mises stress plot is displayed.

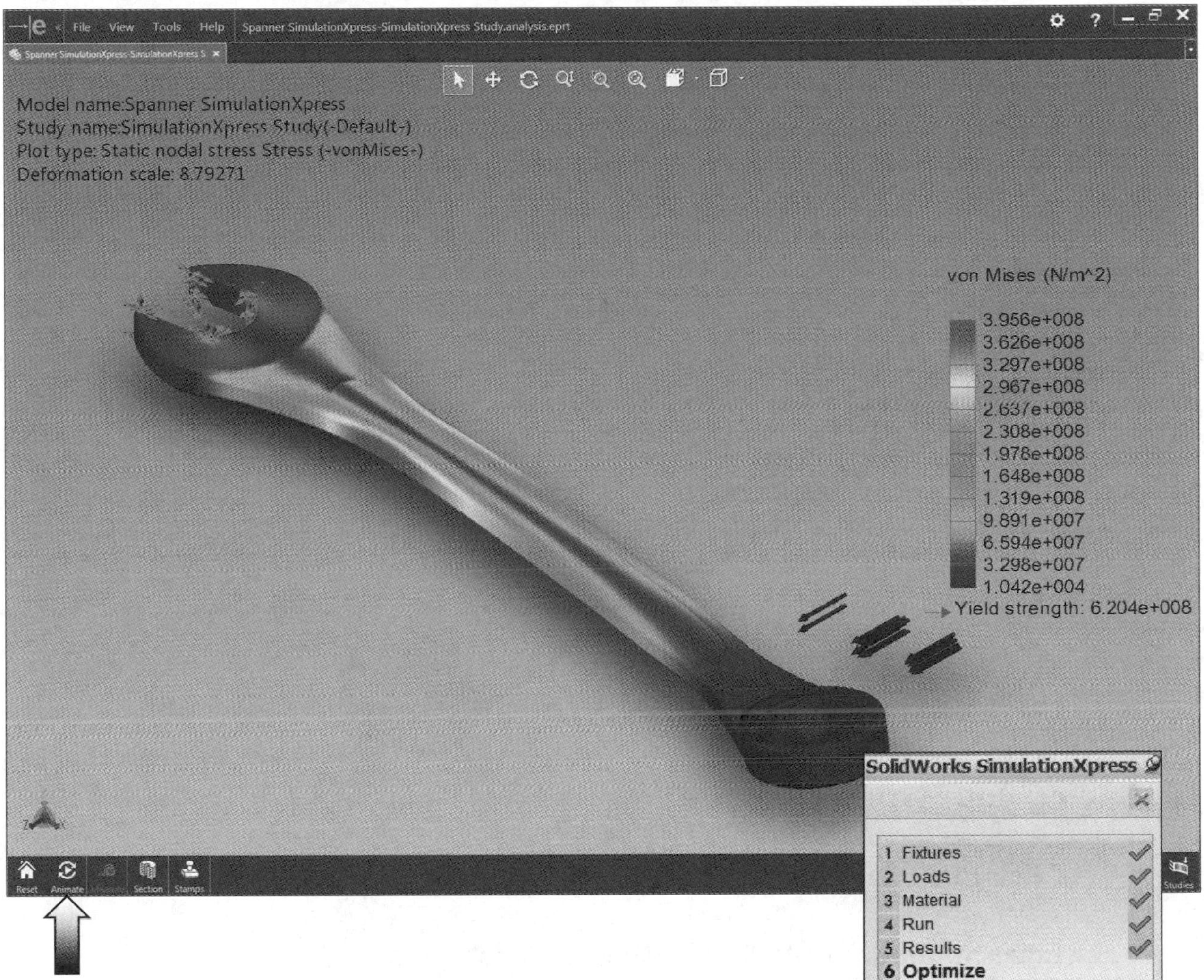

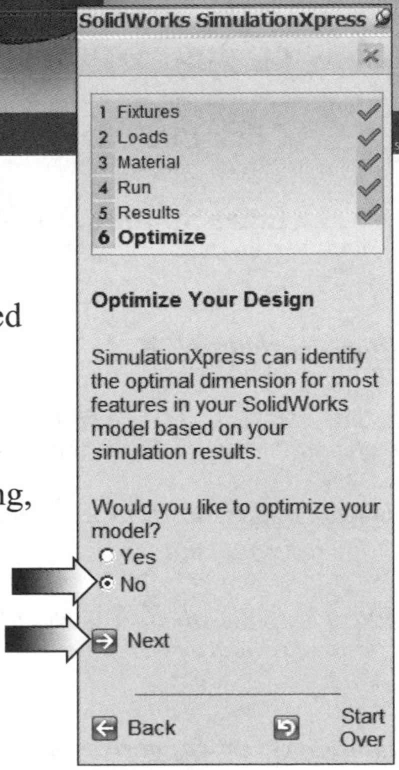

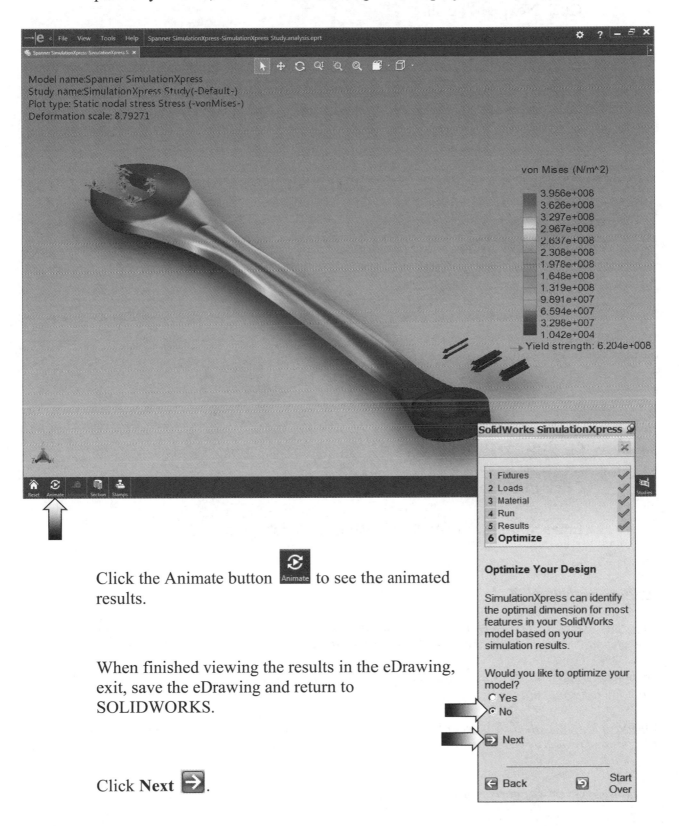

Click the Animate button (Animate) to see the animated results.

When finished viewing the results in the eDrawing, exit, save the eDrawing and return to SOLIDWORKS.

Click **Next**.

Click **No** under **Optimize Your Design.** (Only if the analysis fails then the optimization step is needed.)

Click **Next** again.

At this point, you are prompted that the analysis has been completed. All 5 steps on the SimulationXpress property tree (right side) have the check marks in front of them.

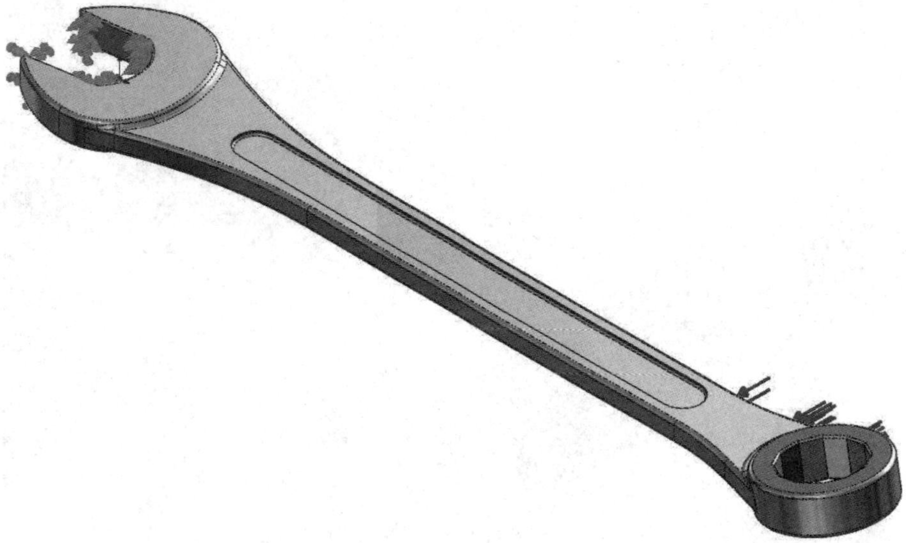

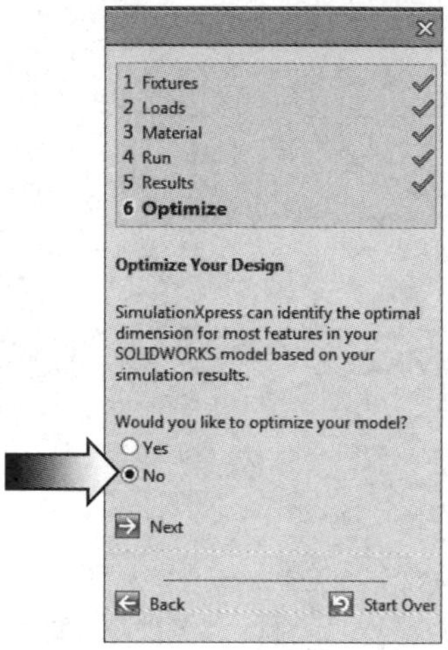

11. Saving your work:

Click **File / Save As**.

Enter **Spanner Study** for file name and click **Save**.

Isotropic, Orthotropic & Anisotropic Materials:

Isotropic Material: If its mechanical properties are the same in all directions. The elastic properties of an Isotropic material are defined by the Modulus of Elasticity (EX) and Poisson's Ratio (NUXY).

Orthotropic Material: If its mechanical properties are unique and independent in the directions of three mutually perpendicular axes.

Anisotropic Material: If its mechanical properties are different in different directions. In general, the Mechanical properties of the anisotropic materials are not symmetrical with respect to any plane or axis.

SimulationXpress supports Isotropic materials only.

Questions for Review

SimulationXpress

1. SimulationXpress can be accessed from the Tools pull down menu.
   a. True
   b. False

2. System Of Units SI (Joules) is the only type that is supported in SimulationXpress.
   a. True
   b. False

3. The material of the part can be selected from the built-in library or input directly by the user.
   a. True
   b. False

4. SimulationXpress supports Isotropic, Orthotropic, and Anisotropic materials.
   a. True
   b. False

5. Restraints/Fixture are used to anchor certain areas of the part so that it will not move during the analysis.
   a. True
   b. False

6. Only one surface/face should be used for restraint in each study.
   a. True
   b. False

7. The elements size (mesh) can be adjusted to a smaller value for more accurate results.
   a. True
   b. False

8. The types of results reported are:
   a. Stress Distribution
   b. Deformed Shape
   c. Deformation
   d. Factor of Safety
   e. All of the above

1. TRUE 2. FALSE
3. TRUE 4. FALSE
5. TRUE 6. FALSE
7. TRUE 8. E

Exercise 1: SimulationXpress: Force

1. Open the existing part: **Extrude Boss & Extrude Cut.**

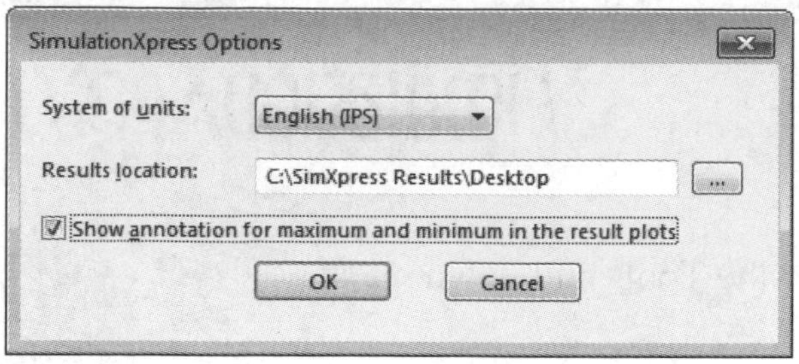

2. Set Unit to: **English (IPS)**.

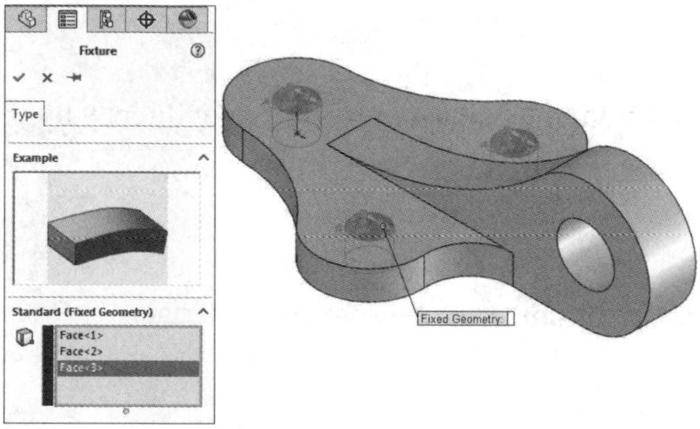

3. Apply Restraint: **to 3 Holes.**

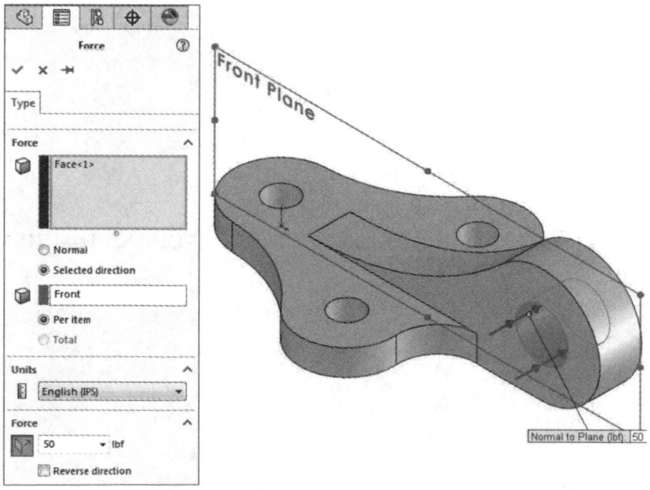

4. Apply Force of **50 Lbs** to the side hole, normal to the **Front** Plane.

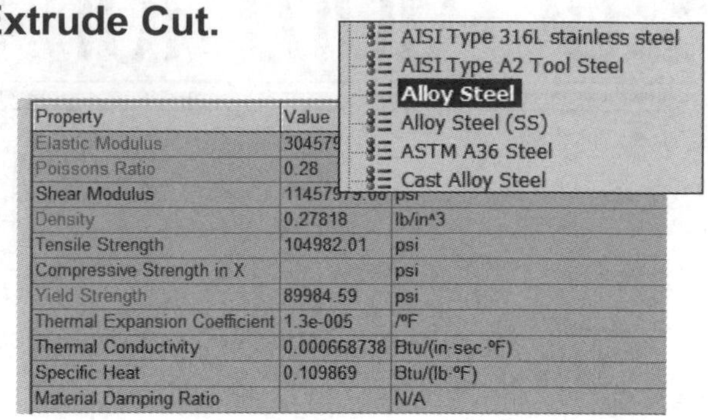

5. Select Material: **Alloy Steel**.

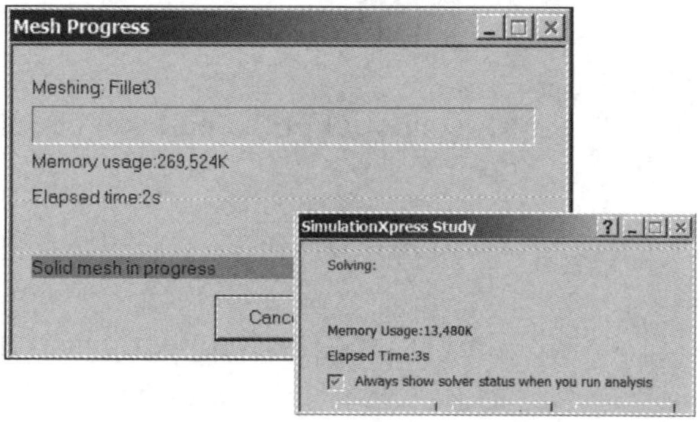

6. Run the Analysis.

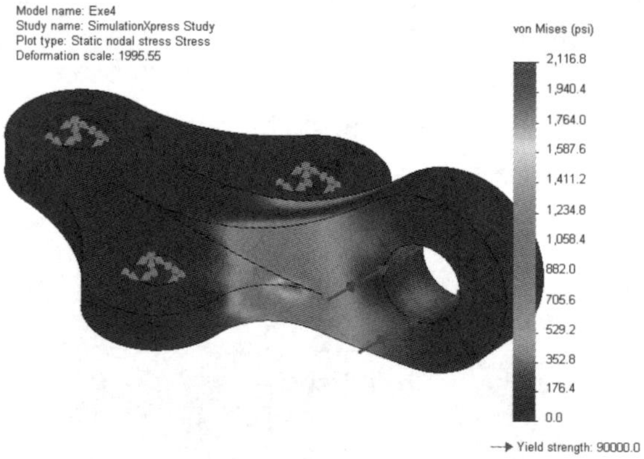

7. Check the von Mises Stress results.

8. Save a copy as: **Simulation_Force**.

Exercise 2: SimulationXpress: Pressure

1. Open the existing part: **Bottle_SimulationXpress**.

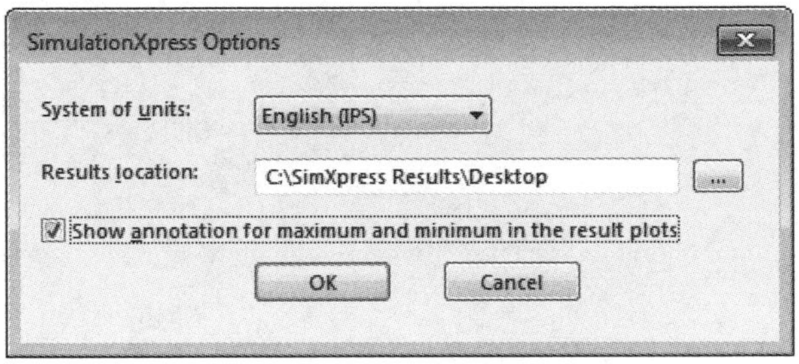

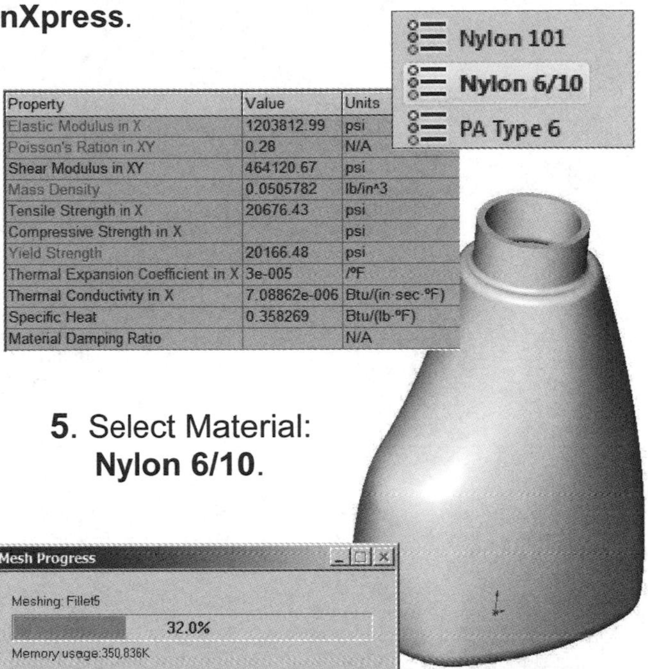

Property	Value	Units
Elastic Modulus in X	1203812.99	psi
Poisson's Ration in XY	0.28	N/A
Shear Modulus in XY	464120.67	psi
Mass Density	0.0505782	lb/in^3
Tensile Strength in X	20676.43	psi
Compressive Strength in X		psi
Yield Strength	20166.48	psi
Thermal Expansion Coefficient in X	3e-005	/°F
Thermal Conductivity in X	7.08862e-006	Btu/(in·sec·°F)
Specific Heat	0.358269	Btu/(lb·°F)
Material Damping Ratio		N/A

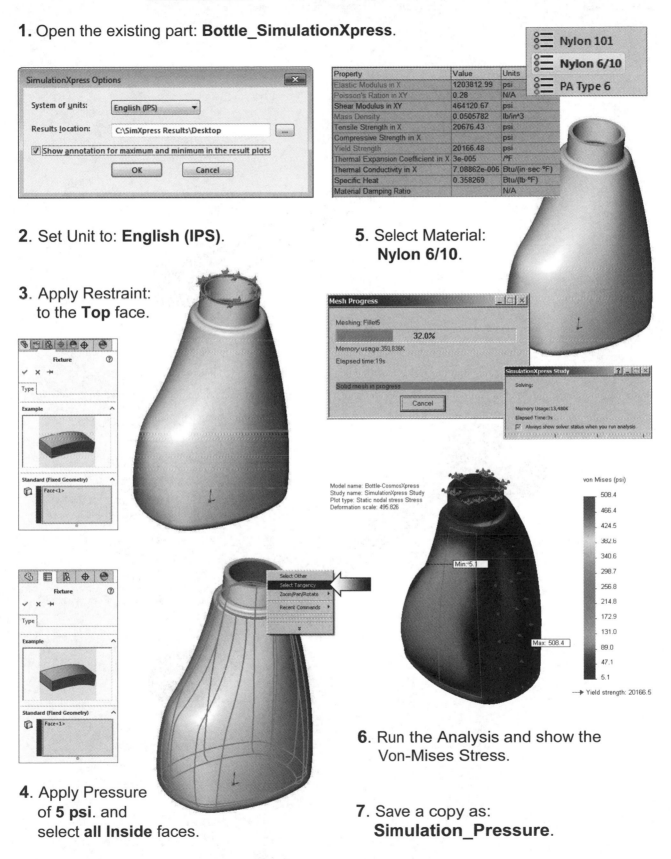

2. Set Unit to: **English (IPS)**.

5. Select Material: **Nylon 6/10**.

3. Apply Restraint: to the **Top** face.

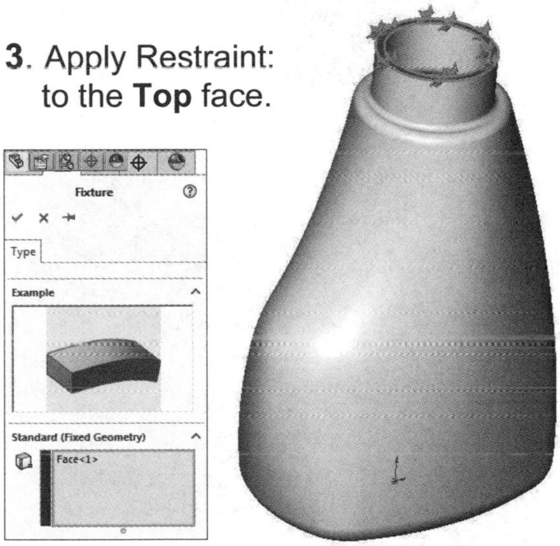

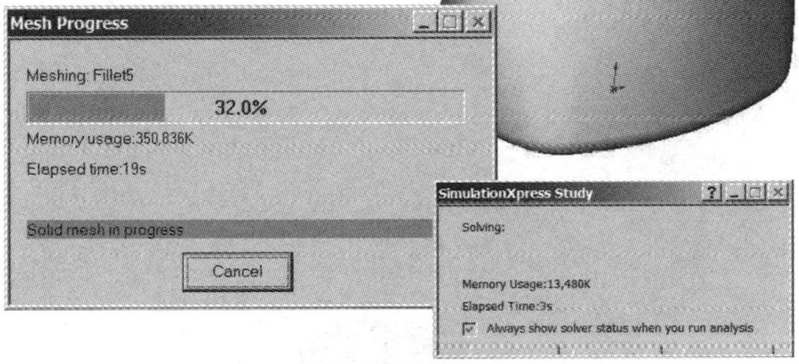

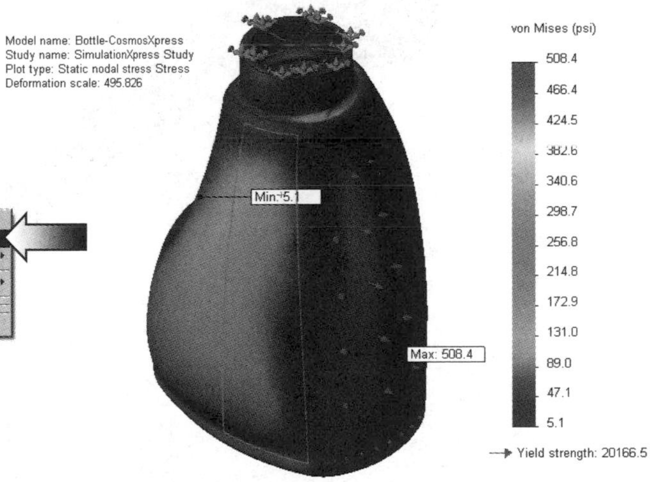

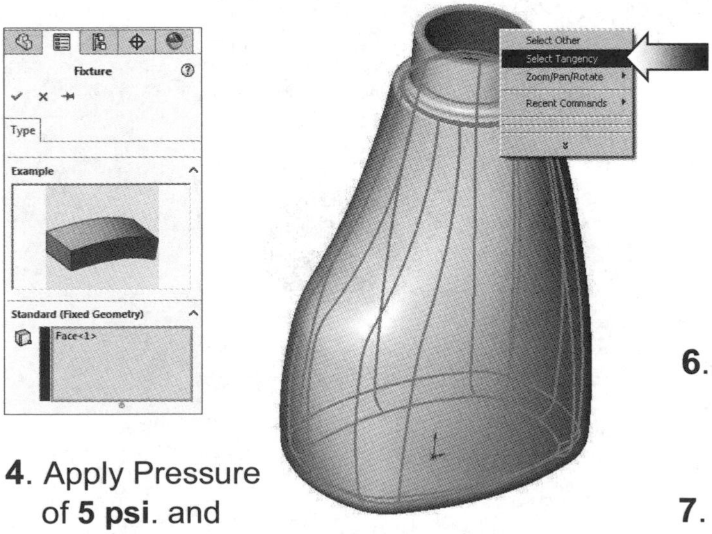

4. Apply Pressure of **5 psi**. and select **all Inside** faces.

6. Run the Analysis and show the Von-Mises Stress.

7. Save a copy as: **Simulation_Pressure**.

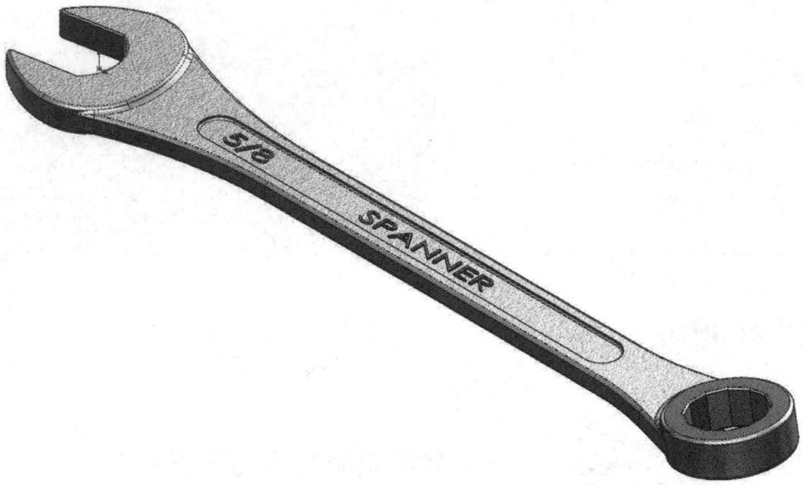
5/8
SPANNER

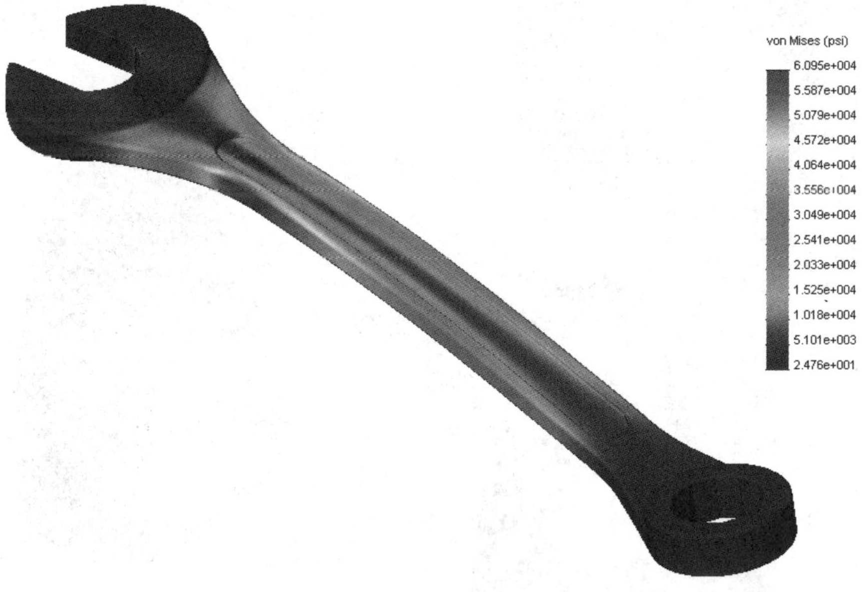
von Mises (psi)
6.095e+004
5.587e+004
5.079e+004
4.572e+004
4.064e+004
3.556e+004
3.049e+004
2.541e+004
2.033e+004
1.525e+004
1.018e+004
5.101e+003
2.476e+001

CHAPTER 13

Sheet Metal Parts

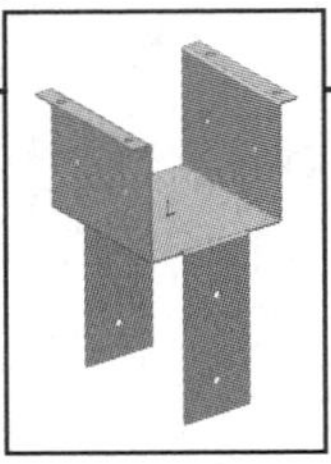

Sheet Metal Parts
Post Cap

This chapter discusses the introduction to designing sheet metal parts.

Create a sheet metal part in the folded stage and add the sheet metal specific flange features such as:

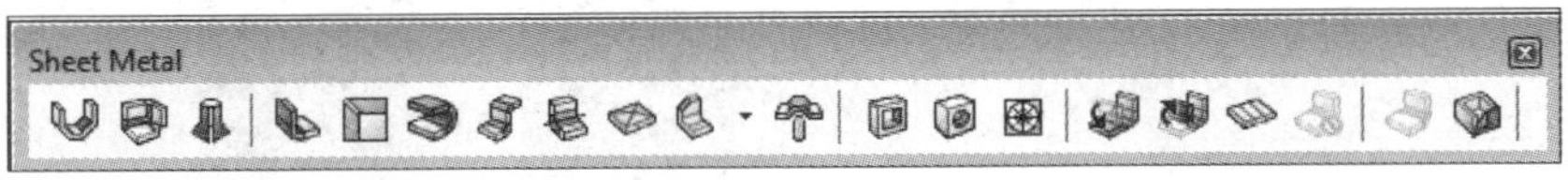

* Base Flange.
* Edge Flanges.
* Sketch Bends.
* Cut with Link to Thickness.
* Normal Cuts.

There are several options for specifying the setback allowance, or the length difference between the fold and the flat patterns of a sheet metal part.

* **Bend Table:** You can specify the bend allowance or bend deduction values for a sheet metal part in a bend table. The bend table also contains values for bend radius, bend angle, and part thickness.

* **K-Factor:** Is a ratio that represents the location of the neutral sheet with respect to the thickness of the sheet metal part.
  Bend allowance using a K-Factor is calculated as follows:
  $BA=\Pi(R + KT)\ A/180$

* **Use Bend Allowance:** Enter your own bend value based on your shop experience.

* **Bend allowance Calculations:** The following equation is used to determine the total flat length when bend allowance values are used: $L_t = A + B + BA$

* **Bend Deduction Calculations:** The following equation is used to determine the total flat length when bend deduction values are used: $L_t = A + B - BD$

Post Cap
Sheet Metal Parts

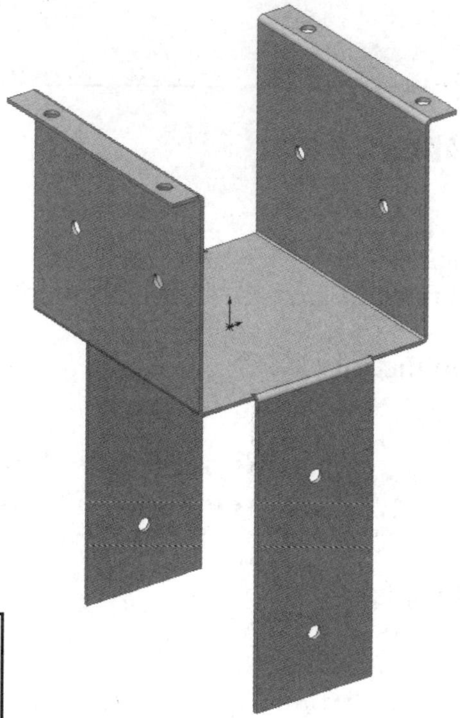

View Orientation Hot Keys:

Ctrl + 1 = Front View
Ctrl + 2 = Back View
Ctrl + 3 = Left View
Ctrl + 4 = Right View
Ctrl + 5 = Top View
Ctrl + 6 = Bottom View
Ctrl + 7 = Isometric View
Ctrl + 8 = Normal To Selection

Dimensioning Standards: **ANSI**

Units: **INCHES** – 3 Decimals

Tools Needed:

Insert Sketch	Line	Circle
Dimension	Add Geometric Relations	Base Flange
Edge Flange	Sketch Bend	Flat Pattern

1. Starting with the base profile:

Select Right plane from FeatureManager tree and insert a **new sketch**.

Sketch the profile below using the **Line** tool.

Add dimensions and relations as indicated to fully define the sketch.

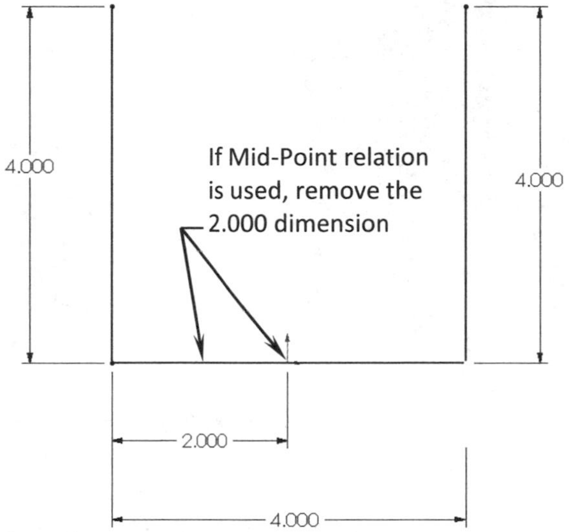

> **Gauge Tables**
>
> To enable the Gauge Tables while in the Base Flange mode, click: **Browse / Installed Dir. / SolidWorks / Lang / English / Sheet Metal Gauge Tables.**

2. Extruding the Base Flange:

Click or select **Insert / Sheet Metal / Base Flange**.

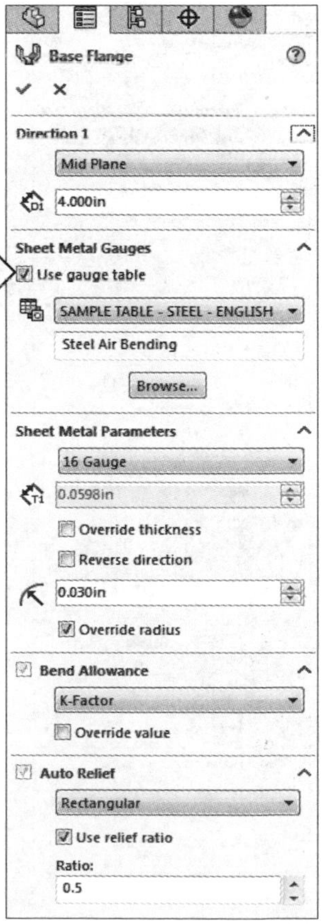

Direction 1: **Mid-Plane**.

Extrude Depth: **4.00 in**.

Use Gauge Table: **Sample Table - Steel**.

Mat'l Thickness: **16 Gauge (.0598)**.

Bend Radius: **.030 in**. (Override Radius).

Click **OK**.

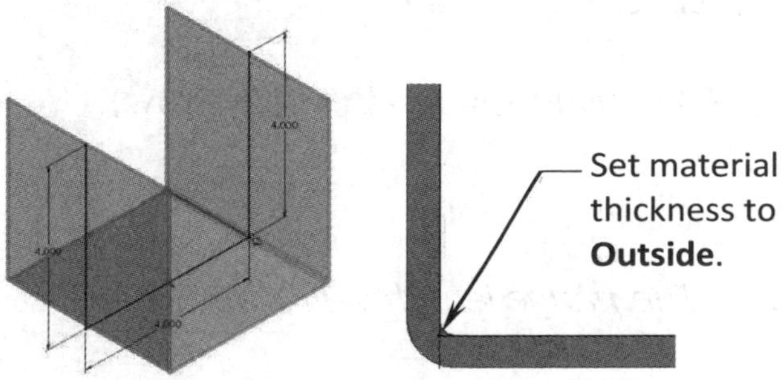

3. Creating an Edge Flange:

Hold the **Control** key and select the **2 edges** as indicated.

Click or select **Insert / Sheet Metal / Edge Flange**.

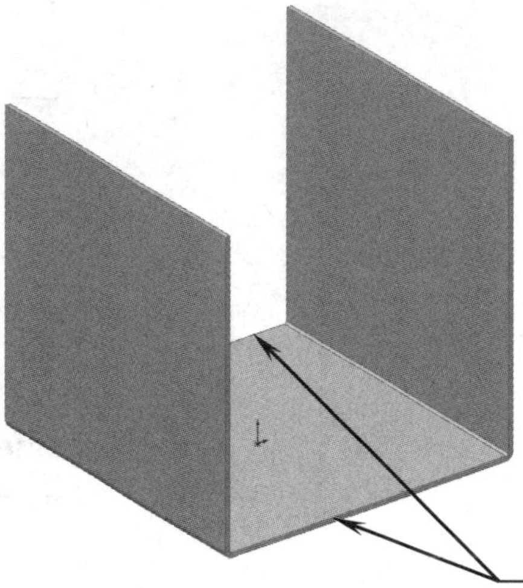

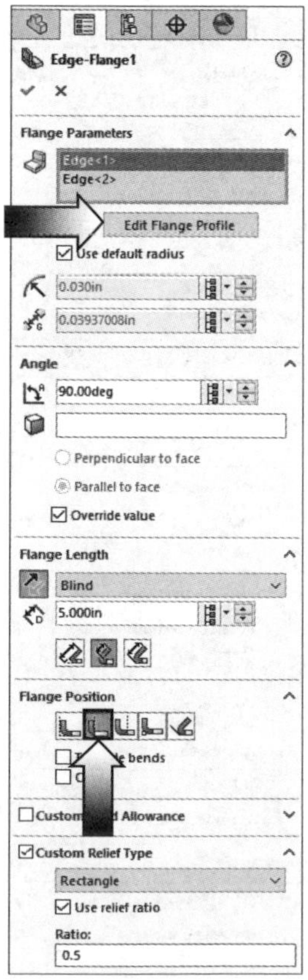

Drag the cursor downwards and click anywhere to lock the direction of the 2 Edge Flanges.

Select **Material Outside** under the Flange Position.

Enter **90°** for Angle (default).

Flange Length: **Blind**.

Enter **5.000"** for depth.

Relief Type: **Rectangle**.

Relief Ratio: **.500**.

Click **Edit Flange Profile** (arrow).

** The flange length will be modified in the next steps.*

4. Editing the Edge Flange Profile:

Move the Profile Sketch dialog out of the way.

Drag the 2 outer lines inward as noted.

Add the dimensions shown below to fully define the sketch.

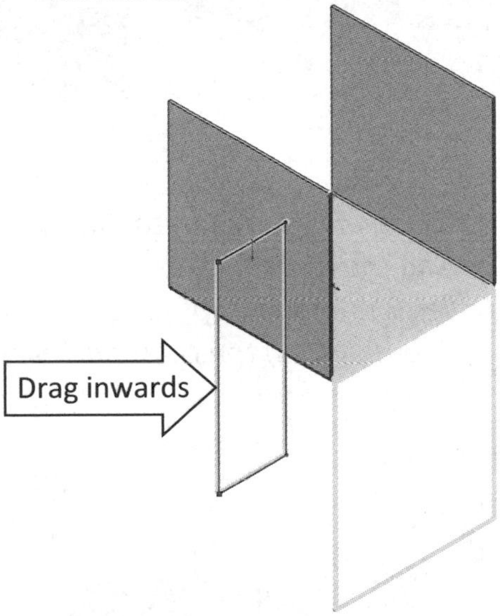

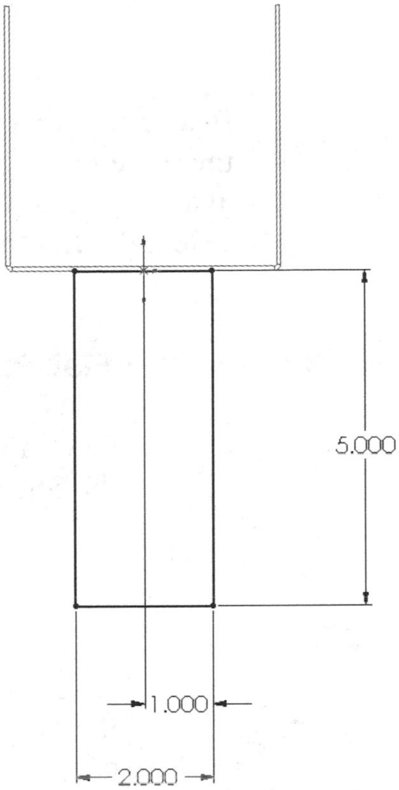

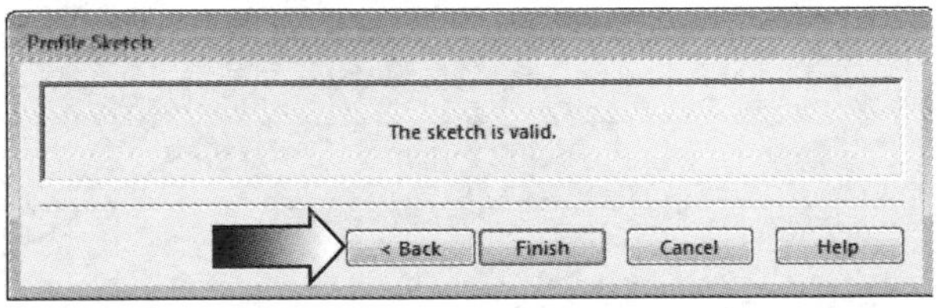

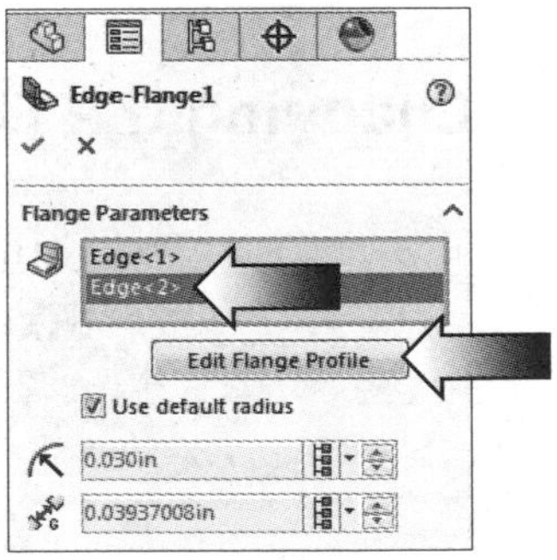

When the 1st edge flange is fully defined, click the **Back** button (arrow) to switch back to the previous screen.

Select the **Edge 2** under the Flange Parameters section and click the **Edit Flange Profile** once again (arrow).

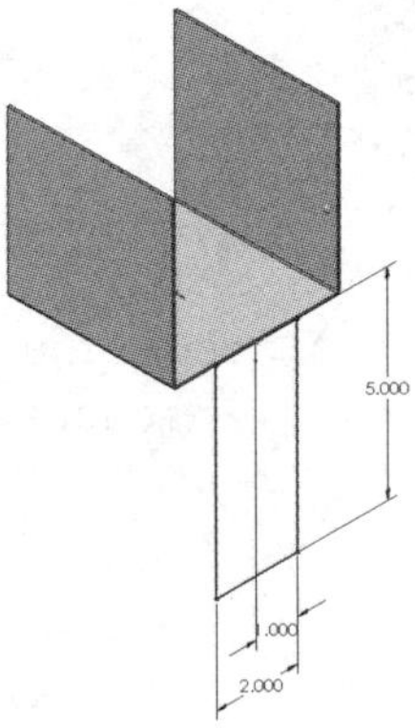

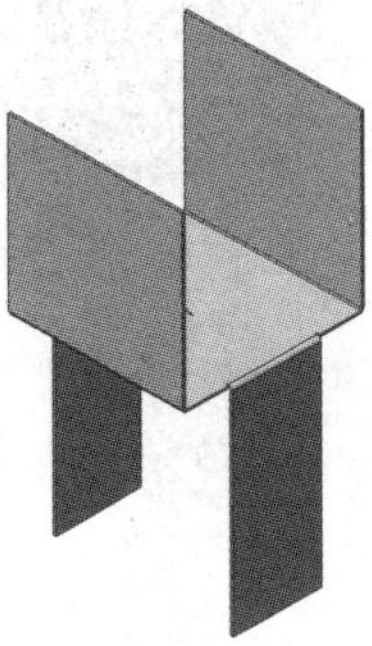

Drag the 2 vertical lines inward similar to the step above, and add the same dimensions as the 1st flange to ensure the 2 flanges are exactly the same size (Collinear relations can also be used for the width and height of the 2nd flange).

Click finished Finish after the 2nd sketch is fully defined.

Zoom in on the upper end of the edge flange to see the relief details.

Beside the Rectangular relief, there are two other types available: Obround and Tear. Select the **Rectangular** type.

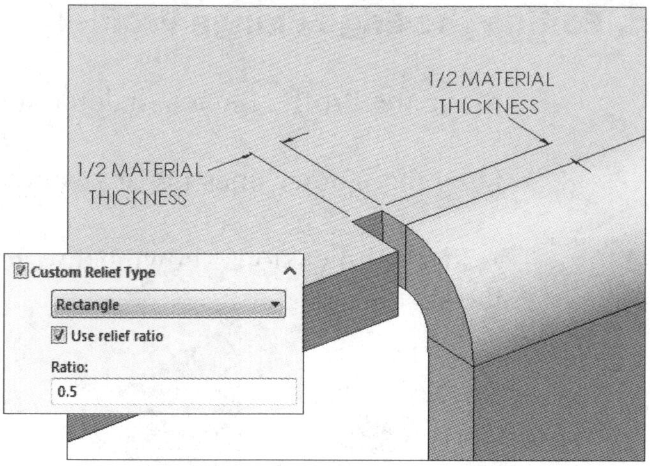

Auto-Relief is added automatically after exiting the Sketch.
Relief Width and Depth are defaulted to one-half (0.5) the Material Thickness.

5. Viewing the Flat Pattern:

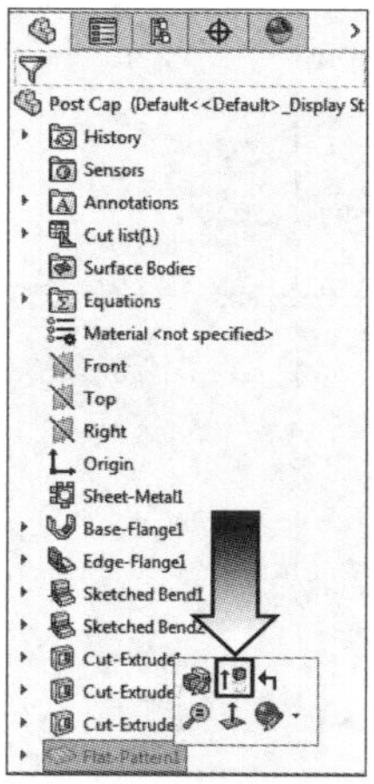

Click **Flatten**  on the Sheet Metal toolbar:

The flat pattern of the part is displayed.

An alternative to viewing the flat pattern is to right click on the Flat-Pattern1 feature and select **Suppress** (arrow).

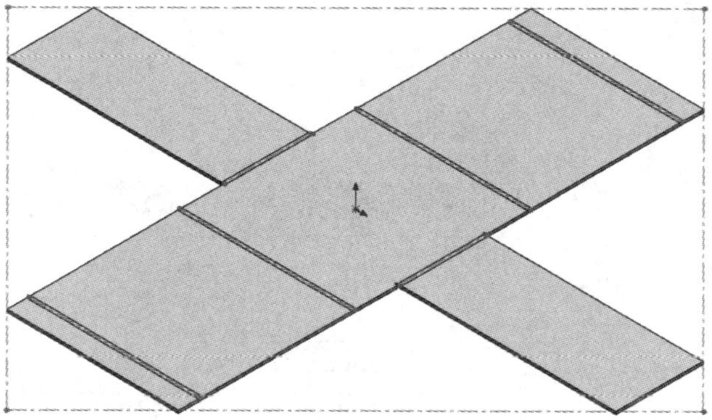

6. Changing the Fixed face:

Edit the **Flat-Pattern** feature and select the <u>face</u> indicated to use as the **Fixed** face.

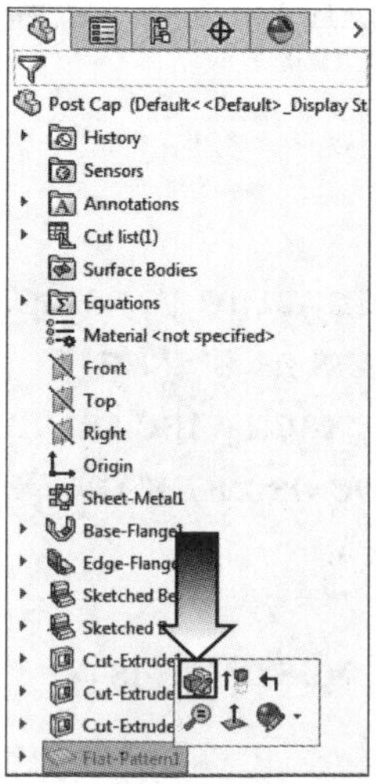

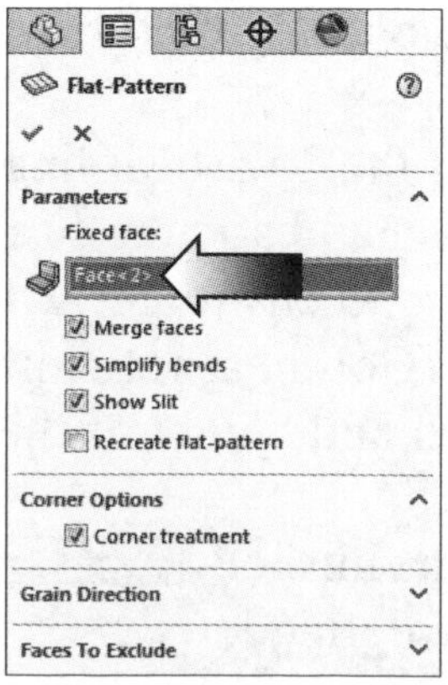

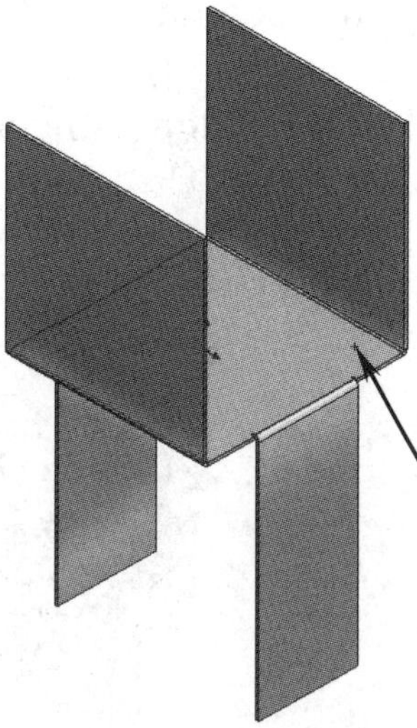

Select the Fixed face

7. Creating a Sketch Bend:

Select the side face and open a **new Sketch**.

Sketch a **Line** starting at one edge and Coincident with the other.

Add dimensions to fully define the sketch.

Sketch Bends

This command adds Bends or Tabs to the Sheet metal part with the sketch lines.

Only sketch lines are allowed, but more than one line can be used in a same sketch to create multiple bends.

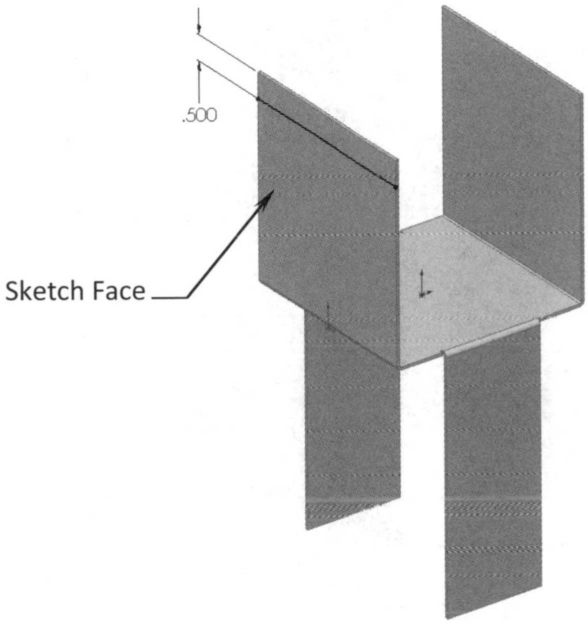

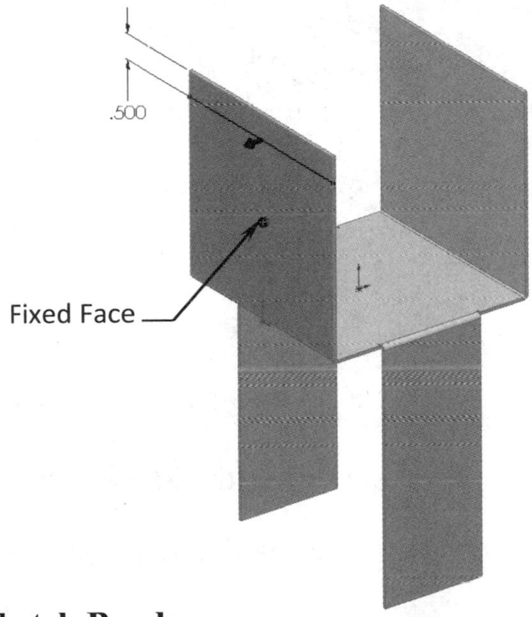

Click or select **Insert / Sheet Metal / Sketch Bends**.

For Fixed Face select the **lower portion** of the surface.

For Bend Position select **Bend Centerline** (default).

For Bend Angle enter **90.00** deg.

Enable **Use Default Radius**.

Click **OK**.

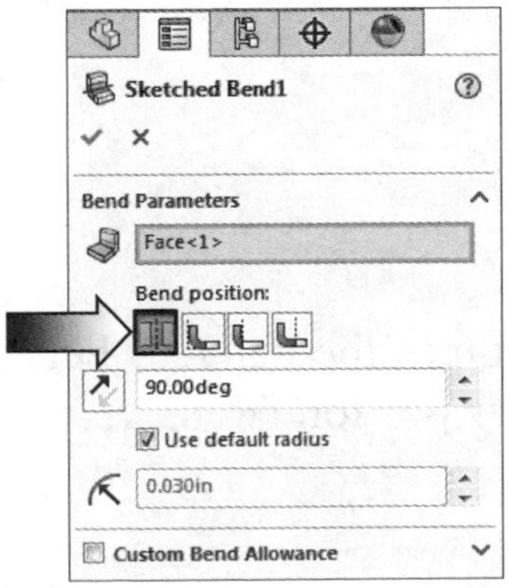

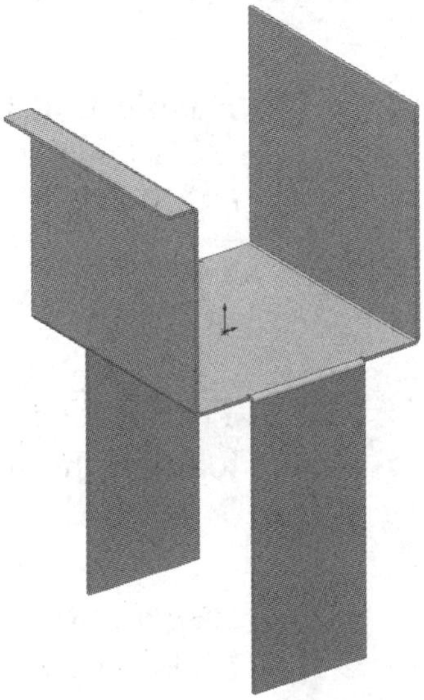

8. Creating the 2nd Sketch Bend:

Select the <u>face</u> on the right and open a **new sketch** or select **Insert / Sketch**.

Sketch a **Line** as shown and add the spacing dimension to fully define the sketch.

Click or select **Insert / Sheet Metal / Sketch Bend**.

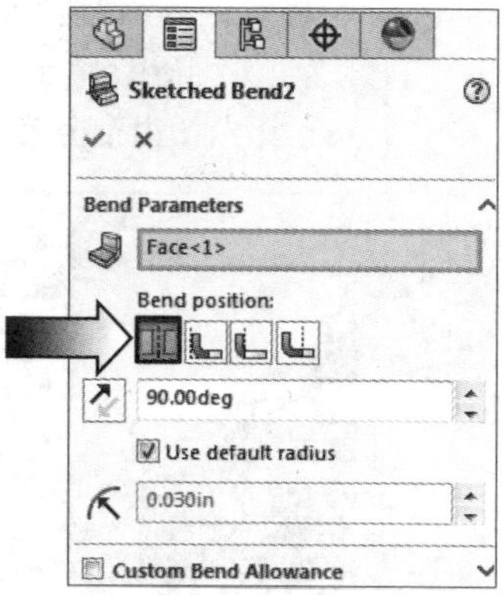

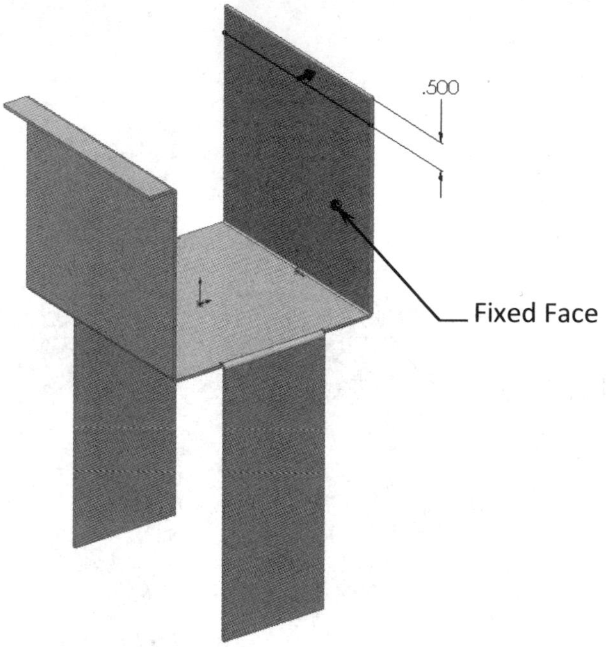

Select the **lower portion** of the surface as the Fixed Face.

For Bend Position, select **Bend Centerline** (default).

For Bend Angle, enter **90.00** deg (default).

Enable **Use Default Radius**.

Click **OK**.

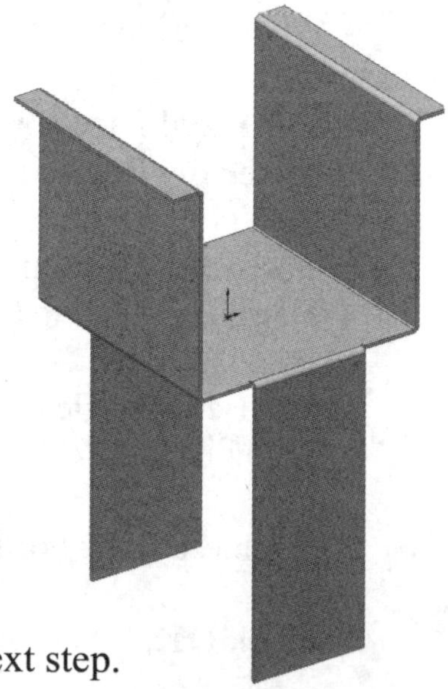

The upper portion of the flange is bent outward 90°, leaving the lower portion fixed.

Inspect your model against the image shown here.
Make any adjustments needed before going to the next step.

9. Adding holes:

Open a **new sketch** on the side face as indicated  or select: **Insert / Sketch**.

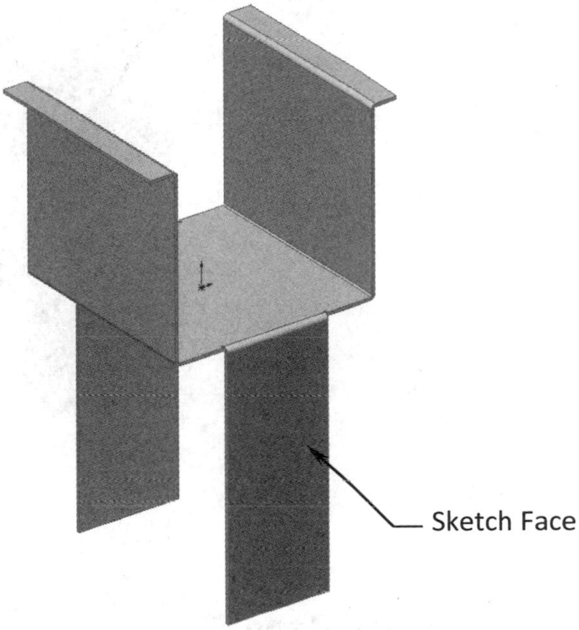

Sketch two circles  on the face.

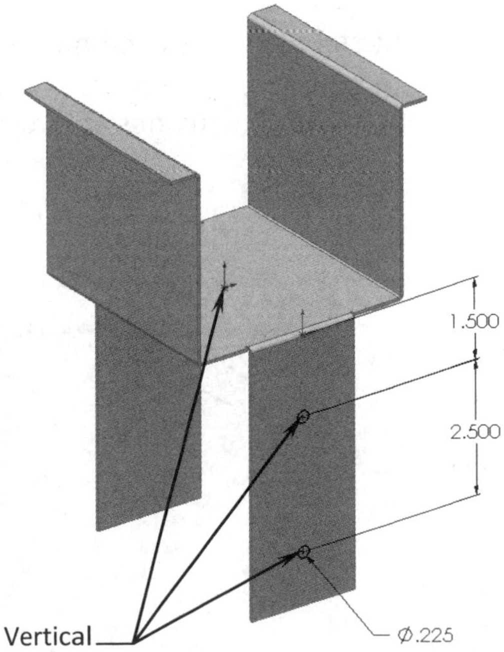

Add dimensions as shown to size and to locate the circles.

Add **Vertical** relations between the centers of the circles and the Origin, to fully define the sketch.

Switch to the **Sheet Metal** tool tab.

Click or select **Insert / Cut / Extrude**.

End Condition: **Through All**.

Enable **Normal Cut** (default).

Click **OK**.

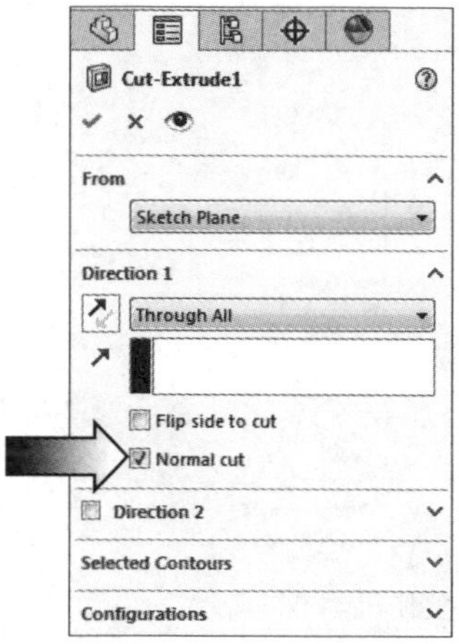

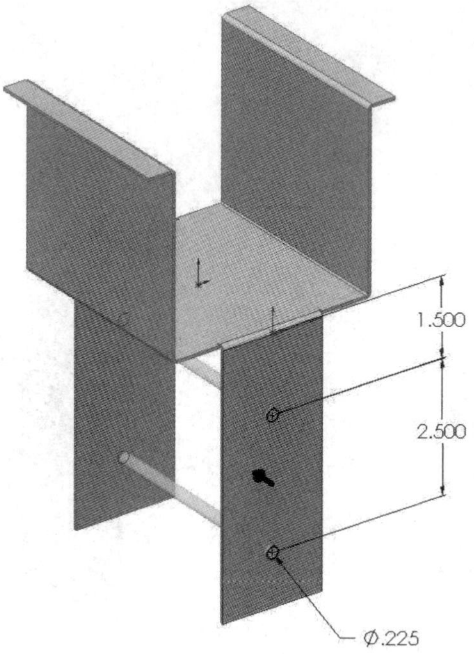

10. Adding holes on the Sketch Bend Flanges:

Select the face as noted and open a **new sketch** or select **Insert / Sketch**.

Sketch **2 Circles** and mirror them as noted.

Add the dimensions and relations needed to fully define the sketch.

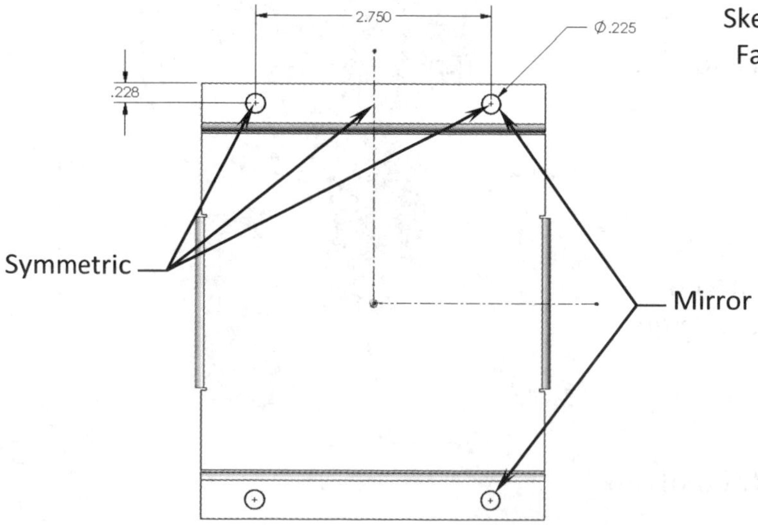

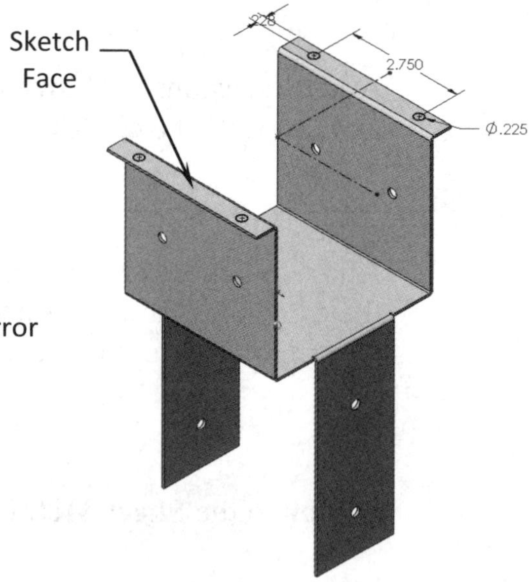

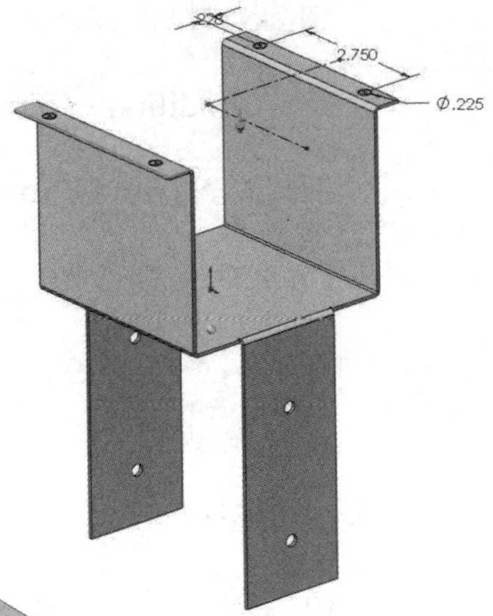

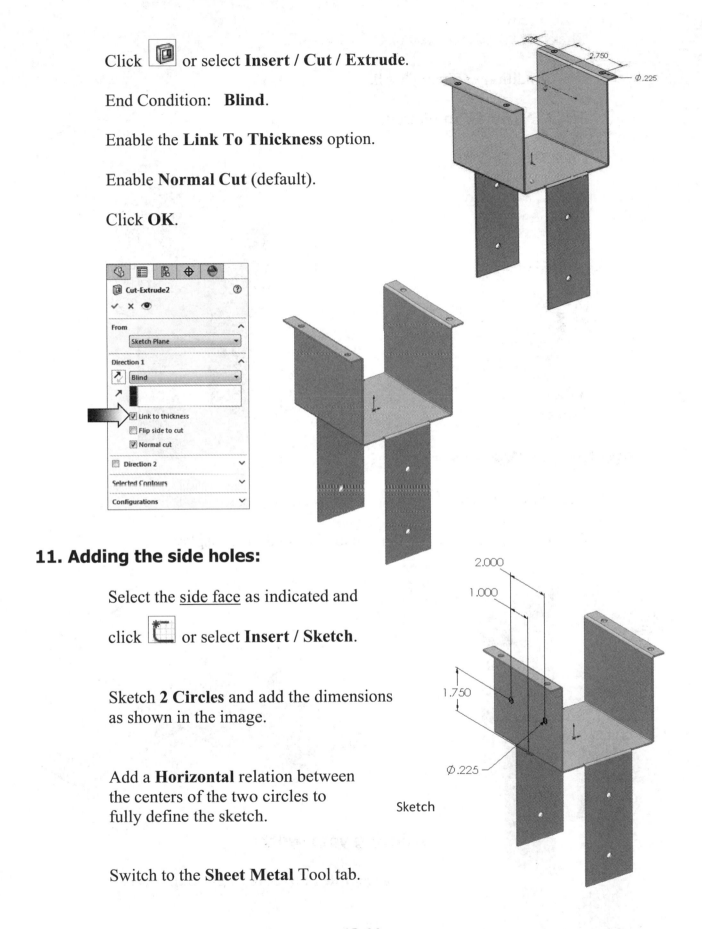

Click  or select **Insert / Cut / Extrude**.

End Condition: **Blind**.

Enable the **Link To Thickness** option.

Enable **Normal Cut** (default).

Click **OK**.

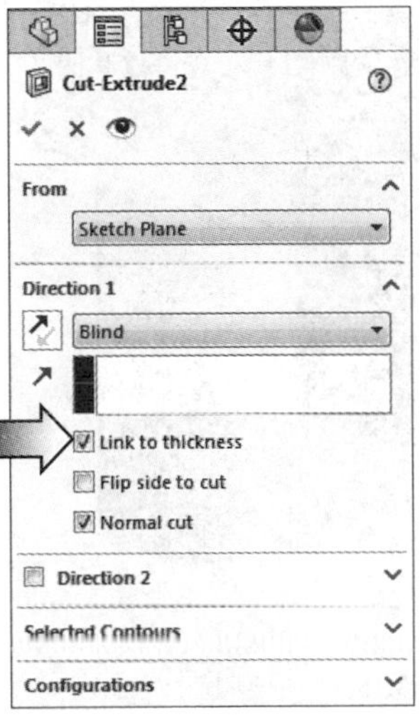

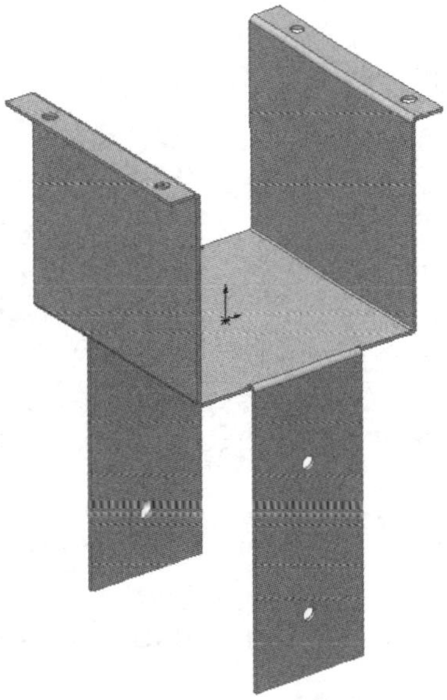

11. Adding the side holes:

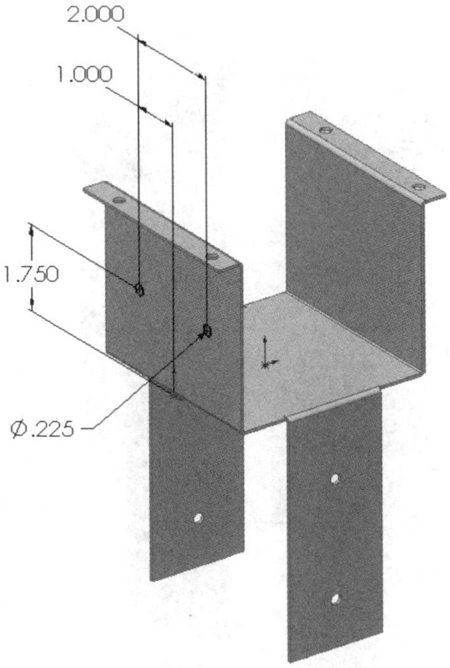

Select the side face as indicated and

click  or select **Insert / Sketch**.

Sketch **2 Circles** and add the dimensions as shown in the image.

Add a **Horizontal** relation between the centers of the two circles to fully define the sketch.

Sketch

Switch to the **Sheet Metal** Tool tab.

Click  or select **Insert / Cut / Extrude**.

End Condition: **Through All**.

Enable **Normal Cut** (default).

Click **OK**.

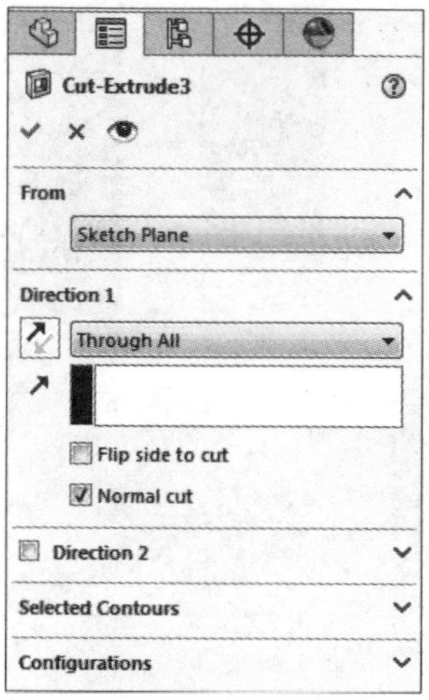

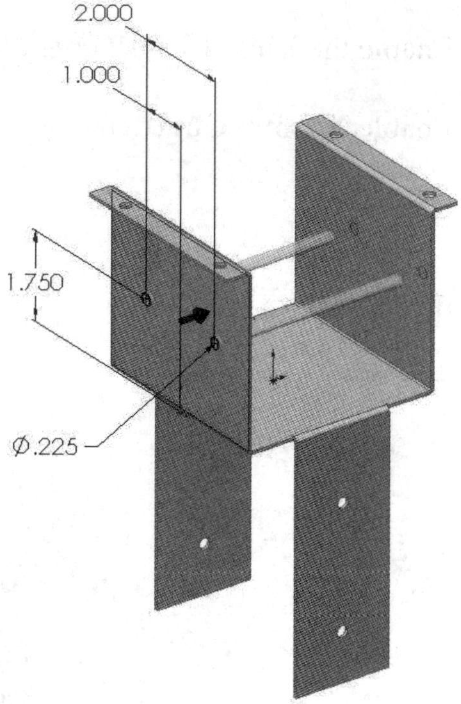

12. Making the Flat Pattern:

Click **Flat Pattern**  on the Sheet Metal toolbar.

The sheet metal part is flattened with the bend lines displayed.

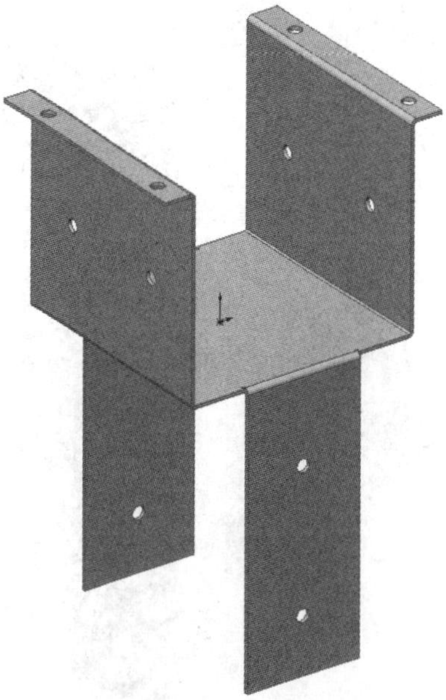

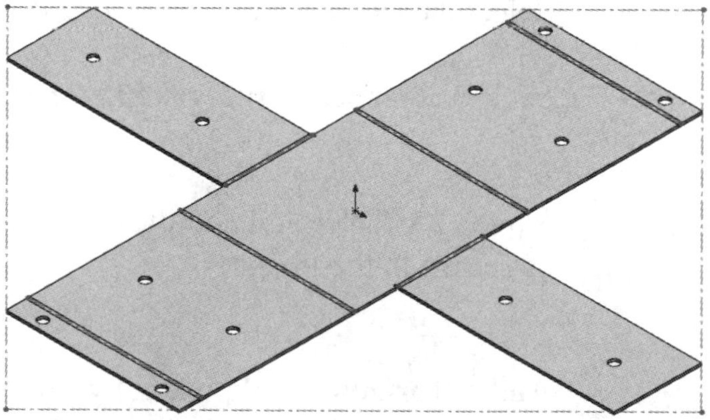

13. Saving your work:

Select **File / Save As / Post Cap / Save**.

Using Sheet Metal Costing

Use SOLIDWORKS sheet metal Costing tools to determine the cost of a sheet metal part. Costing provides a comprehensive breakdown and comparison of manufacturing and material costs for sheet metal parts.

Costing Template Editor for Sheet Metal Parts

You can create and edit costing templates for sheet metal parts or bodies from the Costing Template Editor.
You can specify rates and costs for the procedures required to manufacture a sheet metal part or body in the sheet metal template. You can include customized information in the template, such as material cost and thicknesses, cost of manufacturing, and manufacturing setup costs.
You can determine how manufacturing operations affect the cost of your design. For example, you can set up templates for vendors that use different manufacturing operations.

You can specify the file location for Costing templates in Tools > Options > System Options > File Locations. In Show folders for, select Costing templates to add or delete a location. The default Costing template folder is ***installation directory/lang/language/ Costing templates***.

1. Opening an existing sheet metal part:

Click **File / Open**.

Browse to the Training Files folder, locate and open the **Sheet Metal Costing** part document.

Click **NO** on the Feature-Works dialog box to close it.

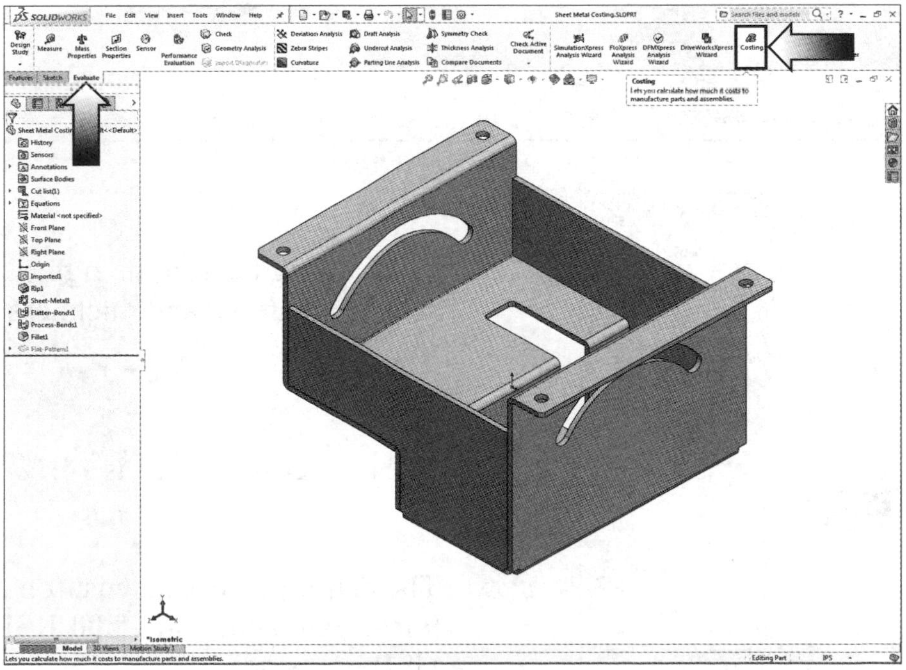

2. Inputting the information:

Switch to the **Evaluate** tab and click **Costing** .

Click the **CostingManager** tab (arrow) on the left side tree to see how the Costing tool categorizes each operation required for manufacturing the part.

Set the **Material** to **Steel AISI 304** and the **Thickness** to **1.90mm** (.075in). The sheet metal part is evaluated based on the selected material and processing costs as set in the default template.

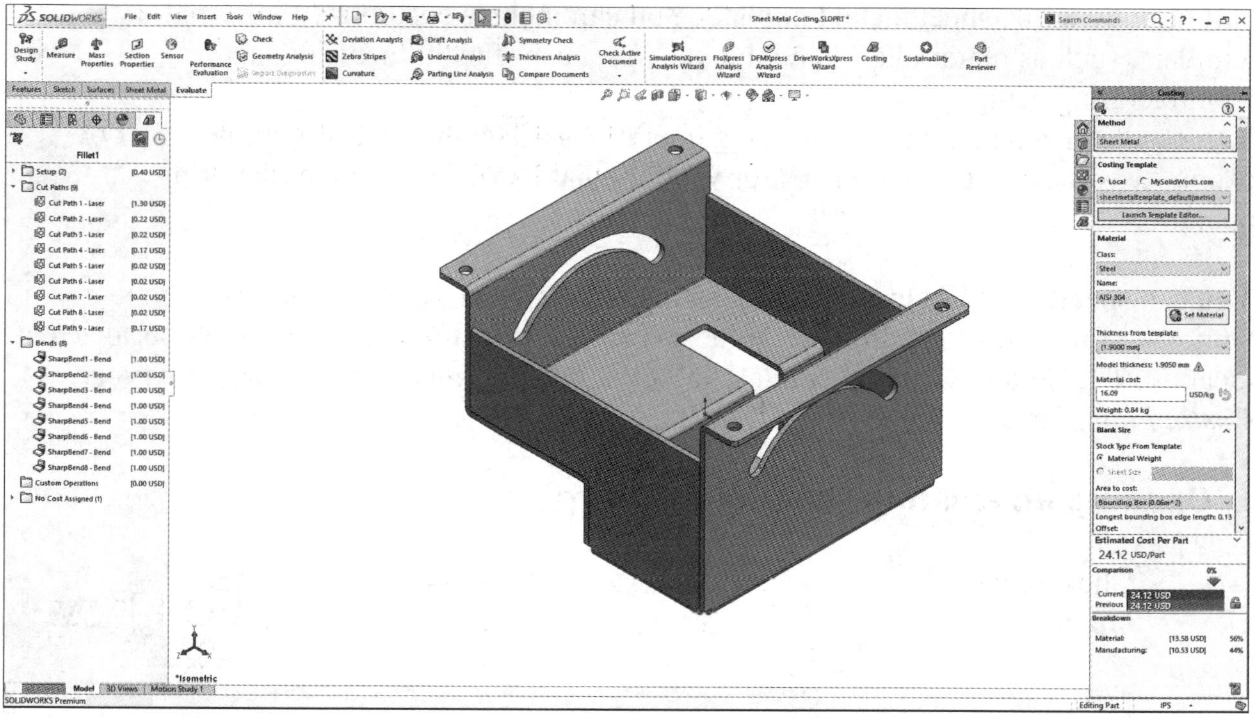

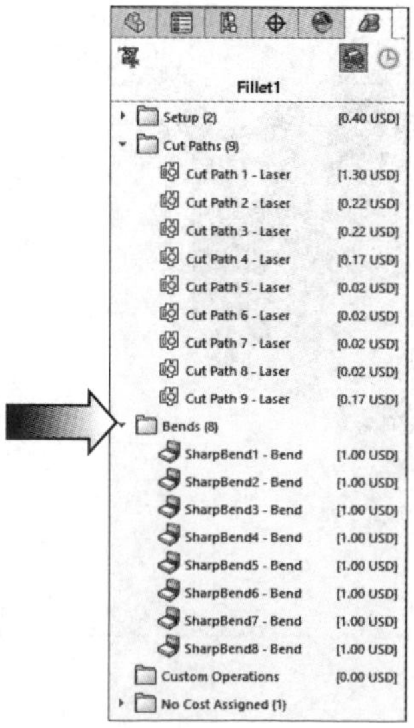

The Costing tab (on the right) indicates the cost for the selected material and thickness is roughly **$16.09 USD per kg**.

The total cost per part is **24.12 USD**.

The major processing cost (on the left) is **$2.16 USD** for 9 laser cut paths and **$8.00 USD** for eight bends ($1.00 USD per bend).

3. Setting the Baseline:

The **Set Baseline** is used to set a baseline cost for comparison. If you change the design later on, the cost is compared to the baseline cost. When you set a baseline cost, any changes to the part are considered Current and the difference is displayed. While the baseline price is set, the part is rotated, flattened, and refolded because the software is capturing images for the Costing report.

Click the **Set Baseline** button  at the lower right, under Comparison.

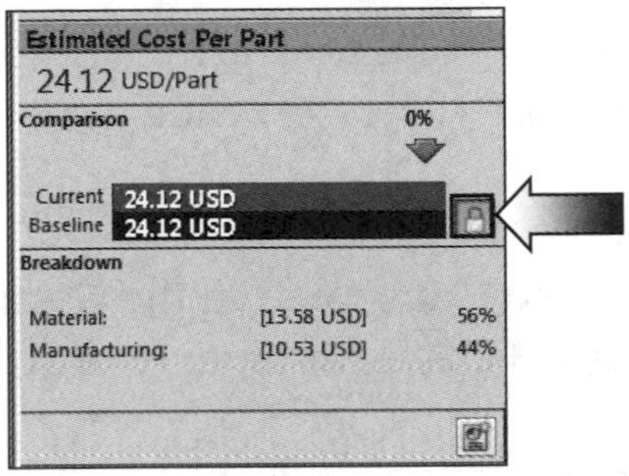

The final cost is based on the values for manufacturing steps.
The values that we are using here are for samples only.

Change the material to:
Aluminum Alloy 6061

Use the same thickness (1.90mm or .075in).

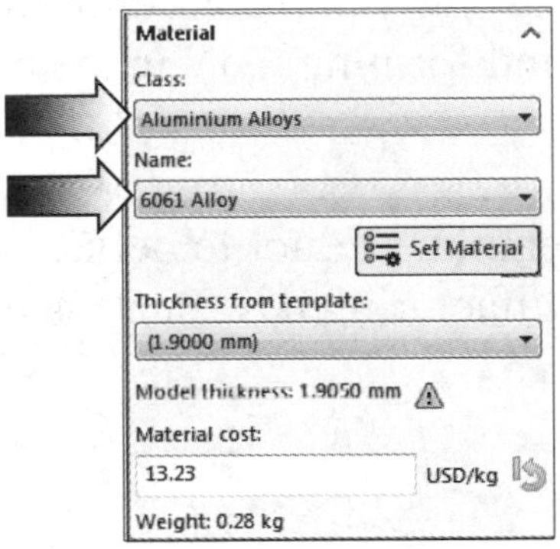

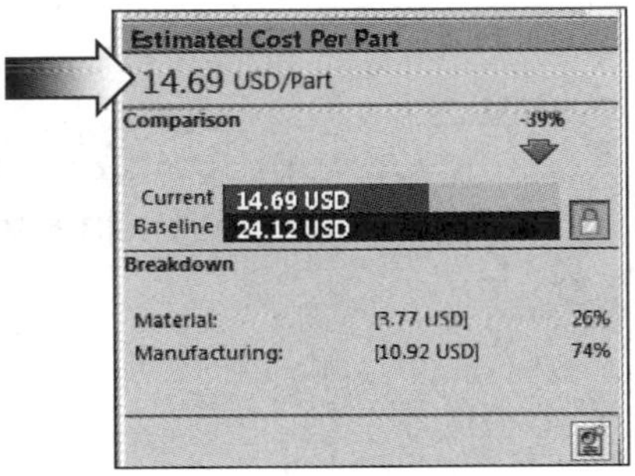

SOLIDWORKS Costing recalculates the cost based on the change in material.

The recalculated cost is now almost half of what it was for Steel ($14.69 instead of $24.12).

4. Saving your work:

Click **File / Save As**.

Enter **Sheet Metal Costing** for the name of the document and click **Save.**

- **General**
  Use the General screen in the Costing Template Editor to set the units and currency options.

- **Material**
  Use the Material screen in the Costing Template Editor to set the materials you need to manufacture the sheet metal part.

- **Thickness**
  Use the Thickness screen in the Costing Template Editor to set the thickness and cost values for each class and material combination.

- **Cut**
  Use the Cut screen in the Costing Template Editor to define the cost of cutting methods based on length or stroke.

- **Bend**
  Use the Bend screen in the Costing Template Editor to define the cost of bending methods based on regular bends or hem bends.

- **Library Features**
  Use the Library Features screen in the Costing Template Editor to define the cost of library features, punch features, and forming tools in the part.

- **Custom**
  Use the Custom screen in the Costing Template Editor to define additional operations that contribute to a part's manufacturing cost, such as powder coating.

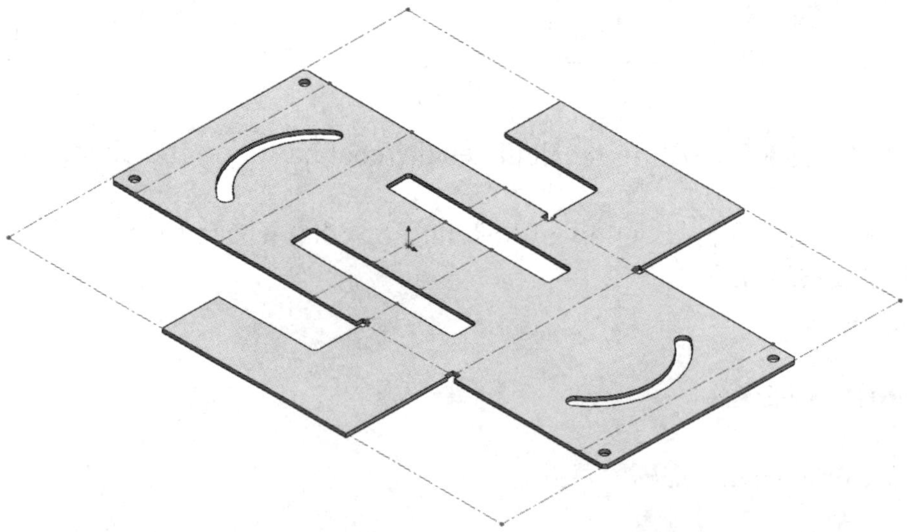

Questions for Review

Sheet Metal Parts

1. A sheet metal part can have multiple thicknesses.
    a. True
    b. False

2. An Edge Flange can be mirrored just like any other feature.
    a. True
    b. False

3. A sheet metal part can be created right from the beginning using the Base flange option.
    a. True
    b. False

4. Auto relief option is not available when extruding a Base Flange.
    a. True
    b. False

5. When the Sketched Bend option is used to create a bend, you will have to specify at least two parameters:
    a. A fixed side and a sketched line.
    b. Fixed side and a bend angle value.
    c. A bend radius and a bend angle value.

6. The only time when the K-Factor option can be changed to Bend Table is when extruding the Base Flange.
    a. True
    b. False

7. A sheet metal part can be designed from a flat sheet and other bends can be added later using Sketched Bend, Edge Flange, etc.
    a. True
    b. False

8. Link-to-Thickness option allows all sheet metal features in a part to have the same wall thickness and they can all be changed at the same time.
    a. True
    b. False

1. FLASE 2. TRUE
3. TRUE 4. FALSE
5. TRUE 6. FALSE
7. TRUE 8. TRUE

Sheet Metal Parts

Sheet Metal Parts

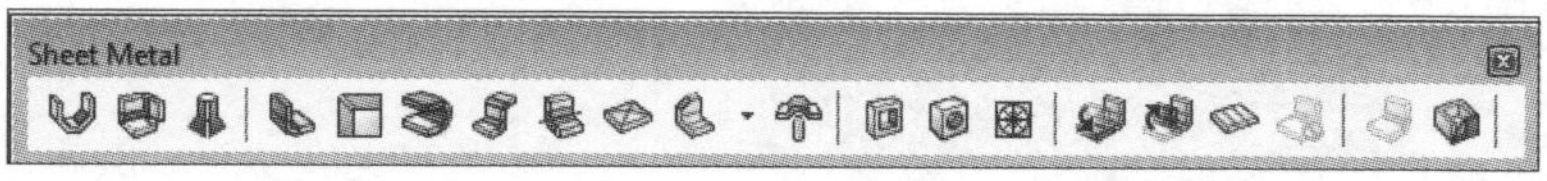

Sheet metal parts can be created using one of the following methods:

* Create the part as a solid and then insert the sheet metal parameters such as rips, bend radius, material thickness, bend allowance, and cut relief so that the part can be flattened.

* Create the part as a sheet metal part from the beginning by using the Base Flange command to extrude the first feature.

* Sheet metal parameters can be applied onto the sheet metal part during the extrusion or after the fact.

This chapter will guide you through the design development of a sheet metal part as well as the use of the sheet metal and forming tool commands to create a louver form tool:

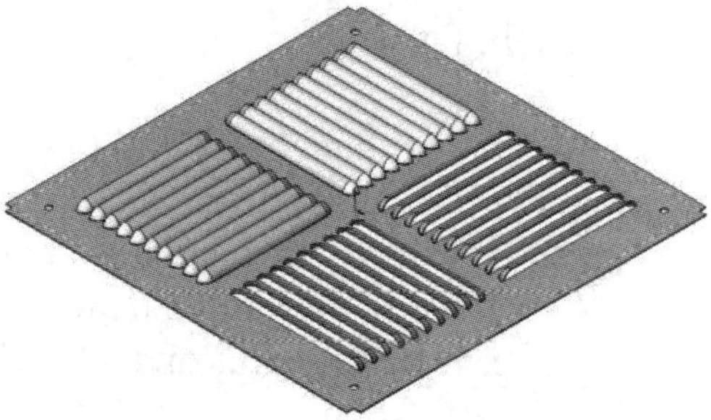

* Creating the parent feature with the Base Flange command
* Using the Miter Flange command
* Create a sheet metal part in the flat or folded stage
* Create revolved features
* Accessing the Design Library
* Using the forming tool to create the louvers
* Create a linear pattern of features
* Create a circular pattern of features
* Create a pattern of patterned features
* Flatten the sheet metal part

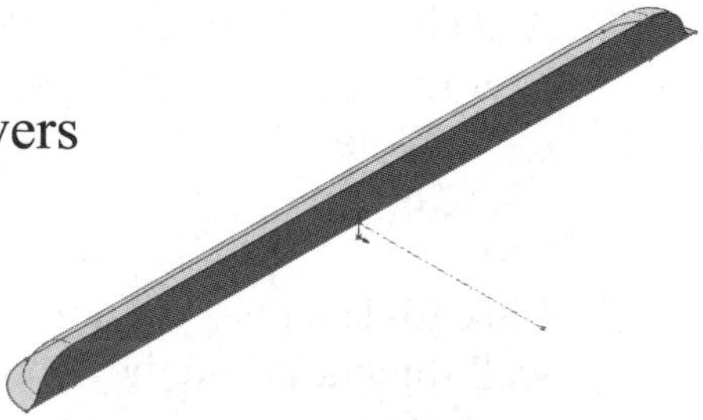

Vents
Sheet Metal Parts

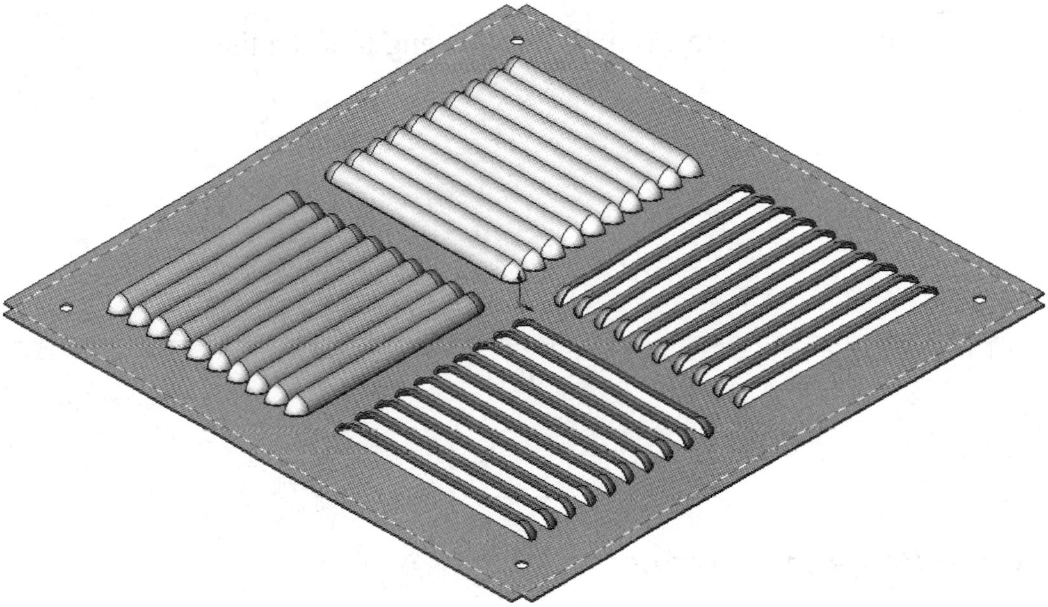

Dimensioning Standards: **ANSI**

Units: **INCHES** – 3 Decimals

Tools Needed:

Insert Sketch	Line	Dimension
Rectangle	Add Geometric Relations	Linear Pattern
Base Flange	Miter Flange	Circular Pattern
Flat Pattern	Extruded Cut	Design Library

1. Sketching the first profile:

Select <u>Front</u> plane from the FeatureManager tree.

Click on the Sketch toolbar or select **Insert / Sketch**.

Sketch a horizontal **Line** <u>below the Origin</u>.

Add the dimension **18.00in** for the length of the line.

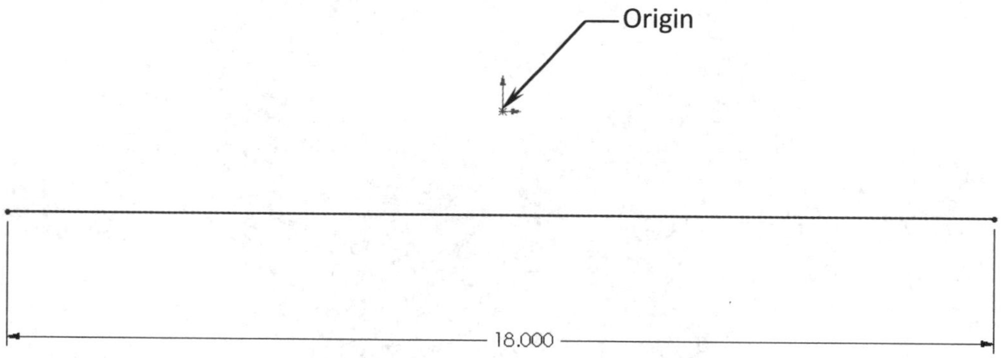

2. Adding a Midpoint relation:

Click from the Sketch toolbar OR select **Tools / Relations /Add**.

Hold the Control key, select the **origin point** and the **line** as noted.

Select **Midpoint** from the Add Relation dialog box.

Click **OK**.

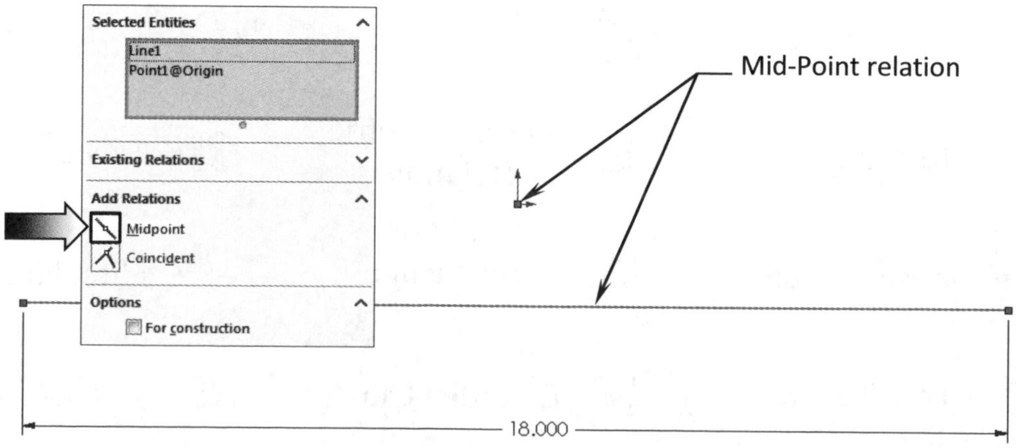

3. Extruding the Base-Flange:

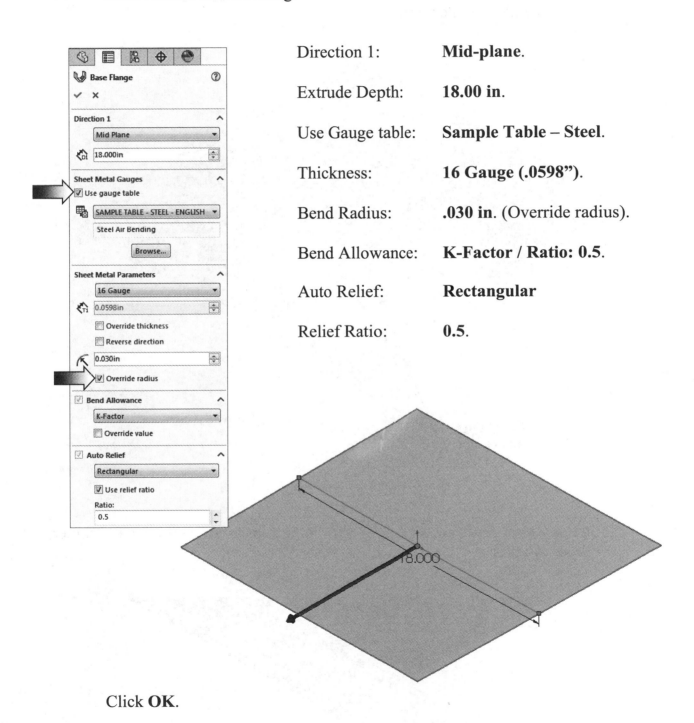

Click (Base Flange) from the Sheet Metal toolbar or select:
Insert /Sheet Metal / Base Flange.

Enter / select the following:

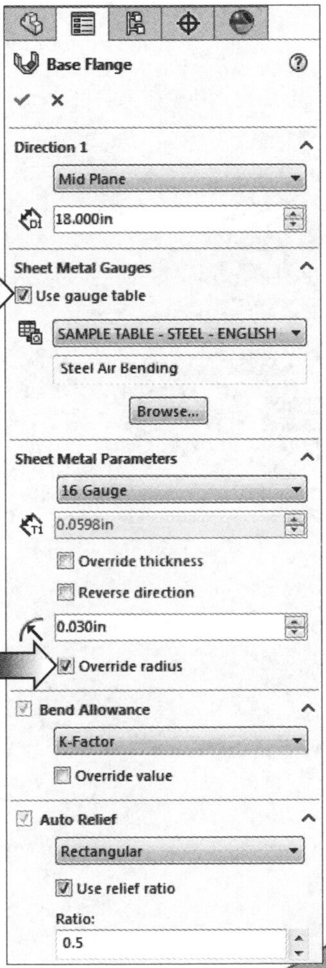

Direction 1:	**Mid-plane**.
Extrude Depth:	**18.00 in**.
Use Gauge table:	**Sample Table – Steel**.
Thickness:	**16 Gauge (.0598")**.
Bend Radius:	**.030 in**. (Override radius).
Bend Allowance:	**K-Factor / Ratio: 0.5**.
Auto Relief:	**Rectangular**
Relief Ratio:	**0.5**.

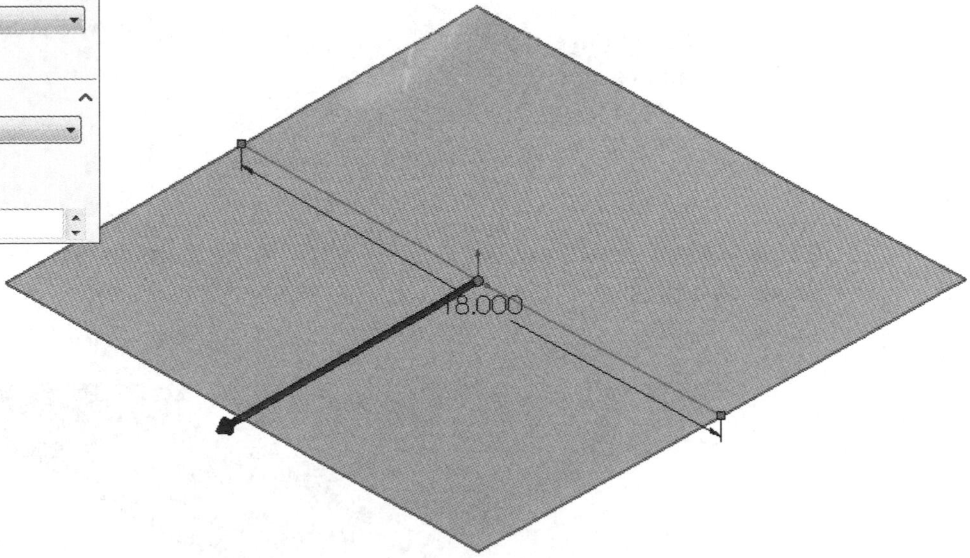

Click **OK**.

4. Creating the Miter-Flanges:

Hold the Control key, select the <u>edge</u> and the <u>vertex</u> as indicated below, and click  or select: **Insert / Sketch**.

SOLIDWORKS automatically creates a new plane <u>normal</u> to the selected edge and coincident to the vertex.

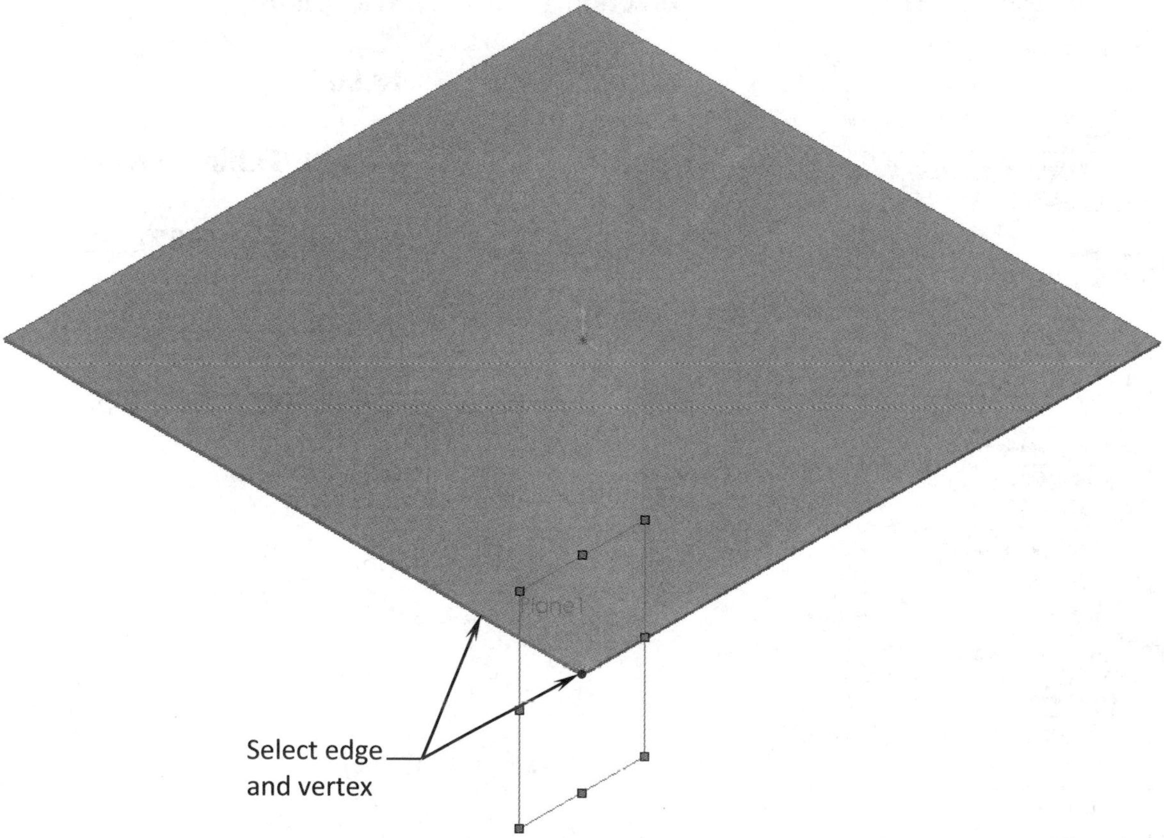

Click **Zoom-to-Area**  from the View toolbar and zoom in on the corner as shown below.

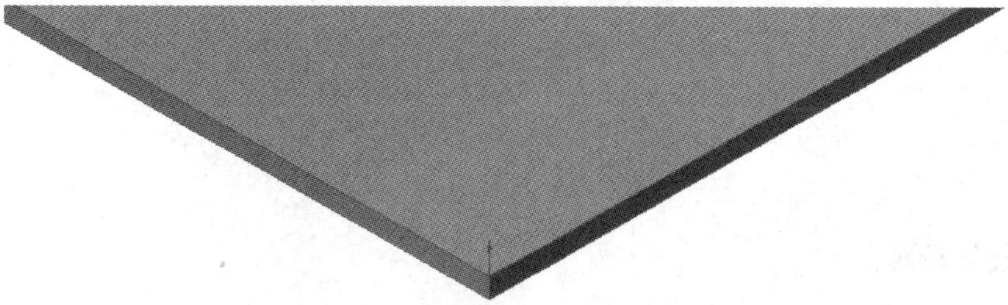

Sketch a vertical **Line** starting at the upper corner.

Add a dimension to make the length of the line **.500 in**.

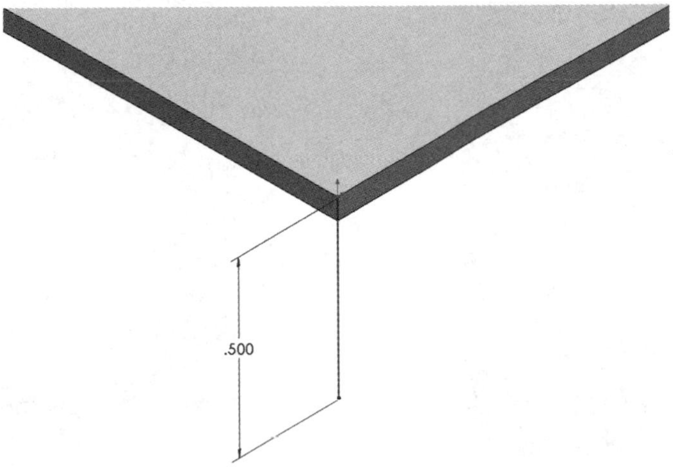

Click  **(Miter-Flange)** from the Sheet Metal toolbar or select **Insert / Sheet Metal / Miter Flange**.

Select the **4 upper edges** as indicated.

Choose **Material Outside** under Flange Position.

Set Gap Distance to **.010 in**.

Click **OK**.

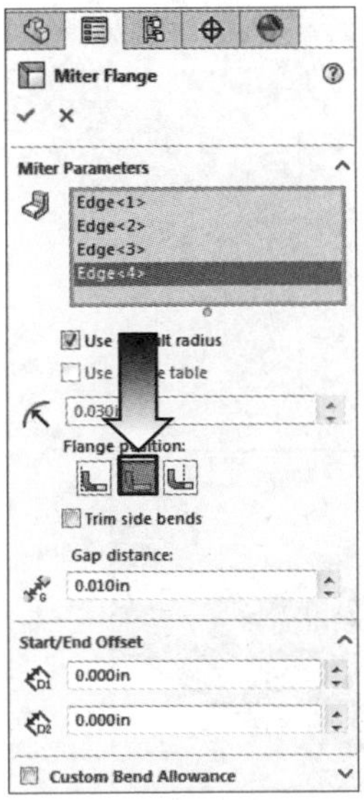

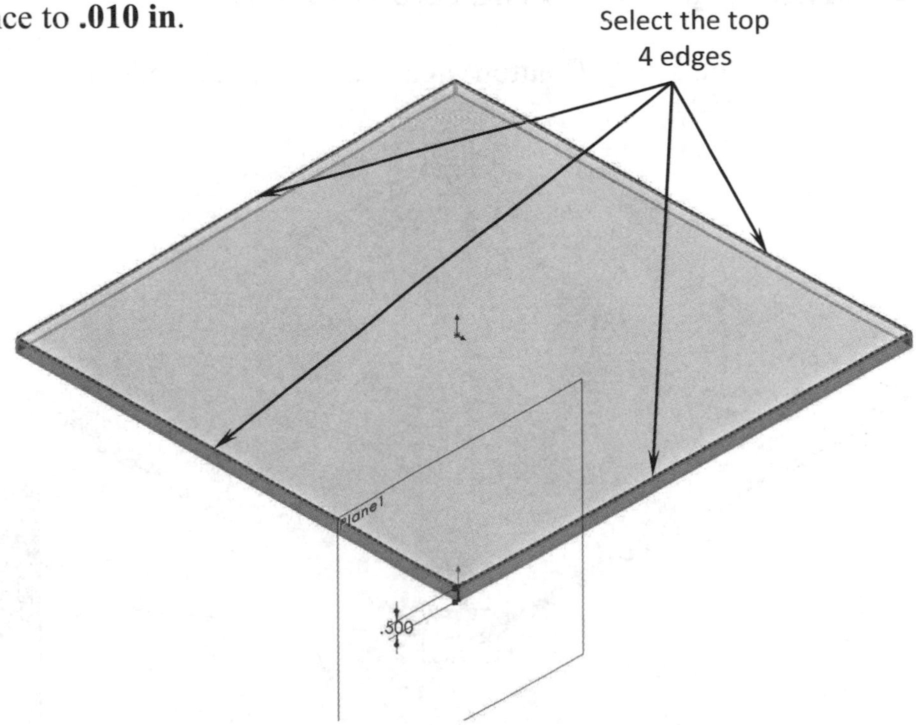

5. Flattening the part:

The Flat Pattern can be toggled at any time during or after the part is created.

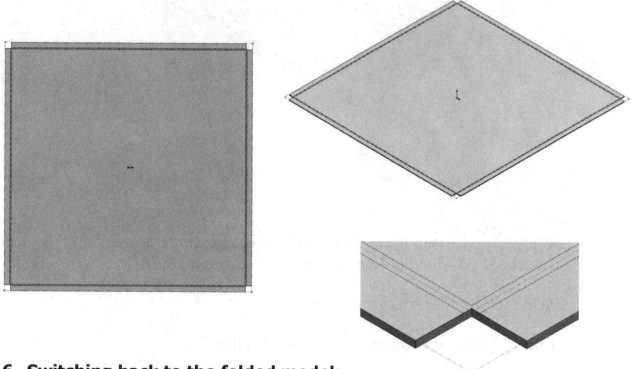

Click  (Flatten) from the Sheet Metal toolbar to flatten the part.

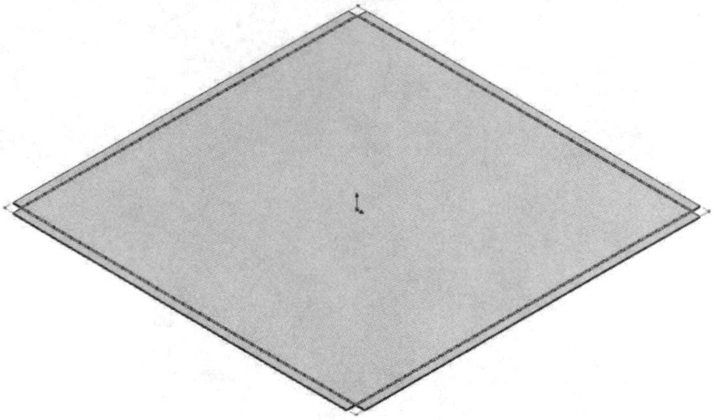

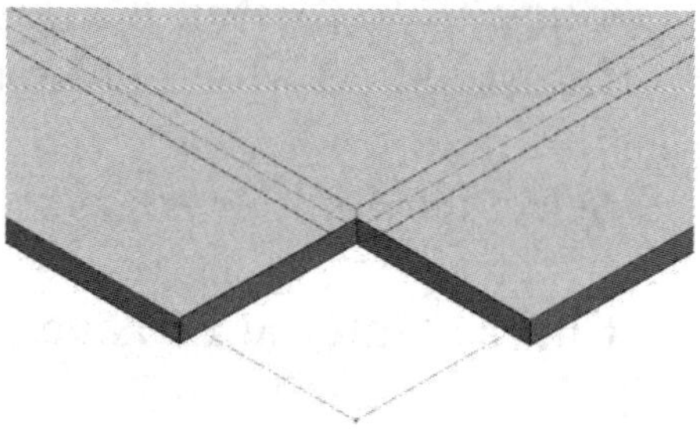

6. Switching back to the folded model:

Click  (Flatten) again to return the model back to the folded stage.

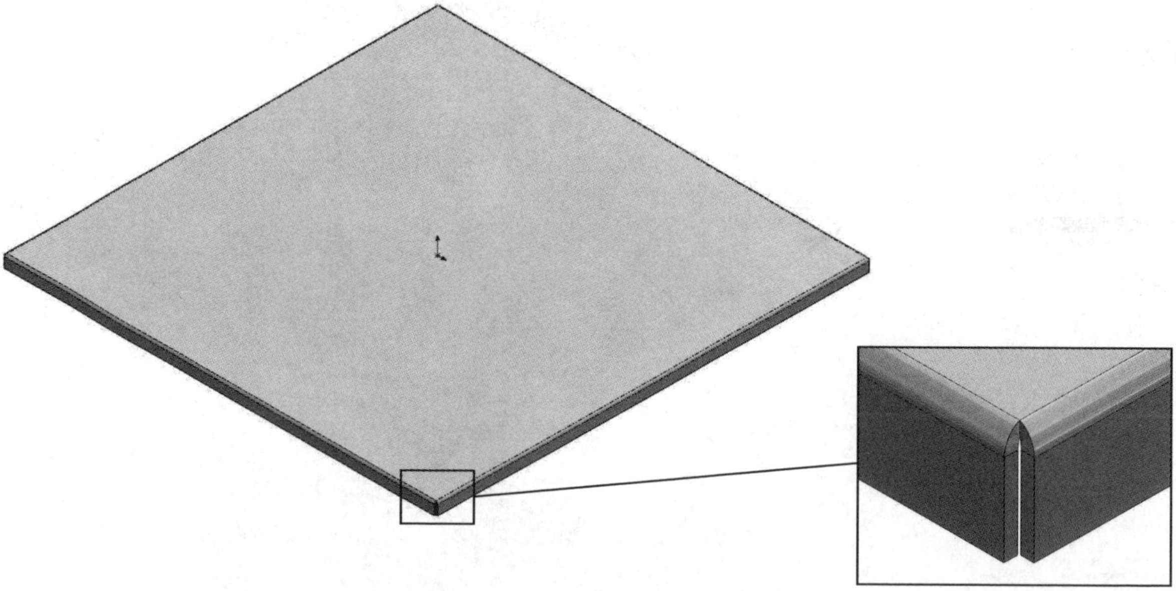

7. Saving your work: Save your model as **Sheet Metal Vent.sldprt**

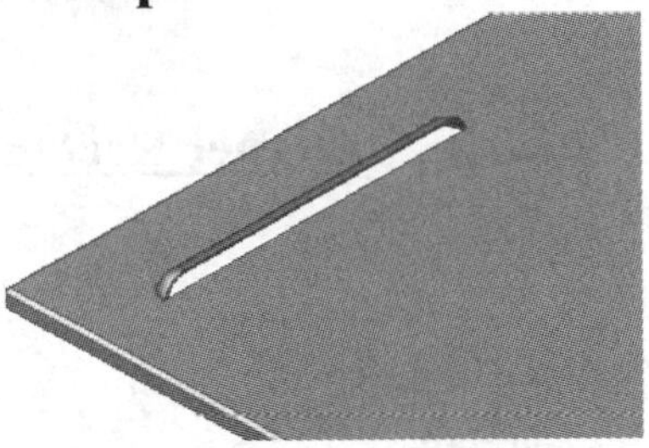

---

8. Creating a new Forming Tool – The Louver:

Start a new Part file: click **File / New / Part / OK.**

Set Units to **Inches – 3 Decimals** (Tools/Options/Document Properties/ Units).

9. Sketching on the TOP reference plane:

Select the Top plane and click **Insert / Sketch** from the Sketch toolbar.

Select the **Center Rectangle** from the Sketch-Tools toolbar.

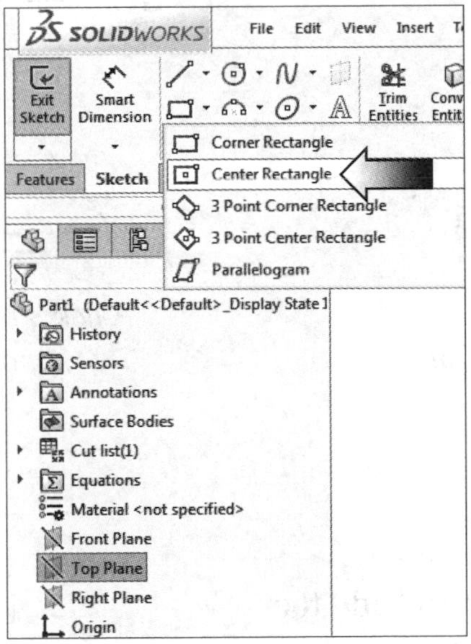

Sketch a **Center Rectangle** that is centered on the Origin, as shown below.

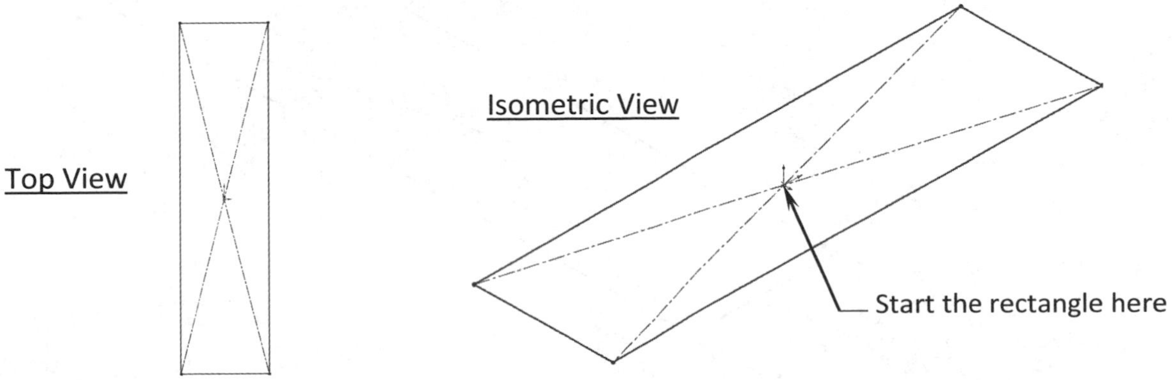

10. Other Rectangle options:

Corner Rectangle

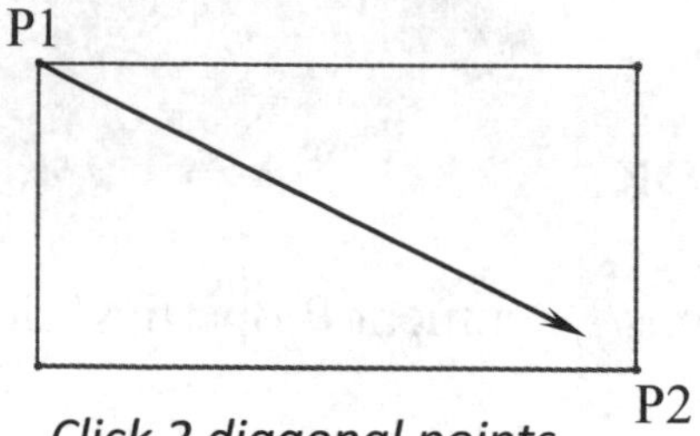

Click 2 diagonal points.

3-Point Corner Rectangle

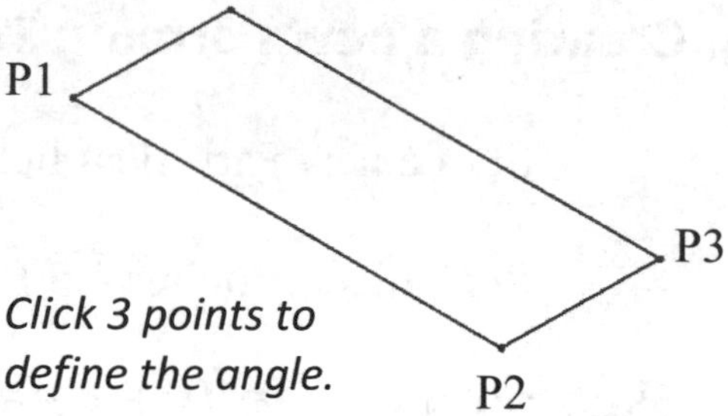

Click 3 points to define the angle.

3-Point Center Rectangle

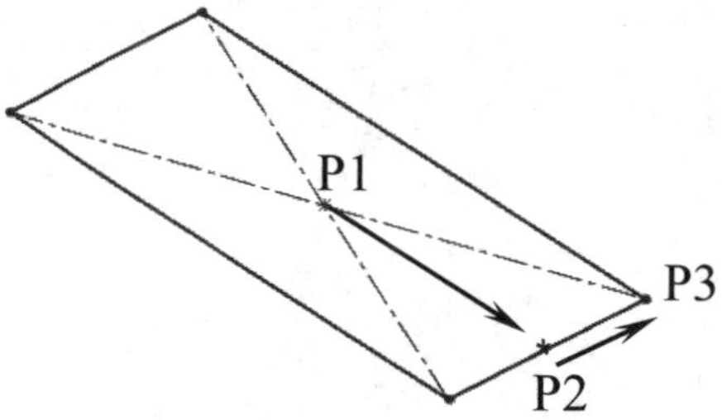

Start at Center point and click 2 other points to define the angle.

Parallelogram

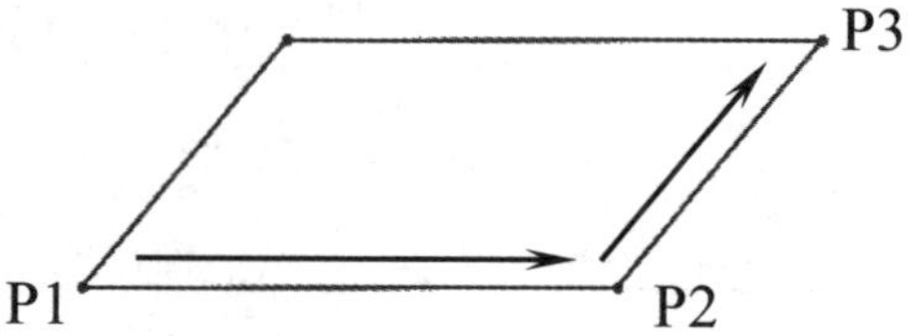

Click 3 points to define the parallelogram.

11. Adding dimensions:

Click **Smart Dimension** and add the dimensions shown below.

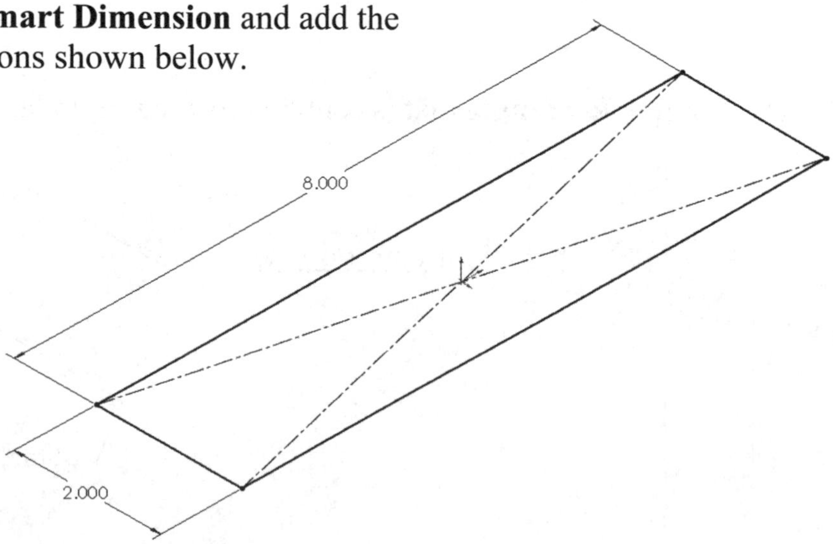

12. Extruding the Base:

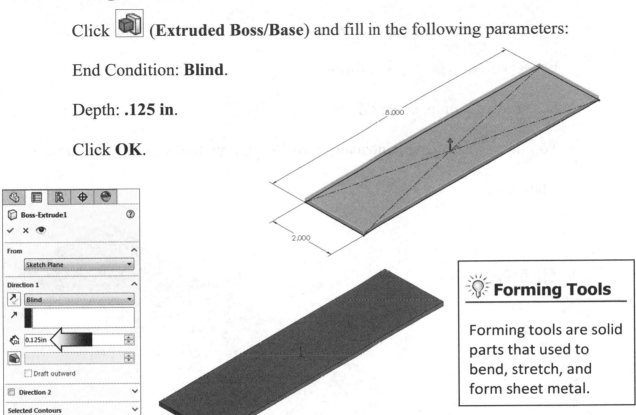

Click (**Extruded Boss/Base**) and fill in the following parameters:

End Condition: **Blind**.

Depth: **.125 in**.

Click **OK**.

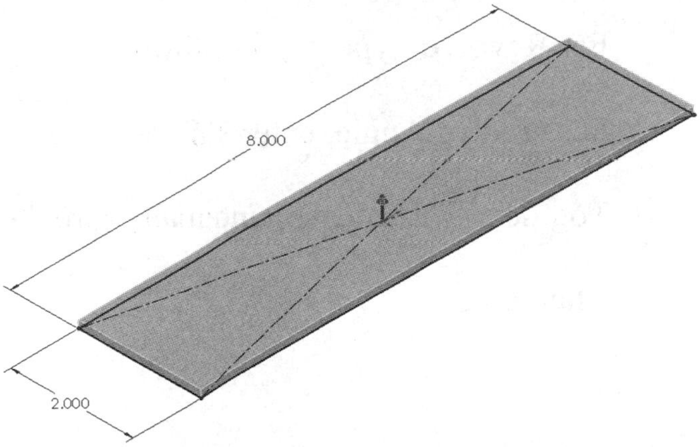

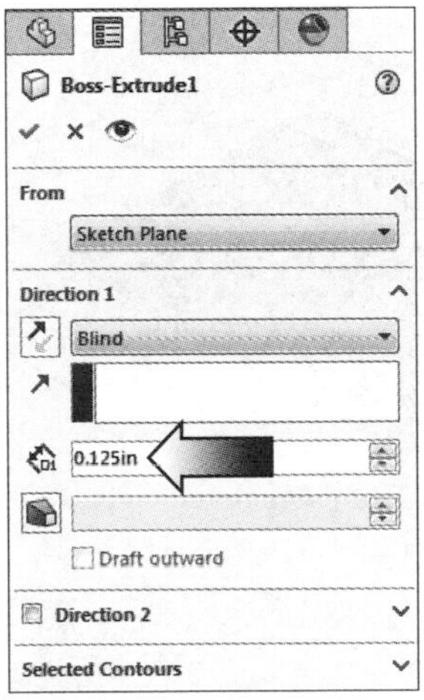

Forming Tools

Forming tools are solid parts that used to bend, stretch, and form sheet metal.

13. Building the Louver's body:

Select the <u>upper face</u> of the part and open a **new sketch**.

Change to the Top View Orientation (Ctrl+5).

Sketch the profile of the Louver-Forming tool and add dimensions shown below:

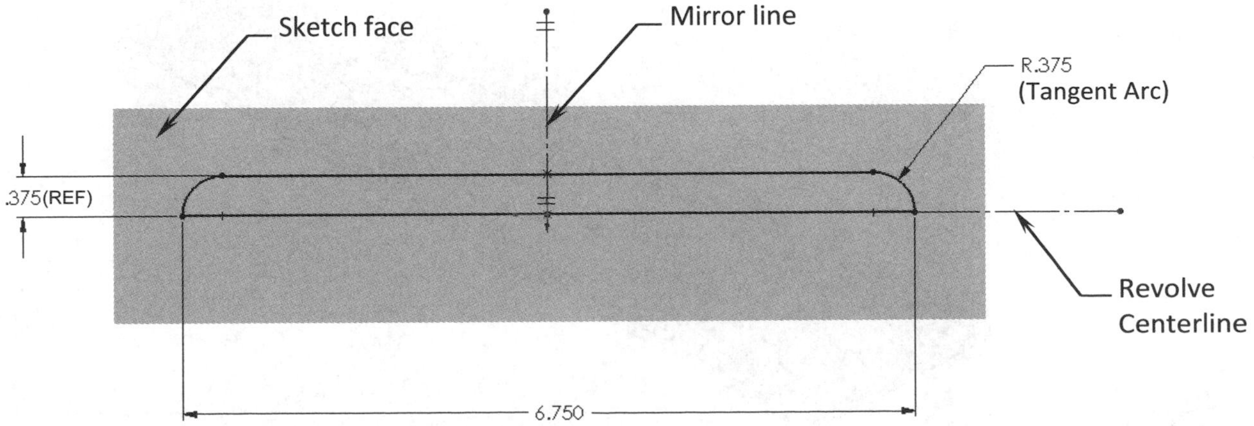

14. Revolving the louver body:

Select the **horizontal centerline** and click  (**Revolve Boss/Base**).

For Revolve Type, select **Blind**.

For revolve Angle, enter **90°**.

Toggle  (Reverse) and make sure the preview looks like the one below.

Click **OK**.

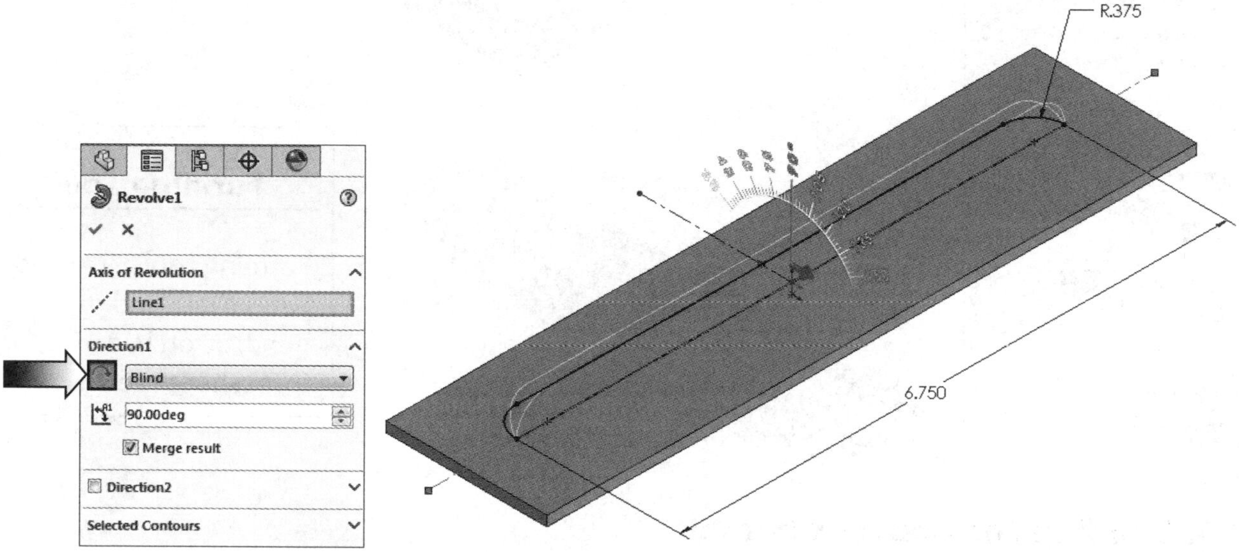

Rotate the model and inspect the model from different angles to ensure it is protruded to the correct side.

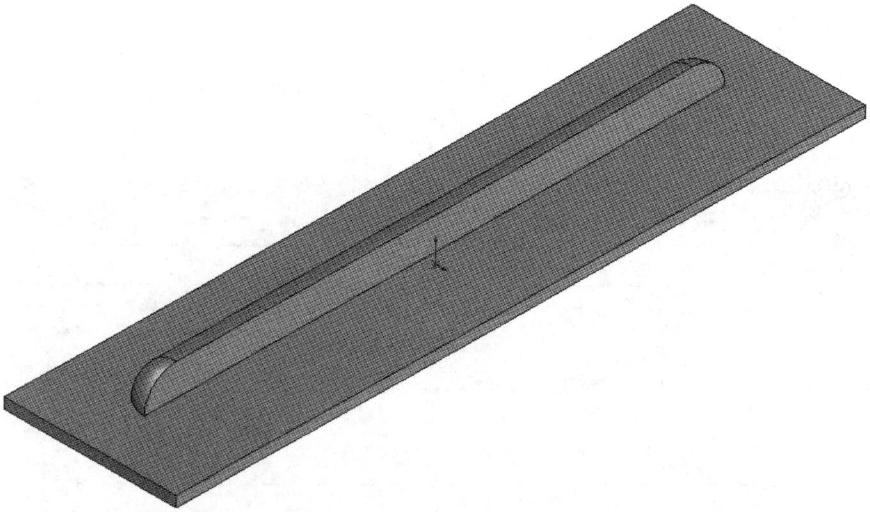

15. Adding a fillet at the base:

Click **Fillet** and enter **.125 in**. for Radius size.

Select the **edge** as indicated.

Select **Tangent Propagation** checkbox (default); the system applies the same fillet to all connected tangent edges.

Click **OK**.

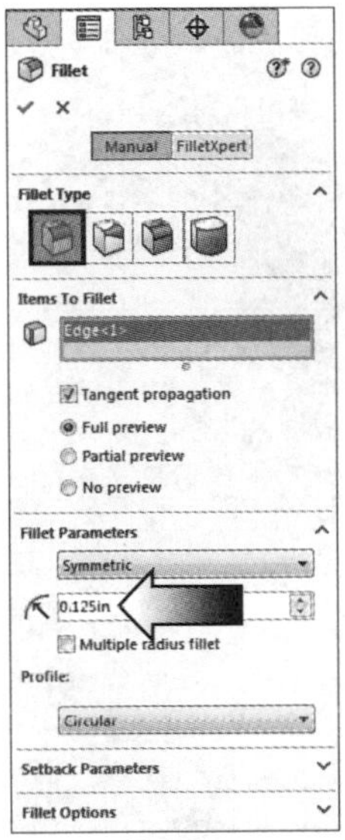

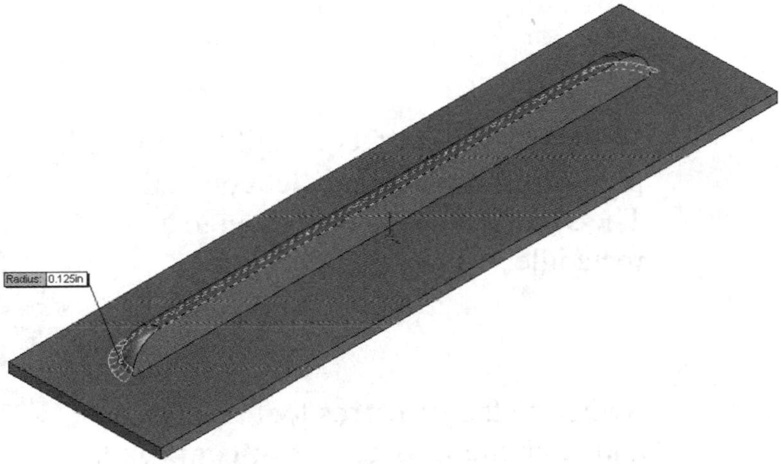

Rotate the model to see the resulting fillet from the backside.

The new fillet should run around the back side but not the front.

16. Removing the base:

The rectangular plate was created so that a fillet can be added between the two features. We no longer need it at this point.

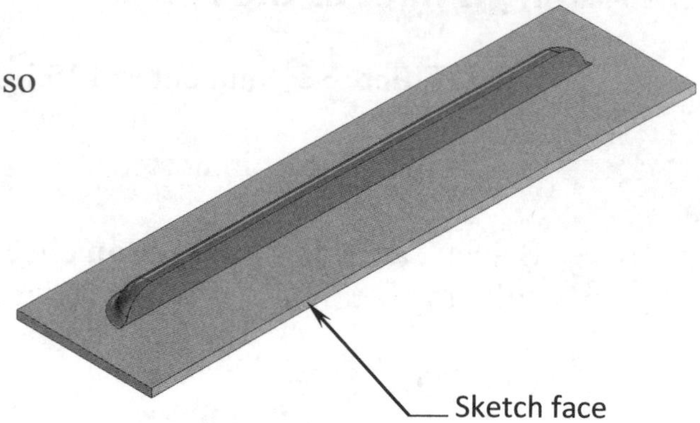

Select the side face of the part and open a **new sketch**.

While the side surface is still highlighted, press the **Convert Entities** command. The selected face is converted to a rectangle.

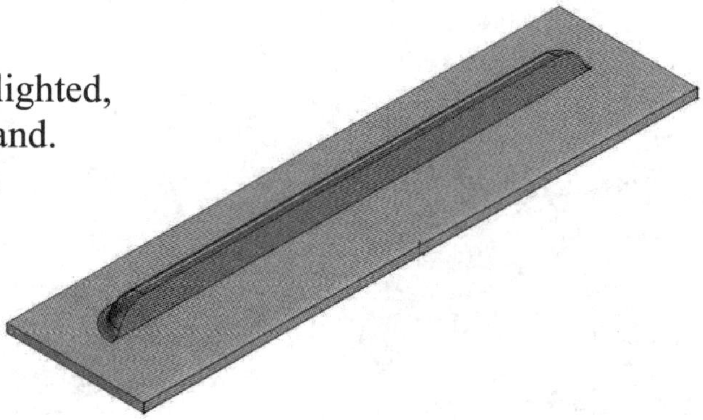

Switch to the **Features** tool tab and click the **Extruded Cut** command.

For End Condition select **Through All**.

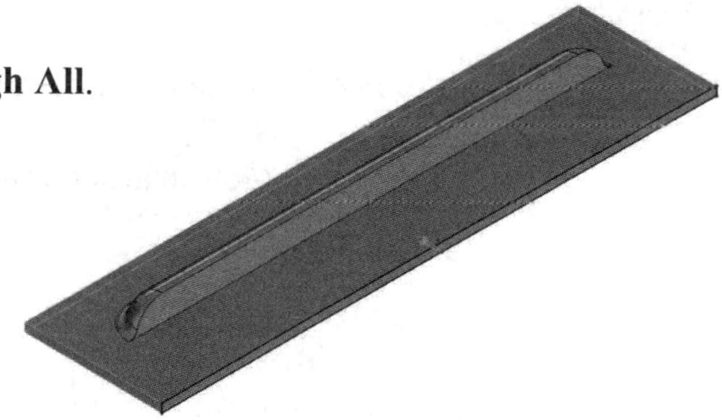

Click **OK**.

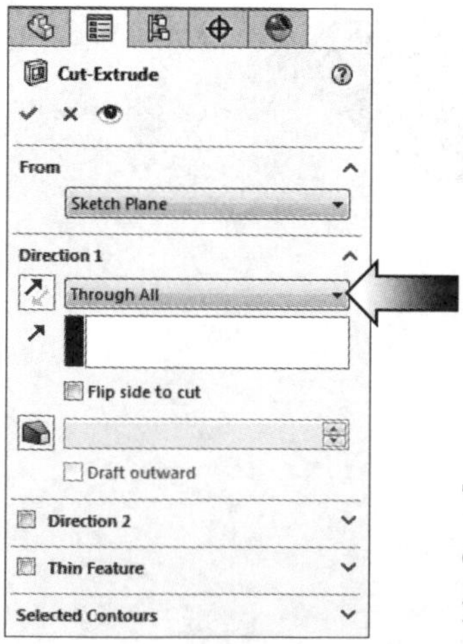

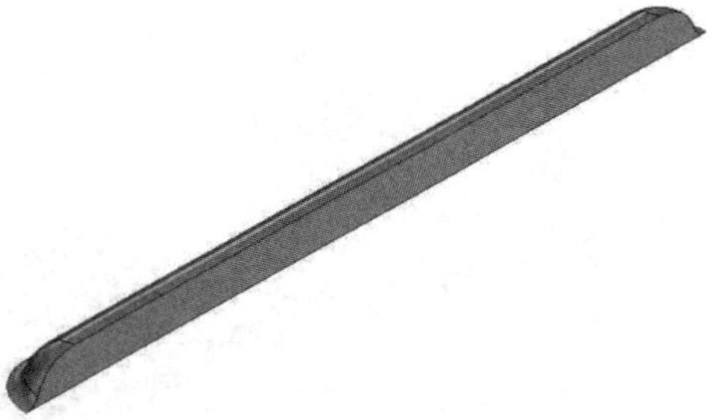

The base is removed and only the form tool portion is kept.

17. Creating the Positioning Sketch:

The Positioning-Sketch displays the preview of the Form tool while it is being dragged from the Design Library. Its sketched entities can be dimensioned to position the Formed feature.

Select the bottom face of the part and open a **new sketch** 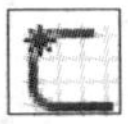.

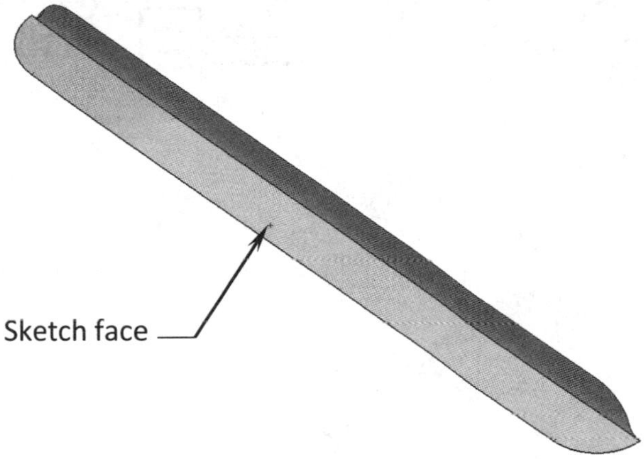

Select the bottom face of the part and click **Convert Entities** .

Add a **Centerline**  as shown to assist with positioning the tool when it is placed on a sheet metal part.

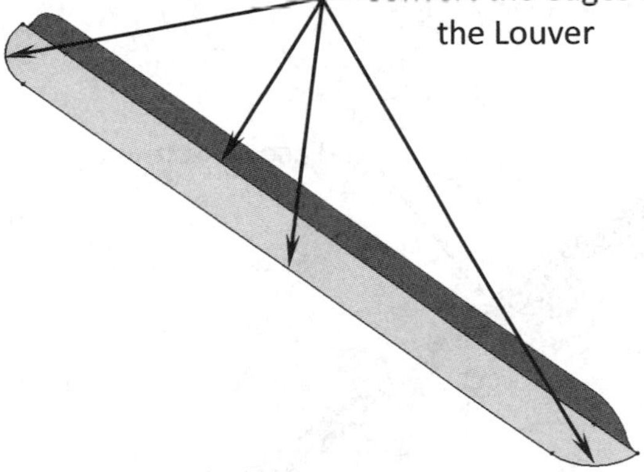

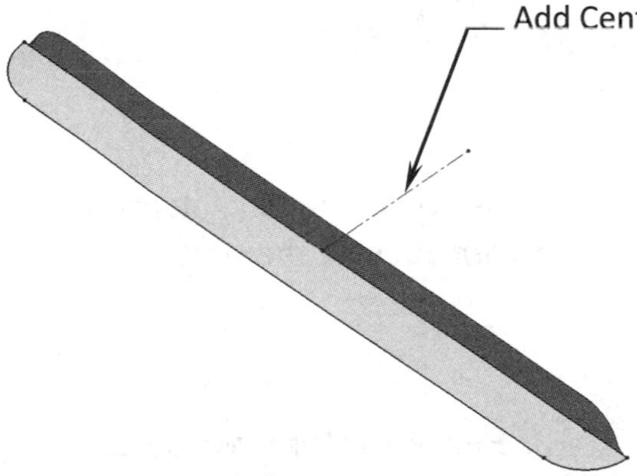

Exit the sketch or click .

Rename the sketch to **Position Sketch** from the FeatureManager tree.

18. Establishing the Stop and Remove faces:

In order for the forming tools to work properly, a set of Stopping Face and Removing Faces will have to be established prior to saving as a forming tool.

Change to the Sheet Metal tool tab and click the **Forming Tool** command.

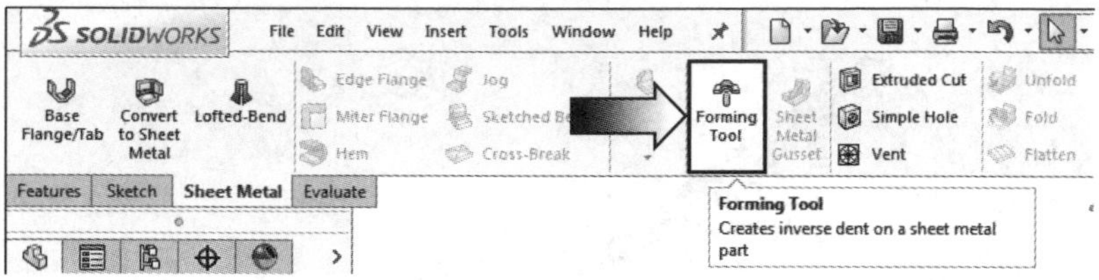

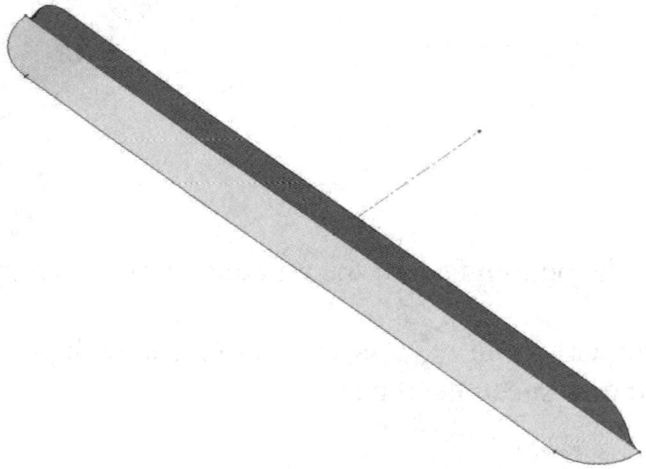

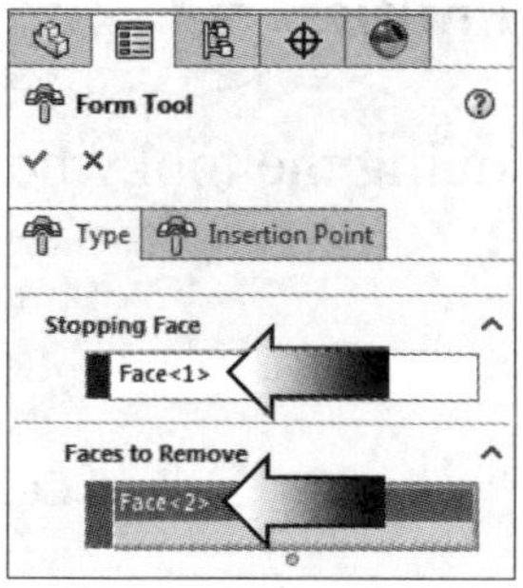

For **Stopping Face**, select the **bottom face** of the part.

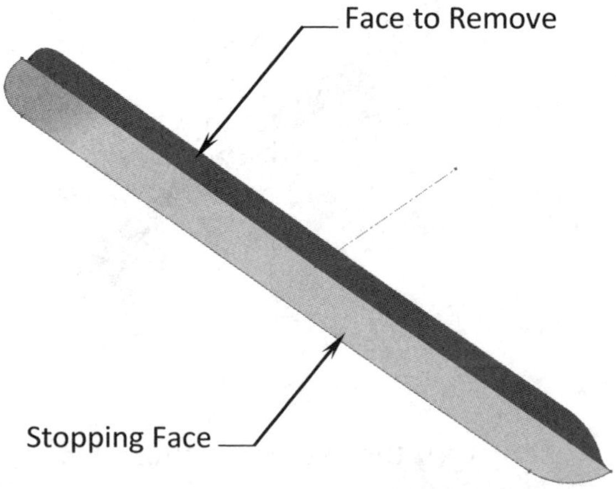

For **Faces to Remove**, select the **face** on the right side of the part.

Click **OK**.

19. Saving the Forming Tool:

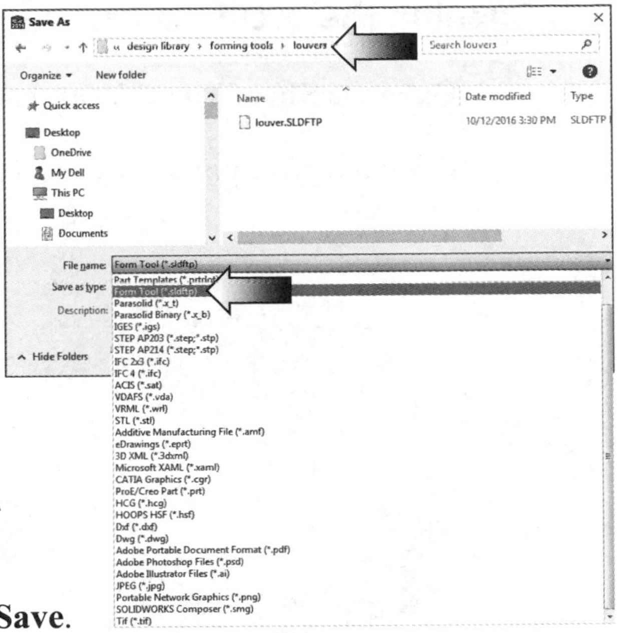

Click: **File / Save As.**

Enter **Long Louver** for the name of the file.

Select **Form Tool** (*.sldftp) in the Save As Type.

Browse to the following directories: **Program Data\ SolidWorks\ SolidWorks 2022\ Design Library\ Forming Tools\Louvers**.

Expand the Louver folder and click **Save**.

NOTE: *The Design Library is a Hidden Folder (Explorer / Organize / Folder and Search Options / View / Show Hidden Files, Folders and Drives).*

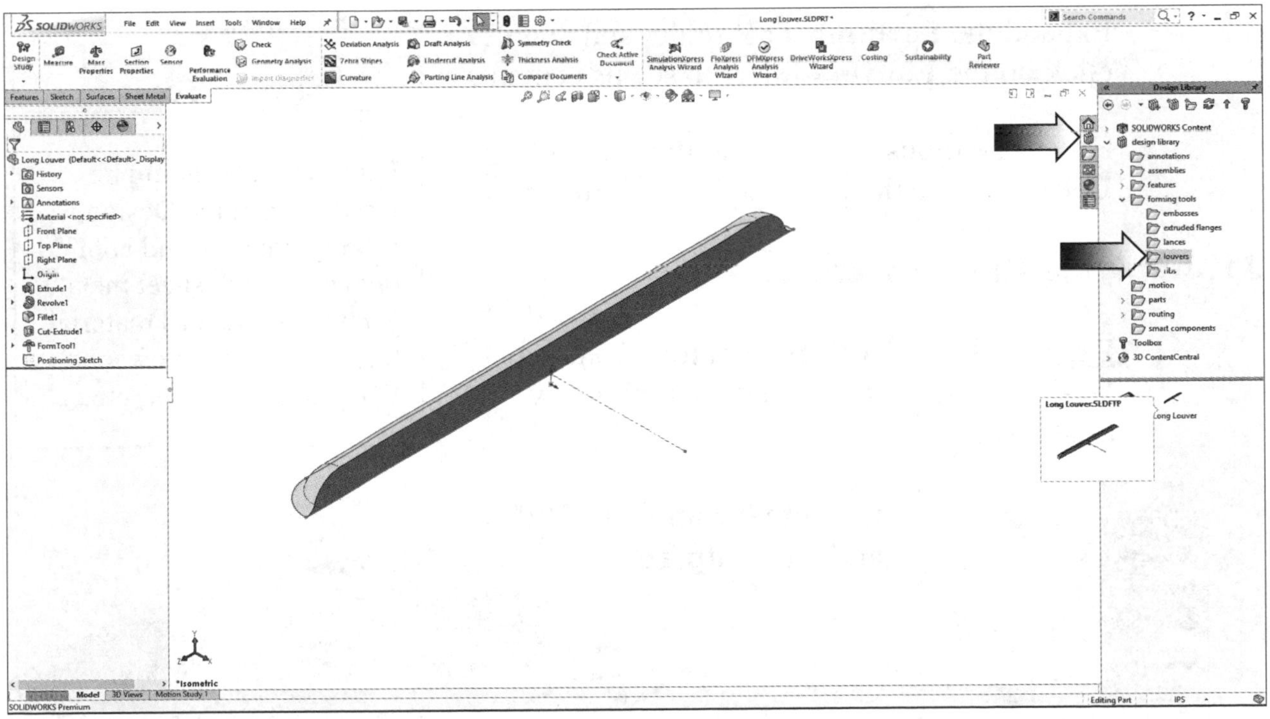

After the forming tool is saved, it can be accessed through the Task Pane by dragging and dropping it from the Design Library folder.

20. Opening the previous part:

Open the **Sheet Metal Vent** that was saved earlier **(Alt+Tab)**.

Click the push-pin to lock the Task Pane in place (arrow).

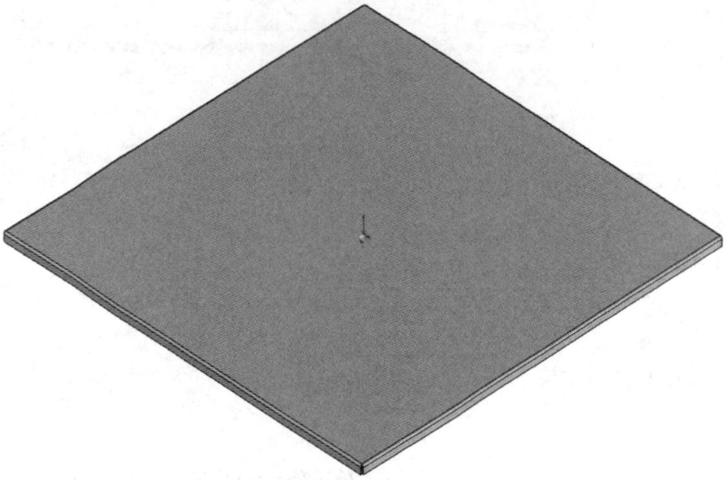

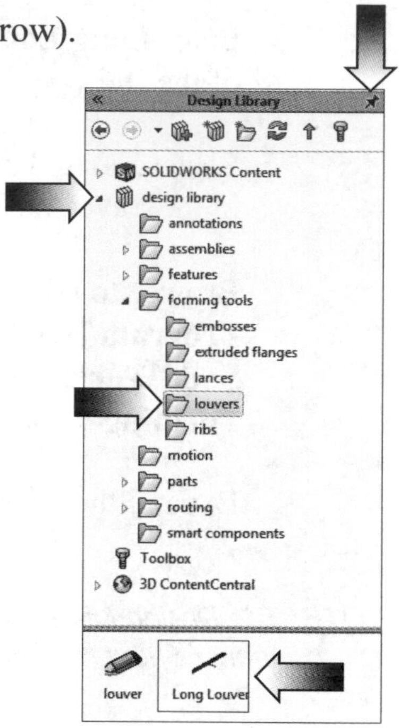

Expand the **Design Library** folder.

Expand the **Forming Tool** folder and double click on the **Louvers** folder to see its content.

Hover the mouse cursor over the name **Long Louver** to see the preview of the Forming tool.

> **Forming Tools**
>
> Forming tools should be inserted from the Design Library window and applied onto parts with sheet metal parameters such as material thickness, bend allowance, fixed face, cut relief, etc…

21. Applying the form tool:

Drag the **Long Louver** form tool from the Design Library and drop it approximately as shown.

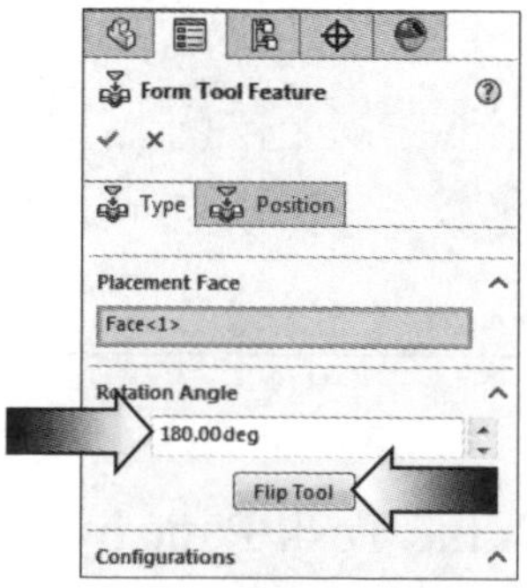

Change the angle to **180°** and click **Flip Tool**.

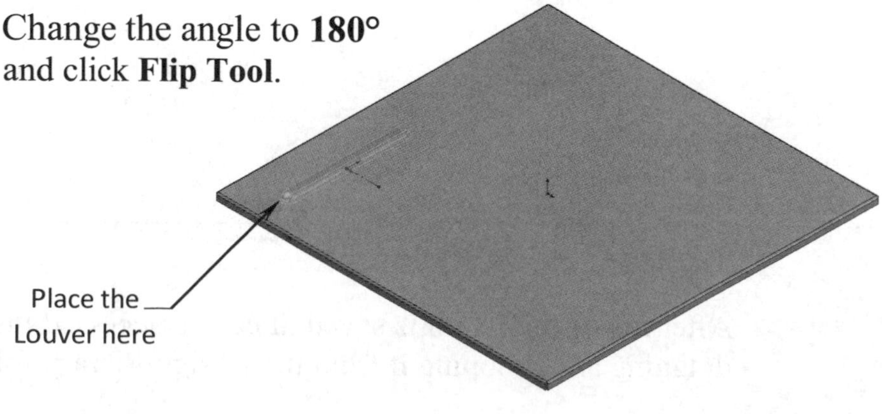

22. Positioning the form tool:

Click the **Position** tab (arrow).

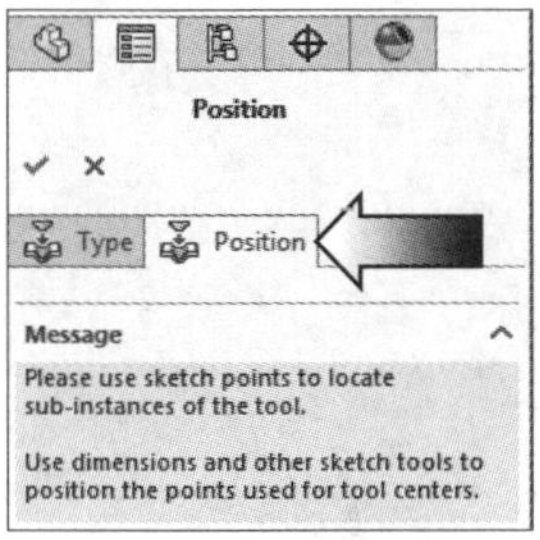

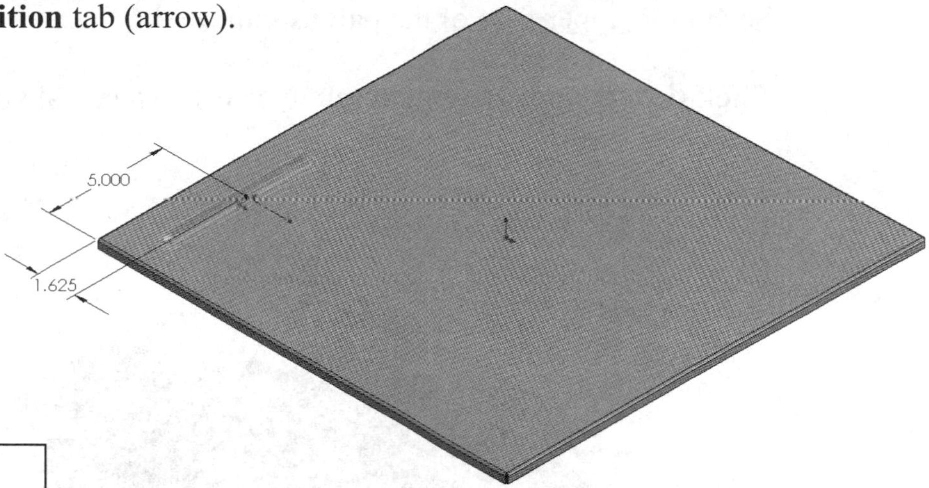

Push / Pull

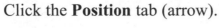

While dragging the Forming Tool from the Design Library, (still holding the mouse button) press the TAB key to reverse the direction from push to pull.

Add the locating dimensions to the outer edges of the model to fully define this sketch.

Use the outer edges of the sheet metal part when adding the dimensions.

Click **OK**.

The Louver feature is formed.

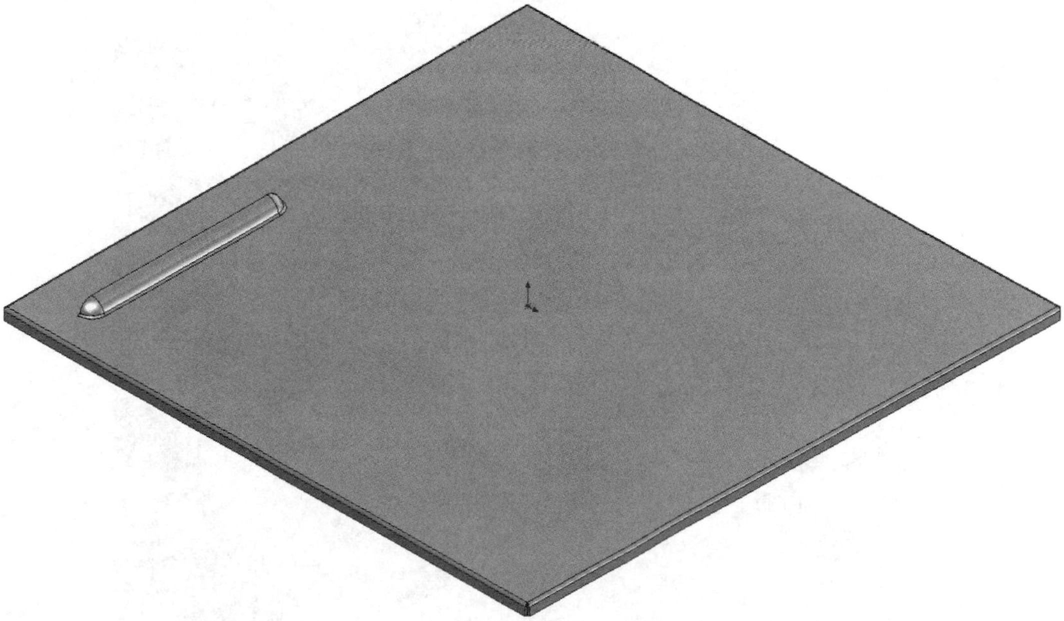

23. Adding a mounting hole:

Select the upper face of the part as indicated.

Click  to open a **new sketch** or select **Insert / Sketch**.

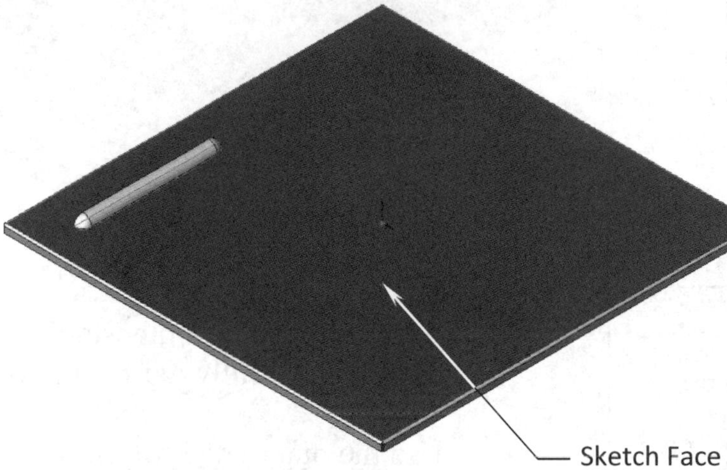

Sketch a **Circle** on the left side of the louver.

Add dimensions to fully define the sketch.

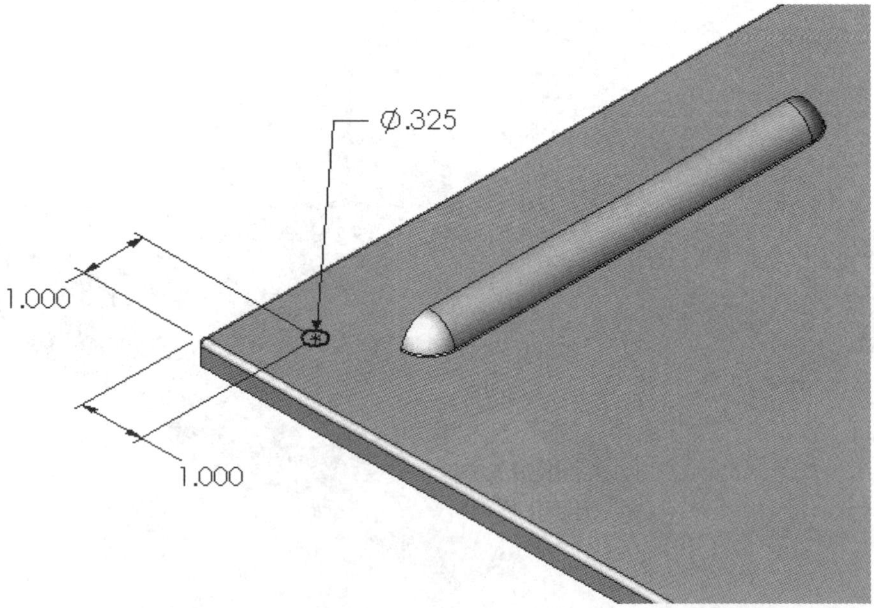

24. Extruding a cut:

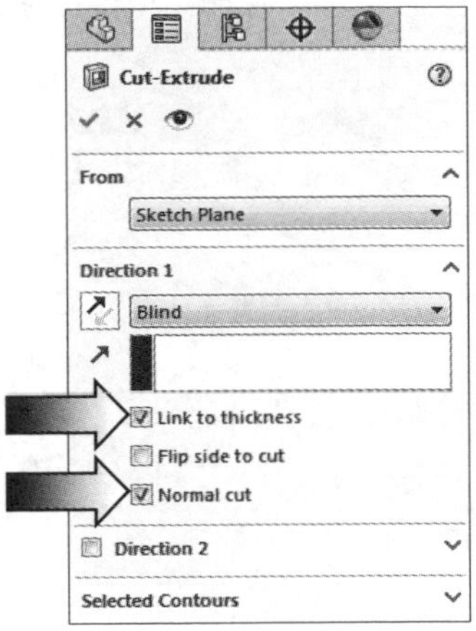

Click or select **Insert / Cut / Extrude**.

End Condition: **Blind**.

Link to Thickness: **Enabled**.

Normal Cut: **Enabled**.

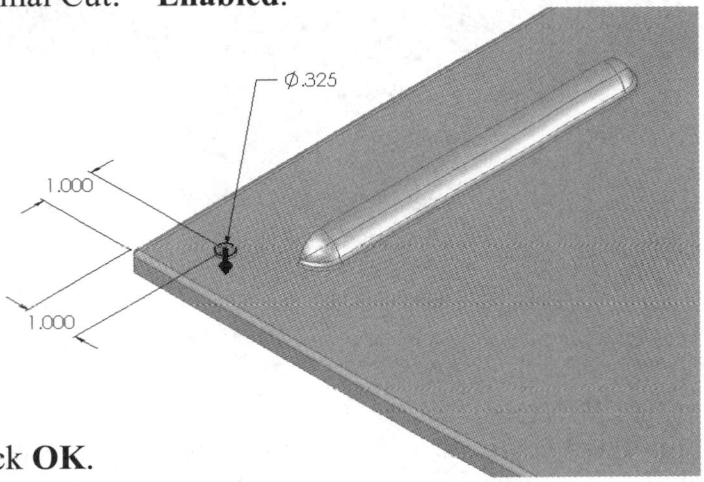

Click **OK**.

25. Creating a Linear Pattern:

Click or select **Insert / Pattern Mirror / Linear Pattern**.

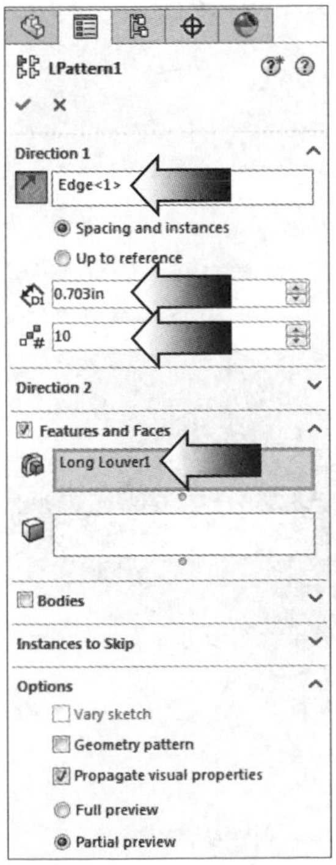

Select the **bottom edge** as Pattern Direction.

Enter **.703 in**. as Spacing.

Enter **10** as Number of Instances.

Select the **Long Louver** as Features to Pattern.

Click **OK**.

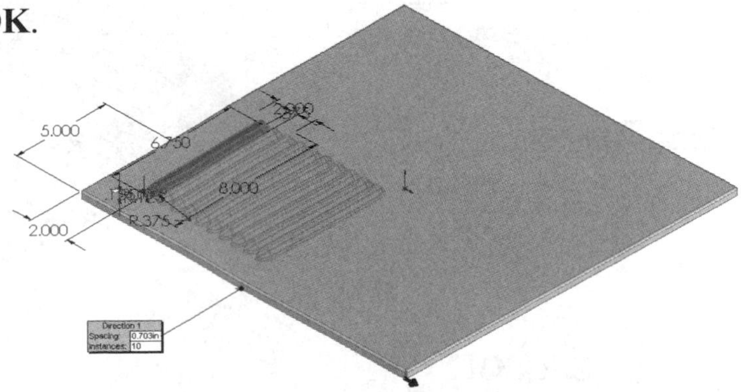

The completed Linear Pattern.

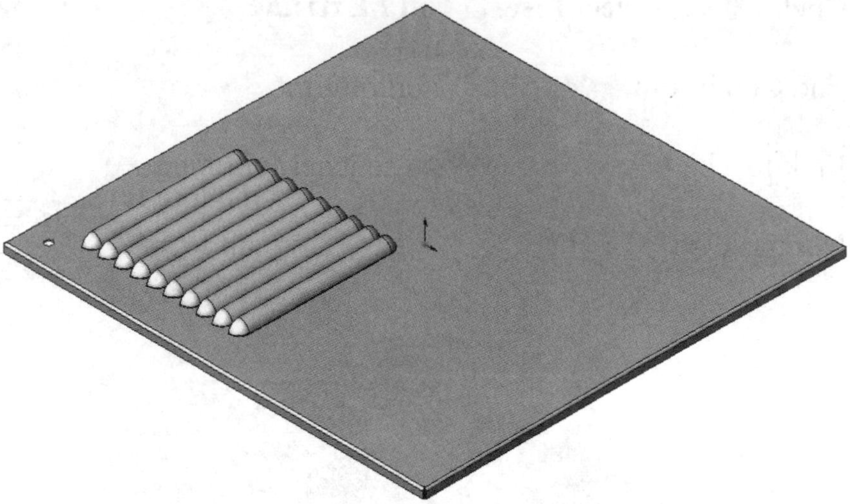

26. Creating an Axis:

An axis can be created at any time so features can be arrayed around it. In this case, an axis in the center of the part will be made and used as the center of the next Circular Pattern.

Click or select **Insert / Reference Geometry / Axis**.

Click the **Two Planes** option .

Select **Front** and **Right** planes from FeatureManager tree.

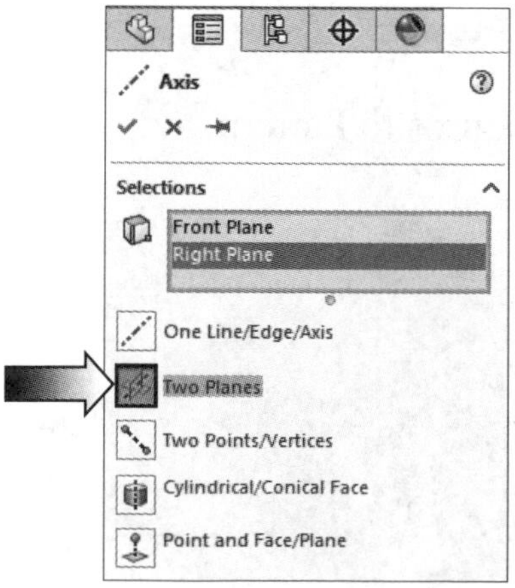

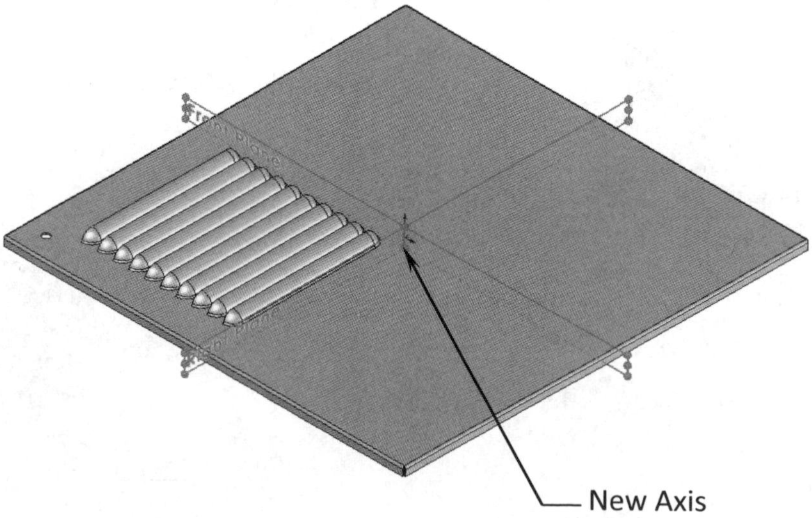

Click **OK**.

27. Creating a Circular Pattern:

Click 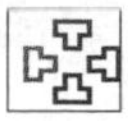 or select **Insert / Pattern Mirror / Circular Pattern**.

Select the new **Axis** for Pattern Axis.

Enter **360 deg.** for Pattern Angle.

Enter **4** for Number of Instances.

For Feature to Pattern select the **Cut-Extrude1**, **LPattern1** and the **Long Louver** from the Feature tree.

Click **OK**.

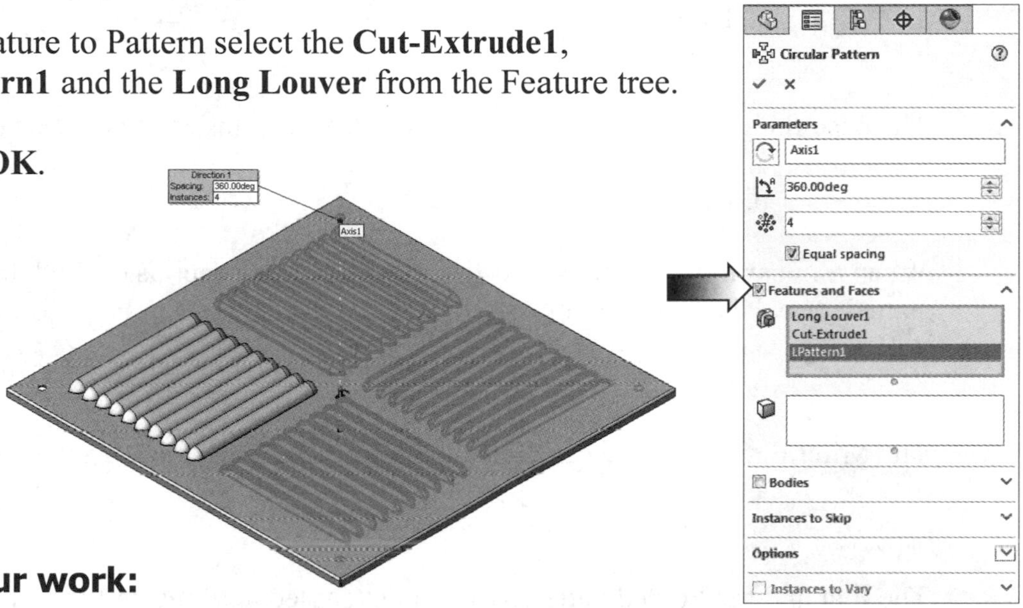

28. Saving your work:

Save the model using the same file name and override the previous document.

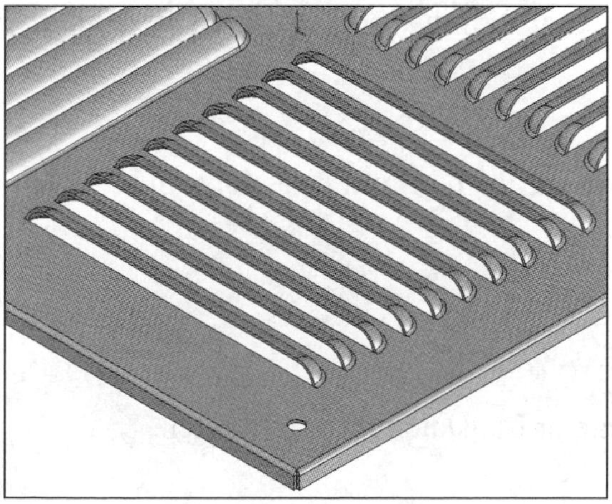

Finished Part (Folded)

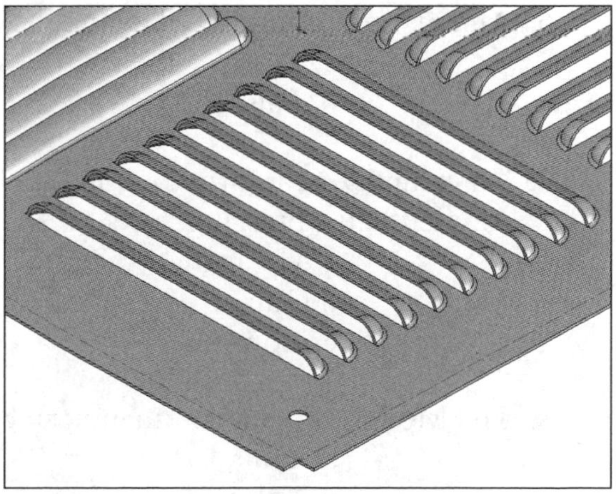

Flat Pattern*

** After the formed features are created, they will retain their formed shapes even when toggled back and forth between Folded or Flattened.*

Questions for Review

Sheet Metal Parts

1. The mid-point relation can be used to center a line onto the Origin.
   a. True
   b. False

2. The base flange command can also be selected from Insert / Sheet Metal / Base Flange.
   a. True
   b. False

3. When a linear model edge is selected and the sketch pencil is clicked, the system creates a plane Normal To Curve automatically.
   a. True
   b. False

4. The Miter Flange feature can create more than one flange in the same operation.
   a. True
   b. False

5. The Flat and the Folded patterns cannot be toggled until the part is completed and saved.
   a. True
   b. False

6. An existing forming tool cannot be edited or changed; forming tools are fixed by default.
   a. True
   b. False

7. When applying a form tool onto a sheet metal part, the push or pull direction can be toggled when pressing:
   a. Up arrow
   b. Tab
   c. Control

8. The Modify Sketch command can be used to rotate or translate the entire sketch.
   a. True
   b. False

9. A formed feature(s) cannot be copied or patterned.
   a. True
   b. False

1. TRUE 2. TRUE
3. TRUE 4. TRUE
5. FALSE 6. FALSE
7. B 8. TRUE
9. FALSE

CHAPTER 14

Sheet Metal Forming Tools

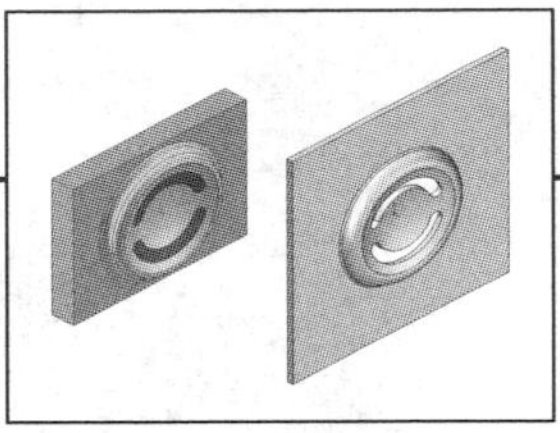

Sheet Metal Forming Tools

Forming tools act as dies that bend, stretch, or otherwise form sheet metal features.

SOLIDWORKS includes some sample forming tools to get you started.
They are stored in: ***Installation_Directory/Data/Design Library/FormingTools/folder_name***.

Some types of form features, such as louvers and lances, create openings on sheet metal parts. To indicate which faces of the form tool will create the openings, the system changes the color of these faces to **red,** the stopping faces to **blue** and the rest to **yellow**.

The user can only insert (drag & drop) forming tools from the **Design Library** window and apply them only to sheet metal parts. The Design Library window gives you quick access to the parts, assemblies, library features, and form tools that are used most often.

Users can create their own forming tools and apply them to sheet metal parts to create form features such as louvers, lances, flanges, and ribs.

The Design Library window has several default folders. Each folder contains a group of palette items displayed as Thumbnail Graphics.

Design Library can include:

* Parts (.sldprt)
* Assemblies (.sldasm)
* Sheet Metal Forming Tools (.sldftp)
* Library Features (.sldlfp)

In this 1st half of the chapter we will learn how to create and save a forming tool.

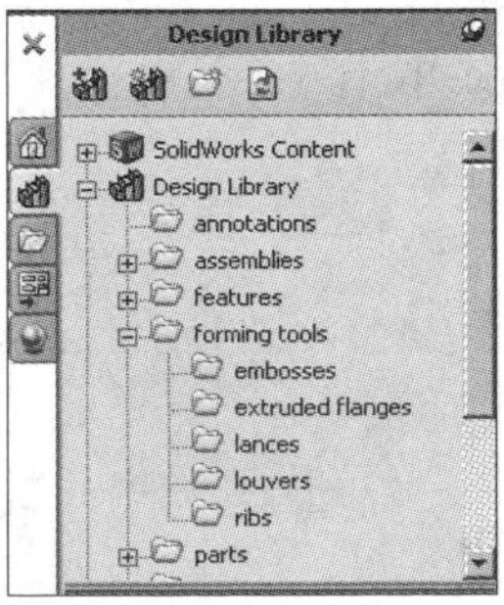

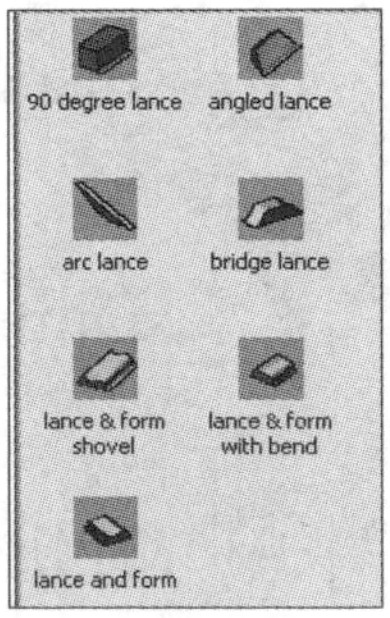

Button with Slots
Sheet Metal Forming Tools

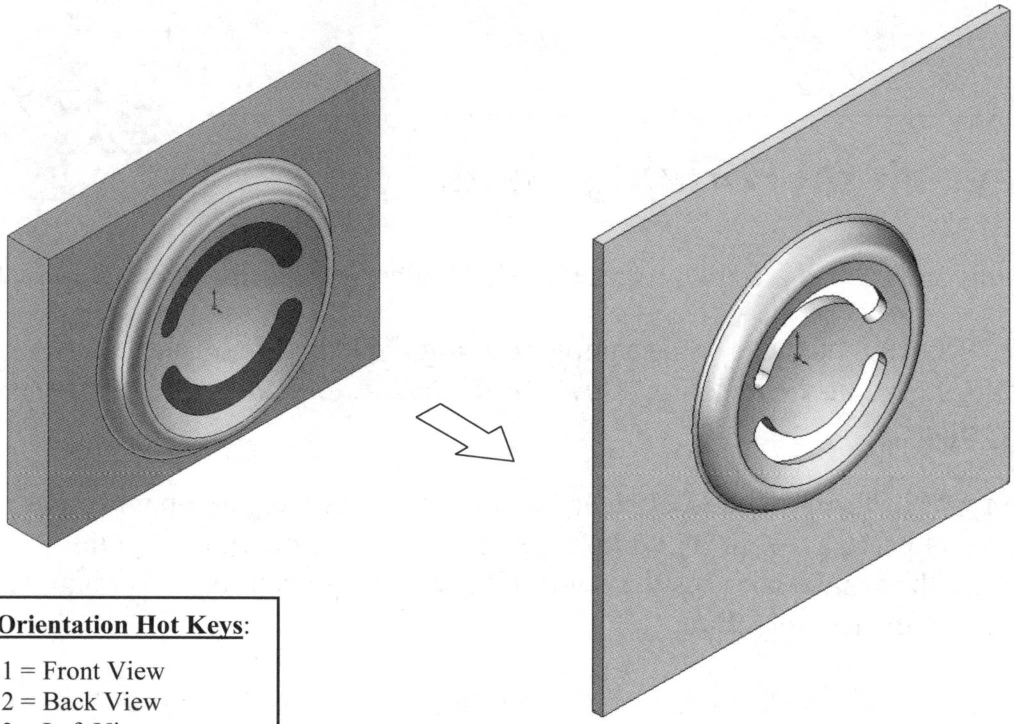

View Orientation Hot Keys:

Ctrl + 1 = Front View
Ctrl + 2 = Back View
Ctrl + 3 = Left View
Ctrl + 4 = Right View
Ctrl + 5 = Top View
Ctrl + 6 = Bottom View
Ctrl + 7 = Isometric View
Ctrl + 8 = Normal To Selection

Dimensioning Standards: **ANSI**

Units: **INCHES** – 3 Decimals

Tools Needed:

Insert Sketch	Boss/Base Revolve	Convert Entities
Dimension	Add Geometric Relations	Extrude Cut
Split Line	Fillet/Round	Forming Tool

1. Creating the base block:

Select the Right plane and insert a **new Sketch** .

Sketch a **Center-Rectangle**  that is centered on the origin.

Add the width and height dimensions shown.

(The sketch should be fully defined at this point.)

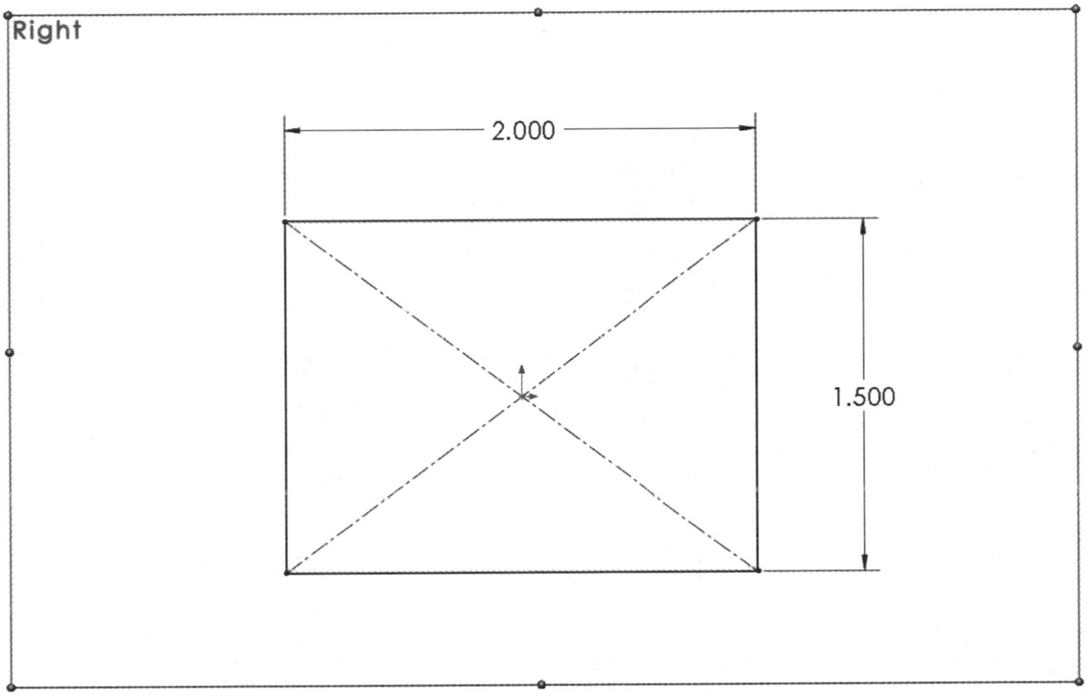

2. Extruding the base:

Click **Extruded Boss/Base** .

Enter the following:

Direction 1: **Blind**.

Reverse Direction.

Depth: **.250in**.

Click **OK**.

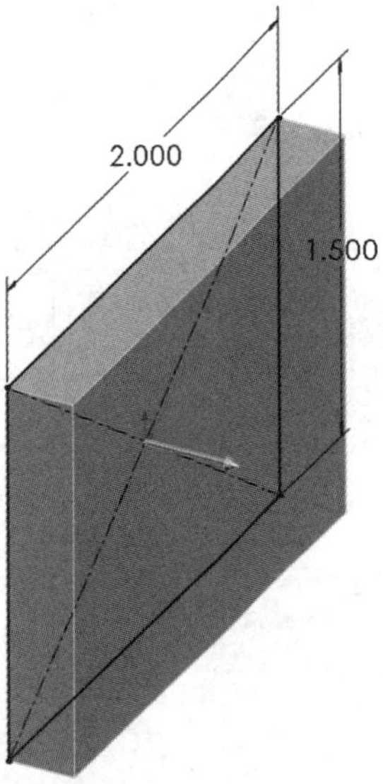

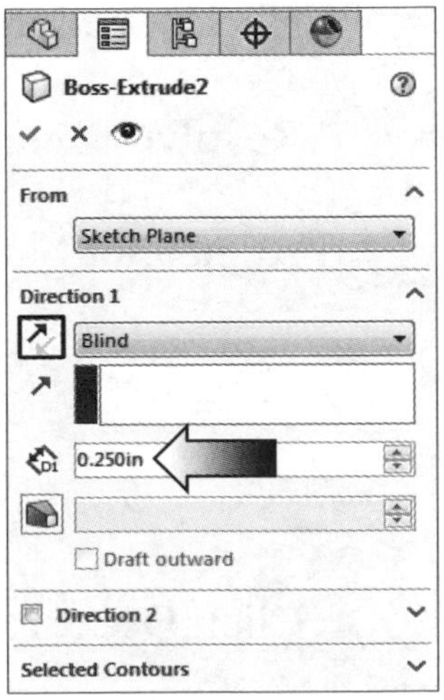

3. Creating the forming tool body:

Select the Front plane and open a **new sketch** .

Sketch the profile shown below.

Add dimensions and relations to fully define the sketch.

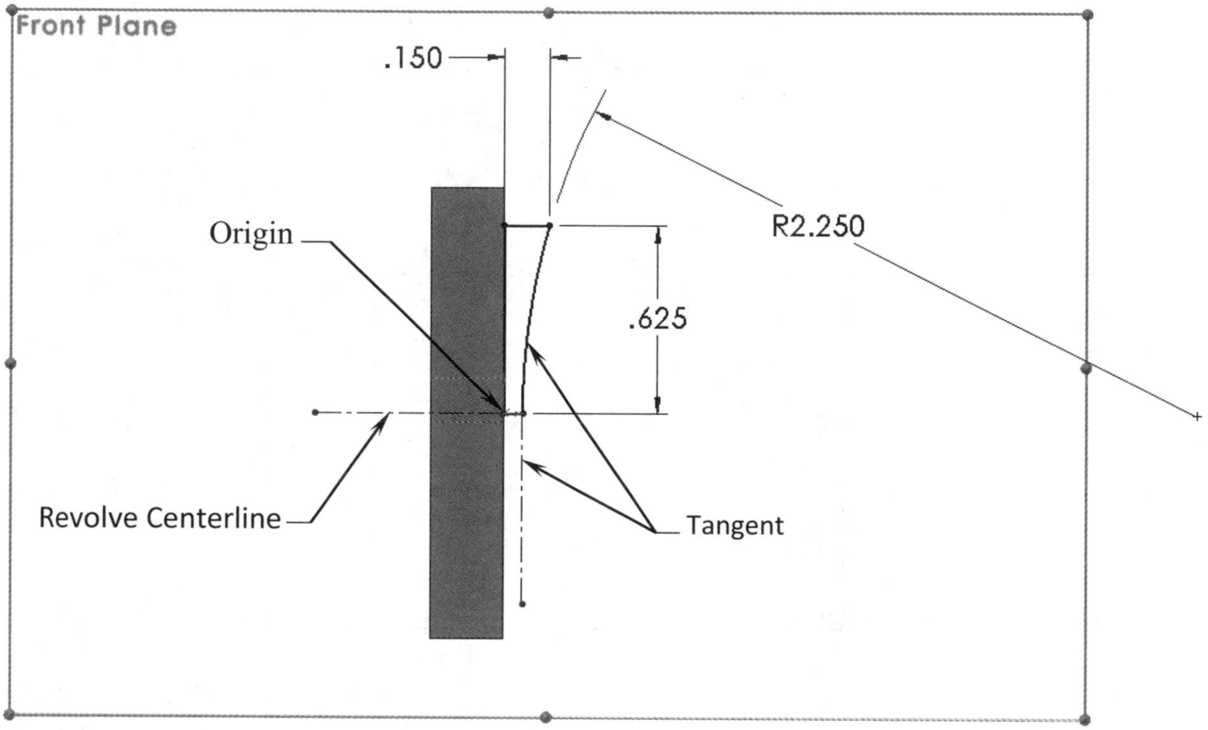

4. Revolving the body:

Click **Revolved Boss/Base** .

Revolve **Blind**.

Angle: **360°**

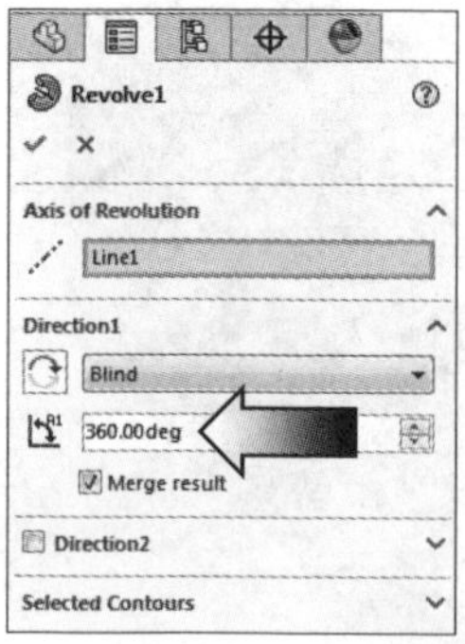

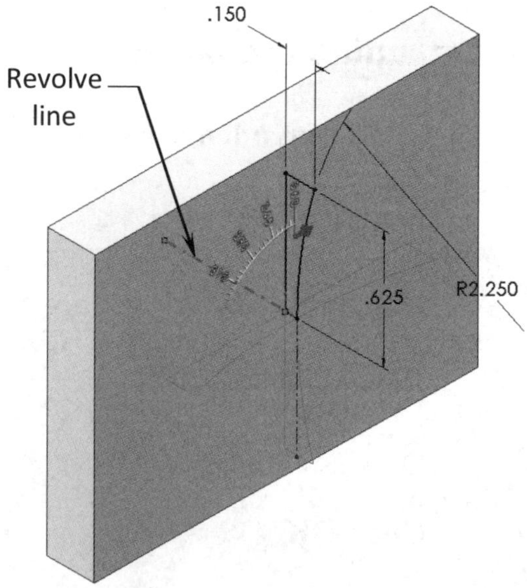

Click **OK**.

5. Adding a fillet:

Click **Fillet** and enter **.080 in**. for Radius.

Select the **edge** as indicated.

Click **OK**.

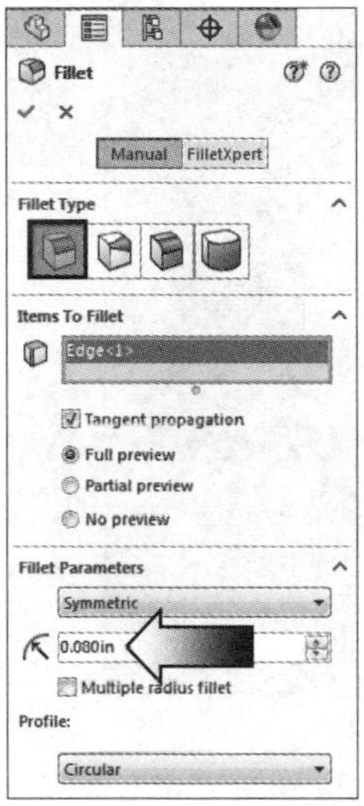

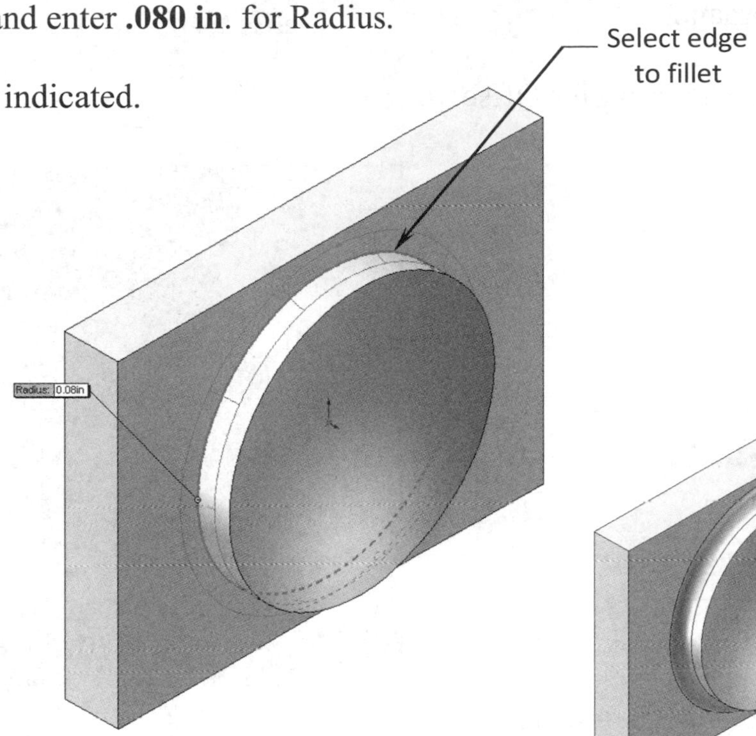

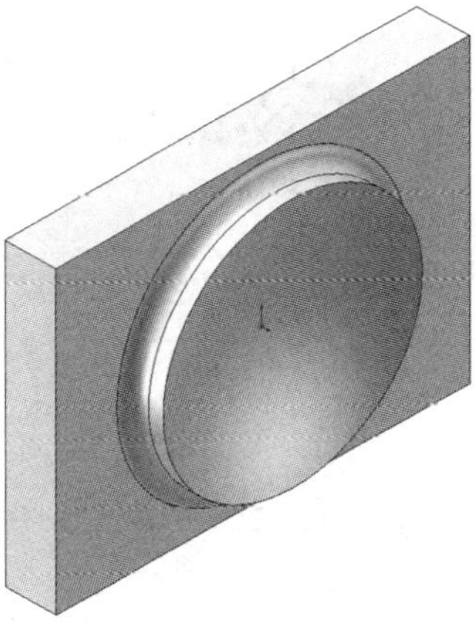

6. Sketching the 1st slot profile:

We will take a look at two different methods to create the slots.
For the 1st Arc Slot let us try the **Offset Entities** option.

Select the <u>face</u> as indicated and open a **new sketch**. Draw a **Center-Point-Arc** following the 3-click as shown below.

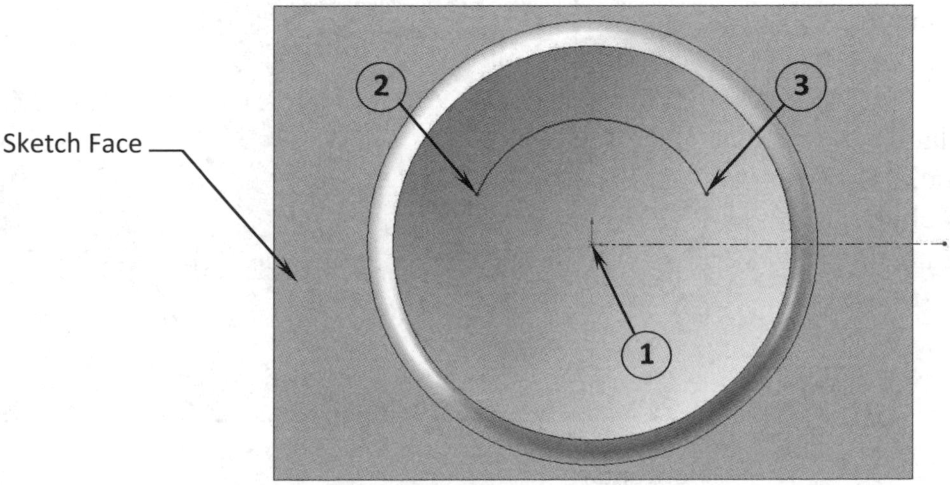

While the arc is still highlighted, click the **Offset Entities** command.

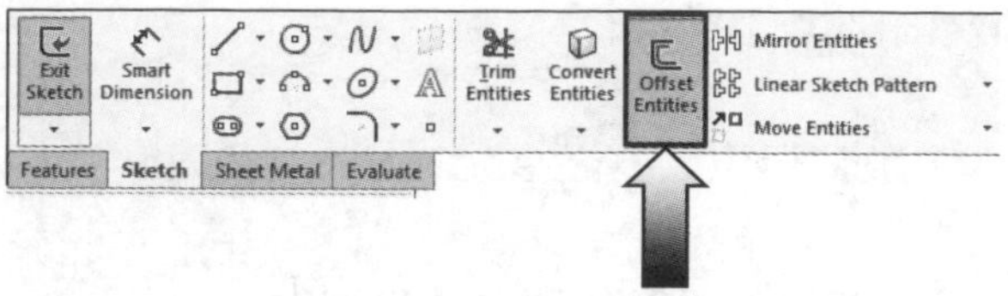

Enter **.0625 in**. for Offset Distance.

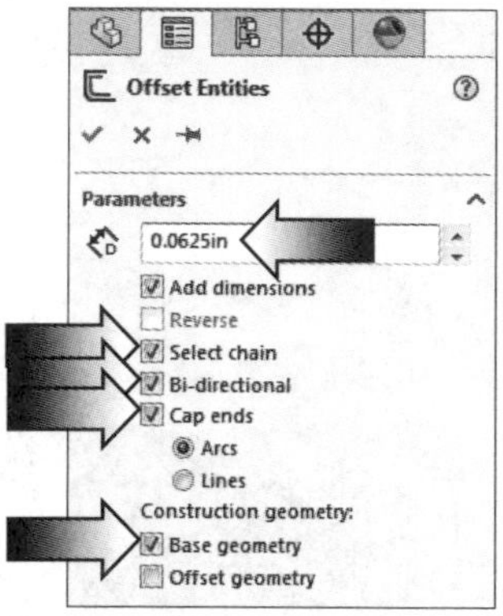

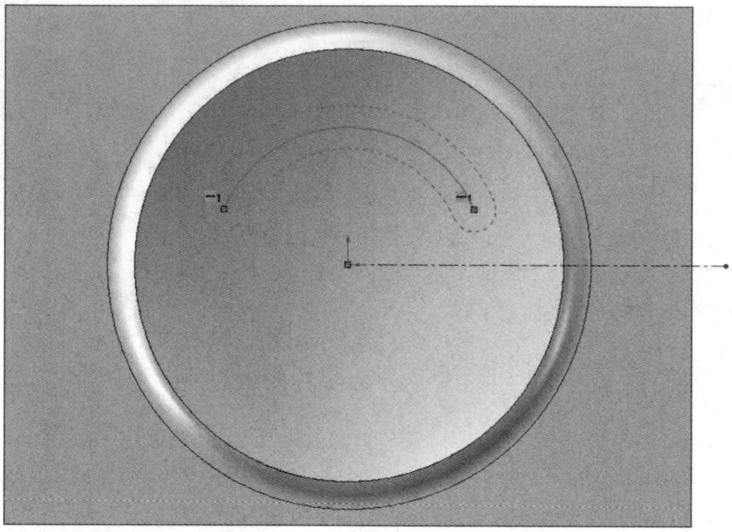

Enable the following:

* **Add Dimensions** * **Select Chain** * **Bi-Directional**

* **Cap End / Arcs**. * **Base Geometry** (to convert the arc to construction)

Click **OK**.

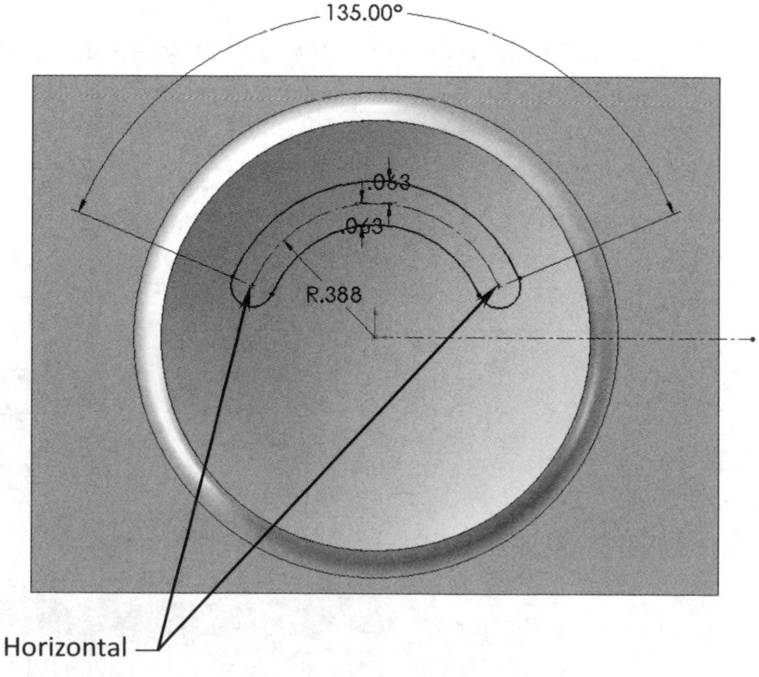

Add the dimensions and a horizontal relation shown to fully define the sketch.

To create the angular dimension: First click the origin and then click the centers of the 2 arcs.

Note:
We will use the Arc-Slot command to create the 2nd slot in step 8.

7. Creating the 1st Split Line:

Click **Split Line**  from the **Curves** toolbar or select: **Insert / Curve / Split Line.**

Select the face as indicated to split.

Click **OK**.

> **Split Line**
>
> The Split Line command projects an entity to the face and divides it into multiple faces.

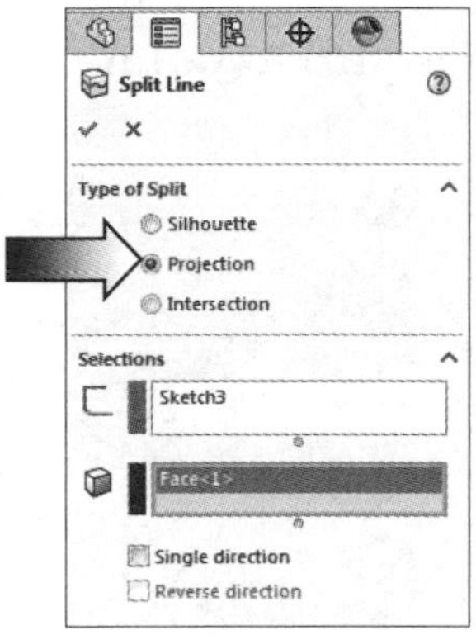

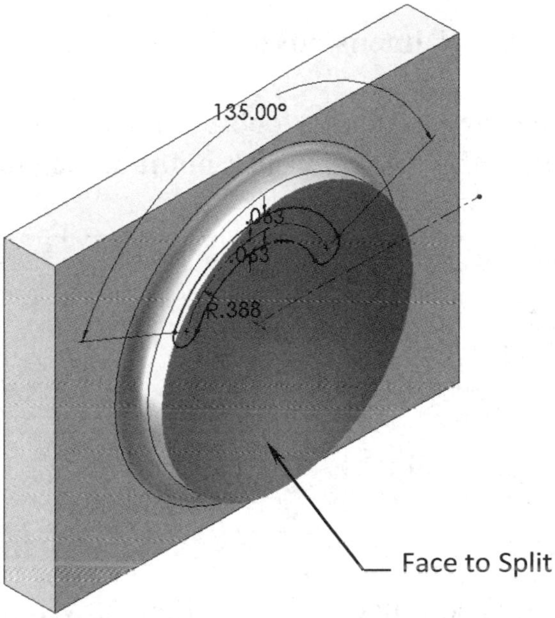

The selected face is split into a new, separate surface.

This new surface can now be used as Faces-to-Remove when the form tool is inserted onto a sheet metal part.

The Faces-to-Remove option specifies what features/area will get a through cut.

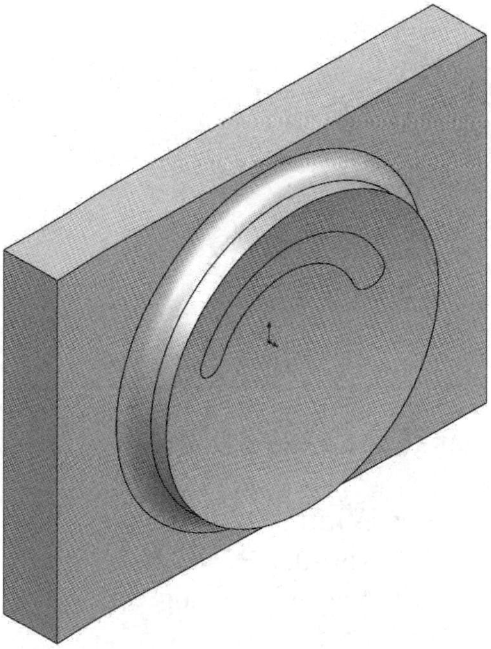

8. Creating the 2nd slot profile:

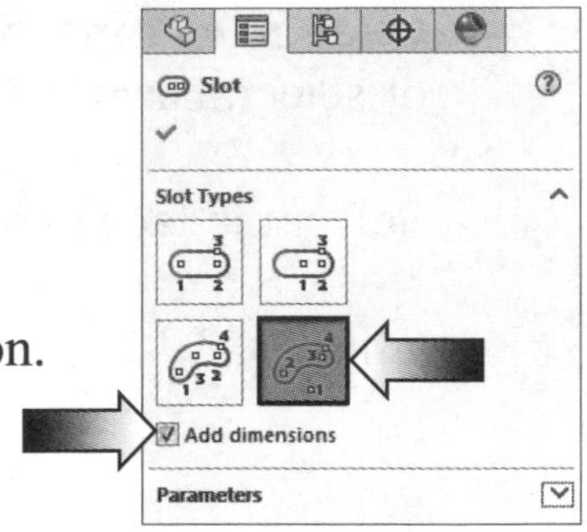

This time we will try another method to create the 2nd curved slot. (The Copy & Paste option also works well.)

Select the Face as noted* and open a **new sketch**.

Click **Straight-Slot** and select the **Center Arc Slot** option.

Enable the **Add Dimensions** checkbox.

Start at the Origin and click **point 1**, move outward and click **point 2**, move the cursor to the other side and click **point 3**, then drag down (or upward) to **point 4**.

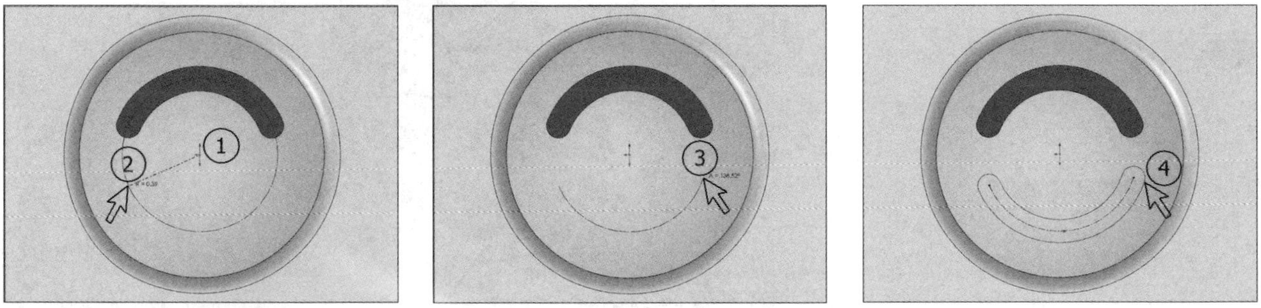

The arc-slot is completed with the Radius, Width, and Angular dimensions.

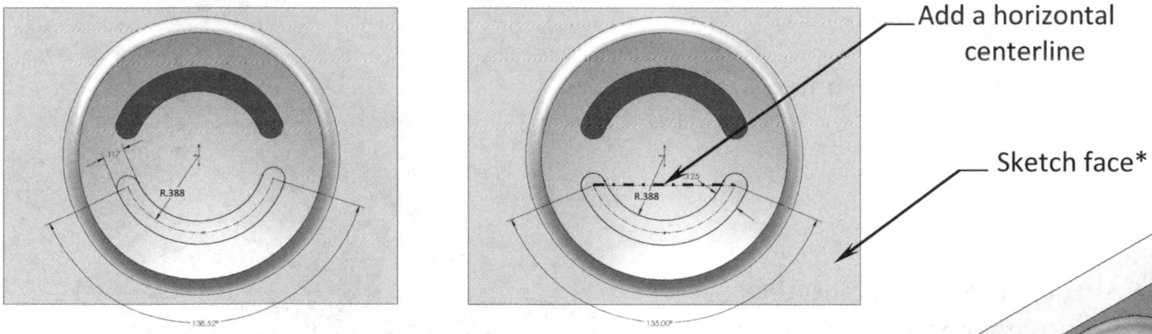

Change the values of the dimensions to:

* Width: **.125 in.**
* Radius: **R.388 in.**
* Angle: **135 deg.**

Create the **2nd split line** as shown in step 8.

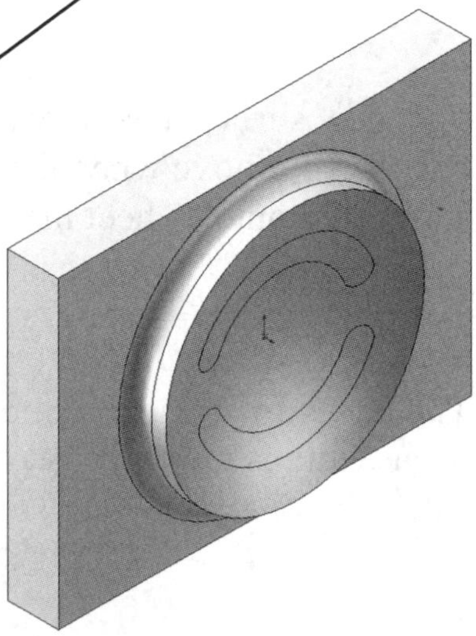

9. Adding more fillets:

Click **Fillet** and enter **.062 in**. for radius.

Select the **edge** as indicated.

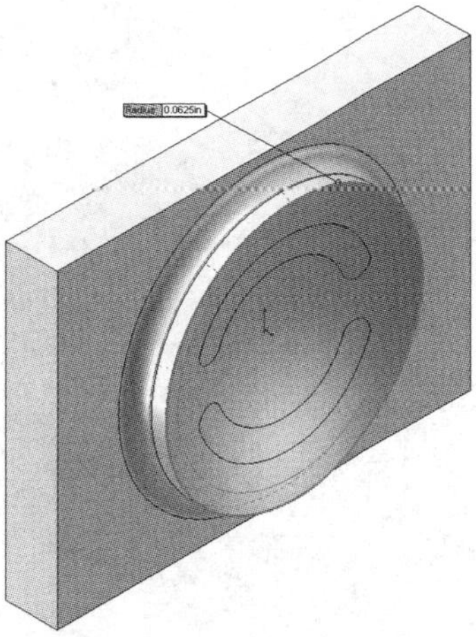

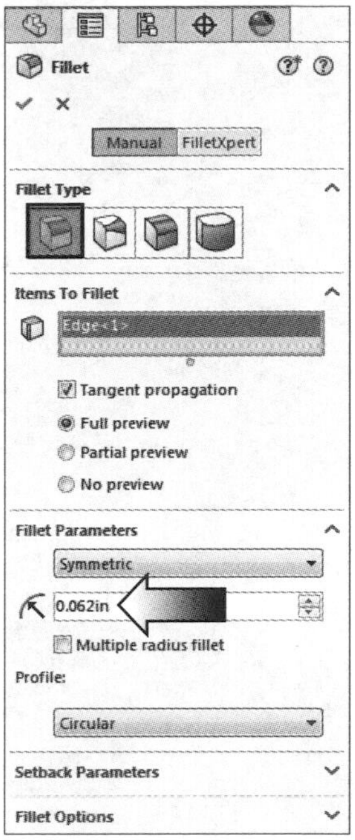

Click **OK**.

10. Inserting the Forming Tool feature:

Select **Insert / Sheet Metal / Forming tool** from the pull-down menu.

Select the **Stopping-Face** and the **Faces-to-Remove** as indicated.

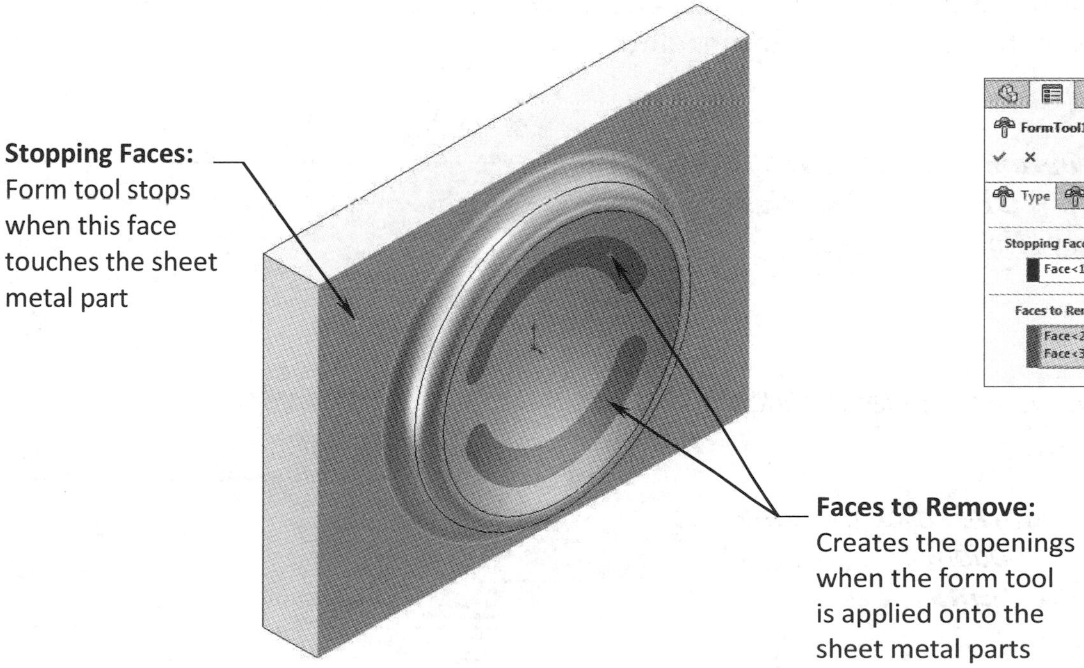

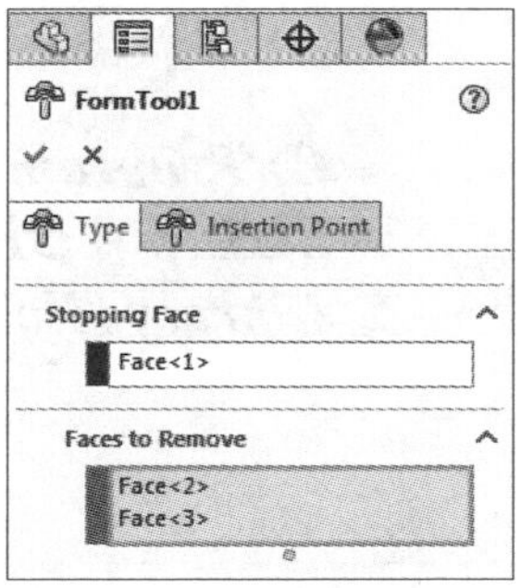

Click **OK**.

The completed form tool.

11. Saving the Forming tool:

Click **File / Save As / Button w-Slots**, change the Save as Type to **Form Tool** and save the part in the following directories:

C:/Program Data/SOLIDWORKS/ SOLIDWORKS 2022/Design-Library /Forming Tools.

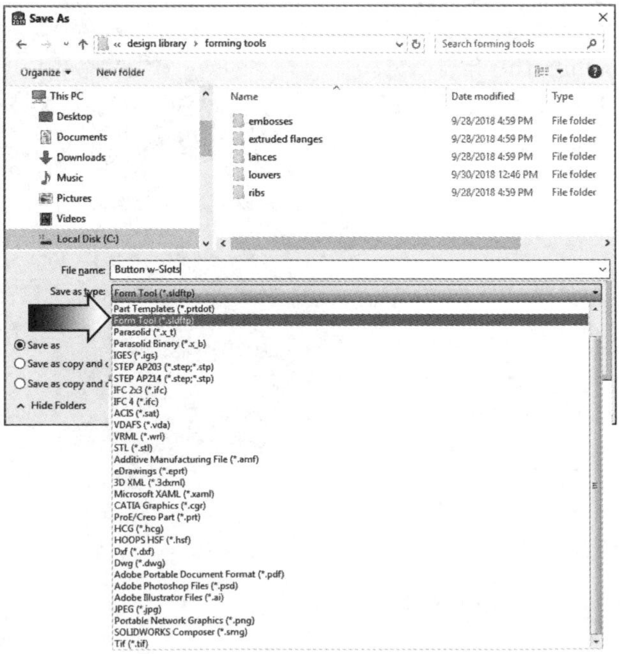

Note:
If the Design Library Folder is Hidden, go to:

Windows Explorer, *click the* ***View*** *tab and enable the option:*
Show Hidden Items

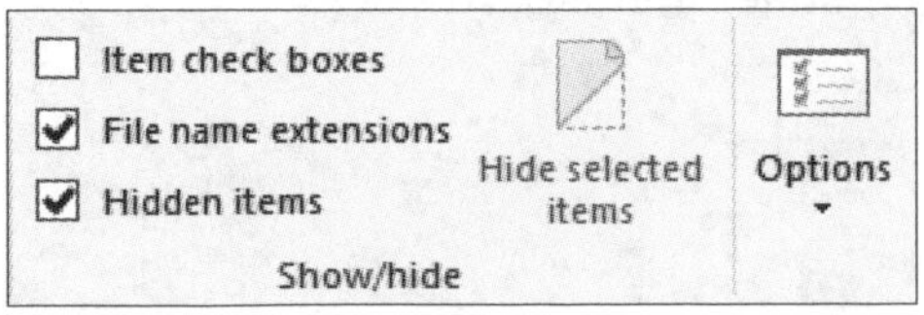

Note:

The Sheet metal forming tools can also be saved by dragging and dropping from the FeatureManager tree to a folder (i.e. Forming Tools) inside the Design Library.

The file name, file type, and path can be selected to save the forming tool at this time.

12. Applying the new forming tool: (Optional)

Create a sheet metal part using the drawing below and test out your new forming tool.

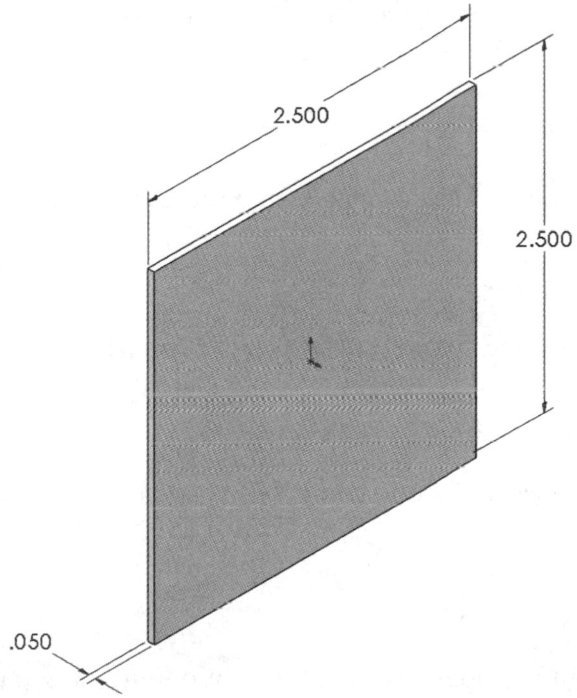

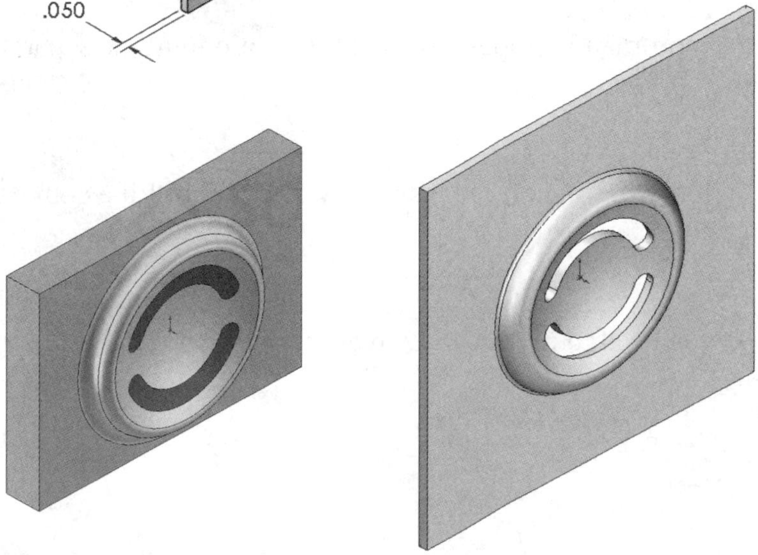

Questions for Review

Sheet Metal Forming Tools

1. Forming tools can bend or stretch sheet metal parts.
   a. True
   b. False

2. Forming tools can be stored in the Design Library window using the file extension:
   a. slddrw
   b. sldftp
   c. dwg

3. Forming tools can be dragged and dropped from the Design Library window.
   a. True
   b. False

4. Forming tools can be used to formed surfaces and solid parts as well.
   a. True
   b. False

5. The Red color on the face(s) of the forming tool creates openings on the sheet metal parts.
   a. True
   b. False

6. The Split Line command divides a selected face into multiple separate faces.
   a. True
   b. False

7. Only one single closed sketch can be used with the Split Line command.
   a. True
   b. False

8. The _____ key is used to reverse the direction of the forming tool (push/pull):
   a. Shift
   b. Tab
   c. Alt

1. TRUE 2. B
3. TRUE 4. FALSE
5. TRUE 6. TRUE
7. TRUE 8. b

Designing Sheet Metal Parts

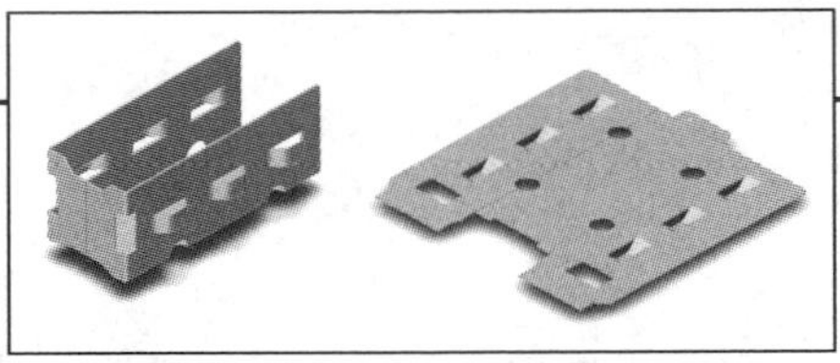

Designing Sheet Metal Parts
Mounting Tray

Sheet metal components are normally used as housings or enclosures for parts or to strengthen and support other parts.

A Sheet Metal part can be created as a single part or it can also be designed in the context of an assembly that has enclosed components.

Forming tools are dies that can bend, stretch, or form sheet metal.

In SOLIDWORKS, forming tools are applied using the "Positive Half" (the raised side) to form features.

When inserting a forming tool, its direction can be reversed using the TAB key (Push or Pull).

The Sheet Metal part can be flattened either by using the Unfold  or Flattened  button, and drawings can be made to show views of the bent or flattened part. Bend lines are also visible in the drawing views.

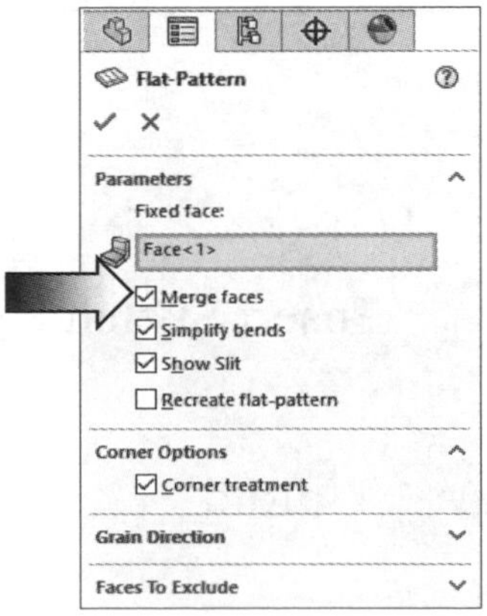

By default, only the Bend-Lines are visible at all times but not the Bend-Regions. To show the Bend Regions, right click the Flat Pattern1 icon at the bottom of the FeatureManager tree, then select Edit Feature and clear the Merge Faces check box.

In this 2nd half of the chapter, besides learning how to create a sheet metal part, we will also learn how to apply the form tool that was created earlier in the 1st half of the lesson.

Mounting Tray
Designing Sheet Metal Parts

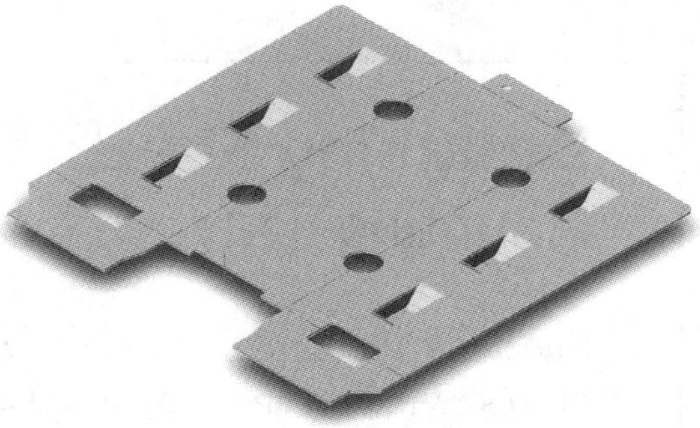

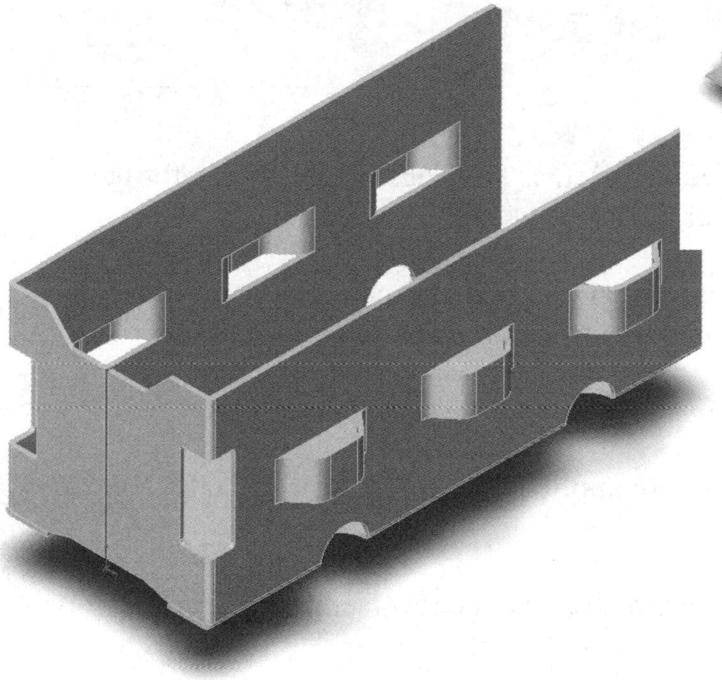

Dimensioning Standards: **ANSI**

Units: **INCHES** – 3 Decimals

Tools Needed:

Insert Sketch	Base Flange	Edge Flange
Unfold	Fold	Extruded Cut
Linear Pattern	Flattened	Break Corner

1. Starting with the base sketch:

Select the Front plane from FeatureManager Tree.

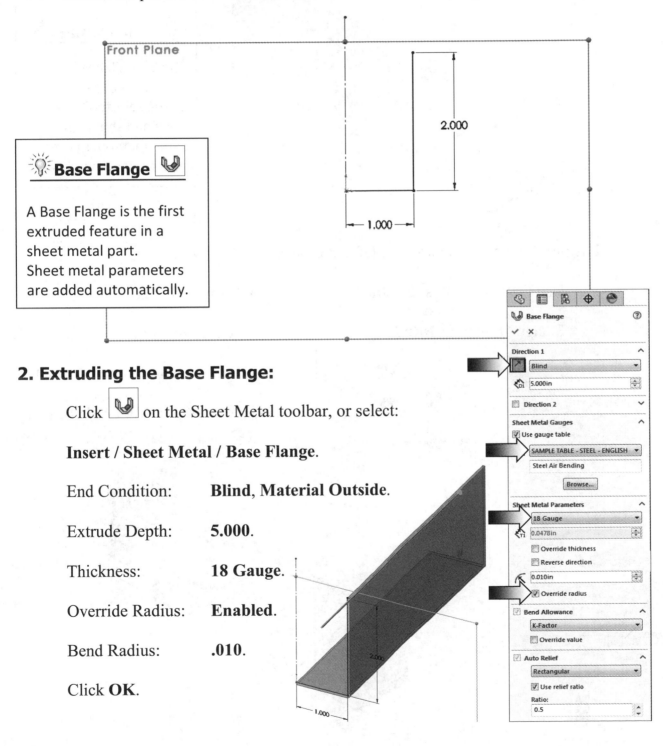

Click or select **Insert / Sketch**.

Sketch the profile shown below and add dimensions to fully define the sketch.

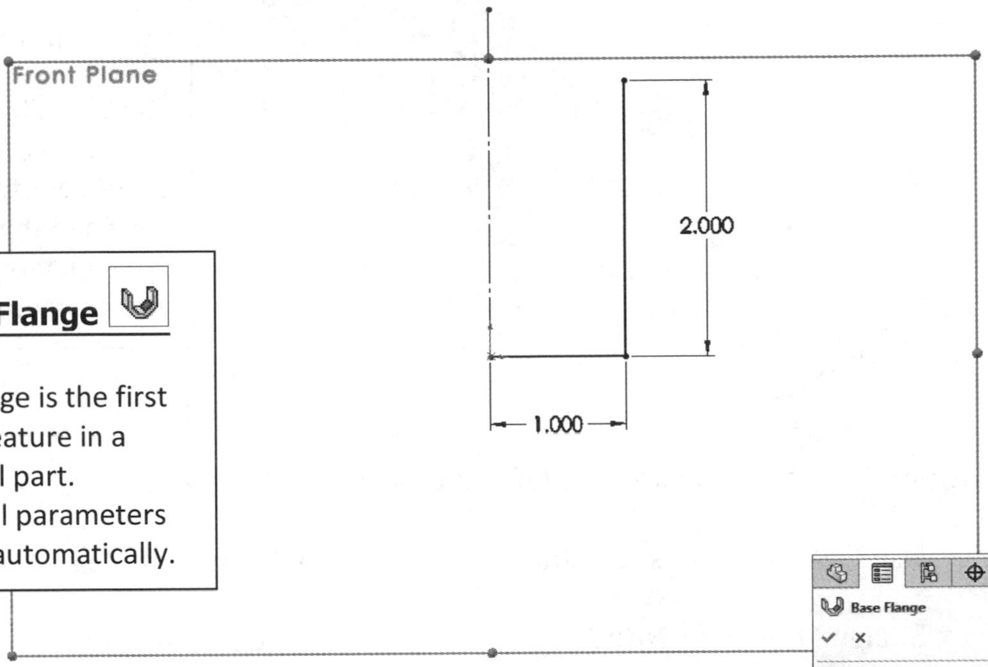

Base Flange

A Base Flange is the first extruded feature in a sheet metal part.
Sheet metal parameters are added automatically.

2. Extruding the Base Flange:

Click on the Sheet Metal toolbar, or select:

Insert / Sheet Metal / Base Flange.

End Condition: **Blind**, **Material Outside**.

Extrude Depth: **5.000**.

Thickness: **18 Gauge**.

Override Radius: **Enabled**.

Bend Radius: **.010**.

Click **OK**.

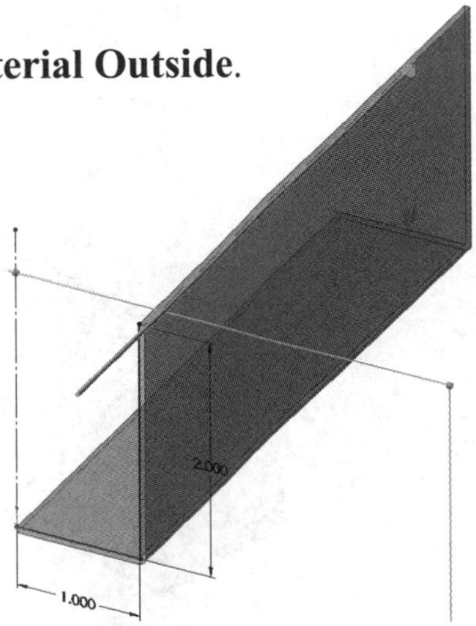

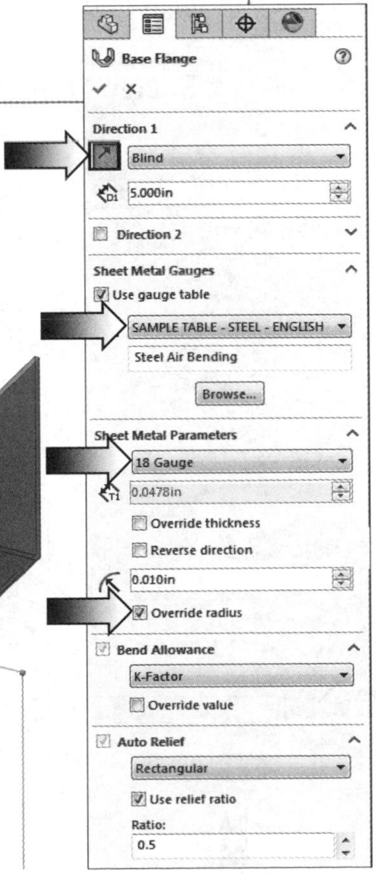

3. Creating an Edge Flange:

Select the **outer edge** as shown and click  on the Sheet Metal toolbar or select: **Insert / Sheet Metal / Edge Flange**.

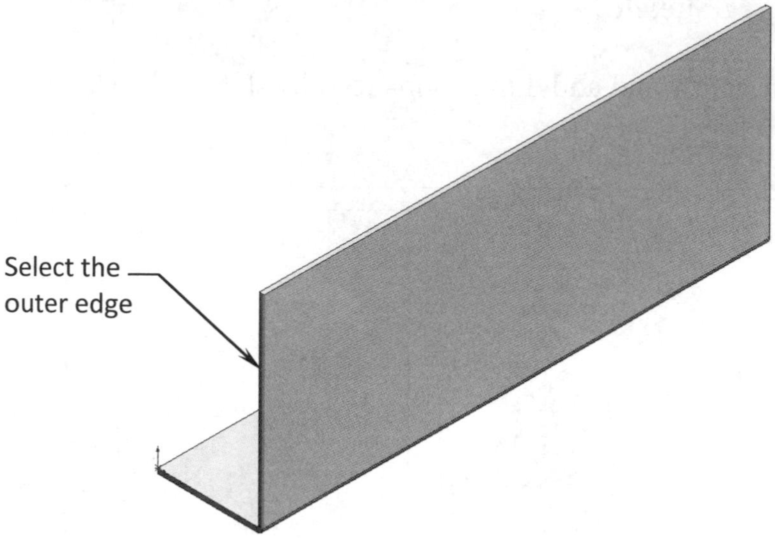

Edge Flange

The Edge Flange command adds a flange to the selected linear edge and shares the same material thickness as the sheet metal part.

Position the flange towards the left side and set the following:

* Use Default Radius: **Enabled**.

* Flange Direction: **Blind**.

* Bend Angle: **90deg**.

* Flange Depth: **.985**.

* Use **Inner Virtual Sharp** (arrow).

* Flange Position: **Material Outside**.

Click **OK**.

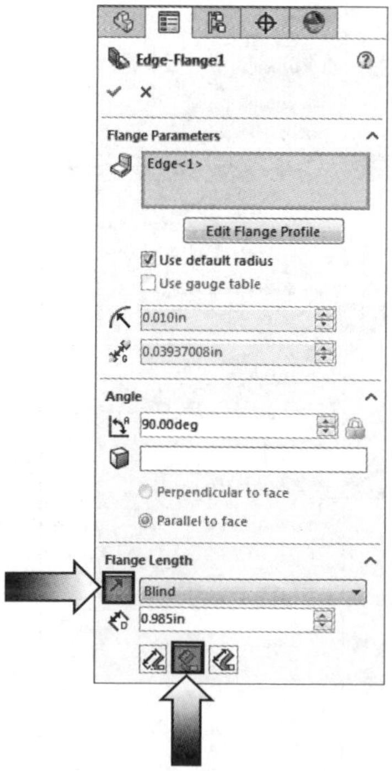

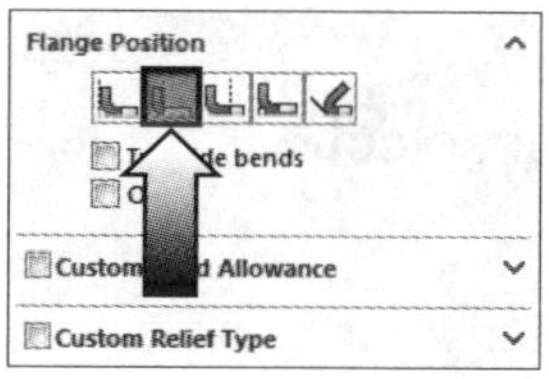

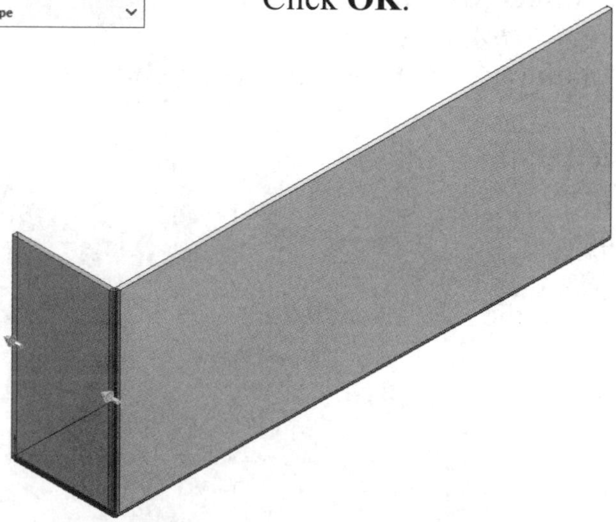

4. Adding cut features:

Select the face as shown and insert a **new sketch** .

Sketch the profile and add dimensions.

All horizontal dimensions are measured from thc centerline.

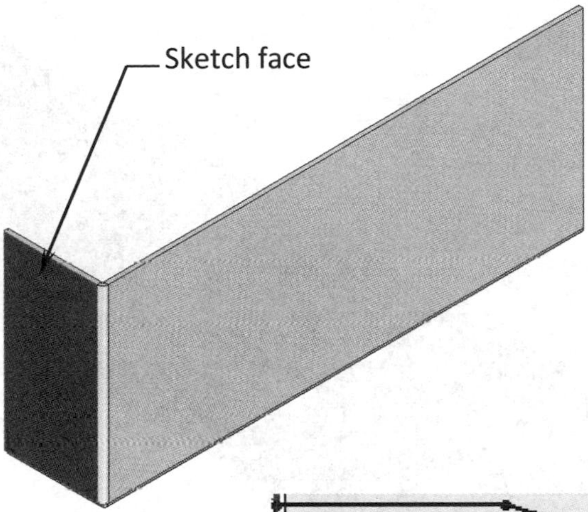

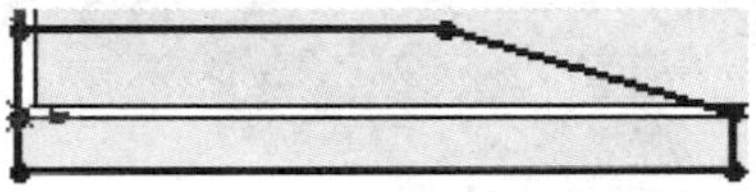

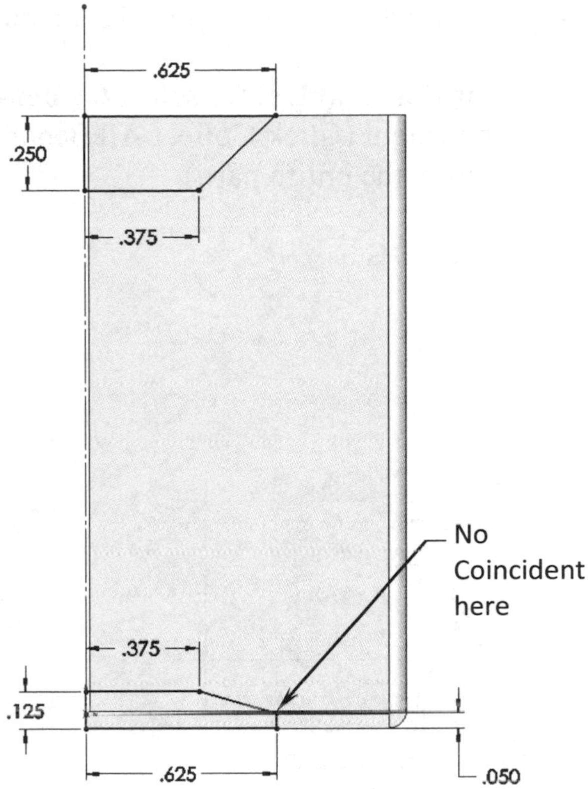

5. Extruding a cut:

Click or select **Insert / Cut / Extrude**.

End Condition: **Blind**.

Extrude Depth: **.165 in.**

Click **OK**.

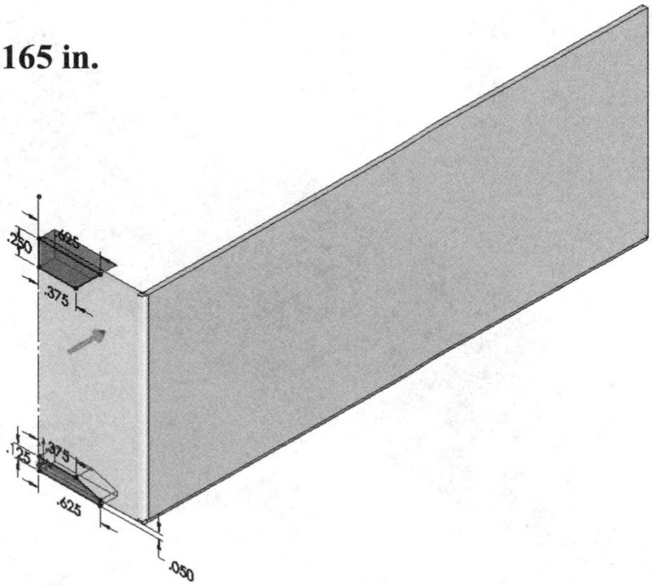

6. Using the Unfold command:

Click on the Sheet metal toolbar or select:
Insert / Sheet Metal / Unfold.

For Fixed face select the **right side face** of the model.

For Bends to Unfold select the **bend radius** as indicated. (Click Collect All Bends if you want to flatten the entire part.)

Click **OK**.

> **Unfold**
>
> When adding cuts across a bend, the Unfold command flattens one or more bend(s) in a sheet metal part.

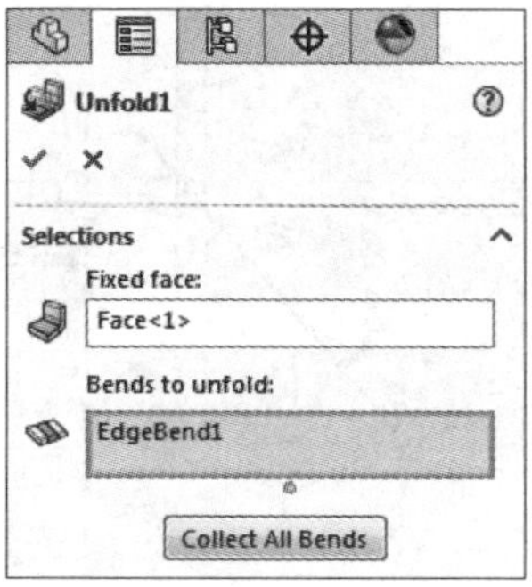

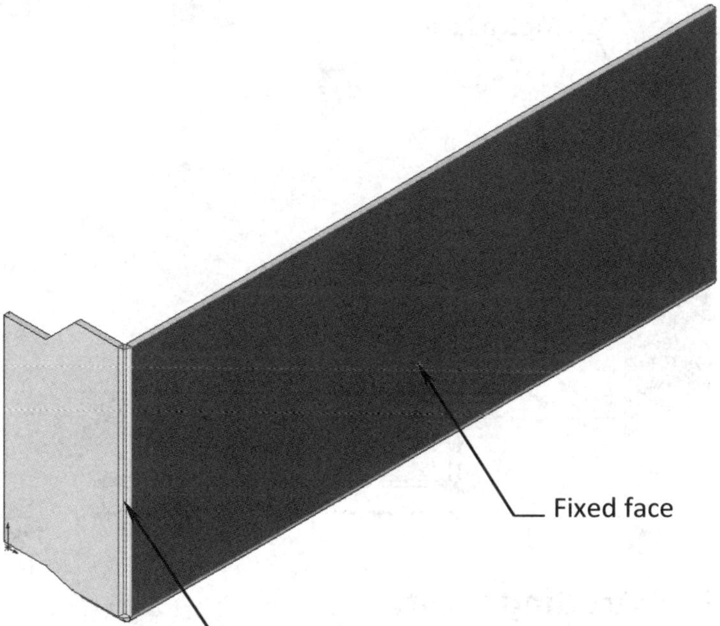

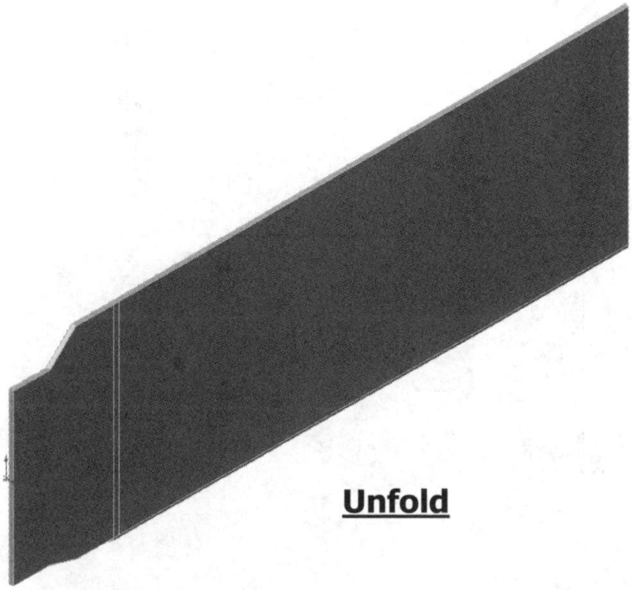

Unfold

7. Creating a Rectangular Window:

Select the face as indicated and insert a **new sketch** .

Sketch a **Corner Rectangle** and a vertical **Centerline** as shown below.

Add dimensions, relation and **4 Sketch Fillets**.

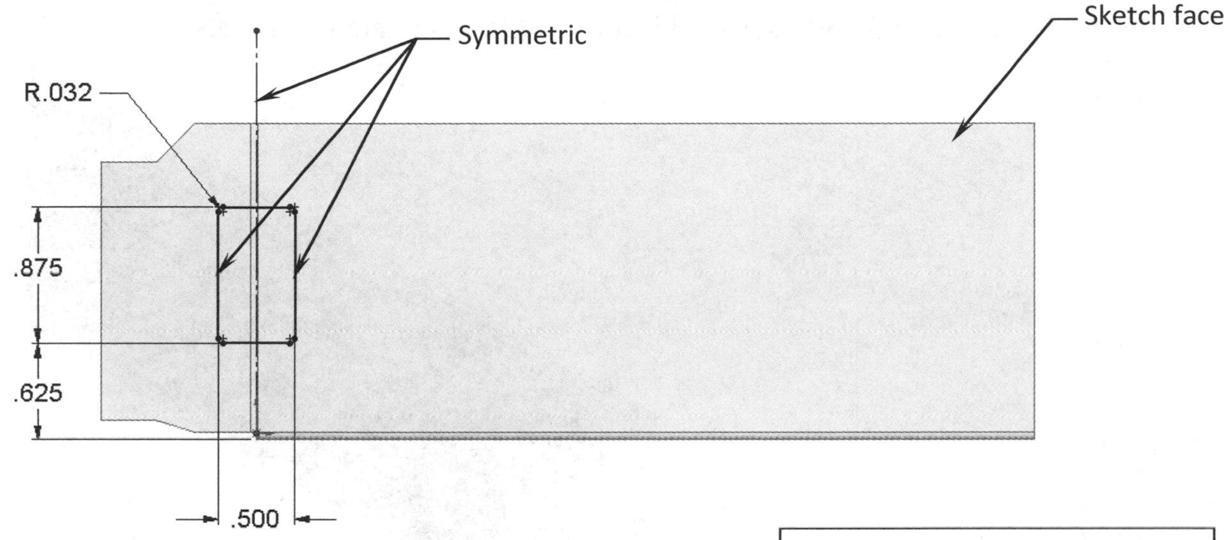

8. Extruding a Cut:

Click or select **Insert / Cut / Extrude**.

End Condition: **Blind**.

Link to Thickness: **Enabled** (default) .

Normal Cut: **Enabled**.

Click **OK**.

Link To Thickness

The Link-to-Thickness option is only available in sheet metal parts.

When this option is enabled, the extruded depth is automatically linked to the part thickness.

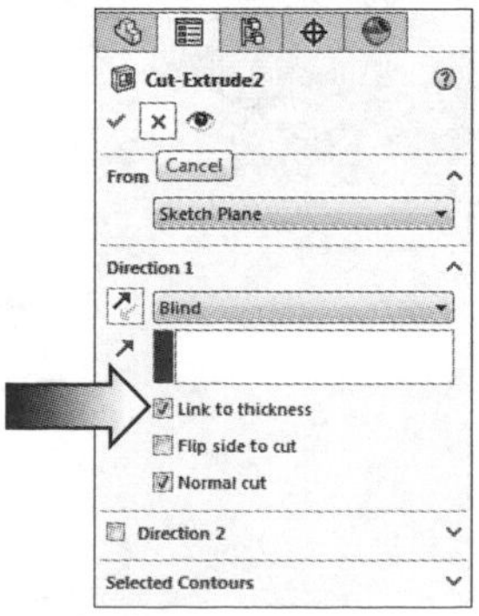

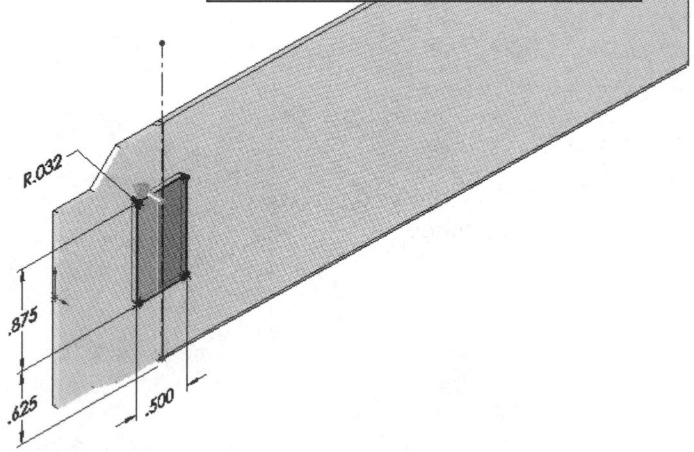

9. Using the Fold command:

> **Fold**
>
> The Fold command returns the bends to their folded state.

Click  on the Sheet Metal toolbar or select:
Insert / Sheet Metal / Fold.

For Fixed face select the **right face** as noted.

Select the **bend radius** as indicated for Bends to Fold.
(If Collect-All-Bends was selected last time, click it again this time.)

Click **OK**.

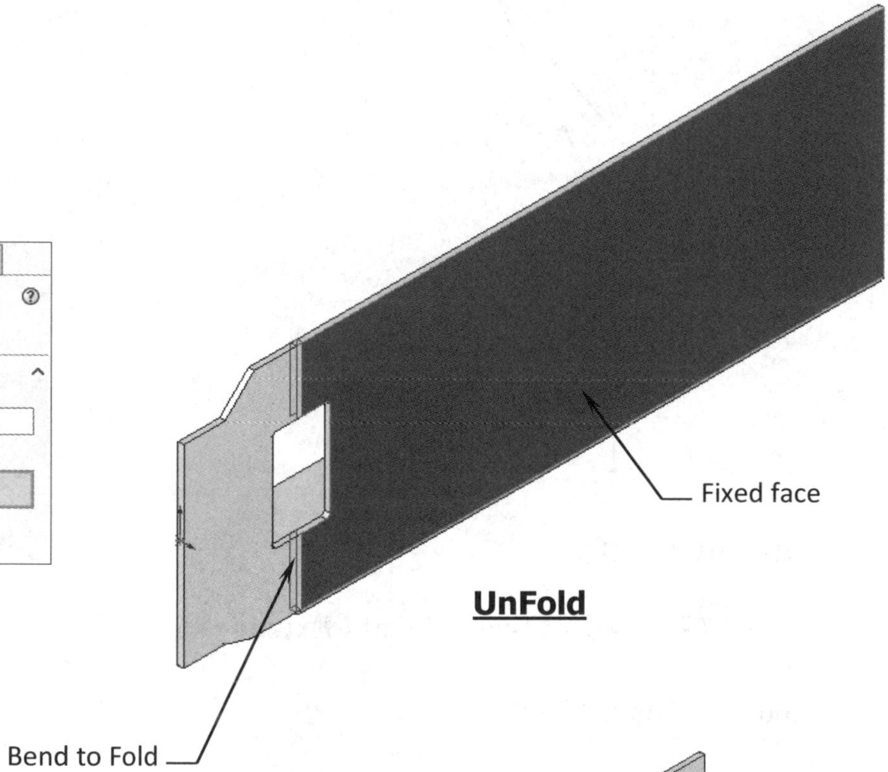

UnFold

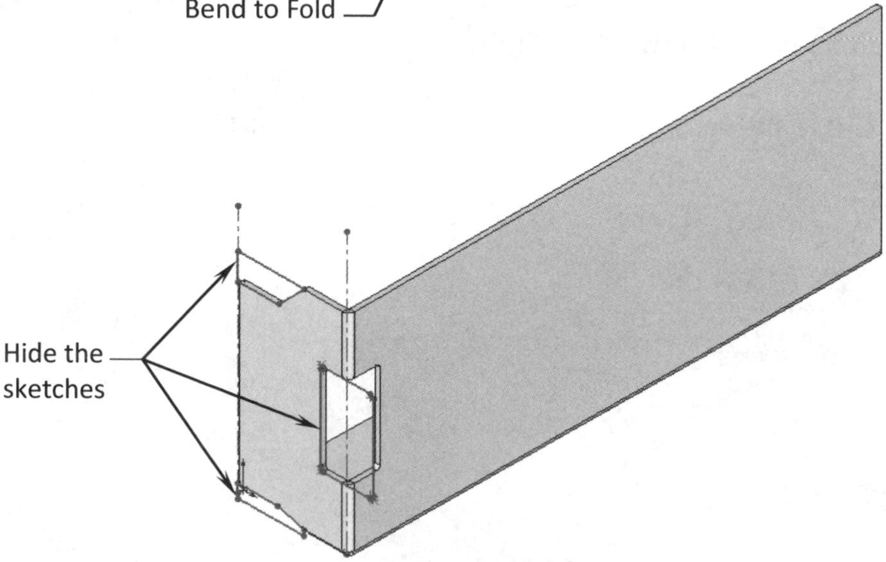

Fold

10. Unfolding multiple bends:

Click  or select **Insert / Sheet Metal / Unfold**.

For Fixed face select the bottom face of the model.

For Bends-To-Unfold select the **faces** of the 2 bends as shown (or click Collect All Bends).

Click **OK**.

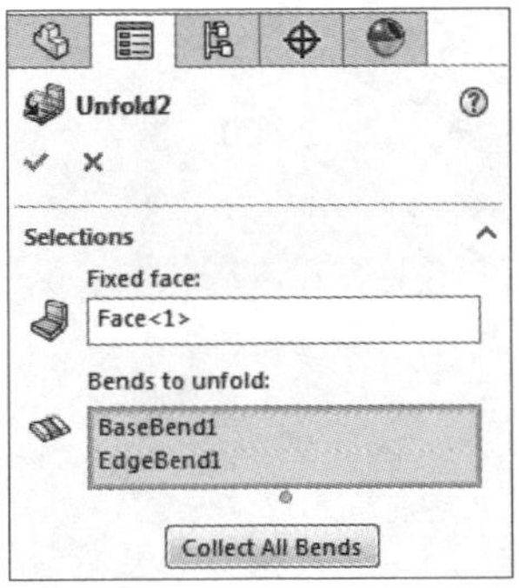

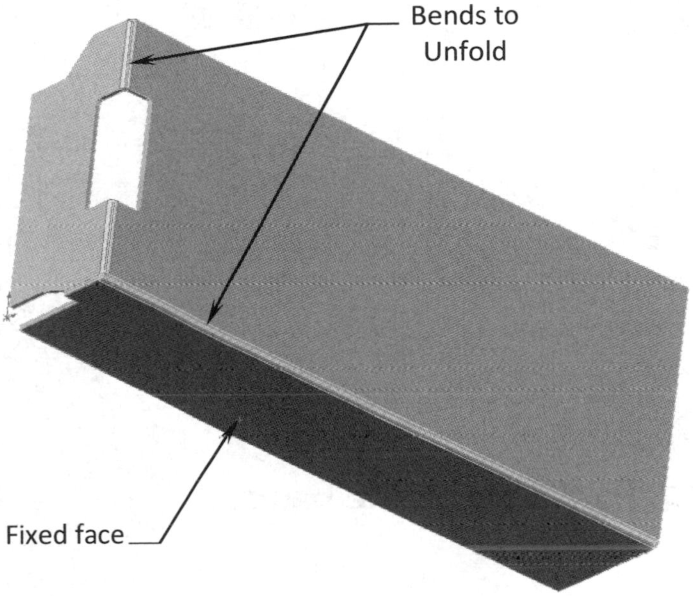

11. Adding more Cuts:

Select the **upper face** as noted and insert a **new sketch** .

Sketch **2 Circles** and add dimensions to size and position them.

Add an **Equal** and a **Vertical** relation between the circles.

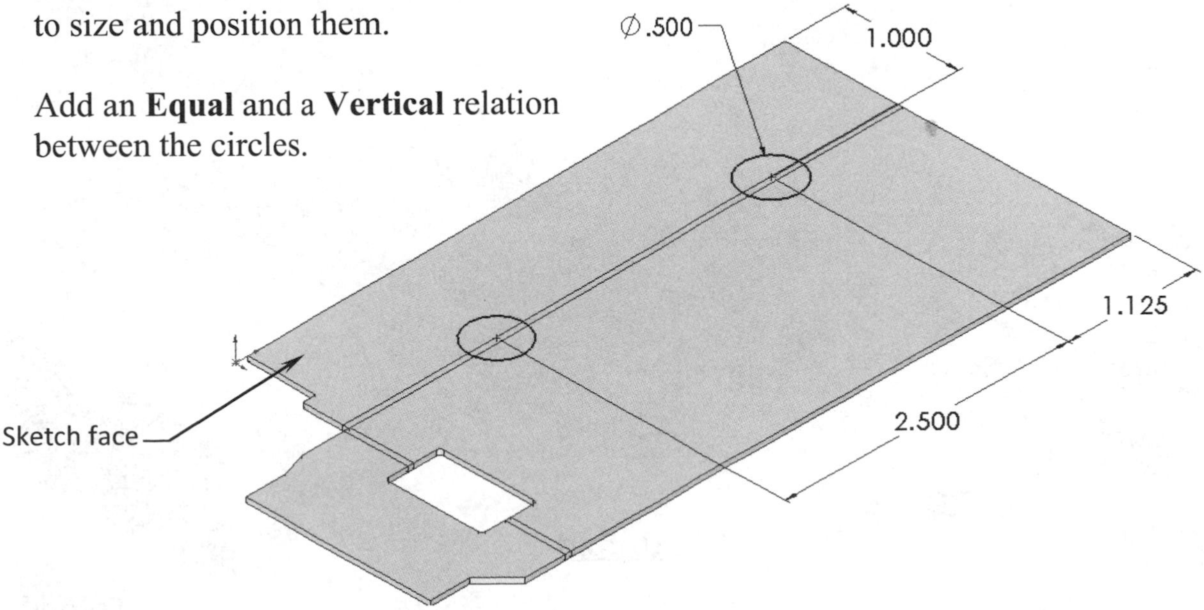

Click  or select **Insert / Cut / Extrude**.

End Condition: **Blind**.

Link To Thickness: **Enabled**.

Click **OK**.

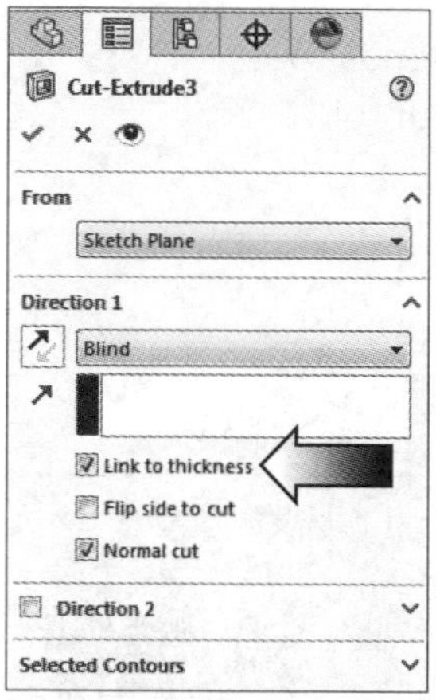

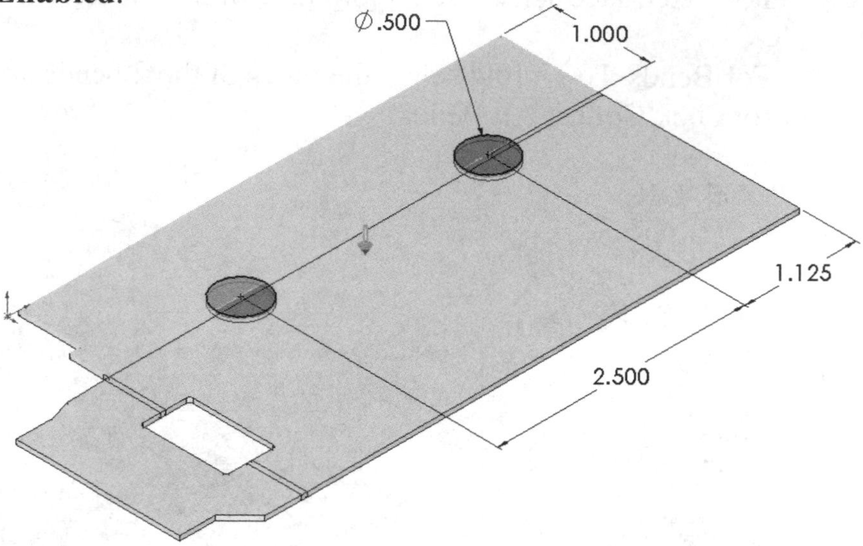

12. Folding multiple bends:

Click  on the Sheet Metal toolbar or select: **Insert / Sheet Metal / Fold**.

The Fixed face is still selected by default.

Under Bends to Fold, click **Collect-All-Bends**.

Click **OK**.

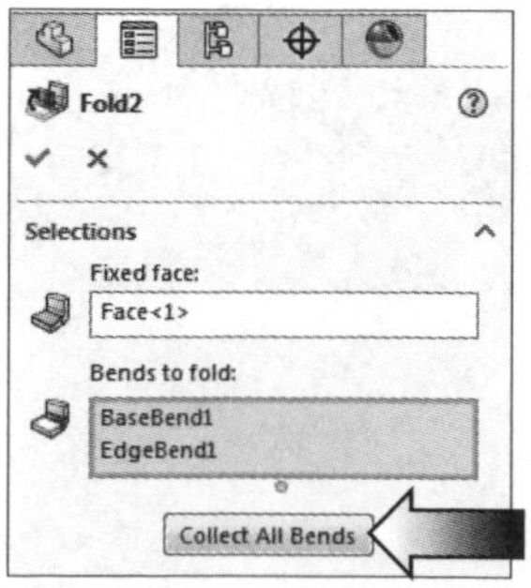

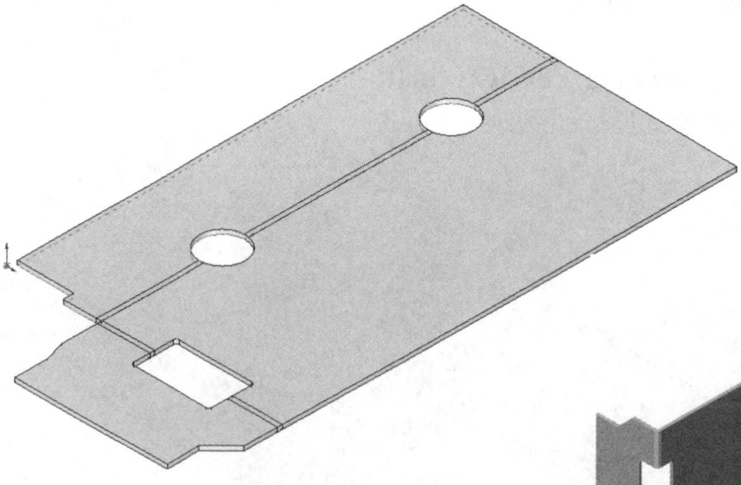

Unfold

Folded

13. Inserting a Sheet Metal Forming Tool:

Click the **Design Library** icon  and click the push pin  to lock it.

> **Forming Tools**
>
> Forming tools are dies that can bend, stretch, or form Sheet Metal and they must be dragged from the Design Library window onto the part.

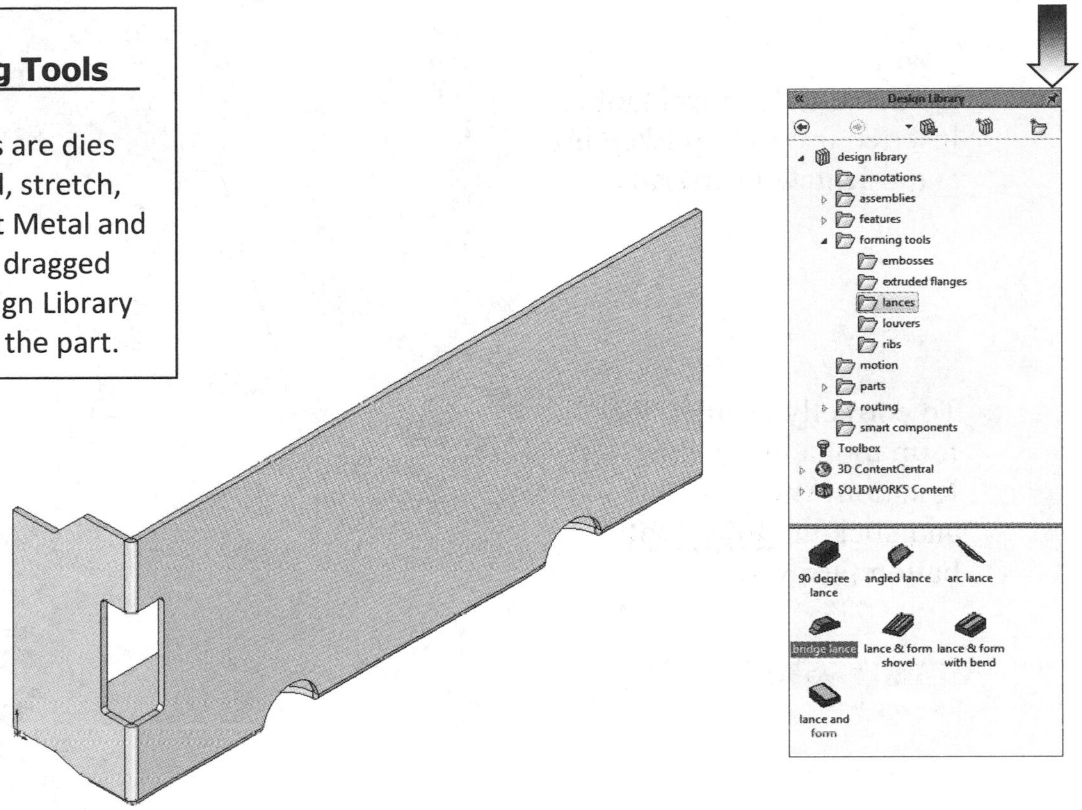

Expand the **Design Library** folder .

Click on the **Forming Tools** folder.

Click on the **Lances** folder.

Locate the **Bridge Lance** form tool.

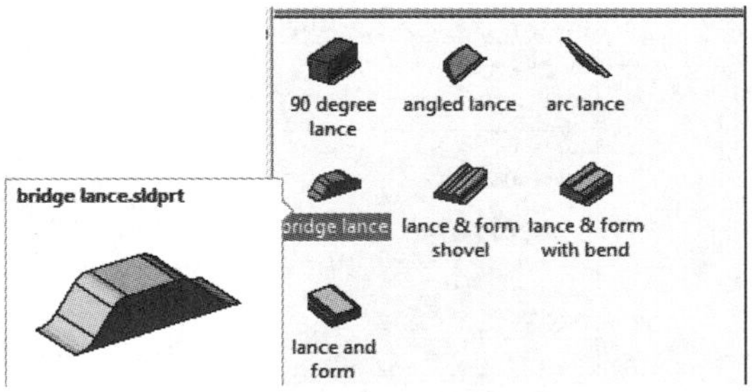

Hover the mouse cursor over the Bridge Lance icon to see its preview graphics.

SOLIDWORKS includes some sample forming tools to get you started. These form tools can be customized and used in different sheet metal parts.

Drag the Bridge Lance* from the Task Pane and drop it on the sheet metal part approximately as shown.

By default, this form tool is inserted inwards (pushed in) and orientated vertically.

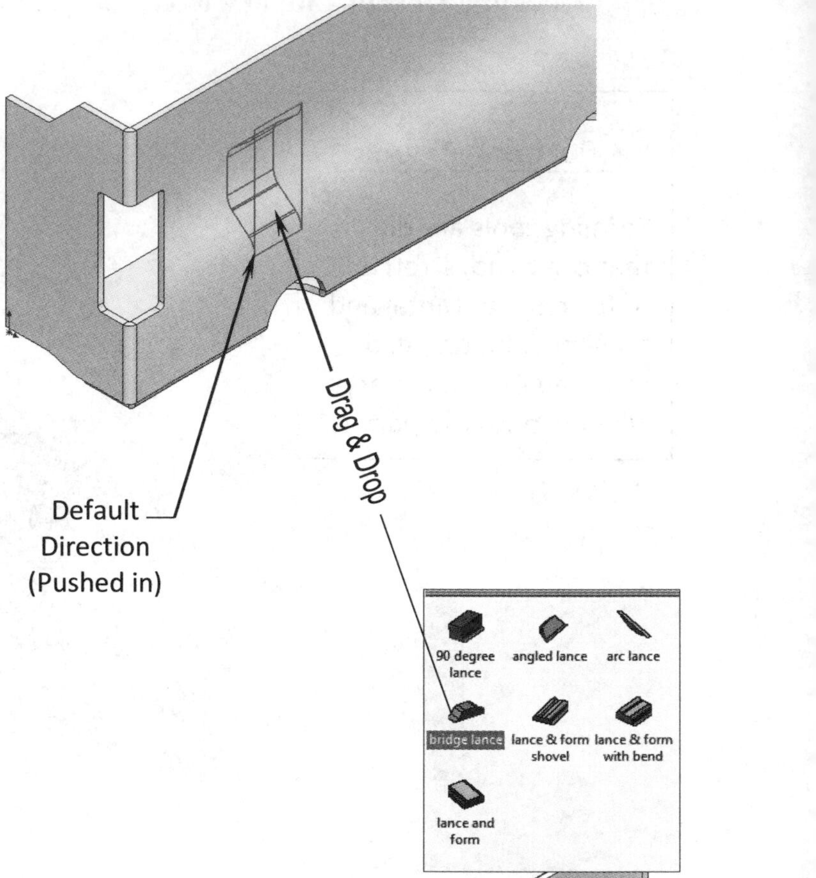

To correctly position the form tool, change the Rotation Angle to **180°** and click the **Flip Tool** button (arrows).

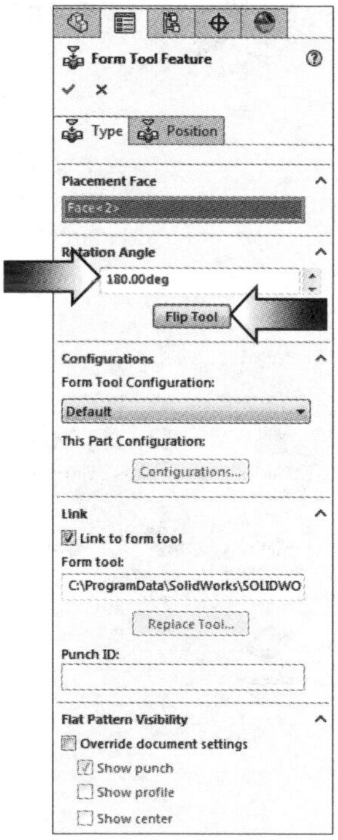

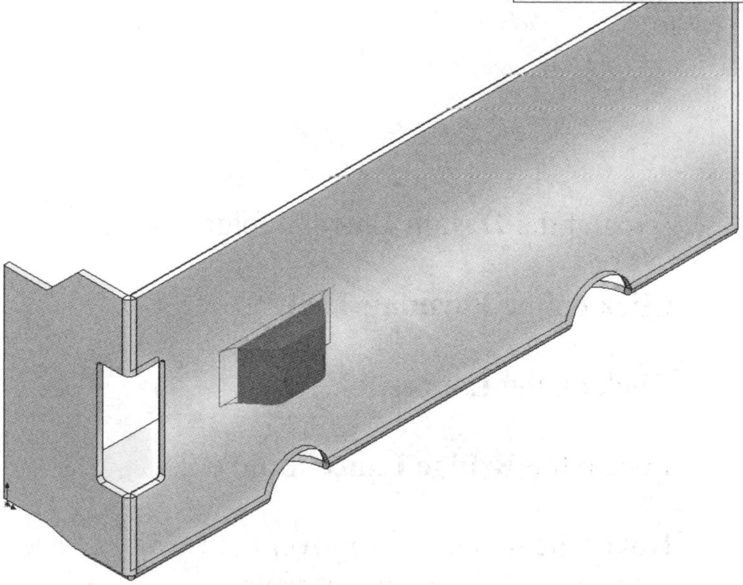

** If the Bridge Lance fails to form, double click on its icon to open the actual part, and re-save it in the same location but using the new Forming Tool extension (.sldftp)*

14. Locating the Bridge Lance:

Click the **Position** tab (arrow).

Add the dimensions shown below to correctly position the formed feature.

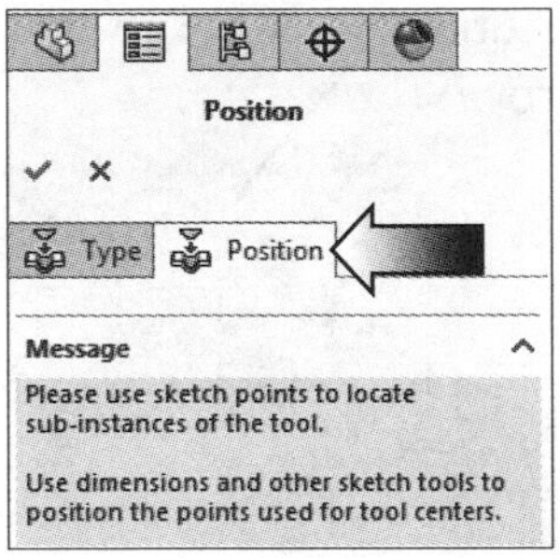

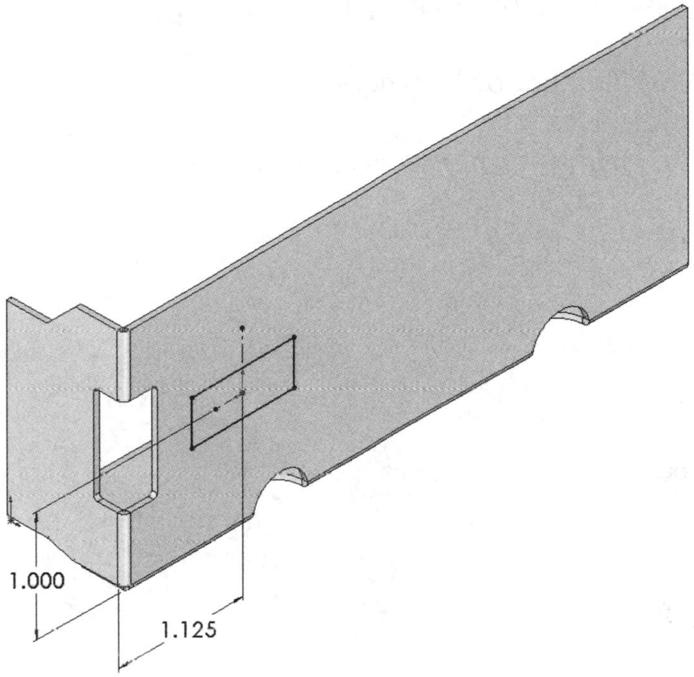

Dimensions <u>from the centerlines</u> to the outer-most edges of the part.

Click 

.

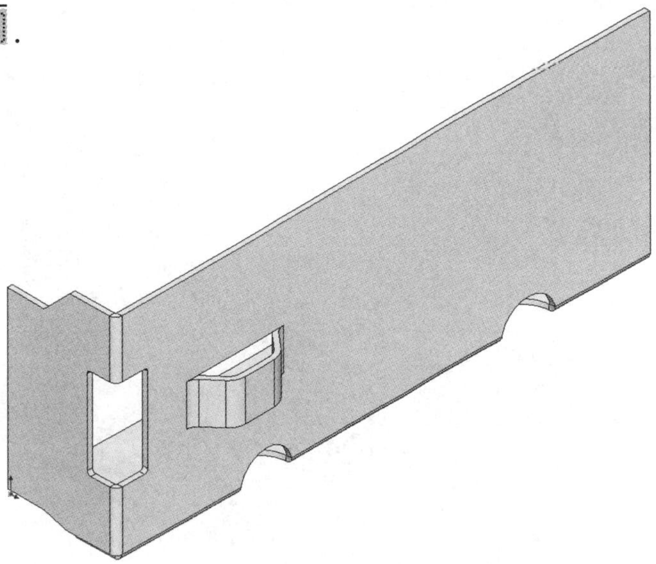

Unpin the Design Library tree  to put it away temporarily.

15. Creating the Linear Pattern of the Bridge Lance:

Click **Linear Pattern** or select **Insert / Pattern Mirror / Linear Pattern**.

Select the top **horizontal edge** of the part as Pattern Direction.

Enter **1.500** in. for Instance Spacing.

Enter **3** for Number of Instances.

Select the **Bridge-Lance** as Features to pattern.

Click **OK**.

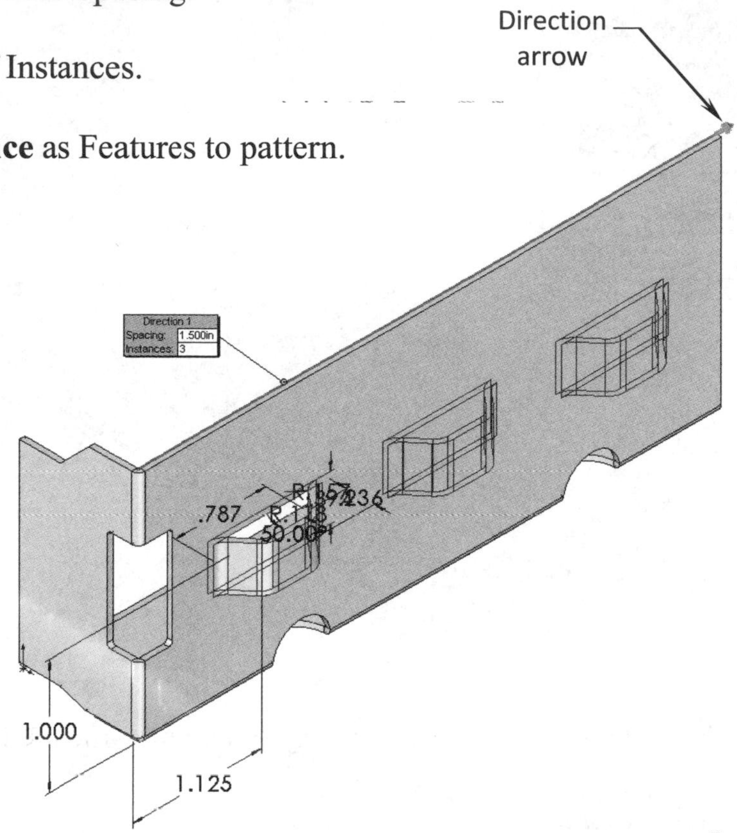

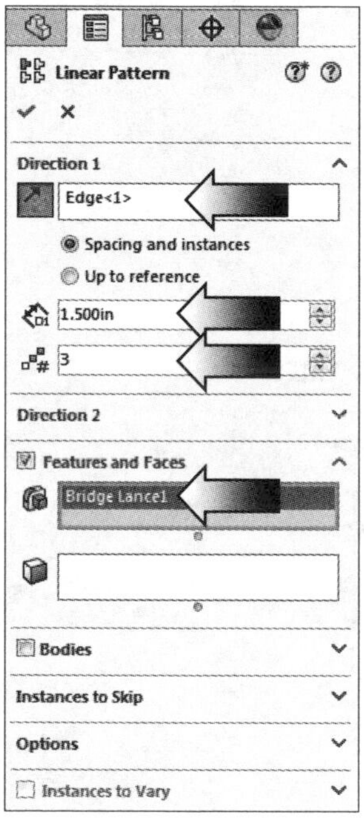

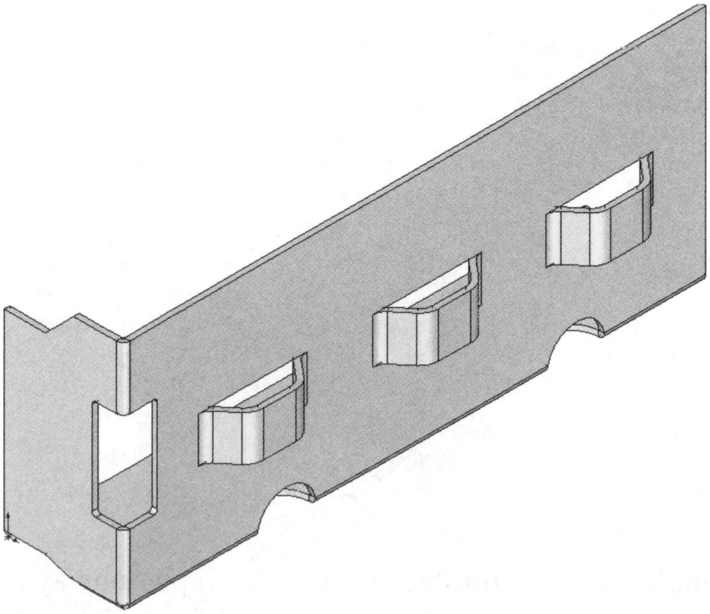

16. Mirroring the body:

Rotate the part and select the face as indicated for mirror face.

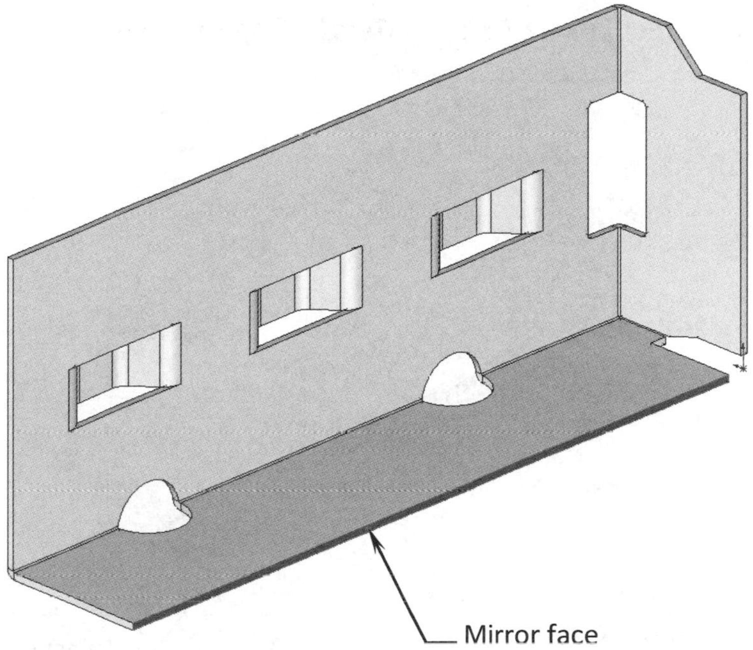

Click **Mirror**  or select **Insert / Pattern Mirror / Mirror**.

Expand the **Bodies to Mirror** section and select the body as shown.

Click **OK**.

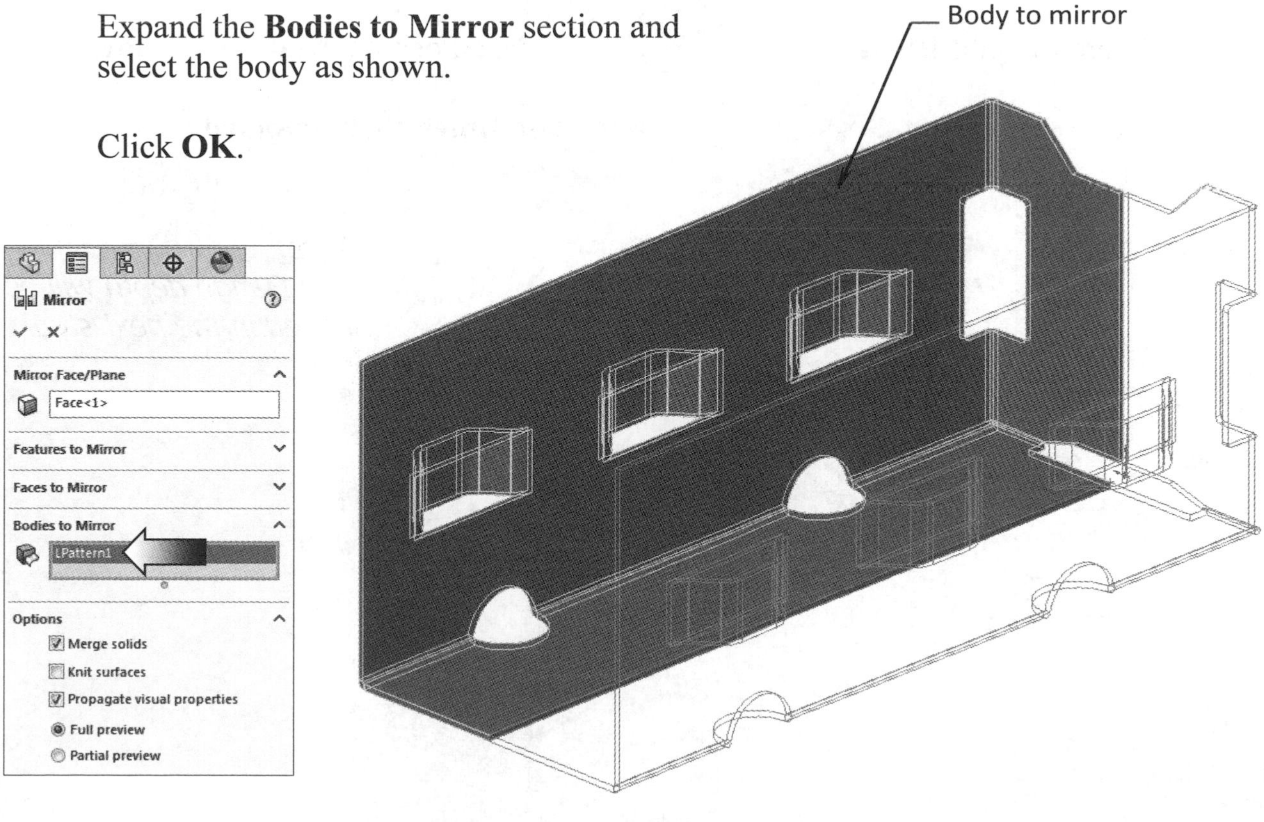

17. Adding the rear Edge Flange:

Select the <u>edge</u> as indicated.

Click  or select **Insert / Sheet Metal / Edge Flange**.

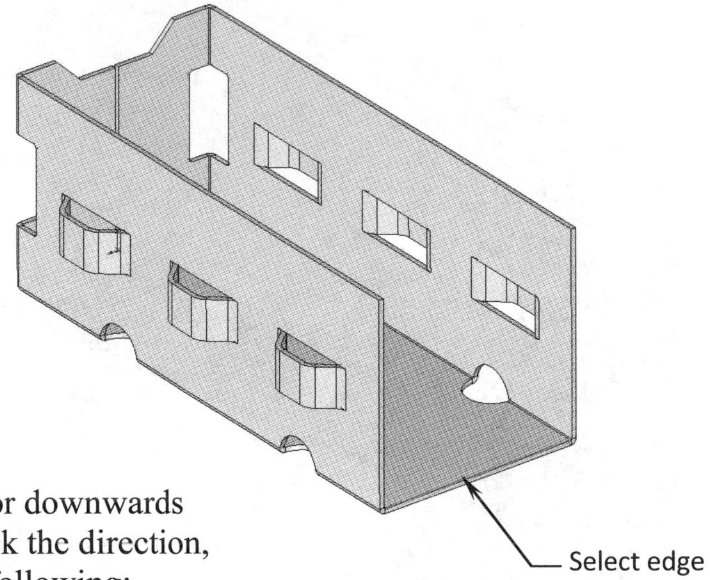

Move the cursor downwards and click to lock the direction, then enter the following:

Use Default Radius: **Enabled.**

Flange Length: **Blind.**

Bend Angle: **90deg**.

Flange Position: **Material Inside**.

Use **Inner Virtual Sharp**.

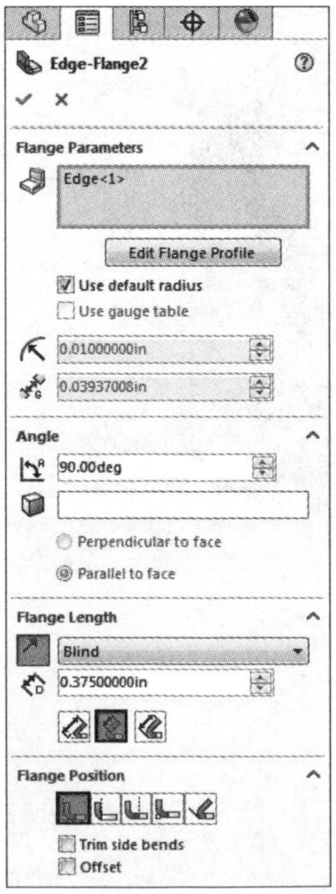

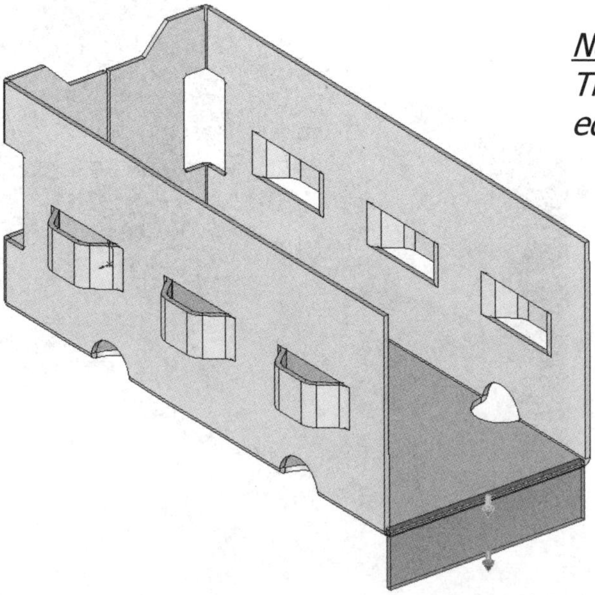

<u>Note:</u>
The Flange depth will be edited in the next step.

18. Resizing the Edge Flange:

Click the **Edit Flange Profile** button (arrow).

The 2D sketch of the flange is activated; its shape and size can now be modified.

Drag the 2 outer-most vertical lines inward (pictured).

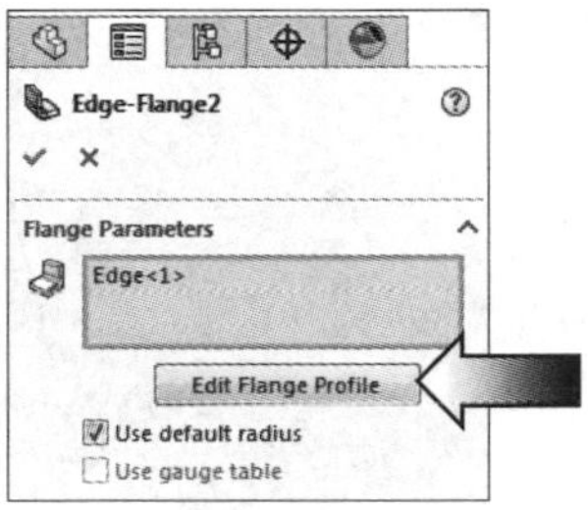

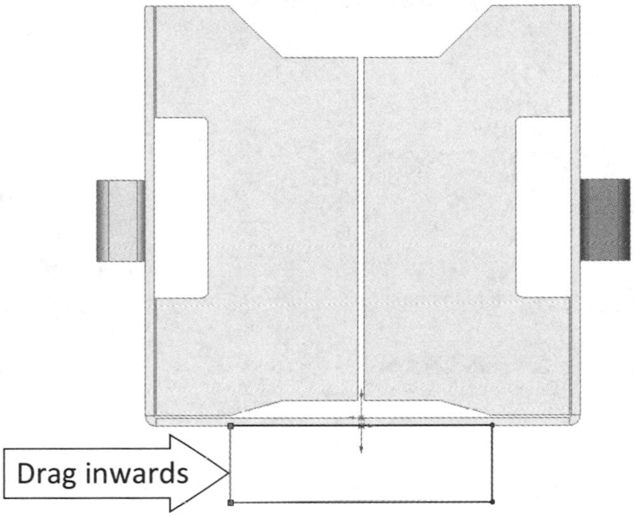

Sketch **2 Circles** and add the Dimensions shown.

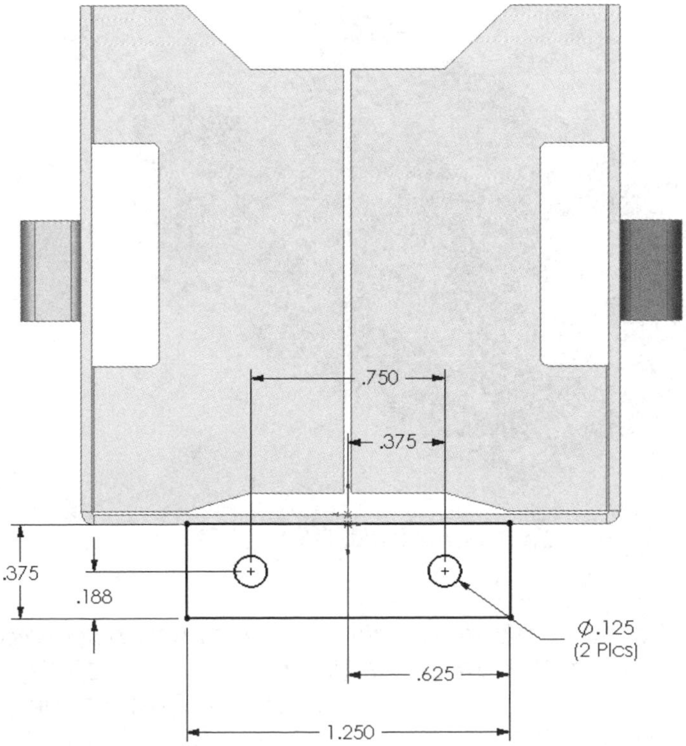

Click [exit sketch icon] to **exit** the sketch,

or click **Rebuilt** [rebuild icon].

19. Adding Chamfers:

Click  or select **Insert / Sheet Metal / Break-Corner**.

Break Type: **Chamfer**.

Enter **.060** for chamfer depth.

Select the **4 Edges** as shown.

Click **OK**.

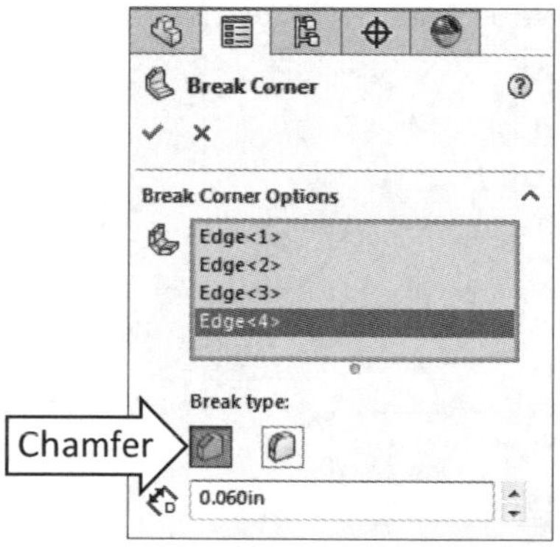

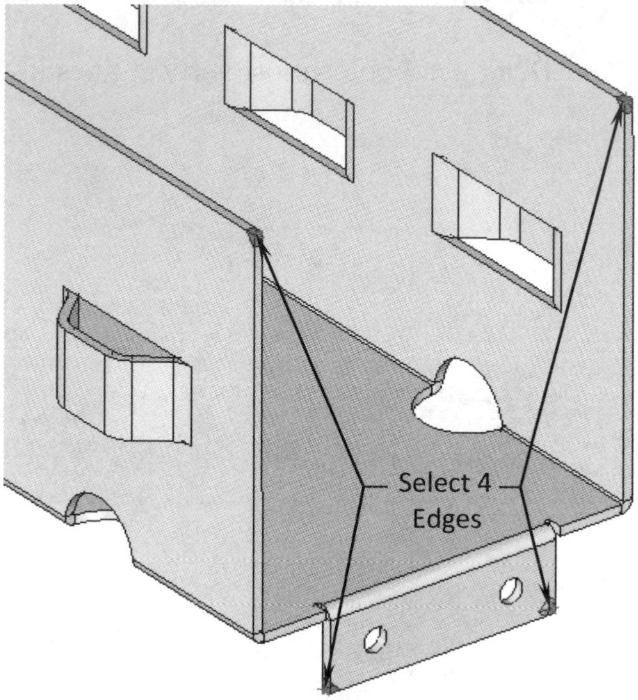

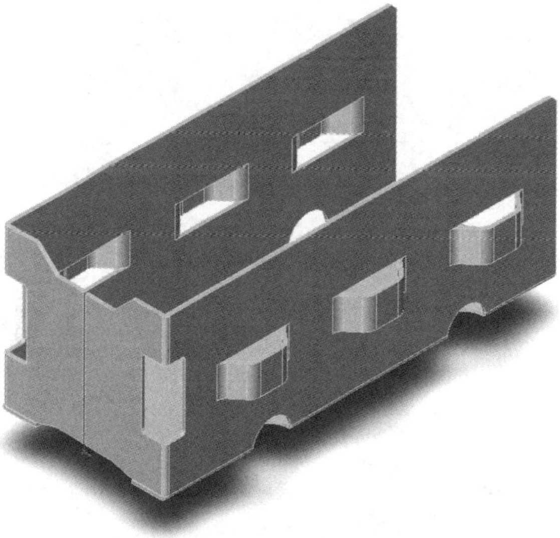

Front Isometric

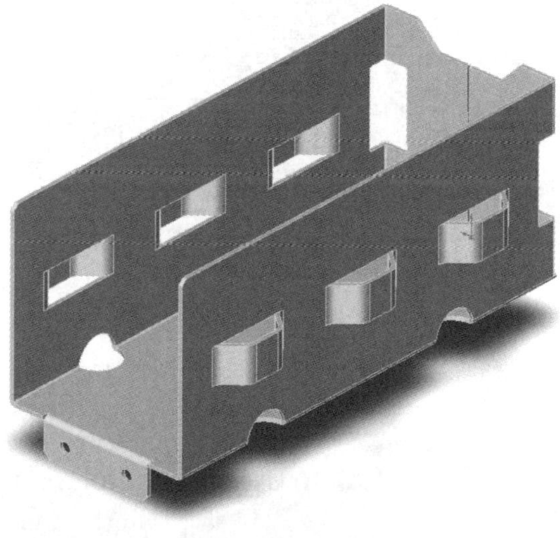

Back Isometric
See note below

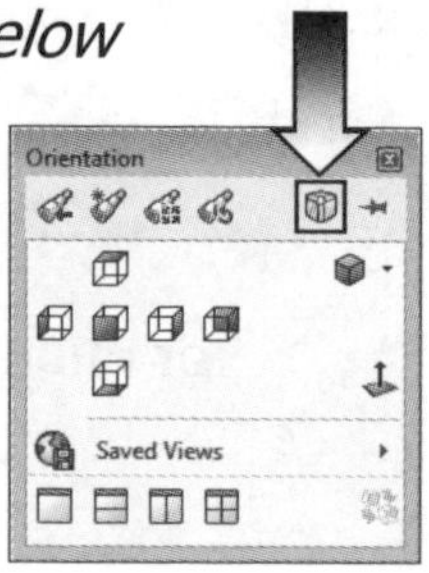

Rotate Option 1: *Shift + Up or Down Arrow twice to rotate 90° each time.*
Rotate Option 2: *Set the View rotation to 15 degrees – Press the **Right** arrow 12 times and the **Down** arrow 4 times.*
OR: *use the new option **View Selector** (Space Bar) and click one of the faces (or projected faces) on the cube to rotate the model to that orientation.*

20. Switching to the Flat Pattern:

Click  or select **Insert / Sheet Metal / Flattened***.

Verify that the part is flattened properly and there are no rebuild errors.

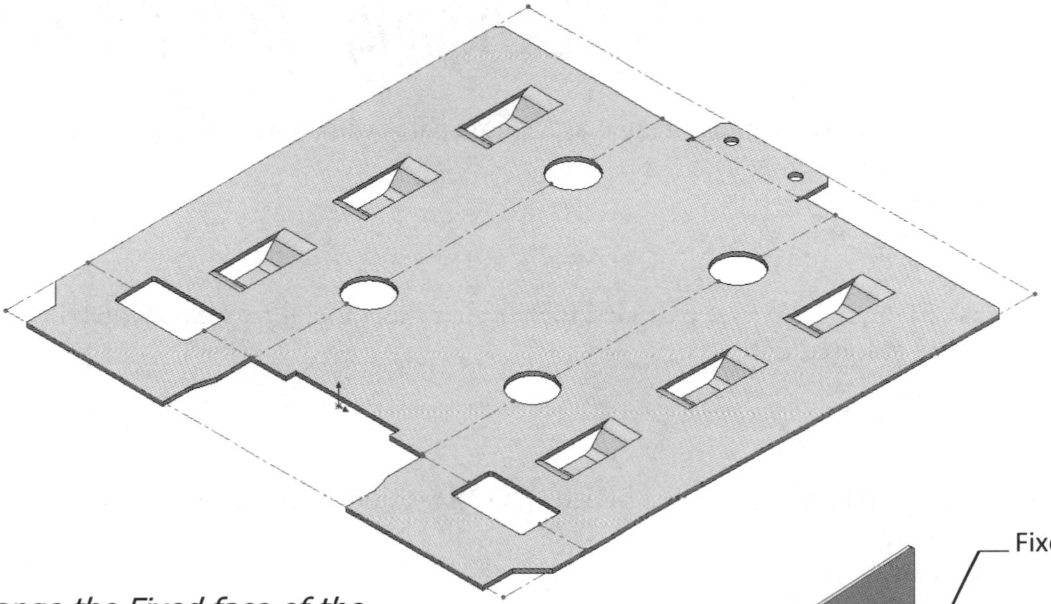

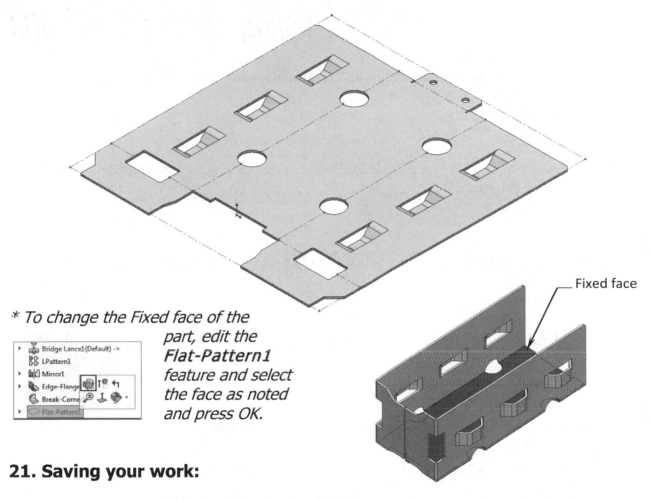

** To change the Fixed face of the part, edit the* ***Flat-Pattern1*** *feature and select the face as noted and press OK.*

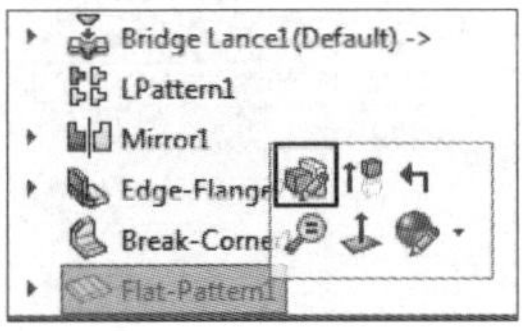

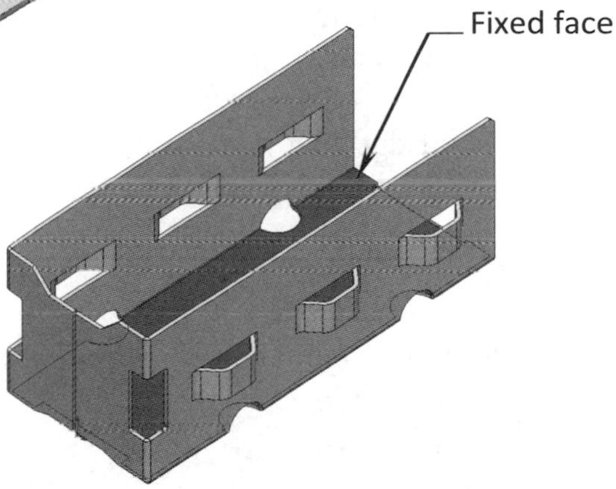

21. Saving your work:

Select **File / Save As / Mounting Tray / Save**.

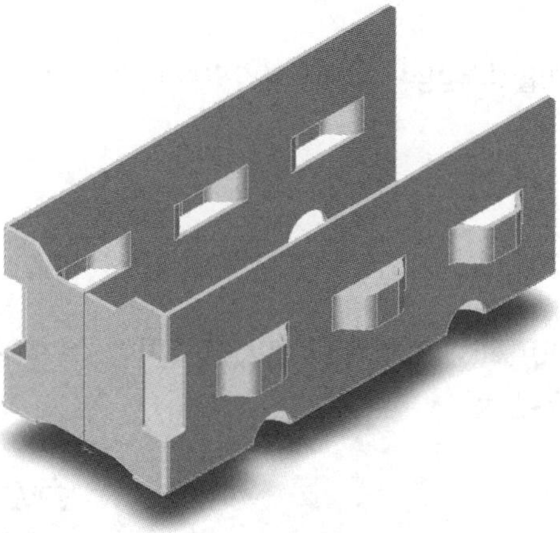

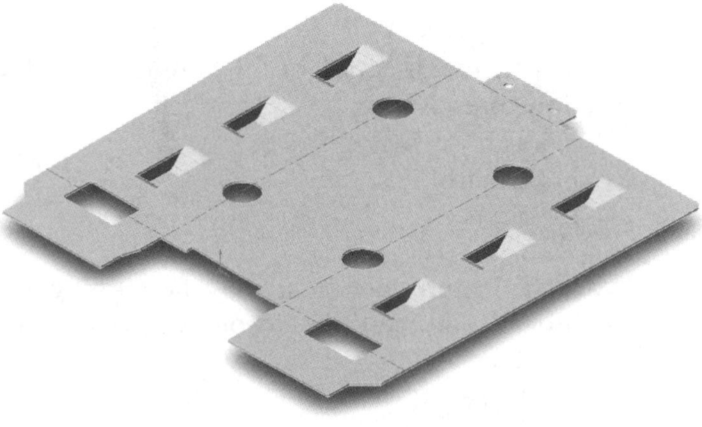

Questions for Review

Designing Sheet Metal Parts

1. A Sheet Metal part can be created as a single part or in context of an assembly with enclosed components.
   a. True
   b. False

2. A Base Flange is the first extruded feature in a sheet metal part. Sheet metal parameters are added automatically.
   a. True
   b. False

3. A sheet metal part designed in SOLIDWORKS can have multiple wall thicknesses.
   a. True
   b. False

4. The Edge Flange command adds a flange to the selected linear edge and shares the same material thickness of the sheet metal part.
   a. True
   b. False

5. Only one bend can be flattened at a time using the Unfold command.
   a. True
   b. False

6. Forming tools have to be inserted from the Feature Palette window.
   a. True
   b. False

7. To reverse the direction of the forming tool while being dragged from the Feature Palette window, press:
   a. Tab
   b. Control
   c. Shift

8. After the features are created by the forming tools, their sketches can only be moved or re-positioned, and their dimension values cannot be changed.
   a. True
   b. False

1. TRUE 2. TRUE
3. FALSE 4. TRUE
5. FALSE 6. TRUE
7. A 8. TRUE

CHAPTER 15

Sheet Metal Conversions

Sheet Metal Conversions
From IGES to SOLIDWORKS

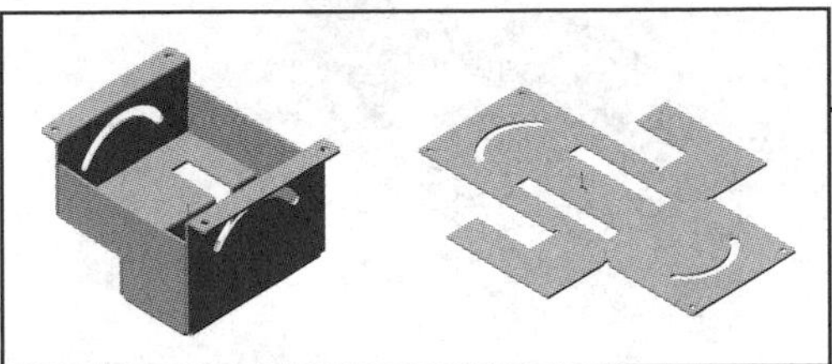

Parts created from other CAD systems and saved as IGES (or Initial Graphics Exchange Specification) can be imported and converted into SOLIDWORKS Sheet Metal.

When importing other CAD formats into SOLIDWORKS, the software recognizes them as follows:

* If there are blank surfaces, they are imported and added to the Feature-Manager design Tree as surface features.
* If the attempt to knit the surfaces into a solid succeeds, the solid appears as the base feature (named **Imported1**) in a new part file.
* If the surfaces represent multiple closed volumes, then one part is generated for each closed volume.
* If the attempt to knit the surfaces fail, the surfaces are grouped into one or more surface features (named **Surface-Imported1...**) in a new part file.
* If you import a **.dxf** or **.dwg** file, the **DXF/DWG import wizard** appears to guide you through the import process.

The imported parts must be of uniform thickness to fold and unfold properly.

After the part is opened in SOLIDWORKS, there are several methods to convert it to a sheet metal part, but the sheet metal parameters such as Rip, Fixed face or edge, Bend radius, etc., must be added before the Flat Pattern can be created.

The converted part appears on the Feature Manager Design tree; it contains the features Sheet Metal1, Flatten Bend1, and Process Bend1.

The sheet metal part can now be Flattened and Folded by toggling the Suppression state of the Process Bends.

Sheet Metal Conversions
From IGES to SOLIDWORKS Flat Pattern

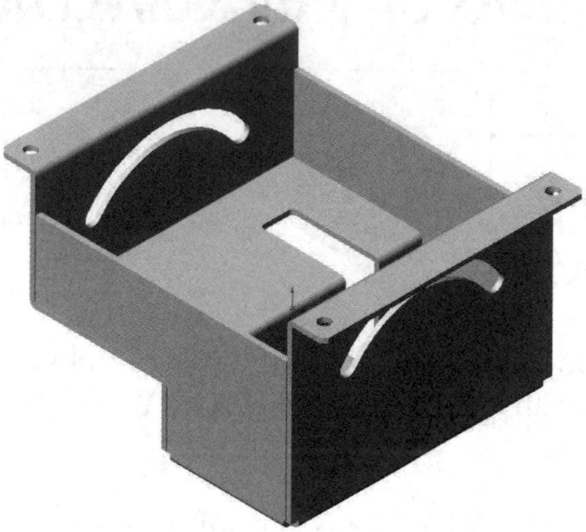

View Orientation Hot Keys:

Ctrl + 1 = Front View
Ctrl + 2 = Back View
Ctrl + 3 = Left View
Ctrl + 4 = Right View
Ctrl + 5 = Top View
Ctrl + 6 = Bottom View
Ctrl + 7 = Isometric View
Ctrl + 8 = Normal To Selection

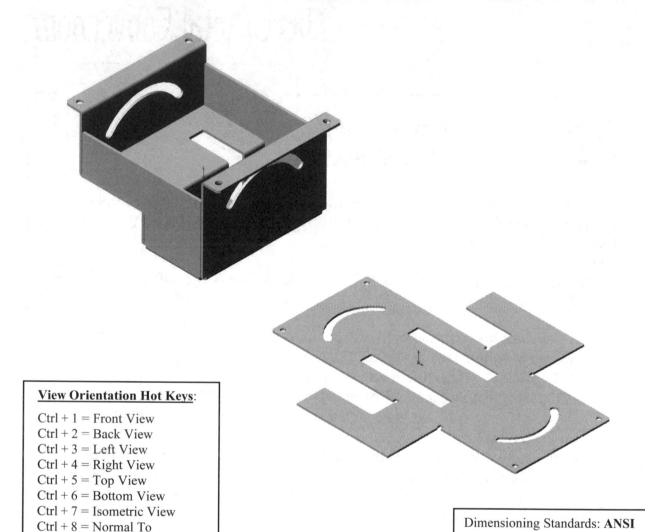

Dimensioning Standards: **ANSI**
Units: **INCHES** – 3 Decimals

Tools Needed:

- Convert to Sheet Metal
- Insert Bend
- Flat Pattern
- Sheet Metal Gusset
- Flatten Bend
- Process Bends

1. Opening an IGES document:

Go to **File / Open**. Change Files of Type to **IGES**.

Browse to the Training Files folder and open: **Sheet Metal Conversion**.

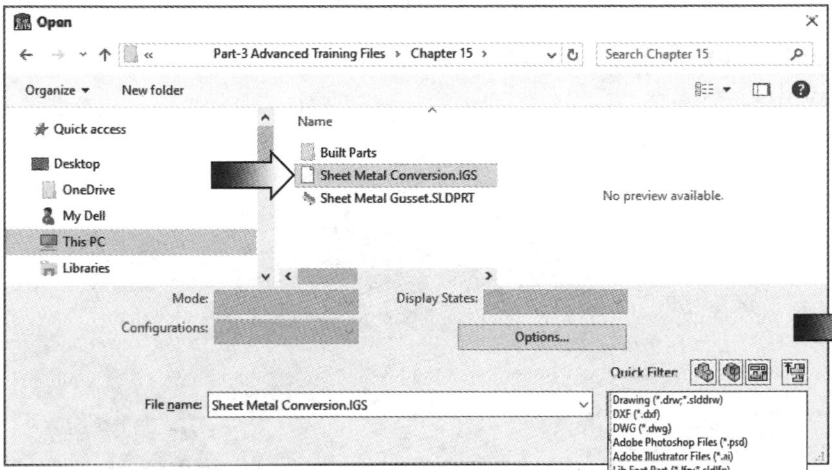

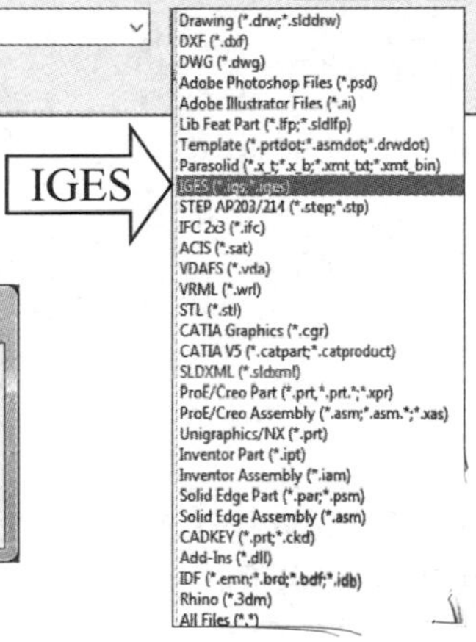

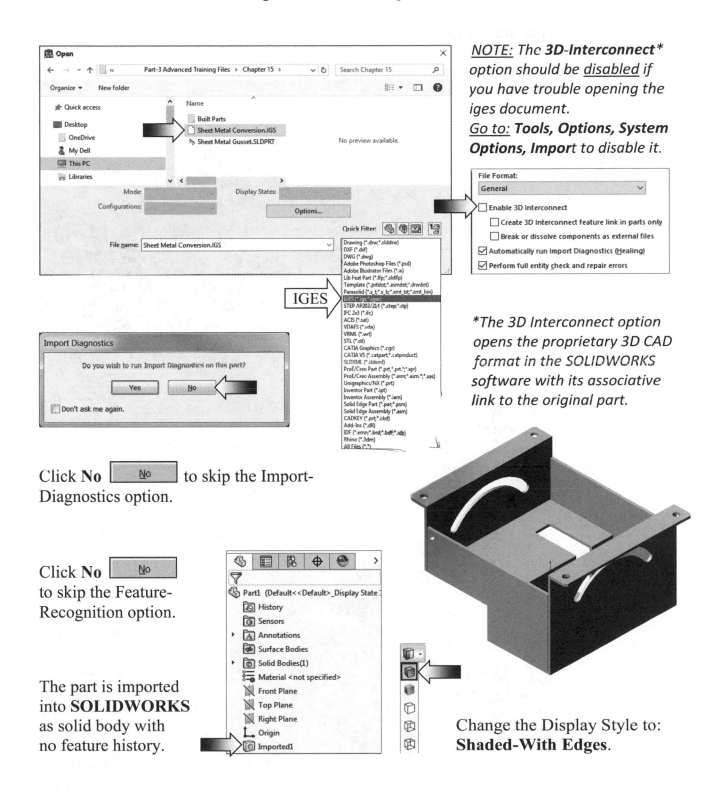

*<u>NOTE:</u> The **3D-Interconnect*** option should be <u>disabled</u> if you have trouble opening the iges document.*
*<u>Go to:</u> **Tools, Options, System Options, Import** to disable it.*

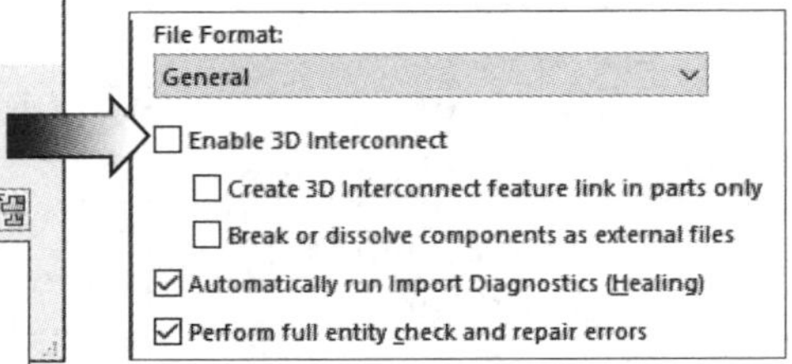

**The 3D Interconnect option opens the proprietary 3D CAD format in the SOLIDWORKS software with its associative link to the original part.*

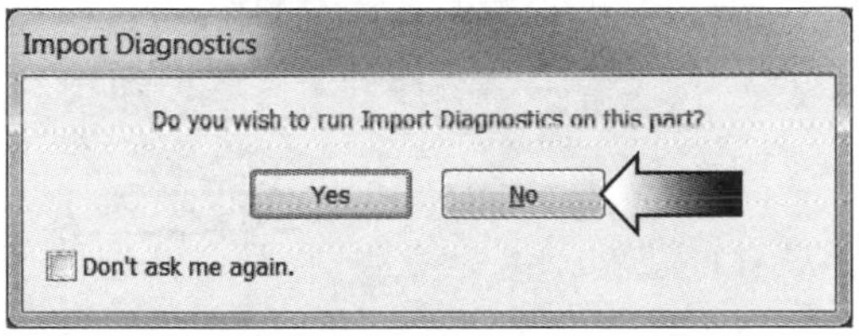

Click **No** [No] to skip the Import-Diagnostics option.

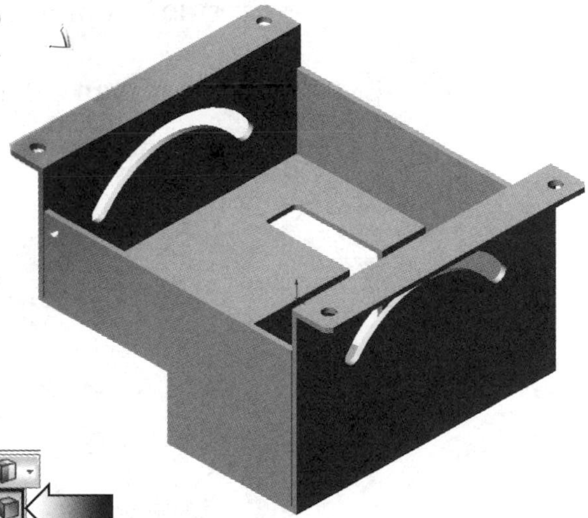

Click **No** [No] to skip the Feature-Recognition option.

The part is imported into **SOLIDWORKS** as solid body with no feature history.

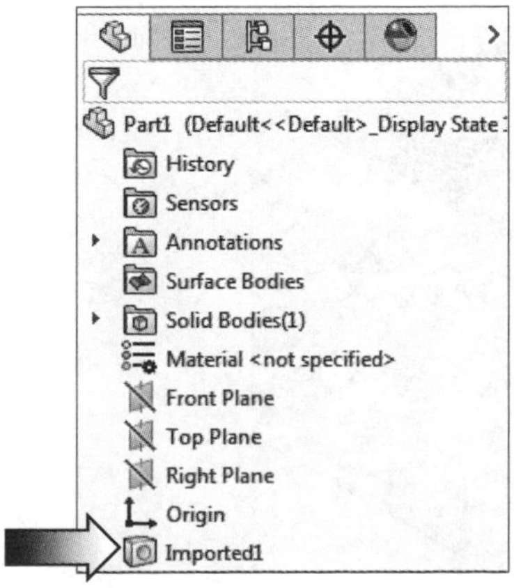

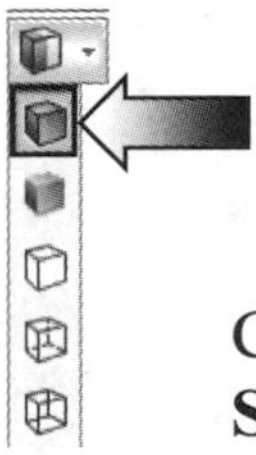

Change the Display Style to: **Shaded-With Edges**.

2. Creating the Rips:

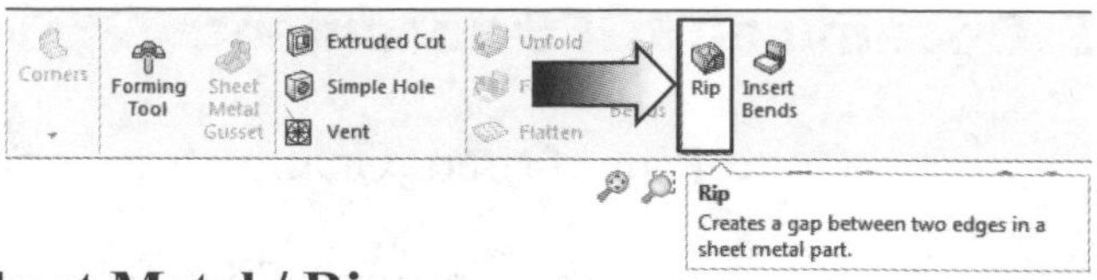

Click **Rip** on the sheet metal toolbar or select **Insert / Sheet Metal / Rip**.

Select the **inner edge** as shown.

The 2 arrows indicate that the Rip command is going to cut both walls.

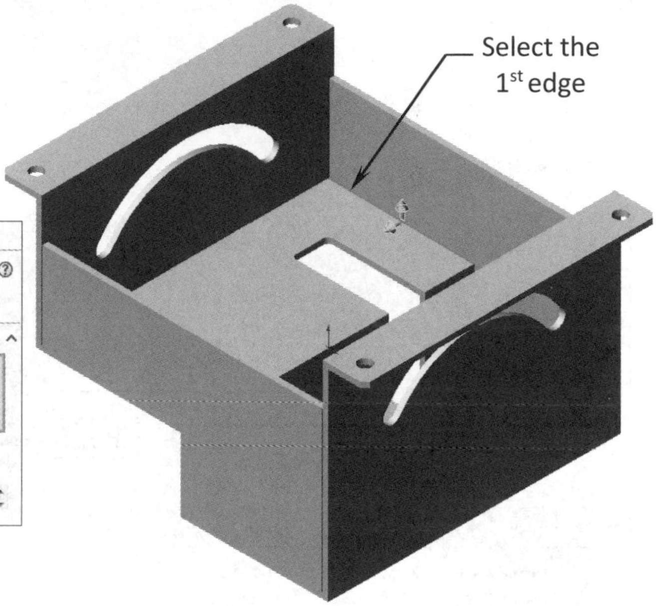

RIP

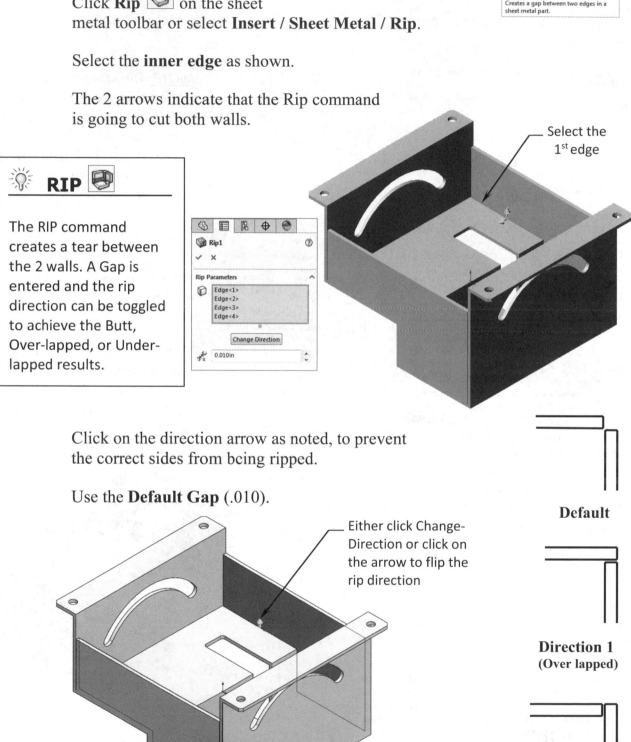

The RIP command creates a tear between the 2 walls. A Gap is entered and the rip direction can be toggled to achieve the Butt, Over-lapped, or Under-lapped results.

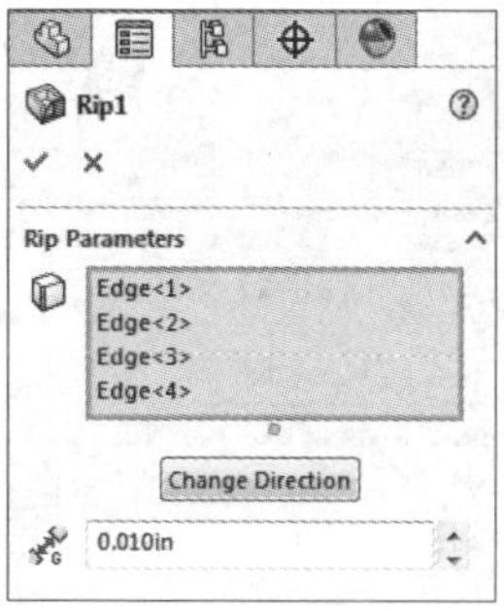

Click on the direction arrow as noted, to prevent the correct sides from bcing ripped.

Use the **Default Gap** (.010).

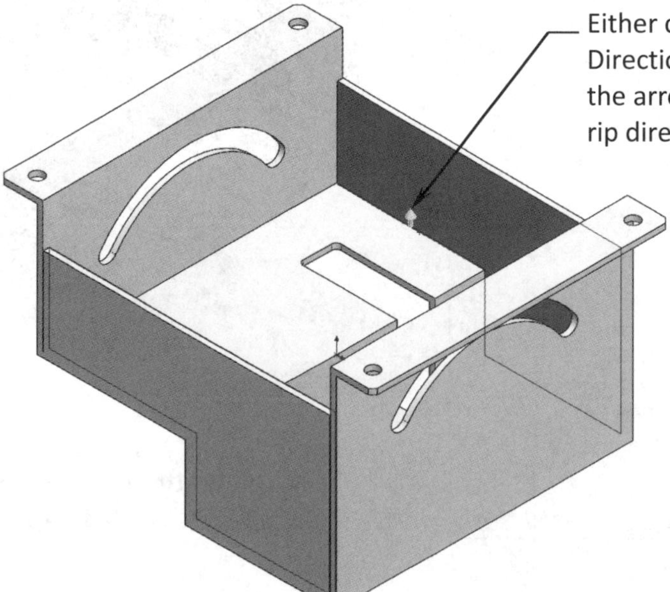

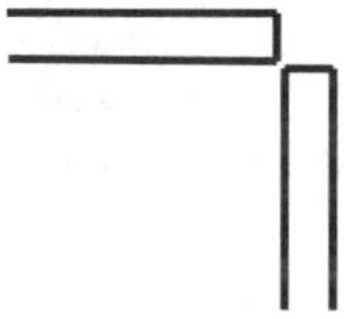

Default

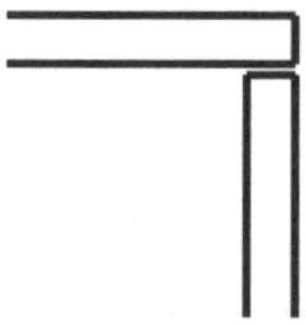

Direction 1
(Over lapped)

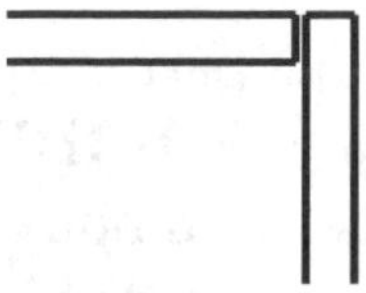

Direction 2
(Under lapped)

Select a total of **4 edges** (2 on each side) as indicated.

Click **OK**.

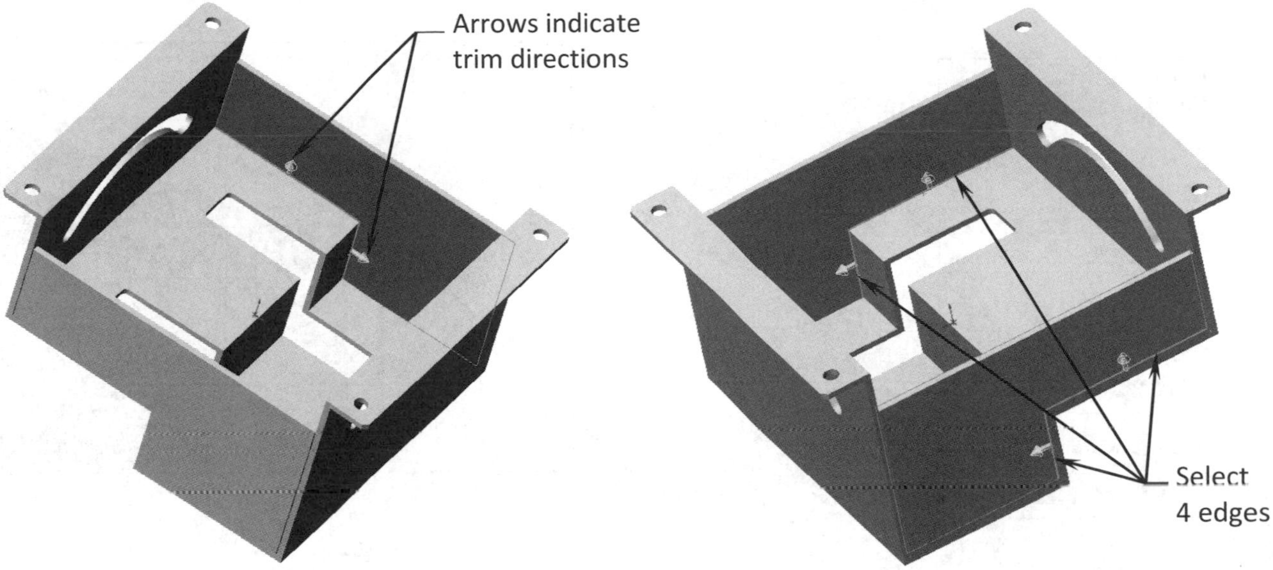

3. Inserting the Sheet Metal Parameters:

Click **Insert Bends** command or select **Insert / Sheet Metal / Bends**.

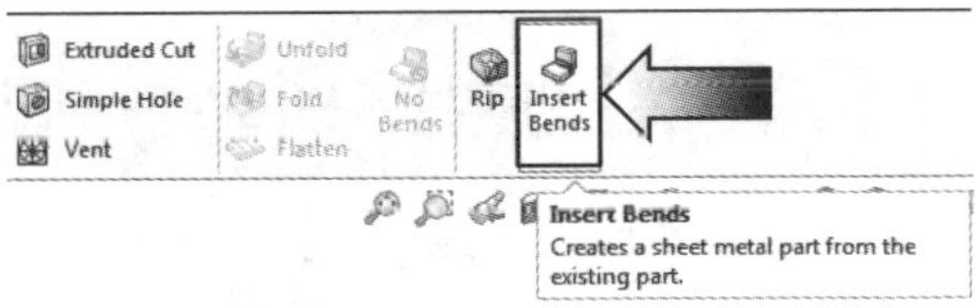

Select the inside **face** to use as the Fixed Face.

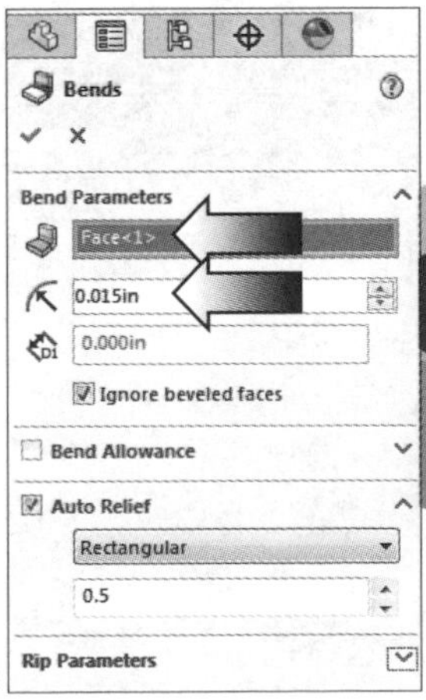

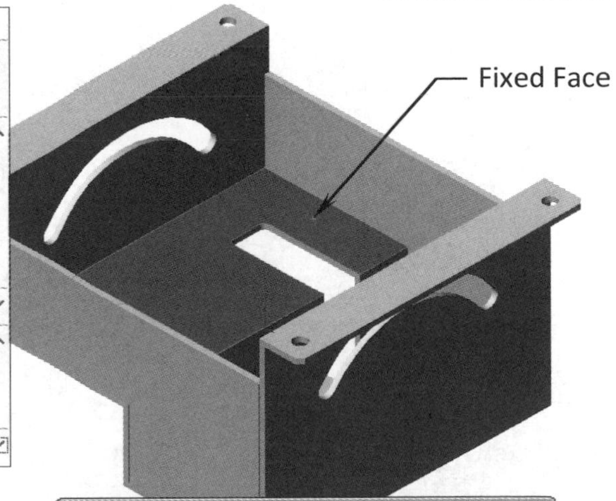

Enter **.015 in.** for inside Bend Radius.

Click **OK**.

A message pops up indicating some Auto Relief Cuts were added.

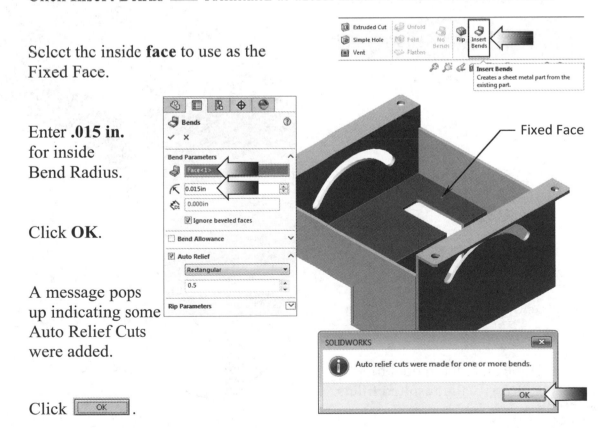

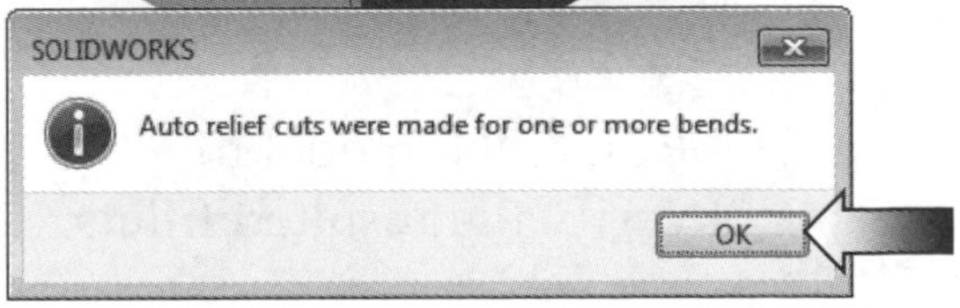

Click OK.

4. Adding Fillets:

Click **Break Corner / Corner Trim** command and select the **Fillet** option.

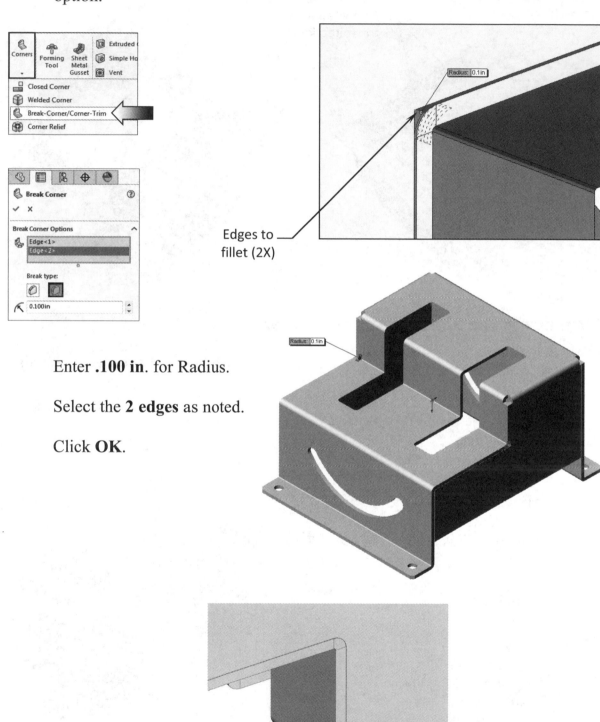

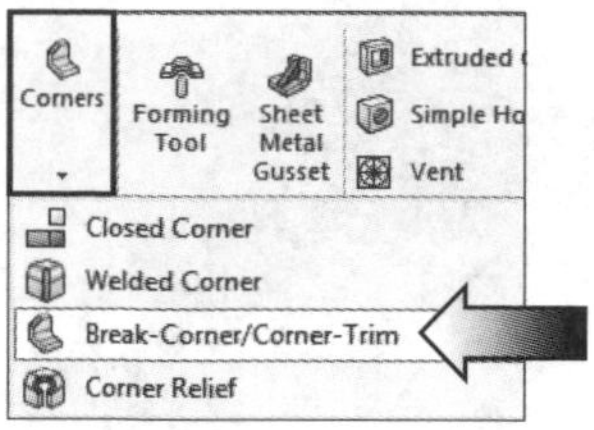

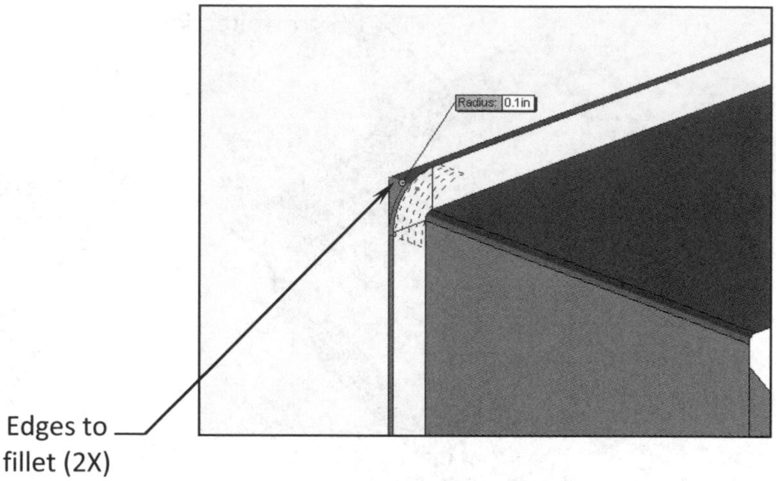

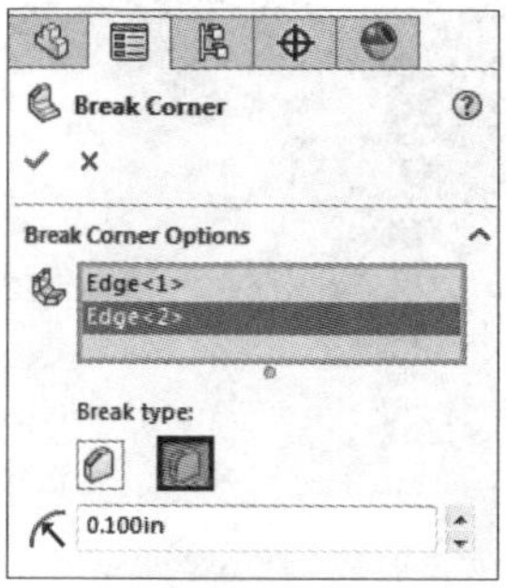

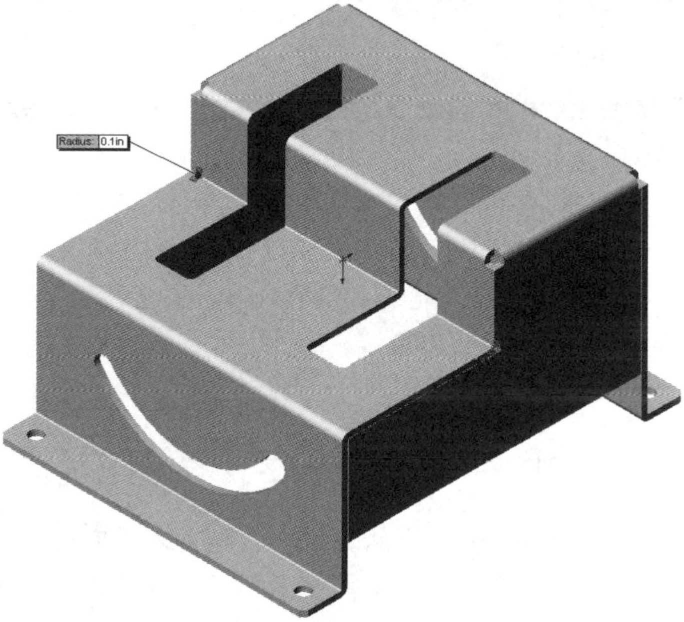

Enter **.100 in**. for Radius.

Select the **2 edges** as noted.

Click **OK**.

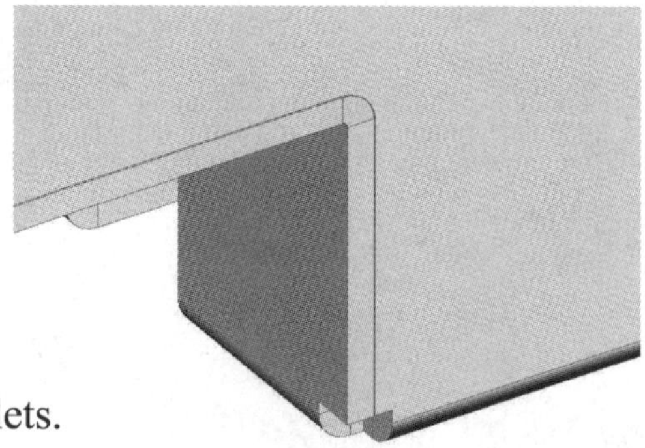

Rotate the model to verify the resulted fillets.

5. Switching to the Flat pattern:

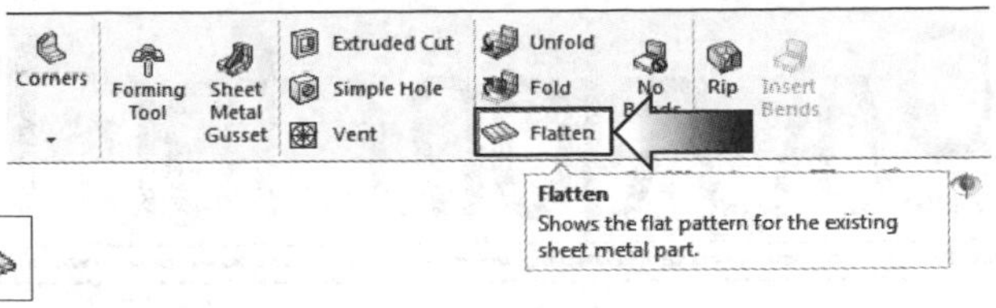

To examine the part in the flattened view, click the **Flatten**  command on the Sheet Metal toolbar (arrow).

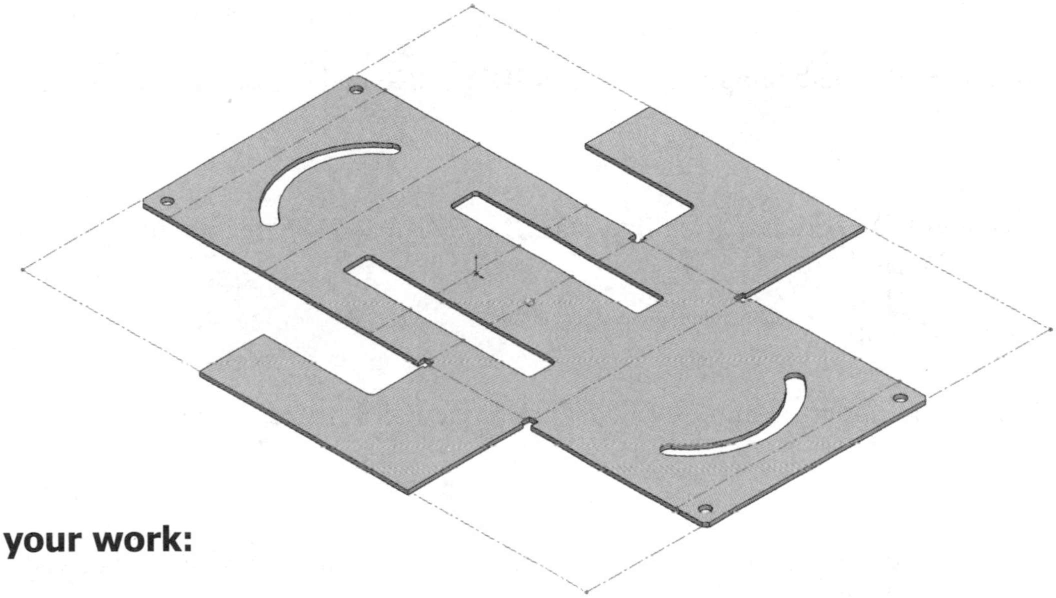

6. Saving your work:

Click **File / Save As**.

Enter **Sheet Metal Conversion** for the file name.

Click **Save**.

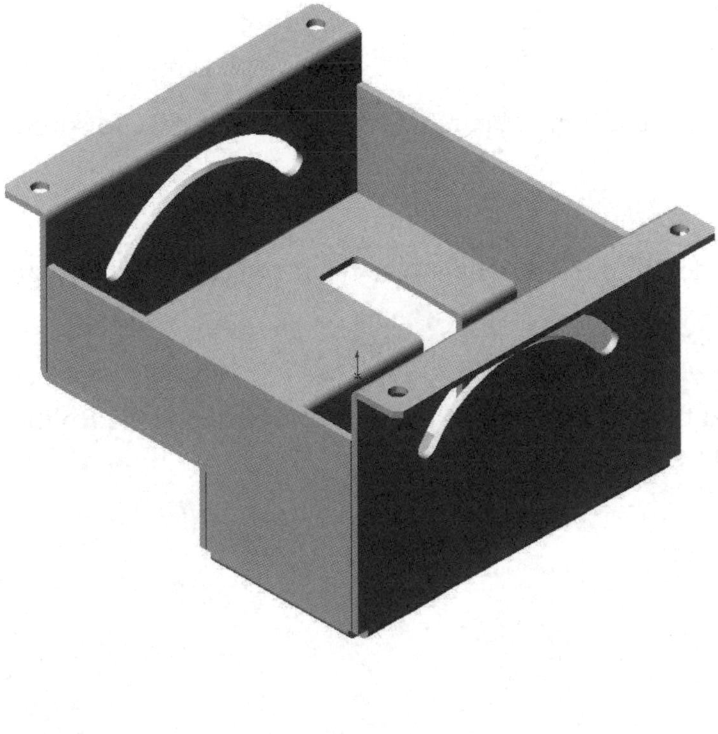

Questions for Review

Sheet Metal Conversions

1. An IGES file can be imported into SOLIDWORKS and converted into a sheet metal part.
   a. True
   b. False

2. DXF and DWG are imported into SOLIDWORKS as 2D Sketches, using the DXF/DWG Import-Wizard.
   a. True
   b. False

3. After being imported into SOLIDWORKS, the IGES file can be flattened instantly.
   a. True
   b. False

4. The imported parts must be of uniform thickness to fold and unfold properly.
   a. True
   b. False

5. The Rip feature removes 1 material thickness based on the side of the direction arrow that you select.
   a. True
   b. False

6. When applying the sheet metal parameters, you do not have to specify a fixed face.
   a. True
   b. False

7. The width and depth of the relief cuts are fixed and cannot be changed.
   a. True
   b. False

8. The Folded and the Flat pattern can be toggled by moving the Rollback Line up or down.
   a. True
   b. False

1. TRUE 2. TRUE
3. FALSE 4. TRUE
5. TRUE 6. FALSE
7. FALSE 8. TRUE

Sheet Metal Gussets

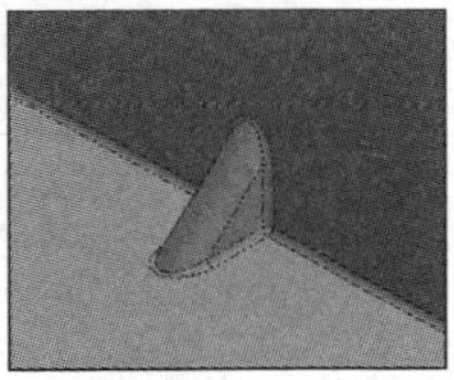

Sheet Metal gussets can be created in SOLIDWORKS with specific indents that go across bends. This exercise will guide us through the creation of a gusset in a sheet metal part.

1. Opening a sheet metal part document:

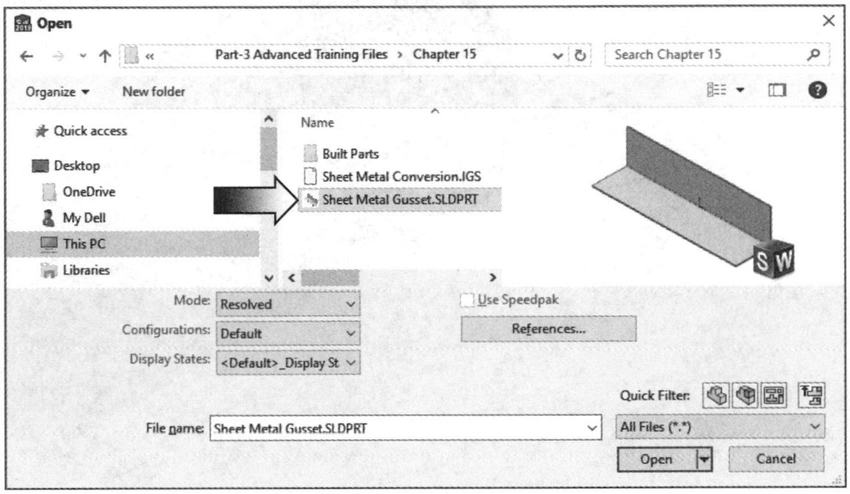

Click **File / Open**.

Browse to the Training files folder and open a part document named: **Sheet-Metal Gusset**.

2. Creating a gusset:

Click the **Sheet Metal Gusset** command from the Sheet Metal tool tab (arrow).

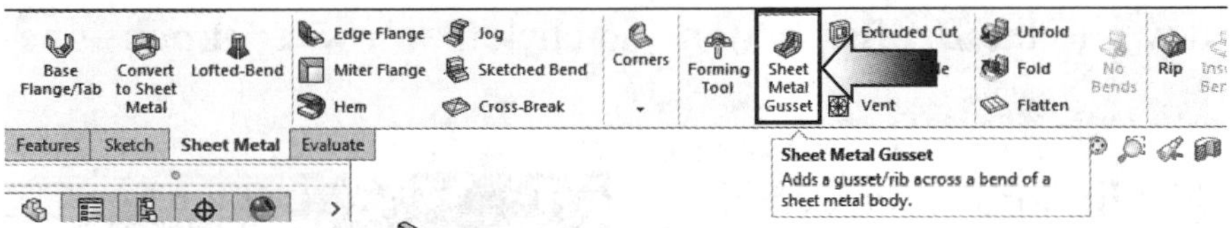

Under the **Position** section, select the **2 Faces** as indicated in the balloon number 1 for **Supporting Faces**.

Select the **horizontal edge** in the middle of the 2 supporting faces as noted in the balloon number-2 to be used as a reference to align the gusset.

Select the **Vertex** as noted in the balloon number 3 to be used for dimensioning.

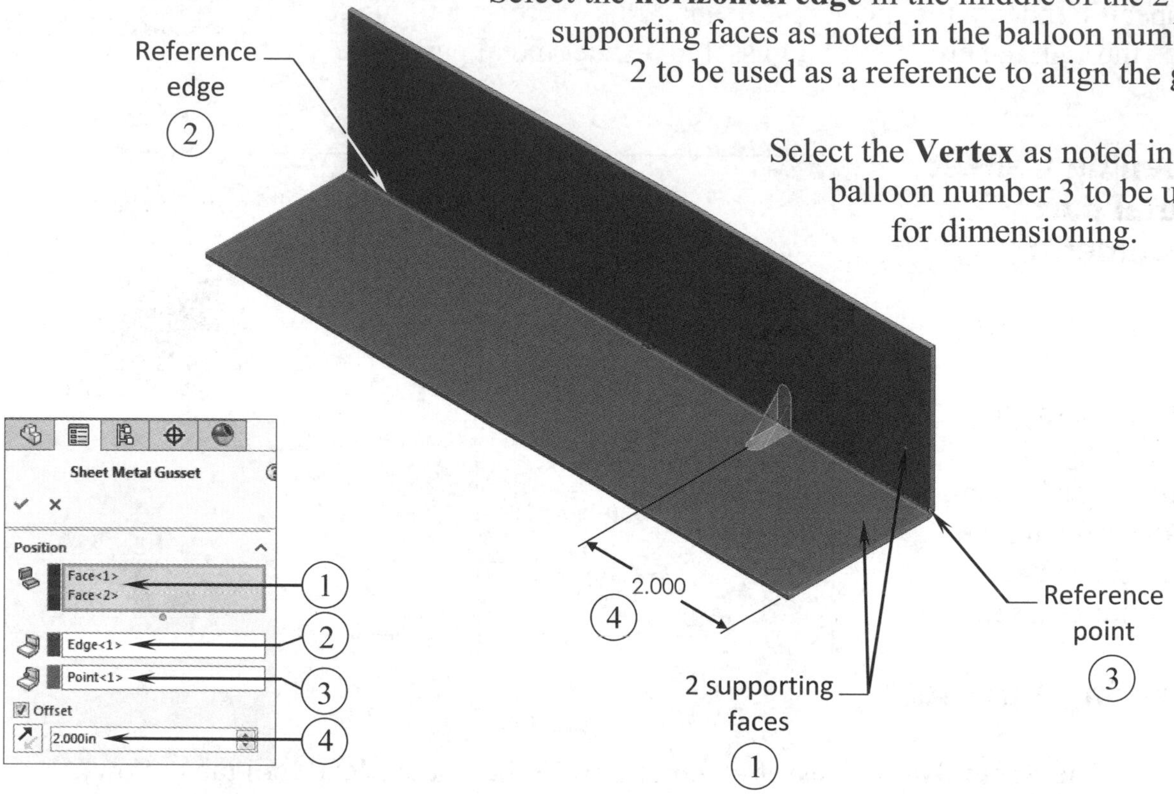

Move down to the **Profile** section and click the **Profile Dimensions** option (arrow).

Enter **.500 in**. for Profile Length.

Enter **.500 in**. for Profile Height.

Click the **Rounded Gusset** button to create a gusset with a rounded edge.

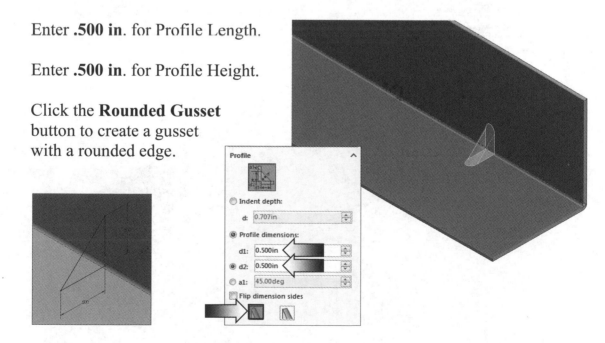

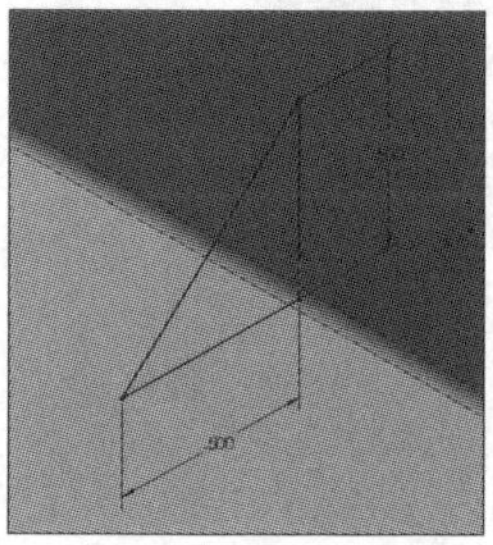

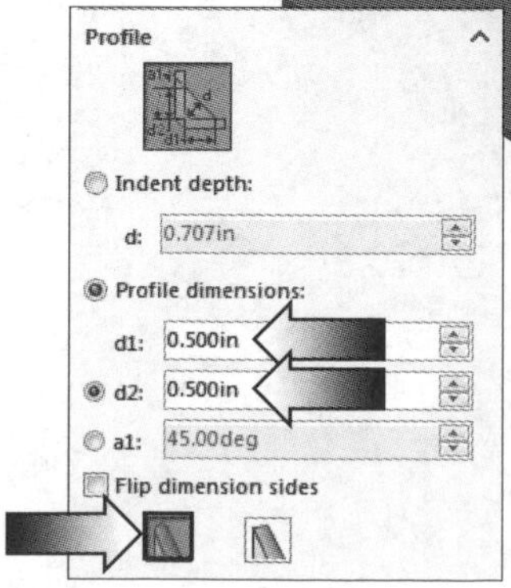

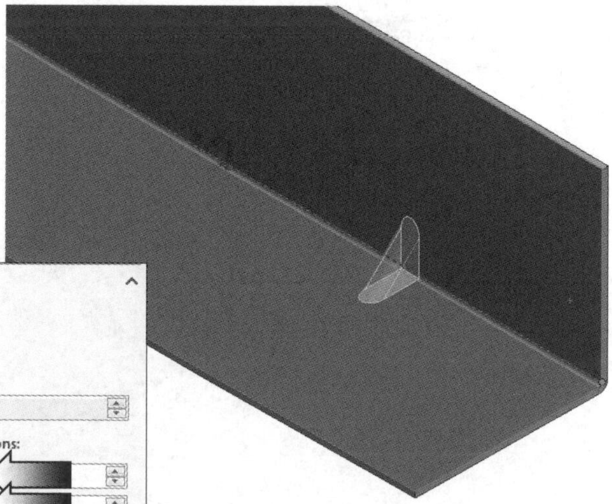

Scroll down to the Dimensions section and enter the following:

* Indent Width Dimension: **.1875 in**.

* Indent Thickness Dimension: **.060 in**.

* Inner Corner Fillet: **.030 in**.

* Outer Corner Fillet: **.090 in**.

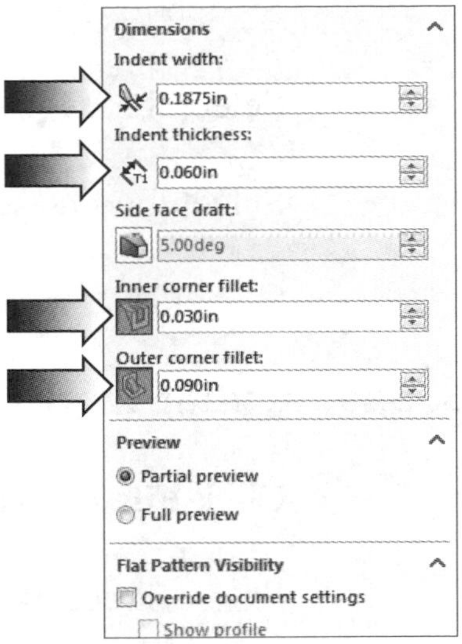

Enable the **Full Preview** option in the Preview section, if needed.

Click **OK** to create the gusset.

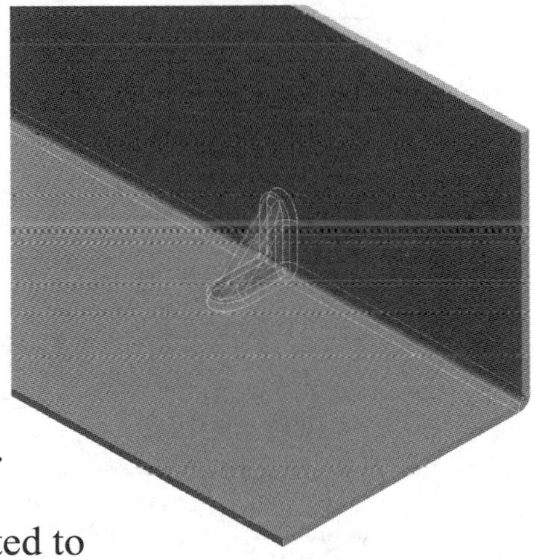

3. Viewing the resulted gusset:

Zoom in on the new gusset to see its details. Also rotate the view to see the indent from the back side.

Click on the feature itself to see its dimensions.

The gusset has a built-in sketch that can be edited to change to a custom profile if needed.

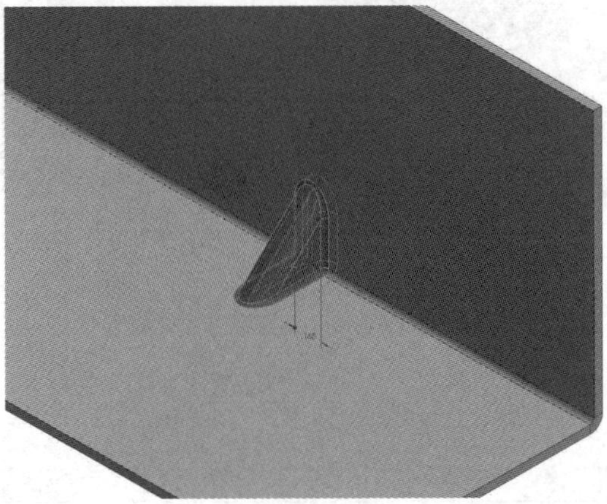

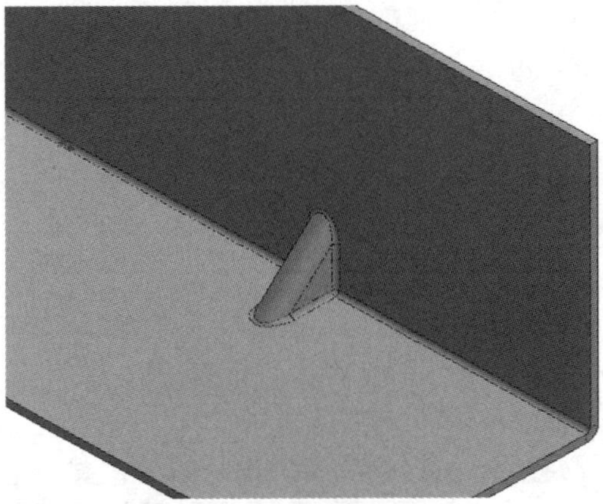

4. Mirroring the gusset:

Switch to the Features tool tab and click the **Mirror** command.

Select the **Right** plane from the Feature tree for Mirror-Plane.

Select the **Gusset1** also from the Feature tree for Features-to-Mirror.

Click **OK**.

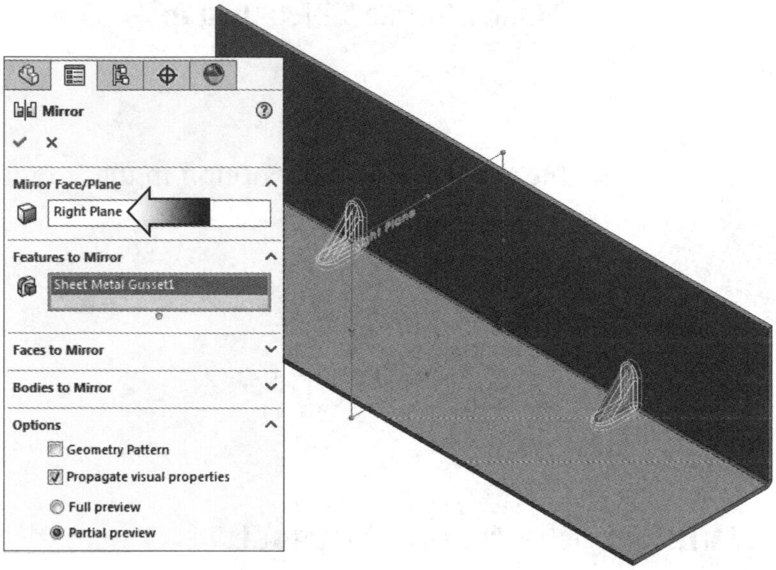

5. Saving your work:

Click **File / Save As**.

Enter: **Sheet Metal Gusset** for the name of the file.

Click **Save**.
Click **Yes** to replace the old file with the new when prompted.

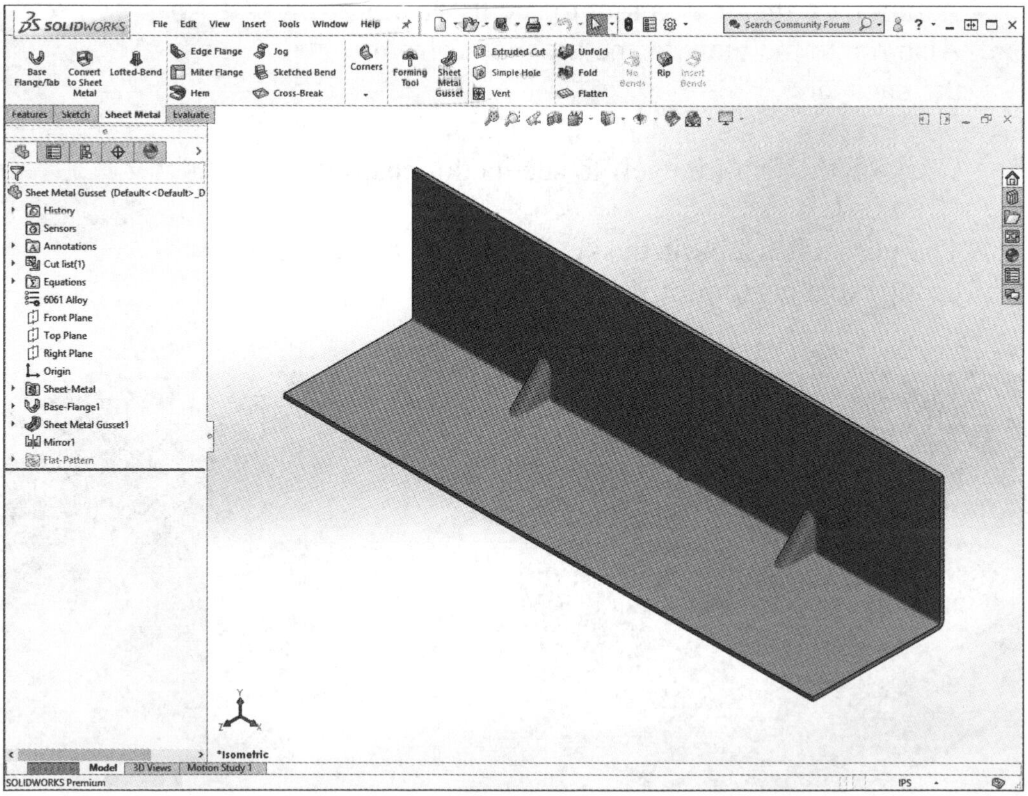

Flat Pattern Stent

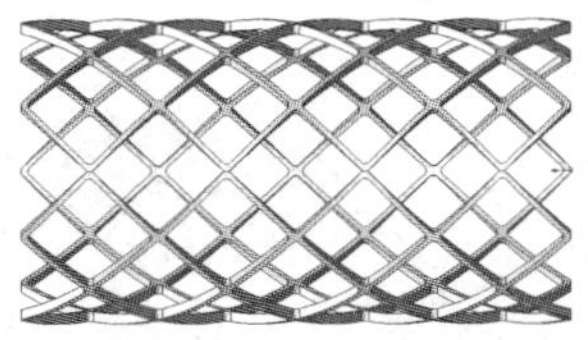

Flat Pattern Stent
A Different Approach

Using the built-in Sheet Metal features in SOLIDWORKS you can flatten or roll solid models such as wire mesh screens, grill meshes, or stent patterns.

When designing a sheet metal part, the material setback is something we must keep in mind: The Bend allowance and bend deduction calculations are methods you can choose to determine the flat length of sheet stock to give the desired dimension of the bent part. This lesson uses the default settings of the K-Factor to calculate the bend allowance (BA=P(R + KT) A/180).

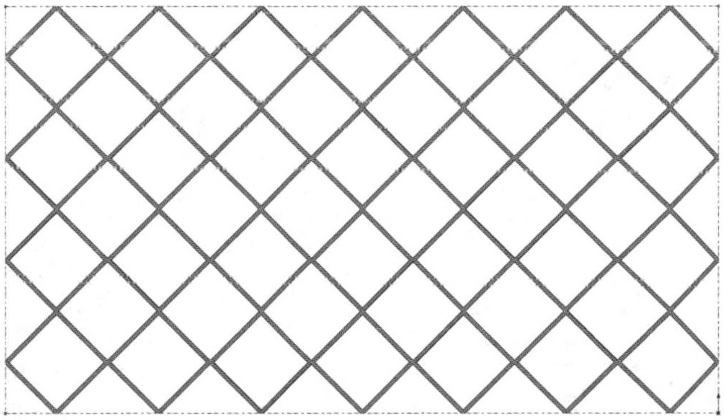

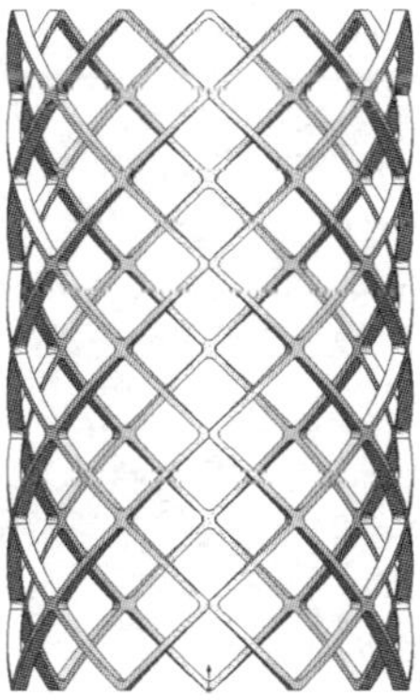

K-Factor is a ratio that represents the location of the neutral sheet with respect to the thickness of the sheet metal part. When you select K-Factor as the bend allowance, you can specify a K-Factor bend table. The SOLIDWORKS application also comes with a K-Factor bend table in Microsoft Excel format.

There are several known methods to create these types of patterns; this lesson will walk you through the use of rolling and unrolling a cylinder and its pattern, using the Sheet Metal functions in SOLIDWORKS.

Flat Pattern Stent
A Different Approach

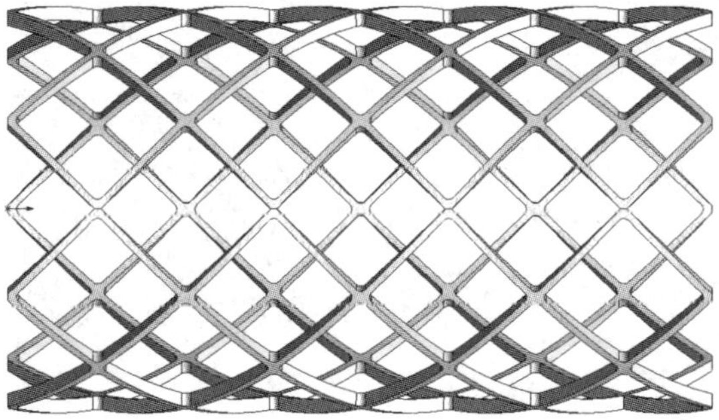

Dimensioning Standards: **ANSI**

Units: **INCHES** – 3 Decimals

Tools Needed:

Revolve Boss-Base	Flatten	Unfold
Fold	Extruded Cut	Fillet / Round

1. Starting with a new part document:

Click **File / New / Part**, set the units to **Inches**, **3 decimal** places.

Select the Front plane and open a **new sketch**.

Create the sketch and add the dimensions / relation shown below.
(The dimensions are scaled up for ease of modeling purposes.)

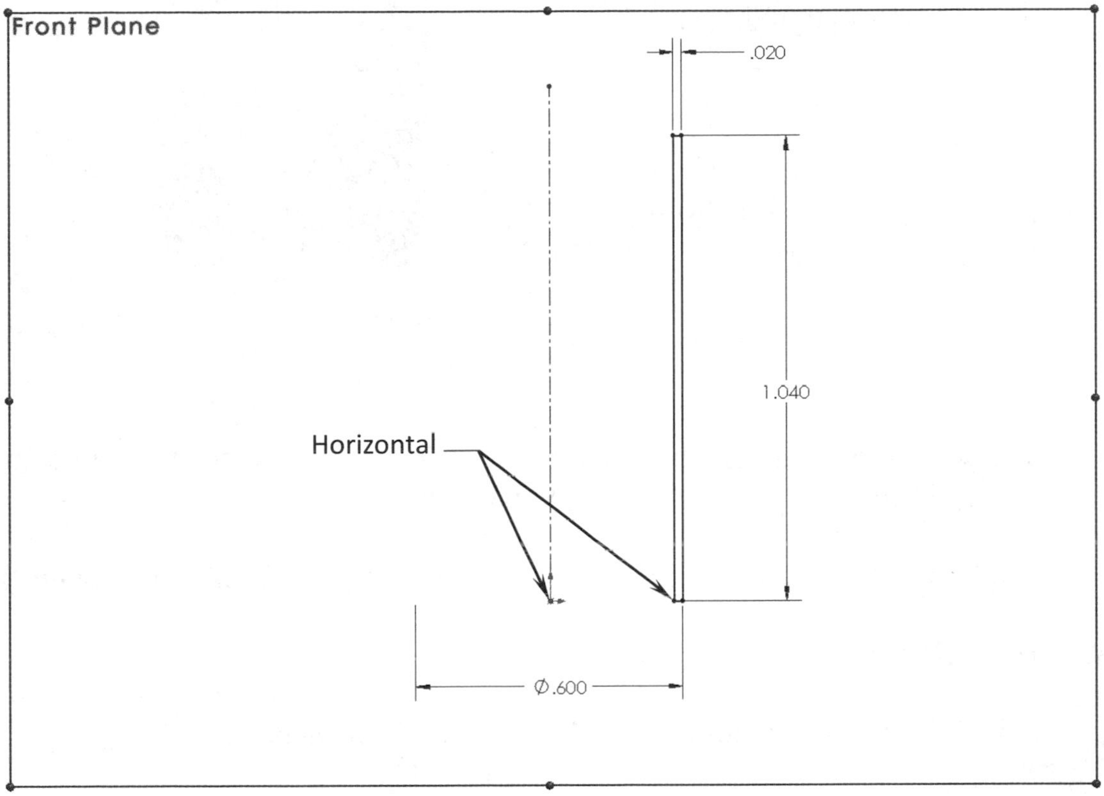

Click **Revolve / Boss-Base** .

The revolve centerline is selected automatically.

Use the default **Blind** type.

Enter **359.9deg** for angle.

Click **OK**.

(The gap is needed to flatten the sheet metal part later on.)

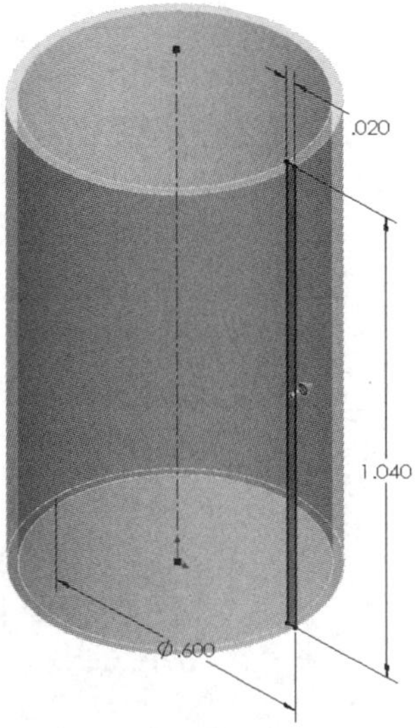

2. Converting to Sheet Metal :

Click the **Insert-Bends** command on the Sheet Metal toolbar.

For Fixed Edge/Face select the <u>edge</u> on the left side as noted.

Enter **0** for bend radius.

Use the default K-Factor and Auto Relief settings.

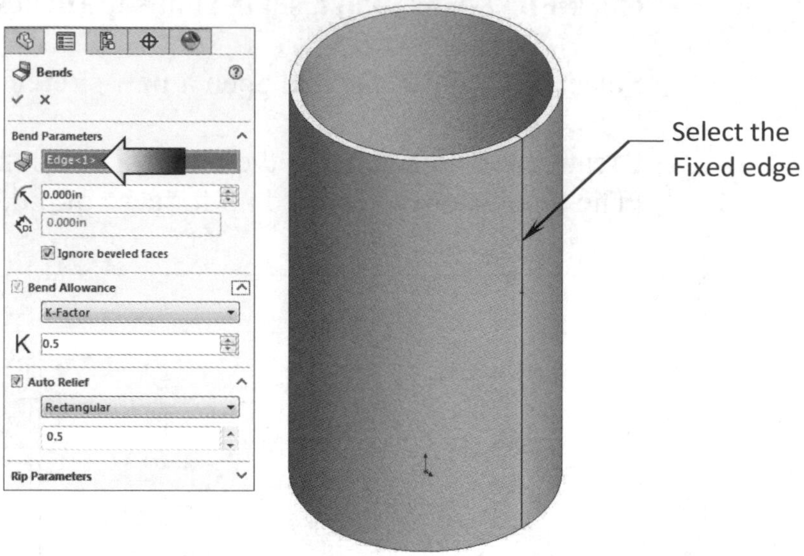

Click **OK**.

The solid model is converted to a Sheet Metal part. Press the **Flatten** command on the Sheet Metal toolbar to see its flat pattern.

Click the **Flat Pattern** command again to roll the part back to its default stage.

3. Unfolding the part:

Select the **Unfold** command  from the Sheet Metal toolbar.

Select the same <u>edge</u> to keep as the Fixed Edge/Face.

Click the **Collect-All-Bends** button (arrow).

Click **OK**.

The part is flattened but this time new features can be added and they will roll back when the part is folded.

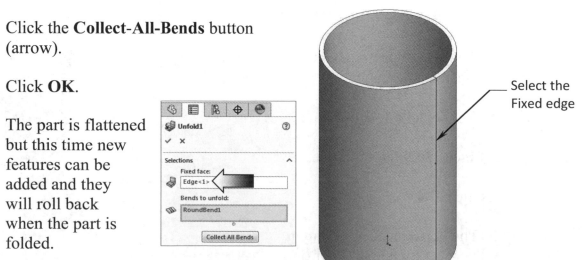

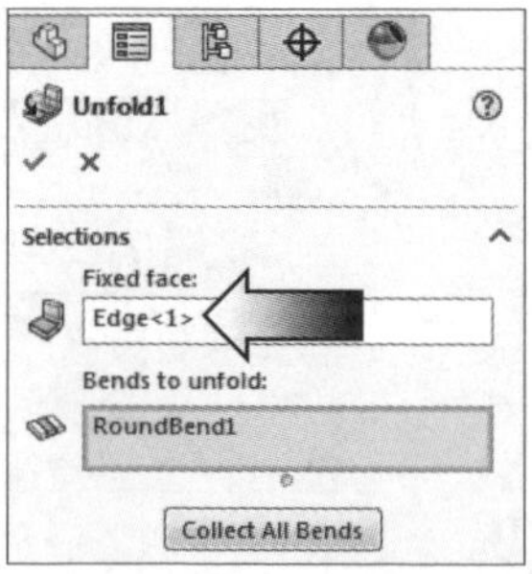

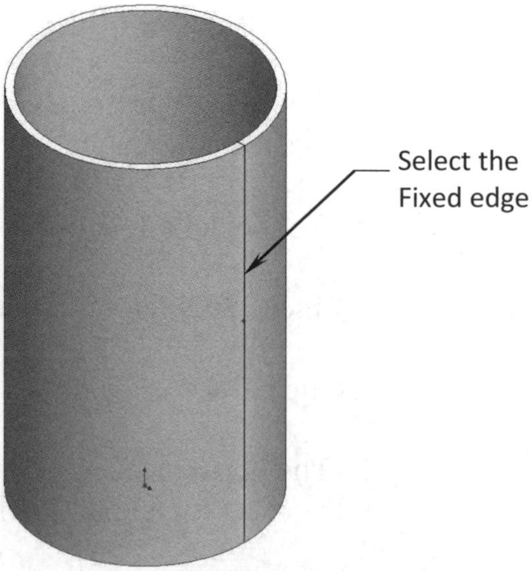

4. Adding the sketch pattern:

Select the <u>face</u> as noted and open a **new sketch**.

Sketch a couple of **squares** (notice the upper square is slightly larger than the lower one by .006").

Add the dimensions and relations shown in the image to fully define the sketch.

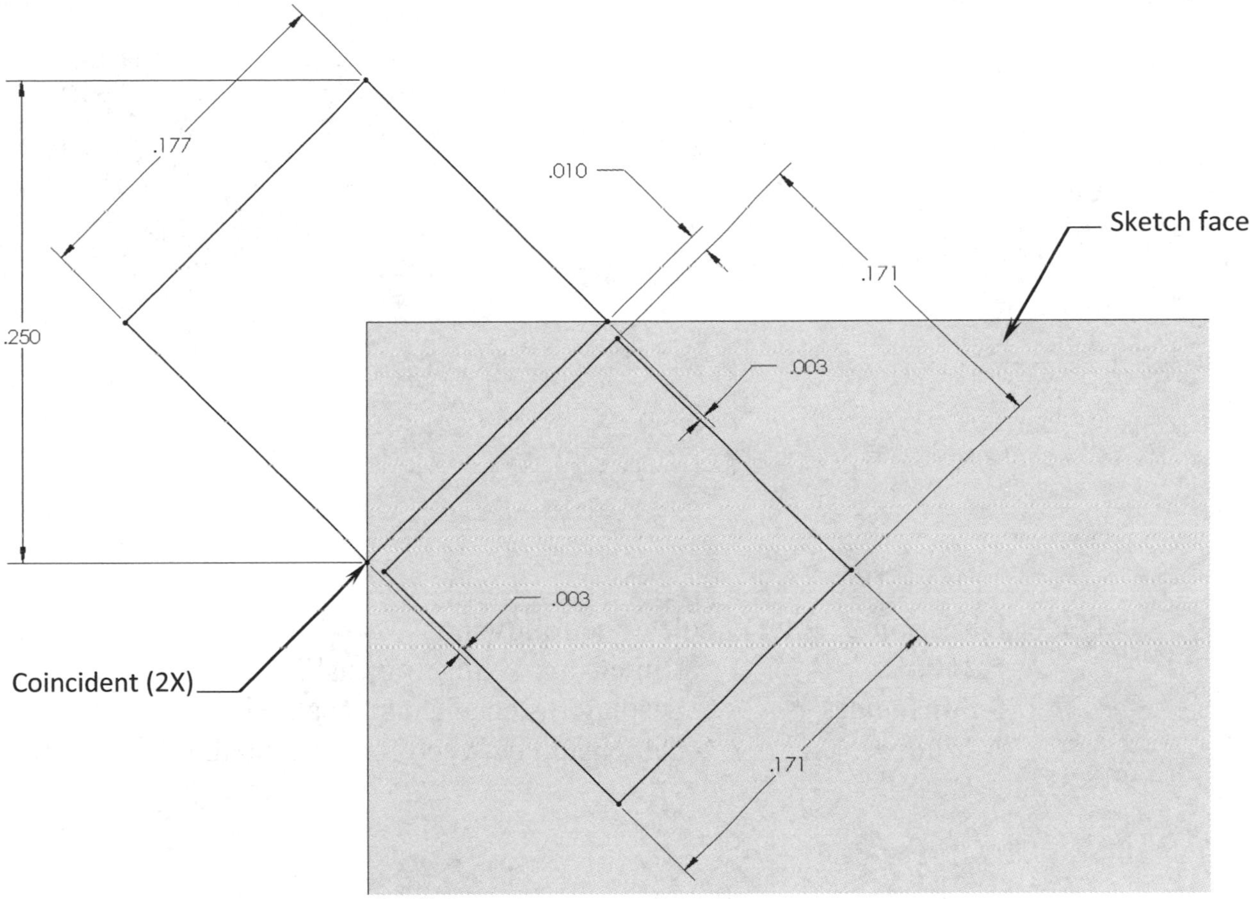

Additional relations such as **Parallel**, **Equal**, or **Perpendicular** can also be used to help eliminate some of the redundant dimensions.

We are going to use the Linear Sketch Pattern command to repeat the two squares several times, so there are a few things to keep in mind:

* Pre-select the entities to pattern. *Auto add spacing dimensions.
* Pattern along 2 directions, use angle (0deg and 270deg for directions).

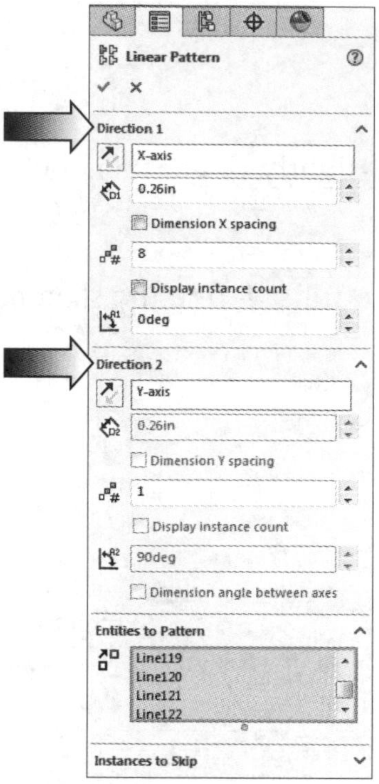

Select both squares and click the **Linear-Sketch-Pattern** command from the Sketch toolbar, or select it from the drop-down menus: **Tools / Sketch Tools**.

Under **Direction 1**, enter / select the following:

- ***.260in** * Dimension X Spacing enabled.
- * **8** Instances * Display Instance Count enabled.
- * **0deg**

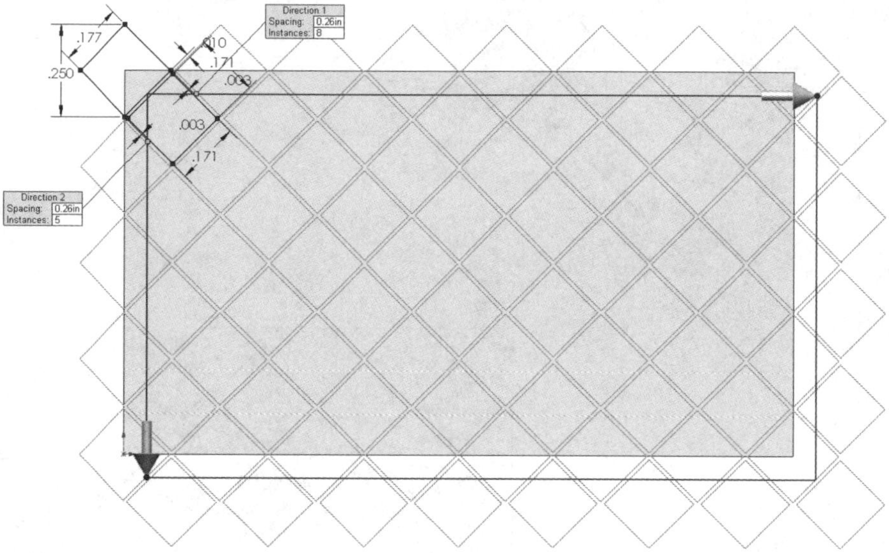

Under Direction 2, enter or select the following:

- ***.260in** * Dimension Y Spacing enabled.
- * **5** Instances * Display Instance Count enabled.
- * **270deg** * Dimension between Axes enabled.

Add a vertical relation between the 2 end points as indicated.

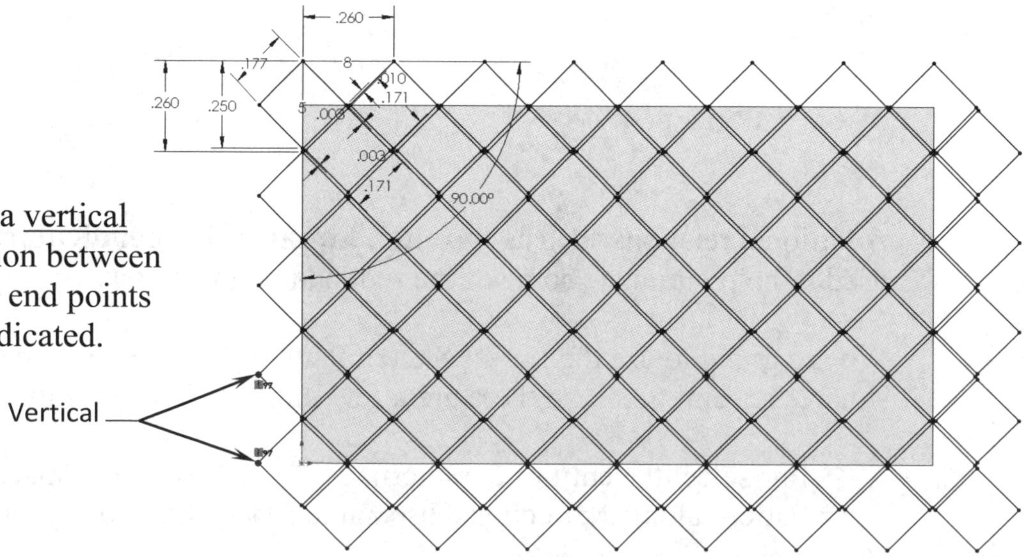

5. Creating a cut with Link to Thickness:

Select the **Extruded Cut** command from the Sheet Metal toolbar.

Us the default **Blind** option and enable the **Link To Thickness** checkbox.

Click **OK**.

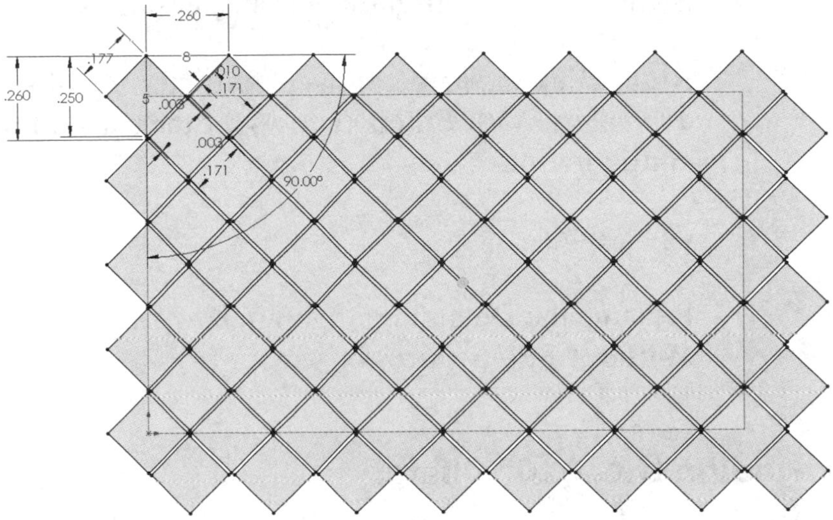

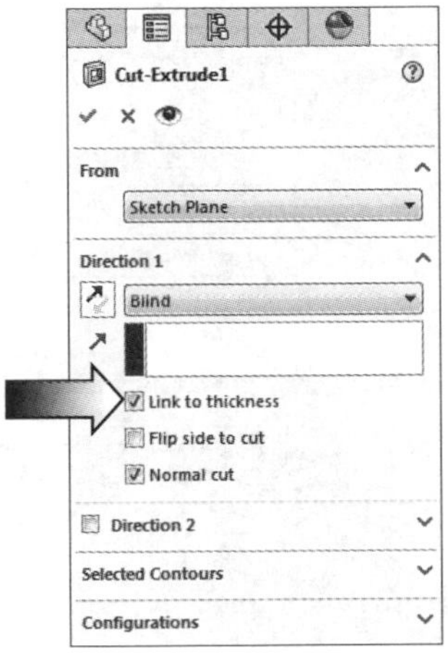

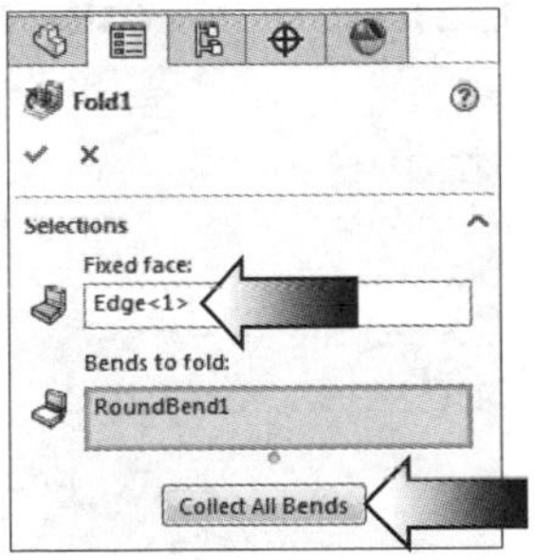

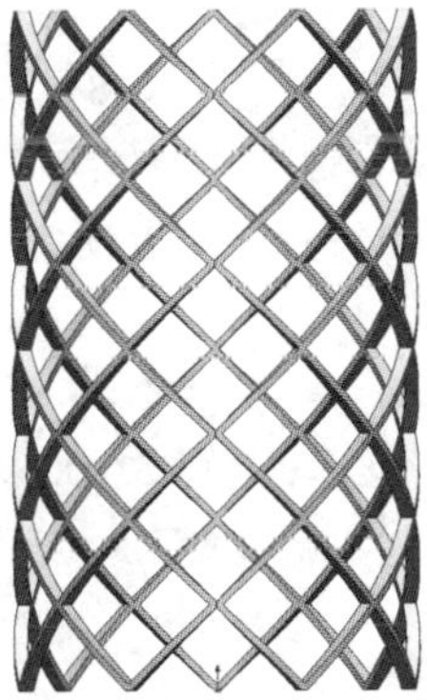

6. Folding the part:

Click the **Fold** command from the Sheet Metal toolbar.

Select the small vertical edge as noted for Fixed Face/Edge.

Click the **Collect-All-Bends** button.

Click **OK**.

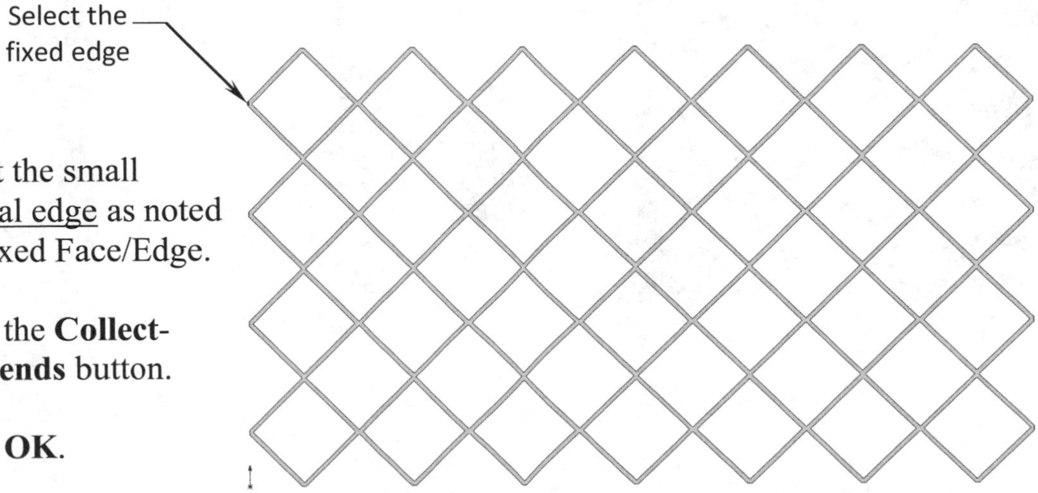

7. Creating a new configuration:

Before adding the fillets to all sharp corners, we will need to create a configuration so that the fillets can be added and captured in a separate configuration.

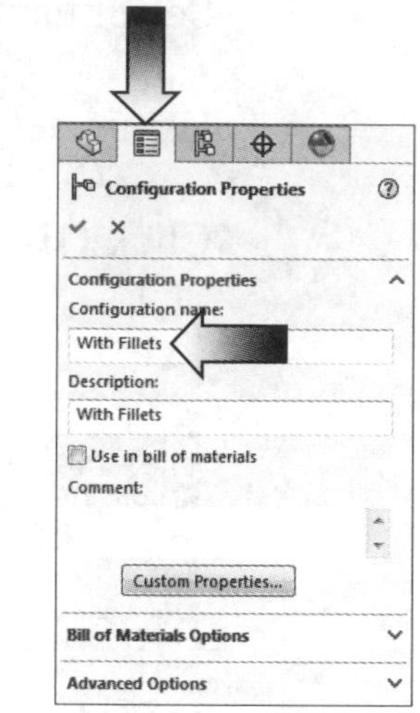

Switch to the ConfigurationManager (arrow).

Right-click the name of the part (on the top of the tree) and enter: **With Fillets** (arrow) for the name of the new configuration.

Click **OK**.

Rename the Default configuration to:
Without Fillets (arrow).

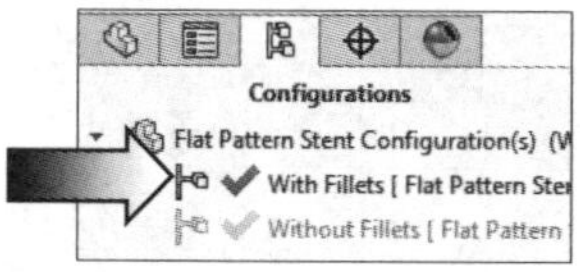

8. Adding the .010" fillets:

Click the **Fillet**  command from the Features toolbar.

Use the default **Constant Radius** option.

Enter **.010in** for radius.

Select all edges except for the ones at the two ends as indicated.

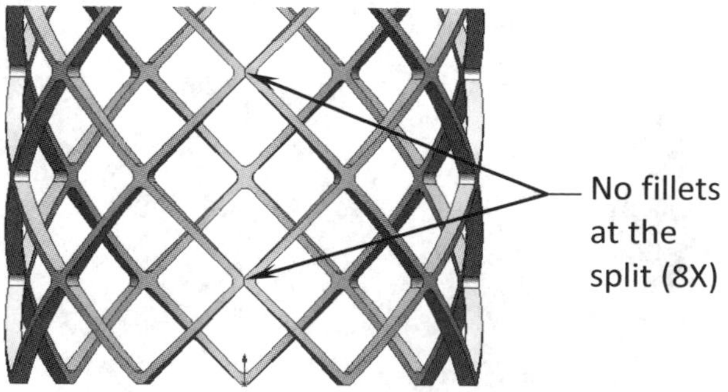

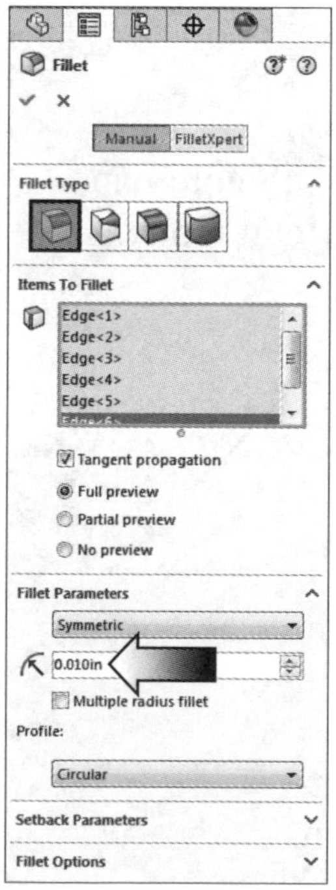

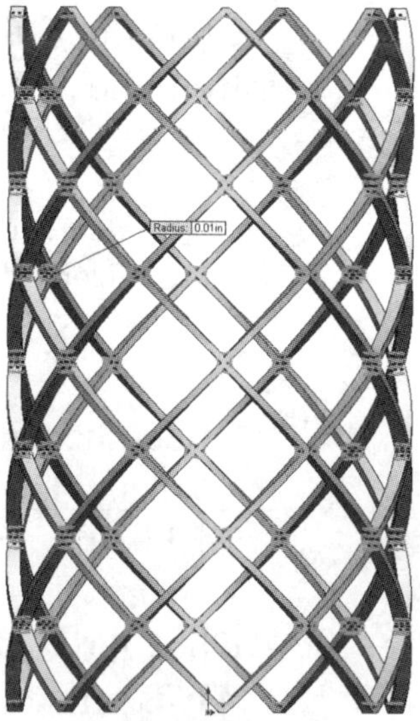

Click **OK**.

9. Switching to Flatten mode:

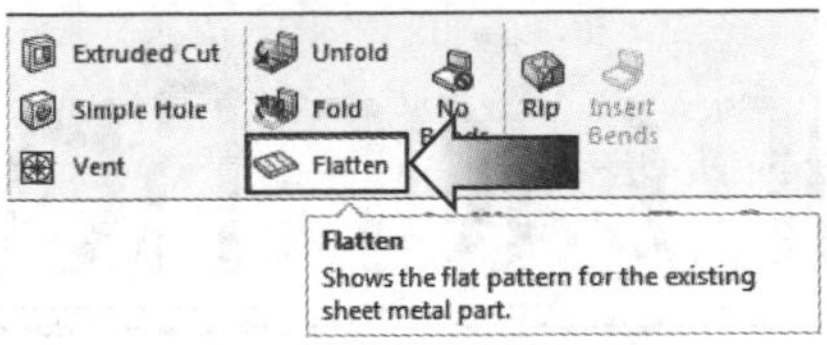

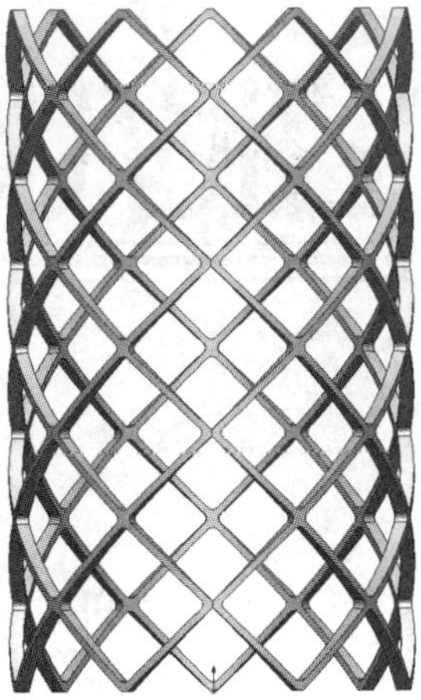

Click **Flatten**  on the Sheet Metal toolbar.

The part is flattened and the cut pattern is unrolled with it.

Use the Flatten command to flatten the sheet metal part to check its dimensions, get a printout from it, or export it as a DXF or DWG to use in manufacturing.

Use the Fold and Unfold commands to flatten the part and add new features, so that these features can roll or unroll with the part.

10. Switching configuration:

Switch back to the **Without Fillets** configuration by double-clicking on its name.

Click the **Flatten** command again to verify the pattern.

At this point, the pattern can be exported or a drawing can be made from it for inspection or documentation purposes.

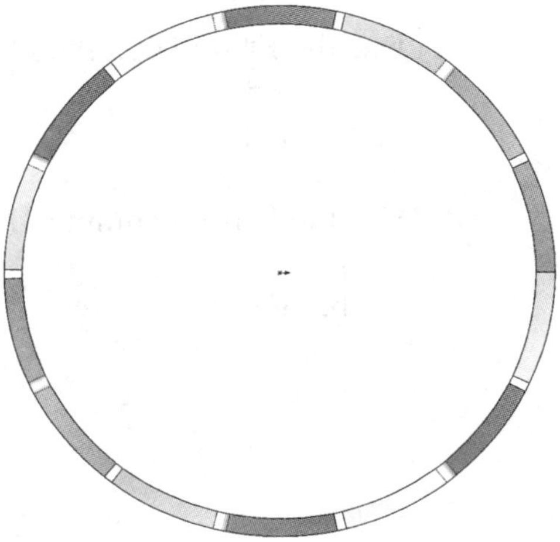

11. Saving your work:

Click **File / Save As**.

Enter **Flat Pattern Stent** for the name of the file.

Click **Save**.

Questions for Review

Flat Pattern Stents

1. Using SOLIDWORKS, a cylinder or a cone can be unrolled into a Sheet Metal flat pattern.
   a. True
   b. False

2. There must be a gap or a slit along the length of the cylinder for it to flatten.
   a. True
   b. False

3. A sheet metal part can have more than one thickness.
   a. True
   b. False

4. A sheet metal part must have one uniform thickness.
   a. True
   b. False

5. The Link to Thickness option links the depth-of-cut to the thickness of the part.
   a. True
   b. False

6. The K-Factor value is locked to .5; this ratio cannot be changed.
   a. True
   b. False

7. Use the Flatten command to flatten the part and add new features.
   a. True
   b. False

8. Use the Unfold command to flatten the part and add new features.
   a. True
   b. False

1. TRUE 2. TRUE
3. FALSE 4. TRUE
5. TRUE 6. FALSE
7. FALSE 8. TRUE

Exercise: Stent Example - Sheet Metal Approach

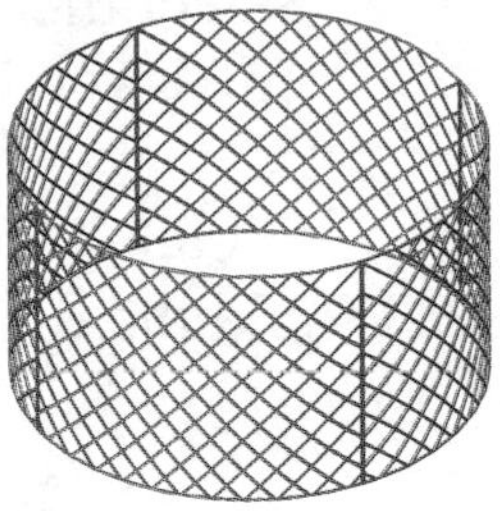

There are many known methods for creating the shapes of stents. This exercise will show the one that uses the combination of Patterns, Ribs and Combine Common options to create the model shown above. (The dimensions in the model are scaled up for visual purposes.)

1. Creating the main sketch:

Select the Front plane and open a **new Sketch**.

Sketch a **Line** centered on the origin and **2 Centerlines** as shown.

Add the dimensions to fully define the sketch.

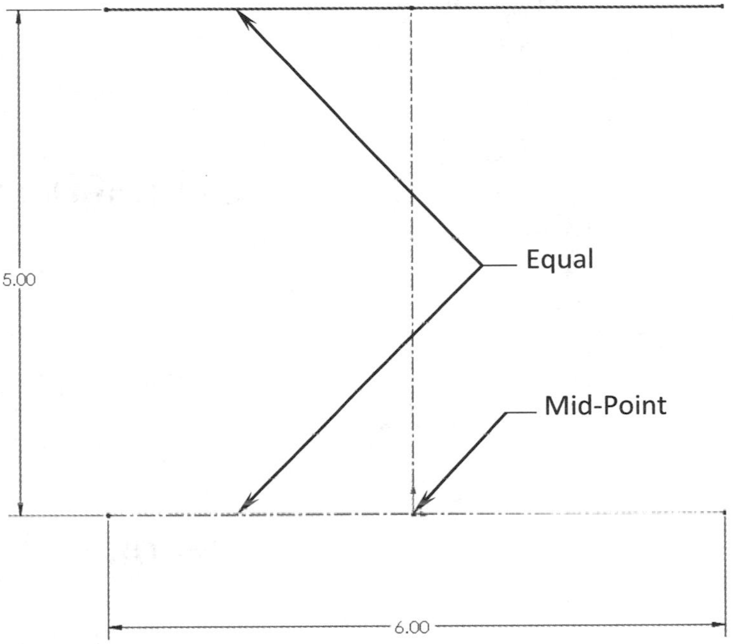

2. Revolving the main body:

From the Features toolbar, click **Revolve Boss-Base**.

Set Direction 1 to: **Mid-Plane**.

Set Revolve Angle to: **90deg**.

Set the Thickness under Thin Feature to: **.040in**.

Click **OK**.

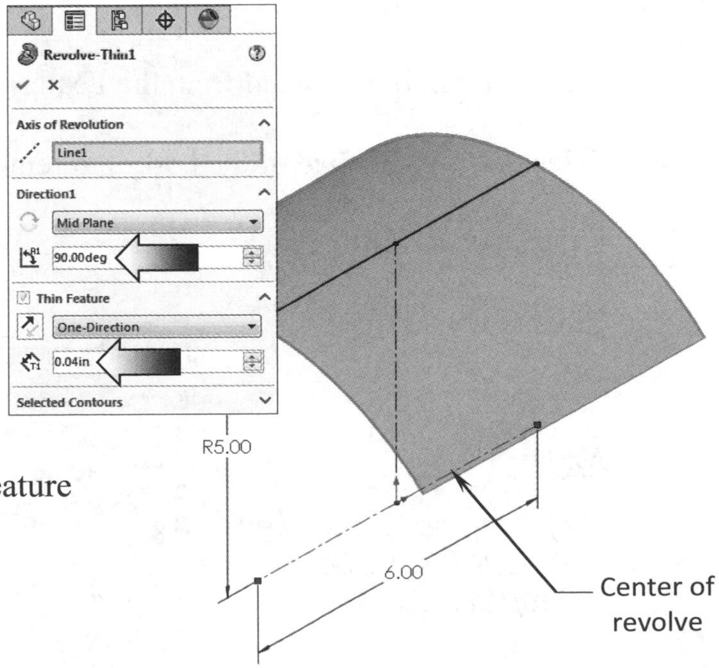

3. Creating the 2nd body:

Select the <u>Top</u> plane and open a **new sketch**.

Sketch a **Center Rectangle** and add the dimension shown.

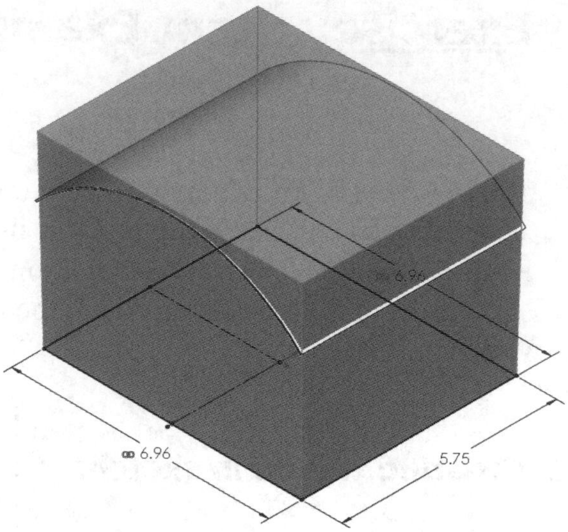

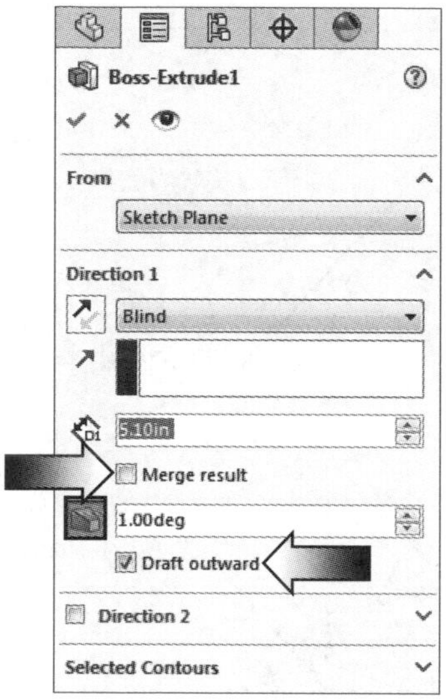

Click **Extruded Boss-Base**:

Blind: **5.10in**

Merge Result: **Cleared** (arrow).

Draft: **1deg** (arrow).

Draft Outward **Enabled**.

Click **OK**.

4. Shelling the body:

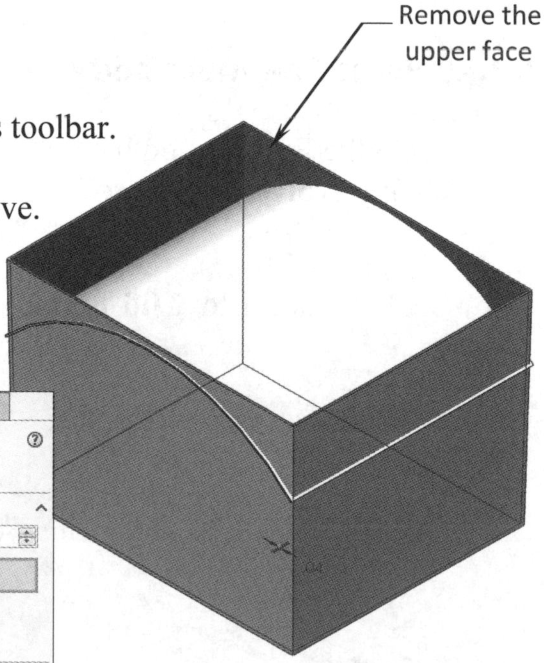

Click the **Shell** command from the Features toolbar.

Select the **upper face** of the **body2** to remove.

Enter **.040in.** for thickness.

Click **OK**.

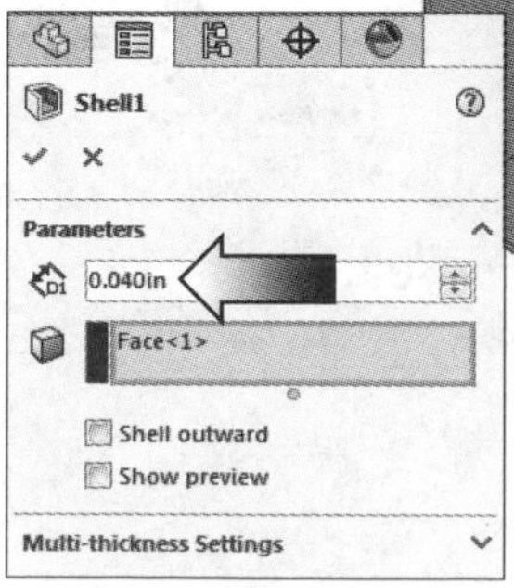

Note:

Only the Body2 is shelled. The Body1 is set below the top surface by.100".

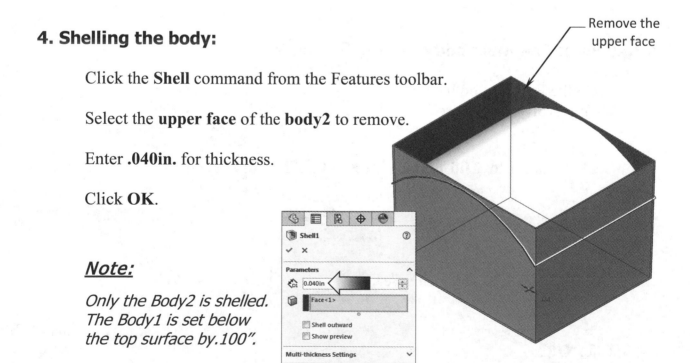

5. Creating an offset plane:

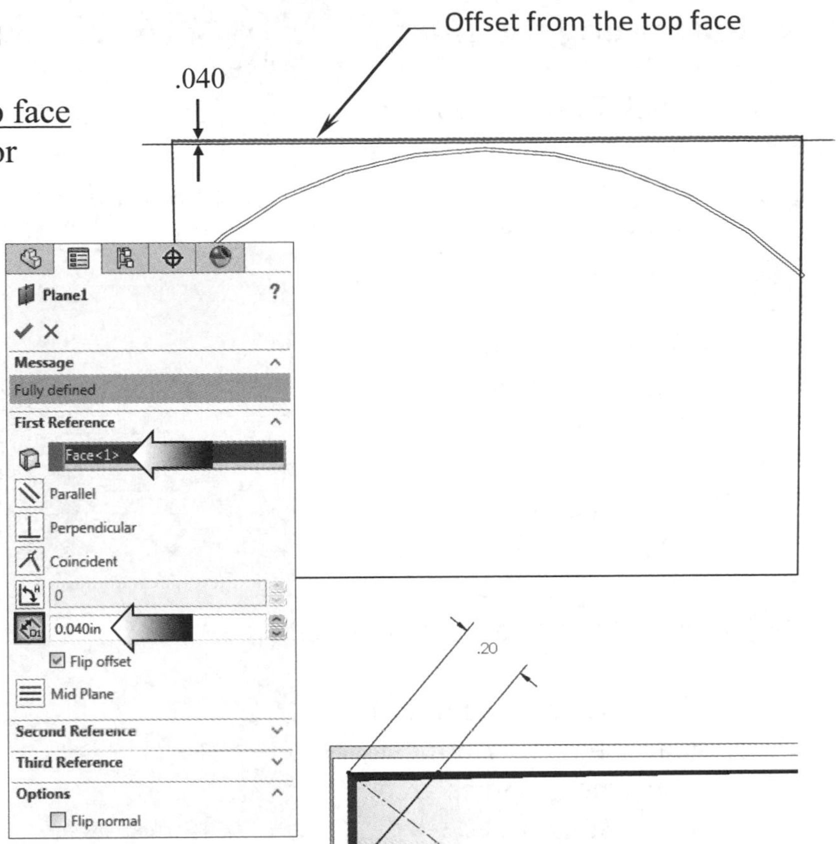

Create a plane from the <u>top face</u> of the part. Enter **.040in**. for the **Offset Distance**.

Click the **Flip** checkbox to place the new plane <u>**below**</u> the surface.

Click **OK**.

6. Creating the 1st Rib:

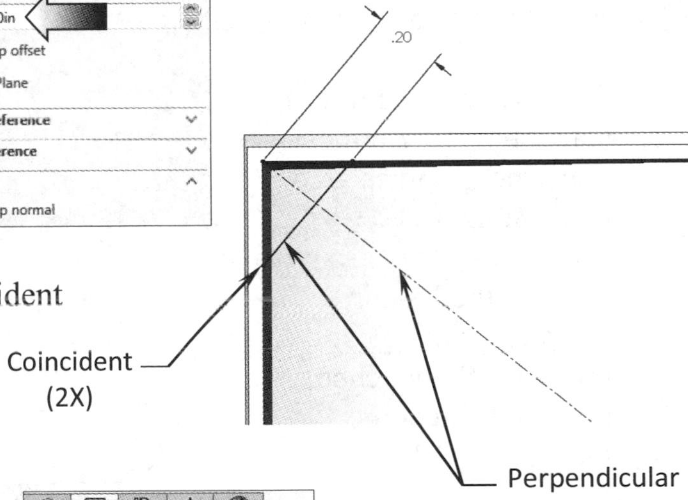

Open a **new sketch** on the <u>new plane</u>.

Add a **line** across the walls, near the upper left corner of the part.

Sketch a **centerline** that is coincident to the 2 diagonal corners.

Add the dimension and relations as noted.

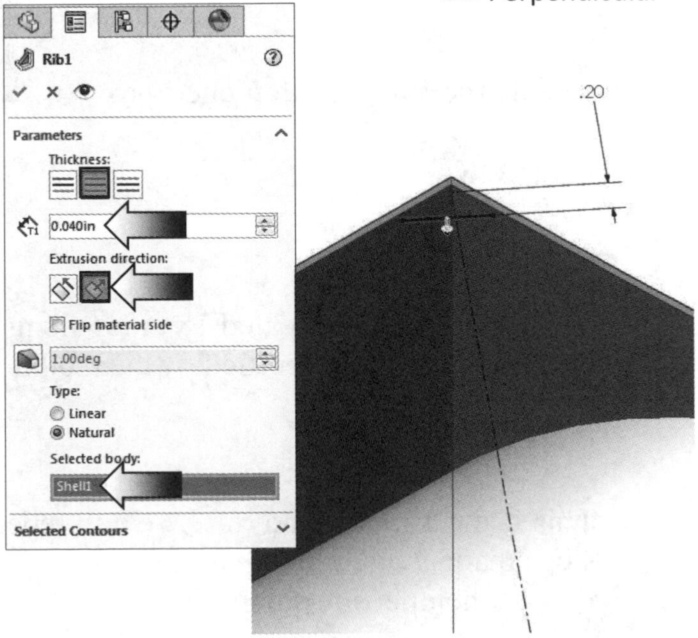

Click the **Rib** command from the Features toolbar.

Set the thickness to **Mid Plane**.

Set the wall to: **.040in**.

Click the **Normal To Sketch** button (arrow).

Under the Selected Body section, click the **Shell1** body (arrow).

Click **OK**.

7. Patterning the Rib:

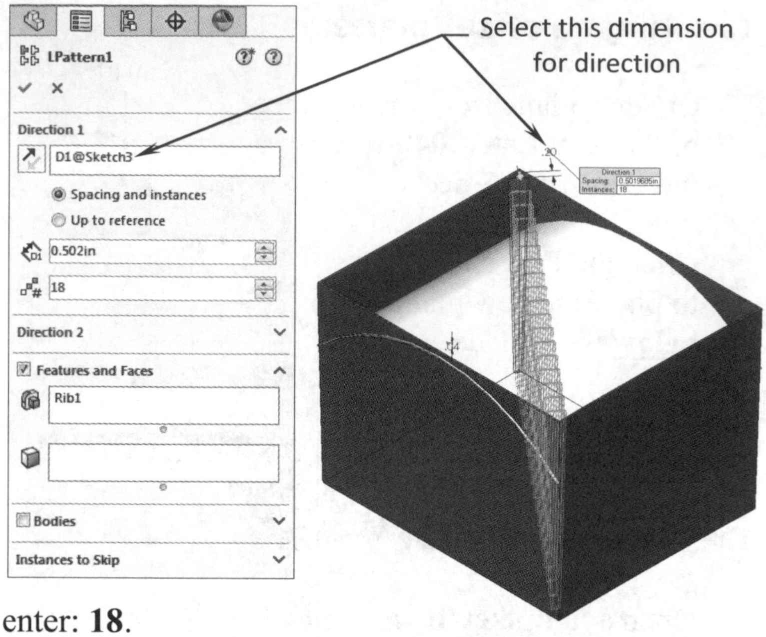

Click **Linear Pattern**.

Click the **front face** of the rib to see its dimensions.

For Pattern Direction: select the **.200in** dimension.

For Spacing, enter: **.502**in.

For Number of Instances, enter: **18**.

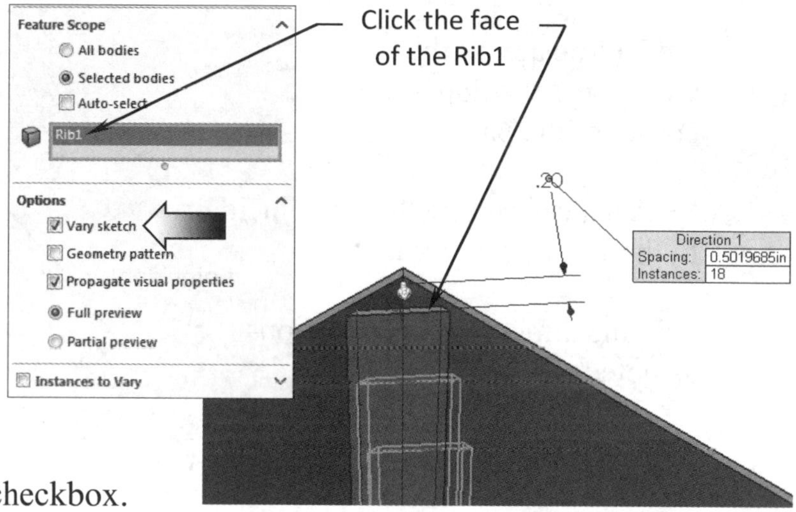

For Features to Pattern, select the **Rib1** either from the graphics area or from the tree.

Expand the Feature-Scope section, clear the Auto-Select check box, and select the face of the Rib as noted.

Enable the **Vary Sketch** checkbox.

Click **OK**.

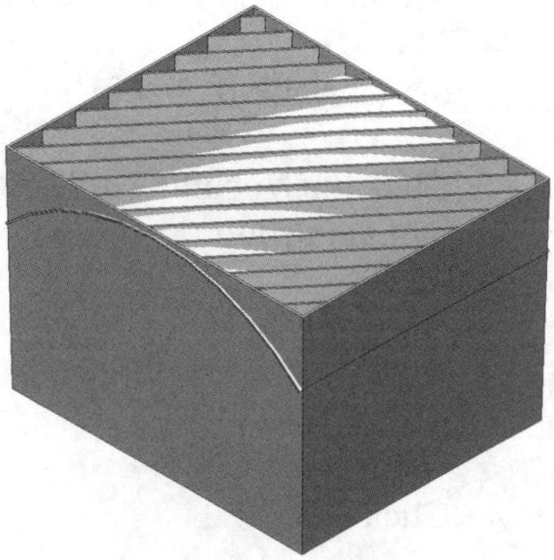

The rib feature is repeated 18 times along the direction that was specified by the .200in dimension.

This is the 1st set of the ribs. We will repeat steps 6 and 7 again to create similar ribs on the opposite side.

8. Creating the 2nd Rib:

Select the upper face of the part and open a **new sketch**.

Sketch a **Line** across the walls as shown.

Add a **Centerline** that is coincident to the 2 diagonal corners.

Add the dimension and relations as indicated.

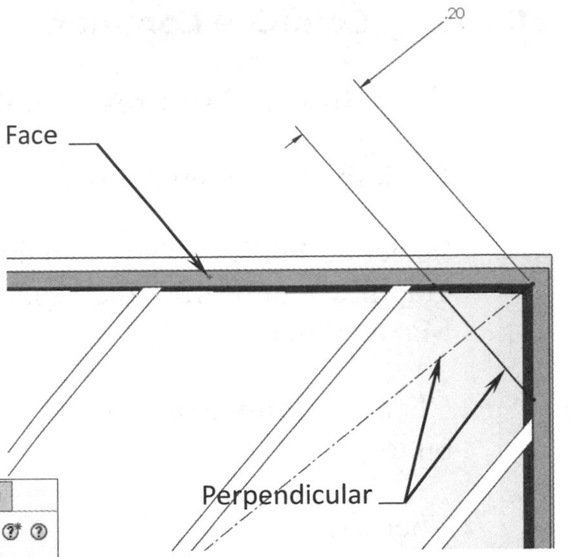

Create another rib using the same settings as the first one.

Click **OK**.

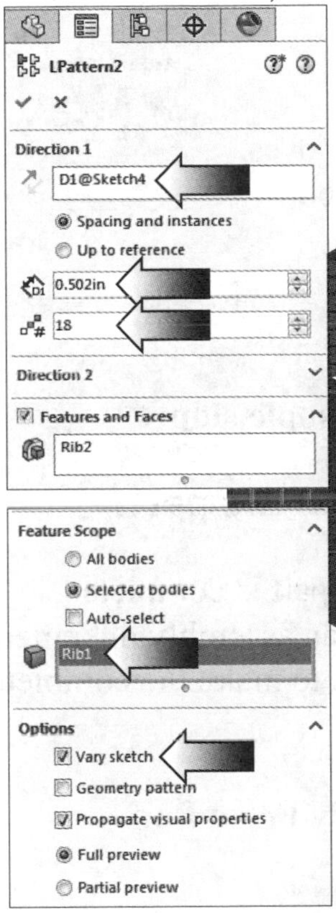

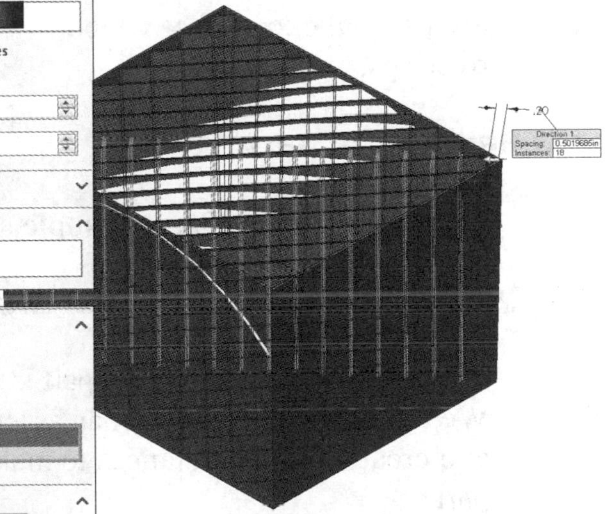

9. Patterning the 2nd set of the ribs:

Click **Linear Pattern**.

Double-click the front face of the rib to see its dimensions.

For Pattern Direction: click the **.20in.** dimension.

For Spacing, enter: **.502in**.

For Number of Instances, enter: **18**.

For Features to Pattern, select the **Rib1**.

Clear the Auto-Select checkbox in the Feature-Scope section and select the face of the Rib. Also enable **Vary Sketch**.

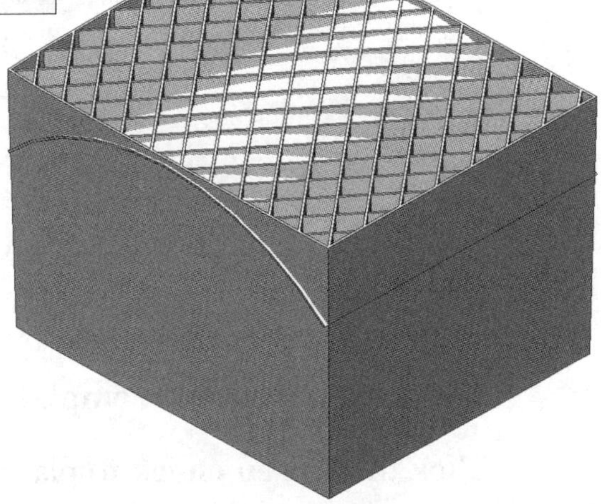

Click **OK**. The Rib is repeated 18 times.

10. Using Combine Common:

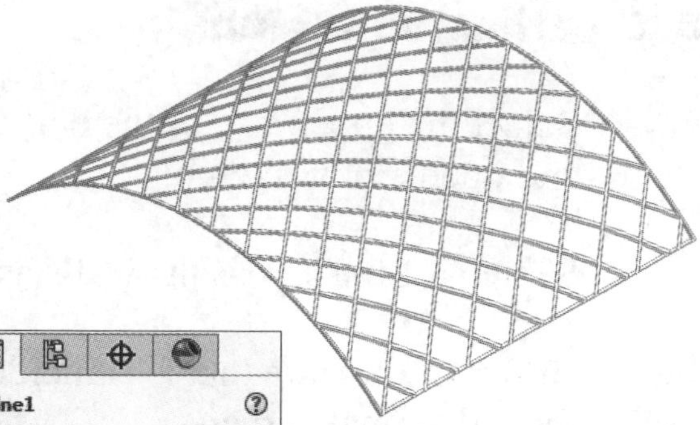

Select **Insert / Features / Combine.**

Click the **Common** option.

Select **all bodies** either from the graphics area or from the Feature Manager tree.

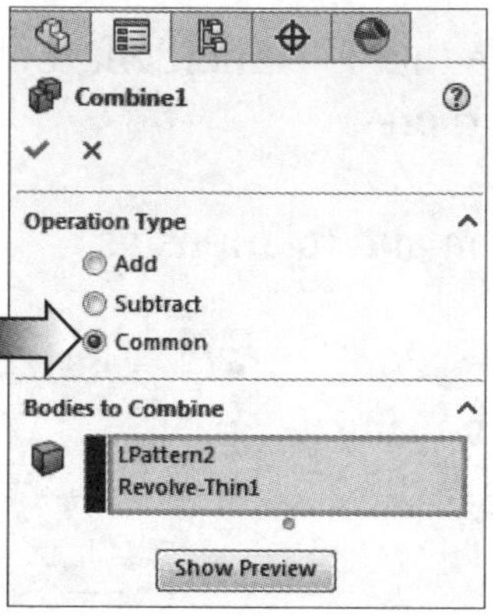

Click the **Show Preview** button.

Click **OK**.

The Combine-Common removes all material except that which overlaps.

11. Saving the part:

<u>**Save**</u> the model as **Stent Sample.sldprt**

12. Making an assembly from the part:

The first one-quarter of the part is completed. We are going to place it in an assembly document and create 3 more instances to make the complete part.

Click **File / Make Assembly From Part**.

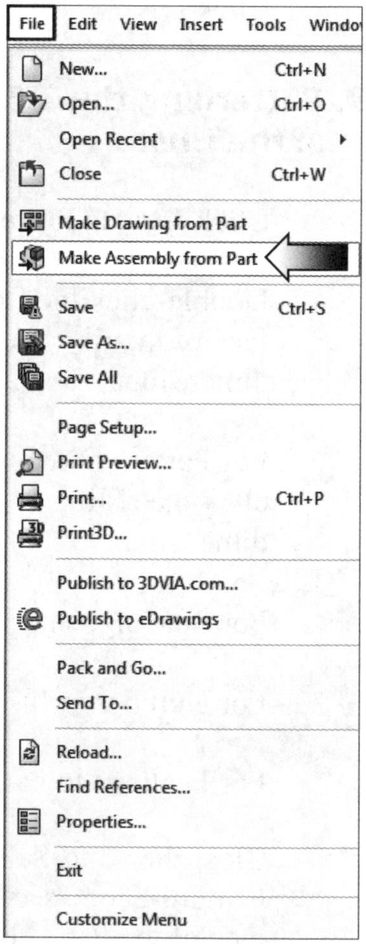

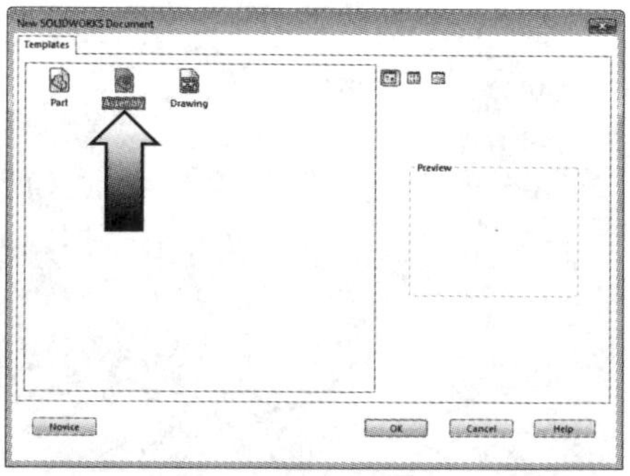

Select the **Assembly Template** and click **OK.**

Click the **Green check** to place the component on the **Origin.**

13. Creating a Circular Component Pattern:

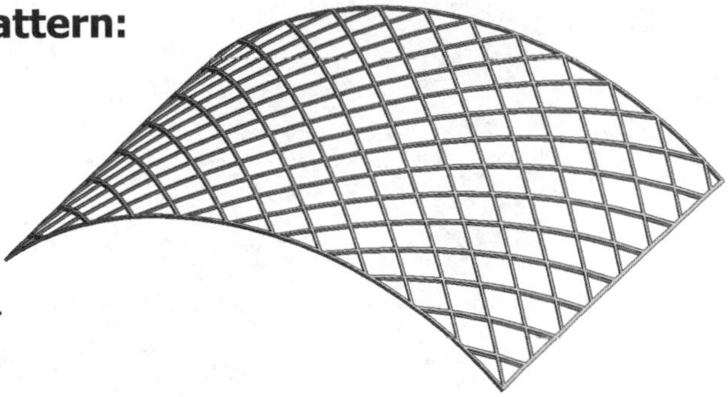

Enable the <u>Temporary Axis</u> under the **View / Hide/Show** pull-down menus.

On the Assembly tool tab, select: **Circular Component Pattern** (arrow).

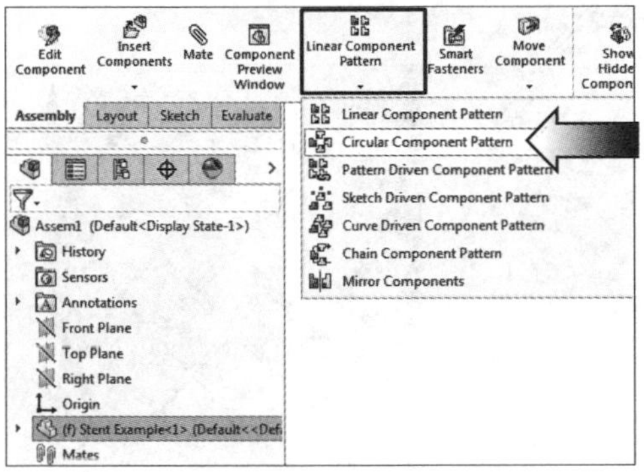

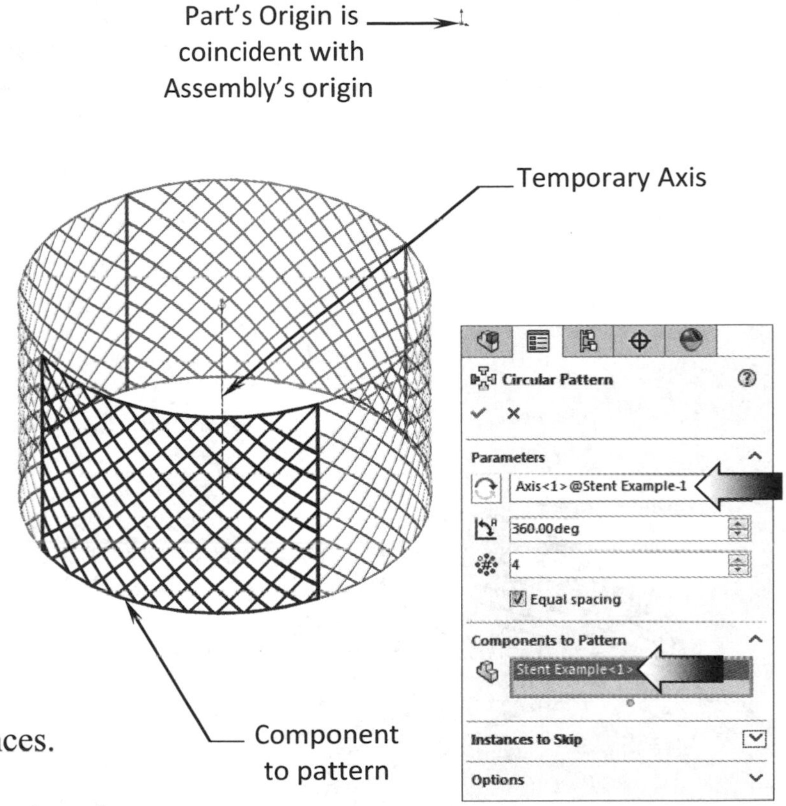

For Pattern Axis, select the **Temporary Axis**.

Enable the **Equal Spacing** checkbox.

Enter **4** for number of instances.

For Components to Pattern, select the component in the graphics area.

Click **OK**.

14. Saving the assembly:

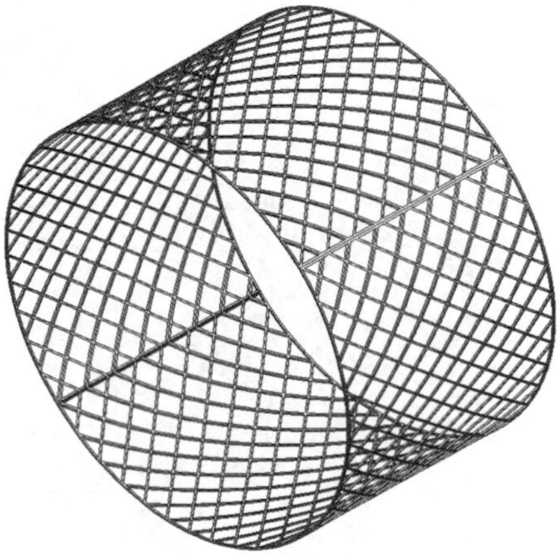

Click **File / Save As**.

Enter **Stent Sample Assembly.sldasm** for the name of the file.

Click **Save**.

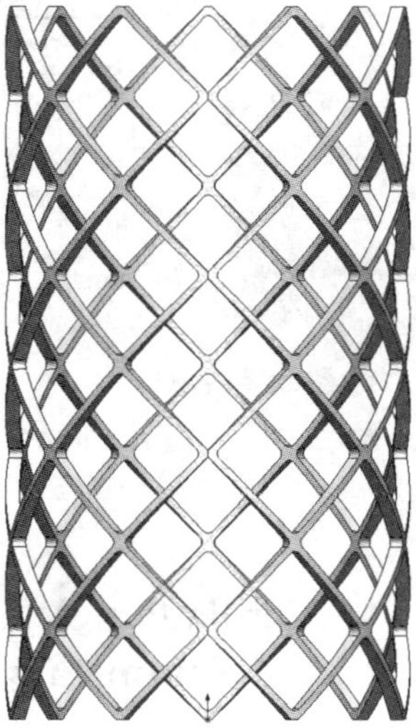

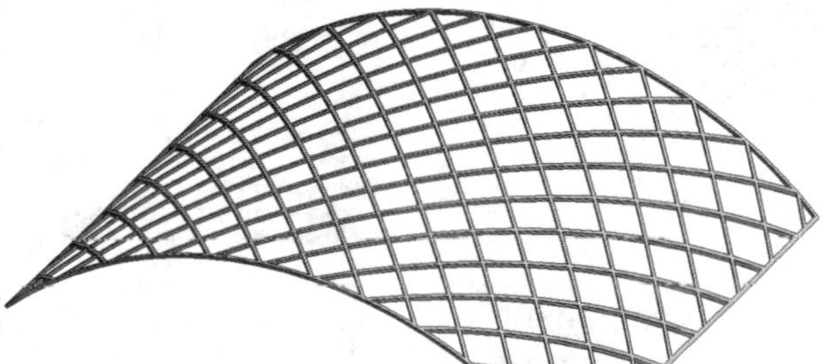

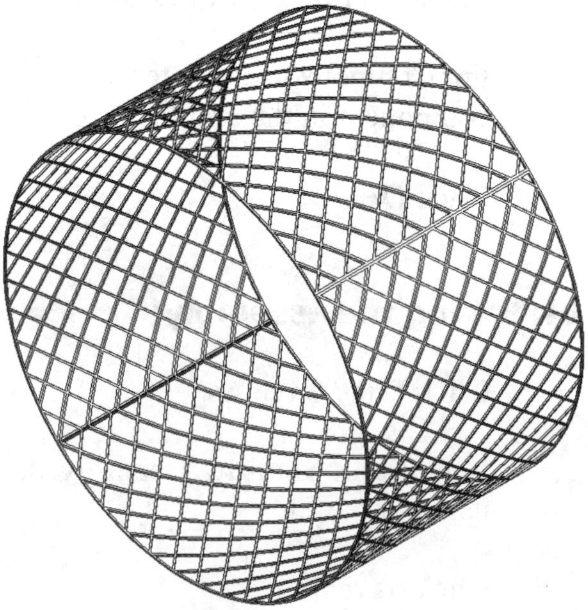

CHAPTER 16

Working with Sheet Metal STEP Files

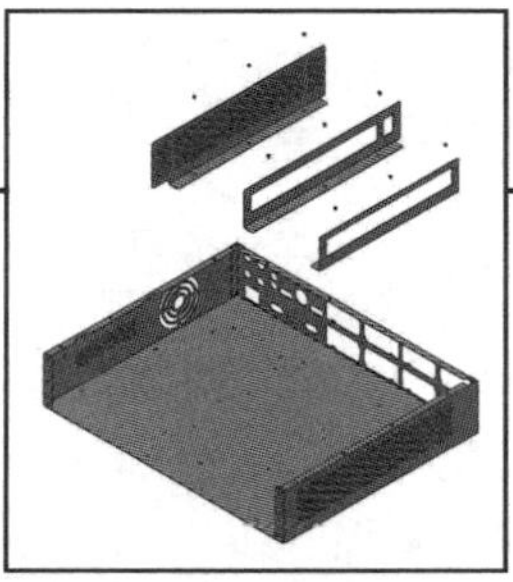

Working with Sheet Metal STEP Files

STEP file extension is short for: **ST**andard for the **E**xchange of **P**roduct data.

The STEP translator supports import and export of body, face, and curve colors from STEP AP214 files.

The STEP AP203 standard does not have any color implementation.
The STEP translator imports STEP files as SOLIDWORKS part or assembly documents.

The STEP translator exports SOLIDWORKS part or assembly documents to STEP files. You can select to export individual parts or subassemblies from an assembly tree, limiting export to only those parts or subassemblies.

If you select a subassembly, all of its components are automatically selected. If you select a component, its ascendants are partially selected, preserving the assembly structure.

This lesson discusses one of the methods to convert an Assembly STEP file to SOLIDWORKS Sheet Metal parts.

After the components are converted, some of the Assembly Features such as the Hole Series and Hole Wizards are used to add the new holes in the assembly mode, then the Fasteners are inserted automatically using the Smart Fasteners feature (requires SOLIDWORKS Toolbox).

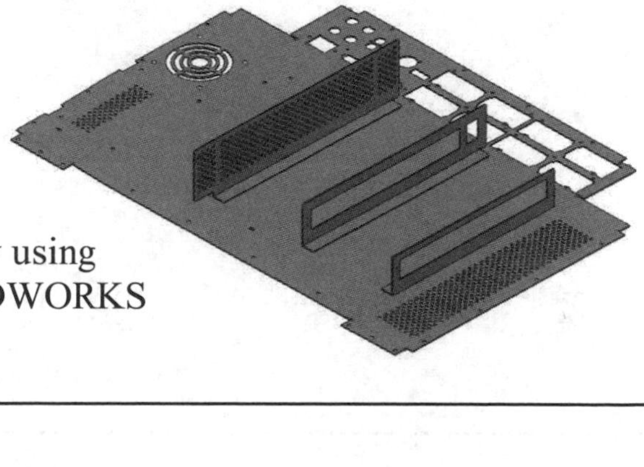

Working with
Sheet Metal STEP Files

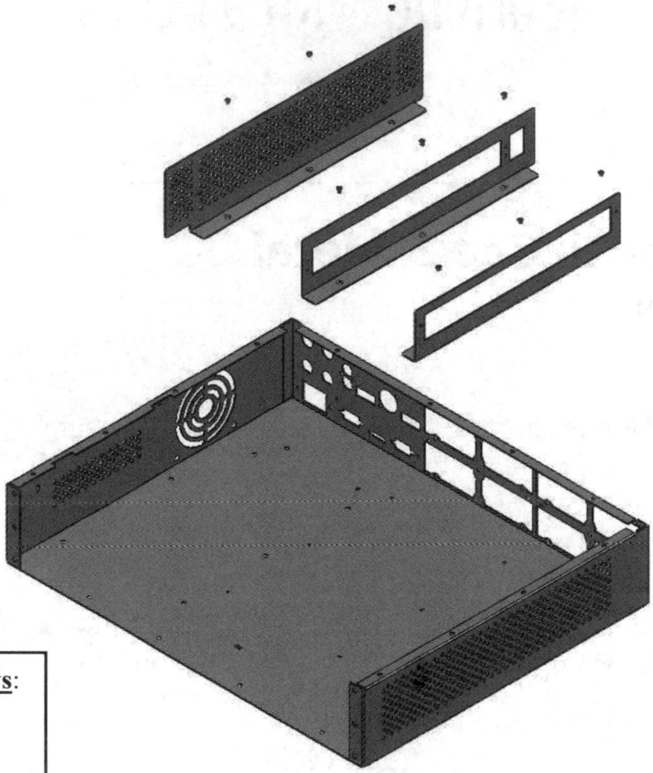

View Orientation Hot Keys:

Ctrl + 1 = Front View
Ctrl + 2 = Back View
Ctrl + 3 = Left View
Ctrl + 4 = Right View
Ctrl + 5 = Top View
Ctrl + 6 = Bottom View
Ctrl + 7 = Isometric View
Ctrl + 8 = Normal To Selection

Dimensioning Standards: **ANSI**

Units: **INCHES** – 3 Decimals

Tools Needed:

Dimension	Insert Bends	Flat Pattern
Hole Series	Hole Wizard	Smart Fasteners

Sheet Metal **STEP Files** and **Smart Fasteners**

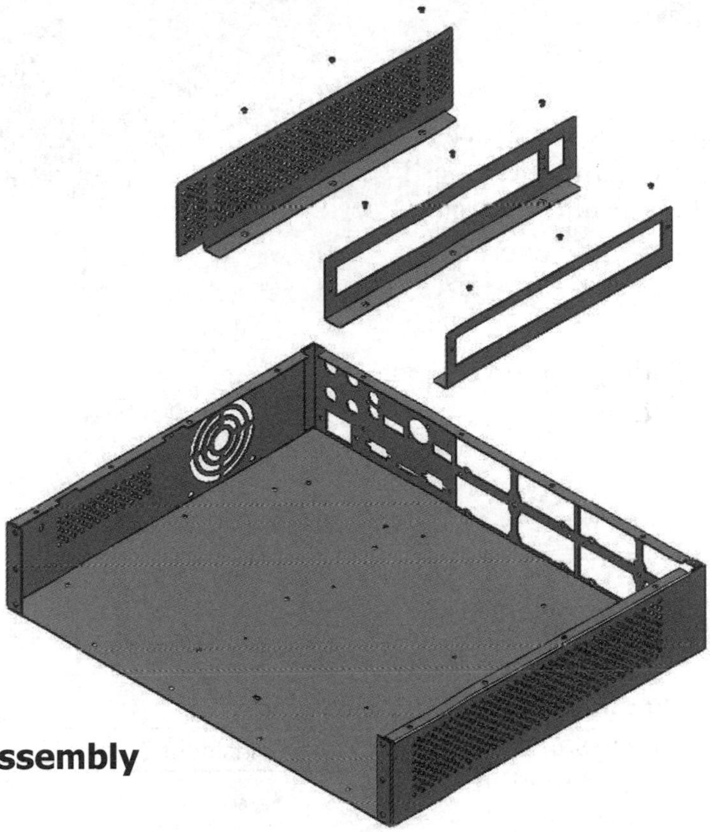

1. Opening an Assembly Step File:

Select to **File / Open**.

Browse to the Training Files folder, change the Files of Type to **STEP** and open a STEP document named: **SM-Assembly.step**.

The part files from the STEP document will appear as SOLIDWORKS documents on the FeatureManager tree without any model history.

*<u>NOTE:</u> The **3D-Interconnect** option should be <u>disabled</u> if you have trouble opening the iges document.*
*<u>Go to:</u> **Tools, Options, System Options, Import** to disable it.*

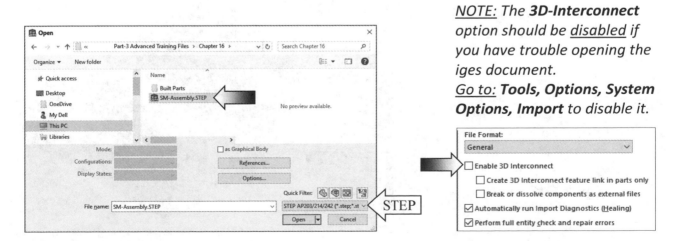

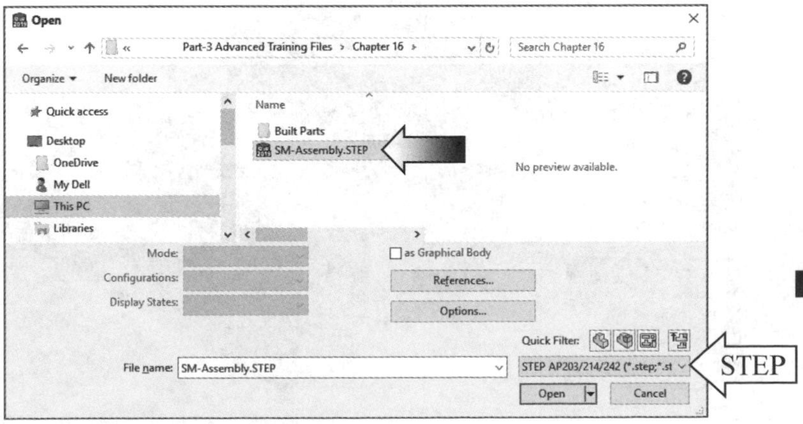

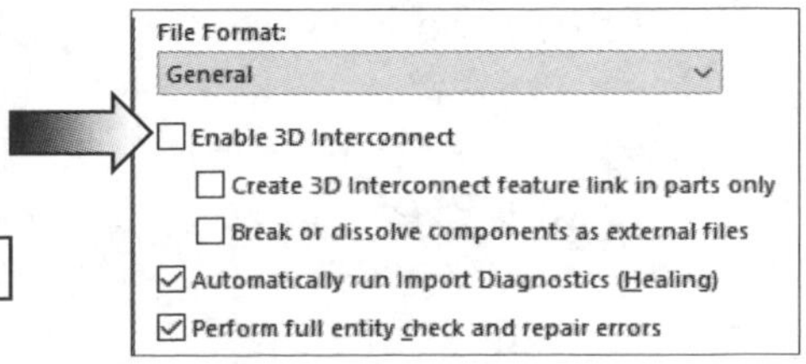

There are 4 components in this assembly and they have not yet been constrained.

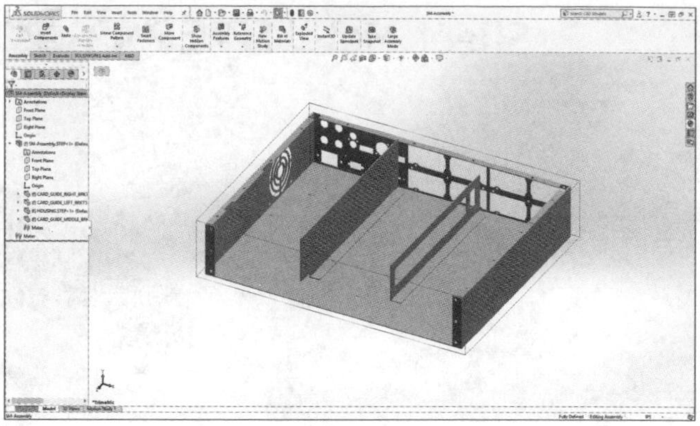

The Housing will be used as the Fixed Component and the 3 Card Guides will be left un-constrained for the purpose of this exercise.

Change the Shading option to: **Shaded with Edges** (arrow).

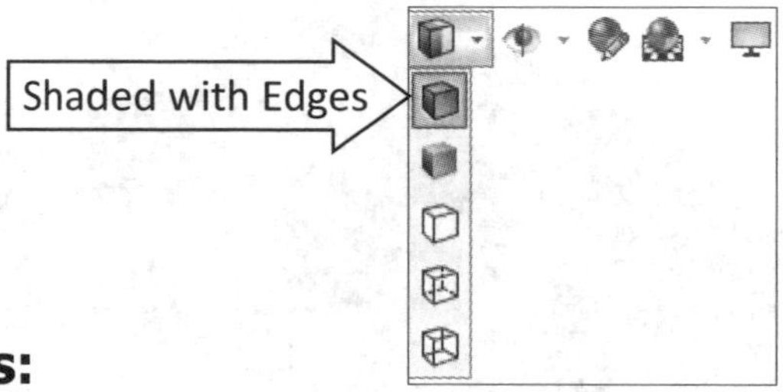

2. Mating the components:

In order to mate the Housing component to the assembly's Origin, we will need to align the 3 Front, Top, and Right planes.

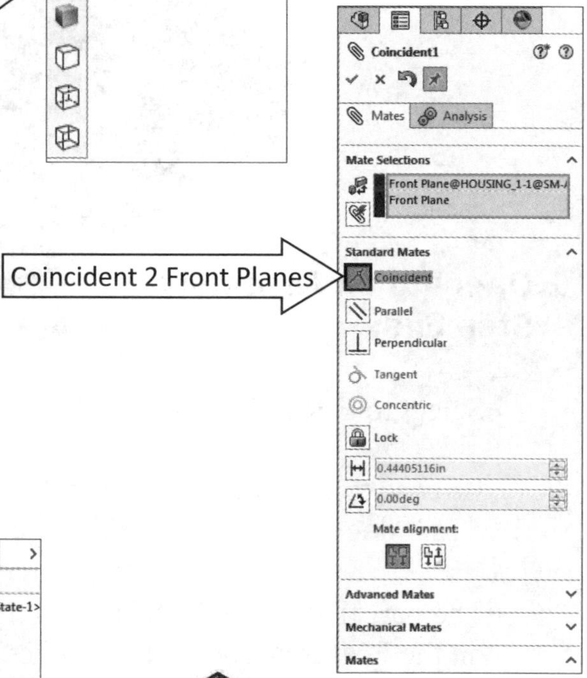

Click the **Mate** command  from the Assembly toolbar.

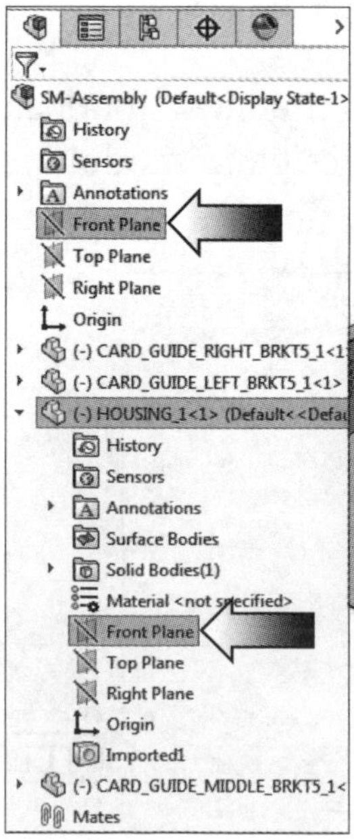

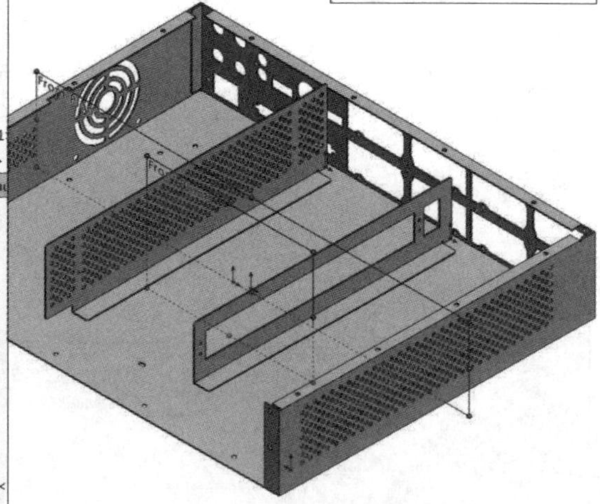

Select the **Front** plane of the Housing and the **Front** plane of the Assembly.

Select **Coincident** from the list.

Click **OK**.

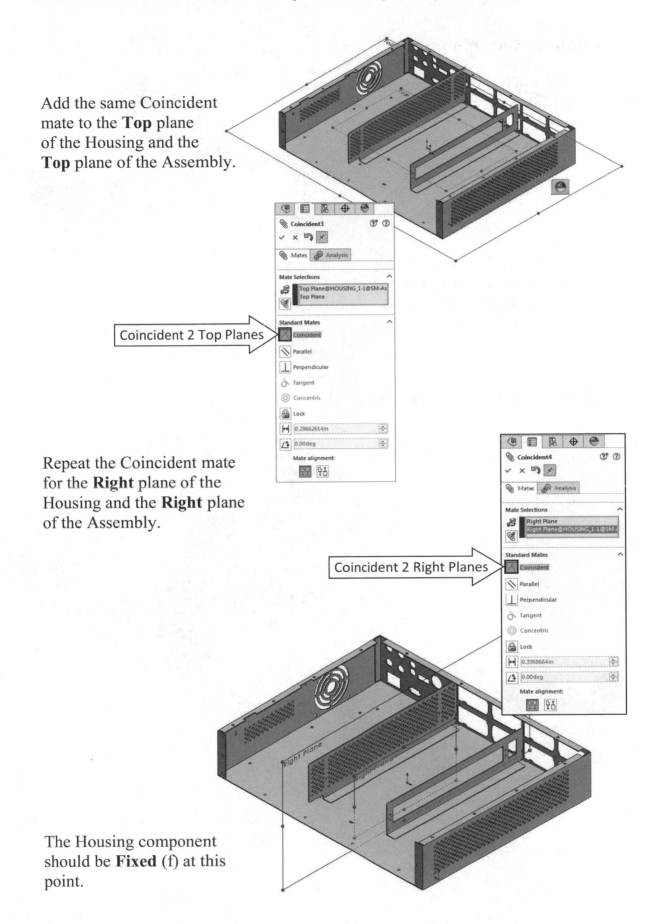

Add the same Coincident mate to the **Top** plane of the Housing and the **Top** plane of the Assembly.

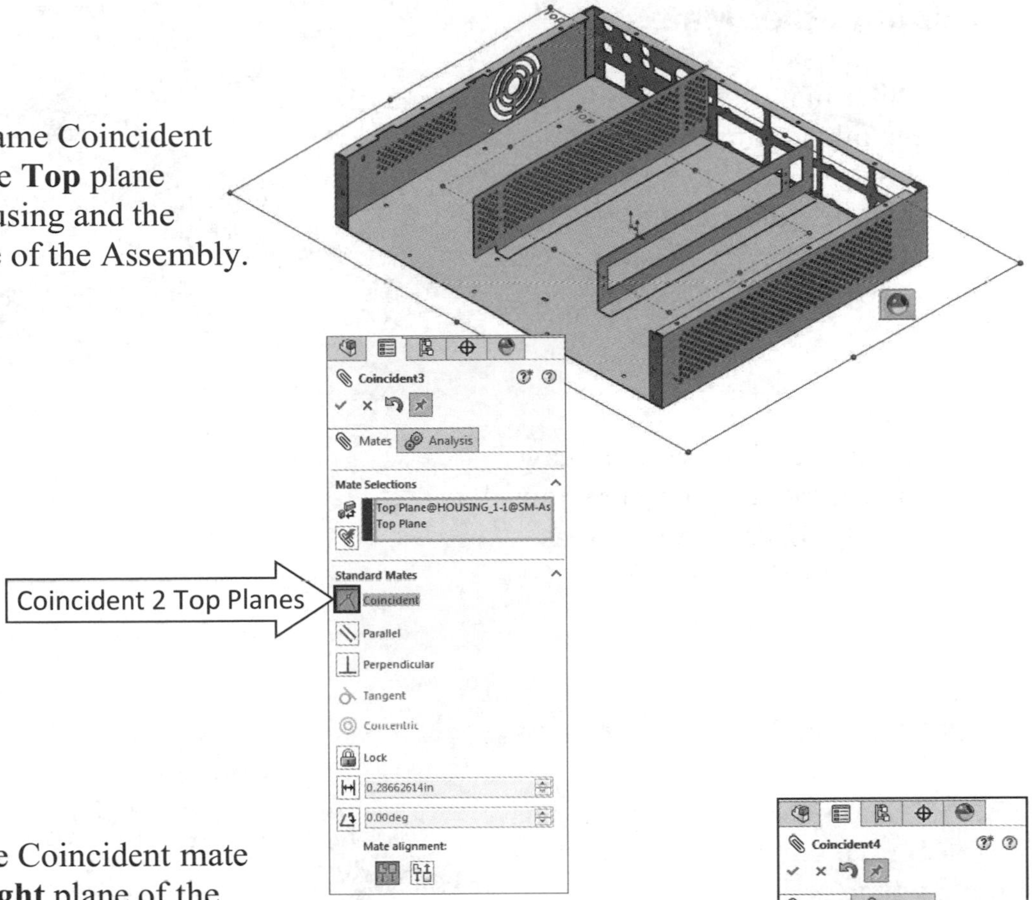

Repeat the Coincident mate for the **Right** plane of the Housing and the **Right** plane of the Assembly.

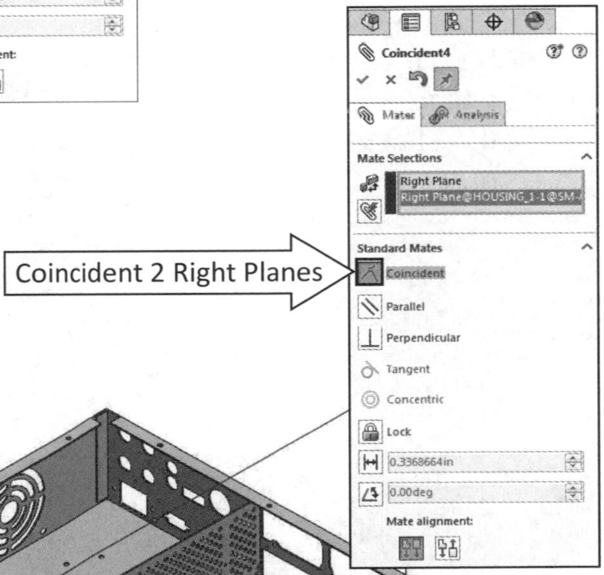

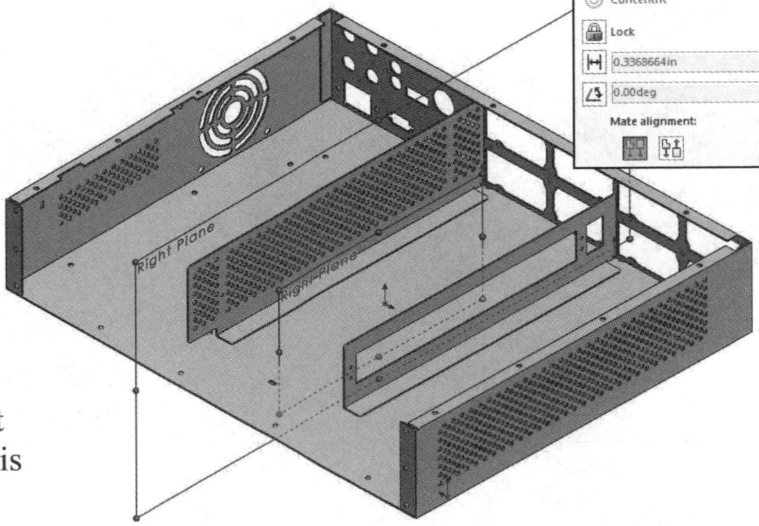

The Housing component should be **Fixed** (f) at this point.

3. Adding other Mates:

Other mates can be added to constrain the other components, but later on they will need to be suppressed so that the final sheet metal components can be flattened properly.

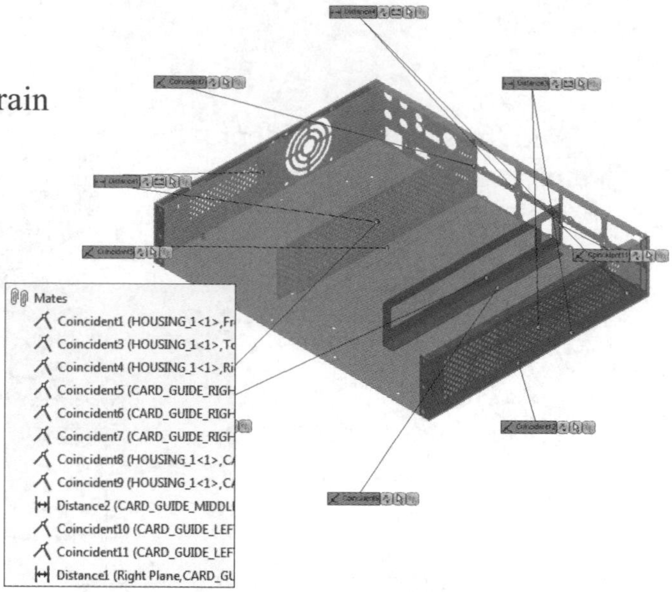

In this exercise we will leave the components un-constrained to help focus in other areas.

4. Examining the components:

The imported components have **Sharp corners** all around, which is not suitable for the Sheet Metal processes.

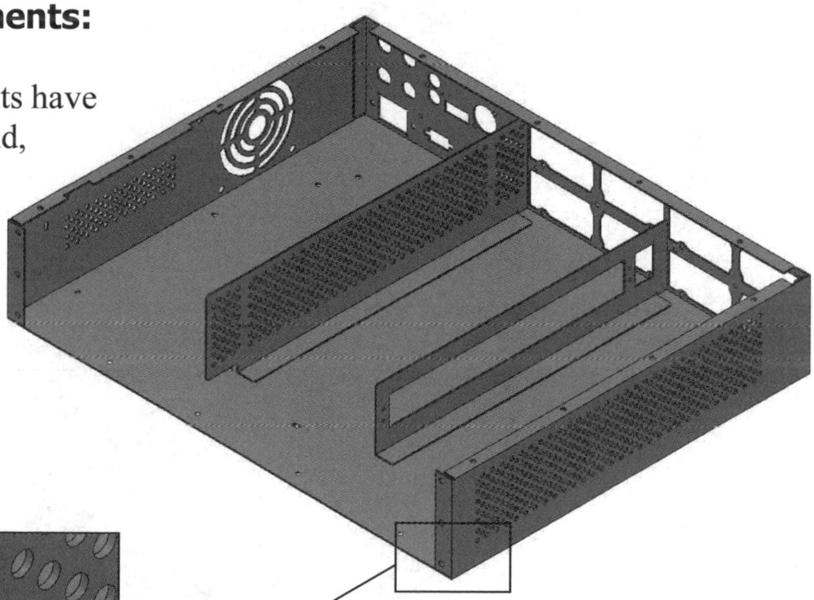

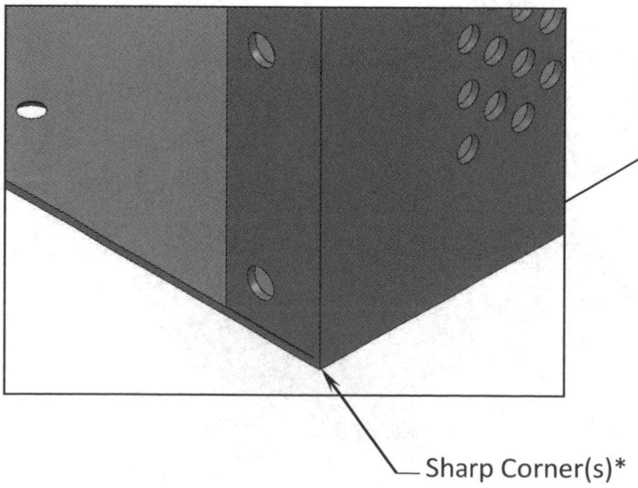

Sheet Metal parameters* must be added to fully convert the imported part into a SOLIDWORKS Sheet Metal part.

5. Adding the Sheet Metal tool tab:

If the Sheet Metal tool tab is not yet enabled on the CommandManager, do the following to add it:

Right-click on the Assembly tool tab and select the **Customize Command-Manager** option (arrow).

Click the **New Tab** and pick: **Sheet Metal** (arrow) and click **OK**.

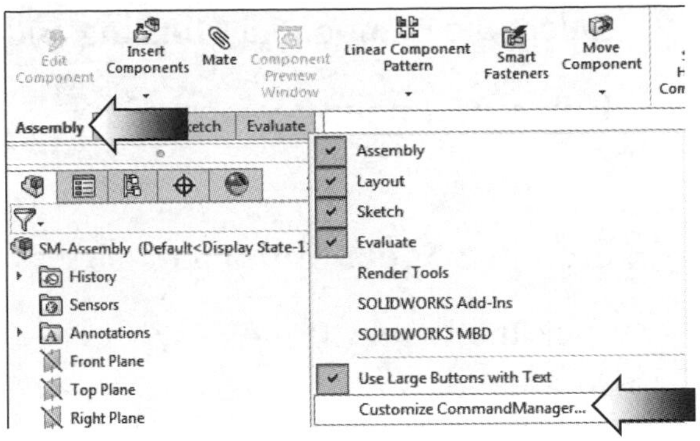

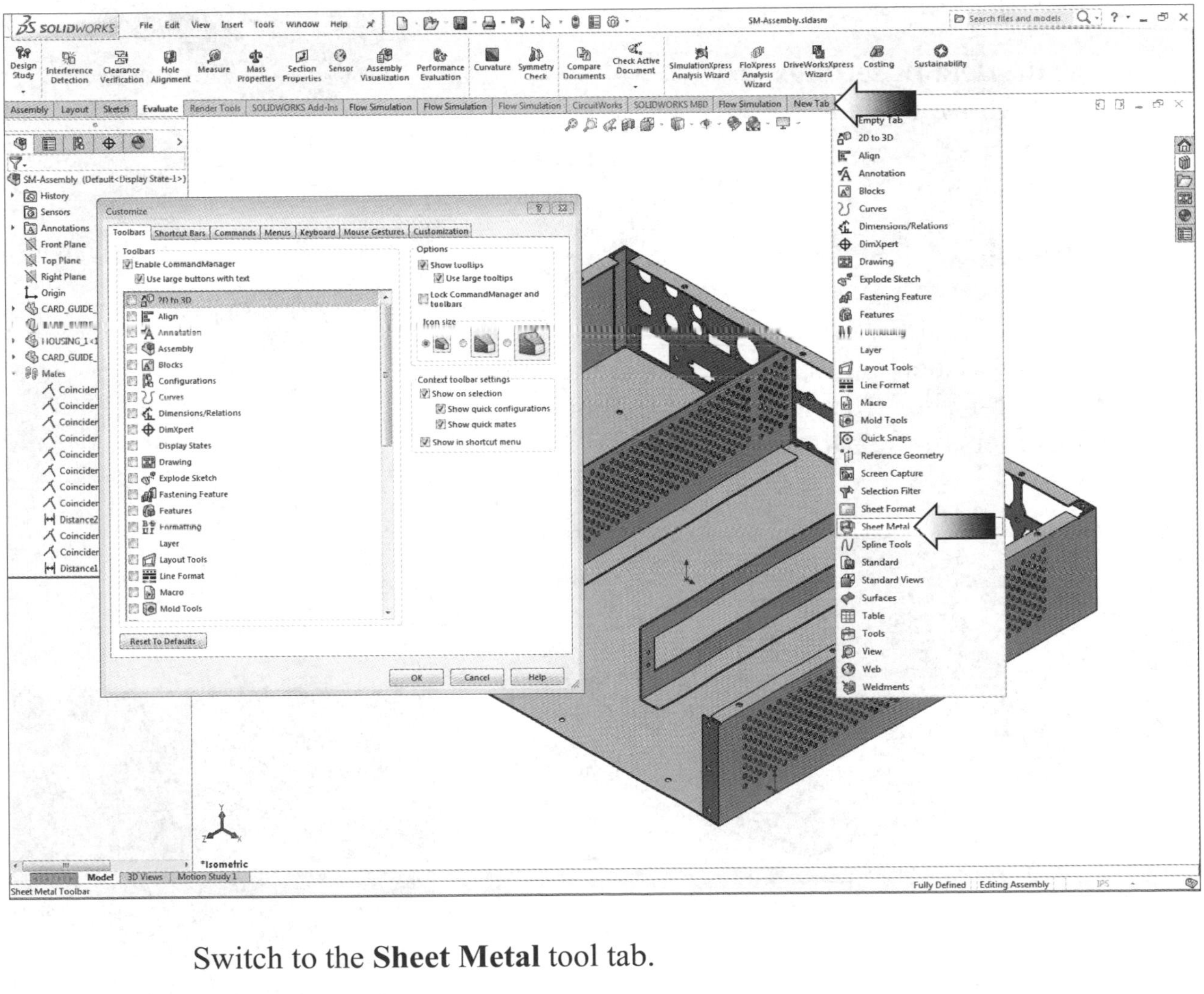

Switch to the **Sheet Metal** tool tab.

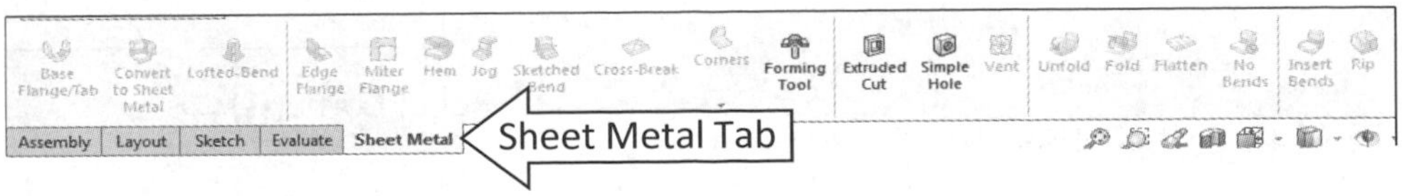

6. Inserting Sheet Metal parameters*:

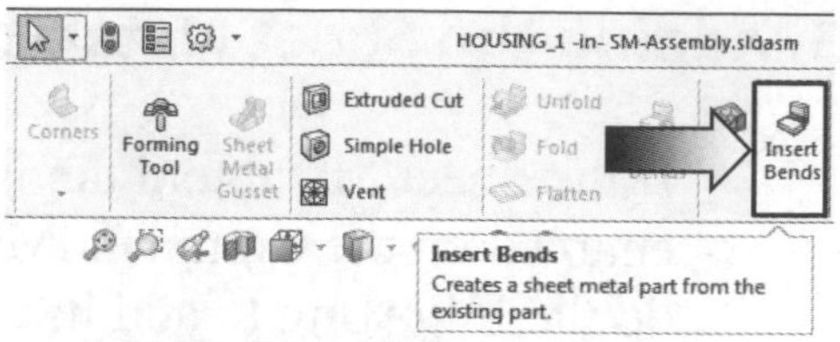

Select the component **Housing** and click the **Edit Component** command.

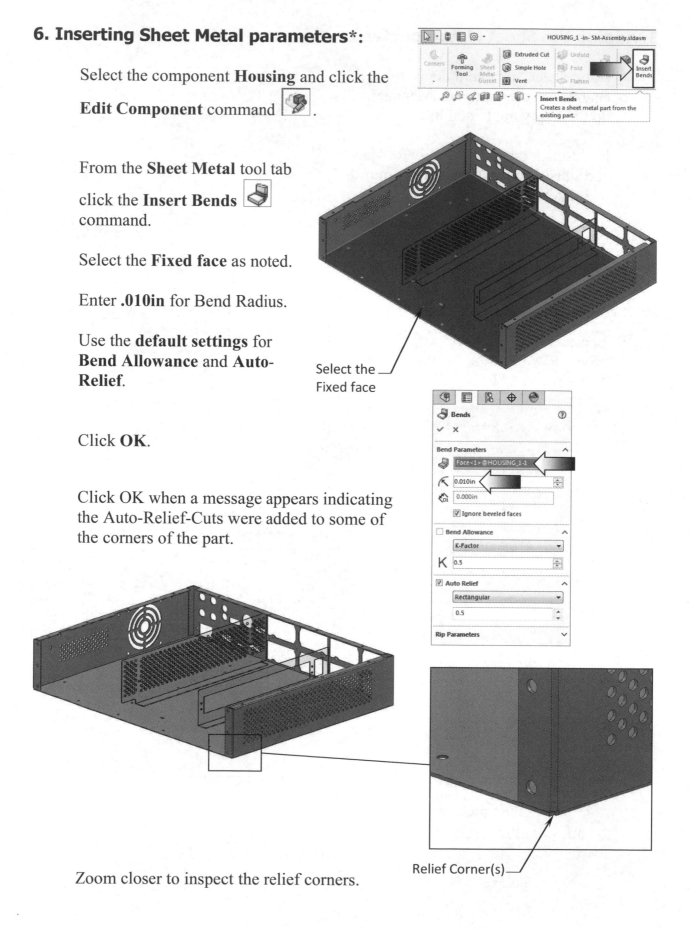

From the **Sheet Metal** tool tab click the **Insert Bends** command.

Select the **Fixed face** as noted.

Enter **.010in** for Bend Radius.

Use the **default settings** for **Bend Allowance** and **Auto-Relief**.

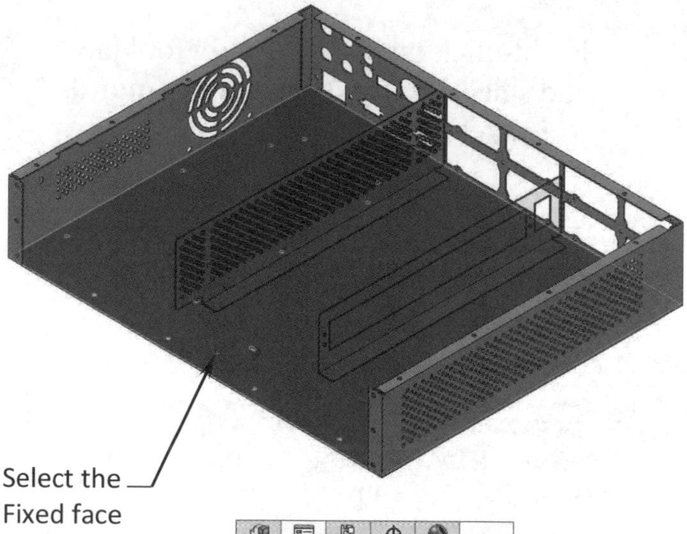

Click **OK**.

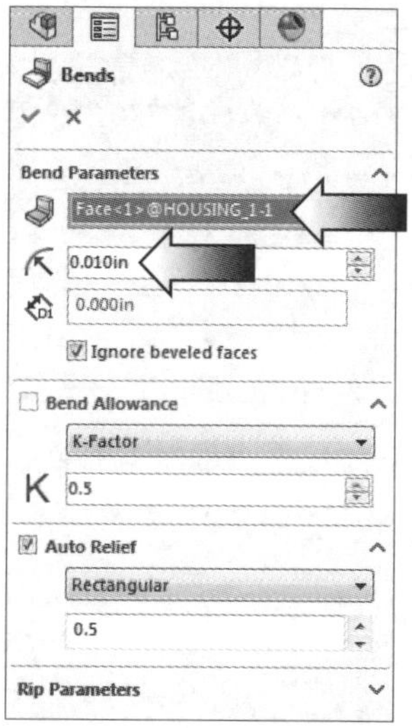

Click OK when a message appears indicating the Auto-Relief-Cuts were added to some of the corners of the part.

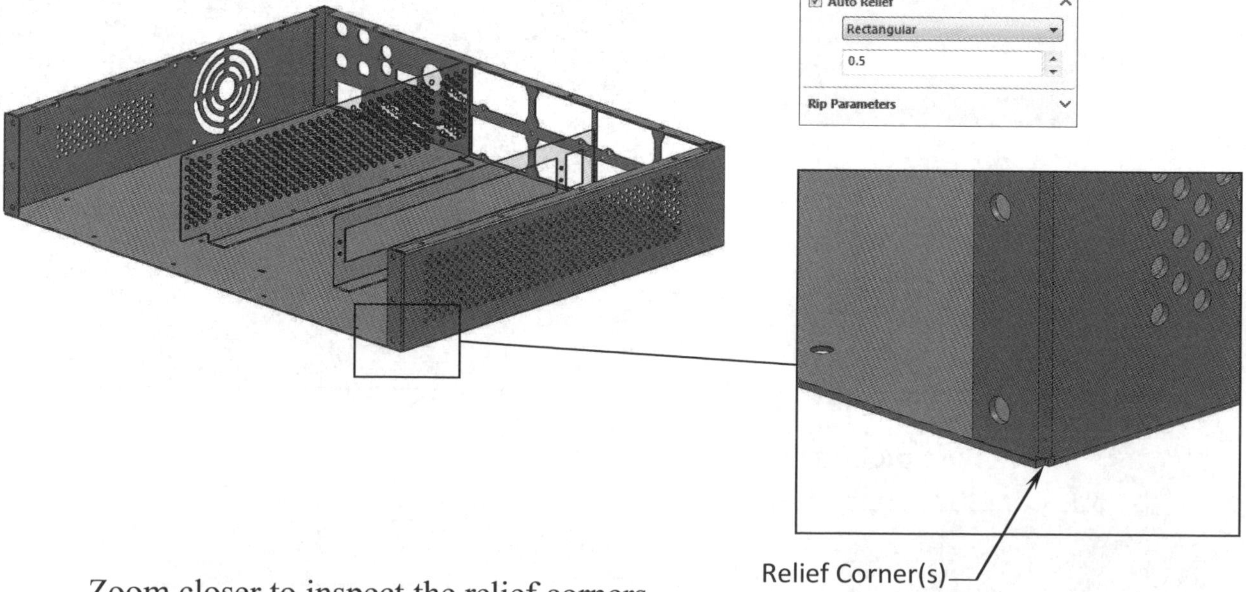

Zoom closer to inspect the relief corners.

7. Viewing the Flat Pattern:

From the **Sheet Metal** tool tab, click the **Flatten** command.

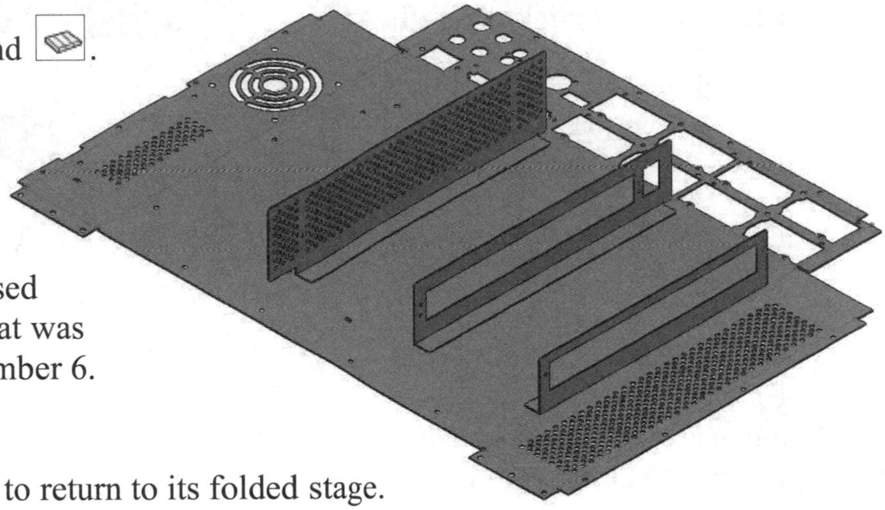

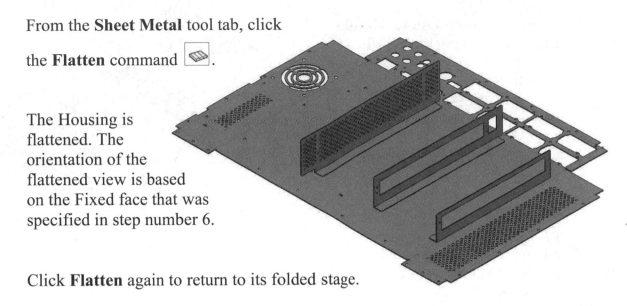

The Housing is flattened. The orientation of the flattened view is based on the Fixed face that was specified in step number 6.

Click **Flatten** again to return to its folded stage.

Click-off the **Edit Component**  command.

8. Converting the 2nd component:

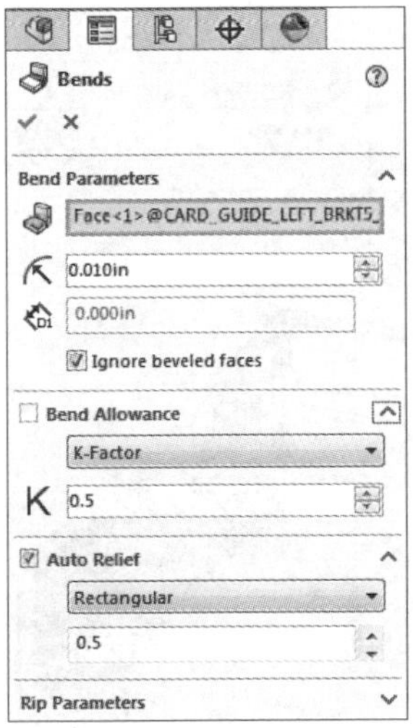

Select the **Card Guide Left** as shown and click the **Edit Component**  command.

From the Sheet Metal tool tab, click **Insert Bends** .

Select the **Fixed face** as noted.

Enter **.010"** for Bend Radius. Keep all other default parameters.

Click **OK**.

Click off the **Edit Component**  command.

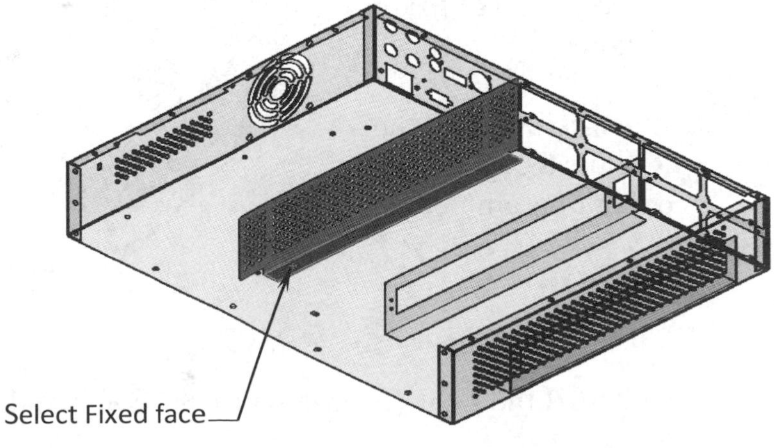

9. Converting the 3rd component:

Select the **Card Guide Middle** as shown and click the **Edit-Component** command.

From the **Sheet Metal** tool tab click **Insert Bends**.

Select the **Fixed face** as noted.

Enter **.010"** for Bend Radius and use the **default settings** for the Bend Allowance and K-Factor.

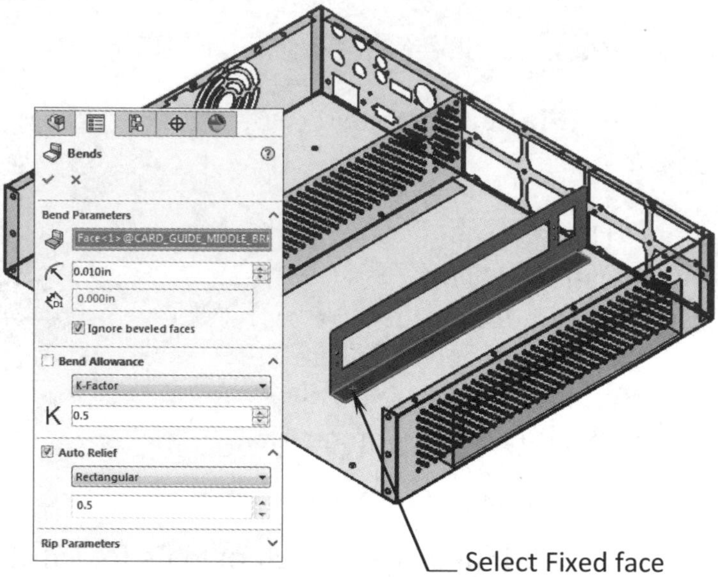

Click **OK**.

Click-off the **Edit Component** command.

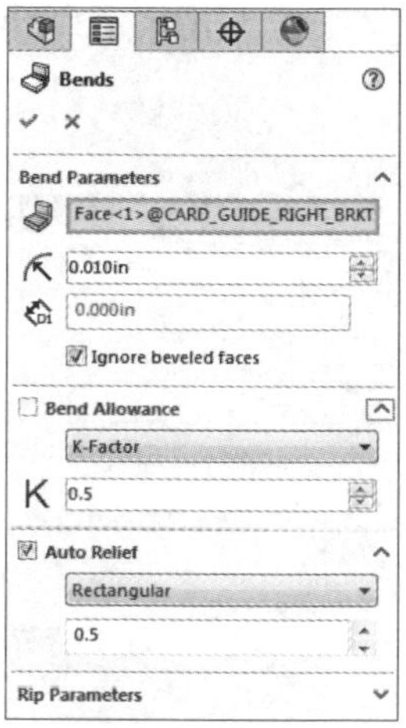

10. Converting the 4th component:

Select the **Card Guide Right** as shown and click the **Edit Component** command.

From the Sheet Metal tool tab, click:

Insert Bends.

Select the **Fixed face** as noted.

Enter **.010"** for Bend Radius. Keep all other default parameters the same.

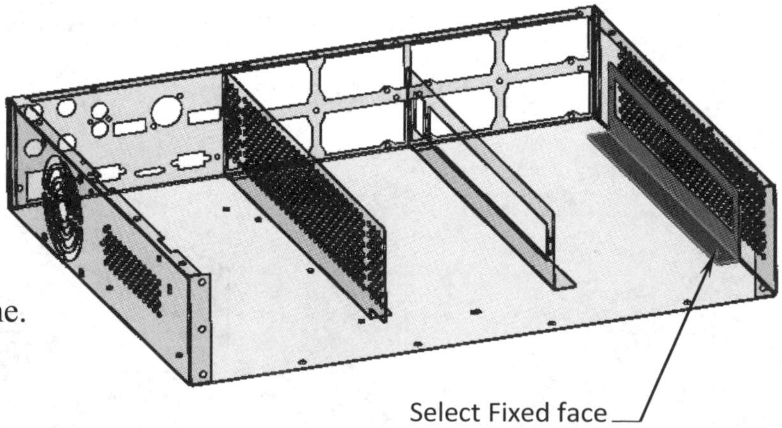

Click **OK**.

Click-off the **Edit Component** command.

11. Using the Hole-Series:

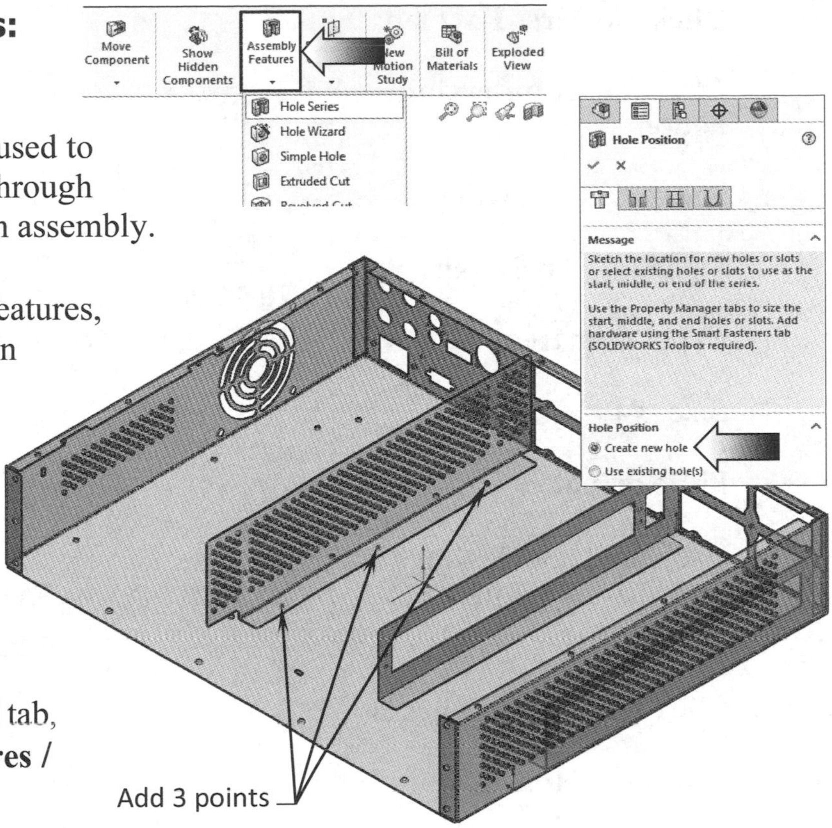

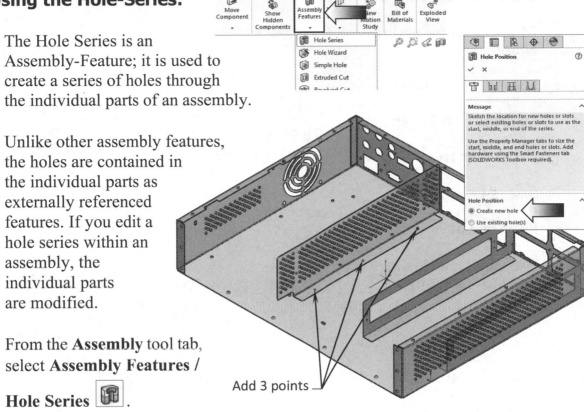

The Hole Series is an Assembly-Feature; it is used to create a series of holes through the individual parts of an assembly.

Unlike other assembly features, the holes are contained in the individual parts as externally referenced features. If you edit a hole series within an assembly, the individual parts are modified.

From the **Assembly** tool tab, select **Assembly Features / Hole Series** .

From the FeatureManager, click: **Create New Hole** (arrow).

The mouse cursor changes to the **Sketch Point** command.

Add **3 Sketch Points** approximately as shown.

Each point is the center of a hole.

Add an **ALONG Z** relation (vertical) between the 3 points.

Add the dimensions as indicated to fully define the positions of the points.

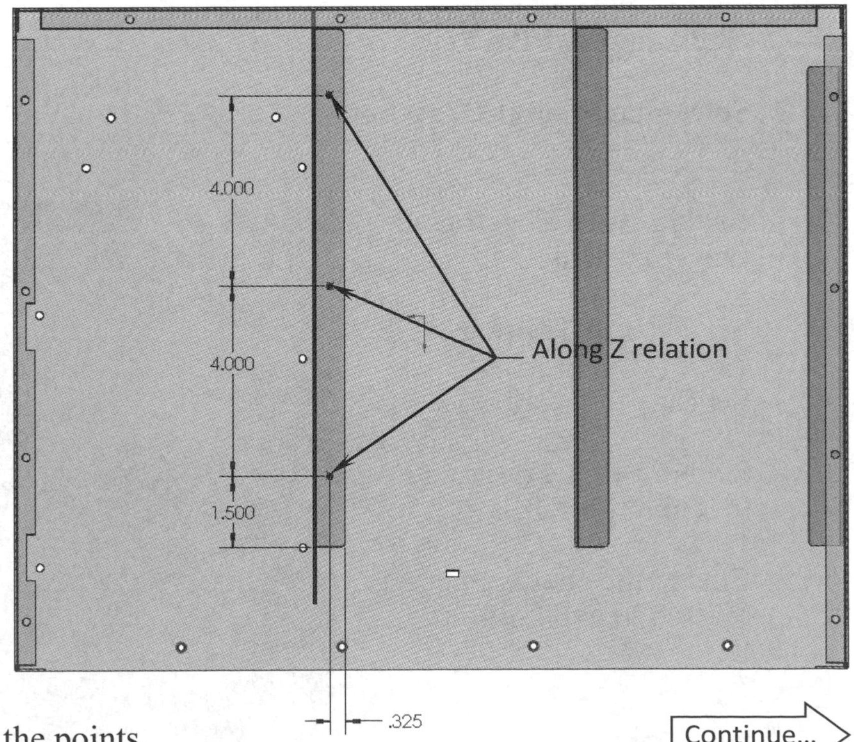

Continue...

Click the **First Part** tab.

Click the **Countersink** option.

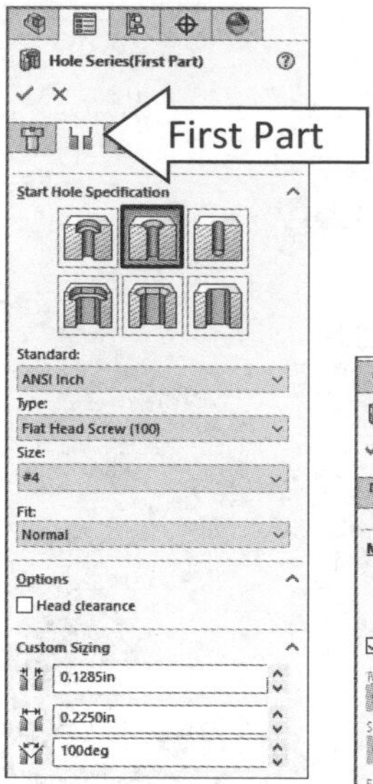

Select the following:

Standard: **Ansi Inch**

Type: **Flat Head Screw**

Size: **#4**

Fit: **Normal**

Use the default settings for Custom Sizing.

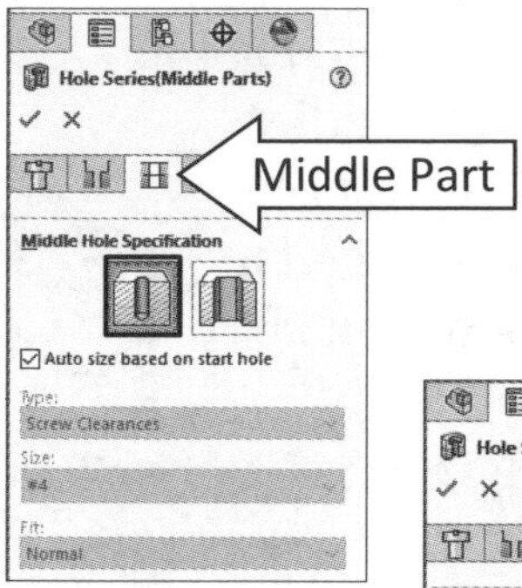

Click the **Middle Part** tab.

Select the **Hole** button.

Enable the check box:
Auto Size based on Start Hole.

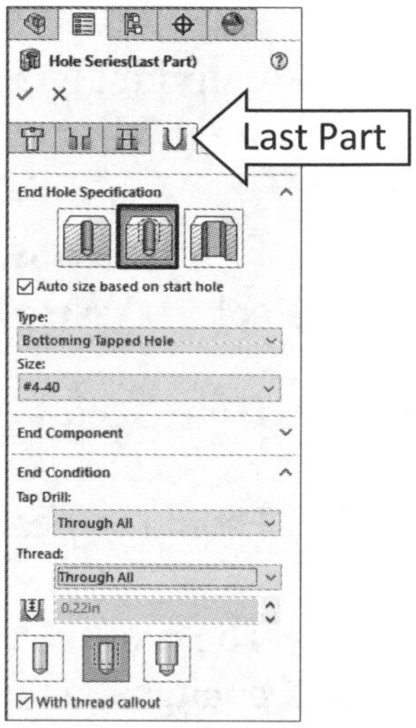

Click the **Last Part** tab.

Select the **Straight Tap** button.

Enable **Auto Size Based On Start Hole**.

Set Type to **Tapped Hole**.

Set Size to **#4-40**.

Set both End Conditions to **Through All**.

Enable the checkbox:
With Thread Callout.

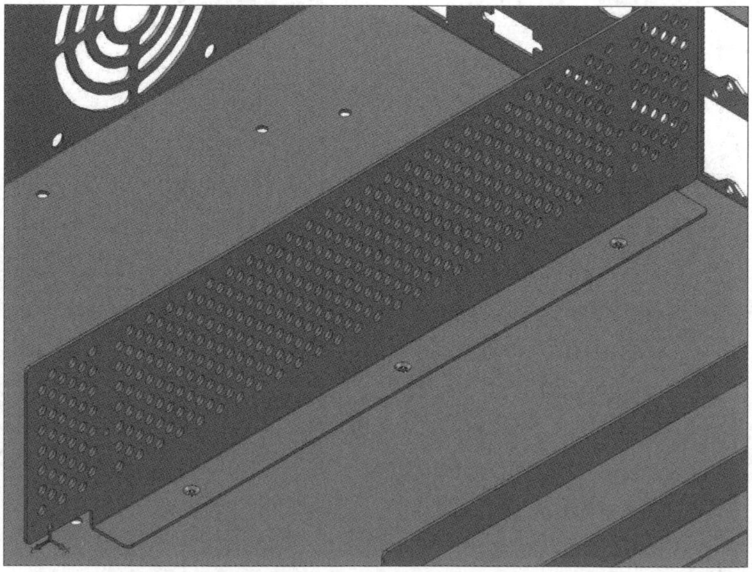

Click **OK**.

12. Using the Hole Wizard:

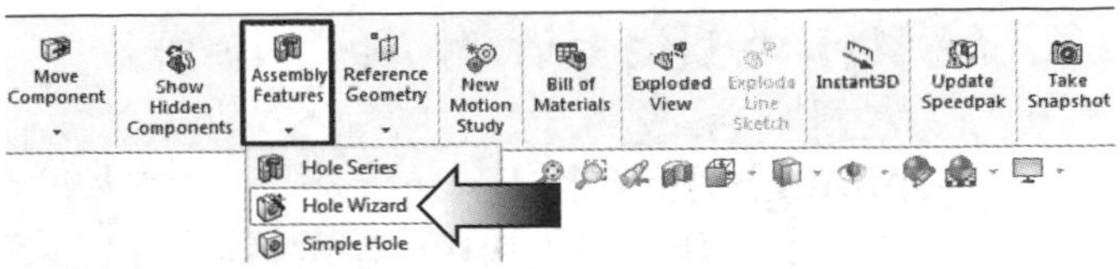

Hole wizard is an Assembly Feature, which creates these types of holes:

- Counterbore
- Countersink
- Hole
- Straight Tap
- Tapered Tap
- Legacy

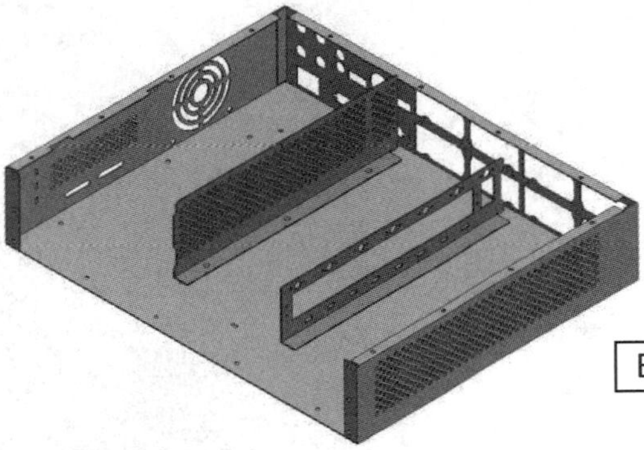

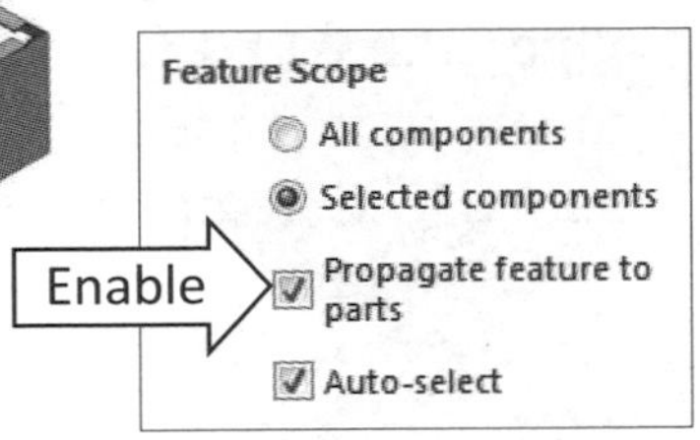

From the **Assembly** tool tab, click **Assembly Features / Hole Wizard**.

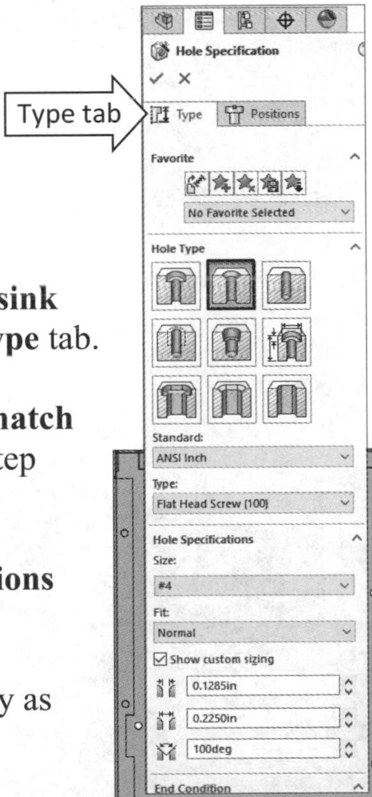

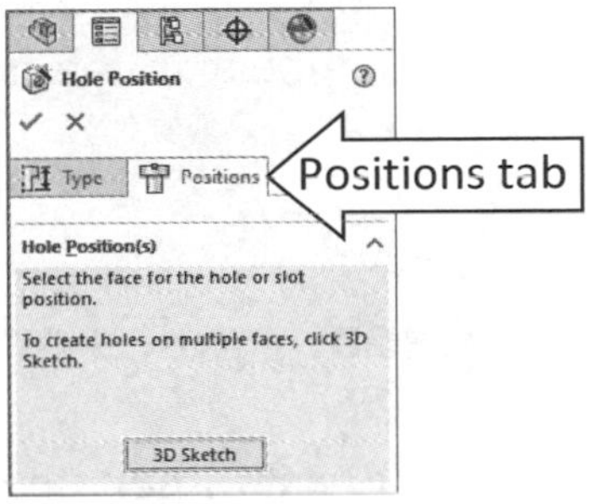

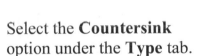

Select the **Countersink** option under the **Type** tab.

Set the options **to match** the last 3 holes in step number 11.

Switch to the **Positions** tab (arrow).

Click approximately as shown to create **3 points**.

Each point represents the center for that hole.

Add the relation and dimensions as indicated.

Click **OK**.

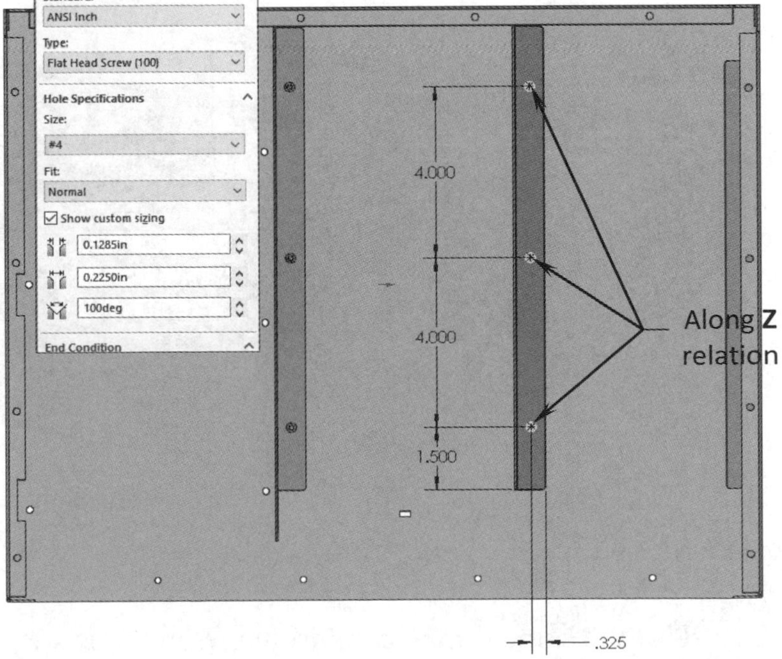

13. Verifying the two hole types:

Although the 6 holes were created with 2 different hole options, they are exactly identical.

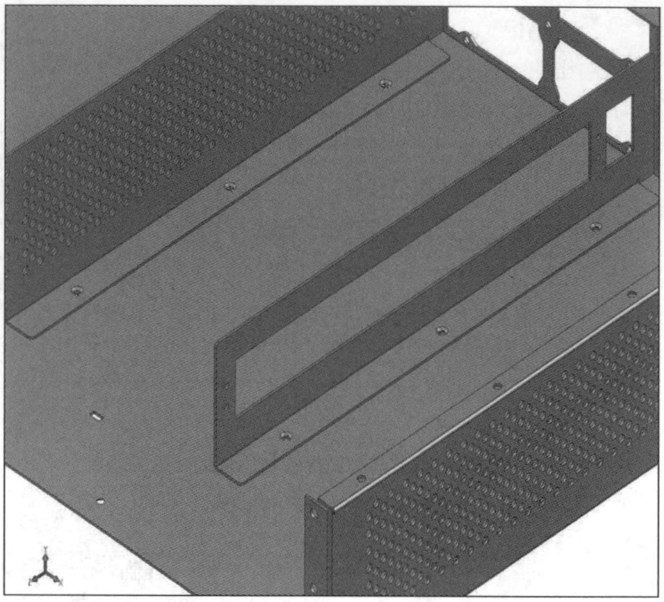

Open the **Card Guide Middle** to verify that the holes are actually there on the part.

The unique option in step 12 (propagate feature to parts) allows these Assembly Features to appear in the part mode as well.

14. Adding holes on the Card Guide Right:

Repeat either step number 11 (Hole Series) or step number 12 (Hole Wizard) and create 3 more holes for the last Card Guide.

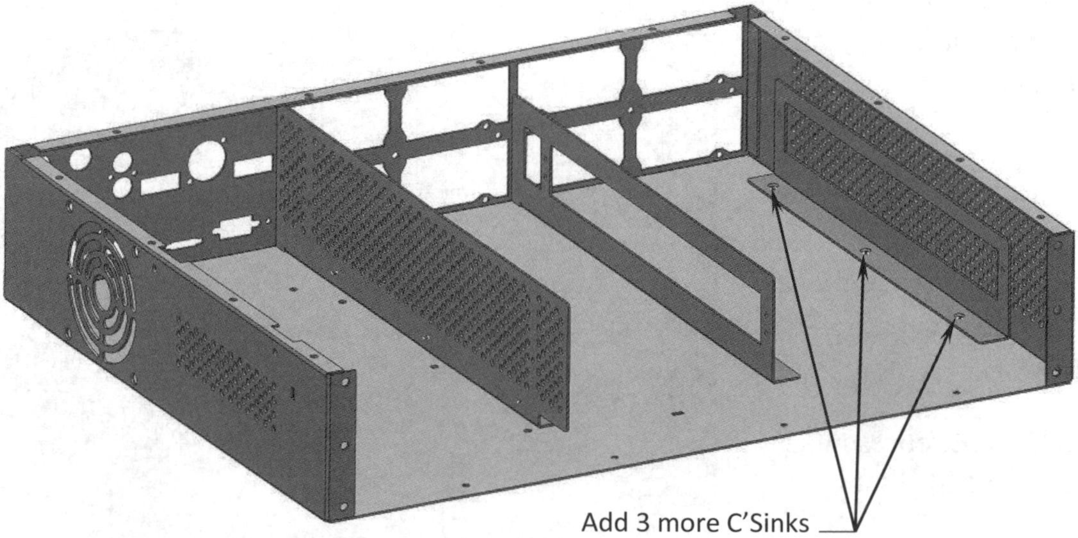

Use the same dimensions from the previous step to position the holes.

15. Adding the Smart Fasteners:

Click **Smart Fasteners** from the Assembly tool tab.

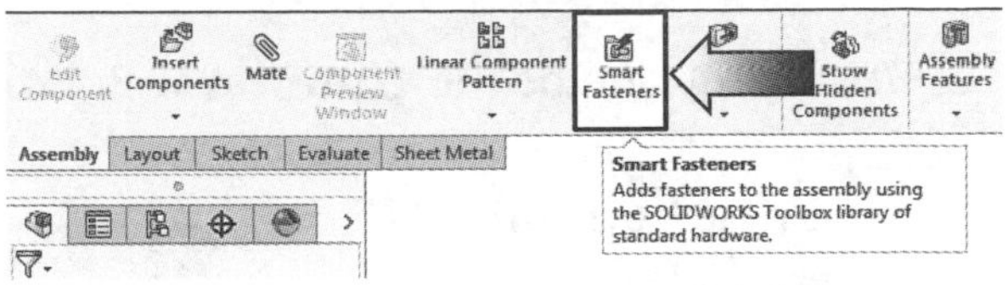

An error message appears inquiring for SOLIDWORKS Toolbox to be activated. (Requires SOLIDWORKS Professional or SOLIDWORKS Premium.)

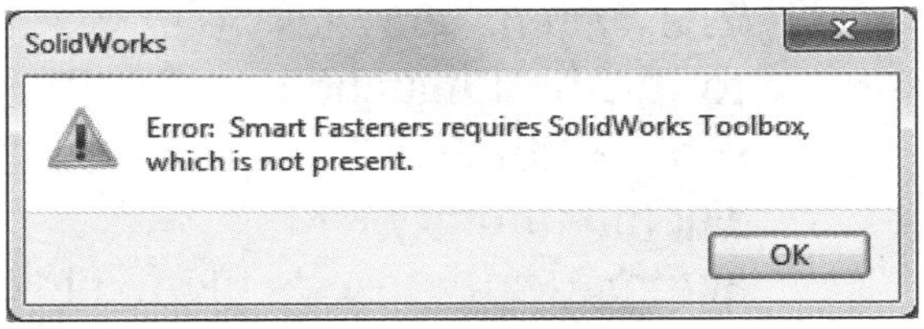

To activate Toolbox, select the following:

Tools / Options / Add Ins.

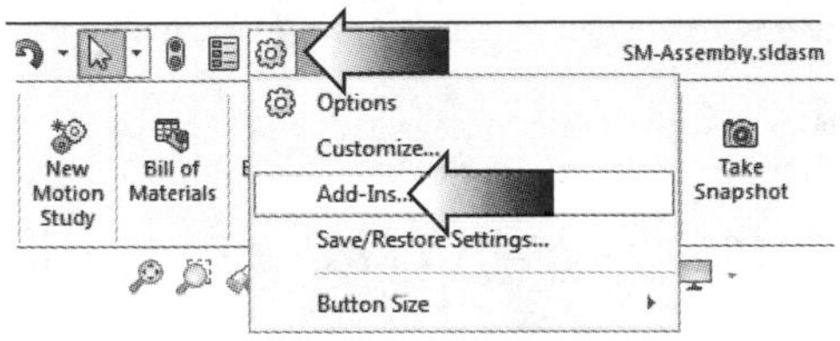

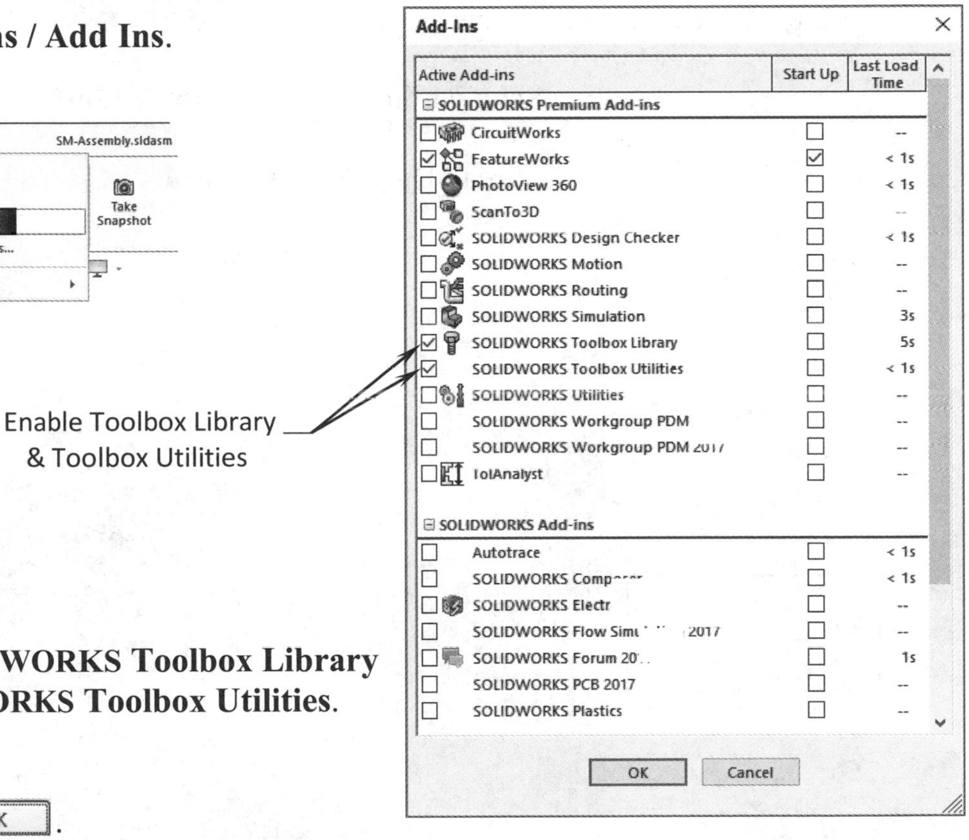

Enable **SOLIDWORKS Toolbox Library** and **SOLIDWORKS Toolbox Utilities.**

Click **OK** OK.

Click the **Smart Fasteners** button once again.

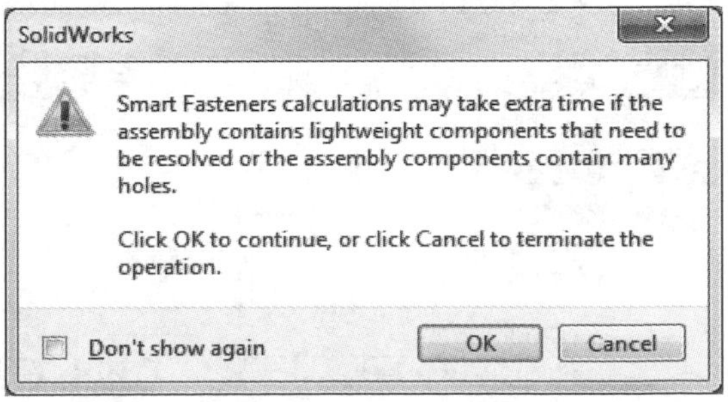

Another message pops up indicating that the Smart Fasteners calculation may take extra time; click **OK** OK.

Select one of the **Countersink** holes in the graphics area.

Click **Populate All** (arrow).

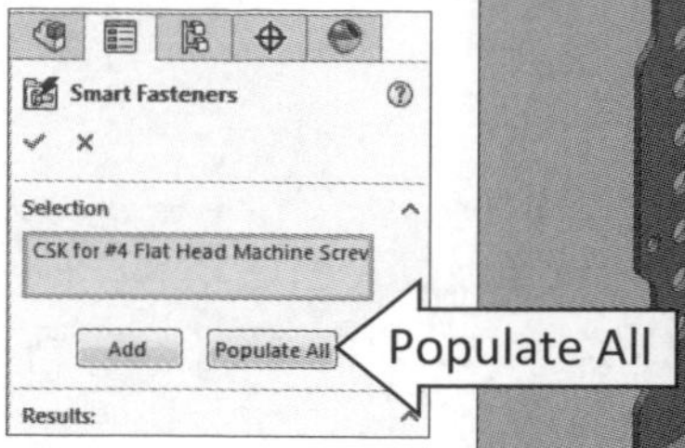

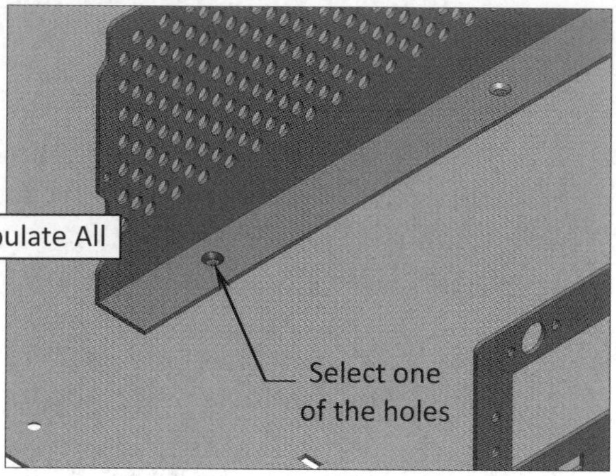

The system searches for the best matched screws from its Toolbox library and automatically inserts them into each hole.

Set the following properties:

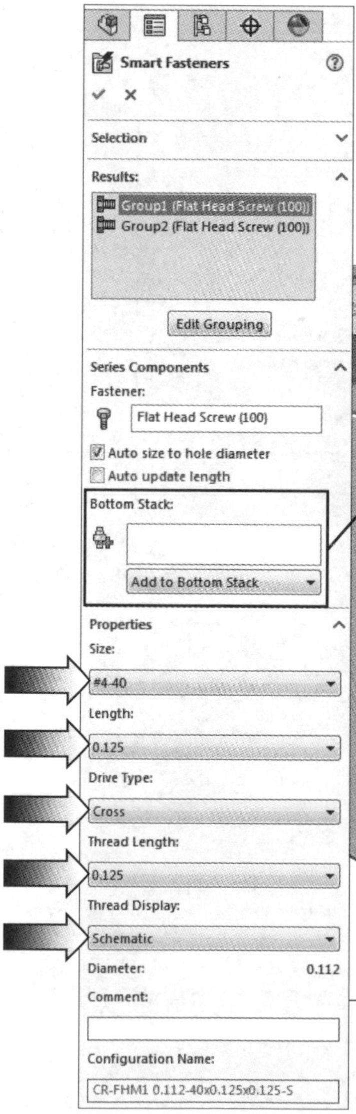

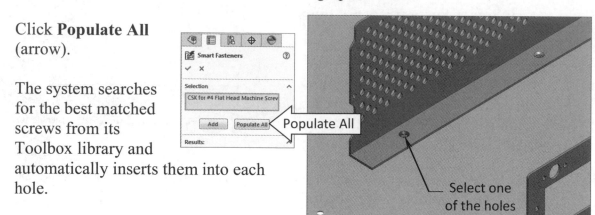

Size: **#4-40** Length: **.125in*** Drive Type: **Cross**

Thread Display: **Schematic**

* *Change the screw length to* ***.250in*** *if adding a washer and a nut.*

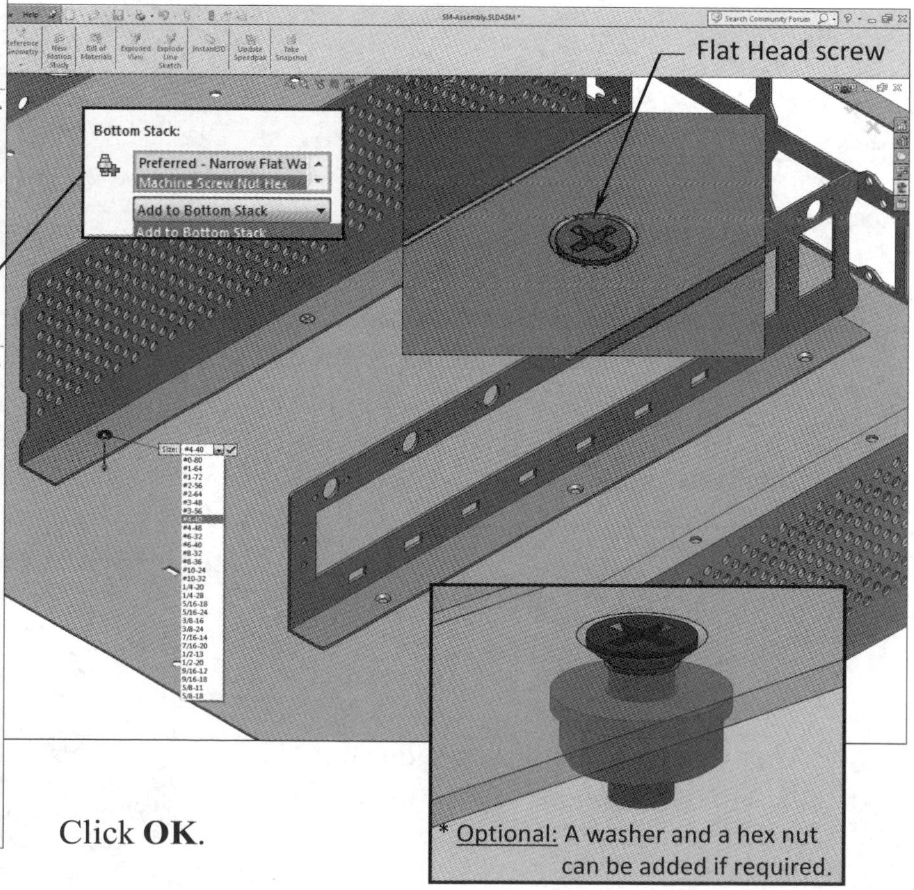

Click **OK**.

16. Creating an Exploded View:

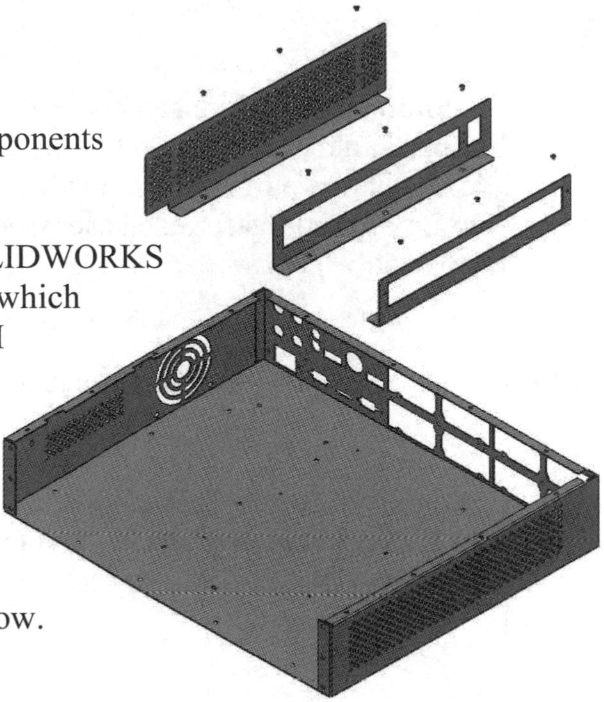

Option 1

Create an exploded view with all 4 components shown in folded stage as shown.

When an exploded view is created, SOLIDWORKS also creates an animated configuration, which can be played back and saved as an AVI file format.

Option 2

Create a 2nd exploded view with all 4 parts as shown in the Flatten view below.

NOTES:

Edit each component in order to switch from the Folded to flatten stage.

Configurations can also be used to capture the flat pattern of each component.

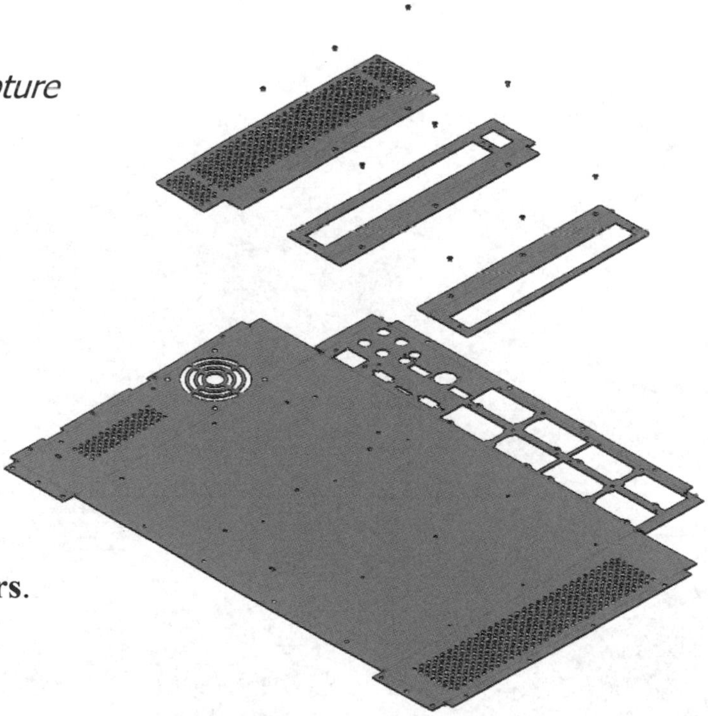

17. Saving your work:

Click **File / Save As**.

For the name of the file enter:
SM_Assembly_Smart Fasteners.

Click **Save**.

Close all documents.

Adding Parts to the Toolbox Library

Customized parts or fasteners can be added to existing Toolbox folders.
For parts that are stored in the shared library, administrators can control who can add or change parts by creating a Toolbox password and setting permissions for Toolbox functions for each user (see Toolbox Permission in the SOLIDWORKS Help section).

1. Starting the Toolbox Settings Utility:

From the Windows desktop select the following:
Start / All Programs / SOLIDWORKS 20XX / SOLIDWORKS Tools 20XX/ Toolbox Settings (arrow).

Select the option number 2 (arrow):
Customize your hardware.

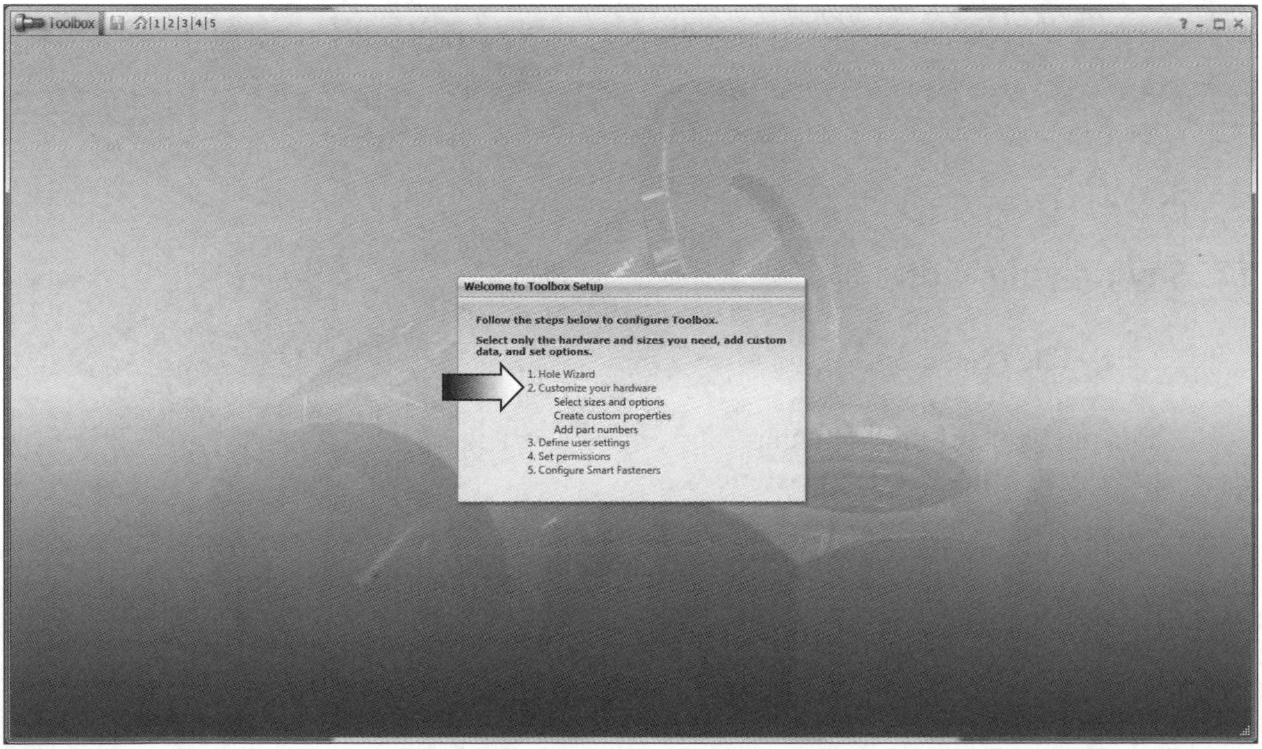

2. Adding a part to a folder:

From the upper left side of the tree, select the following:

ANSI Inch
Bolts and Screws
Countersunk Head

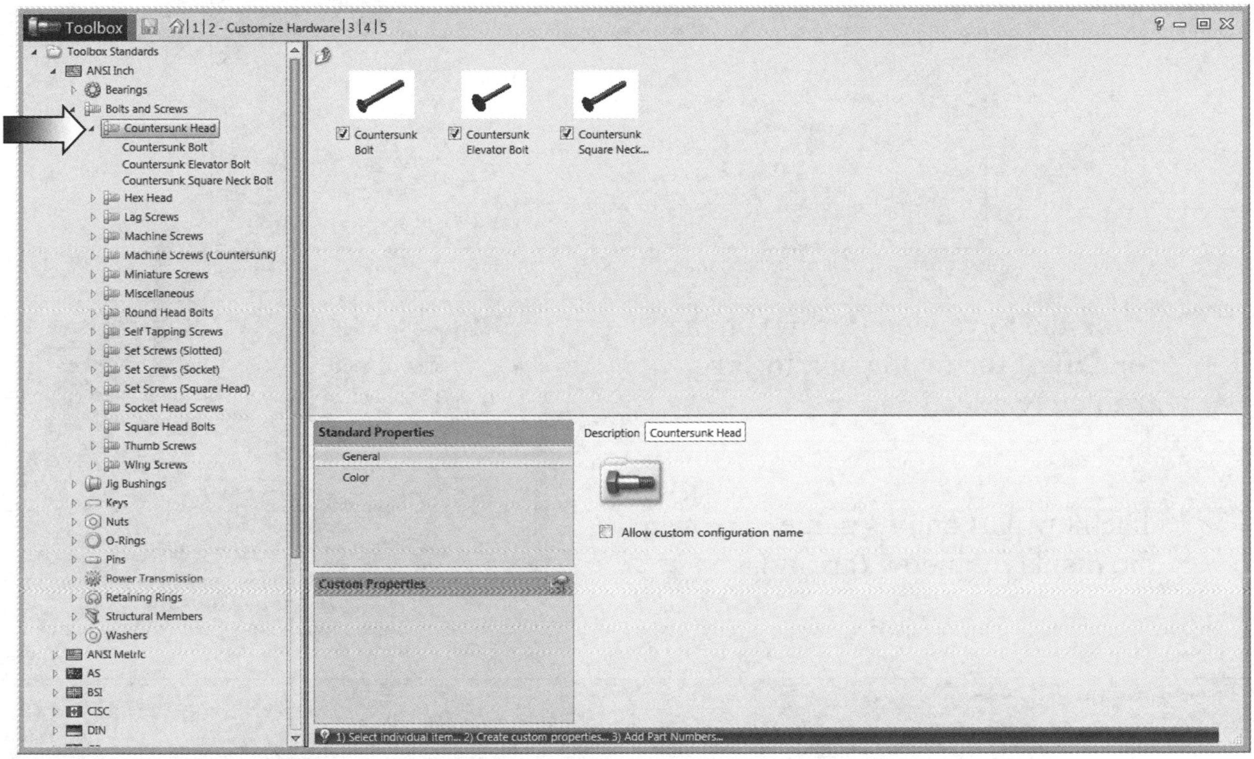

In this folder there are 3 existing screw / bolt types: **Countersunk Bolt**, **Countersunk Elevator Bolt**, and **Countersunk Square Neck Bolt**.

Right-click the **Countersunk Head** folder and select **Add File** (arrow).

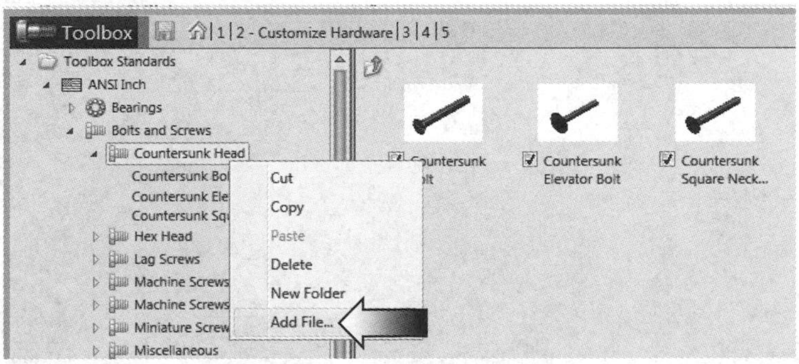

Browse to the Training Files folder and select the document named:
C-Sink .250-20x1x0.75-Bolt.sldprt and open it.

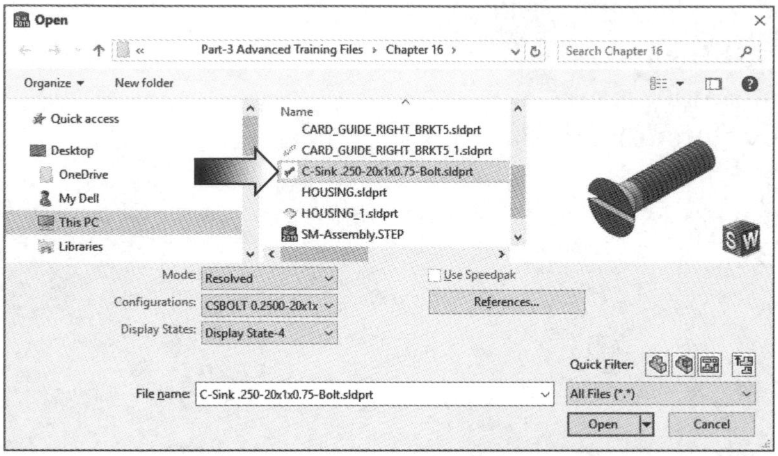

Click the Save icon (arrow) on the top left of the dialog box to **save** the newly added part.

The new part and its name appear in the display window (arrow).

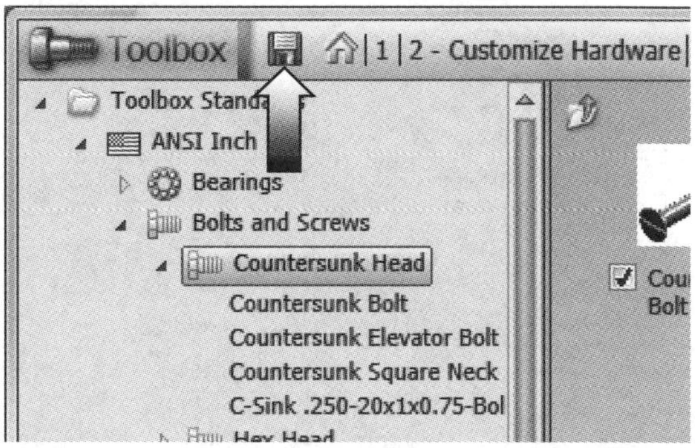

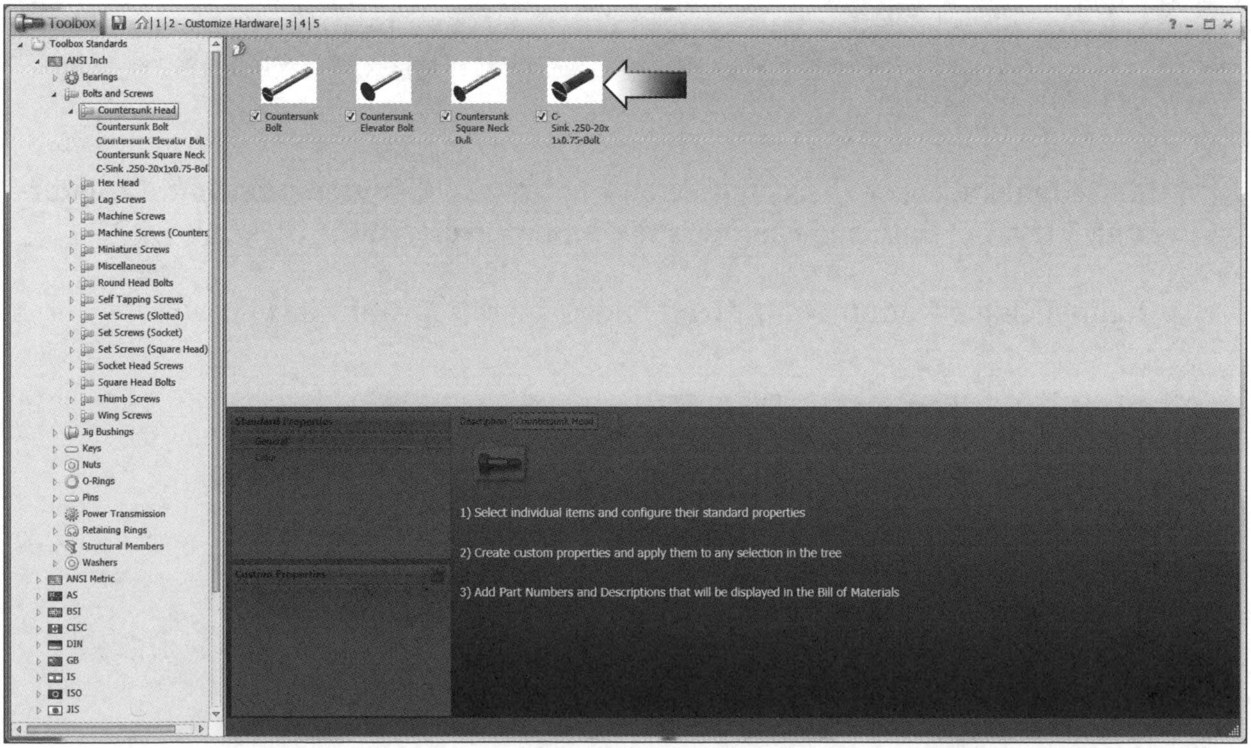

3. Activating Toolbox:

Launch the SOLIDWORKS application.

Select: **Options / Add-Ins** (arrow).

Enable the 2 options:

* **SOLIDWORKS Toolbox Library**
* **SOLIDWORKS Toolbox Utilities**

Click **OK**.

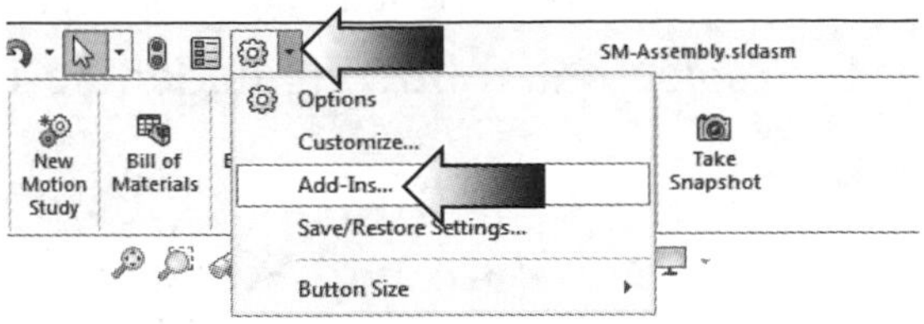

From the **Task Pane** (on top right hand corner) click the **Design Library** folder and then expand the **Toolbox** folder (arrow).

NOTE: Toolbox is only available in SOLIDWORKS Professional and SOLIDWORKS Premium.

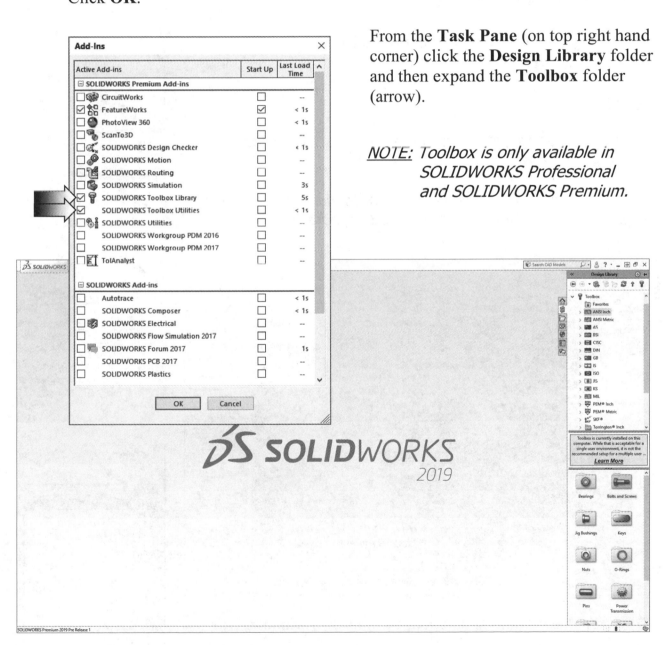

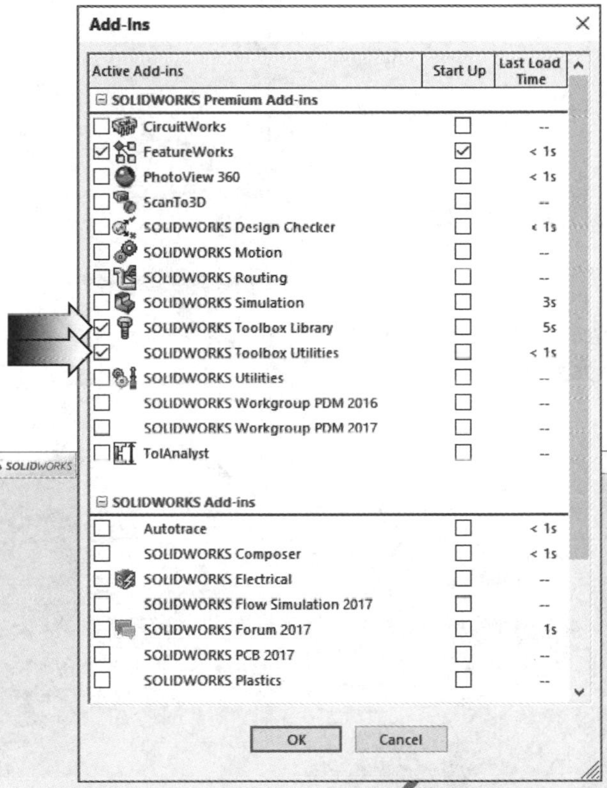

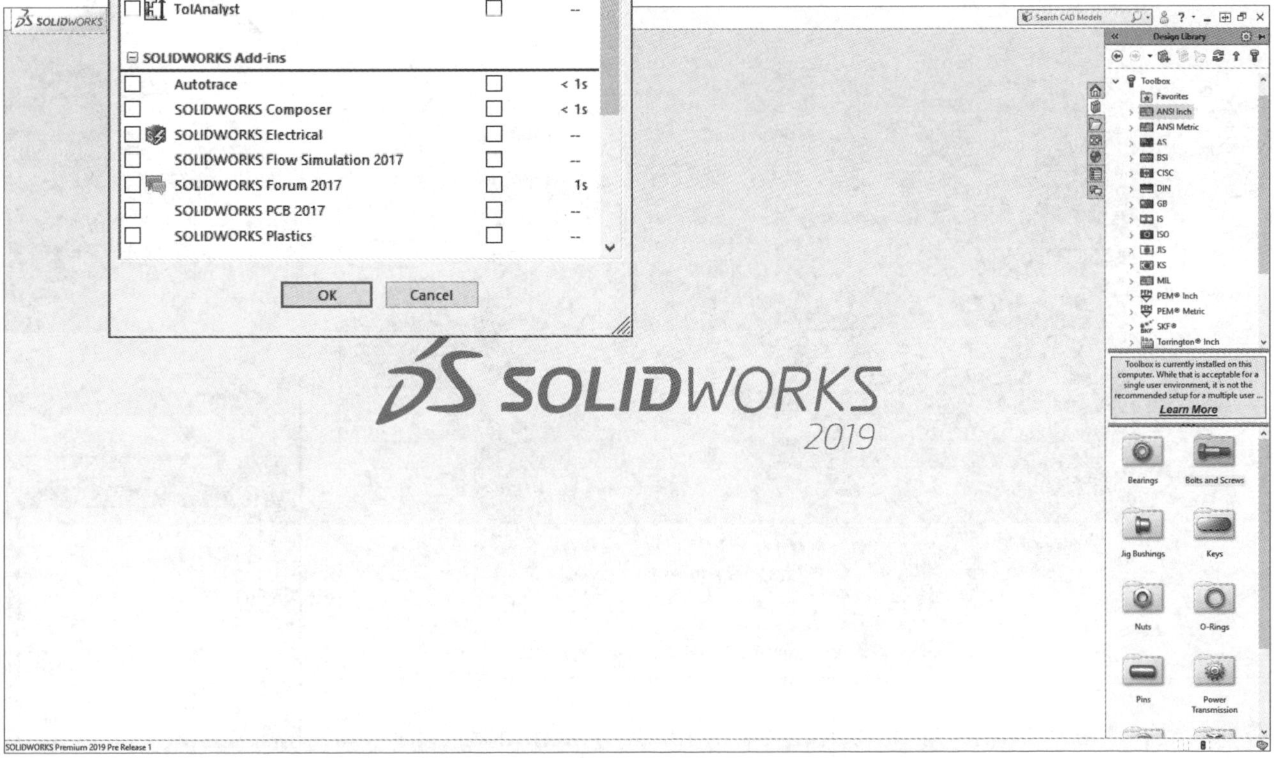

4. Locating the new part:

Select the following under the Toolbox folder:

* **ANSI Inch**
* **Bolts and Screws**
* **Countersunk Head**

Locate the new part **C-Sink .250-.20x1x0.75-Bolt** on the lower right side of the Task Pane (arrow).

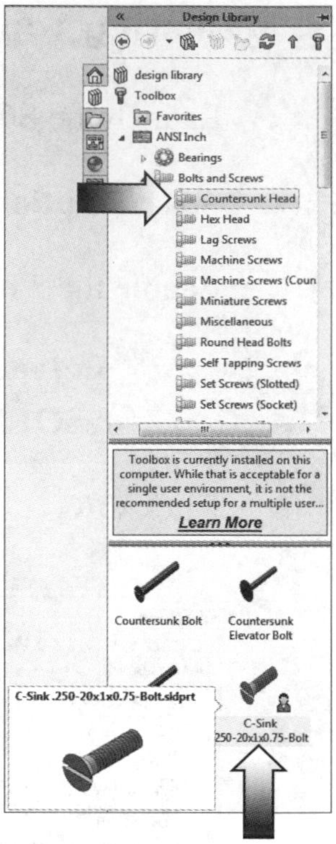

5. Viewing the new part:

Right-click on the new part and select:
Configure Part.

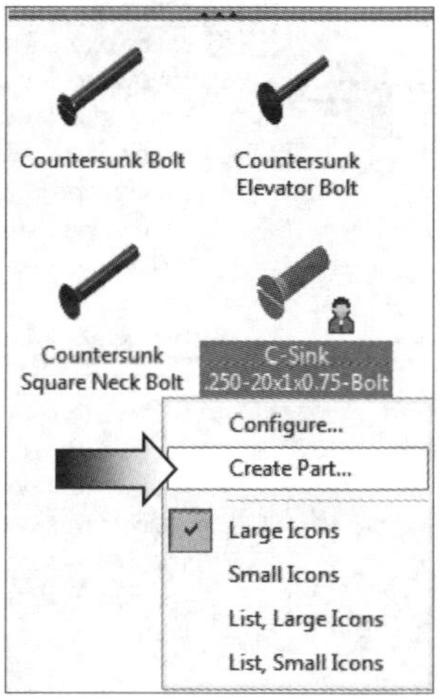

NOTE:

There is no configuration created for this custom screw; its length will have to be modified manually.

Existing Toolbox parts will have several configurations such as Size, Length, Thread Display, etc., to choose from.

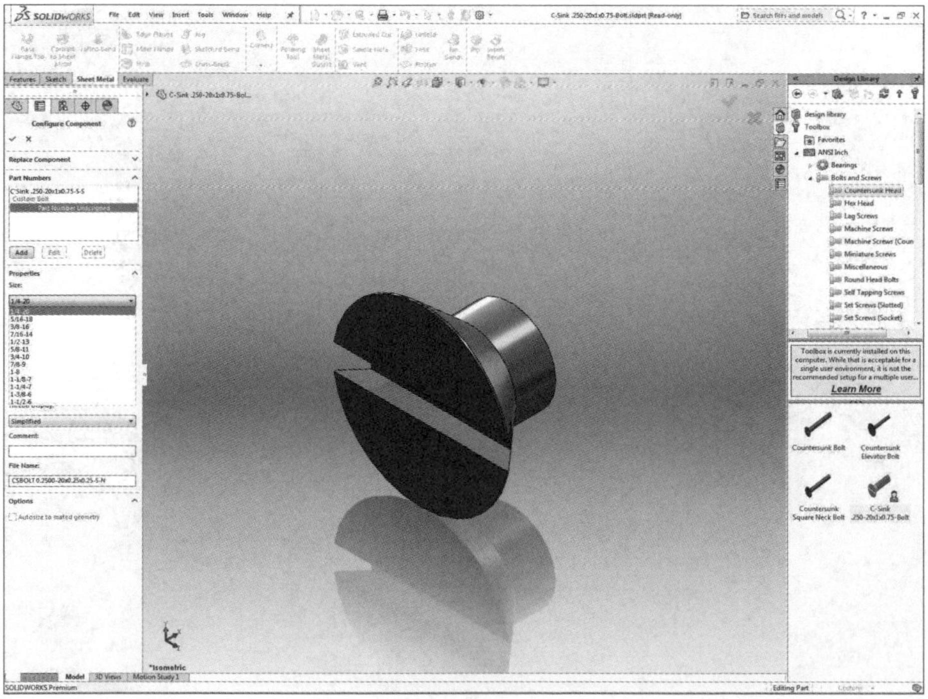

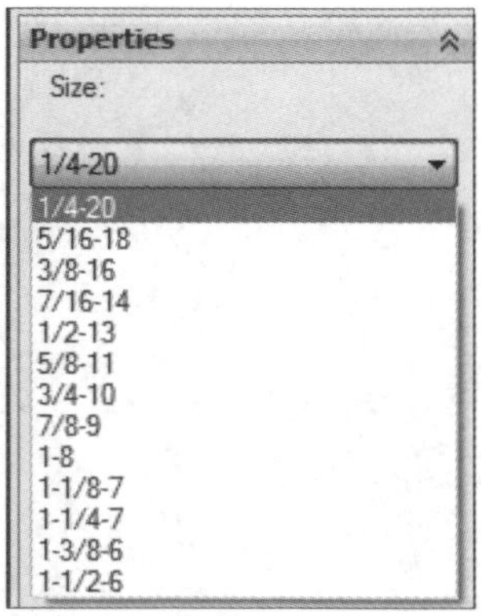

The new Toolbox part appears in its own window.

6. Adding a Part Number and Description:

Click the **Edit** button (arrow).

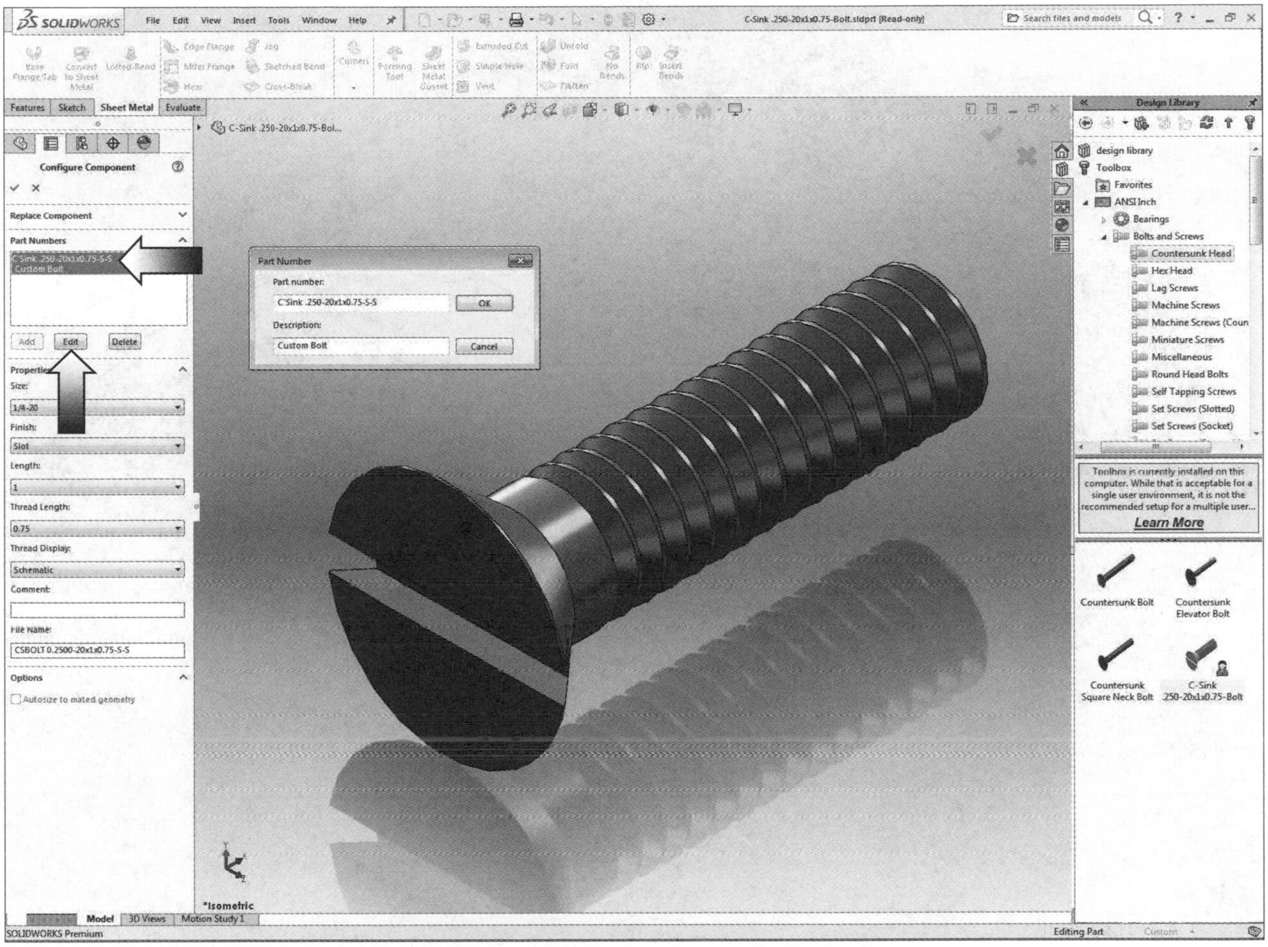

Enter a new **Part Number** (if applicable).

Enter **Custom Bolt** as Description.

Save and close your documents.

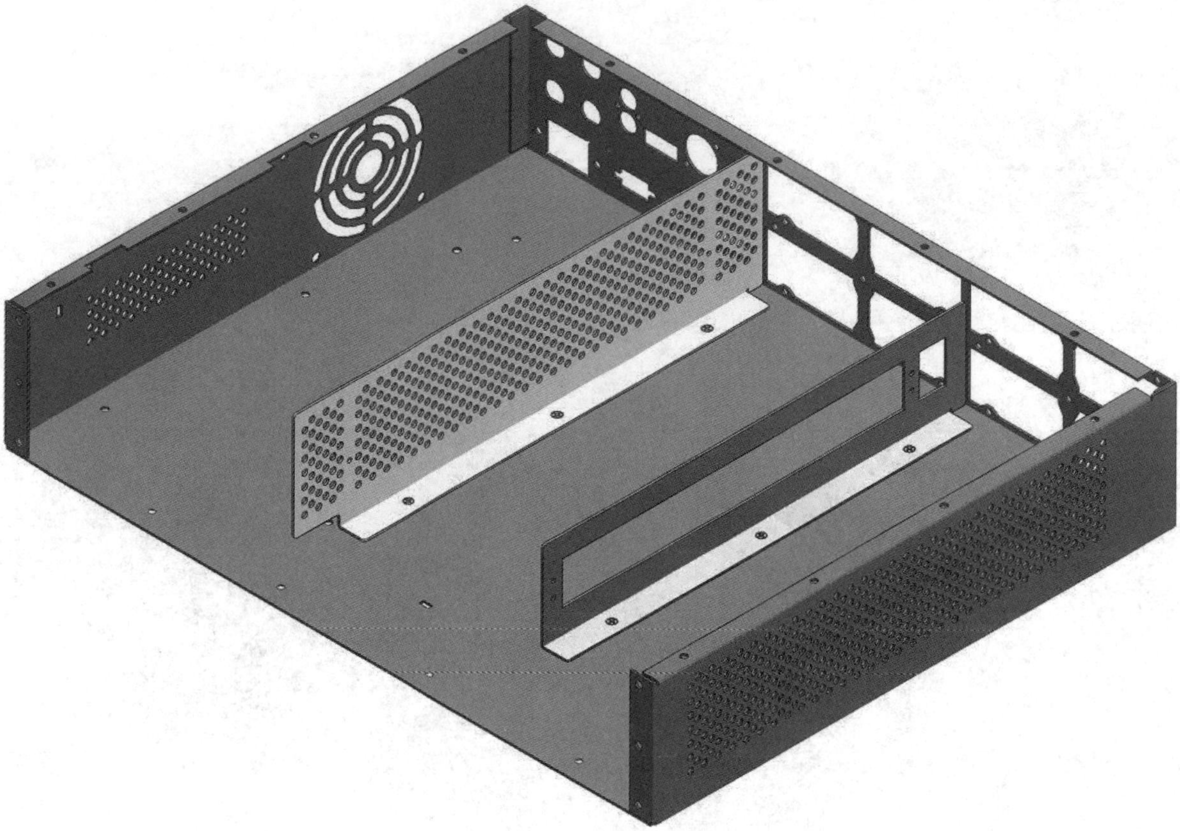

CHAPTER 17

Advanced Weldments

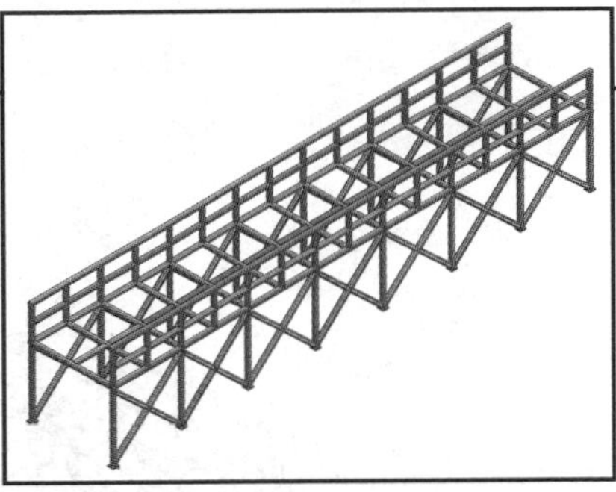

Advanced Weldments
Weldments Platform

Weldments can be used to create a structure as single multibody part. Either a 2D or 3D sketch can be used to define the framework, then create structural members containing groups of sketch segments.

Other Weldments tools can be used to add objects such as gussets and end caps. Weld gaps and beads can also be added to the model. Drawings can be made to document the design, including tables of cut materials, cut length, and weld bead totals.

Weldments are normally created in groups. A group is a collection of related segments in a structural member. A group is configured to affect all its segments without affecting other segments or groups in the structural member.

The 2 types of groups are:
Contiguous: A continuous contour of segments joined end-to-end. You can control how the segments join to each other. The end point of the group can optionally connect to its beginning point.

Parallel: A discontinuous collection of parallel segments. Segments in the group cannot touch each other.

* You can define a group in a single plane or in multiple planes.
* A group can contain one or more segments.
* A structural member can contain one or more groups.
* After you define a group, you can operate on it as a single unit. Use the Structural Member PropertyManager to specify the corner treatment for the segments in the group.

Advanced Weldments
Weldments Platform

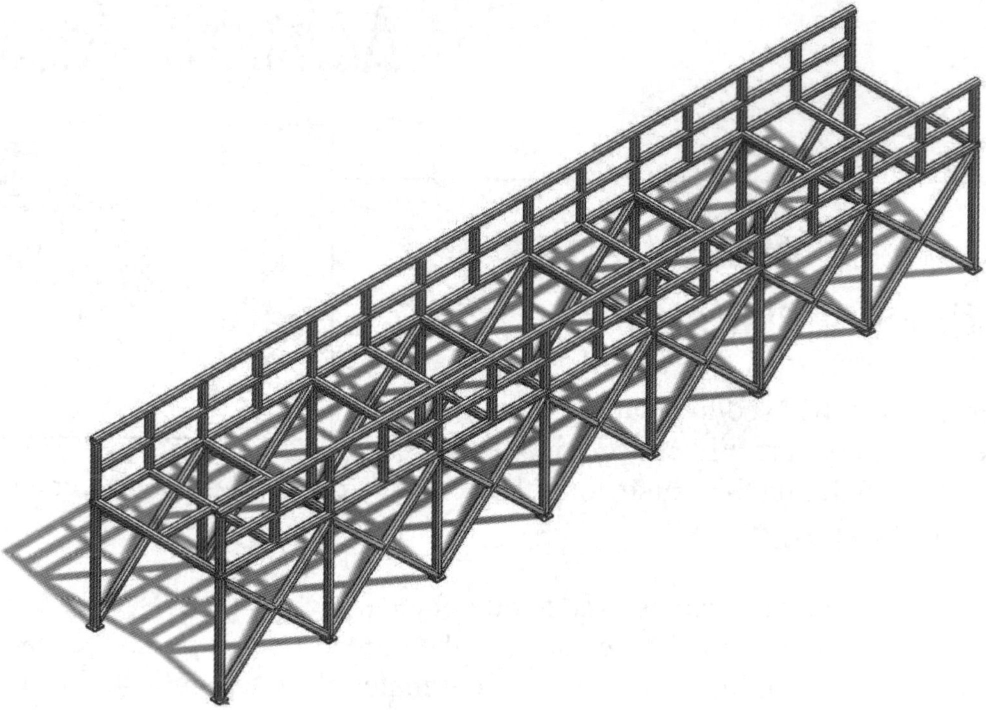

View Orientation Hot Keys:

Ctrl + 1 = Front View
Ctrl + 2 = Back View
Ctrl + 3 = Left View
Ctrl + 4 = Right View
Ctrl + 5 = Top View
Ctrl + 6 = Bottom View
Ctrl + 7 = Isometric View
Ctrl + 8 = Normal To Selection

Dimensioning Standards: **ANSI**
Units: **INCHES** – 3 Decimals

Tools Needed:

- 3D Sketch
- Extruded Boss-Base
- Linear Pattern
- Structural Member
- Trim/Extend
- End Trim Type

1. Opening a part document:

Click **File / Open**.

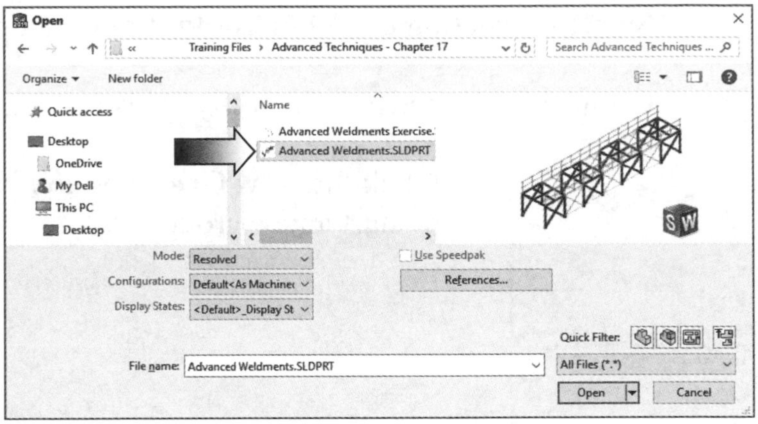

Browse to the Training Folder and open a part document named: **Advanced Weldments**.

2. Adding the Weldments toolbar:

Right-click one of the tool tabs and enable the **Weldments** toolbar.

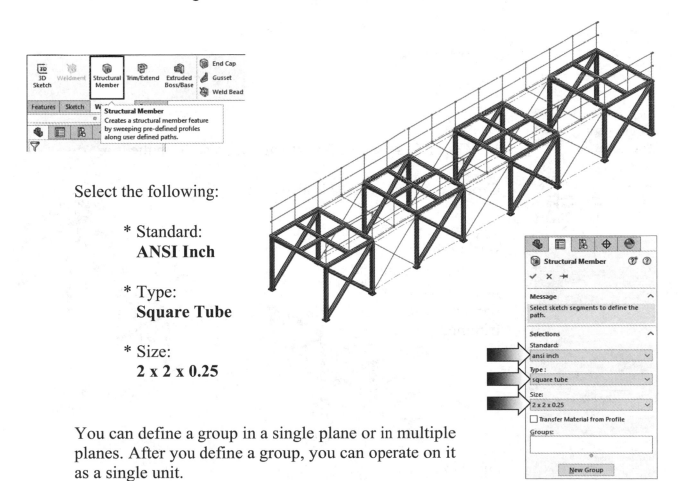

Select the **Structural Member** command. This adds a Weldments feature on the FeatureManager tree.

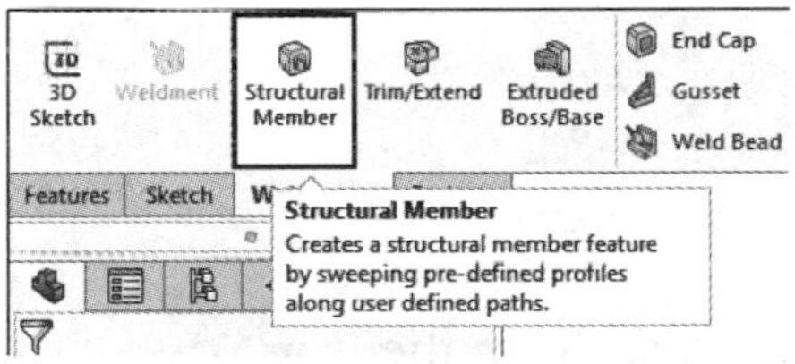

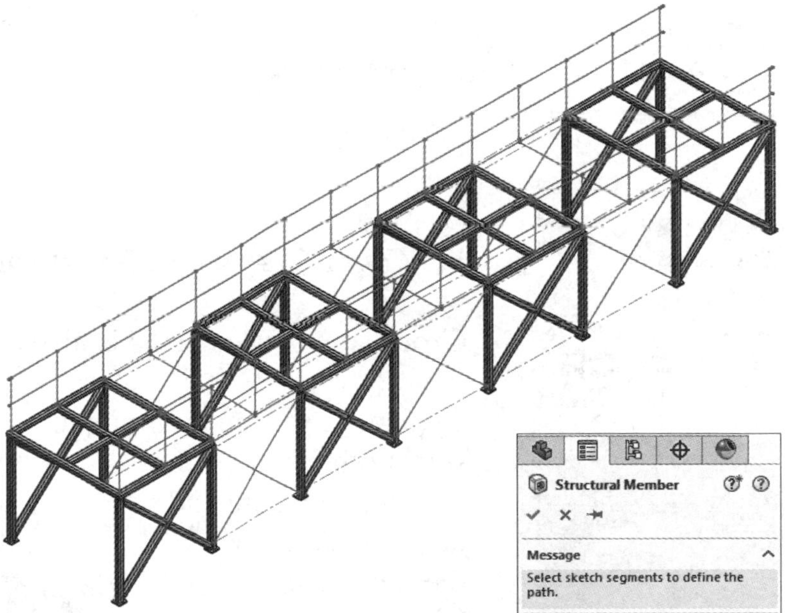

Select the following:

* Standard: **ANSI Inch**

* Type: **Square Tube**

* Size: **2 x 2 x 0.25**

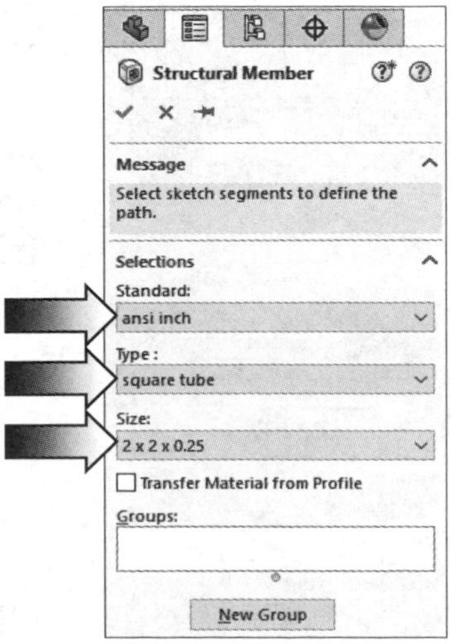

You can define a group in a single plane or in multiple planes. After you define a group, you can operate on it as a single unit.

3. Adding the Structural Members:

Start by selecting the **3 horizontal lines** on top as indicated.

Group 1 is created automatically. This is a Parallel group.

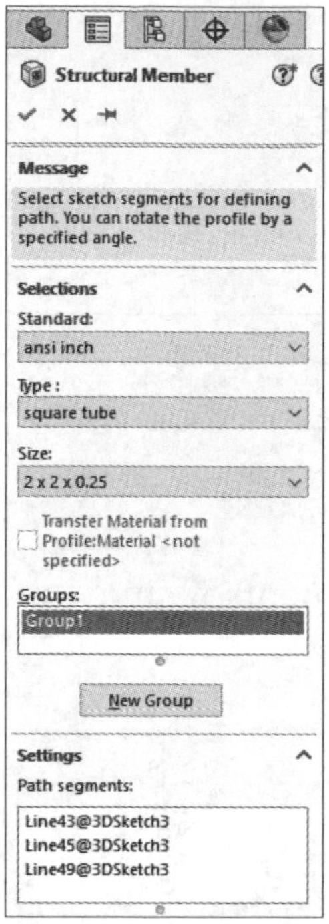

Click the **New Group** button (arrow) to start a new group.

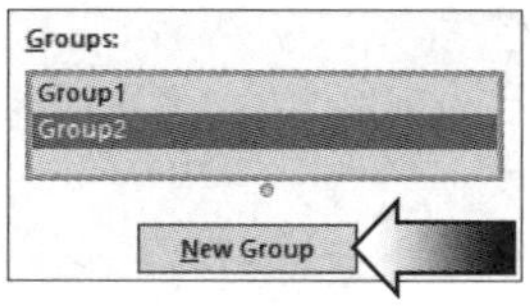

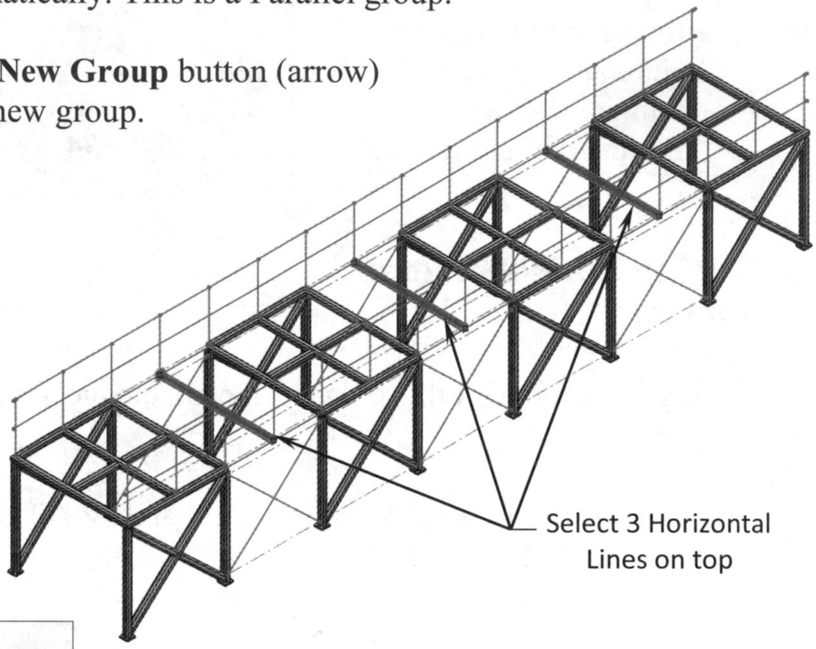

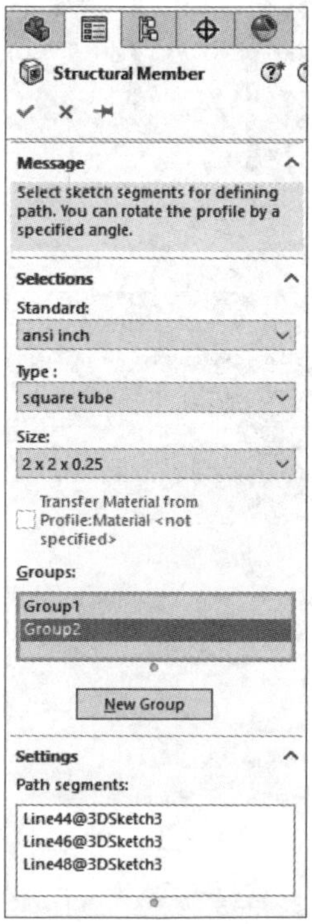

For Group 2, select the **3 vertical lines** also from the top.

Group 2 is created. This is also a Parallel group.

Typically, we can continue to select other groups, but we will stop here and take a look at how the tubes are trimmed.

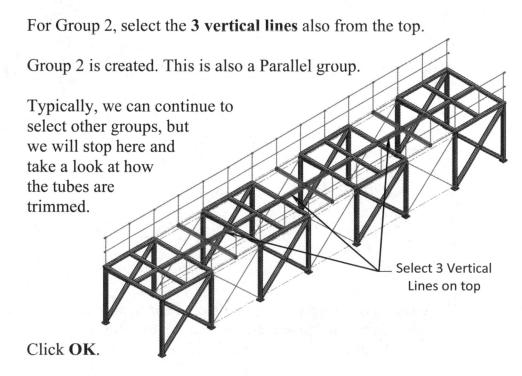

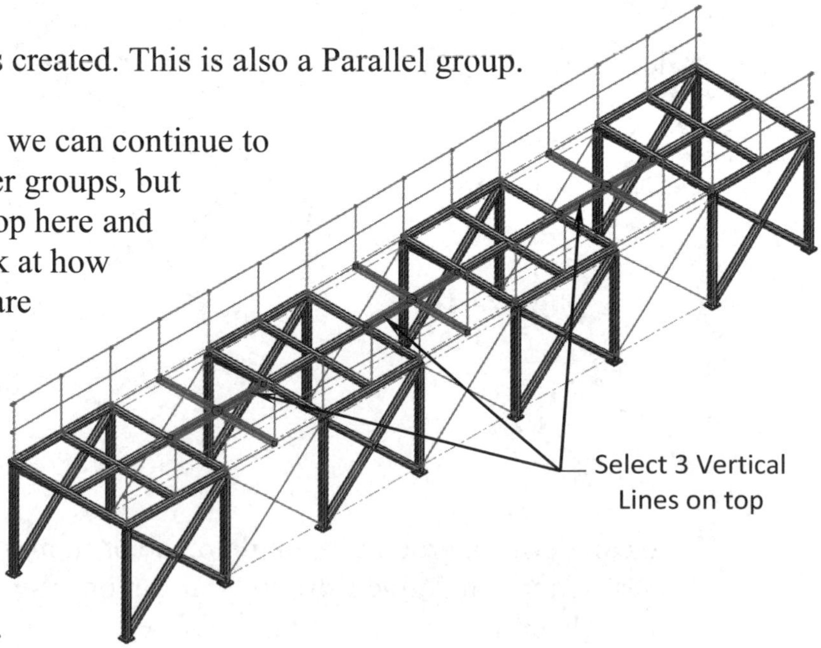

Click **OK**.

Zoom in on the first two sections of the platform frame.

Notice the new tubes were trimmed automatically.

If the Structural Members are created separately, manual trim is needed.

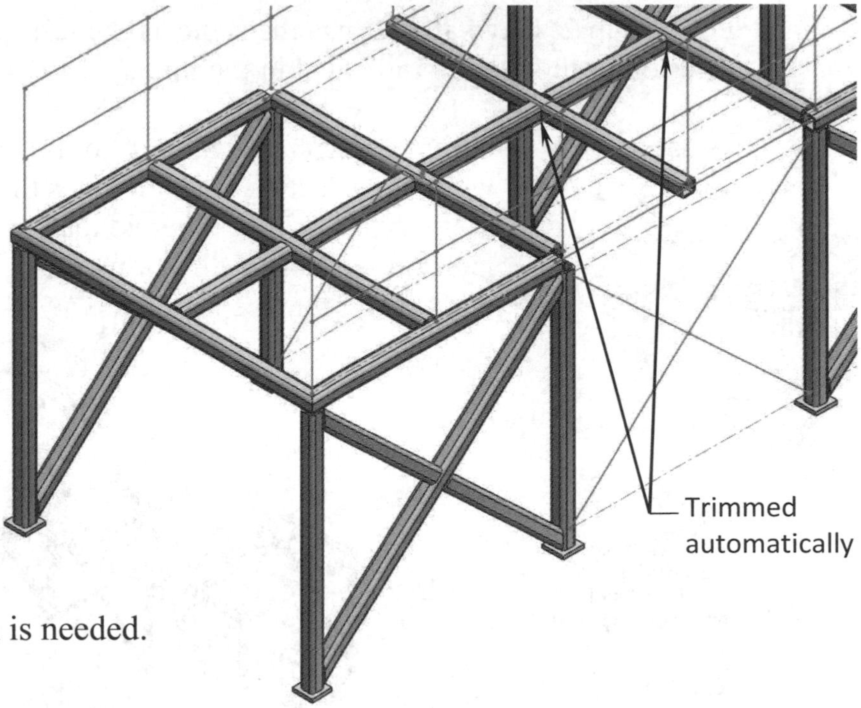

4. Adding more Structural Members:

Click the **Structural Member** command  again.

Keep all parameters the same as the last group and select the **6 parallel lines** on both sides as noted.

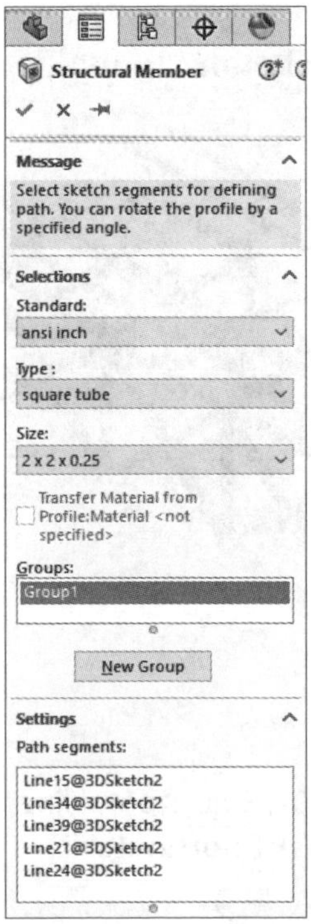

The new Group 1 is created. This is also a Parallel group.

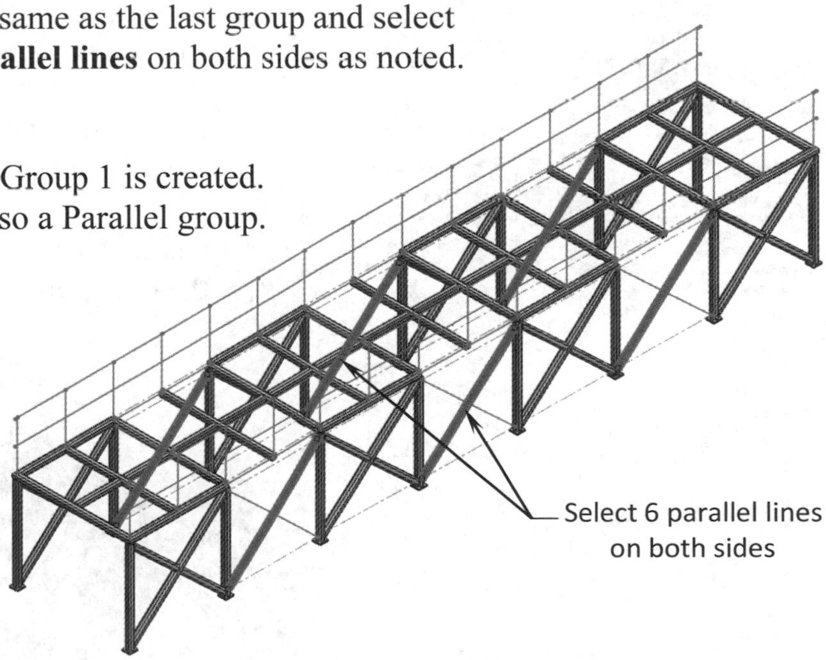

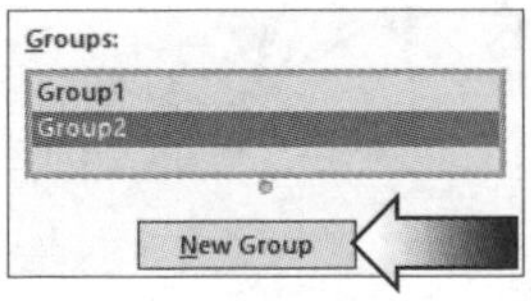

Click the **New Group** button to start a new group.

For Group 2, leave all settings the same as the last group and select the **6 parallel lines** on both sides as indicated in the image.

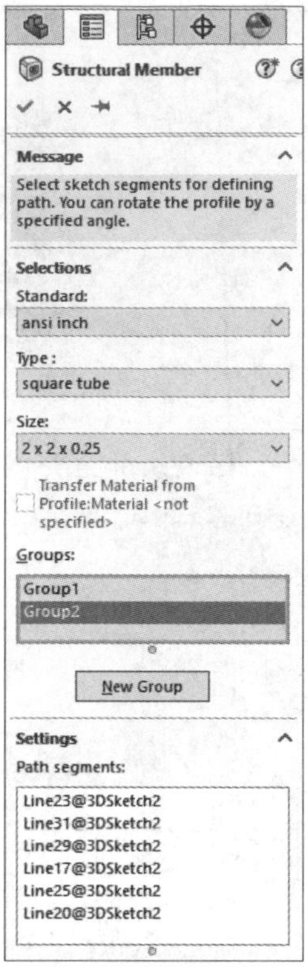

Group 2 is created. Similar to the other groups, we were selecting only the lines that are parallel to each other so that later on we can modify or work with each group separately.

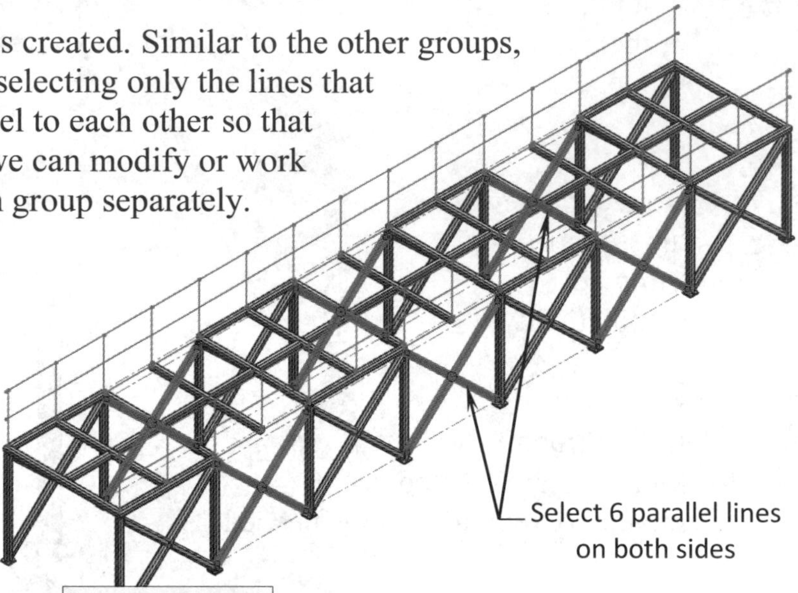

Click the **New Group** button.

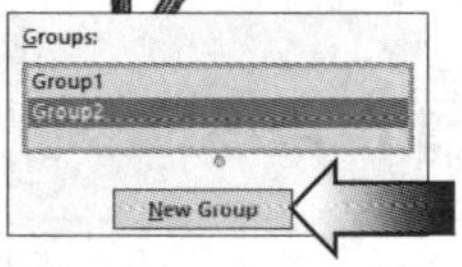

For the new Group 2, select the **10 horizontal lines** for the rails.

Group 2 is created but the weldment platform is becoming very busy, making it hard to see the other lines.

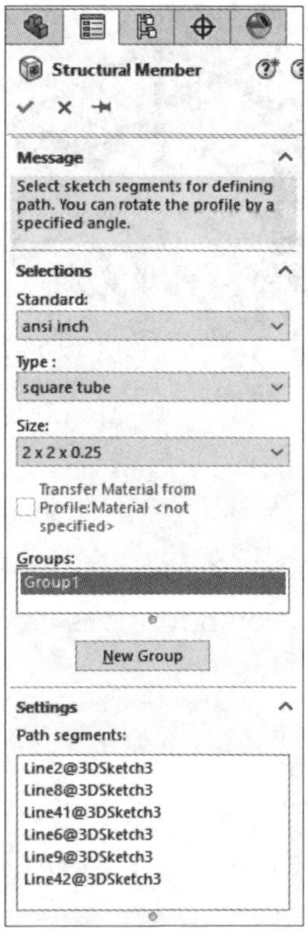

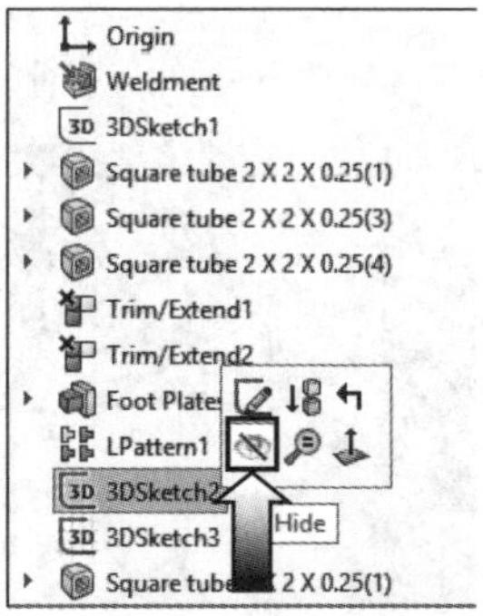

For clarity, we will hide the Sketch2 to make the other lines easier to see.

Click **OK** to exit the command.

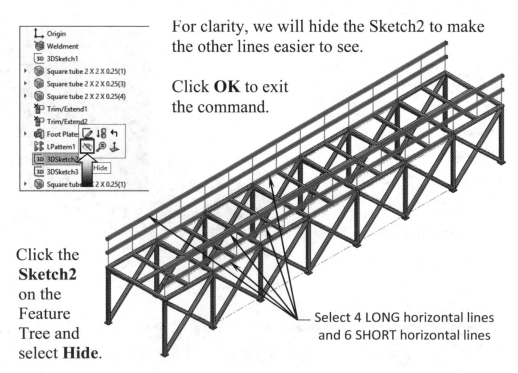

Click the **Sketch2** on the Feature Tree and select **Hide**.

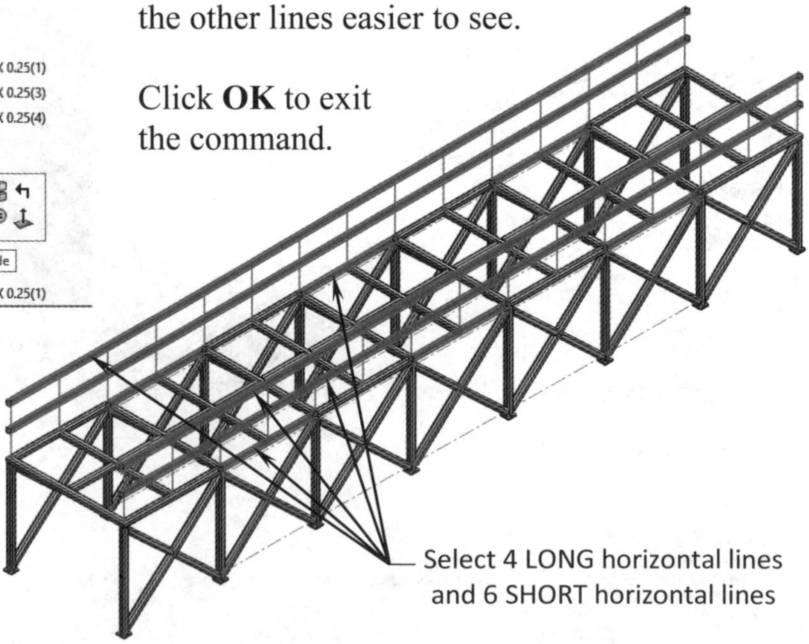

5. Adding the vertical Structural Members to the rails:

Click the **Structural Member** command  again.

For the new Group 1, select the **30 vertical lines** as indicated.

The new Group 1 is created with all parallel lines in it. Again, by grouping the lines, they become one unit. Editing will be much easier in the future.

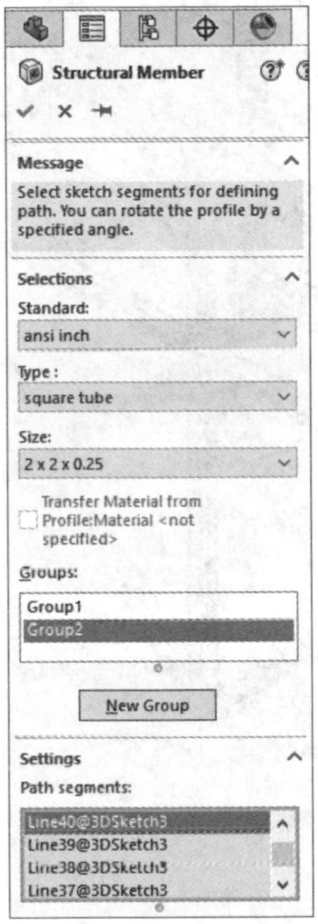

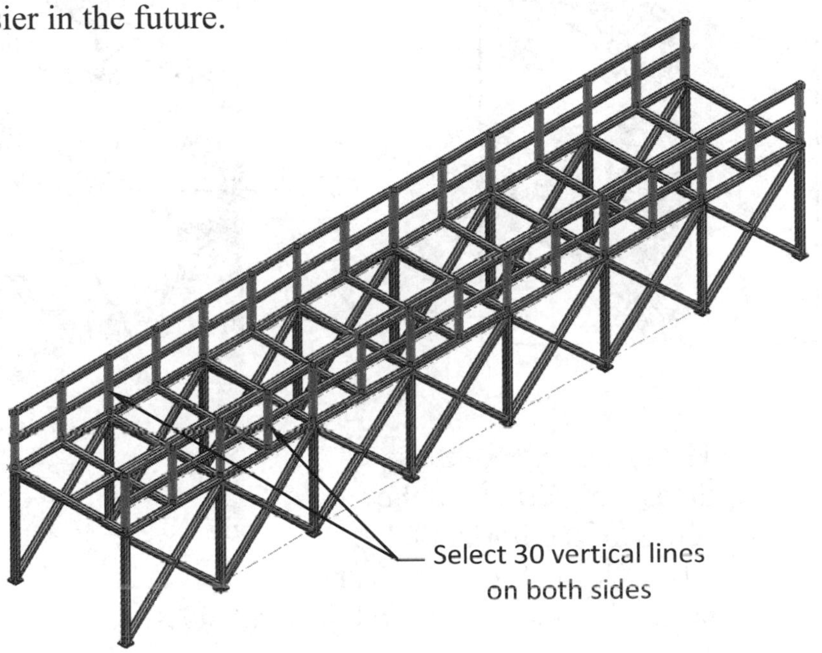

Click **OK** to exit the command.

Click on the **Sketch3** and select **Hide** (arrow).

Inspect your model against the image shown here.

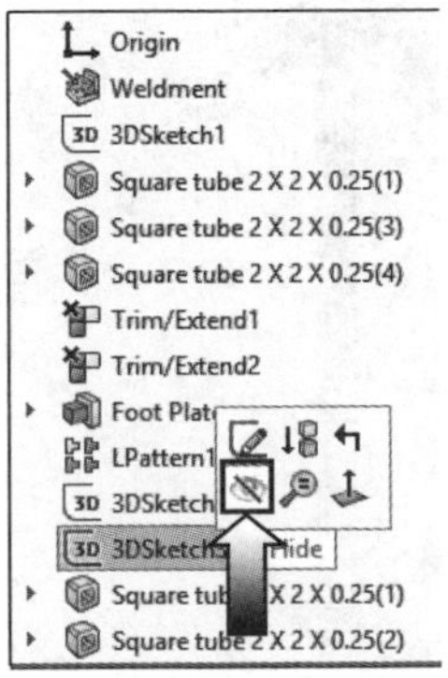

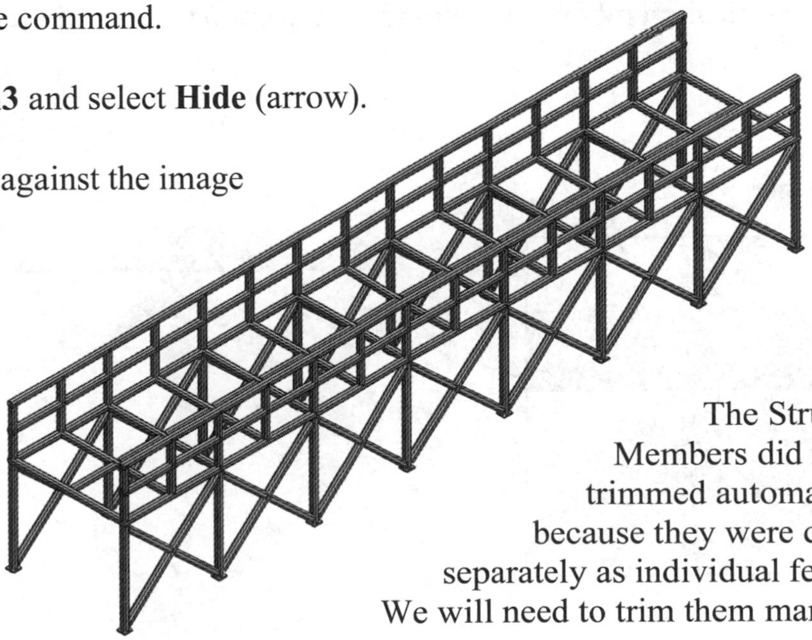

The Structural Members did not get trimmed automatically because they were created separately as individual features. We will need to trim them manually.

6. Viewing the overlapped areas:

Zoom in on the area shown in the enlarged view to see the overlapped issues.

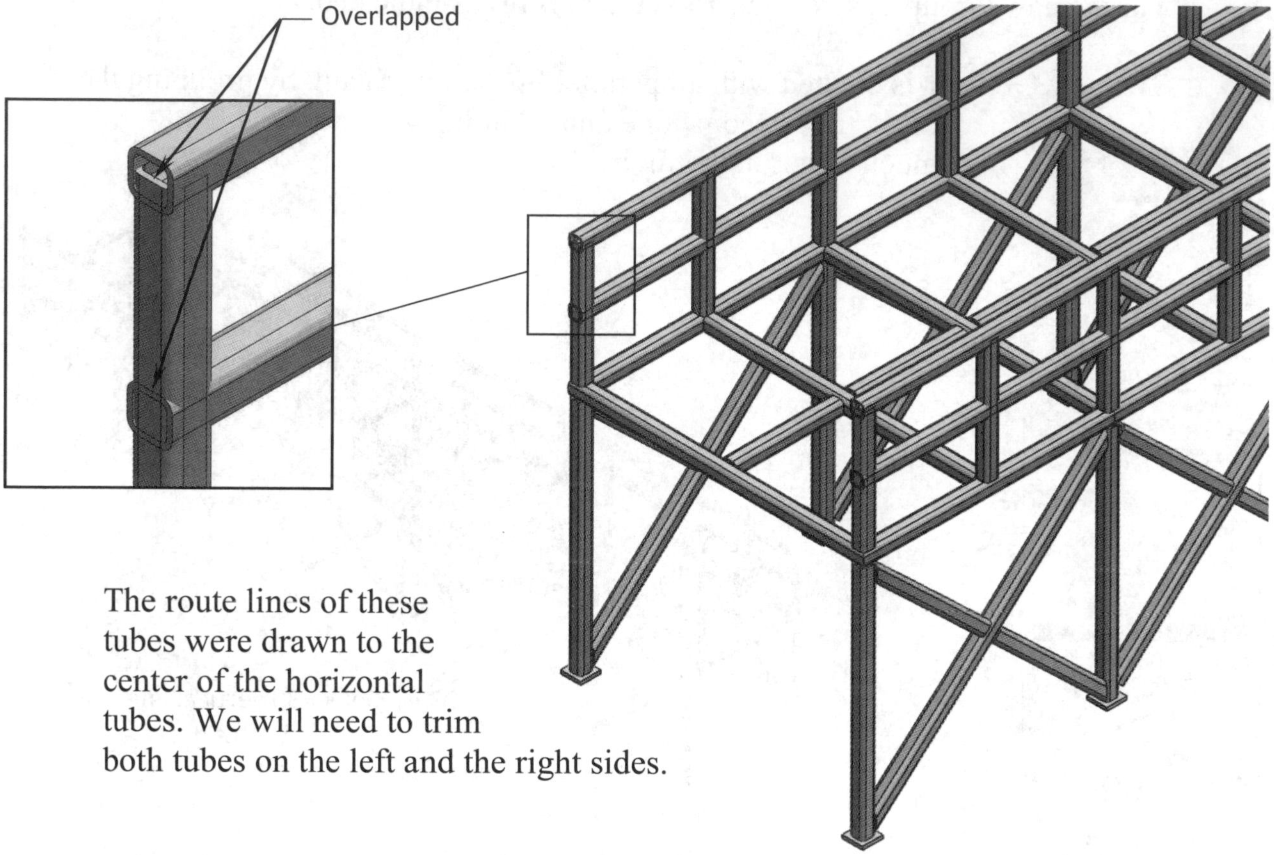

The route lines of these tubes were drawn to the center of the horizontal tubes. We will need to trim both tubes on the left and the right sides.

The cross-tubes are also overlapped with the vertical members. We will need to trim/extend all of them, front and back sides.

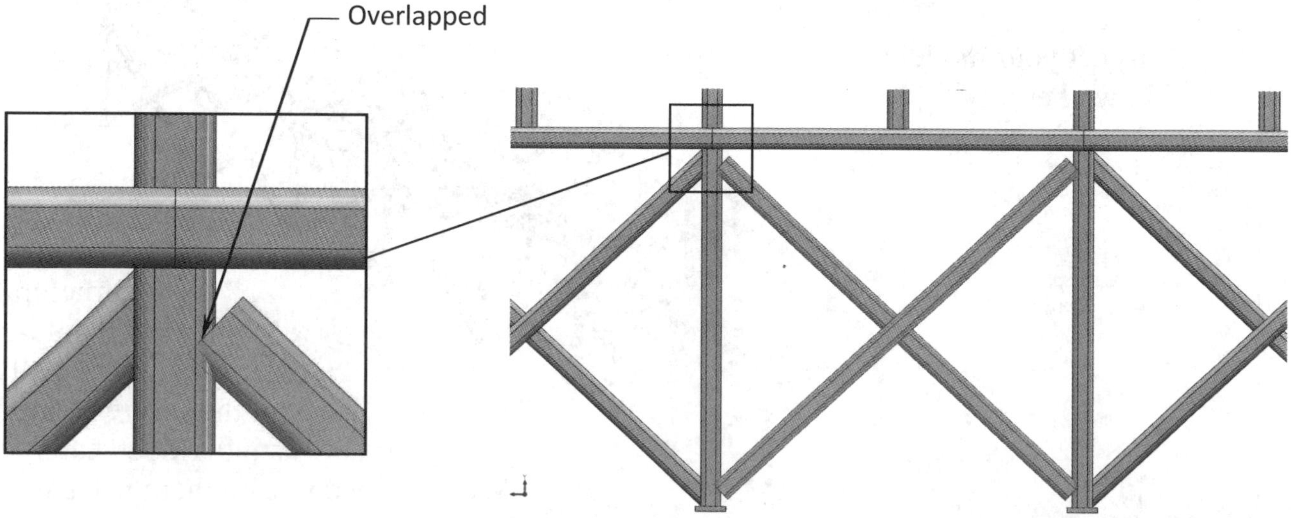

7. Trimming the overlapped:

Click the **Trim/Extent** command.

Use the default **End Trim** option.

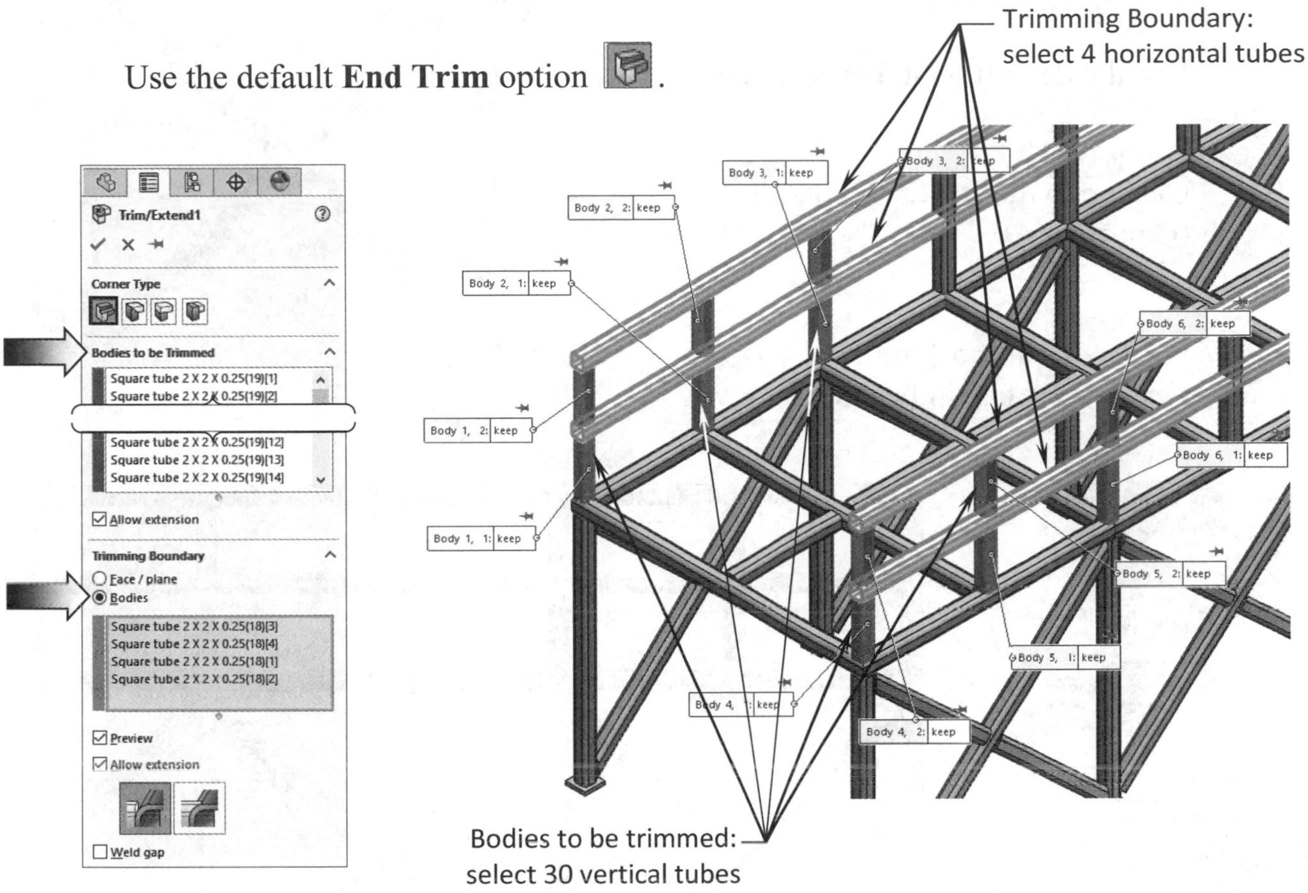

For Bodies to be Trimmed, select the **30 vertical members** as indicated.

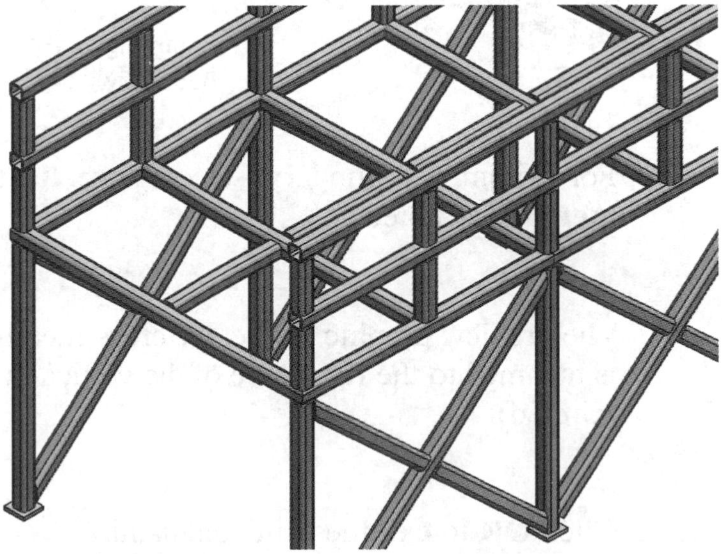

For Trimming Boundary, click the **Bodies** button and select the **4 horizontal tubes** as noted.

The preview graphics shows the vertical members are trimmed to match the faces of the horizontal bodies.

Click **OK** to exit the Trim.

8. Trimming the cross-members:

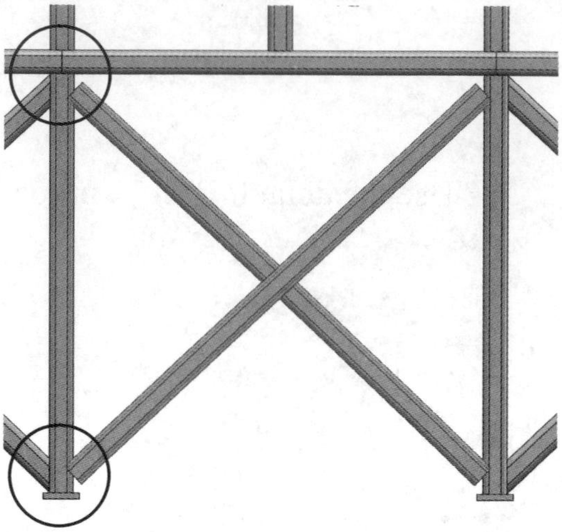

Click the **Trim/Extent** command .

Use the default **End Trim** option .

Change to the **Right** orientation (Control + 4).

For Bodies to be Trimmed, select the **Cross-member** indicated.

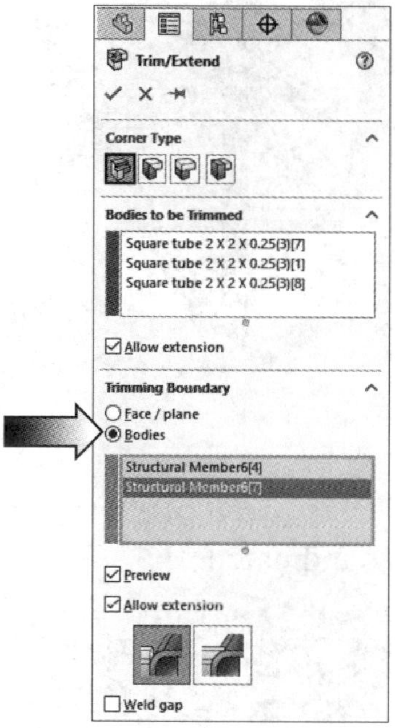

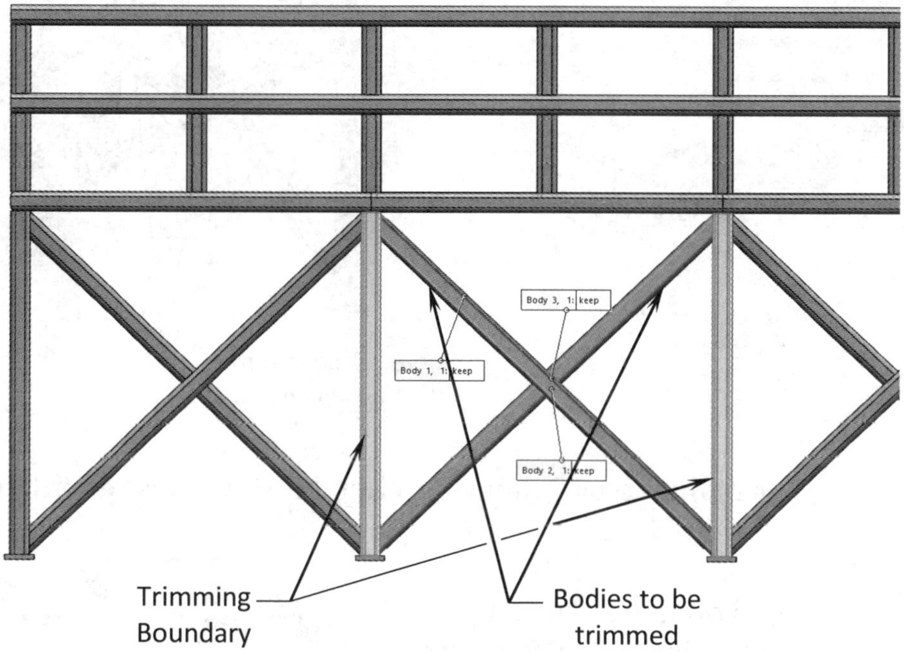

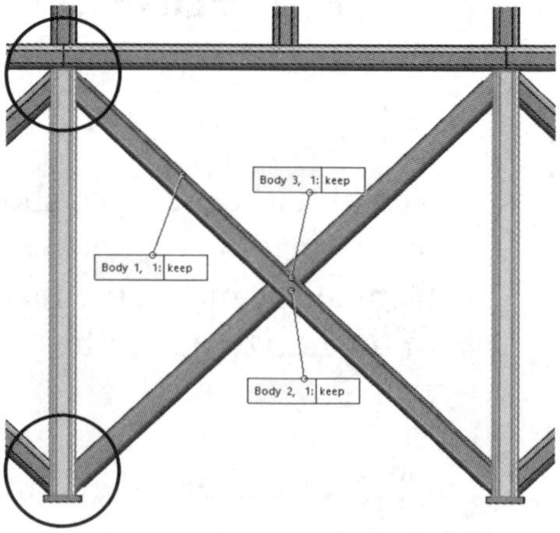

For Trimming Boundary, select the **vertical member** as noted.

The preview graphic shows the cross-member is trimmed to the right face of the vertical tube (circled).

Click **OK** to exit the trim command.

9. Trimming the other structural members:

There are several other structural members throughout the platform that still need to be trimmed (circles).

Change to the **Right** orientation (Control + 4).

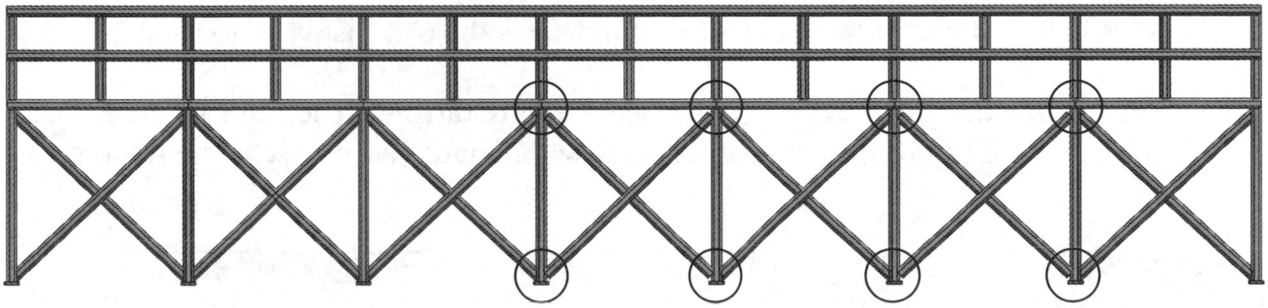

Click the **Trim/Extend** command and trim the tubes shown in the circles.

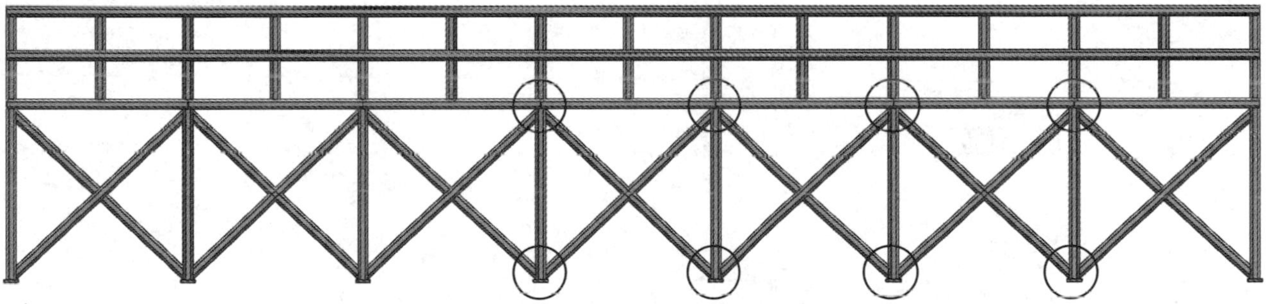

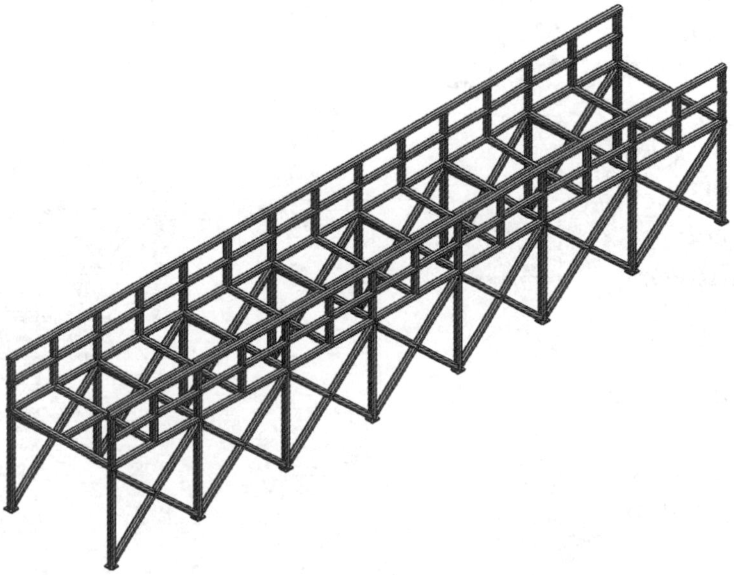

10. Updating the Cut list:

SOLIDWORKS updates the cut list automatically when new features are added, such as: extruded foot plates or 3D bounding boxes, edit existing features, or rebuilds the model. The model's custom properties and internal supporting data are also updated, preventing custom property errors.

This Icon in front of the Cut List  indicates the Cut List needs updating.

This Icon in front of the Cut List  indicates the Cut List is up to date.

Right-click the **Cut List** icon and select **Update** (arrow). The Cut List items that were created as groups are automatically sorted and placed in separate folders.

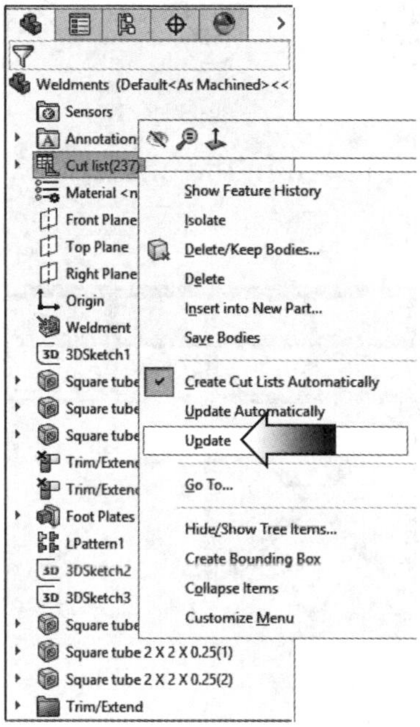

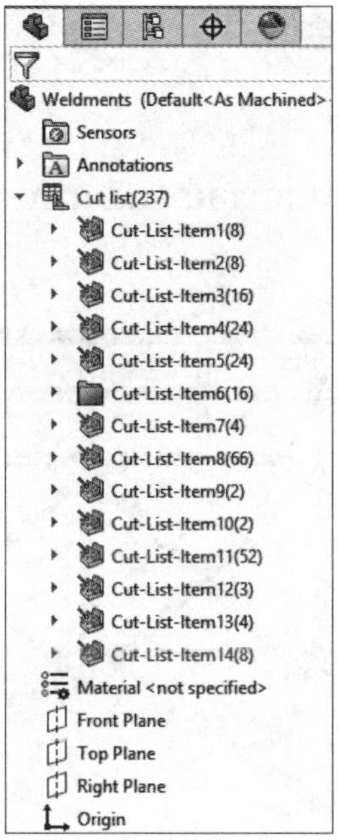

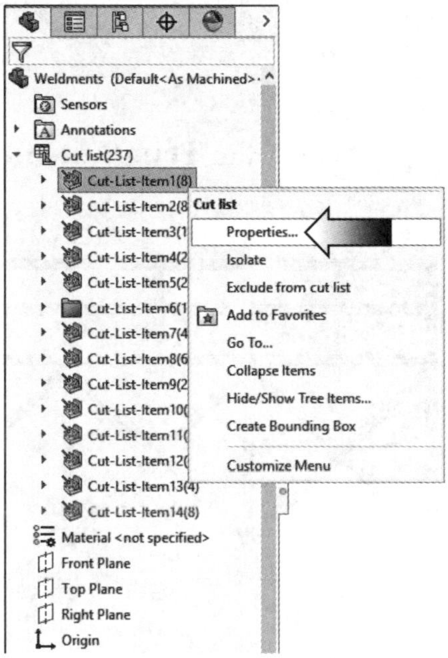

To view the Cut List details, right click on one of the Cut List-Items, and select: **Properties** (arrow).

Select one of the items such as Angle, Length, Description, etc. to see the details of the item.

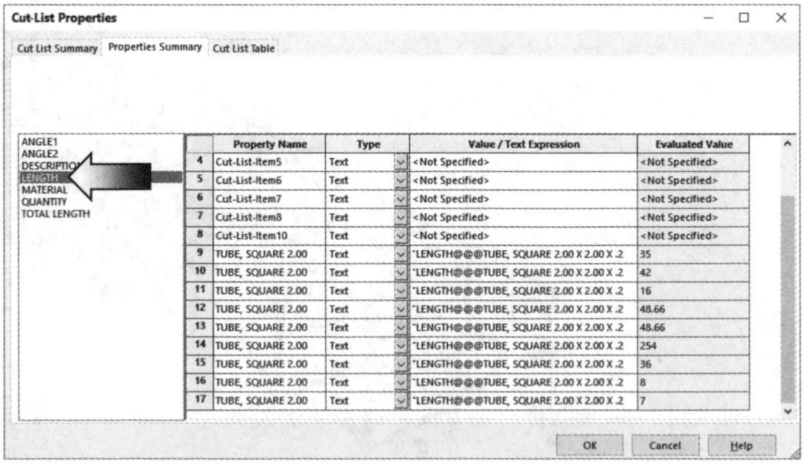

Click **OK** to exit the Cut List Properties.

11. Saving your work:

Select File / **Save As**.

Enter **Weldments Platform.sldprt** for the name of the file.

Click **Save**.

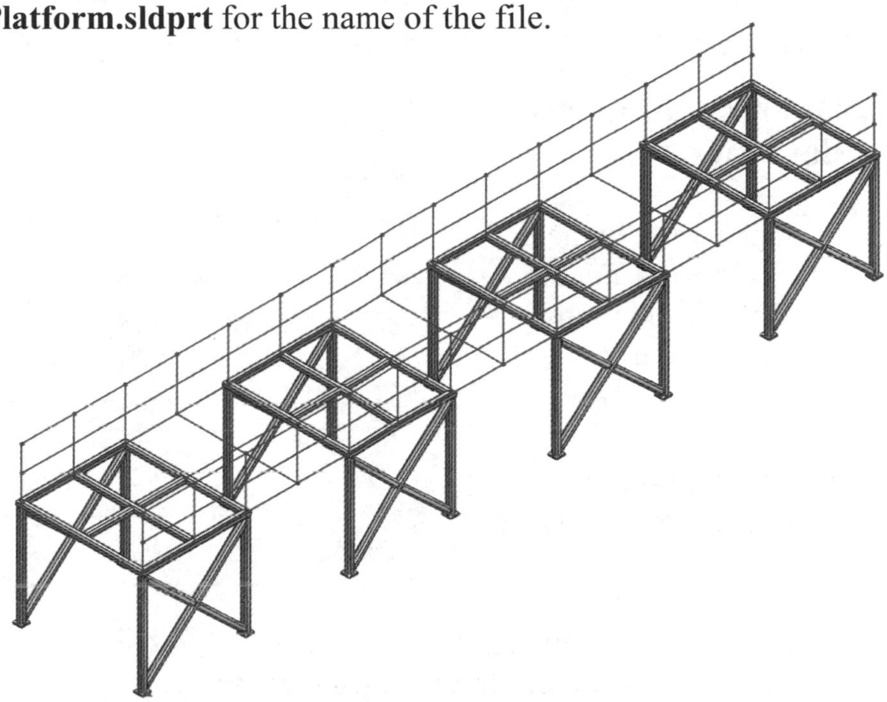

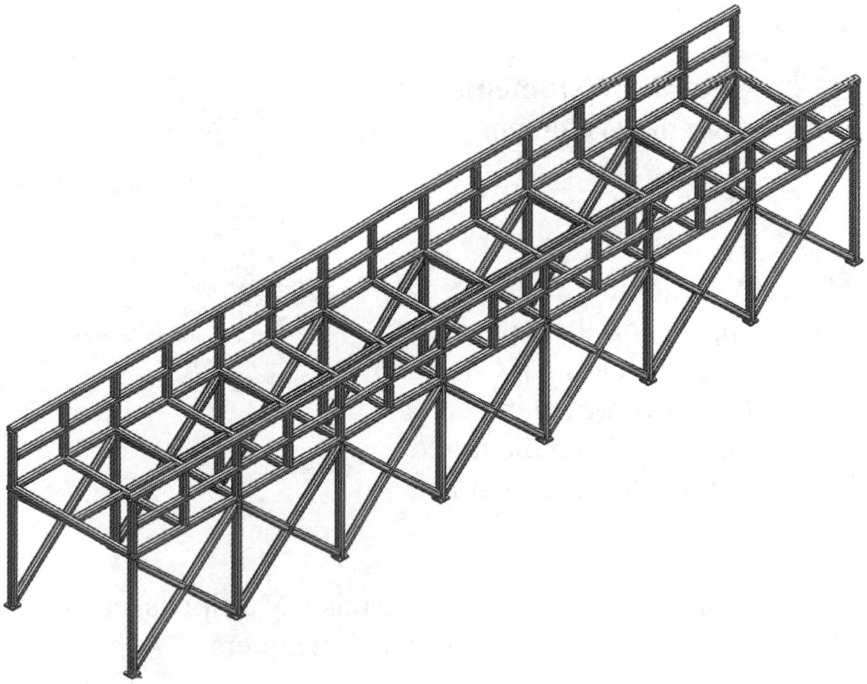

Using Weldments – Structural Members

The options in Weldments allow you to develop a weldment structure as a single multibody part. The basic framework is defined using 2D or a 3D sketch, and then structural members like square or round Tubes are added by sweeping the tube profile along the framework. Gussets, end caps, and weld-beads can also be added using the tools on the **Weldments** toolbar.

1. Opening an existing document:

Open a file named **Weldment Frame** from the Training Files folder.

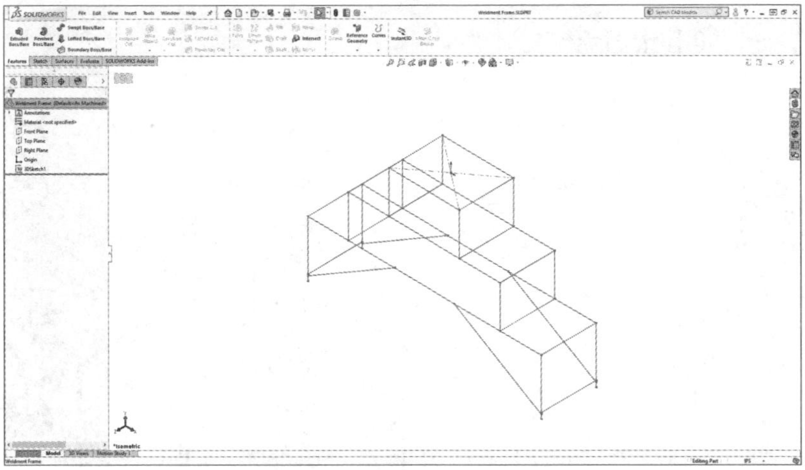

2. Enabling the Weldments toolbar:

Right-click the **Evaluate** tab and select the **Weldments** toolbar from the list (arrow).

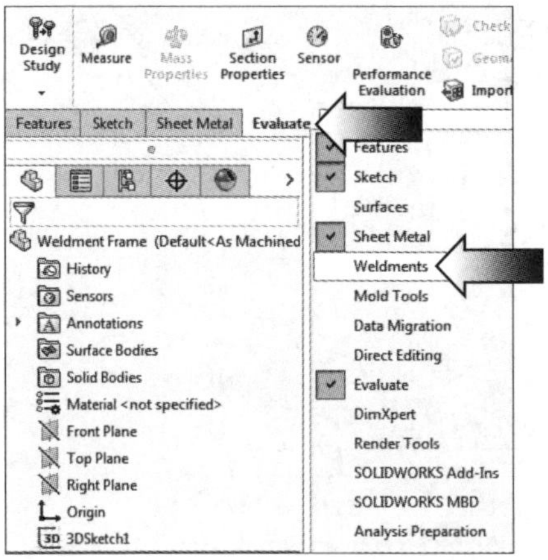

Click the **Weldments** button from the Weldments toolbar.

A Weldment Feature  appears on the feature tree with a Weldment Cut List (arrows), which indicates the items from the model to include in this cut list.

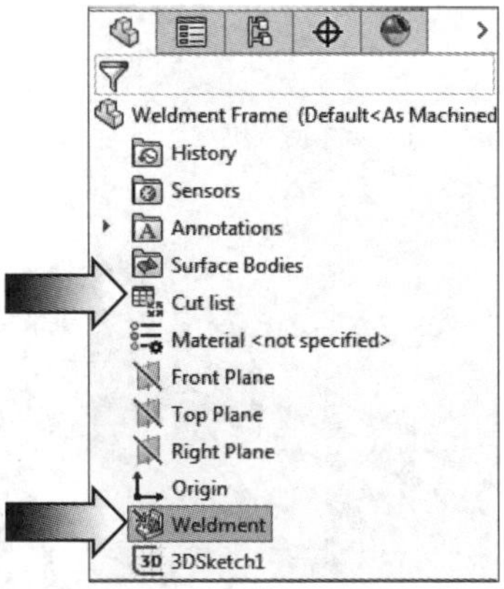

A single 3D Sketch is created for the purpose of this exercise. Multiple sketches (2D & 3D) can be used to design the weldments structural members.

3. Adding Structural Members:

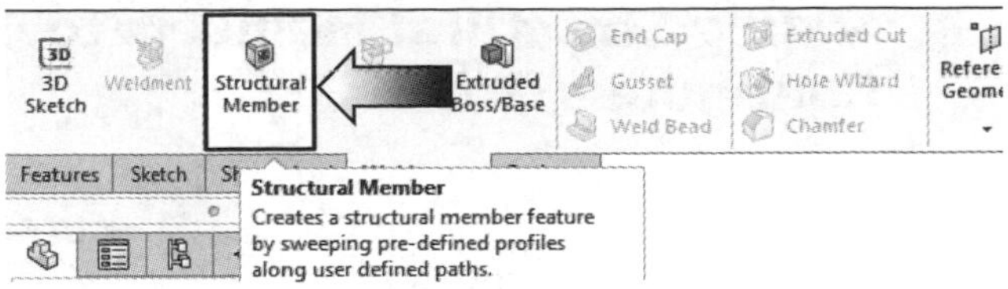

Click the **Structural Member** button from the Weldments toolbar.

Select the following:

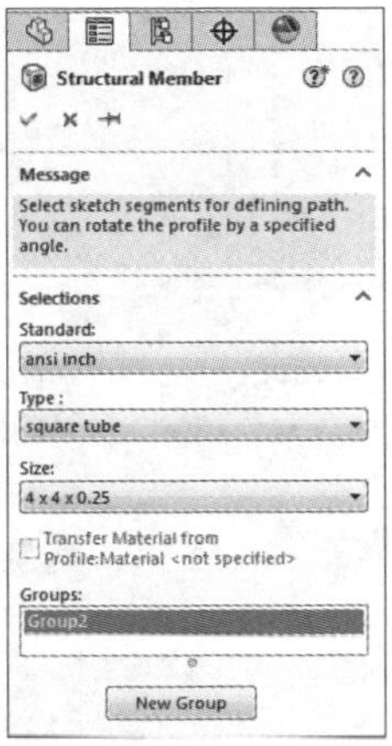

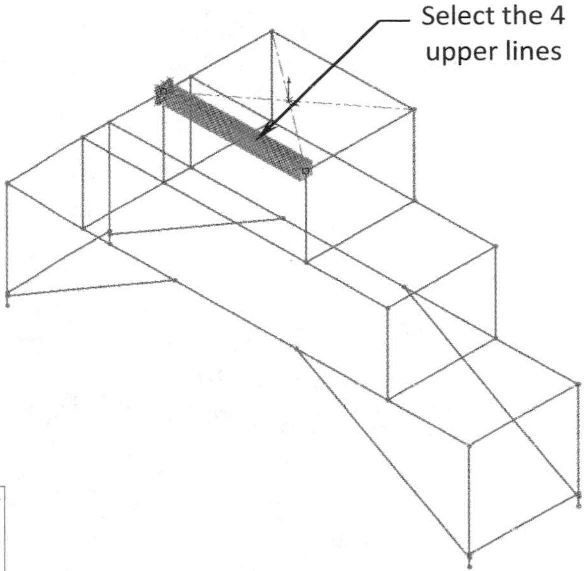

* **Ansi Inch.**
* **Square Tube**
* **4 X 4 X 0.25**

Click the **4 lines** on the top of the frame.

Select the **MITER** under Apply Corner-Treatment.

Click **OK**.

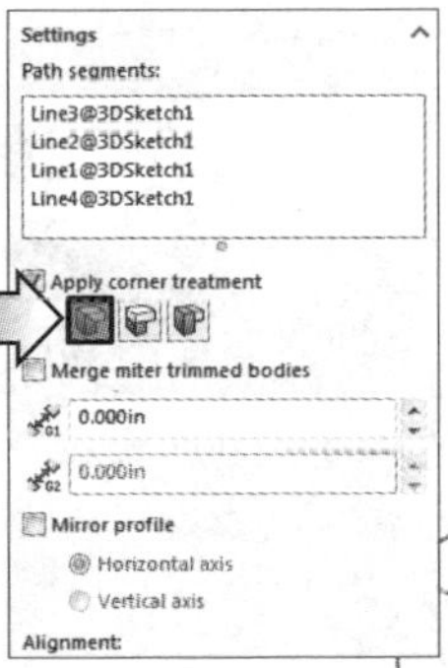

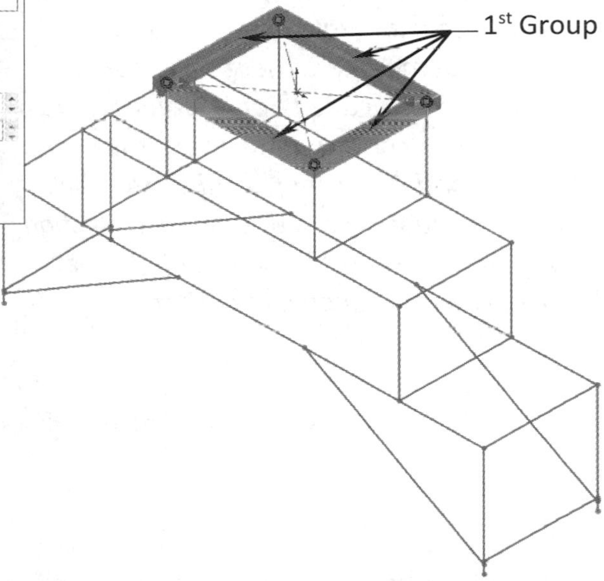

By default, the profile of the tube is automatically centered on the end of each line.

Try out all 3 options for corner treatments: End Miter, End Butt1, and End Butt2.

Switch back to the **End Miter** option when finished.

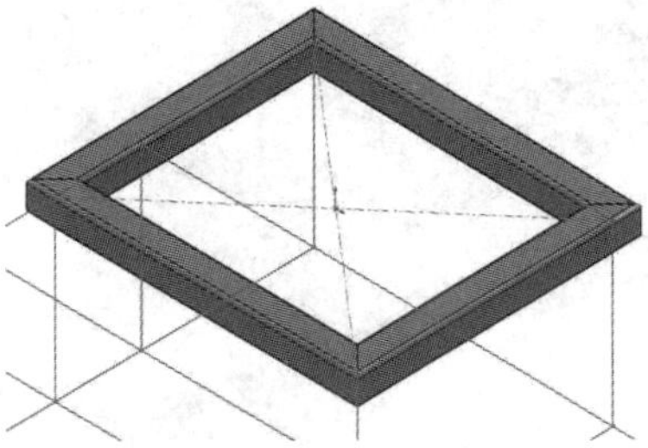

End Miter

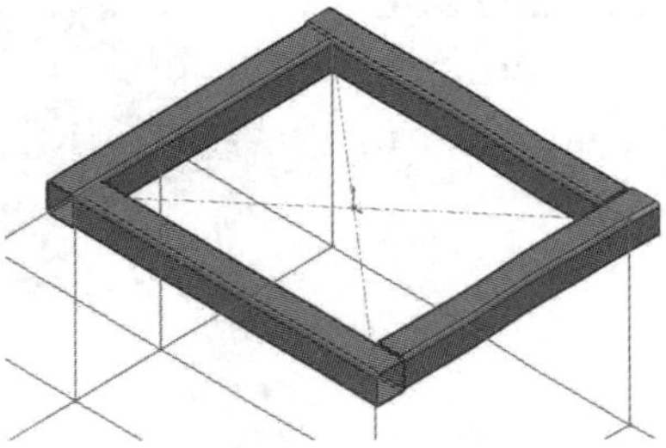

End Butt1
(Overlapped)

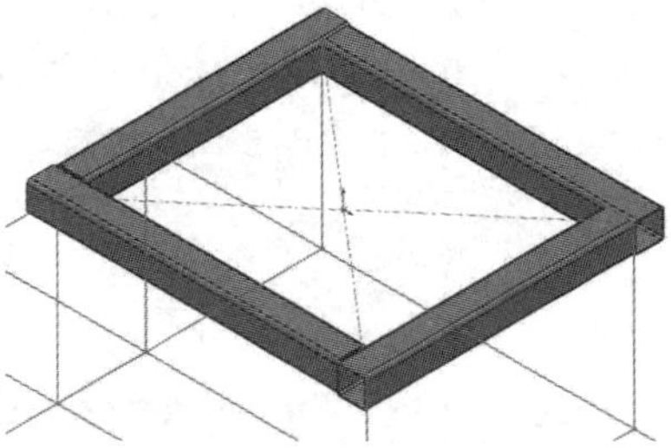

End Butt2
(Under-lapped)

4. Adding Structural Members to Contiguous Groups*:

Repeat the previous step and add another 4 square tubes to the 2nd group as shown.

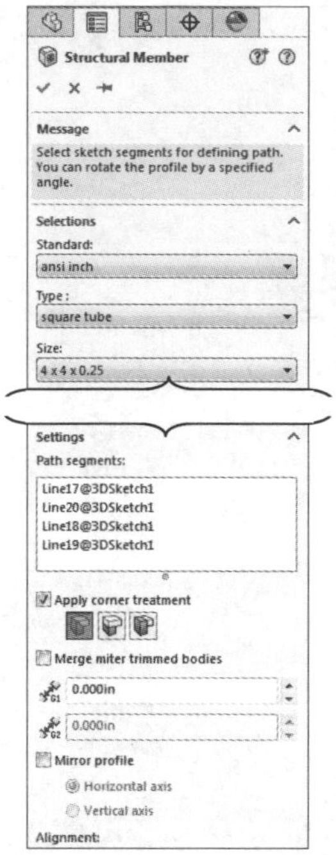

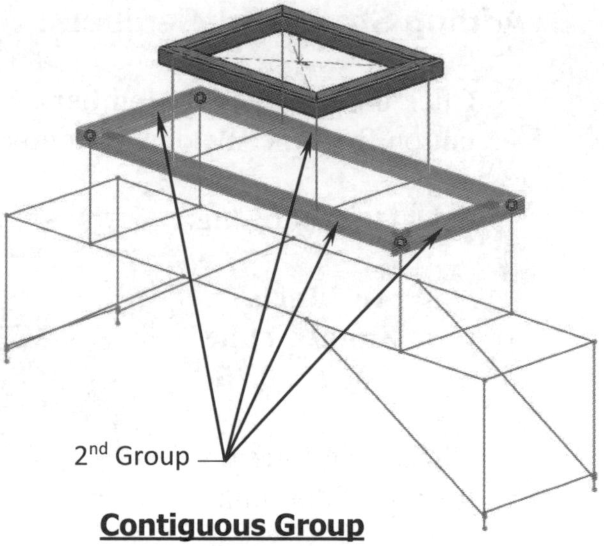

Contiguous Group

Use these same settings:

*** Ansi Inch.**
*** Square Tube**
*** 4 X 4 X 0.25**

** A group is a collection of related segments in a structural member. There are 2 types of groups, one is called Contiguous, where a continuous contour of segments is joined end-to-end. The other is called Parallel, which includes a discontinuous collection of parallel segments. Segments in the group cannot touch each other.*

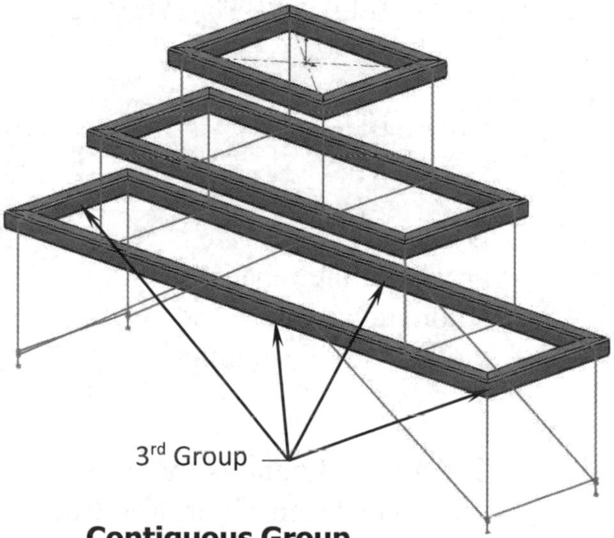

Contiguous Group

Repeat the same step for the 3rd group.

Follow the same procedure and add the same size tubing to the vertical members as noted for the 4th group.

Note: *Select the exact same vertical tubes on both sides (total of 12).*

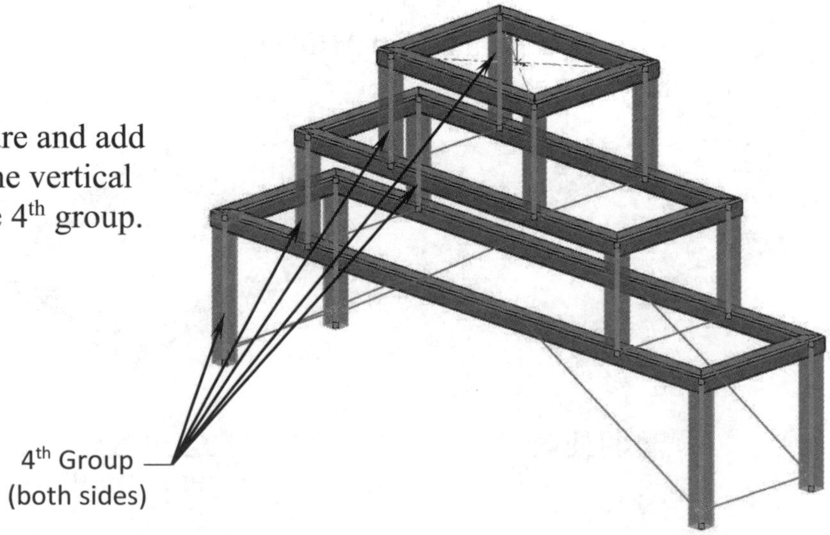

5. Adding Structural Members to the Parallel Groups:

Repeat the previous step and add the same structural members to the 5th group as indicated.

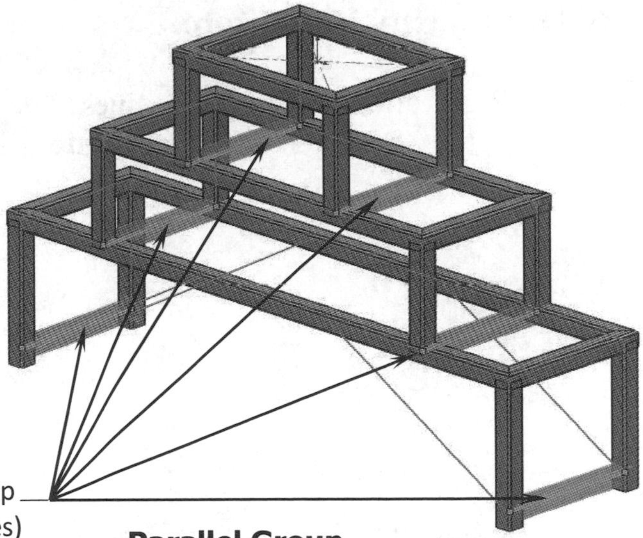

5th Group (both sides)

Parallel Group

> ### Groups
>
> You can define a group in a single plane or in multiple planes. A 3D sketch is best suited for weldment designs since all entities can be drawn and controlled in the same sketch.

Create the same type of structural members for the 6th and 7th group, which has only 2 lines in each group…

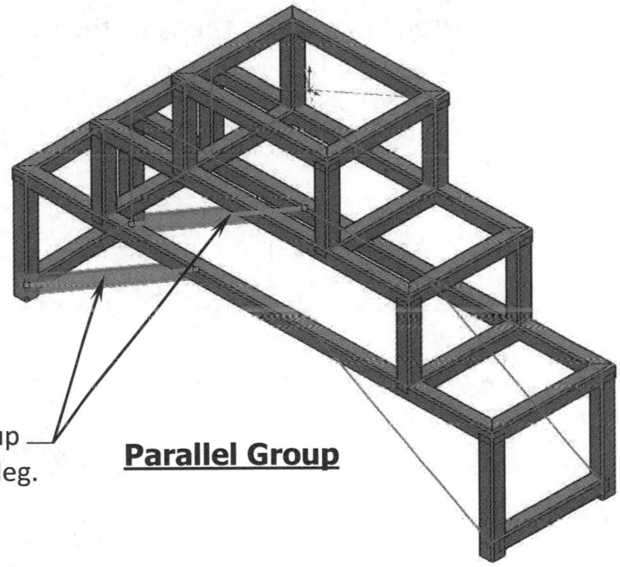

6th Group rotate 30 deg.

Parallel Group

Rotate the profile to **30 deg.** for the 6thth group and **60 deg.** for the 7th group.

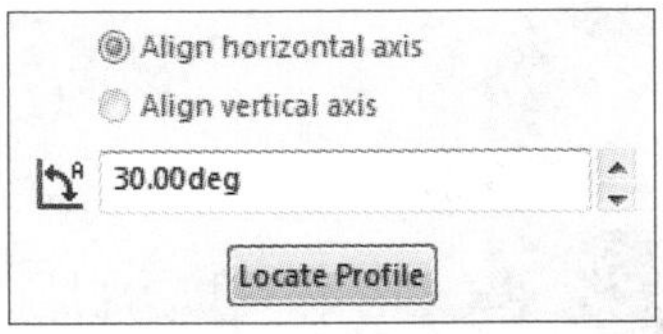

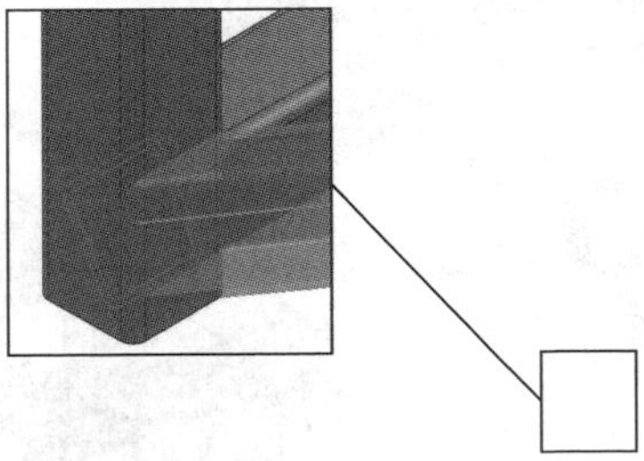

There are several over-lapped areas that need trimming; we will look into that in the next steps.

Parallel Group

7th Group rotate 60 deg.

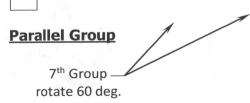

6. Hiding the 3D sketch:

Right-click on one of the lines in the 3D sketch and select **Hide** .

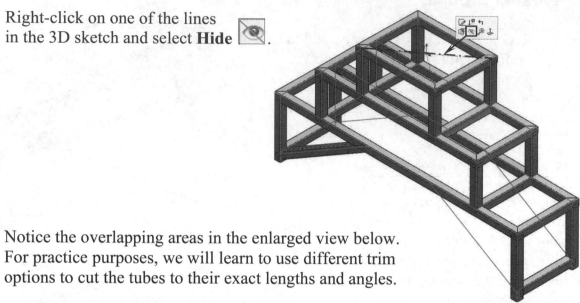

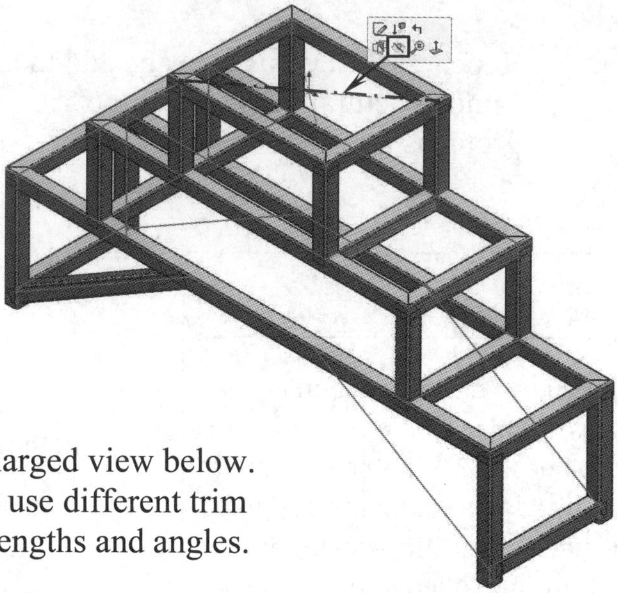

Notice the overlapping areas in the enlarged view below. For practice purposes, we will learn to use different trim options to cut the tubes to their exact lengths and angles.

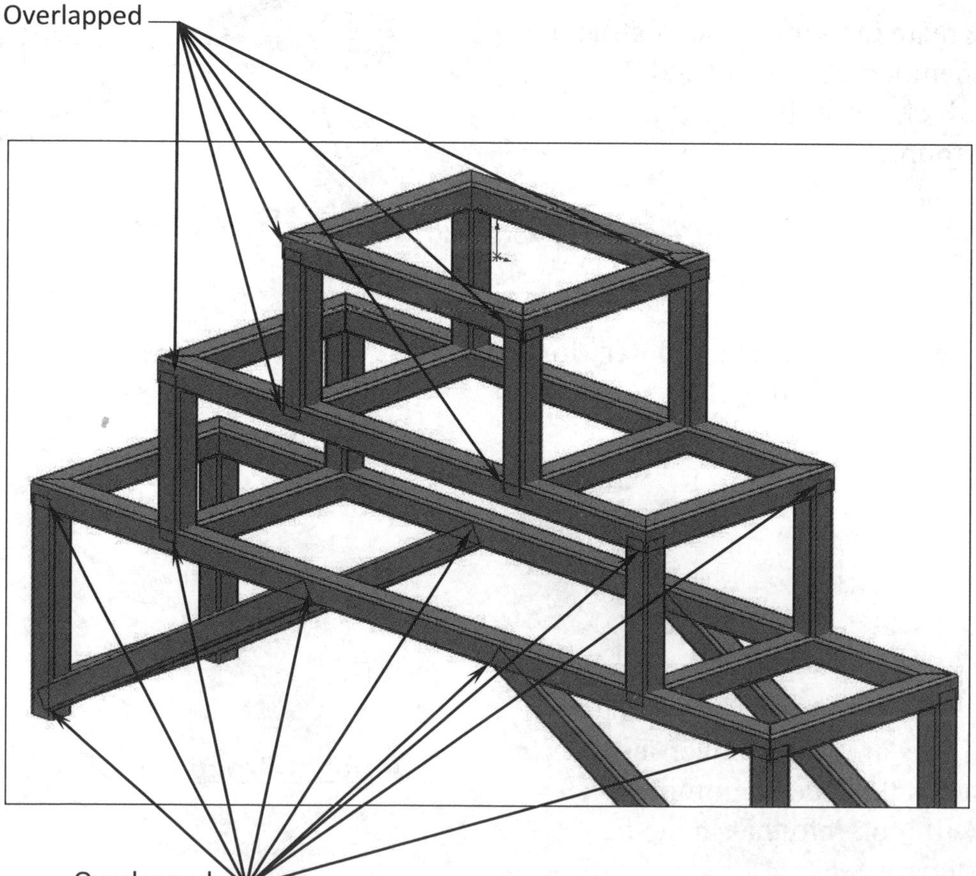

7. Trimming the Structural Members:

Click **Trim/Extend** on the Weldments toolbar.

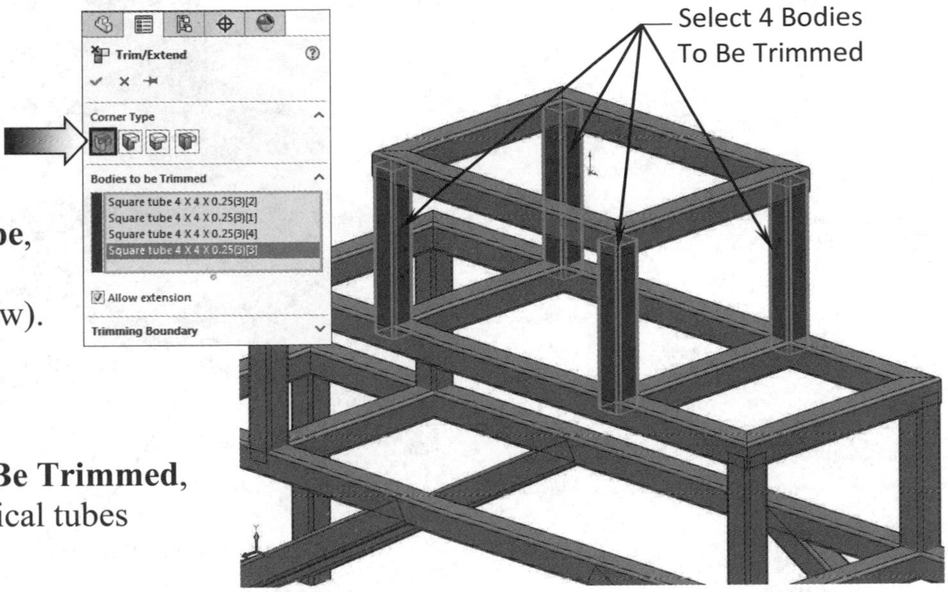

For **Corner Type**, use the default **End Trim** (arrow).

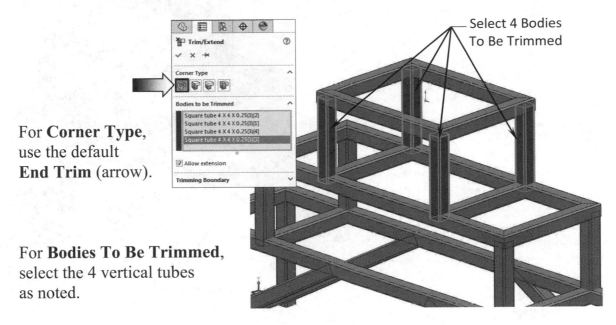

For **Bodies To Be Trimmed**, select the 4 vertical tubes as noted.

For **Trimming Boundary**, select the **Body** option (arrow).

For **Trimming Bodies**, select the **6 horizontal tubes** as indicated.

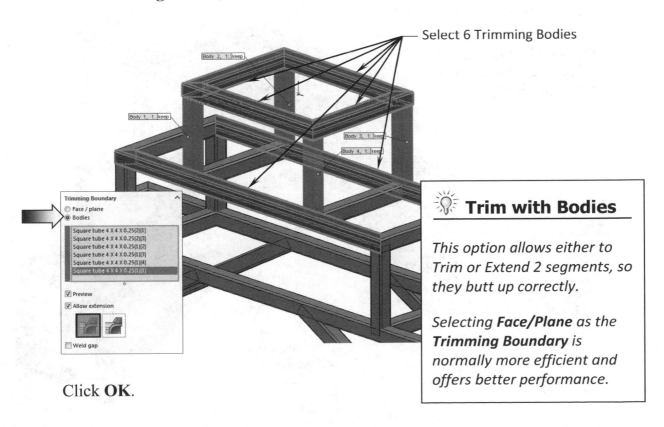

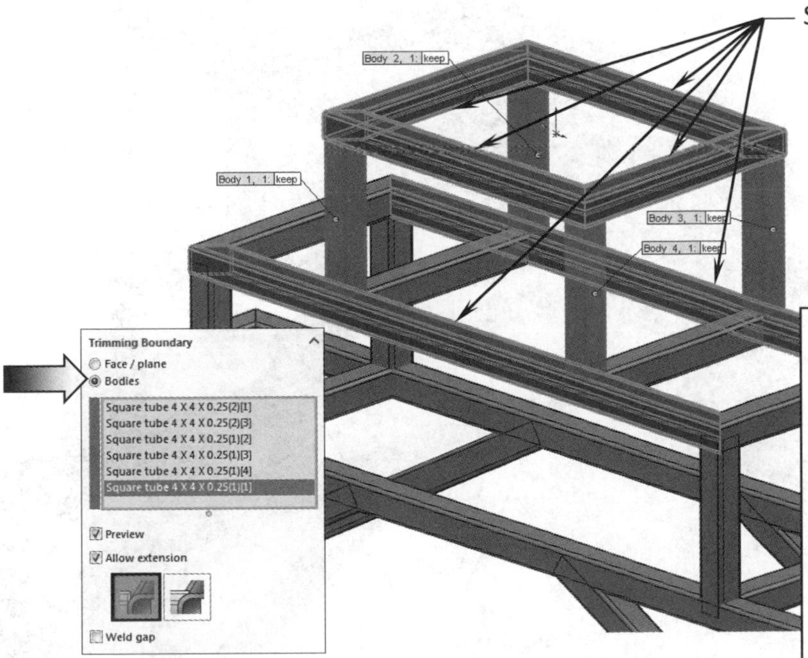

Trim with Bodies

This option allows either to Trim or Extend 2 segments, so they butt up correctly.

*Selecting **Face/Plane** as the **Trimming Boundary** is normally more efficient and offers better performance.*

Click **OK**.

8. Trimming the Parallel Groups:

Select the **Trim/Extend** command once again from the Weldments toolbar.

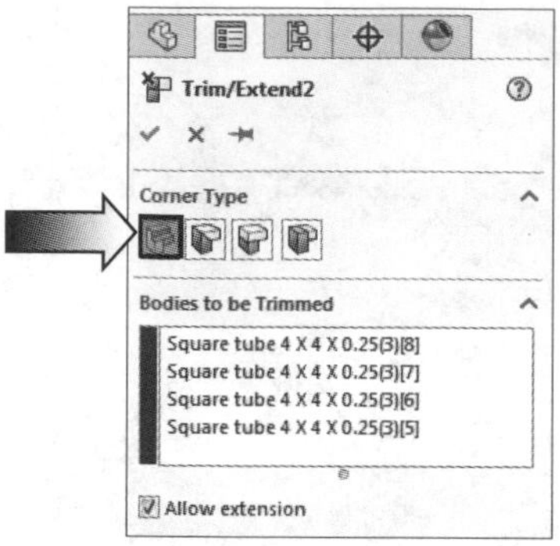

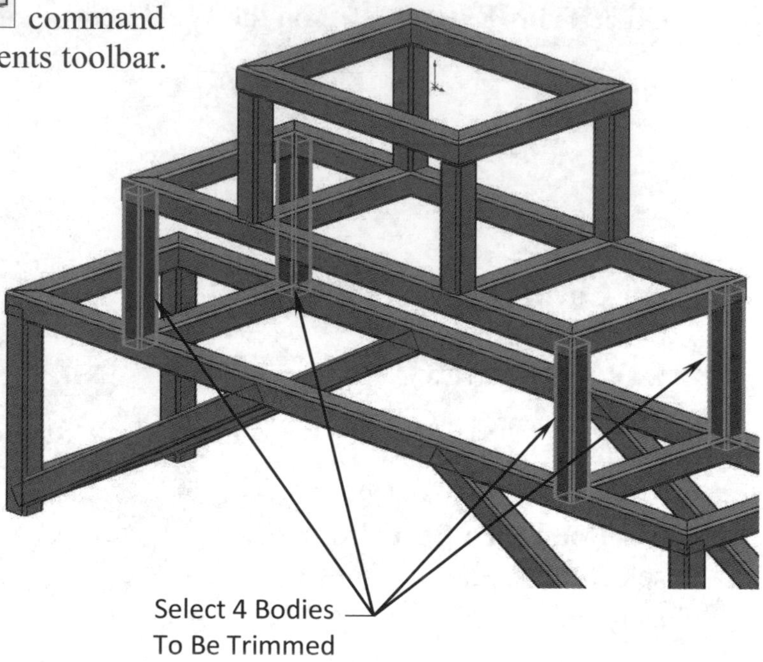

For **Bodies To Be Trimmed**, select the next 4 vertical tubes as noted.

For **Trimming Boundary**, select the **Body** option again (arrow).

For **Trimming Bodies**, select the **6 horizontal tubes** as shown.

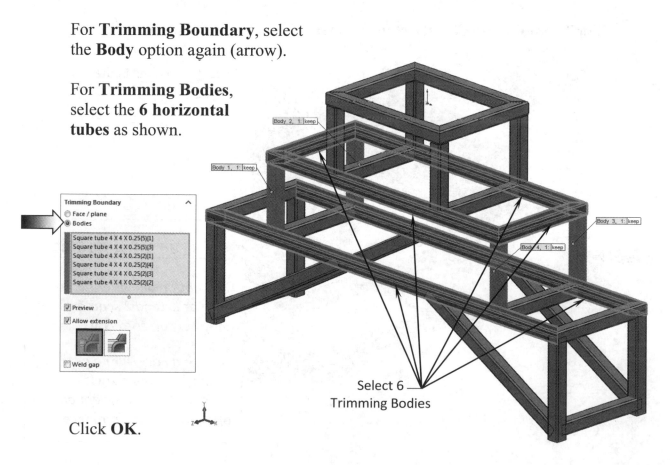

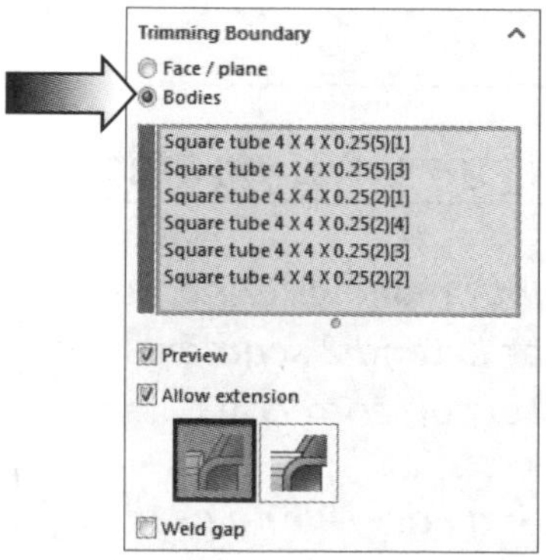

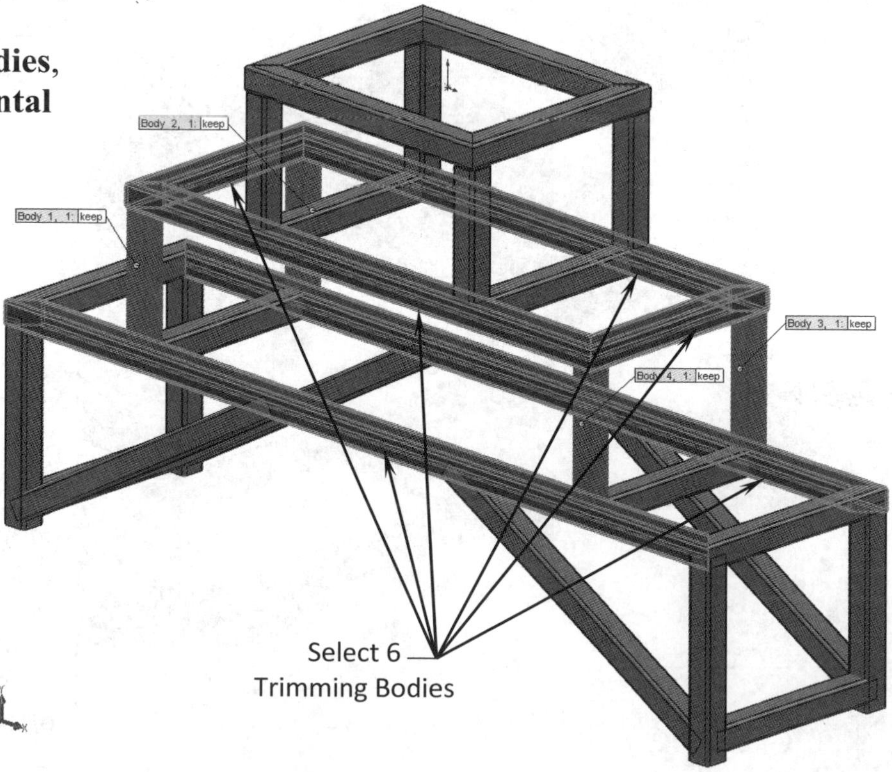

Click **OK**.

9. Trimming the next Parallel Groups:

Select the **Trim/Extend**  command from the Weldments toolbar.

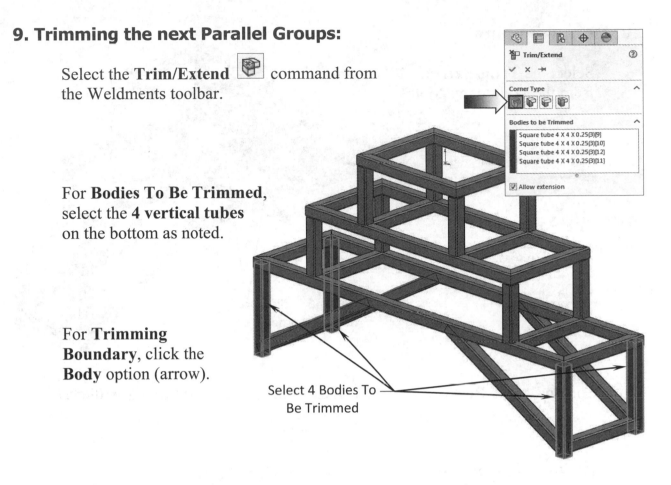

For **Bodies To Be Trimmed**, select the **4 vertical tubes** on the bottom as noted.

For **Trimming Boundary**, click the **Body** option (arrow).

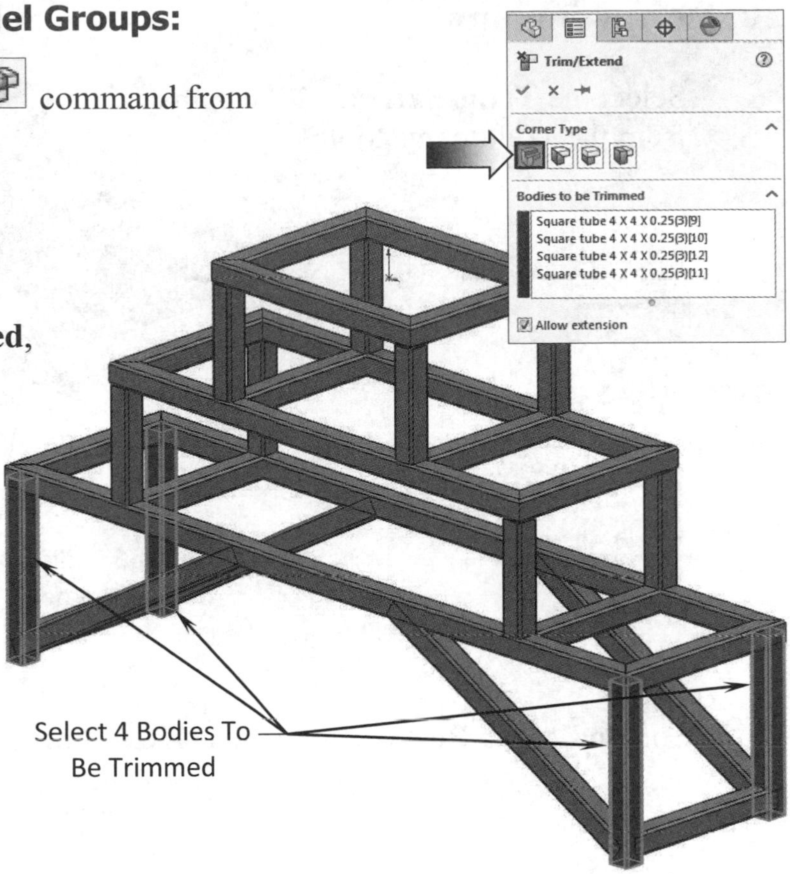

For **Trimming Bodies**, select the **4 horizontal tubes** as shown.

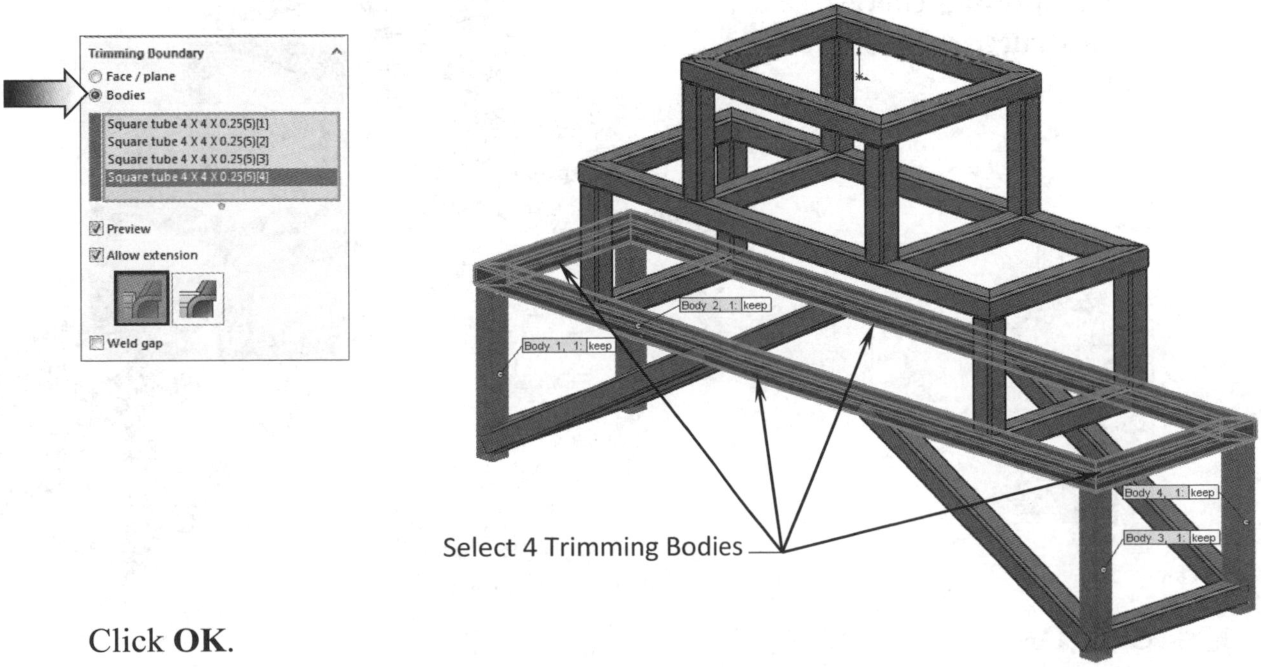

Click **OK**.

10. More Trimming:

Select the **Trim/Extend**  command from the Weldments toolbar.

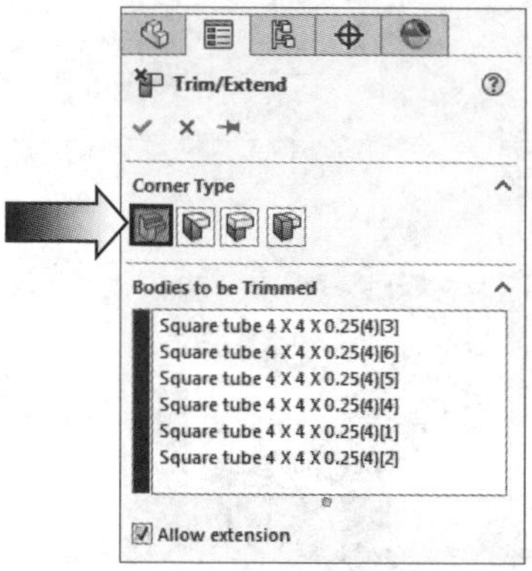

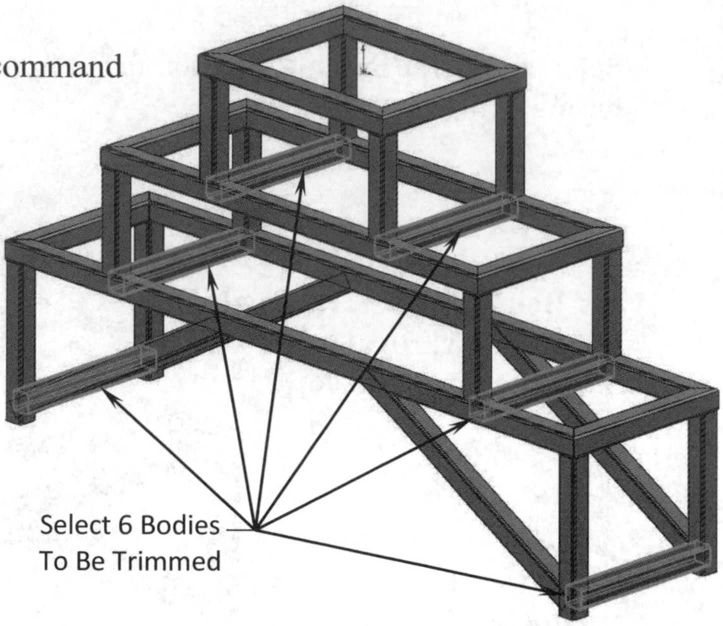

For **Bodies To Be Trimmed** select the **6 short horizontal tubes** as indicated.

For **Trimming Boundary**, click the **Body** option (Arrow).

For **Trimming Bodies**, select the **8 structural members** as indicated.

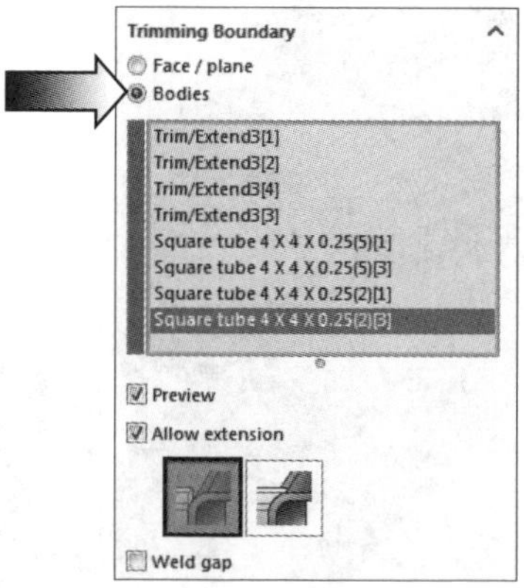

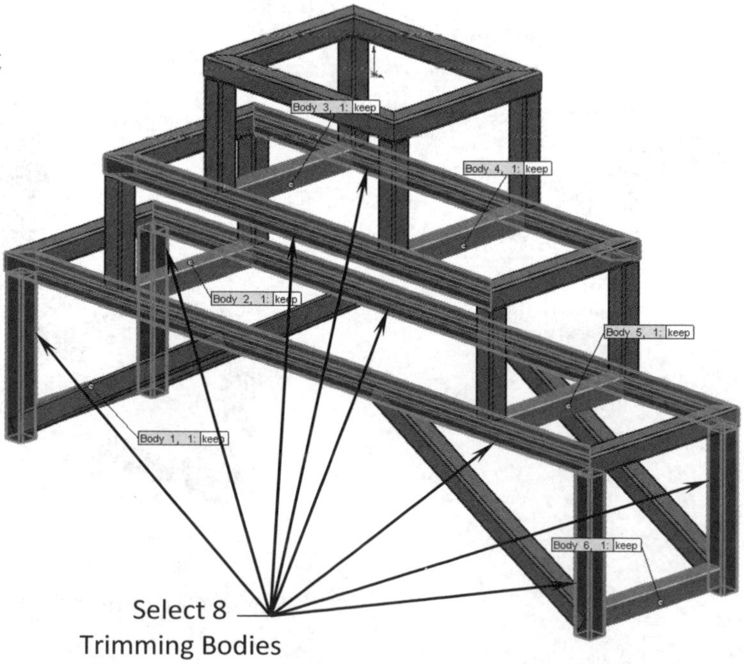

Click **OK**.

11. Trimming with Face/Plane:

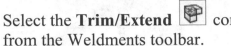

Select the **Trim/Extend**  command from the Weldments toolbar.

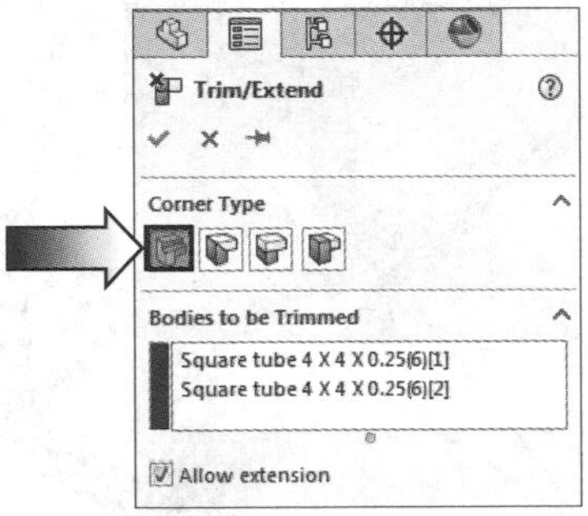

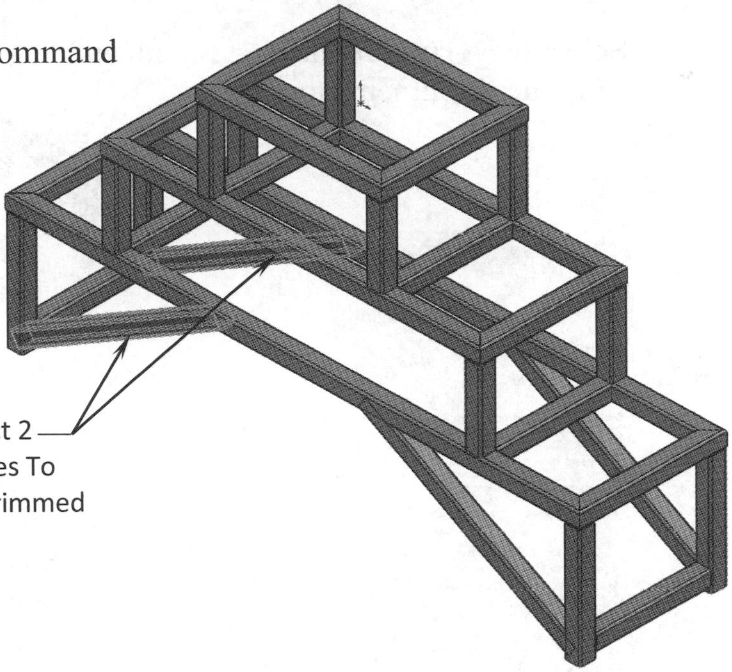

For **Bodies To Be Trimmed** select the 2 structural members as noted.

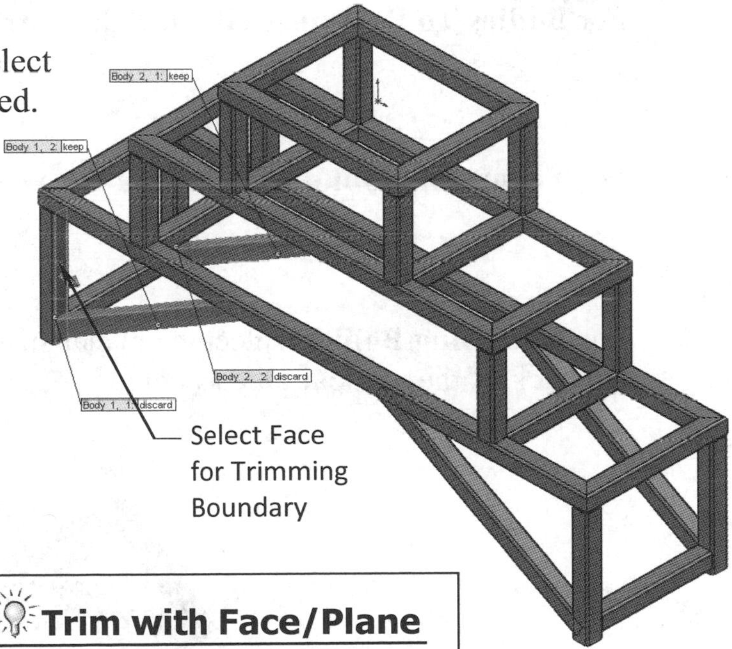

For **Trimming Boundary**, click the **Face / Plane** option (arrow).

Select the planar surface as noted, for **Trimming Bodies**.

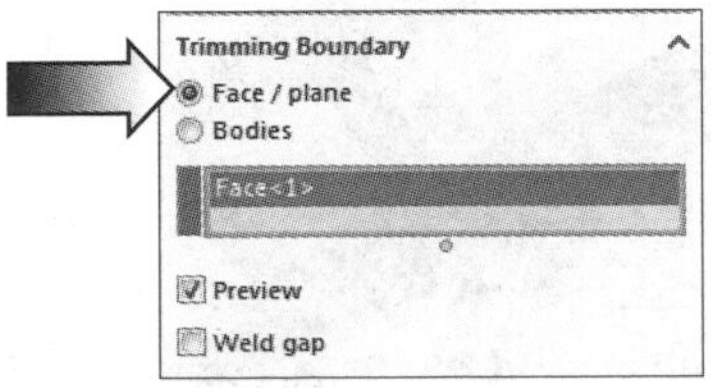

Click **OK**.

Trim with Face/Plane

This option allows a planar face(s) as a trimming boundary to trim one or more solid bodies.

*Selecting **Face/Plane** as the **Trimming Boundary** is normally more efficient and offers better performance.*

12. More Trimming with Face/Plane:

Select the **Trim/Extend**  command from the Weldments toolbar.

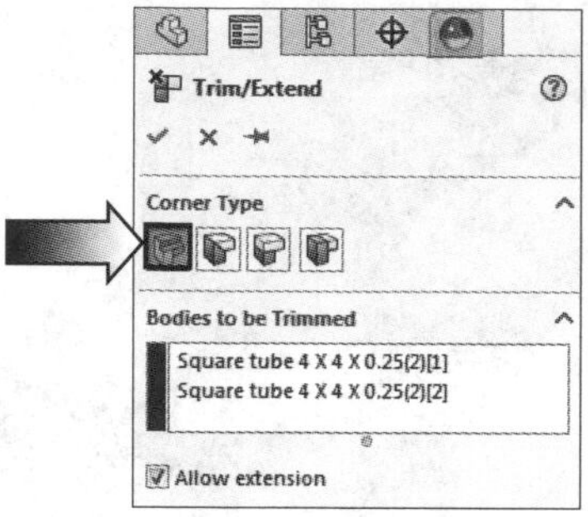

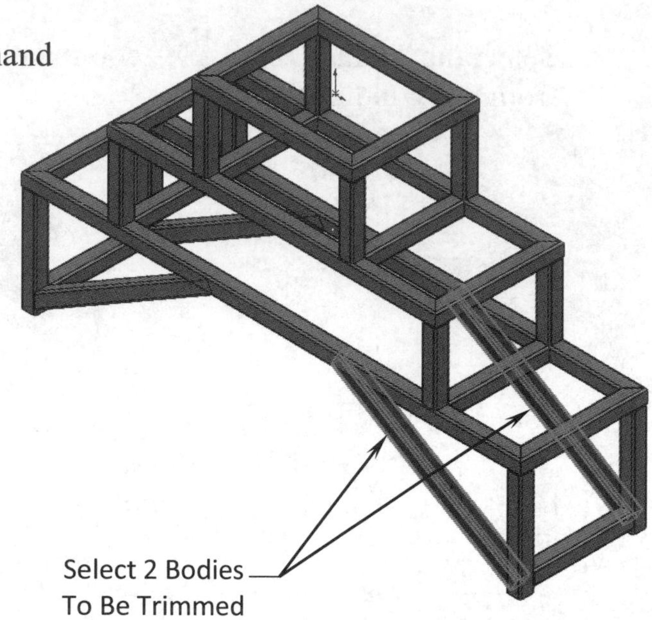

For **Bodies To Be Trimmed** select the **2 structural members** as indicated.

For **Trimming Boundary**, click the **Face / Plane** option (Arrow).

For **Trimming Bodies**, select the planar surface on the back of the vertical tube as noted.

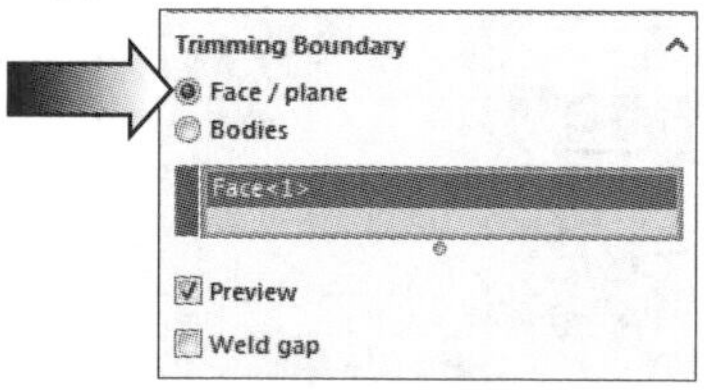

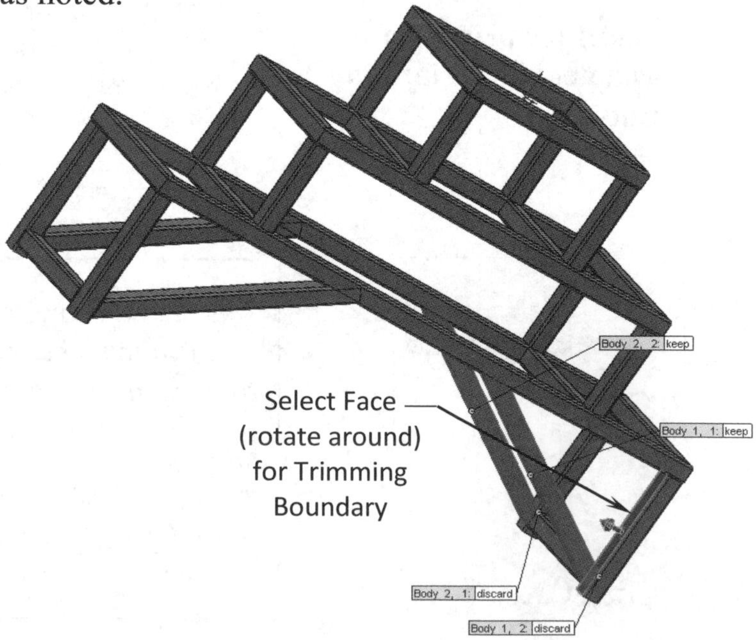

Click **OK**.

13. Trimming the last 4 structural members:

Select the **Trim/Extend**  command from the weldment toolbar.

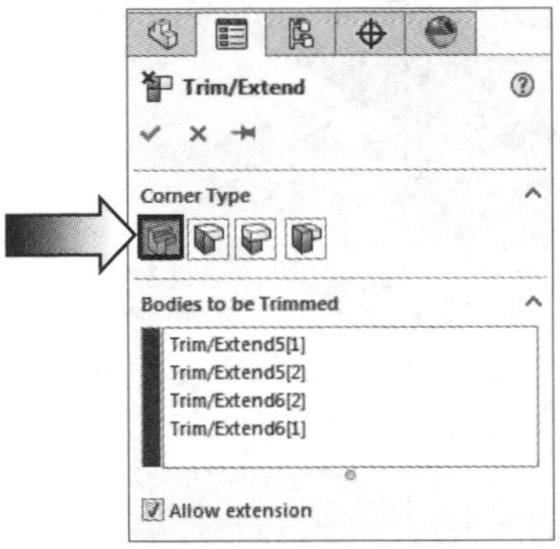

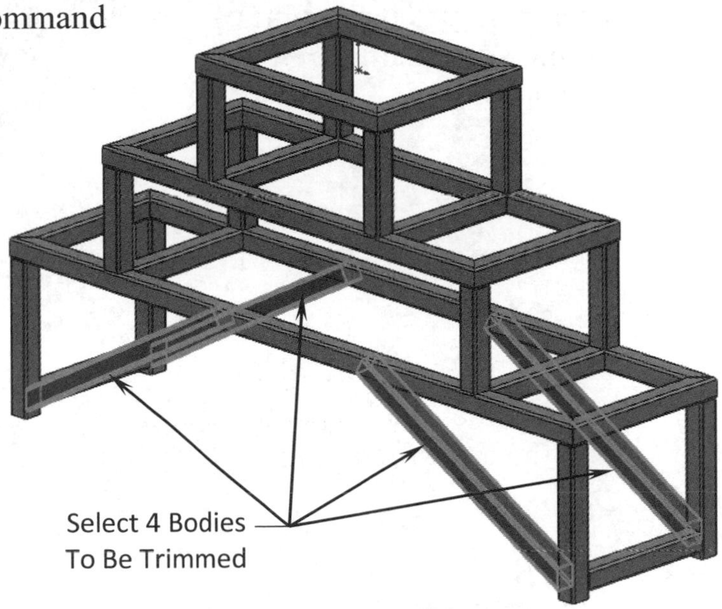

For **Bodies To Be Trimmed** select the **4 structural members** as shown.

For **Trimming Boundary**, click the **Body** option (arrow).

For **Trimming Bodies**, select the **2 horizontal tubes** as noted.

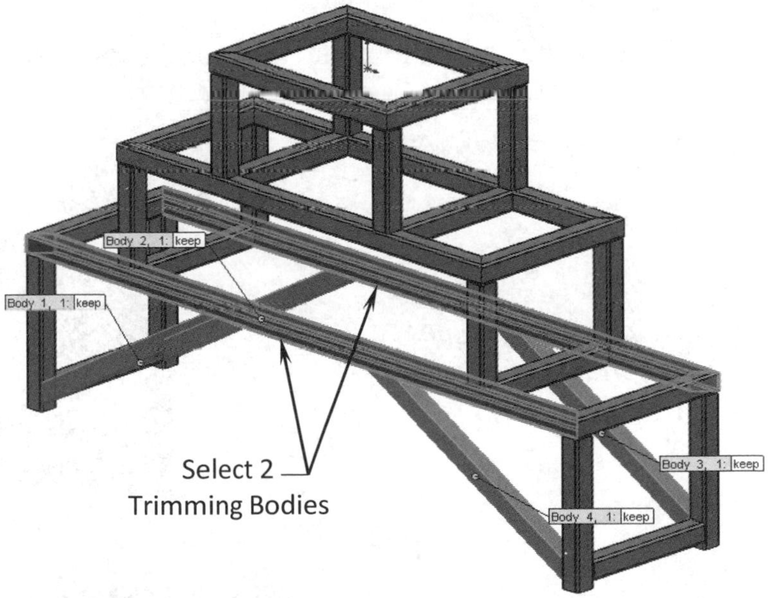

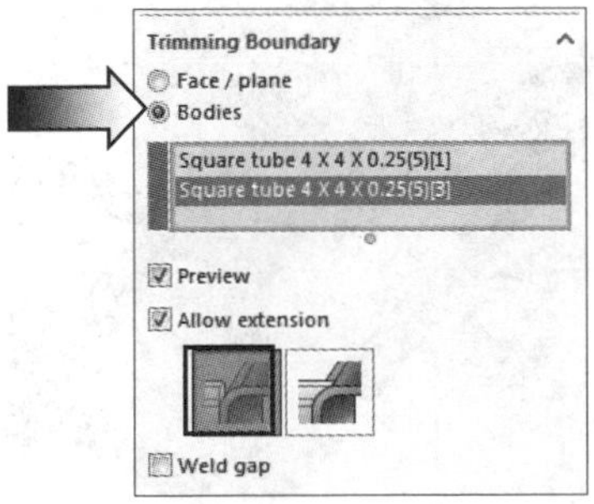

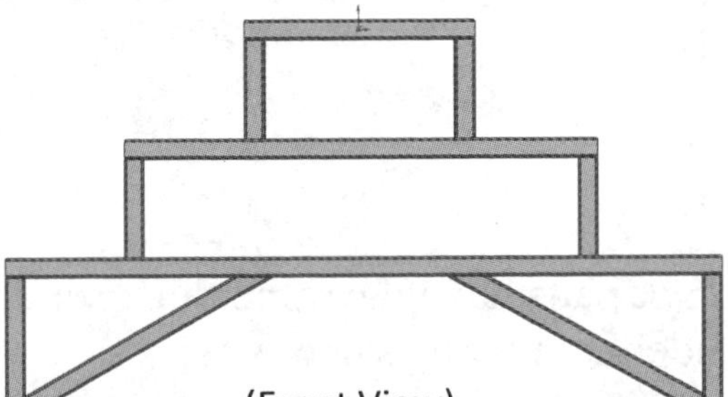
(Front View)

Click **OK**.

14. Adding the foot pads:

Insert a new sketch on the bottom surface of one of the 4 legs.

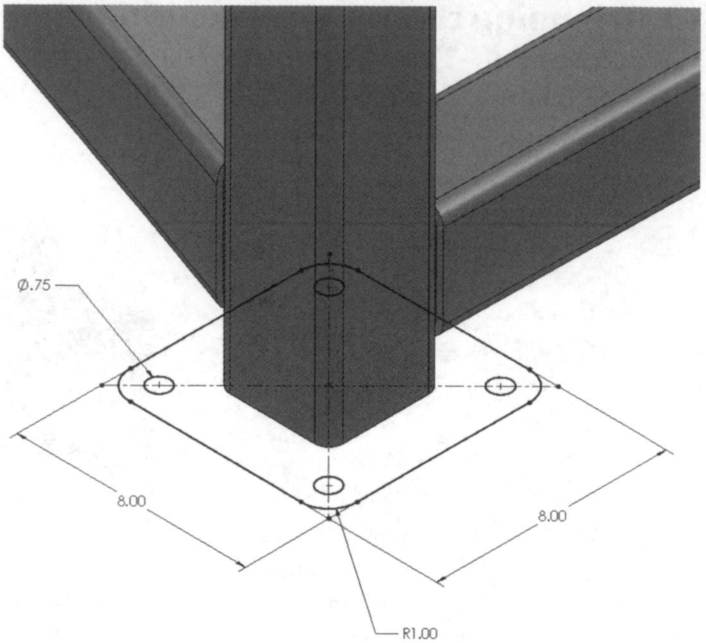

Sketch the profile as shown.

The 4 circles are concentric with the corner radiuses.

Add the dimensions and relations needed to fully define the sketch.

Mirror the sketch to make a total of 4 foot pads.

Note:
Add a couple of centerlines as shown prior to making the mirror.

Add centerlines to mirror

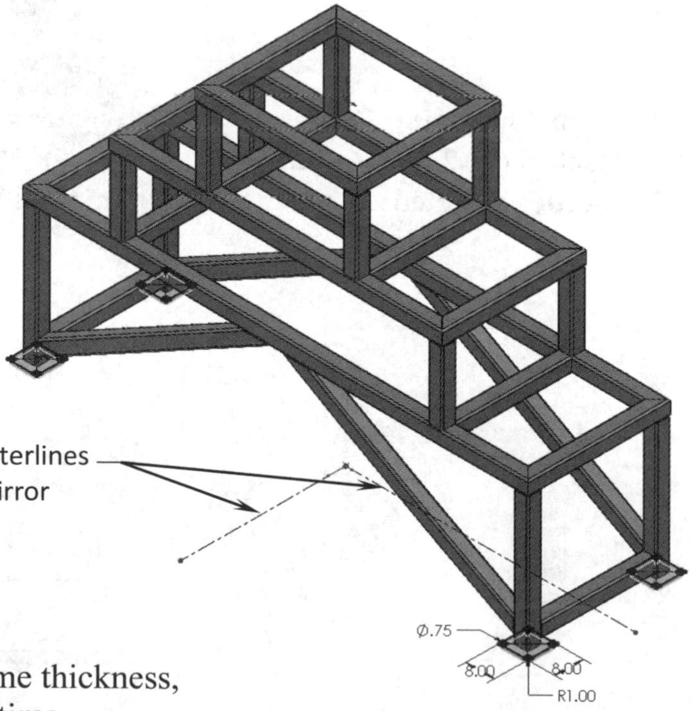

Since the 4 foot pads have the same thickness, we can extrude them at the same time.

Click **Extruded Boss/ Base.**

Enter the following:

* Type: **Blind**

* Depth: **1.000**

Click **OK**.

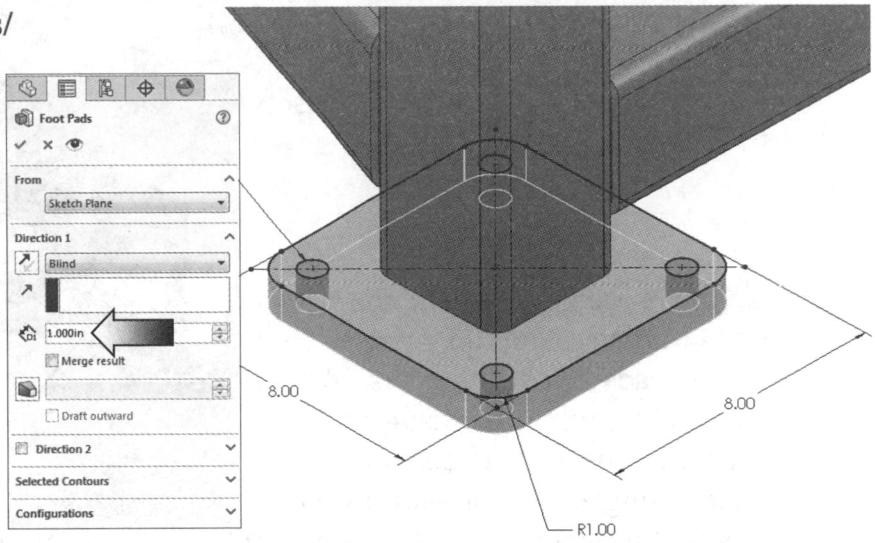

15. Adding the Gussets:

Rotate and zoom to an orientation that looks similar to the view below.

Click the **Gusset** command.

Enter the following:

For **Supporting Faces**, select the **2 faces** as noted.

* Distance1: **4.00 in.**

* Distance2: **4.00 in.**

* Distance3: **.500 in.**

* Thickness: **.500 in.** (Both Sides)

* Location: **Midpoint**

Click **OK**.

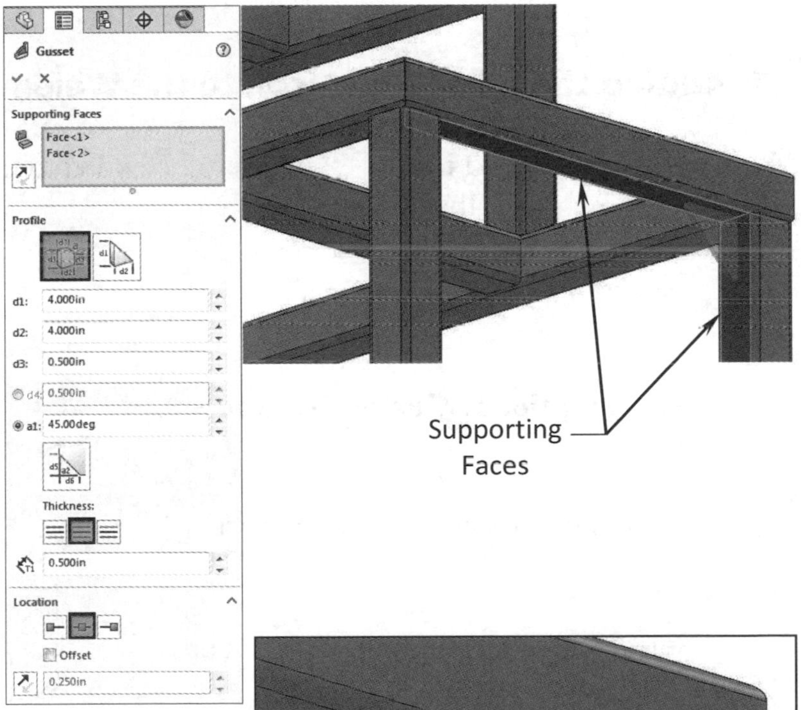

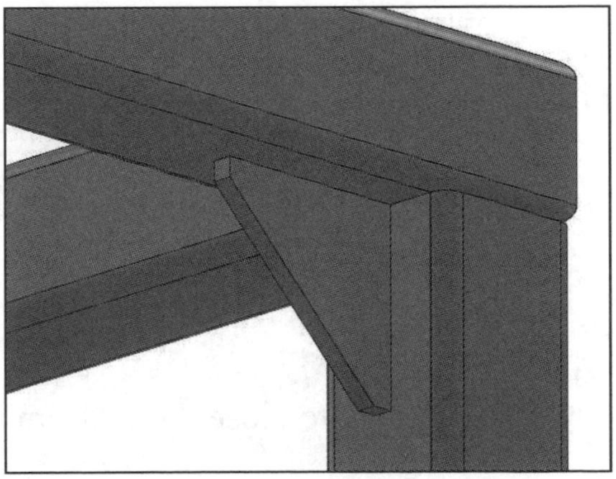

16. Adding more Gussets:

Repeat the step 15 and add a gusset to each corner of the frame.

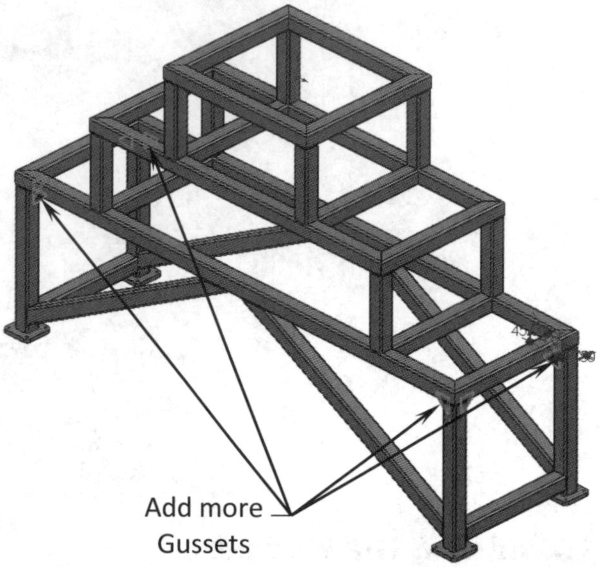

Next, we are going to add the weld beads around the gussets. Weld beads can be added as full length, intermittent, or staggered fillet weld beads between any intersecting weldment entities such as structural members, plate weldments, or gussets.

17. Adding the Fillet-Bead icon to the Weldments toolbar:

The **Fillet Bead** icon  needs to be added to the Weldments toolbar.

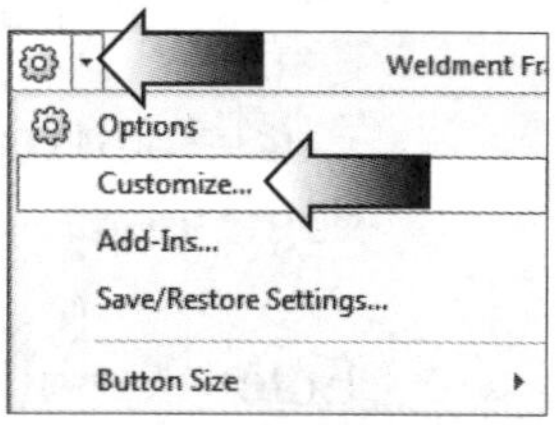

To add the missing icon:

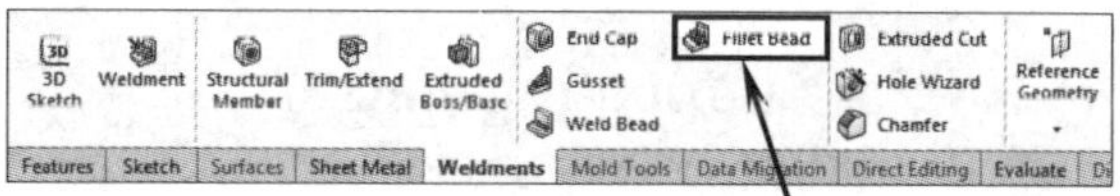

Select **Options / Customize** (arrows).

Click the **Commands** tab (arrow).

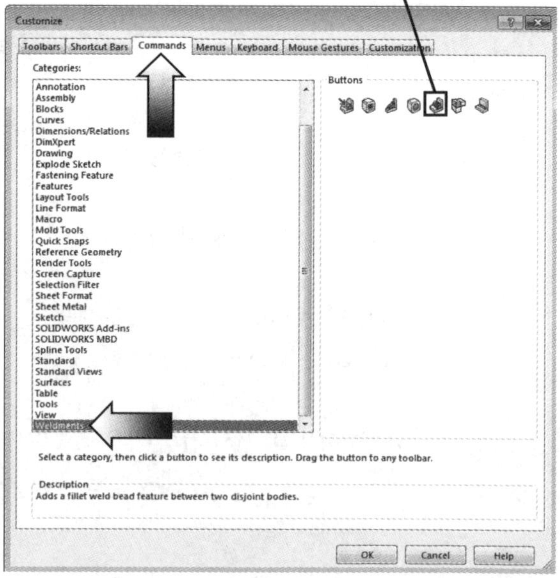

Select the **Weldments** option under **Categories** (arrow).

Drag the **Fillet Bead** icon and drop it onto the Weldments toolbar as noted.

Click **OK** to close the Customize dialog.

18. Adding the Fillet Beads:

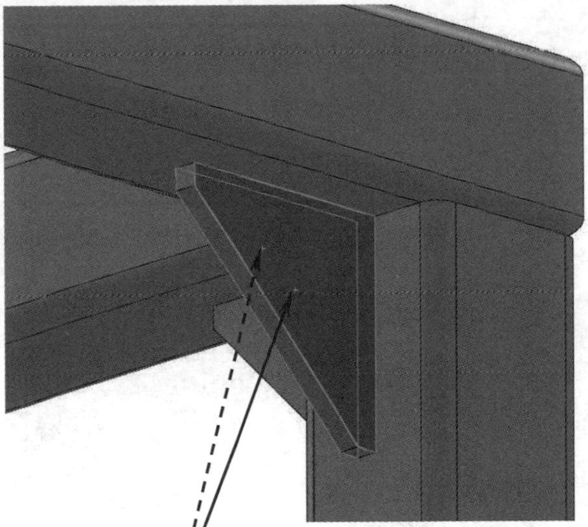

Click **Fillet Bead**  on the Weldments toolbar.

From the Weld Bead properties tree, enter the following:

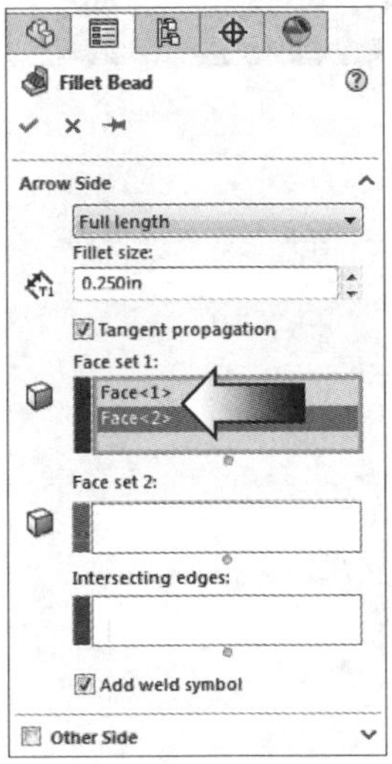

* Bead Type: **Full Length**

* Fillet Size: **.250 in.**

* Tangent Prop: **Enabled**

* Face Set1: **Select the 2 faces** as noted.

Select **2 Faces** (front & back) for Face Set1

For **Face Set2**, select the next **2 faces** as indicated.

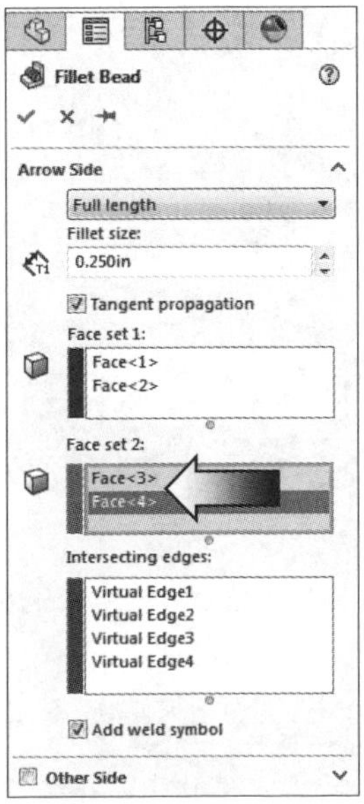

***Intersecting Edges**: Highlights edges where **Face Set1** and **Face-Set2** intersects.*

*(You can also right-click an edge and select **Delete** to remove from the weld bead.)*

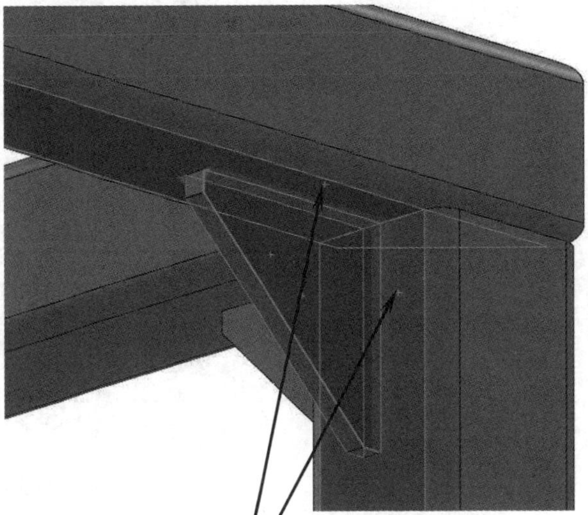

Select **2 Faces**

Enable the **Other Side** check-box and apply the <u>same settings</u> to the back end of the gusset.

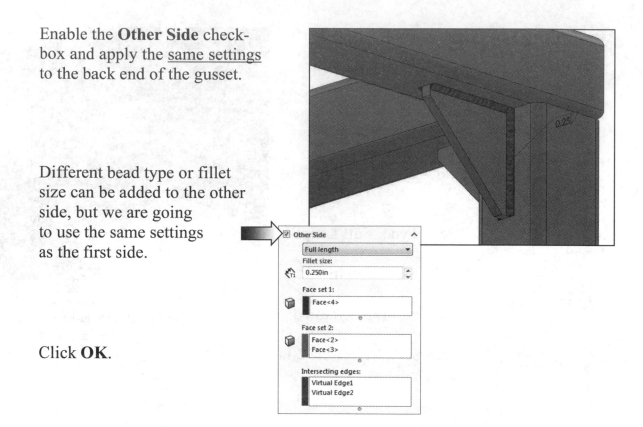

Different bead type or fillet size can be added to the other side, but we are going to use the same settings as the first side.

Click **OK**.

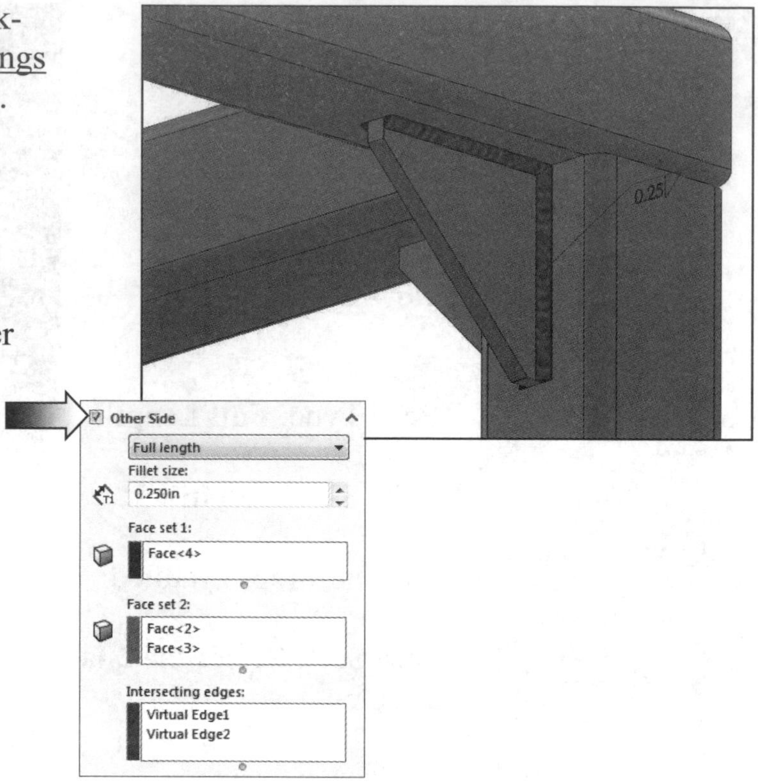

A bead call out is added automatically (see example below).

<u>*Example*</u>: *0.25 = Length of the leg of the fillet bead.*
0.375 = Length of each bead segment.
0.7 = Distance between the start of each bead.

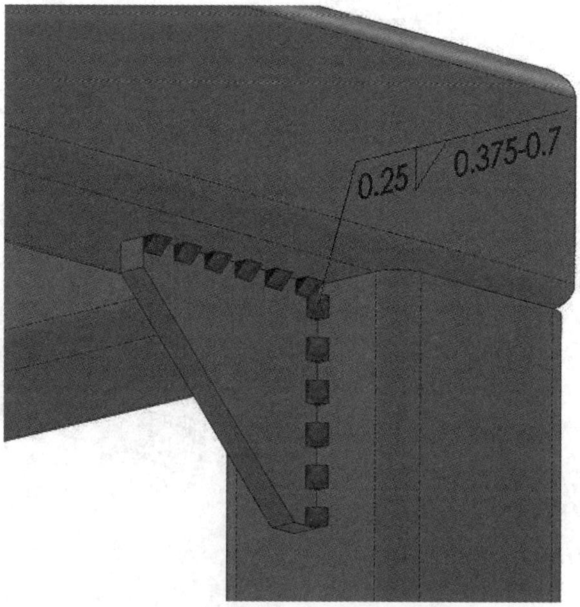

Intermittent or Staggered

19. Adding more Fillet Beads:

Repeat step 17 and add a set of fillet beads to each gusset that was created earlier.

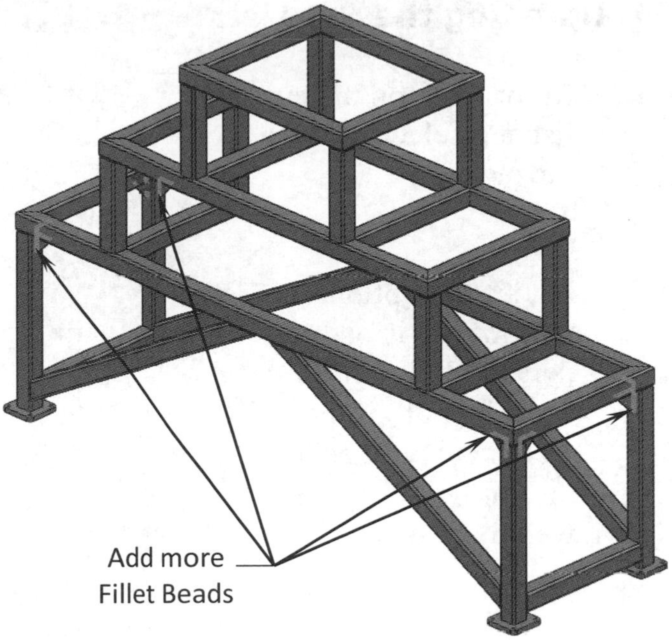

When adding the fillet beads, try using the different types of beads: Full Length, Intermittent, and Staggered to see the different results and callouts.

20. Viewing the Weldment Cut List:

Locate the **Cut List** on the FeatureManager tree and click the Plus (+) sign to expand.

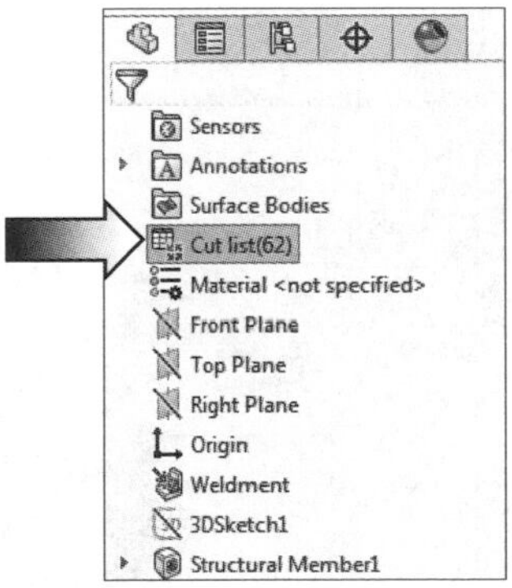

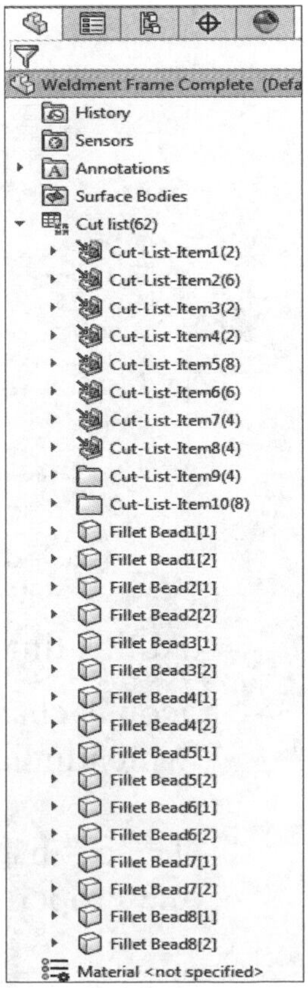

The Cut List needs to be updated every time something is added to the model.

The icon  in front of the cut list indicates that it needs updating and the icon  indicates the list is up to date.

The current list displays all items in the order that they were created. We will update the list in the next step.

21. Updating the Cut List:

Right-click on the cut list and select **Update** (arrow).

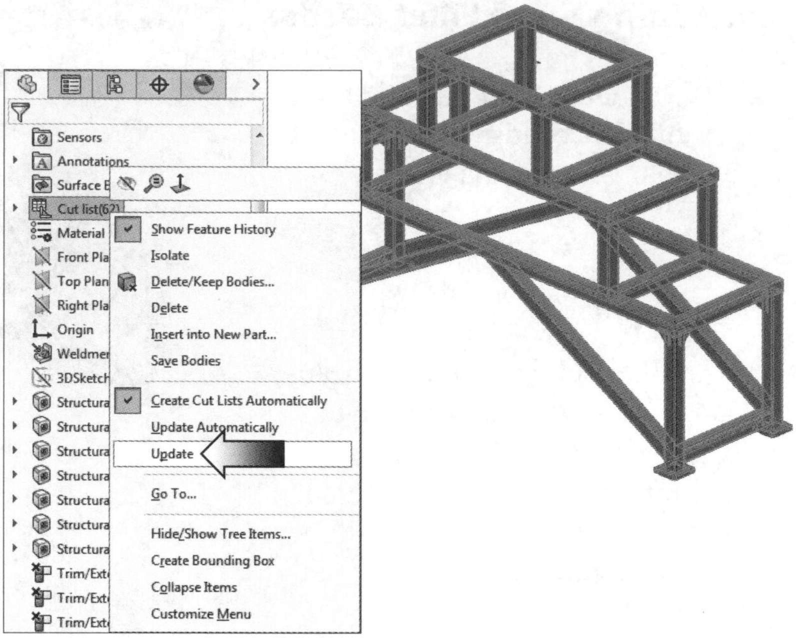

Notice the option **Automatic** is on by default. This option organizes all of the weldment entities in the cut list for the new weldment parts.

Although the cut list is generated automatically, you can manually specify when to update the cut list in a weldment part document.

This enables you to make many changes, and then update the cut list once. However, the cut list updates automatically when you open a drawing that references the list.

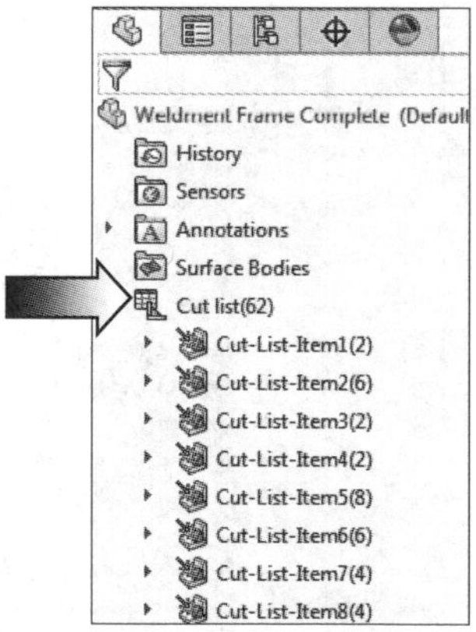

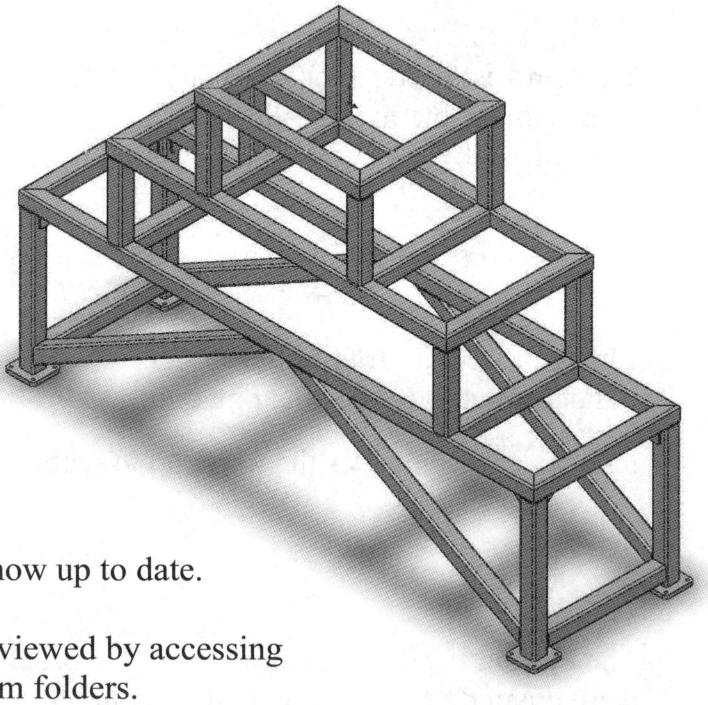

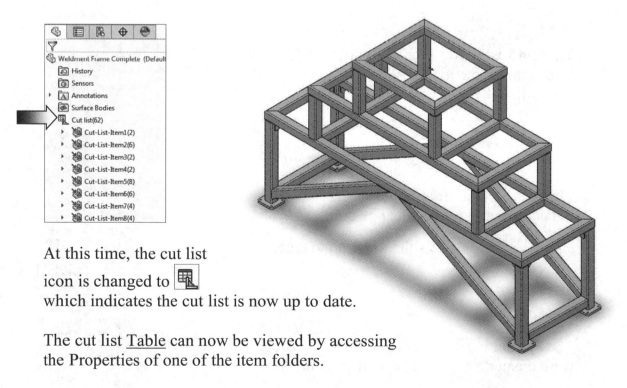

At this time, the cut list icon is changed to  which indicates the cut list is now up to date.

The cut list Table can now be viewed by accessing the Properties of one of the item folders.

22. Creating a drawing (OPTIONAL):

A drawing that includes the cut list can be generated. (Refer to the Part-1 Basic-Tools textbook for more information on how to create a detail drawing using SOLIDWORKS.)

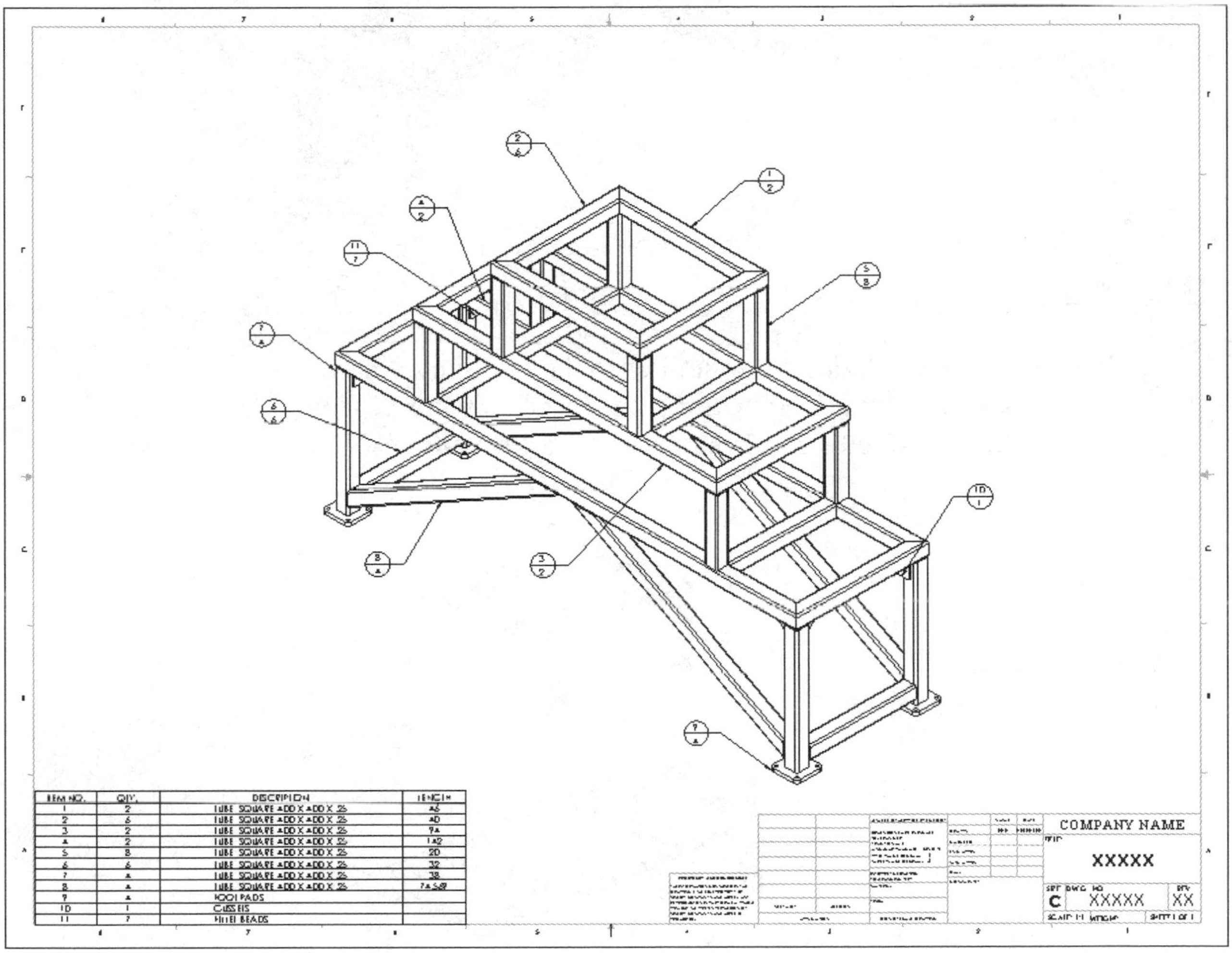

23. Saving your work:

Click **File / Save As**.

Enter **Weldments Frame** for the name of the file.

Click **Save**.

Replace the existing file when prompted.

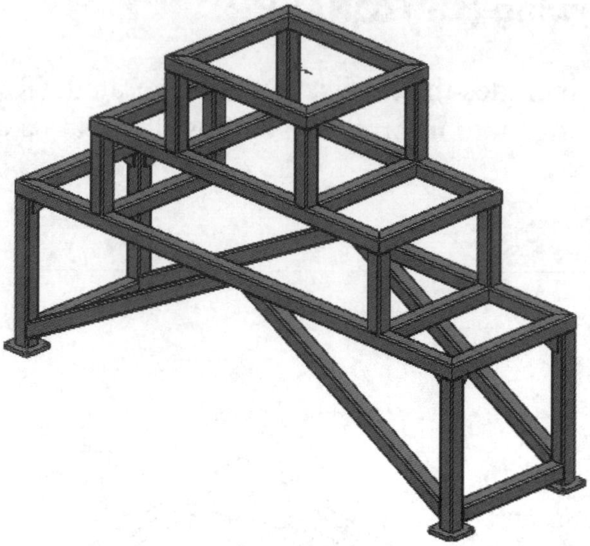

Optional:

To create the exploded view similar to the one shown here:
Either use the Move/Copy command (Insert / Features/Move-Copy)
or use the Exploded View command (Insert /Exploded View).

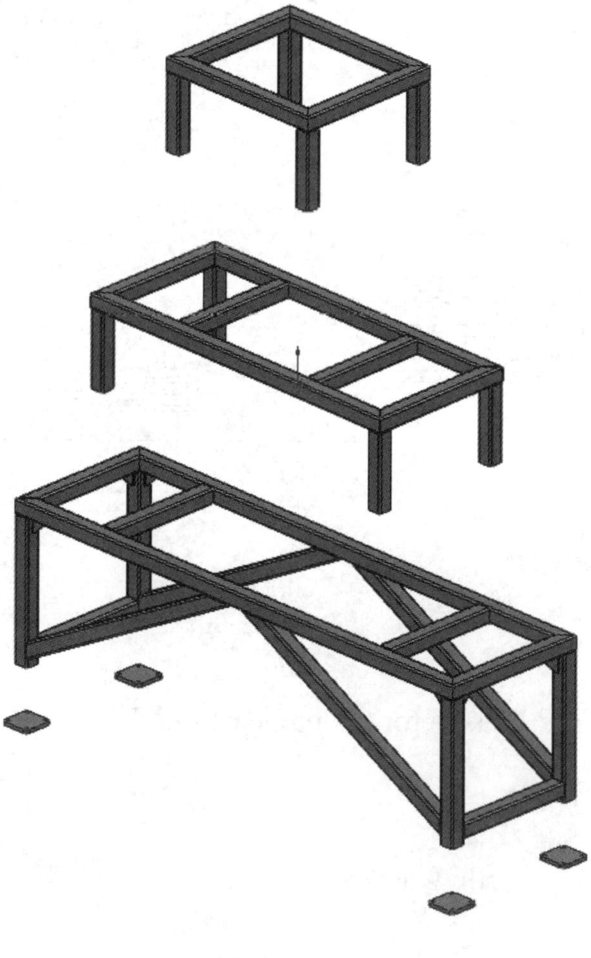

CHAPTER 18

Creating a Core and Cavity

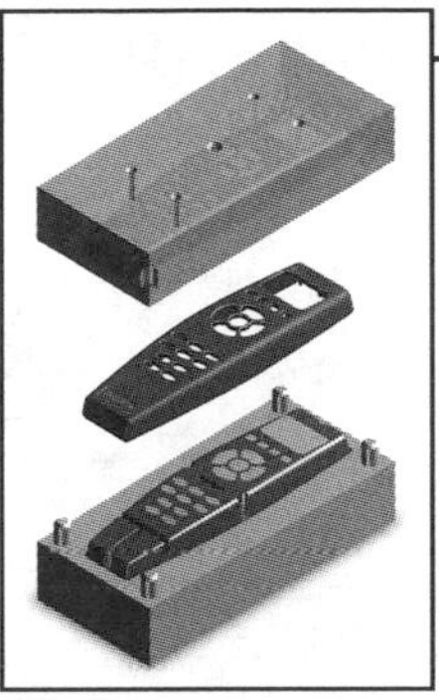

Creating a Core and Cavity
Linear parting Lines

A mold is usually designed in SOLIDWORKS using a sequence of intergraded tools that control the mold creation process. Using the finished model, these mold tools can be used to analyze and correct deficiencies in the part.

The process usually follows these steps: Draft analysis, Undercut Detection, Parting Lines, Shut-Off Surfaces, Parting Surfaces, Interlock Surfaces (Ruled Surfaces), and Tooling Split.

The Parting Lines lie along the edge of the molded part, between the core and the cavity surfaces. They are used to create the Parting Surfaces and to separate the surfaces.

The Shut-Off Surfaces are created after the Parting Lines. A shut-off surface closes up a through hole by creating a surface patch along the Edges that form a continuous loop, or a parting line you previously created, to define a loop.

After the Parting Lines and the Shut-Off Surfaces are determined, the Parting Surfaces are created. The Parting Surfaces extrude from the parting lines and are used to separate the mold cavity from the core.

After a parting surface is defined, the Tooling Split tool is used to create the core and cavity blocks from the model. To create a tooling split, at least three surface bodies are needed in the Surface Bodies folder, a Core, a Cavity and a Parting surface.

With most mold parts, the interlock surfaces need to be created. The interlock surfaces help prevent the core and cavity blocks from shifting, and are located along the perimeter of the parting surfaces. Usually they have a 5-degree taper.

Creating a Core and Cavity
Linear Parting Lines

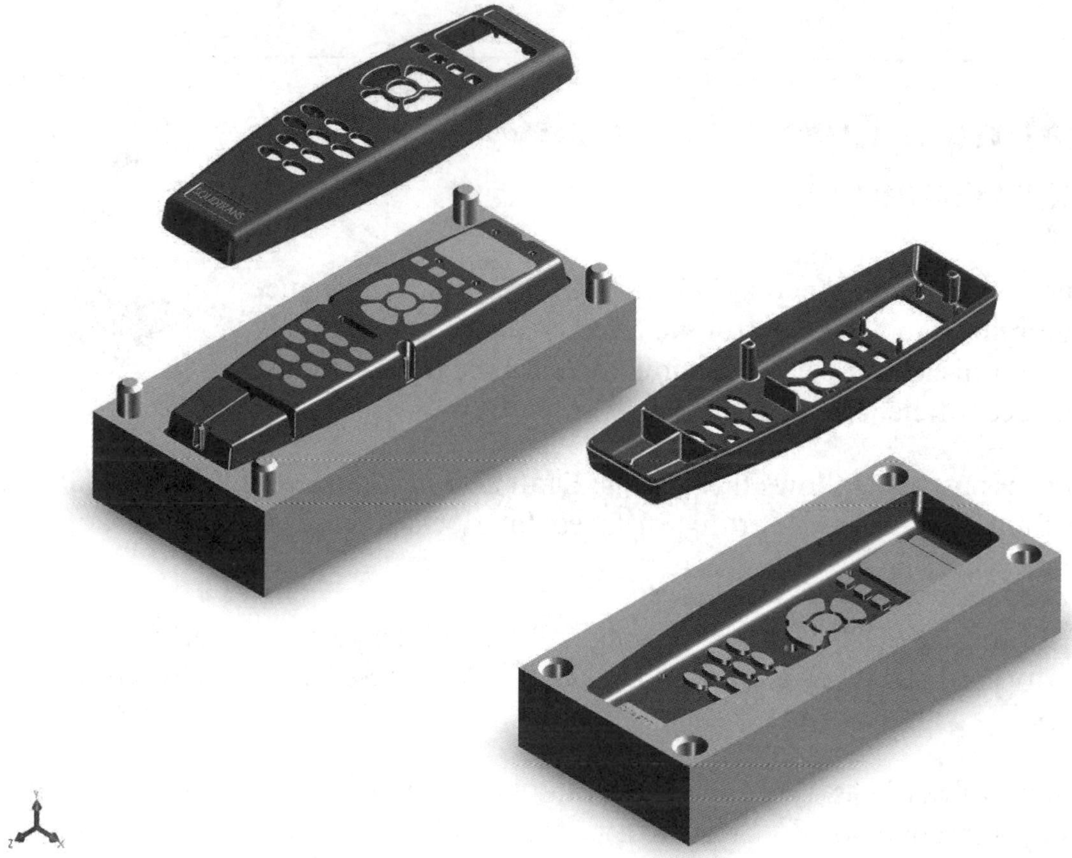

Dimensioning Standards: **ANSI**
Units: **INCHES** – 3 Decimals

Tools Needed:

Parting Lines	Parting Surfaces	Shut-Off Surfaces
Tooling Split	Planes	2D Sketch

1. Opening an existing Parasolid document:

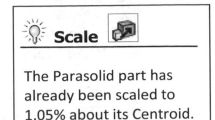

> **Scale**
>
> The Parasolid part has already been scaled to 1.05% about its Centroid.

Select **File / Open.**

Browse to the Training Files folder, change the Files of Type to **Parasolid**.

Select **Remote Control.x_b** and click **Open.**

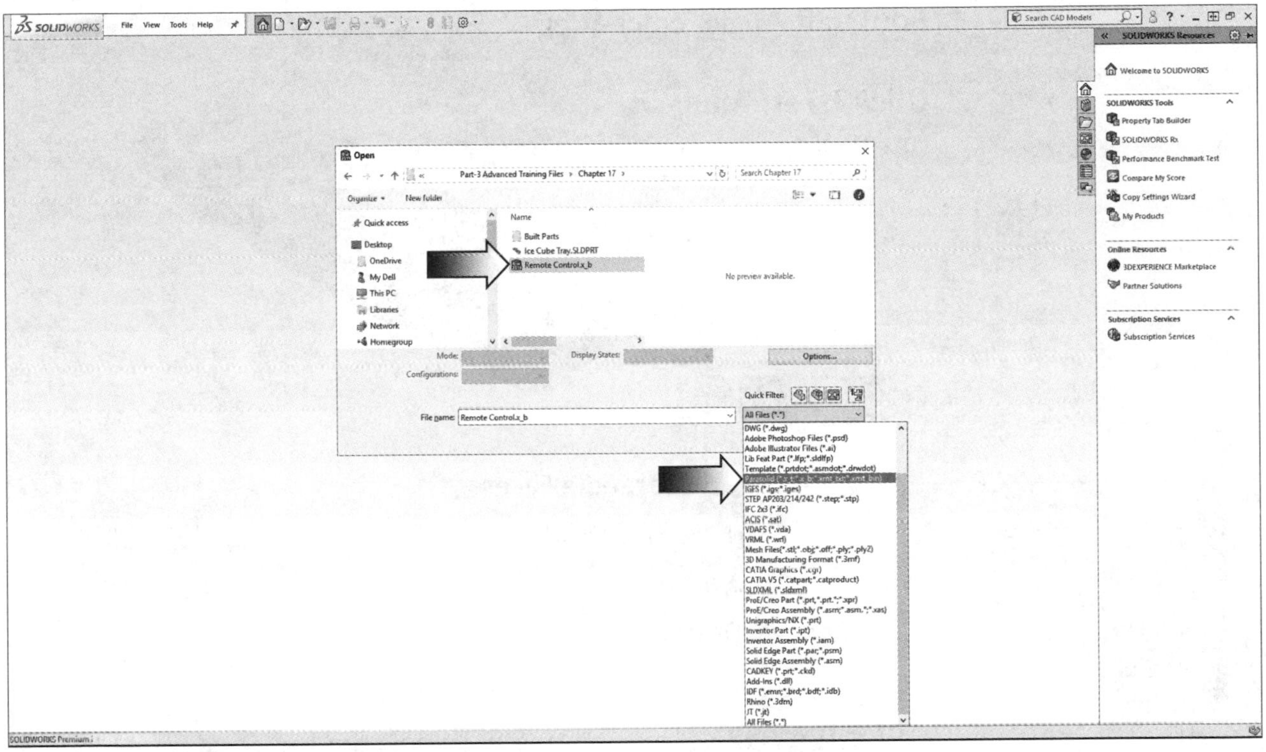

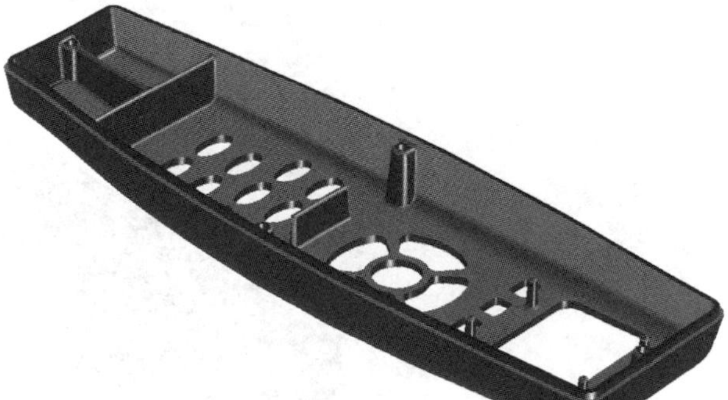

Click **NO** on the Import Diagnostics dialog and close it.

2. Creating the Parting Lines:

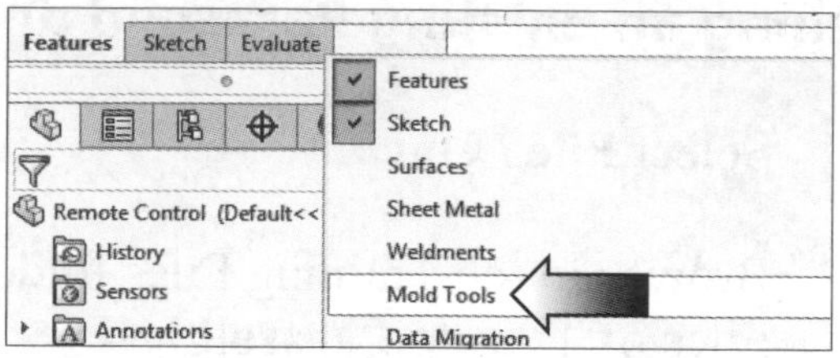

Right-click the **Evaluate** tab and enable the **Mold Tools** option.

Click **Parting Line** on the Mold Tools toolbar.

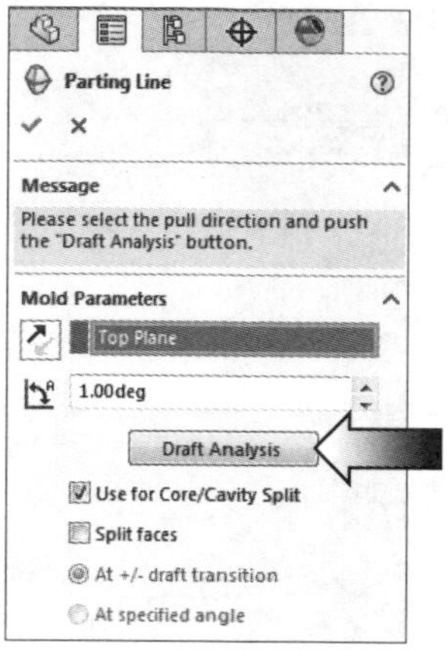

For Direction of Pull, select the **Top** plane from the Feature tree.

For Draft Angle, enter **1deg**.

Click **Draft Analysis** (arrow).

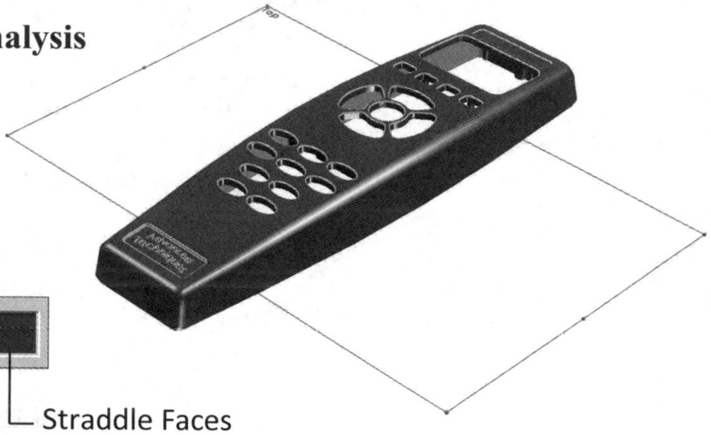

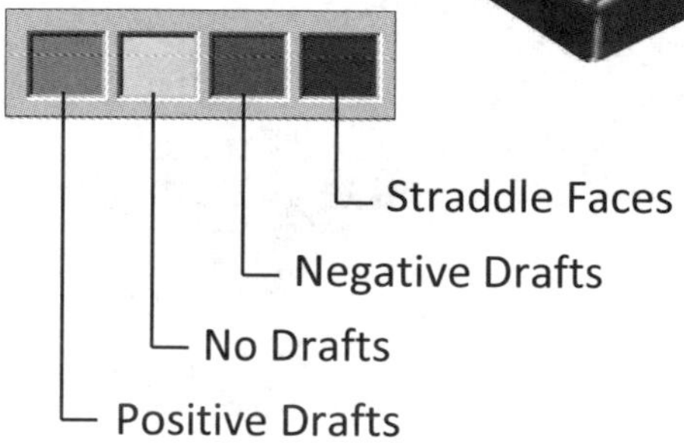

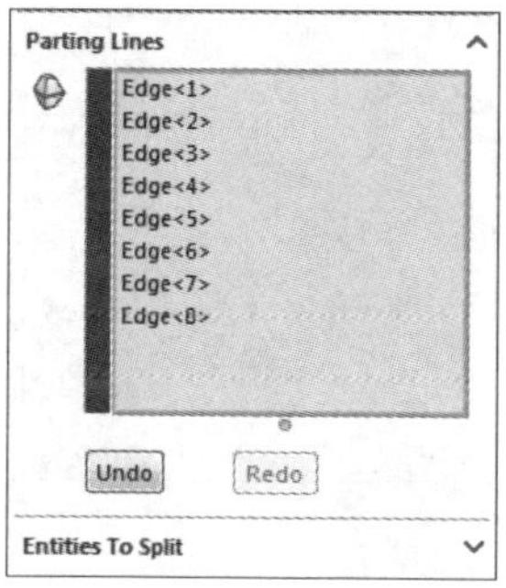

SOLIDWORKS automatically selects the lower edges of the part where the red surfaces meet the green, and places them in the Parting Lines section.

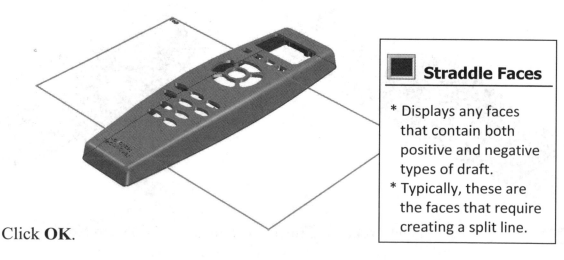

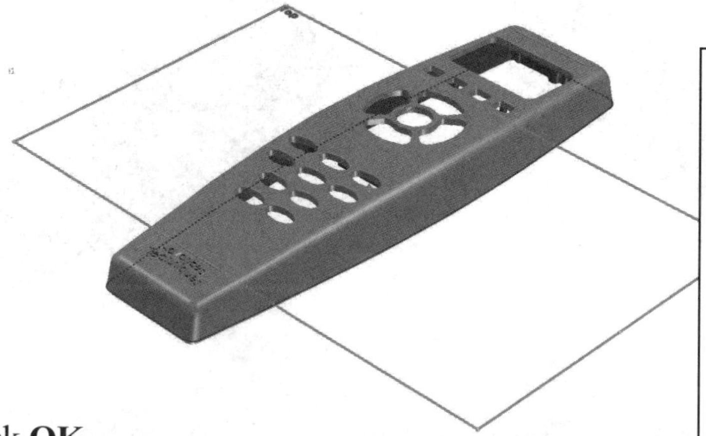

Straddle Faces

* Displays any faces that contain both positive and negative types of draft.
* Typically, these are the faces that require creating a split line.

Click **OK**.

3. Creating the Shut-Off Surfaces:

Click [icon] or select **Insert / Molds / Shut-Off Surfaces**.

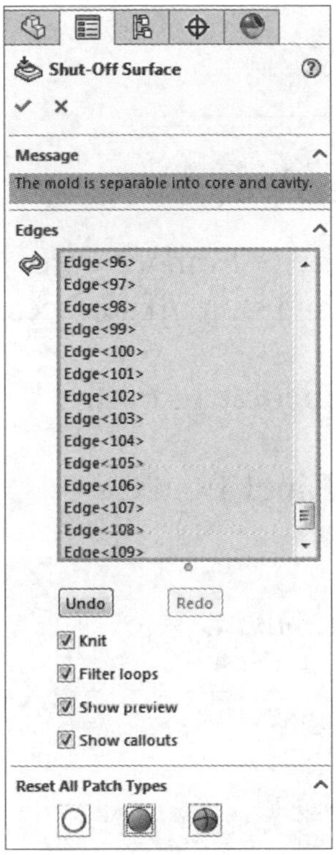

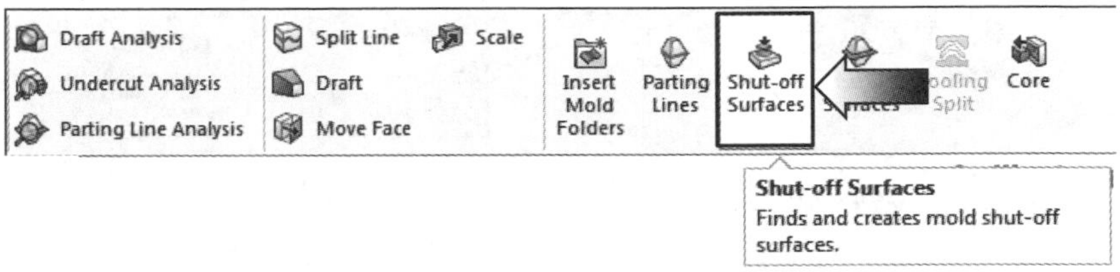

SOLIDWORKS automatically selects the edges of all through openings and labels them as Loop/Contacts.

Keep the Patch Type at **Contact**.

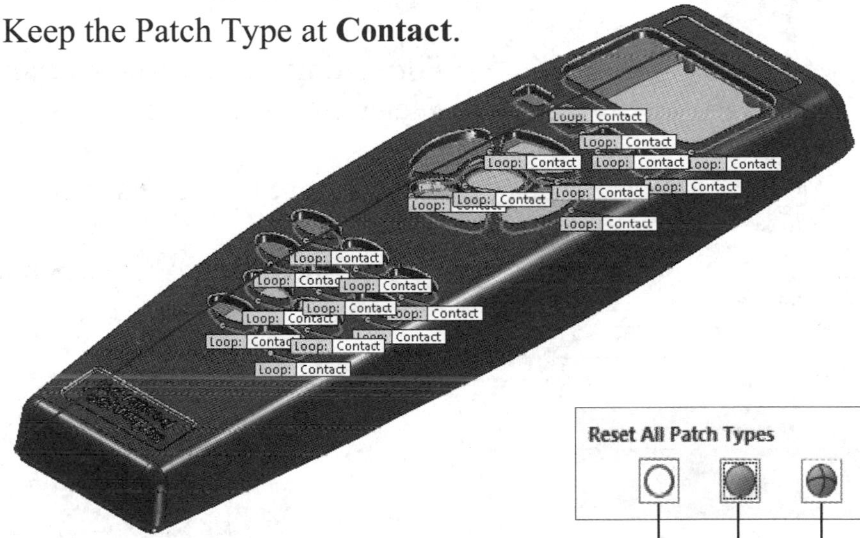

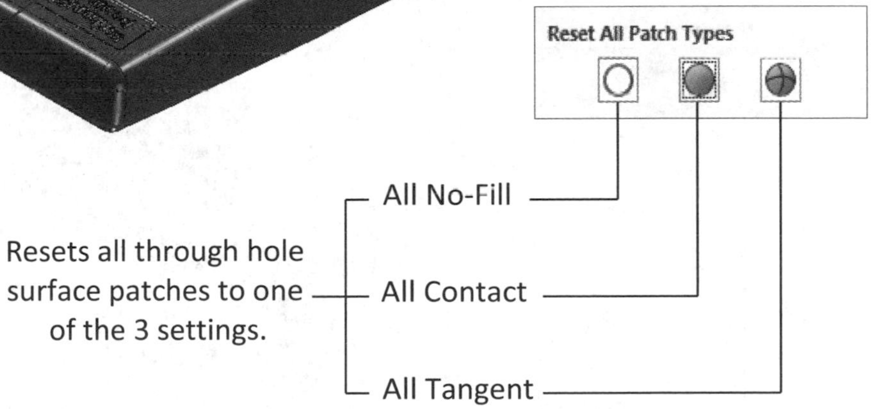

Patch Types

Only one Shut-Off Surface feature is allowed in a model. Therefore, within the one feature, you must assign a fill type of **Contact**, **Tangent**, or **No Fill** to every hole.

Click **OK**.

4. Creating the Parting Surfaces:

Click or select **Insert / Molds / Parting Surfaces**.

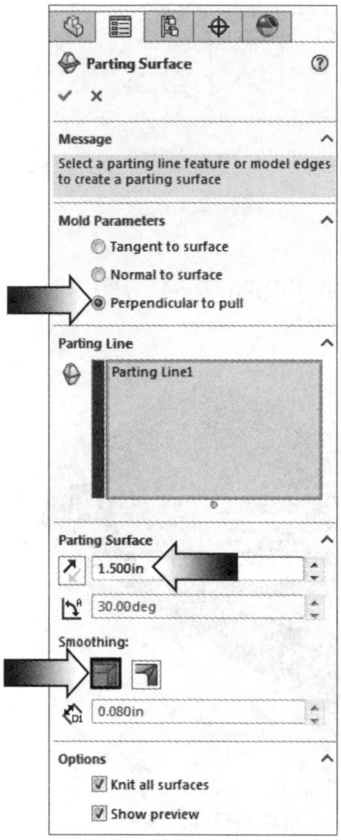

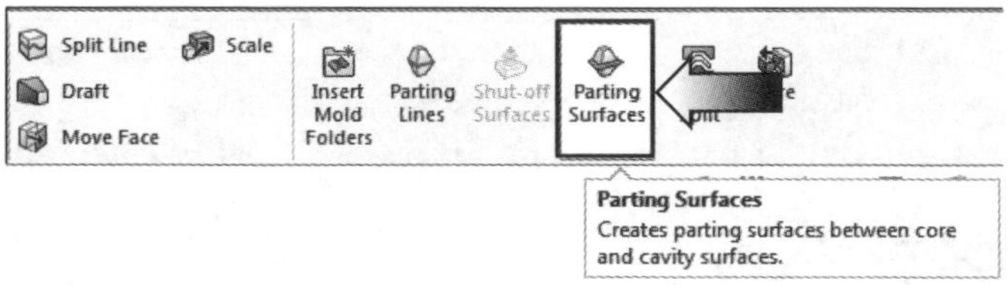

The Parting surfaces split the mold cavity from the core. It gets created after the parting lines and shut off surfaces.

For Mold Parameters, select **Perpendicular to Pull**.

For Parting Line, select the **Parting Line1** from the FeatureManager Tree.

For Parting Surface Distance, enter **1.500 in**.

Select **Sharp Edges** under Smoothing section.

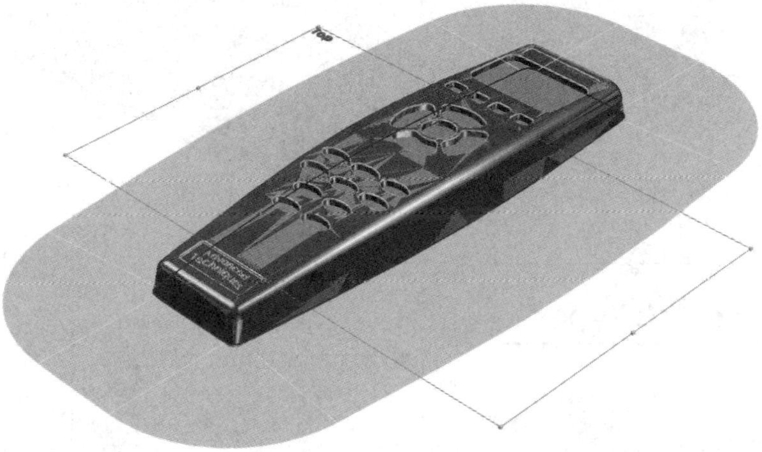

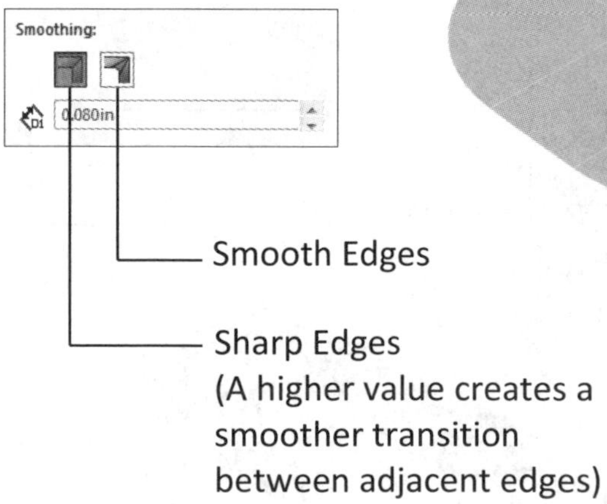

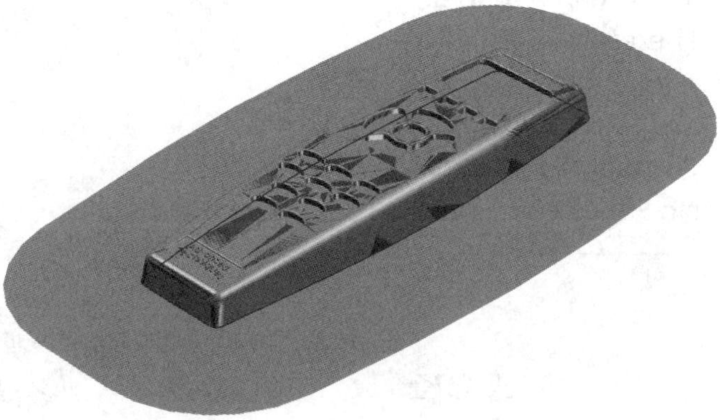

Click **OK**.

5. Sketching the profile of the mold-blocks:

Select the <u>Top</u> plane and open a **new sketch**  .

Sketch a **Center Rectangle** 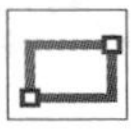 centered on the Origin.

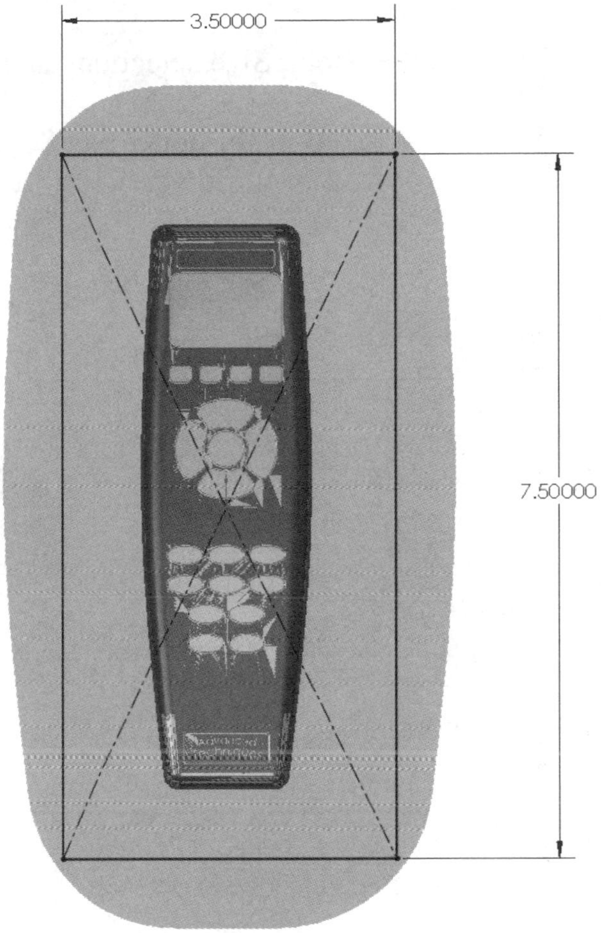

Add the width and the height dimensions.

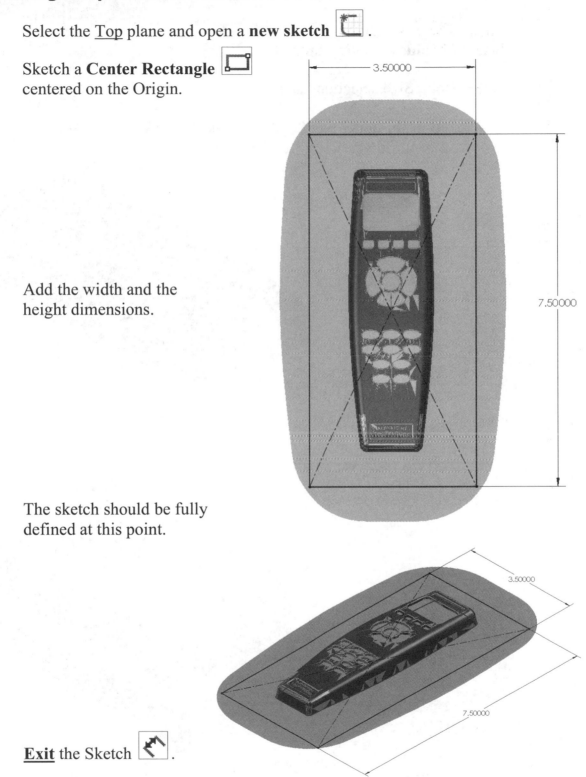

The sketch should be fully defined at this point.

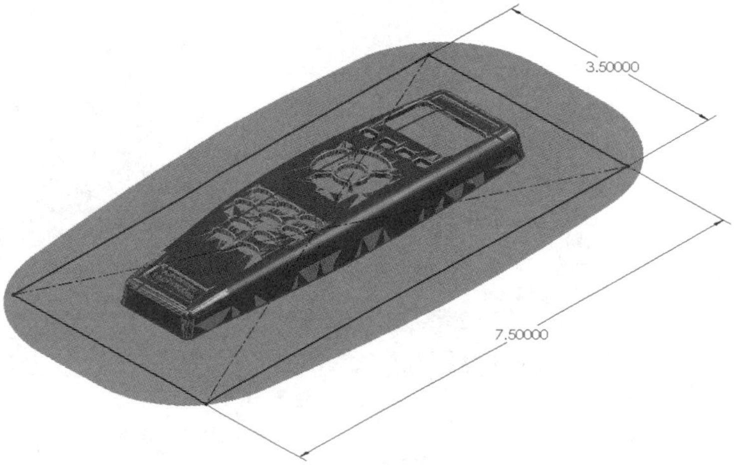

<u>Exit</u> the Sketch  .

(The next step is to create the upper and lower blocks using the Tooling Split command. This is only available when the Sketch is off.)

6. Creating the Tooling Split:

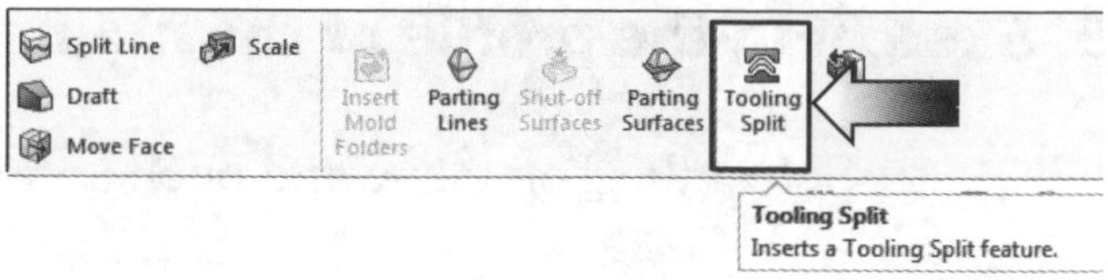

Click [icon] or select:
Insert / Molds / Tooling Split.

In the Block Size selection, enter:

1.500 in for upper block.
1.50 in for lower block.

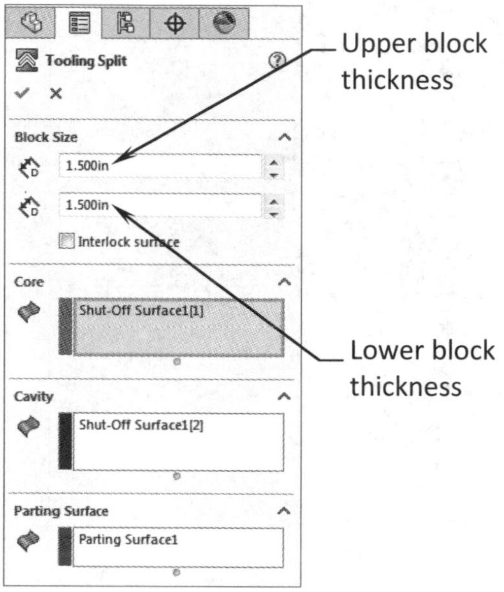

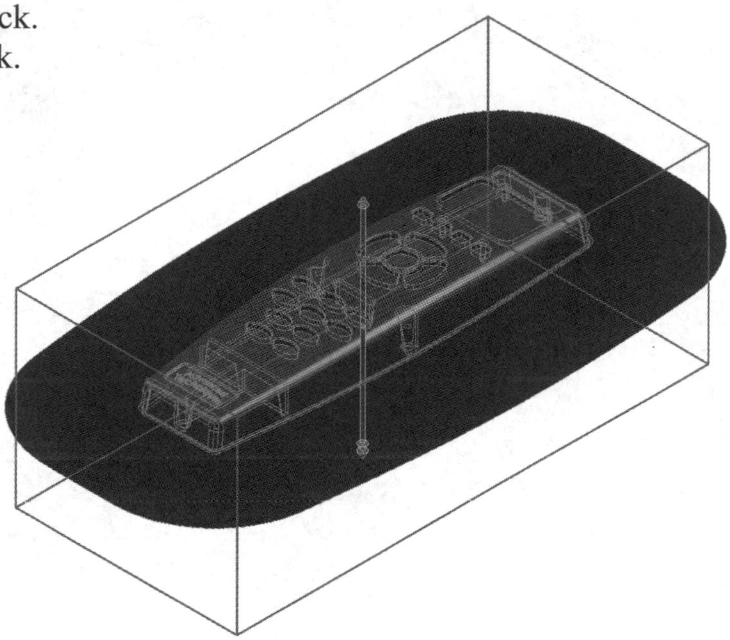

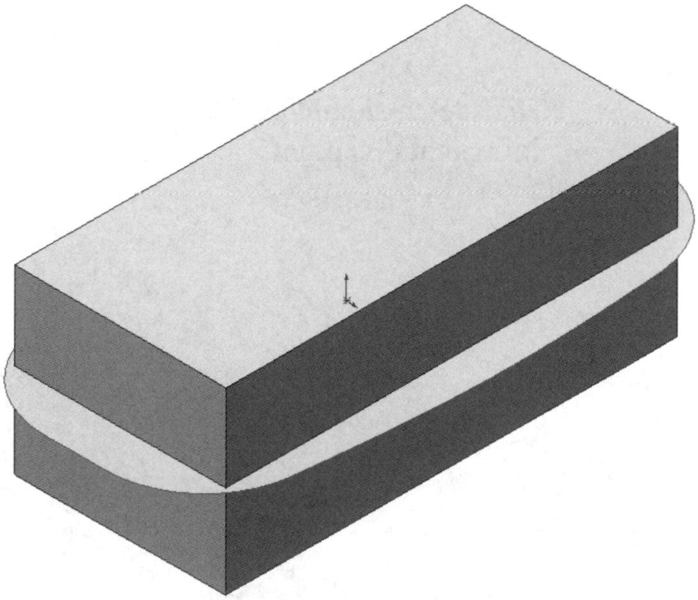

The Cavity and the parting surfaces options should already be filled.

Click **OK**.

* *The Interlock Surface surrounds the perimeter of the parting lines in a slight tapered direction. It helps seal the mold to prevent resins from leaking, prevents shifts, and maintains alignment between the tooling entities.*

7. Hiding the Solid Bodies:

From the FeatureManager Tree, expand the **Surface Bodies** folder. There are 3 groups of surfaces in this folder.

Click the **Surface Bodies** folder and select **Hide** .

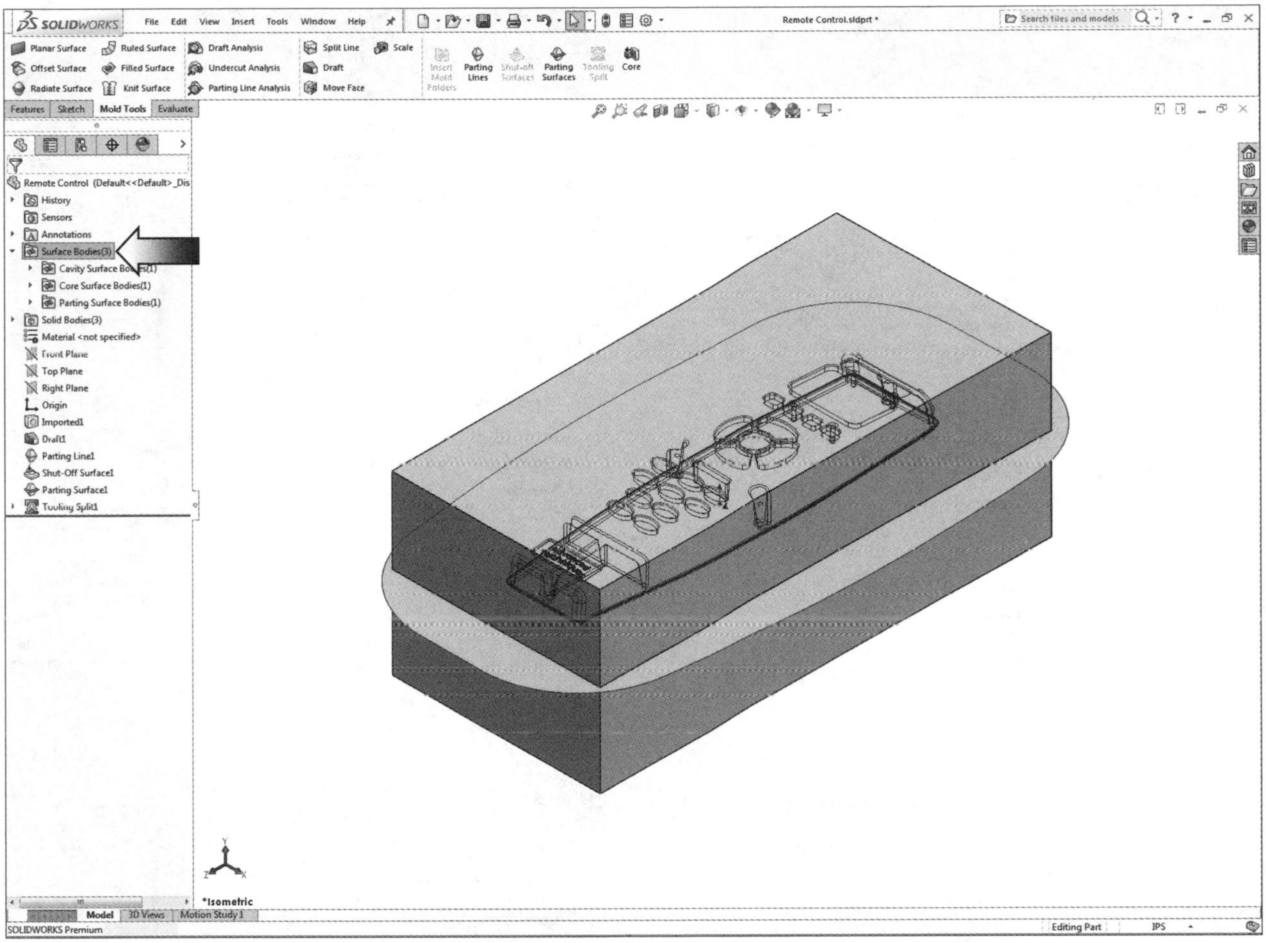

The 3 surfaces that were created in the previous steps are temporarily removed from the graphics display.

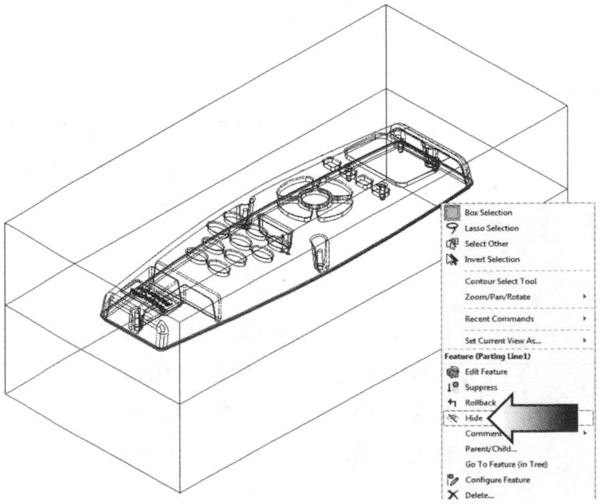

Change to the Wireframe mode to see the inside details of the blocks.

Right-click on the blue parting line and hide it (arrow).

8. Saving the bodies as part files:

Expand the **Solid Bodies** folder, right-click on **Tooling Split [1]** and select: **Insert into New Part**.

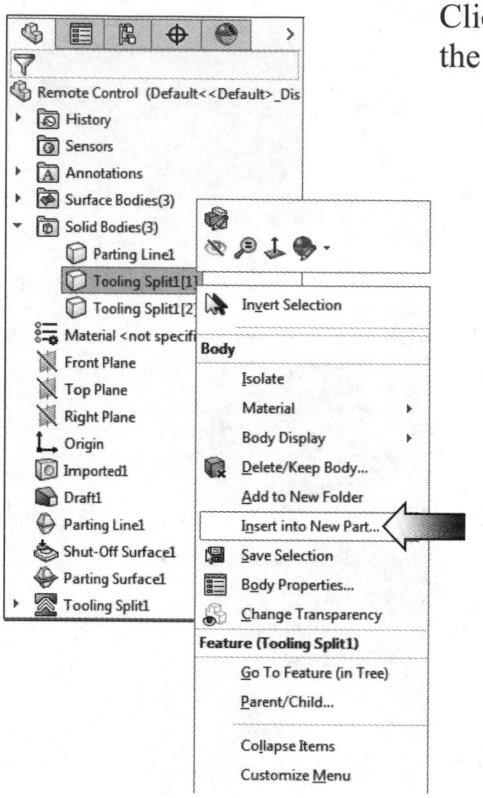

Click the **Browse**  button and select a location to save the document.

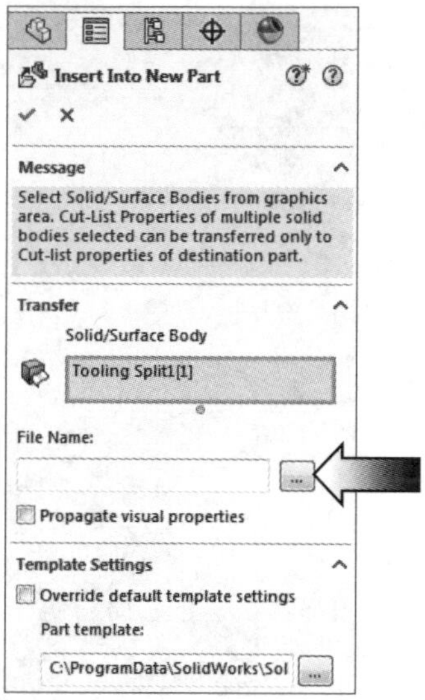

Enter: **Core_Remote_Control**, for the name of the file.

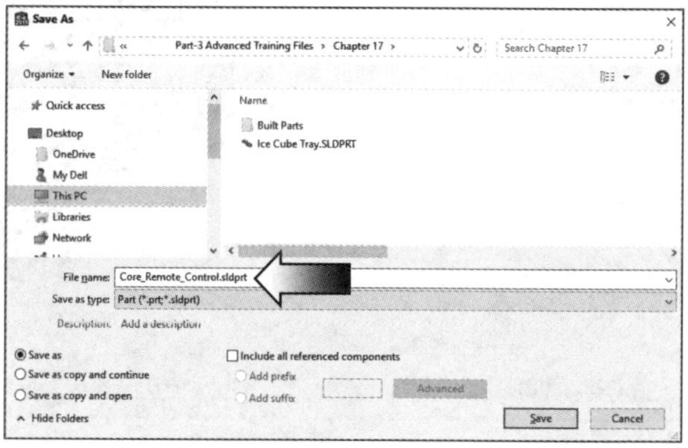

Click **Save**.

Hold the **Control** key and push the **Tab** key to switch back to the main part.

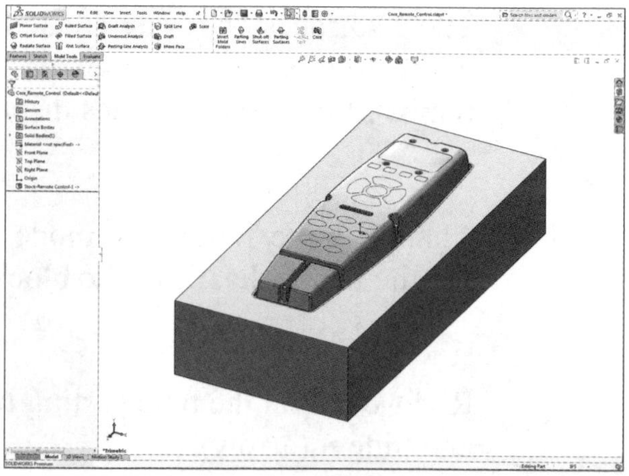

Repeat the last step to save the **Tooling Split[2]**, enter: **Cavity_Remote_Control** for the name of the 2nd block.

9. Separating the 2 blocks:

Select **Insert / Feature / Move-Copy**.

Select the **upper block** in the graphics area.

Click the **vertical arrow** to define the explode direction.

Under the Translate section, enter **4.00in** and press **Enter**.

Click **OK**.

The upper block moves 4 inches upward from its original position.

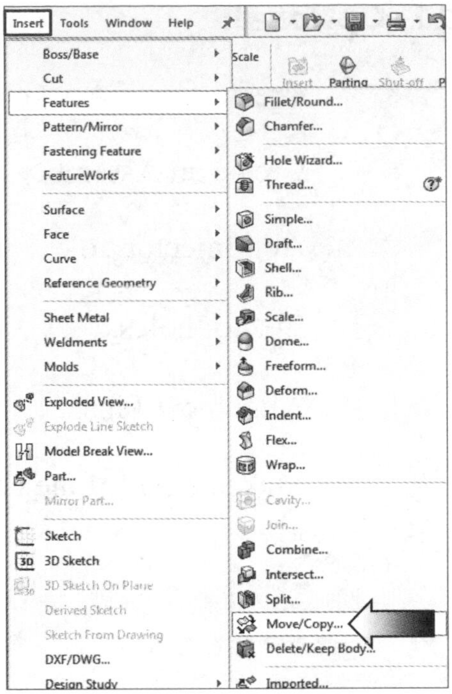

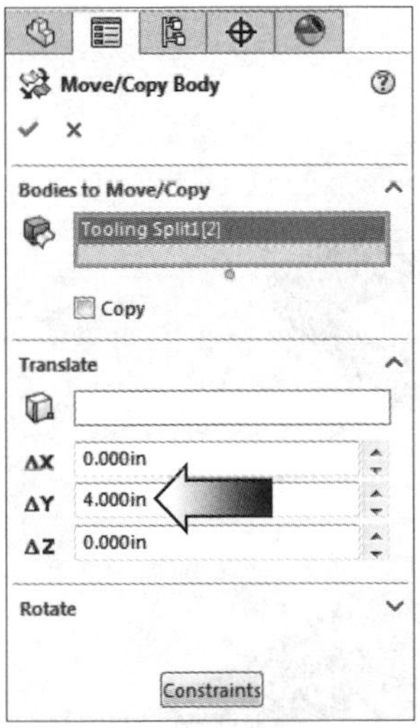

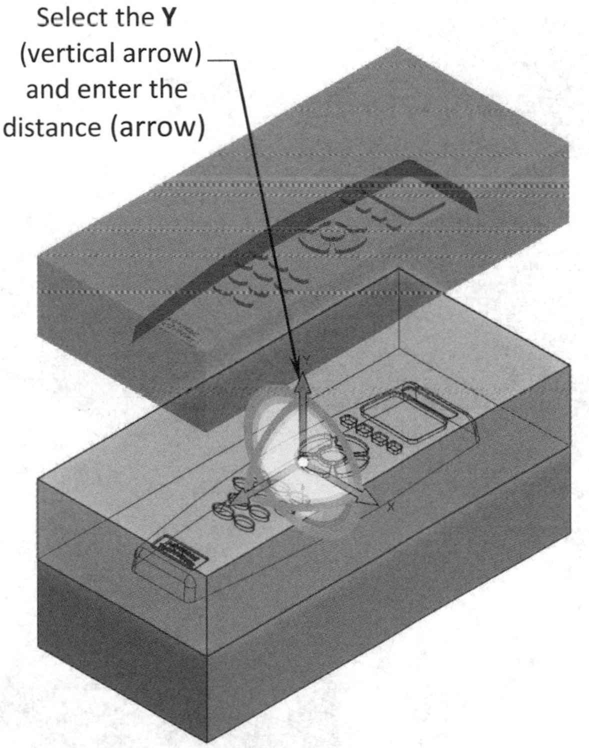

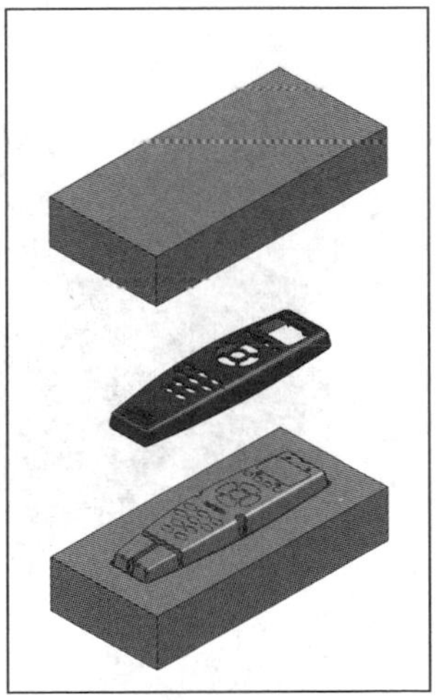

Repeat the same step to move the lower block downward (use -4.00" for distance).

10. Saving your work:

Save a copy of your work as **Remote Control Tooling**.

11. Optional:

Create a new Assembly document and assemble the 3 components.

Create an Assembly Exploded View as a separate configuration.

Add Injector hole.

Ejector holes.

Alignment Pins.

Make copies of the components and create an exploded view as shown.

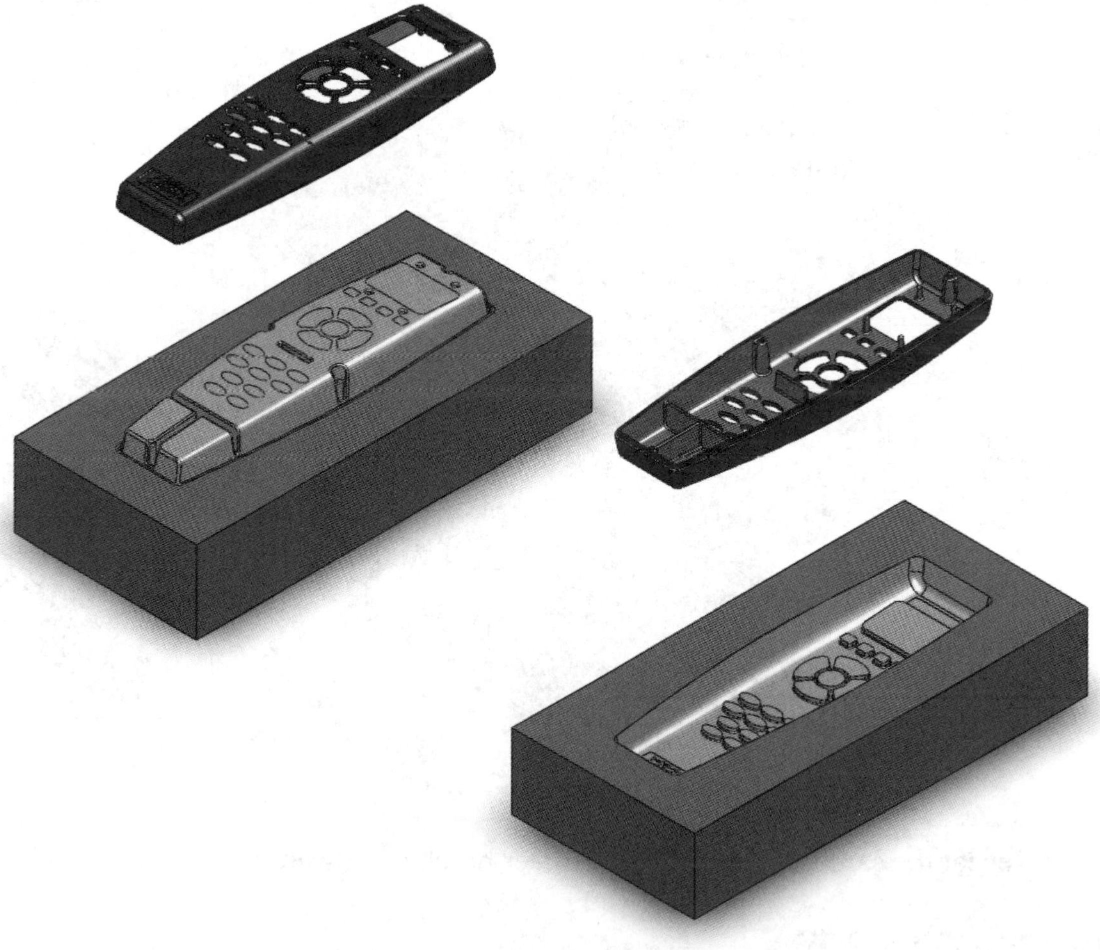

Exercise: Linear Parting Lines

1. Opening a part document:

Open a part document named: **Ice Cube Tray.sldprt** from the training files folder.

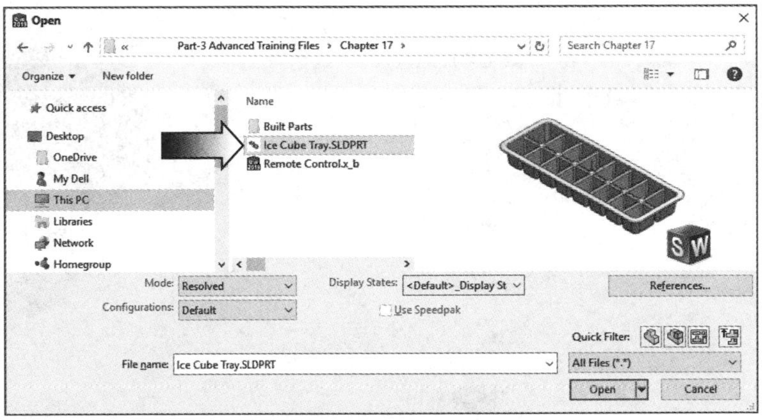

2. Enabling the Mold Tools:

Right-click one of the tool tabs and select the **Mold Tools** (arrow).

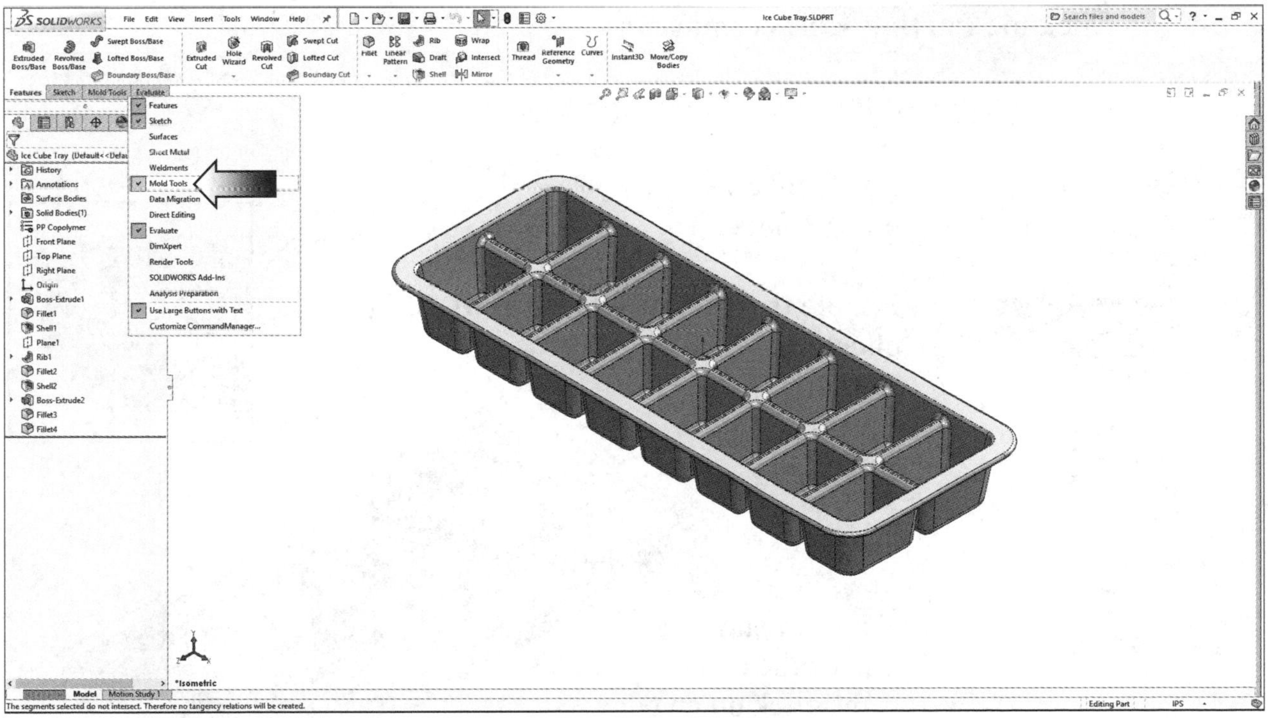

A material has already been assigned to the part (PP Copolymer) but a new material will be assigned to the mold blocks after they are created.

3. Applying Scale:

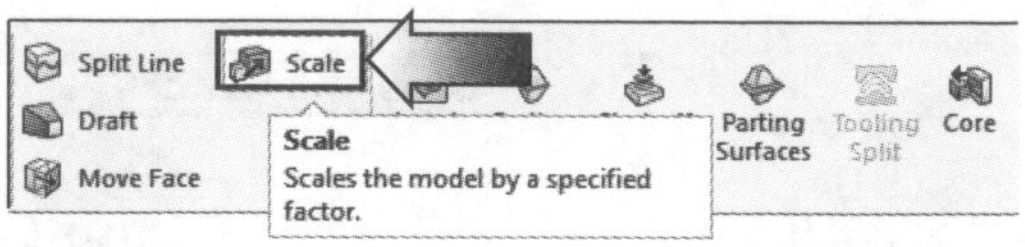

Switch to the **Mold Tools** tab.

Click the **Scale** command and enter **1.02** (2% larger).

Use the default **Centroid** option.

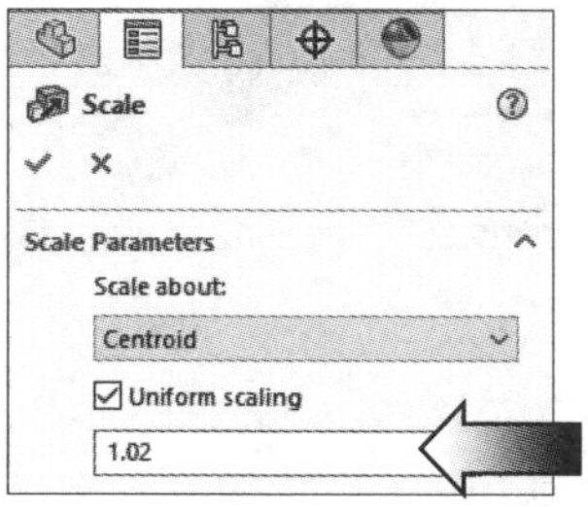

Click **OK**.

4. Creating the parting line:

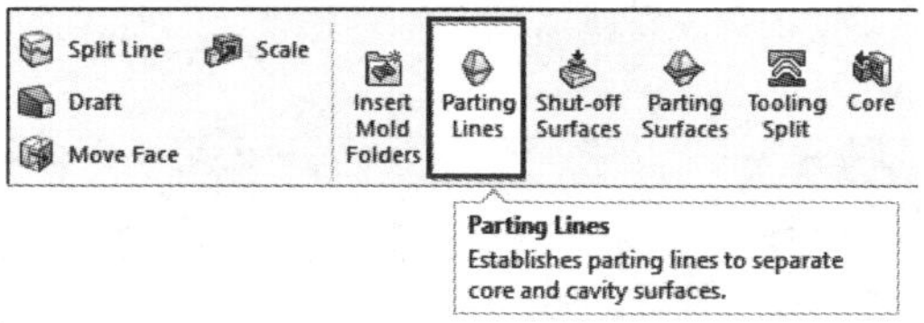

Click the **Parting Lines** command.

For Direction of Pull, select the **Top** plane.

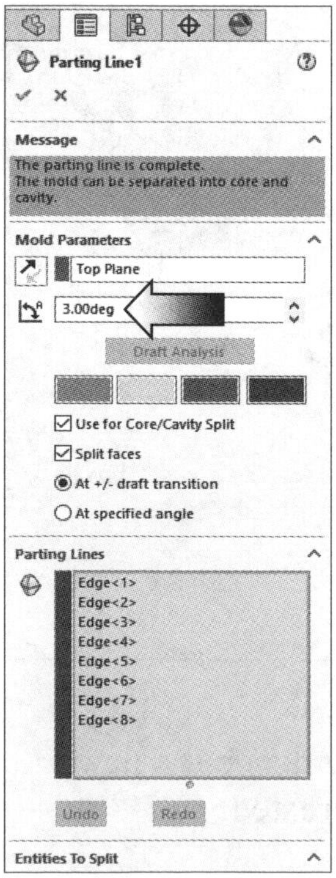

For Draft Angle, enter **3.00deg**.

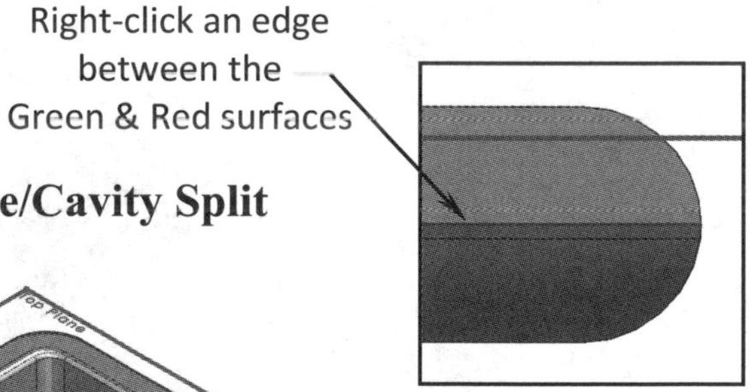

Enable the **Use for Core/Cavity Split** checkbox.

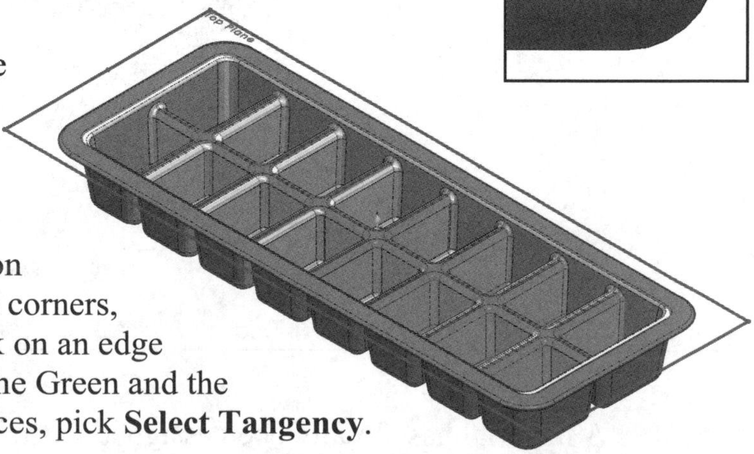

Enable the **Split-Faces** box.

Zoom in on one of the corners, right-click on an edge <u>between</u> the Green and the Red surfaces, pick **Select Tangency**.

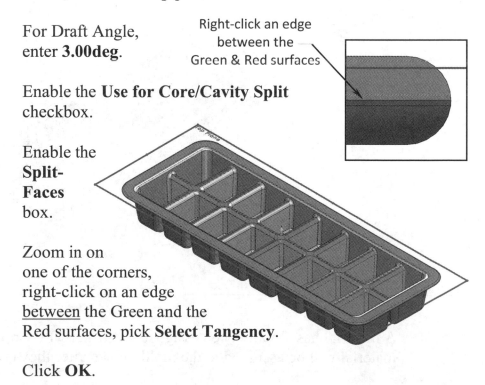

Click **OK**.

5. Creating the parting surface:

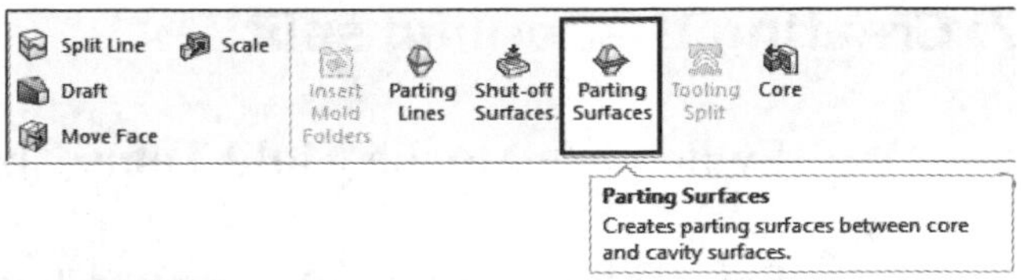

Click the **Parting Surfaces** command.

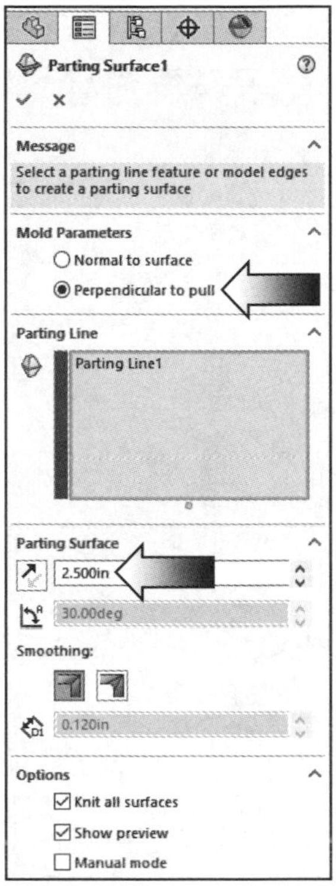

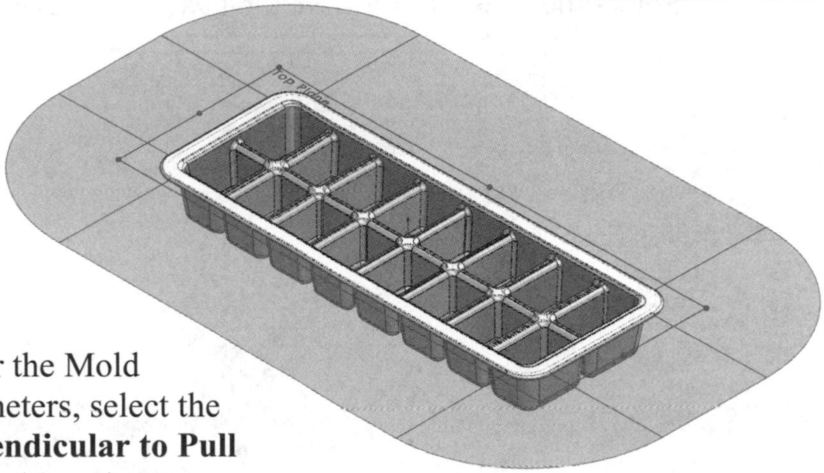

Under the Mold Parameters, select the **Perpendicular to Pull** option (arrow).

For Parting Surface, enter **2.500in** for Distance (arrow).

Enable the **Knit All Surfaces** checkbox.

Click **OK**.

6. Sketching the mold block:

Open a **new sketch** on the Parting Surface.

Sketch a **Center-Rectangle** that is centered on the Origin.

Add the dimensions shown to fully define the sketch.

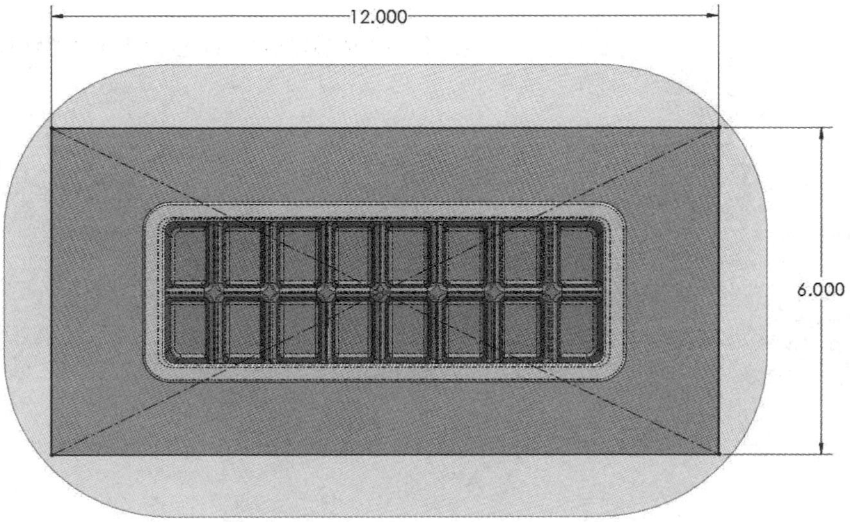

Exit the sketch.

7. Creating the tooling split:

Switch back to the **Mold Tools** tab.

Select the **new sketch** from the FeatureManager tree and click **Tooling Split**.

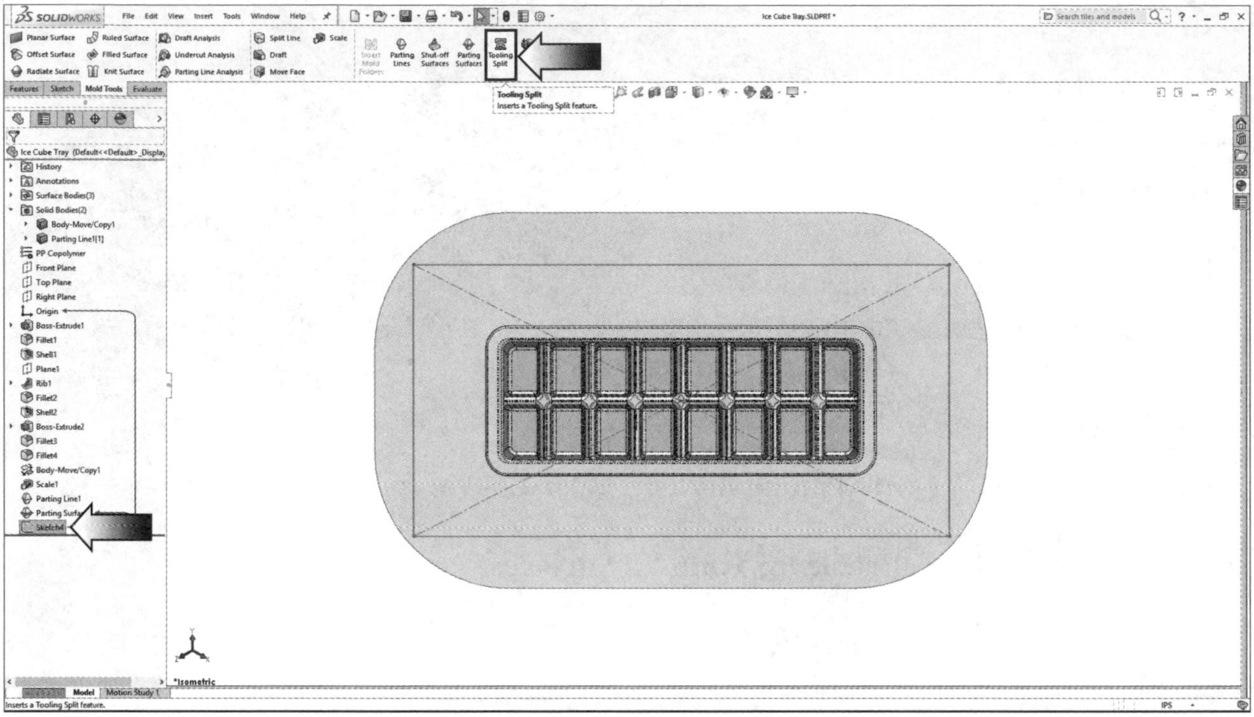

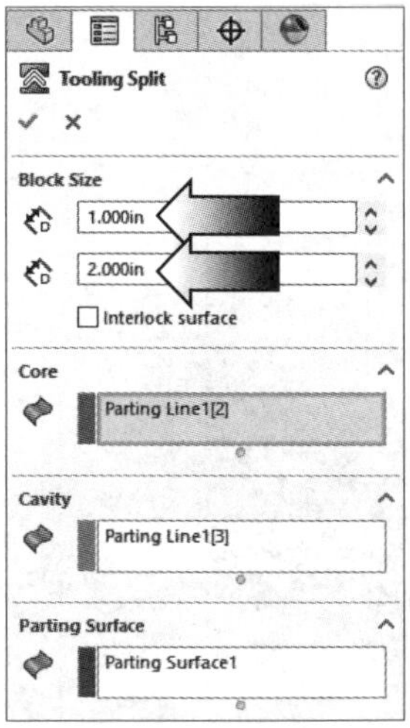

For Upper Block Size, enter **1.000in**.

For Lower Block Size, enter **2.000in**.

The reference surfaces are automatically placed in their correct locations.

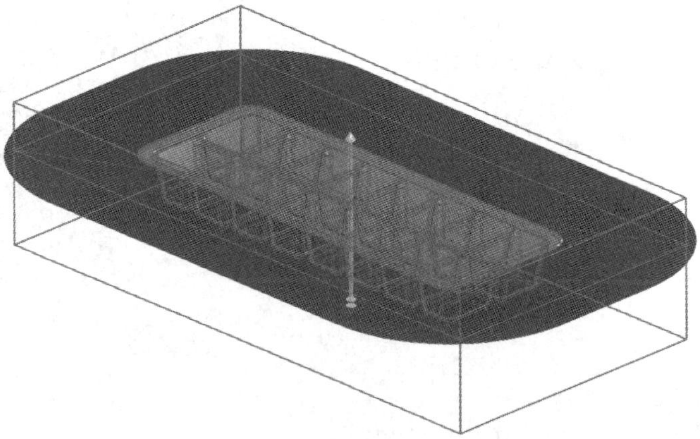

Click **OK**.

8. Hiding the reference surfaces:

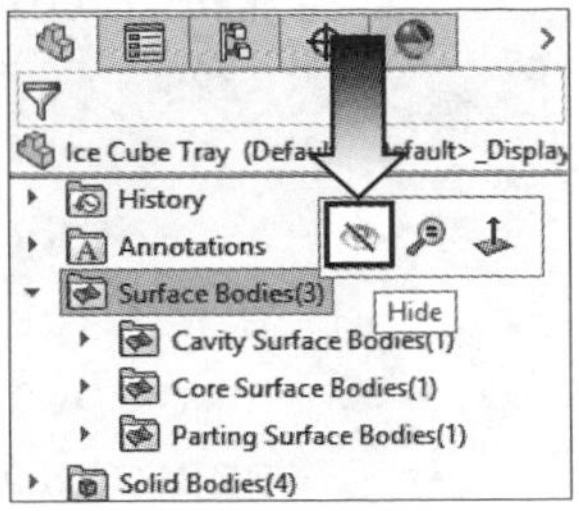

Locate the **Surface Bodies** folder near the top of the FeatureManager tree.

Click the Surface Bodies folder and select **Hide** (arrow).

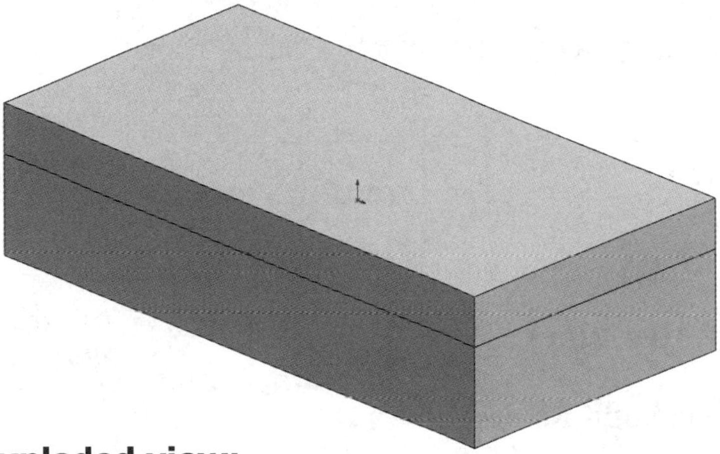

9. Creating an exploded view:

Select **Insert / Exploded View** (arrow).

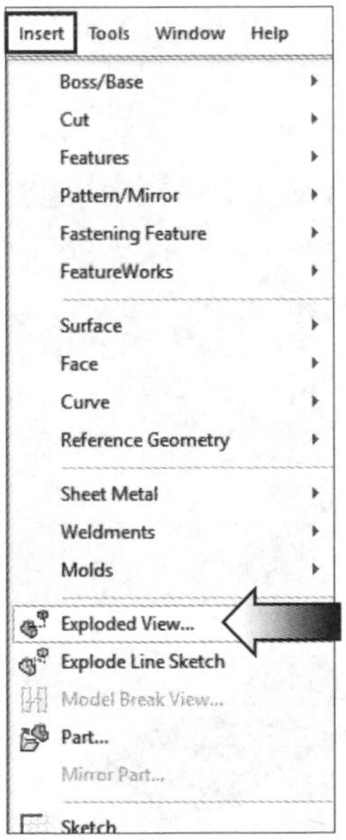

Select the **upper mold block** and drag the **Y** arrowhead upwards approximately **6 inches**.

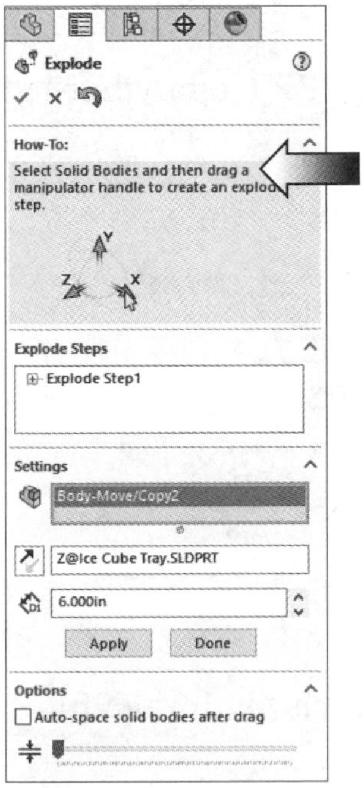

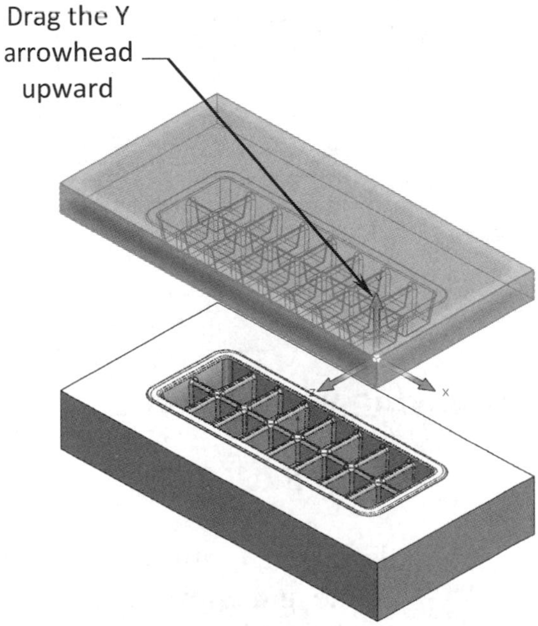

The Explode Step1 is created and stored under the Explode Steps section.

Select the **lower mold block** and drag the **Y** arrowhead downwards, approximately **6.5 inches**.

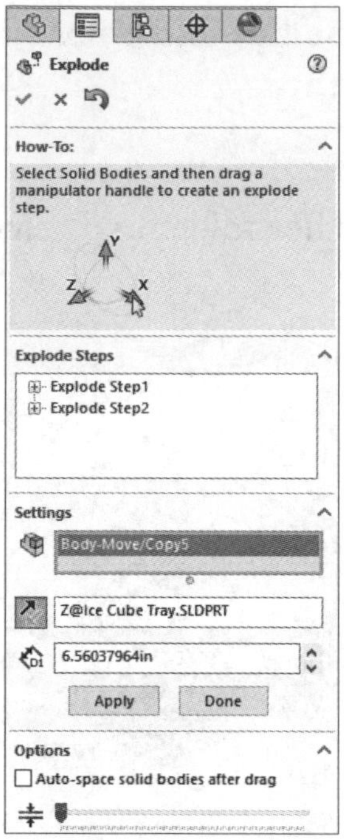

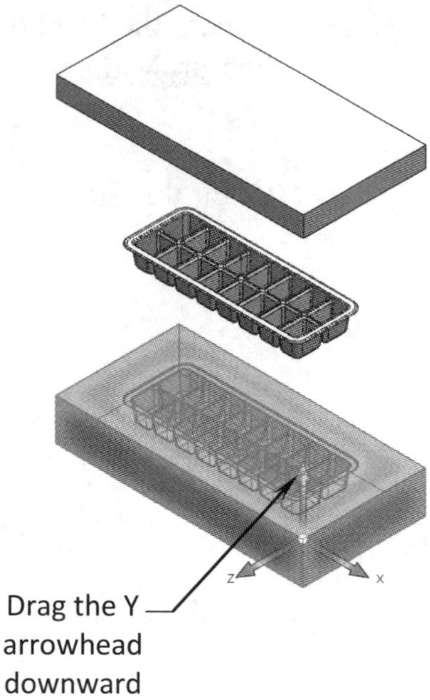

The Explode Step2 is created.

Click **OK**.

10. Assigning materials:

Expand the **Solid Bodies** folder.

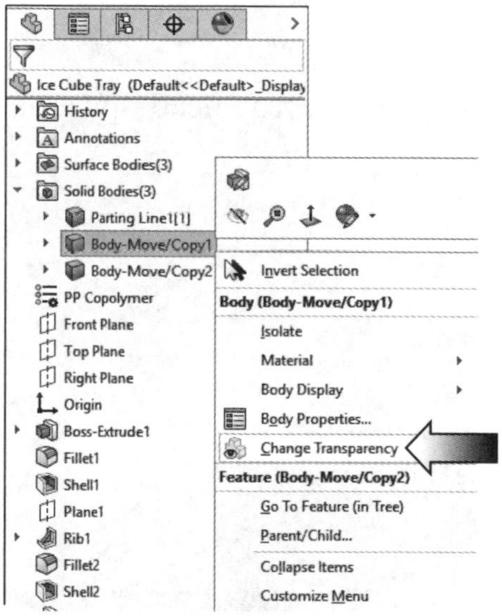

Right-click the **Body-Move/Copy1** (the upper mold block) and select: **Change Transparency**.

Right-click the same body and change the material to **Plain Carbon Steel**.

Assign the same material to the lower mold block.

Assign **PP Copolymer** to the plastic part.

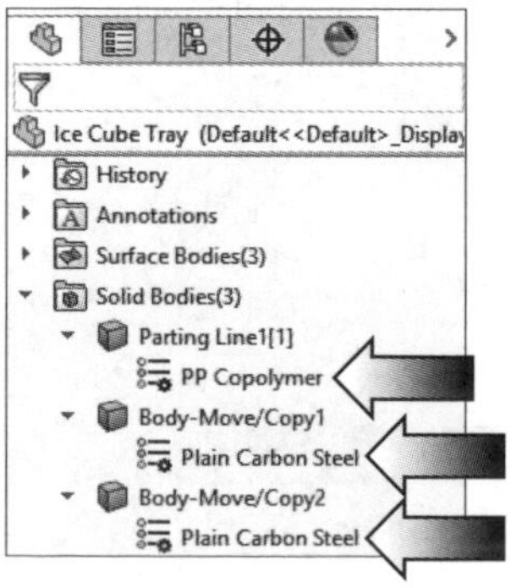

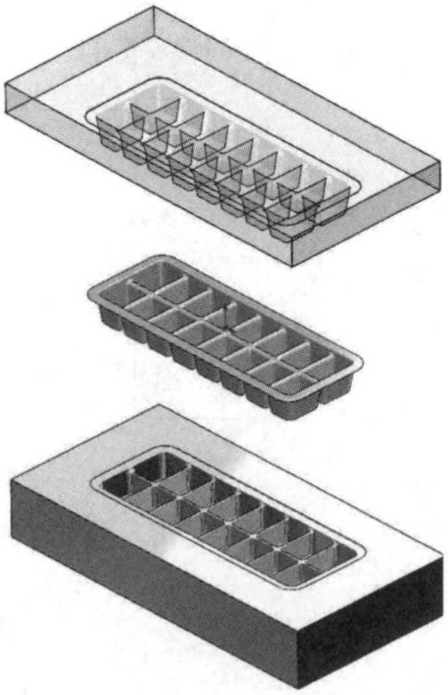

11. Saving your work:

Click **File / Save As**.

Enter **Ice Cube Tray (Completed).sldprt** for the file name and click **Save**.

Questions for Review

Creating a Core and Cavity

1. Using the finished model, the mold tools can be used to analyze and correct the deficiencies such as undercuts, draft angles, shut-off surfaces, etc.
   a. True
   b. False

2. The Parting Lines are used to create the Parting Surfaces and to separate the surfaces.
   a. True
   b. False

3. A shut-off surface closes up a through hole by creating a surface patch along the edges that form a continuous loop.
   a. True
   b. False

4. The Parting Surfaces extrude from the parting lines and are used to separate the mold cavity from the core.
   a. True
   b. False

5. To create a tooling split, what surface bodies are needed for this operation?
   a. The Core
   b. The Cavity
   c. The Parting Surface
   d. All of the above

6. The Interlock surfaces help prevent the core and cavity blocks from shifting and are located along the perimeter of the parting surfaces.
   a. True
   b. False

7. The solid bodies can be hidden or shown just like any other features in SOLIDWORKS.
   a. True
   b. False

1. TRUE 2. TRUE
3. TRUE 4. TRUE
5. D 6. TRUE
7. TRUE

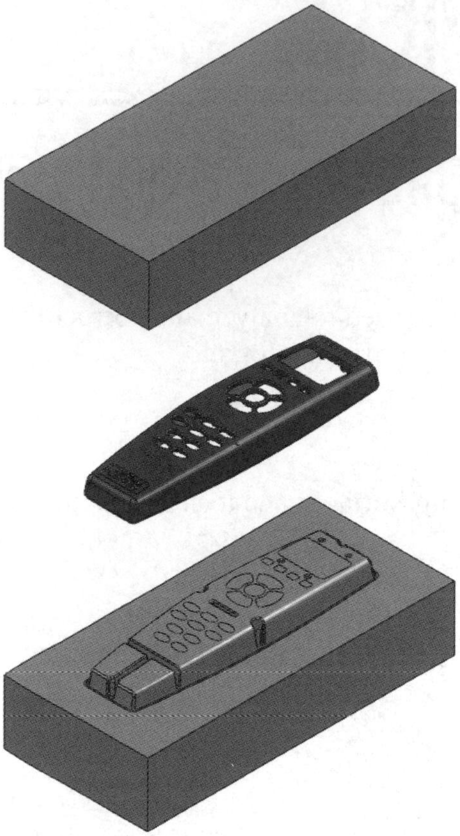

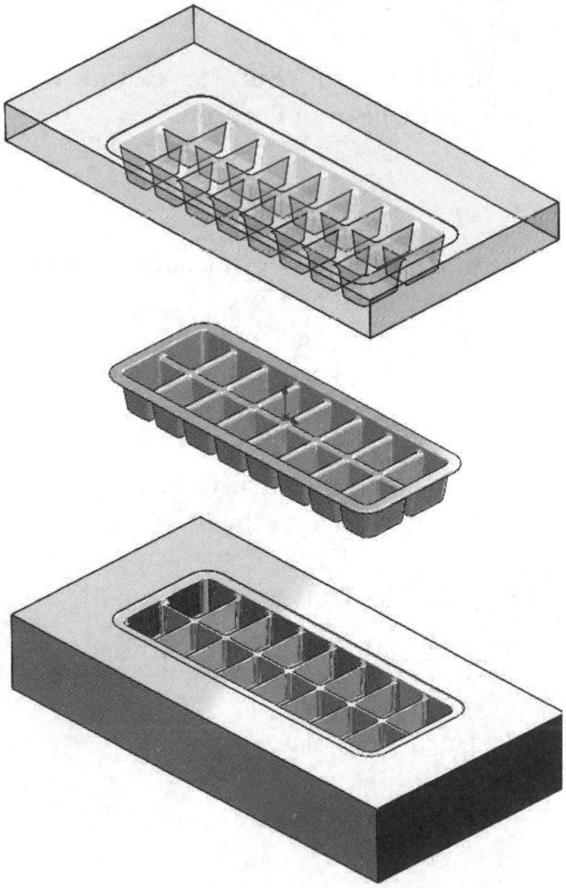

CHAPTER 19

Non-Planar Parting Lines

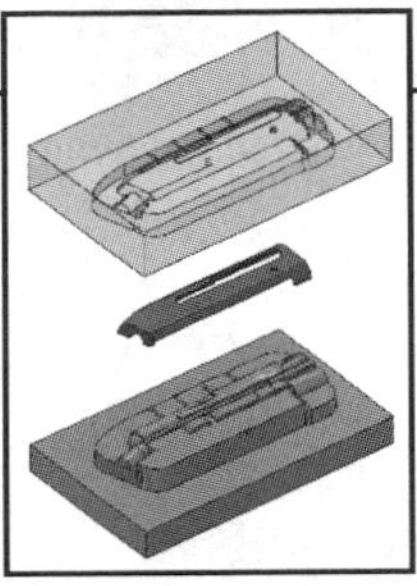

Non-Planar Parting Lines
Mold-Tooling Design

Using SOLIDWORKS a mold is created by following a sequence of integrated tools that control the mold creation process.

The mold tools are used to analyze and correct deficiencies such as draft angles or undercuts with the plastic models to be molded.

The mold tools span from initial analysis to creating the tooling split. The result of the tooling split is a multibody part containing separate bodies for the molded part, the core, and the cavity, plus other optional bodies such as side cores.

The multibody part file maintains your design intent in one convenient location. Changes to the molded part are automatically reflected in the tooling bodies.

The mold design process is as follows:

* **Draft Analysis:** Examines the faces of the model for sufficient draft, to ensure that the part ejects properly from the tooling.
* **Undercut Analysis:** Identifies trapped areas that prevent the part from ejecting.
* **Parting Line Analysis:** Analyzes transitions between positive and negative draft to visualize and optimize possible parting lines.
* **Parting Lines:** Creates a parting line from which you create a parting surface.
* **Shut-off Surfaces:** Creates surface patches to close up through holes in the molded part.
* **Parting Surfaces:** Extrude from the parting line to separate mold cavity from core. You can also use a parting surface to create an interlock surface.
* **Ruled Surface:** Adds draft to surfaces on imported models. You can also use the Ruled Surface tool to create an interlock surface.
* **Tooling Split:** Creates the core and cavity bodies, based on the steps followed earlier.

Non-Planar Parting Lines
Mold-Tooling Design

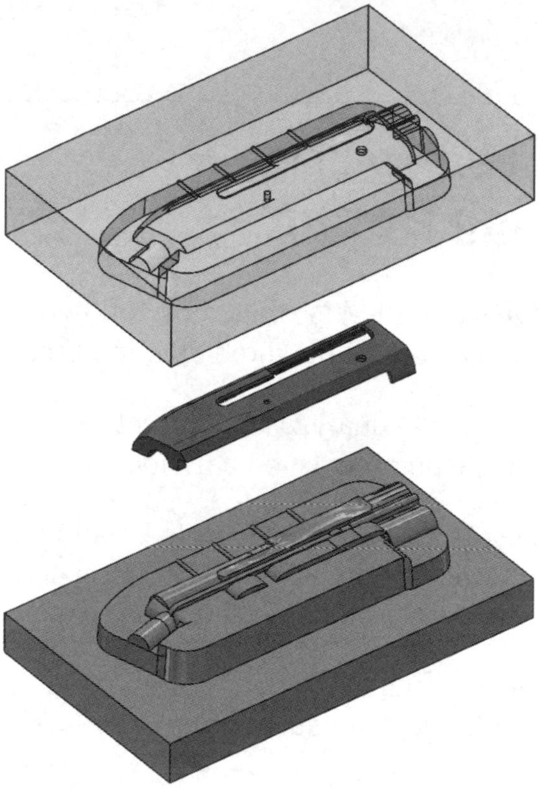

View Orientation Hot Keys:
Ctrl + 1 = Front View
Ctrl + 2 = Back View
Ctrl + 3 = Left View
Ctrl + 4 = Right View
Ctrl + 5 = Top View
Ctrl + 6 = Bottom View
Ctrl + 7 = Isometric View
Ctrl + 8 = Normal To Selection

Dimensioning Standards: **ANSI**
Units: **INCHES** – 3 Decimals

Tools Needed:

Parting Lines	Tooling Split	Planar Surface
Shut-Off Surfaces	Ruled Surface	Knit Surface
Parting Surfaces	Filled Surface	Trim Surface

1. Opening an existing part document:

Click **File / Open**.

Browse to the Training Files folder and open a part document named: **Receiver.sldprt**.

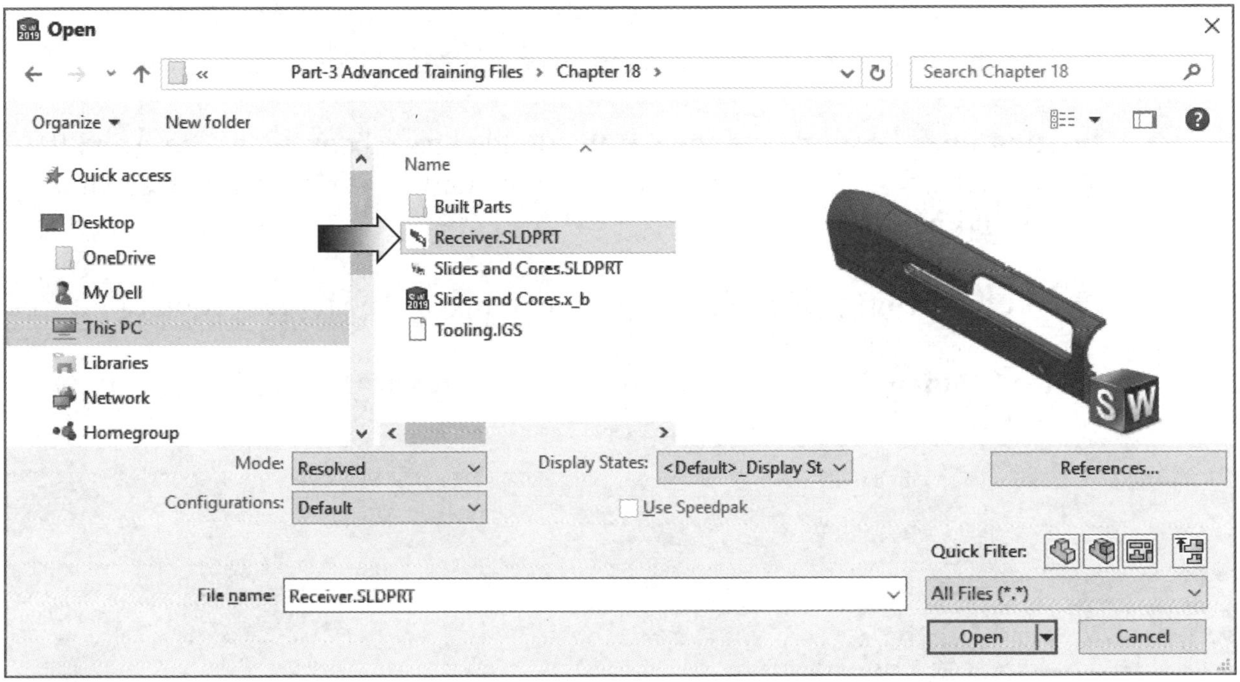

2. Enabling the Mold Tools toolbar:

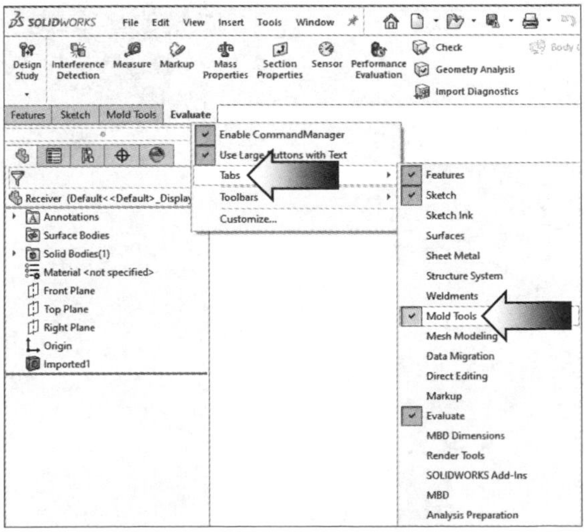

Right-click one of the existing tool tabs and enable the **Mold Tools** checkbox (arrows).

For clarity, only keep the following toolbars enabled:

* **Features**
* **Sketch**
* **Mold Tools**
* **Evaluate**

Disable the other toolbars.

NOTE: *This model has already been scaled to accommodate the mold shrinkage. The draft angles also have been added to all features. The next step is to add the parting lines.*

3. Creating the Parting Lines:

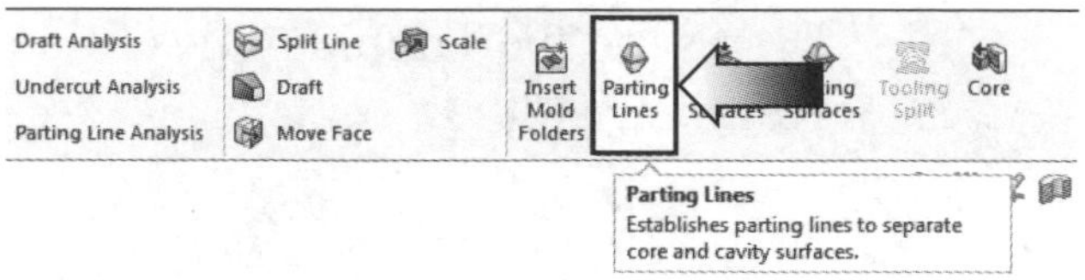

Switch to the new **Mold Tools** tab (arrow).

Select the **Parting Lines** command from the **Mold Tools** tab (arrow).

Parting lines lie along the edge of the molded part, between the core and the cavity surfaces. They are used to create the parting surfaces and to separate the two mold halves.

Select the **Front** plane from the FeatureManager tree for Direction of Pull.

Enter **1.00deg** for Draft Angle and click the **Draft Analysis** button (arrow).

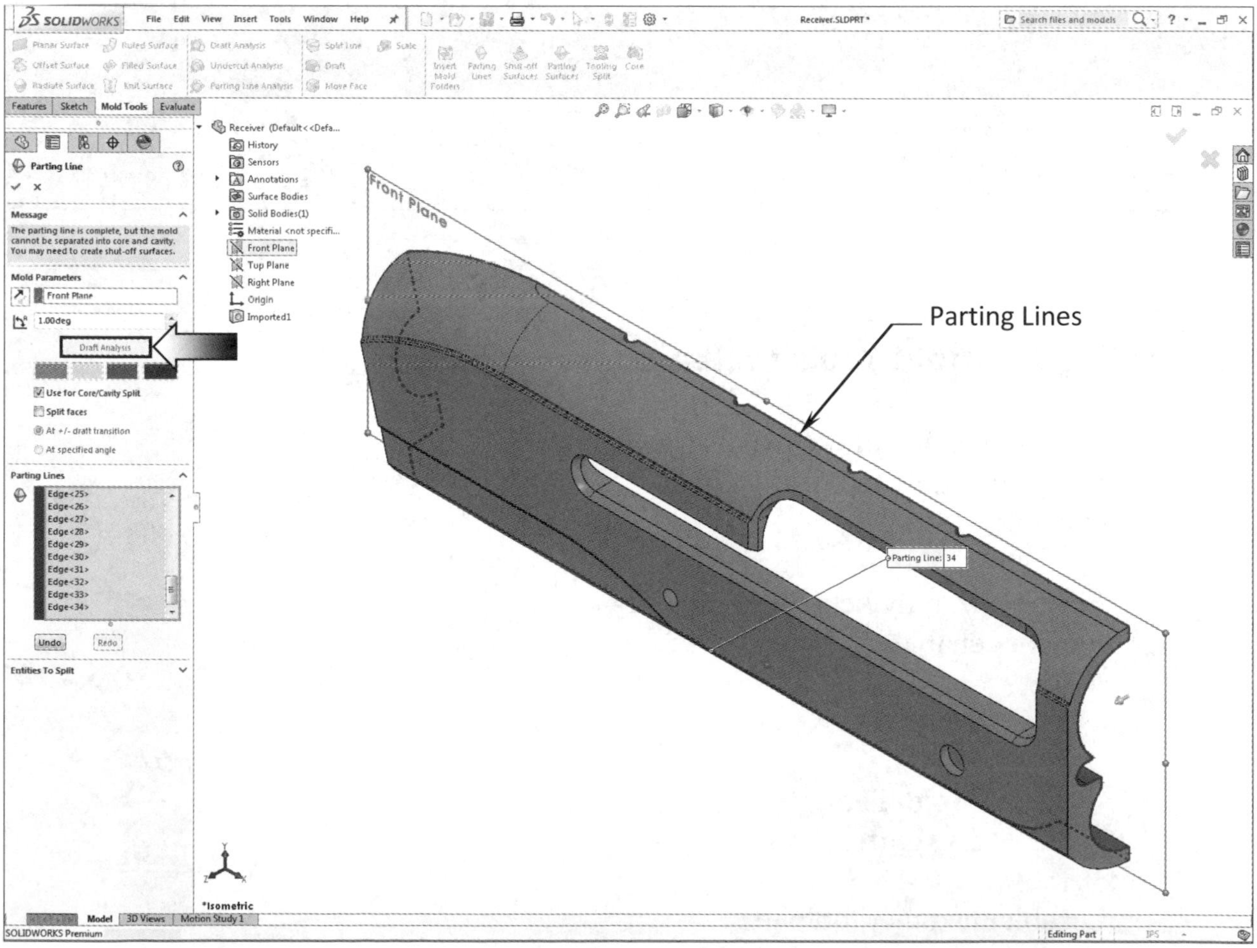

SOLIDWORKS automatically selects the edges of the model that border the two halves of the mold. The parting lines will be used to separate the surfaces between the core and the cavity.

The **Green** surfaces on the model represent the positive draft surfaces on the Cavity half, and the Red surfaces are negative draft surfaces on the Core half.

Click **OK**.

4. Creating the Shut-Off Surfaces:

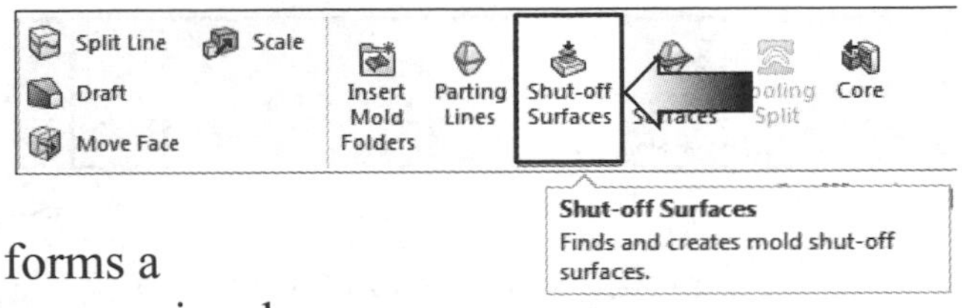

A shut-off surface closes up a through hole by creating a surface patch along the edges that forms a continuous loop, or a parting line you previously created to define a loop.

Click the **Shut-Off Surfaces** command on the **Mold Tools** tab (arrow).

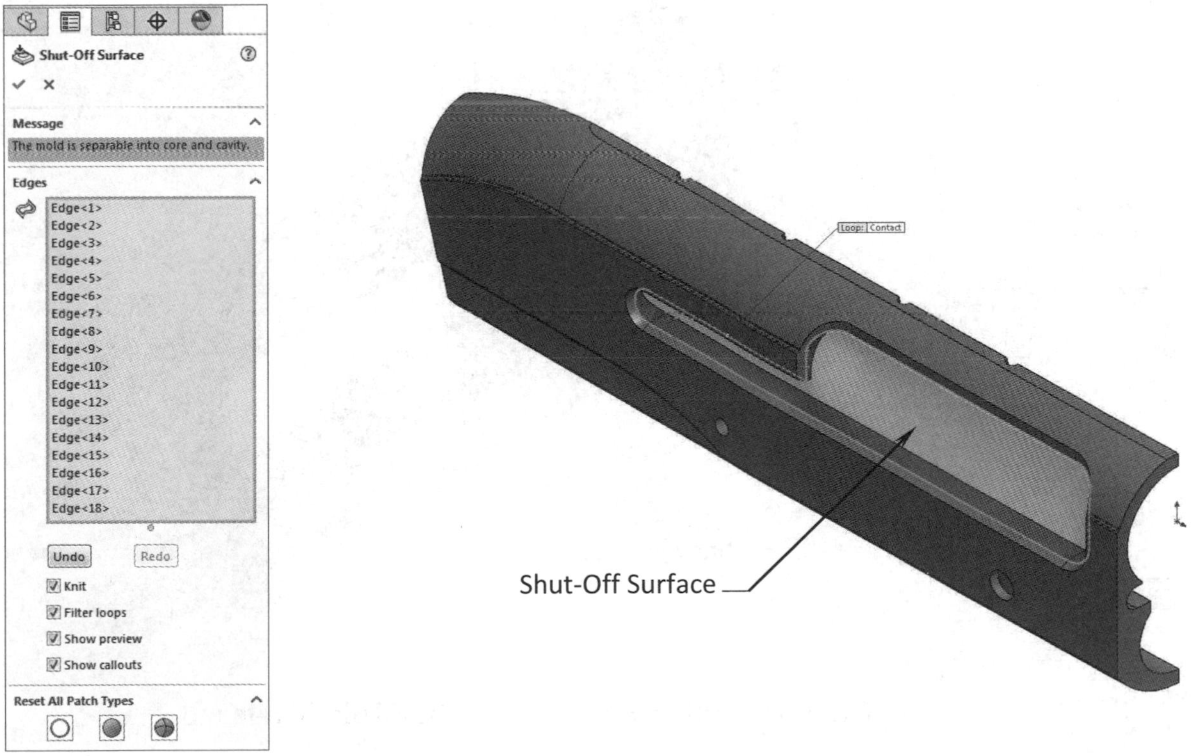

SOLIDWORKS searches for any through holes and automatically creates a surface patch along the edges that form a continuous loop.

A "Green Message" appears on the Feature tree indicating the mold is separable into core and cavity.

Enable the **Knit** checkbox (arrow).

Click **OK**.

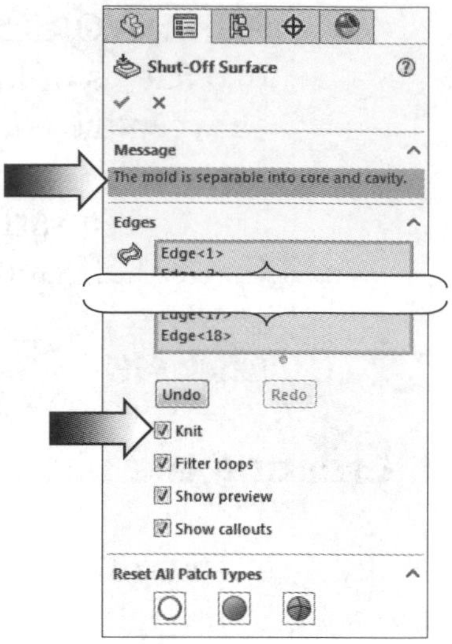

5. Creating the Parting Surfaces:

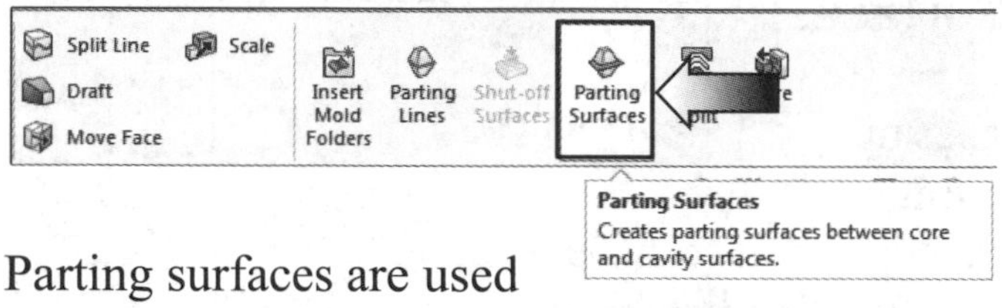

Parting surfaces are used to separate the mold cavity from the core. They must be created right after the Shut-Off Surfaces. (The Shut-Off Surface is not shown for clarity.)

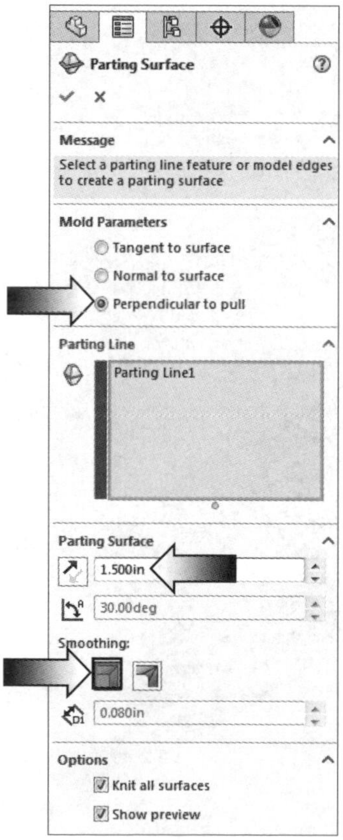

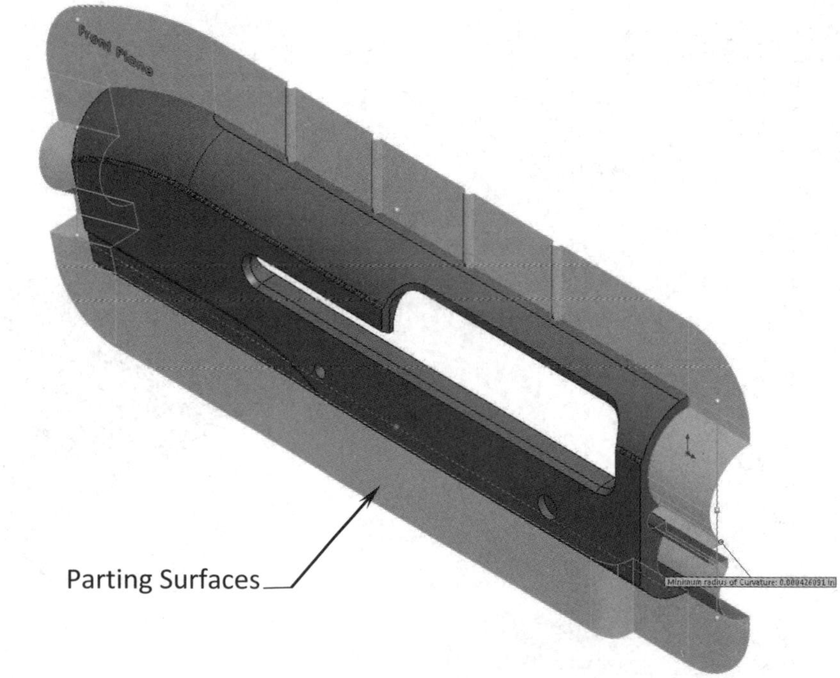

Click **Parting Surfaces** from the **Mold Tools** tab.

Select the **Perpendicular to Pull** option (arrow).

Enter **1.00in** for **Distance** (arrow).

Use the default **Sharp*** option for smoothness (arrow).

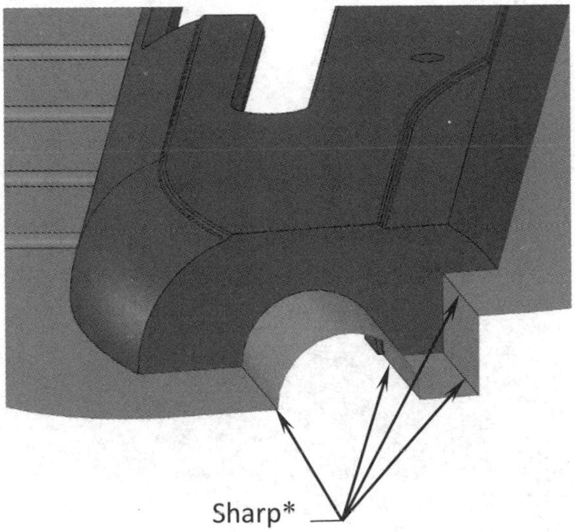

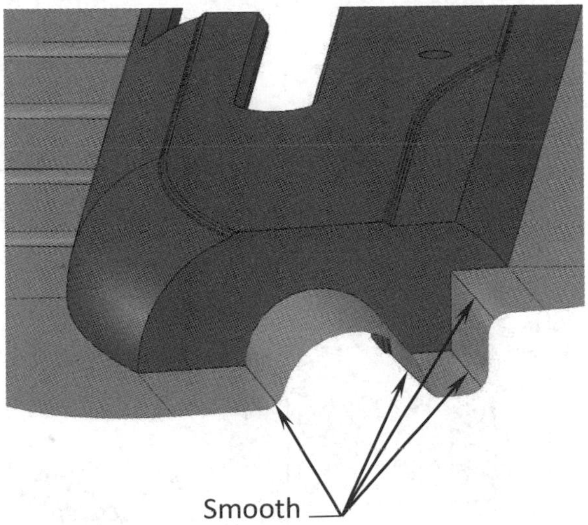

6. Creating a Ruled Surface:

To help prevent the core and cavity blocks from shifting, you can add an interlock surface. It is created along the perimeter of parting surfaces prior to inserting a tooling split in a mold part. The interlock surface can be created manually or automatically.

Because of the sudden changes in the parting surface geometry, the interlock surfaces will need to be created manually.

The Ruled Surface command is used to create the tapered surfaces that form the interlocks.

Select the **Rule Surface** command from the **Mold Tools** tab (arrow).

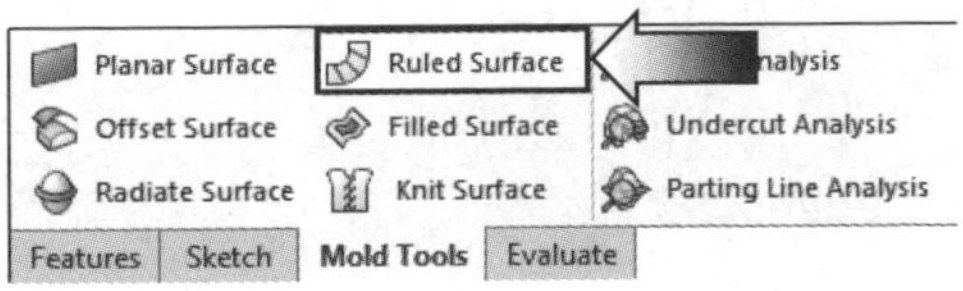

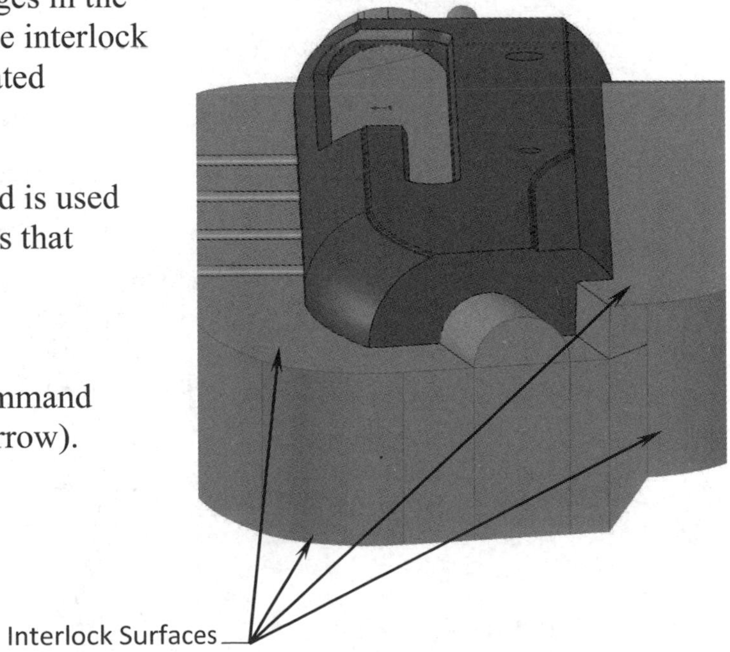

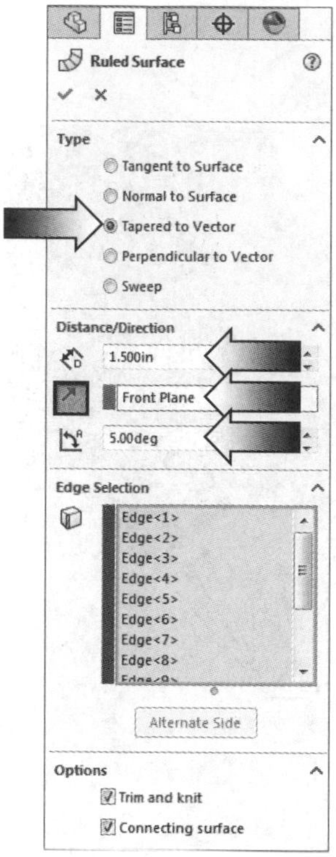

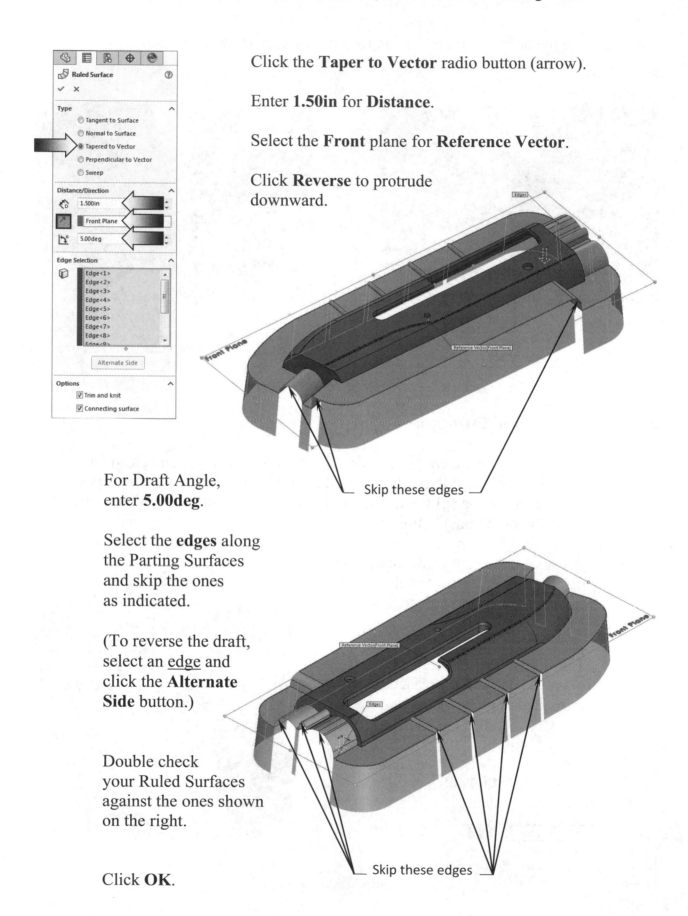

Click the **Taper to Vector** radio button (arrow).

Enter **1.50in** for **Distance**.

Select the **Front** plane for **Reference Vector**.

Click **Reverse** to protrude downward.

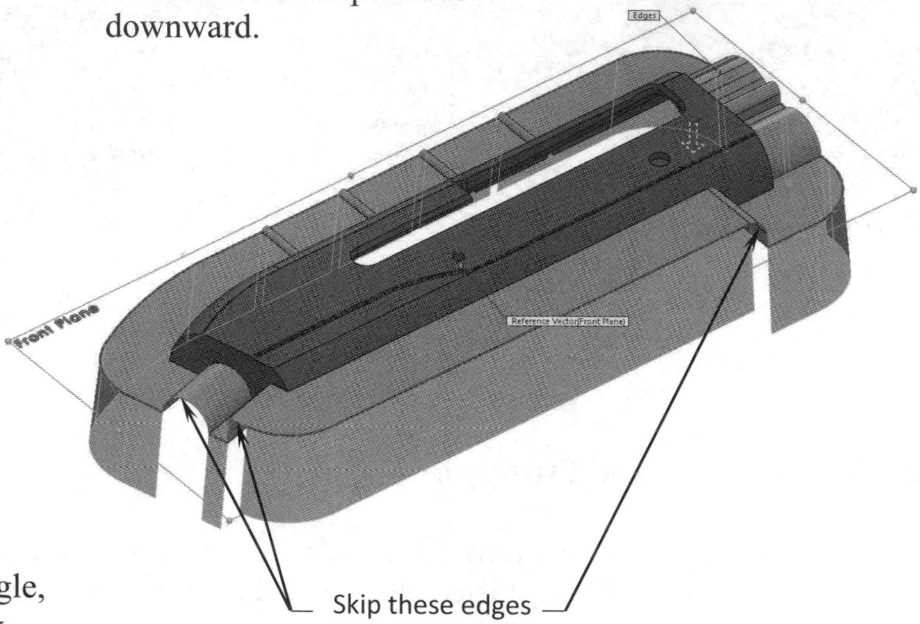

For Draft Angle, enter **5.00deg**.

Select the **edges** along the Parting Surfaces and skip the ones as indicated.

(To reverse the draft, select an edge and click the **Alternate Side** button.)

Double check your Ruled Surfaces against the ones shown on the right.

Click **OK**.

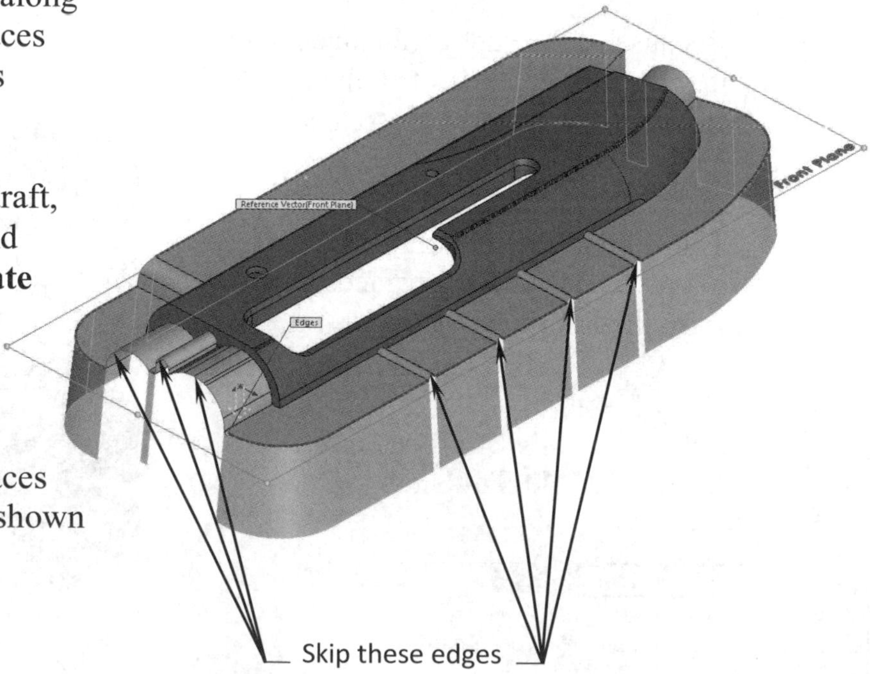

7. Creating the patches:

We will use a combination of Lofted Surface and Filled Surface commands to create the patches and fill the openings in the ruled surfaces.

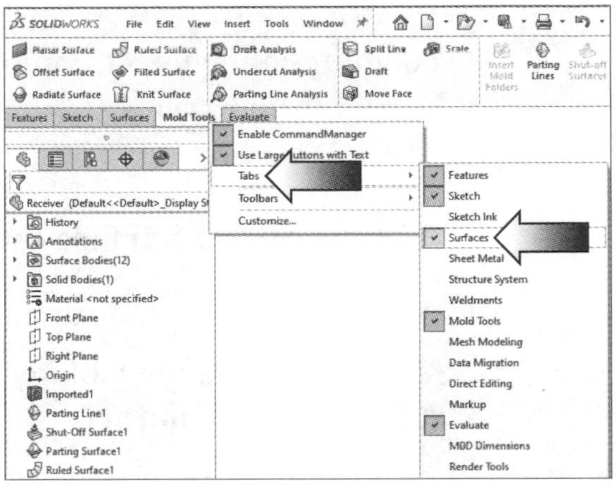

Right-click one of the tool tabs and enable the **Surfaces** toolbar (arrow).

Select **Lofted Surface** from the **Surfaces** tool tab.

Rotate the model to a position similar to the one shown here.

Select the **2 edges** of the opening as noted.

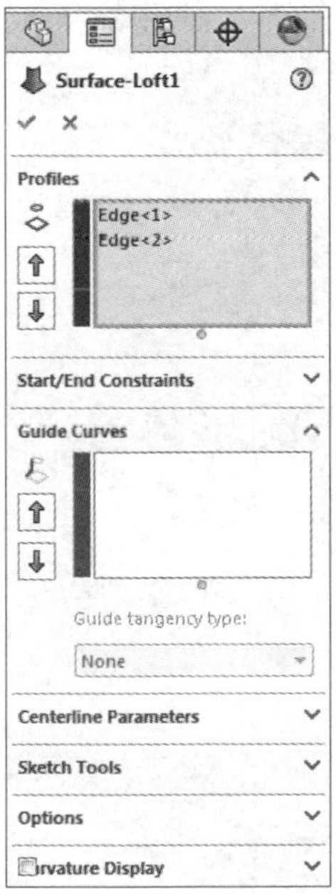

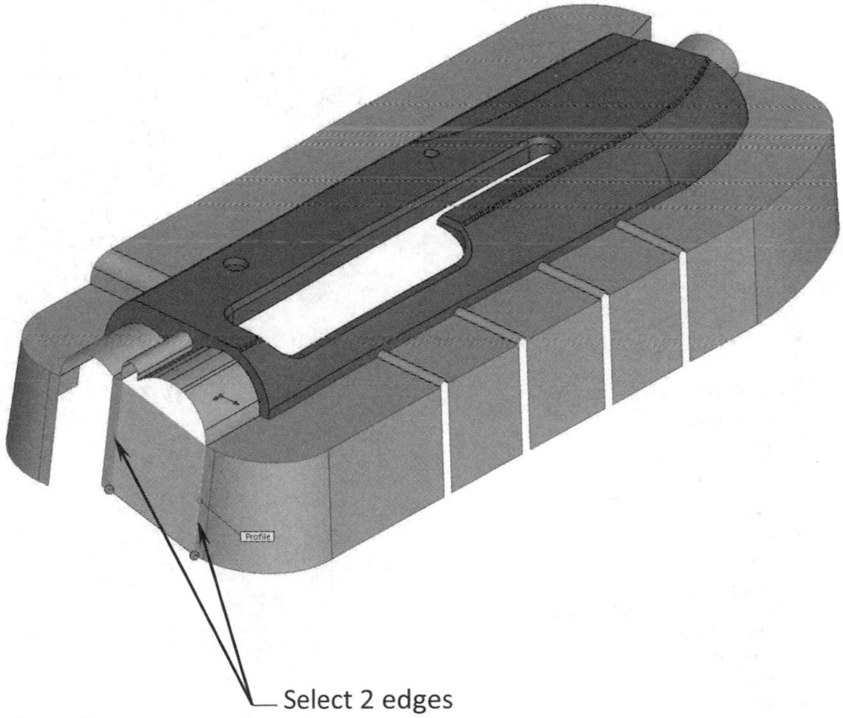

A preview of a lofted surface appears filling the lower portion of the opening.

Click **OK**.

8. Patching with Filled Surface:

The Filled Surface command creates a surface patch with any number of sides, within a boundary defined by existing model edges, sketches, or curves, including composite curves.

Zoom in on the upper portion of the opening. We will patch it up with the Filled Surface command.

Click the **Filled Surface** command from the Surfaces toolbar.

Keep the Curvature Control at the default: **Contact**.

Select all remaining edges on the upper portion of the same opening.

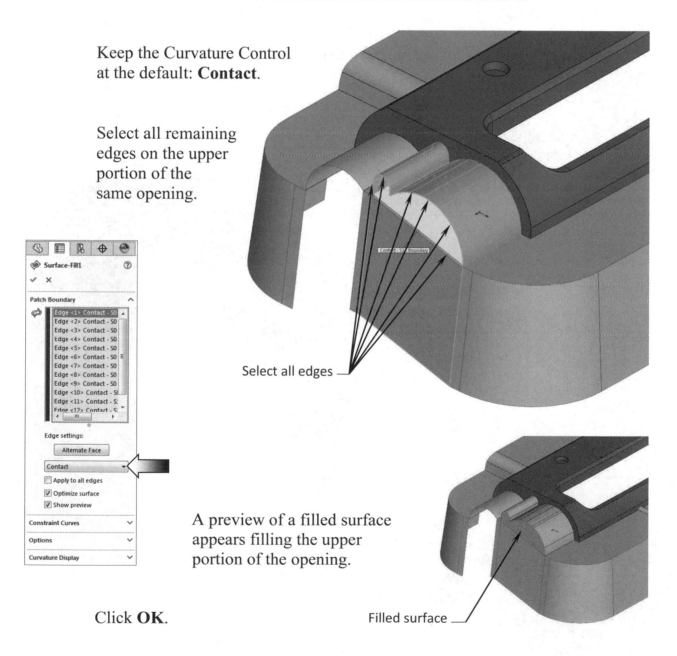

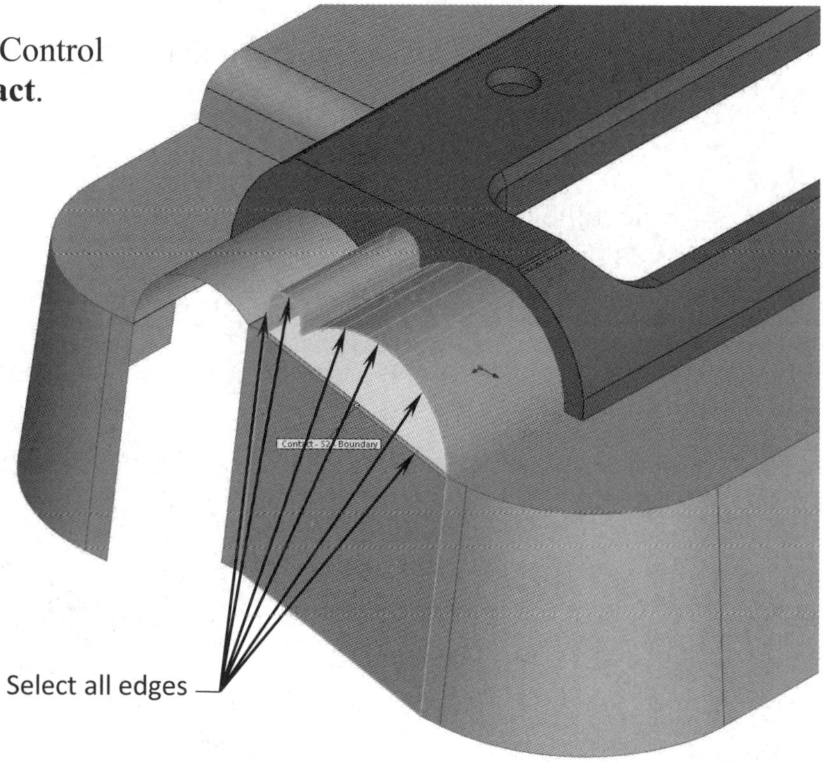

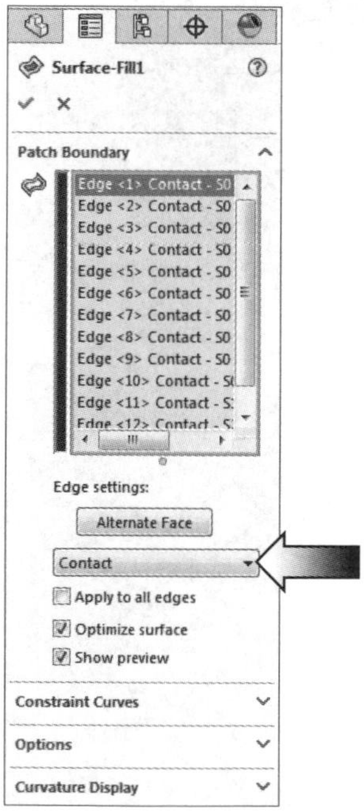

A preview of a filled surface appears filling the upper portion of the opening.

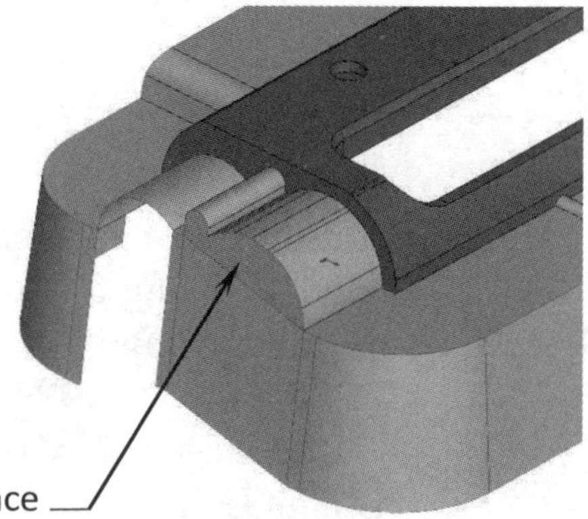

Click **OK**.

9. Patching all openings:

Repeat the steps 7 and 8 and patch / fill the rest of the openings in the model.

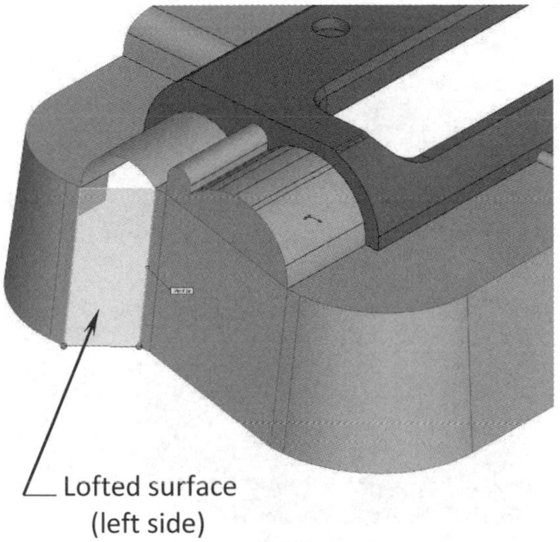

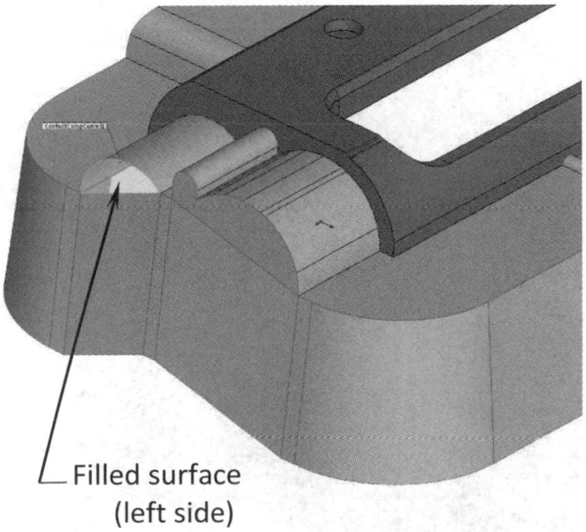

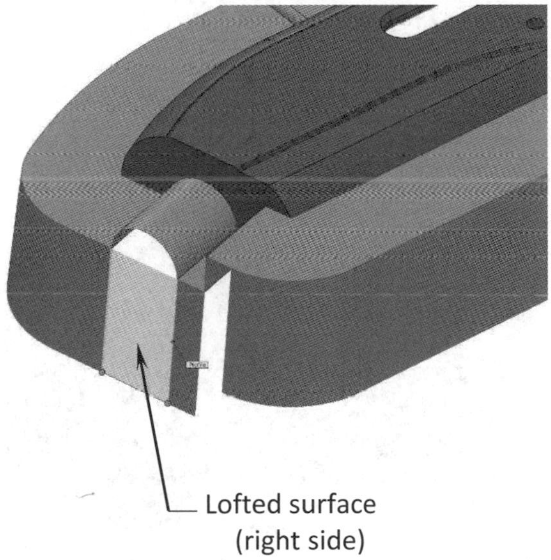

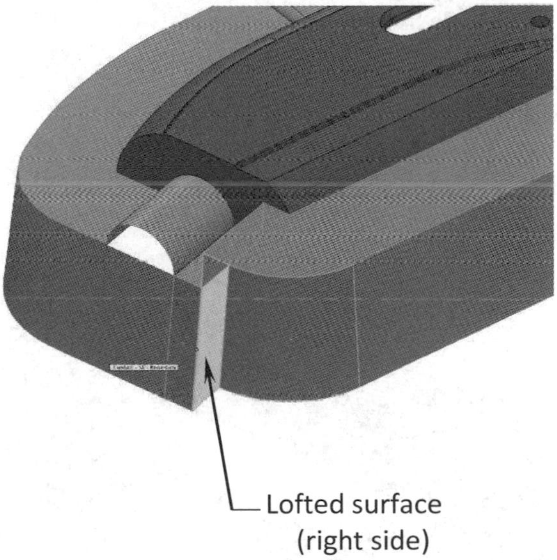

The Filled Surface command requires the boundary to be closed in order to fill or patch that area. The Curve Through Reference Points command can be used to create the missing entities in those openings.

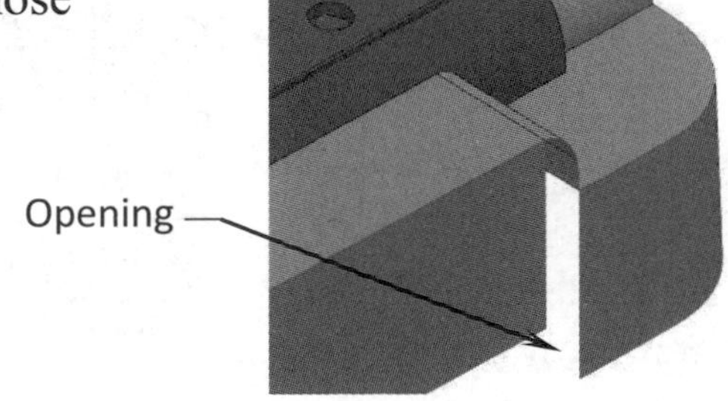

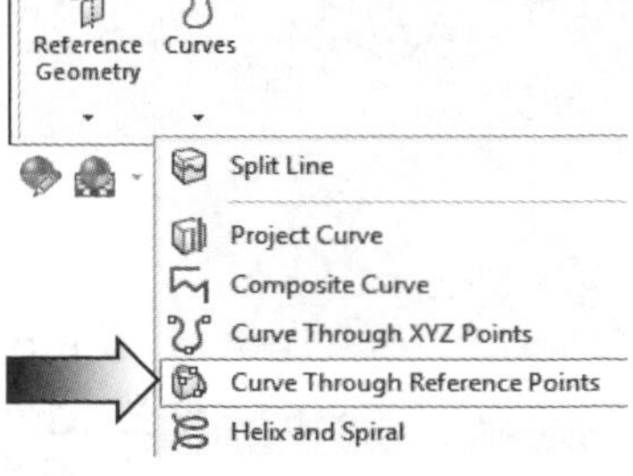

From the **Surfaces** toolbar, select **Curves / Curve Through Reference Points**.

Select the **2 vertices** as indicated and the preview of a curve appears.

Click **OK**.

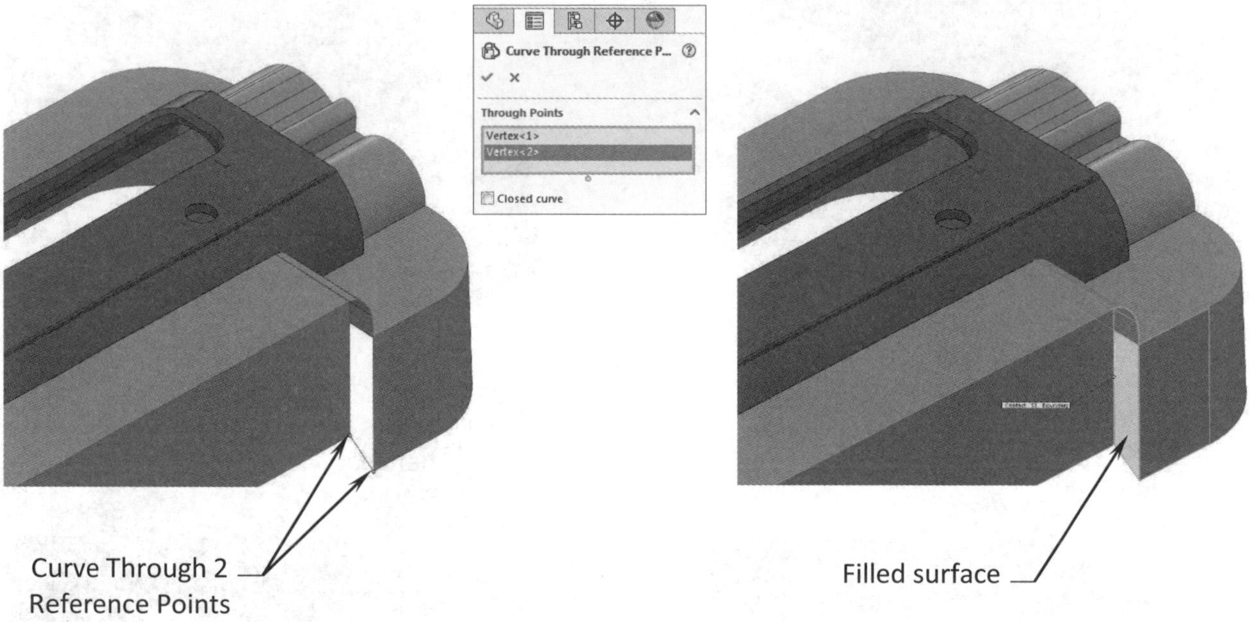

Continue with creating the patches to fill the rest of the openings.

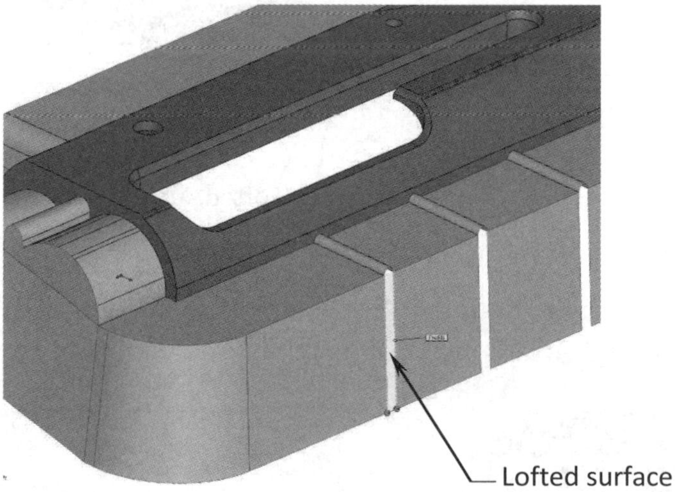

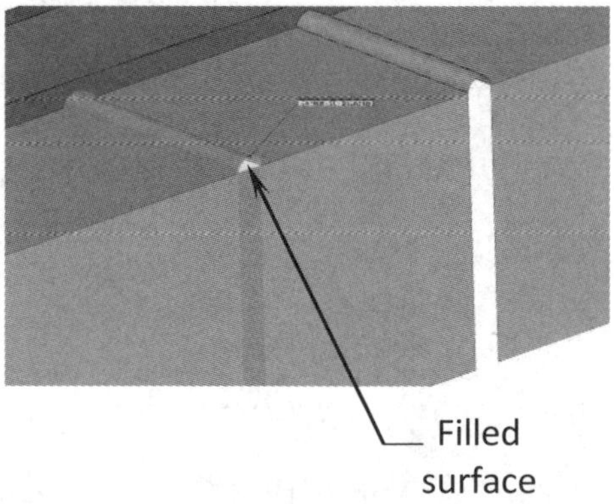

After all openings are patched, expand the Surface Bodies folder and show all surfaces.

10. Knitting the surfaces:

To help select multiple surfaces more easily, we will need to knit all of the ruled surfaces and the patched surfaces into a single surface.

Click the **Knit Surface** command (arrow).

Expand the Surface Bodies folder and select all surfaces inside this folder (arrow).

Click **OK**.

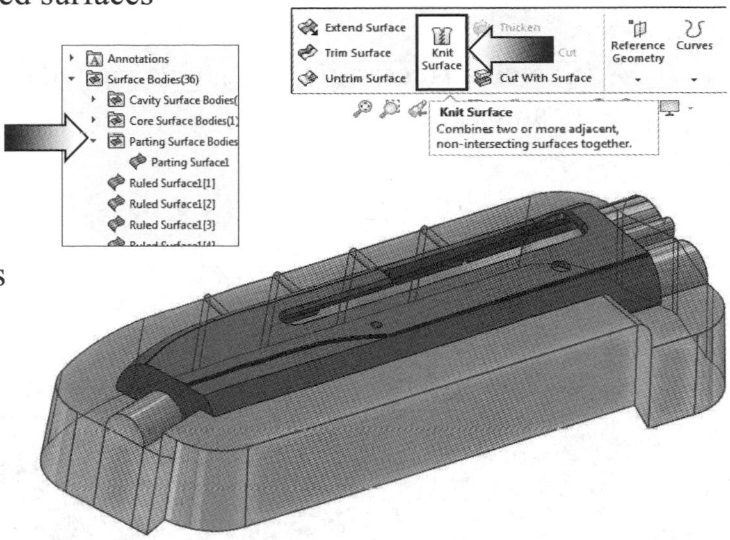

11. Creating a new plane:

The bottom of the interlock surface is not flat. We will need to create a plane and a planar surface and use them to trim the bottom.

Click **Reference Geometry / Plane**, from the **Surfaces** tool tab.

Select the **Front** plane for **First Reference**.

Click the **Offset Distance** button and enter **.750in** for **Distance**.

Place the new plane below the Front plane and click **OK**.

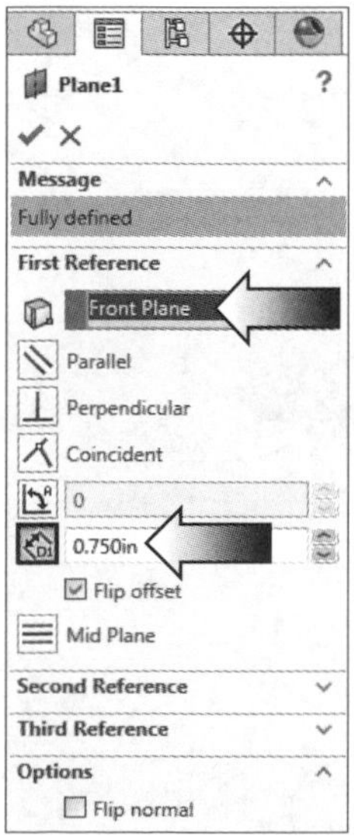

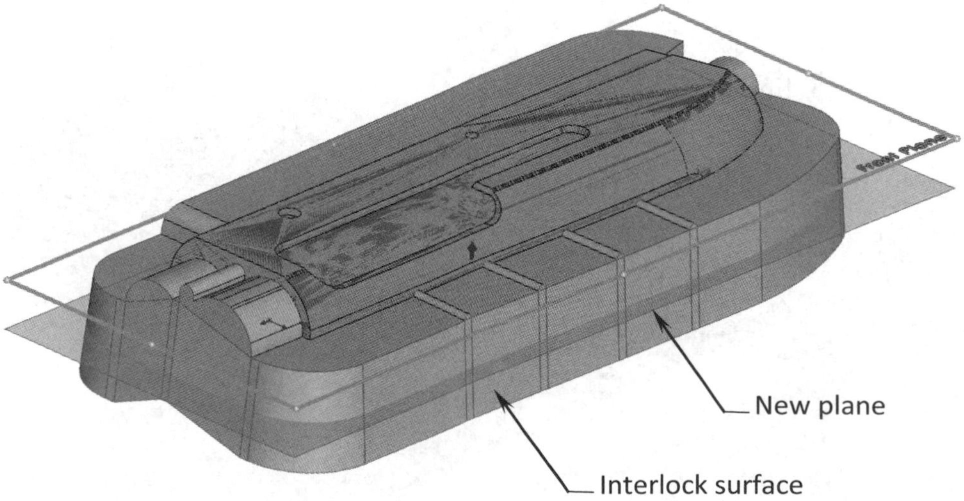

12. Creating a new sketch:

Select the new plane1 and open a **new sketch**.

Sketch a **Rectangle** around the model and add the dimensions shown to fully position it.

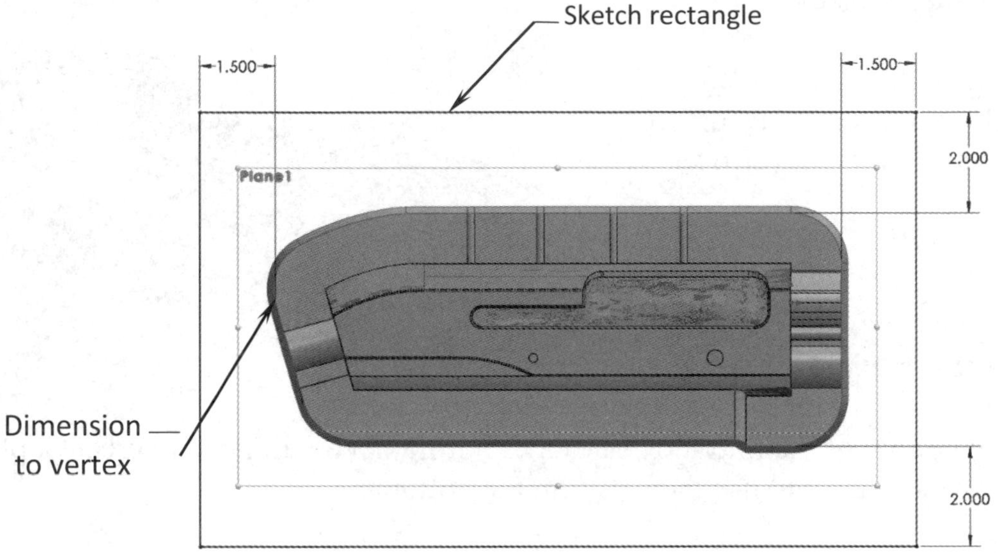

While the sketch is still active click the **Planar Surface** command from the Surfaces tool tab (arrow).

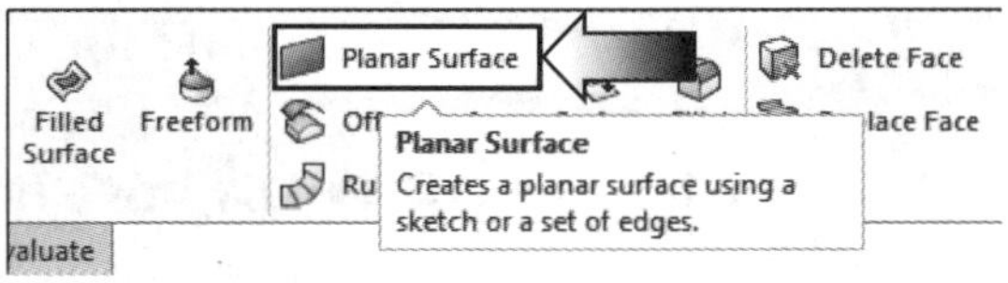

The rectangular sketch is converted into a planar surface.

Click **OK**.

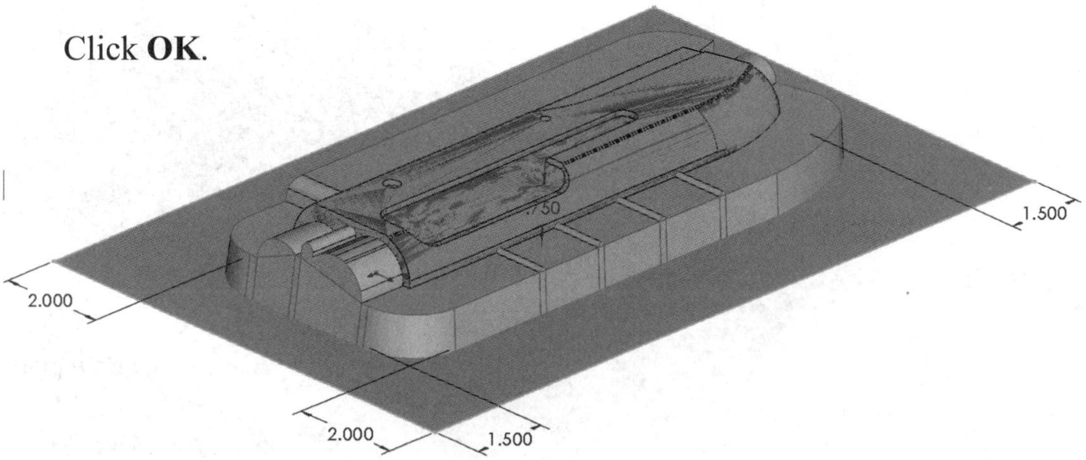

13. Trimming the bottom of the ruled surface:

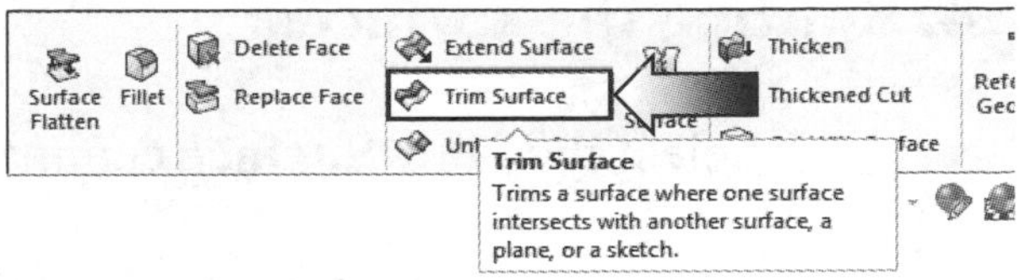

Select the **Trim Surface** command from the **Surfaces** tool tab.

Click the **Mutual** trim option (arrow). Select the **Planar Surface** and <u>all</u> of the surfaces along the perimeter of the Interlock Surface.

Click the **Remove Selections** options and select the <u>inside</u> of the Interlock Surface plus all of its surfaces along the perimeter as indicated.

Click **OK**.

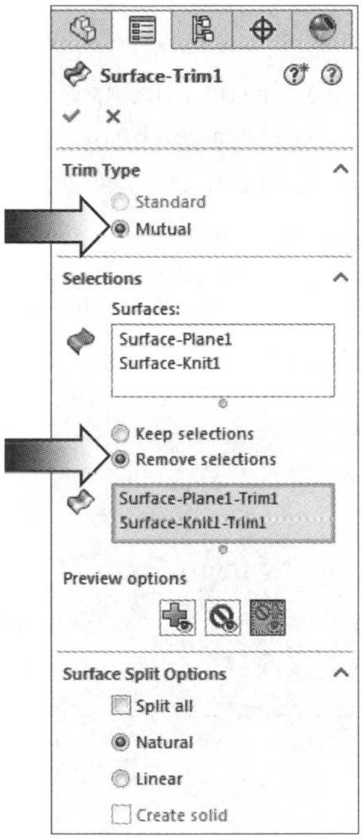

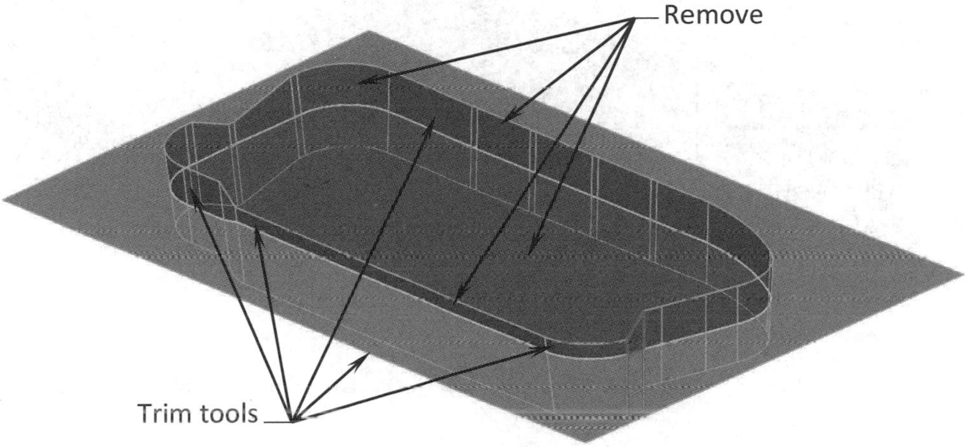

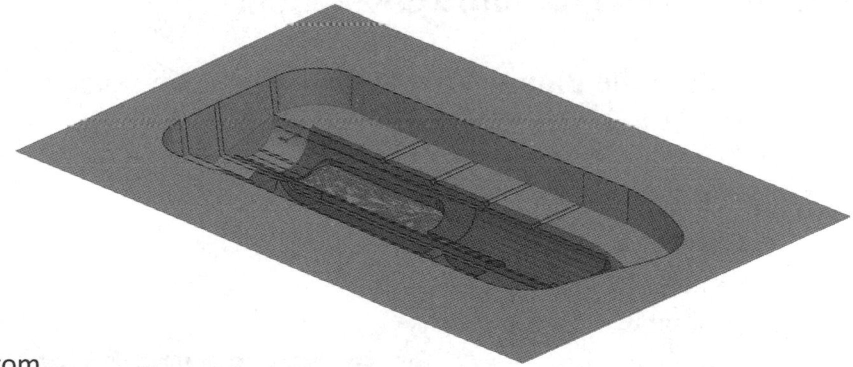

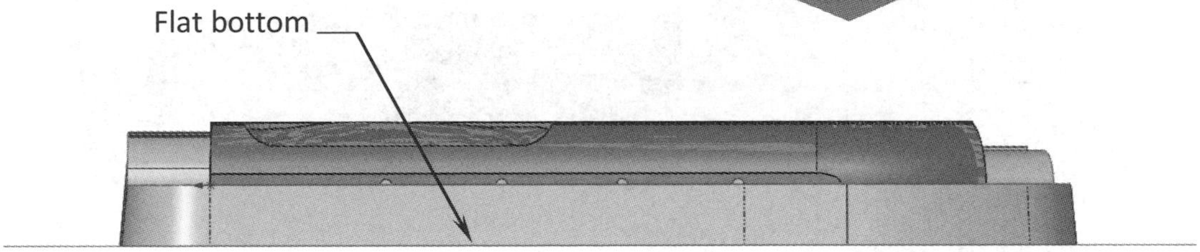

The resulting Trimmed Surface. Change to the side view to verify the trim. The bottom of the interlock surface should be flat at this point.

14. Knitting the surfaces:

Select the **Knit Surface** command from the Surfaces tool tab.

Expand the **Surfaces Bodies** folders. Select the **Parting Surface** and the **Surface-Trim1** from the Surface Bodies folder.

Click **OK**.

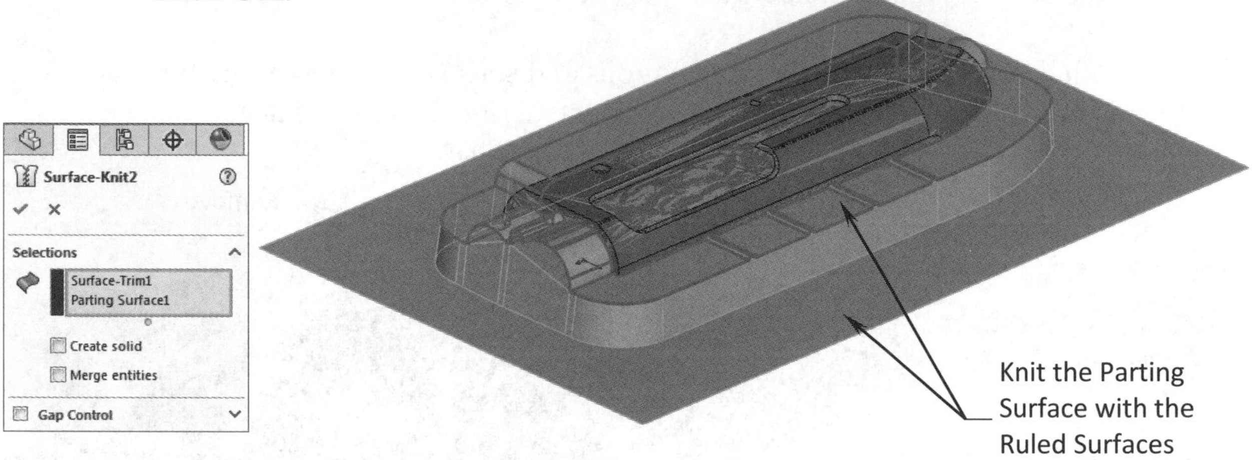

(Skip this step if SOLIDWORKS knits the surfaces automatically.)

15. Creating the tooling split sketch:

Select the **Planar Surface** and open a new sketch.

Locate the **Sketch1** under the Planar-Surface and Show it.

Create an offset of **.125in** (inside) from Sketch1.

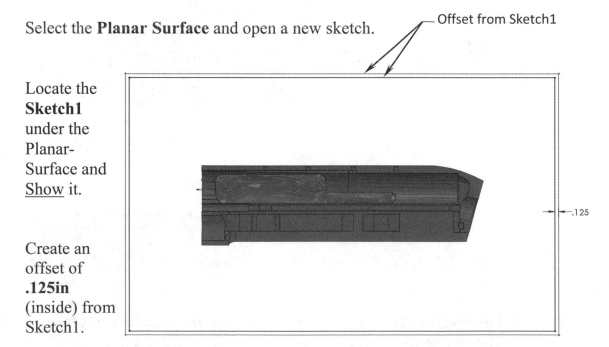

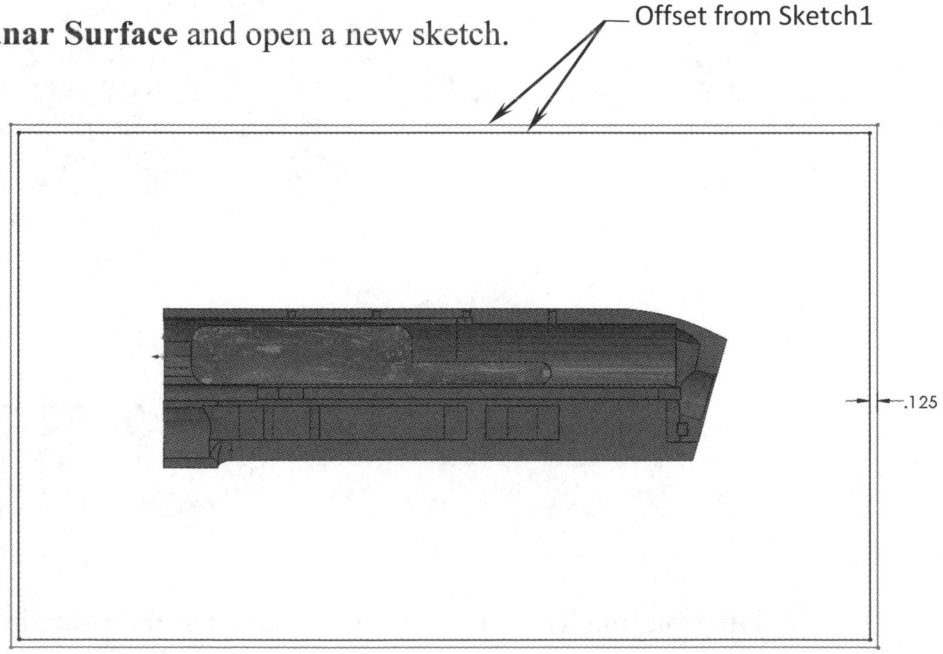

Exit the sketch and click **Tooling Split**. The Tooling Split properties appears.

For Block Size, enter the following:

Upper block: 2.500in
Lower block: 1.500in

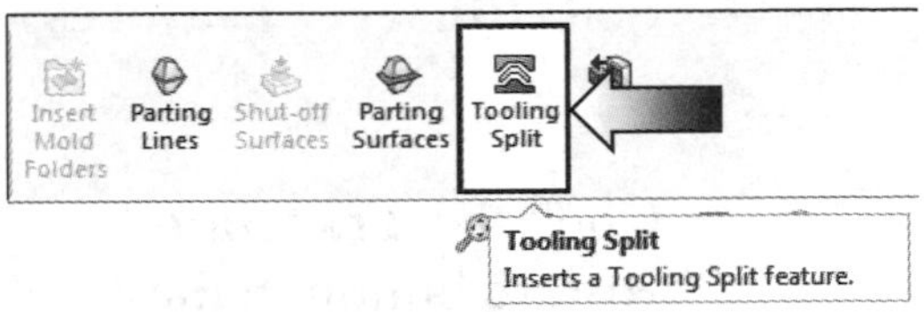

Click **OK**.

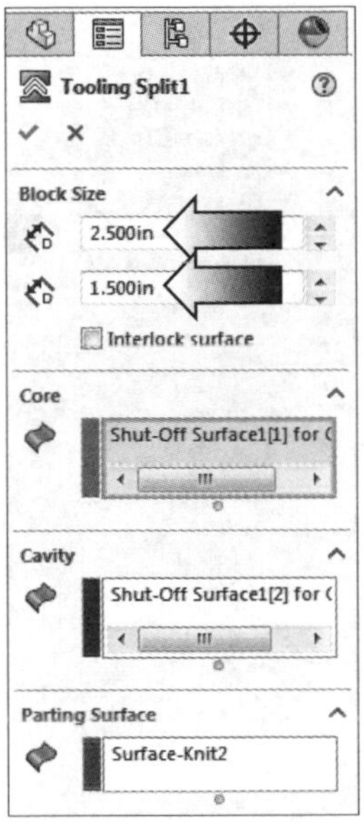

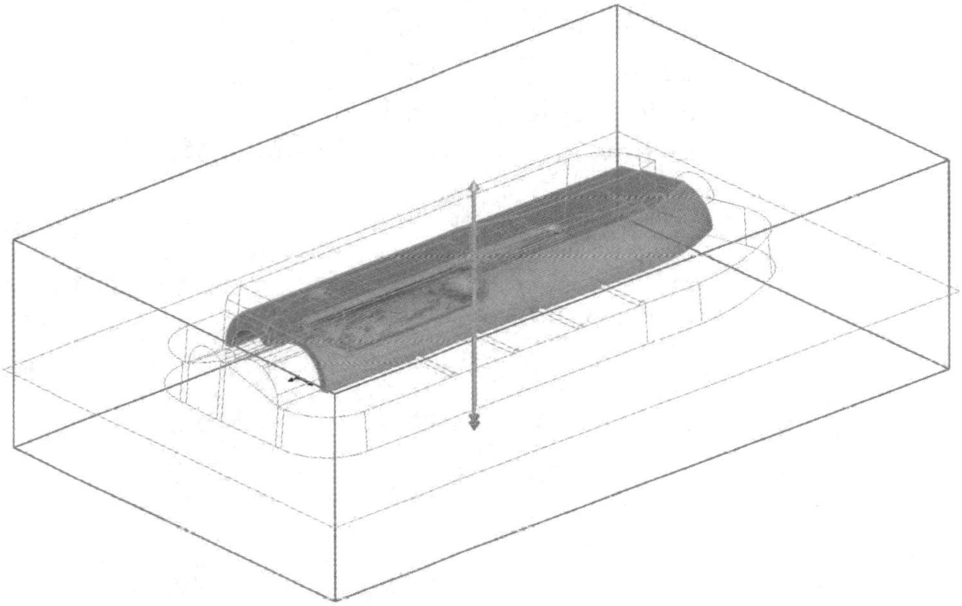

There are 3 solid bodies on the FeatureManager tree: the Original part, the Upper Mold Block, and the Lower Mold Block.

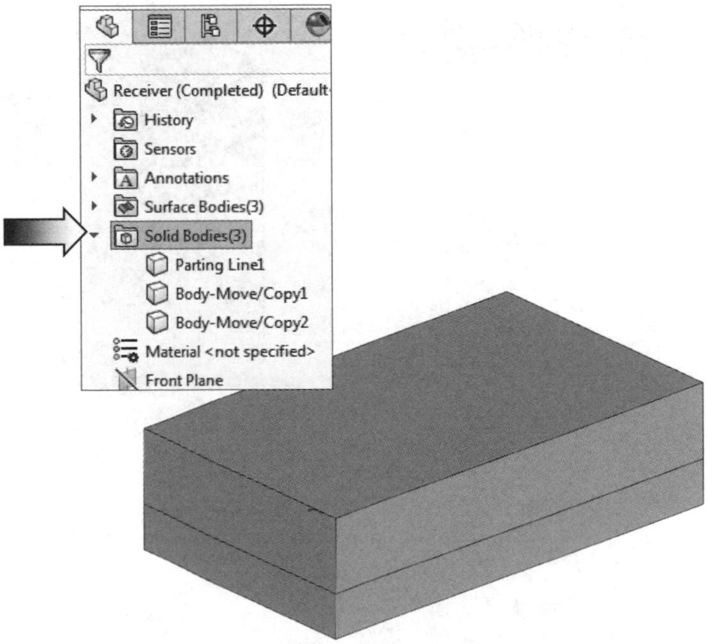

We will separate them in the next step.

16. Separating the solid bodies:

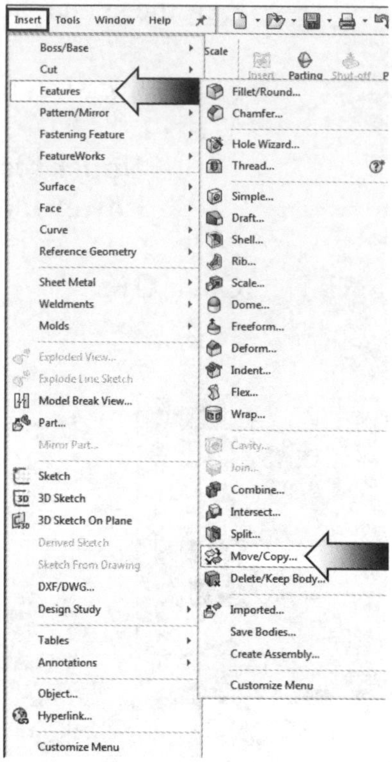

Select **Insert / Features / Move-Copy**.

Click the **Translate / Rotate** button in the Options section.

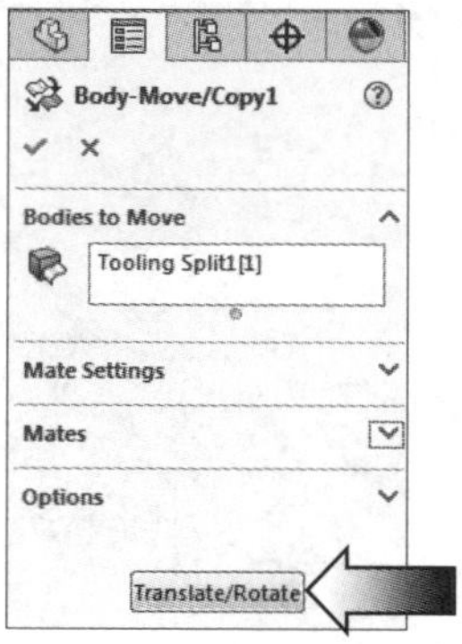

Select the <u>bottom half</u> of the mold (the core) for Bodies to Move/Copy.

Enter **-6.500in** for Delta Z distance. The core half is moved outward.

Click **OK**.

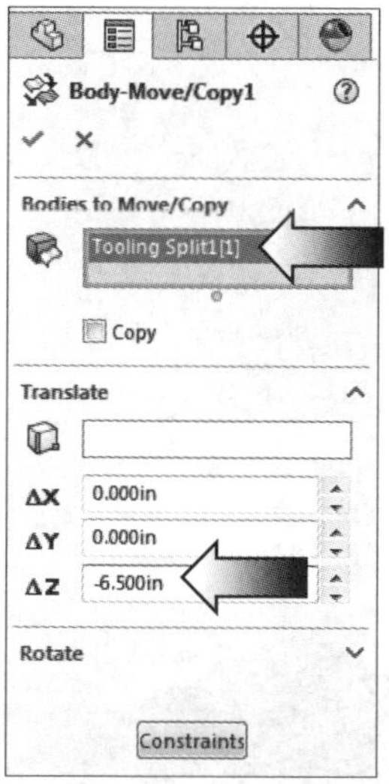

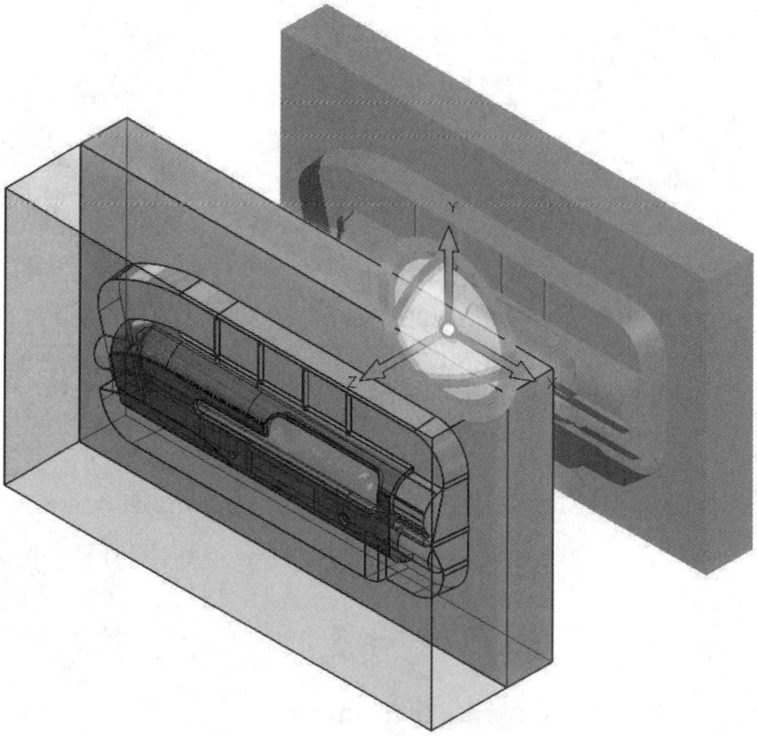

Repeat step number 16 and move the left half of the mold (the cavity).

Move the cavity block about the **Delta Z** direction; use a distance of **8.250in**.

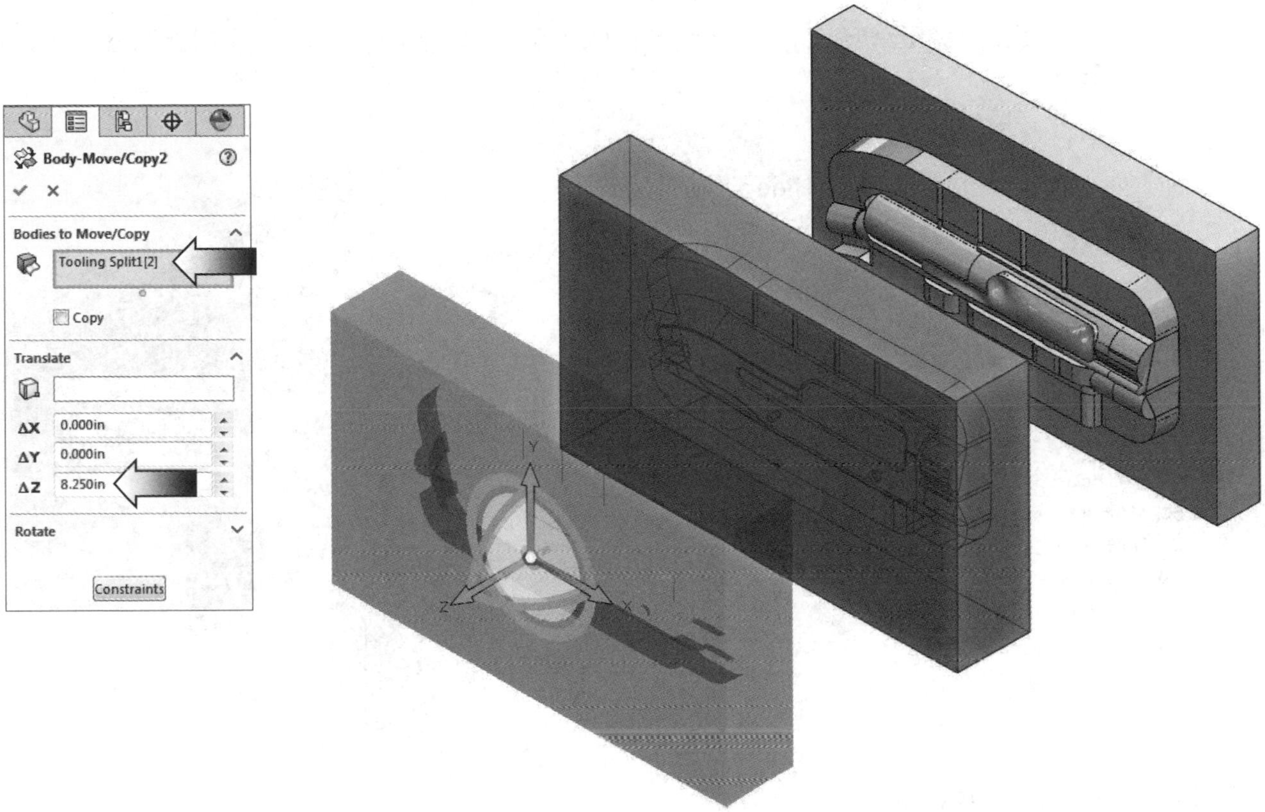

17. Making the body transparent:

From the FeatureManager tree, Expand the Solid Bodies folder.

Right-click on the Cavity body and select: **Change-Transparency** (arrow).

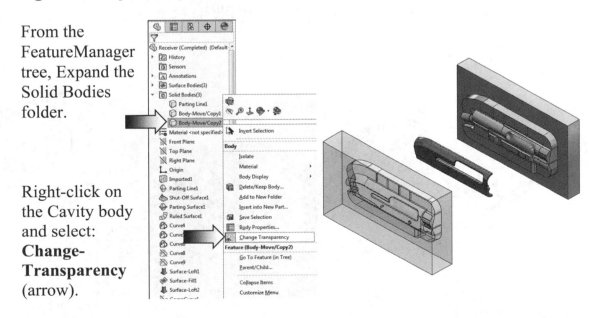

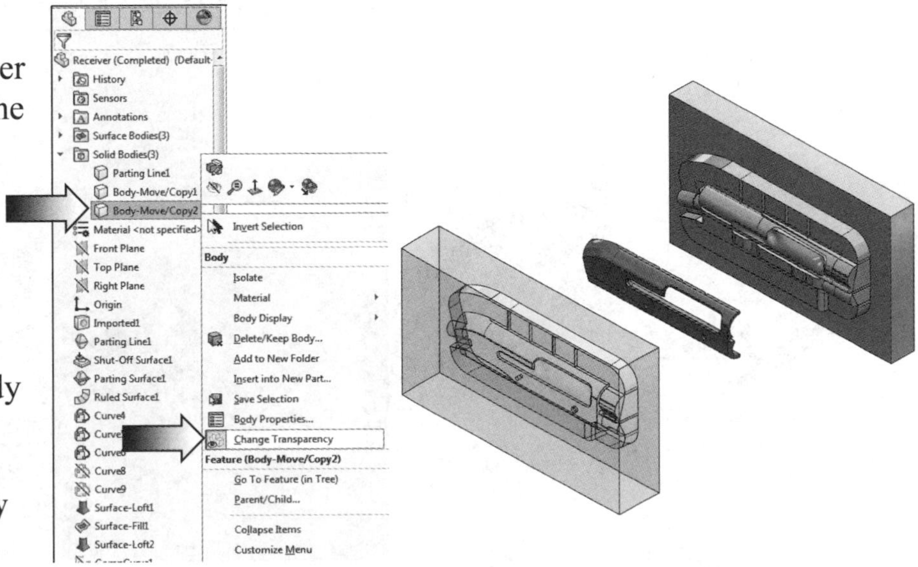

18. Hiding the surface bodies:

The Core, Cavity, and other surfaces are still visible in the graphics making it difficult to see the molded part. We will need to hide them.

From the FeatureManager tree, right click the Surface Bodies folder and select: **Hide**.

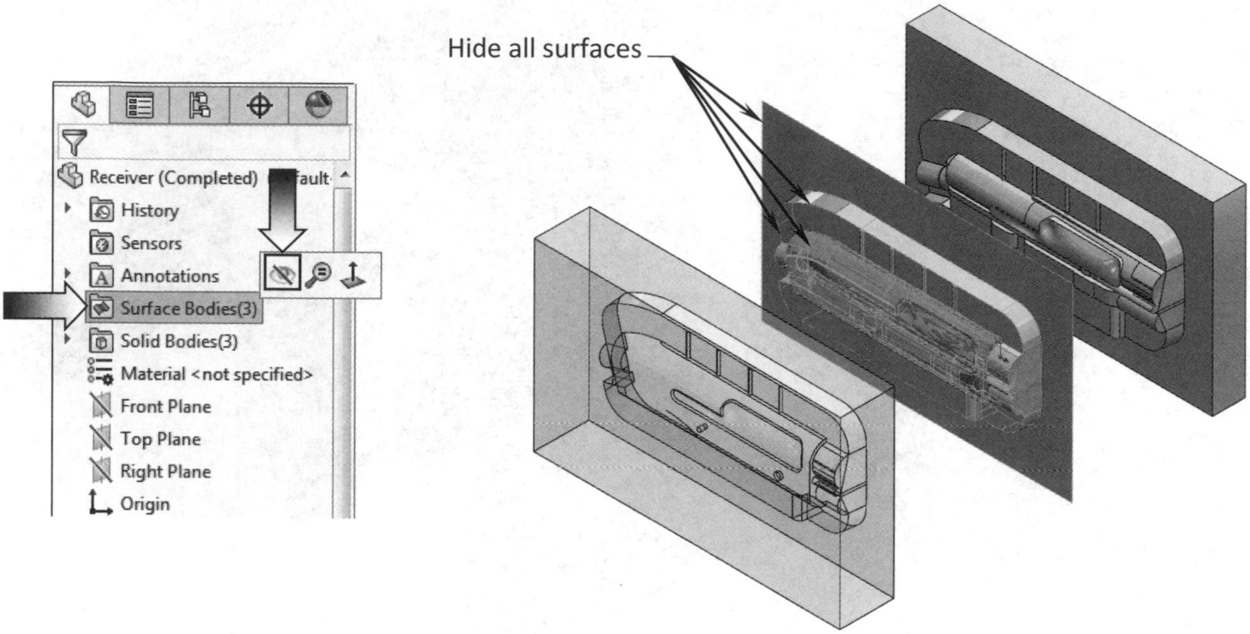

19. Saving your work:

Select **File / Save As**.

Enter **Mold Manual Creation** for the file name.

Click **Save**.

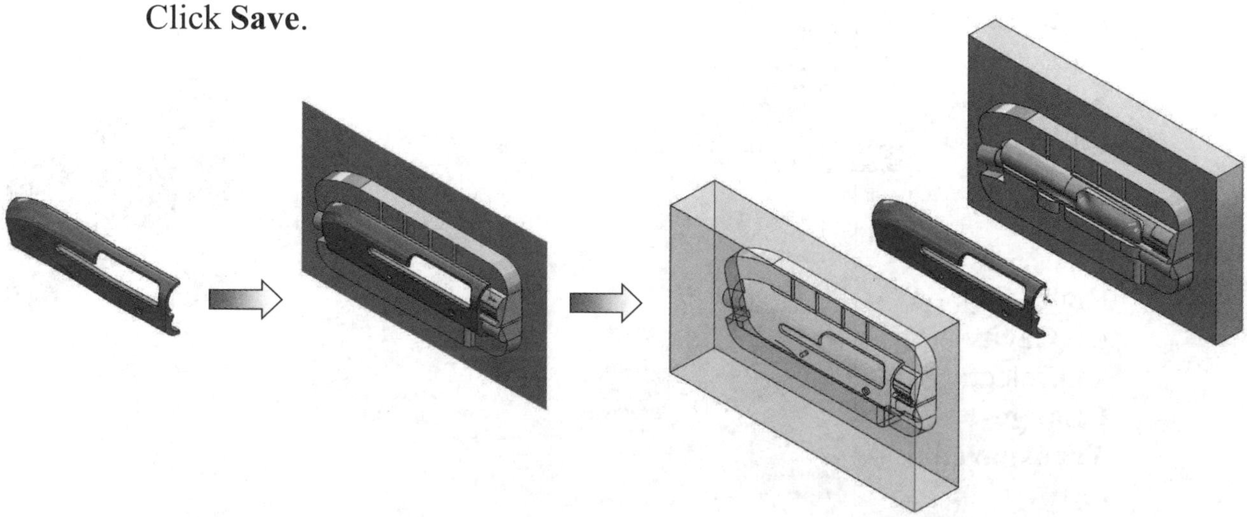

Creating Slides and Cores

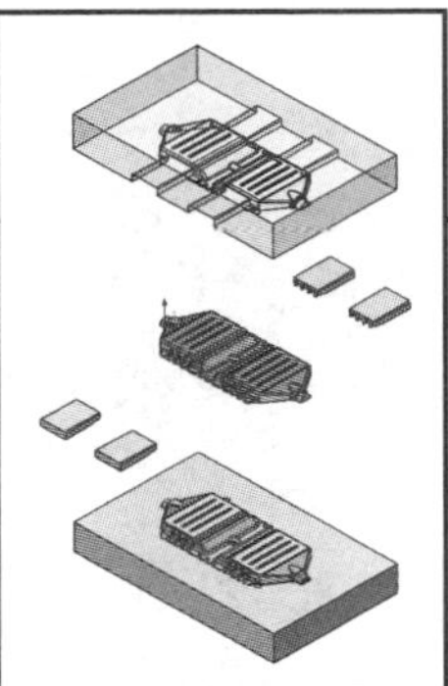

Mold Tooling Design
Creating Slides and Cores

Many times undercuts cannot be avoided in plastic part designs. Features such as slots, triggers, locks, and latches, etc. are often seen in plastic parts even though they may be more expensive to design and manufacture.

The Undercut Analysis tool can assist you with finding and visualizing trapped areas that may prevent the part from ejecting from the mold. These areas will need additional tooling such as lifters and cores to allow the part to be released from the tooling using the primary direction of pull.

A Slide or Core is a piece of tooling that slides out of the mold from the side, perpendicular to the direction that the part is ejected from the mold.

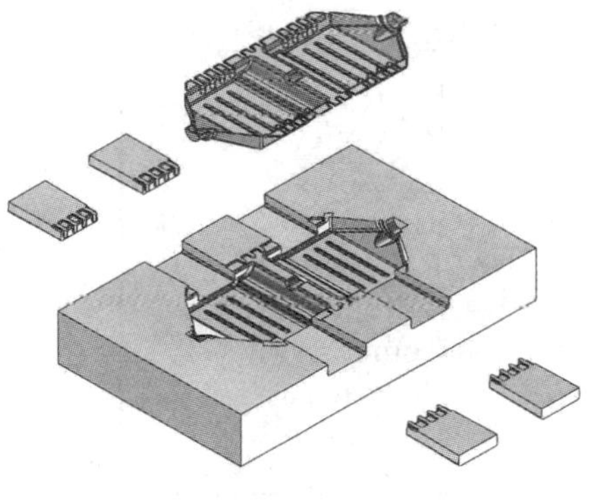

A sketch is used to define the shape and location of the core and it is normally drawn directly on the sides of the core or cavity block. Generally the sketch plane is parallel or perpendicular to the direction in which the side core travels away from the plastic part.

The Core command is used to extrude the sketch into a separate solid body and at the same time, it is subtracted from the core or cavity body. The Core command works like the Split command; they both divide a body into two or more bodies.

Additional tooling like Lifters, Core Pins, and Ejector Pins may be required to complete the mold design, but this lesson will discuss the use of the standard mold tools to automate the creation of the Core and Cavity as well as the use of the Core command.

Creating Slides and Cores
Mold Tooling Design

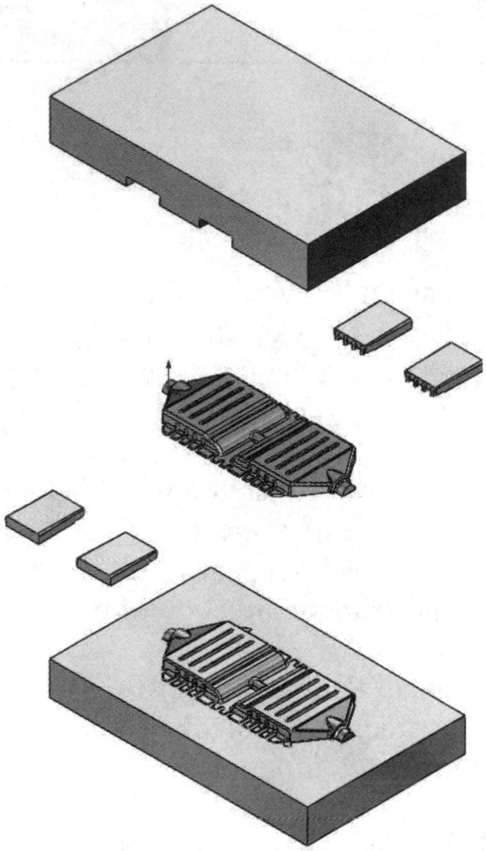

View Orientation Hot Keys:

Ctrl + 1 = Front View
Ctrl + 2 = Back View
Ctrl + 3 = Left View
Ctrl + 4 = Right View
Ctrl + 5 = Top View
Ctrl + 6 = Bottom View
Ctrl + 7 = Isometric View
Ctrl + 8 = Normal To Selection

Dimensioning Standards: **ANSI**
Units: **INCHES** – 3 Decimals

Tools Needed:

 Scale

 Parting Lines

 Parting Surfaces

 Tooling Split

 Extruded Boss/Base

 Core

1. Opening a part document:

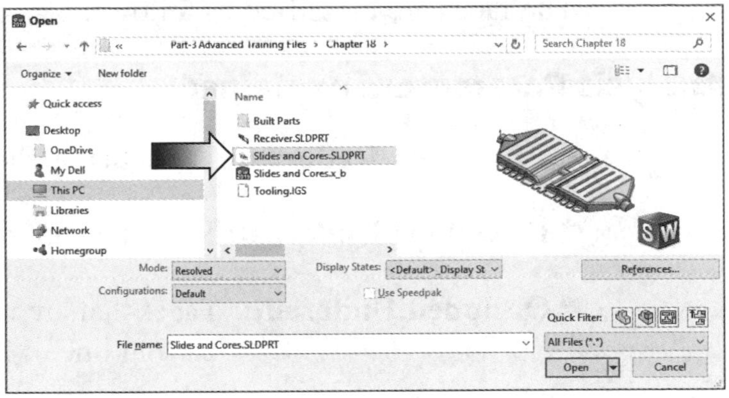

Select **File / Open.**

Browse to the Training Files folder and open a part document named:
Slides and Cores.sldprt

One of the first things to do is to examine the model for potential problems that might prevent the core and cavity from separating.

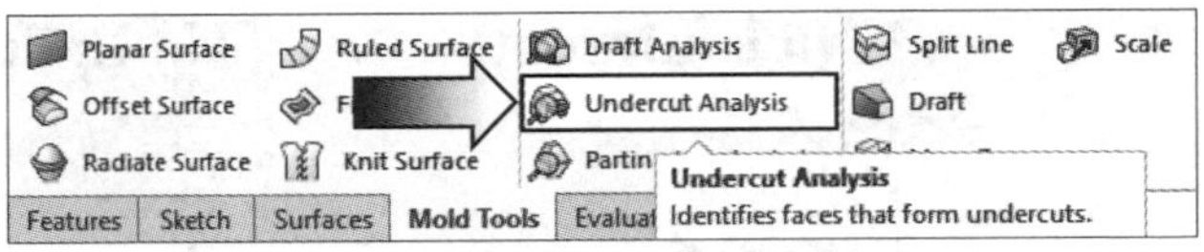

2. Analyzing the undercuts:

The Undercut Analysis tool finds trapped areas in a model that cannot be ejected from the mold. These areas require a side core. When the main core and cavity are separated, the side core slides in a direction perpendicular to the motion of the main core and cavity, enabling the part to be ejected.

Undercut Analysis works only on solid bodies, not surface bodies.

Right-click on one of the tool tabs and enable the **Mold Tools**.

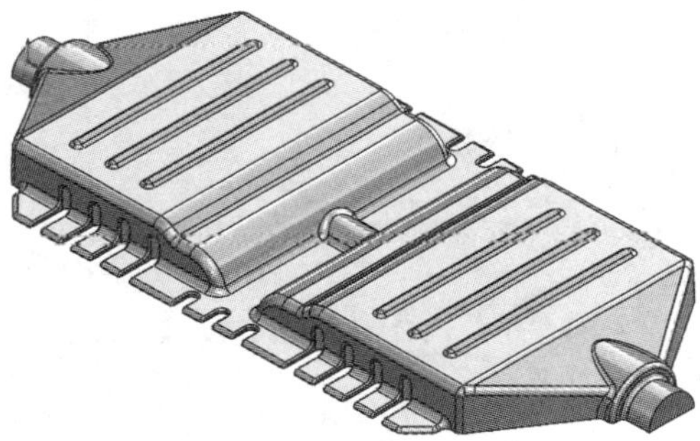

Select the **Undercut Analysis** command.

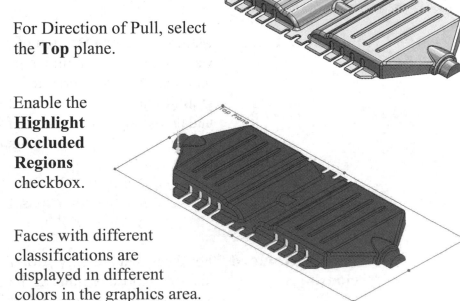

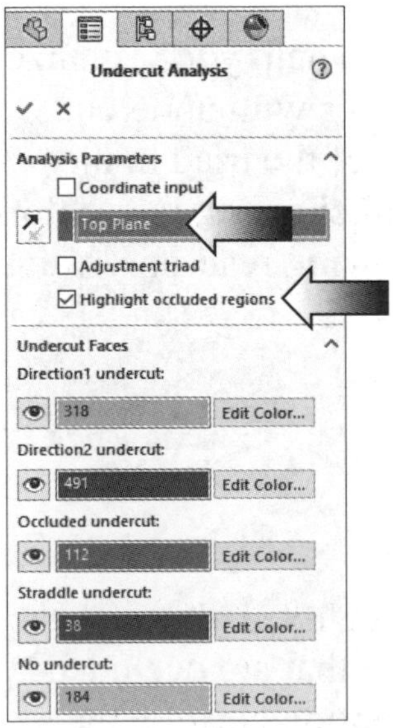

For Direction of Pull, select the **Top** plane.

Enable the **Highlight Occluded Regions** checkbox.

Faces with different classifications are displayed in different colors in the graphics area.

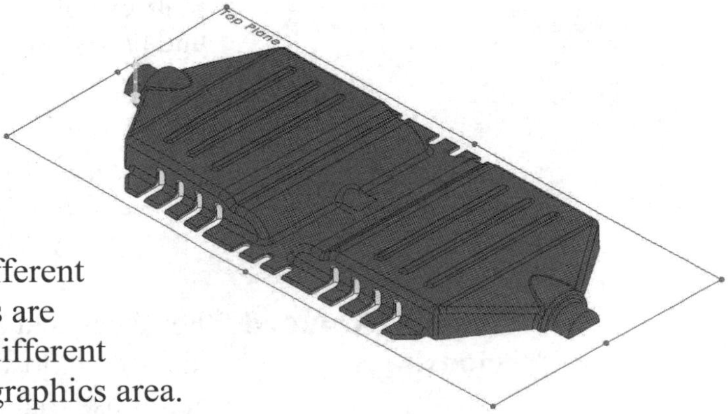

The faces are classified as follows:

Analysis Parameters explained:

* **Direction1 Undercut:** Faces that are not visible from above the parting line.

* **Direction2 Undercut:** Faces that are not visible from below the parting line.

* **Occluded Undercut:** Faces that are not visible from above or below the parting line.

* **Straddle Undercut:** Faces that have undercuts in both directions.

* **No Undercut:** No Undercut.

---

* **Direction of Pull:** (Not required if you select a Parting Line below.) Select a planar face, a linear edge, or an axis to define the draw direction, or select Coordinate input and set conditions along the X, Y, and Z axes.

* **Parting Line:** Faces above the parting line are evaluated to determine if they are visible from above the parting line. Faces below the parting line are evaluated to determine if they are visible from below the parting line. This identifies depressions in the wall of the part that require a side core, and also helps you to identify sections of the parting line that you can modify to avoid the need for side cores.

* **Adjustment Triad:** Manipulates the direction of pull to help you visualize ways to avoid or minimize problems with undercut regions. When you drag the rings of the triad in the graphics area, the direction of pull changes, face colors update dynamically, and the following read-only values appear in the PropertyManager:

  Angle with X axis
  Angle with Y axis
  Angle with Z axis

* **Highlight Occluded Regions:** For faces that are only partially occluded, the analysis identifies those regions of the face that are occluded and those that are not. With this option cleared, the

analysis identifies the entire face as being occluded.

*** Undercut Faces:** Faces with different classifications are displayed in different colors in the graphics area. The results update in real time when you change the direction of pull.

Click **Cancel** [X] to close the Undercut Analysis. A side core will be created to release the trapped areas in the mold.

3. Scaling the part:

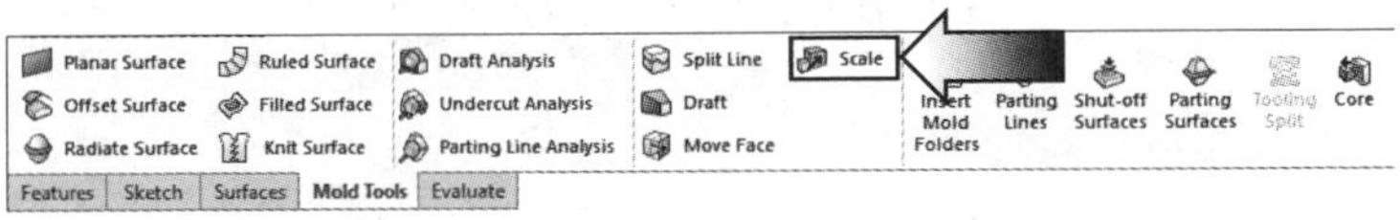

The Scale feature scales only the geometry of the model. It does not scale dimensions, sketches, or reference geometry. To temporarily restore the model to its un-scaled size, you can roll-back or suppress the Scale feature.

Use the Scale tool to account for the shrink factor when plastic cools.
For odd shaped parts and glass filled plastics, you can specify nonlinear values.

Click the **Scale** command on the **Mold Tools** toolbar.

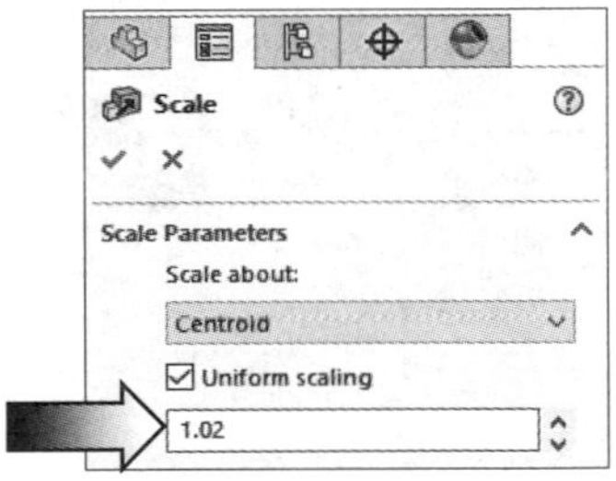

For Scale About, use the default **Centroid** option.

Enable the **Uniform Scaling** checkbox.

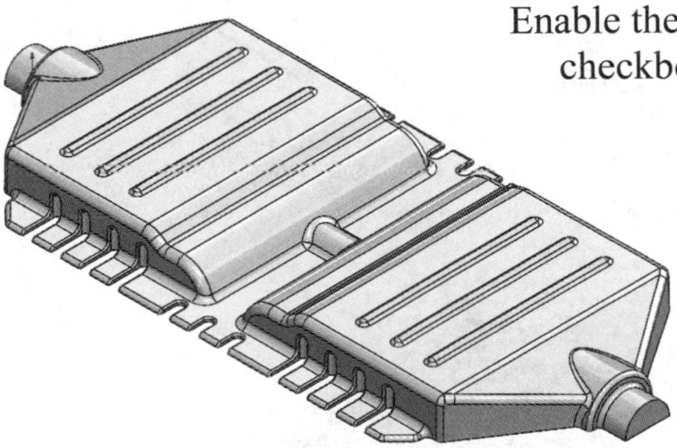

For Scale Factor, enter **1.02%** (arrow) (2% larger).

Click **OK**.

A Scale feature is like any other features listed in the FeatureManager design tree: it manipulates the geometry, but it does not change the definitions of features created before it was added.

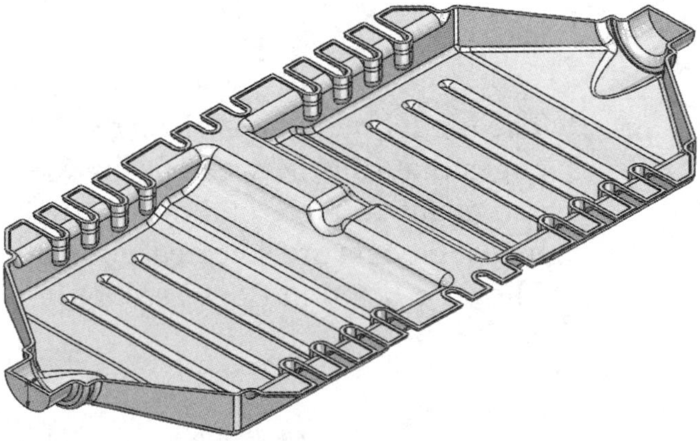

4. Creating the parting lines:

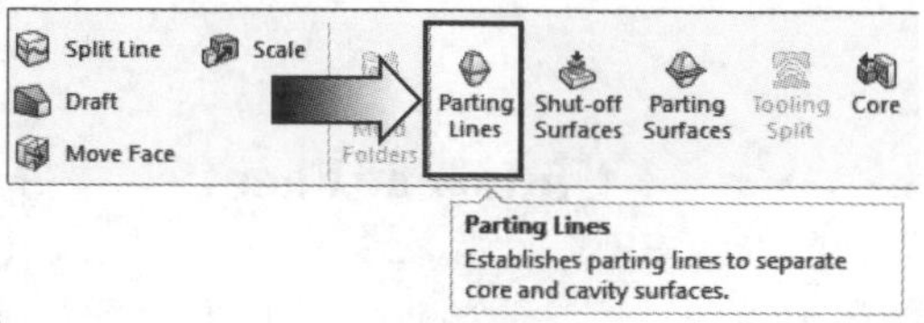

The parting lines lie along the edge of the molded part, between the core and the cavity surfaces. They are used to create the parting surfaces and to separate the surfaces.
The parting line is created after the model is scaled and proper draft is applied.

Select the **Parting Lines** command from the Mold Tools toolbar (arrow).

For Direction of Pull, select the **Top** plane (arrow).

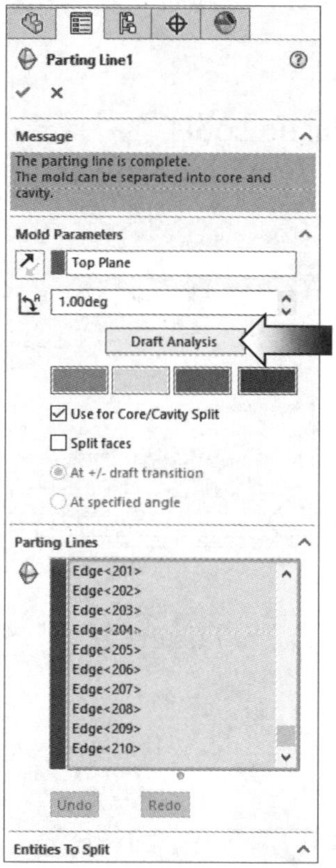

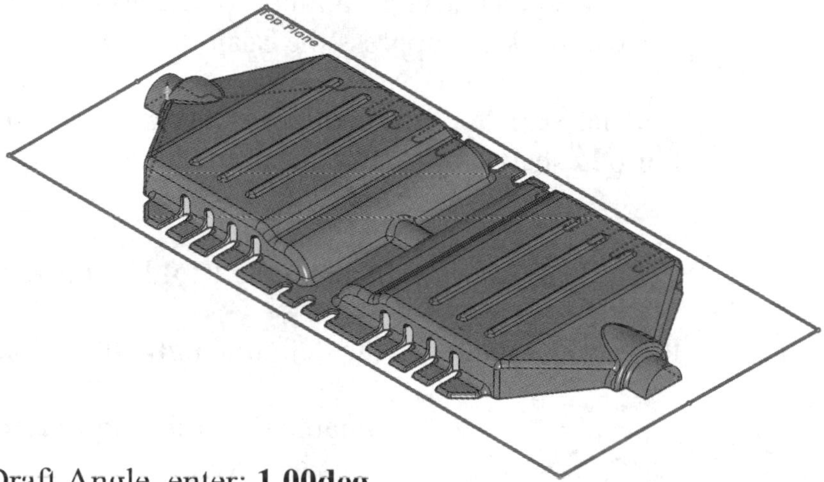

For Draft Angle, enter: **1.00deg**.

Click the **Draft Analysis** button (arrow).

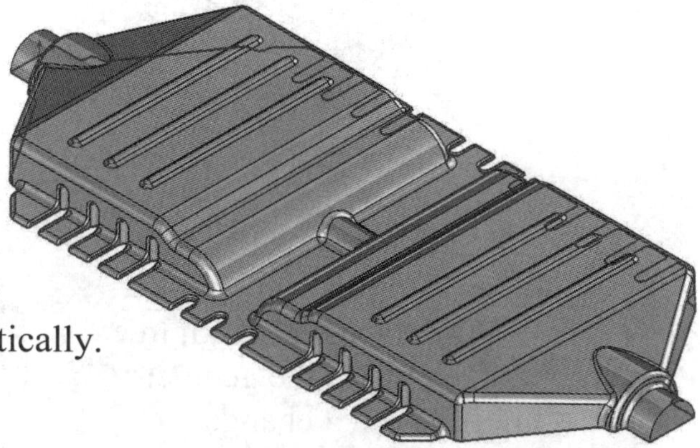

The model includes a chain of edges that runs between positive and negative faces and the parting line segments are selected automatically. They are listed in Edges section.

Click **OK**.

5. Creating the parting surfaces:

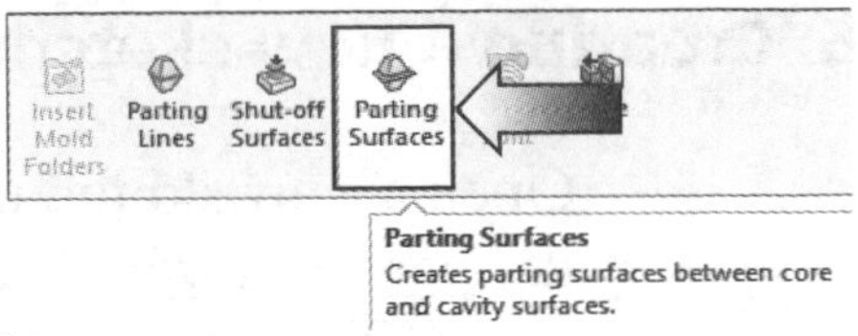

The Parting Surfaces split the mold cavity from the core. Create the parting lines and shut off surfaces before creating parting surfaces.

The Shut-Off surfaces is used to shut-off the openings of thc through holes in the plastic parts. Since this model does not have any through holes, we can skip this step and move on to creating the parting surfaces.

Select the **Parting Surfaces** command (arrow).

For Mold Parameters, select: **Perpendicular to Pull** (arrow).

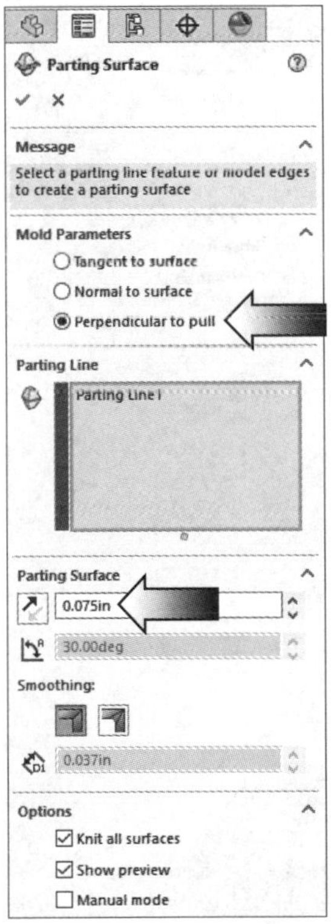

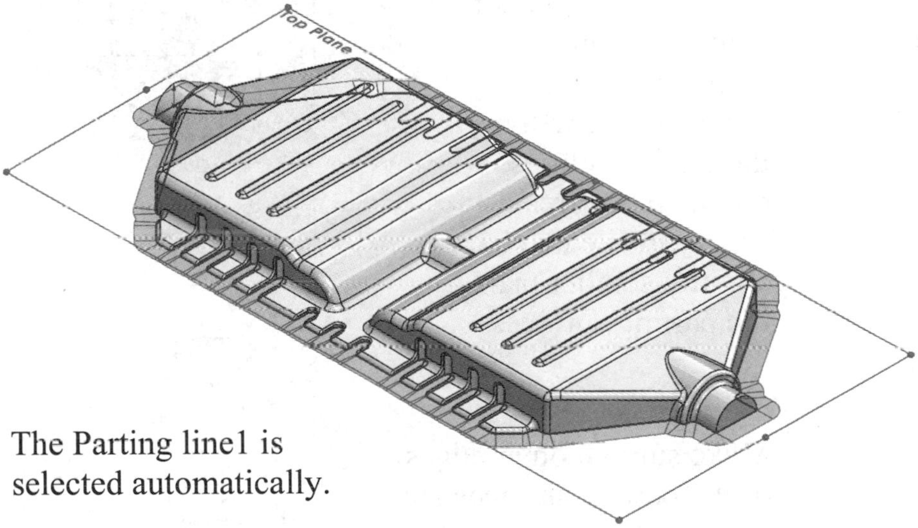

The Parting line1 is selected automatically.

We will use a small distance to prevent the parting surface from overlapping with itself. Additional materials will be added to the mold blocks later on when finalizing the mold block sizes.

For Distance, enter **.075in.**; also enable the **Knit All Surfaces** checkbox.

Click **OK**.

6. Creating a new sketch:

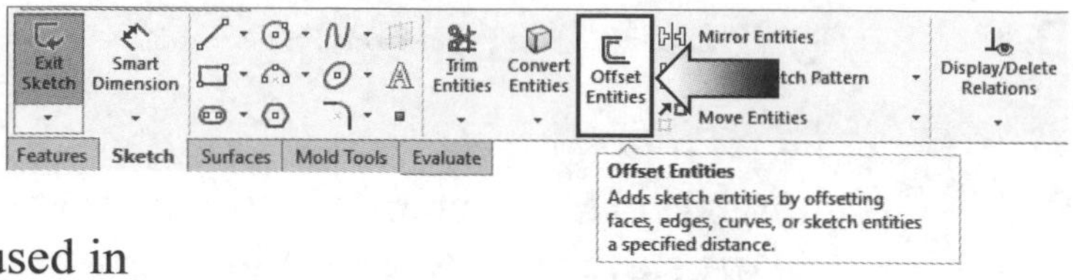

Open a **new sketch** on the parting surfaces as noted. This sketch will be used in the next step to create the tooling split.

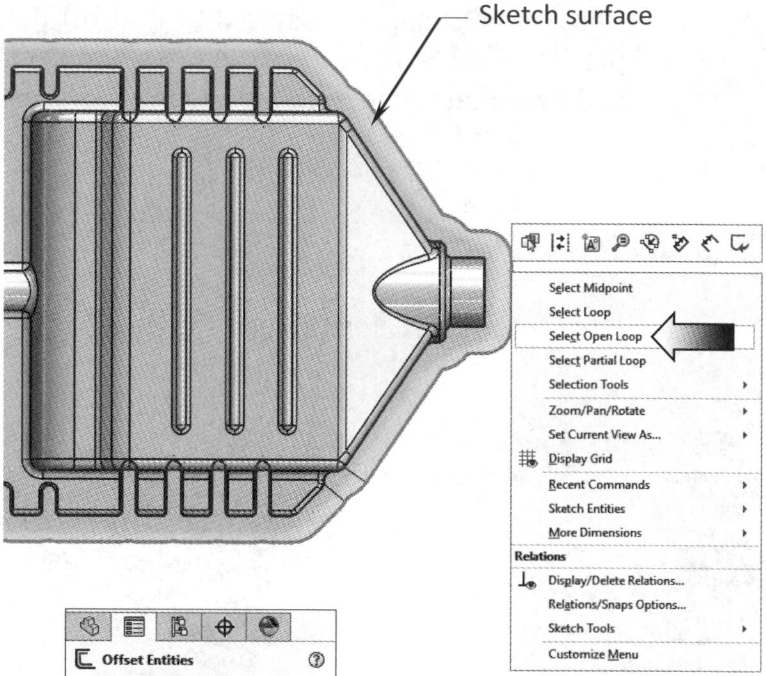

Right-click on one of the outer edges of the parting surfaces and pick: **Select Open Loop** (arrow).

The sketch that is used for tooling split should be smaller than the parting surfaces. We will use the Offset Entities command to create a new profile that is slightly smaller than the parting surfaces.

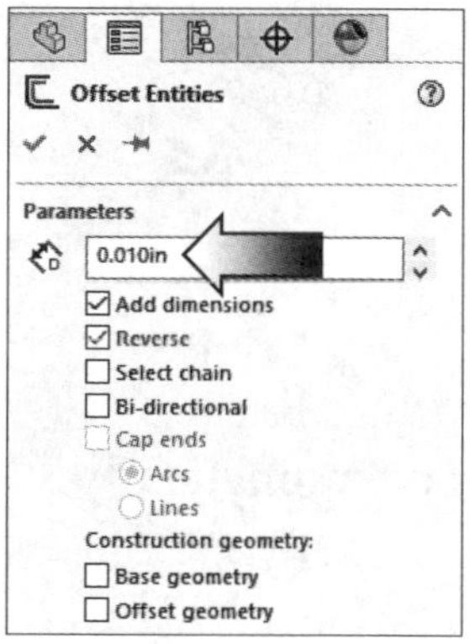

Make sure all outer edges of the parting surfaces are still highlighted; click the **Offset Entities** command.

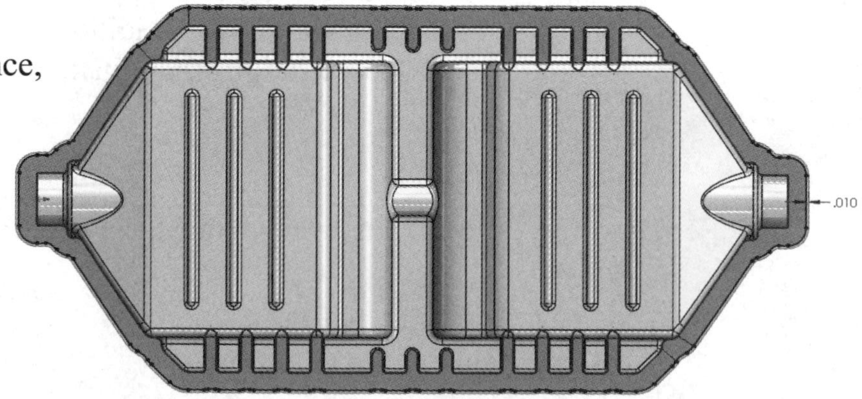

For Offset Distance, enter **.010in**. and click **Reverse** to place the entities on the inside of the parting surfaces.

Click **OK**.

7. Creating the tooling split:

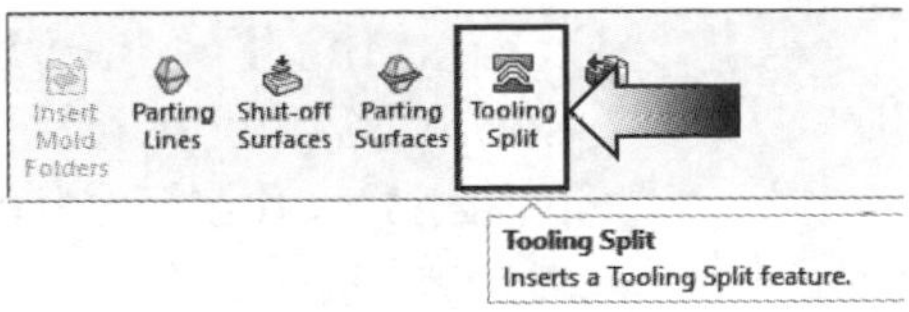

The Tooling Split tool uses the parting line, the shut off surfaces, and the parting surfaces information to create the core and cavity and allows the user to specify the block sizes.

Exit the Sketch and select **Tooling Split** (arrow).

For Upper Block size, enter: **.600in**

For Lower Block size, enter: **.500in**

Click **OK**.

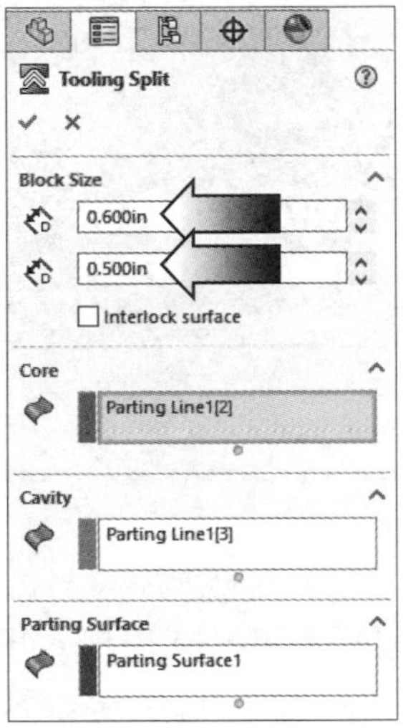

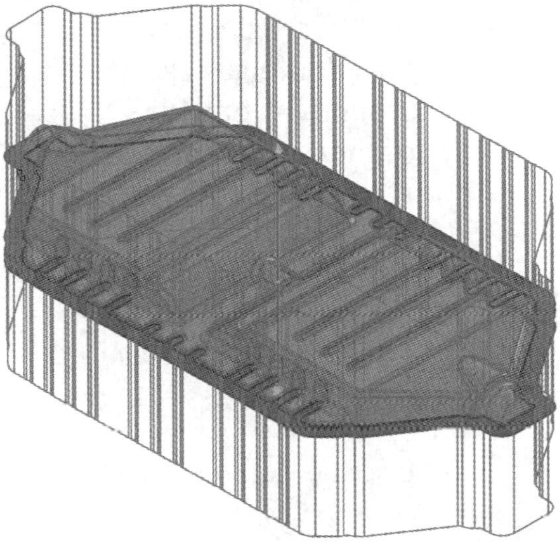

8. Finalizing the upper block size:

For clarity, right-click the **Surface Bodies** folder and select **Hide**. Also right-click and hide **Parting Line1**.

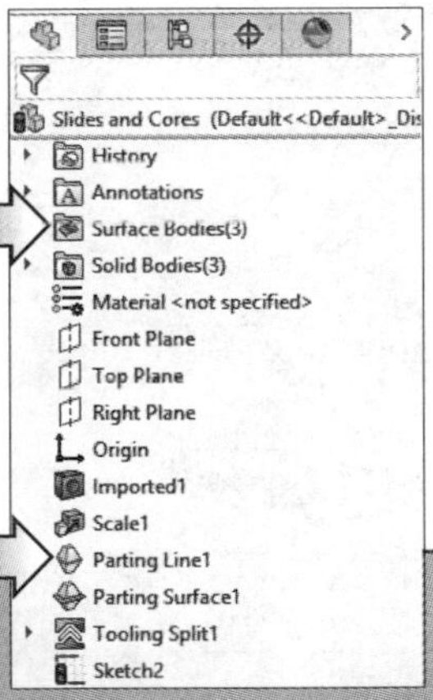

Select the topmost surface of the upper block and open a **new sketch**.

Sketch a **Corner Rectangle** as shown on the right.

Add the **dimensions** as indicated to fully define the sketch.

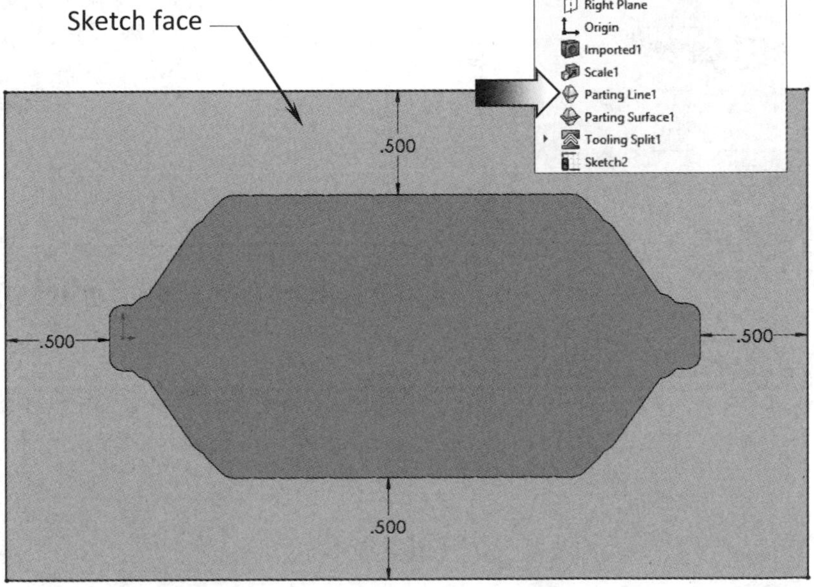

Expand the **Tooling Split1** on the FeatureManager tree, select the **Sketch1** and click **Convert Entities**.

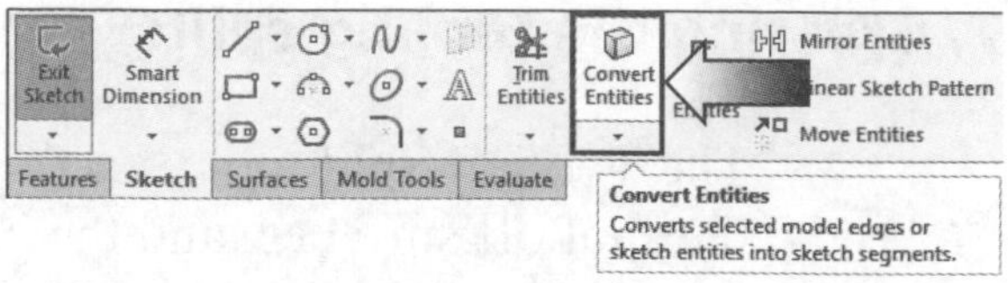

The Corner Rectangle and the Converted Entities formed a closed contour and will be used to add the additional material around the outside of the upper block.

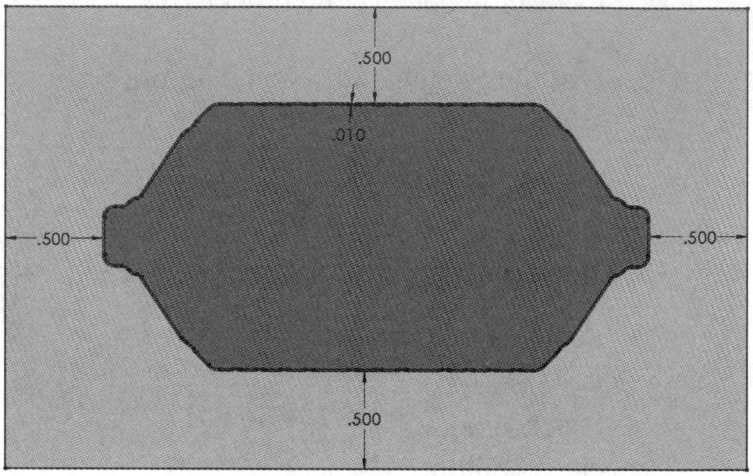

Switch to the Features toolbar and select: **Extruded Boss-Base**.

Use the default **Blind** type and a depth of **.600in**. Click **Reverse** direction.

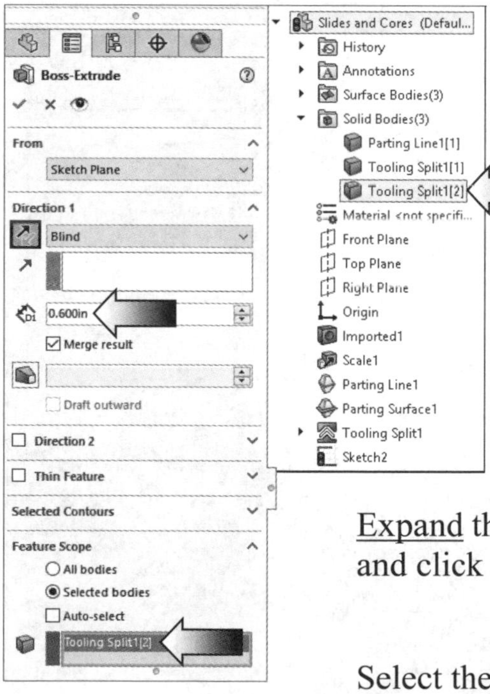

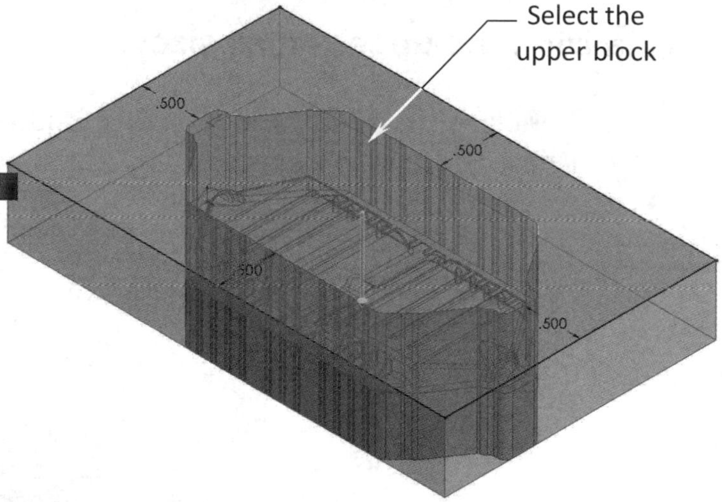

Expand the **Feature Scope** section and click the **Selected Bodies** option.

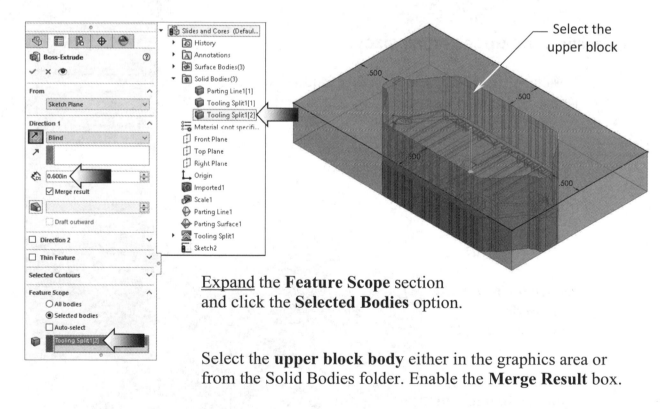

Select the **upper block body** either in the graphics area or from the Solid Bodies folder. Enable the **Merge Result** box.

Click **OK**.

9. Finalizing the lower block size:

Switch to the **Isometric** view (Control+7).

Hold the **Shift** key and press the **Up Arrow** twice to rotate to the reverse isometric view **180º** (the default angle is 90º when rotating with the Shift and Arrow keys).

Open a **new sketch** on the top surface of the lower block.

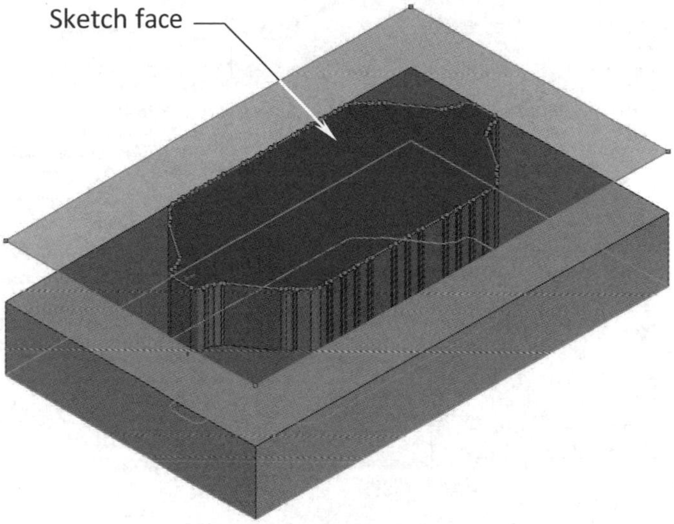

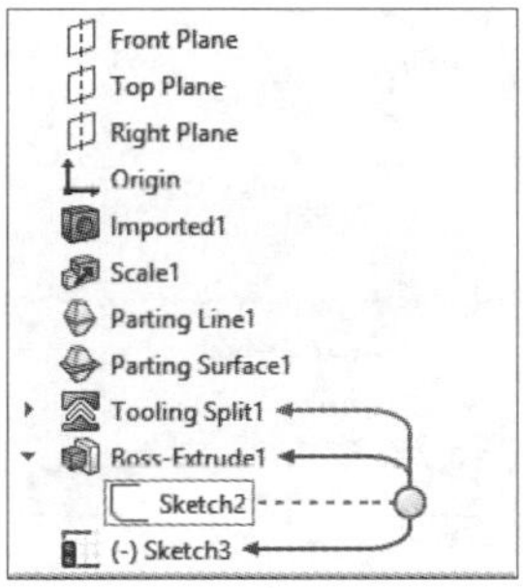

Expand the Boss-Extrude1 and select the **Sketch2**.

Click **Convert-Entities** to copy the entire sketch.

Switch to the **Features** toolbar and select: **Extruded Boss-Base**.

Use the default **Blind** type and a depth of **.500in**.

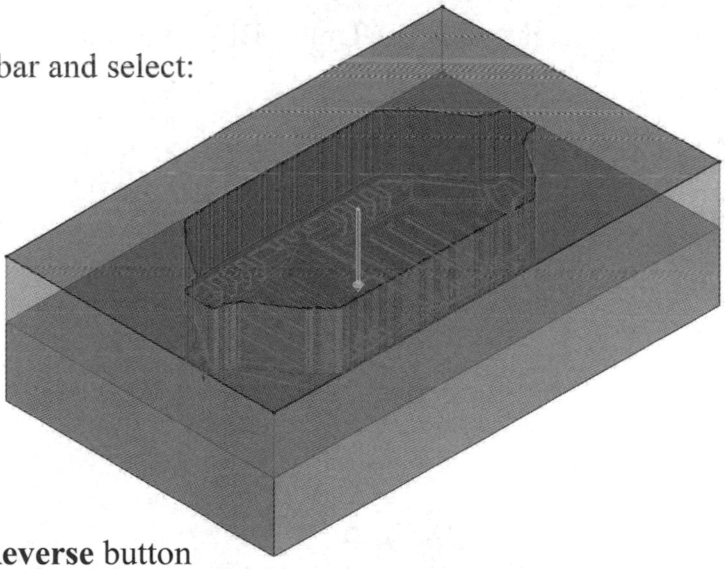

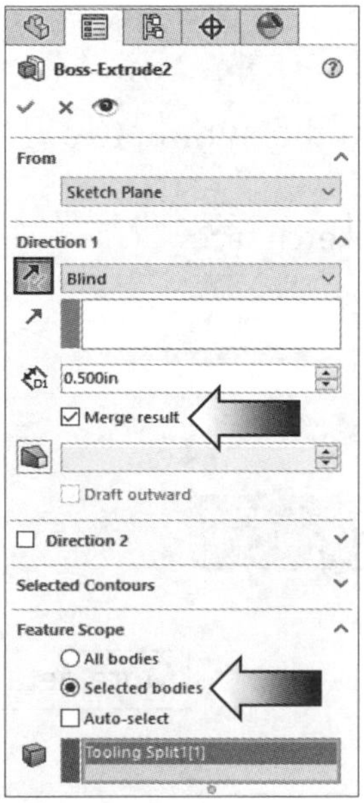

Enable the **Reverse** button and the **Merge Result** checkbox.

Expand the **Feature Scope** section and click the **Selected Bodies** option. Select the **lower block** body from the graphics area.

Click **OK**.

10. Renaming the bodies and assigning materials:

Expand the **Solid Bodies folder** and rename the solid bodies as shown below.

Right-click **each solid body** and assign the material as noted.

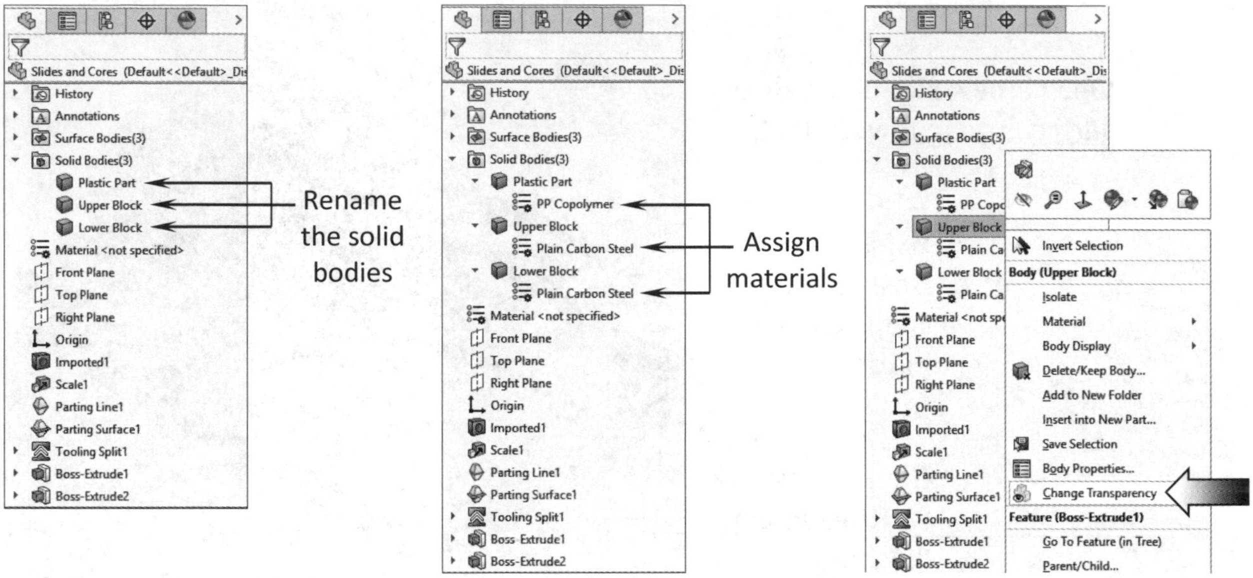

Right-click the Upper Block body and select: **Change Transparency** (arrow).

11. Creating the front slide core:

We will create a couple of slide cores to capture the undercut features in the part.

Open a **new sketch** on the face indicated.

Sketch the profile shown. Use the mirror function to maintain the symmetric relations between the sketch entities.

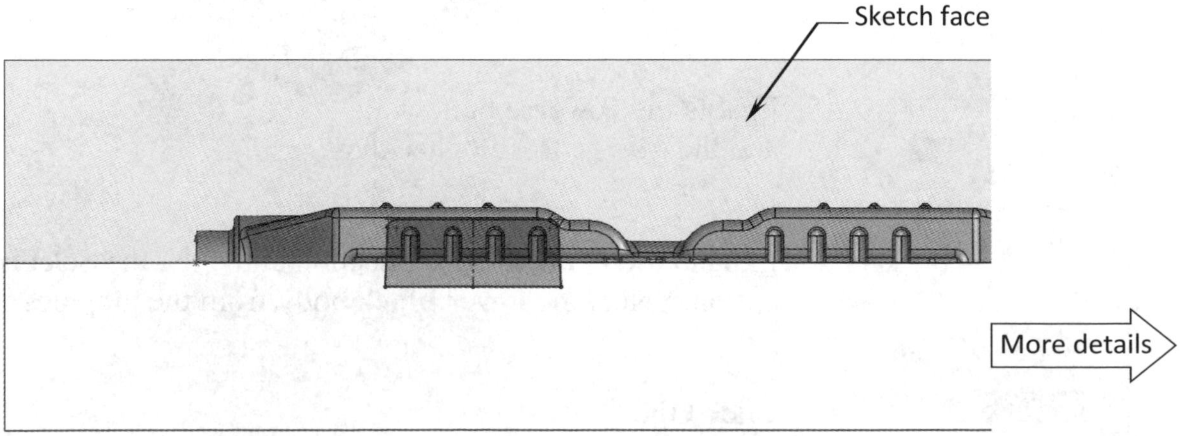

Notice the sketch is overlapped with the lower block by .073in? We will select only the upper block when making the core.

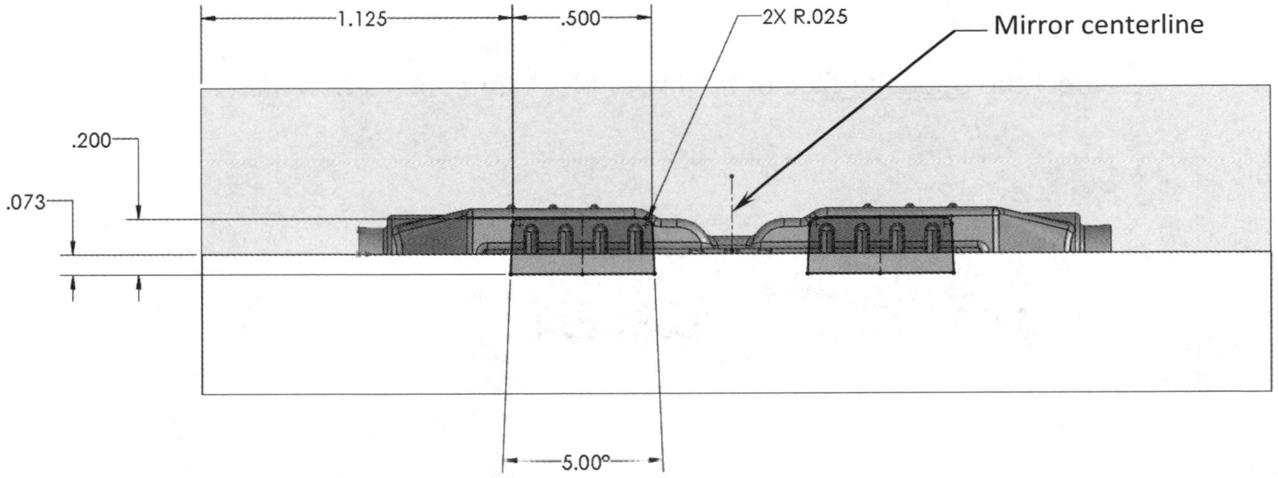

Mirror the sketch and add the dimensions shown to fully define it.

Exit the sketch and select the **Core** command on the **Mold Tools** toolbar.

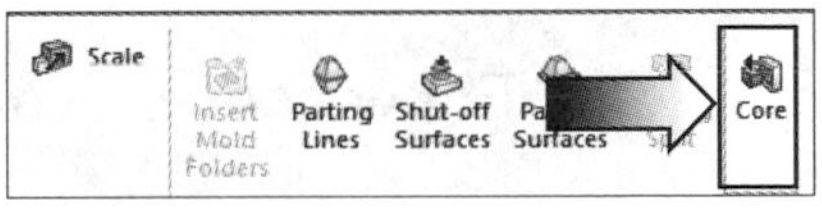

The upper block (Boss-Extrude1) should be selected automatically.

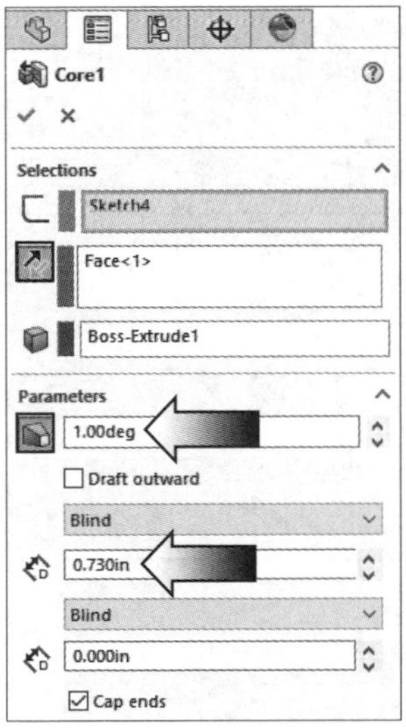

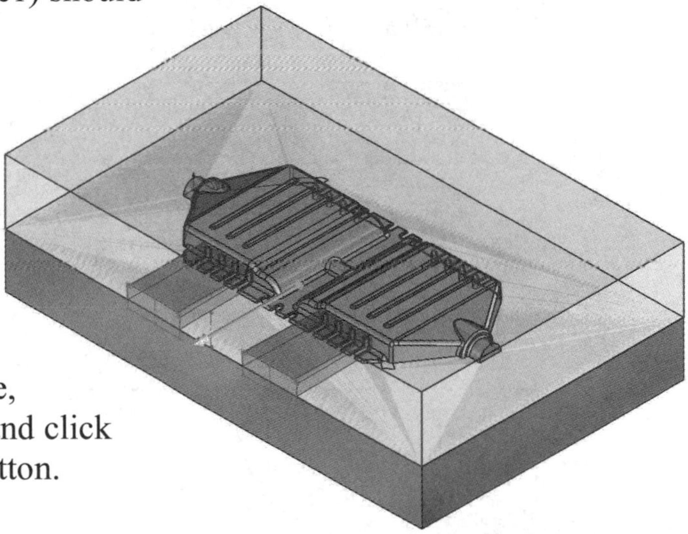

For Draft Angle, enter **1.00deg** and click the **Reverse** button.

Use the default **Blind** type and enter **.730in** for depth.

The extrude direction is towards the plastic part.

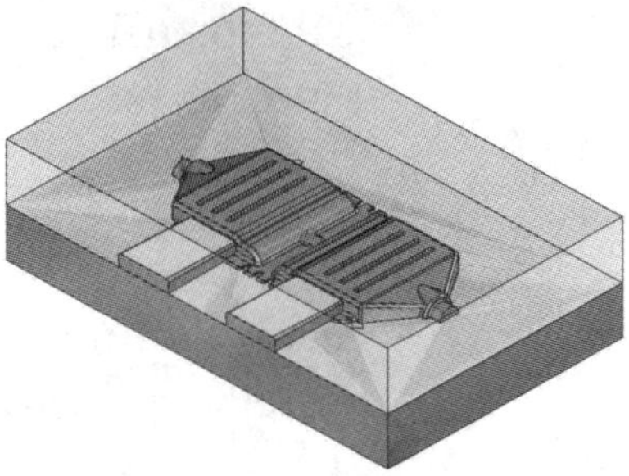

Click **OK**.

12. Creating the back slide core:

We will create an exploded view after both slide cores are created.

Select the opposite face of the lower block and open a **new sketch**.

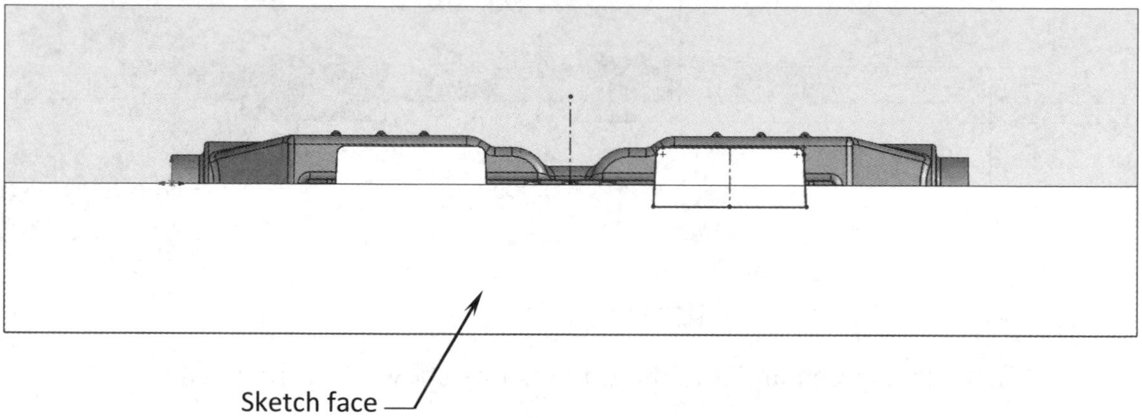

Either convert, copy & paste the last sketch, or re-create it once again.

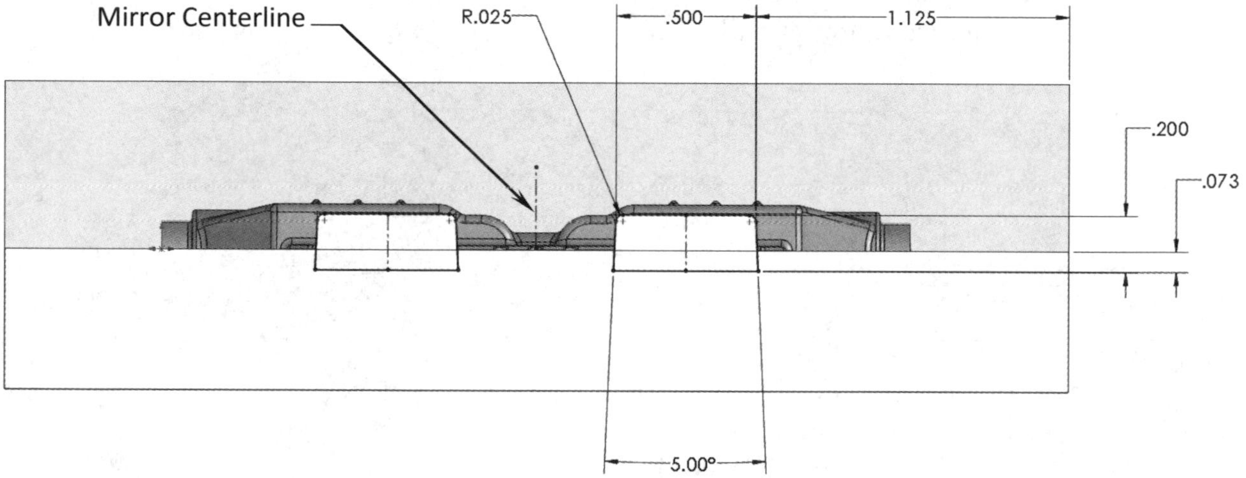

Avoid snapping any endpoints of the profile as it may cause the sketch to become over defined.

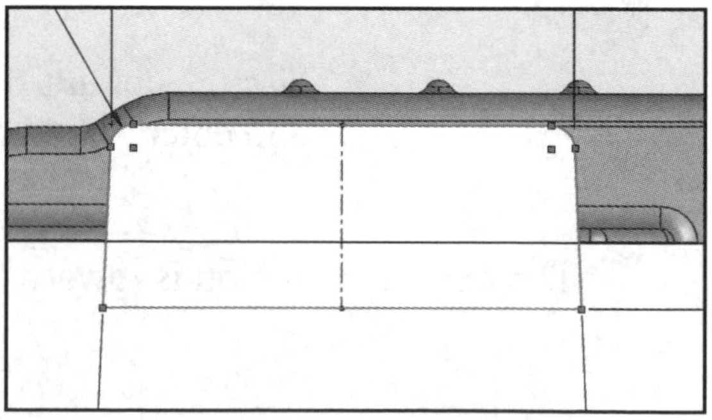

Verify that the sketch is fully defined and positioned correctly.

Exit the sketch and select the **Core** command on the **Mold Tools** toolbar.

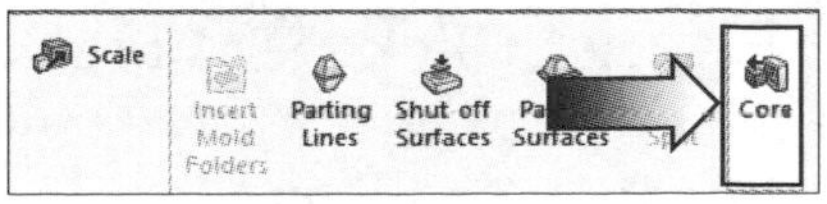

The lower block (Boss-Extrude2) should be selected automatically.

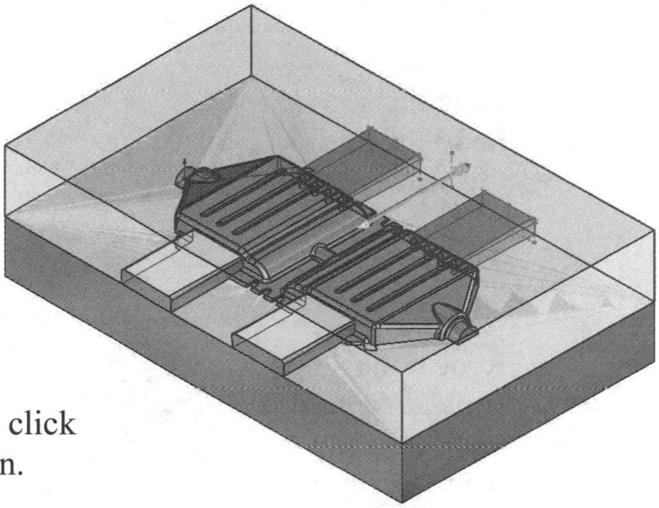

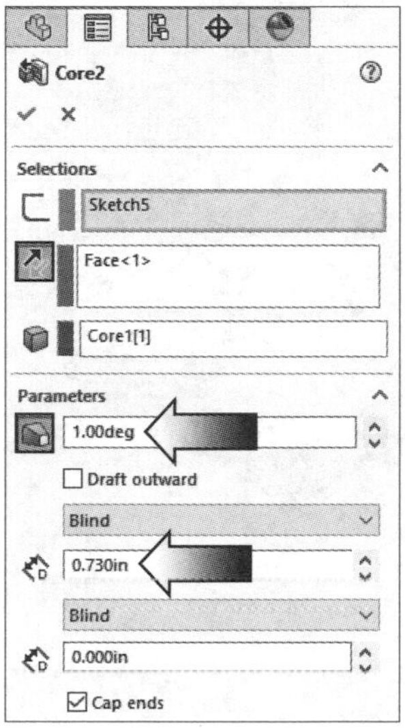

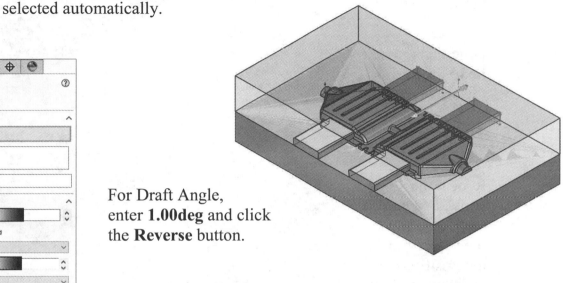

For Draft Angle, enter **1.00deg** and click the **Reverse** button.

Use the default **Blind** type and enter **.730in** for depth.

Verify the extrude direction is towards the part.

Click **OK**.

13. Creating an exploded view:

An exploded view in a multibody part shows the solid bodies spread out but positioned to show how they fit together.

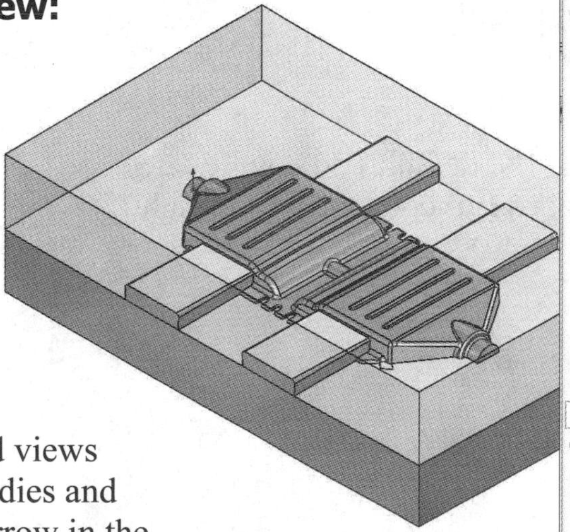

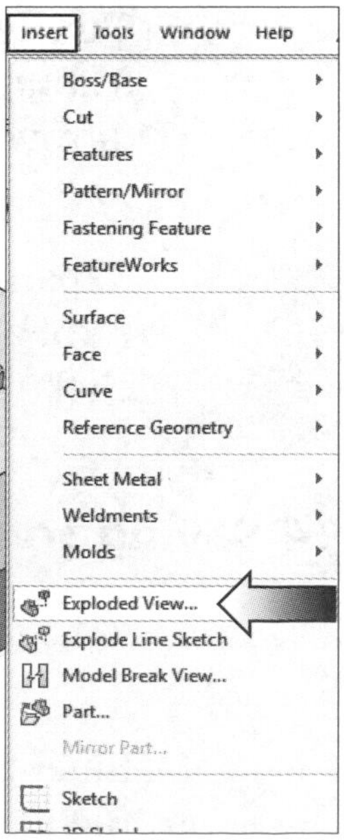

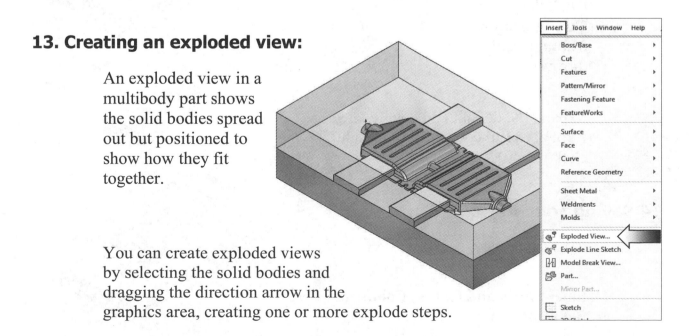

You can create exploded views by selecting the solid bodies and dragging the direction arrow in the graphics area, creating one or more explode steps.

Click [icon] or select **Insert, Exploded View**.

Select the **upper block** from the graphics area.

Drag the **Y arrow-head** upward towards the vertical direction approx. **3.500 inches**.

Repeat the step above and explode the lower block and the 4 core blocks to the positions approximately as shown.

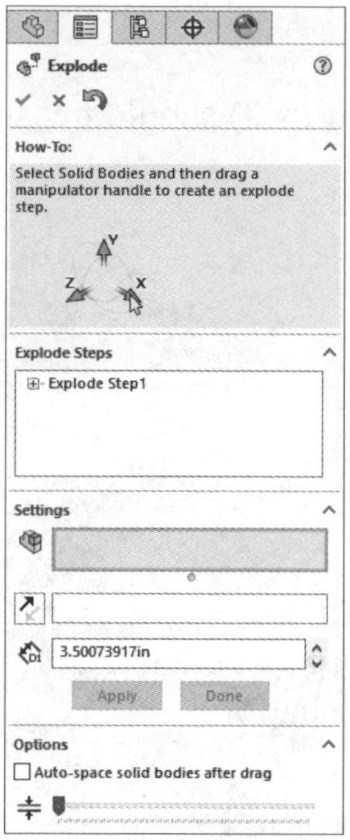

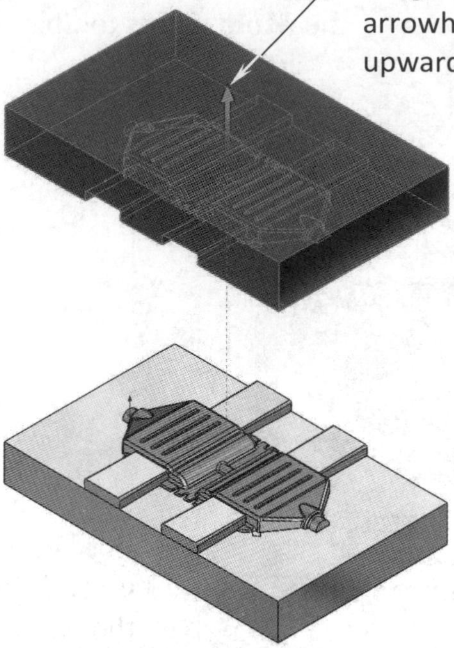

Click **OK** when the exploded view is completed.

Click the **Configuration-Manager** tab (arrow).

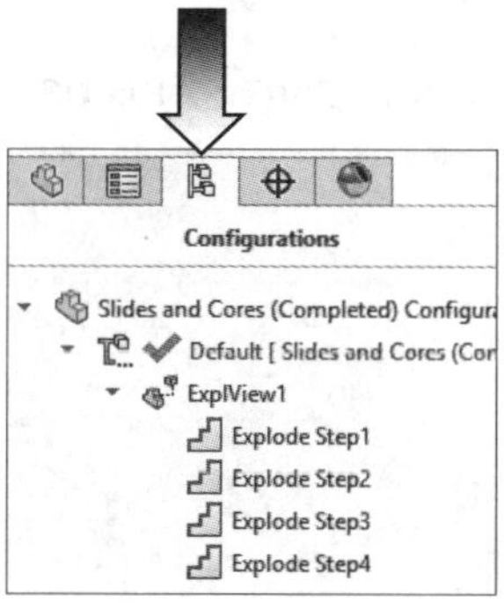

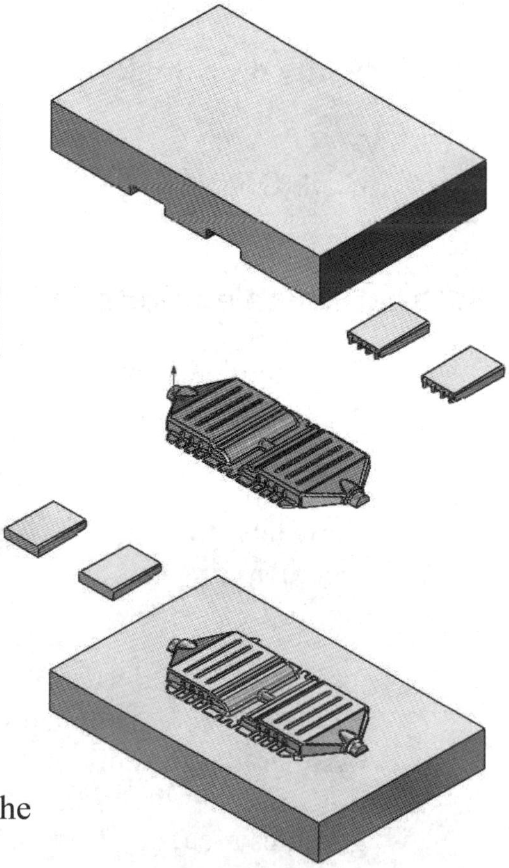

Expand the default and double-click **ExplView1** to collapse it (and right-click to edit).

14 Saving your work:

Select **File, Save As**.

Enter **Slides and Cores (Completed)** for the file name and press **Save**.

CHAPTER 20

Top Down Assembly – Part 1

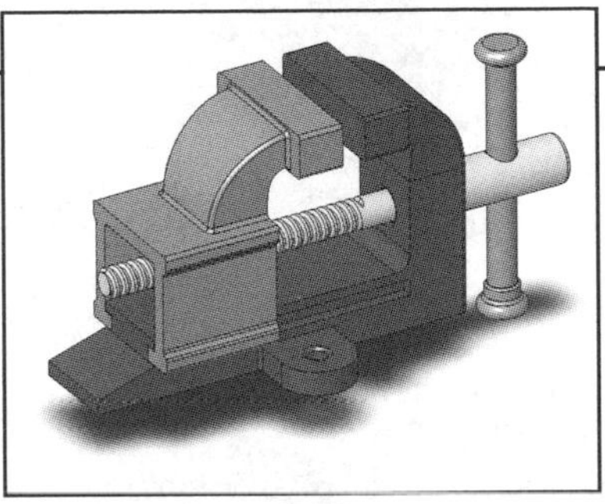

Top Down Assembly - Part 1
Miniature Vise

This chapter will guide us through the use of special techniques for creating new parts in the context of an assembly or Top Down Assembly Design.

Using the existing geometry of other components such as their locations, sketch entities, and model edges to construct the new components is referred to as In-Context Assembly. This option greatly helps you capture your design intent and reduce the time it takes to do a design change, and having the parts update within themselves based on the way they were created.

While working in the top down assembly mode, every time a face or a plane is selected for a new sketch, the system automatically creates an INPLACE mate to reference the new part.

The Inplace mates can be suppressed so that components can be moved or re-positioned and the Inplace mates can also be deleted as well; new mates can be added to establish new relationships with other components.

When a part is being edited in the Top Down Assembly mode, the Edit-Component icon  is selected, and the component's color changes to Blue (or Magenta depending on the color settings in the system options).

Upon the successful completion of this lesson, you will have a better understanding of the 2 assembly methods used in SOLIDWORKS: the traditional Bottom Up assembly (where parts are created separately, then inserted into an assembly document and mated together) and the dynamic Top-Down assembly (where parts can be created together in the context of an assembly).

Miniature Vise
Top Down Assembly – Part 1

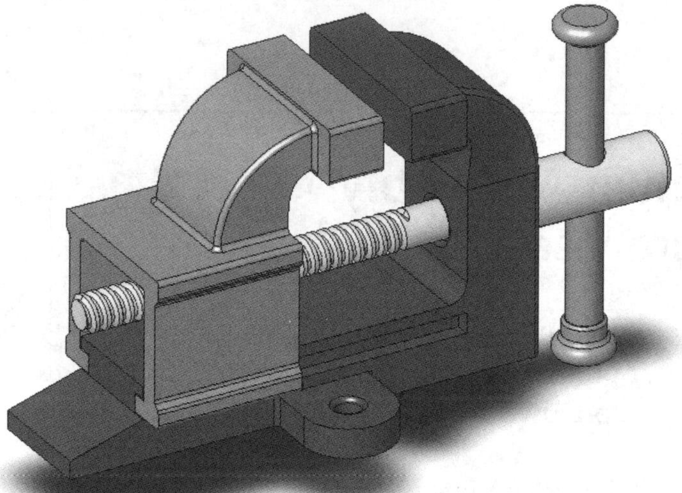

View Orientation Hot Keys:

Ctrl + 1 = Front View
Ctrl + 2 = Back View
Ctrl + 3 = Left View
Ctrl + 4 = Right View
Ctrl + 5 = Top View
Ctrl + 6 = Bottom View
Ctrl + 7 = Isometric View
Ctrl + 8 = Normal To Selection

Dimensioning Standards: **ANSI**

Units: **INCHES** – 3 Decimals

Tools Needed:

Insert Sketch	Rectangle	Circle
Dimension	Add Geometric Relations	Sketch Mirror
Offset Entities	Planes	Fillet/Round
Base/Boss Extrude	Loft	Edit Component

1. Starting a new assembly template:

Select **File / New / Assembly**.

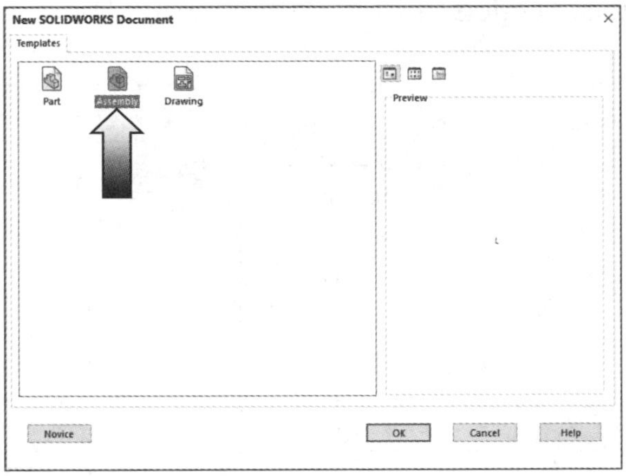

Click **Cancel**  to exit the **Begin Assembly** mode.

<u>**Save**</u> the new assembly document as **Mini Vise.sldasm**.

2. Creating the Base part:

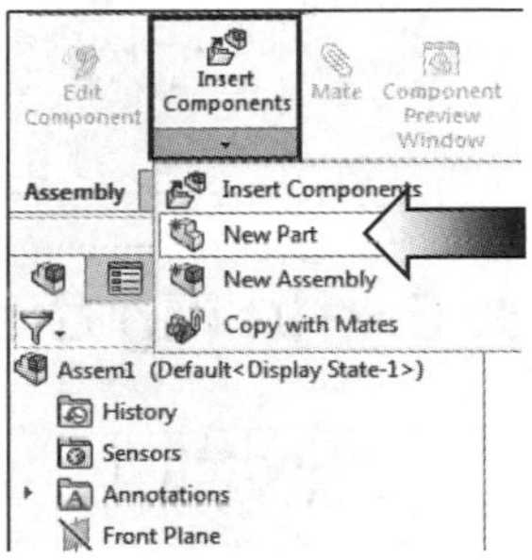

Select **Insert, Component, New Part.**

Select the <u>Front</u> plane from the FeatureManager tree. An Inplace1 mate is created referencing the new part.

A new component is created using a default name **[Part1^Assembly]<1>**.

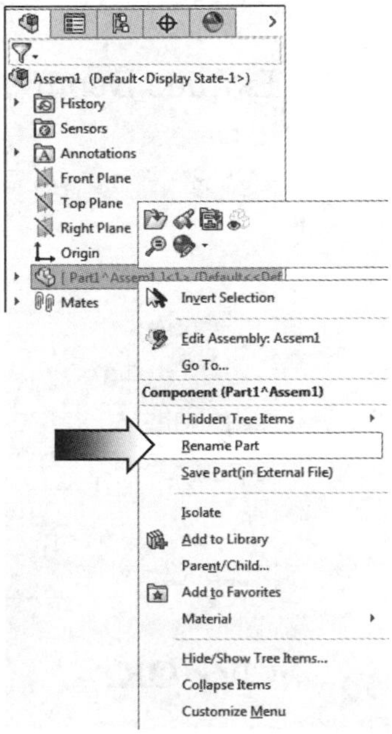

To rename the part, right click on the default name and select **Rename Part**.

Enter: **Base** as the name of the first component.

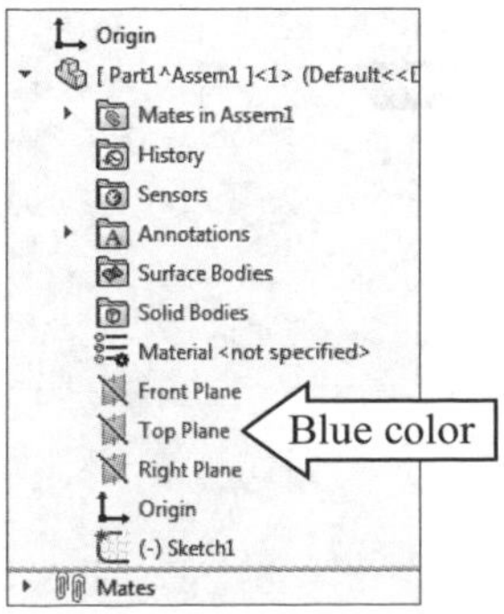

The color of the new component changes to the default Blue color.

To change the part's color, go to: **Tools/ Options / System Options / Colors / Assembly / Edit Part.**

A **new sketch** is created automatically when a new component is inserted.

Sketch the profile shown below; keep the **Origin** at the lower right corner.

Add the dimensions or relations needed to fully define the sketch.

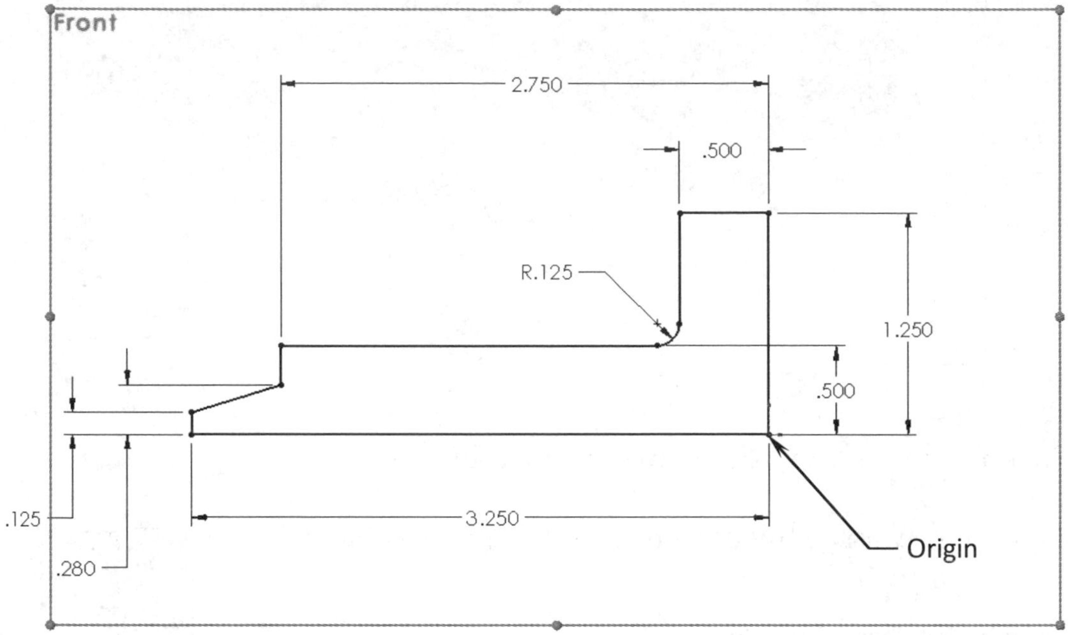

3. Extruding the Base:

Switch to the **Features** tab and click  or select **Insert, Boss-Base, Extrude**.

Direction 1: **Mid-Plane**.

Extrude Depth: **.750 in.**

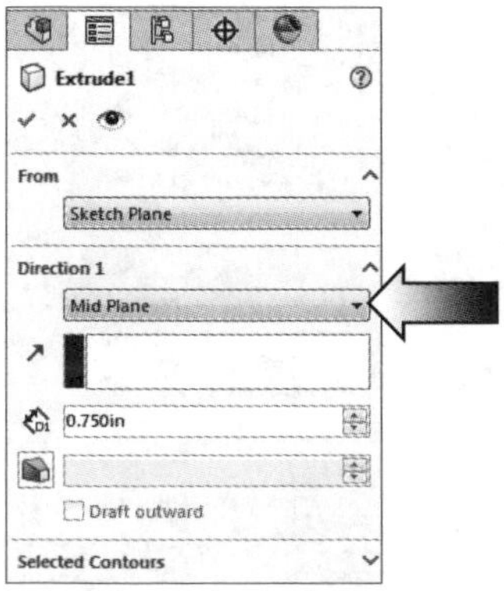

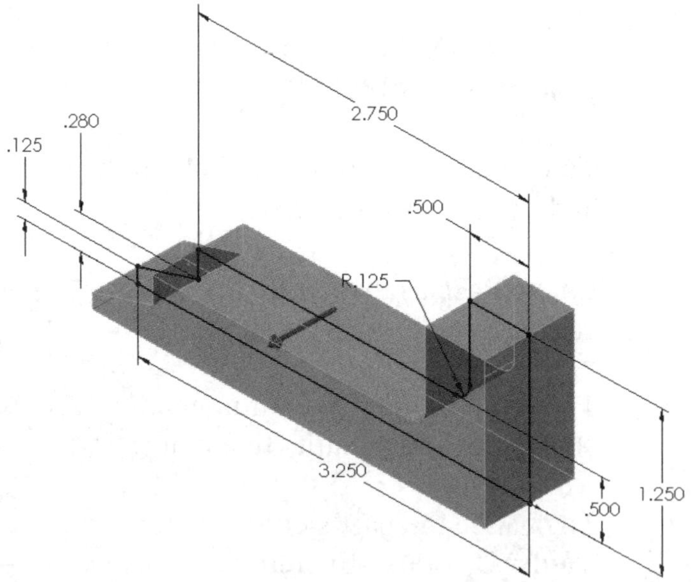

Click **OK**.

4. Adding the side flanges:

Select the <u>bottom face</u> of the base and open a **new sketch** .

Sketch the profile below; use the Mirror  option when sketching to keep the sketch entities symmetrical with the Centerline.

Add the dimensions shown. (Hold the **Shift** key when adding the .625 dim.)

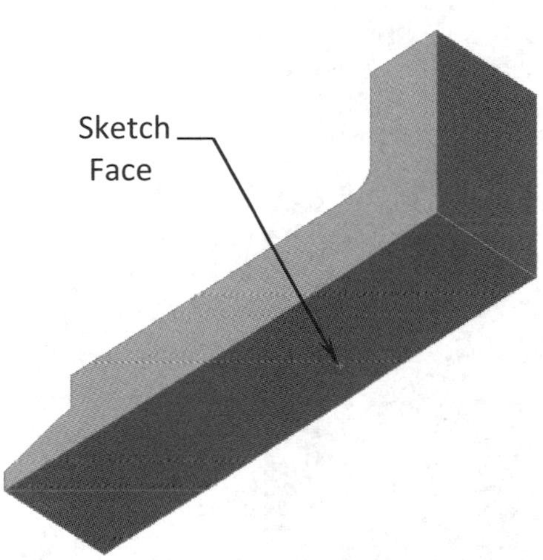

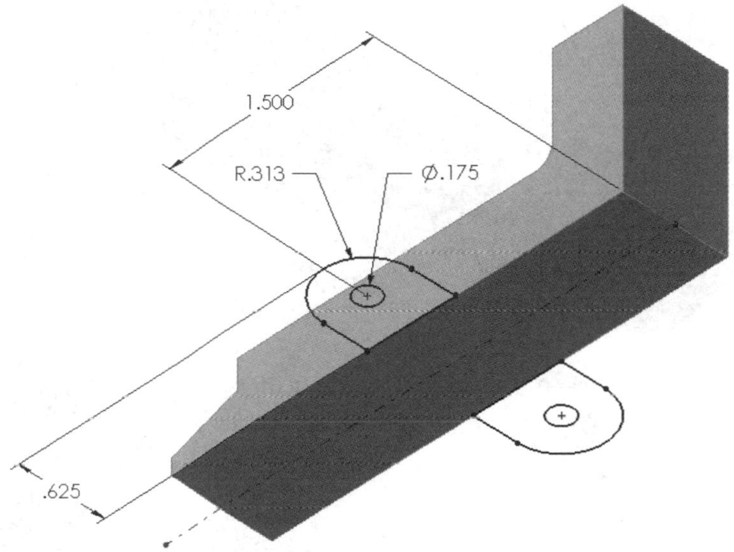

5. Extruding the Flanges:

Switch to the **Features** tab and click  or select **Insert, Boss-Base, Extrude**.

Direction 1: **Blind** (Reverse)   Depth: **.200 in.**

Enable **Reverse Direction**.

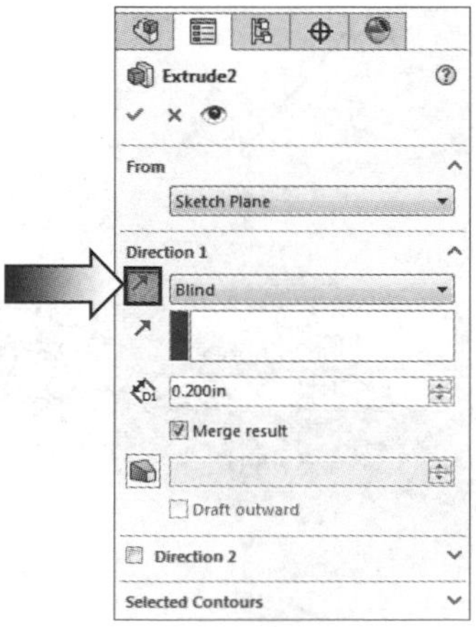

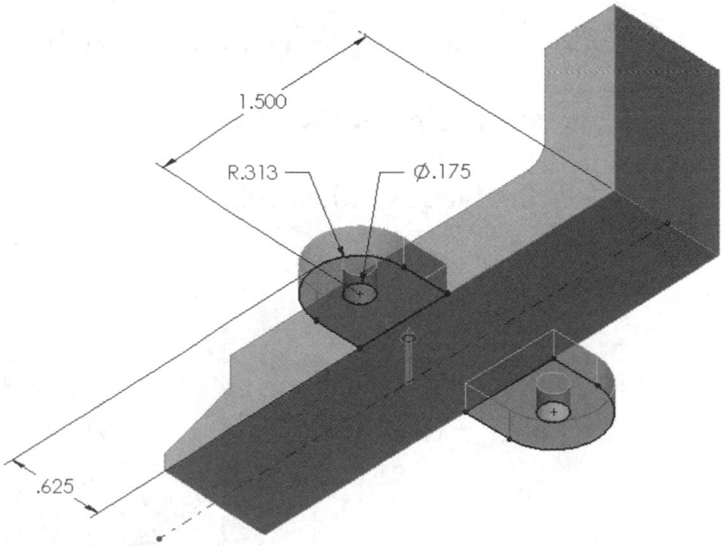

Click **OK**.

6. Adding the side cuts:

Select the face as indicated and click or select **Insert, Sketch.**

Sketch a **Centerline** starting at the Origin and click **Dynamic Mirror**.

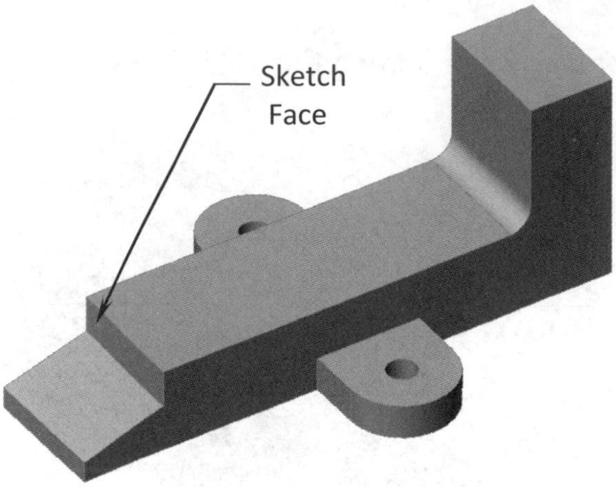

Sketch a **Rectangle** and add the dimensions/relations shown below.

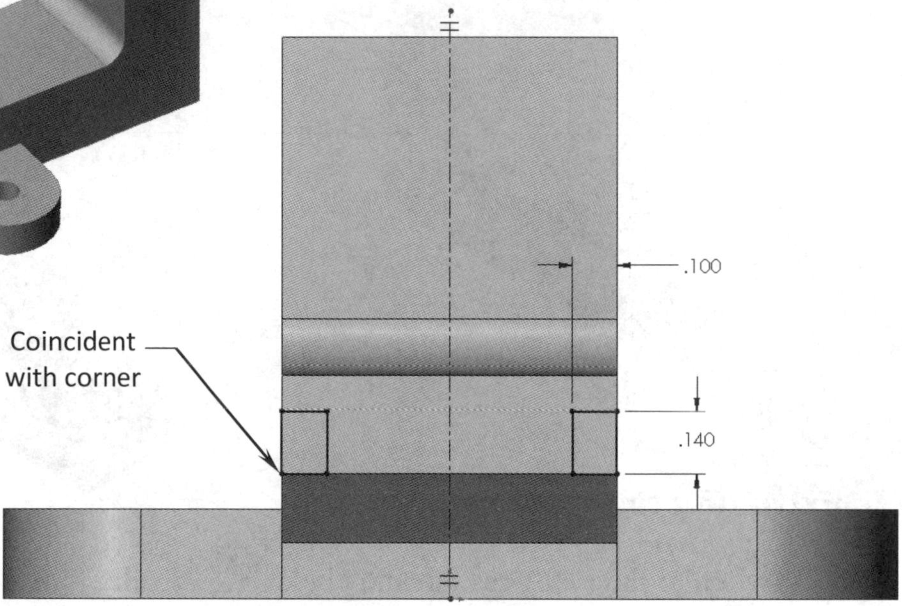

7. Extruding the side cuts:

Click or select **Insert, Cut, Extrude**.

Direction 1: **Up-To-Surface**.

Select the face as indicated.

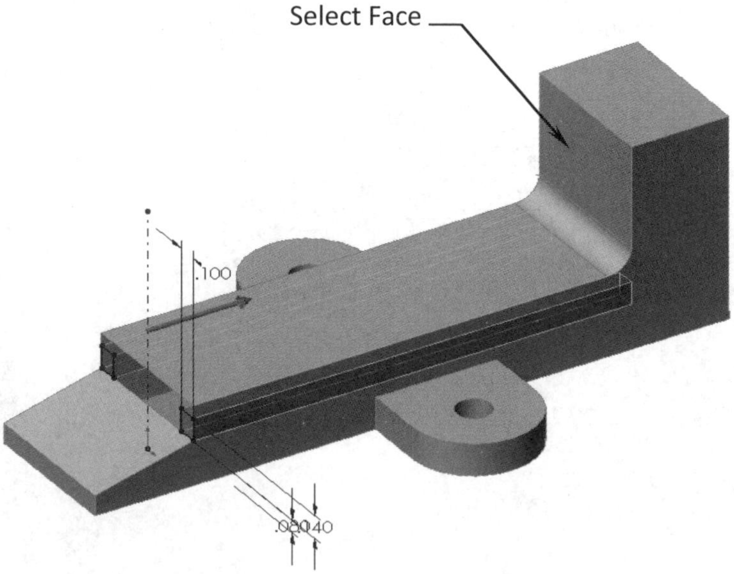

Click **OK**.

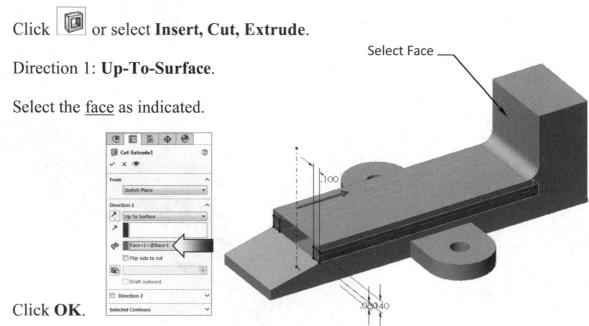

8. Creating an offset distance plane:

Select the face as shown and click  or select **Insert, Reference Geometry, Plane**.

Click **Offset Distance** option.

Enter **.150 in**. (the new plane is created away from the face).

Click **OK**.

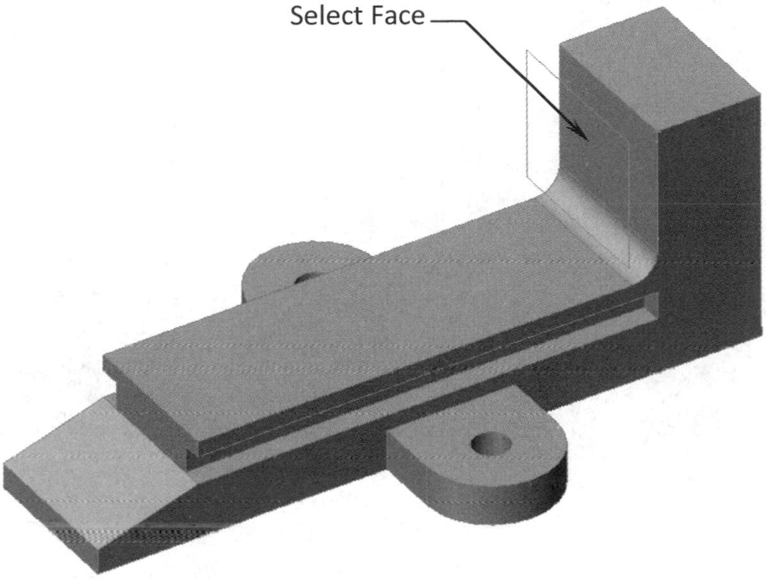

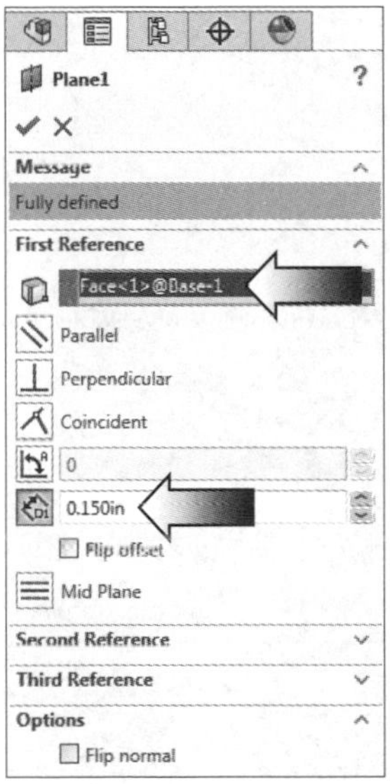

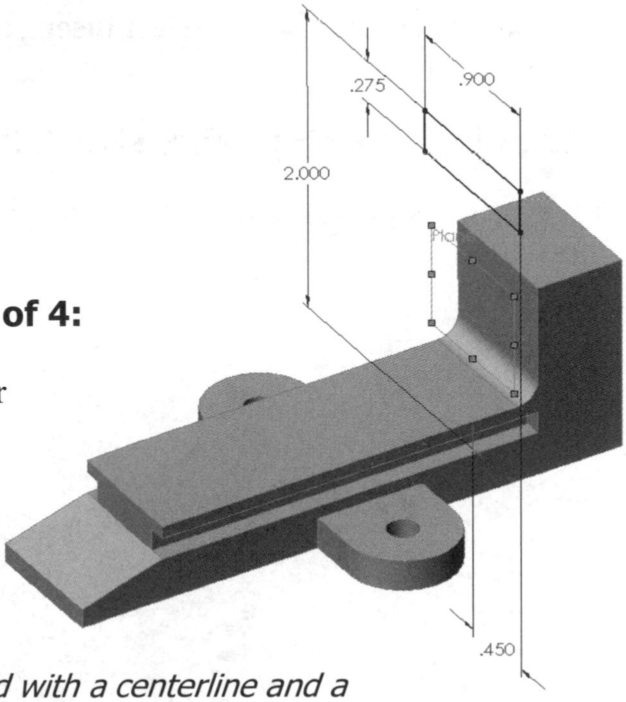

9. Creating the Fixed Jaw, sketch 1 of 4:

Select the new plane and click  or select **Insert, Sketch**.

Sketch a **Rectangle** approximately 2 inches above the origin.

NOTE: The dimension .450 can be replaced with a centerline and a symmetric relation.

More...

Add the dimensions shown below to fully position the sketch.
(Sketch a verical centerline and use Symmetric relation to center the rectangle will also work well.)

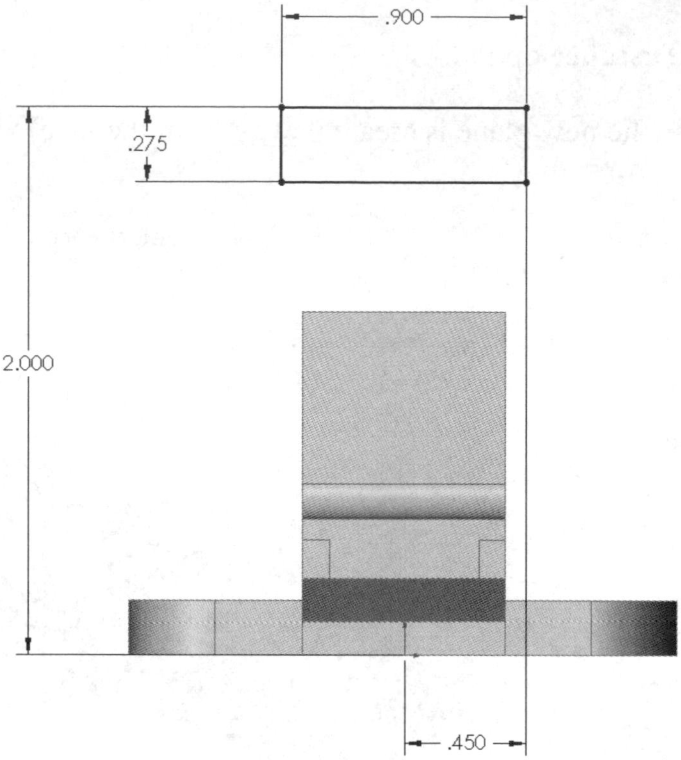

Exit the sketch or select **Insert, Sketch**.

10. Creating the 2nd profile, sketch 2 of 4:

Select the face as indicated and click or select **Insert, Sketch**.

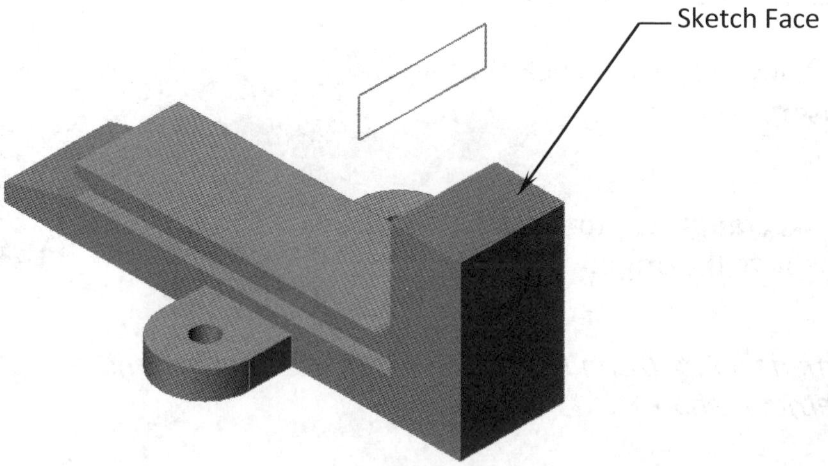

Hold the **Control** key and select the **4 edges** as indicated (or simply select the rectangular face and click the Convert Entities command).

Click **Convert Entities** on the Sketch-Tools toolbar.

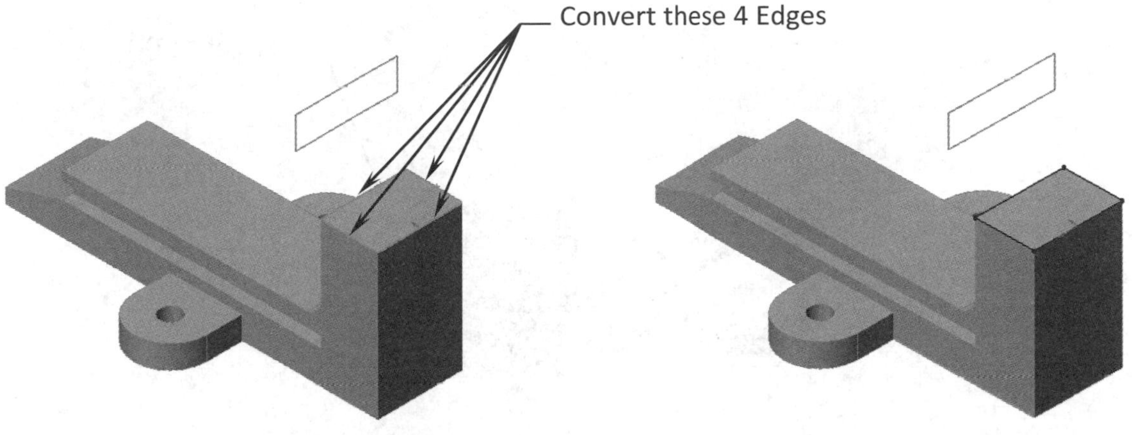

The 4 selected edges are converted into a new 2D rectangle.

Exit the sketch or select **Insert, Sketch**.

11. Creating the 3D Guide Curves:

Select **3D Sketch** from the Sketch tool tab or select **Insert, 3D Sketch**.

Sketch a **3-Point-Arc** approximately as shown and add the **Coincident** relation between the ends of the arc and the corners of the rectangles.

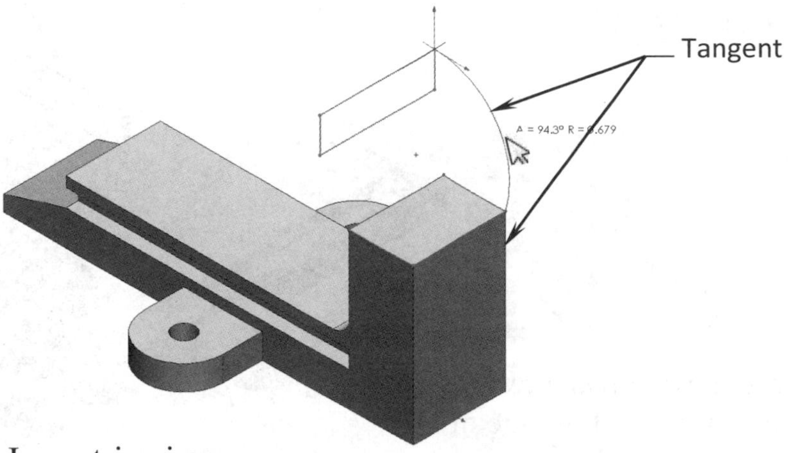

Remain in the Isometric view while sketching the 3-Point-Arc.

Add a **Perpendicular** relation between the endpoint of the arc and the upper face of the part as noted.

Repeat the last step and create the **other 3 arcs** the same way.

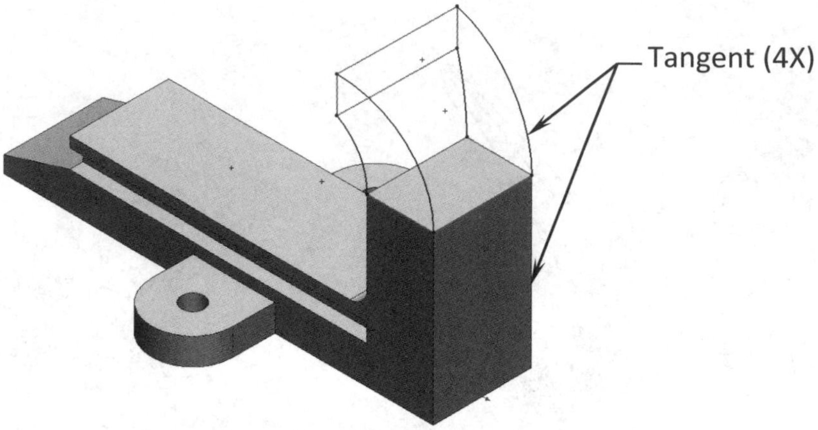

Exit the 3D Sketch  or press Control + Q.

12. Creating the Fixed Jaw loft:

Click  or select **Insert, Boss-Base, Loft.**

Select the **2 sketch profiles** as labeled (Profile 1 and Profile 2).
(Click at or near the ends of thc rcctangles. SOLIDWORKS will select the nearest endpoints automatically.)

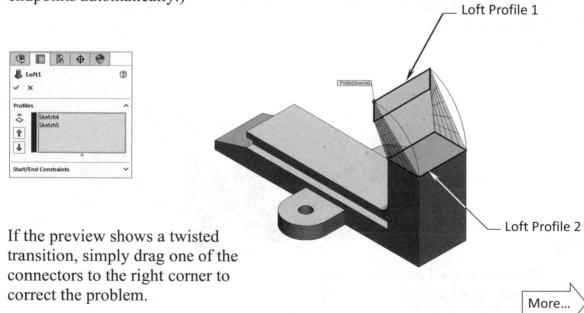

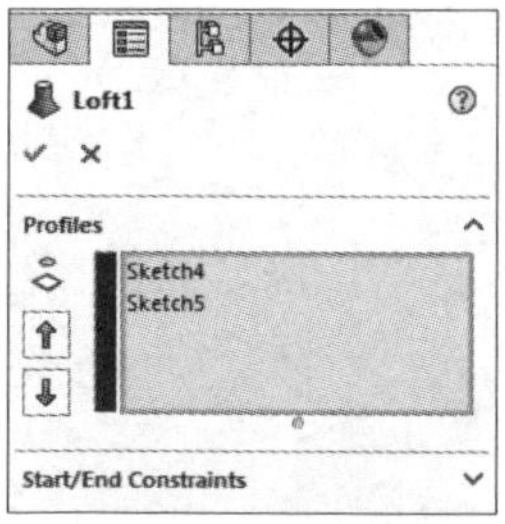

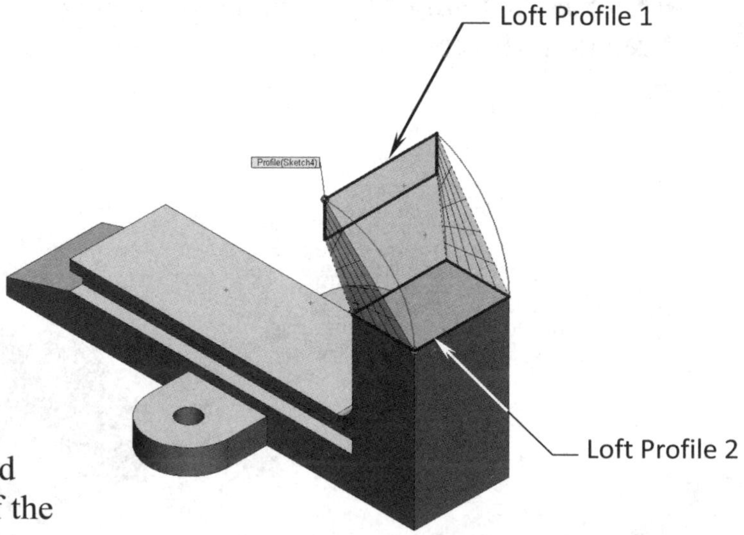

If the preview shows a twisted transition, simply drag one of the connectors to the right corner to correct the problem.

More...

Expand the **Guide Curve** section and select **one of the guide curves** in the 3D-Sketch.

Because this sketch has multiple entities that are not connected with one another, you will have to click the OK button (the check mark) on the SelectionManager after selecting each arc (arrow).

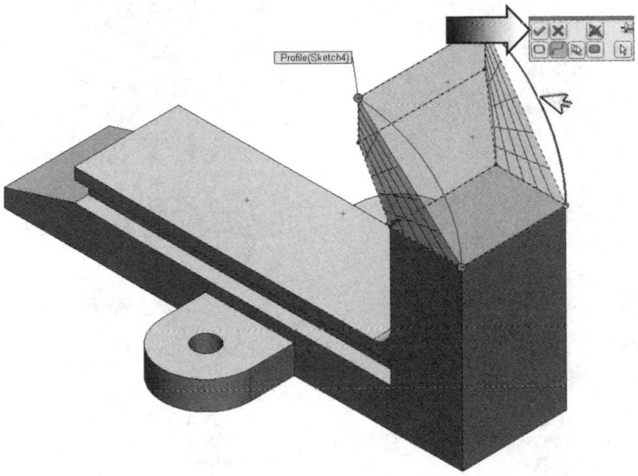

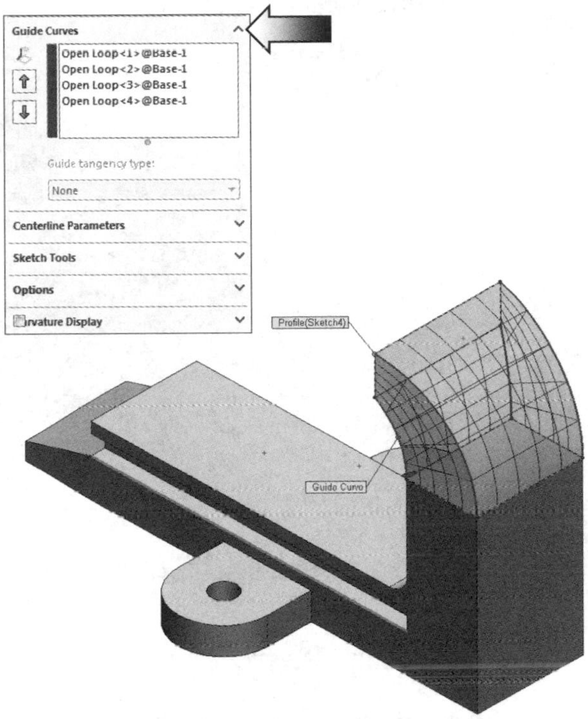

Click **OK**.

The resulted loft with four guide curves.

13. Creating the Fixed Jaw Clamp:

Select the Front plane from the FeatureManager tree and click [sketch icon] or select: **Insert, Sketch**.

Sketch a **Rectangle** and add the dimensions and relations as indicated. *(Add a horizontal centerline and a midpoint relation between the right endpoint and the model edge.)*

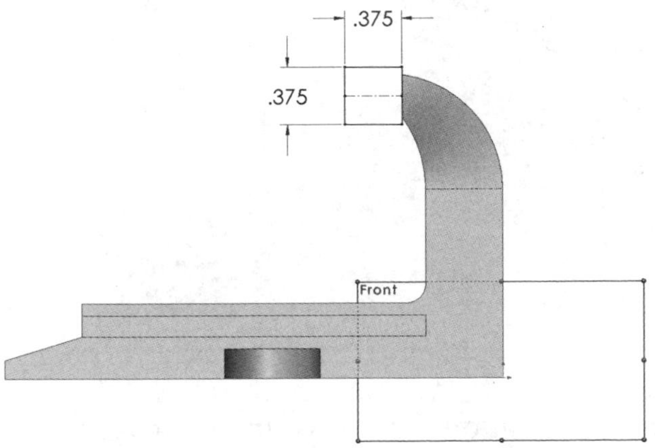

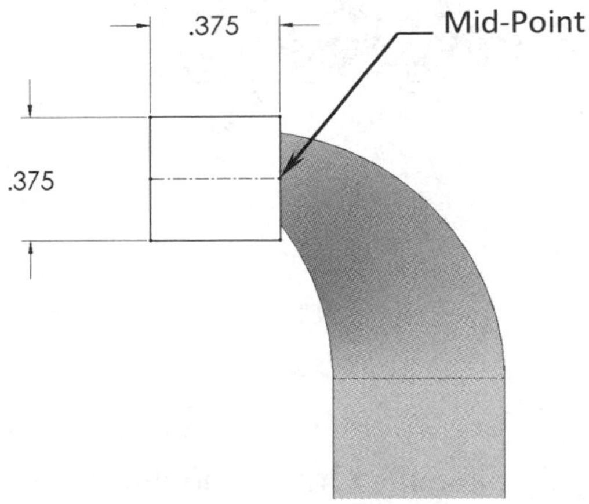

14. Extruding the Fixed Jaw Clamp:

Click or select **Insert, Boss-Base, Extrude.**

Direction 1: **Mid-Plane**.

Extrude Depth: **1.250 in**.

Click **OK**.

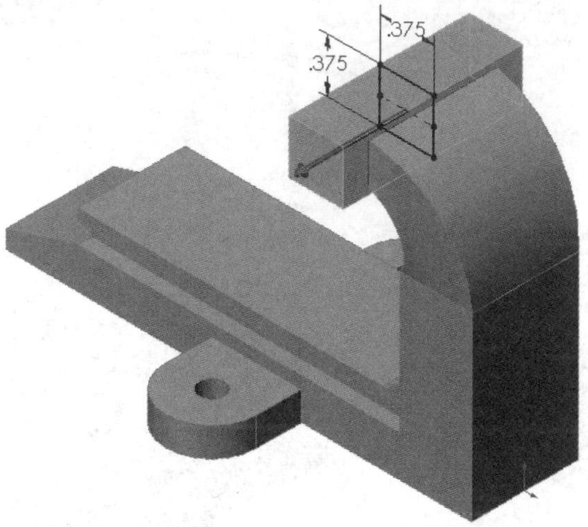

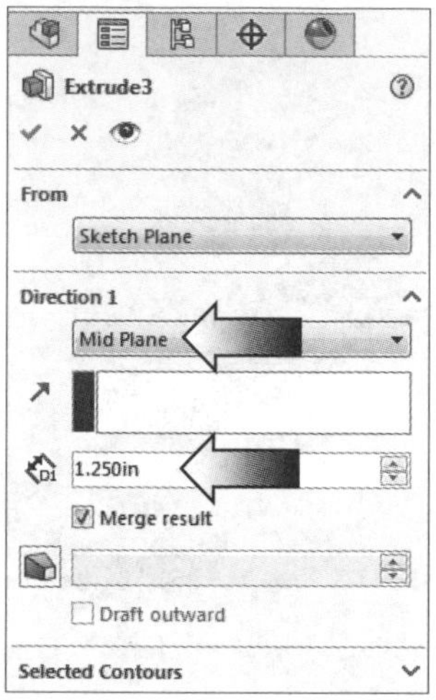

15. Creating the Lead Screw Hole:

Select the <u>face</u> as indicated

and open a **new sketch** .

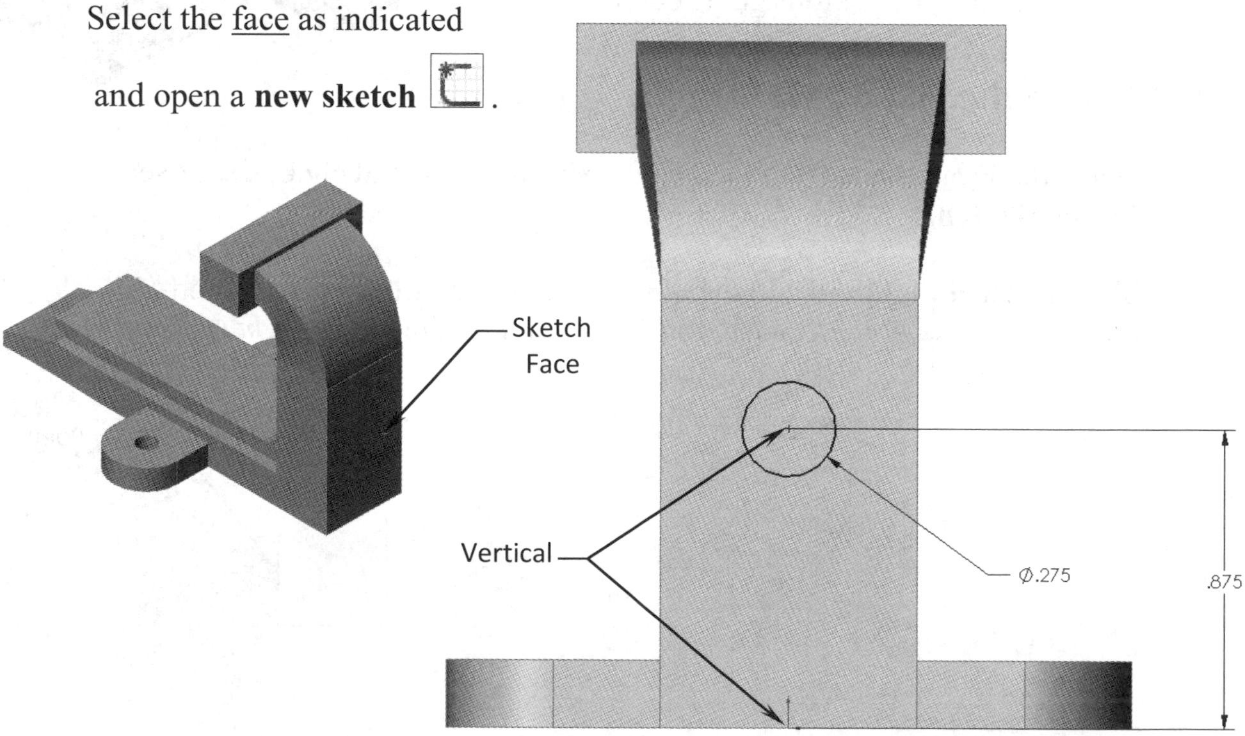

Sketch a **Circle** and add the dimensions and relations as shown in the image.

16. Extruding the Hole:

Click  or select **Insert, Cut, Extrude**.

Direction 1: **Through All**.

Click **OK**.

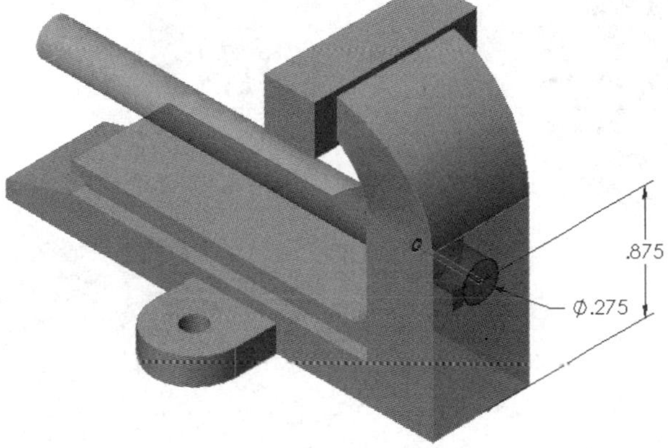

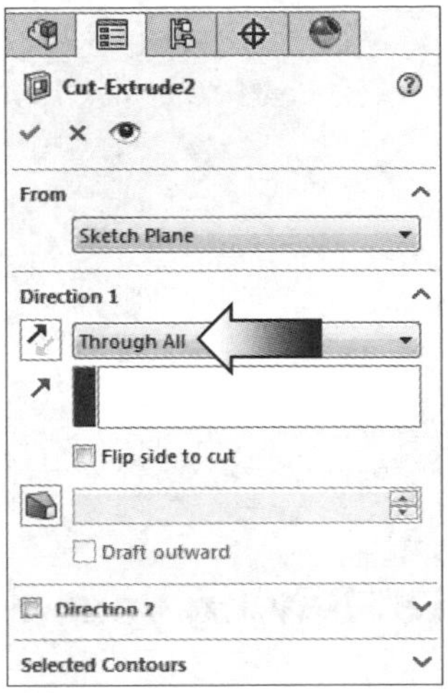

17. Adding Fillets:

Click Fillet  or select **Insert, Features, Fillets-Rounds**.

Enter **.032 in**. for Radius.

For Edges to Fillet, select the edges as indicated.

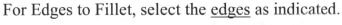

Click **OK**.

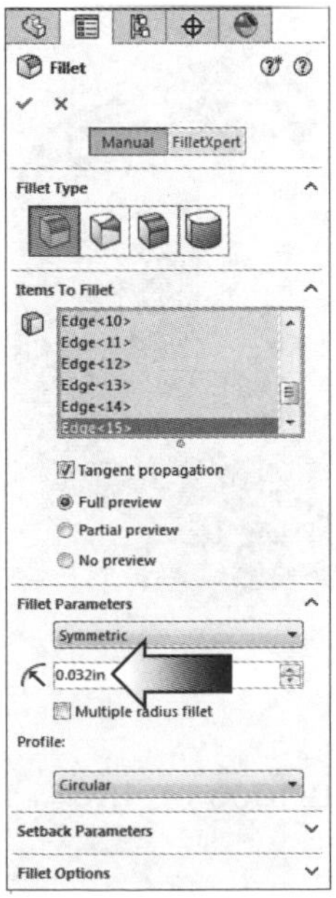

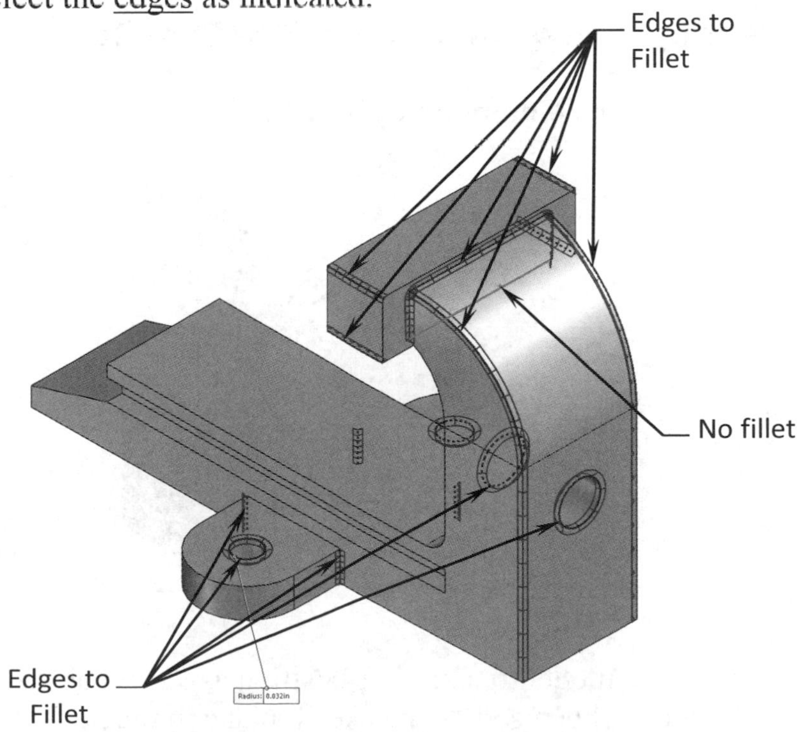

The Base component is shown in Front and Back Isometric views.

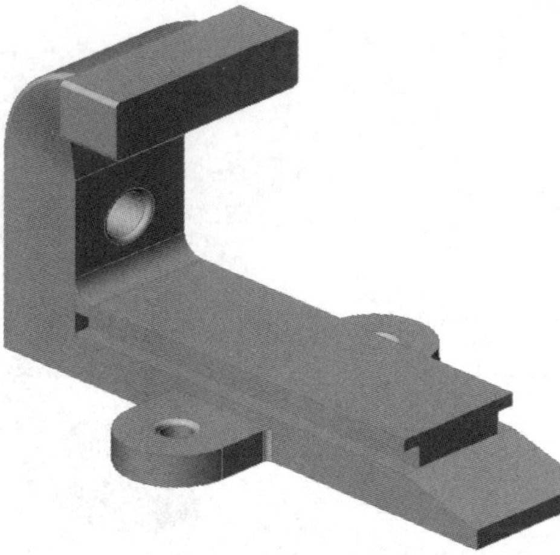

18. Saving your work:

Select **File, Save As**.

Enter **Base** for the file name and press **Save**.

Click  to <u>exit</u> the **Edit Component** mode.

19. Creating a new component: The Slide Jaw

Select **Insert, Component, New Part**.

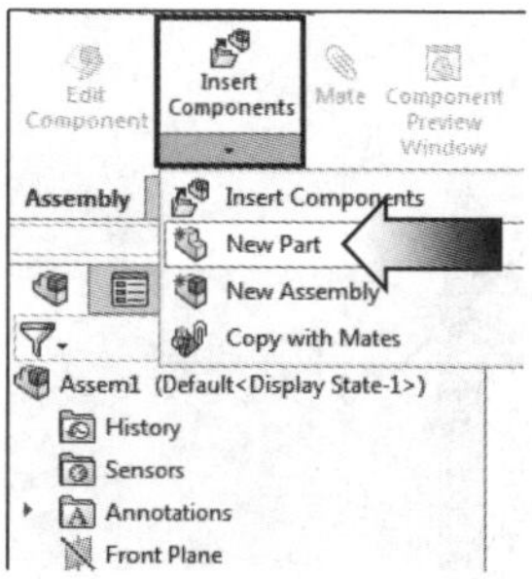

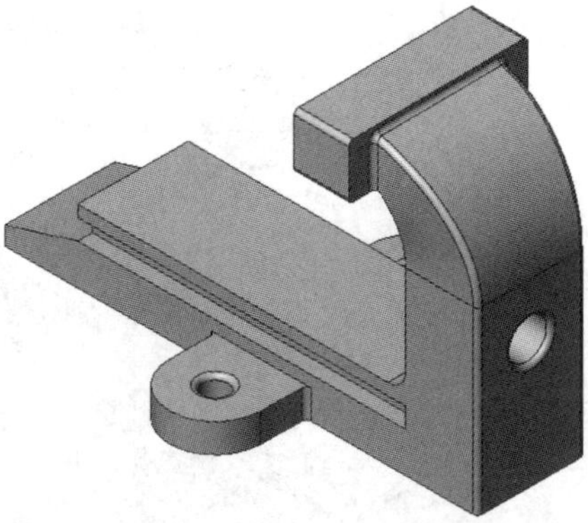

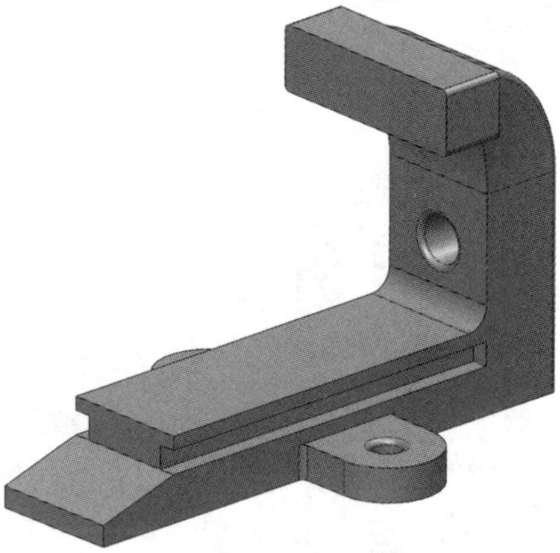

Rotate the model to a similar position as shown above; the planar surface on the left side will be used as the sketch plane in the next step.

Select the face indicated as sketch plane for the new component (Inplace2). A new part and a new sketch are created in the FeatureManager tree.

Rename the new component to **Slide Jaw**.

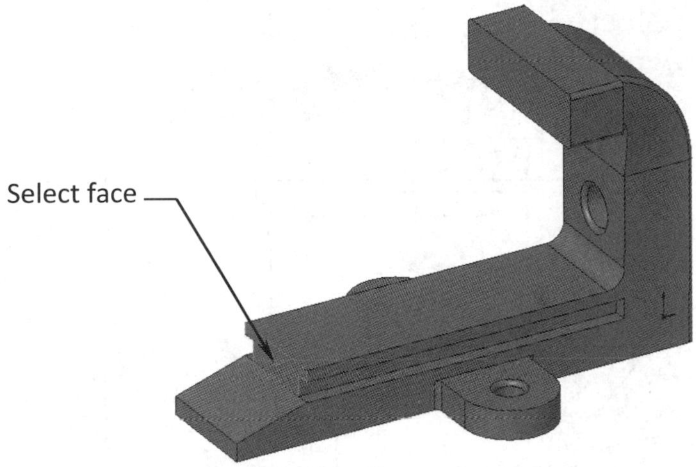

A new part is created in the FeatureManager tree and a **new sketch** is activated.

An **Inplace** mate 

is also created for the new component to reference its location.

20. Using the Offset Entities command:

Select the **4 edges** of the model (as shown) and click **Offset Entities**.

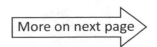

Enter **.010 in**. for offset value. This offset distance between 4 the lines and the model edges will remain locked and get updated at the same time when the value is changed.

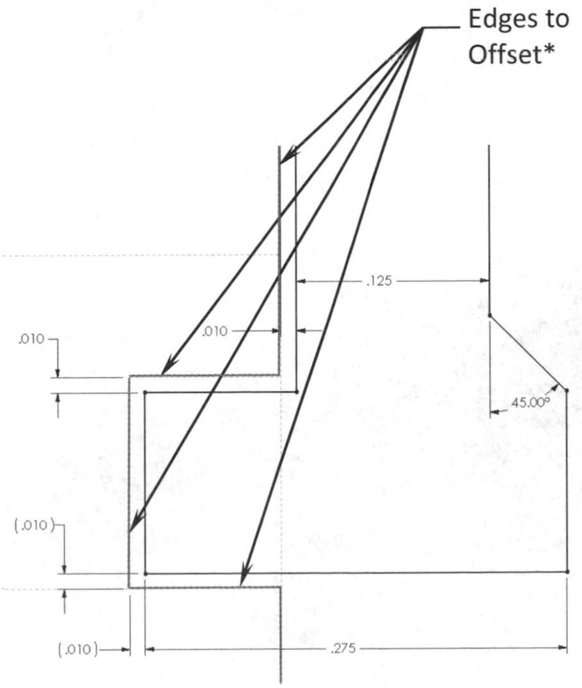

Offset Entities

* The geometry of a model such as edges, faces, and other sketch entities can be offset or converted to create the new geometry for the new part.
* The offset entities can be set to one direction or bidirectional.
* An On-Edge relation is created for each converted sketch entity.

More on next page

Sketch the rest of the profile and add the dimensions or relations needed to fully define the sketch.

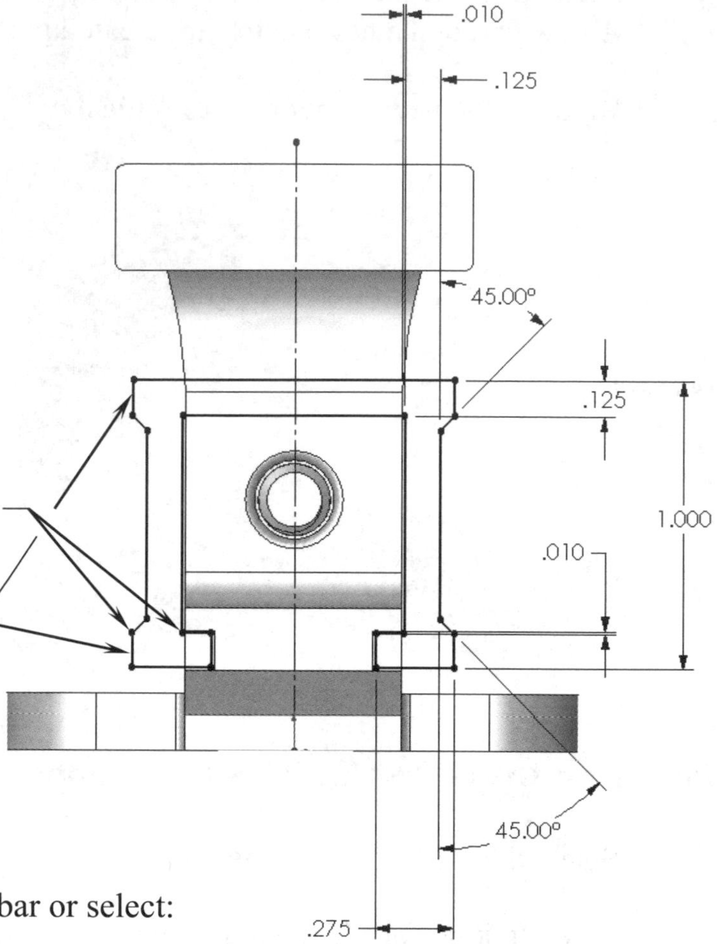

Note:

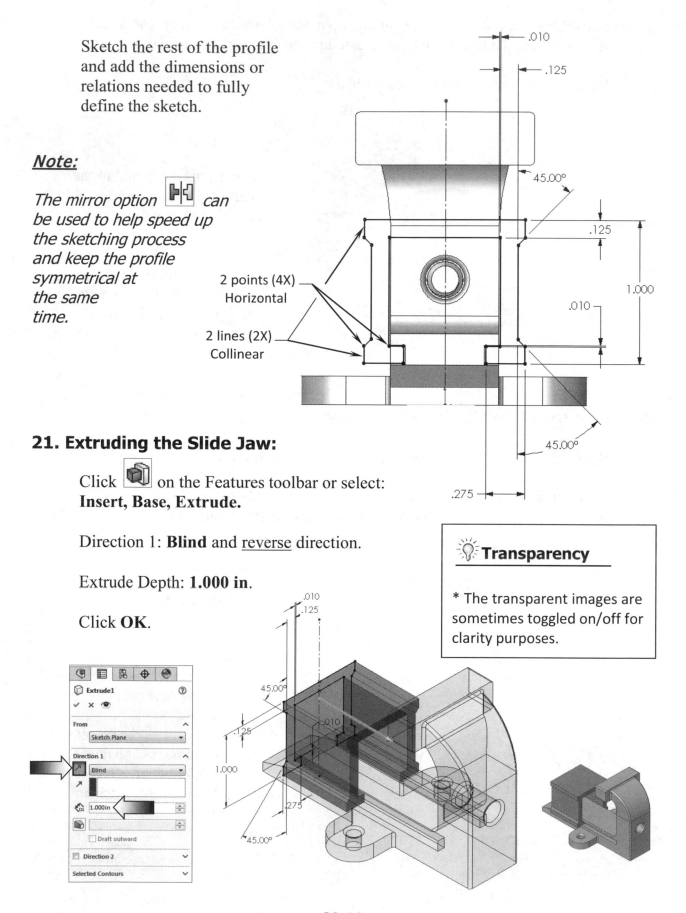

The mirror option [mirror icon] can be used to help speed up the sketching process and keep the profile symmetrical at the same time.

21. Extruding the Slide Jaw:

Click [extrude icon] on the Features toolbar or select: **Insert, Base, Extrude.**

Direction 1: **Blind** and <u>reverse</u> direction.

Extrude Depth: **1.000 in**.

Click **OK**.

> **Transparency**
>
> * The transparent images are sometimes toggled on/off for clarity purposes.

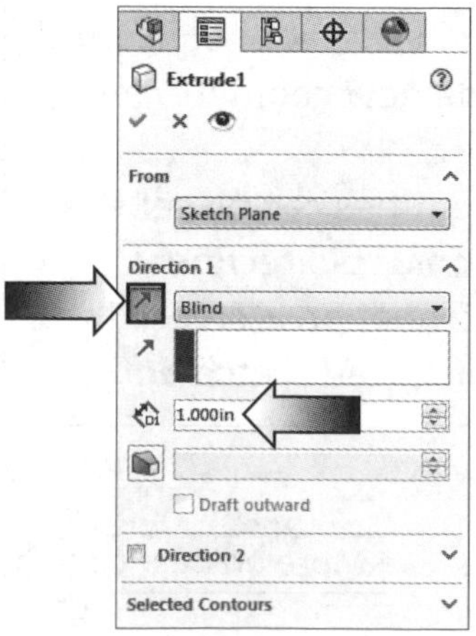

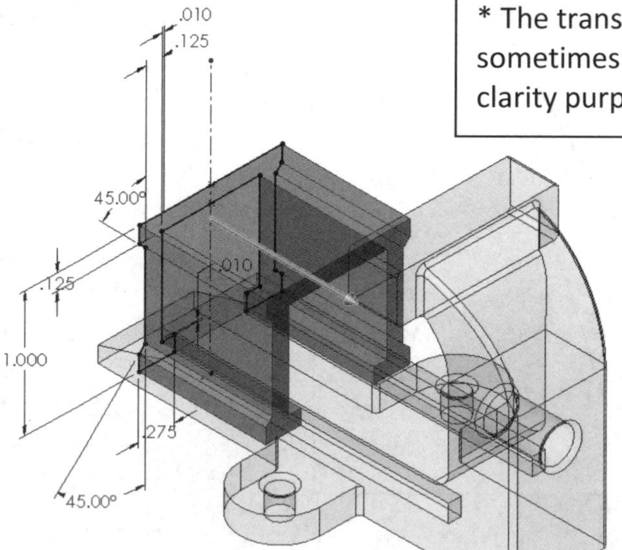

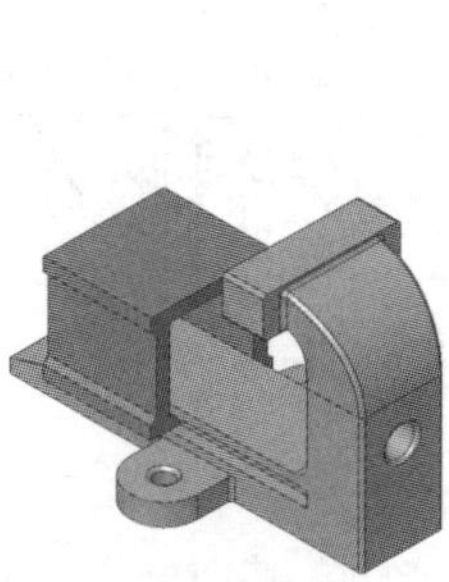

22. Adding the support wall:

Select the face indicated and open a **new sketch** or select **Insert, Sketch.**

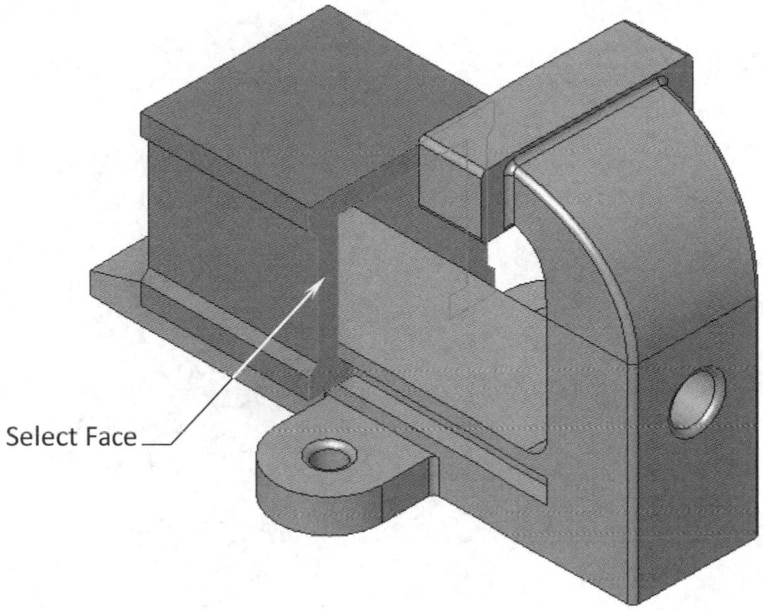

Sketch the profile and add the dimensions and relation as shown below to fully define the sketch.

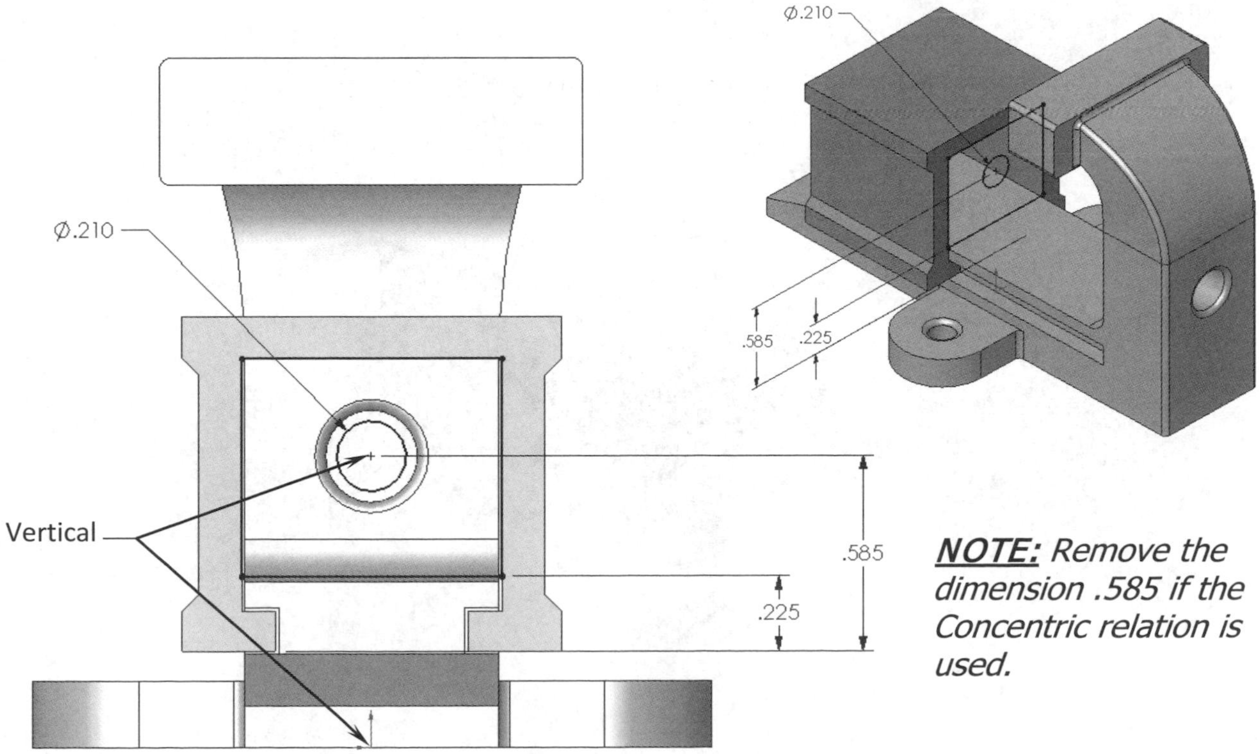

NOTE: *Remove the dimension .585 if the Concentric relation is used.*

23. Extruding the Support Wall:

Click  on the Features toolbar or select **Insert, Base, Extrude.**

Direction 1: **Blind** and click the reverse direction arrow.

Extrude Depth: **.375 in**.

Click **OK**.

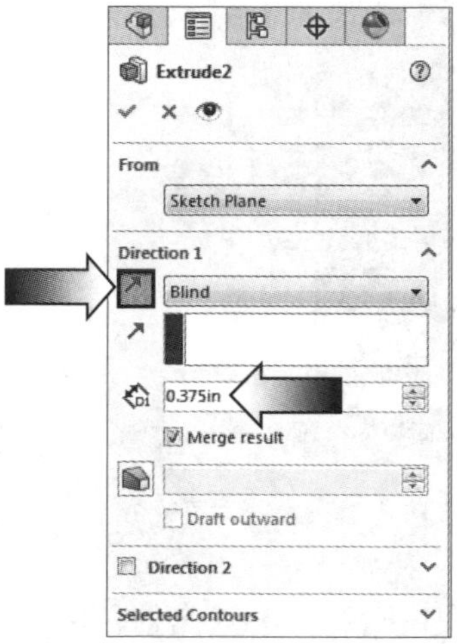

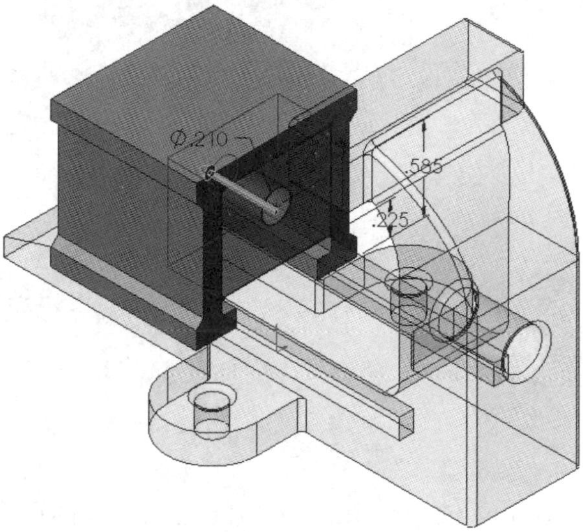

The Support Wall is built with a guide hole.

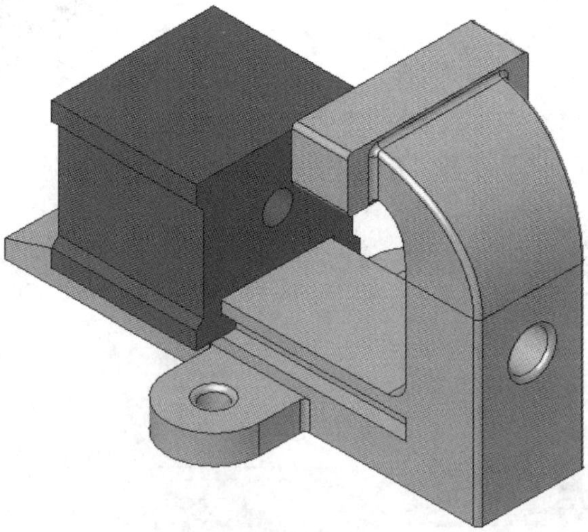

24. Creating a new work plane:

Select the face as indicated and click or select:
Insert, Reference Geometry, Plane.

Enter **.150 in**. for Offset Distance and place the new plane on the **outside**.

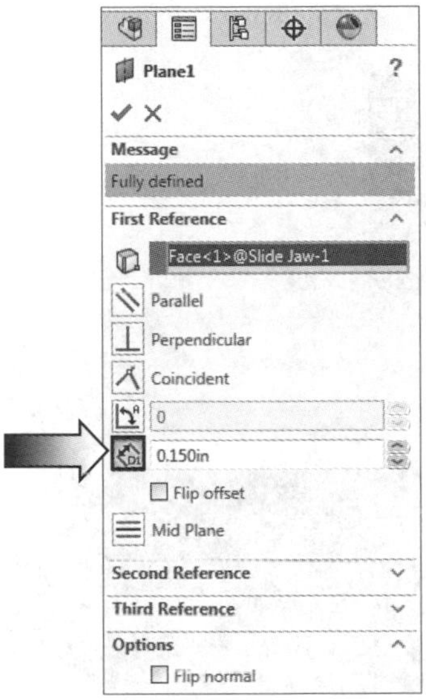

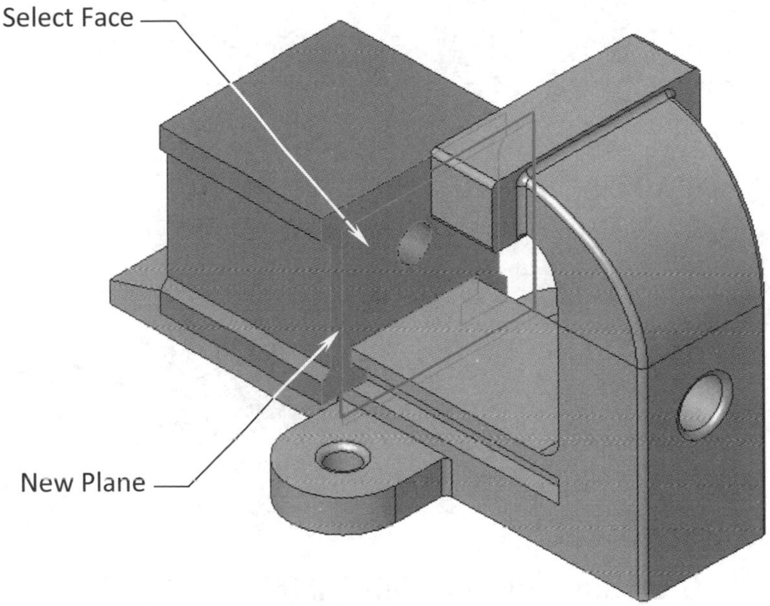

25. Creating the Slide Jaw 1st sketch:

Open a **new sketch** on the new plane or select **Insert, Sketch.**

Sketch a **Rectangle** and add the dimensions shown to fully define the sketch (Sketch a vertical center-line and add a Symmetric relation to center the rectangle will also work fine).

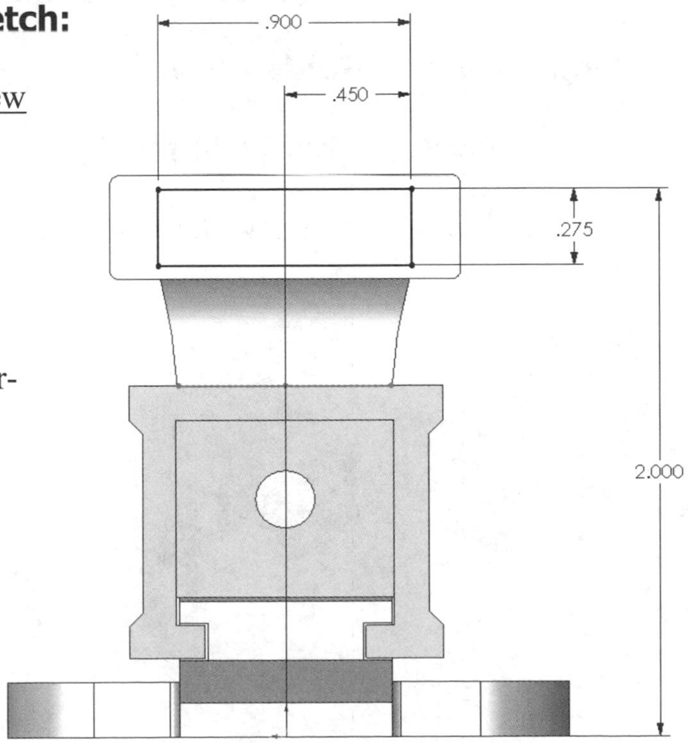

Exit the sketch or select:
Insert, Sketch.

26. Creating the Slide Jaw 2nd sketch:

Select the face indicated and open a new sketch or select: **Insert, Sketch.**

Sketch a **Rectangle** as shown.

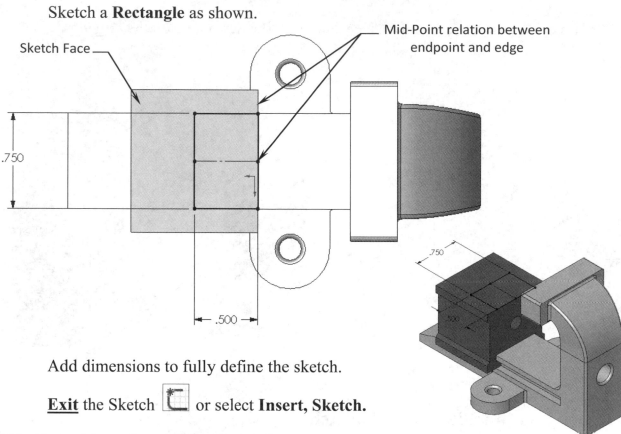

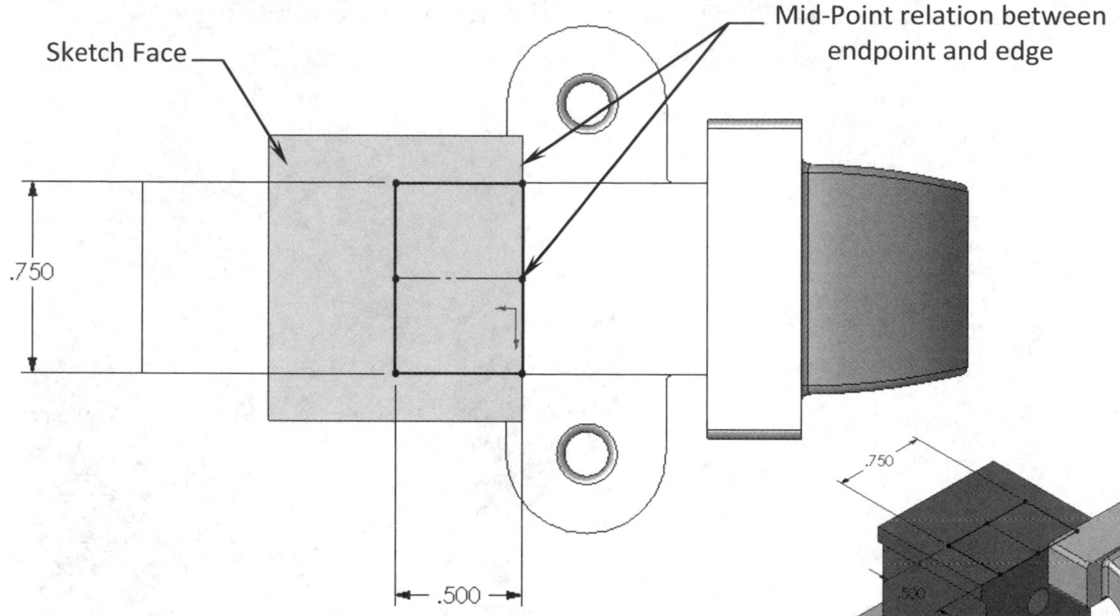

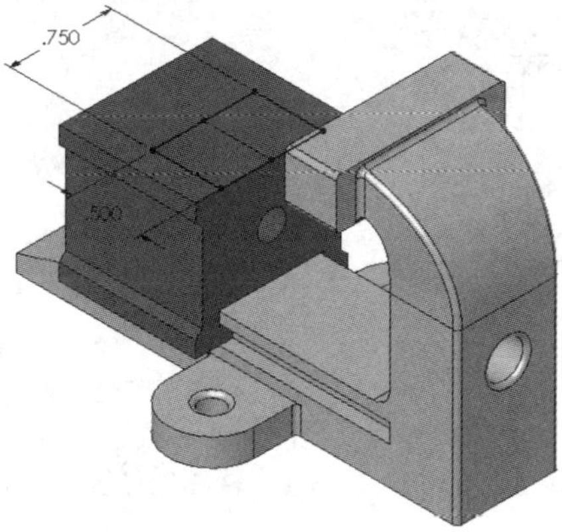

Add dimensions to fully define the sketch.

Exit the Sketch or select **Insert, Sketch.**

27. Sketching the Guide Curve:

Select the Right plane of the part from the FeatureManager tree.

Click to open a new sketch or select: **Insert, Sketch.**

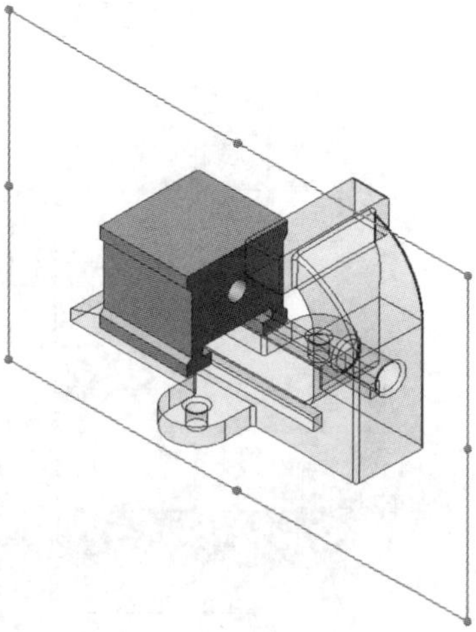

Guide Curves

* Guide curves are used to control the profile from twisting as the sketch is swept along the path.

* Guide curves are also used in Loft to control the transitions between the profiles.

Sketch either a **CenterPoint Arc**  or a **3-Point Arc** 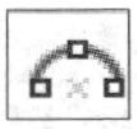 that connects the two sketches.

Add the relations shown below to fully define the sketch.

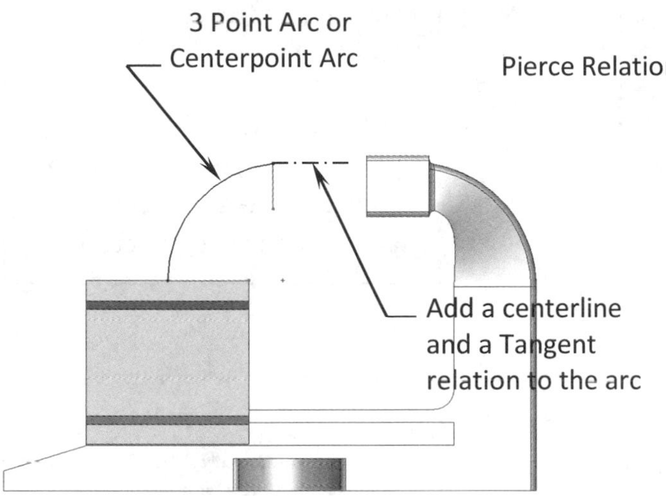

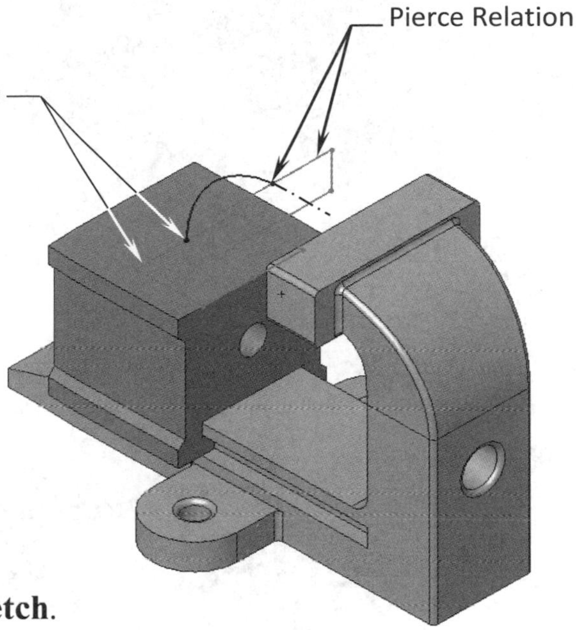

Exit the sketch  or Select **Insert, Sketch.**

28. Creating the Slide Jaw Loft:

Click  on the Features toolbar or select **Insert, Boss-Base, Loft.**

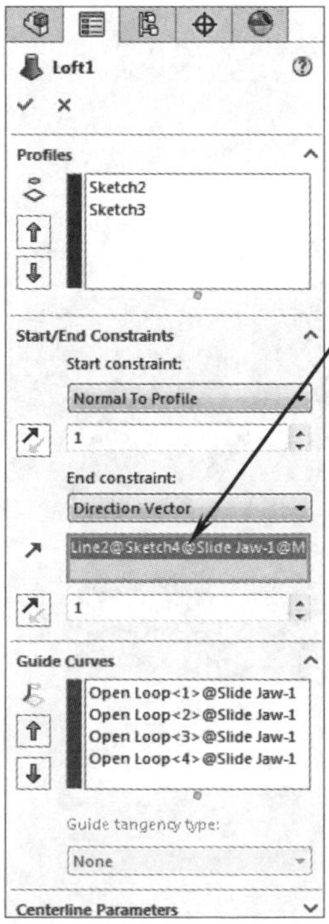

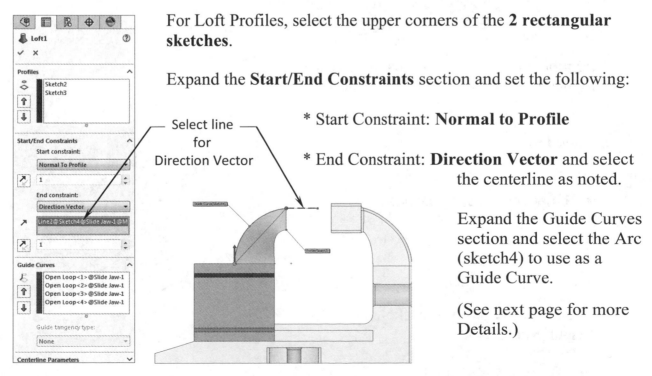

For Loft Profiles, select the upper corners of the **2 rectangular sketches**.

Expand the **Start/End Constraints** section and set the following:

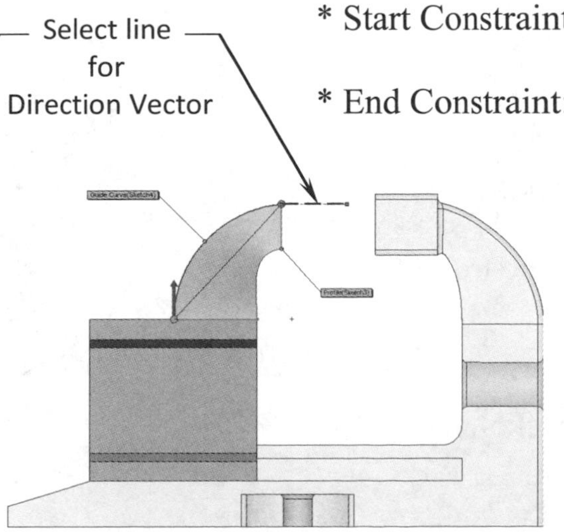

* Start Constraint: **Normal to Profile**

* End Constraint: **Direction Vector** and select the centerline as noted.

Expand the Guide Curves section and select the Arc (sketch4) to use as a Guide Curve.

(See next page for more Details.)

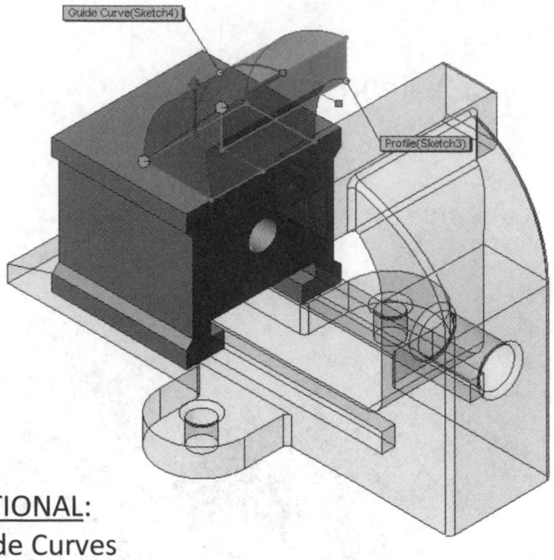

> ### Start/End Constraints
>
> * The Start constraint and End constraint option applies a constraint to control tangency to the start and end profiles.
>
> * The Direction Vector option applies a tangency constraint based on a selected entity used as a direction vector.

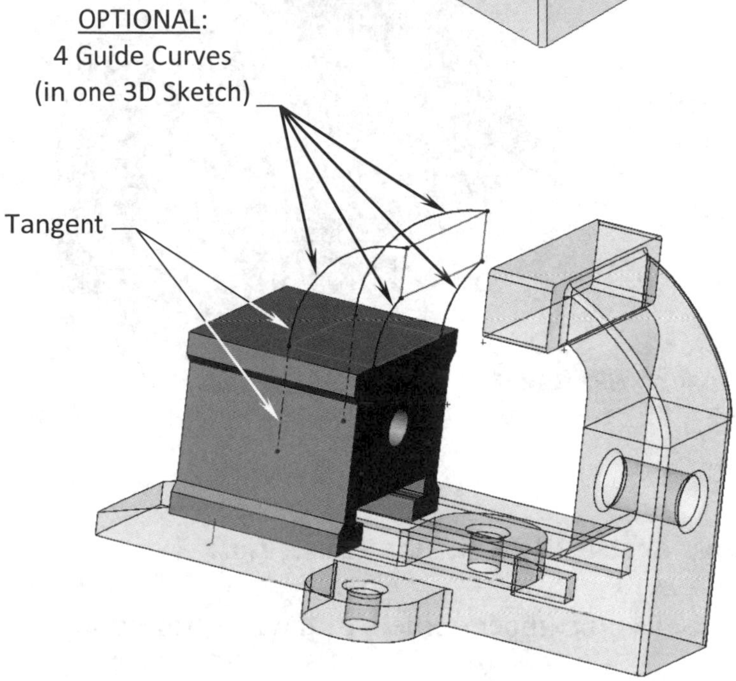

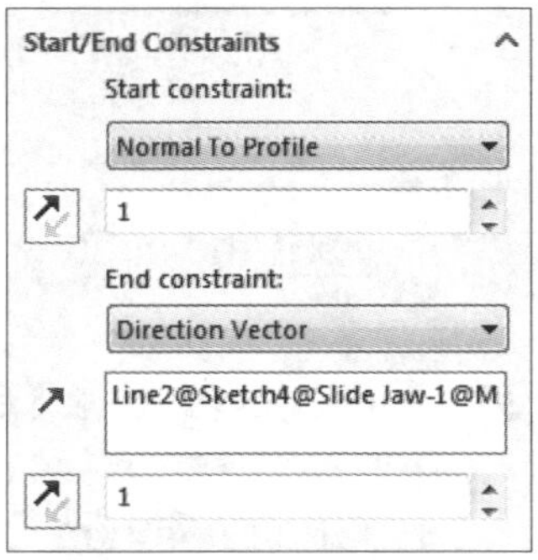

(Optional: Four guide curves can be used to increase accuracy of the loft if needed.)

Click **OK**.

29. Creating the Clamp block:

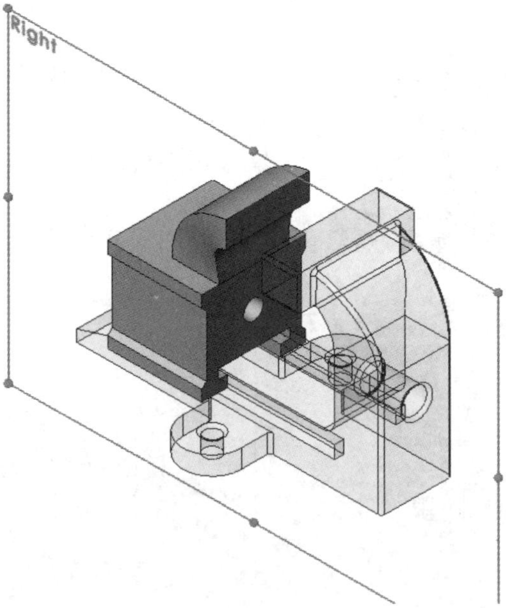

Select the part's Right plane from the FeatureManager tree.

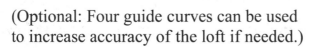

Click to open a **new sketch** or select: **Insert, Sketch.**

Sketch a **Rectangle** and add the dimensions shown.

Add a **Centerline**  in the center of the rectangle and position it on the Mid-Point of the vertical edge.

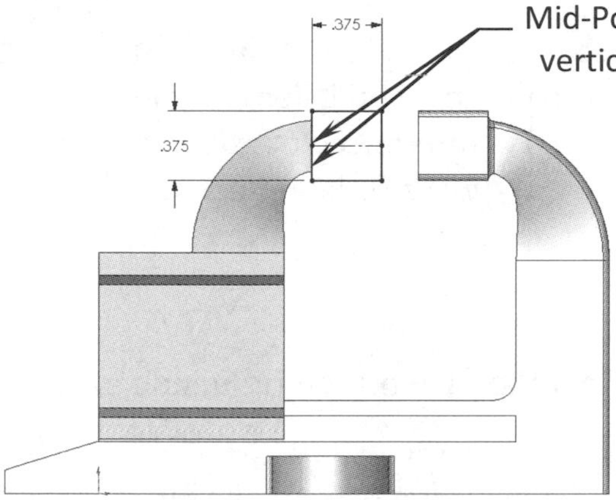

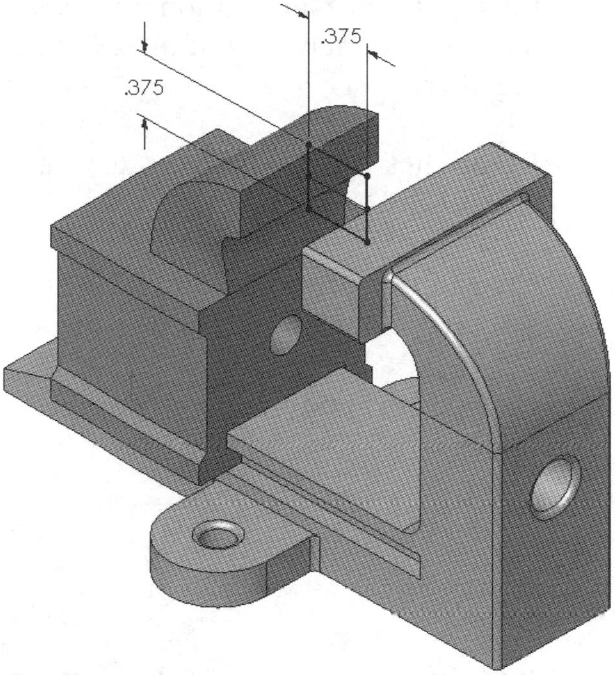

30. Extruding the Clamp block:

Click  or select **Insert, Boss-Base, Extrude.**

Direction 1: **Mid-Plane**.

Extrude Depth: **1.250 in**.

Merge Result: **Enabled**.

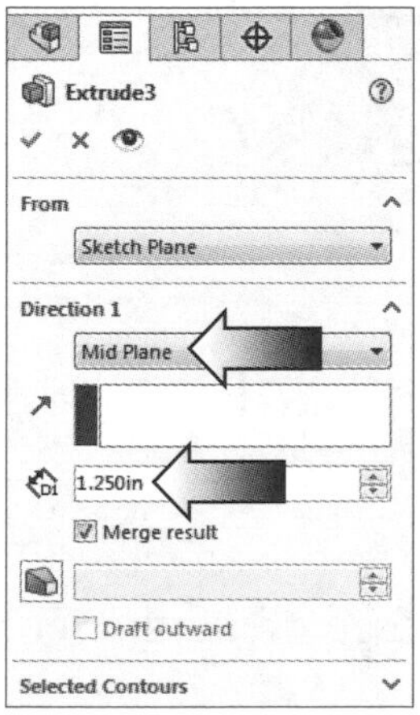

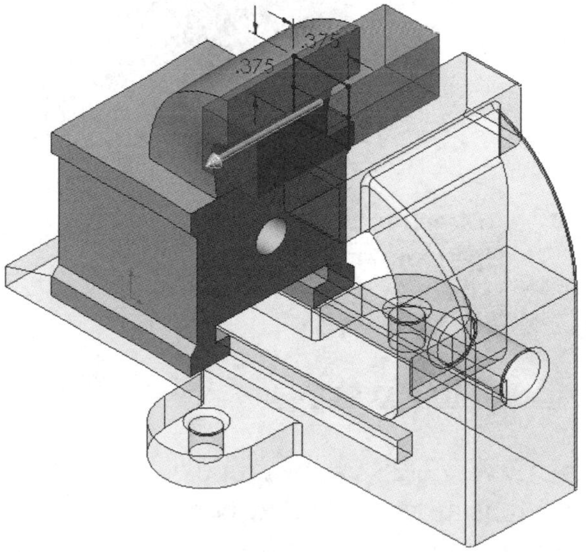

Click **OK**.

31. Which option is better?

Instead of using the Mid-Plane extrude type, the **Up-To-Surface** option can be used to link the length dimensions of the 2 Clamp Jaws together.

Right-click on the last Extruded feature and select **Edit Feature**.

Change **Direction 1** from Mid-Plane to **Up-To-Surface** and select the face on the left side.

Up-To-Surface

* Extends the feature from the sketch plane to the selected surface.
* When the driving surface is changed in length, the referenced extruded feature will also be reflected.

Change **Direction 2** to **Up-To-Surface** and select the face on the right side as indicated.

Click **OK**.

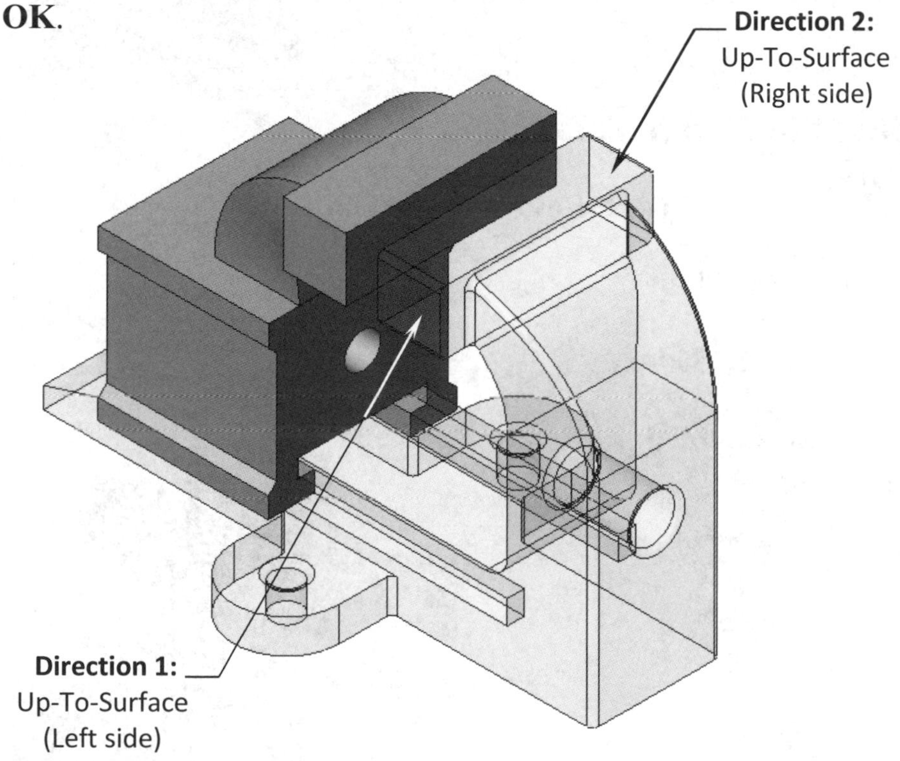

External References:

SOLIDWORKS creates an external relation every time a face of another component is used as an extrude end-condition.
If the reference face is moved or rotated, the related feature will also move or rotate with it.

32. Adding fillets:

Click  or select **Insert, Features, Fillet-Round.**

Enter **.032** for radius value.

Select the **edges** as shown.

Click **OK**.

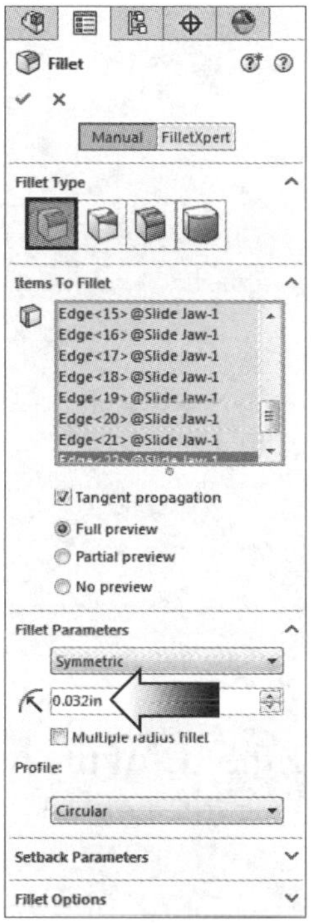

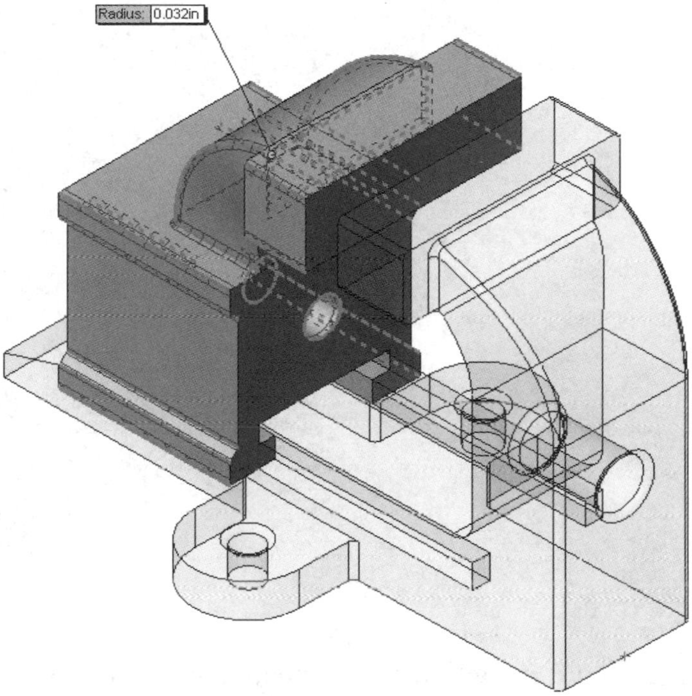

Inspect the fillets from the Front and Back Isometric views. Make any corrections if needed.

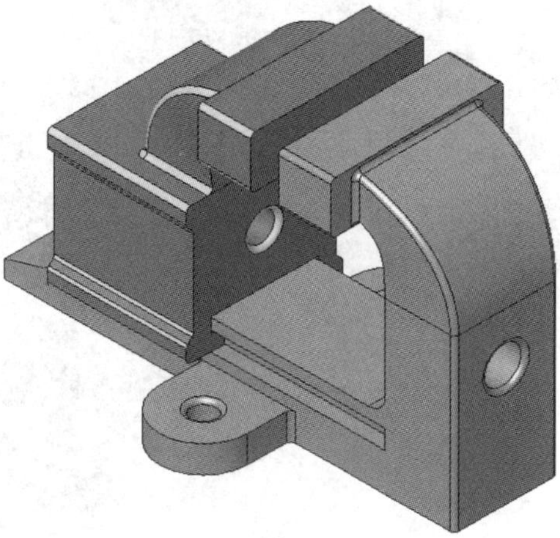

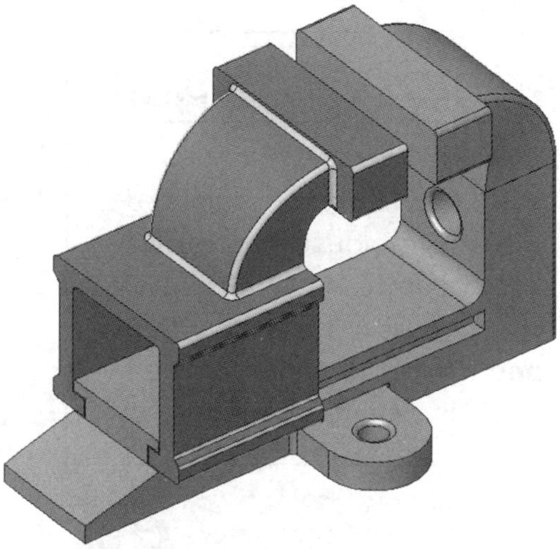

33. Creating the internal threads:

Starting with the sweep path.

Select the face indicated and open a **new sketch** or select: **Insert, Sketch.**

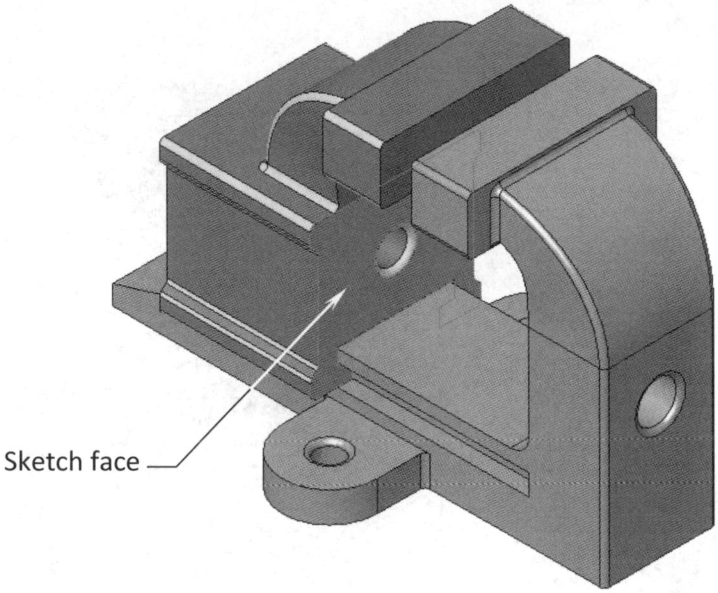

Sketch a **Circle** that is Concentric with the hole. (Converting the ID of the hole is another good way to link the diameter of the circle to the hole's diameter.)

Add a **Ø.210** dimension to fully define.

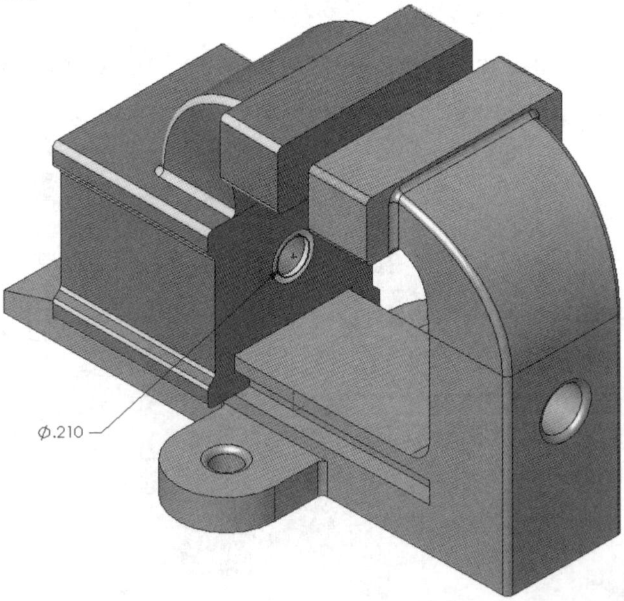

Wake up Center Points

* With the Circle tool selected, hover the mouse cursor over the circumference of the hole; the 4 quadrant points appear, and the center-point of the circle is visible for snapping.

Select **Insert, Curve, Helix-Spiral.**

Enter the following parameters:

Defined by:	**Pitch and Revolution**.
Pitch:	**.080 in**.
Revolution:	**5.000**.
Starting Angle:	**90.00 deg**.
Reverse Direction:	**Enabled**.
Clockwise:	**Selected**.

Click **OK**.

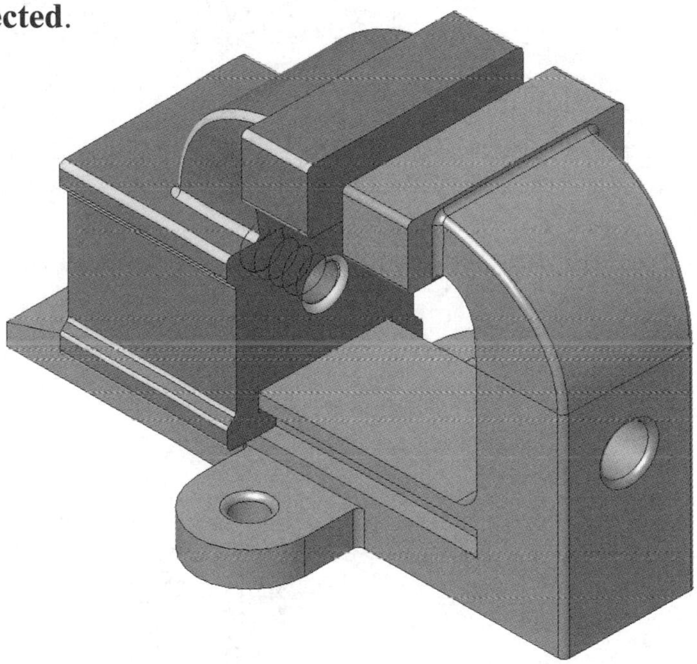

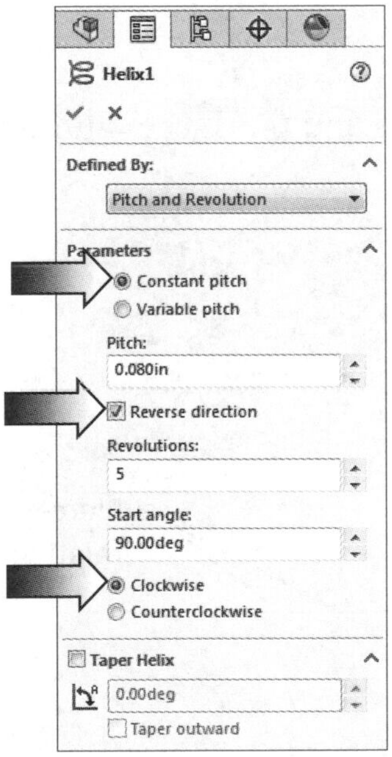

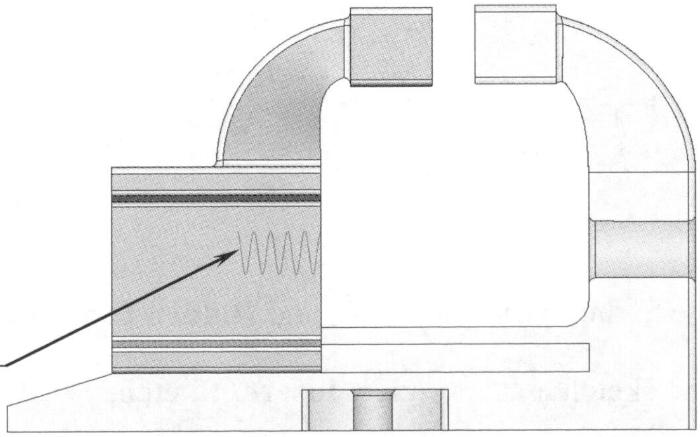

Click **View / Hide-Show / Curves** if the helix is not visible

Sketching the Sweep Profile:

Select the Right plane of the part and open a **new sketch** or select: **Insert, Sketch**.

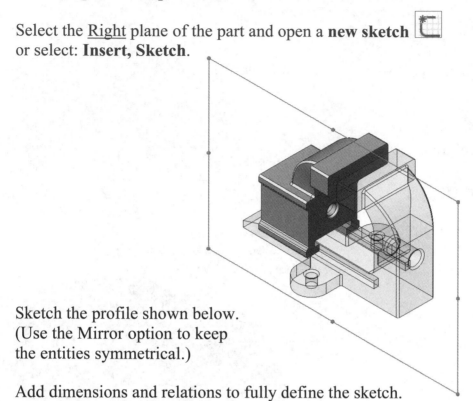

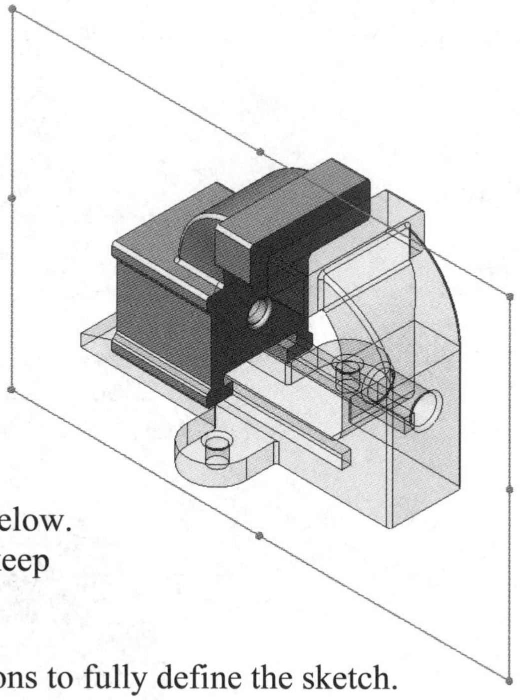

Sketch the profile shown below. (Use the Mirror option to keep the entities symmetrical.)

Add dimensions and relations to fully define the sketch.

Add a **Pierce** relation between the endpoint of the Centerline and the 1st revolution of the helix.

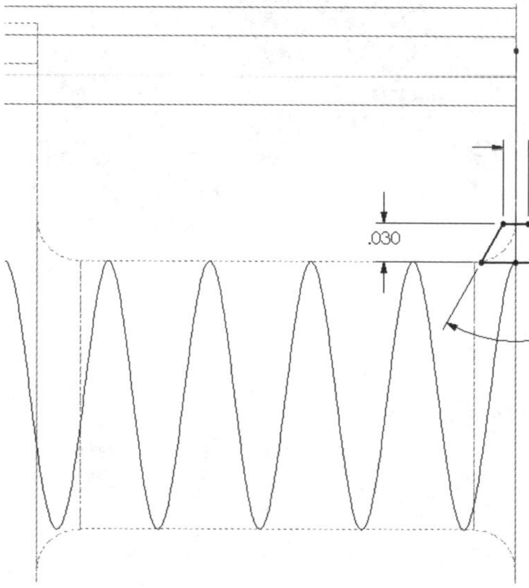

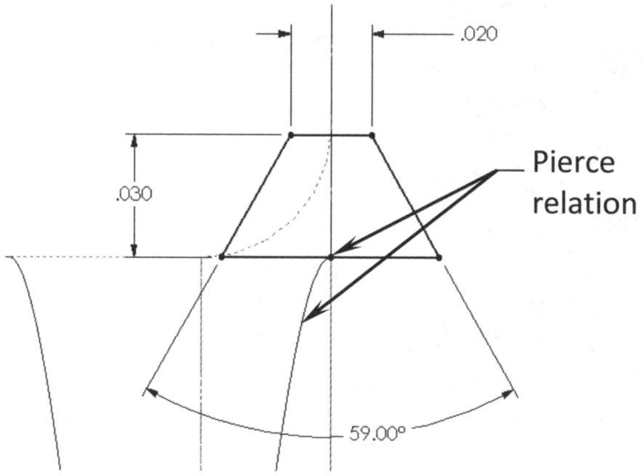

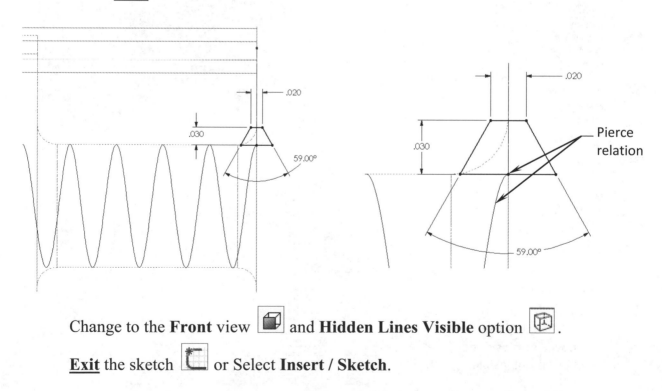

Change to the **Front** view and **Hidden Lines Visible** option.

Exit the sketch or Select **Insert / Sketch**.

34. Sweeping the thread Profile along the Helix:

Click  or select **Insert, Cut, Sweep.**

For Sweep Profile, select the **Thread Profile**.

For Sweep Path, select the **Helix**.

Click **OK**.

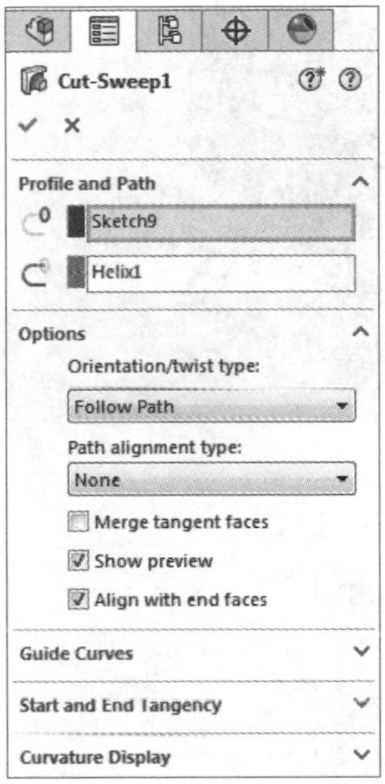

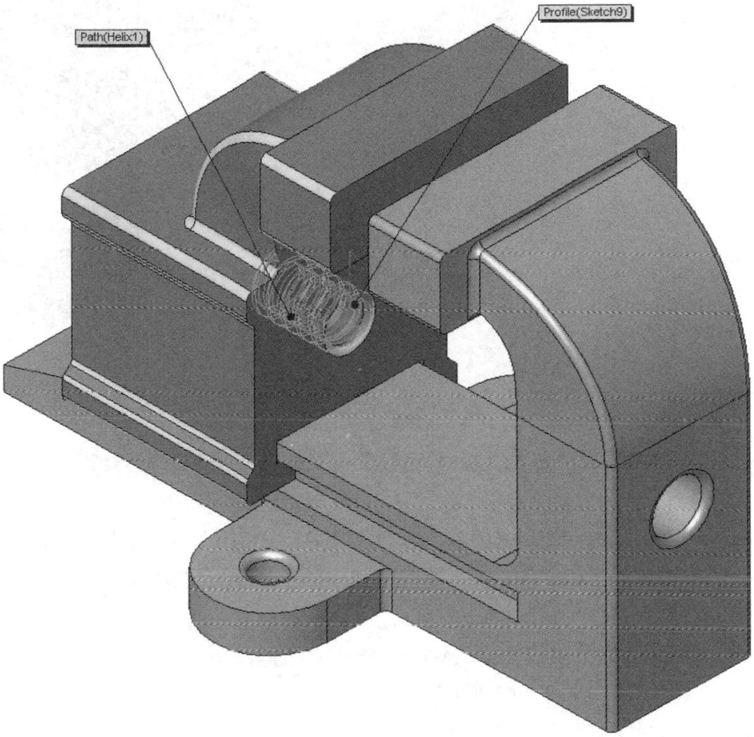

35. Creating a Section View:

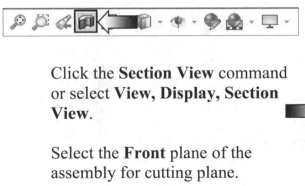

Click the **Section View** command or select **View, Display, Section View**.

Select the **Front** plane of the assembly for cutting plane.

Verify the details of the threads.

Click the **Section View** icon again to turn it off.

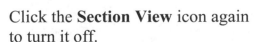

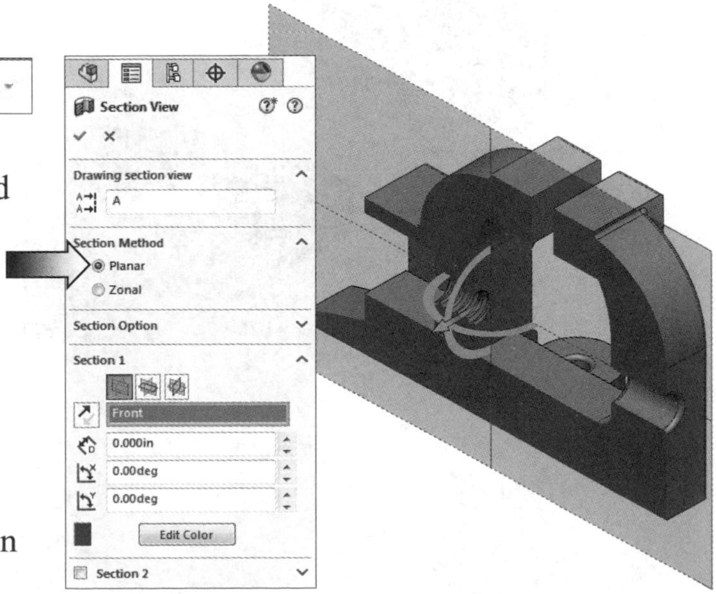

36. Saving your work.

Save your work once again using the same file name: **Mini-Vise.sldasm**

Overwrite the old file when prompted.

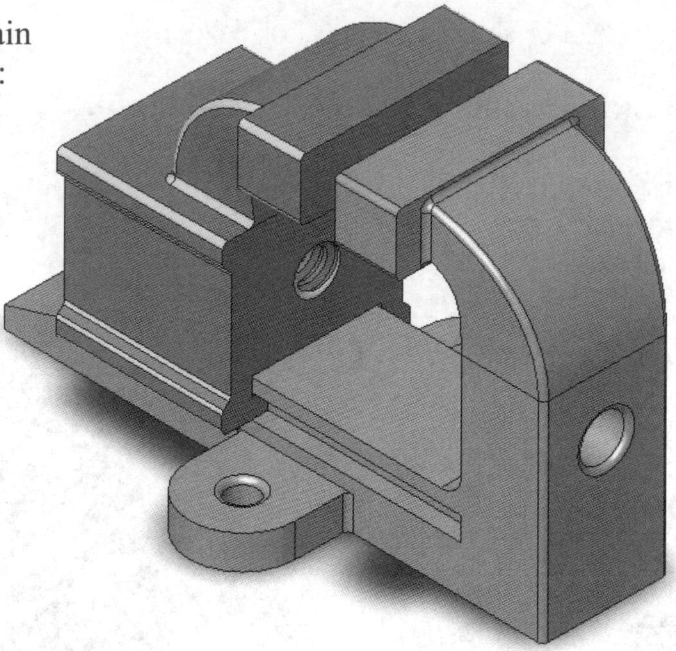

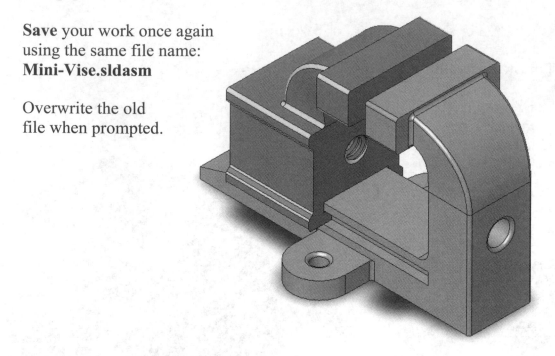

37. Assembly Exploded view (Optional):

Create the additional components: Lead Screw, Crank Handle, and Crank Knob using the Top Down Assembly method.

Create an assembly exploded view as shown (details on next page).

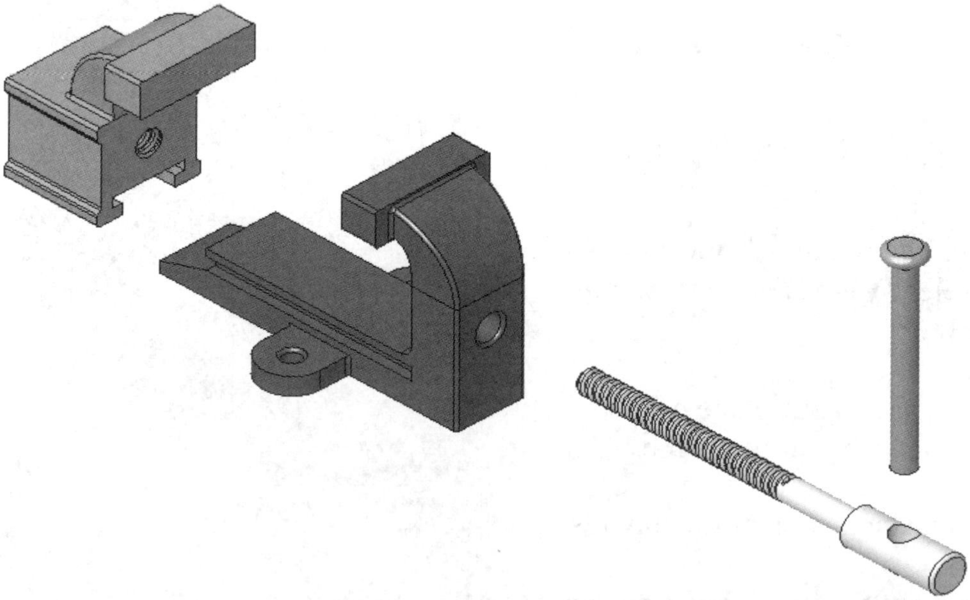

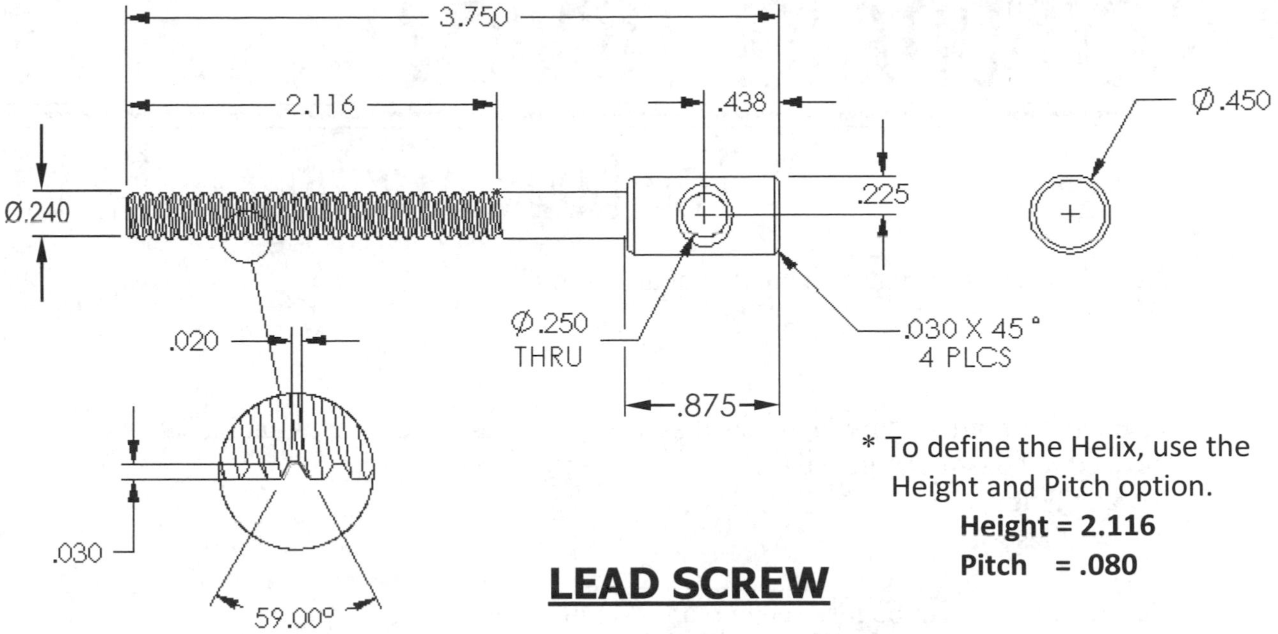

* To define the Helix, use the Height and Pitch option.
Height = 2.116
Pitch = .080

LEAD SCREW

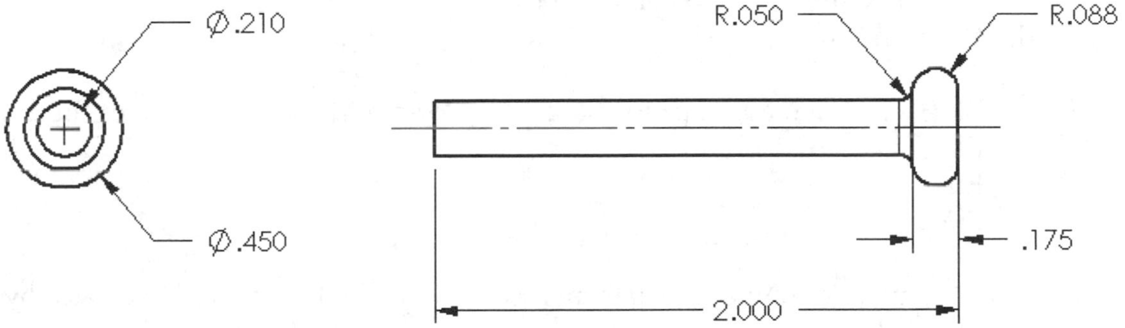

CRANK HANDLE

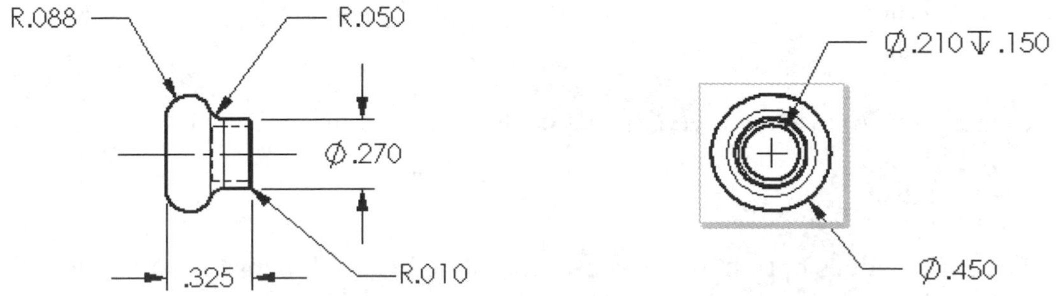

CRANK KNOB

Questions for Review

Top Down Assembly – Part 1

1. New parts can be created in context of an assembly.
   a. True
   b. False

2. Geometry of other components such as sketch entities, model edges, and locations etc., can be used to construct a new part.
   a. True
   b. False

3. Part documents can be inserted into an assembly using:
   a. Insert menu
   b. Windows Explorer
   c. Drag and drop from an open window
   d. All of the above

4. The suffix (f) next to the first part's name in the FeatureManager tree stands for:
   a. Fail
   b. Fixed
   c. Float

5. When inserting new components into an assembly, the Inplace mates are created by the user.
   a. True
   b. False

6. Either in the part or assembly mode, the guide curves are used to help control the profiles from twisting, as they are swept along the path.
   a. True
   b. False

7. Centerpoint Arcs are drawn from its center, then radius, and angle.
   a. True
   b. False

8. The Link Values option allows a user to link only two dimensions at a time.
   a. True
   b. False

1. TRUE 2. TRUE
3. D 4. B
5. FALSE 6. TRUE
7. TRUE 8. FALSE

CHAPTER 21

Top Down Assembly – Part 2

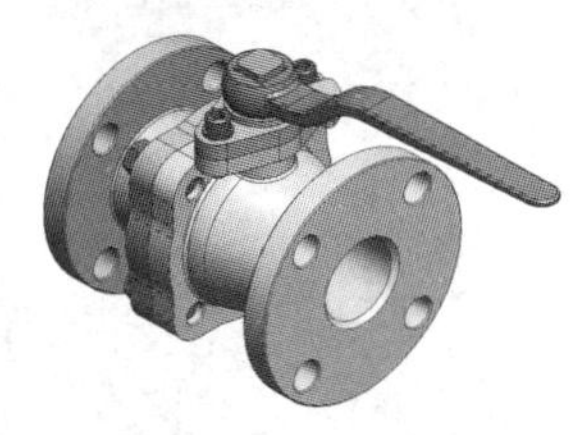

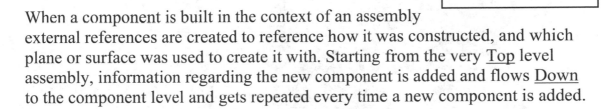

Top Down Assembly – Part 2
Water Control Valve

When a component is built in the context of an assembly external references are created to reference how it was constructed, and which plane or surface was used to create it with. Starting from the very Top level assembly, information regarding the new component is added and flows Down to the component level and gets repeated every time a new component is added.

For example: The mounting holes in the second part can be converted from the first, so that the hole diameters and the location dimensions are the same for both parts. When the holes in the 1st part are changed, the holes in the 2nd part would also change. Thus the sketch of the holes in the 2nd part is defined in the assembly, not by sketching and dimensioning them as in the part mode.

Using the Top Down assembly design, one of the better approaches is to use the geometry of the existing parts to create the new. This way several parts can be controlled and changed at the same time.

There are many advantages for creating parts in Top Down mode, and just to mention a few: not only is this method much quicker than the others due to the ability to use existing geometry to reference the new parts, but because all the parts are always visible in the assembly to help develop the Form of the new part and how it is supposed to Fit with other parts, it is more predictable how it is going to Function. Interference, friction and or clearance fits can be created and controlled within the very same screen.

However, there are a few things to consider when designing in Top Down mode:
* External references are created to the geometry that the new part is referenced to, and that means:
  * When changes occurred, the assembly updates all of its internal parts, and if drawings were made from these parts earlier; they will get updated as well.

Top Down Assembly – Part 2
Water Control Valve

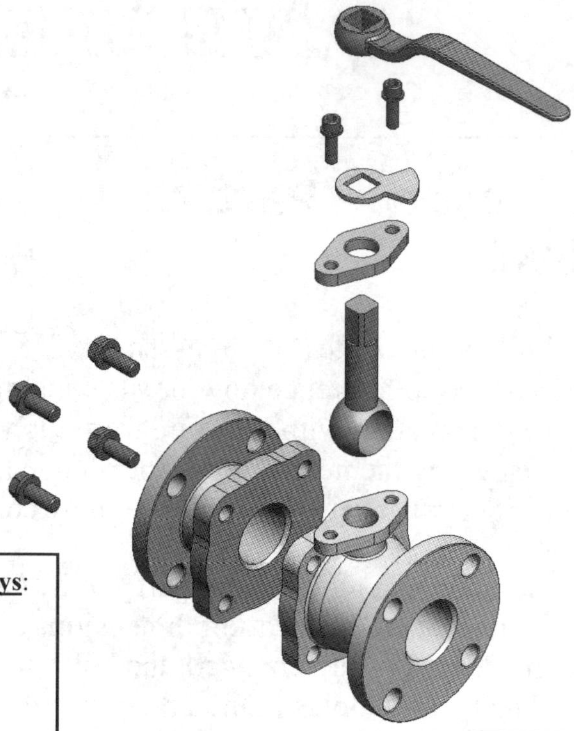

View Orientation Hot Keys:

Ctrl + 1 = Front View
Ctrl + 2 = Back View
Ctrl + 3 = Left View
Ctrl + 4 = Right View
Ctrl + 5 = Top View
Ctrl + 6 = Bottom View
Ctrl + 7 = Isometric View
Ctrl + 8 = Normal To Selection

Dimensioning Standards: **ANSI**

Units: **INCHES** – 3 Decimals

Tools Needed:

Insert Sketch	Line	Circle
Add Geometric Relations	Sketch Fillet	Trim
Dimension	Centerline	Fillet/Round
Base/Boss Revolve	Extruded Boss/Base	Edit Component

1. Starting with a new Assembly Template:

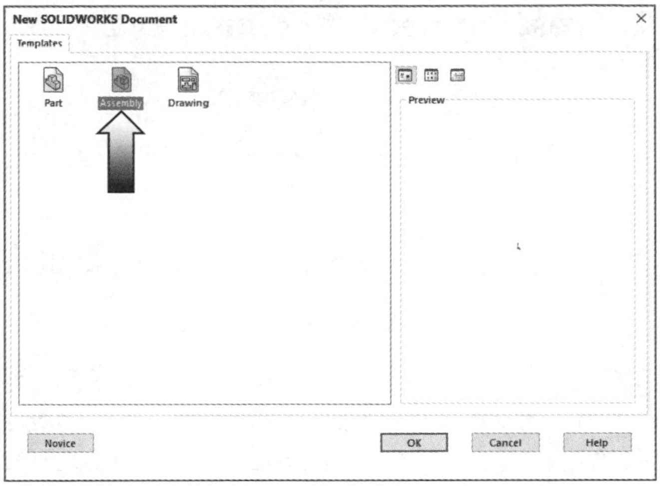

Click **File / New**.

Select an **Assembly** template either from the Template or the Tutorial tab.

(Click the **Advance** button at the lower left corner of this dialog box if you do not see the similar templates.)

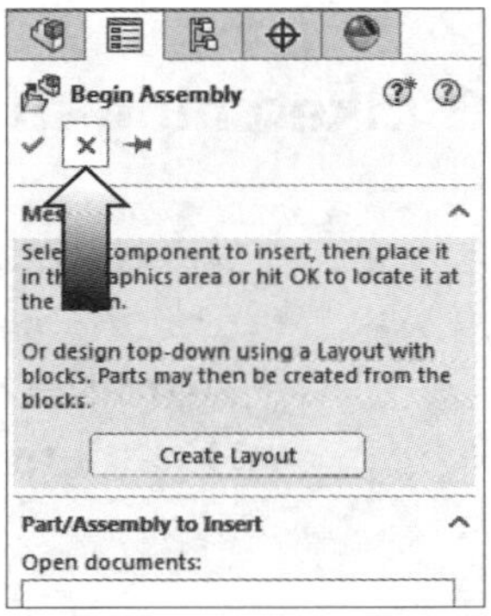

The **Begin Assembly** dialog appears on the left side; click **Cancel**. We are going to use a different approach to create the new components.

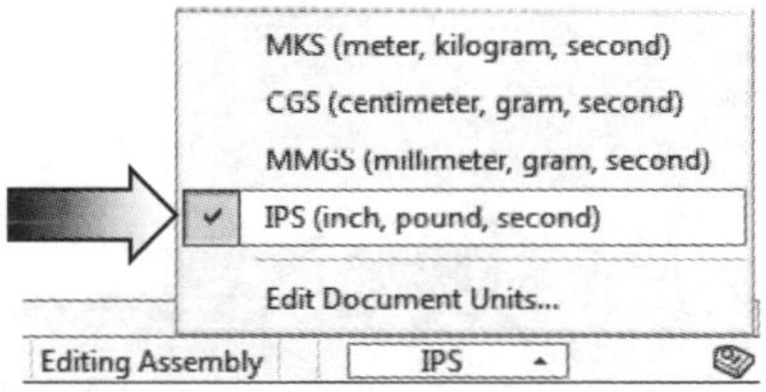

At the bottom right of the screen, set the Units to **IPS** (Inch, Pound, Second) 3 decimal places.

From the **Assembly tab**, click the **drop arrow** below the Insert Components command and select: **New Part**.

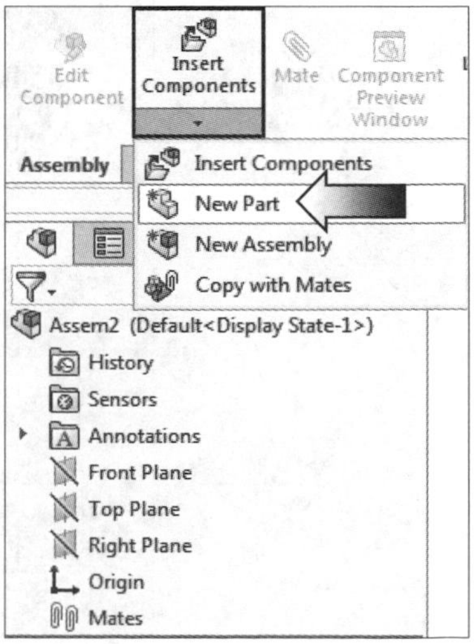

Creating components in context of an assembly will require a few additional steps:

a/. A new part is inserted into an Assembly and the Part's name is entered.
b/. A plane is selected at this time to reference the new part.
c/. The Edit Component command is activated and the Sketch mode is enabled for the plane selected in step b.
d/. The active part will change to the blue color by default.

2. Creating the 1st component:

When the symbol  appears next to your mouse cursor, select the **Front** plane on the Feature tree. An Inplace mate is created to reference the new part.

Press **Control** + **1** to switch to the Front orientation.

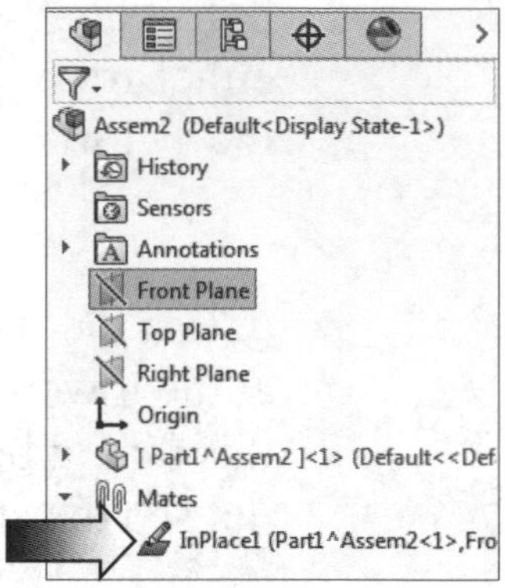

NOTE: *To automatically rotate normal to the sketch plane, go to System* ***Options / Sketch****, and enable the checkbox:* ***Auto Rotate View Normal to Sketch Plane…***

3. Sketching the Base Profile:

Sketch the profile above the origin.

Add the dimensions shown. The diameter dimensions are shown as Virtual Diameters.

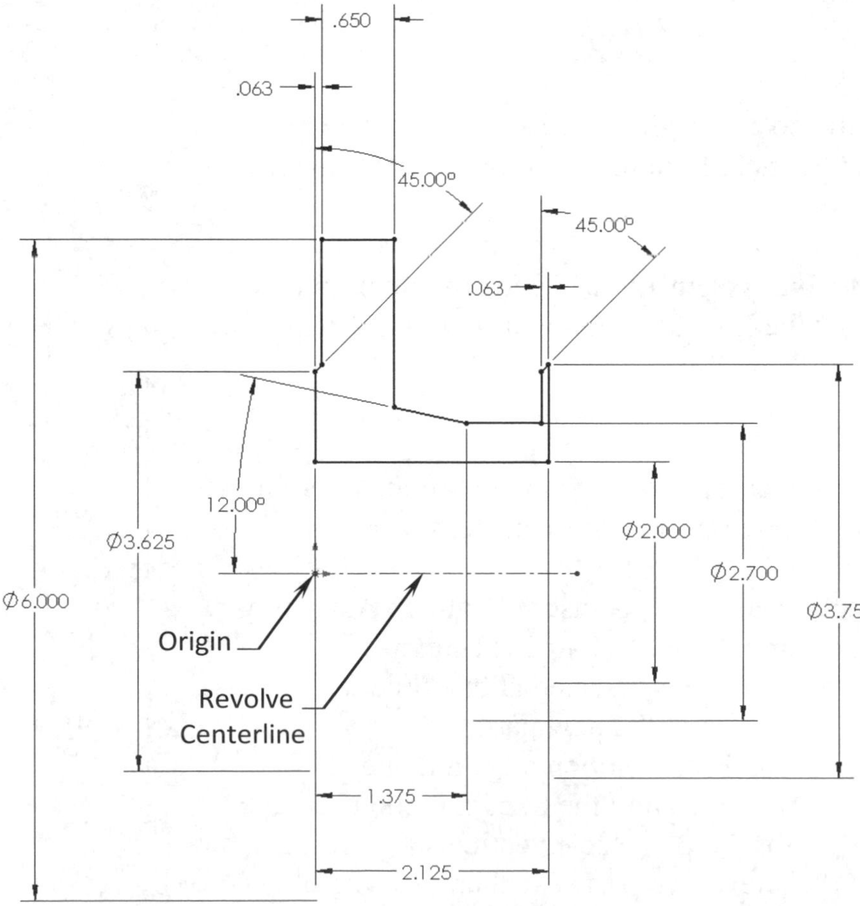

Click **Revolve** .

The centerline should be selected automatically.

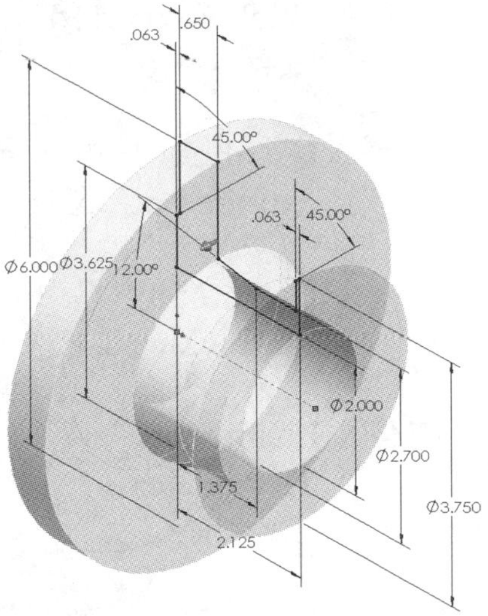

Use the default **Blind** option and revolve the sketch one complete revolution.

Click **OK**.

4. Adding the Inlet Flange:

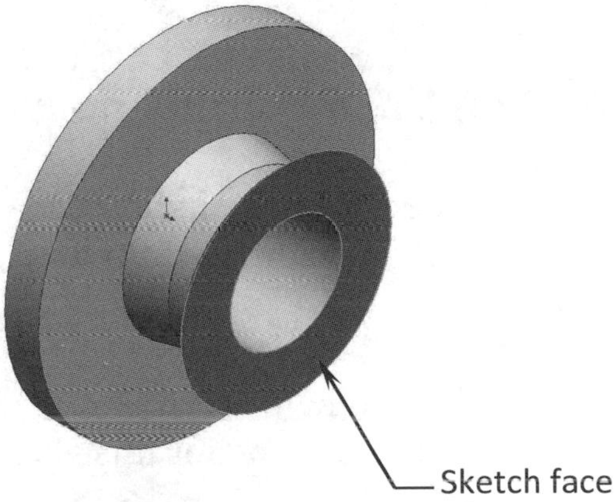

Select the face indicated and open a **new sketch**.

Sketch the profile shown below. Use the Dynamic Mirror option to help speed up the sketching process.

Add the dimensions and relations needed to fully define the sketch.

The number of instances (4X, 8X) are added to help clarify the sketch; you do not have to add them.

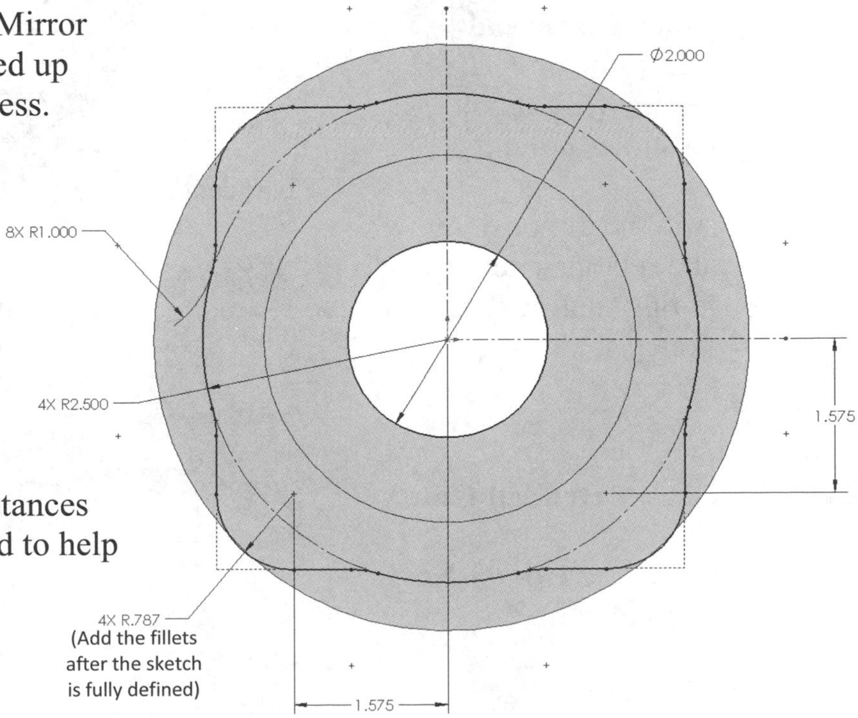

Click **Extruded Boss-Base**.

Use the **Blind** extrude option.

Enter **.650"** for thickness.

Enable the **Merge Result** checkbox.

Click **OK**.

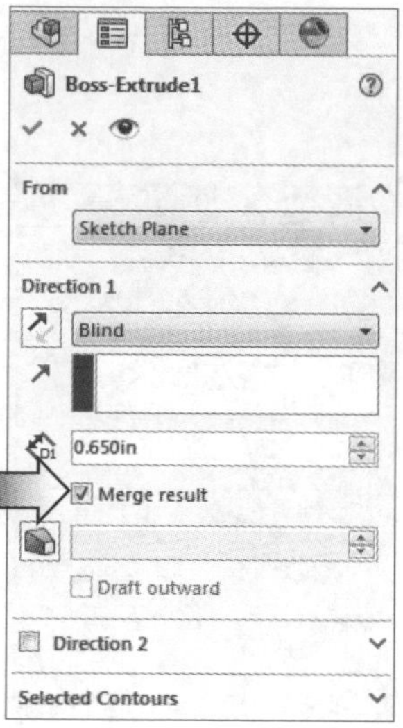

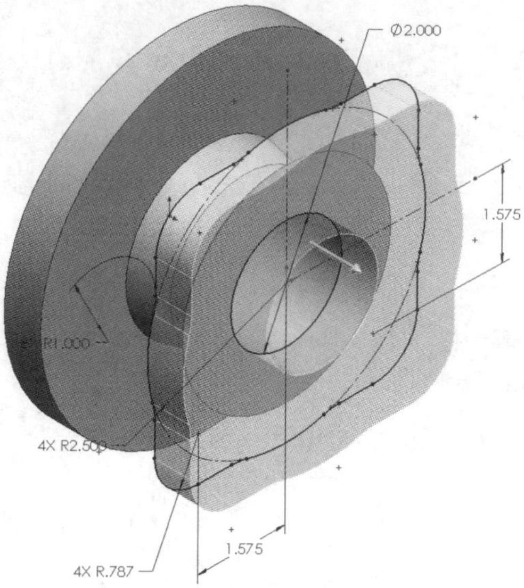

5. Adding the mounting holes:

Select the face indicated and open a **new sketch**.

Sketch a couple of **Centerlines** to help locate the center and directions for this sketch.

Add a **Circle** and either mirror it or circular pattern it 4 times around.

Add the dimensions and relations needed to fully define the sketch.

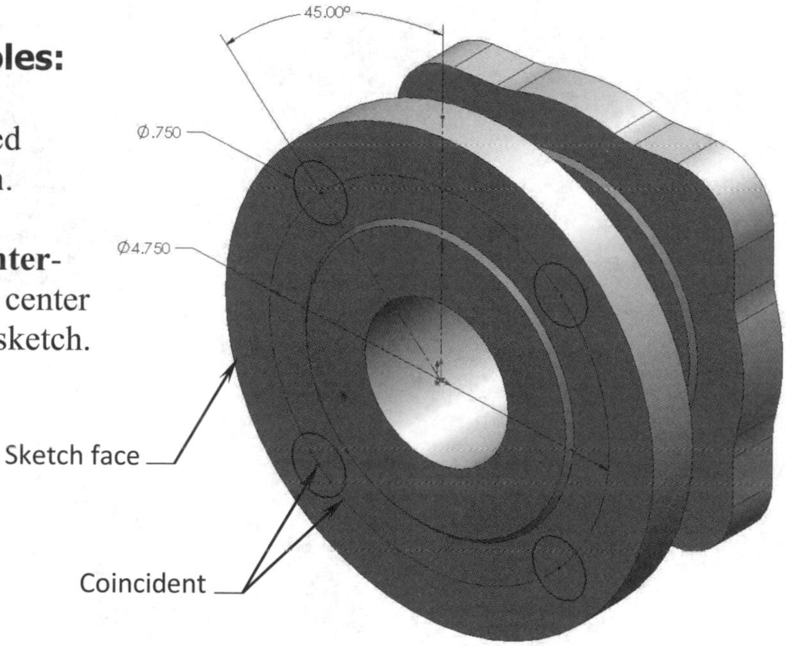

Click **Extruded Cut**.

Select the **Up-To-Next** extrude option.

Click **OK**.

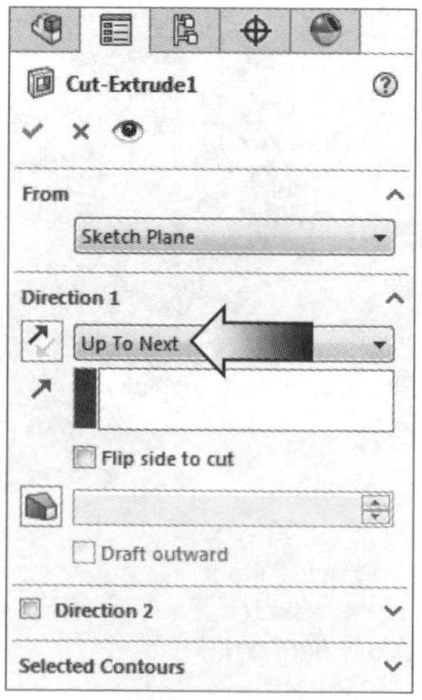

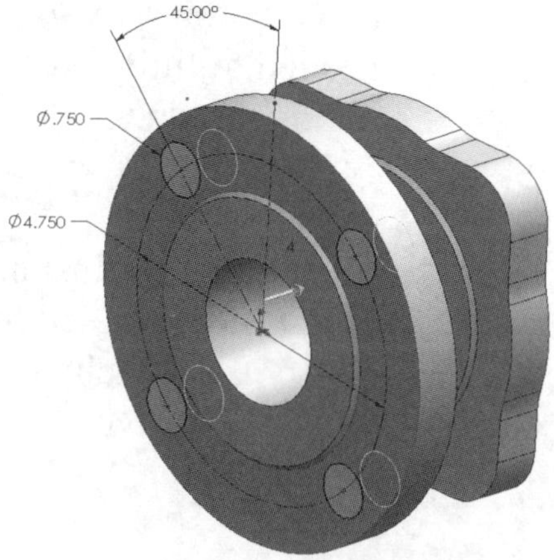

6. Adding other mounting holes:

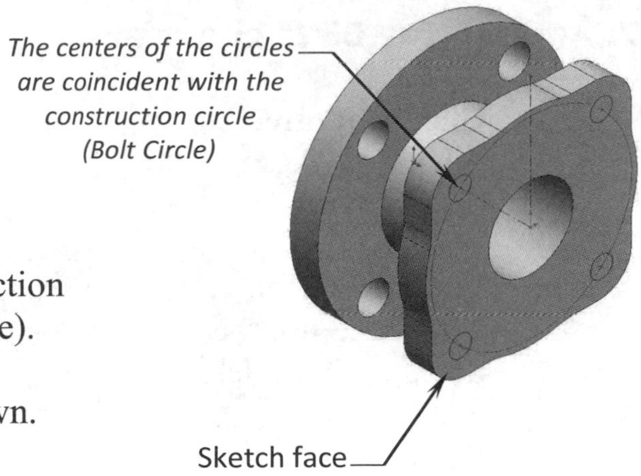

Select the face as noted and open a **new sketch**.

Sketch a **Circle** and **convert** it to **construction** (click the For-Construction checkbox on the FeatureManager tree).

Add a couple of **Centerlines** as shown.

Sketch a smaller **Circle** that is coincident with the construction circle and the endpoint of the centerline.

Use the **Circular-Sketch-Pattern** option to array the small circle 4 times around.

Add the dimensions and relations needed to fully define this sketch.

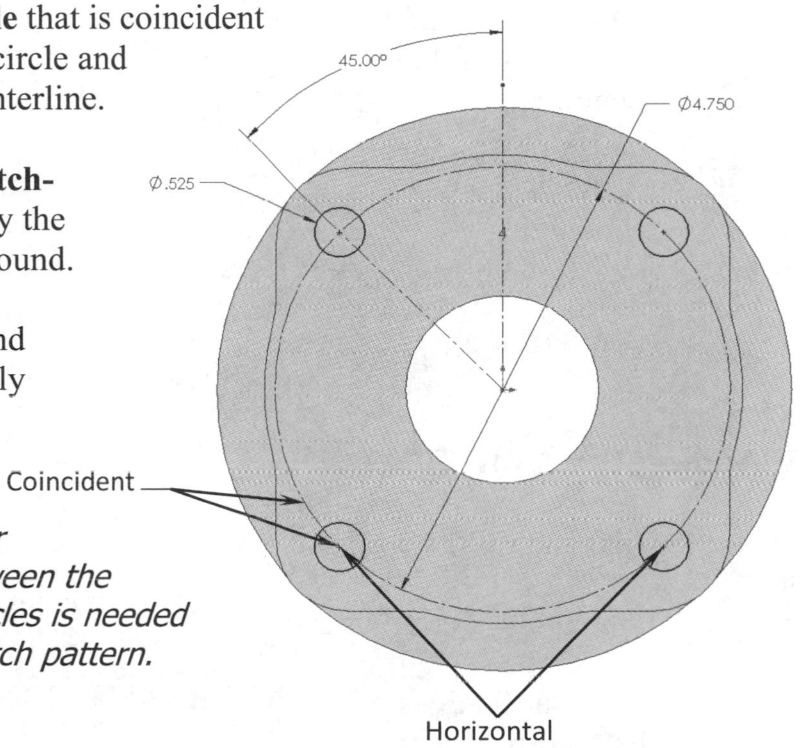

NOTE: *An additional Vertical or Horizontal relation between the centers of the small circles is needed when using the 2D sketch pattern.*

Click **Extruded Cut**.

Use the **Up-To-Next** extrude option to ensure the cut only goes through the thickness of the flange.

Click **OK**.

Rotate the view to inspect the cut result.

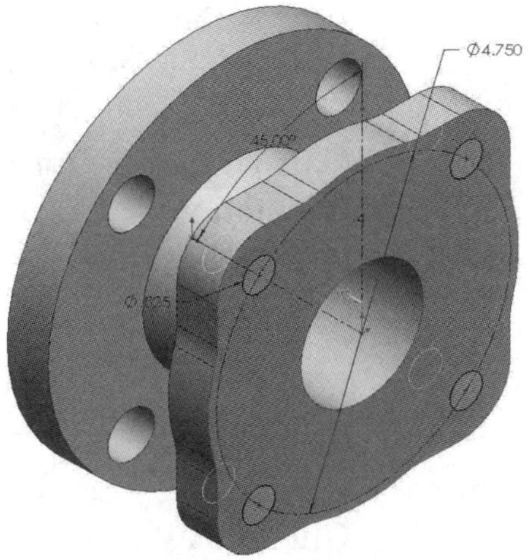

7. Adding the .032" chamfers:

Click **Chamfer** (under the Fillet drop-down).

Enter **.032"** for Depth and use the default **45°** angle.

Select the edges of the **8 holes** and the 2 edges of the round flange.

To deselect an edge simply click it once again.

Click **OK**.

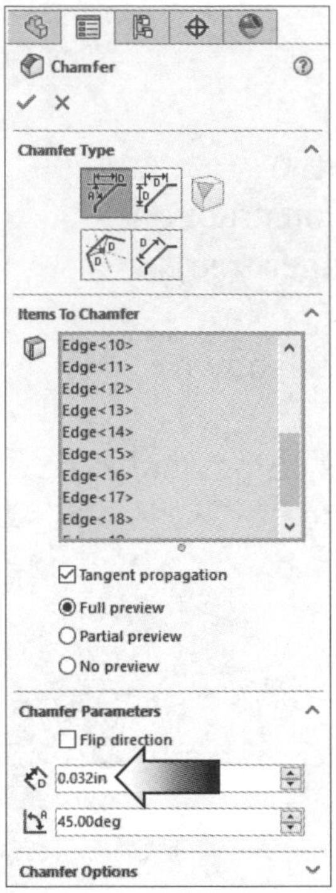

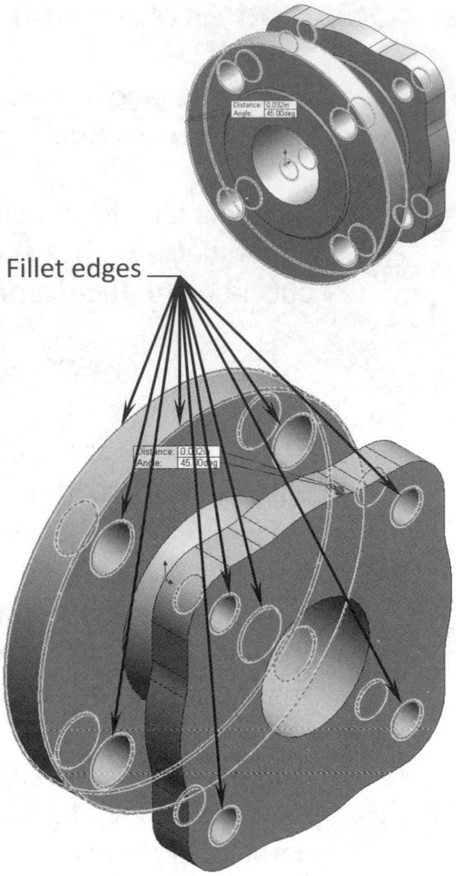

8. Adding the .060" fillets:

Click the **Fillet** command.

Enter **.060"** for radius size.

Select the **3 edges** of the transition body.

Enable the **Full Preview** checkbox.

Change to the **Top** orientation (Control + 5) to verify the selection.

Click **OK**.

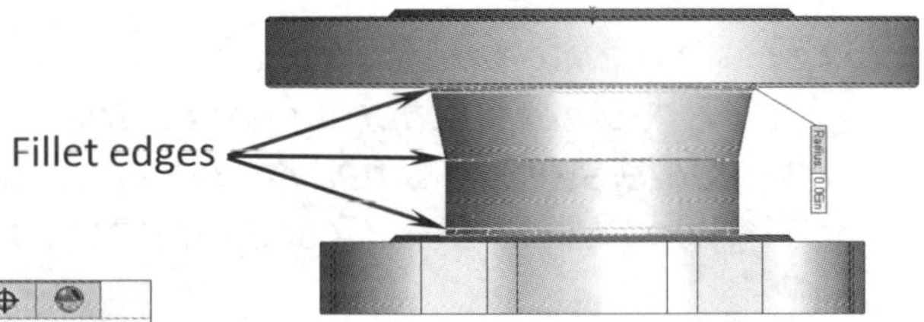

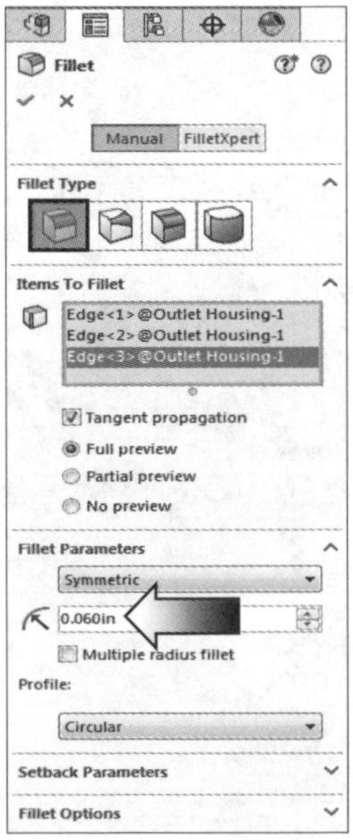

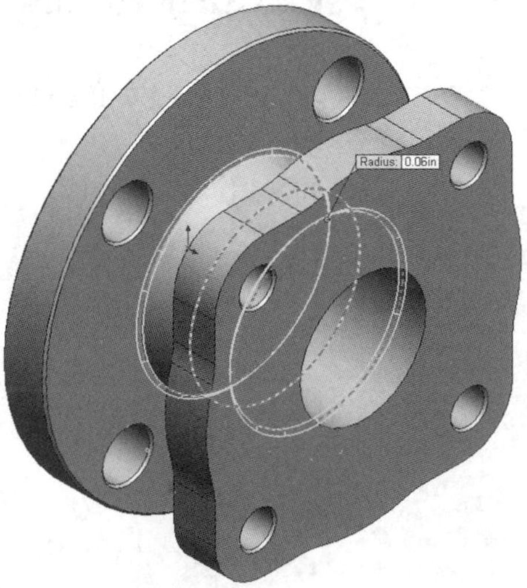

9. Adding the .125" chamfers:

Click **Chamfer** once again.

Enter **.125"** for Depth and use the default **45°** angle.

Select the **2 edges** of the center hole.

Selecting the face of the hole would get the same result as selecting the 2 edges.

Click **OK**.

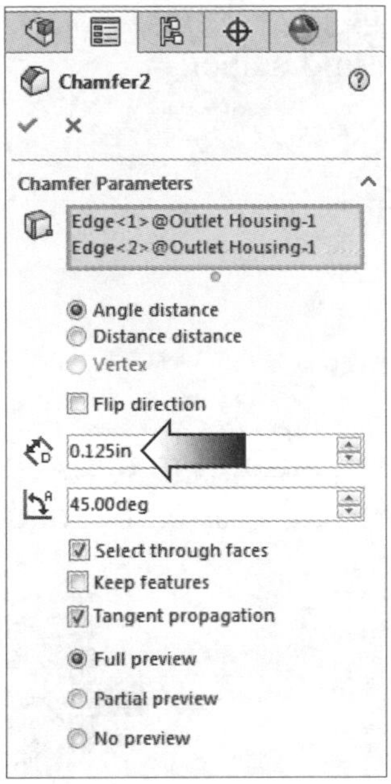

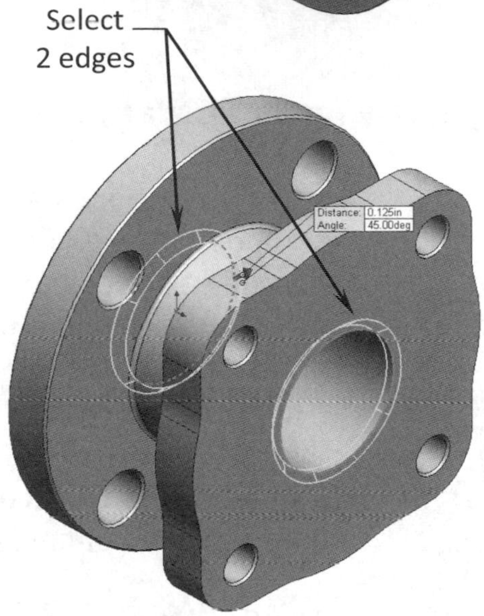

10. Exiting the Edit Component mode:

Click-off the **Edit Component** button to return to the Edit Assembly mode.

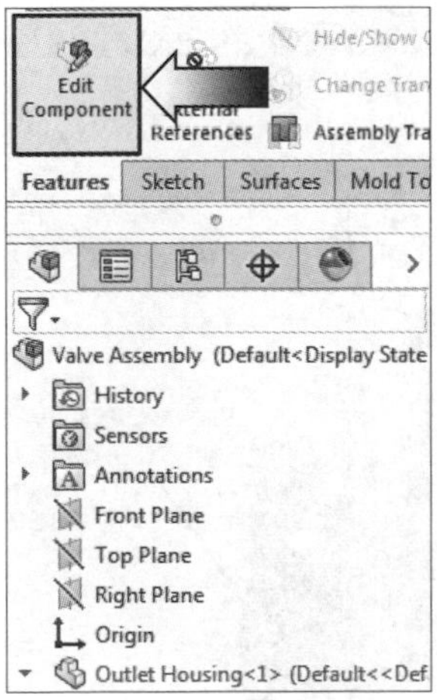

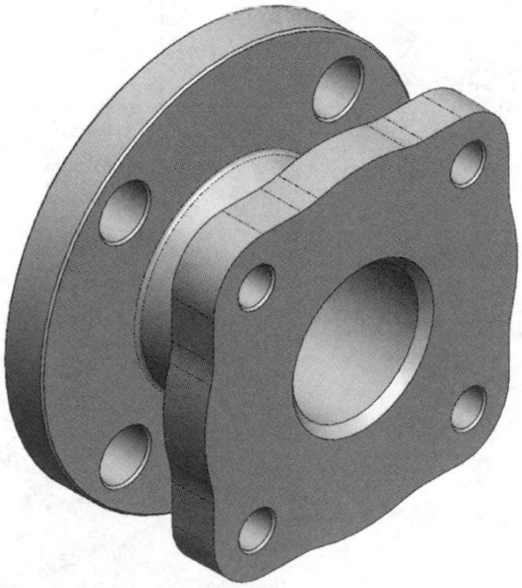

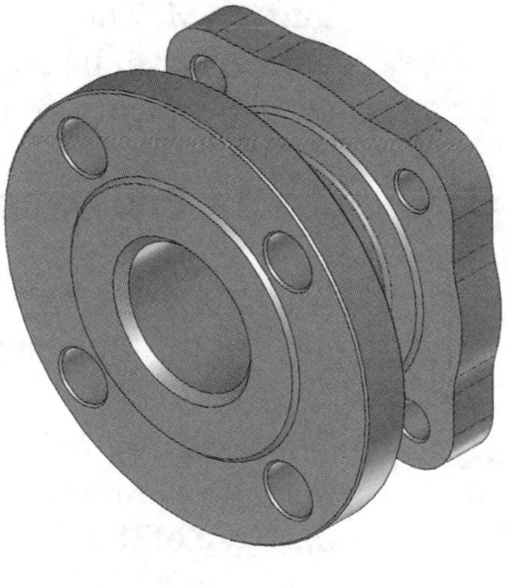

When the Edit Component command is <u>not active</u>, the part's color changes back to its default color (gray).

11. Renaming the component:

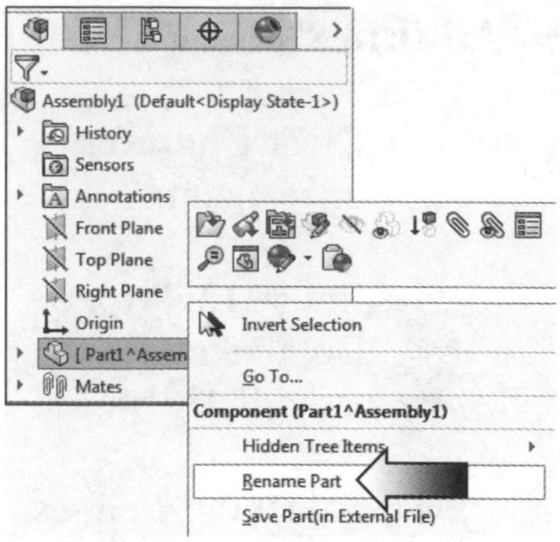

Right-click the name of the part (Part1) from the FeatureManager and select: **Rename Part** (arrow).

Enter **Outlet Housing** and press enter.

12. Saving as Virtual Component:

Virtual components are quite useful in the Top-Down Assembly mode. These components are saved internally, or embedded in the assembly document, instead of as separate part or sub-assembly documents.

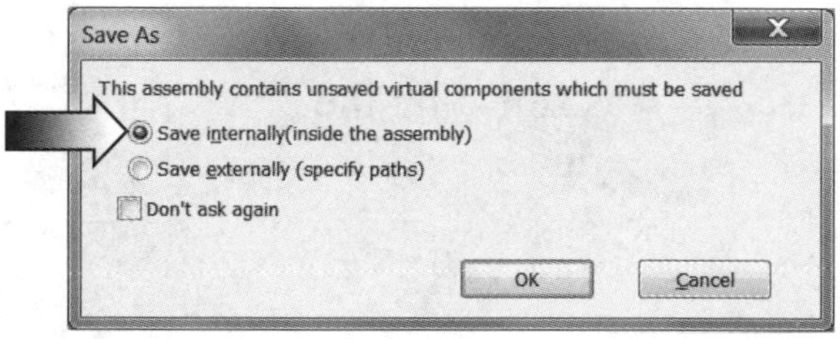

Click **File, Save As**.

Enter: **Water Control Valve** for the file name and press **Save**.

The Save As Virtual Component dialog appears; click the **Save Internally** option (Inside the Assembly) and click **OK**.

When the parent assembly is opened, all virtual components are also loaded into RAM. The virtual components can then be opened so that the detail drawings can be generated from them, or they can simply be saved as external part documents to share with others.

13. Creating the 2nd component:

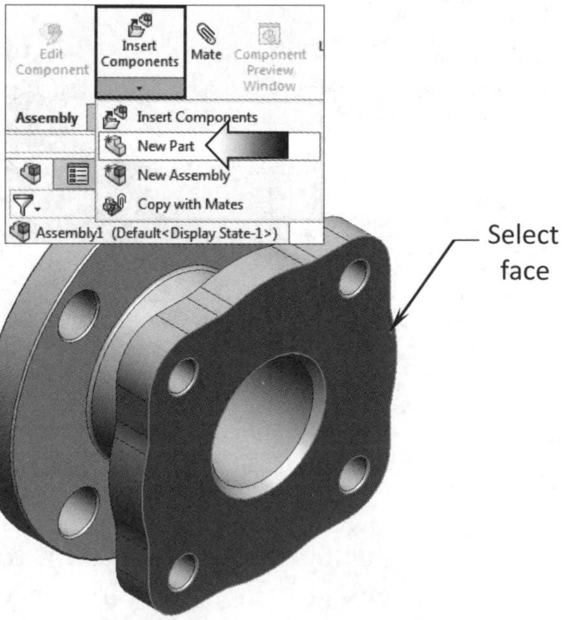

Click the **New Part** command under the Insert Components drop down.

When the symbol  appears next to your mouse cursor, click the Face of the flange as indicated.

At this point, another Inplace mate is created for the new part.

A new (Part2) component is created on the FeatureManager tree.

The Outlet Housing changes to transparent (inactive).

The **Edit Component**  command is activated.

A **new Sketch** is also enabled automatically.

Right-click one of the **outer edges** and pick **Select Tangency**.

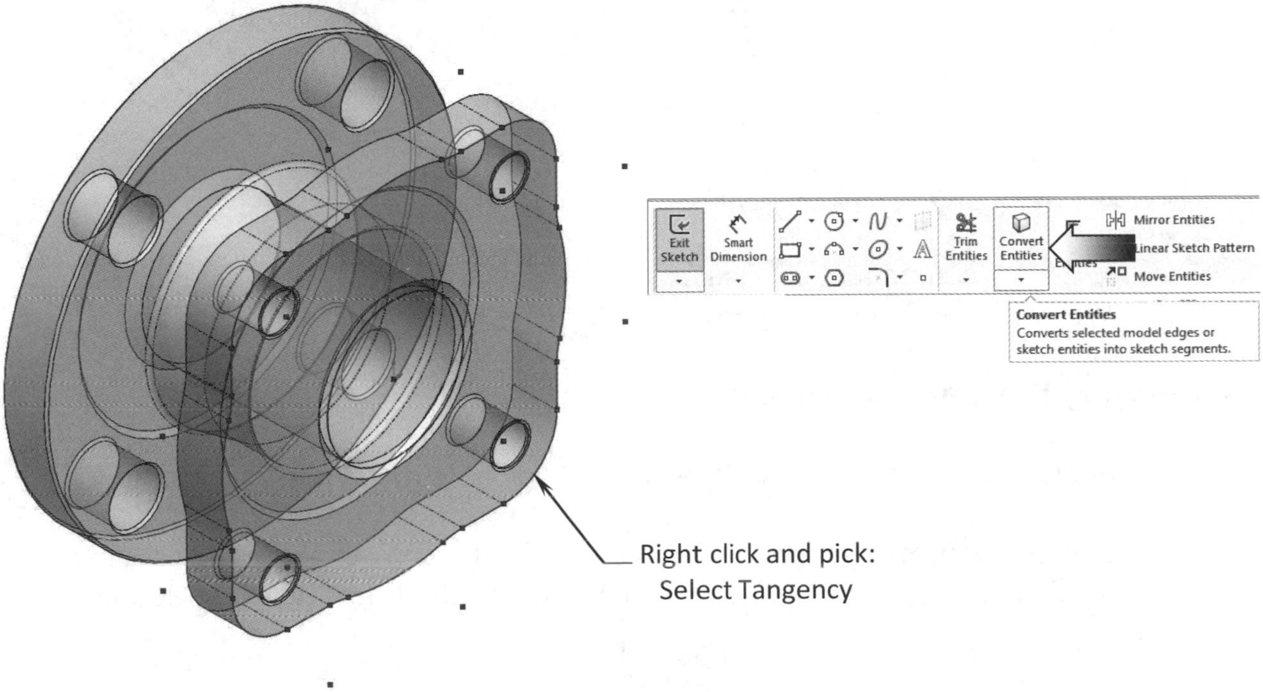

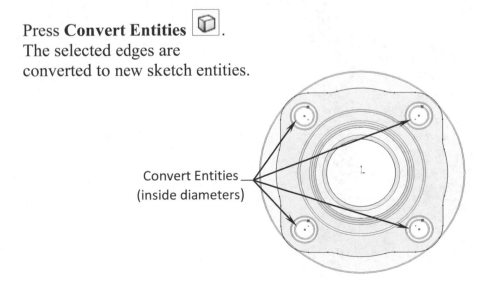

Press **Convert Entities** .
The selected edges are
converted to new sketch entities.

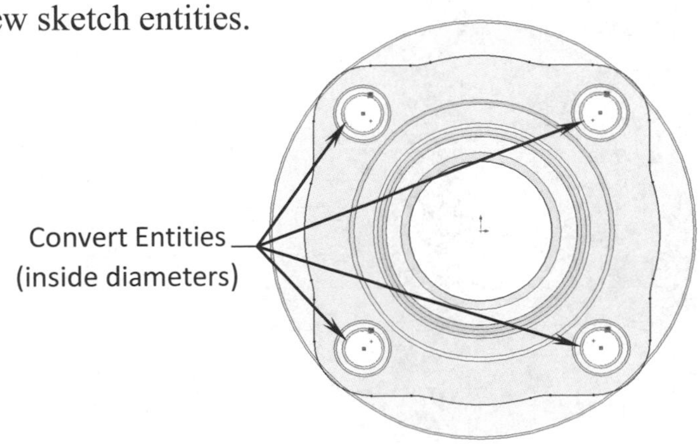

Also convert the **circular edges** of the 4 holes.

Click **Extruded Boss-Base.**

Use the default **Blind** extrude option.

Enter **.650"** for thickness.

Click **OK**.

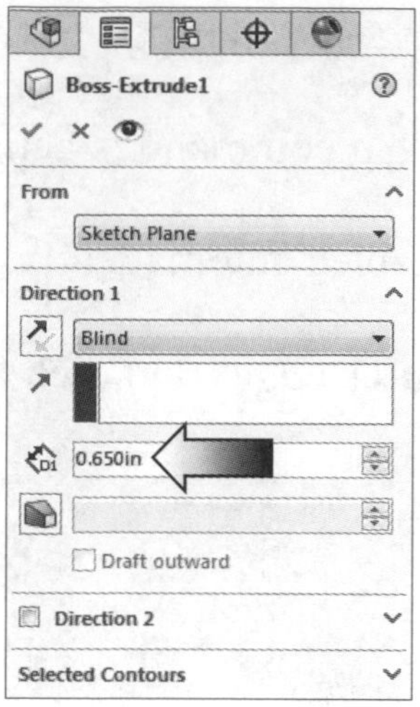

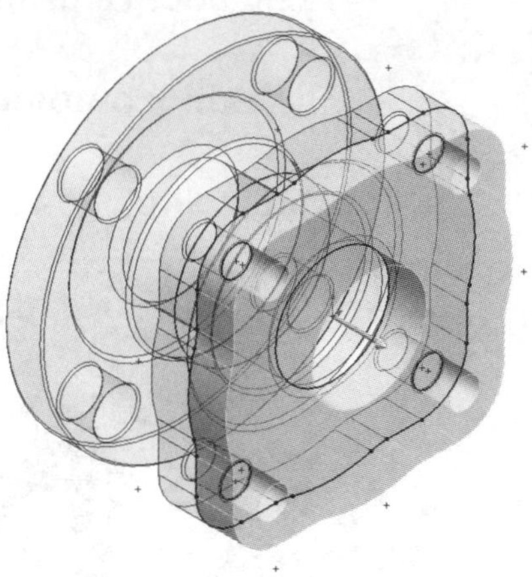

The new flange is created by converting the geometry of the 1st component.

If the 1st component is changed the 2nd component will also get updated. We will take a look at some of the changes toward the end of this chapter.

14. Creating the transition body:

Select the component's <u>Right</u> plane and open a **new sketch.**

Sketch the profile shown below and add the dimensions and any relations needed to fully define this sketch. (Add the R2.00 fillet after fully defined.)

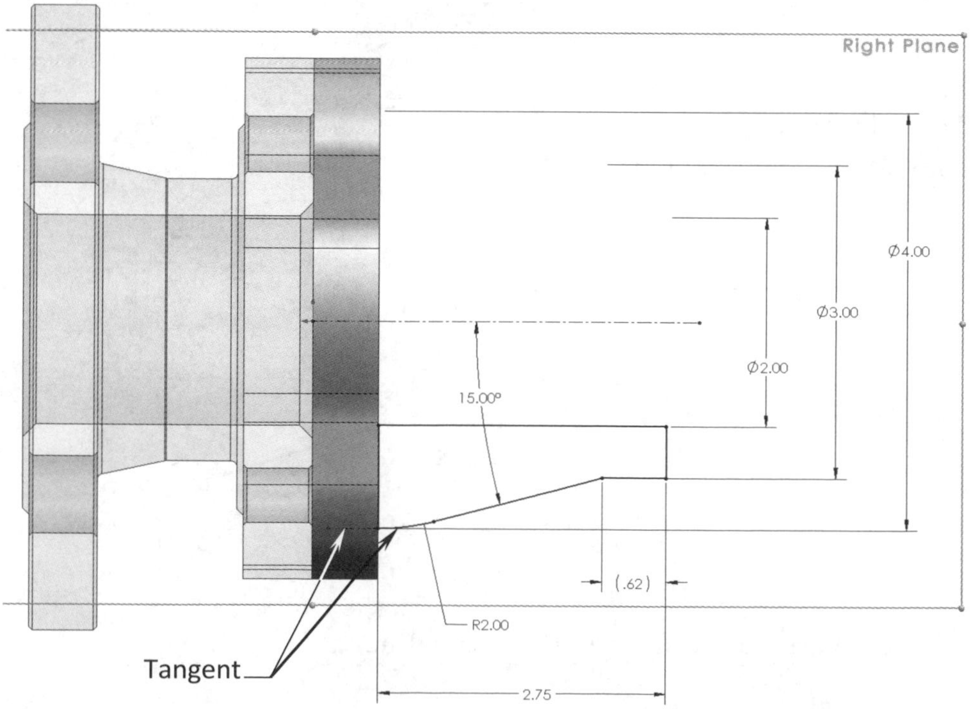

Click **Revolve Boss-Base**.

The revolve centerline is selected by default.

Use the default settings:

* **Blind**
* **360deg**

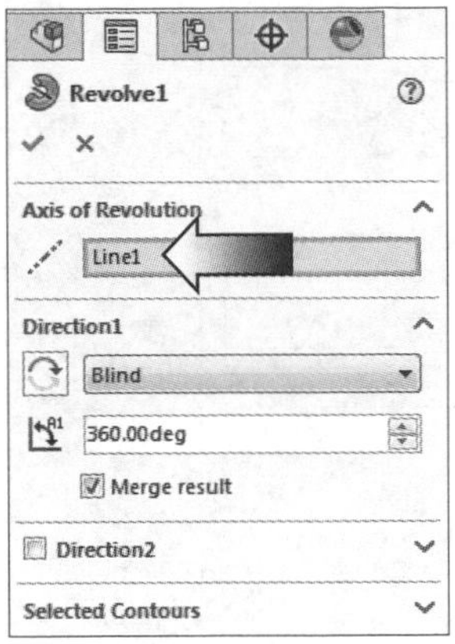

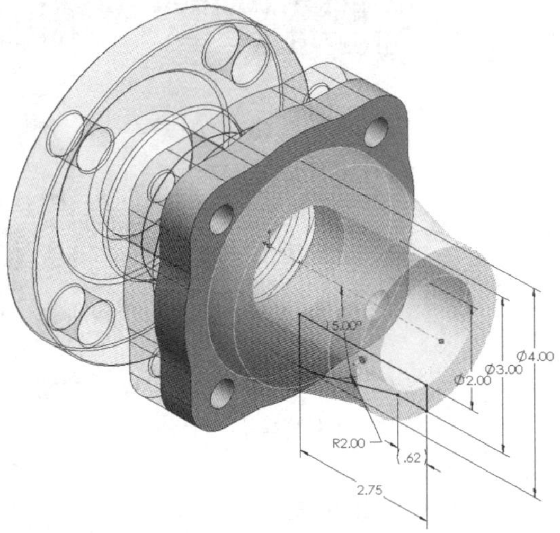

Click **OK**.

15. Adding another mounting flange:

Select the <u>face</u> as indicated and open a **new sketch**.

Hold the control key and select the **circular edge** of the round flange <u>and</u> the **4 edges** of its mounting holes (arrow).

Click **Convert Entities**.

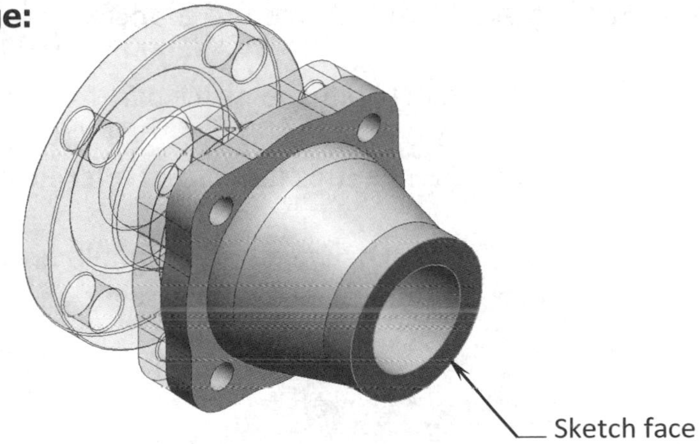

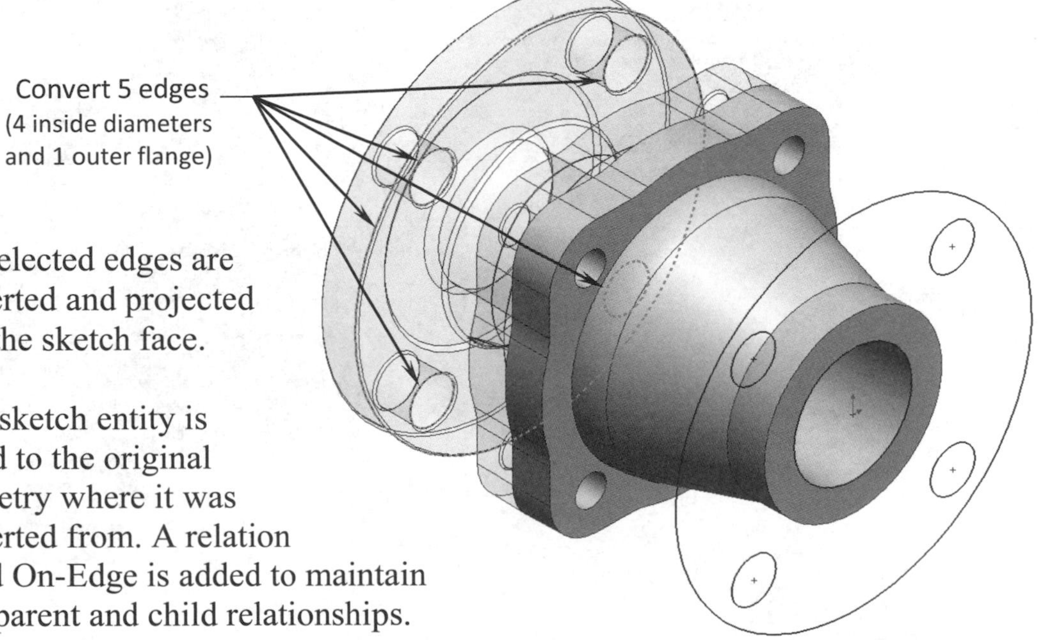

The selected edges are converted and projected onto the sketch face.

Each sketch entity is linked to the original geometry where it was converted from. A relation called On-Edge is added to maintain their parent and child relationships.

Click **Extruded Boss-Base**.

Use the default **Blind** option.

Enter **.650"** for depth.

Extrude direction is outward.

Click **OK**.

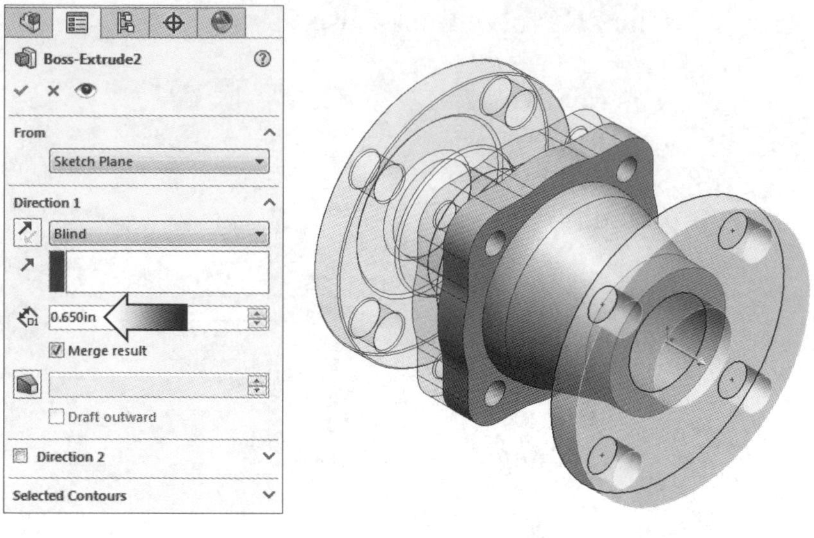

16. Adding an Offset-Distance plane:

Select the Top plane of the part from the FeatureManager tree.

From the Features tab, click **Reference Geometry / Plane**.

The Offset Distance option should be selected by default; enter **1.625"** for distance, and place the new plane above the Top plane.

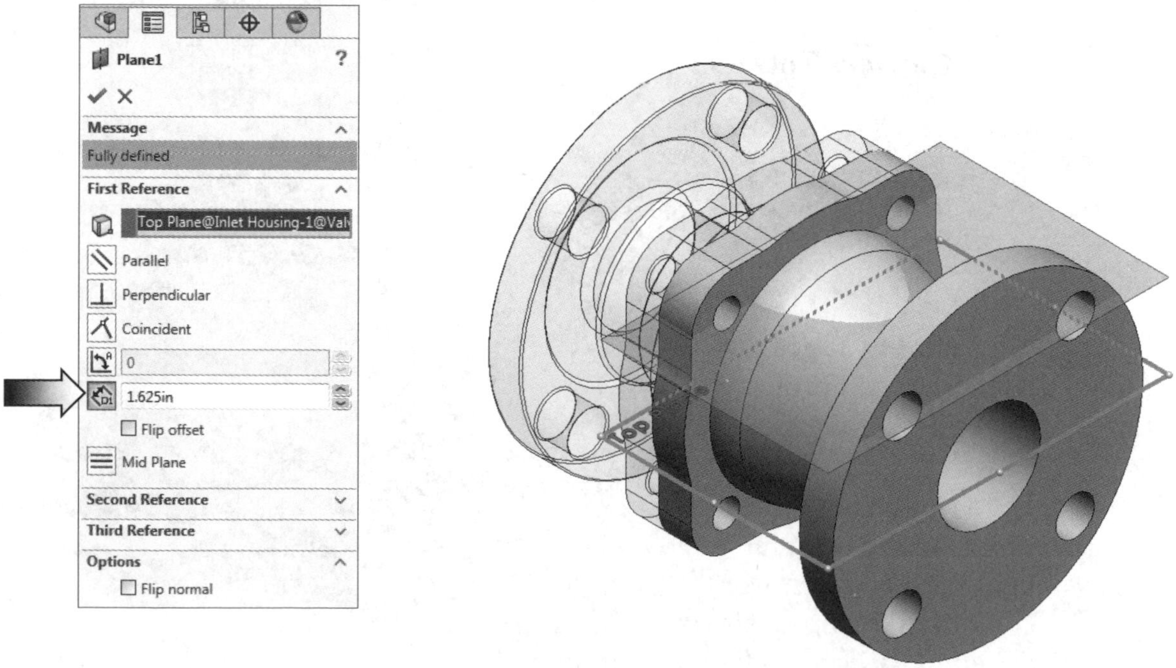

Click **OK**.

17. Adding a circular boss:

Select the new Plane1 and open a **new sketch**.

Sketch a **Circle** approx. as shown.

Add the **1.50"** diameter dimension and the **1.00"** locating dimension.

Add the **horizontal** relation as noted to fully define the sketch.

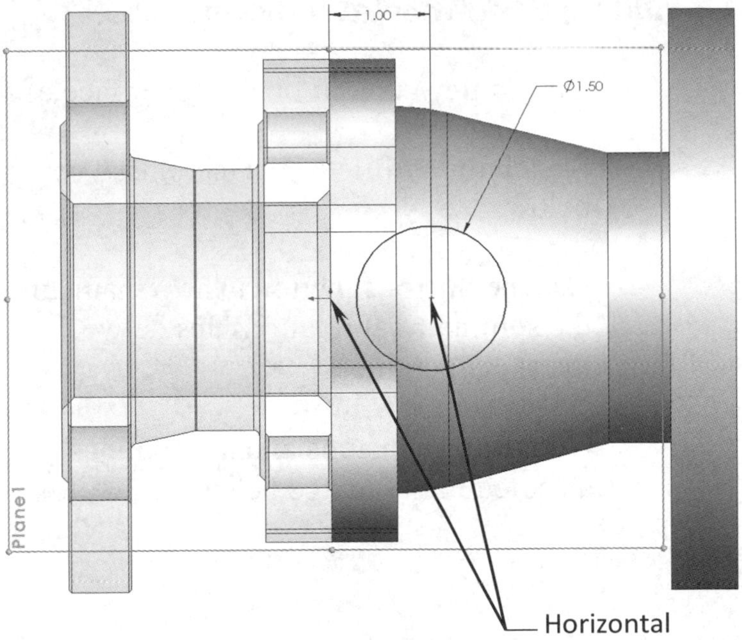

Click **Extruded Boss-Base**.

Use the default **Blind** option.

Enter: **.875"** for depth.

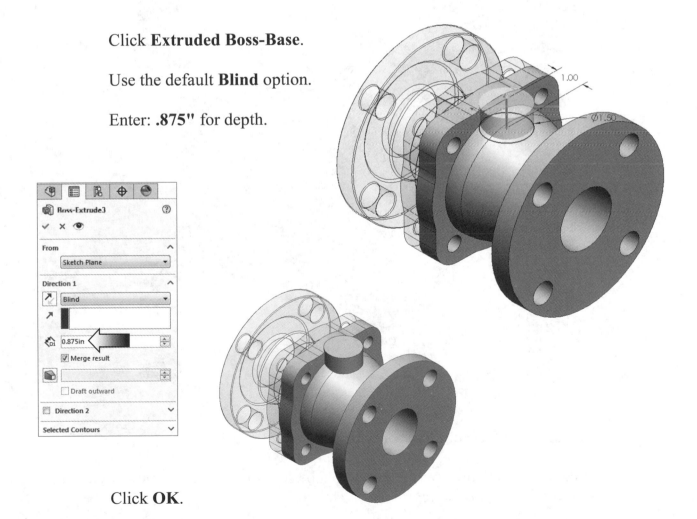

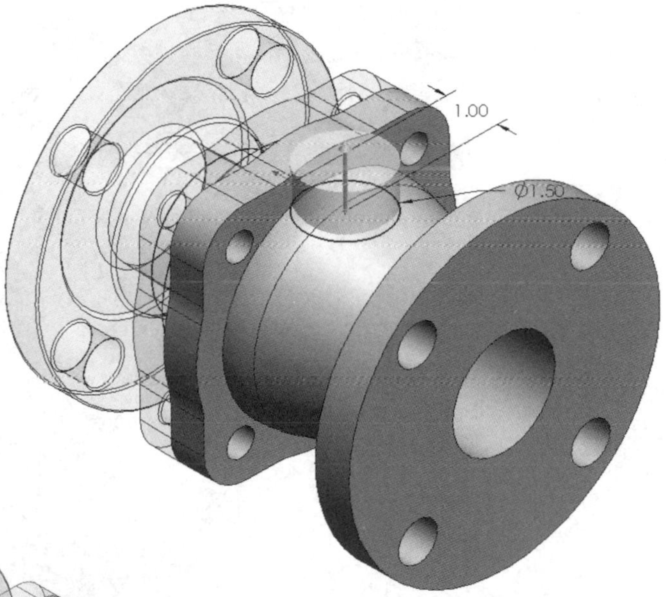

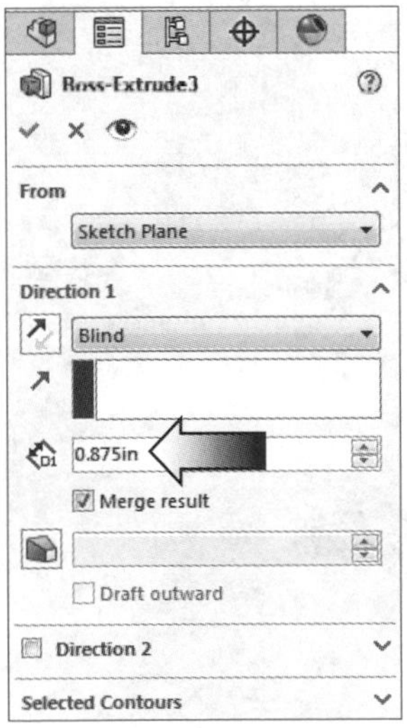

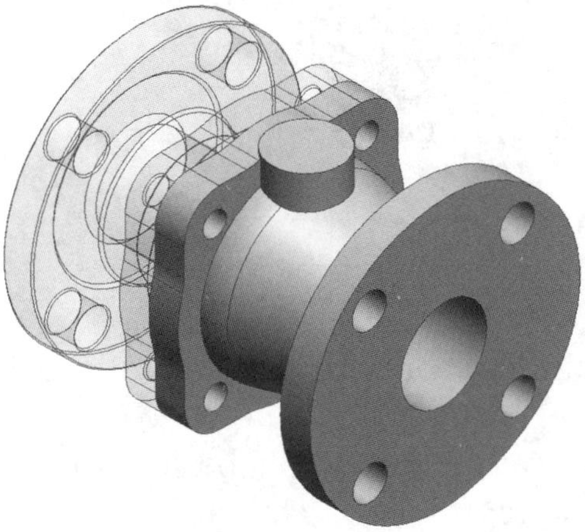

Click **OK**.

18. Adding a thermostat valve mount:

Open a **new sketch** on the upper face of the boss.

Sketch the profile of the thermostat shown below.

Use the mirror function to help maintain the symmetrical relationships between the sketch entities.

Add the dimensions and the relations as noted in the image below.

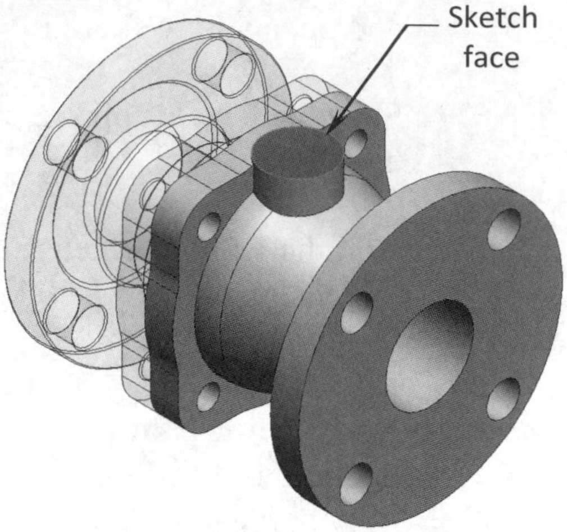

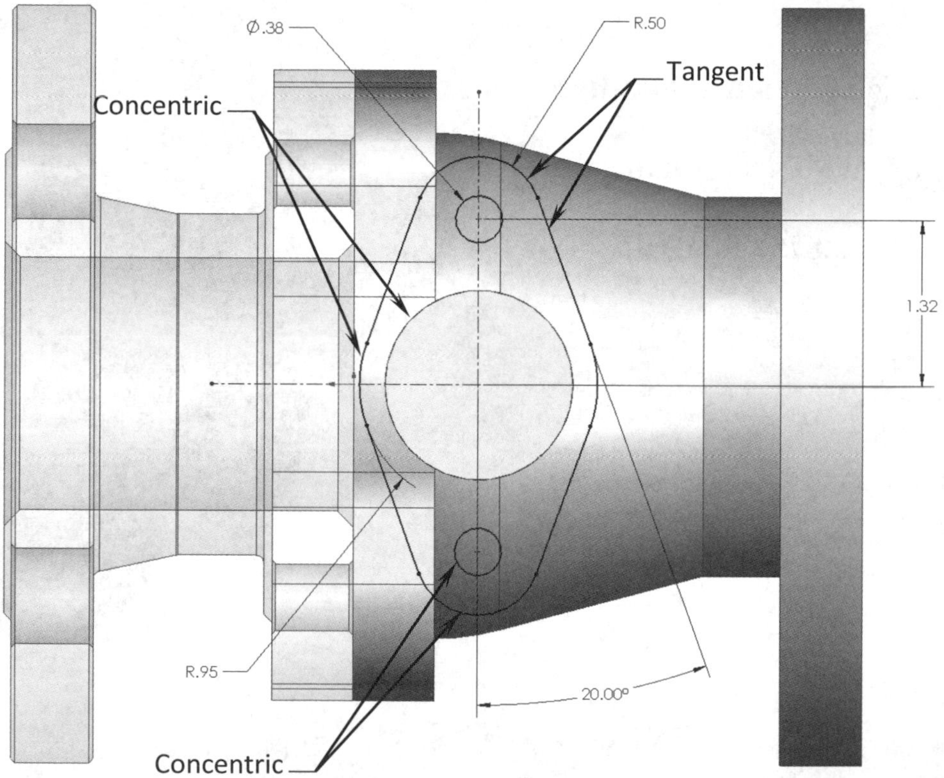

The centers of the circles should be vertical with each other.

If the mirror feature was not used, then be sure to add the symmetric relations to fully define this sketch.

Click **Extruded Boss-Base**.

For **Direction 1**: Use **Blind** and the depth of **.250"**.

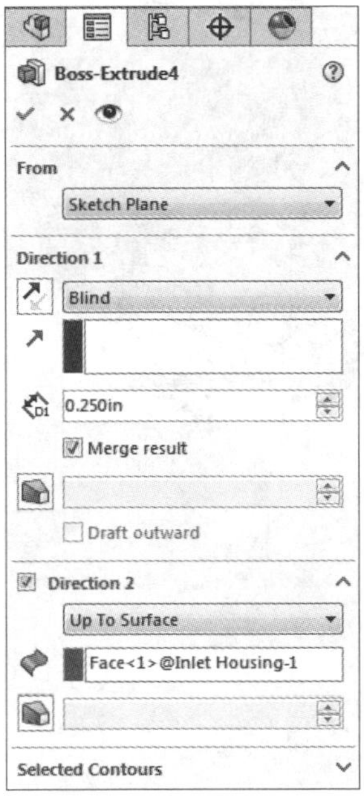

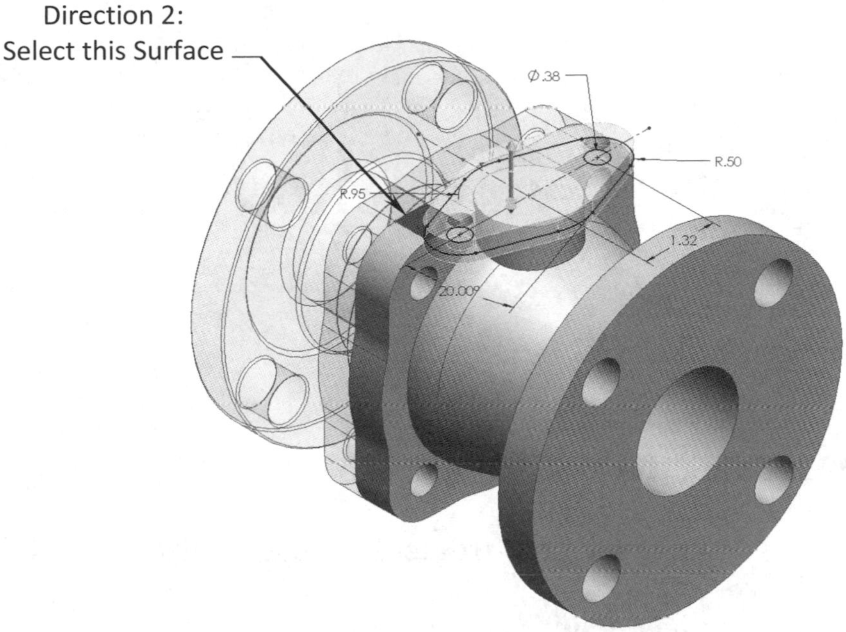

For **Direction 2**: Use **Up-To-Surface** and select the <u>planar face</u> as indicated.

Click **OK**.

19. Adding a hole:

Open a **new sketch** on the <u>upper face</u> of the thermostat.

Sketch a **Ø1.00 circle** and add a **concentric** relation to center it.

Make use of the "Wake-up the Entities Snap Mode".

(With the Circle command selected, hover the mouse cursor over the circular edge of the radius; the appropriate snap-entities will appear to allow snapping to the existing geometry.)

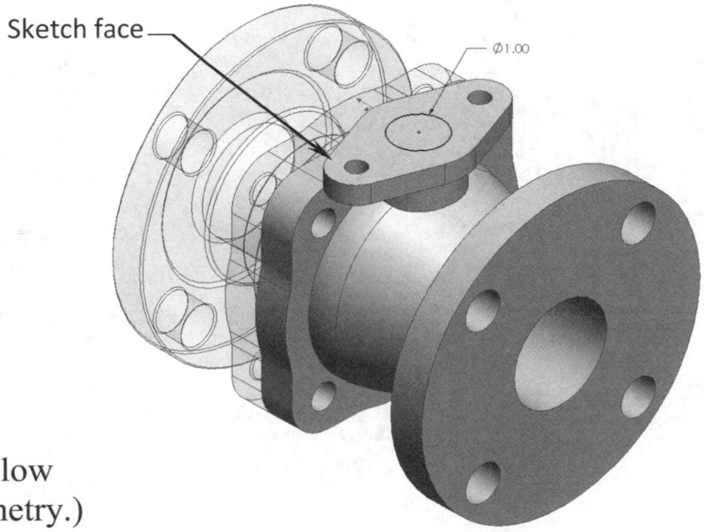

Click **Extruded Cut**.

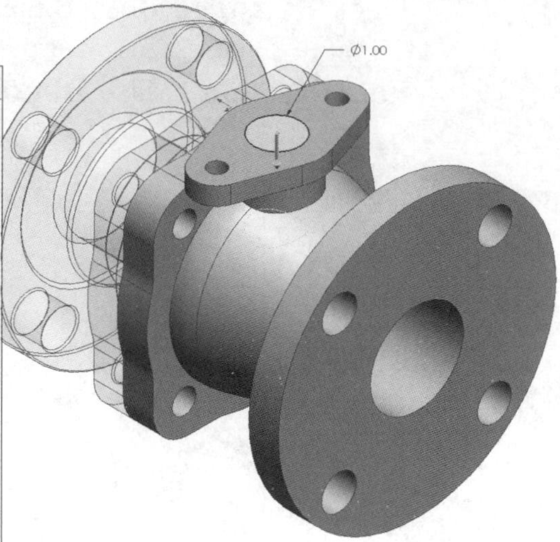

Click the **Reverse** direction button.

Select the option **Up-To-Next** from the list.

Click **OK**.

20. Adding the .080" fillets:

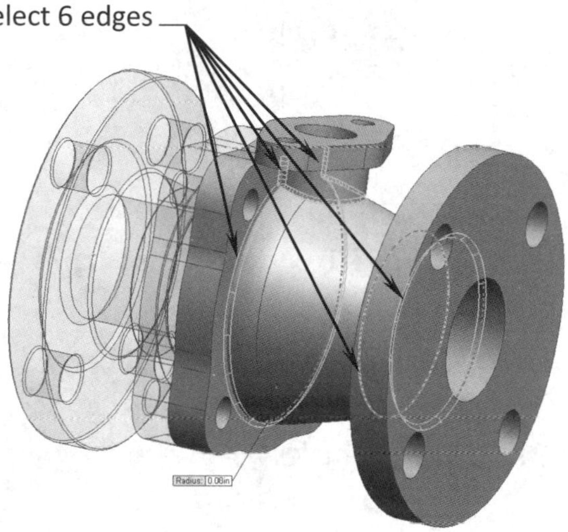

From the **Features** tool tab, click **Fillet**.

Enter **.080"** for radius size.

Select the **6 edges** as indicated.

The **Tangent Propagation** checkbox should be selected.

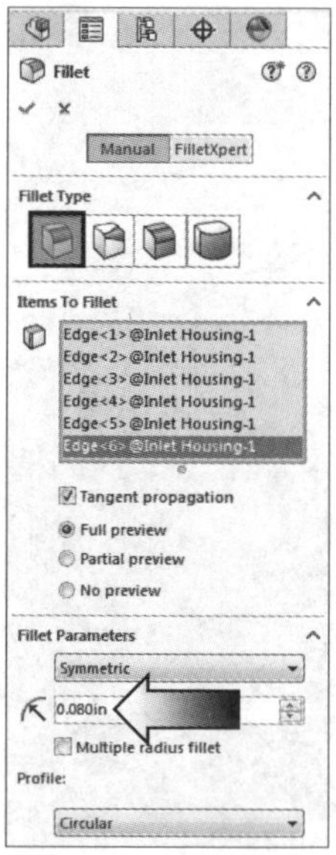

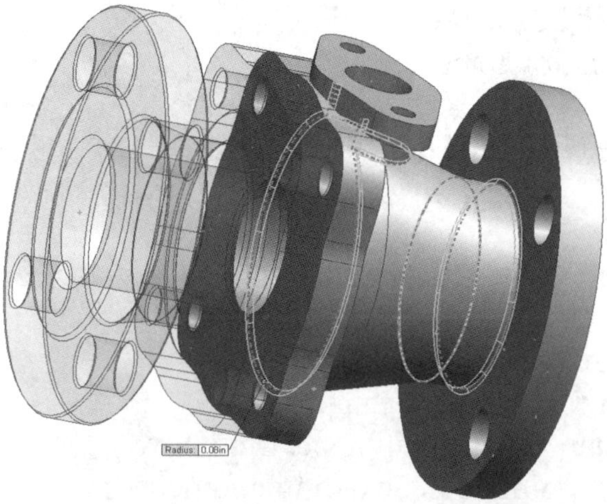

Click **OK**.

21. Adding the .032" chamfers:

Click **Chamfer** under the Fillet drop-down arrow.

Enter **.032"** for depth.

Use the default **45deg** angle.

Select the **edges** of the holes as shown in the preview image (23 edges total).

The same result can be achieved by selecting the inner faces of the holes.

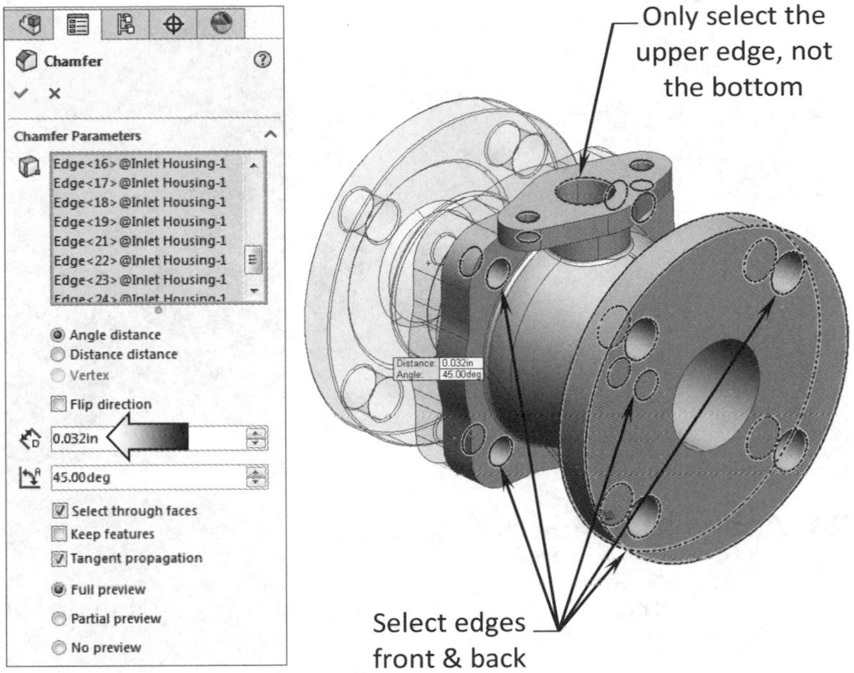

Click **OK**.

22. Adding the .125" chamfers:

Click **Chamfer** once again.

Enter **.125"** for depth.

Use the same **45deg** angle.

Select the **2 edges** of the center hole. (Selecting the face of the hole would get the same result.)

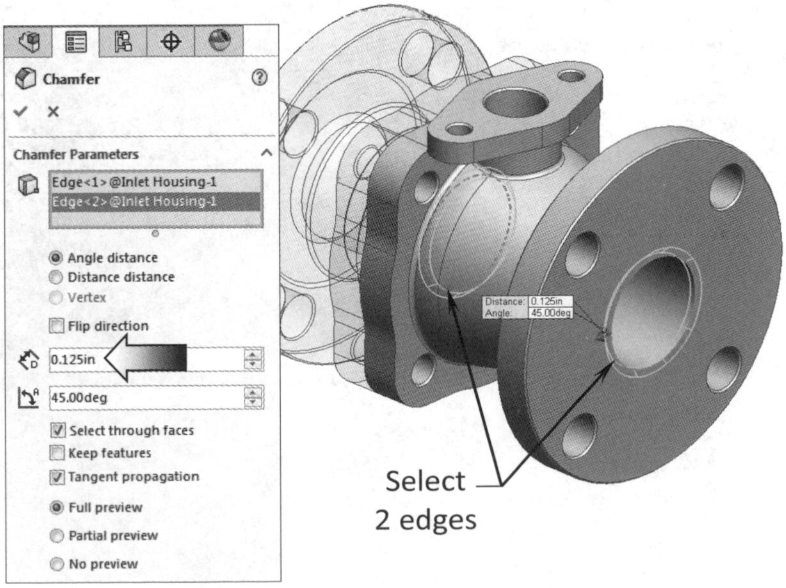

Click **OK**.

23. Exiting the Edit Component mode:

On the Assembly toolbar, click-off the **Edit Component** command (arrow).

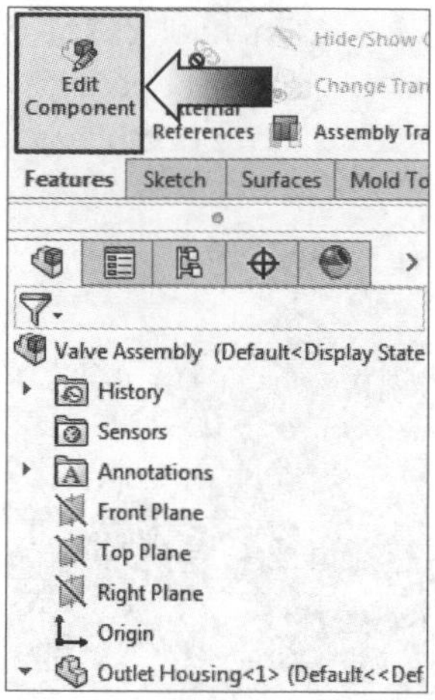

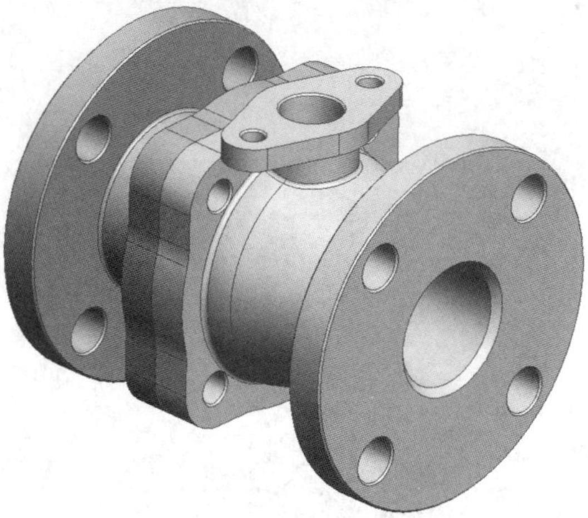

24. Applying dimension changes:

Expand the component Outlet Housing and double-click on the feature **Revolve1**.

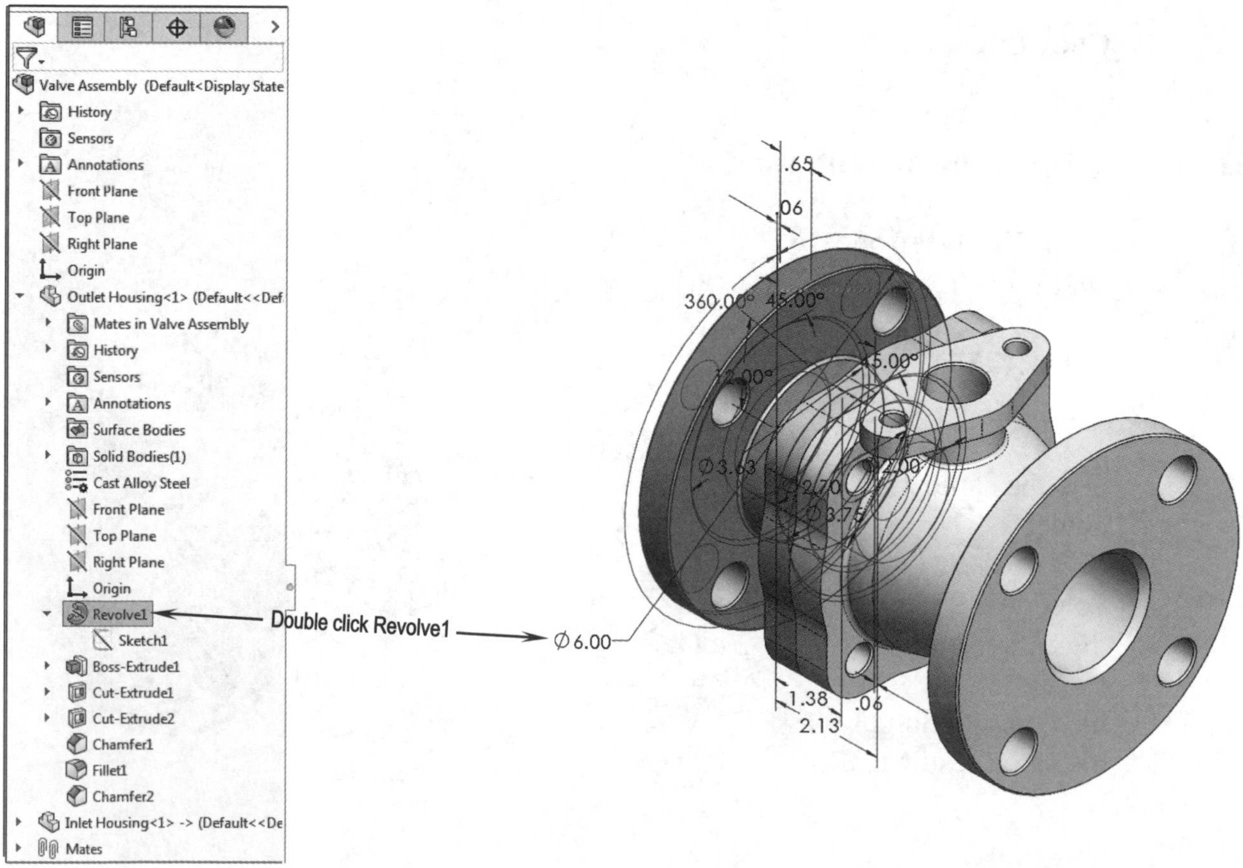

Change the flange diameter **from 6.00" to 7.00"**

Click the **Rebuild** (the stop light) to execute the change*.

Notice the dimension change also updates the flange diameter on the right.

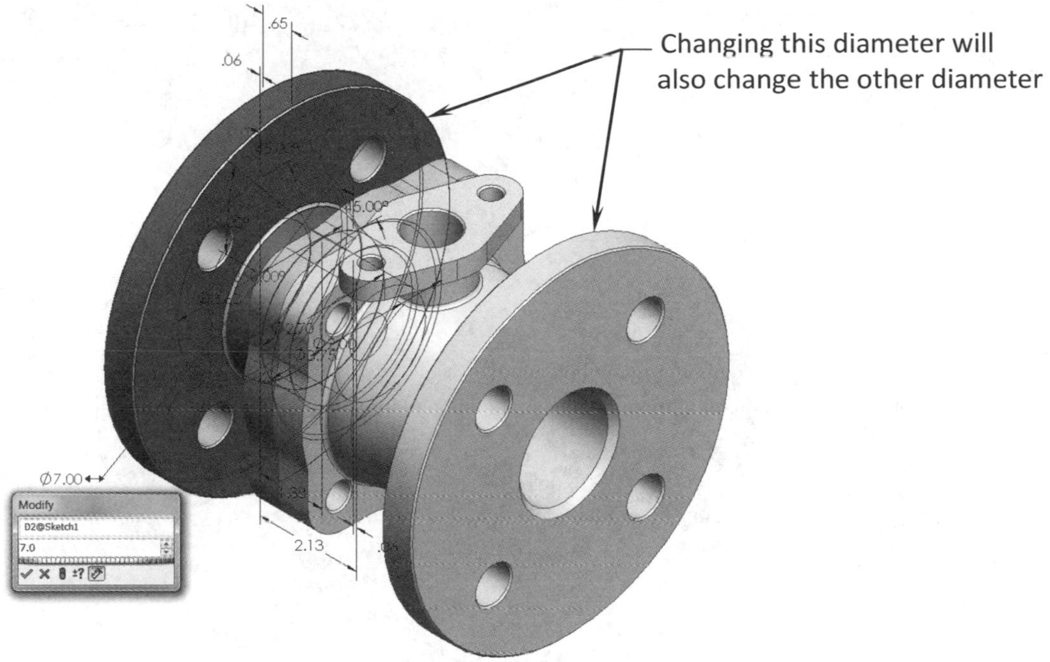

* Press undo to switch the dimension back to its original value, or double-click the same dimension, re-enter the previous value and click Rebuild again. Both cylindrical flanges should be 6.00” in diameter.

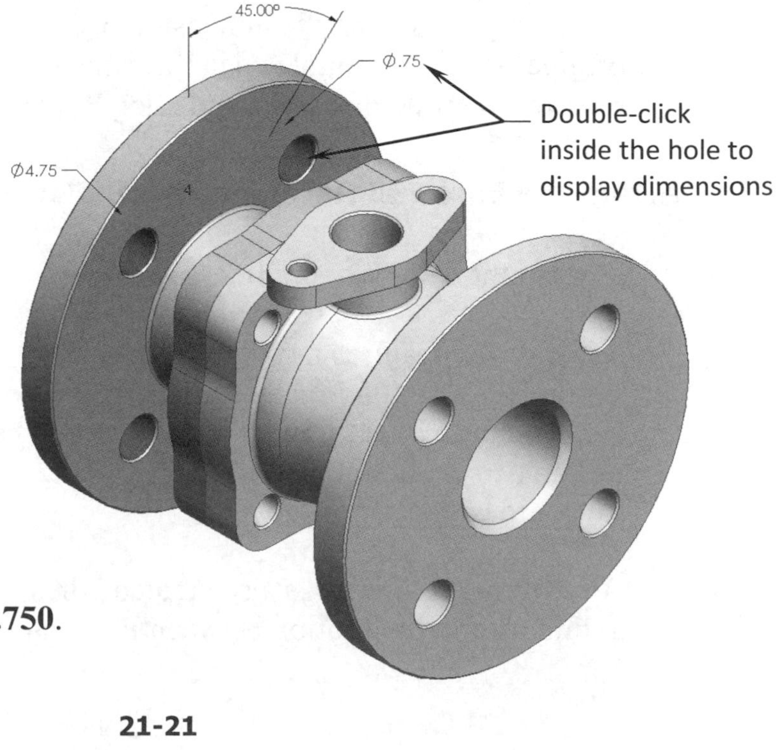

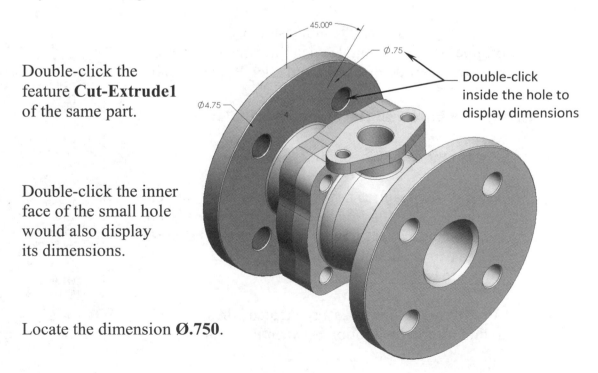

Double-click the feature **Cut-Extrude1** of the same part.

Double-click the inner face of the small hole would also display its dimensions.

Locate the dimension **Ø.750**.

Change the hole diameter **from .750" to .500"**.

Click the **Rebuild** (the stop-light) to execute the change*.

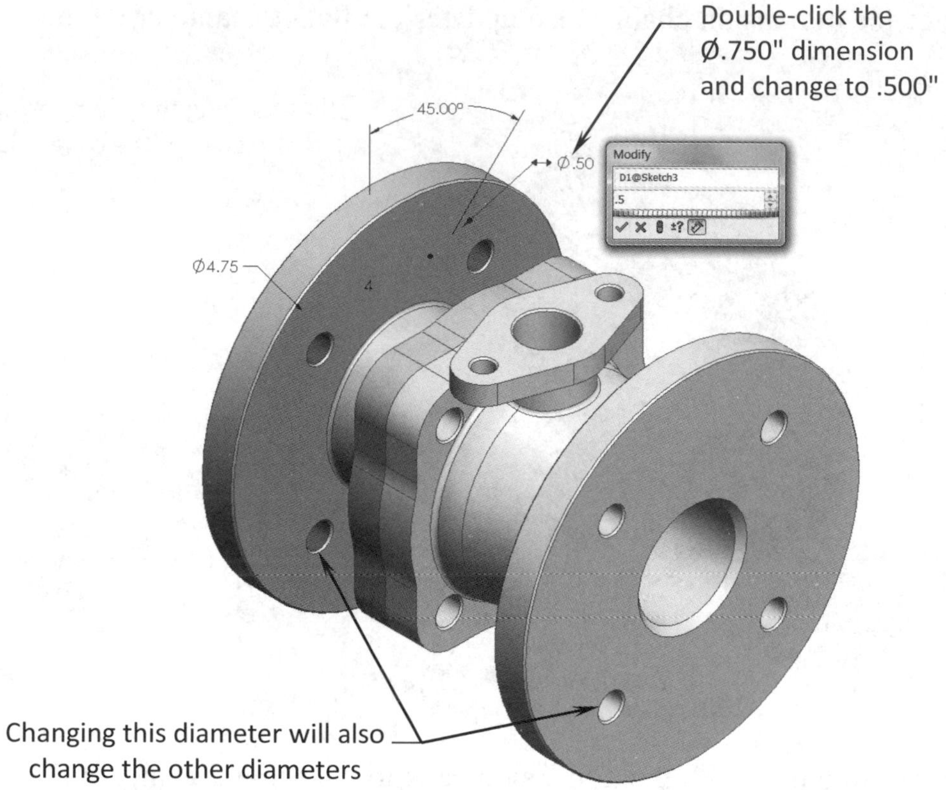

Notice the dimension change also updates the hole diameters on the right.

* Press undo to switch the dimension back to its original value, or double click the same dimension, re-enter the previous value and click Rebuild again.

25. Viewing the External Reference Symbols:

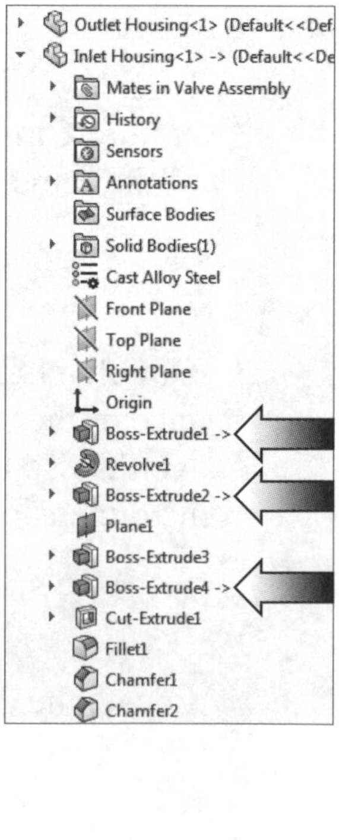

Expand the 2nd part, the **Inlet Housing**.

Some of the features have the External (->) reference symbols next to their names. These references were created automatically when we converted the entities of Part1 to create the new sketch for Part2. They are called On-Edge relations.

An external reference is also created when we add a dimension or a relation between Part1 and Part2.

26. Inserting other components:

Due to the length of this lesson, we are going to insert and assemble the rest of the components that belong to this assembly.

Browse to the Training Files folder and insert the components as labeled below.

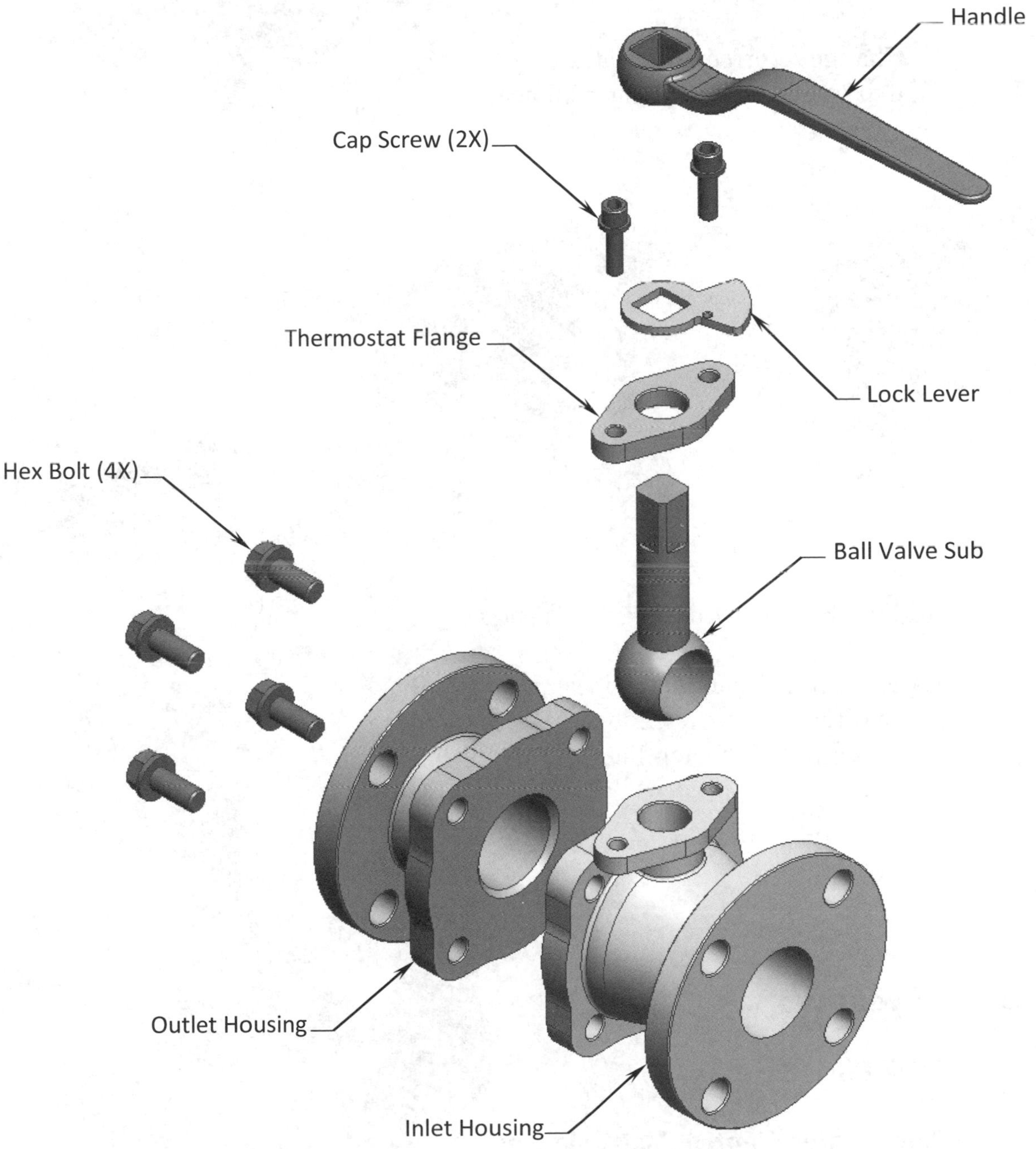

Create the mates that are appropriate for each component. The non-moving components will get 3 mates, and the moving components will need only two.

27. Optional:

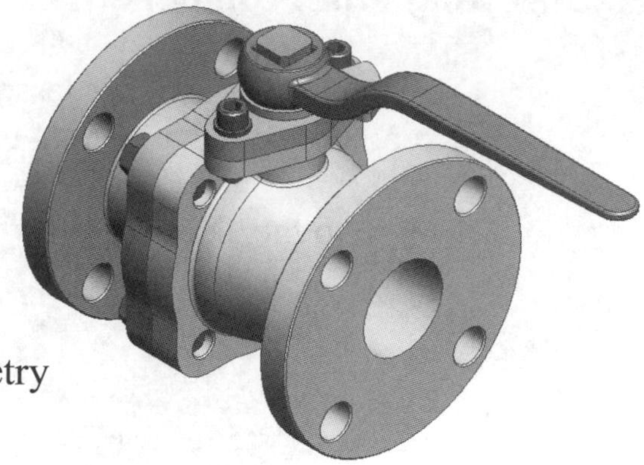

a/. Create a section view to verify how the components were mated.

b/. To center the 2 components, it is best to use the Width mate option.

c/. Change / correct any mates or geometry that would cause the interferences.

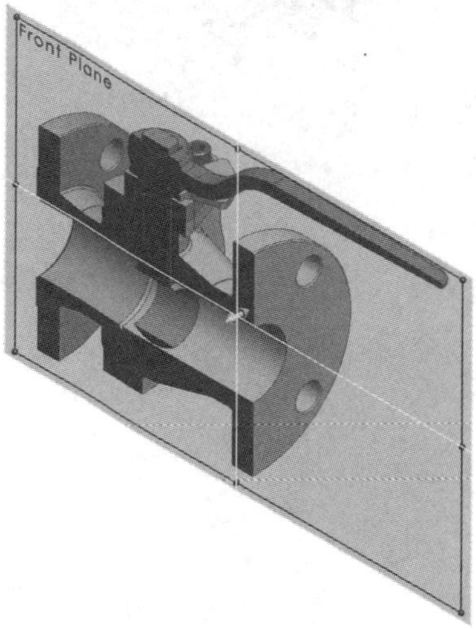

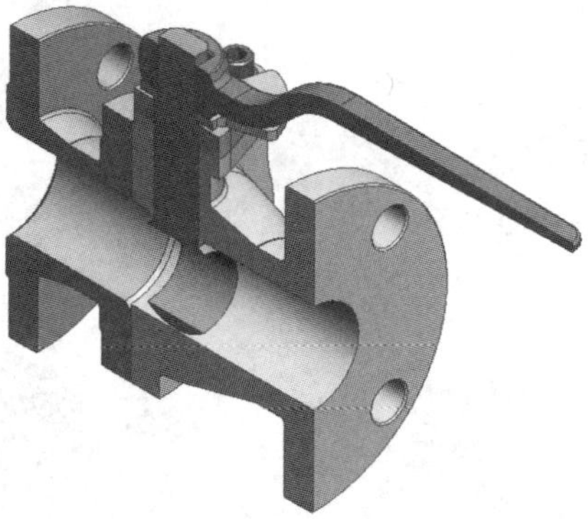

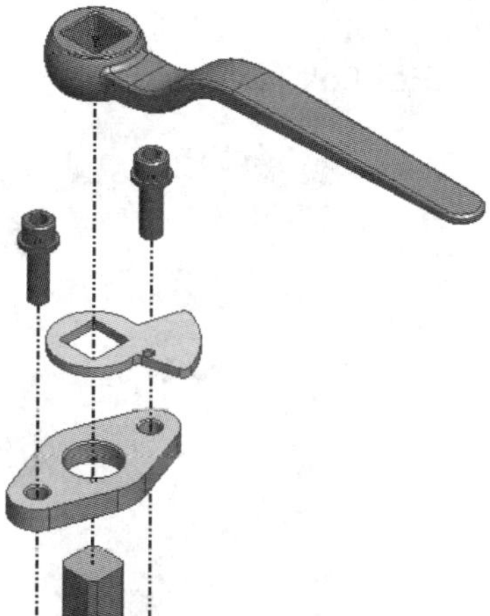

Add the Explode-Line Sketch as shown.

When adding the explode lines, pay attention to the direction arrows, flip or reverse them before completing each line.

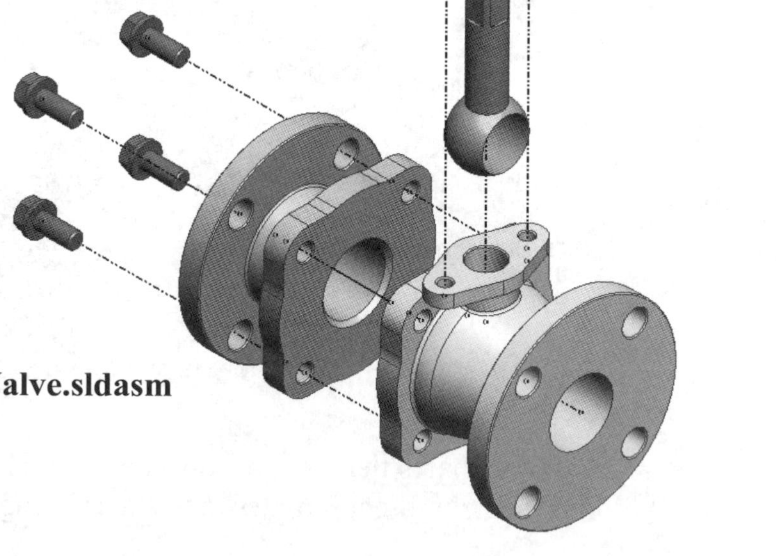

28. Saving your work:

Click **File / Save As**.

Enter **Water Control Valve.sldasm** for the name of the file.

Press **Save**.

Questions for Review

Top Down Assembly – Part 2

1. In Top Down mode, when a plane is selected to sketch the new part, SOLIDWORKS will create an Inplace mate to reference the new part.
   a. True
   b. False

2. After the plane is selected, SOLIDWORKS will also activate the Edit Component command and the Sketch mode at the same time.
   a. True
   b. False

3. The option Auto-Rotate Normal to the Sketch is not available to set as the default.
   a. True
   b. False

4. The Convert Entities command can only be used when the Sketch pencil is turned off.
   a. True
   b. False

5. The Virtual part is saved / embedded inside an assembly document.
   a. True
   b. False

6. When editing a part in Top Down mode, both the current part and other parts can be filleted at the same time.
   a. True
   b. False

7. The Edit Component command should be left active prior to inserting a new part.
   a. True
   b. False

8. When a dimension is changed, any reference geometry of other parts should also change.
   a. True
   b. False

1. TRUE 2. TRUE
3. FALSE 4. FALSE
5. TRUE 6. FALSE
7. FALSE 8. TRUE

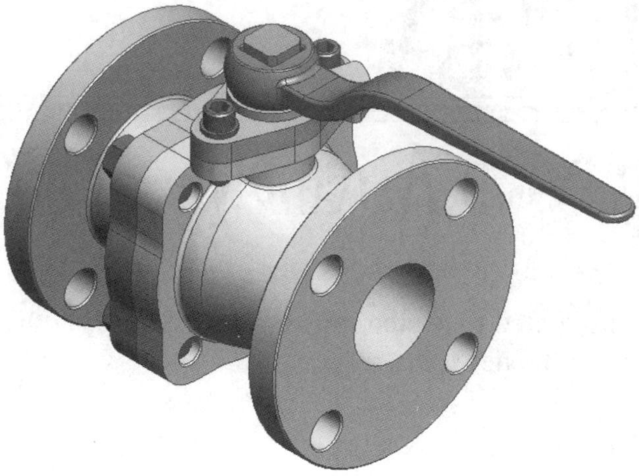

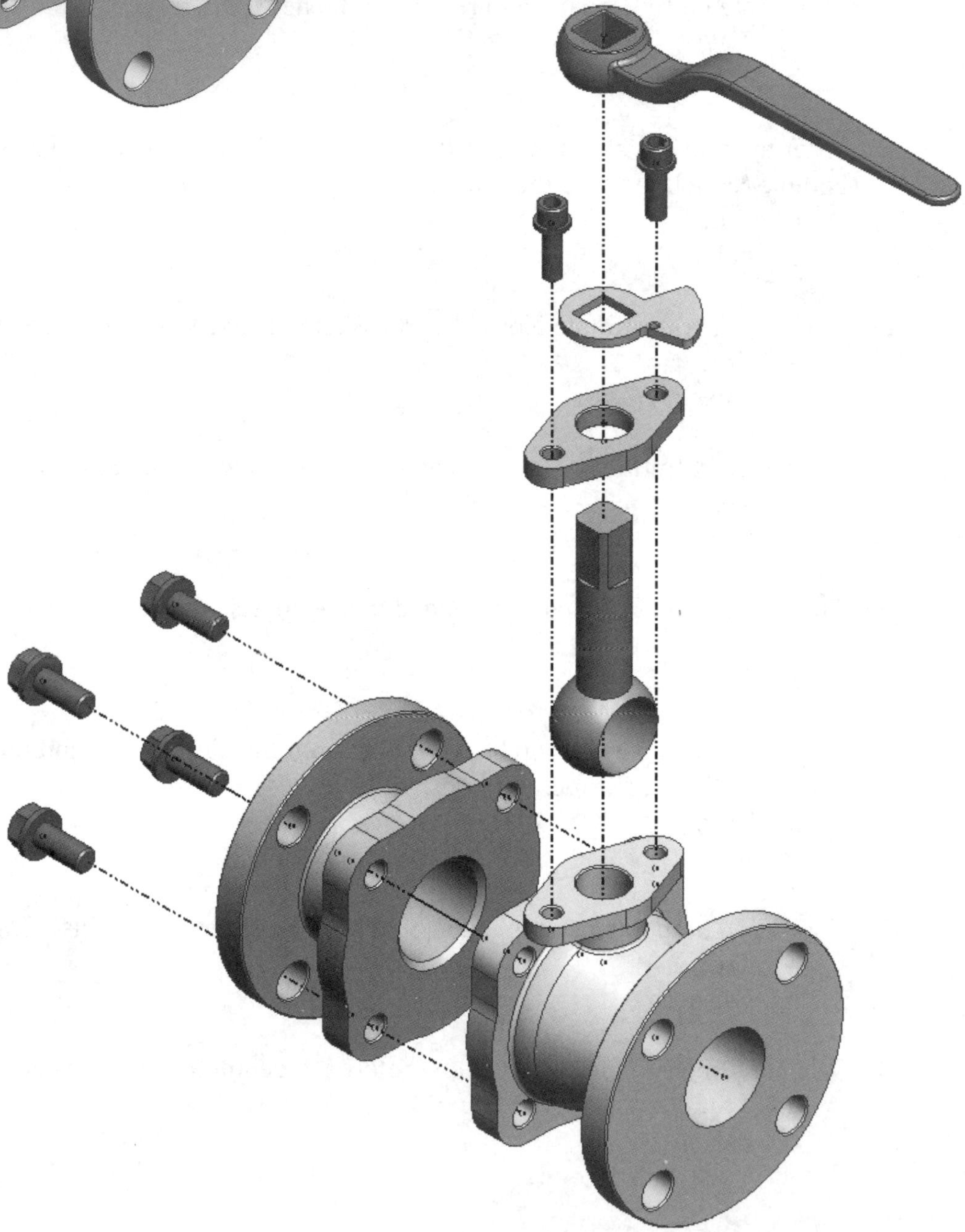

CHAPTER 22

External References & Repair Errors

External References & Repair Errors

An ***external reference*** is created when one component is dependent on another component for its solution. If the original document is changed, the dependent document will also change.

In an assembly, you can create an ***in-context*** feature on one component that references the geometry of another component. This in-context feature has an external reference to the other component. If you change the geometry on the referenced component, the associated in-context feature changes accordingly.

The External Symbols:

-> .... External Reference

(+) .... Over Defined

X .... Reference Broken

? .... Out Of Context

***** .... Reference Locked

-> External Reference:
The part itself or some of its entities are depending on the geometry of other parts for their solutions.

? Out Of Context:
The part or its features are not solved, not up-to-date or disconnected from its assembly.

(+) **Over Defined**:
The Dimensions or Relations of the sketch are conflicting, redundant dimensions or wrong relations were used.

* **Reference Locked**:
Lock the external references on a part, the existing references no longer update and the part will not accept any new references from that point.

X Reference Broken:
The references between the part and the others are broken. Changes done to the part will not affect the others.

External References & Repair Errors

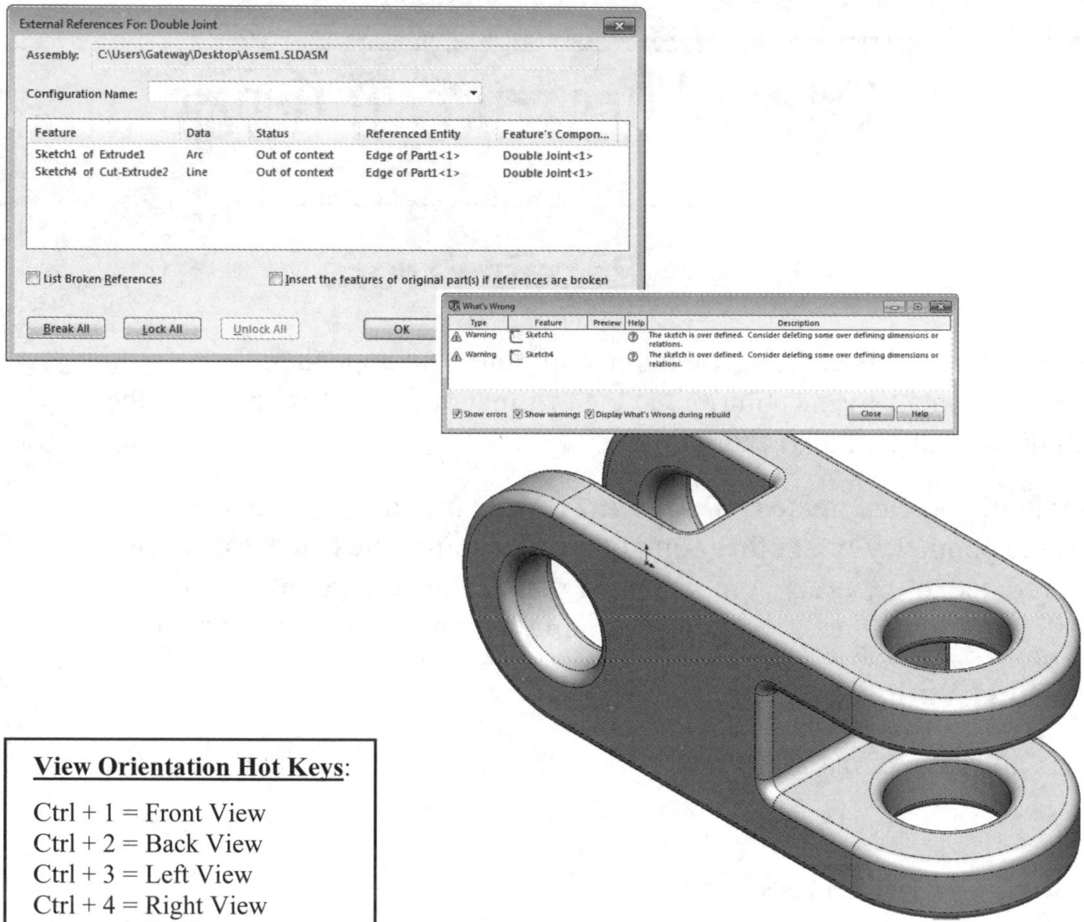

View Orientation Hot Keys:

Ctrl + 1 = Front View
Ctrl + 2 = Back View
Ctrl + 3 = Left View
Ctrl + 4 = Right View
Ctrl + 5 = Top View
Ctrl + 6 = Bottom View
Ctrl + 7 = Isometric View
Ctrl + 8 = Normal To Selection

Dimensioning Standards: **ANSI**
Units: **INCHES** – 3 Decimals

External Reference Symbols:

->	External Reference	?	Out of Context
(+	Over Defined	*	External Reference Locked
X	External Reference Broken		Display/Delete Relations

Understanding & Removing External References

1. Opening a part document:

Go to: Training Files folder

Open a part named: **Double Joint**.

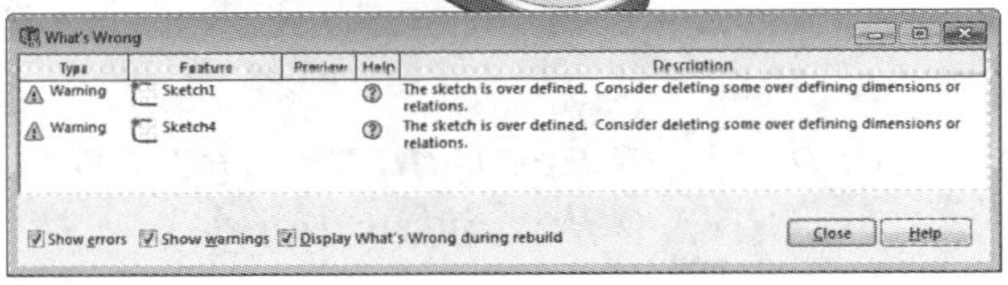

The **What's Wrong** dialog appears displaying the current errors. Close it. We need to look at the external references first.

2. Listing the External References:

Right-click on the part's name and select **External References**.

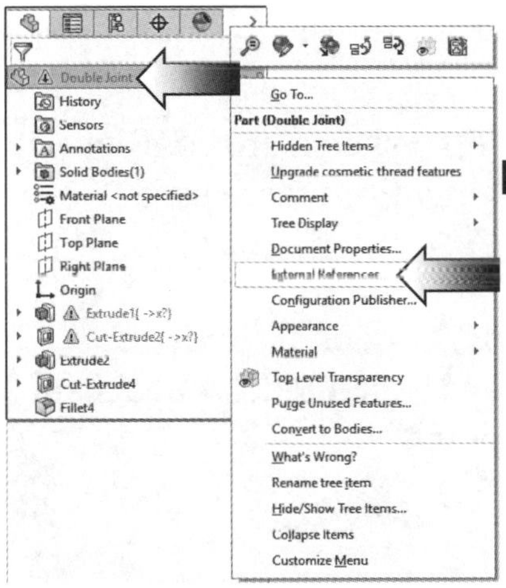

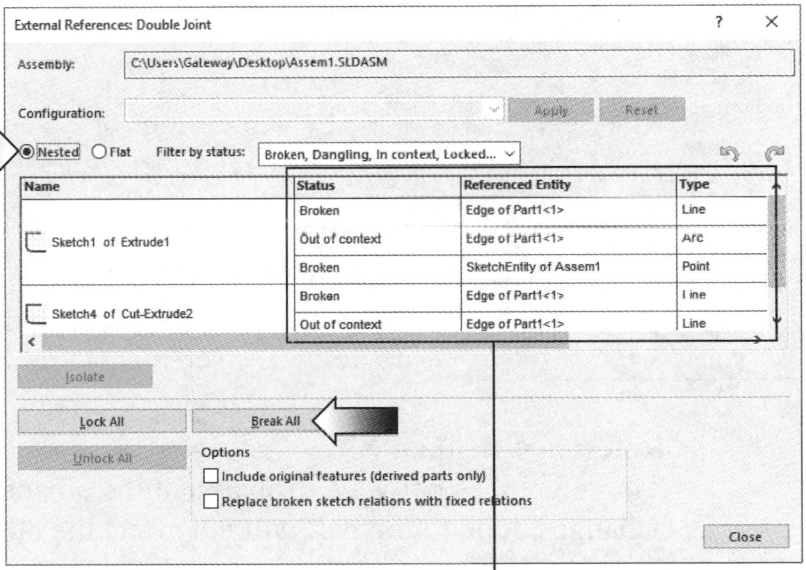

Enable the option: **Nested** (arrow).

Several Lines and Arcs were converted from other parts in an assembly; deleting these references will cause those entities to become Dangling

3. Removing External References:

Click **Break All** to remove the external references.

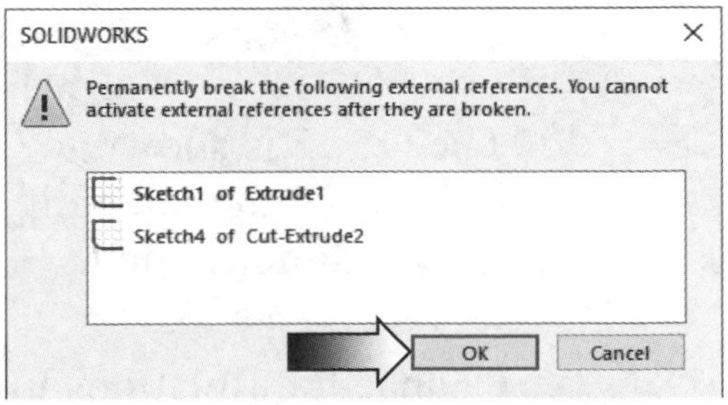

Click **Close** 

and **OK**.

(Click Continue Ignore Error to close the message.)

4. Understanding the External Symbols:

-> .... External Reference
? .... Out Of Context
(+) .... Over Defined
* .... Reference Locked
X .... Reference Broken

(From the FeatureManager tree, expand the first two features to see their sketches.)

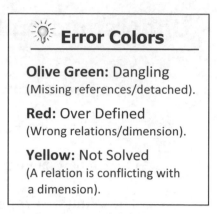

> **Error Colors**
>
> **Olive Green:** Dangling
> (Missing references/detached).
>
> **Red:** Over Defined
> (Wrong relations/dimension).
>
> **Yellow:** Not Solved
> (A relation is conflicting with a dimension).

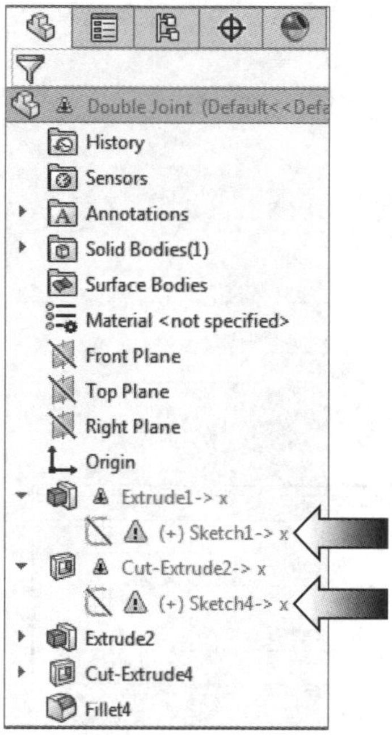

a. External Reference ->:
The model itself or some of its entities are depending on the geometry of other parts for their solutions.

b. Out Of Context ?:
The model or its features are not solved, not up-to-date, or disconnected from its assembly.

c. Over Defined (+):
The Dimensions or Relations of the sketch are conflicting; redundant dimensions or wrong relations were used.

d. Reference Locked *:
Lock the external references on a model; the existing references no longer update - and - the model will accept any new references from that point.

e. Reference Broken X:
The references between the part and the others are broken. Changes done to the part will not affect the others.

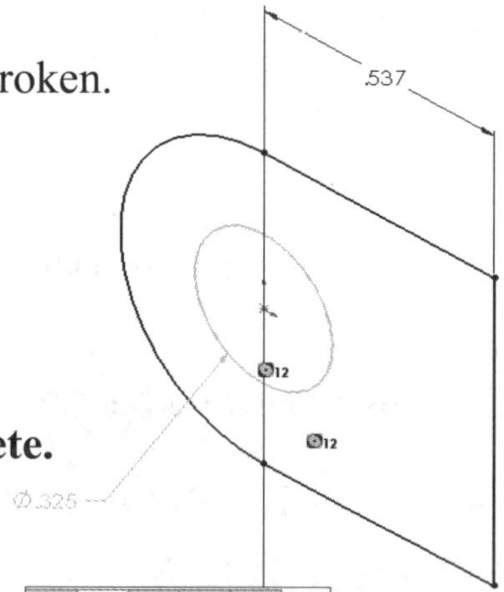

5. Viewing the existing Relations:

Right-click on **Sketch1** and select **Edit-Sketch**.

Click or select **Tools, Relations, Display-Delete.**

The Ø.325 is shown in Olive Green color; this indicates either an entity is missing or the sketch is over dimensioned.

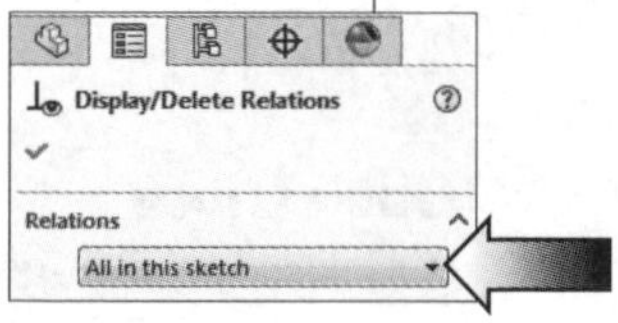

Change the Relations Filter to **All In This Sketch**.

6. Repairing the 1st sketch:

Click **Display / Delete Relations** and delete the **Coradial** relation (the Circle is Coradial with an entity that no longer exists).

Delete all relations that have the External Relations (->X) next to their names.

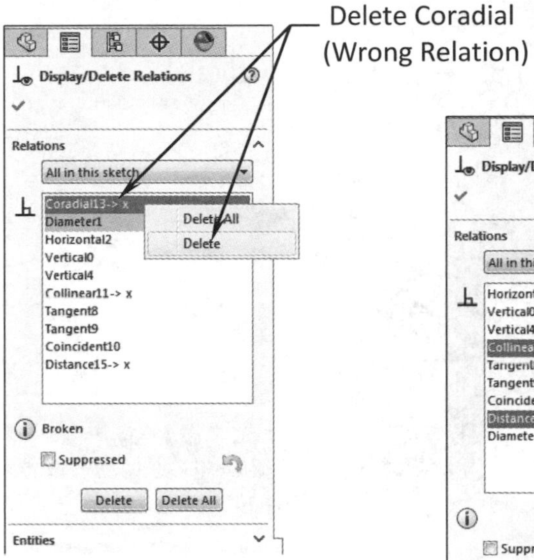

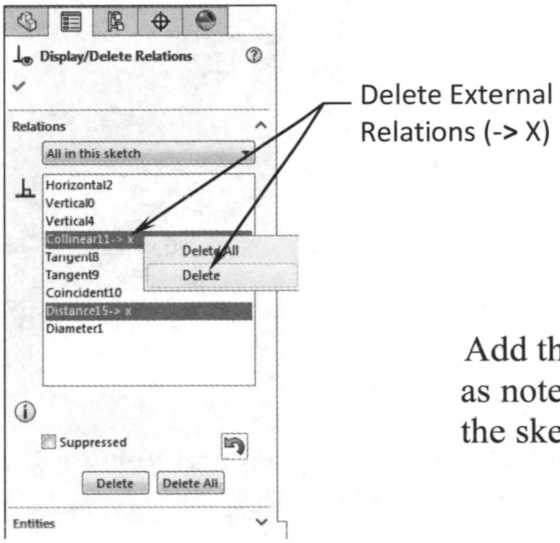

Add the new Dimension as noted to fully define the sketch.

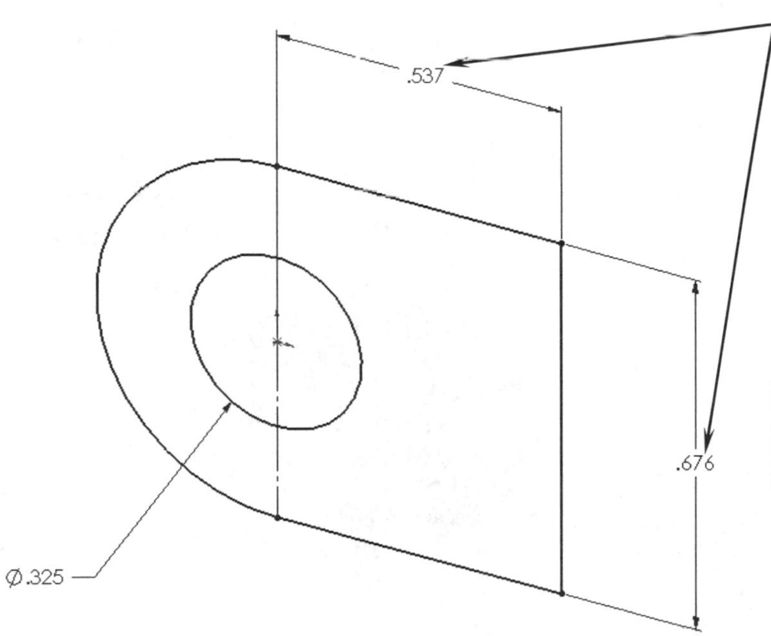

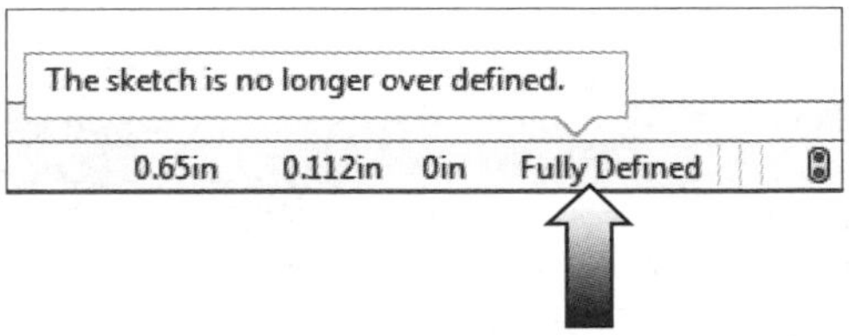

A message on the lower right indicates the sketch is no longer over defined.

Click **OK** and **Exit** the sketch. There are still some other errors in the part.

7. Repairing the 2nd sketch:

Right-click on the 2nd sketch and select **Edit-Sketch** .

Select the **Display / Delete Relations** command once again.

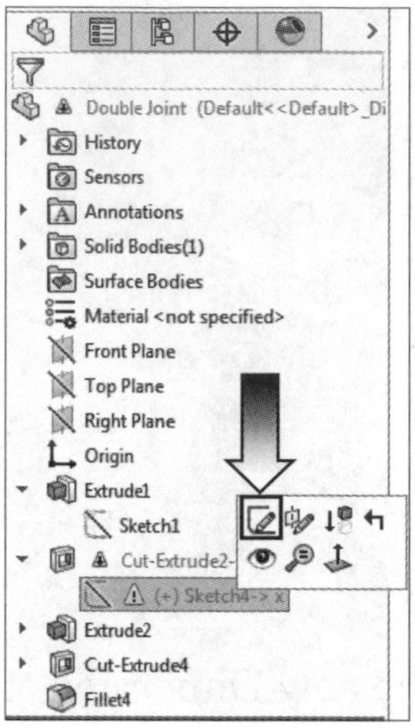

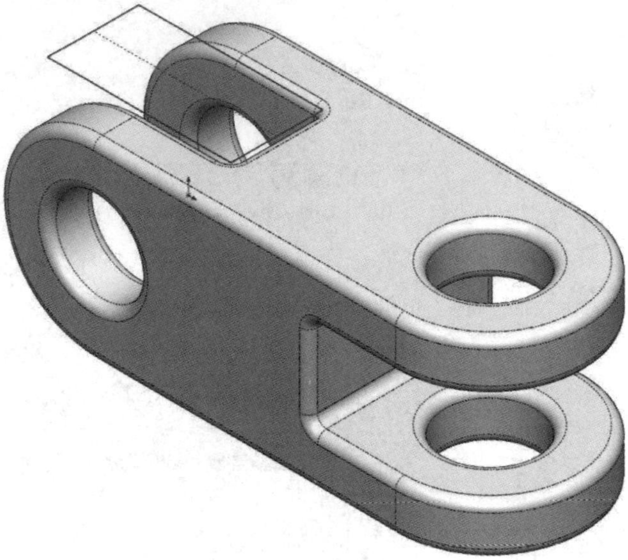

Delete the External Dimensions and Relations that were created in context of other parts.

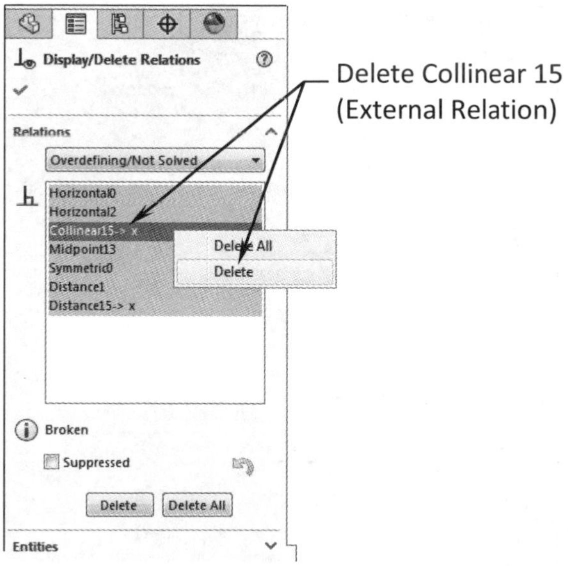

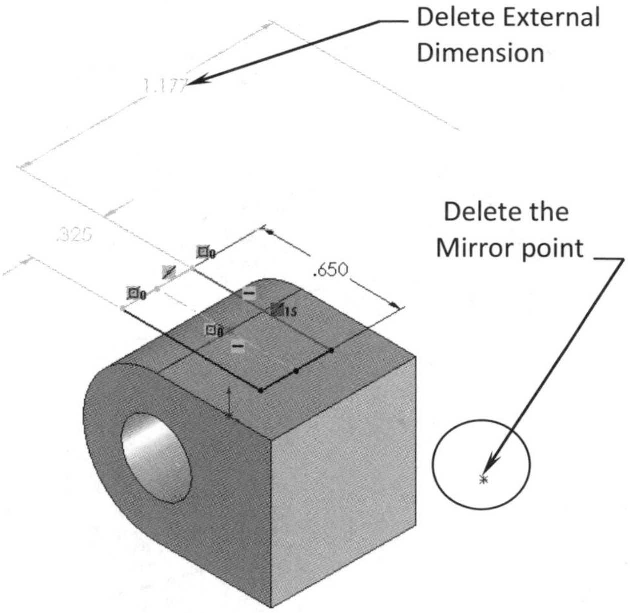

Delete also the two "Mirror-Points" as noted.

<u>Exit</u> the sketch when the message:
Fully Defined status is displayed on the lower right.

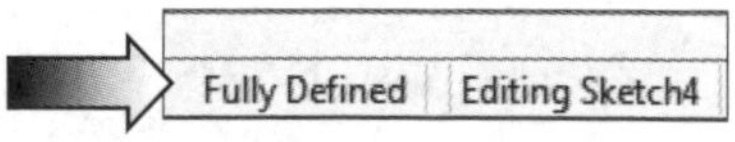

8. Rebuilding the model:

Press **Rebuild**  to re-generate the model.

Verify that the part has no rebuild errors and there should not be any external reference symbols in the FeatureManager tree.

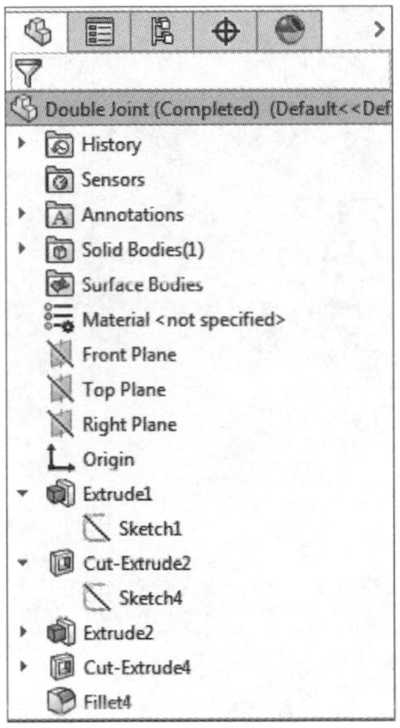

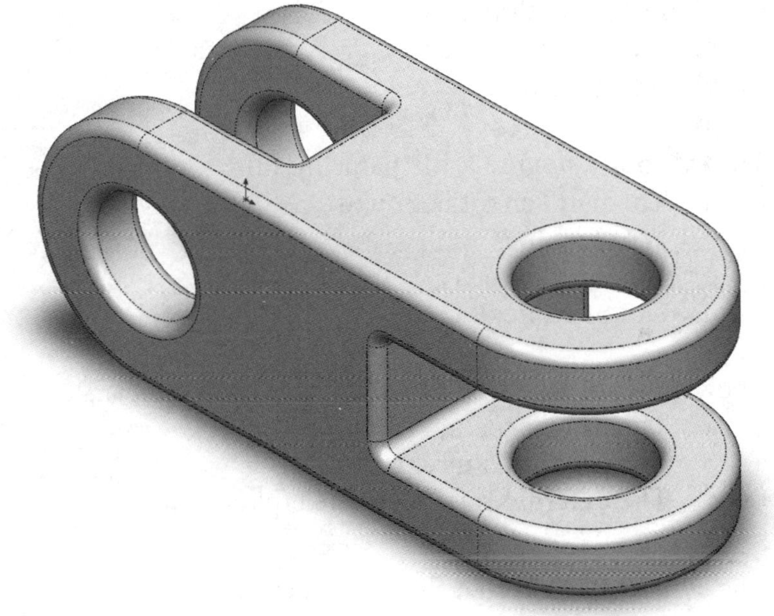

9. Saving your work:

Select **File / Save As.**

Enter **Breaking External References** for the file name.

Click **Save**.

Questions for Review

External References

1. The symbol **->** next to a file name means:
   a. Dangling dimension
   b. External reference
   c. Not solved

2. The symbol **?** next to a file name means:
   a. The part cannot be found
   b. Wrong mates
   c. Wrong relations
   d. Out of context

3. The symbol **X** next to a file or a feature name means:
   a. The part or feature is wrong
   b. The part or feature is deleted
   c. The external references are broken

4. The symbol * next to a file name means:
   a. External references are locked
   b. Select all references
   c. Deselect all references

5. The symbol ***X** next to a feature name means:
   a. The feature is fully defined
   b. The feature is over defined
   c. The feature is under defined
   d. None of the above

6. The Olive-Green color in a sketch means:
   a. The sketch entity is selected
   b. The sketch entity is being copied
   c. The sketch has dangling entities, relations, or dimensions

7. The dangling dimensions can be "re-attached" simply by dragging its handle point to a sketch line or a model edge.
   a. True
   b. False

1.B 2.D
3.C 4.A
5.D 6.C
7.TRUE

Understanding and Repairing Part Errors

When an error occurs SOLIDWORKS will try and solve it based on the settings below:

To pre-set the rebuild action:

A. Click **Options**  or **Tools / Options / General**.

B. Select **Stop**, **Continue**, or **Prompt** for **When rebuild error occurs**, and then click **OK**.

With **Stop** or **Prompt**, the rebuild action stops for each error so you can fix feature failures one at a time.

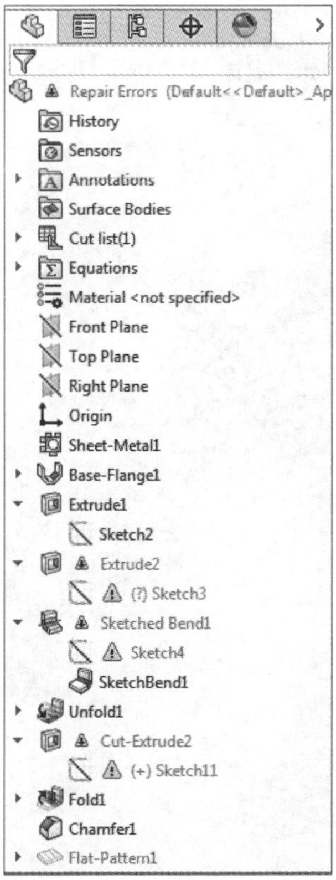

 Indicates an error with the model. This icon appears on the document name at the top of the FeatureManager design tree, and on the feature that contains the error. The text of the part or feature is displayed in **red** color.

 Indicates an error with a feature. This icon appears on the feature name in the FeatureManager design tree. The text of the feature is displayed in **red** color.

 Indicates a warning underneath the node indicated. This icon appears on the document name at the top of the FeatureManager design tree and on the parent feature in the FeatureManager design tree whose child feature issued the error. The text of the feature is displayed in **olive green** color.

 Indicates a warning with a **feature** or **sketch**. This icon appears on the specific feature in the FeatureManager design tree that issued the warning. The text of the feature or sketch is displayed in **olive green** color.

1. Opening a part document:

In the Training File folder, browse to the Repair Errors folder, and open a part document named: **Repair Errors**.

When opening a document that contains errors, the What's Wrong dialog box will appear and display where the errors are located and suggest some solutions in solving them.

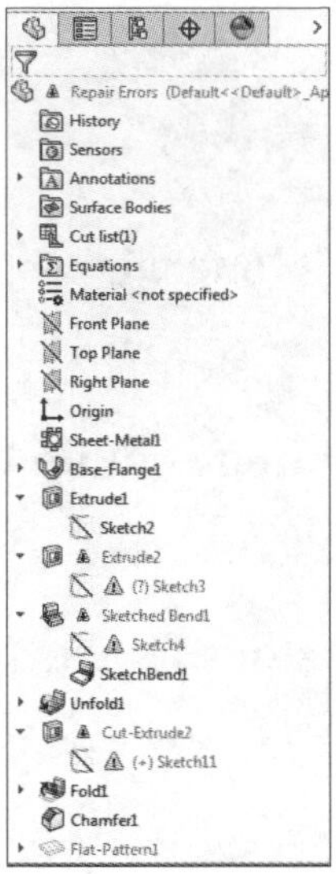

Expand each feature on the FeatureManager tree and hover the pointer over the **Sketch3** (Control+T is the hot key to expand the feature tree).

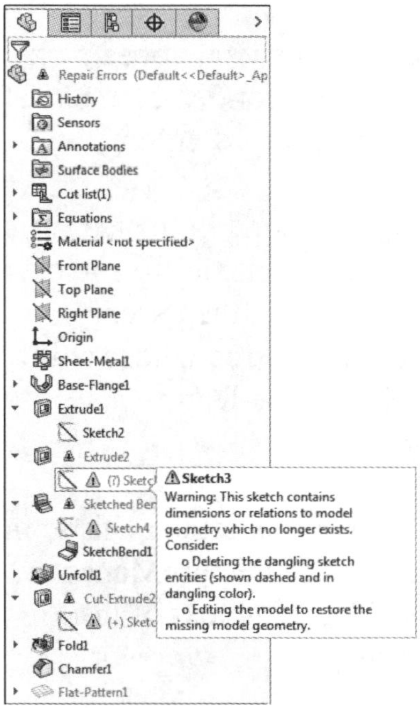

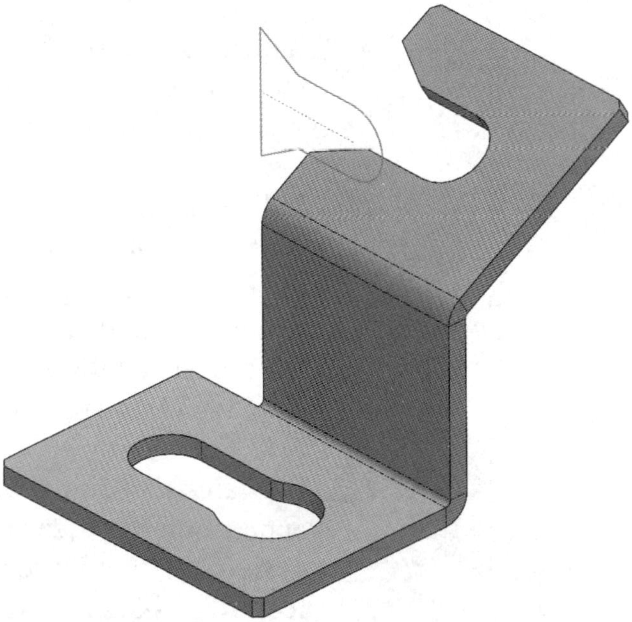

A description about the error or the warning is displayed in the tooltip. This is the same as right clicking on the error and selecting the What's Wrong option

2. Repairing the 1st error:

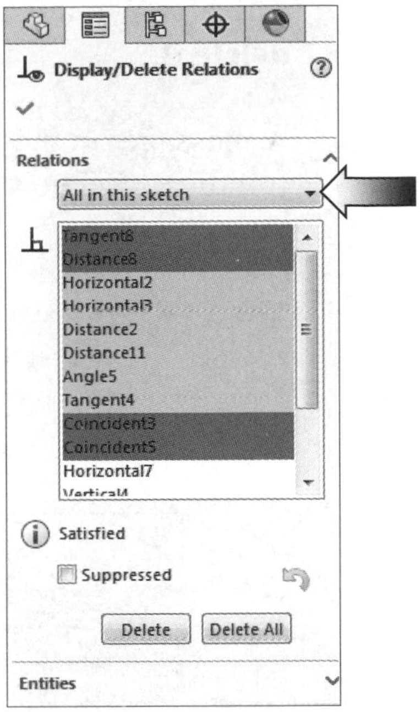

Select the **Sketch3** and click **Edit Sketch**.

Click the **Display/Delete Relation** command .

Change the Filter to **All In This Sketch** (arrow).

There are some dangling coincident relations in this sketch; they are displayed in the **olive green** color.

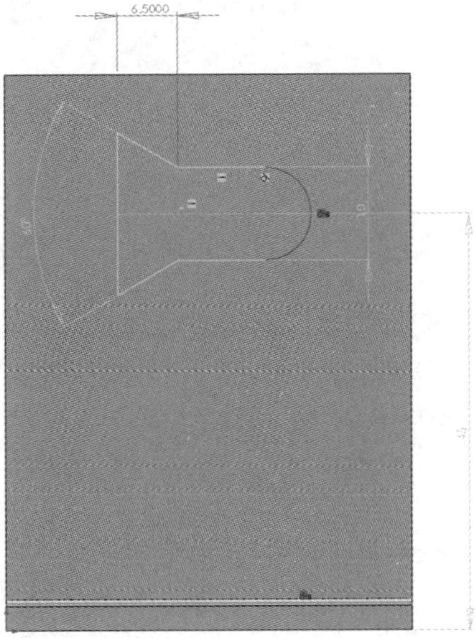

Select the **Coincident3**. One endpoint of a line is coincident to a missing edge. **Delete** this Coincident3.

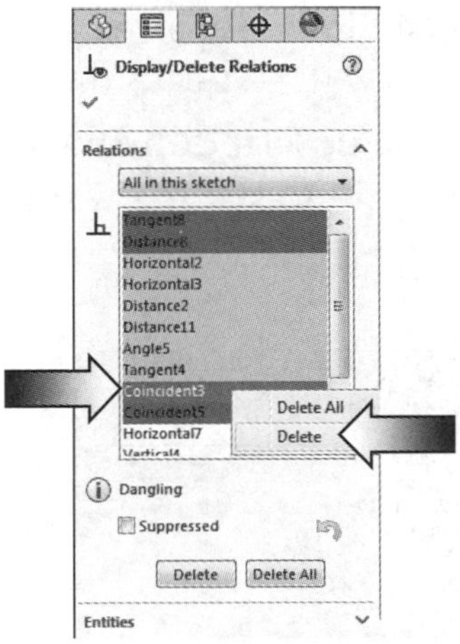

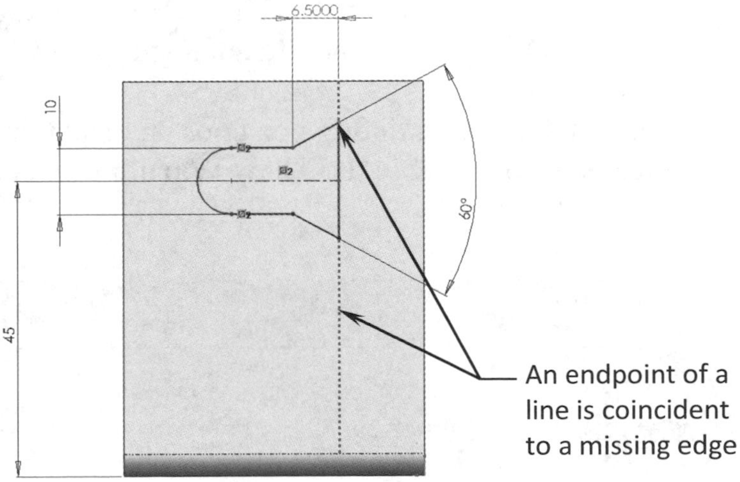

Delete also the **Coincident5**. It is missing the same edge as the previous relation.

The dimension **6.500** is also dangling. It was measured to a missing entity, **delete it**.

A message on the bottom right of the screen appears.
It indicates: **The sketch can now find a valid solution**.

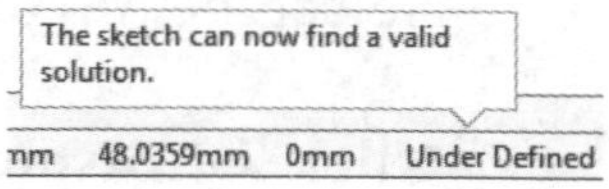

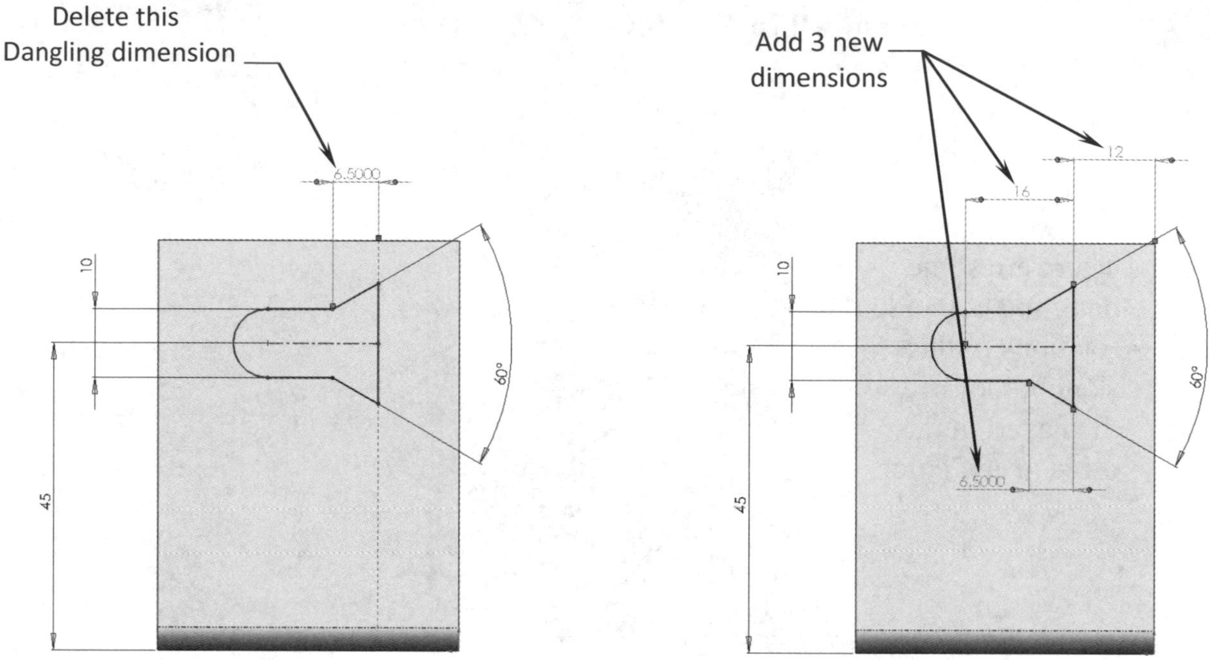

Add **3 new dimensions** shown above to fully define the sketch.
Exit the sketch when completed.

After exiting the sketch, SOLIDWORKS continues to report other errors still remained in the part. The **Sketch4** and **Sketch11** still need to be repaired.

The What's Wrong dialog box pops up displaying the same description about the selected error. Enable the **Show Warnings** checkbox (arrow) to see the warnings after each rebuild.

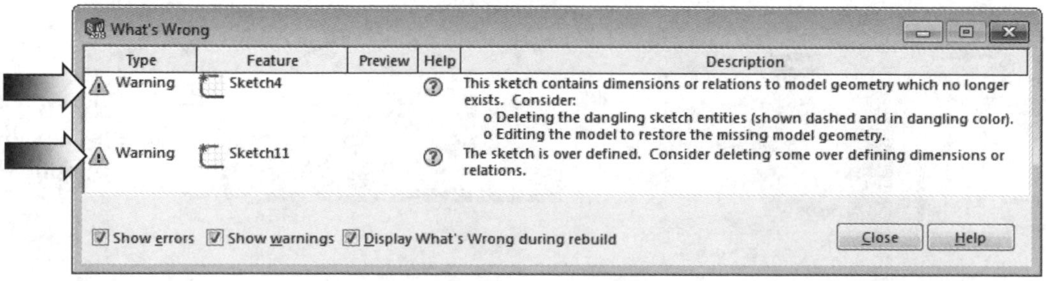

Close the What's Wrong dialog box.

3. Repairing the 2nd error:

Click the **Sketch4** (under the Sketched Bend1 feature) and select: **Edit Sketch**.

Select the **Display / Delete Relations** command once again.

The **Coincident1** is dangling and has the **Olive Green** color. **Delete it**.

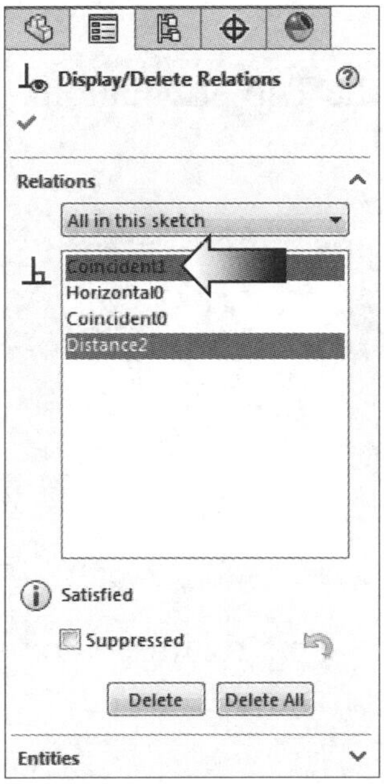

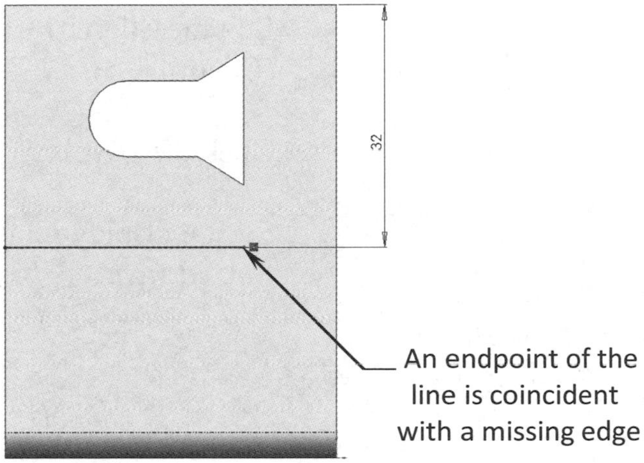

The endpoint of the line changes to **Blue** color. This indicates the sketch is now under defined.

Drag/drop the endpoint of the line until it touches the vertical right edge of the part.

A coincident relation is added automatically.

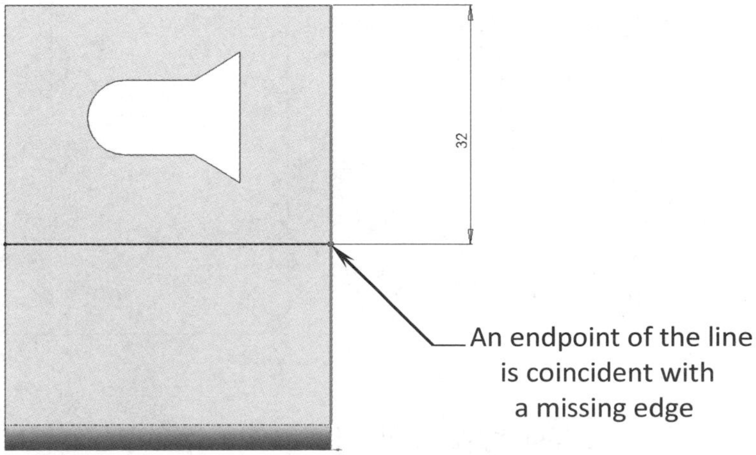

Exit the sketch or press **Control + B** (Rebuild).

SOLIDWORKS continues to report the last error in the model. Click the **Close** button and continue with repairing the last error.

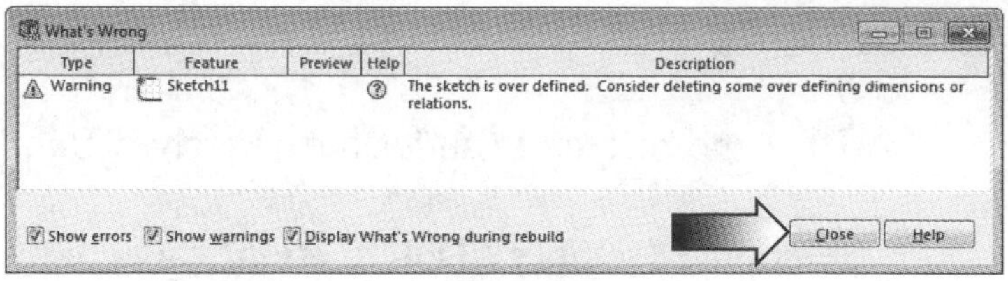

4. Repairing the 3rd error:

The **Sketch11** has a plus sign next to its name. This indicates that the sketch is **Over Defined**.

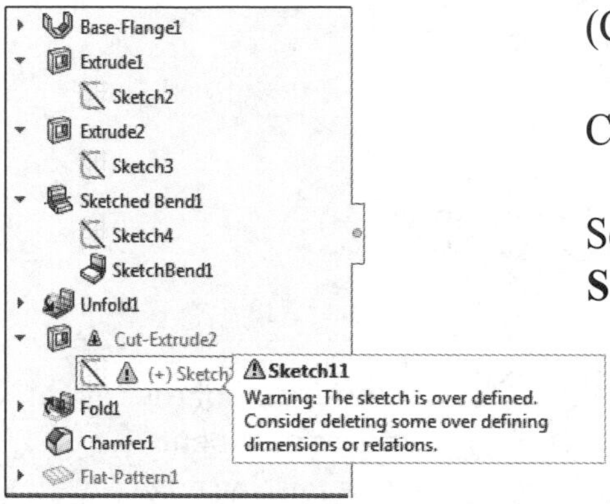

Edit the **Sketch11** and change to the Top orientation (Control + 5).

Click the **Display / Delete Relations** button.

Set the display relations filter to **Over Defining / Not-Solved**.

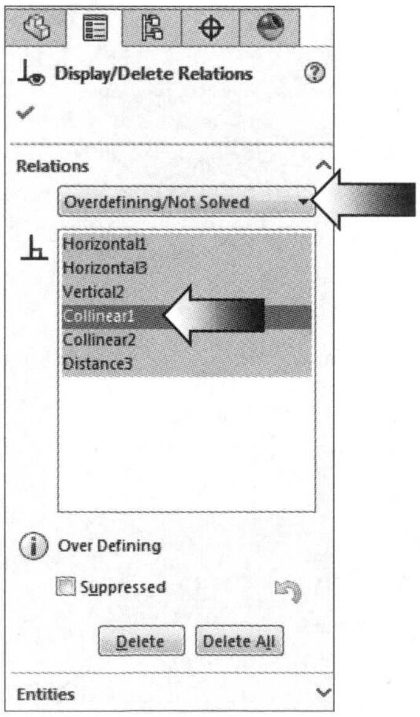

Select the **Collinear1** from the list. This relation shows a **Magenta** color next to its Collinear symbol.

Delete the **Collinear1** from the list.

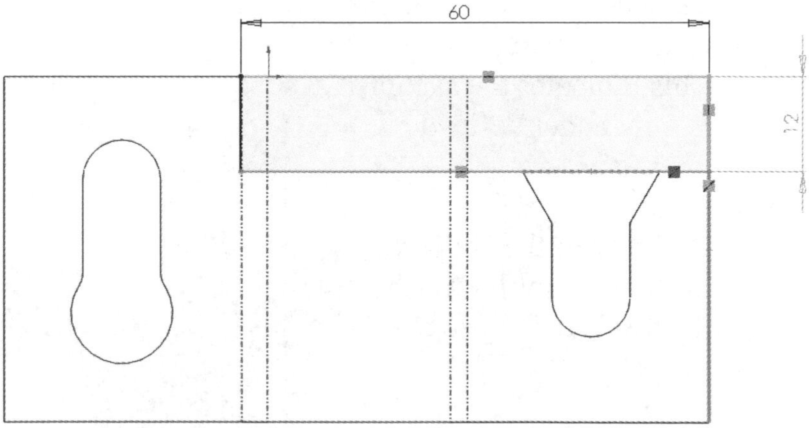

This last step should bring the sketch back to its Fully Defined status.

Exit the sketch or press **Control + B**.

5. Saving your work:

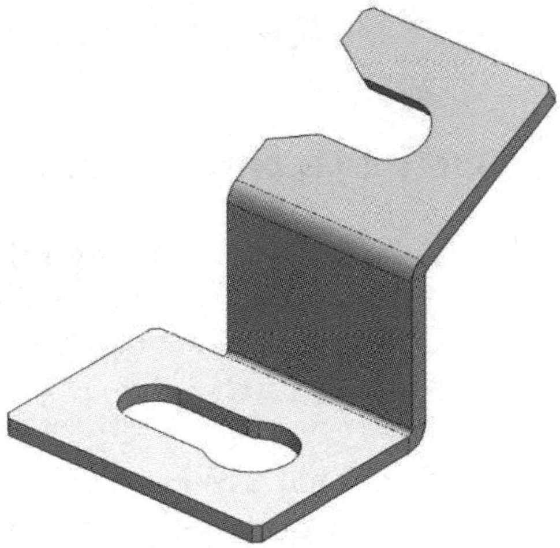

Click **File / Save As**.

Enter **Repair Errors (Completed)** for the name of the file.

Click **Save**.

All errors have been repaired. The Feature-Manager tree is now free of errors.

(Press Control+T to collapse the FeatureManager tree.)

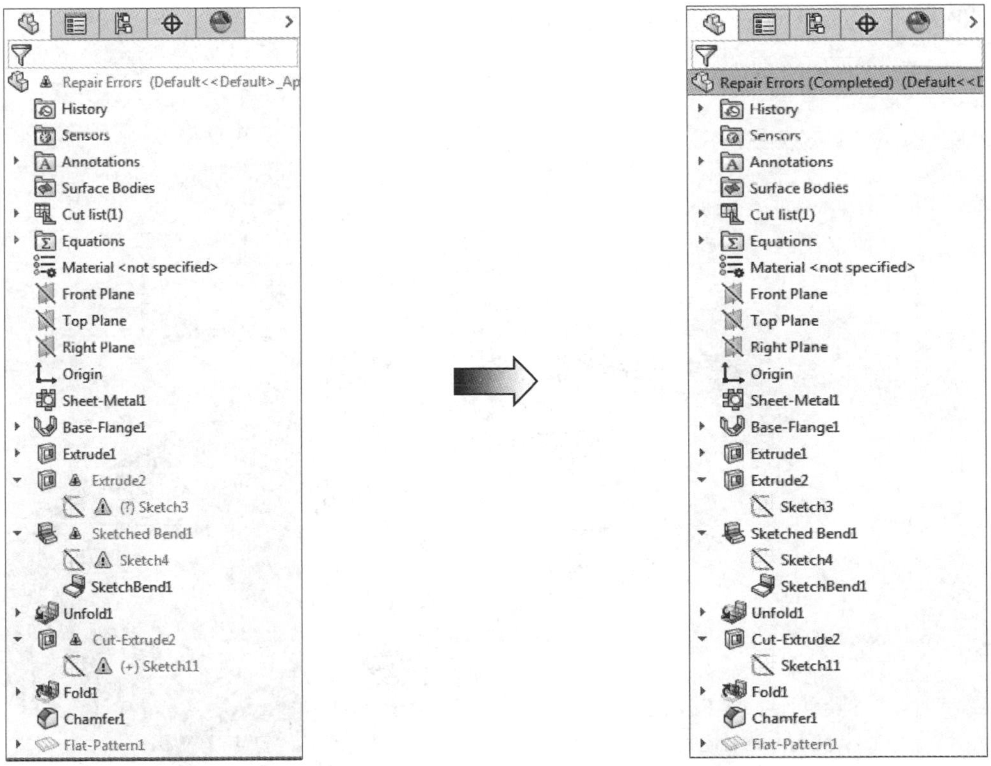

Close all documents.

Repair Errors & External References

1. Opening a part document:

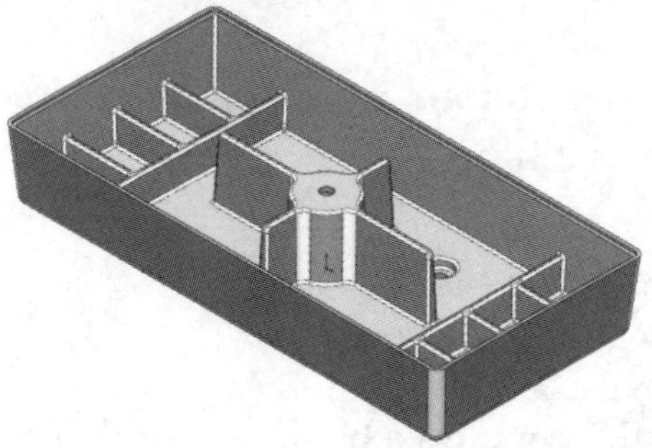

Browse to the Training Files folder.

Open a document named:
Molded Part.

The **What's Wrong** dialog appears displaying the current errors in the model.

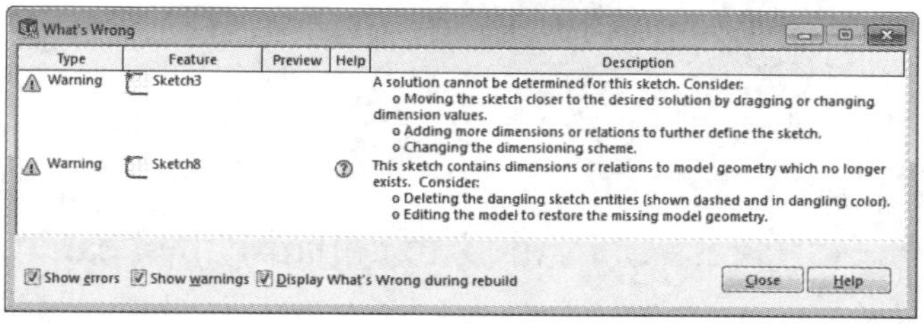

Click **Close**. We are going to take a look at breaking the external references first.

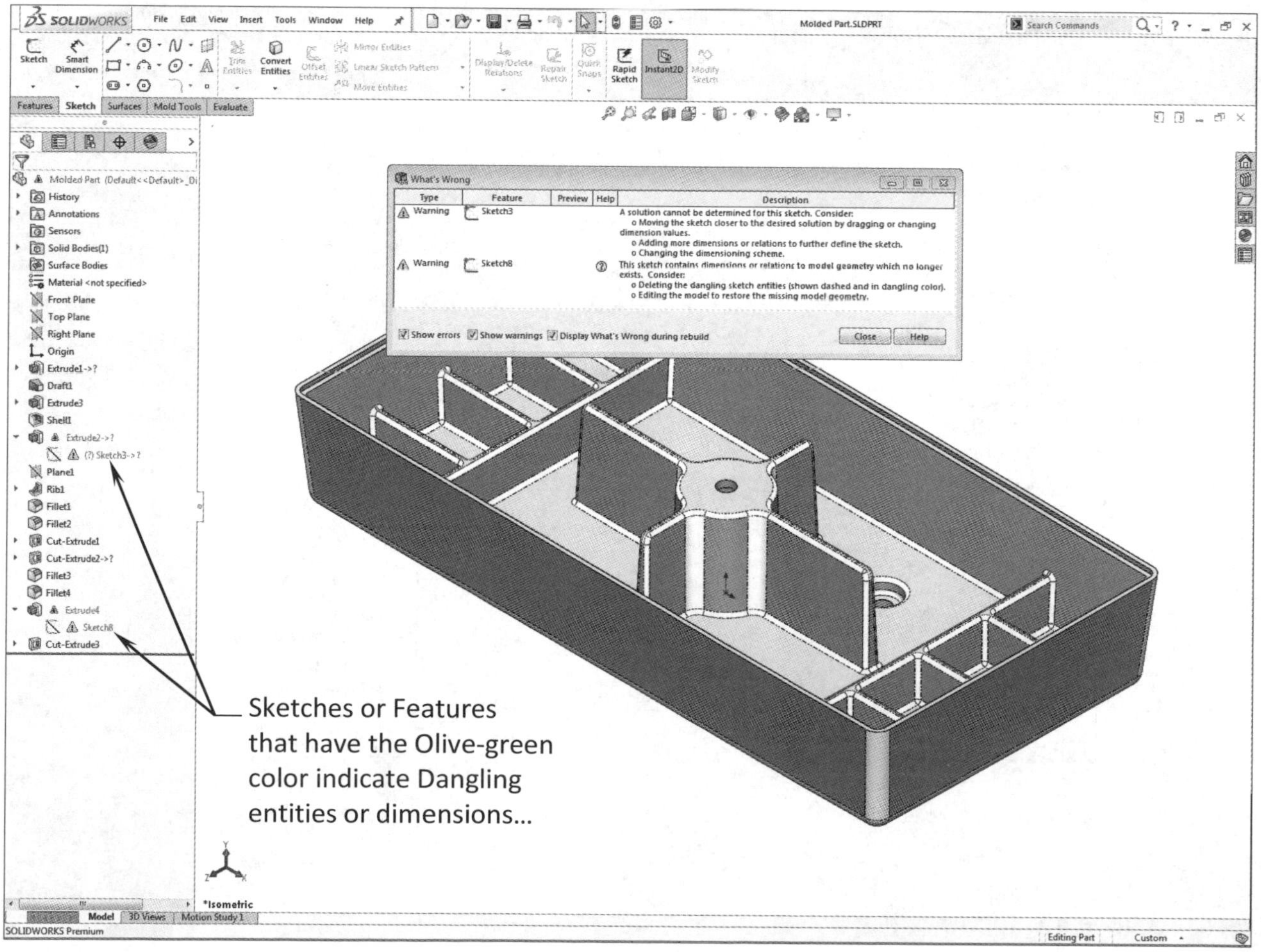
Sketches or Features that have the Olive-green color indicate Dangling entities or dimensions...

2. Breaking all External References:

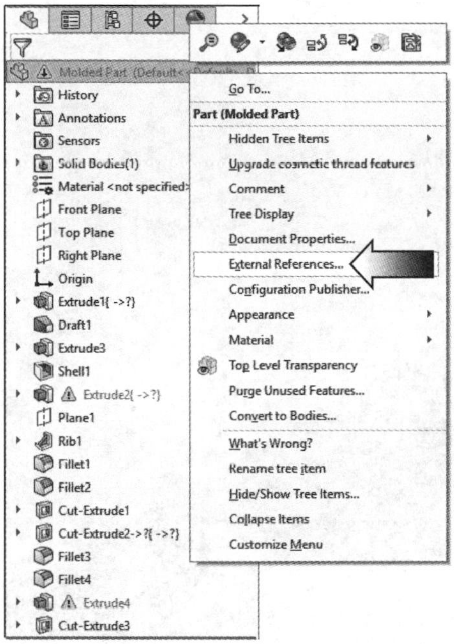

Right-click the part's name and select: **External References**.

All of the Out of Context entities are displayed in the dialog box along with the names of the components to which they were related.

Click **Break All** (Arrow).

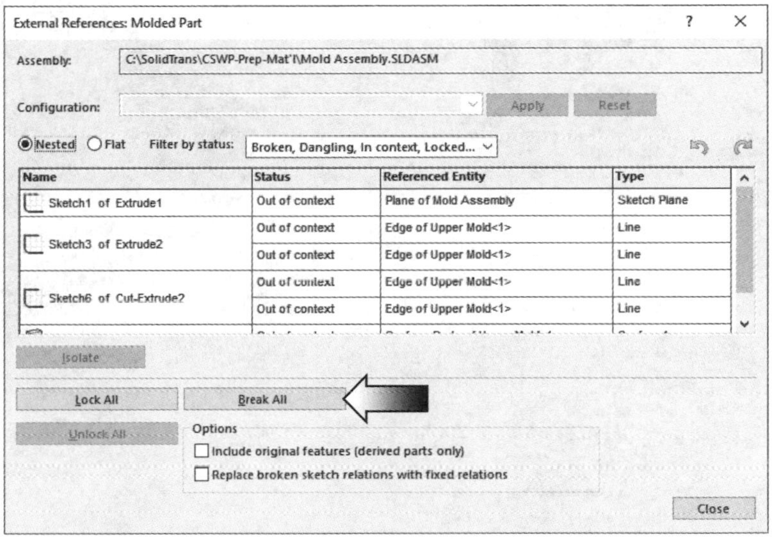

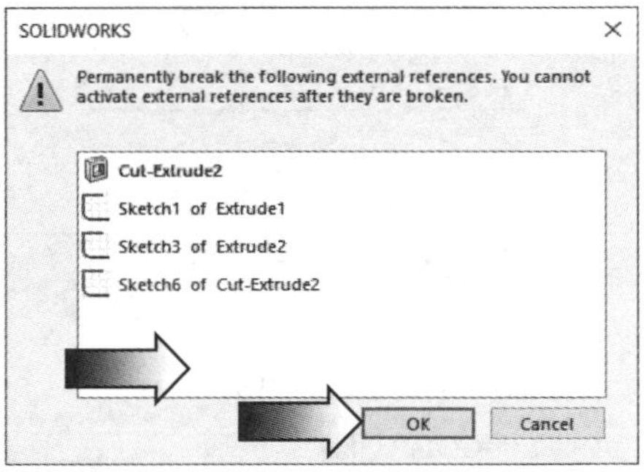

Click **OK** [OK] to confirm the delete of all External References.

Click **Continue** and close the External dialog box.

TIPS:

4 basic steps should be done, in most cases, to repair or replace External References:

1. Break all External references (Right click on the part's name and select External Refs.).

2. Replace the sketch Plane or Face (if missing).

3. Delete or replace any Relation with an External Reference symbol next to it (Display/Delete Relations).

4. Repair or replace the extrude type.

3. Replacing the Sketch Plane:

Expand the **Extrude1** feature to see the **Sketch1** below.

Right-click **Sketch1** and select **Edit Sketch Plane**  .

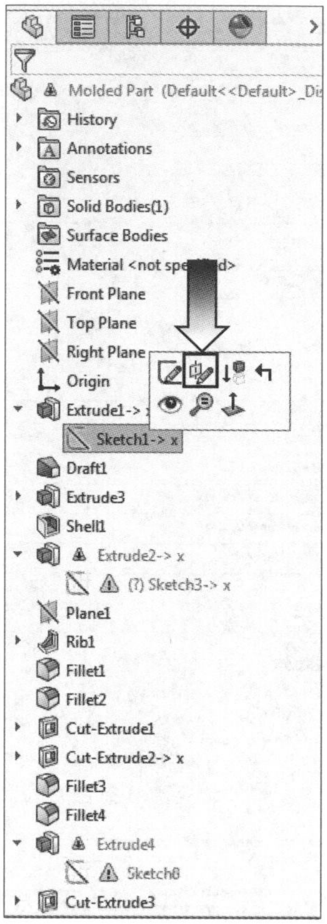

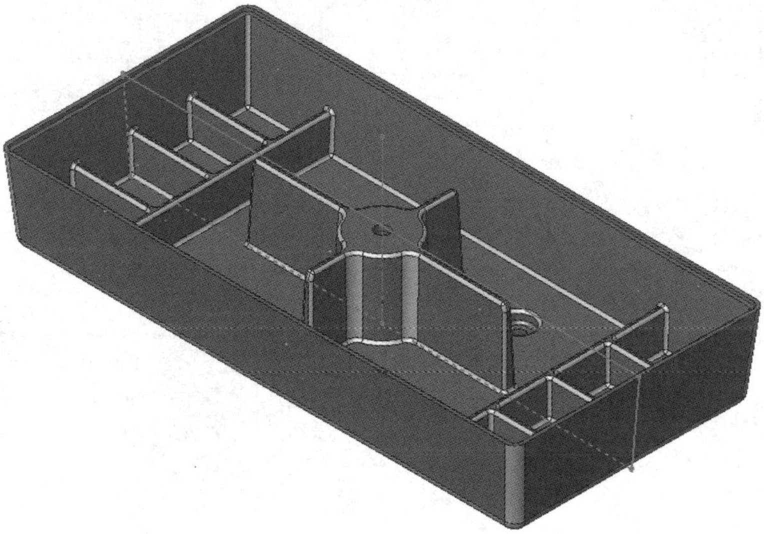

The Sketch Plane is missing; a new plane or face must be selected to replace it.

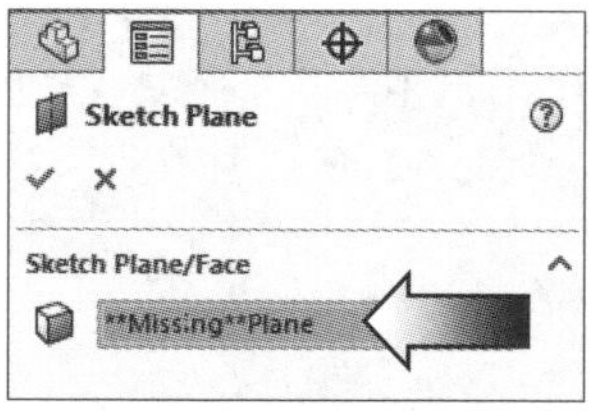

Select the **Front** plane from the FeatureManager tree to replace the missing plane.

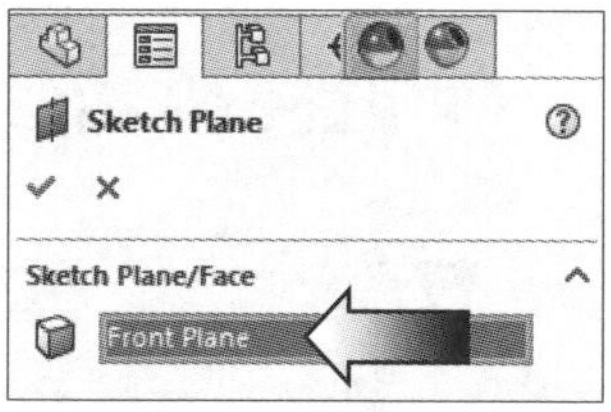

After replacing the plane, click **OK**.

The system displays the warnings on other errors along with the solutions for repairing them.

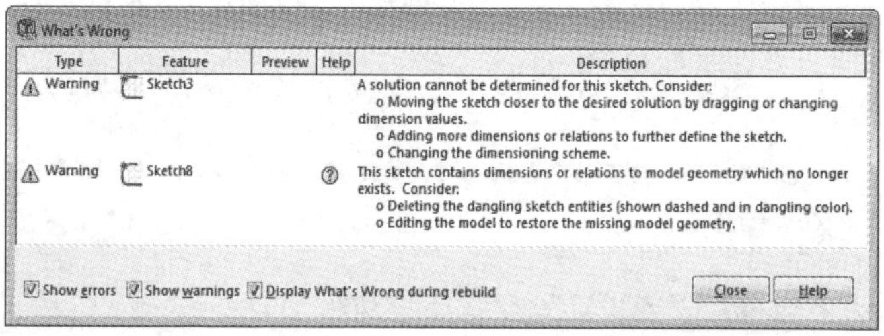

Click **Close** 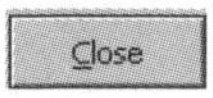 .

4. Repairing the sketch Relations and Dimensions:

Right-click **Sketch3** and select **Edit Sketch**.

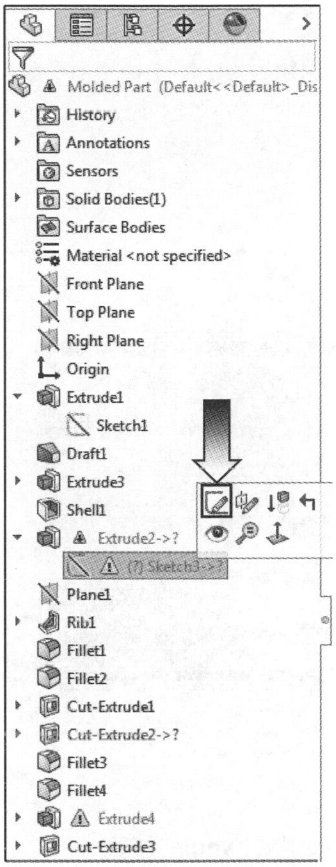

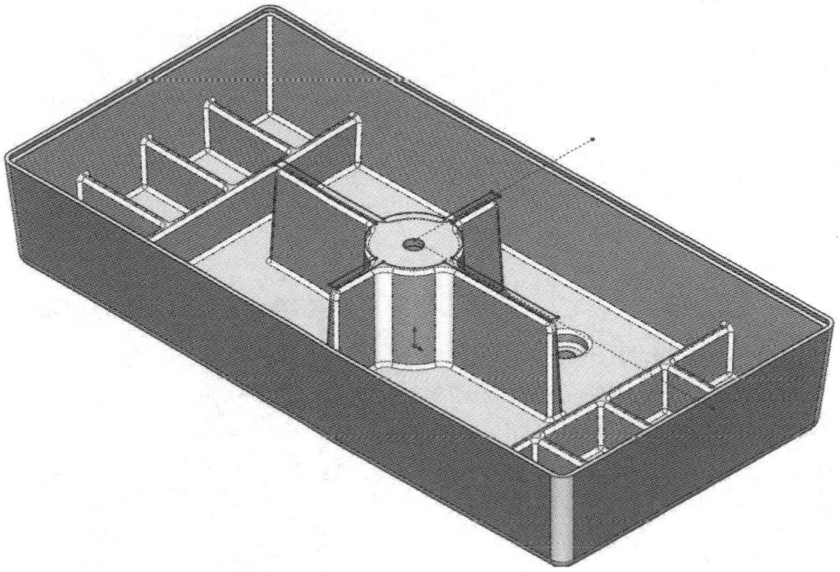

Click the **Display/Delete Relations** command  or select: **Tools / Relations / Display-Delete**.

Delete the 2 geometric relations that have the external symbols next to their names (arrows).

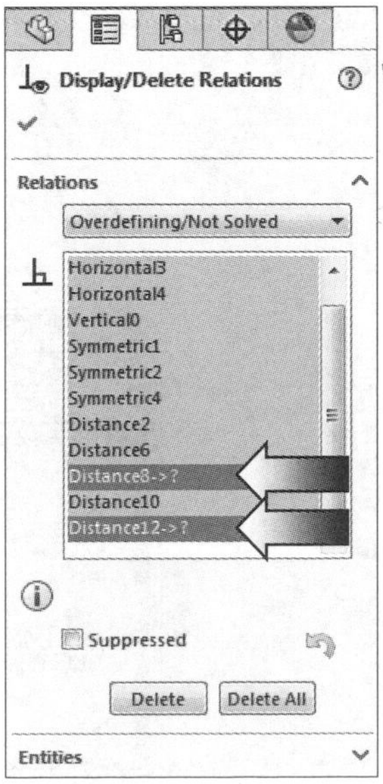

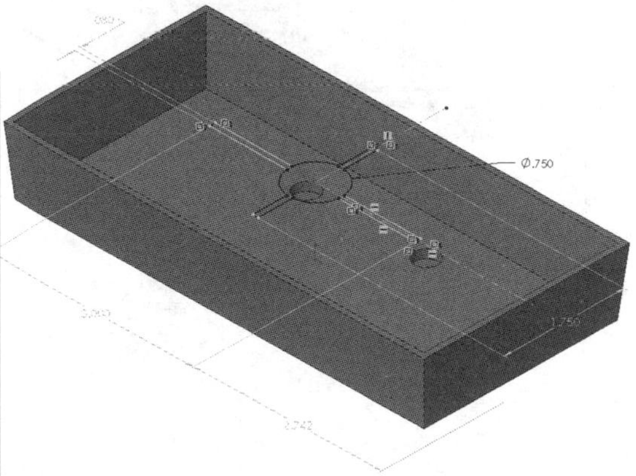

The sketch becomes fully-defined.

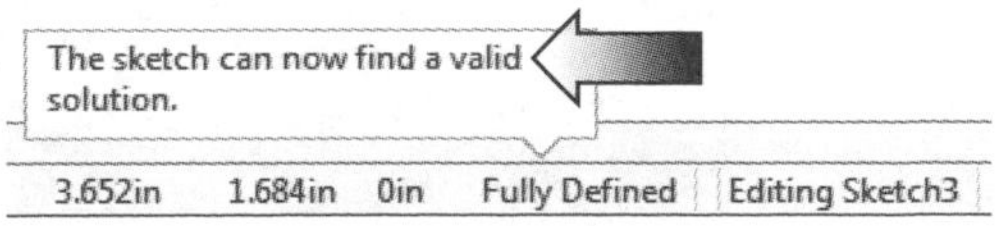

Click **OK**.

Exit the sketch .

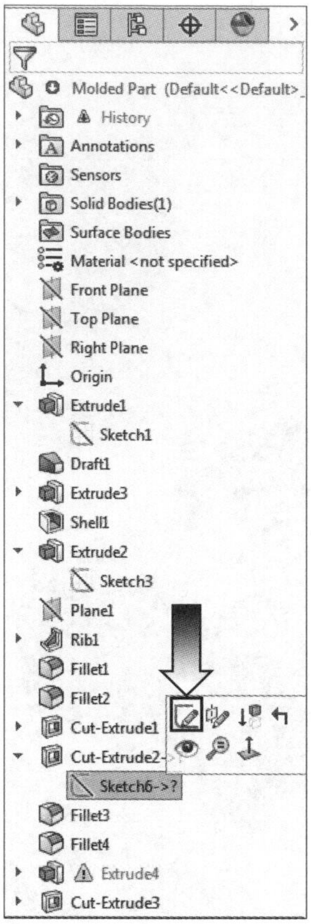

5. Repairing the next sketch errors:

Right-click **Sketch6** (under Cut-Extrude2) and select: **Edit Sketch** .

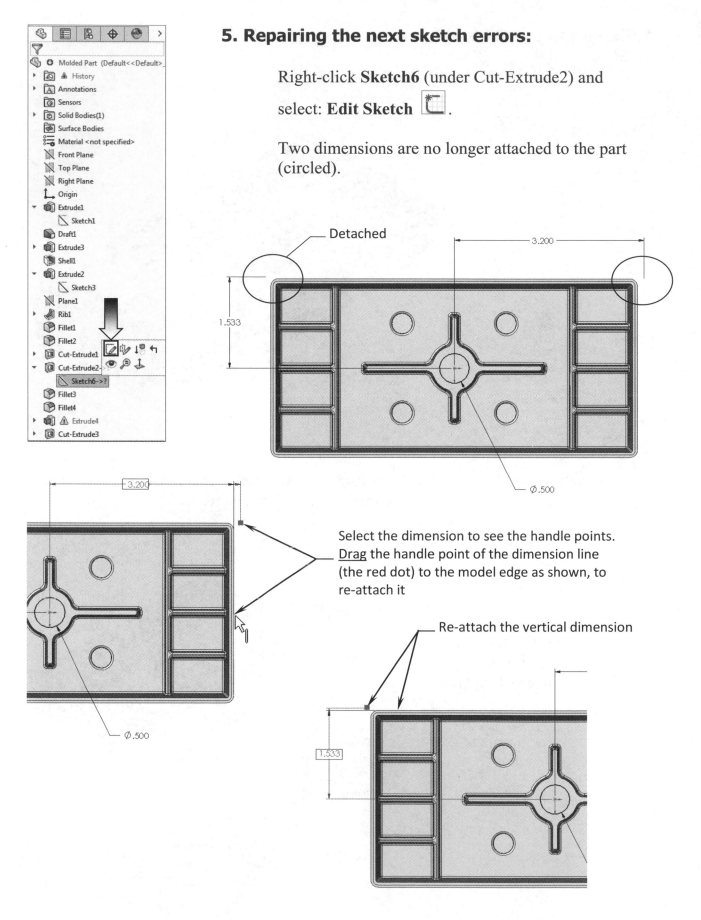

Two dimensions are no longer attached to the part (circled).

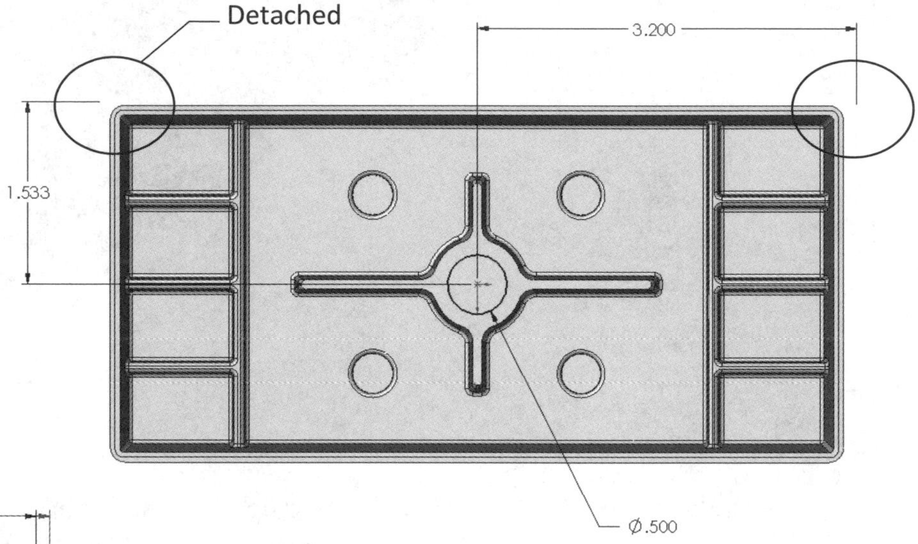

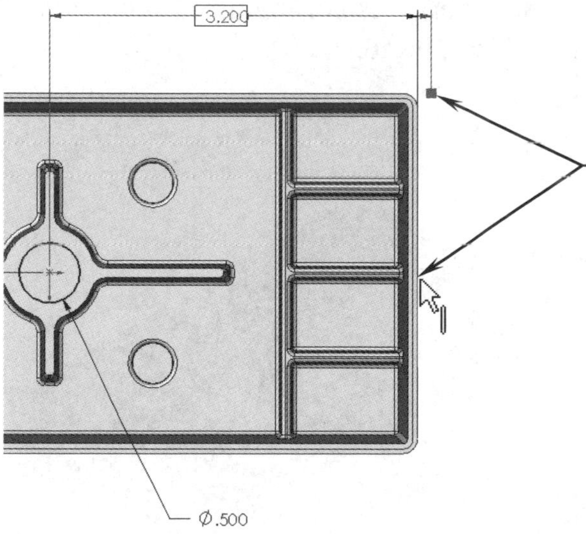

Select the dimension to see the handle points. Drag the handle point of the dimension line (the red dot) to the model edge as shown, to re-attach it

Re-attach the vertical dimension

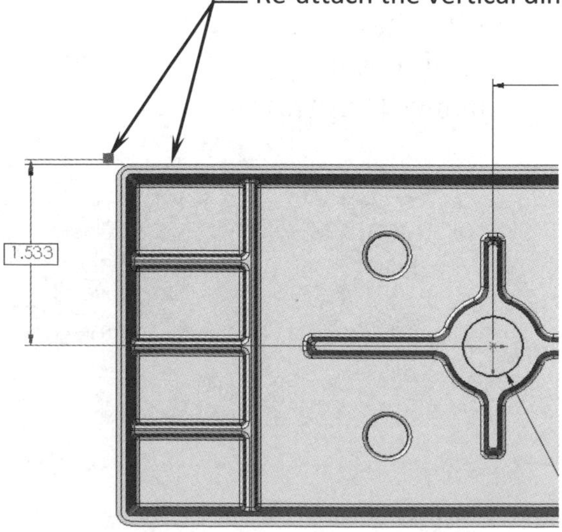

The sketch becomes fully defined. The two dimensions are now attached to the edges of the model and also locate the center of the circle.

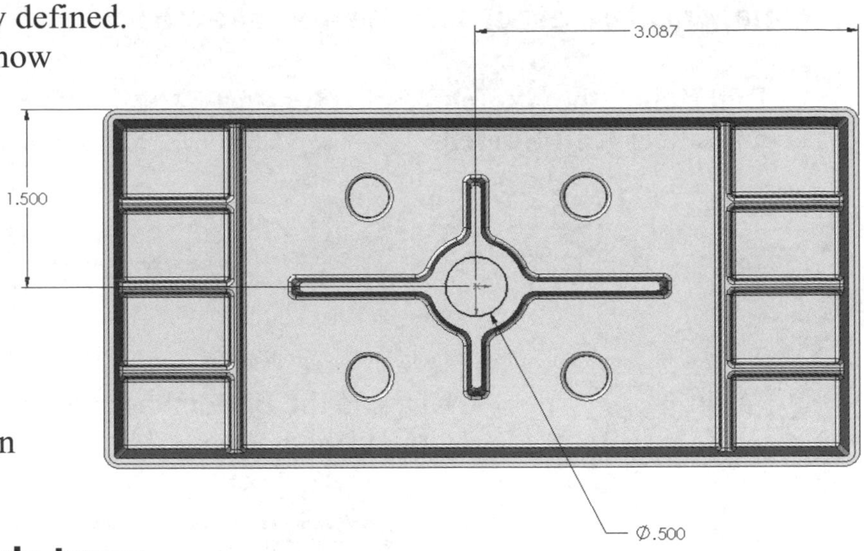

Exit the sketch .

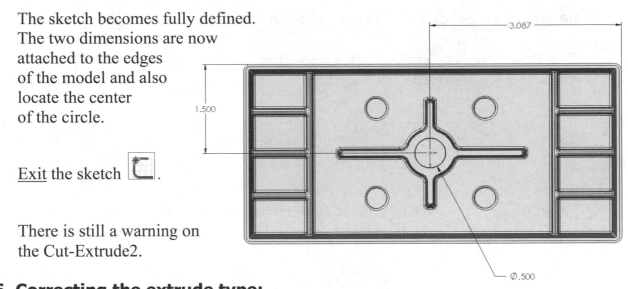

There is still a warning on the Cut-Extrude2.

6. Correcting the extrude type:

Right-click on **Cut-Extrude2** and select **Edit Feature**.

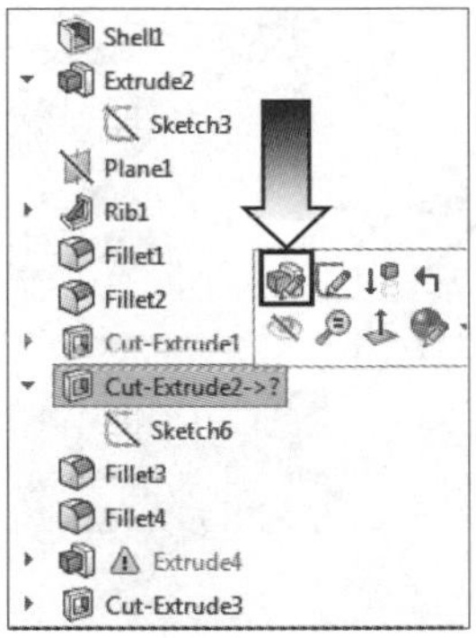

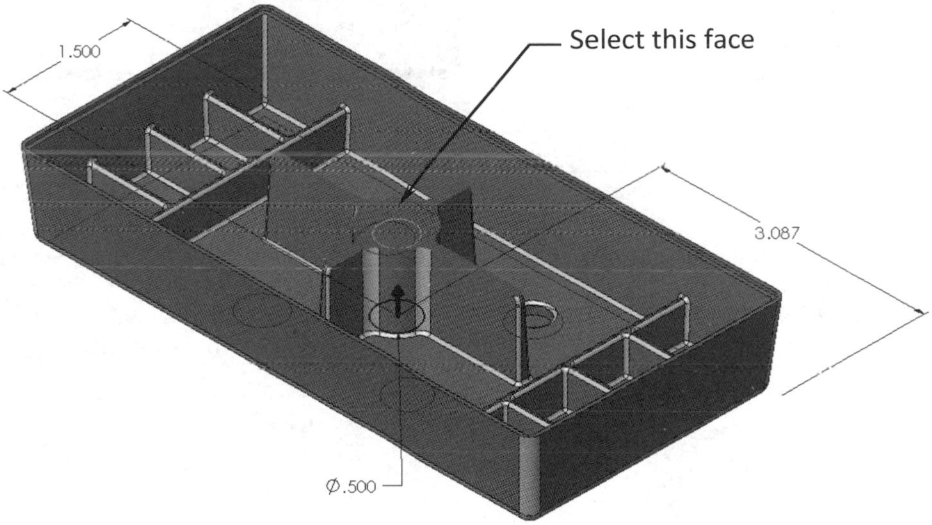

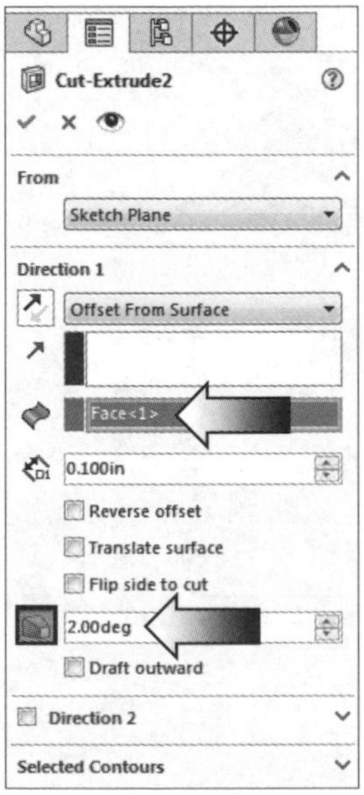

The surface that was used as the end condition option is no longer recognized; a new surface must be selected to replace the missing one.

Select the **face** as indicated.

Leave extrude depth as **.100**in. and the draft angle at **2deg**.

Click **OK**.

7. Repairing the errors in the last sketch:

Right-click the **Sketch8** and select **Edit Sketch**.

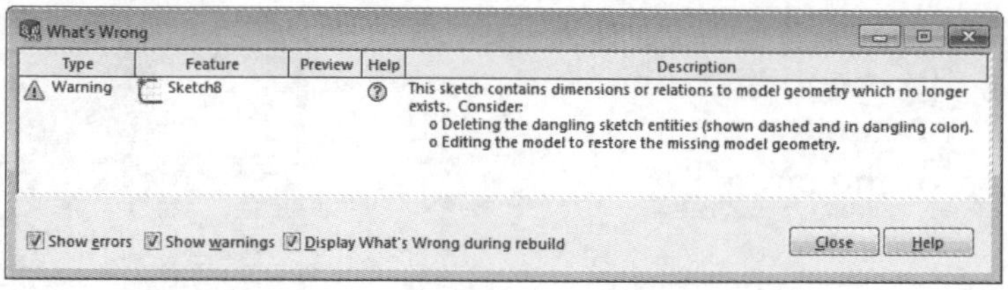

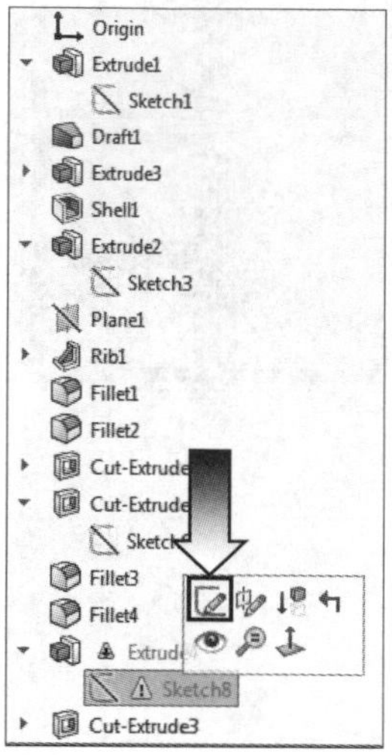

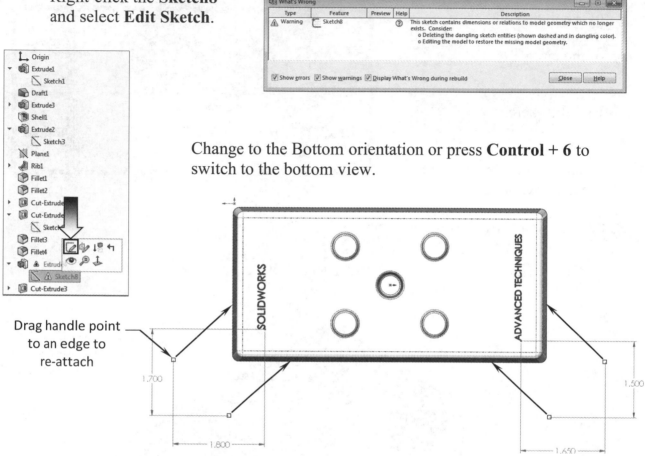

Change to the Bottom orientation or press **Control + 6** to switch to the bottom view.

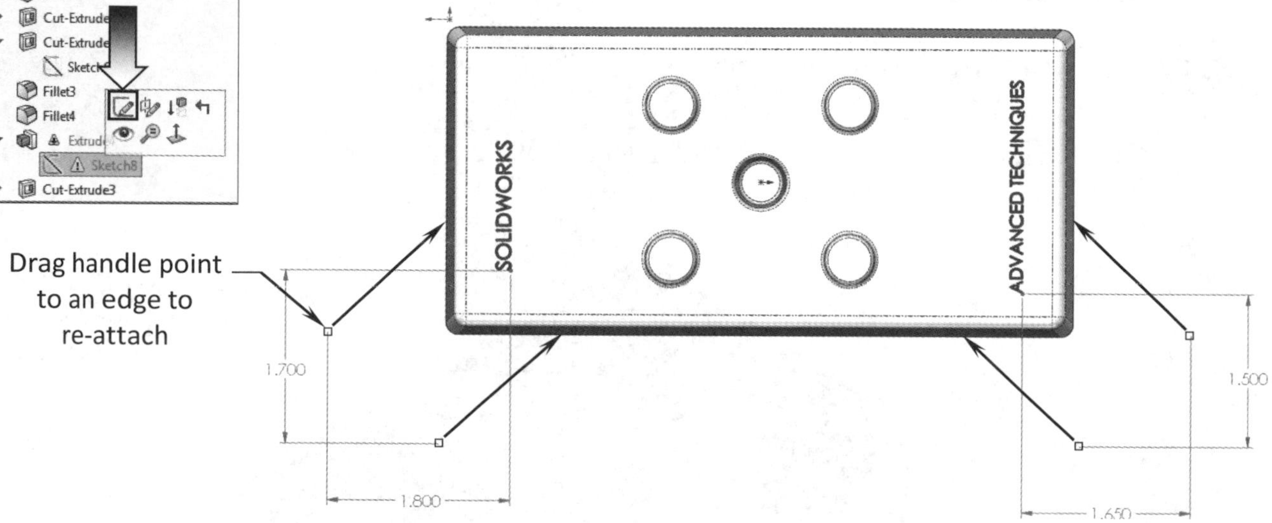

The dimensions that were created from or to other parts became Dangling (Olive-green color) because they are no longer attached to the models.

Select one of the dimensions and <u>drag</u> its handle point (the red dot) to a model edge to reattach it.

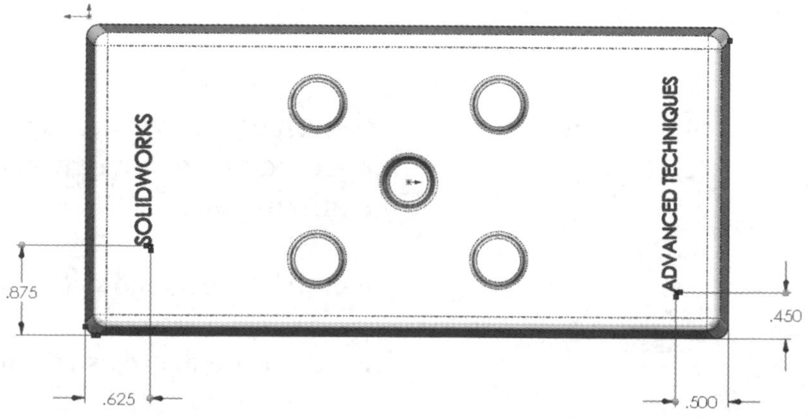

Repeat the same step to re-attach all other dimensions.

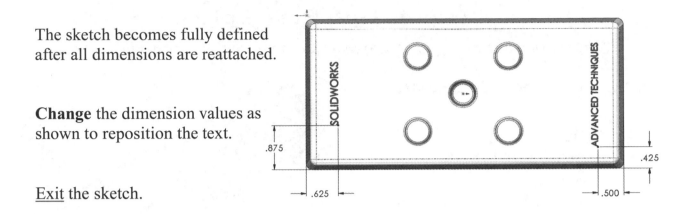

The sketch becomes fully defined after all dimensions are reattached.

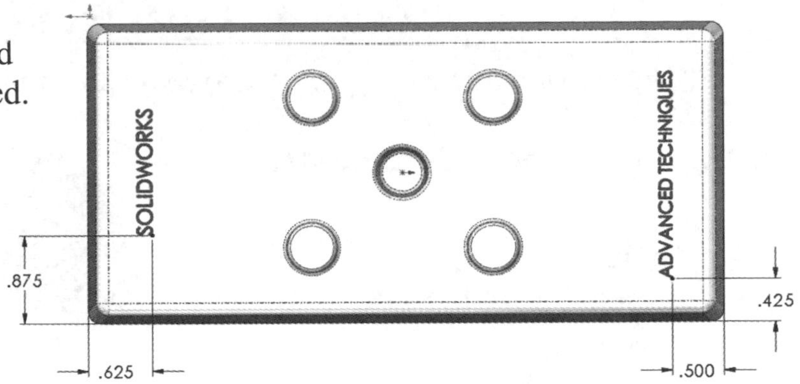

Change the dimension values as shown to reposition the text.

Exit the sketch.

The reference symbols and the error colors on the FeatureManager tree should now be all removed.

8. Saving your work:

Select **File / Save As / Repair Errors / Save**.

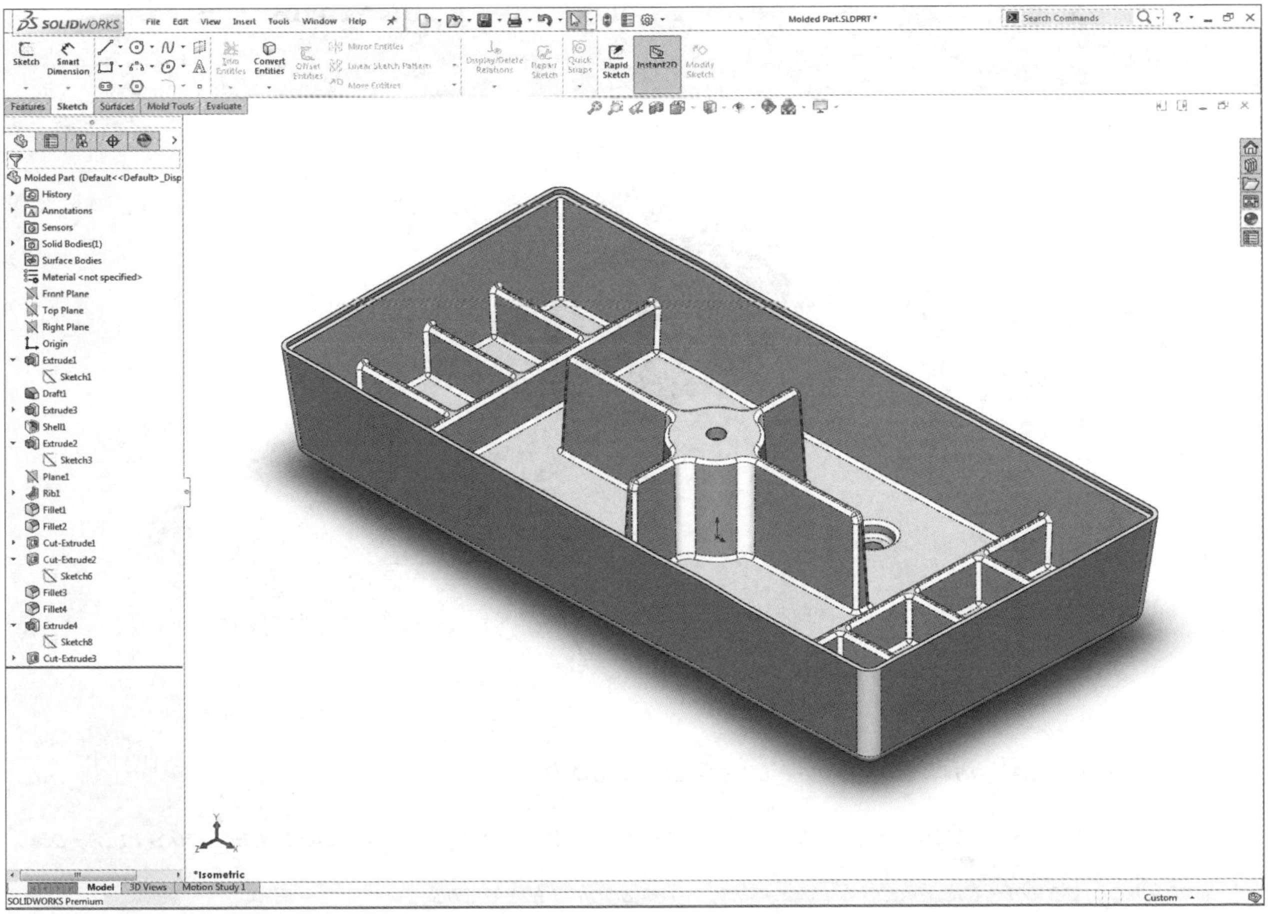

Level 4: Final Exam 1 of 2

(The instructions will be limited so that you can try out your own approaches.)

1. Opening a part document:

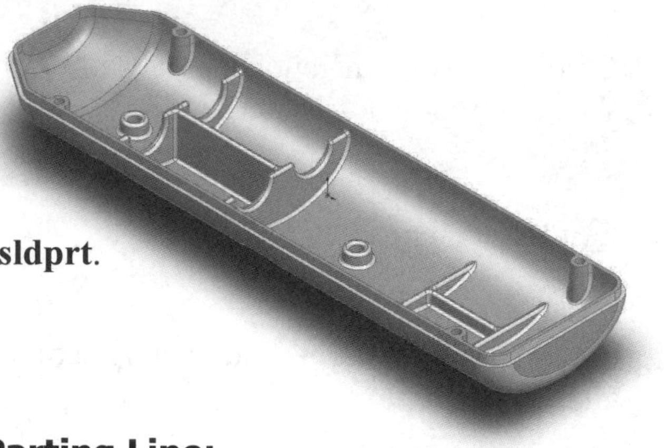

Go to:

Training Files folder
Tooling Design folder
Open: **L4 Final Exam_Handle Mold.sldprt**.

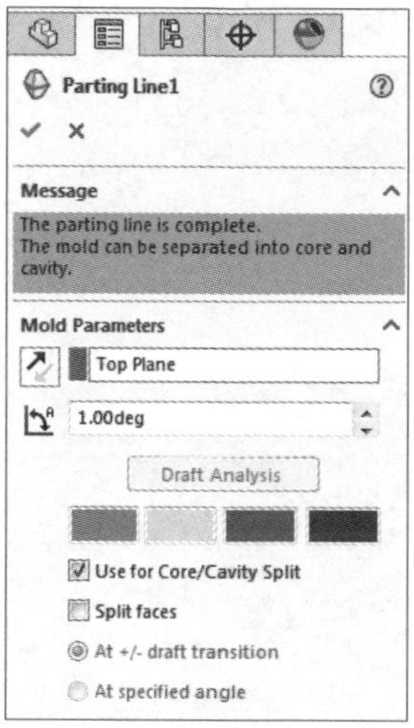

2. Creating a Parting Line:

Direction of Pull = **TOP plane**.

Draft Angle = **1deg**.**

Parting Lines = **All outer edges** as noted.

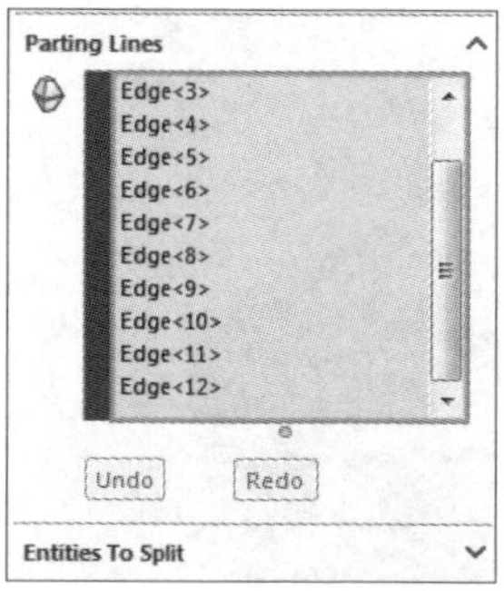

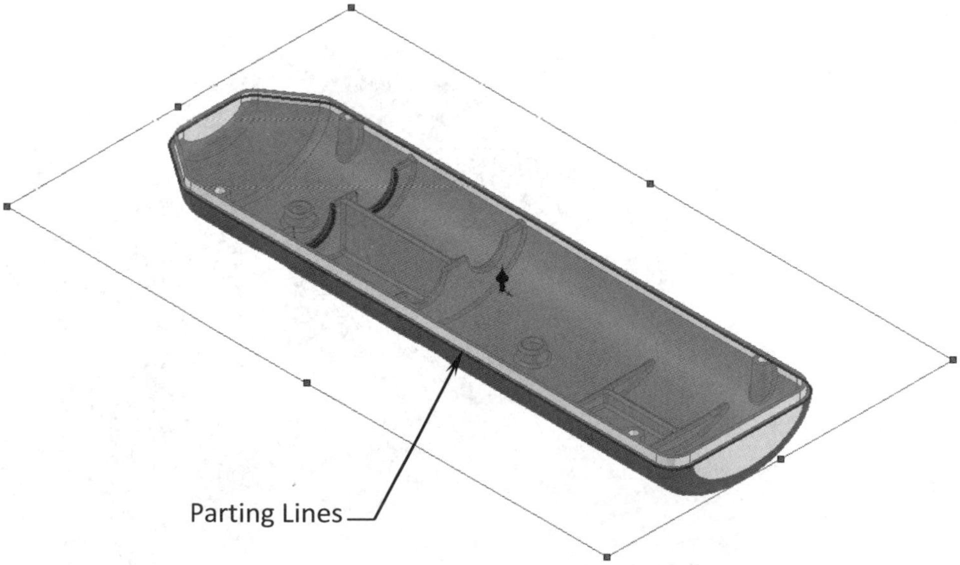

*** <u>Important</u> - Roll back under the Loft feature and:***

a/ Add 1° drafts to all Yellow faces, including the 4 holes; this change will create some errors in the part.

b/ <u>Reorder</u> or <u>recreate</u> the fillets, if necessary, <u>after</u> adding the drafts.

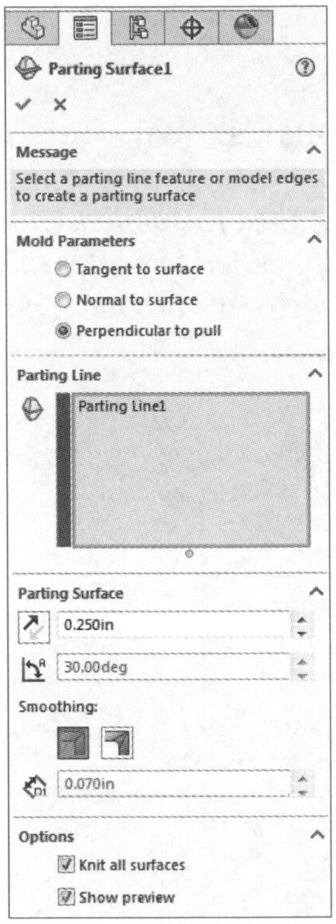

3. Creating a Parting Surface:

Select the **Perpendicular to Pull** option.

For Parting Line: select the **Parting Line1** from the FeatureManager tree.

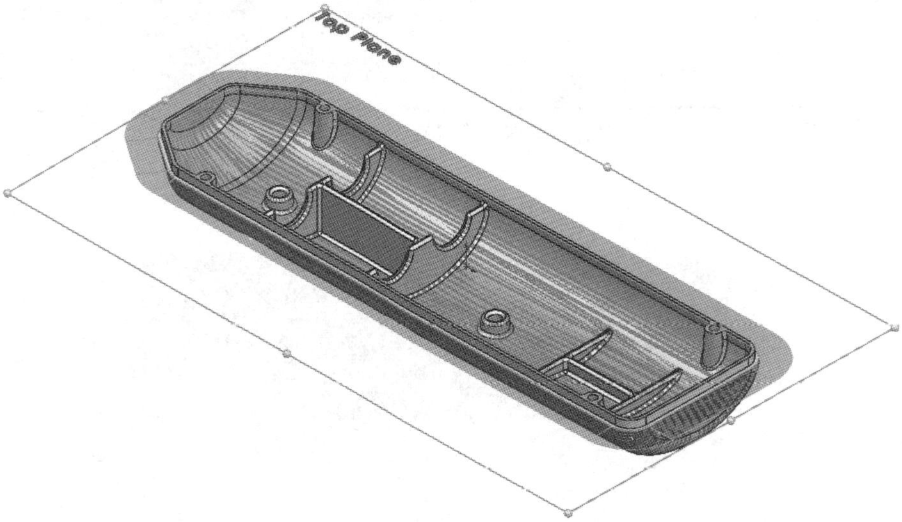

4. Adding an Offset Distance plane:

Use the **Top** reference plane and **.375 in**. distance.

The new plane is placed **above** the Top plane. (This new plane will also be used to create the Interlock Surfaces.)

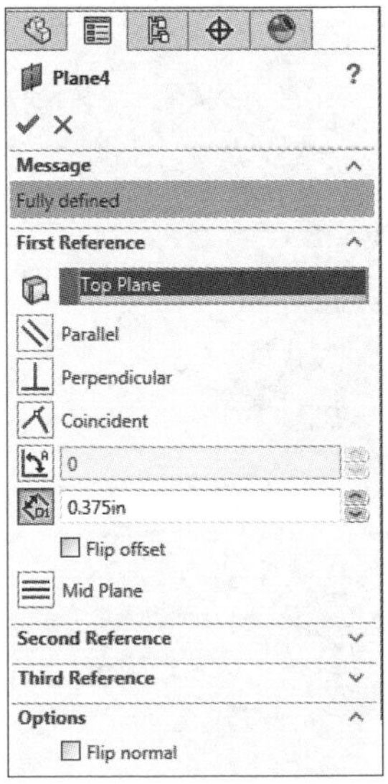

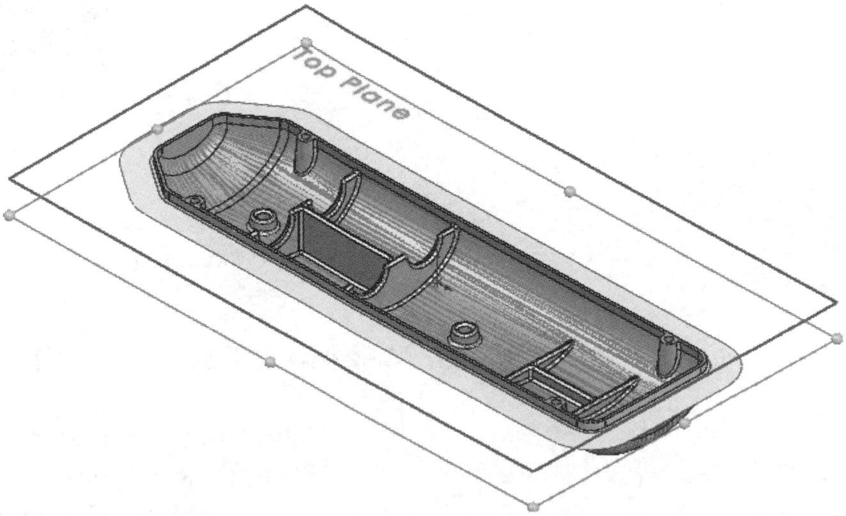

5. Sketching the mold block profile:

Sketch a **Rectangle** on the new plane (Plane1) and add dimensions* shown.

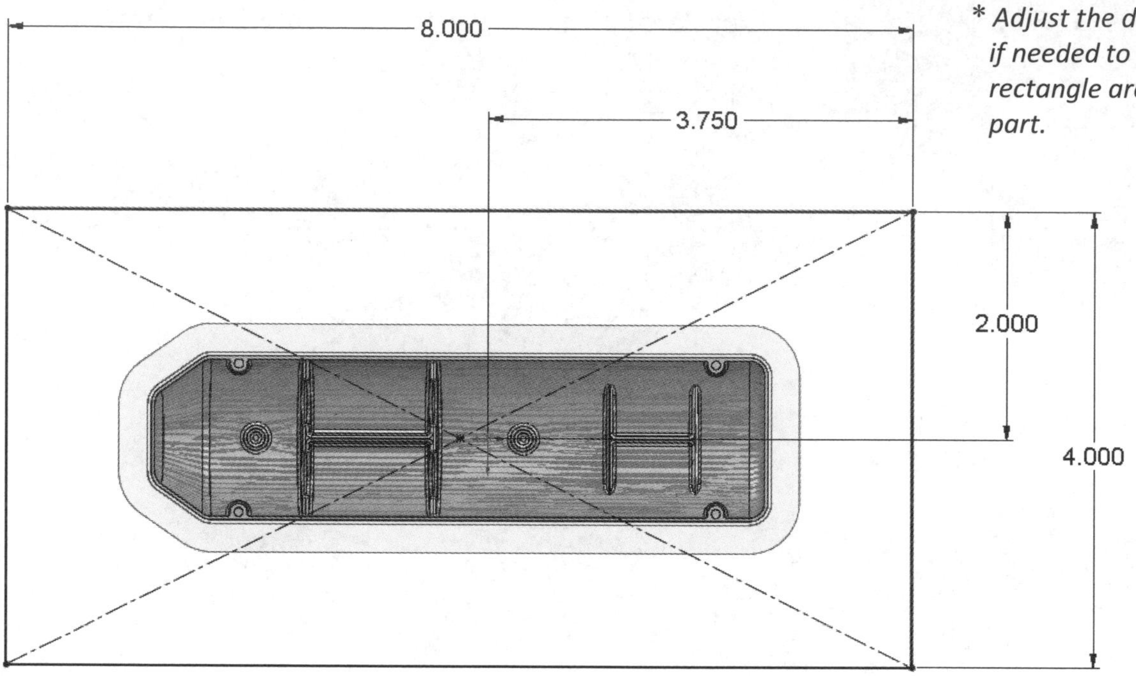

** Adjust the dimensions if needed to center the rectangle around the part.*

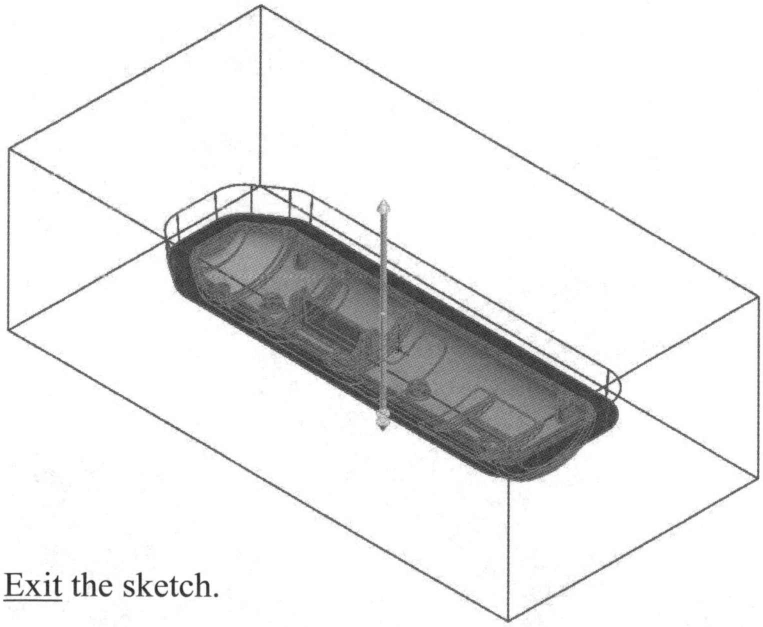

Exit the sketch.

6. Making a Tooling split:

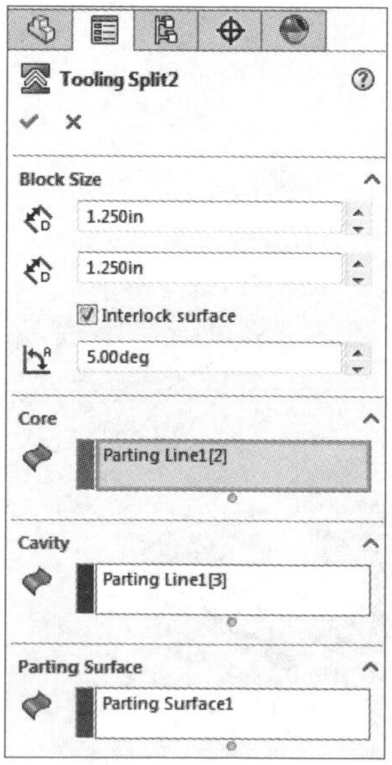

1.250in. (upper block)
1.250in. (lower block)
Use **Interlock Surface** with **5° Draft**.

7. Separating the mold blocks:

Use the **Move/Copy** command to separate the two halves.

** **OPTIONAL:** Make the upper and lower solid bodies transparent for clarity.

8. Saving your work as:

L4 Final Exam_Handle Mold_Completed.

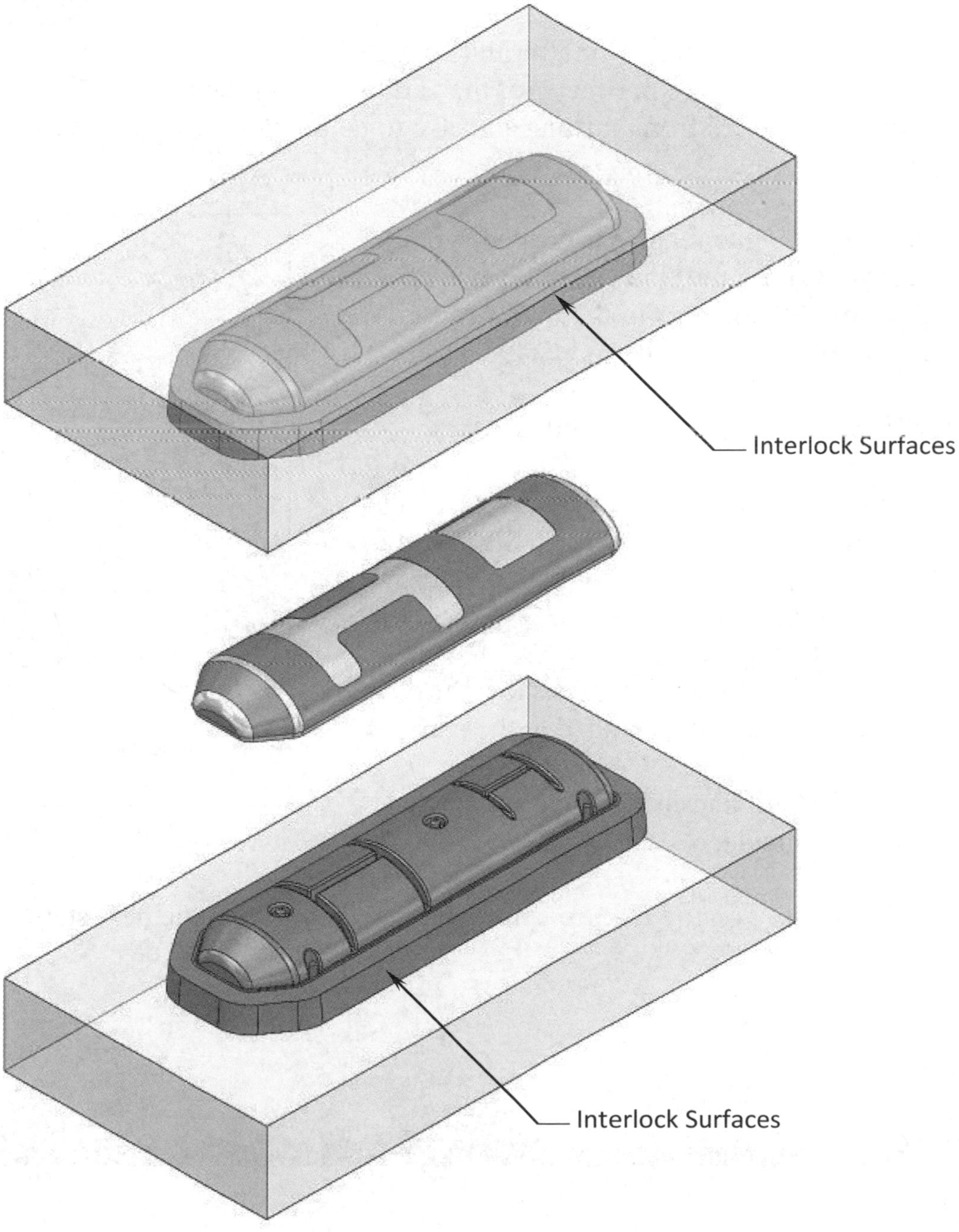

Level 4: Final Exam 2 of 2

(The instructions will be limited so that you can try out your own approaches.)

1. Opening a part document:

Go to:

Training Files folder
Tooling Design folder
Open: **L4 Final Exam_Mouse Mold.sldprt**.

The intent is to create a small wall (Tongue & Groove) that runs around the perimeter of the part to prevent the upper and lower halves from shifting and also to help align them more easily during assembly stage.

After the tongue feature is created, a pair of Core & Cavity mold needs to be made from the part. All surfaces in the model must have a minimum of 3° draft angle.

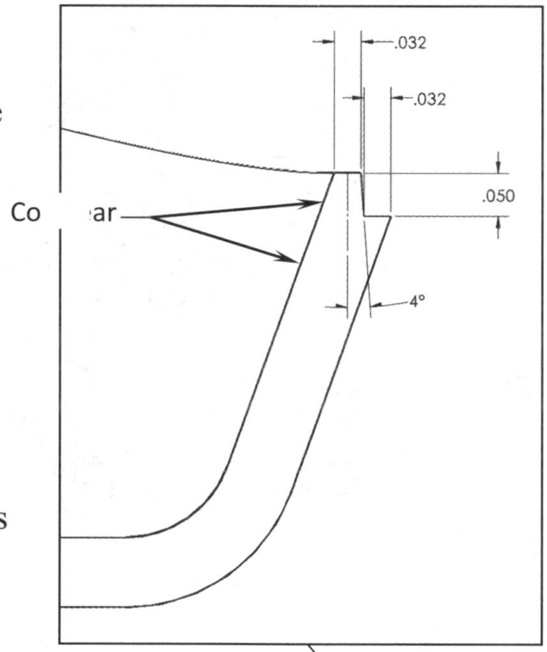

2. Creating a Tongue feature*:

There are several ways to create the Tongue feature. You can use any methods that you like as long as it has all of the required size and location dimensions.

The section and detail views show the details of the Tongue feature added to the part.

It is **.050"** tall and **.032"** wide and has a **4°** draft on outside.

* Create the Tongue feature before making the mold in the next step.

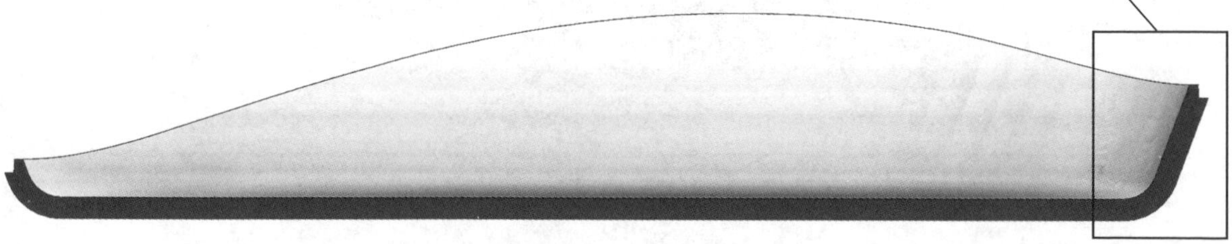

3. Analyzing the Draft Angles:

Switch to the **Evaluate** tab and click **Draft Analysis**.

For Direction of Pull, select the **Top** plane.

For Draft Angle enter **3.00deg**.

Click the **Face Classification** checkbox.

The **Required Draft** section (yellow) must show **0** (zero).

Click **OK**.

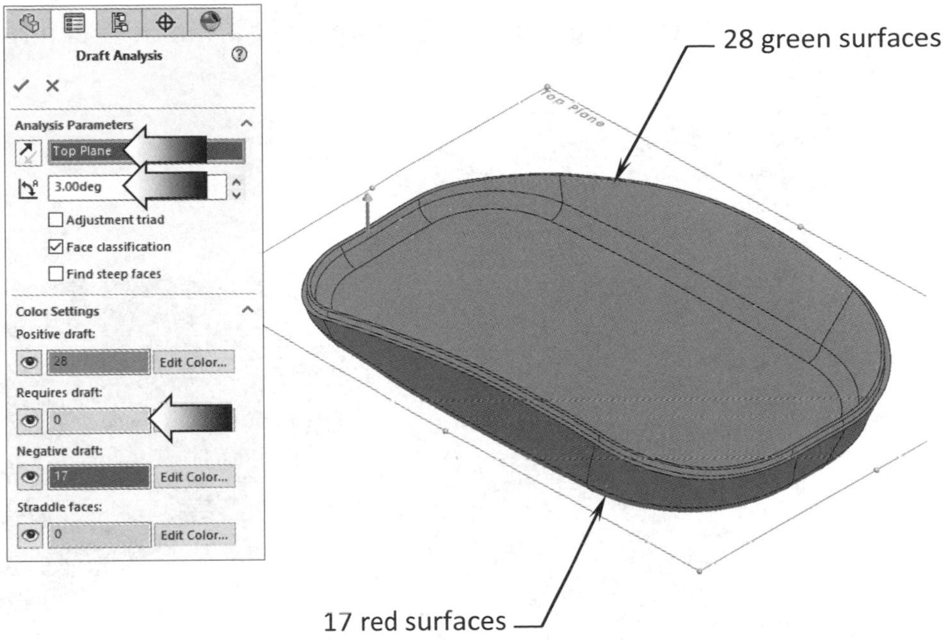

4. Applying the Scale Factor:

Switch to the **Mold Tools** tab.

Click **Scale**.

For Scale Parameter, select the part.

For Scale About, use the default **Centroid**.

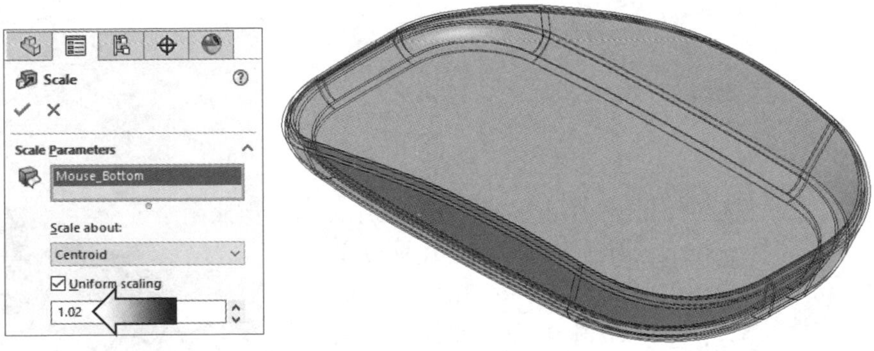

For Scale Factor, enter **1.02** (2% larger) and enable the **Uniform Scaling** checkbox.

Click **OK**.

Set the material to **ABS**.

5. Creating a Parting Line:

Click **Parting Line** on the **Mold Tools** tab.

For Pull Direction, select the **Top** plane from the FeatureManager tree.

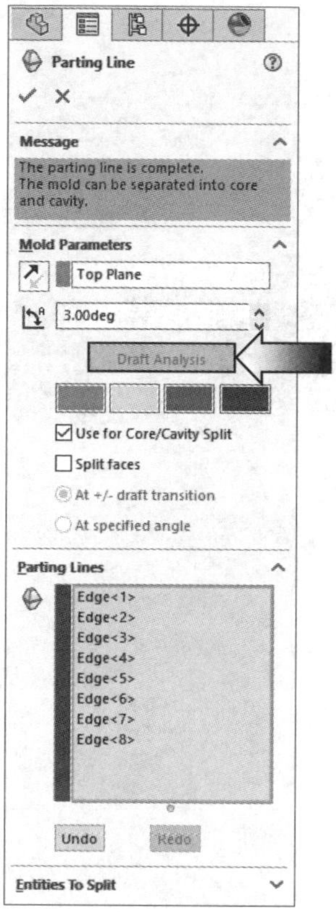

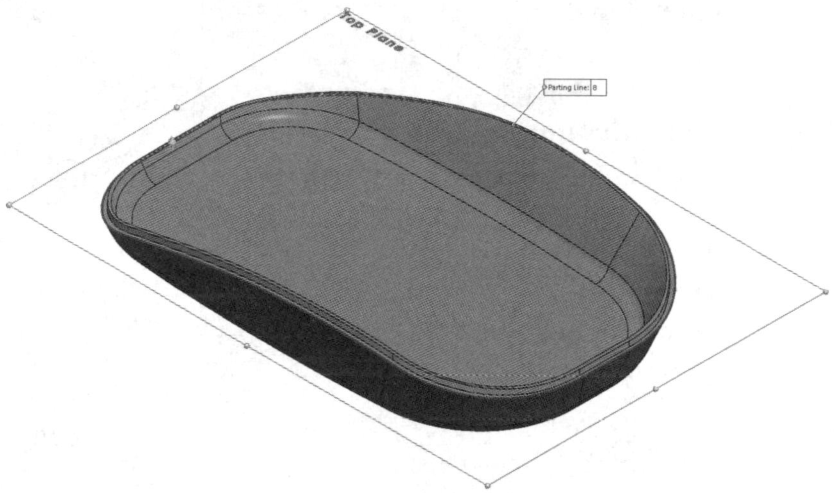

For Draft Angle, enter **3.00deg**.

Click the **Draft Analysis** button to analyze the drafts in the model.

Click **OK**.

Ensure that there are no yellow surfaces in the model at this point before moving forward to the next step. Stop and make any corrections if needed.

A Parting Line is created where the green surfaces (Core) meet the red surfaces (Cavity)

6. Adding a Parting Surface:

A parting surface is used to split the mold into the Core and Cavity halves.

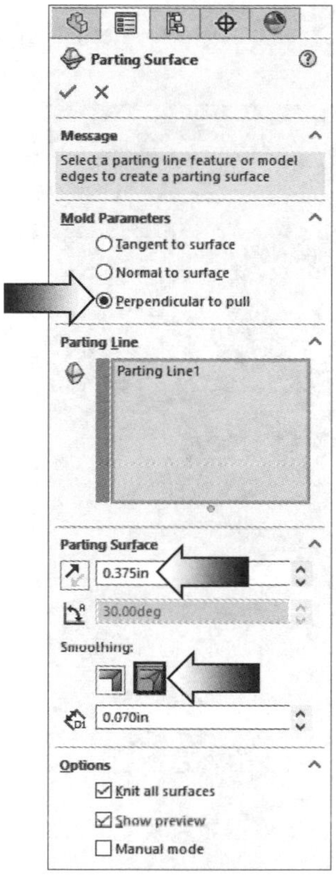

Click **Parting Surface** on the **Mold Tools** tab.

For Mold Parameters, select **Perpendicular to Pull**.

The **Parting Line** should be selected automatically.

For Parting Surface Distance, enter **.375in**.

For Smoothing between adjacent surfaces, select **Smooth.**

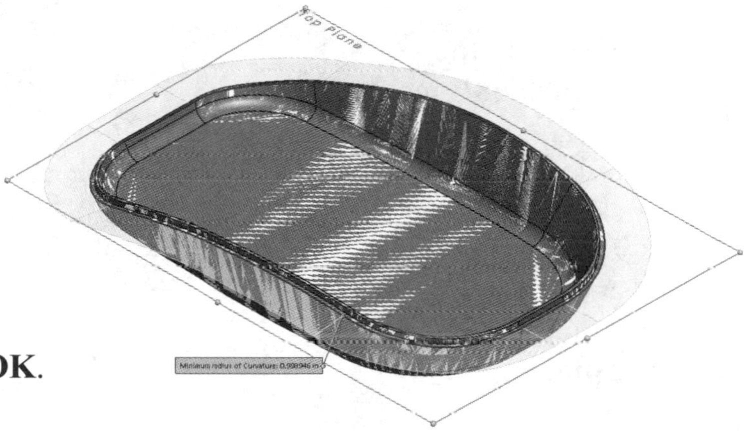

Click **OK**.

7. Creating a new plane:

This plane is used to sketch the profile of the mold block and also to define the height of the Interlock Surface.

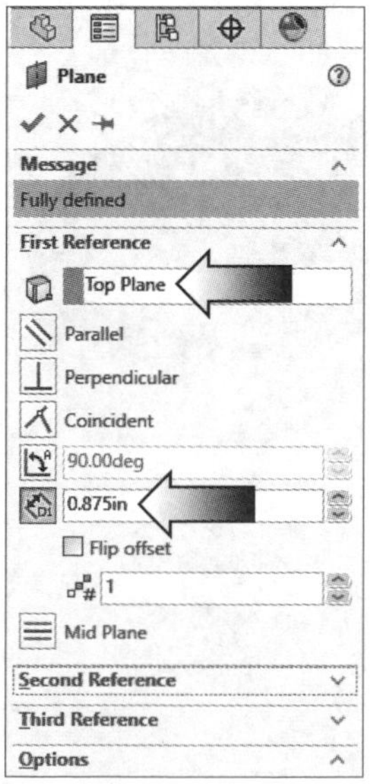

Switch to the **Features** tab and click:
Reference Geometry, Plane.

For First Reference, select the **Top** plane from the tree.

For Offset Distance, enter **.875in**.

Click **OK**. The new plane is placed <u>above</u> the Top plane.

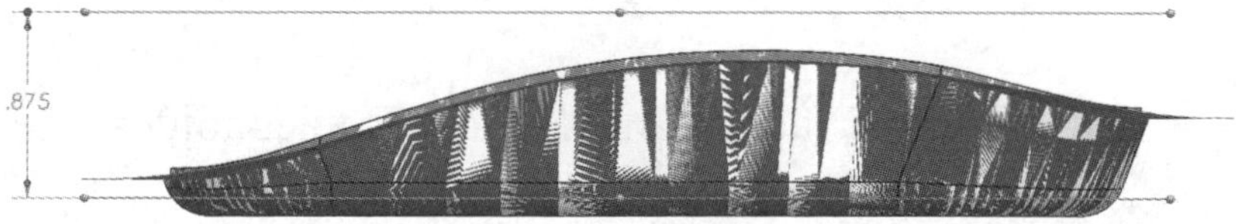

8. Creating a Tooling Split:

Select Plane1 and open a **new sketch**.

Sketch a **Corner Rectangle** around the model.

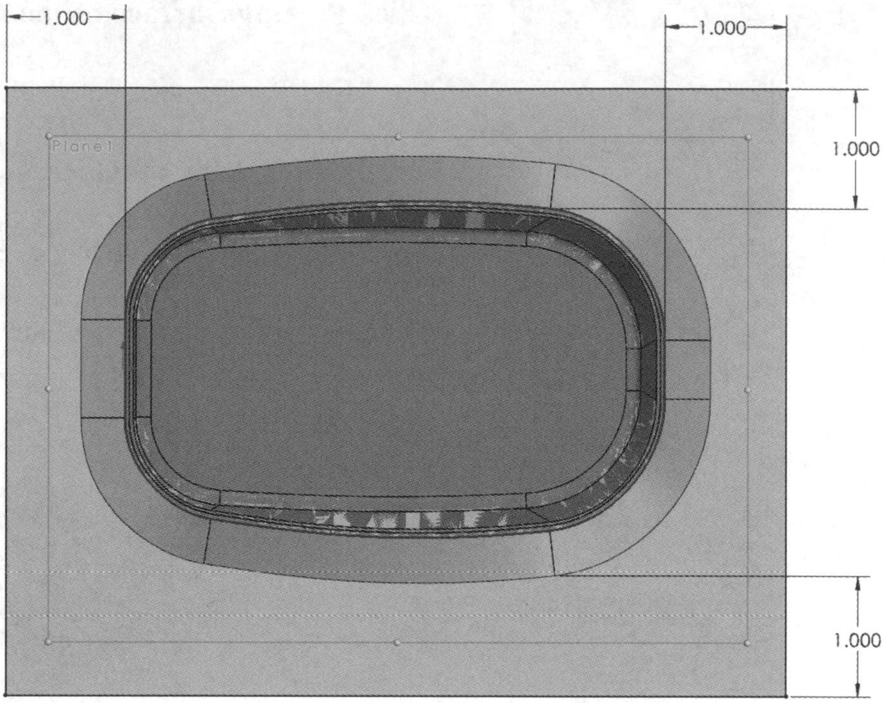

Add the location dimensions shown to fully define the sketch.

Exit the sketch.

Switch to the **Mold Tools** tab.

Select the sketch of the **Rectangle** and click: **Tooling Split**.

For Block Size enter:

.750in.
1.750in.

Enable the **Interlock-Surface** checkbox.

Enter **3°** for Draft Angle.

Click **OK**.

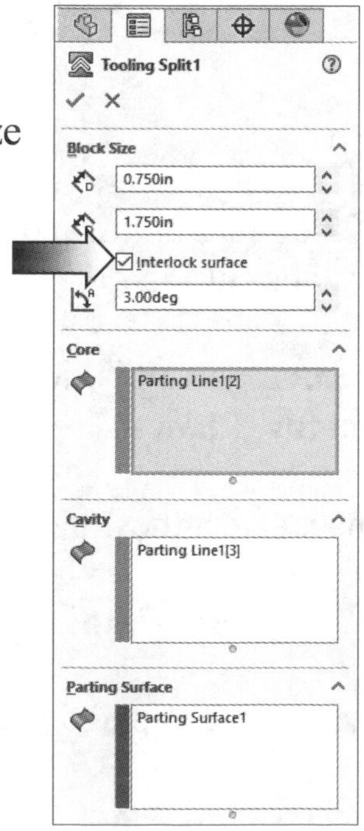

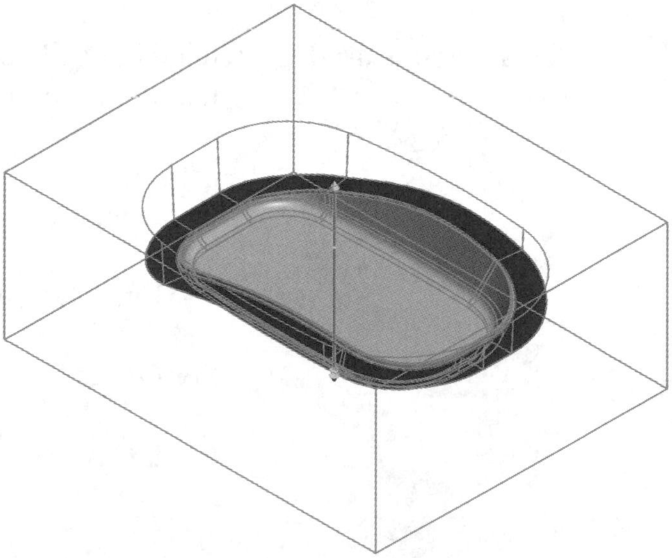

Hide all surfaces in the Surface Bodies folder.
Also hide the Parting Line.

9. Separating the mold blocks:

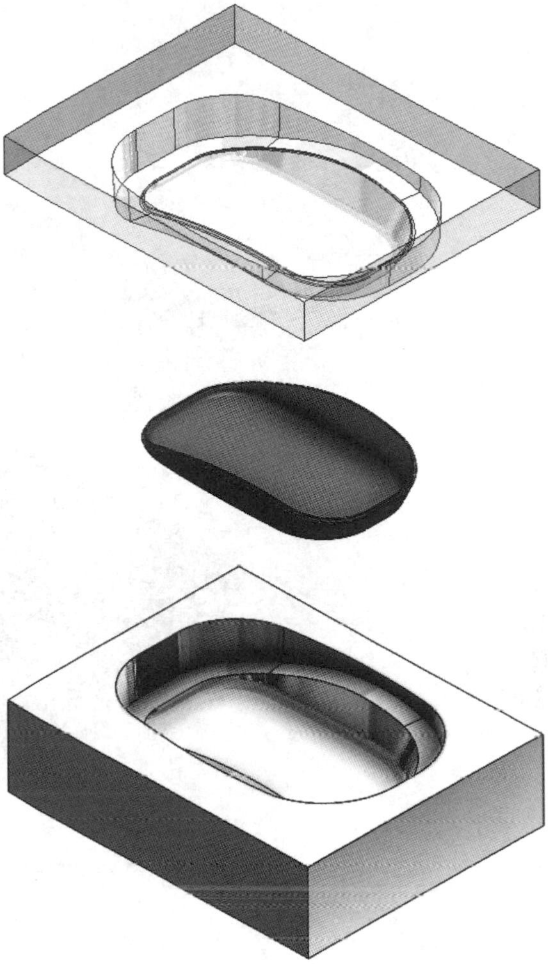

Use the **Move/Copy Bodies** command to separate the 2 mold blocks.

Move the Core Block along the Y direction **4.50in**.

Also move the Cavity Block along the Y direction **-5.50in**.

Rename the 3 solid bodies to:

Core Block
Plastic part
Cavity Block

Change the material of the 2 mold blocks to **Plain Carbon Steel**

The material **ABS** was assigned to the plastic part in step number 4.

10. Saving your work:

Select **File, Save As**.

Enter: **Level 4 Final Exam_Mouse Mold_Completed**

Click **Save**.

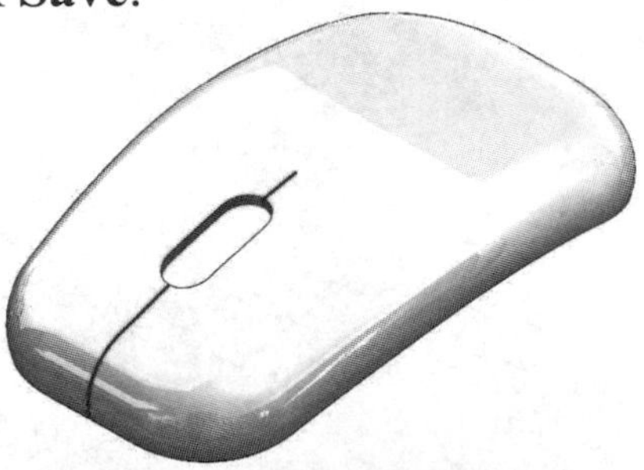

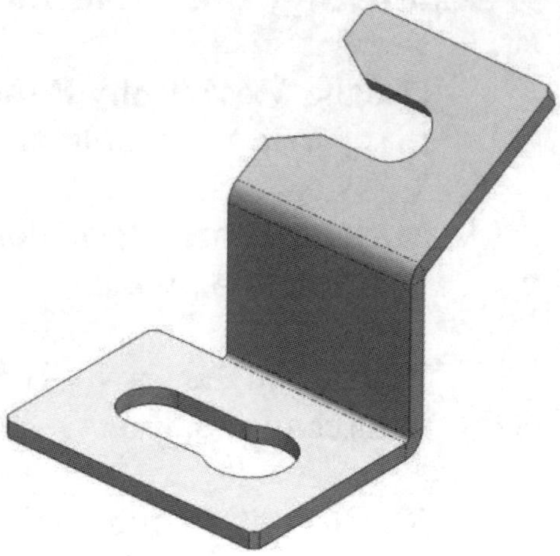

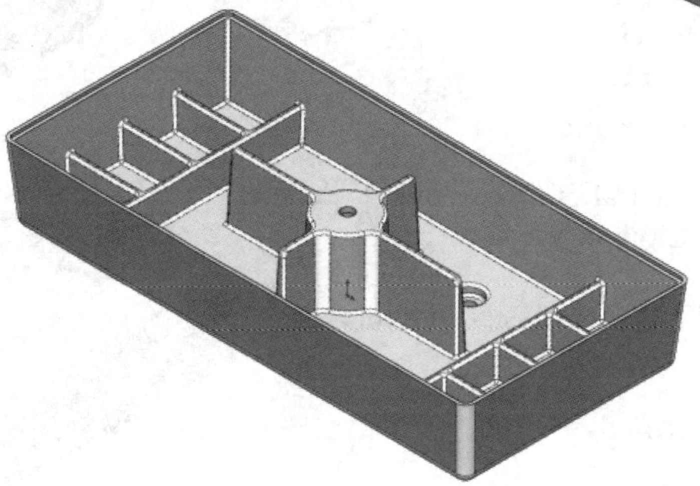

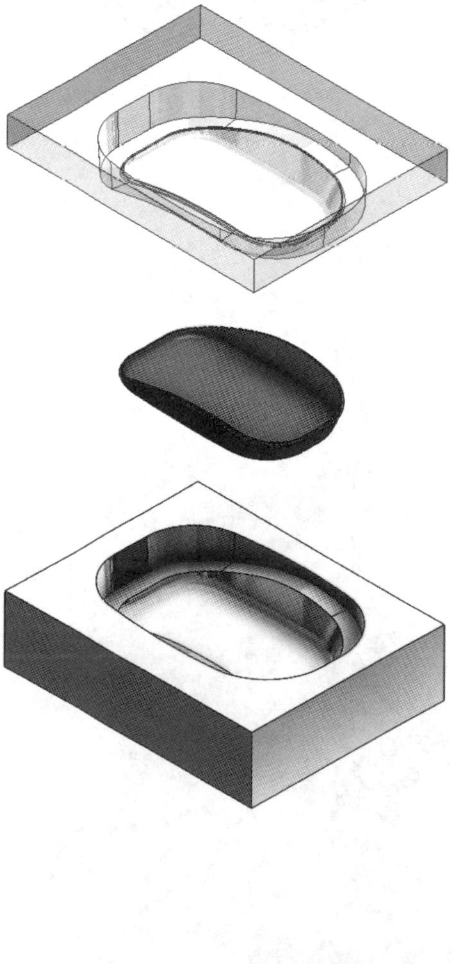

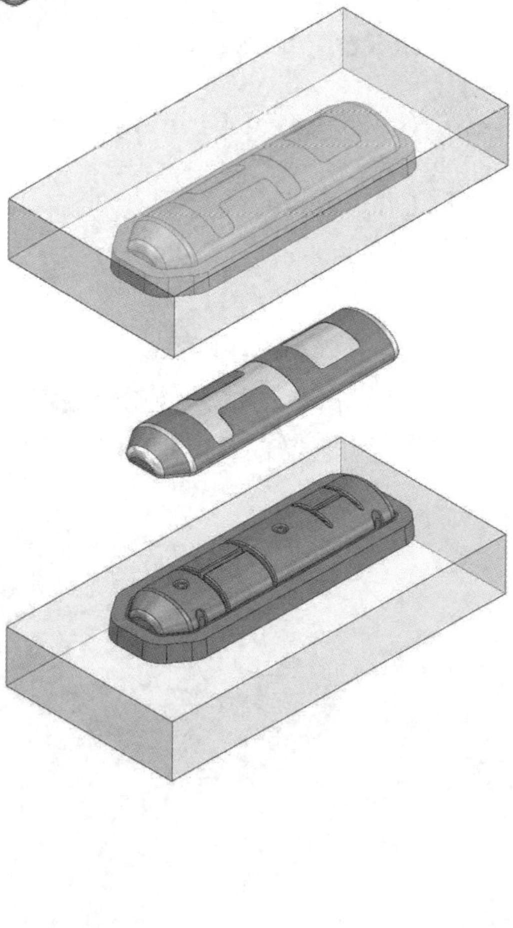

CHAPTER 23

Using Appearances

A knurled feature will create hundreds of extra faces in the model; it will increase the file size and bog down graphical performance. If it is important to show a knurl feature in a model, then there are a couple of techniques that we can use to accommodate it.

This first half of the chapter will guide you through the method of modeling the Raised Diamond Knurls where a sketch profile is swept along a path to create a spiral cut. The spiral cut is then patterned and mirrored to repeat the raised diamond shapes of the knurl.

Modeling Diamond Knurls

1. Opening a part document:

Browse to the Training Files folder and open a part document named: **Knurled Handle**.

This model has a single solid body, a sweep profile, and a sweep path that was previously created.

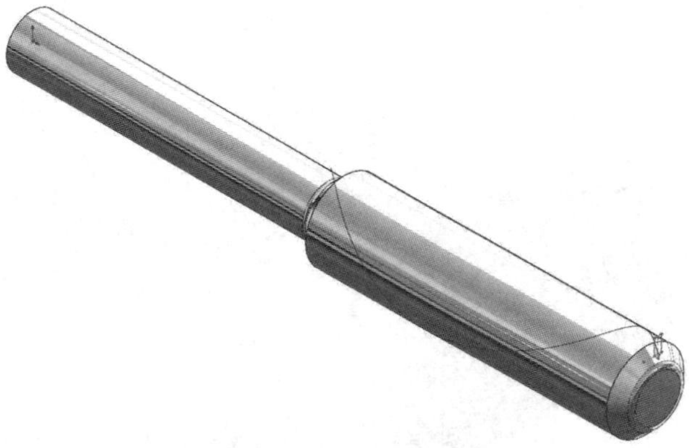

Before

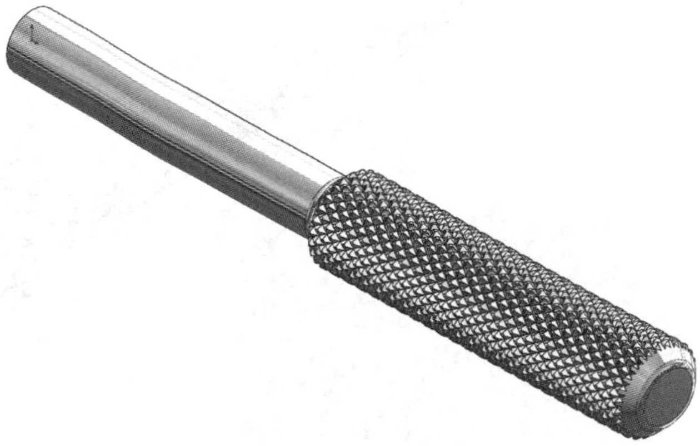

After

2. Creating a swept cut:

Switch to the **Features** tool tab and click the **Swept Cut** command.

For Sweep profile, select **Sketch2** from the FeatureManager tree.

For Sweep path, select the **Helix**.

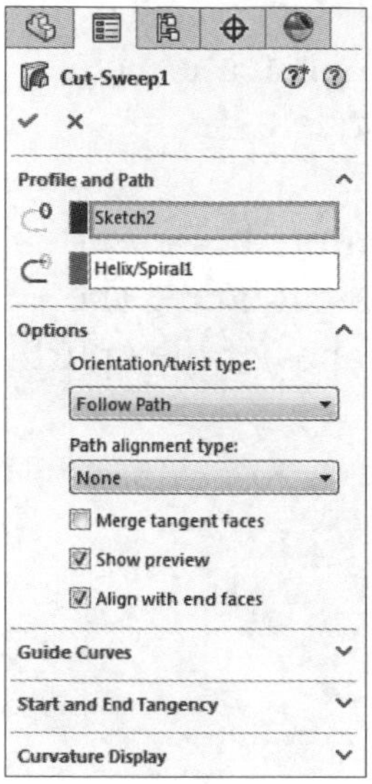

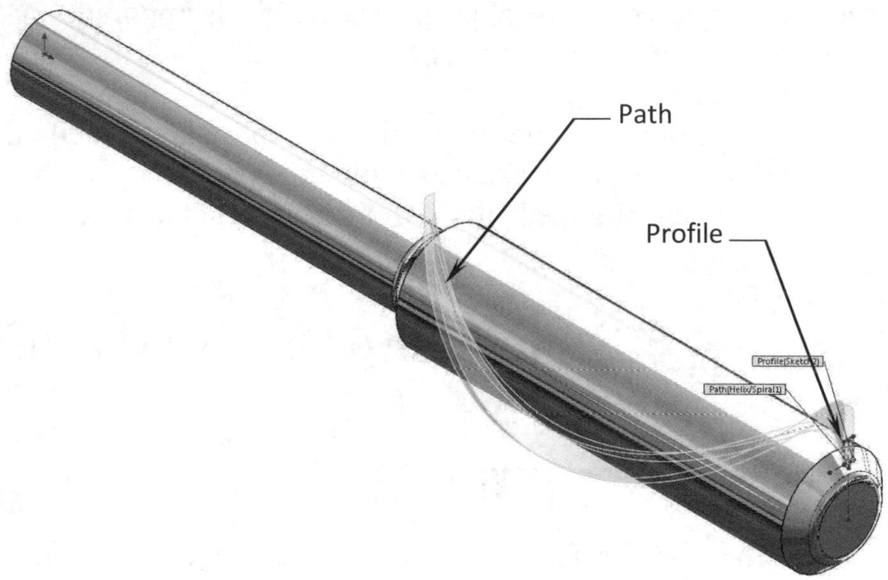

Click **OK**.

Zoom closer to inspect the model and to examine the swept cut feature.

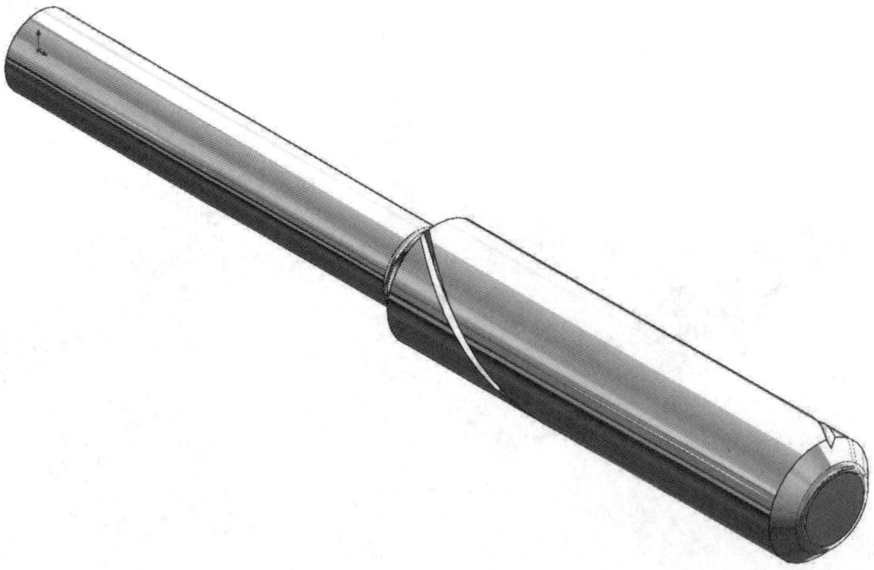

3. Creating a circular pattern:

Click the **Circular Pattern** command below the Linear Pattern drop-down.

For Pattern Direction select the circular edge on the left end of the model.

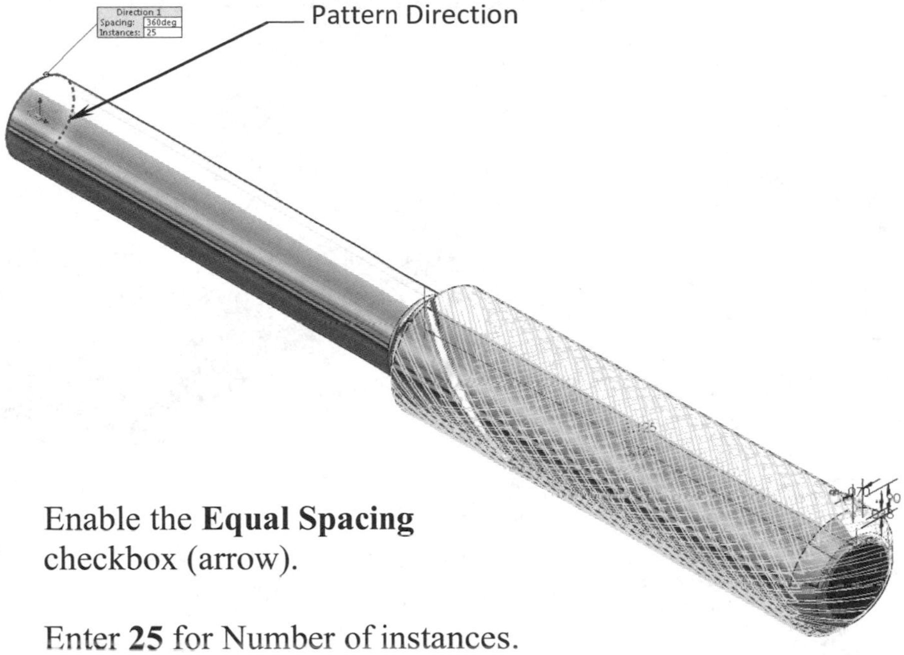

Enable the **Equal Spacing** checkbox (arrow).

Enter **25** for Number of instances.

Select the **Swept Cut** feature either from the Feature tree or Directly from the graphics area.

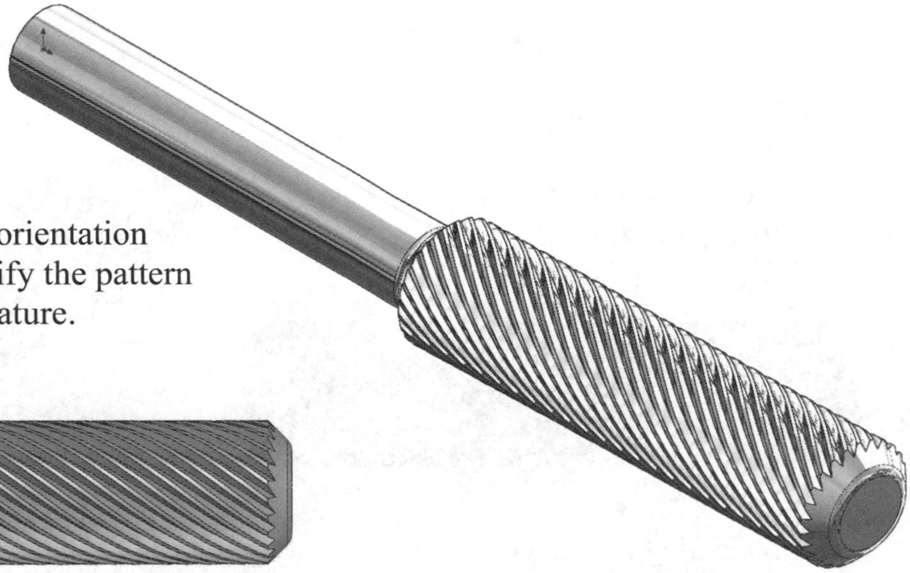

Click **OK**.

Change to the top orientation (Control+5) to verify the pattern of the swept cut feature.

4. Creating a mirror pattern:

Click the **Mirror** command on the Features tool tab.

Expand the FeatureManager tree and select the **Front** plane to use as Mirror Plane.

For Features to Mirror select both the **Cut-Swept1** and the **CirPattern1** features.

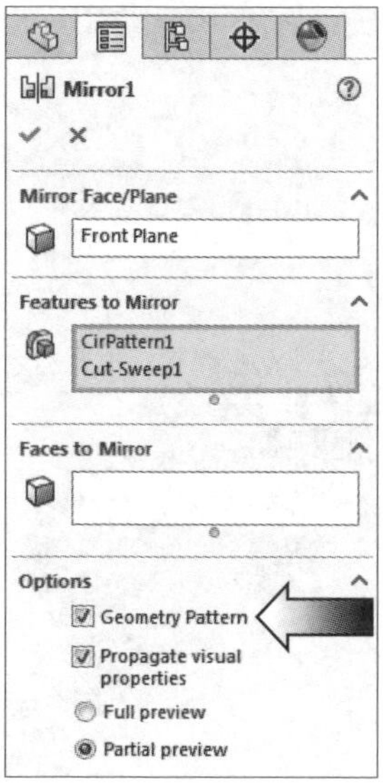

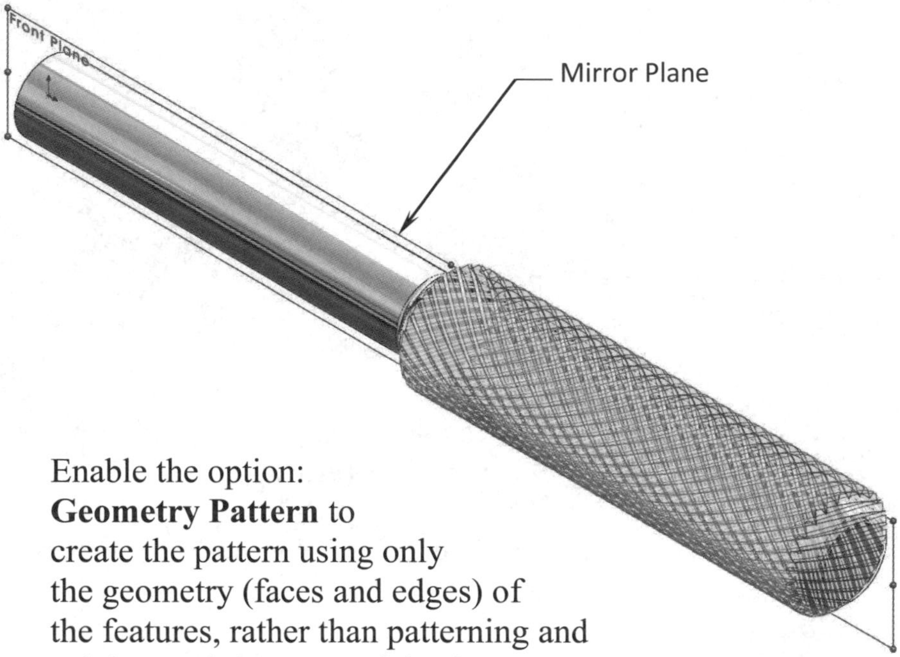

Enable the option:
Geometry Pattern to create the pattern using only the geometry (faces and edges) of the features, rather than patterning and solving each instance of the feature.

Click **OK**.

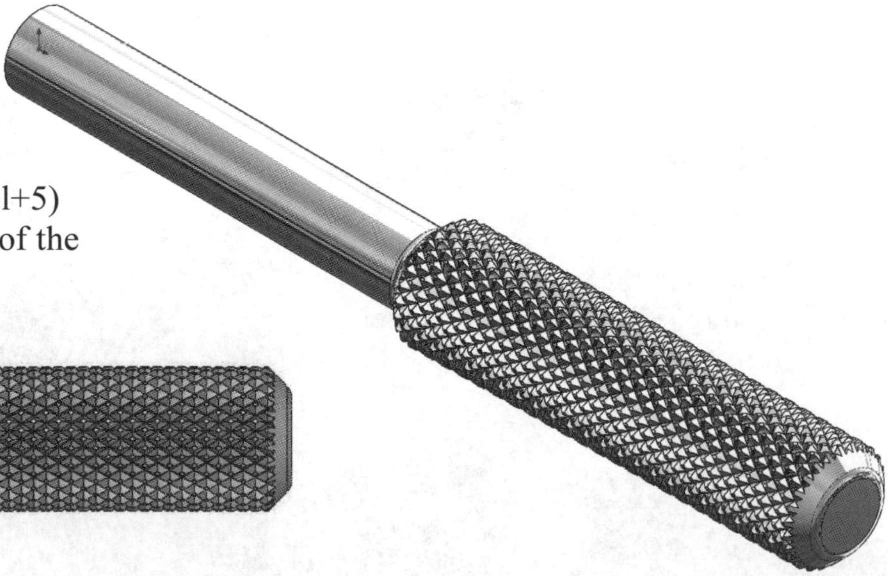

Change to the top orientation (Control+5) to verify the result of the mirror pattern.

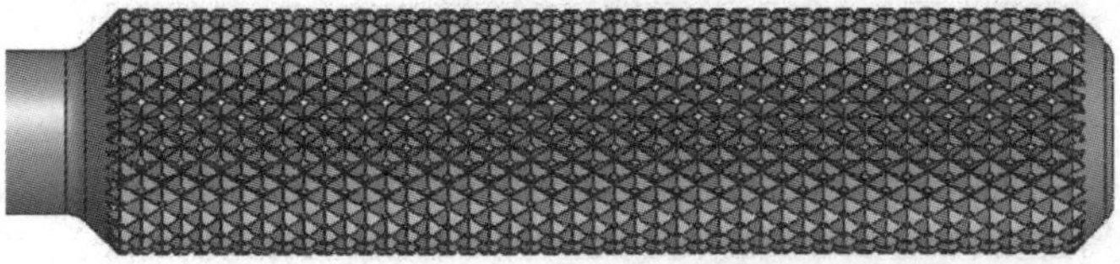

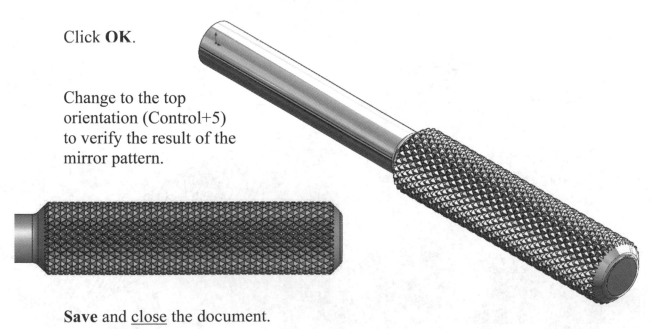

Save and close the document.

Applying the Knurl Appearance

Certain appearances require the use of RealView Graphics for more realistic Representation; knurled, dimpled, or sandblasted are some examples.

RealView gives models a realistic and dynamic representation without the need to render. If your graphics card is RealView-compatible, RealView is enabled by default.

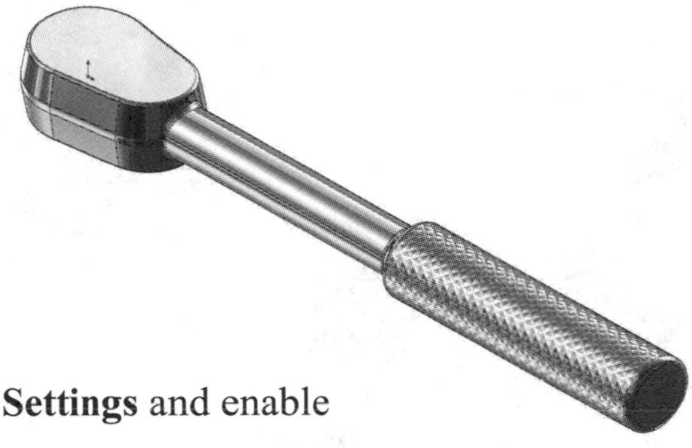

1. Opening a part document:

Open a part document named:
Knurl Appearance.

Click the drop-down arrow next to **View Settings** and enable the **RealView Graphics** option (arrow).

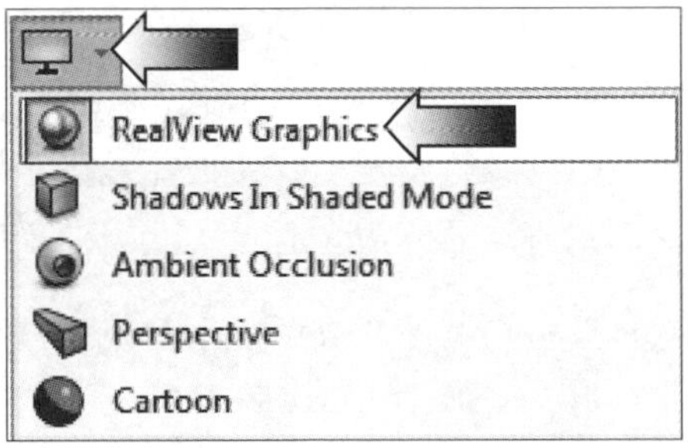

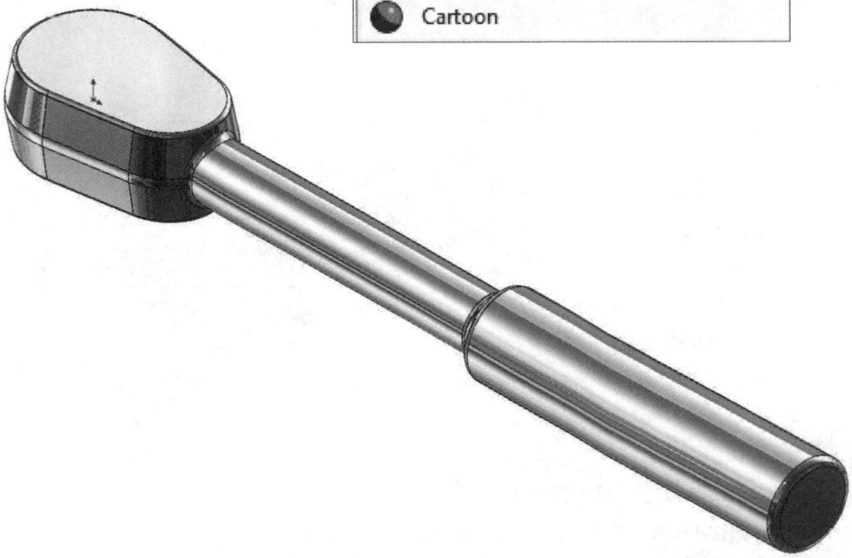

2. Applying the knurl appearance:

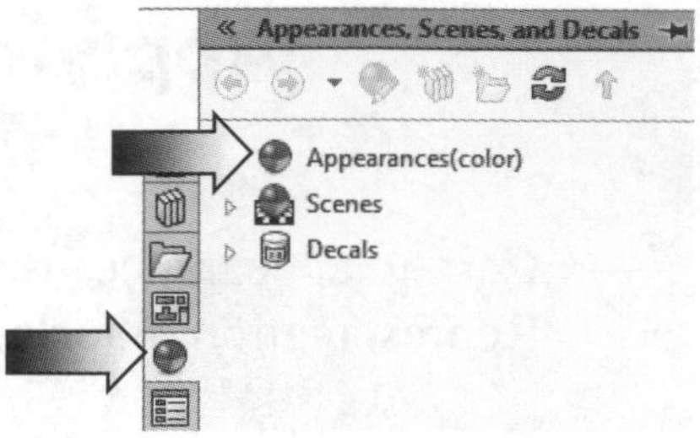

Expand the **Task pane** on the right side of the screen and pin it.

Expand the **Metal** and **Steel** folders. Drag and drop the **Stainless Steel Knurled** appearance onto the surface of the handle.

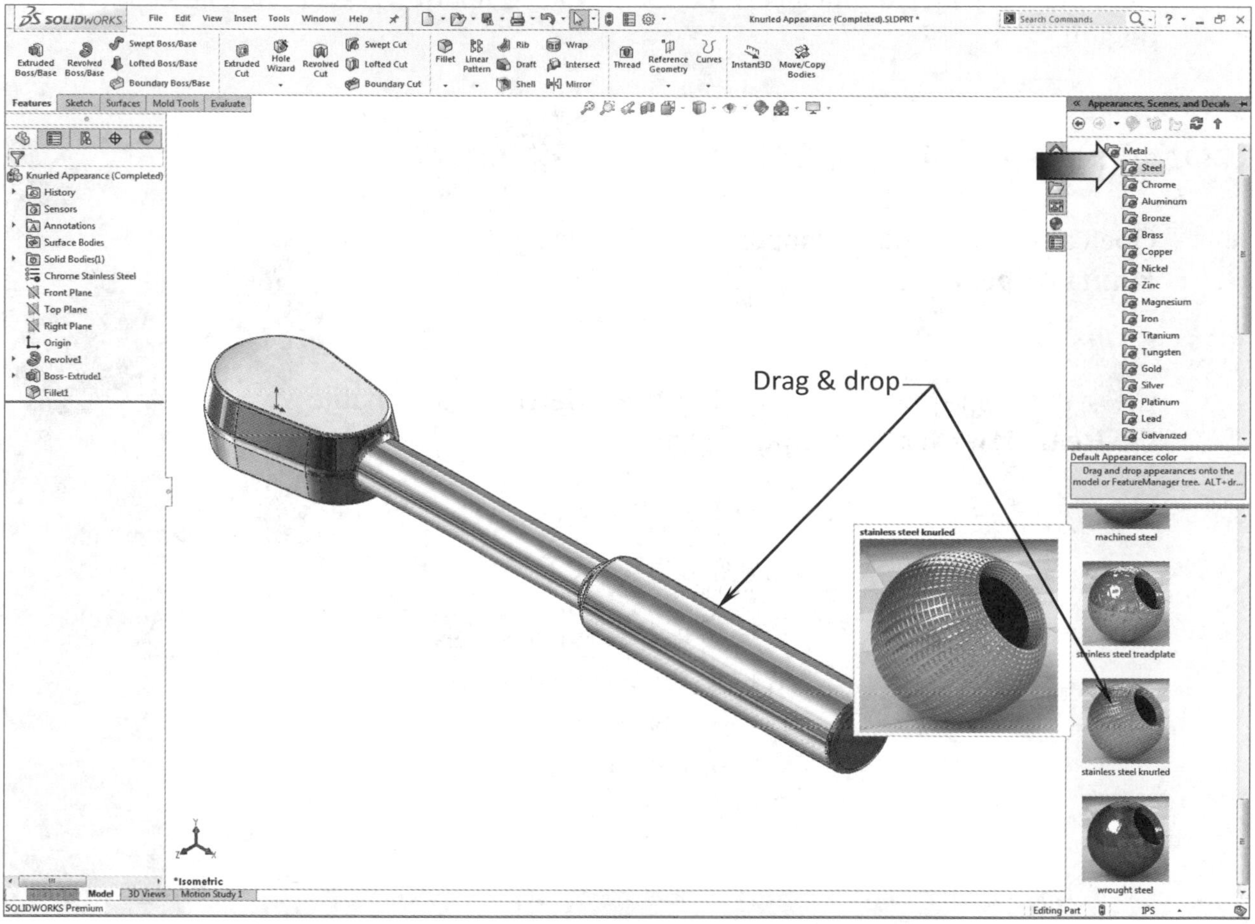

Select the **Apply to Face** option in the pop-up menu.

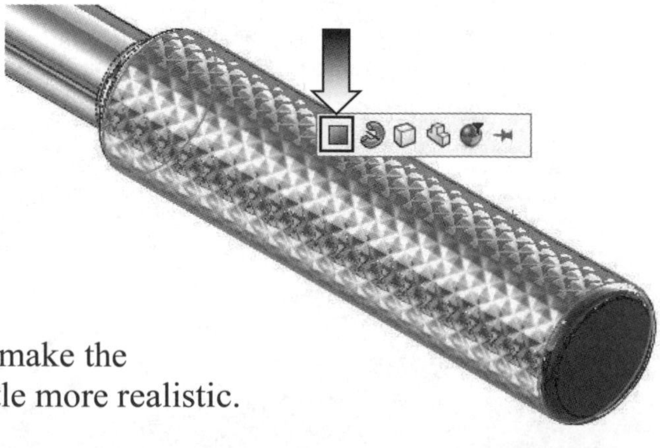

The default knurled appearance is applied to the selected face.

Next, we will modify the settings to make the diamond knurl appearance look a little more realistic.

3. Modifying the knurl appearance:

Right-click the surface of the handle where the knurled appearance was applied and select: **Appearance > Face** edit option (arrow).

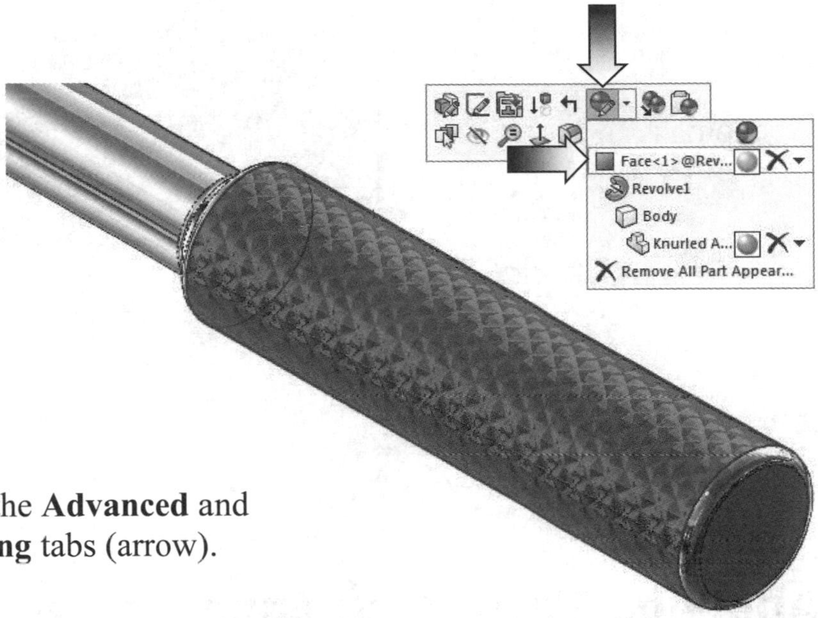

Select the **Advanced** and **Mapping** tabs (arrow).

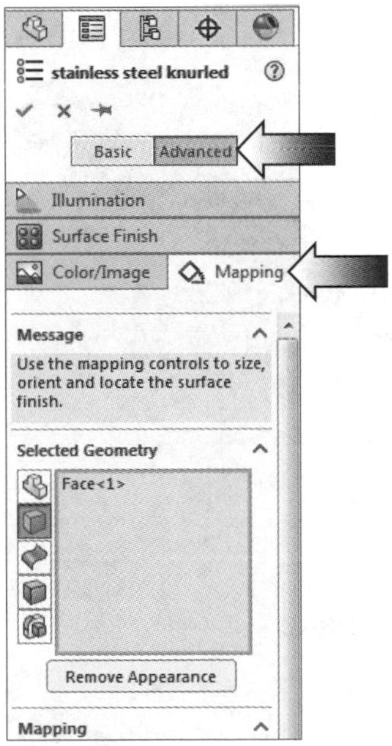

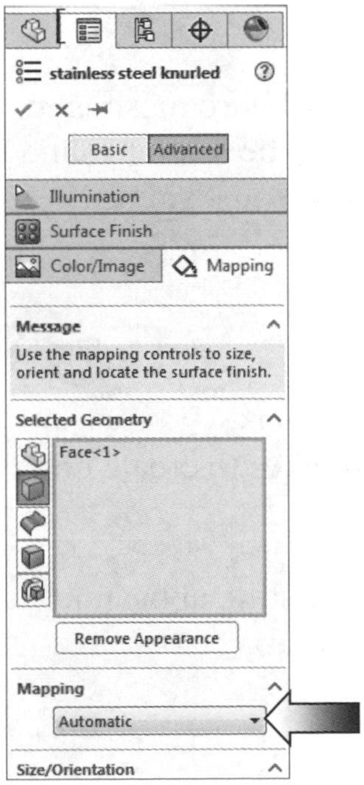

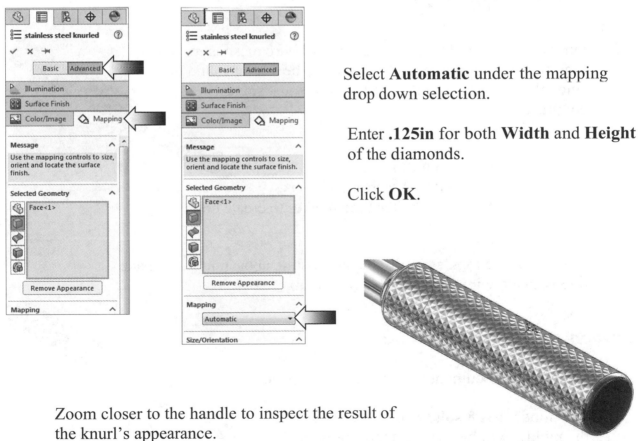

Select **Automatic** under the mapping drop down selection.

Enter **.125in** for both **Width** and **Height** of the diamonds.

Click **OK**.

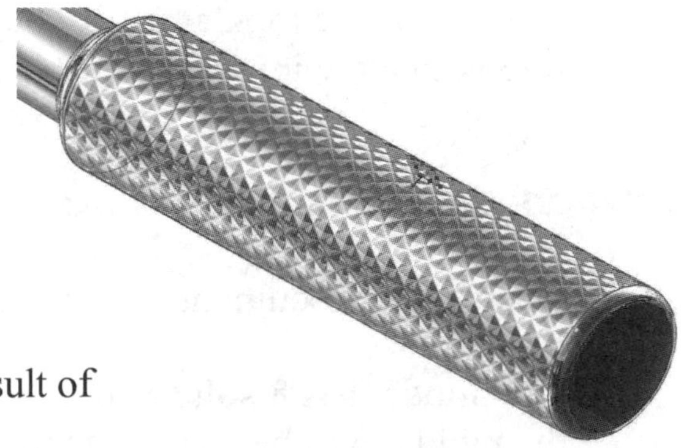

Zoom closer to the handle to inspect the result of the knurl's appearance.

Additionally, change the following:

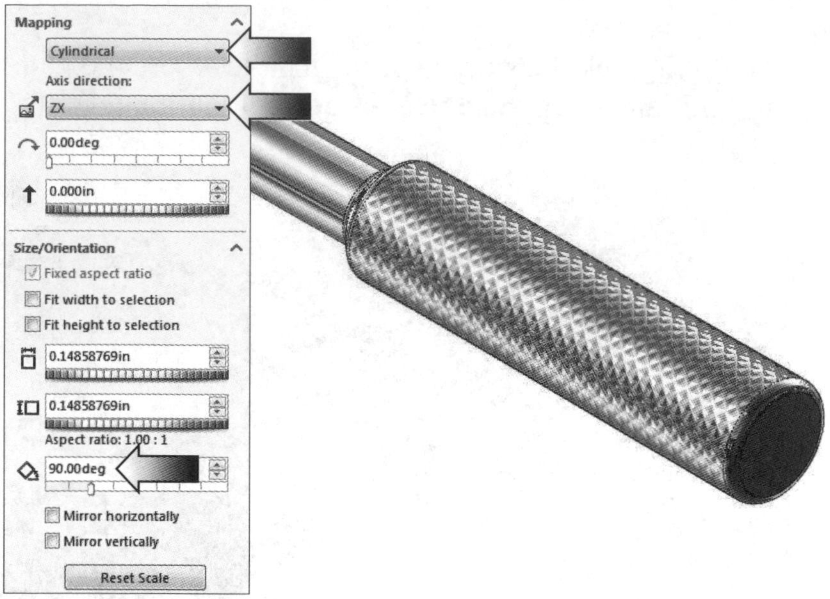

* Mapping Type: **Cylindrical**

* Axis Direction: **ZX**

* Rotation: **90deg**.

Click **OK**.

Save and close the document.

Applying Wire Mesh Appearance

Similar to diamond knurls, modeling 3D wire mesh is just too time consuming; the rebuild time would be too long, and the amount of memory needed for the task is just simply not practical.

Appearance once again can offer some acceptable results without sacrificing your computer performance, or the time it takes to create one.

We will take a look at modifying an existing appearance and make it look like a wire meshed cable (pictured)

1. Opening a part document:

Open a part document named: **Wire Mesh.sldprt**

This model has 8 solid bodies and the sleeve in the middle will be used to apply the wire mesh.

2. Applying the wire mesh:

Expand the **Task Pane** on the right side of the screen and click the pin to lock it.

Expand the following 2 folders: **Plastic > Composite** (arrows).

Drag and drop the **Carbon Fiber Epoxy** appearance to the middle sleeve as noted.

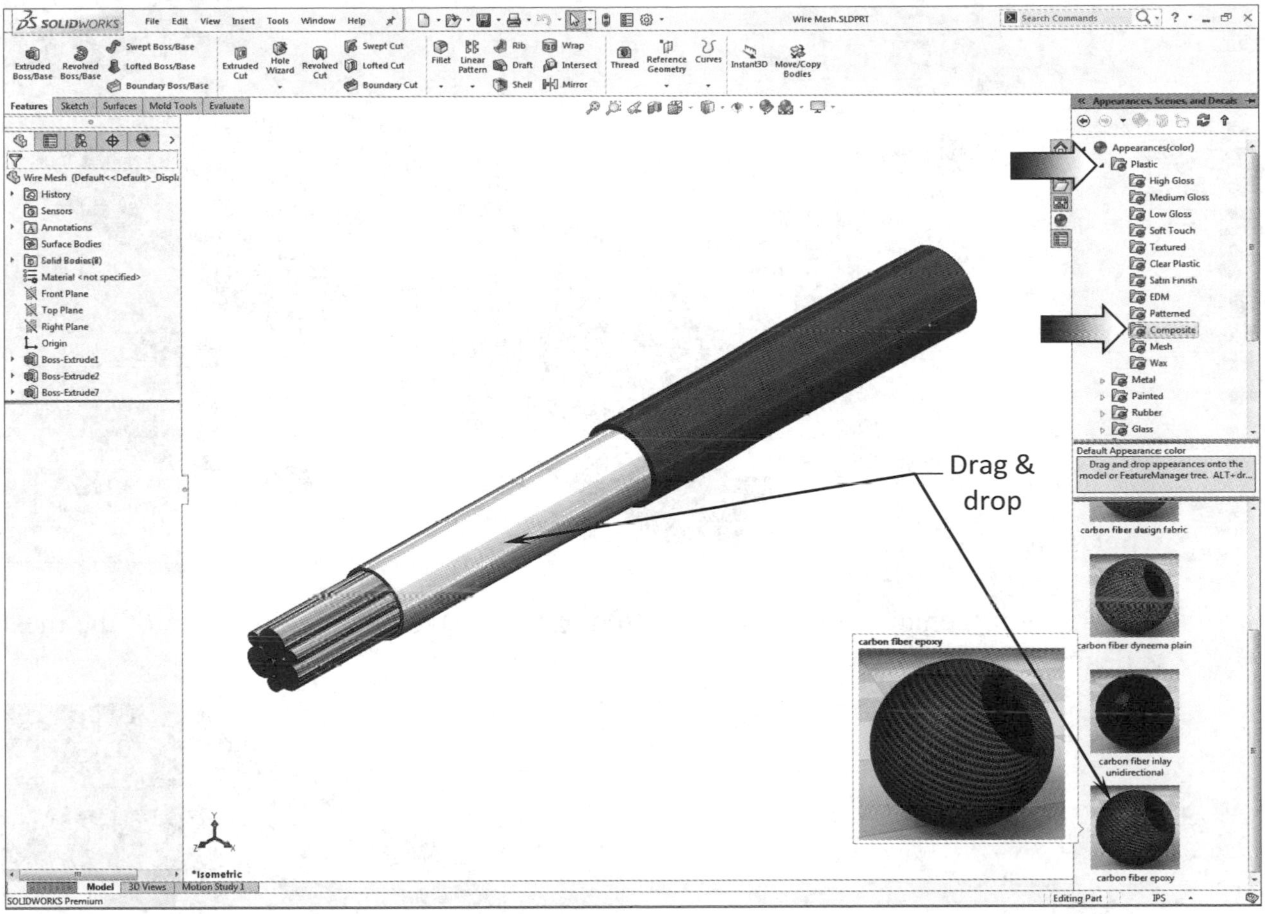

Select the **Advanced** and the **Mapping** tabs and set the parameters shown below.

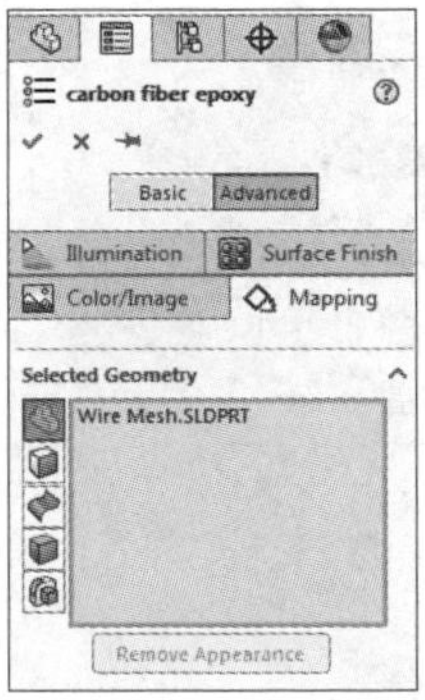

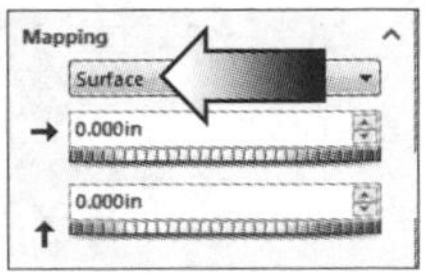

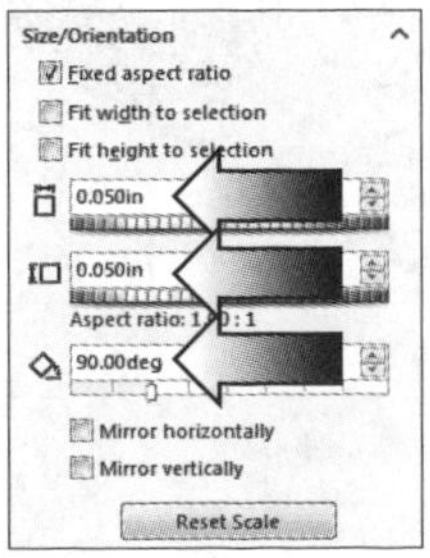

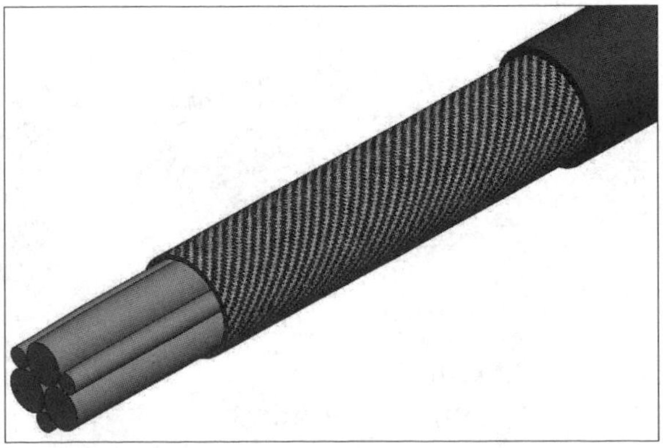

3. Modifying the wire mesh appearance:

Right-click the middle sleeve and select: **Appearance > Edit** (arrow).

Select the **Advanced** and **Mapping** tabs. Set the Mapping type to Surface, and set the **Width** and **Height** to **.075in**, and the **angle** to **25deg**.

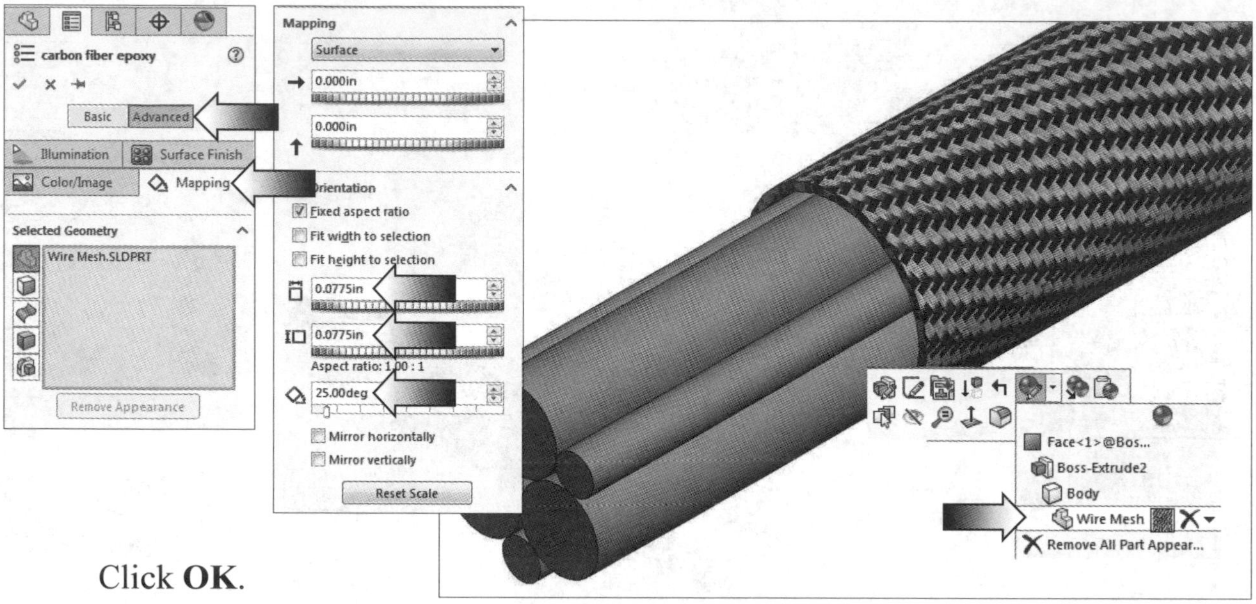

Click **OK**.

Alternatively, change the rotation to **90deg** (arrow) to change the angle of the mesh if desired.

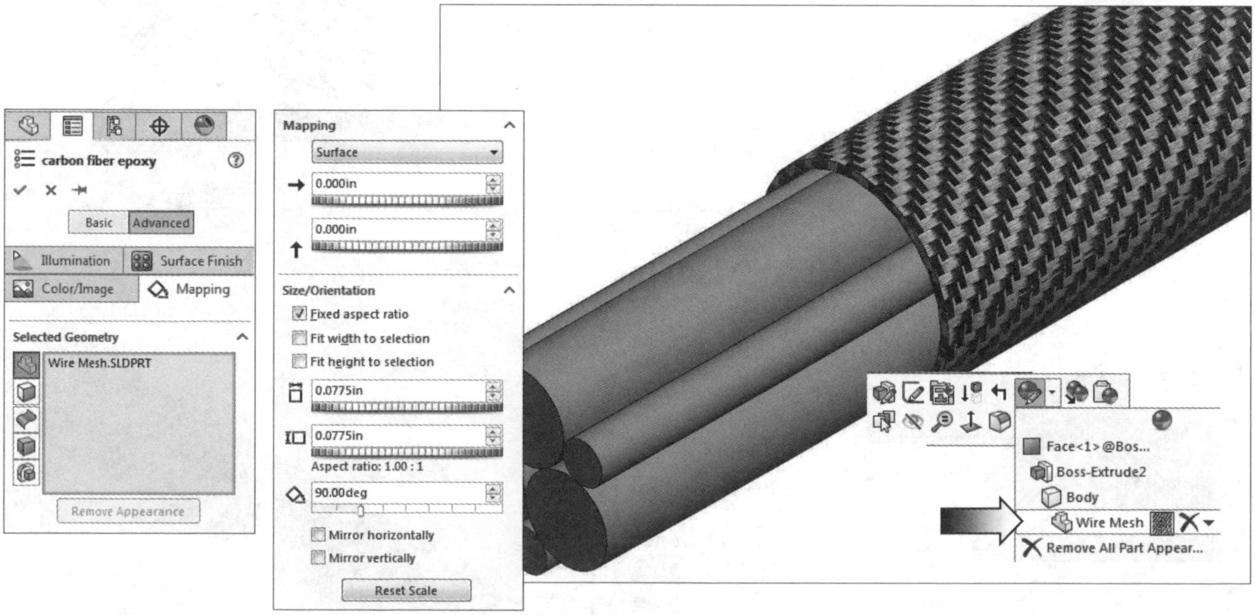

Save and
close the document.

Applying the Car-Paint Appearance

An appearance defines the visual properties of a model, including color and texture. Appearances do not affect physical properties, which are defined by materials.

View2

1. Opening an assembly document:

Open an assembly document named:
Concept Helicopter Assembly.sldasm

There are two **Named Views** saved under the Orientation dialog box.

Press the **Space-Bar** to see them.

View1 is shown.

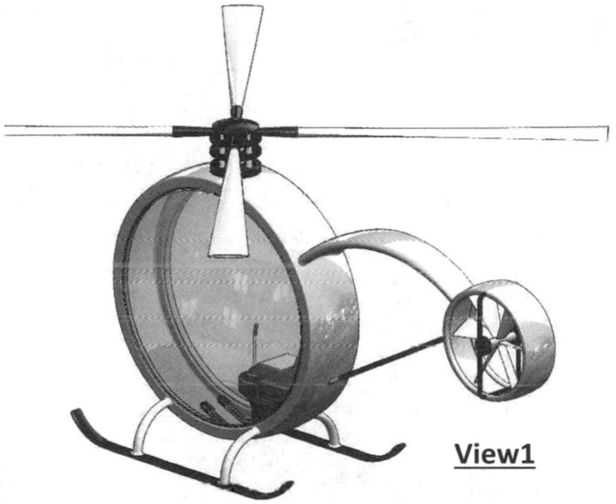

View1

2. Changing the appearance of a body:

Click the outer face of the Main Body and select:

Appearance, Main Body.

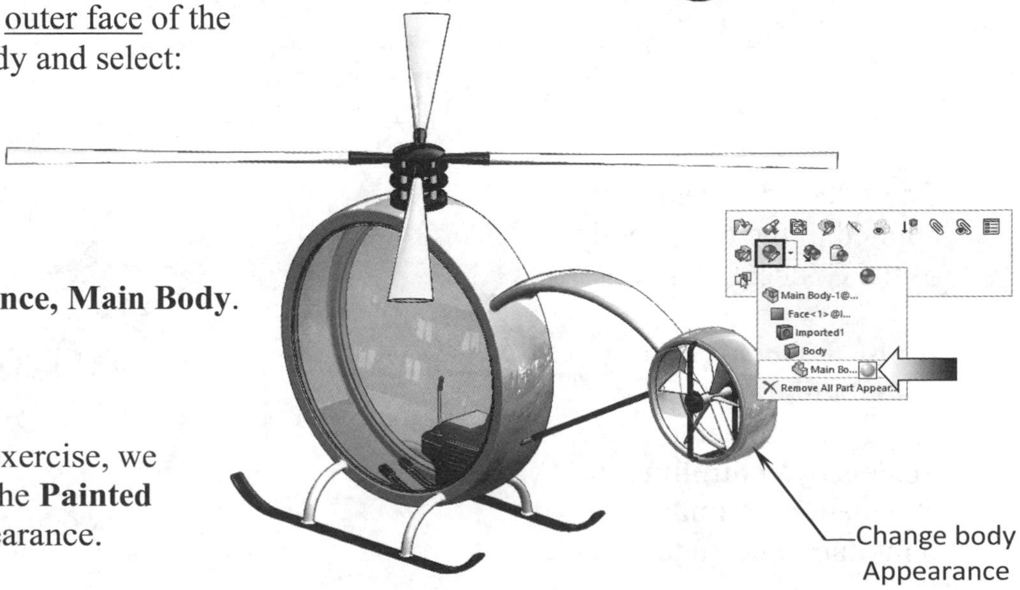

For this exercise, we will use the **Painted Car** appearance.

You can drag/drop an appearance from the Task Pane directly onto the component as shown in the previous chapter.

This exercise shows us an alternative method to see step by step how an appearance is applied to a model.

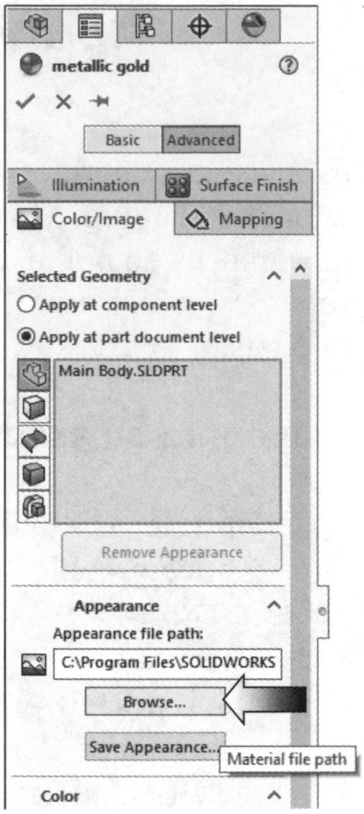

Click the **Browse** button and select the following directories:

C:\Program Files\SOLIDWORKS Corp\SOLIDWORKS (version)\Data\Graphics\Materials\Painted\Car.

Select the **Metallic Gold.p2m** appearance 

Click **Open**.

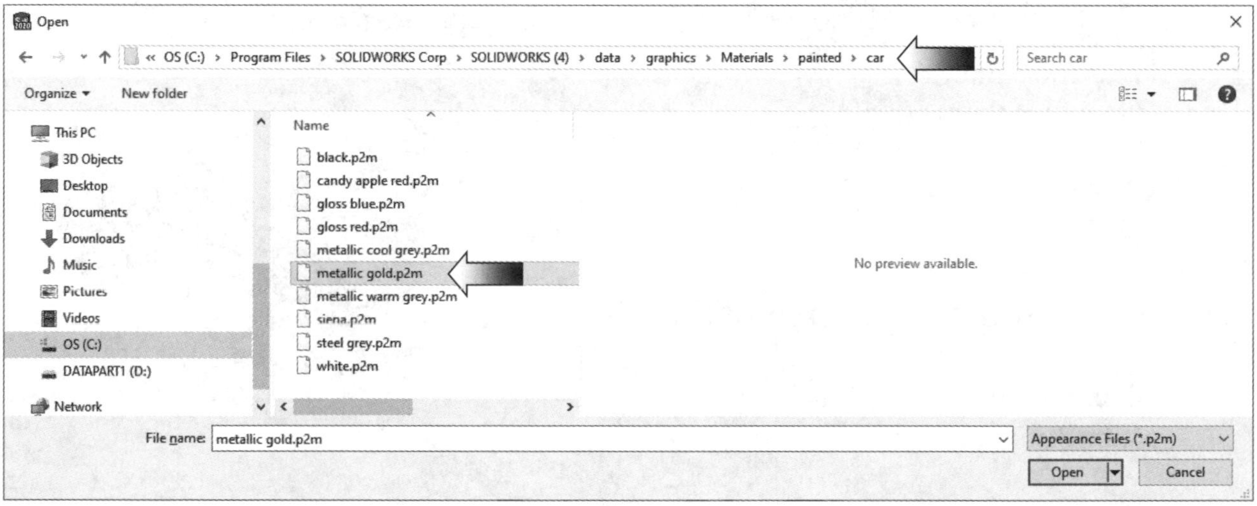

The selected appearance is applied to the entire body of the Concept Helicopter.

Ambient Occlusion is a global lighting method that adds realism to models by controlling the attenuation of ambient light due to occluded areas.

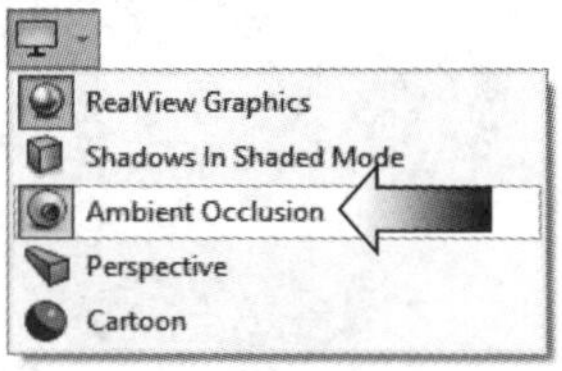

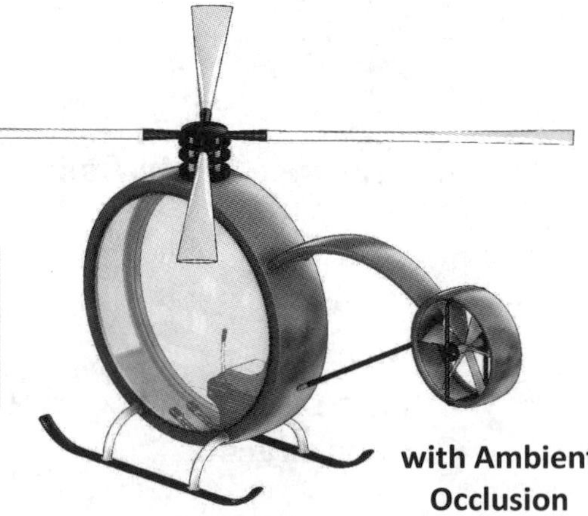

with Ambient Occlusion

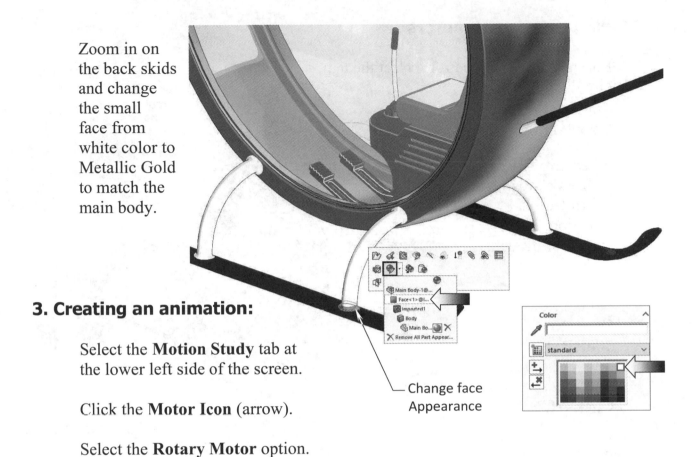

Zoom in on the back skids and change the small face from white color to Metallic Gold to match the main body.

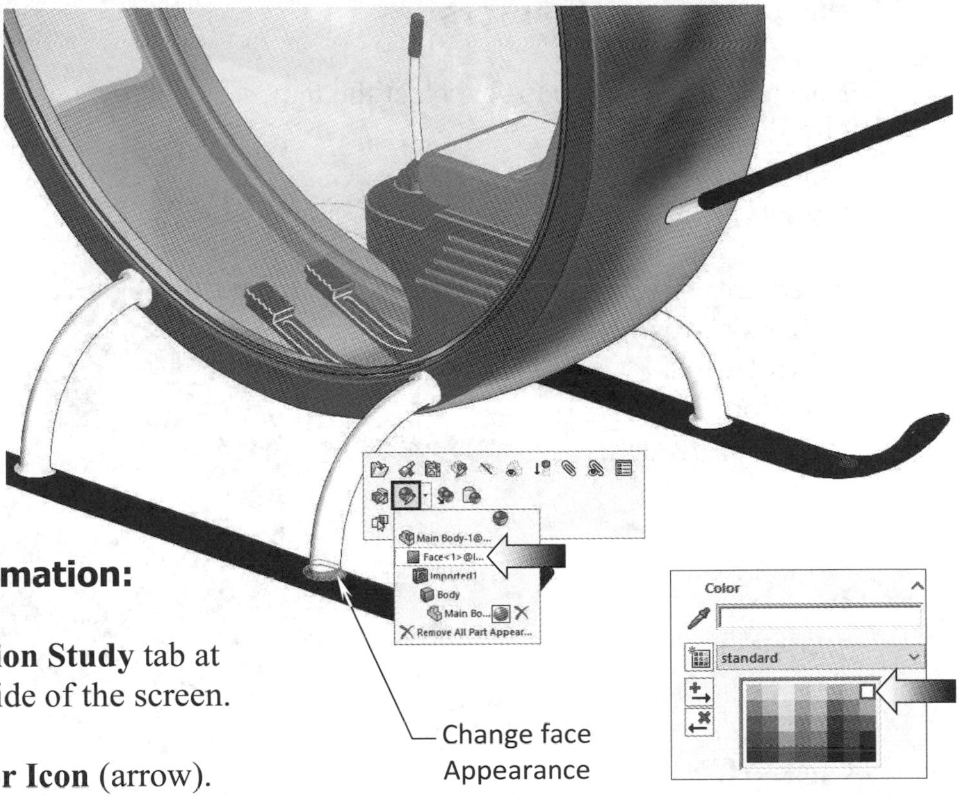

3. Creating an animation:

Select the **Motion Study** tab at the lower left side of the screen.

Click the **Motor Icon** (arrow).

Select the **Rotary Motor** option.

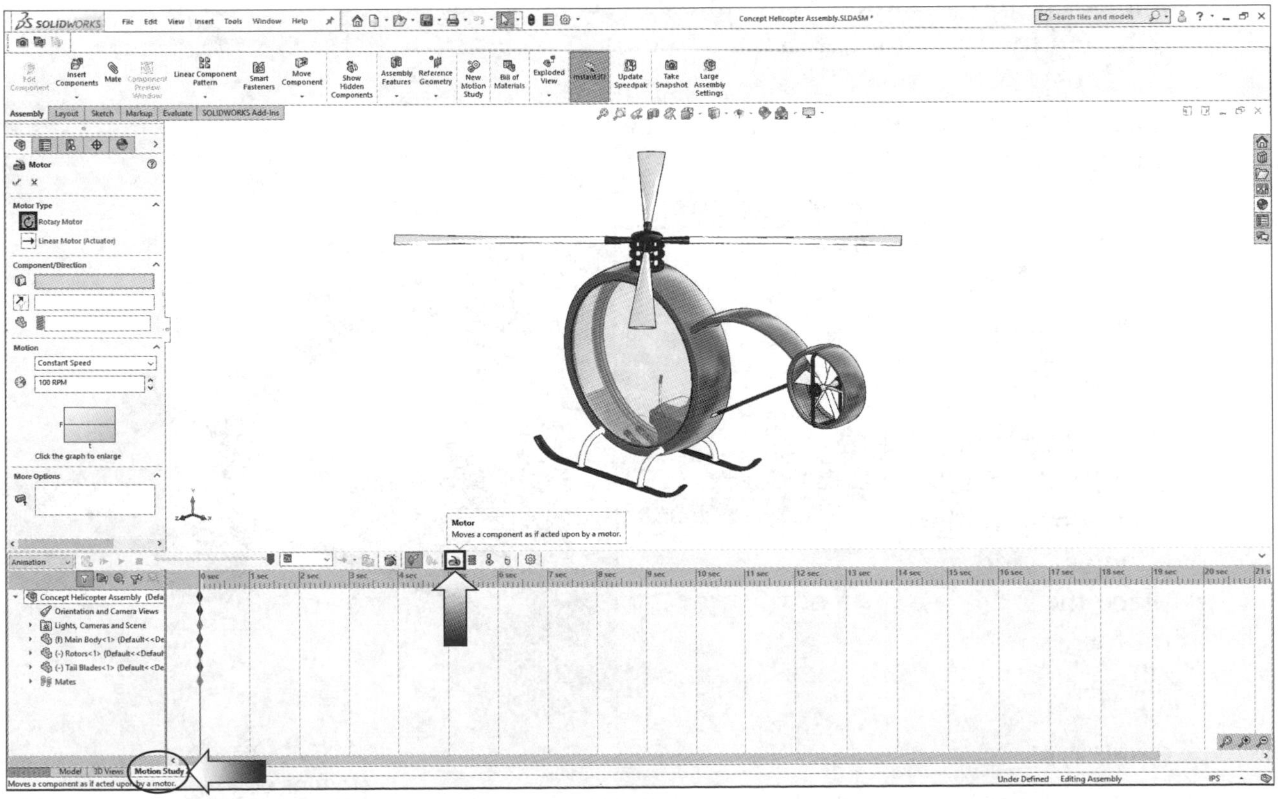

4. Placing the virtual motors:

For virtual motor location, select the upper <u>circular edge</u> as indicated.

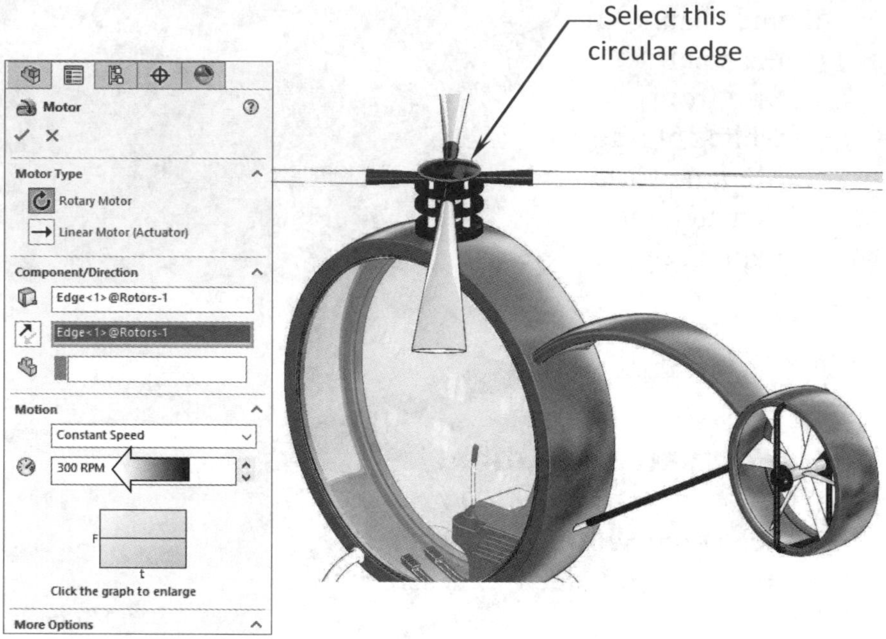

A red arrow appears showing the rotate direction; click reverse if needed.

Set the Constant Speed to: **300RPM**

Click **OK**.

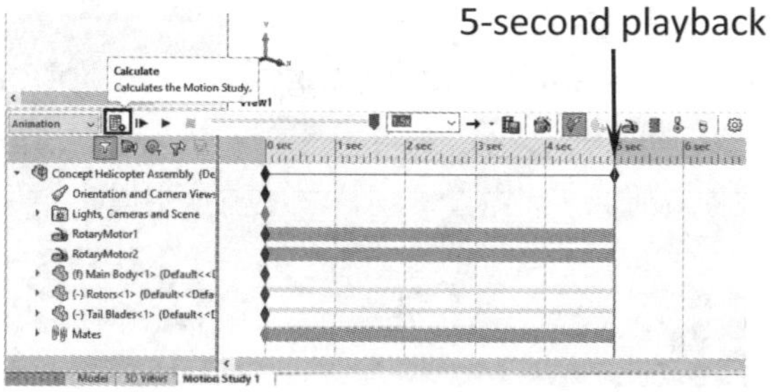

A 5-second playback time is automatically applied to the 1st motor. (The time can be changed by dragging the diamond key.)

Click the **Calculate** button to view the animation.

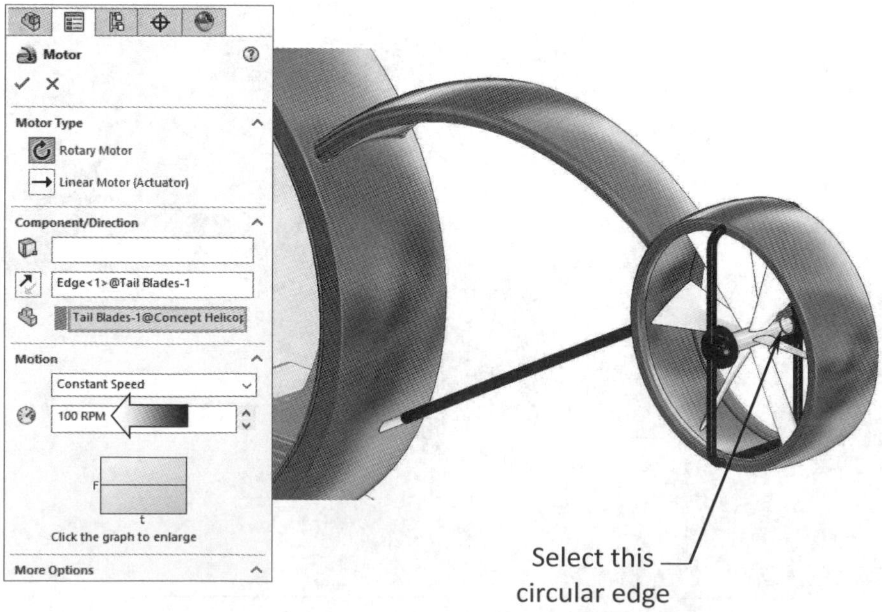

Click the **Motor** button again.

Zoom in on the Tail Blades and select the <u>circular edge</u> as noted to place the 2nd motor.

Click **OK**.

Click **Calculate** to view the rotary motions of the Rotors and the Tail Blades.

To play back continuously either select the **Loop** or **Reciprocate** options from the drop-down list.

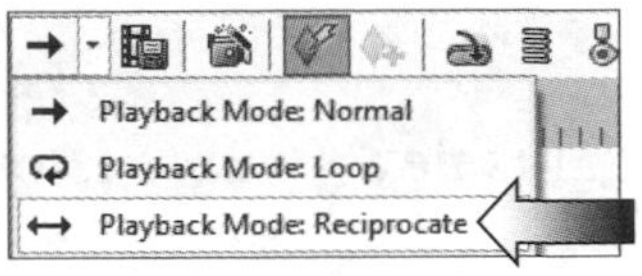

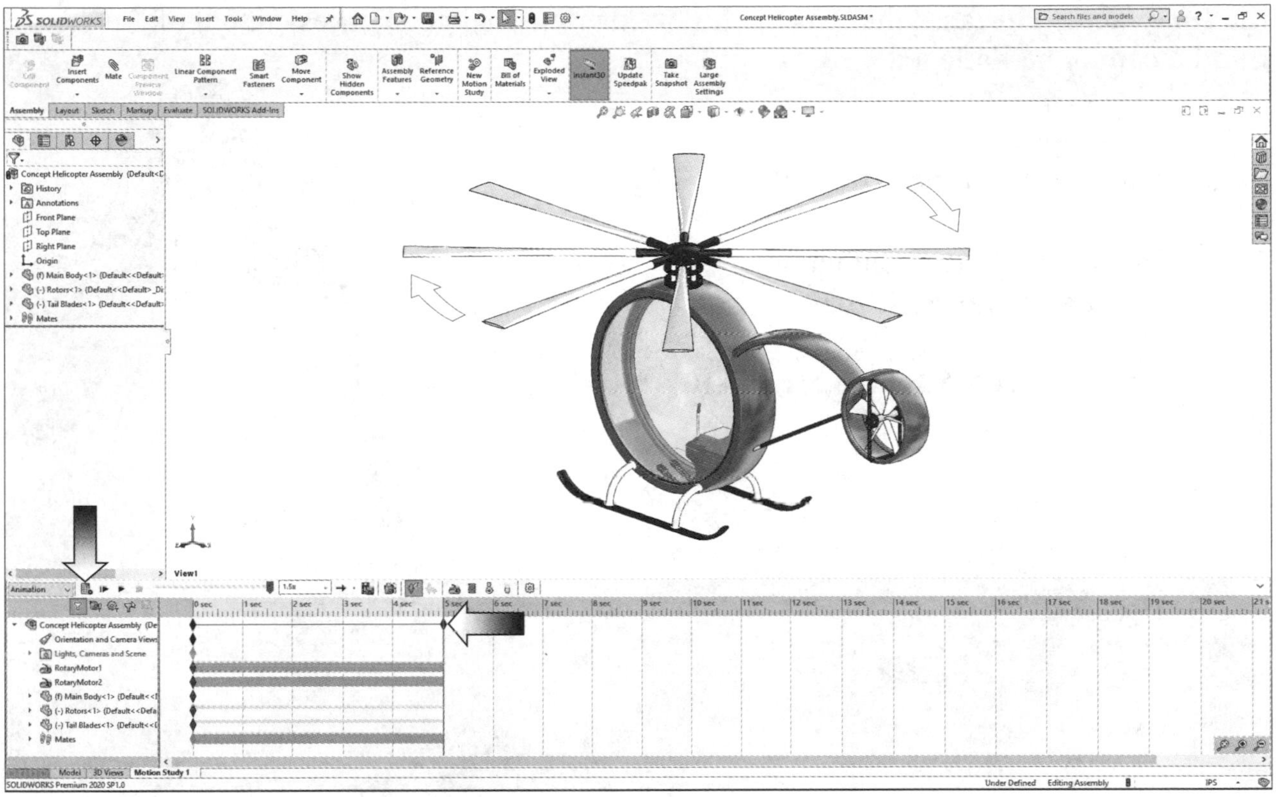

Change the play back speed (arrow) to speed up or to slow down the animation.

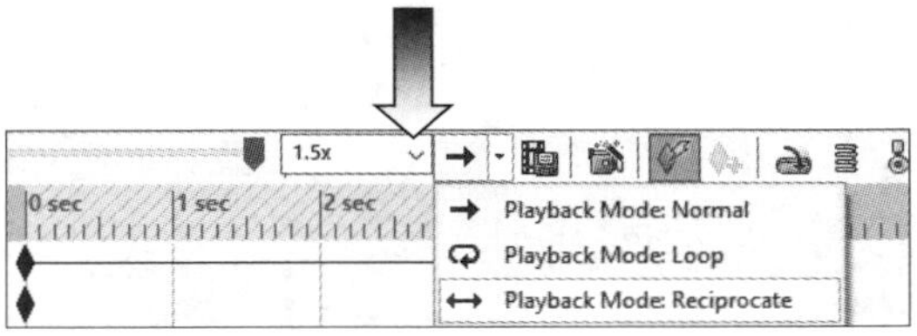

Save and close all documents.

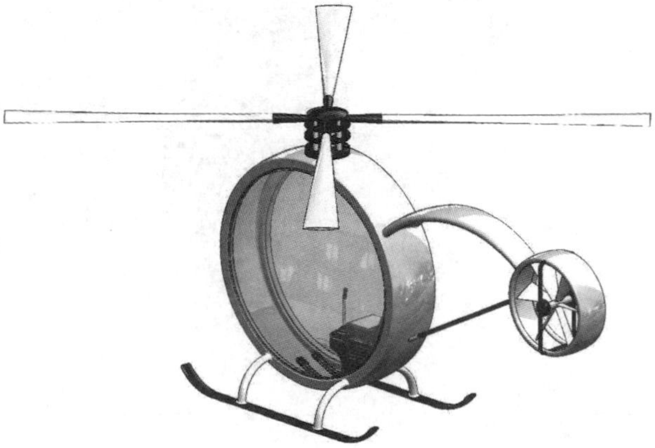

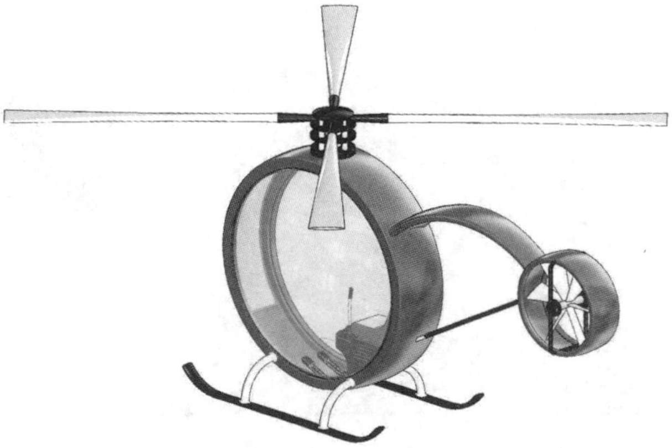

Flatten Surfaces

This new feature is only available in SOLIDWORKS Premium.

You can flatten surfaces and multi-faced surfaces (such as surfaces that are split into multiple faces). Surfaces that have holes or other internal geometries cut out of the middle cannot be flattened.

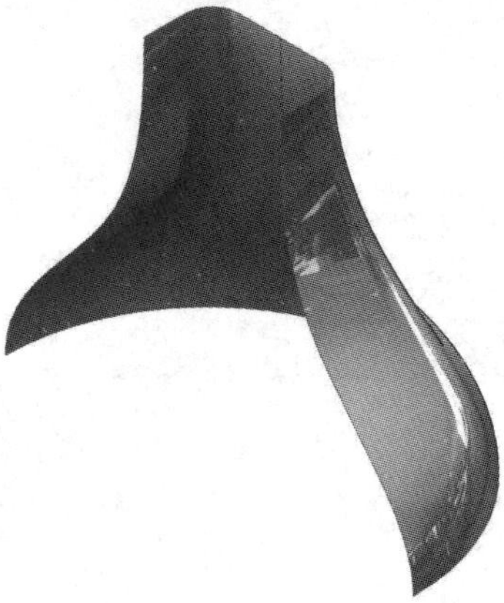

1. Opening a part document:

Open a part document named:
Flatten Surfaces.sldprt

Click **Insert > Surface > Flatten** .

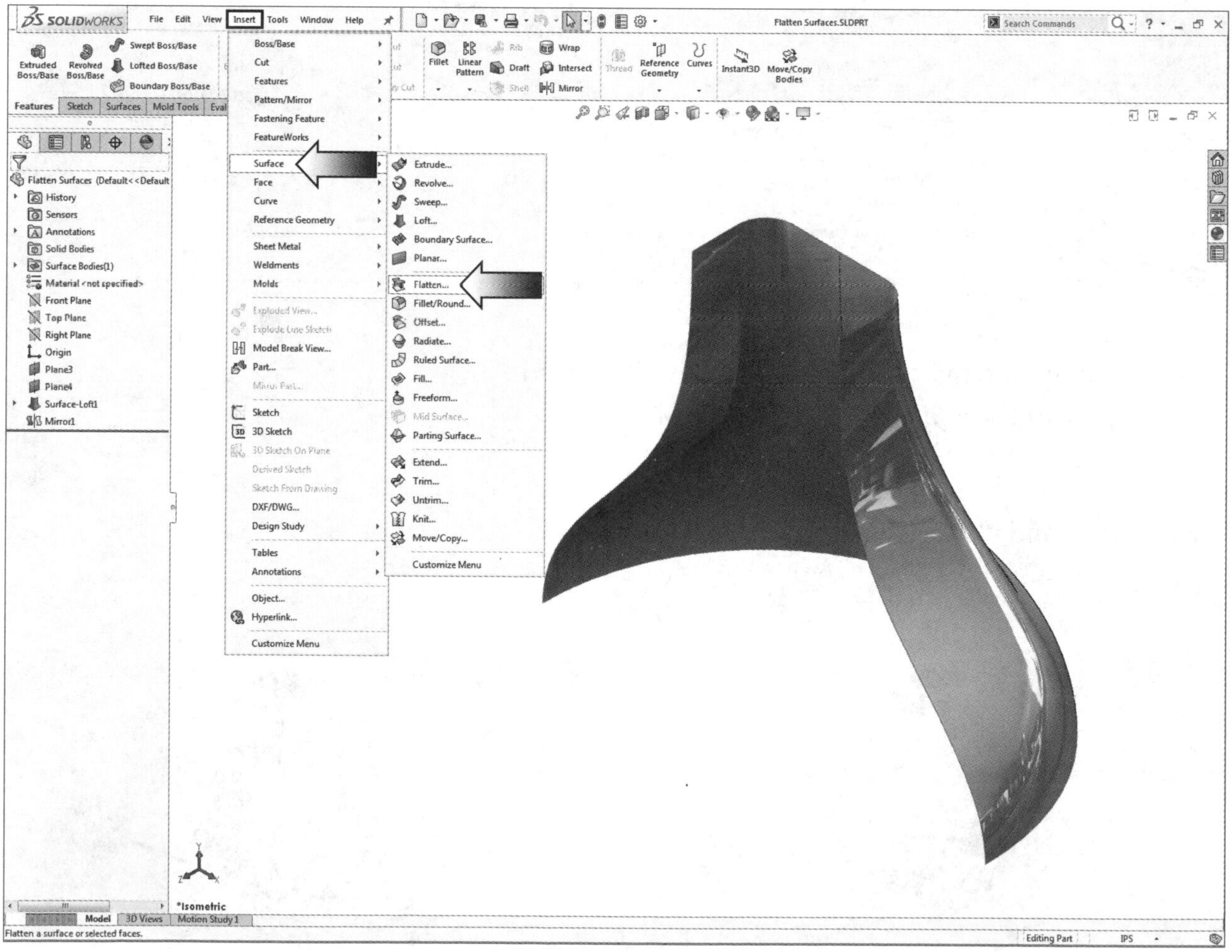

2. Flattening a surface:

Click in the Selections box; select the left and the right surfaces of the model.

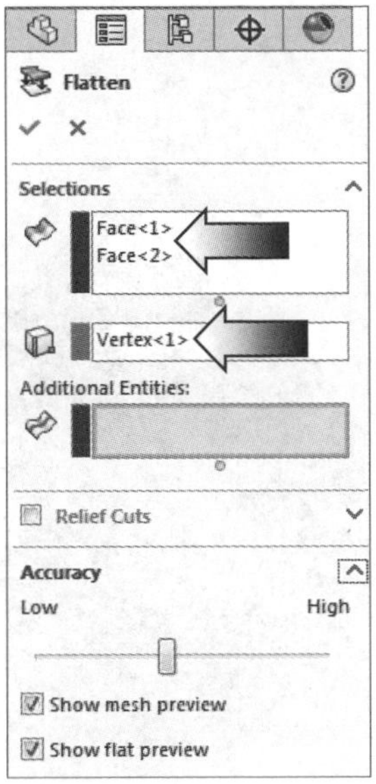

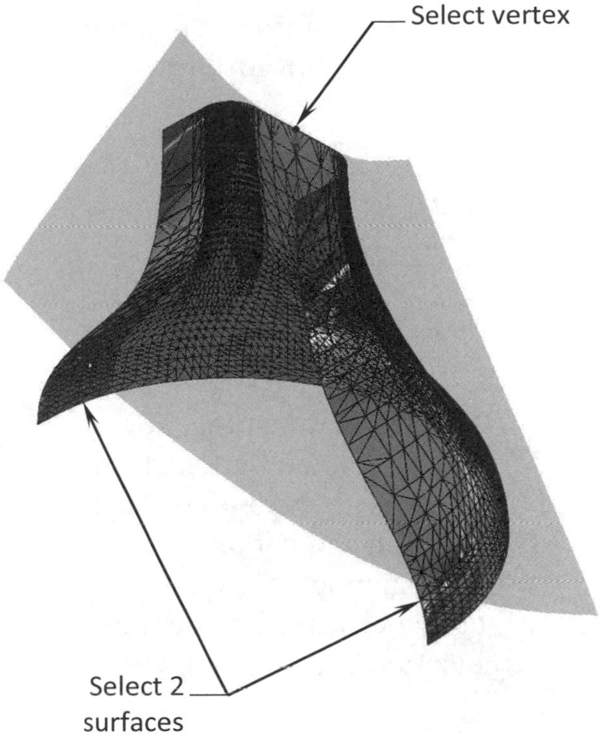

Click in the **Vertex to Flatten From** box and select the vertex as indicated.

Leave the Accuracy slider at its default location.

Click O**K**.

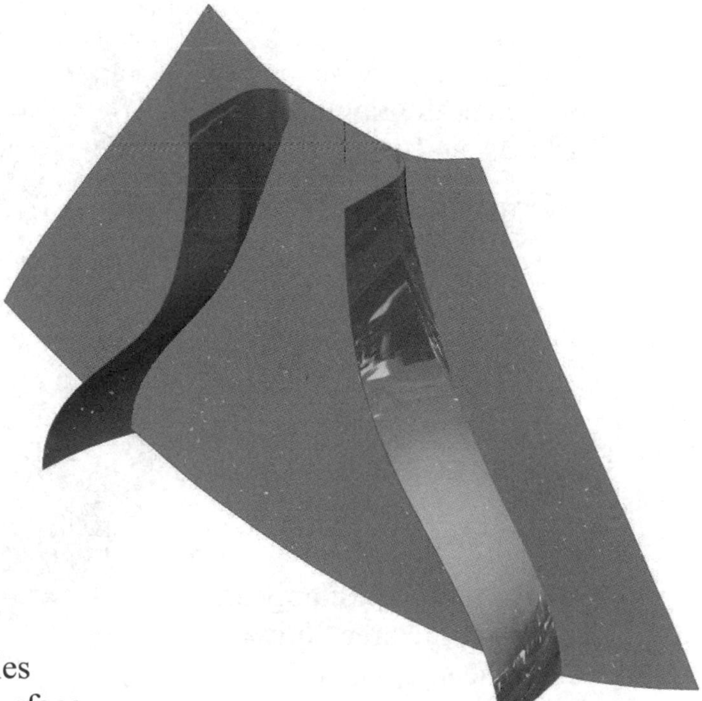

The new flatten surface is created over the original surface. These surfaces can be toggled to show or hide.

Rotate the model to different angles to verify the result of the flatten surface.

3. Viewing the deformation plot:

To view a deformation plot of the flattened surface, **right-click** the surface and click **Deformation Plot**.

The deformation plot shows the areas on the flattened surface with the highest levels of stretch and compression. You can mouse over the surface to see the percent of deviation at any given point.

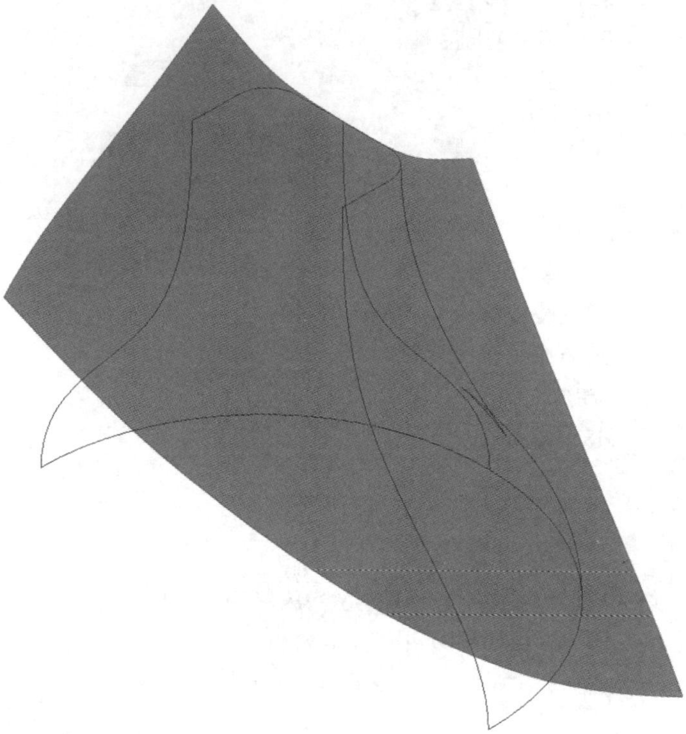

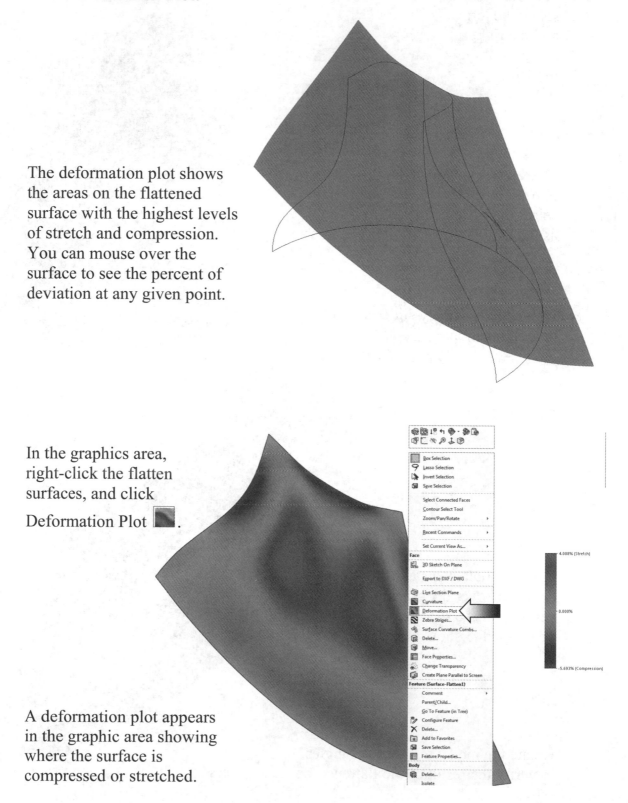

In the graphics area, right-click the flatten surfaces, and click Deformation Plot .

A deformation plot appears in the graphic area showing where the surface is compressed or stretched.

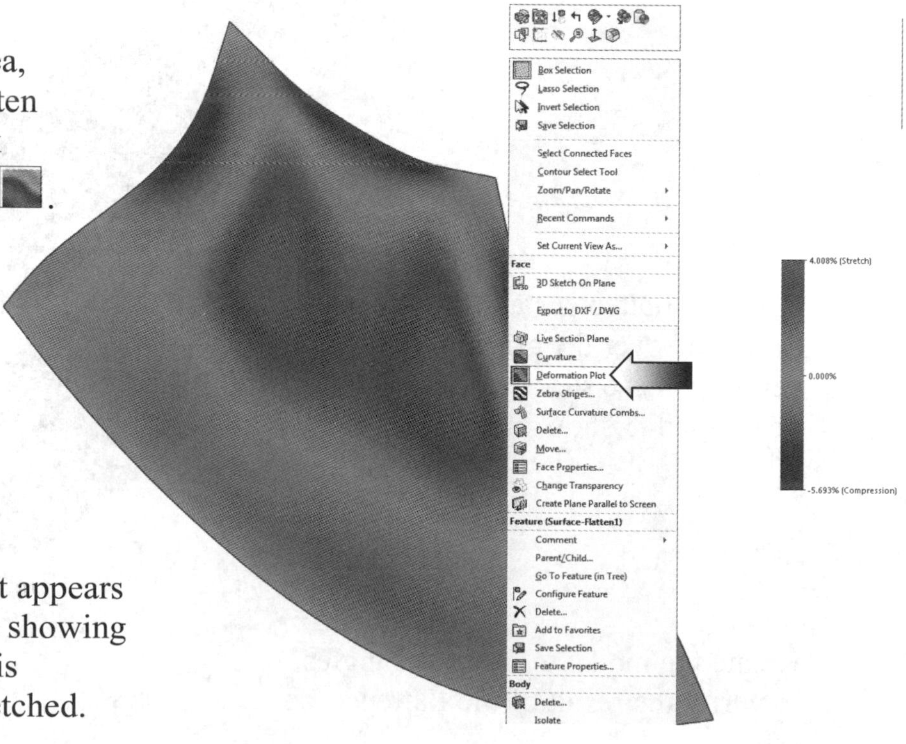

Change to the front orientation and mouse over one of the red areas to see the percentage of the deformation (stretch).

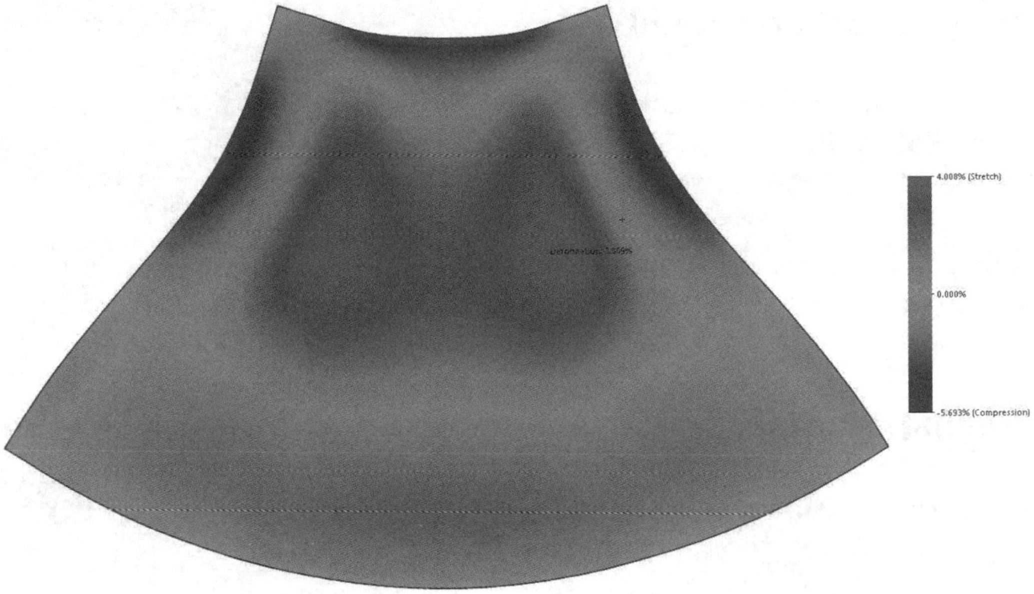

4. Moving the surfaces:

Click **Insert, Surface, Move/Copy**.

In the Translate section, select the **Flatten Surface** to move.

Enter **10.00in** in the Delta X box.

Click **OK**.

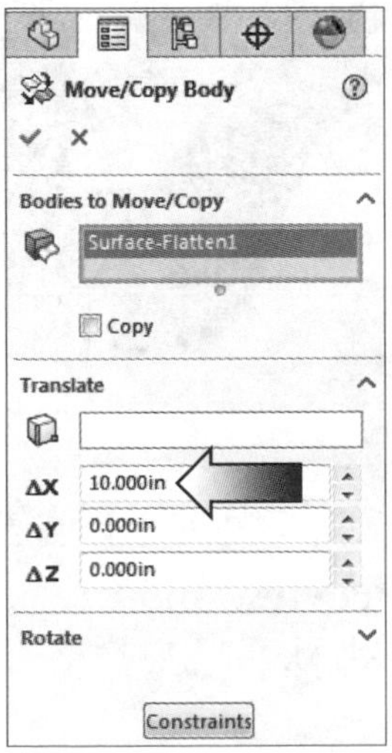

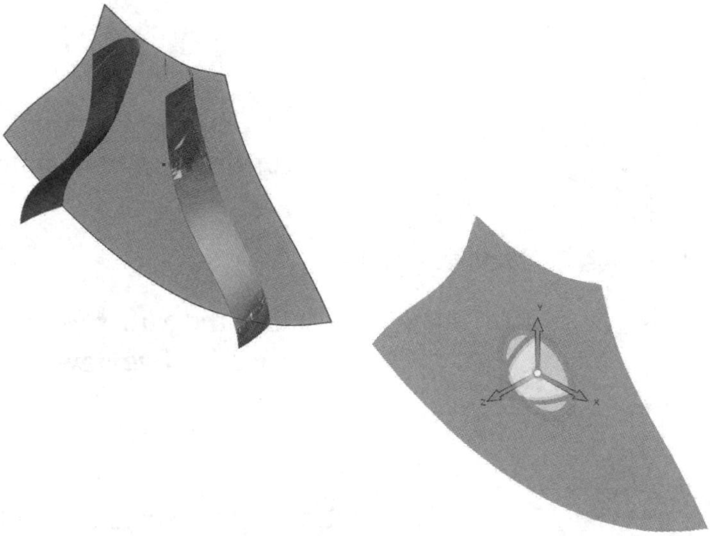

Save and close the document.

Exercise: Flattening a Shoe Sole

1. Opening a part document:

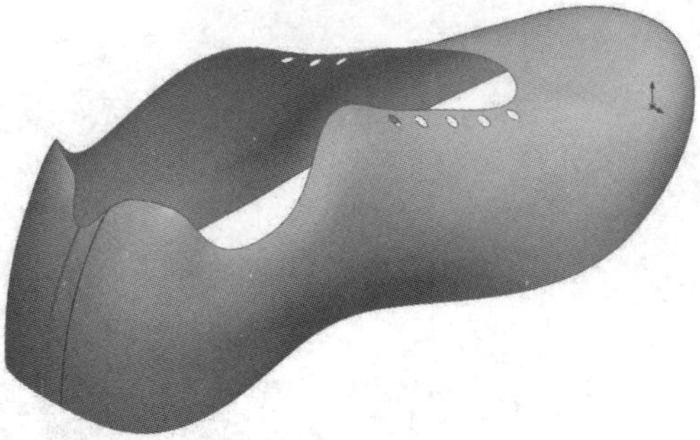

Open a Parasolid document named:
Shoe Sole.x_b

Click **NO** to close the Import Diagnostic dialog box.

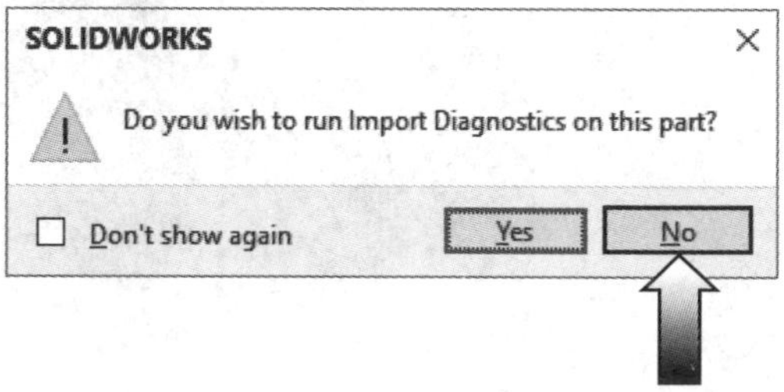

2. Flattening a surface:

Click **Insert, Surface, Flatten** .

For Surface to Flatten, select the **3 surfaces** of the entire sole.

Press **F5** to display the Selection Filters at the bottom left of the screen. Select the **Filter Vertices** button (arrow).

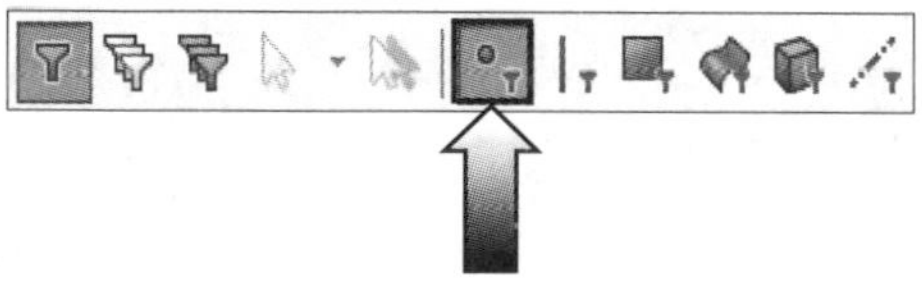

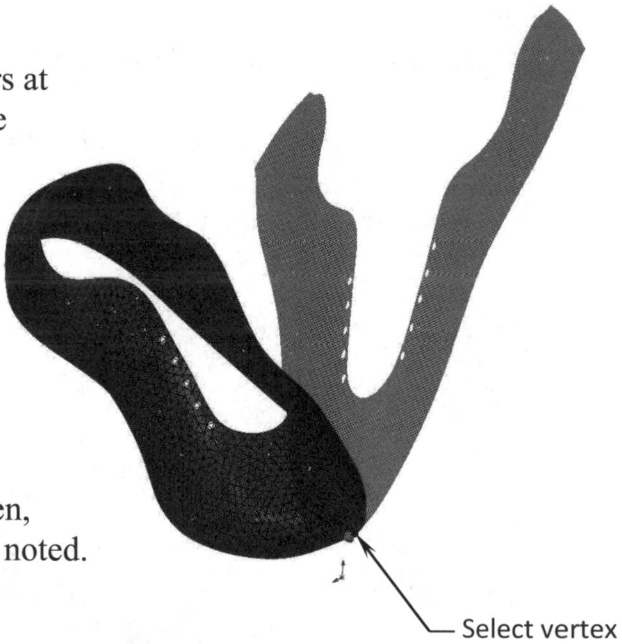

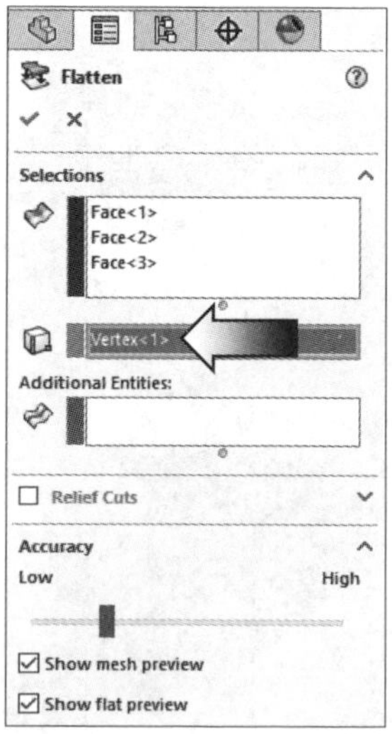

For **Vertex** to Flatten, select the Vertex as noted.

Enable the **Show Mesh Preview** and **Show Flat Preview** checkboxes.

Leave the Accuracy slider at its default value and click **OK**.

Click-off the Filter Vertices button and press F5 again to hide the Selection Filters toolbar.

Click the **Surface-Imported1** on the FeatureManager tree and select **Hide** (arrow).

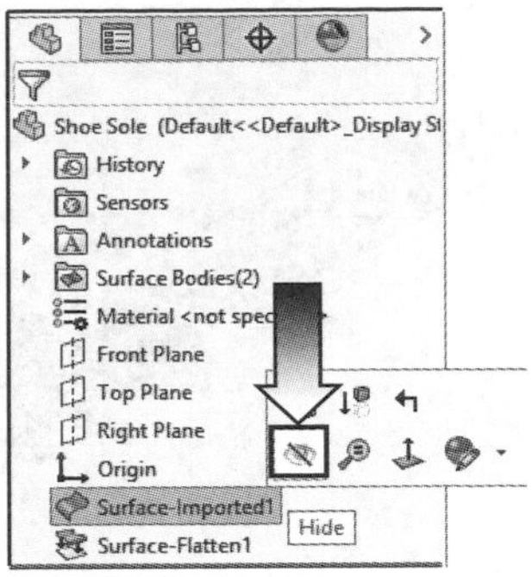

3. Viewing the deformation plot:

Right-click the surface and select **Deformation Plot**.

A deformation plot appears in the graphic area showing where the surface is compressed or stretched.

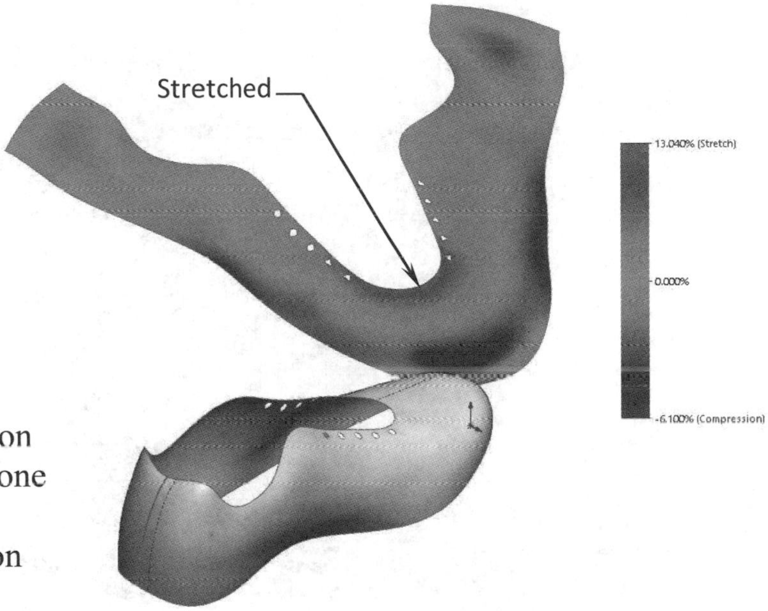

Change to the front orientation (Control+1) and hover over one of the red areas to see the percentage of the deformation (stretch).

4. Saving your work:

Click **File > Save As**.

Enter **Shoe Sole Completed** for the file name.

Press **Save**.

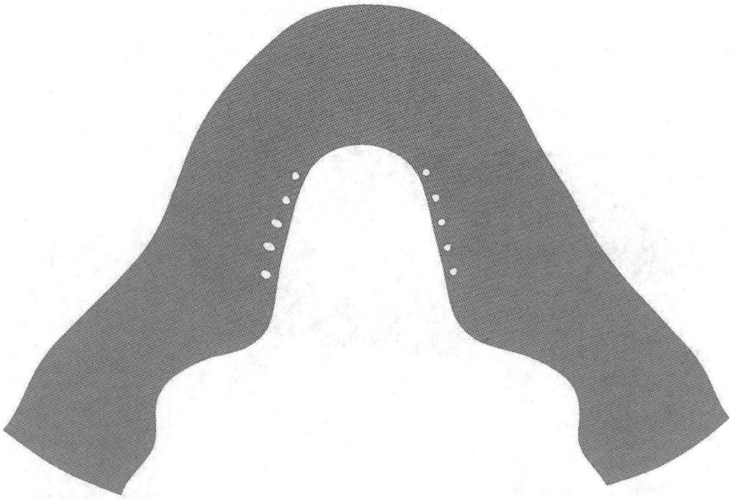

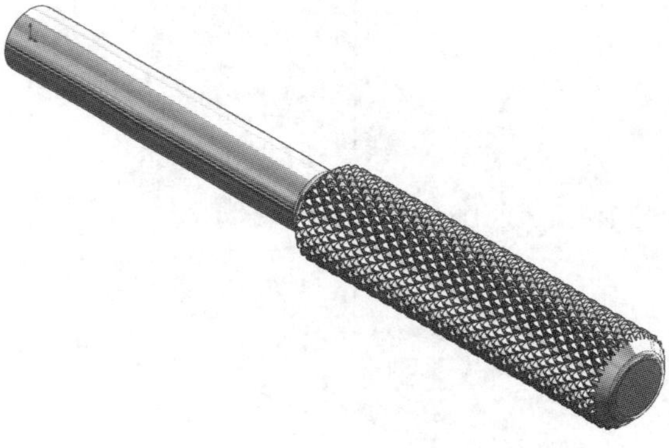

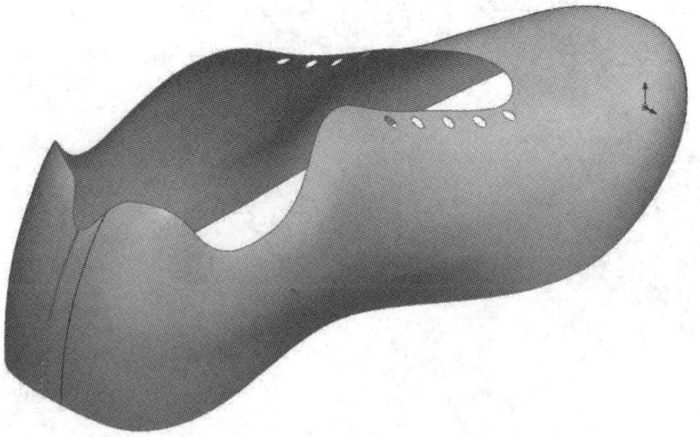

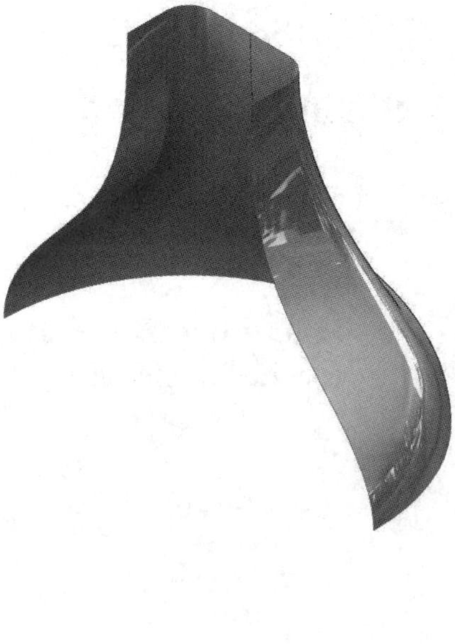

SOLIDWORKS 2022

Certified SOLIDWORKS Professional (CSWP)
Certification Practice for the Mechanical Design Exam

Courtesy of Paul Tran, Sr. Certified SOLIDWORKS Instructor

Certified-SOLIDWORKS-Professional (CSWP)

Certification Practice for the Mechanical Design Exam

Challenge I: Part Modeling & Modifications

Complete this challenge within 70 minutes (1 part)

(The following examples are intended to assist you in familiarizing yourself with the structures of the exams and the method in which the questions are asked.)

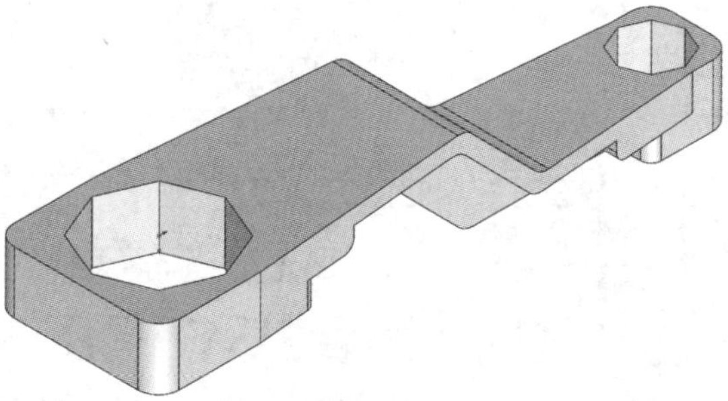

Create this part in SOLIDWORKS 2010 or newer.
Unit: **Inches, 3 decimals**
Drafting Standards: **ANSI**
Origin: **Arbitrary**
Material: **Cast Alloy Steel**
Density: **0.264 lb/in^3**

Save the model after each question as a different file in case it must be reviewed.

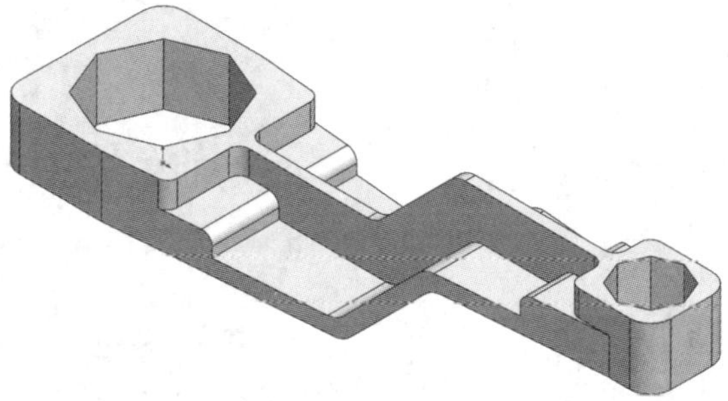

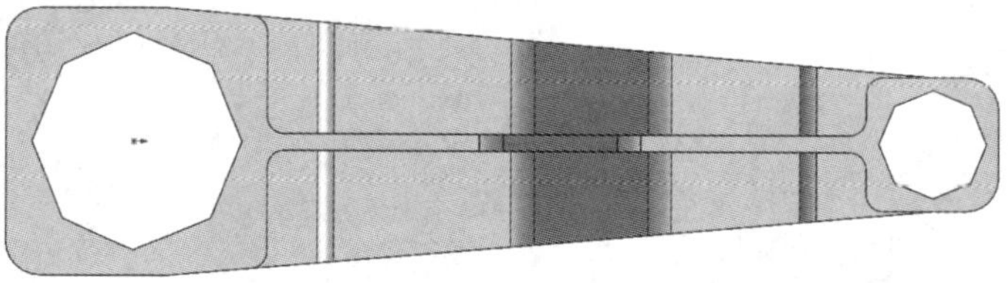

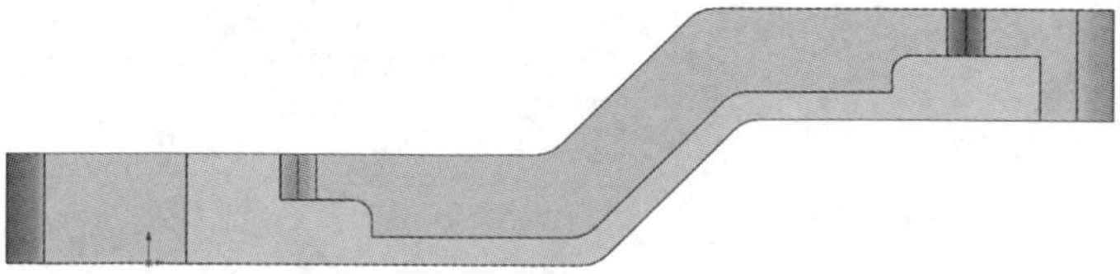

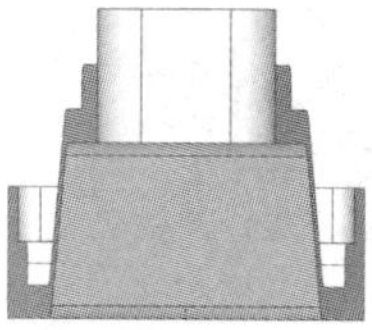

(The CSWP-Segment 1 examines your skills on creating a parametric model, where various methods will be used to create and constrain the geometry of each features.)

1. Creating the base plate:

Select the Front plane and open a **new sketch**.

Sketch the profile shown below.

Keep the Origin at the lower left corner of the profile.

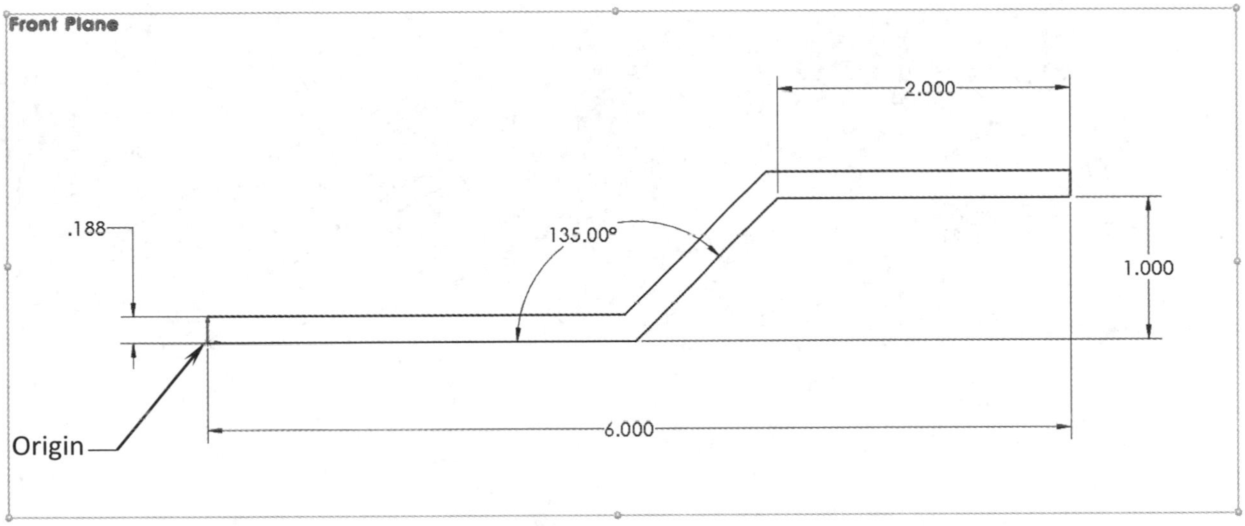

Ensure that all the Horizontal lines are parallel with each other.

The Vertical lines are also parallel with one another.

Add **dimensions** to fully define the sketch (dimensions are in Inches, three decimal places).

The .188" thickness dimension applies to the entire profile.

2. Extruding the sketch:

Switch to the **Features** tool tab.

Select the **Extruded Boss-Base** command.

For Direction 1, select the **Mid Plane** type.

For Extrude Depth, enter **2.000in**

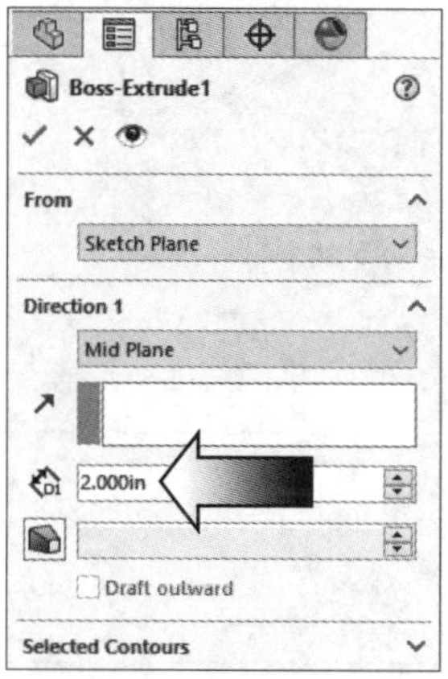

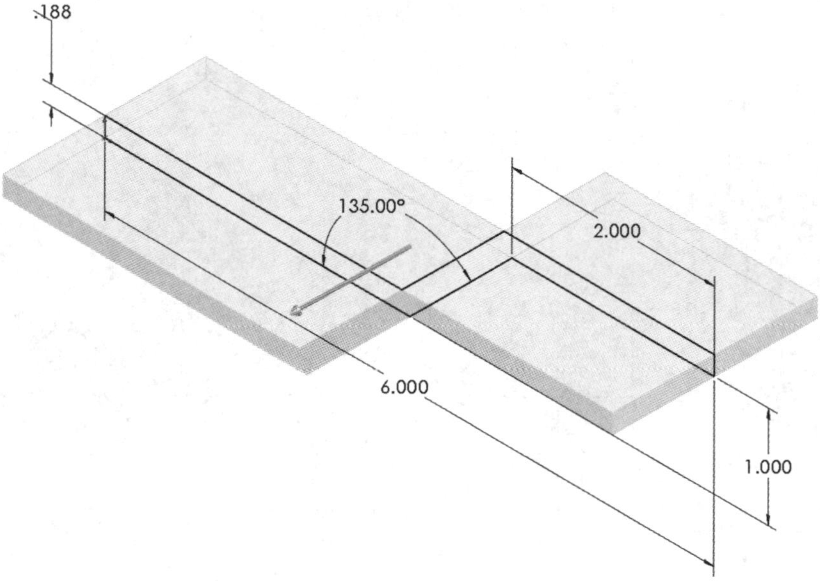

Click **OK**.

3. Making the 1st block:

Open a **new sketch** on the Bottom Face as indicated.

Sketch a **Center Rectangle** similar to the one shown in the image below.

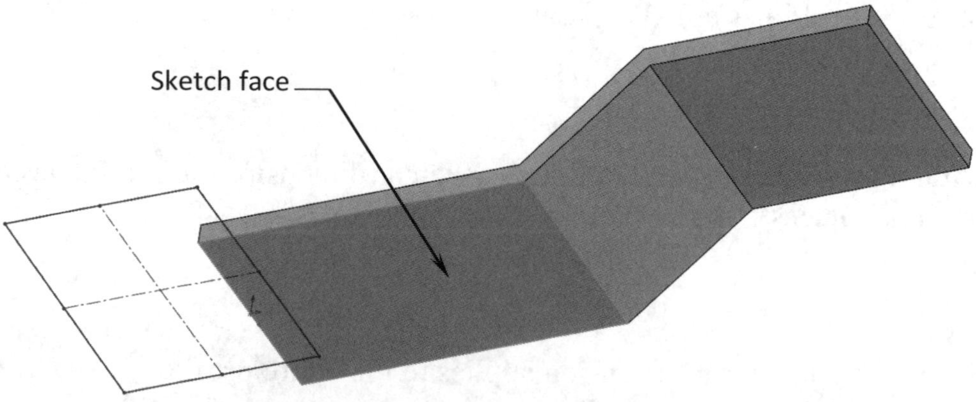

Add the relations shown to fully define the sketch.

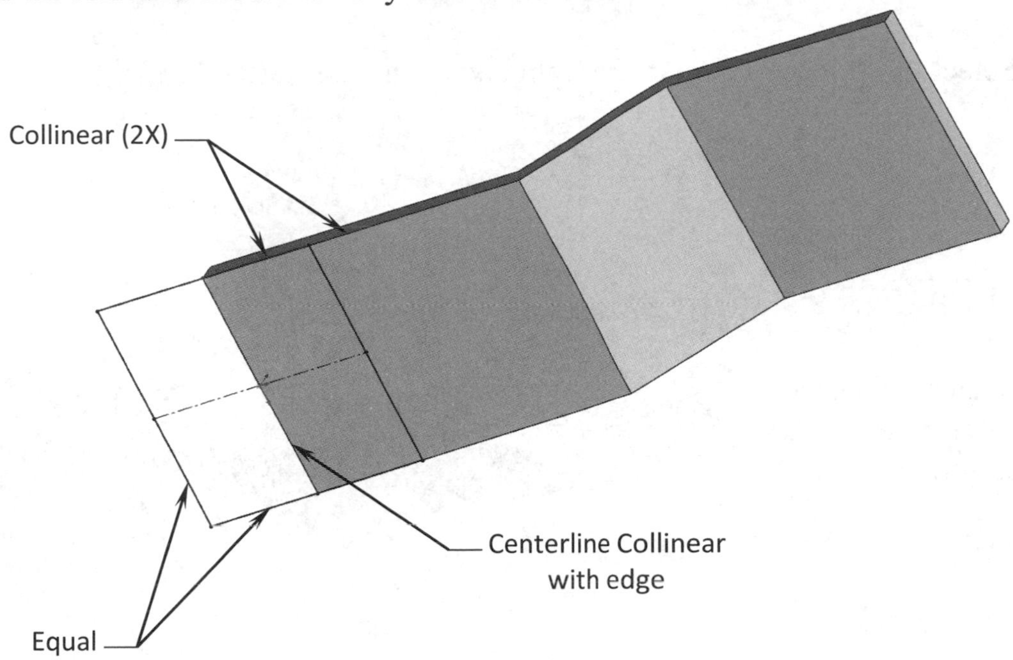

The relations above will make the rectangle the same width as the base (2.00in).

4. Extruding the sketch:

Switch to the **Features** tool tab and select the **Extruded Boss-Base** command.

For Direction 1, select **Blind**.

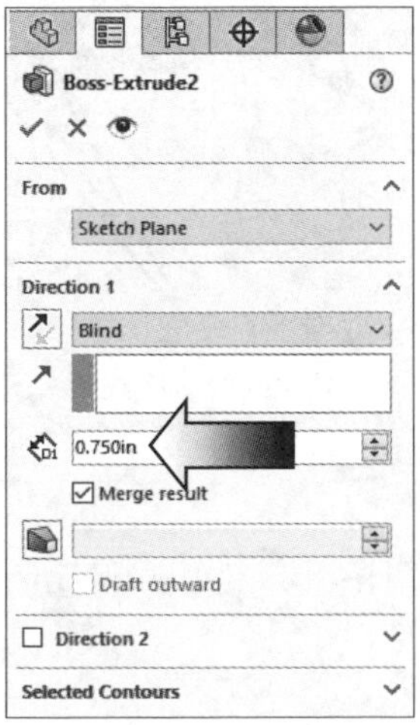

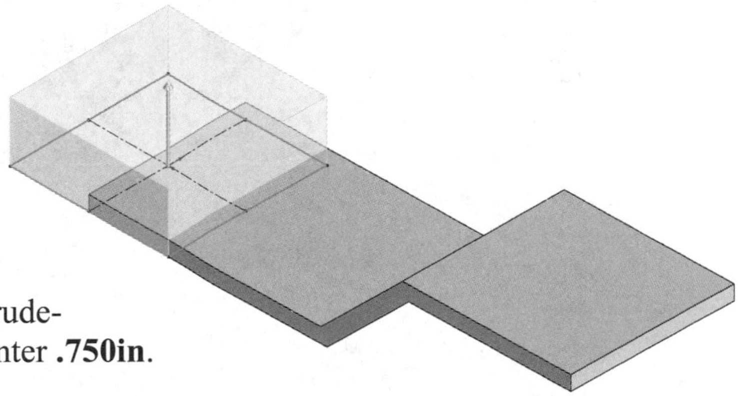

For Extrude-Depth enter **.750in**.

Click **OK**.

5. Making the 2nd block:

Select the Bottom Face on the right and open a **new sketch**.

Sketch a **Center Rectangle** approx. as shown.

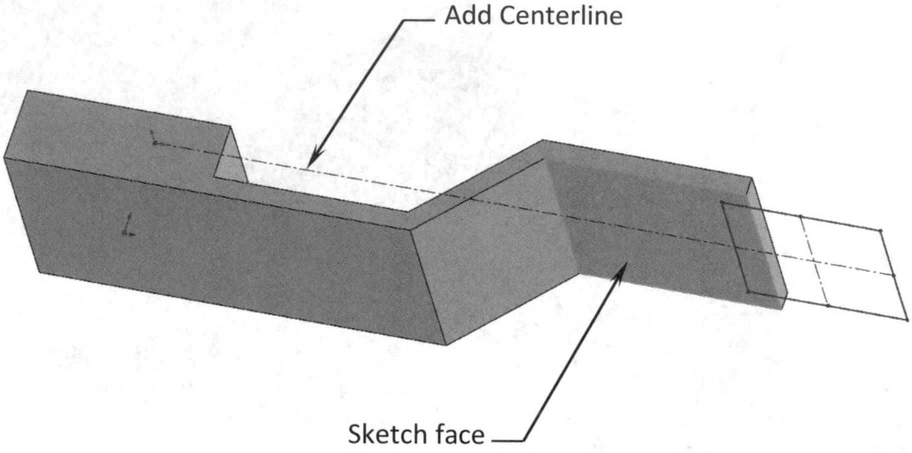

Add a **Centerline** that connects the Origin and the midpoint of the line on the right side of the rectangle.

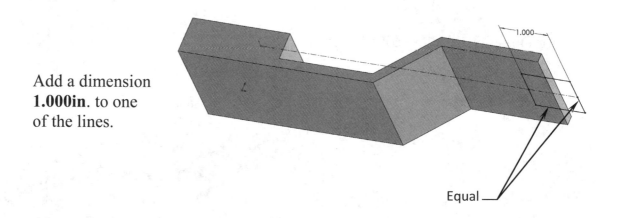

Add a dimension **1.000in**. to one of the lines.

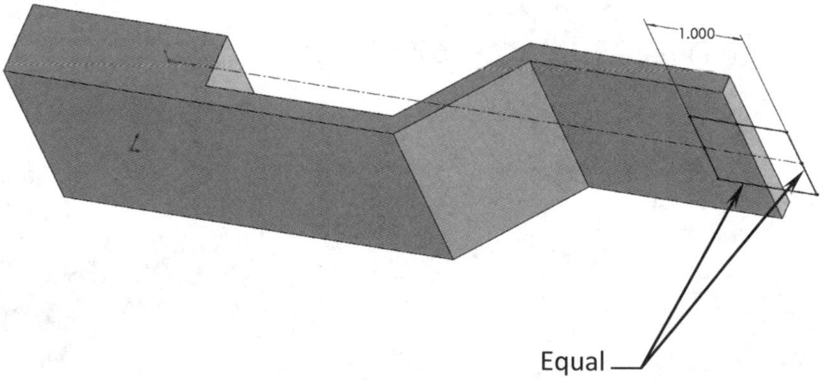

Add an **Equal** relation to one horizontal and one vertical line to fully define the sketch.

6. Extruding the sketch:

Switch to the **Features** tool tab.

Click the **Extruded Boss-Base** command.

For Direction 1, select the **Blind** type.

For Extrude Depth, enter **.750in**

Enable the **Merge Result** checkbox.

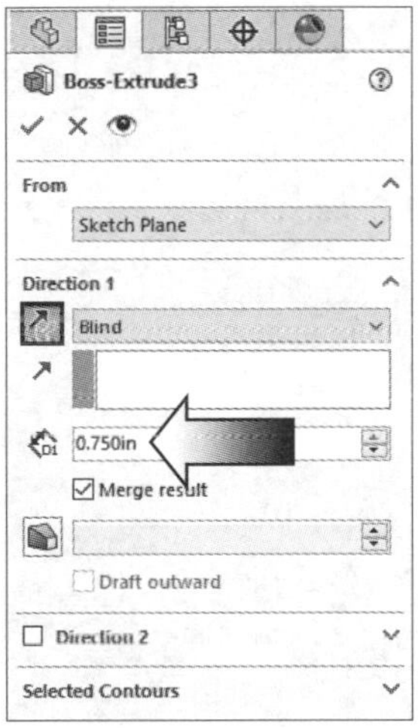

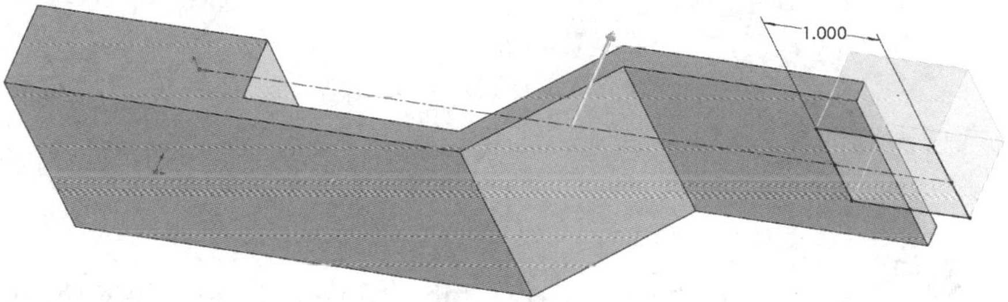

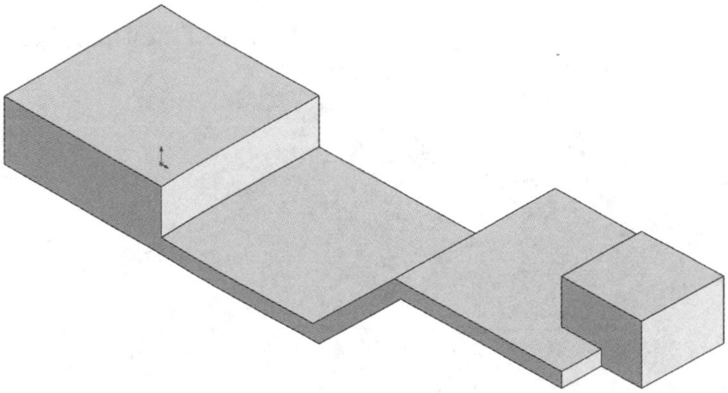

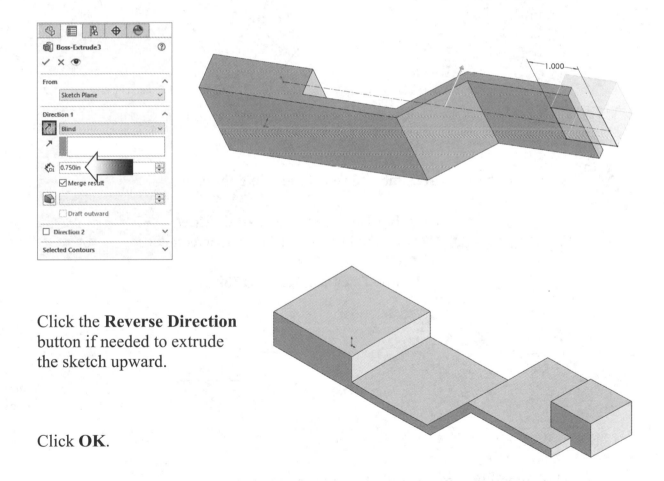

Click the **Reverse Direction** button if needed to extrude the sketch upward.

Click **OK**.

The 1st block and the 2nd block should have the same thickness (.750in.).

7. Adding a Rib feature:

The next step is to add the support rib across the length or the part.

Select the Front plane and open a **new sketch**.

Sketch 3 lines shown below and add a **Parallel** relation as indicated.

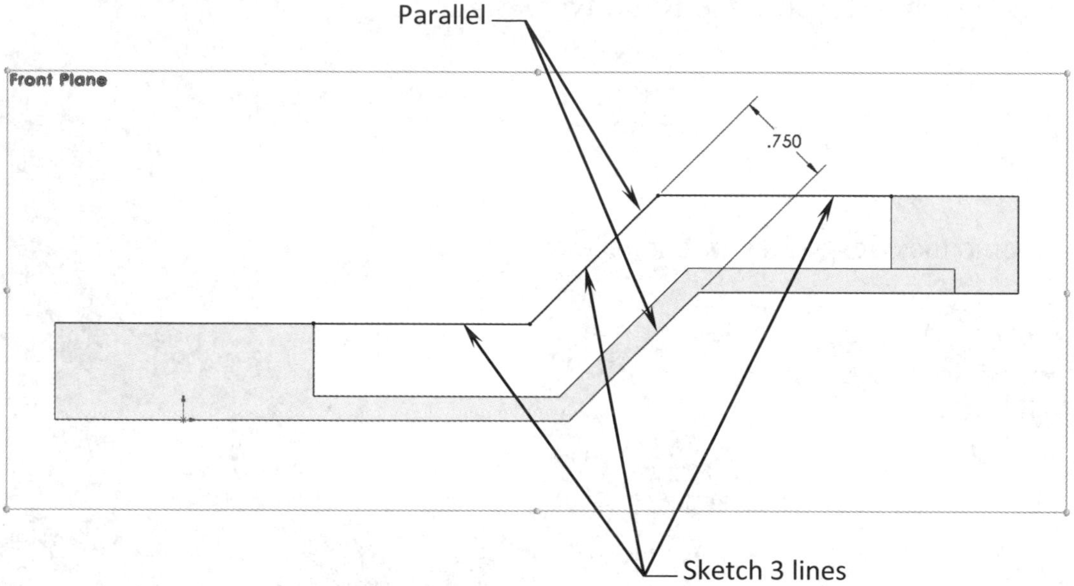

Switch to the **Features** tab and select the **Rib** command.

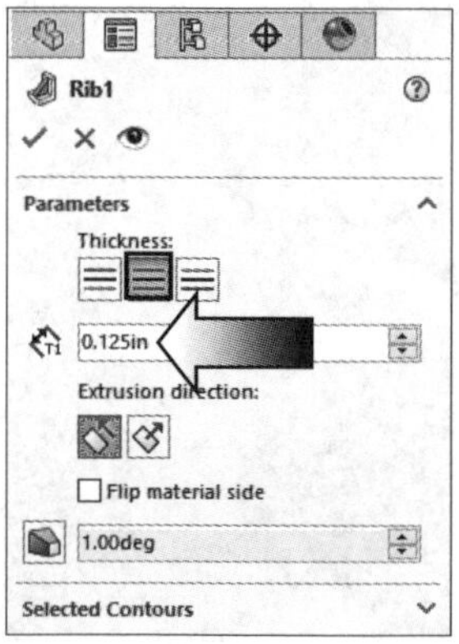

For Direction, select **Both Sides** on the left side of the model as noted.

For Thickness, enter **.125in**.

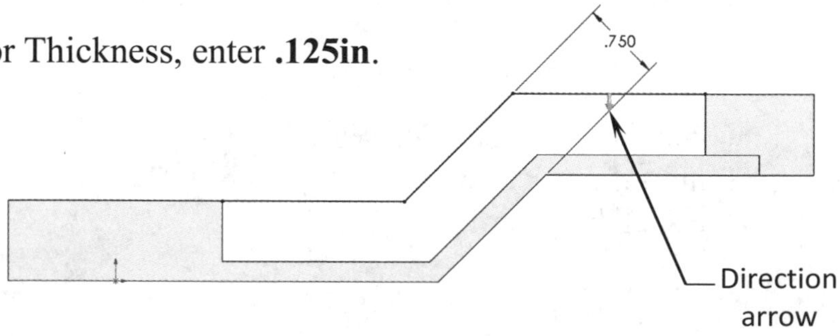

The Direction arrow should be pointing downward.

Click **OK**.

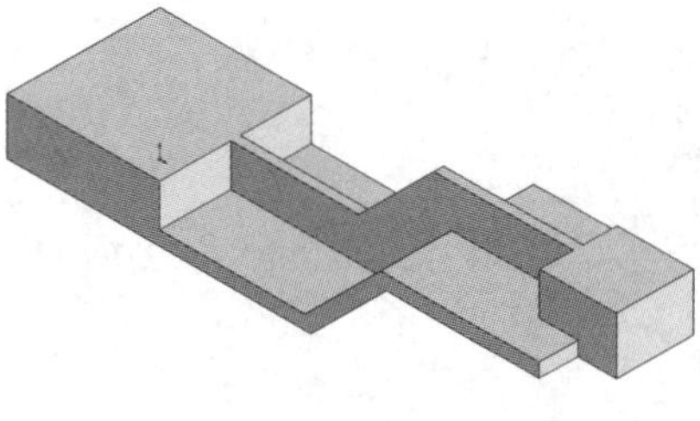

8. Trimming the base:

Open a **new sketch** on the Top plane.

Sketch the profile shown. Use the Mirror Entities option to keep the 2 halves symmetrical about the horizontal centerline.

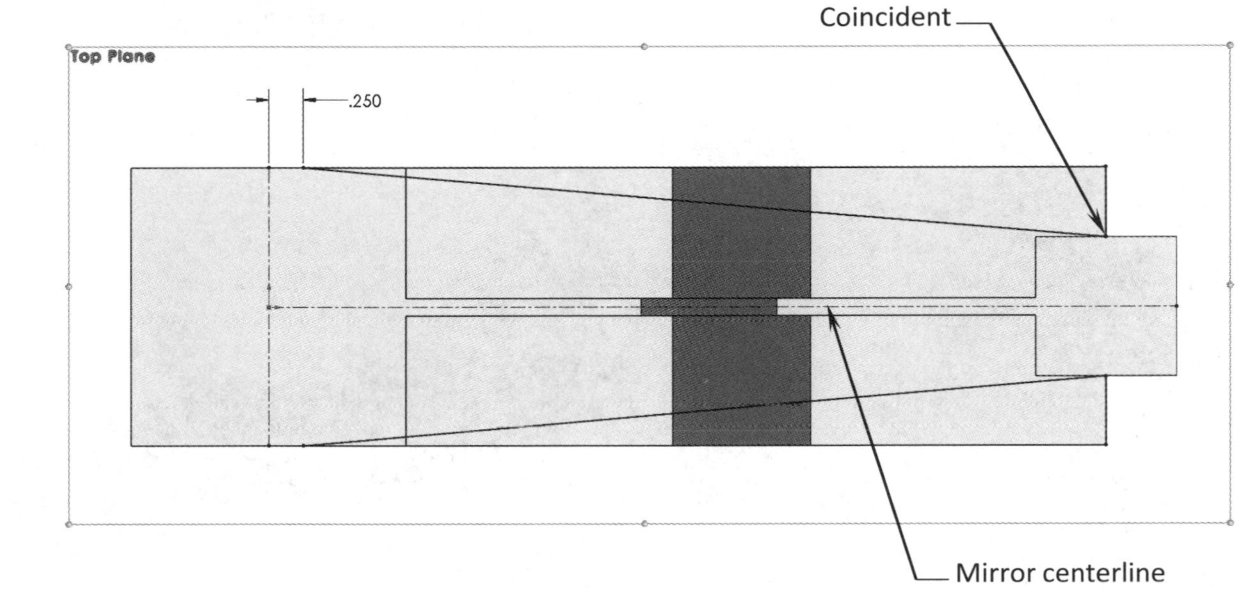

Switch to the **Features** tab and select **Extruded Cut**.

For Direction 1, select **Through All**.

Click the **Reverse** button (arrow).

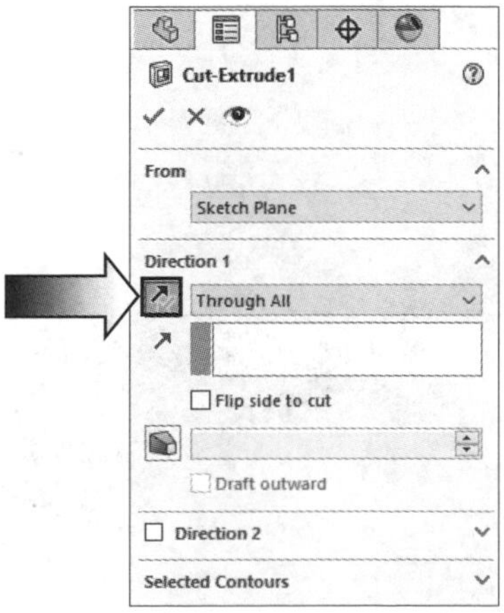

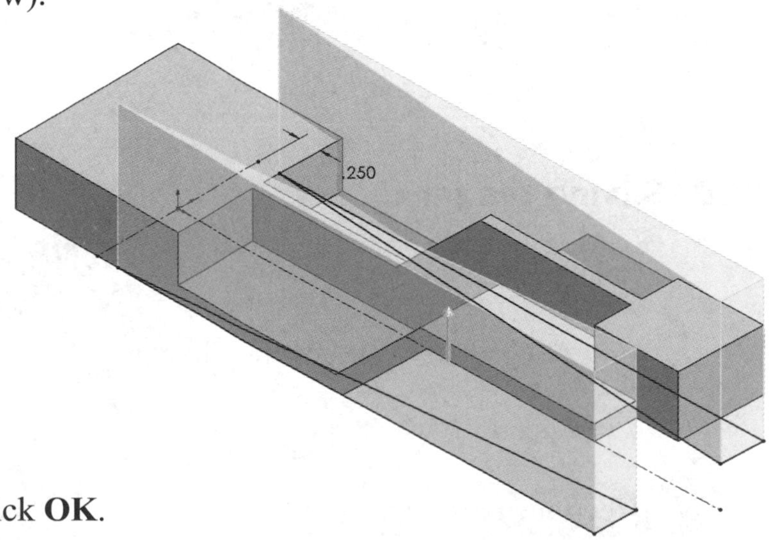

Click **OK**.

9. Calculating the mass properties:

Right-click the Material option and select:
Plain Carbon Steel.

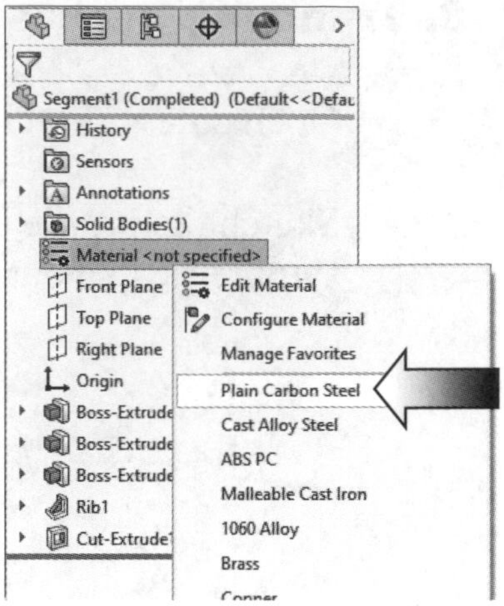

Switch to the Evaluate tab and click **Mass Properties.**

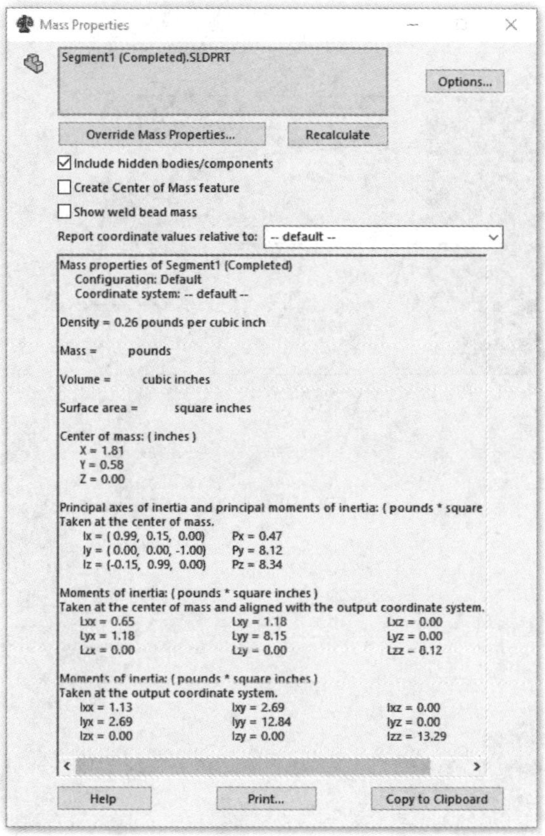

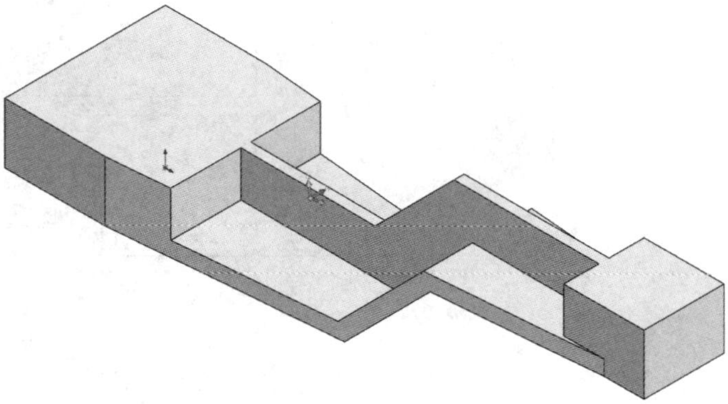

Enter the mass of the part below:

____________ pounds.

10. Saving the part:

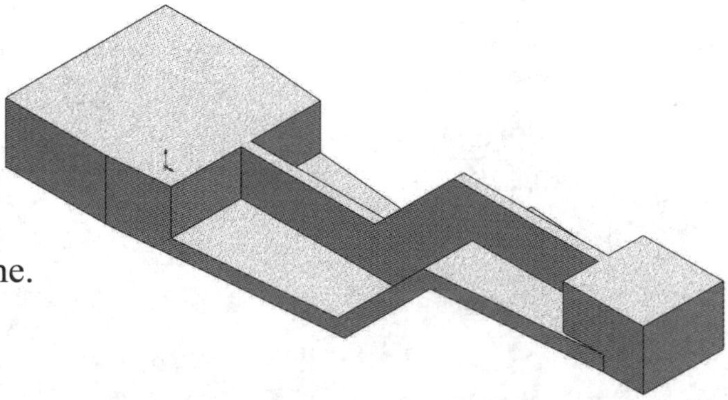

Select **File, Save As.**

Enter **Segment1_Q1** for the file name.

Click **Save.**

Keep the part open.

11. Modifying the dimensions:

Double-click the **Boss-Extrude1** to display its dimensions.

Change the dimension **6.000in** (circled) to **7.000in**.

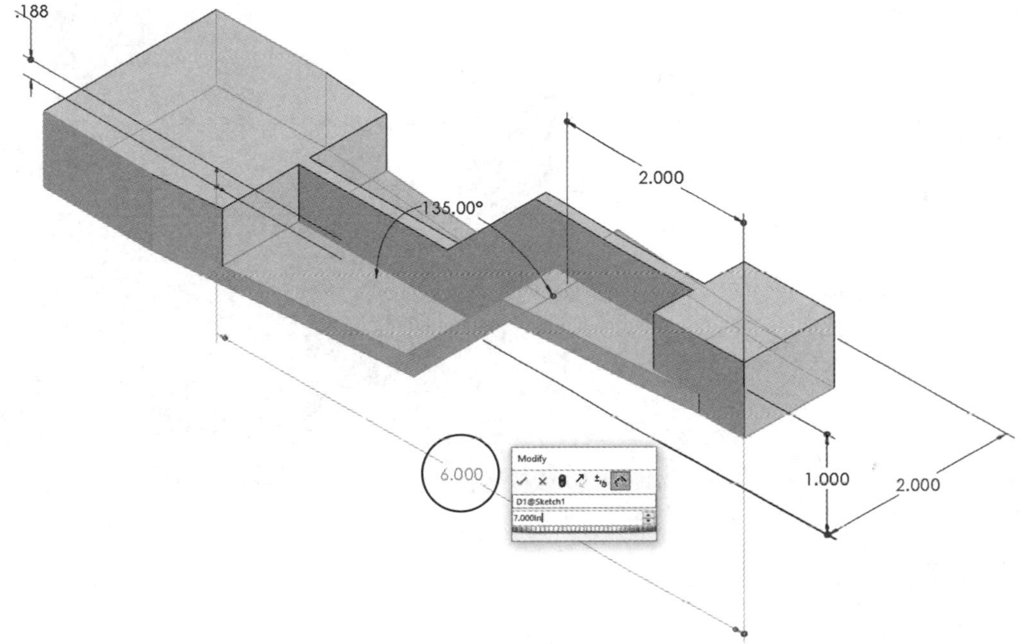

Double-click **Boss-Extrude2** and **Boss-Extrude3** and change both thickness dimensions from **.750in** (circled) to **.625in**.

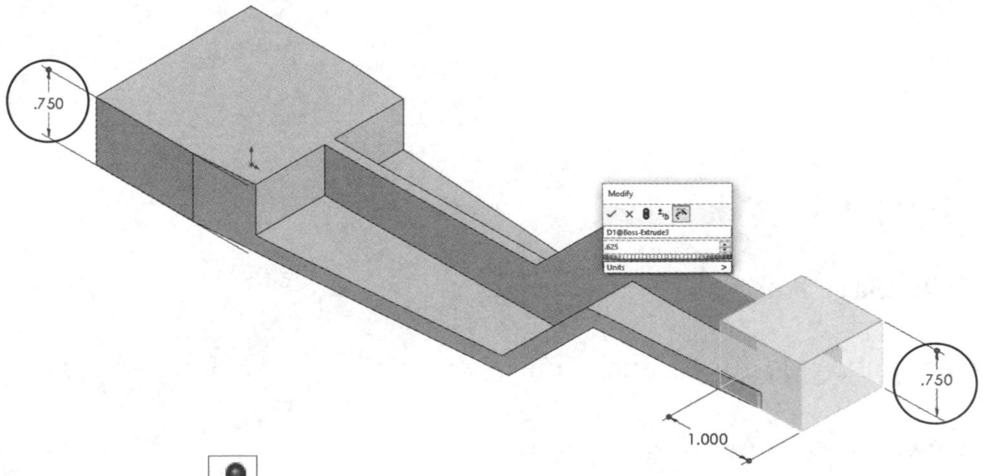

Press **Rebuild** to update the model.

12. Calculating the mass:

There may be several dimensions changes in the actual exam, but we will change only the 3 main ones in this exercise.

Switch to the **Evaluate** tab.

Click **Mass Properties.**

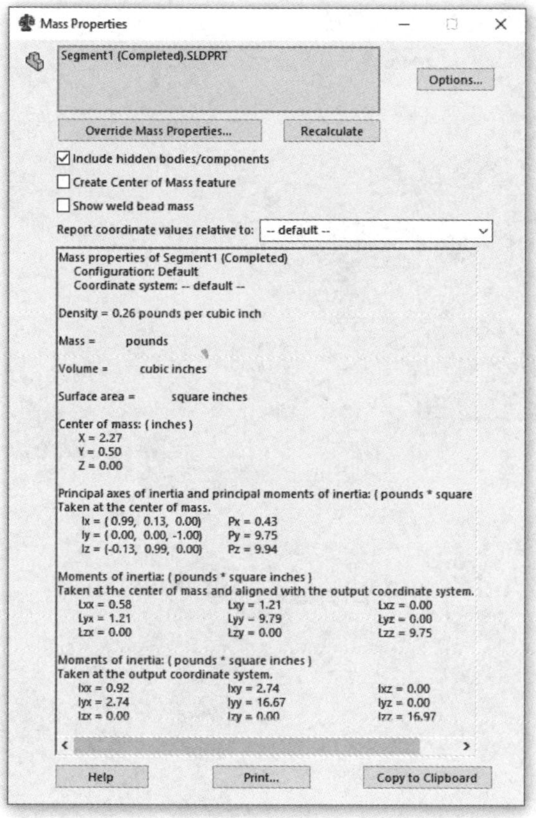

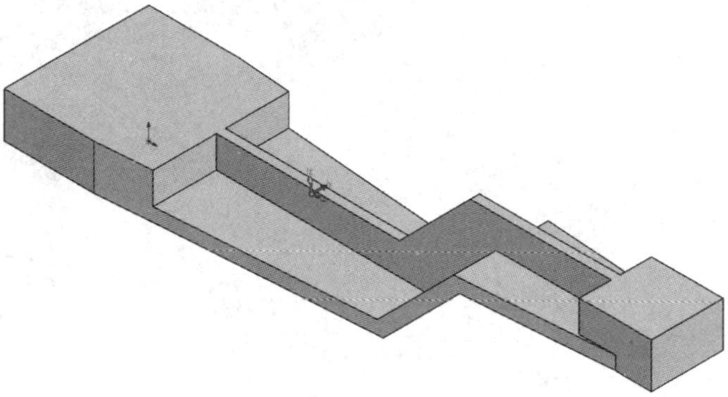

Enter the mass of the part below:

____________ pounds.

13. Saving the part:

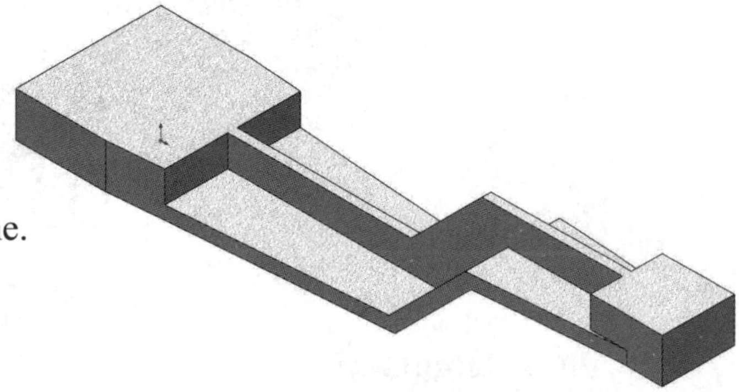

Select **File, Save As**.

Enter **Segment1_Q2** for the file name.

Click **Save**.

Keep the part open.

14. Adding fillets:

Switch to the **Features** tool tab.

Click **Fillet** .

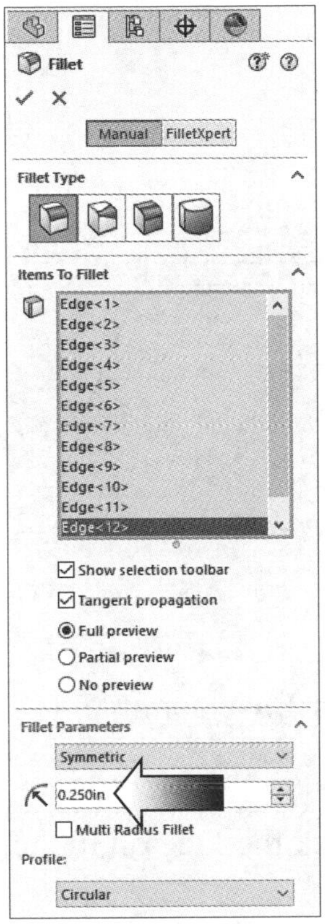

For Fillet Type, use the default **Constant Size** fillet.

Enter **.250in** for radius size.

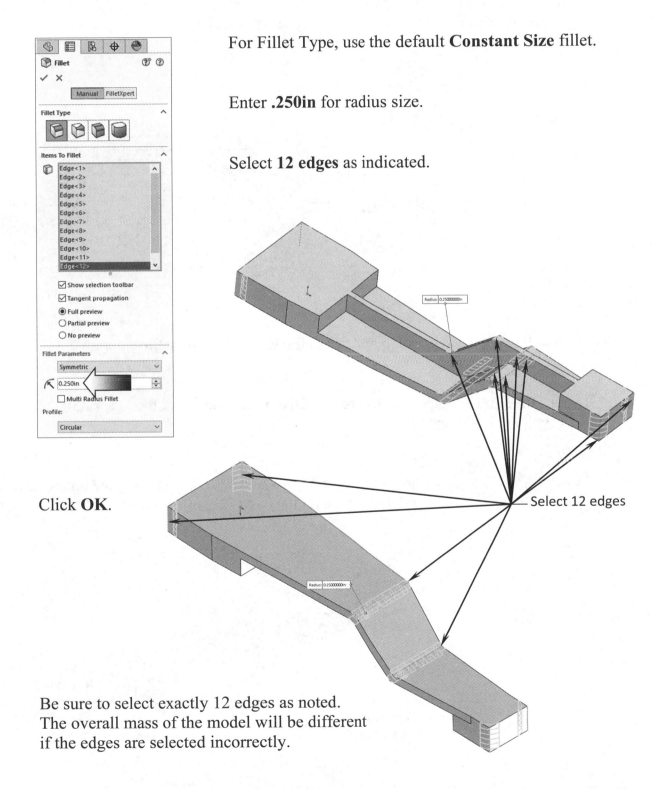

Select **12 edges** as indicated.

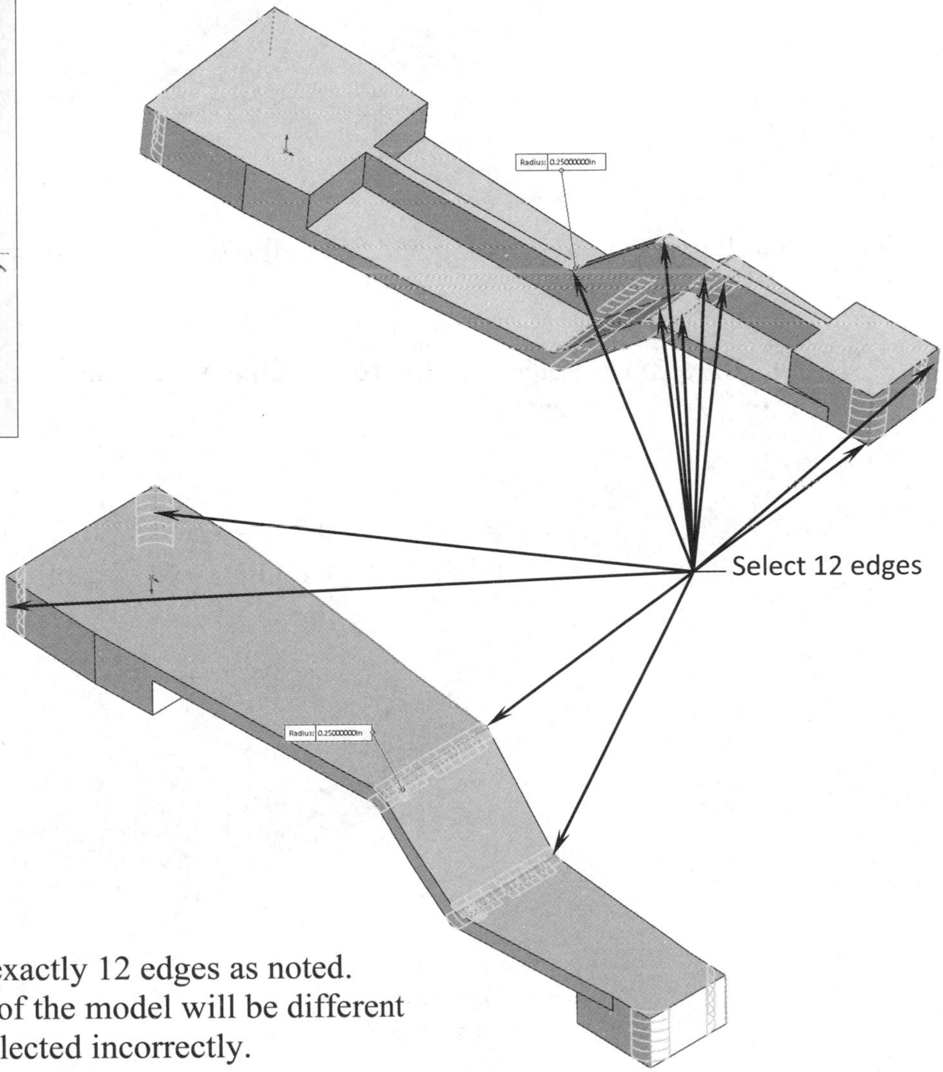

Click **OK**.

Be sure to select exactly 12 edges as noted.
The overall mass of the model will be different if the edges are selected incorrectly.

15. Creating the side tabs:

Select the Front plane and open a **new sketch**.

Sketch **2 Corner Rectangles**. The left and the right corners of the 2 rectangles are **Coincident** to the vertical edges of the square blocks.

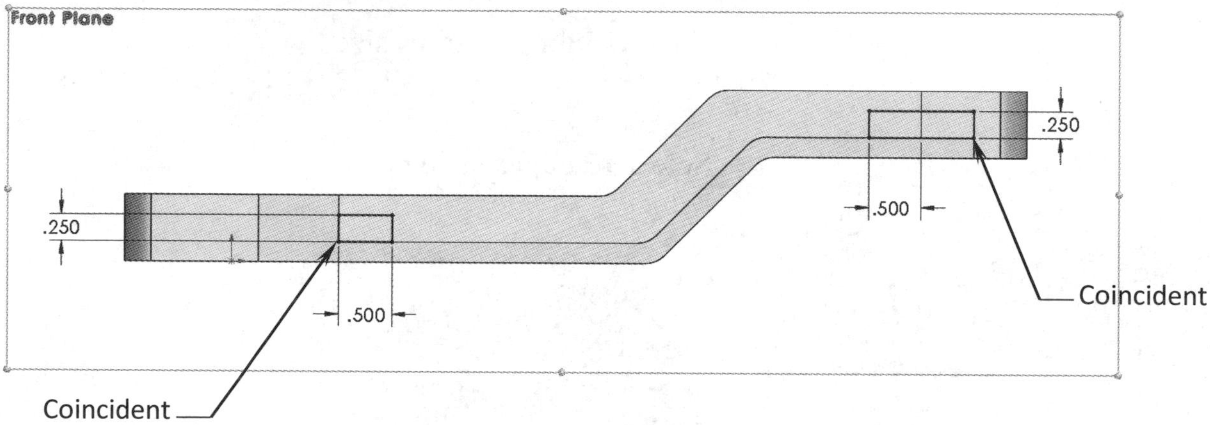

Switch to the **Features** tab and select the **Extruded Boss-Base** command.

For Direction 1, select the **Up-to-Surface** option and click the face on the left side of the part as indicated.

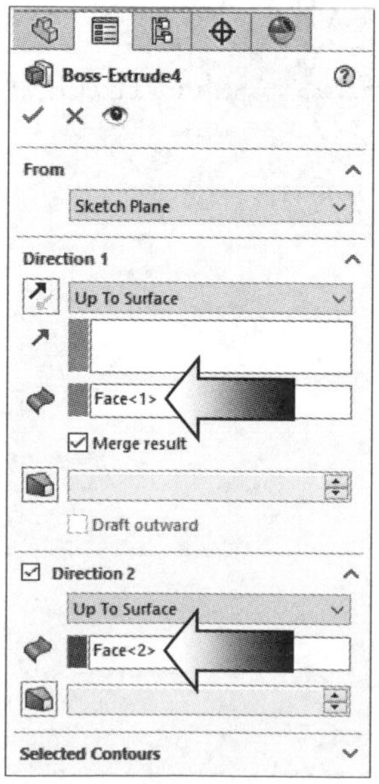

For Direction 2, also select the **Up-to-Surface** option and click the face on the right side of the part.

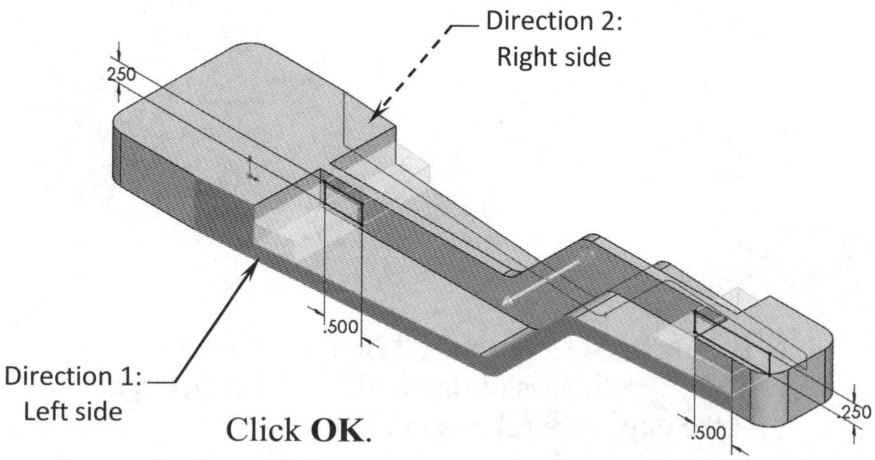

Click **OK**.

16. Adding the 8-sided Polygonal holes:

Select the Top plane and open a **new sketch**.

Sketch **two 8-sided Polygons** and add the **Horizontal** relations as indicated.

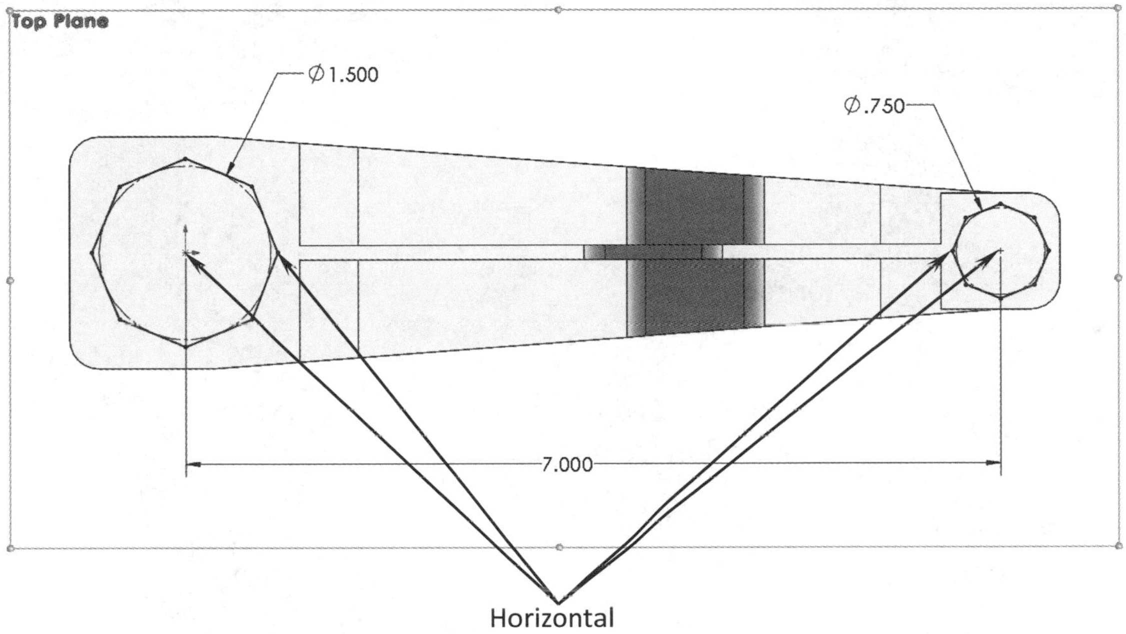

Add the dimensions shown to fully define the sketch.

Switch to the **Features** tool tab and click **Extruded Cut**.

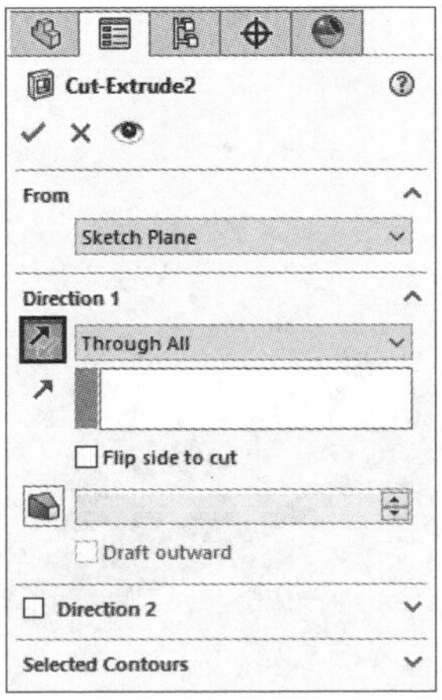

For Direction 1 select **Through All**.

Click **Reverse Direction**.

Click **OK**.

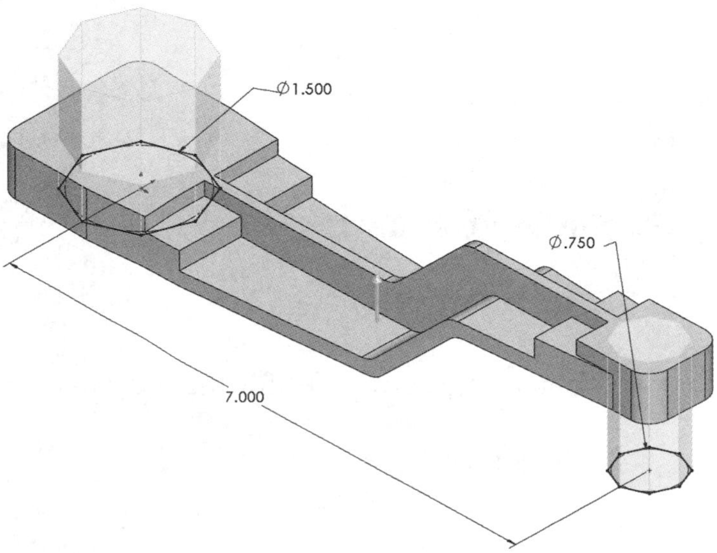

17. Adding fillets:

Click **Fillet** .

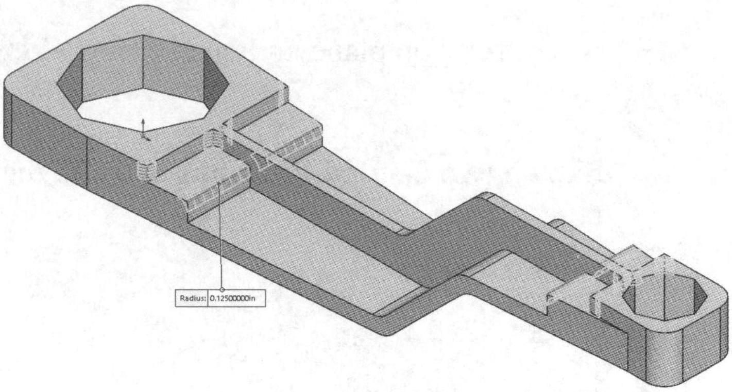

Use the default **Constant Size Radius** option.

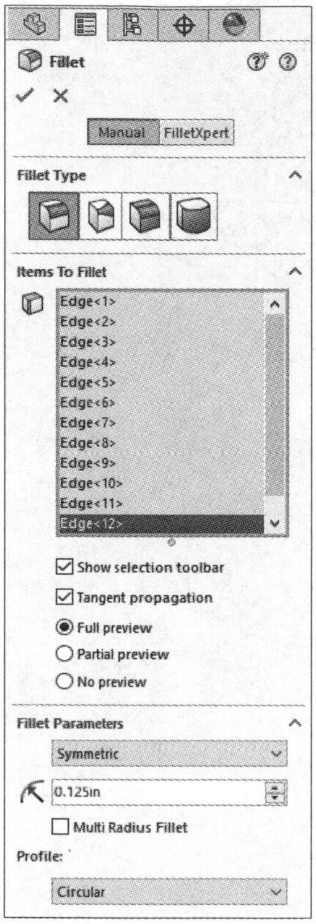

Enter **.125in.** for radius size.

Select the **12 edges** as shown in the images.

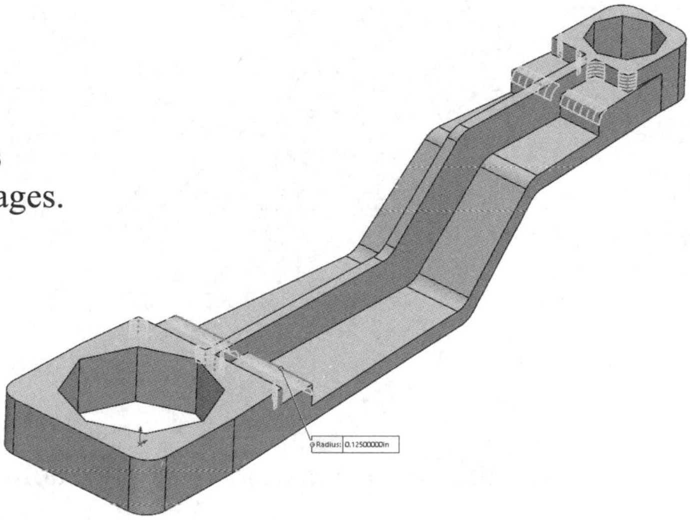

Click **OK**.

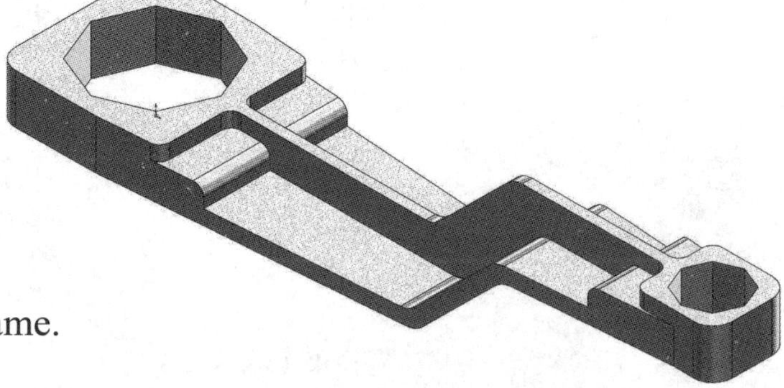

18. Saving the part:

Select **File**, **Save As**.

Enter **Segment1_Q3** for the file name.

Click **Save**.

19. Calculating the final mass:

Switch to the **Evaluate** tab.

Click **Mass Properties**.

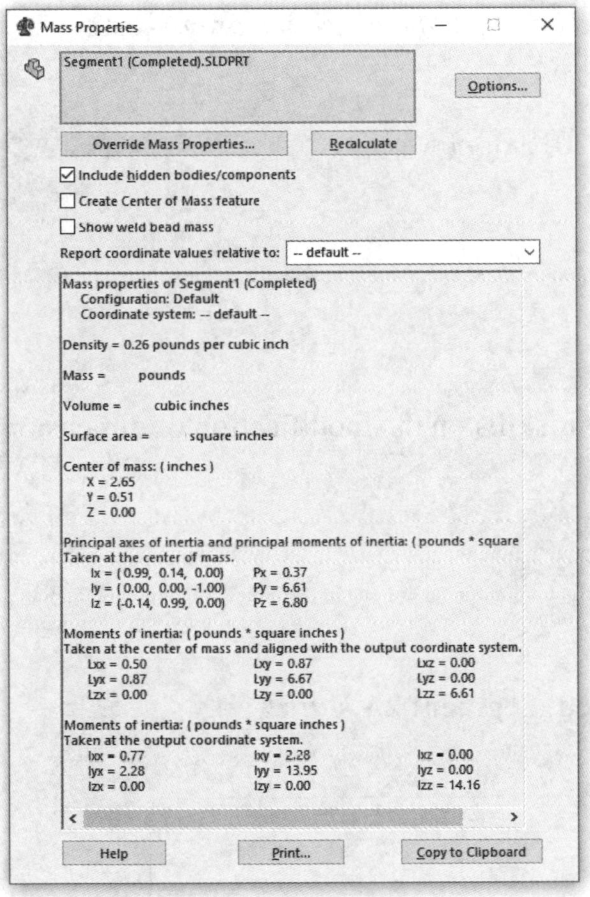

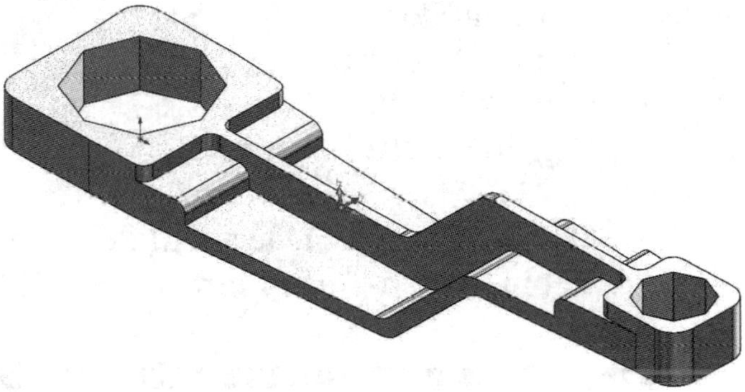

Locate the mass of the part and enter it below:

___________ pounds

Resave the part and close all documents.

Certified-SOLIDWORKS-Professional (CSWP)

Certification Practice for the Mechanical Design Exam

Challenge II: Part Modifications & Configurations

Complete this challenge within 50 minutes (3 parts)

(The following examples are intended to assist you in familiarizing yourself with the structures of the exams and the method in which the questions are asked.)

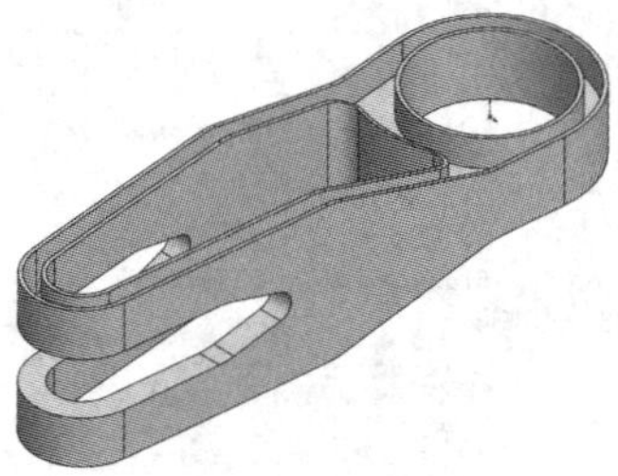

Modify this part using SOLIDWORKS 2010 or newer.
Unit: **Inches, 3 decimals**
Drafting Standards: **ANSI**
Origin: **Arbitrary**
Material: **ABS**
Density: **0.2037 lb/in^3**

(This portion of the test will examine your skills on the modification of dimensions, geometry, and repair errors in a part.)

1. Opening a part document: (1 of 3)

Click **File / Open**.

Locate and open the part document named: **Segment 2A.sldprt**

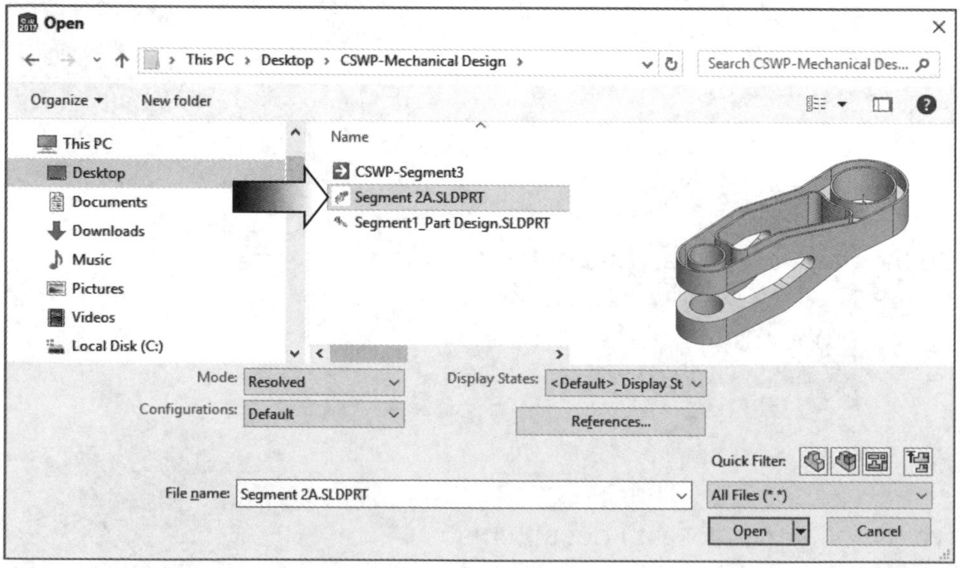

2. Modifying dimensions:

Double-click the feature named **Cut-Extrude1** and change the dimension **4.875** (circled) to **4.000**.

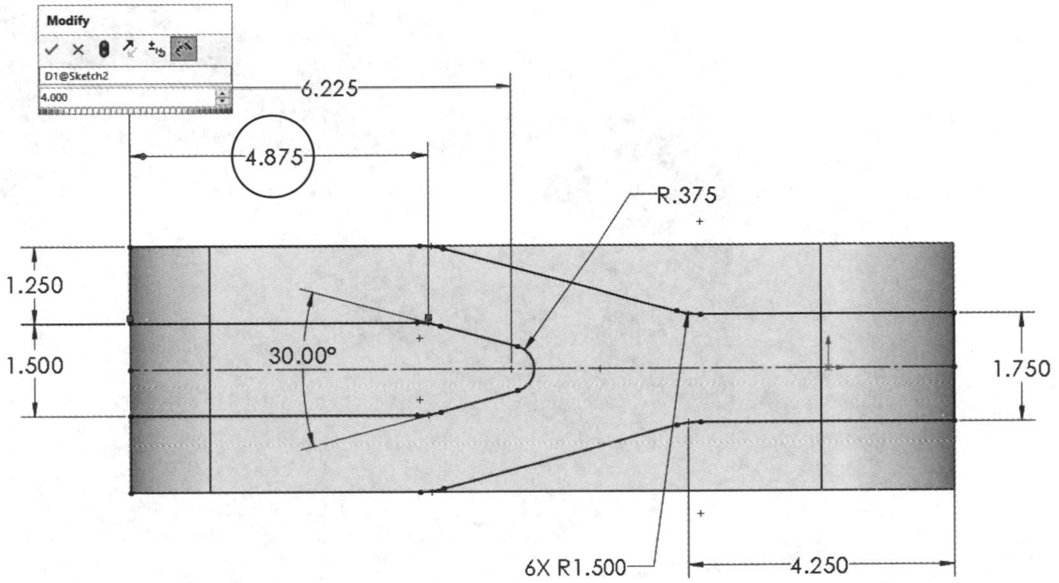

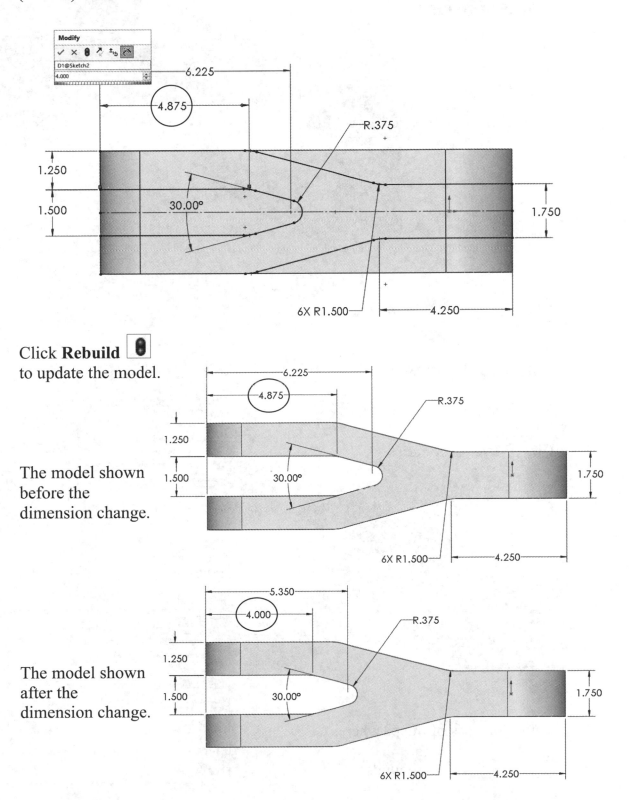

Click **Rebuild** to update the model.

The model shown before the dimension change.

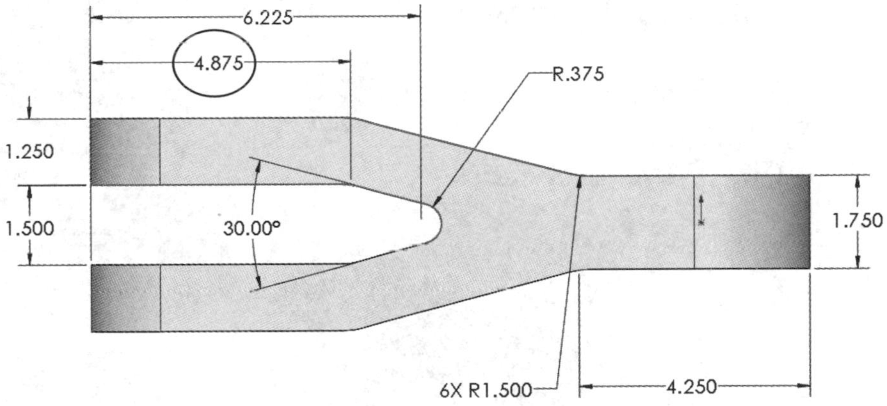

The model shown after the dimension change.

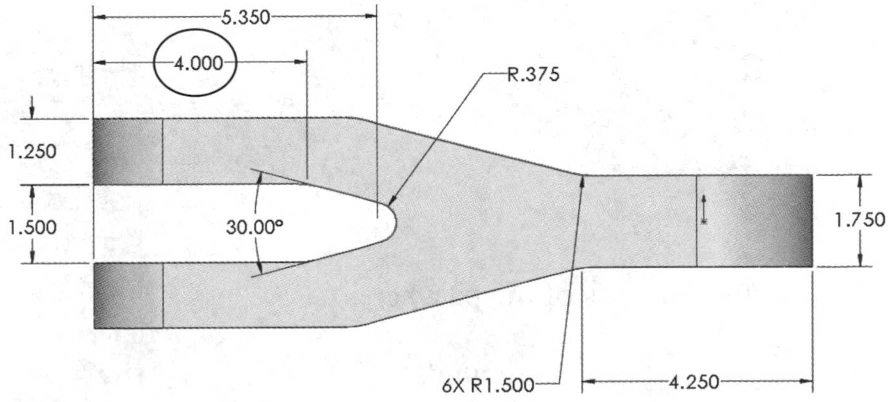

Double-click the **Shell** feature from the FeatureManager tree.

Change the wall thickness dimension from **.090** to **.125**.

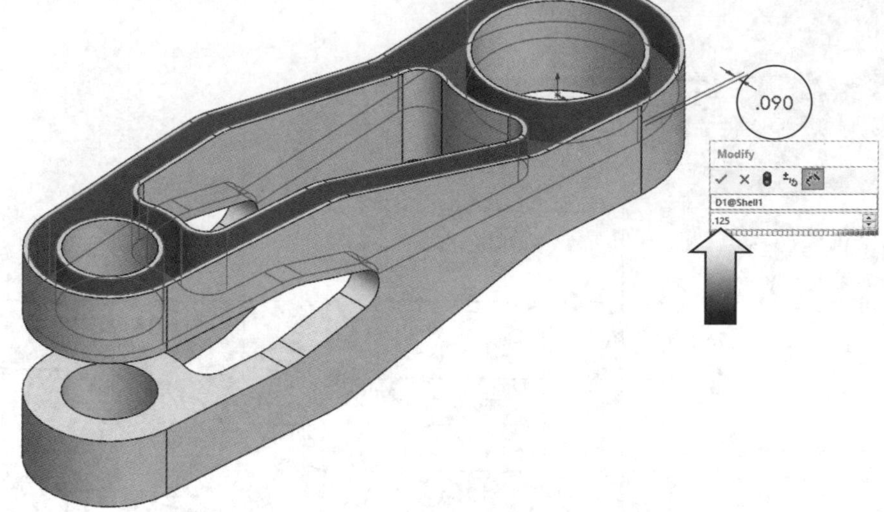

Click **Rebuild** to update the model.

3. Calculating the mass:

Switch to the **Evaluate** tab.

Click **Mass Properties**.

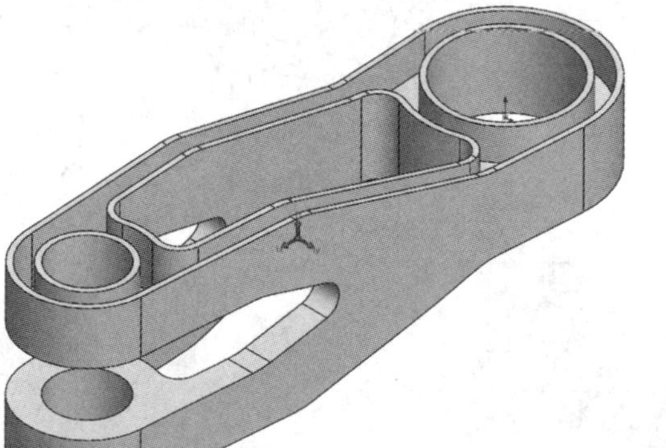

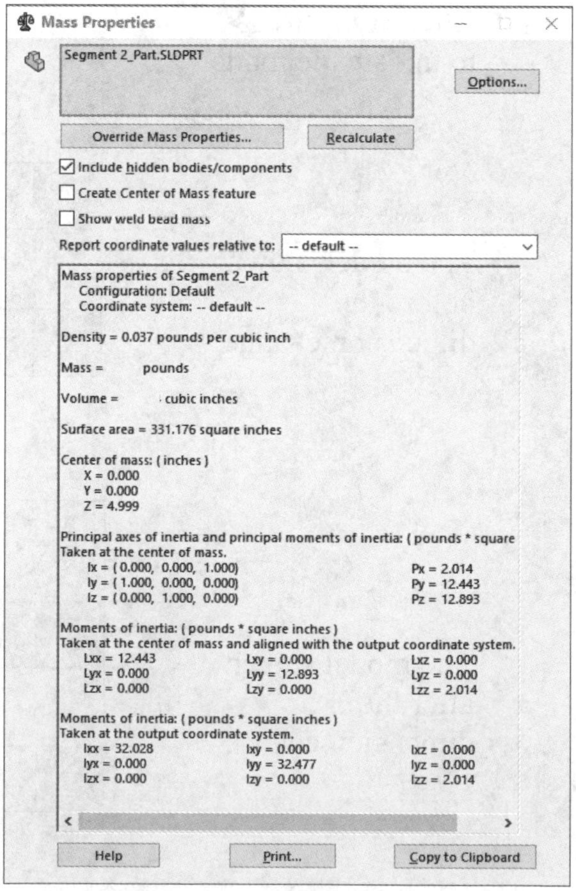

Enter the mass of the part here:

_______________ pounds.

4. Modifying the geometry:

Edit the **Sketch1** under the Boss-Extrude1 feature.

Modify the geometry of the sketch so that it would look like the images on the right.

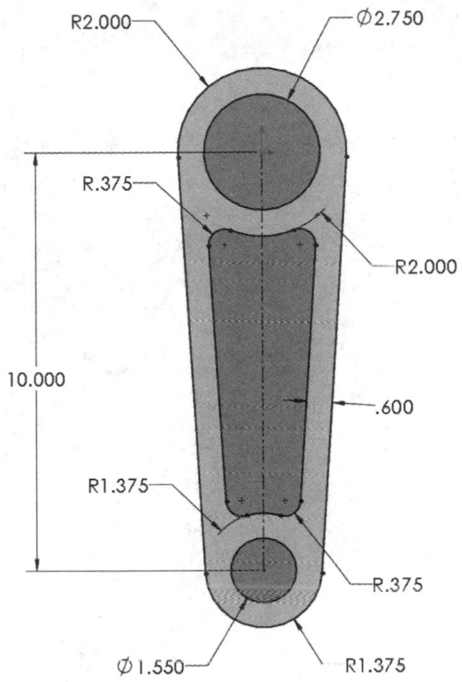

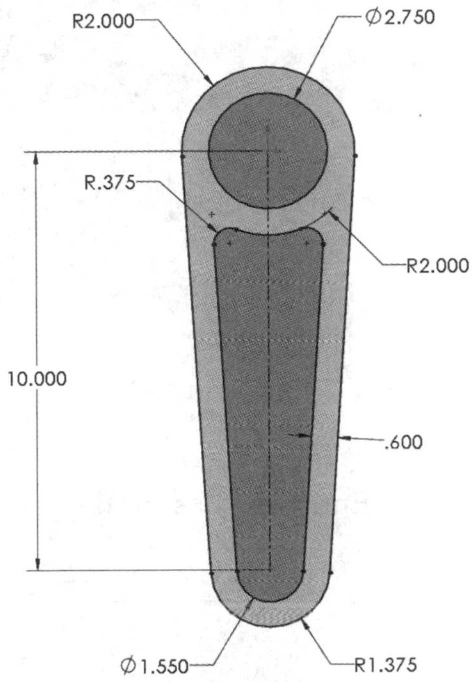

Before the changes

After the changes

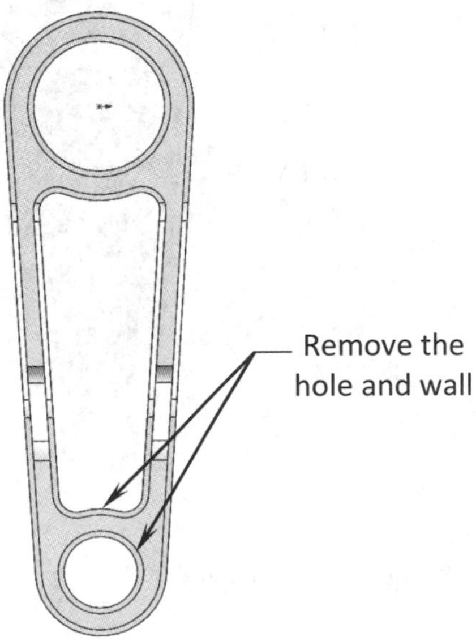

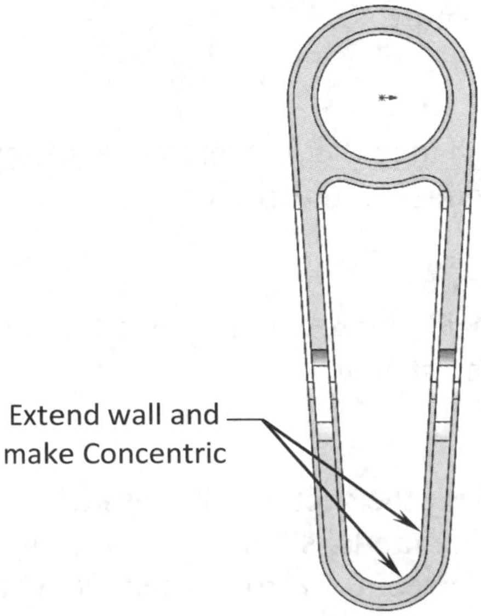

5. Calculating the mass:

Switch to the **Evaluate** tab.

Click **Mass Properties**.

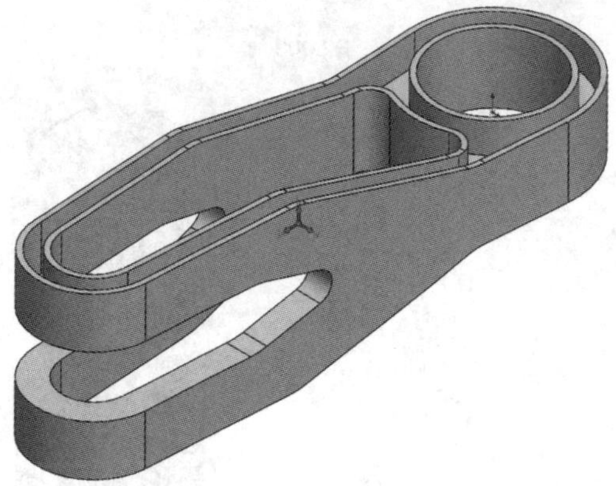

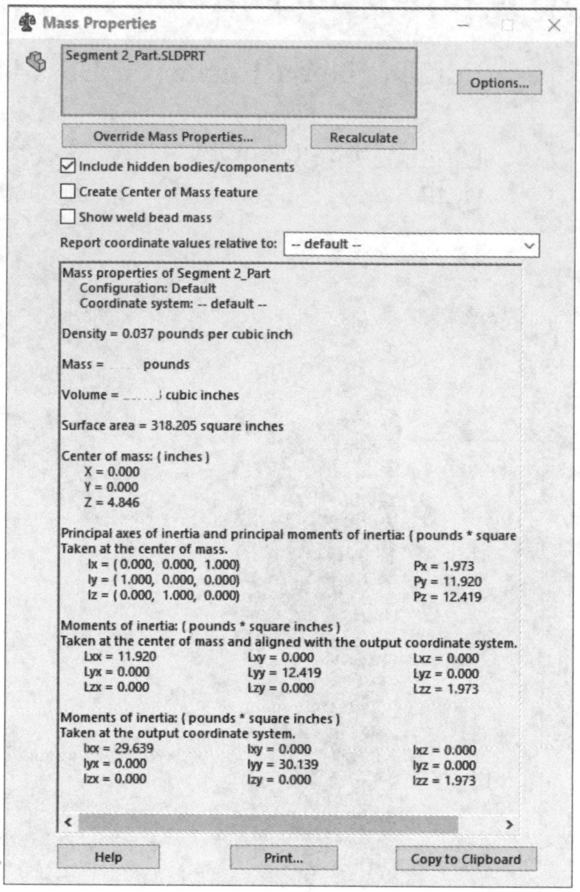

Enter the mass of the part here:

_______________ pounds

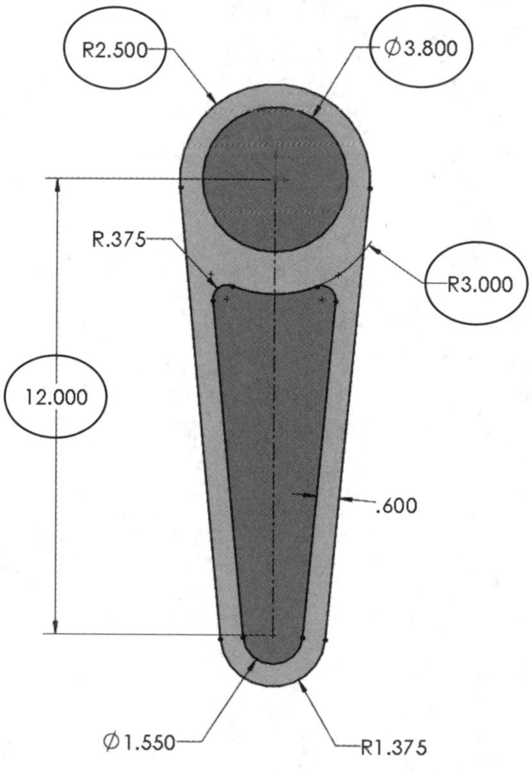

6. Making additional changes:

Edit the **Sketch1** under the Boss-Extrude1 feature.

Change the dimensions to match the values in the circles.

Add any relations needed to fully define the sketch.

Exit the sketch and **Rebuild** the model. The model should not have any errors. Correct the errors when/where they occur.

Edit the **Sketch2** under the Cut-Extrude1 feature.

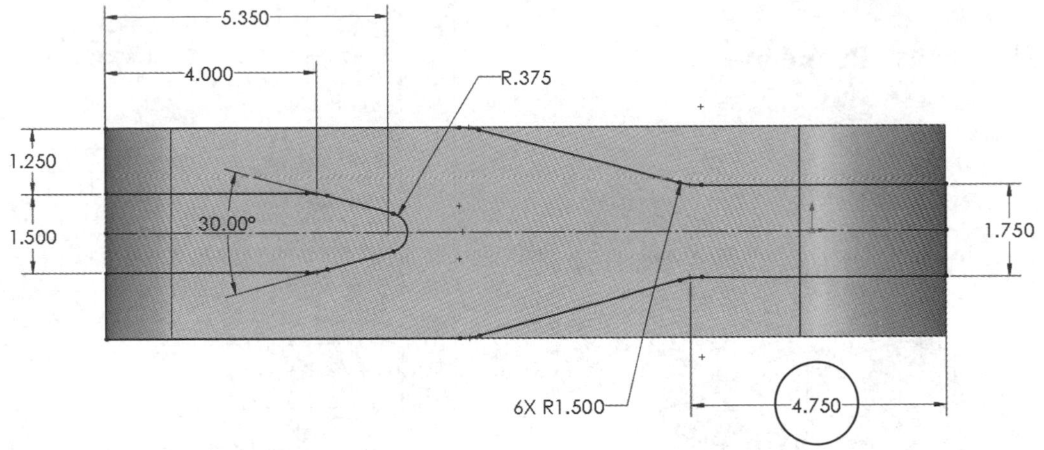

Change the dimension **4.250** to **4.750** (circled).

Click **Rebuild** to update the model.

Double-click the **Shell** feature from the FeatureManager tree.

Change the wall thickness dimension from **.125** to **.100**.

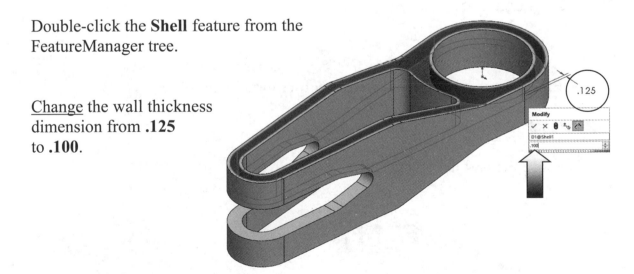

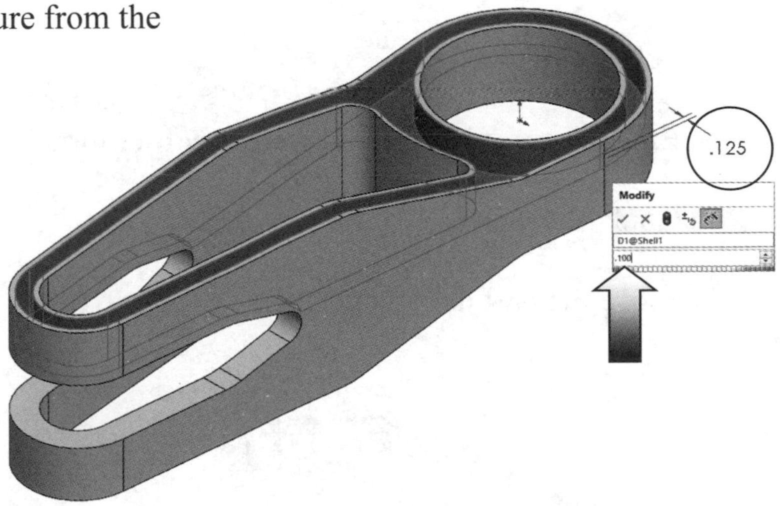

Click **Rebuild** to update the model.
The model should be free of errors. Correct/repair any errors when/where they occur.

7. Calculating the mass:

Switch to the **Evaluate** tab.

Click **Mass Properties**.

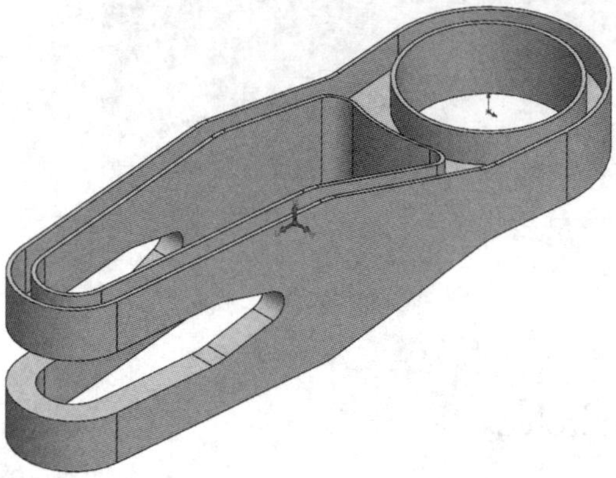

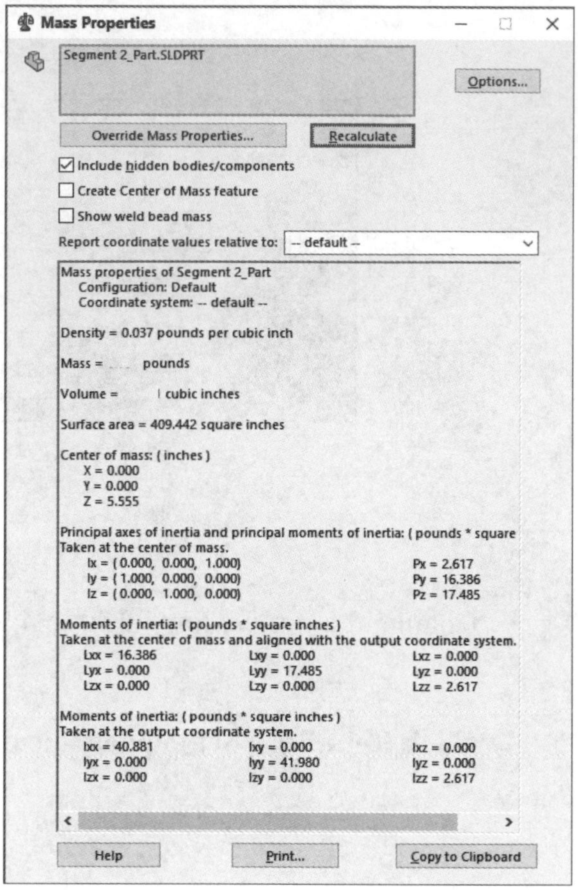

Enter the mass of the part here:

_______________ pounds.

8. Saving the part:

Select **File**, **Save As**.

Enter **Segment 2A (Completed).sldprt** for the file name.

Click **Save**.

Close the part document.

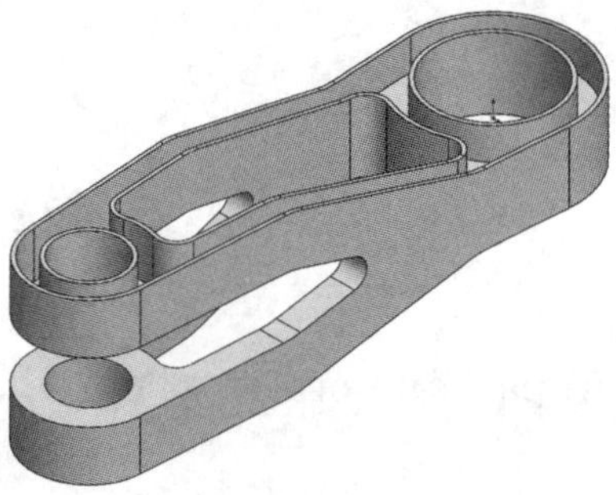

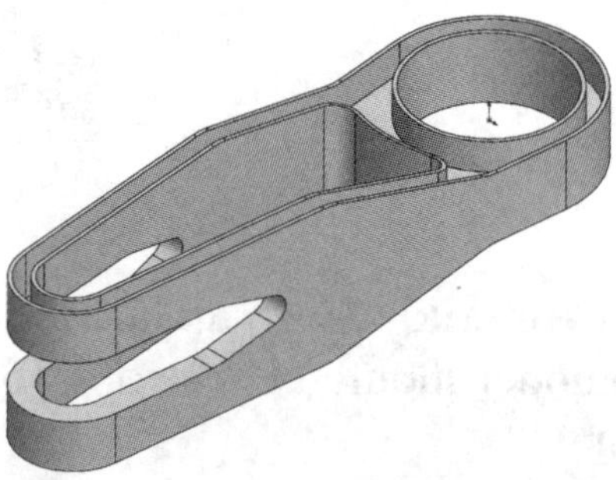

1. Opening a part document: (2 of 3)

Click **File / Open**.

Locate and open the part document named: **Segment 2B.sldprt**

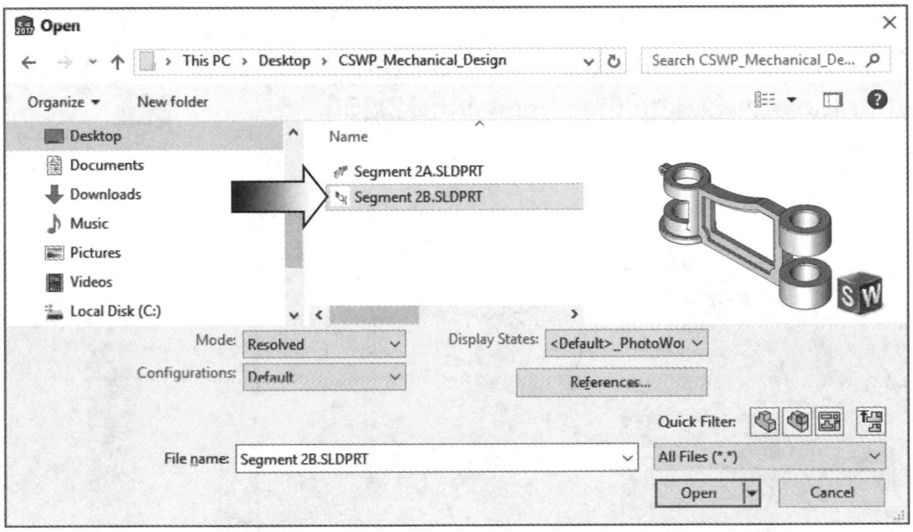

(This portion of the test will examine your skills on the modification of dimensions, geometry, and repair errors in a part.)

2. Modifying dimensions:

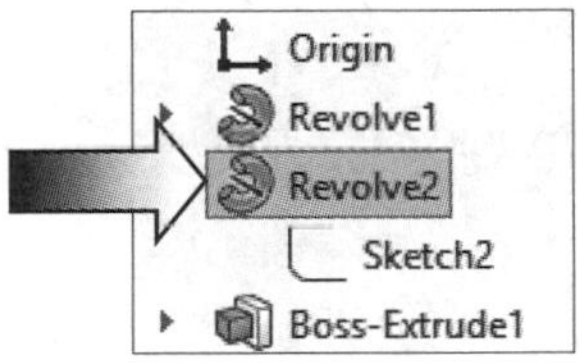

Double-click the feature **Revolve2** from the Feature-Manager tree.

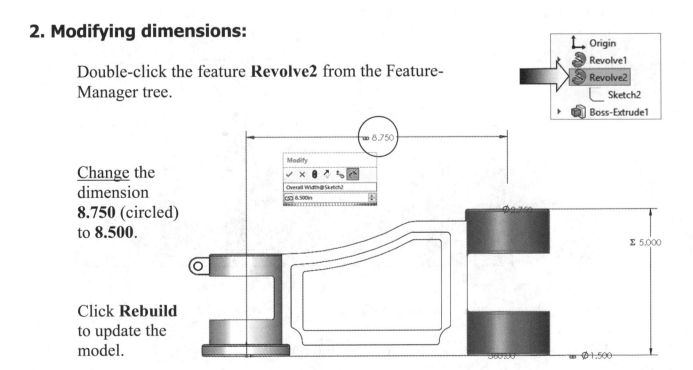

<u>Change</u> the dimension **8.750** (circled) to **8.500**.

Click **Rebuild** to update the model.

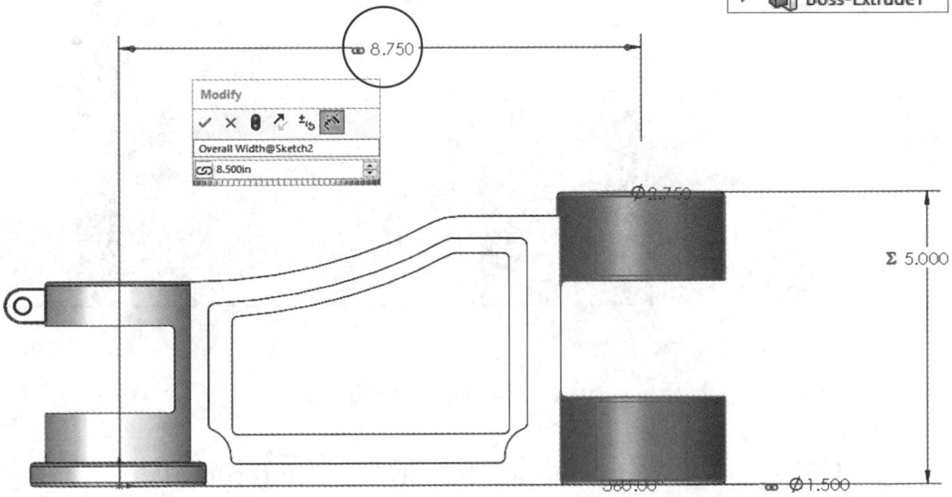

The dimension change triggers some errors in the model. These errors will be repaired after the geometry in the Sketch3 are corrected.

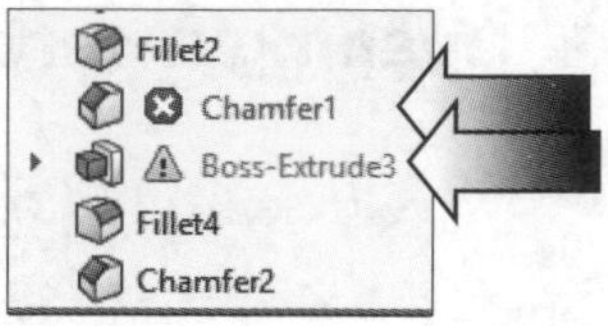

Edit the **Sketch3** under the Boss-Extrude1 feature (arrow).

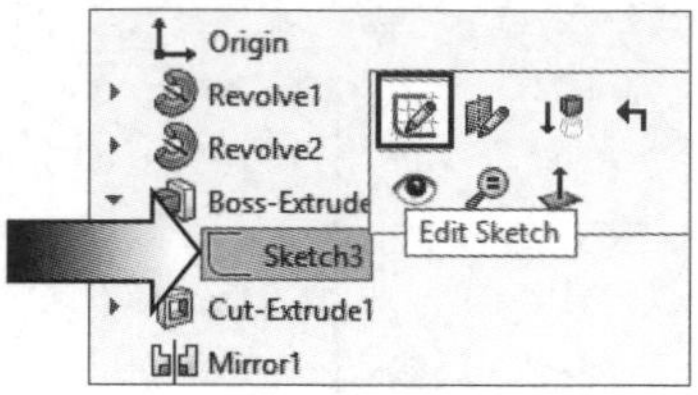

The dimension 12.250 prevents the line from being tangent to the arc. Delete the dimension **12.250** as indicated.

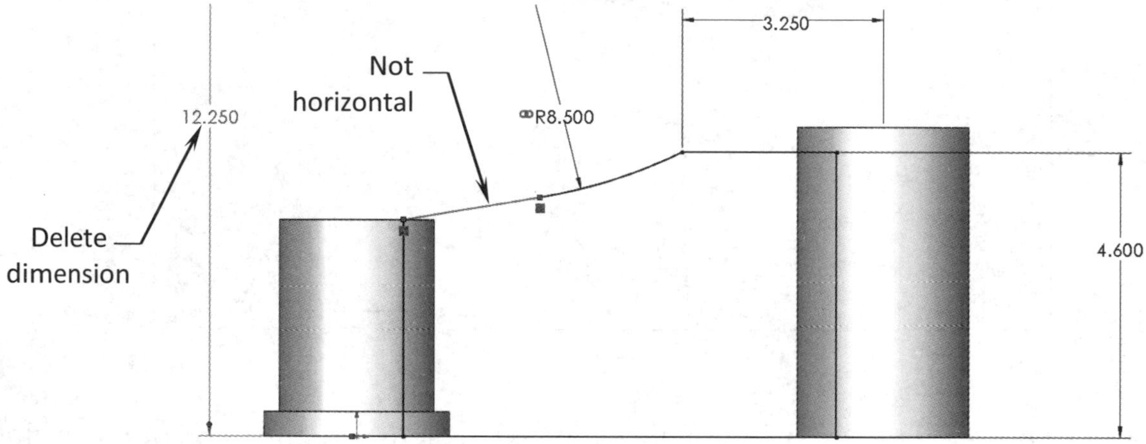

The sketch is no longer over defined.
Add a **Horizontal** relation to the line as noted.

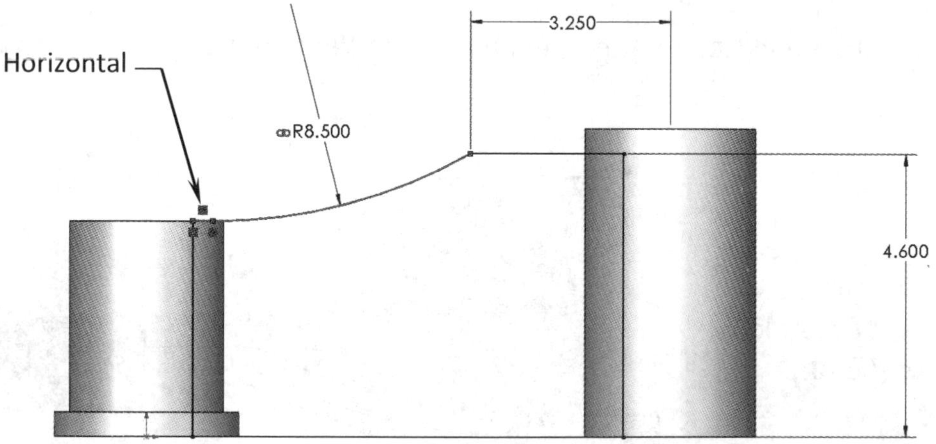

Exit the sketch when it is fully defined.

The errors are automatically corrected and removed.

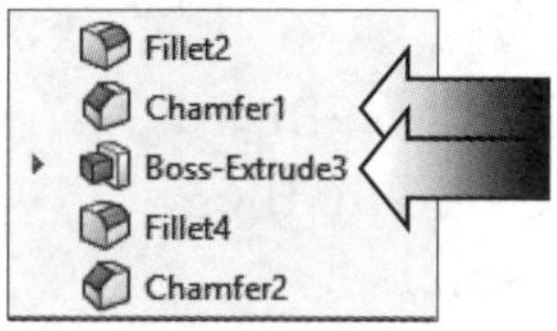

3. Suppressing features:

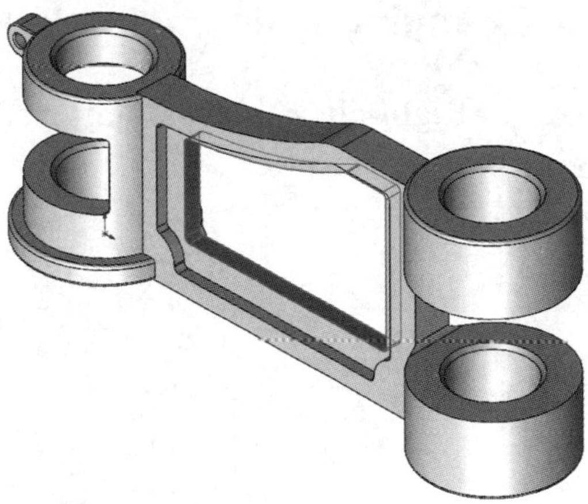

Locate the feature named **Cut-Extrude3** from the FeatureManager tree. It is the cutout feature in the middle of the part.

Click the **Cut-Extrude3** feature and select **Suppress**.

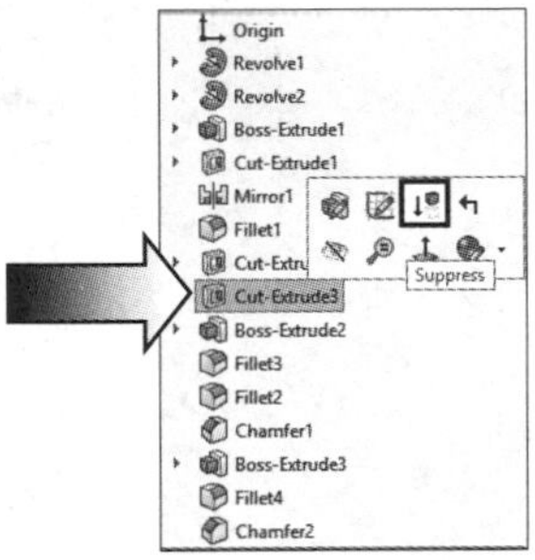

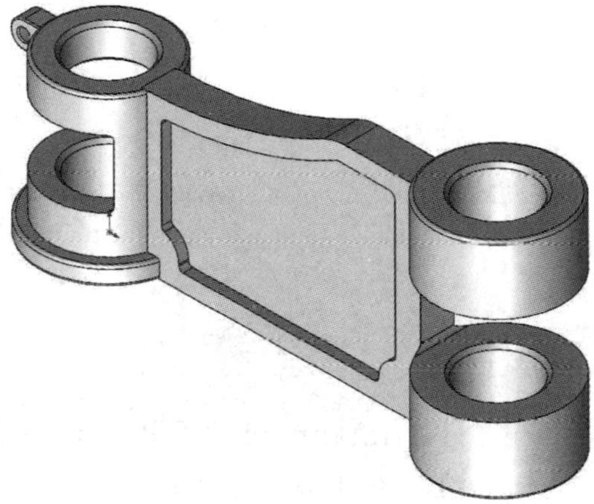

The cutout feature is suppressed.

4. Calculating the mass:

Switch to the **Evaluate** tab.

Click **Mass Properties**.

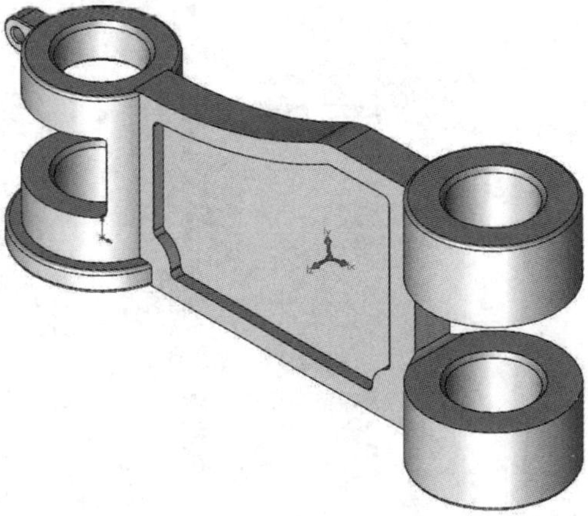

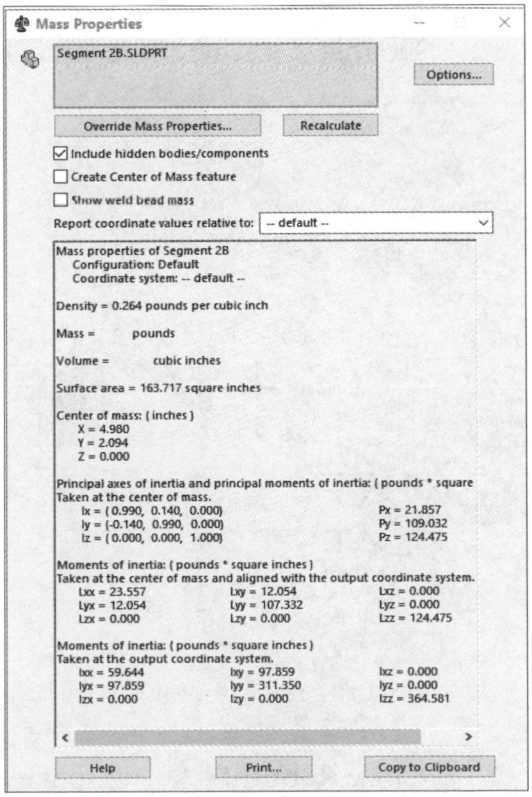

Enter the mass of the part here:

_______________ pounds.

5. Changing a dimension linked to an equation:

Right-click the Annotations folder and select **Display Annotations**.

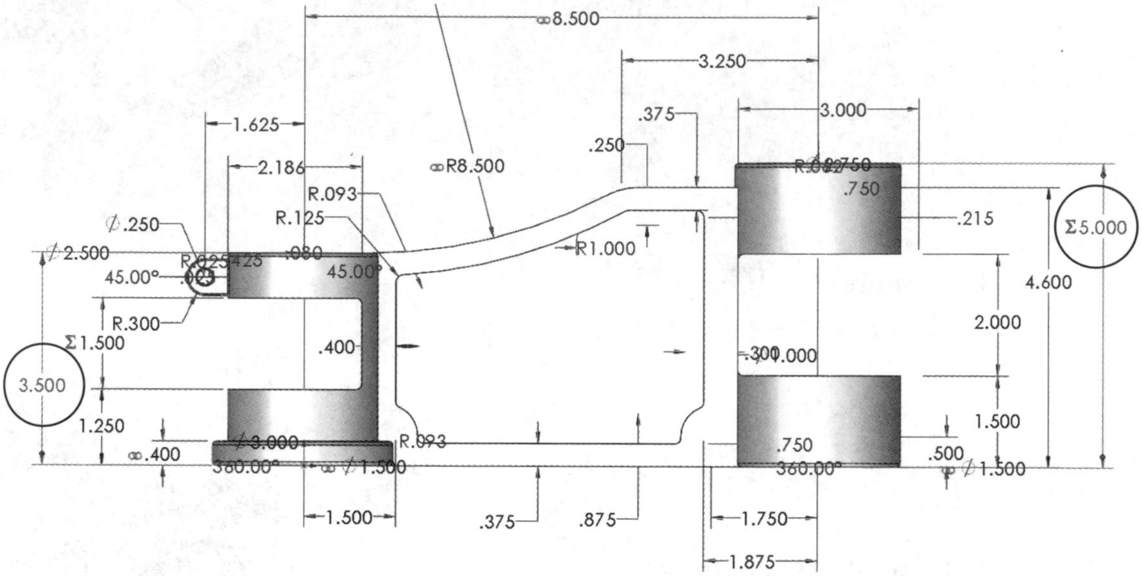

The dimension 3.500 (circled) is tied to the dimension 5.000 (circled) in an equation that was previously created. It is the driving dimension.

Double-click the dimension **3.500** and <u>change</u> it to **4.000**.

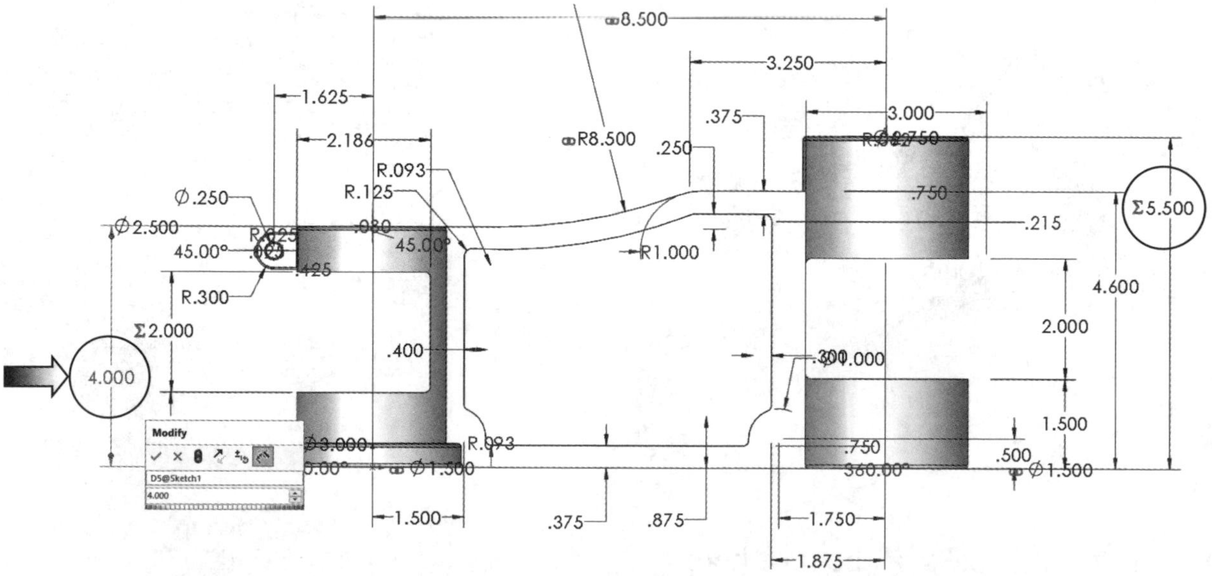

Click **Rebuild** to update the model. <u>Hide</u> all annotations.

6. Modifying the revolved feature:

Select the Top plane and open a **new sketch**.

Select the outer edge of the revolved feature (on the right side) and create an **Offset Entity** using a distance of **.300in**. (smaller). Also **Convert** the outer edge of the same revolved feature into a circle as indicated.

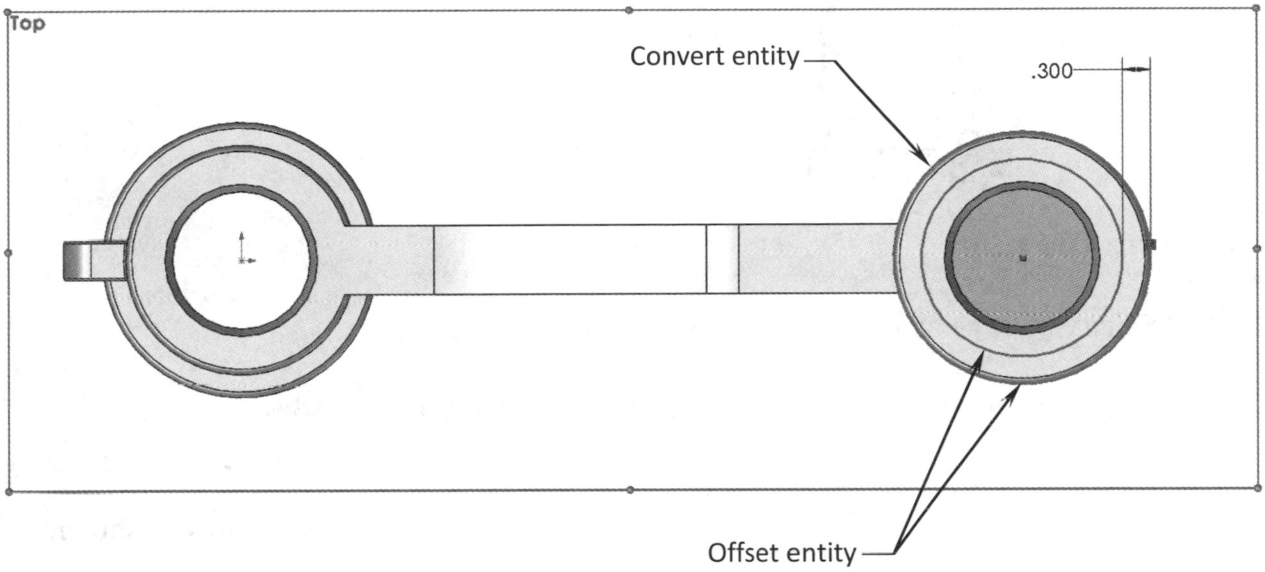

Switch to the **Features** tab and click **Extruded Cut**.

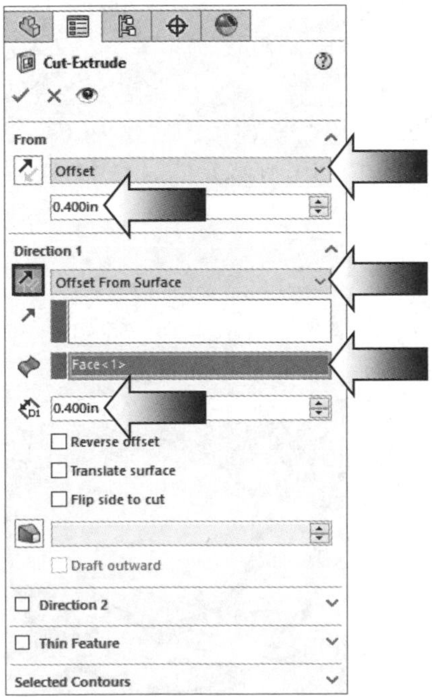

For **Extrude From**, select **Offset** and enter **.400** (arrow).

For **Direction 1**, select **Offset From Surface** (arrow).

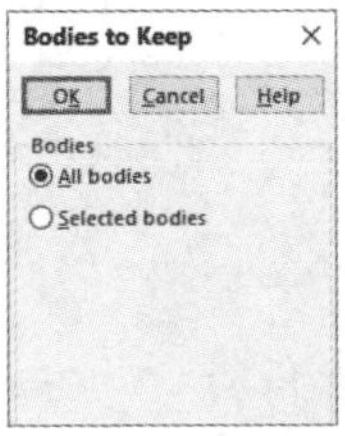

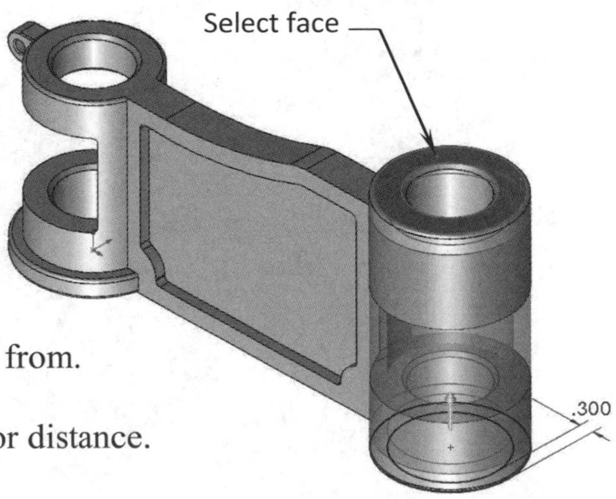

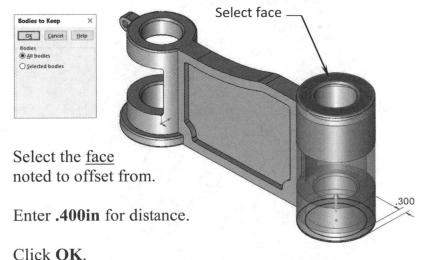

Select the face noted to offset from.

Enter **.400in** for distance.

Click **OK**.

The cut feature causes the part to have 2 disjoined solid bodies. The gap between the 2 bodies must be filled in.

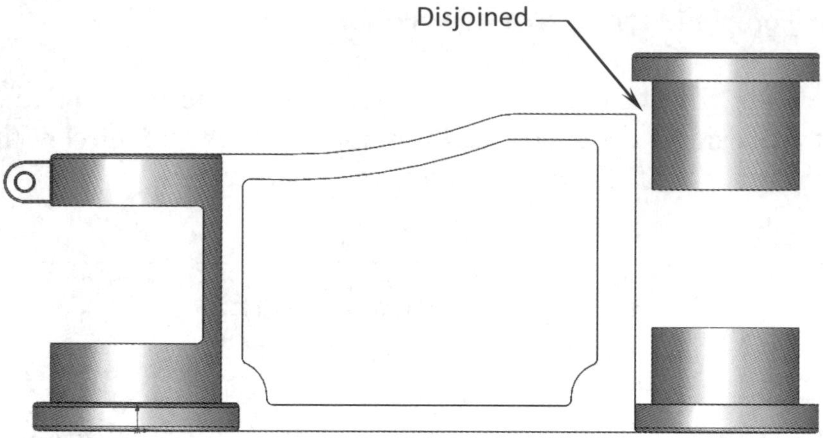

7. Filling the gap:

Select the **face** as indicated and open a new sketch.

Select the outer edge of the cylinder and click **Convert Entities** as noted.

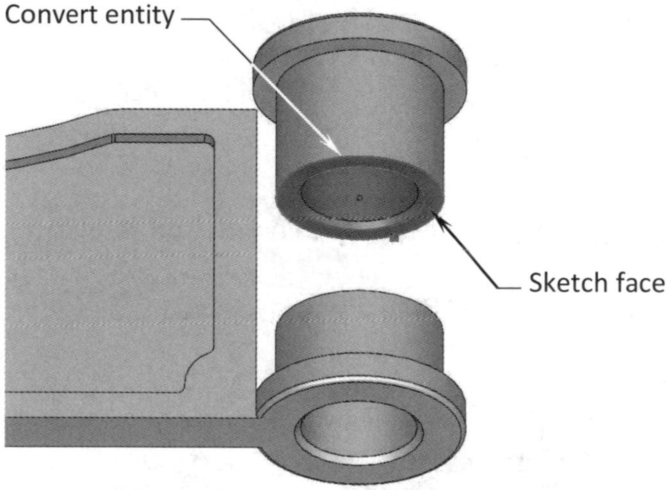

Add 3 additional **lines** as shown below.

Trim the circle and add the **1.500** dimension.

Add **Collinear** relations between the lines and the model edges.

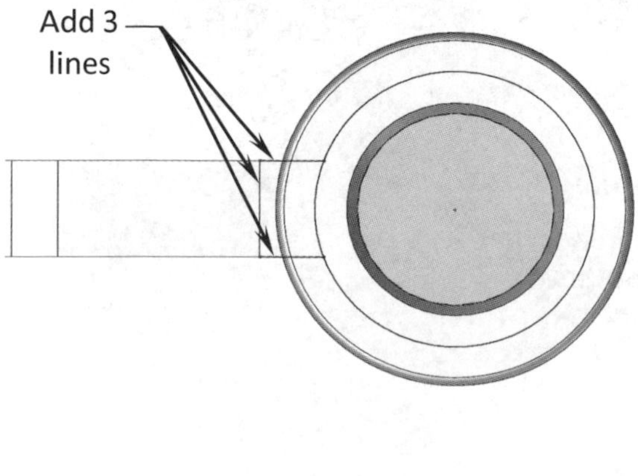

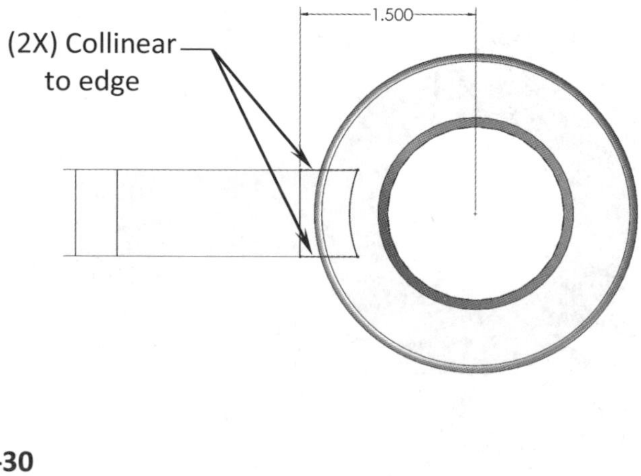

Switch to the **Features** tab and select **Extruded Boss-Base**.

For Direction 1, select **Up-to-Surface** and click the face as noted.

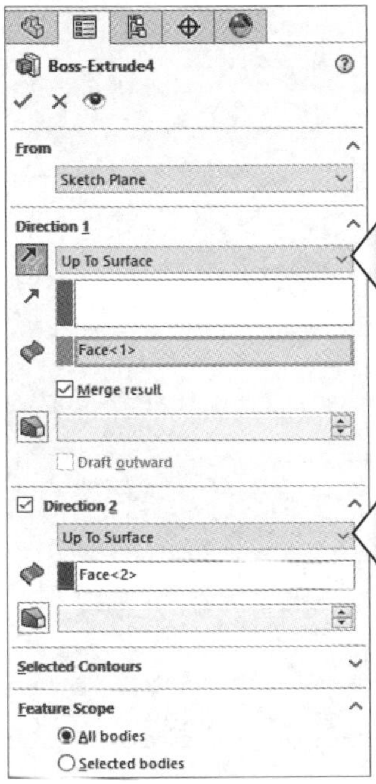

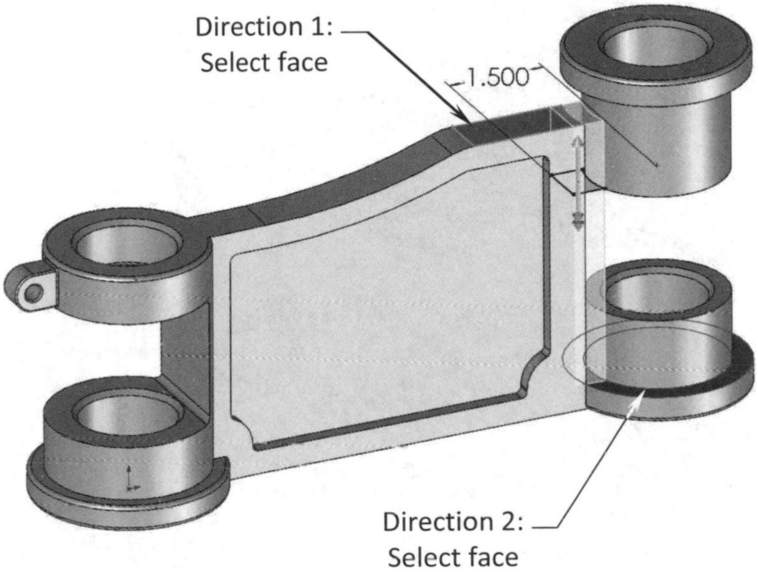

For Direction 2, also select **Up-to-Surface** and click the face indicated.

Click **OK**.

The extruded feature joins the 2 solid bodies to a single body.
Rotate the model and examine the model from different angles.

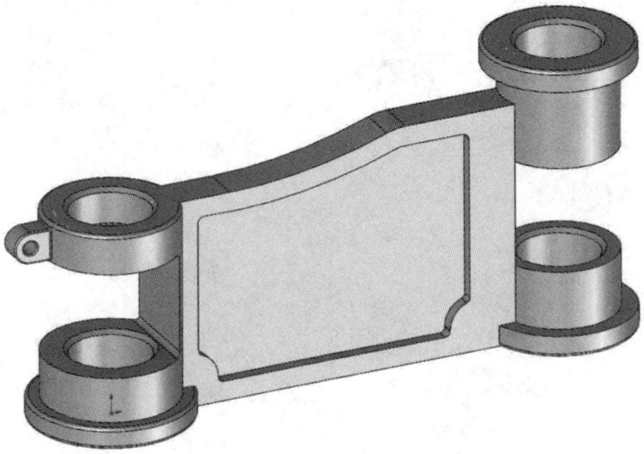

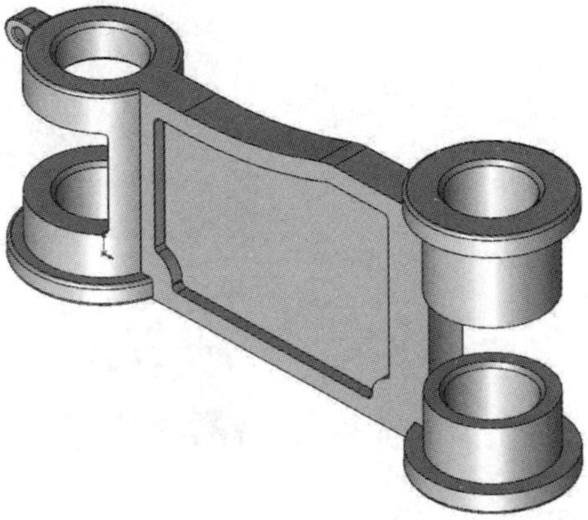

8. Calculating the mass:

Switch to the **Evaluate** tab.

Click **Mass Properties**.

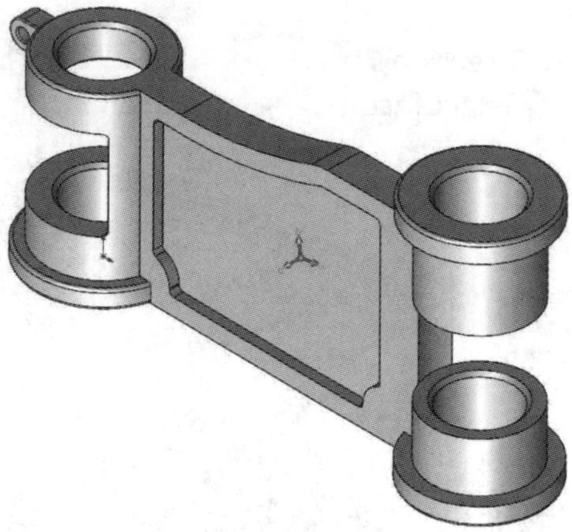

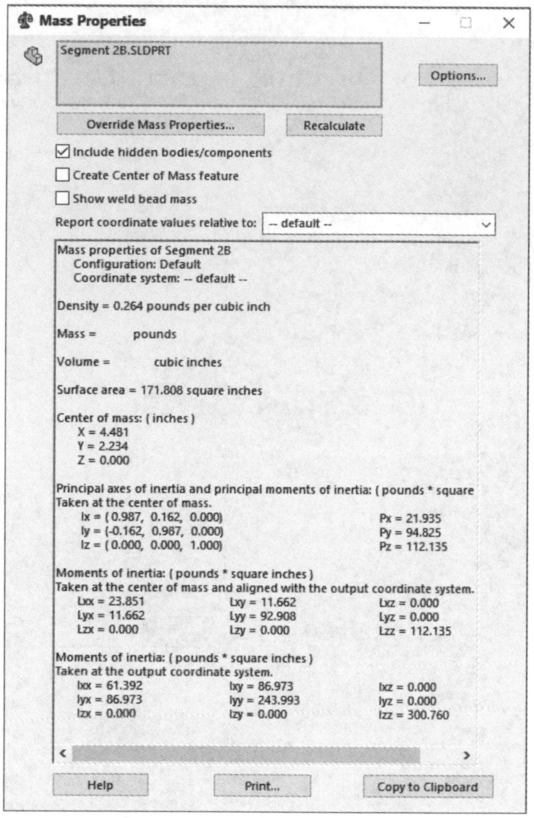

Enter the mass of the part here:

_______________ pounds.

9. Making the final changes:

Change the dimension **8.500** in the **Revolve2** to **9.000** (circled).

Change the dimension **4.000** in the **Revolved1** to **3.400** (circled).

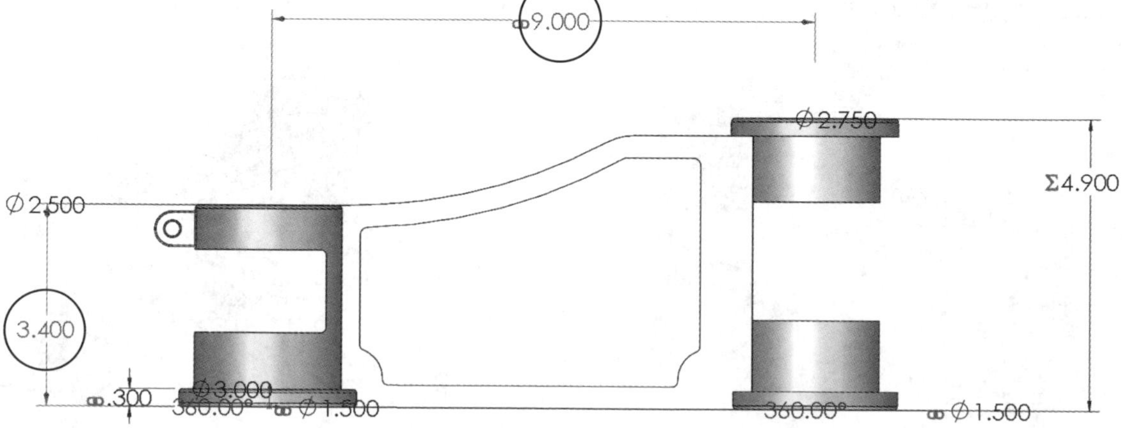

Change the dimension **.400** in the **Revolved1** to **.300** (circled).

Also change both of the dimension **.400** in the **Cut Extrude4** to **.300** (circled).

Click **Rebuild** to update the model.

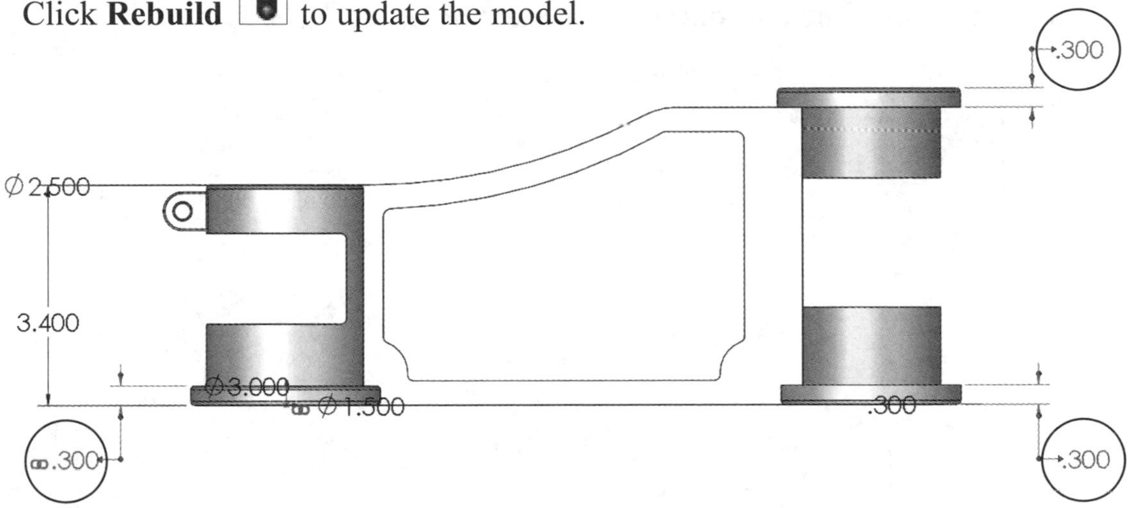

10. Calculating the mass:

Switch to the **Evaluate** tab.

Click **Mass Properties**.

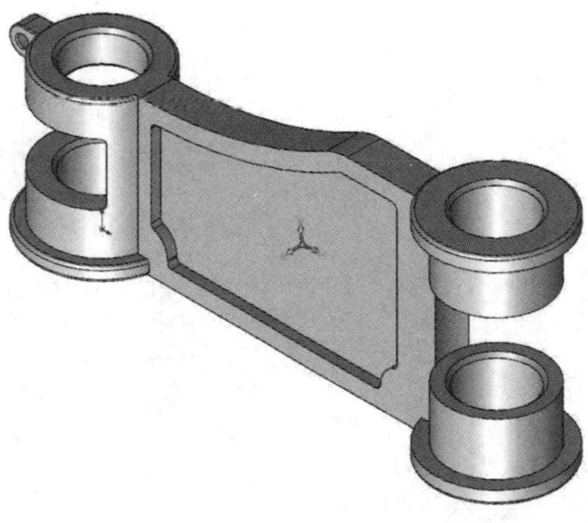

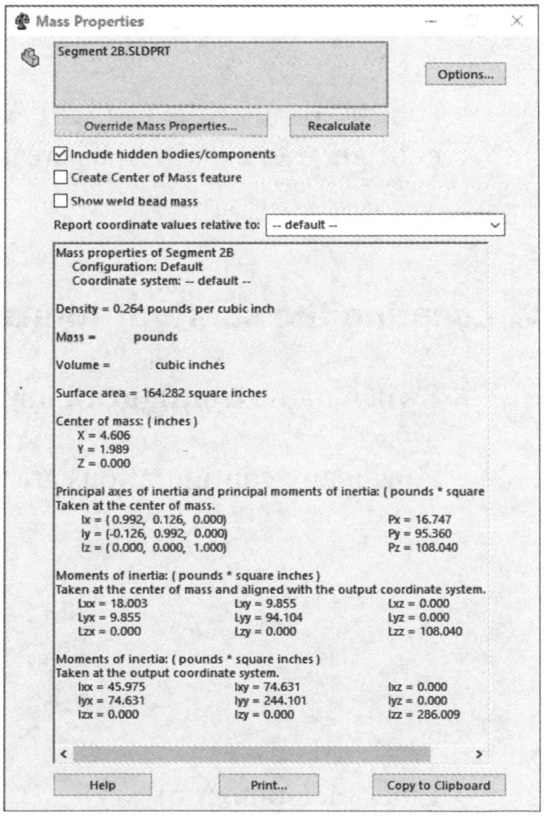

Enter the mass of the part here:

_______________ pounds.

Save you model as **Segment 2B (Completed)** and close the part document.

1. Opening a part document: (3 of 3)

Click **File / Open**.

Locate and open the part document named: **Segment 2C.sldprt**

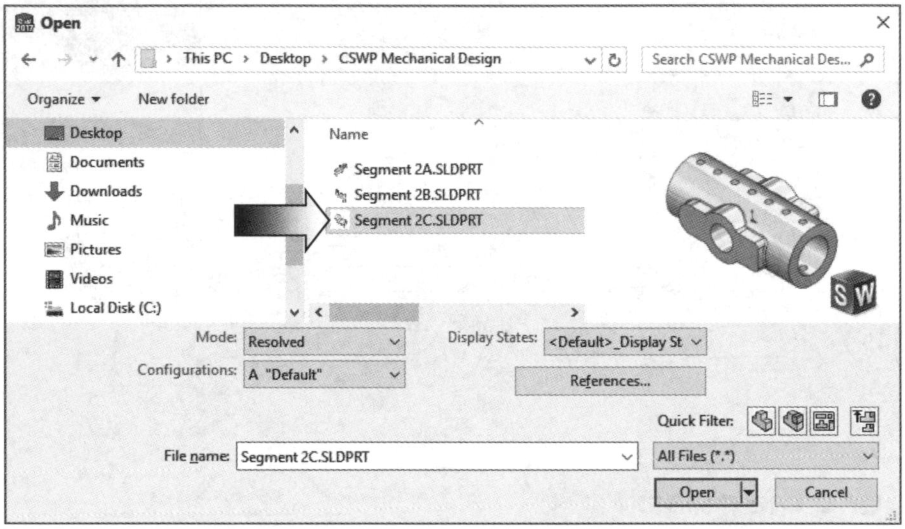

(This portion of the test will examine your skills on the modification of existing configurations, as well as creating new configurations in a part.)

2. Locating the configurations:

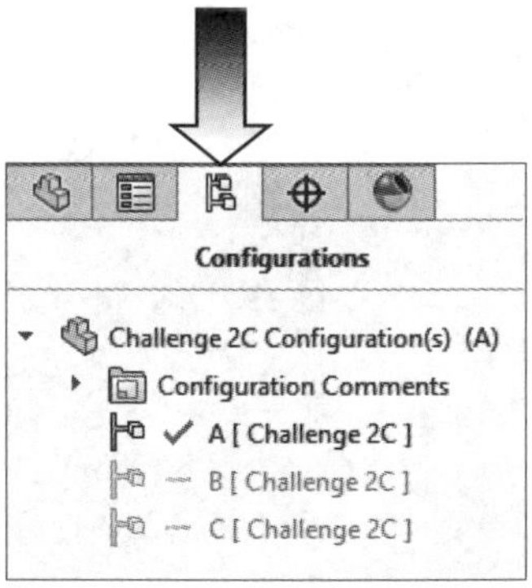

Switch to the **ConfigurationManager** (arrow).

How many configurations are there in this model?

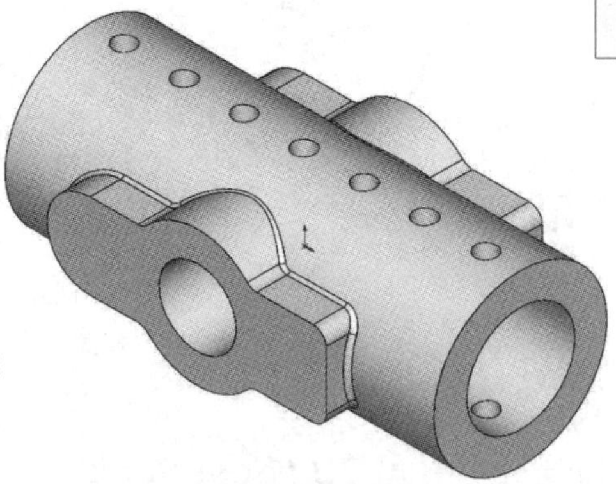

Enter the number of configurations found here:

_______ (1, 2, or 3).

3. Calculating the mass:

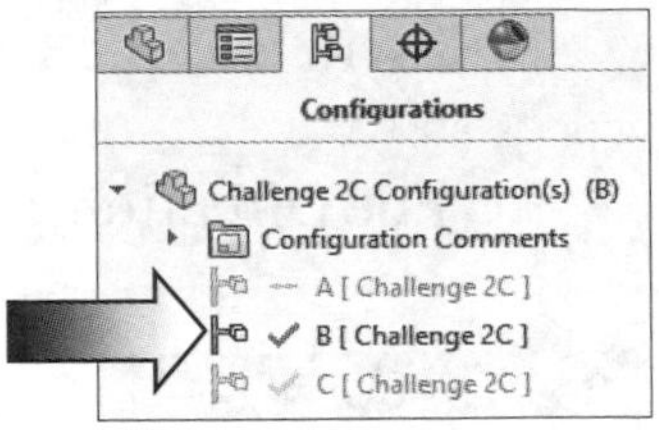

Double-click configuration **B** to activate it. Each configuration has its own material and different feature sizes.

Switch to the **Evaluate** tab.

Click **Mass Properties**.

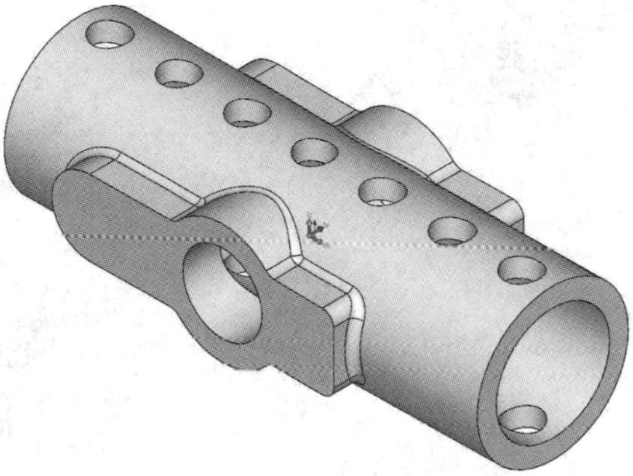

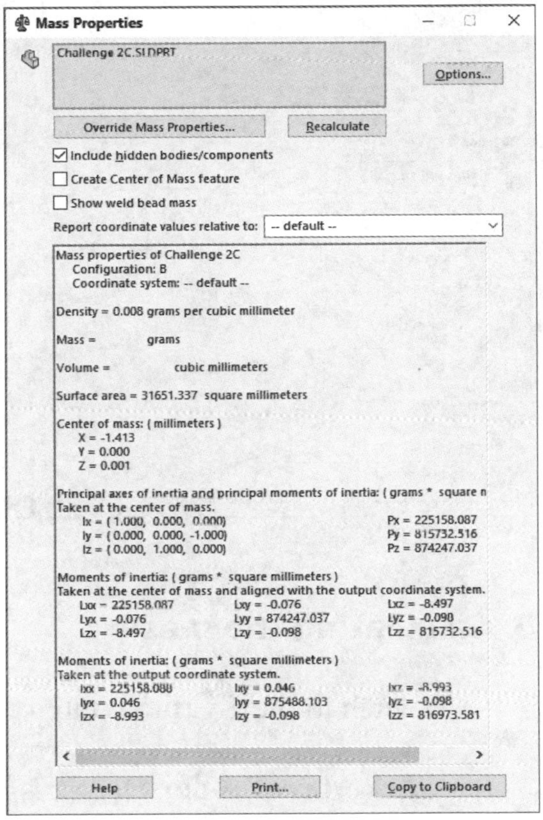

Enter the mass of **Configuration B** here:

_______________ pounds.

4. Adding a hole:

Double-click **Configuration C** to activate it.

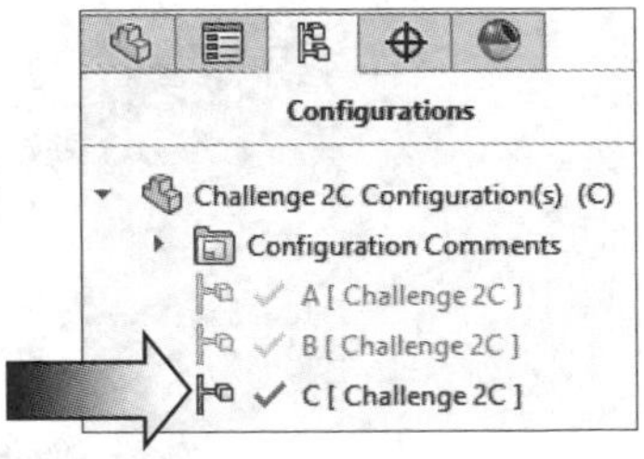

Open a **new sketch** on the face as noted.

Sketch a **circle** and add a diameter dimension to fully define the sketch.

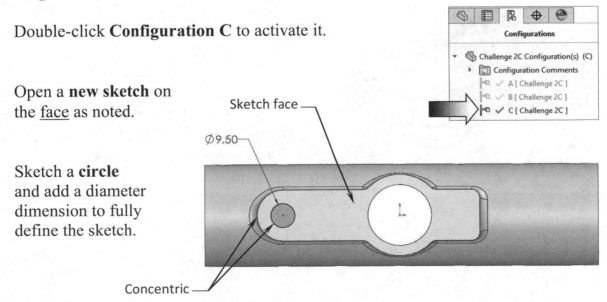

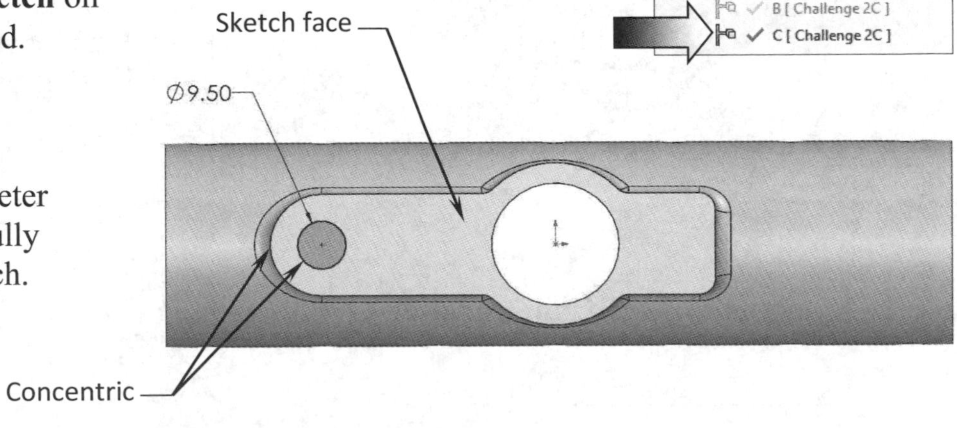

Switch to the **Features** tab and click **Extruded Cut**.

For Direction 1, select **Through All**.

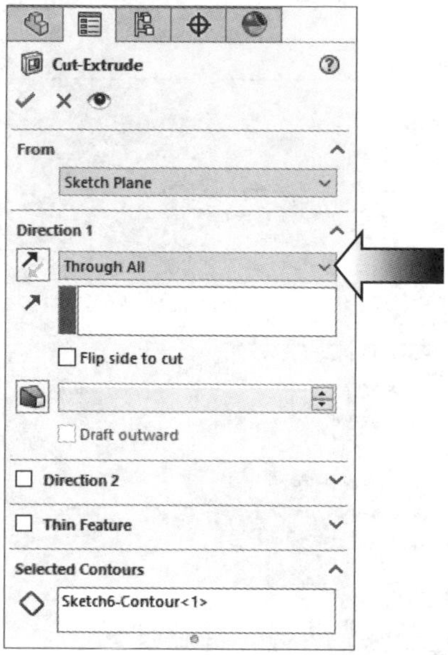

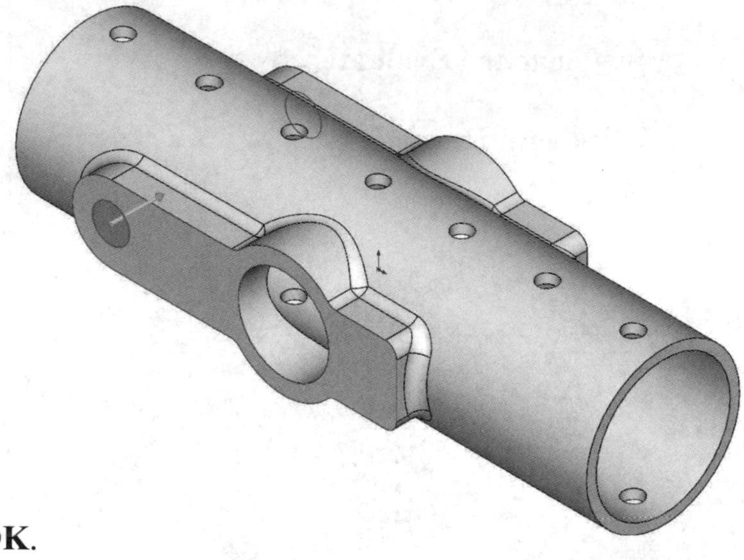

Click **OK**.

5. Calculating the mass:

Switch to the **Evaluate** tab.

Click **Mass Properties**.

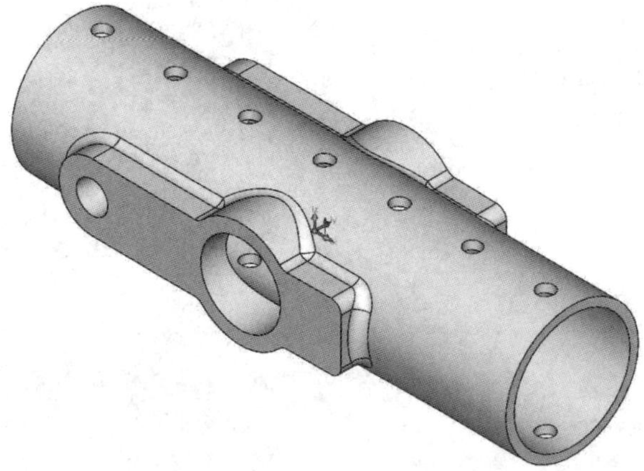

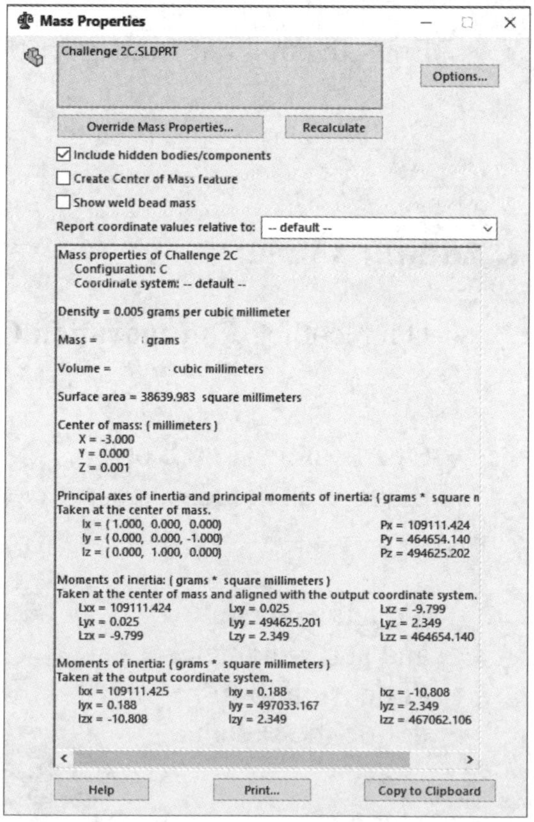

Enter the mass of **Configuration C** here:

_______________ pounds.

6. Creating a new configuration:

Double-click **Configuration A** to activate it.

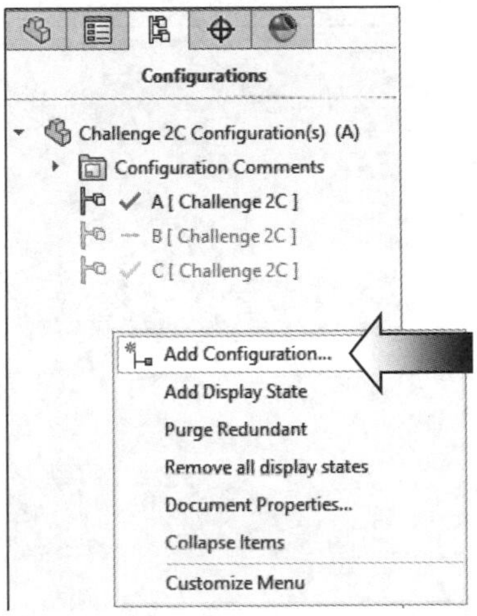

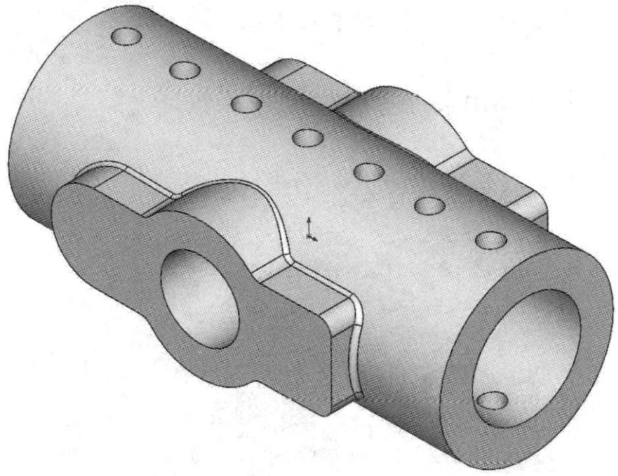

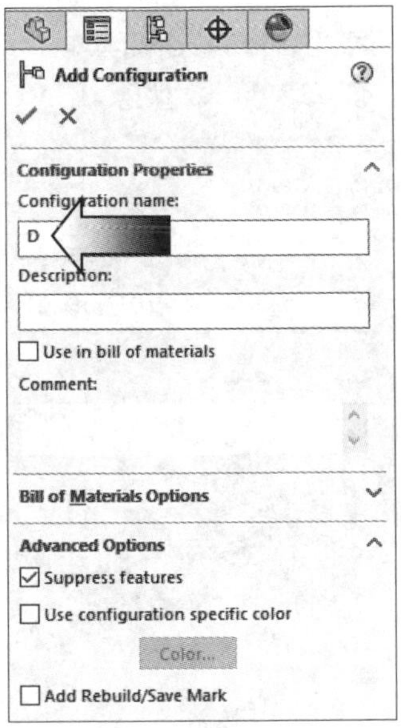

Right-click the name of the part and select: **Add Configuration**.

Enter **D** for the name of the new configuration.

Change the **3 dimensions** to match the values in the circles.

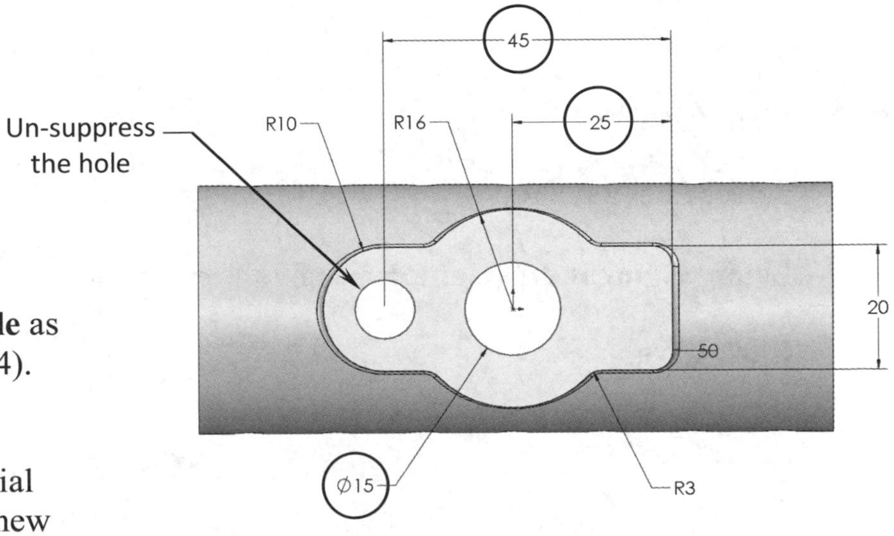

Un-suppress the **hole** as noted (Cut-Extrude4).

Use the same material **1060 Alloy** for the new configuration.

The hole (Cut-Extrude4) should be <u>unsuppressed</u> in the configurations C and D.

The hole (Cut-Extrude4) should be <u>suppressed</u> in the configurations A and B.

7. Calculating the mass:

Remain in **Configuration D**.

Switch to the **Evaluate** tab.

Click **Mass Properties**.

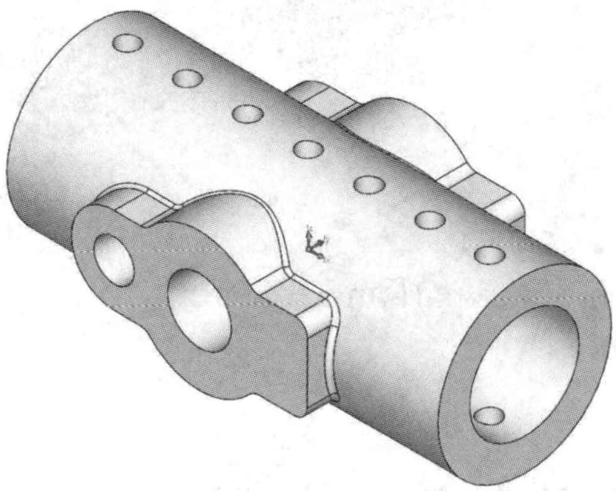

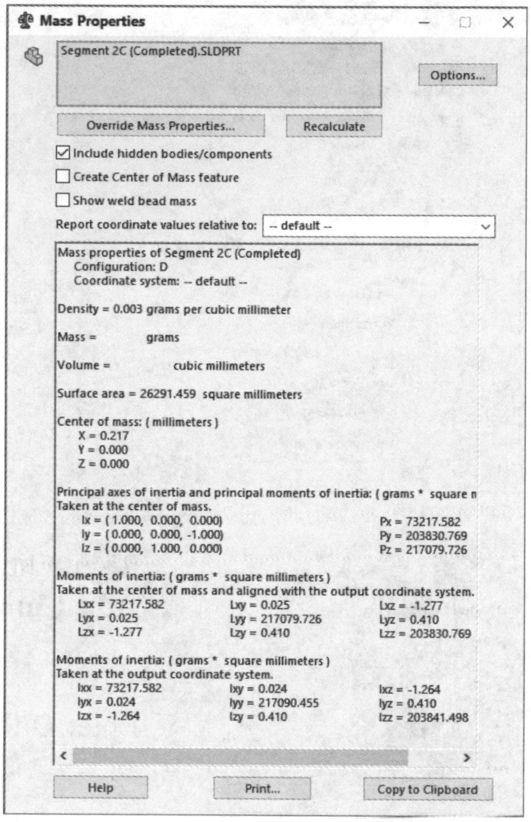

Enter the mass of configuration **D** here:

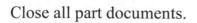

_______________ pounds.

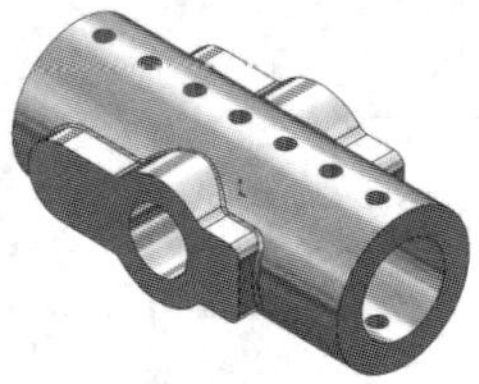

8. Saving your work:

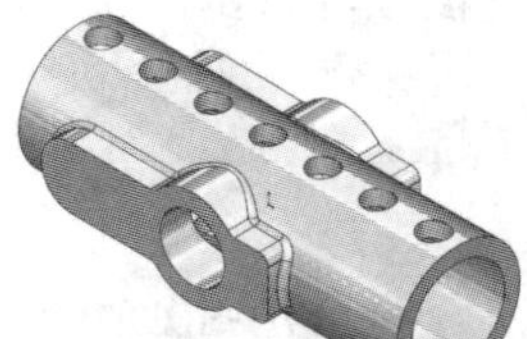

Select **File, Save As**.

Enter: **Segment 2C (Completed)** for the file name.

Click **Save**.

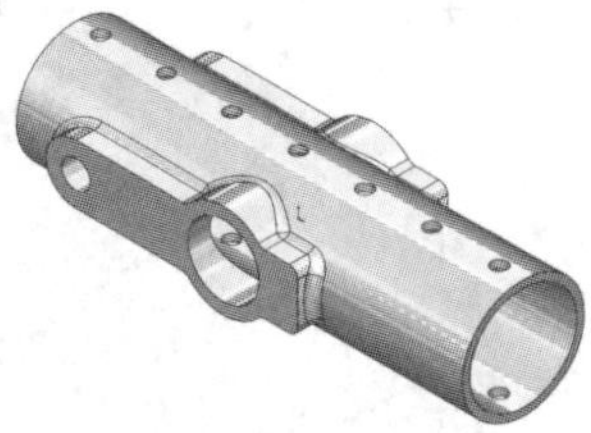

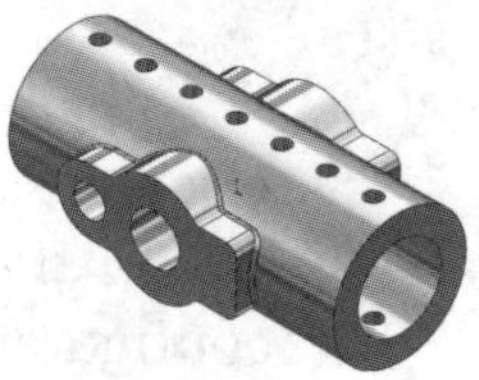

Close all part documents.

Certified-SOLIDWORKS-Professional (CSWP)

Certification Practice for the Mechanical Design Exam

Challenge III: Bottom Up Assembly & Mates

Complete this challenge within 80 minutes

(The following examples are intended to assist you in familiarizing yourself with the structures of the exams and the method in which the questions are asked.)

Modify this part using SOLIDWORKS 2010 or newer
Unit: **Inches, 3 decimals**
Drafting Standards: **ANSI**
Origin: **Arbitrary**
Material: **Already Specified for each part**

(This portion of the test will examine your skills on constraining components using the Bottom Up Assembly approach.)

1. Starting a new assembly document:

Click **File / New**.

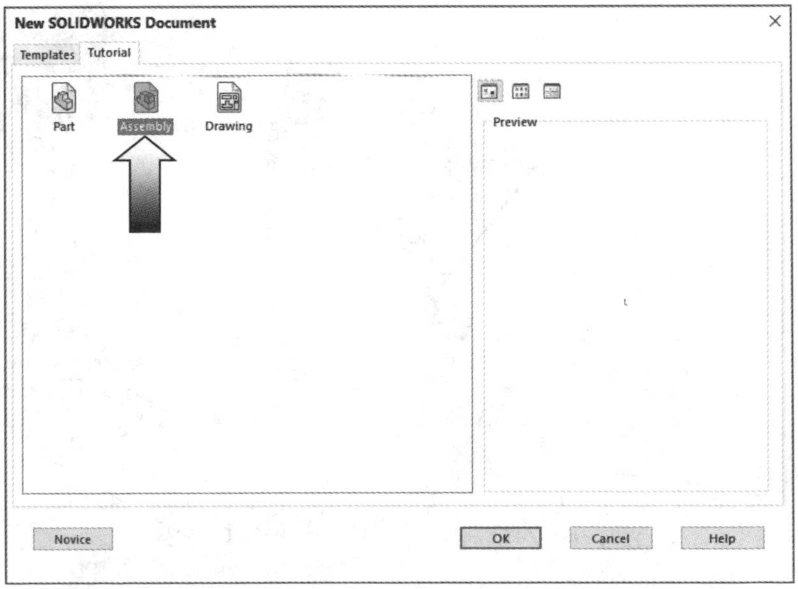

Select an **Assembly Template** and click **OK**.

Change the Drafting Standard to **ANSI** and the Units to **IPS**, 3 decimal places.

You will be asked to demonstrate the use of different mate types to constrain the component in an assembly.

2. Inserting the 1st component:

Change to the **Assembly** tab and select: **Insert Component**.

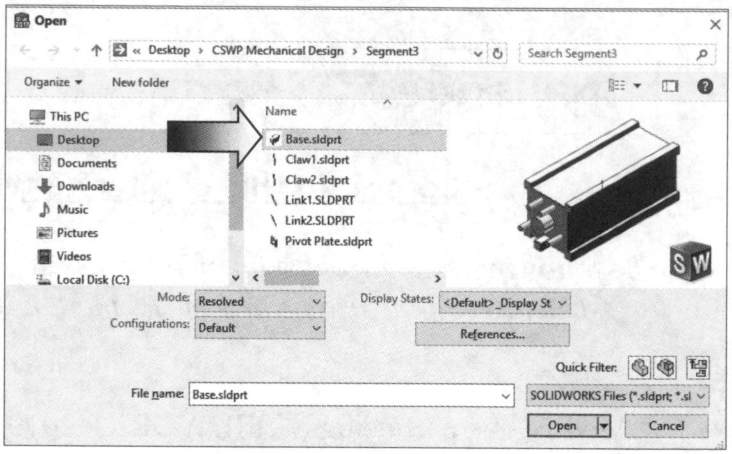

Locate the component named **Base.sldprt** and click **Open**.

Click the **Green check** to place the component on the Origin.

3. Inserting other components:

Insert all other components into the assembly.

Place the components approximately as shown.
Make a copy of the **Link1** and **Link2**.

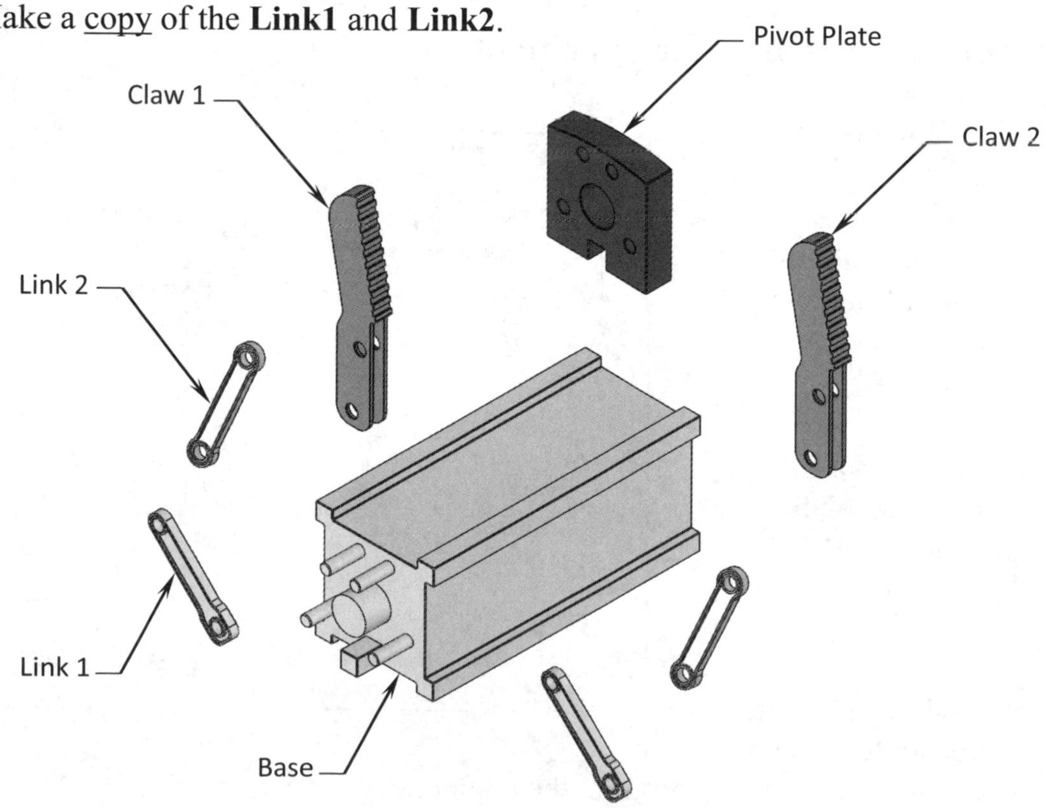

Rotate the Claw and the Link ahead of time to make it easier to see and to mate them to the Base.

Also hide the origins.
Click **View**, **Hide/Show**, **Origins**.

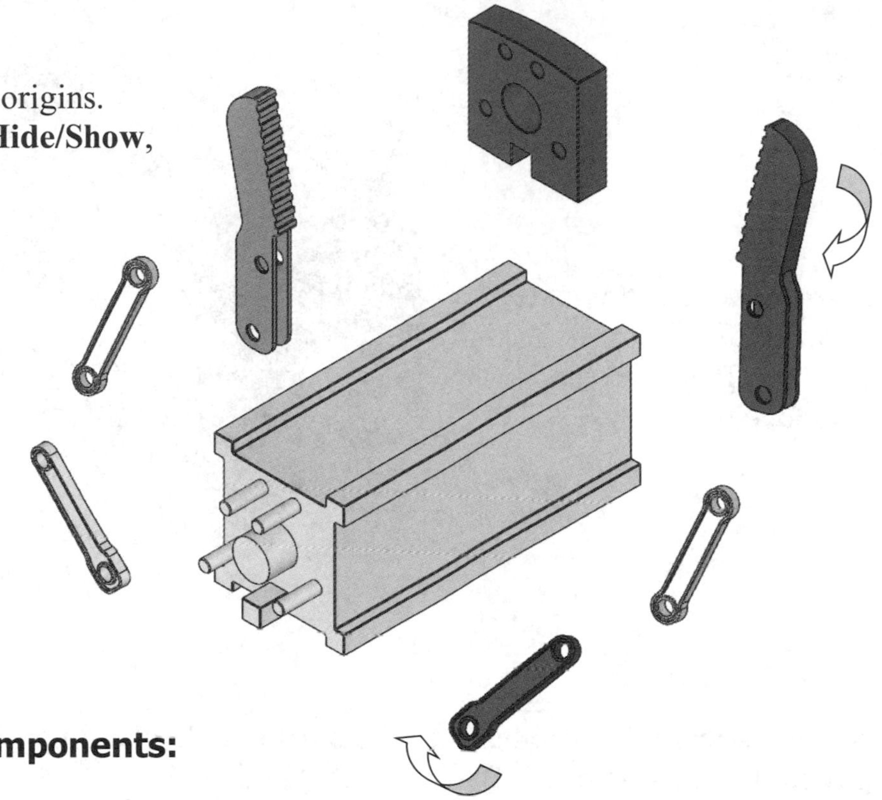

4. Mating the components:

The **F5** function key enables the **Selection Filters**. Use the **Filter Faces** (press the letter **X** shortcut key) to help select the faces more easily.

Hold the **Control** key and select the <u>hole</u> and the <u>shaft</u> as noted.

Release the control key to see the mate shortcuts popup options.

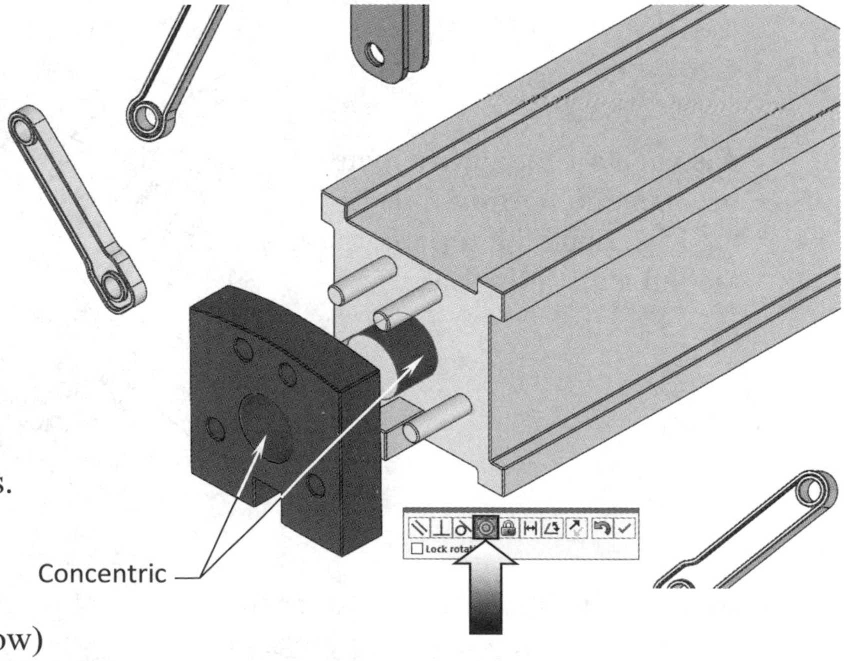

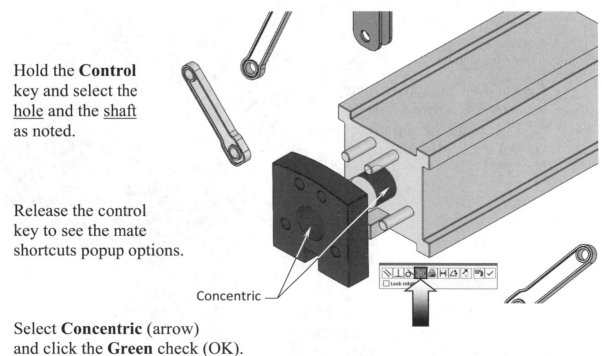

Select **Concentric** (arrow) and click the **Green** check (OK).

Add a **Coincident** mate between the **back face** of the Pivot Plate and the **front face** of the Base.

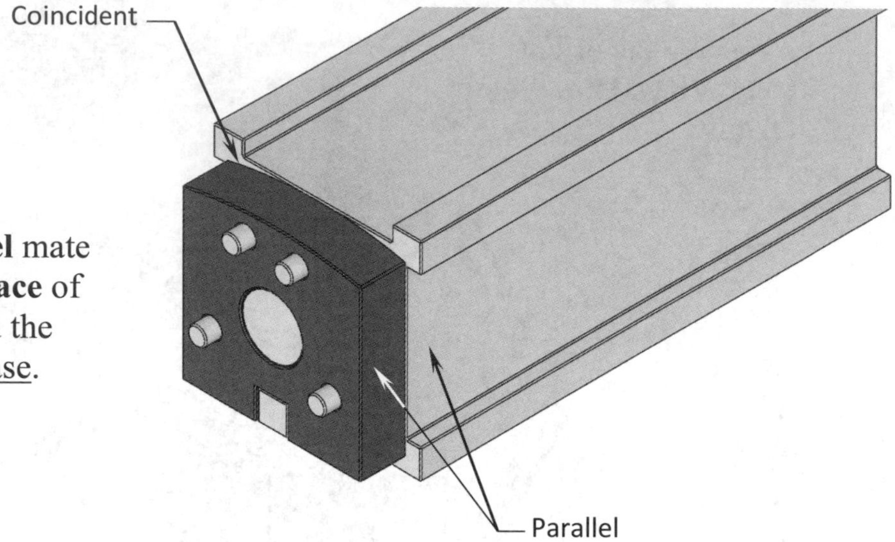

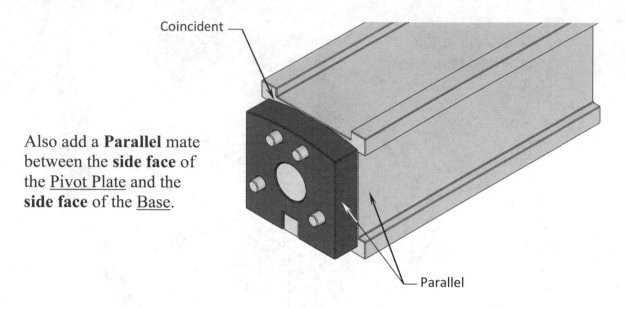

Also add a **Parallel** mate between the **side face** of the Pivot Plate and the **side face** of the Base.

Add a **Concentric** mate between the **bottom hole** of the Link1 and the **bottom pin** on the left side of the Base.

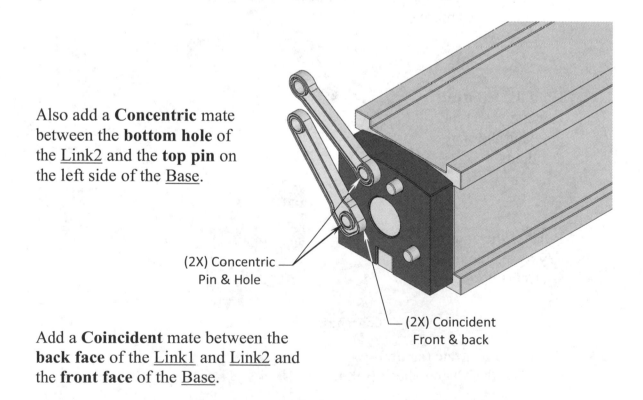

Also add a **Concentric** mate between the **bottom hole** of the Link2 and the **top pin** on the left side of the Base.

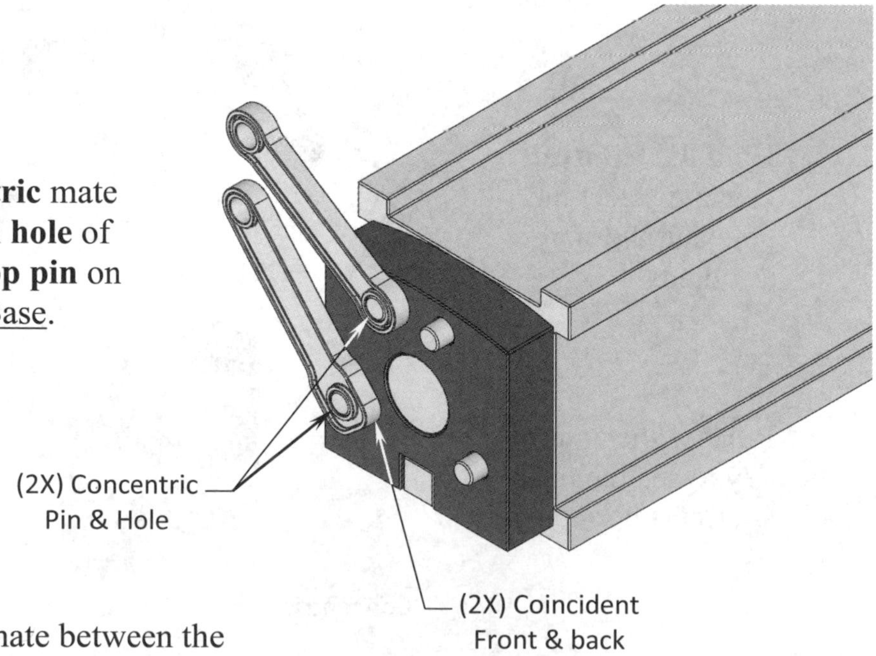

Add a **Coincident** mate between the **back face** of the Link1 and Link2 and the **front face** of the Base.

5. Assembling the 1st Claw:

Zoom closer to the Claw and the Links components.

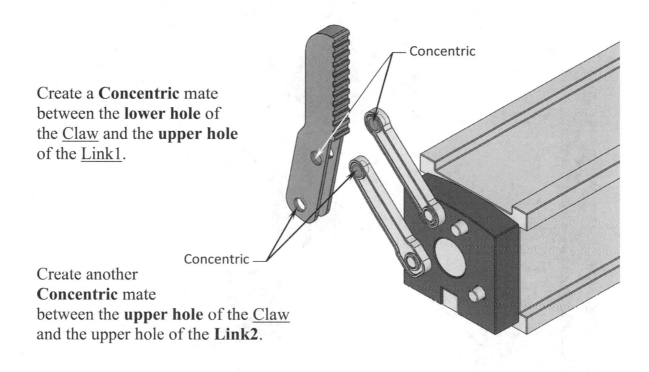

Create a **Concentric** mate between the **lower hole** of the Claw and the **upper hole** of the Link1.

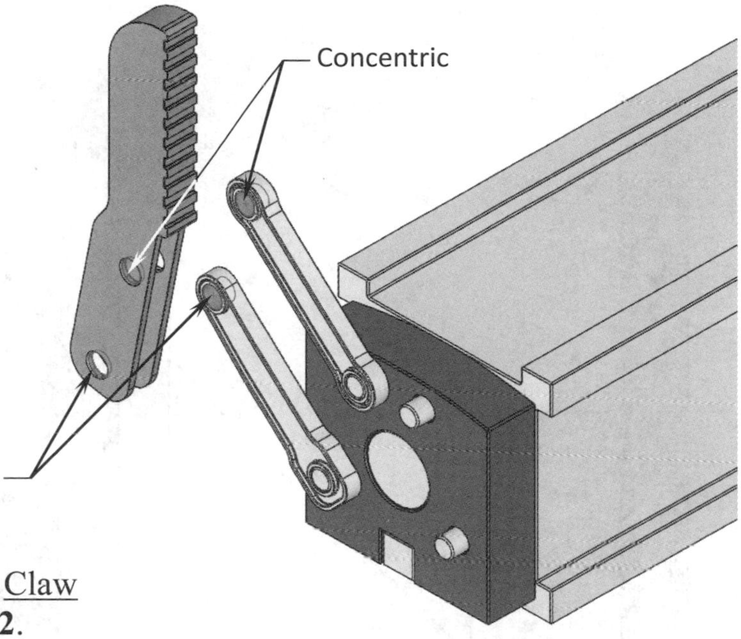

Create another **Concentric** mate between the **upper hole** of the Claw and the upper hole of the **Link2**.

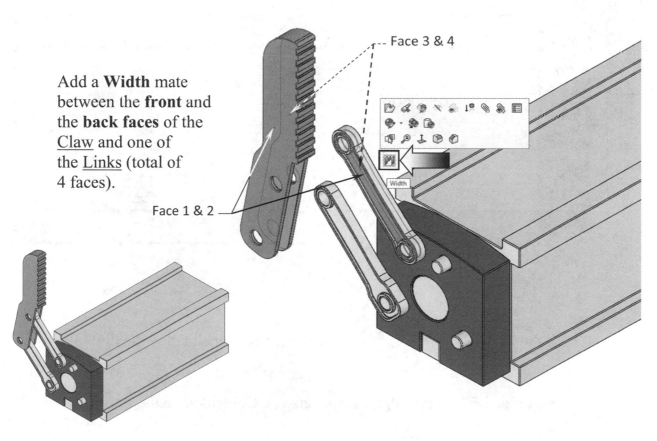

Add a **Width** mate between the **front** and the **back faces** of the Claw and one of the Links (total of 4 faces).

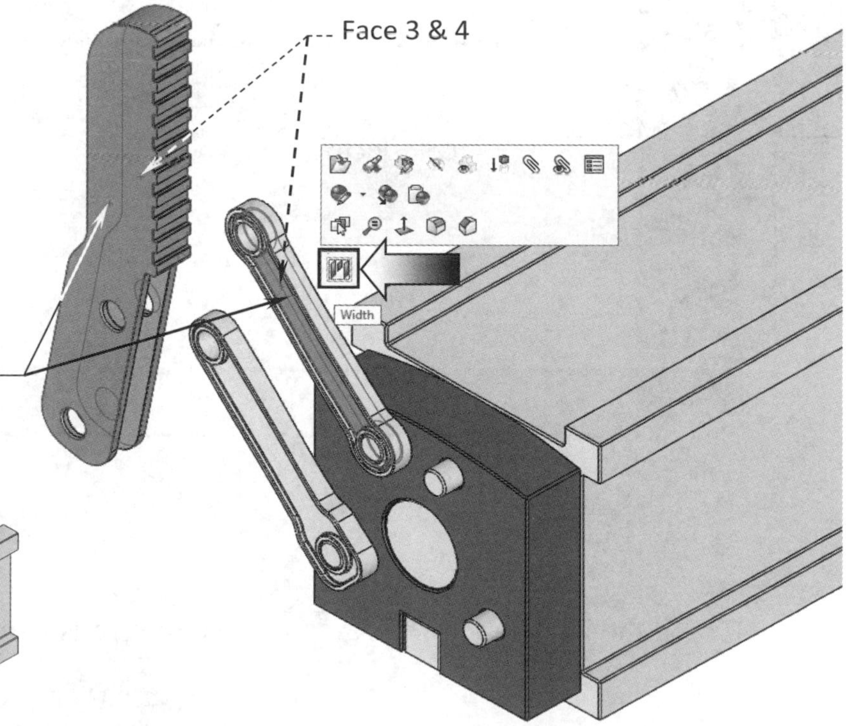

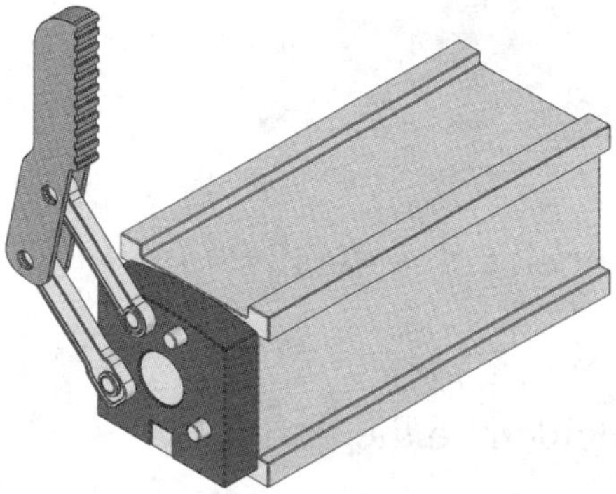

Change to the **Right** orientation (Control+4).

Zoom very close and examine the gaps between the Claw and the 2 Links. The gaps must be exactly the same on both sides.

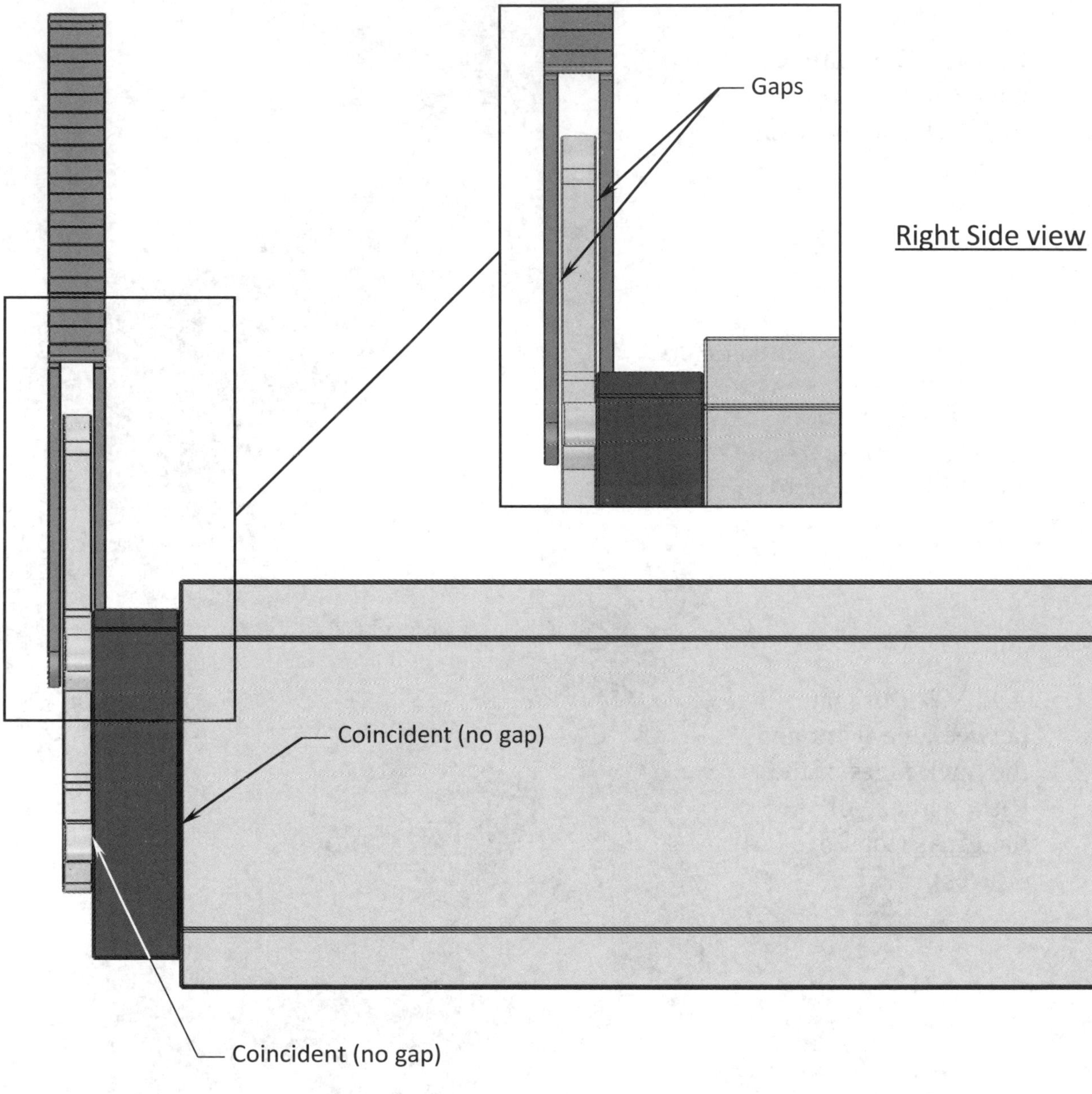

Check the back side of the 2 Links to ensure it is Coincident to the Pivot Plate.

The back side of the Pivot Plate should also be Coincident to the Base.

6. Assembling the 2nd Claw:

Add a **Concentric** relation and a **Coincident** relation between the Link1 and the **Lower Pin** in the Base.

Add the same **Concentric** and **Coincident** relations between the Link2 and the **Upper Pin** in the Base.

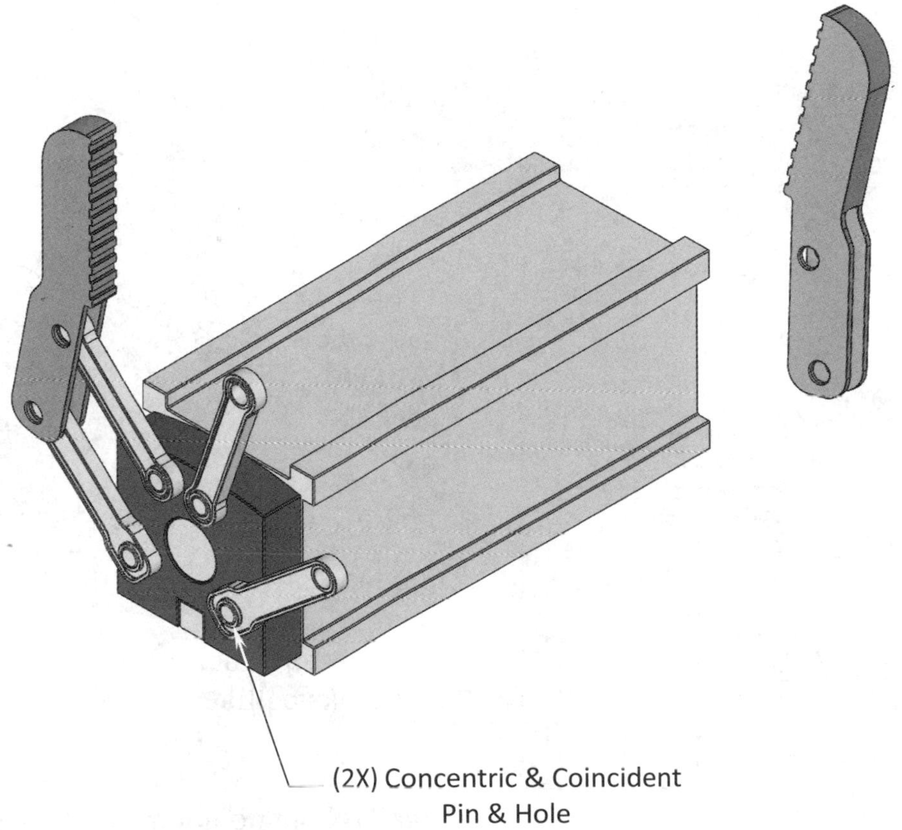

Also add a **Width** mate between the **Claw2** and one of the **Links**. (Use the front & back faces of the Claw2, and the front & back faces of the Link.)

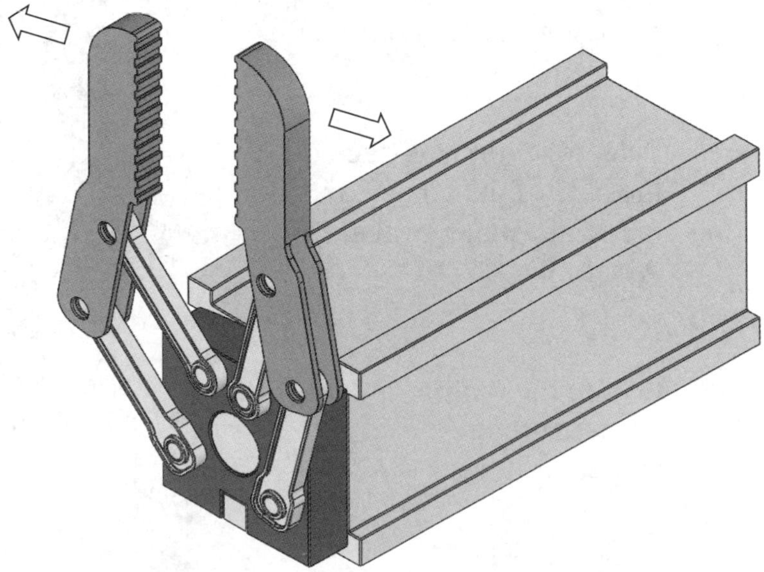

Move the 2 claws back and forth to ensure that they can be moved freely.

7. Detecting Collision:

Switch to the **Assembly** tab and click the **Move Component**  command.

Expand the **Options** section and select the **Collision Detection** option (arrow).

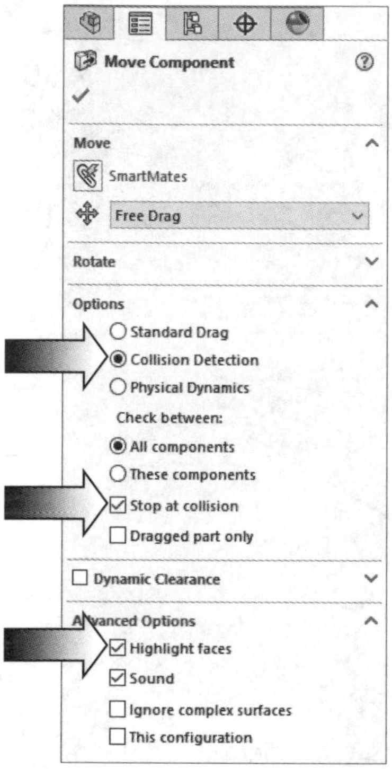

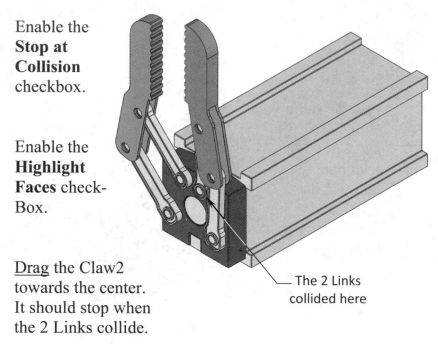

Enable the **Stop at Collision** checkbox.

Enable the **Highlight Faces** check-Box.

Drag the Claw2 towards the center. It should stop when the 2 Links collide.

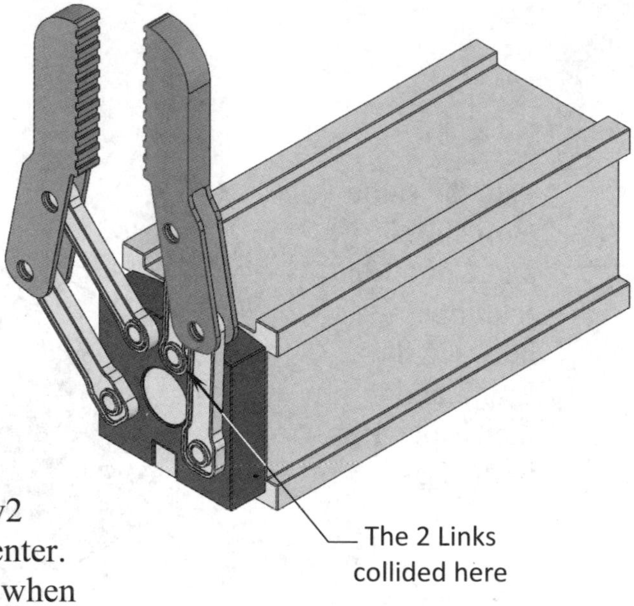

Click **OK** but do not move the Claw2.

8. Measuring the angle:

Switch to the **Evaluate** tab.

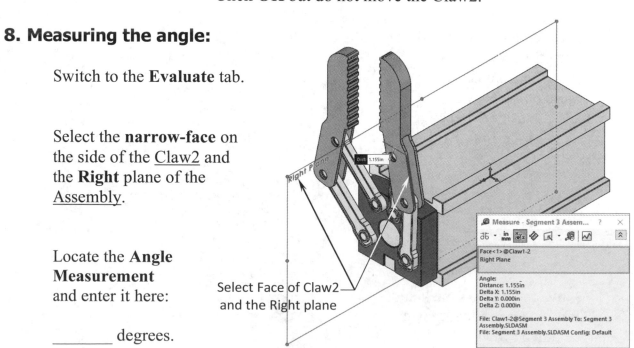

Select the **narrow-face** on the side of the Claw2 and the **Right** plane of the Assembly.

Locate the **Angle Measurement** and enter it here:

_______ degrees.

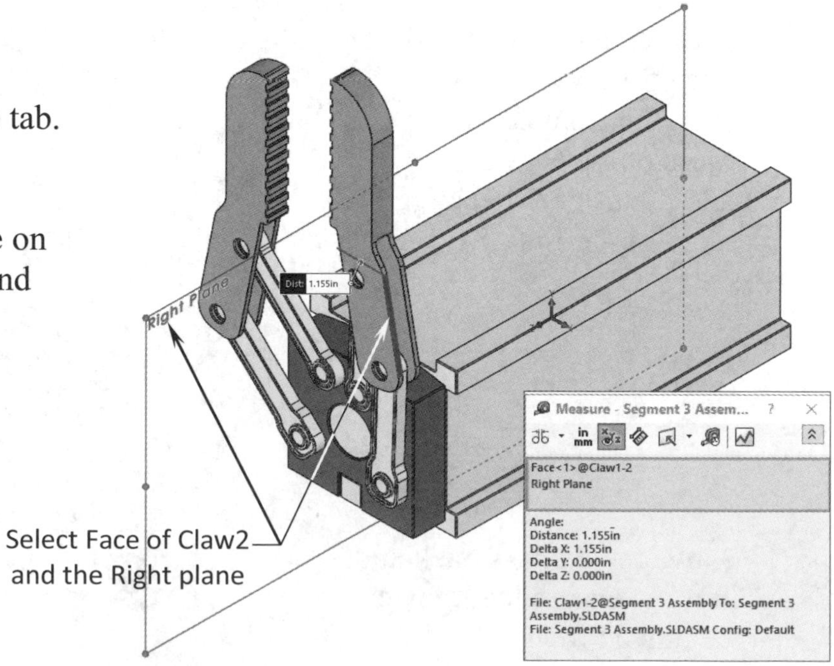

9. Moving components with Collision Detection:

Click **Move Component** .

Select the following options:

* **Collision Detection**
* **These Components** (select the 2 Claws)
* **Stop at Collision**
* **Highlight Faces**

Click the **Resume Drag** button and drag one of the claws until it stops.

It is important to keep the Claws at the collided position for measurement.

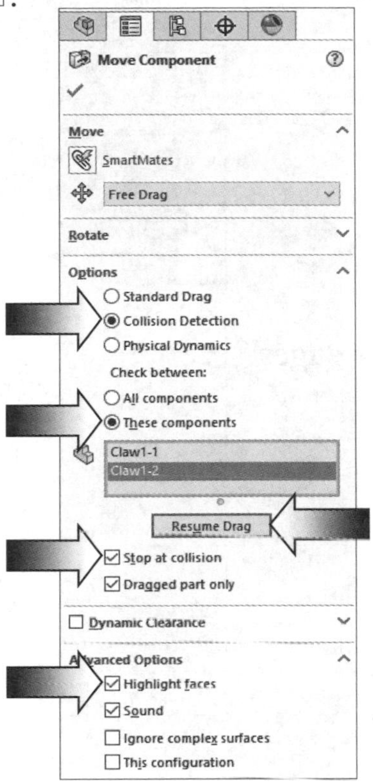

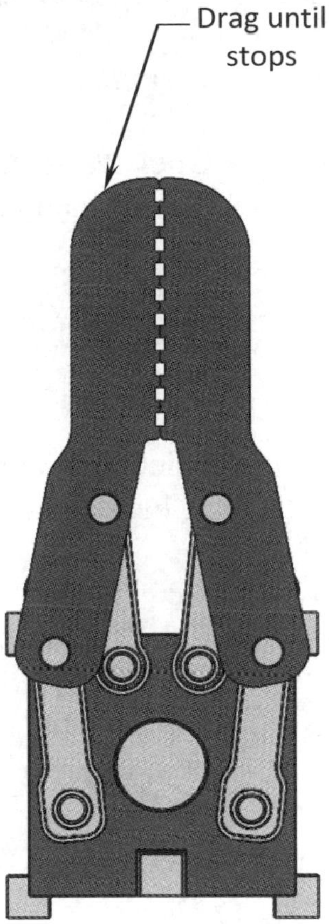

10. Calculating the Center of Mass:

Switch to the **Evaluate** tab.

Click **Mass Properties**.

Enter the **Center of Mass** here:

X = __________

Y = __________

Z = __________

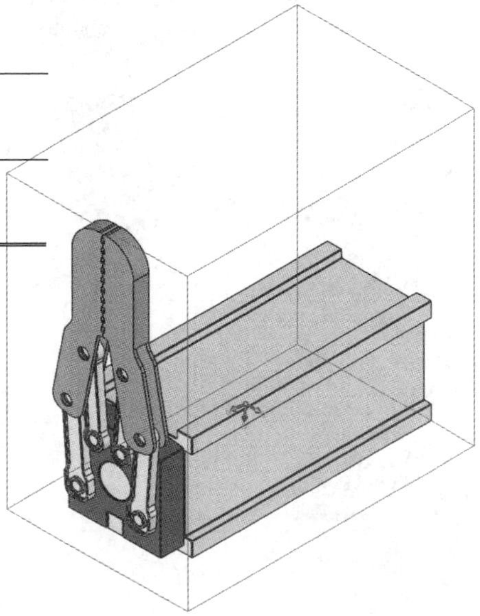

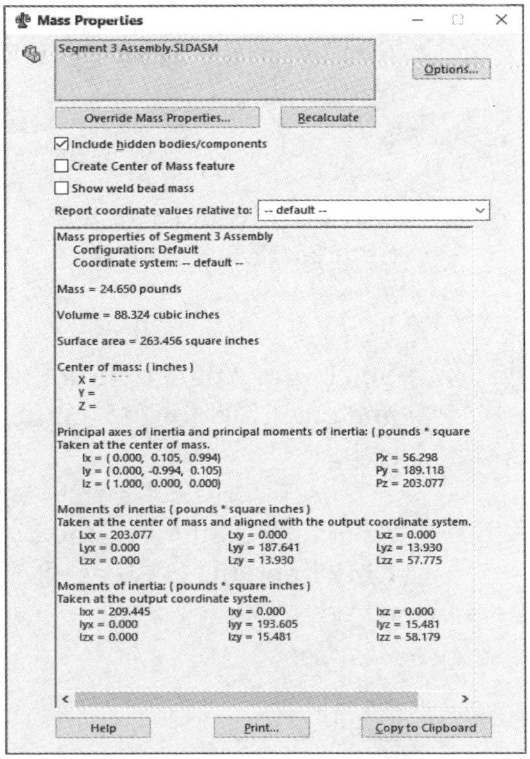

11. Adding a Gear mate:

Change to the **Front** view.

Click **Mate** and expand the Mechanical Mate option.

Click **Gear**.

Select the **2 circular edges** as indicated.

Enter the ratio **0.5in** for the left side and **1.0in** for the right side.

Click **OK**.

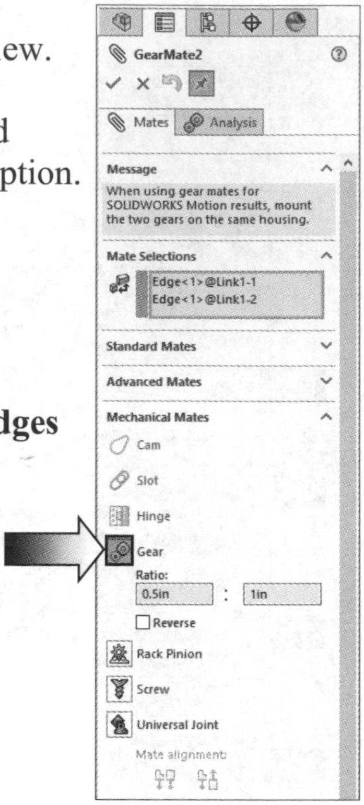

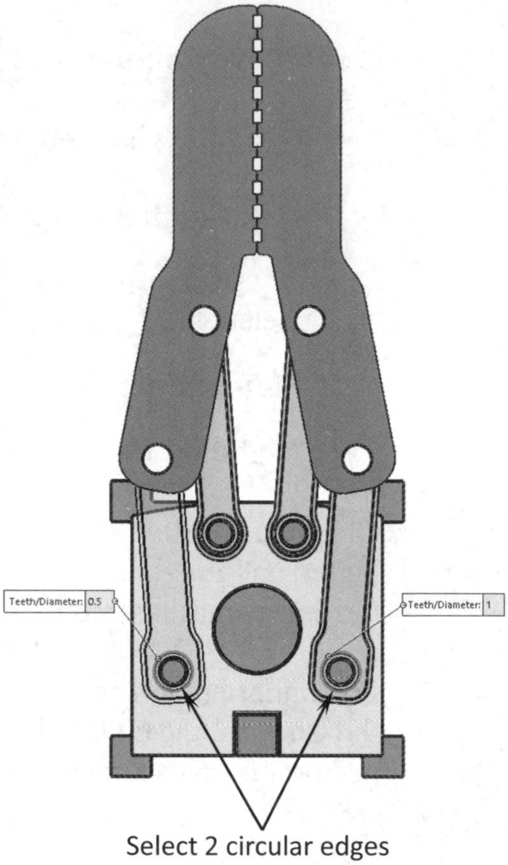

12. Adding an angle mate:

Remain in the **Front** view orientation.

Click **Mate** and select the Standard Mates option.

Select the **Angle** option and enter **30°** for the angle.

Select the 2 edges of the **Link1** and the **Base** as noted.

Click **OK**.

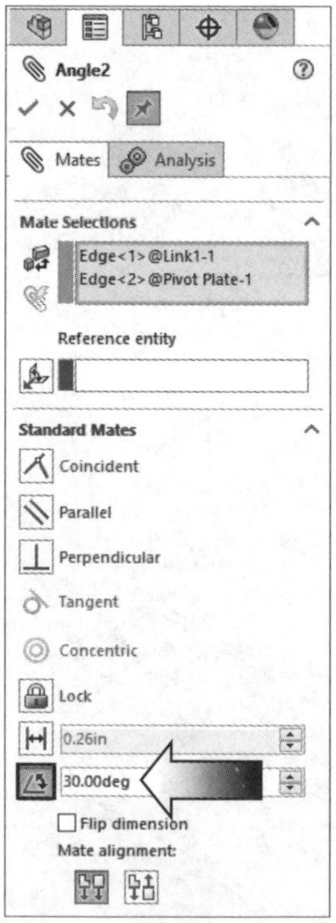

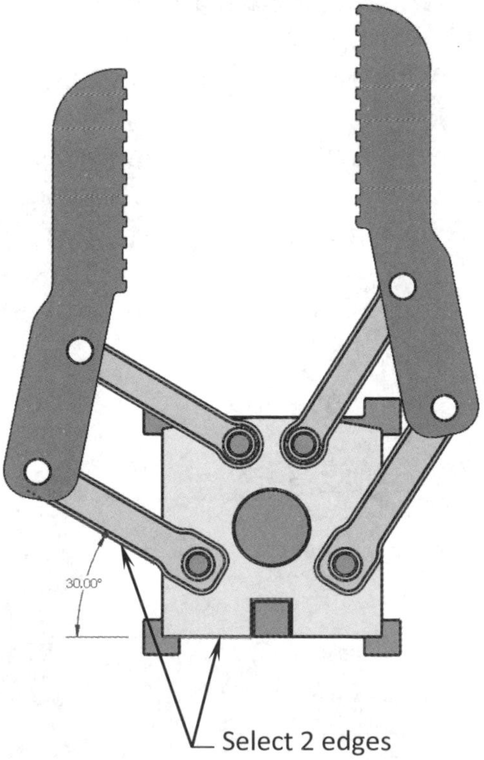

13. Measuring the angle:

Switch to the **Evaluate** tab.

Click **Measure**.

Select the 2 edges of the **Link1** and the **Base** as indicated

Enter the angle below:

__________ deg.

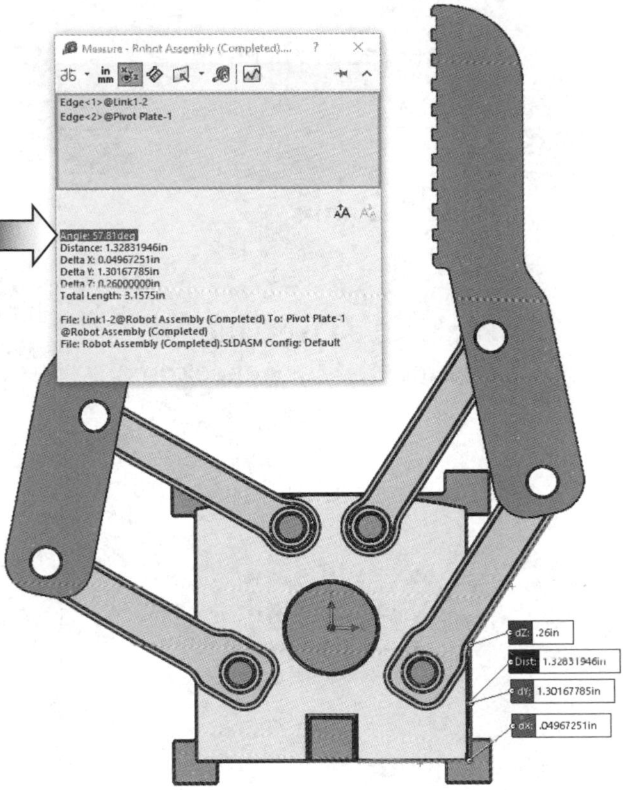

14. Modifying the Mass:

This step will modify the valucs of the Mass, Center of mass, and Moments of Inertia of the model.

Click **Mass Properties**.

Click the **Override Mass Properties** button (arrow).

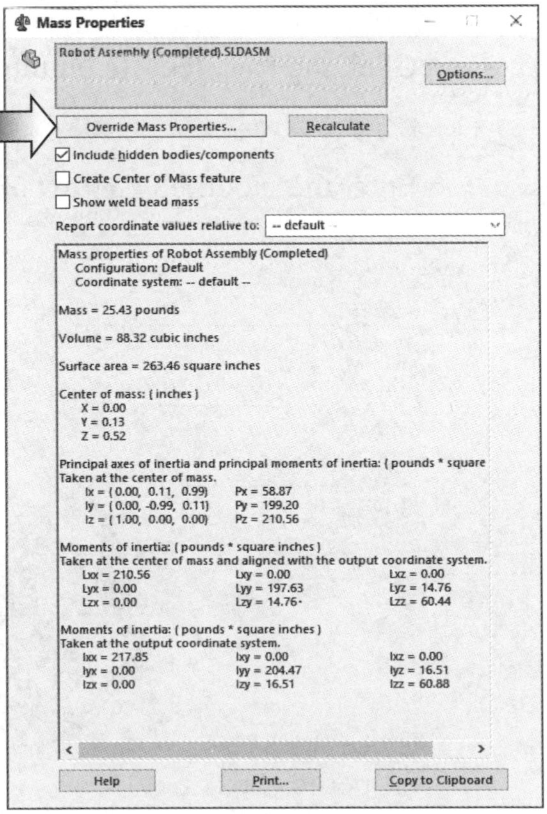

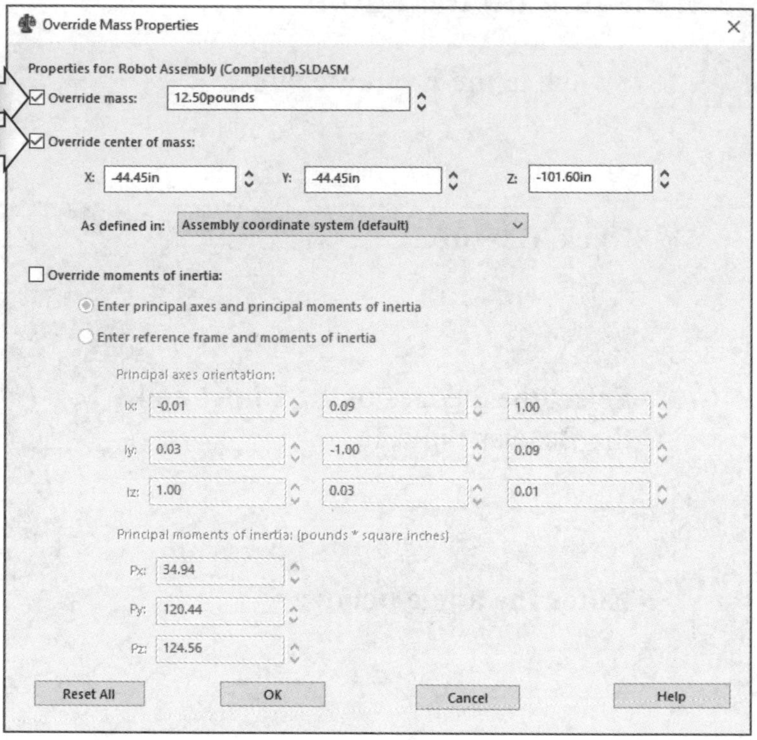

Enable the **Override Mass** checkbox.

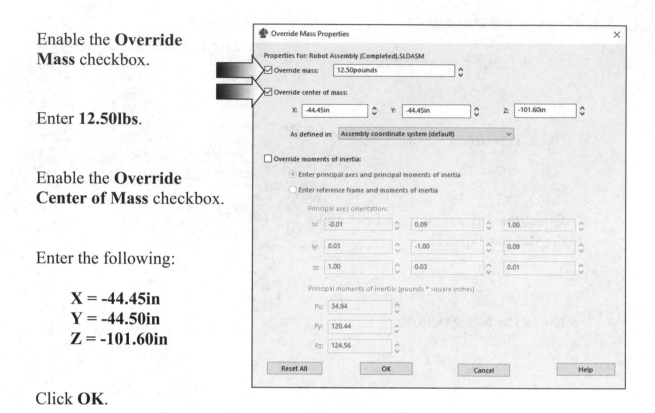

Enter **12.50lbs.**

Enable the **Override Center of Mass** checkbox.

Enter the following:

X = -44.45in
Y = -44.50in
Z = -101.60in

Click **OK.**

15. Calculating the modified Center of Mass:

Click the **Recalculate** button.

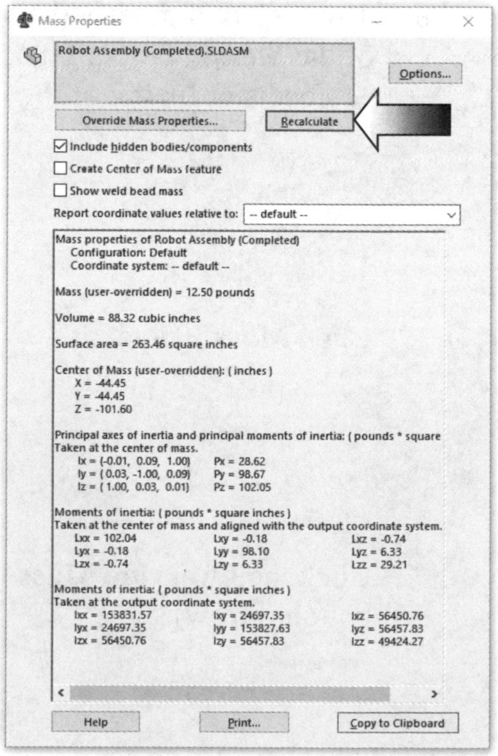

Enter the modified **Center of Mass** below:

X = ___________

Y = ___________

Z = ___________

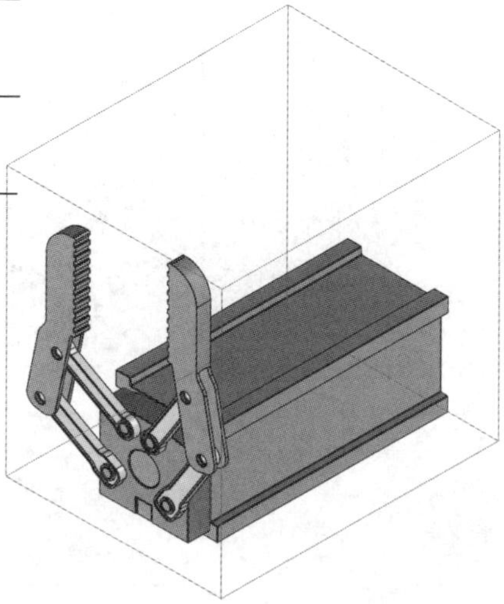

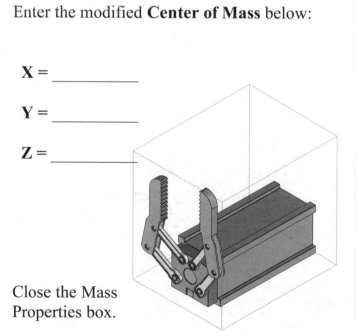

Close the Mass Properties box.

16. Saving your work:

Select **File**, **Save As**.

Enter: **Segment 3C (Completed)** for the file name.

Click **Save**. Close all part documents.

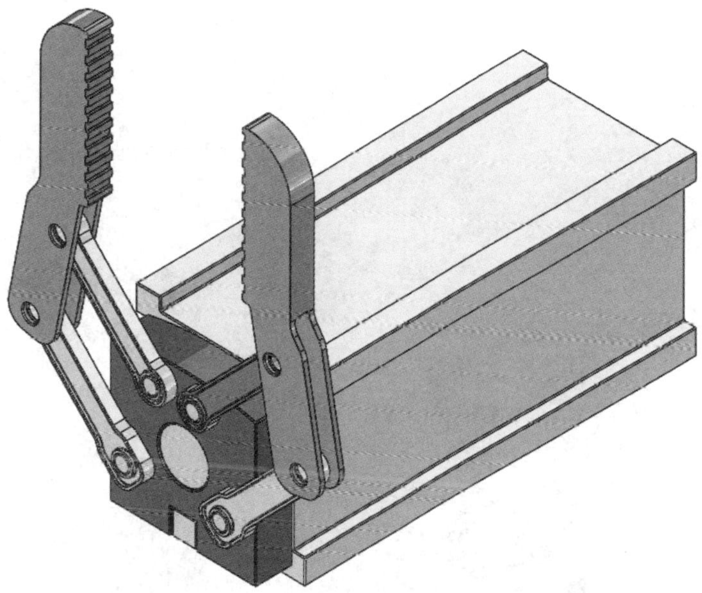

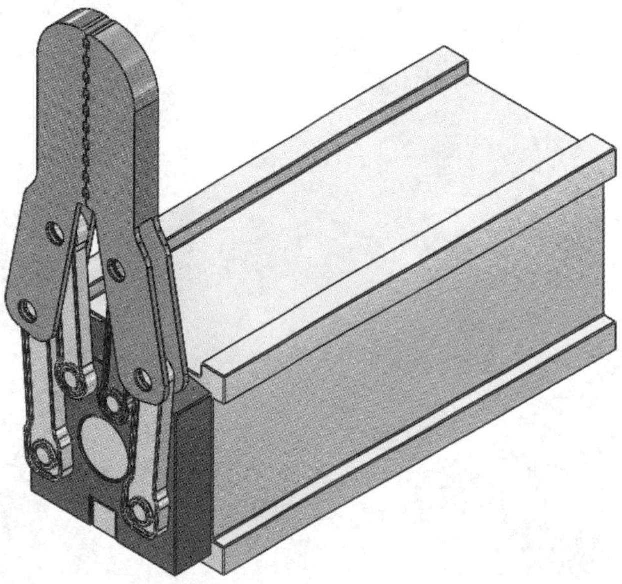

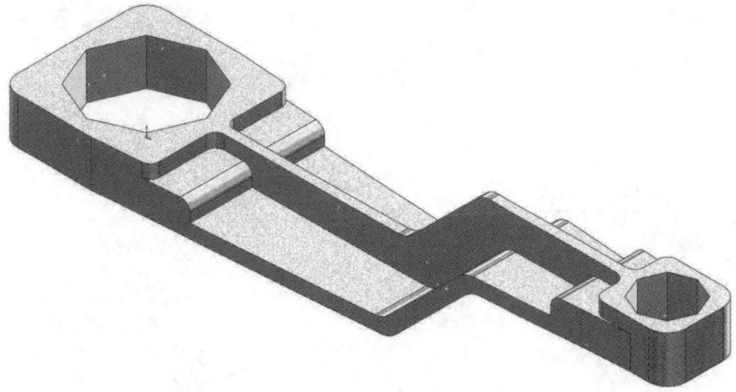

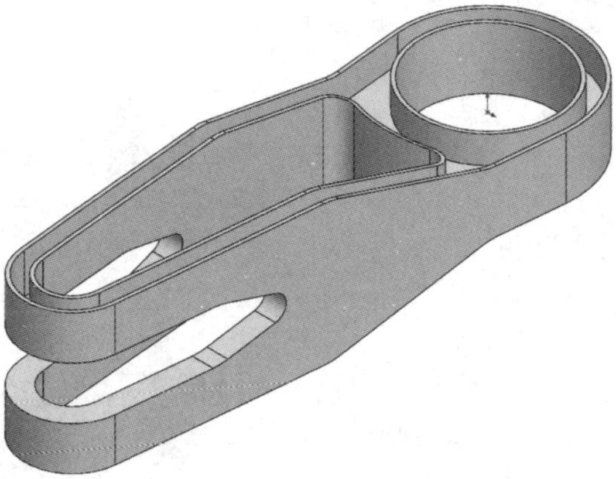

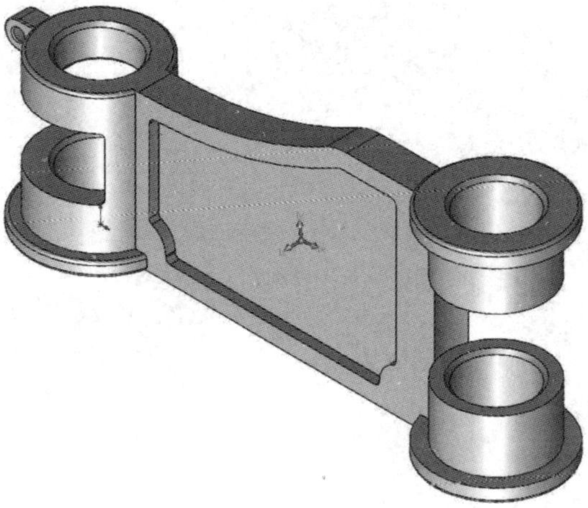

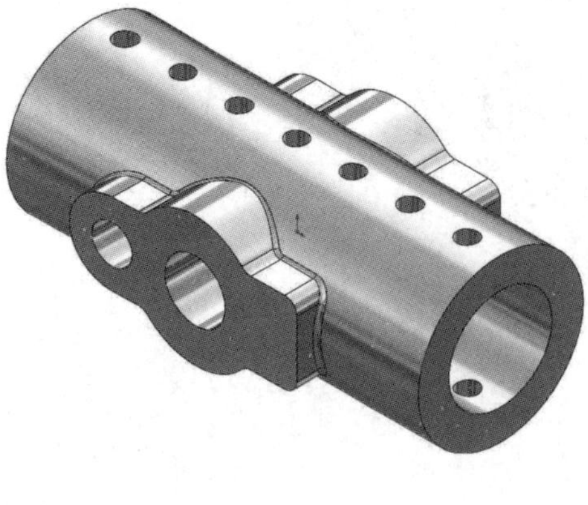

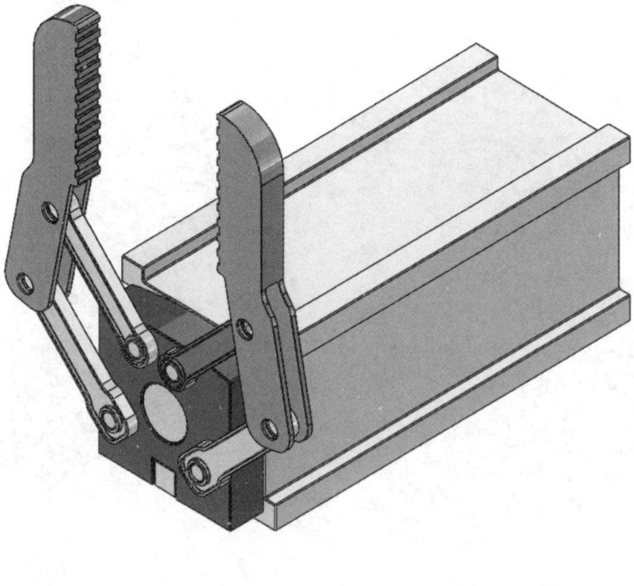

Glossary

Alloys:

An Alloy is a mixture of two or more metals (and sometimes a non-metal). The mixture is made by heating and melting the substances together.
Example of alloys are Bronze (Copper and Tin), Brass (Copper and Zinc), and Steel (Iron and Carbon).

Gravity and Mass:

Gravity is the force that pulls everything on earth toward the ground and makes things feel heavy. Gravity makes all falling bodies accelerate at a constant 32ft. per second (9.8 m/s). In the earth's atmosphere, air resistance slows acceleration. Only on airless Moon would a feather and a metal block fall to the ground together.
The mass of an object is the amount of material it contains.
A body with greater mass has more inertia; it needs a greater force to accelerate.
Weight depends on the force of gravity, but mass does not.

When an object spins around another (for example: a satellite orbiting the earth) it is pushed outward. Two forces are at work here: Centrifugal (pushing outward) and Centripetal (pulling inward). If you whirl a ball around you on a string, you pull it inward (Centripetal force). The ball seems to pull outward (Centrifugal force) and if released will fly off in a straight line.

Heat:

Heat is a form of energy and can move from one substance to another in one of threc ways: by Convection, by Radiation, and by Conduction.

Convection takes place only in liquids like water (for example: water in a kettle) and gases (for example: air warmed by a heat source such as a fire or radiator).
When liquid or gas is heated, it expands and becomes less dense. Warm air above the radiator rises and cool air moves in to take its place, creating a convection current.

Radiation is movement of heat through the air. Heat forms match set molecules of air moving and rays of heat spread out around the heat source.

Conduction occurs in solids such as metals. The handle of a metal spoon left in boiling liquid warms up as molecules at the heated end moves faster and collide with their neighbors, setting them moving. The heat travels through the metal, which is a good conductor of heat.

Inertia:

A body with a large mass is harder to start and also to stop. A heavy truck traveling at 50mph needs more power breaks to stop its motion than a smaller car traveling at the same speed.
Inertia is the tendency of an object either to stay still or to move steadily in a straight line, unless another force (such as a brick wall stopping the vehicle) makes it behave differently.

Joules:

The Joules is the SI unit of work or energy.
One Joule of work is done when a force of one Newton moves through a distance of one meter. The Joule is named after the English scientist James Joule (1818-1889).

Materials:

Stainless steel is an alloy of steel with chromium or nickel.

Steel is made by the basic oxygen process. The raw material is about three parts melted iron and one part scrap steel. Blowing oxygen into the melted iron raises the temperature and gets rid of impurities.

All plastics are chemical compounds called polymers.

Glass is made by mixing and heating sand, limestone, and soda ash. When these ingredients melt they turn into glass, which is hardened when it cools.
Glass is in fact not a solid but a "supercooled" liquid; it can be shaped by blowing, pressing, drawing, casting into molds, rolling, and floating across molten tin to make large sheets.

Ceramic objects, such as pottery and porcelain, electrical insulators, bricks, and roof tiles are all made from clay. The clay is shaped or molded when wet and soft, and heated in a kiln until it hardens.

Machine Tools:

Are powered tools used for shaping metal or other materials, by drilling holes, chiseling, grinding, pressing, or cutting. Often the material (the work piece) is moved while the tool stays still (lathe), or vice versa, the work piece stays while the tool moves (mill).
Most common machine tools are Mill, Lathe, Saw, Broach, Punch press, Grind, Bore and Stamp break.

CNC

Computer Numerical Control is the automation of machine tools that are operated by precisely programmed commands encoded on a storage medium, as opposed to controlled manually via hand wheels or levers, or mechanically automated via cams alone. Most CNC today is computer numerical control in which computers play an integral part of the control.

3D Printing

All methods work by working in layers, adding material, etc. different to other techniques, which are subtractive. Support is needed because almost all methods could support multi material printing, but it is currently only available in certain top tier machines.

A method of turning digital shapes into physical objects. Due to its nature, it allows us to accurately control the shape of the product. The drawbacks are size restraints and materials are often not durable.

While FDM does not seem like the best method for instrument manufacturing, it is one of the cheapest and most universally available methods.

EDM
Electric Discharge Machining.

SLA
Stereo Lithography.

SLM
Selective Laser Melting.

FDM
Fused Deposition Modeling.

SLS
Selective Laser Sintering.

J-P
Jetted Photopolymer (or Polyjet)

Newton's Law:

1. Every object remains stopped or goes on moving at a steady rate in a straight line unless acted upon by another force. This is the inertia principle.
2. The amount of force needed to make an object change its speed depends on the mass of the object and the amount of the acceleration or deceleration required.
3. To every action there is an equal and opposite reaction. When a body is pushed on way by a force, another force pushes back with equal strength.

Polymers:

A polymer is made of one or more large molecules formed from thousands of smaller molecules. Rubber and Wood are natural polymers. Plastics are synthetic (artificially made) polymers.

Speed and Velocity:

Speed is the rate at which a moving object changes position (how far it moves in a fixed time).
Velocity is speed in a particular direction.
If either speed or direction is changed, velocity also changes.

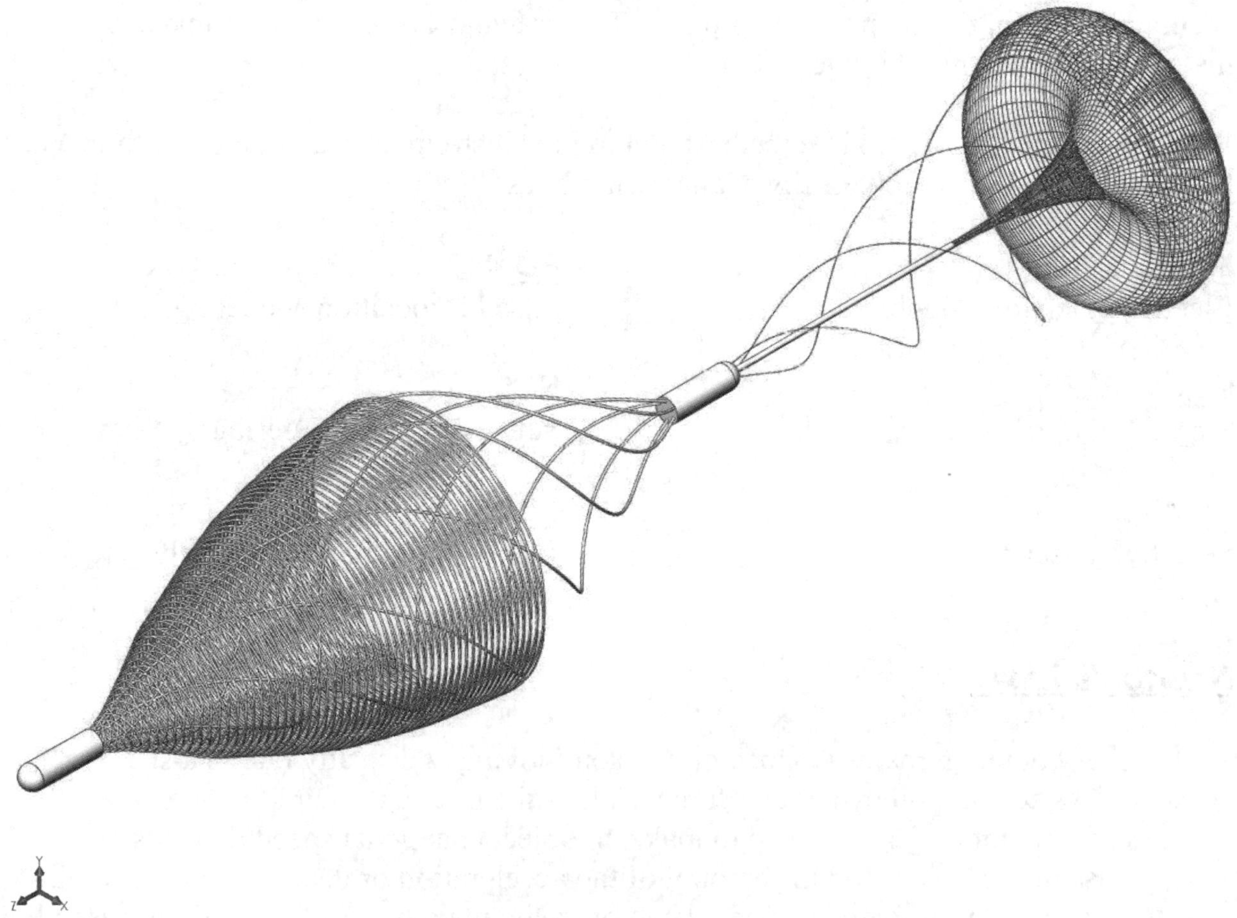

Absorbed
A feature, sketch, or annotation that is contained in another item (usually a feature) in the FeatureManager design tree. Examples are the profile sketch and profile path in a base-sweep, or a cosmetic thread annotation in a hole.

Align
Tools that assist in lining up annotations and dimensions (left, right, top, bottom, and so on). For aligning parts in an assembly.

Alternate position view
A drawing view in which one or more views are superimposed in phantom lines on the original view. Alternate position views are often used to show range of motion of an assembly.

Anchor point
The end of a leader that attaches to the note, block, or other annotation. Sheet formats contain anchor points for a bill of materials, a hole table, a revision table, and a weldment cut list.

Annotation
A text note or a symbol that adds specific design intent to a part, assembly, or drawing. Specific types of annotations include note, hole callout, surface finish symbol, datum feature symbol, datum target, geometric tolerance symbol, weld symbol, balloon, and stacked balloon. Annotations that apply only to drawings include center mark, annotation centerline, area hatch, and block.

Appearance callouts
Callouts that display the colors and textures of the face, feature, body, and part under the entity selected and are a shortcut to editing colors and textures.

Area hatch
A crosshatch pattern or fill applied to a selected face or to a closed sketch in a drawing.

Assembly
A document in which parts, features, and other assemblies (sub-assemblies) are mated together. The parts and sub-assemblies exist in documents separate from the assembly. For example, in an assembly, a piston can be mated to other parts, such as a connecting rod or cylinder. This new assembly can then be used as a sub-assembly in an assembly of an engine. The extension for a SOLIDWORKS assembly file name is .SLDASM.

Attachment point
The end of a leader that attaches to the model (to an edge, vertex, or face, for example) or to a drawing sheet.

Axis
A straight line that can be used to create model geometry, features, or patterns. An axis can be made in a number of different ways, including using the intersection of two planes.

Balloon
Labels parts in an assembly, typically including item numbers and quantity. In drawings, the item numbers are related to rows in a bill of materials.

Base
The first solid feature of a part.

Baseline dimensions
Sets of dimensions measured from the same edge or vertex in a drawing.

Bend
A feature in a sheet metal part. A bend generated from a filleted corner, cylindrical face, or conical face is a round bend; a bend generated from sketched straight lines is a sharp bend.

Bill of materials
A table inserted into a drawing to keep a record of the parts used in an assembly.

Block
A user-defined annotation that you can use in parts, assemblies, and drawings. A block can contain text, sketch entities (except points), and area hatch, and it can be saved in a file for later use as, for example, a custom callout or a company logo.

Bottom-up assembly
An assembly modeling technique where you create parts and then insert them into an assembly.

Broken-out section
A drawing view that exposes inner details of a drawing view by removing material from a closed profile, usually a spline.

Cavity
The mold half that holds the cavity feature of the design part.

Center mark
A cross that marks the center of a circle or arc.

Centerline
A centerline marks, in phantom font, an axis of symmetry in a sketch or drawing.

Chamfer
Bevels a selected edge or vertex. You can apply chamfers to both sketches and features.

Child
A dependent feature related to a previously built feature. For example, a chamfer on the edge of a hole is a child of the parent hole.

Click-release
As you sketch, if you click and then release the pointer, you are in click-release mode. Move the pointer and click again to define the next point in the sketch sequence.

Click-drag
As you sketch, if you click and drag the pointer, you are in click-drag mode. When you release the pointer, the sketch entity is complete.

Closed profile
Also called a closed contour, it is a sketch or sketch entity with no exposed endpoints, for example, a circle or polygon.

Collapse
The opposite of explode. The collapse action returns an exploded assembly's parts to their normal positions.

Collision Detection
An assembly function that detects collisions between components when components move or rotate. A collision occurs when an entity on one component coincides with any entity on another component.

Component
Any part or sub-assembly within an assembly

Configuration
A variation of a part or assembly within a single document. Variations can include different dimensions, features, and properties. For example, a single part such as a bolt can contain different configurations that vary the diameter and length.

ConfigurationManager
Located on the left side of the SOLIDWORKS window, it is a means to create, select, and view the configurations of parts and assemblies.

Constraint
The relations between sketch entities, or between sketch entities and planes, axes, edges, or vertices.

Construction geometry
The characteristic of a sketch entity that the entity is used in creating other geometry but is not itself used in creating features.

Coordinate system
A system of planes used to assign Cartesian coordinates to features, parts, and assemblies. Part and assembly documents contain default coordinate systems; other coordinate systems can be defined with reference geometry. Coordinate systems can be used with measurement tools and for exporting documents to other file formats.

Cosmetic thread
An annotation that represents threads.

Crosshatch
A pattern (or fill) applied to drawing views such as section views and broken-out sections.

Curvature
Curvature is equal to the inverse of the radius of the curve. The curvature can be displayed in different colors according to the local radius (usually of a surface).

Cut
A feature that removes material from a part by such actions as extrude, revolve, loft, sweep, thicken, cavity, and so on.

Dangling
A dimension, relation, or drawing section view that is unresolved. For example, if a piece of geometry is dimensioned, and that geometry is later deleted, the dimension becomes dangling.

Degrees of freedom
Geometry that is not defined by dimensions or relations is free to move. In 2D sketches, there are three degrees of freedom: movement along the X and Y axes, and rotation about the Z axis (the axis normal to the sketch plane). In 3D sketches and in assemblies, there are six degrees of freedom: movement along the X, Y, and Z axes, and rotation about the X, Y, and Z axes.

Derived part
A derived part is a new base, mirror, or component part created directly from an existing part and linked to the original part such that changes to the original part are reflected in the derived part.

Derived sketch
A copy of a sketch, in either the same part or the same assembly that is connected to the original sketch. Changes in the original sketch are reflected in the derived sketch.

Design Library
Located in the Task Pane, the Design Library provides a central location for reusable elements such as parts, assemblies, and so on.

Design table
An Excel spreadsheet that is used to create multiple configurations in a part or assembly document.

Detached drawing
A drawing format that allows opening and working in a drawing without loading the corresponding models into memory. The models are loaded on an as-needed basis.

Detail view
A portion of a larger view, usually at a larger scale than the original view.

Dimension line
A linear dimension line references the dimension text to extension lines indicating the entity being measured. An angular dimension line references the dimension text directly to the measured object.

DimXpertManager
Located on the left side of the SOLIDWORKS window, it is a means to manage dimensions and tolerances created using DimXpert for parts according to the requirements of the ASME Y.14.41-2003 standard.

DisplayManager
The DisplayManager lists the appearances, decals, lights, scene, and cameras applied to the current model. From the DisplayManager, you can view applied content, and add, edit, or delete items. When PhotoView 360 is added in, the DisplayManager also provides access to PhotoView options.

Document
A file containing a part, assembly, or drawing.

Draft
The degree of taper or angle of a face usually applied to molds or castings.

Drawing
A 2D representation of a 3D part or assembly. The extension for a SOLIDWORKS drawing file name is .SLDDRW.

Drawing sheet
A page in a drawing document.

Driven dimension
Measurements of the model, but they do not drive the model and their values cannot be changed.

Driving dimension
Also referred to as a model dimension, it sets the value for a sketch entity. It can also control distance, thickness, and feature parameters.

Edge
A single outside boundary of a feature.

Edge flange
A sheet metal feature that combines a bend and a tab in a single operation.

Equation
Creates a mathematical relation between sketch dimensions, using dimension names as variables, or between feature parameters, such as the depth of an extruded feature or the instance count in a pattern.

Exploded view
Shows an assembly with its components separated from one another, usually to show how to assemble the mechanism.

Export
Save a SOLIDWORKS document in another format for use in other CAD/CAM, rapid prototyping, web, or graphics software applications.

Extension line
The line extending from the model indicating the point from which a dimension is measured.

Extrude
A feature that linearly projects a sketch to either add material to a part (in a base or boss) or remove material from a part (in a cut or hole).

Face
A selectable area (planar or otherwise) of a model or surface with boundaries that help define the shape of the model or surface. For example, a rectangular solid has six faces.

Fasteners
A SOLIDWORKS Toolbox library that adds fasteners automatically to holes in an assembly.

Feature
An individual shape that, combined with other features, makes up a part or assembly. Some features, such as bosses and cuts, originate as sketches. Other features, such as shells and fillets, modify a feature's geometry. However, not all features have associated geometry. Features are always listed in the FeatureManager design tree.

FeatureManager design tree
Located on the left side of the SOLIDWORKS window, it provides an outline view of the active part, assembly, or drawing.

Fill
A solid area hatch or crosshatch. Fill also applies to patches on surfaces.

Fillet
An internal rounding of a corner or edge in a sketch, or an edge on a surface or solid.

Forming tool
Dies that bend, stretch, or otherwise form sheet metal to create such form features as louvers, lances, flanges, and ribs.

Fully defined
A sketch where all lines and curves in the sketch, and their positions, are described by dimensions or relations, or both, and cannot be moved. Fully defined sketch entities are shown in black.

Geometric tolerance
A set of standard symbols that specify the geometric characteristics and dimensional requirements of a feature.

Graphics area
The area in the SOLIDWORKS window where the part, assembly, or drawing appears.

Guide curve
A 2D or 3D curve used to guide a sweep or loft.

Handle
An arrow, square, or circle that you can drag to adjust the size or position of an entity (a feature, dimension, or sketch entity, for example).

Helix
A curve defined by pitch, revolutions, and height. A helix can be used, for example, as a path for a swept feature cutting threads in a bolt.

Hem
A sheet metal feature that folds back at the edge of a part. A hem can be open, closed, double, or tear-drop.

HLR
(Hidden lines removed) a view mode in which all edges of the model that are not visible from the current view angle are removed from the display.

HLV
(Hidden lines visible) A view mode in which all edges of the model that are not visible from the current view angle are shown gray or dashed.

Import
Open files from other CAD software applications into a SOLIDWORKS document.

In-context feature
A feature with an external reference to the geometry of another component; the in-context feature changes automatically if the geometry of the referenced model or feature changes.

Inference
The system automatically creates (infers) relations between dragged entities (sketched entities, annotations, and components) and other entities and geometry. This is useful when positioning entities relative to one another.

Instance
An item in a pattern or a component in an assembly that occurs more than once. Blocks are inserted into drawings as instances of block definitions.

Interference detection
A tool that displays any interference between selected components in an assembly.

Jog
A sheet metal feature that adds material to a part by creating two bends from a sketched line.

Knit
A tool that combines two or more faces or surfaces into one. The edges of the surfaces

must be adjacent and not overlapping, but they cannot ever be planar. There is no difference in the appearance of the face or the surface after knitting.

Layout sketch
A sketch that contains important sketch entities, dimensions, and relations. You reference the entities in the layout sketch when creating new sketches, building new geometry, or positioning components in an assembly. This allows for easier updating of your model because changes you make to the layout sketch propagate to the entire model.

Leader
A solid line from an annotation (note, dimension, and so on) to the referenced feature.

Library feature
A frequently used feature, or combination of features, that is created once and then saved for future use.

Lightweight
A part in an assembly or a drawing has only a subset of its model data loaded into memory. The remaining model data is loaded on an as-needed basis. This improves performance of large and complex assemblies.

Line
A straight sketch entity with two endpoints. A line can be created by projecting an external entity such as an edge, plane, axis, or sketch curve into the sketch.

Loft
A base, boss, cut, or surface feature created by transitions between profiles.

Lofted bend
A sheet metal feature that produces a roll form or a transitional shape from two open profile sketches. Lofted bends often create funnels and chutes.

Mass properties
A tool that evaluates the characteristics of a part or an assembly such as volume, surface area, centroid, and so on.

Mate
A geometric relationship, such as coincident, perpendicular, tangent, and so on, between parts in an assembly.

Mate reference
Specifies one or more entities of a component to use for automatic mating. When you

drag a component with a mate reference into an assembly, the software tries to find other combinations of the same mate reference name and mate type.

Mates folder
A collection of mates that are solved together. The order in which the mates appear within the Mates folder does not matter.

Mirror
(a) A mirror feature is a copy of a selected feature, mirrored about a plane or planar face.
(b) A mirror sketch entity is a copy of a selected sketch entity that is mirrored about a centerline.

Miter flange
A sheet metal feature that joins multiple edge flanges together and miters the corner.

Model
3D solid geometry in a part or assembly document. If a part or assembly document contains multiple configurations, each configuration is a separate model.

Model dimension
A dimension specified in a sketch or a feature in a part or assembly document that defines some entity in a 3D model.

Model item
A characteristic or dimension of feature geometry that can be used in detailing drawings.

Model view
A drawing view of a part or assembly.

Mold
A set of manufacturing tooling used to shape molten plastic or other material into a designed part. You design the mold using a sequence of integrated tools that result in cavity and core blocks that are derived parts of the part to be molded.

Motion Study
Motion Studies are graphical simulations of motion and visual properties with assembly models. Analogous to a configuration, they do not actually change the original assembly model or its properties. They display the model as it changes based on simulation elements you add.

Multibody part
A part with separate solid bodies within the same part document. Unlike the components in an assembly, multibody parts are not dynamic.

Native format
DXF and DWG files remain in their original format (are not converted into SOLIDWORKS format) when viewed in SOLIDWORKS drawing sheets (view only).

Open profile
Also called an open contour, it is a sketch or sketch entity with endpoints exposed. For example, a U-shaped profile is open.

Ordinate dimensions
A chain of dimensions measured from a zero ordinate in a drawing or sketch.

Origin
The model origin appears as three gray arrows and represents the (0,0,0) coordinate of the model. When a sketch is active, a sketch origin appears in red and represents the (0,0,0) coordinate of the sketch. Dimensions and relations can be added to the model origin, but not to a sketch origin.

Out-of-context feature
A feature with an external reference to the geometry of another component that is not open.

Over defined
A sketch is over defined when dimensions or relations are either in conflict or redundant.

Parameter
A value used to define a sketch or feature (often a dimension).

Parent
An existing feature upon which other features depend. For example, in a block with a hole, the block is the parent to the child hole feature.

Part
A single 3D object made up of features. A part can become a component in an assembly, and it can be represented in 2D in a drawing. Examples of parts are bolt, pin, plate, and so on. The extension for a SOLIDWORKS part file name is .SLDPRT.

Path
A sketch, edge, or curve used in creating a sweep or loft.

Pattern
A pattern repeats selected sketch entities, features, or components in an array, which can

be linear, circular, or sketch driven. If the seed entity is changed, the other instances in the pattern update.

Physical Dynamics
An assembly tool that displays the motion of assembly components in a realistic way. When you drag a component, the component applies a force to other components it touches. Components move only within their degrees of freedom.

Pierce relation
Makes a sketch point coincident to the location at which an axis, edge, line, or spline pierces the sketch plane.

Planar
Entities that can lie on one plane. For example, a circle is planar, but a helix is not.

Plane
Flat construction geometry. Planes can be used for a 2D sketch, section view of a model, a neutral plane in a draft feature, and others.

Point
A singular location in a sketch, or a projection into a sketch at a single location of an external entity (origin, vertex, axis, or point in an external sketch).

Predefined view
A drawing view in which the view position, orientation, and so on can be specified before a model is inserted. You can save drawing documents with predcfined views as templates.

Profile
A sketch entity used to create a feature (such as a loft) or a drawing view (such as a detail view). A profile can be open (such as a U shape or open spline) or closed (such as a circle or closed spline).

Projected dimension
If you dimension entities in an isometric view, projected dimensions are the flat dimensions in 2D.

Projected view
A drawing view projected orthogonally from an existing view.

PropertyManager
Located on the left side of the SOLIDWORKS window, it is used for dynamic editing of sketch entities and most features.

RealView graphics
A hardware (graphics card) support of advanced shading in real time; the rendering applies to the model and is retained as you move or rotate a part.

Rebuild
Tool that updates (or regenerates) the document with any changes made since the last time the model was rebuilt. Rebuild is typically used after changing a model dimension.

Reference dimension
A dimension in a drawing that shows the measurement of an item but cannot drive the model and its value cannot be modified. When model dimensions change, reference dimensions update.

Reference geometry
Includes planes, axes, coordinate systems, and 3D curves. Reference geometry is used to assist in creating features such as lofts, sweeps, drafts, chamfers, and patterns.

Relation
A geometric constraint between sketch entities or between a sketch entity and a plane, axis, edge, or vertex. Relations can be added automatically or manually.

Relative view
A relative (or relative to model) drawing view is created relative to planar surfaces in a part or assembly.

Reload
Refreshes shared documents. For example, if you open a part file for read-only access while another user makes changes to the same part, you can reload the new version, including the changes.

Reorder
Reordering (changing the order of) items is possible in the FeatureManager design tree. In parts, you can change the order in which features are solved. In assemblies, you can control the order in which components appear in a bill of materials.

Replace
Substitutes one or more open instances of a component in an assembly with a different component.

Resolved
A state of an assembly component (in an assembly or drawing document) in which it is fully loaded in memory. All the component's model data is available, so its entities can be selected, referenced, edited, and used in mates, and so on.

Revolve
A feature that creates a base or boss, a revolved cut, or revolved surface by revolving one or more sketched profiles around a centerline.

Rip
A sheet metal feature that removes material at an edge to allow a bend.

Rollback
Suppresses all items below the rollback bar.

Section
Another term for profile in sweeps.

Section line
A line or centerline sketched in a drawing view to create a section view.

Section scope
Specifies the components to be left uncut when you create an assembly drawing section view.

Section view
A section view (or section cut) is (1) a part or assembly view cut by a plane, or (2) a drawing view created by cutting another drawing view with a section line.

Seed
A sketch or an entity (a feature, face, or body) that is the basis for a pattern. If you edit the seed, the other entities in the pattern are updated.

Shaded
Displays a model as a colored solid.

Shared values
Also called linked values, these are named variables that you assign to set the value of two or more dimensions to be equal.

Sheet format
Includes page size and orientation, standard text, borders, title blocks, and so on. Sheet formats can be customized and saved for future use. Each sheet of a drawing document can have a different format.

Shell
A feature that hollows out a part, leaving open the selected faces and thin walls on the remaining faces. A hollow part is created when no faces are selected to be open.

Sketch
A collection of lines and other 2D objects on a plane or face that forms the basis for a feature such as a base or a boss. A 3D sketch is non-planar and can be used to guide a sweep or loft, for example.

Smart Fasteners
Automatically adds fasteners (bolts and screws) to an assembly using the SOLIDWORKS Toolbox library of fasteners.

SmartMates
An assembly mating relation that is created automatically.

Solid sweep
A cut sweep created by moving a tool body along a path to cut out 3D material from a model.

Spiral
A flat or 2D helix, defined by a circle, pitch, and number of revolutions.

Spline
A sketched 2D or 3D curve defined by a set of control points.

Split line
Projects a sketched curve onto a selected model face, dividing the face into multiple faces so that each can be selected individually. A split line can be used to create draft features, to create face blend fillets, and to radiate surfaces to cut molds.

Stacked balloon
A set of balloons with only one leader. The balloons can be stacked vertically (up or down) or horizontally (left or right).

Standard 3 views
The three orthographic views (front, right, and top) that are often the basis of a drawing.

StereoLithography
The process of creating rapid prototype parts using a faceted mesh representation in STL files.

Sub-assembly
An assembly document that is part of a larger assembly. For example, the steering mechanism of a car is a sub-assembly of the car.

Suppress
Removes an entity from the display and from any calculations in which it is involved. You can suppress features, assembly components, and so on. Suppressing an entity does not delete the entity; you can unsuppress the entity to restore it.

Surface
A zero-thickness planar or 3D entity with edge boundaries. Surfaces are often used to create solid features. Reference surfaces can be used to modify solid features.

Sweep
Creates a base, boss, cut, or surface feature by moving a profile (section) along a path. For cut-sweeps, you can create solid sweeps by moving a tool body along a path.

Tangent arc
An arc that is tangent to another entity, such as a line.

Tangent edge
The transition edge between rounded or filleted faces in hidden lines visible or hidden lines removed modes in drawings.

Task Pane
Located on the right-side of the SOLIDWORKS window, the Task Pane contains SOLIDWORKS Resources, the Design Library, and the File Explorer.

Template
A document (part, assembly, or drawing) that forms the basis of a new document. It can include user-defined parameters, annotations, predefined views, geometry, and so on.

Temporary axis
An axis created implicitly for every conical or cylindrical face in a model.

Thin feature
An extruded or revolved feature with constant wall thickness. Sheet metal parts are typically created from thin features.

TolAnalyst
A tolerance analysis application that determines the effects that dimensions and tolerances have on parts and assemblies.

Top-down design
An assembly modeling technique where you create parts in the context of an assembly by referencing the geometry of other components. Changes to the referenced components propagate to the parts that you create in context.

Triad
Three axes with arrows defining the X, Y, and Z directions. A reference triad appears in part and assembly documents to assist in orienting the viewing of models. Triads also assist when moving or rotating components in assemblies.

Under defined
A sketch is under defined when there are not enough dimensions and relations to prevent entities from moving or changing size.

Vertex
A point at which two or more lines or edges intersect. Vertices can be selected for sketching, dimensioning, and many other operations.

Viewports
Windows that display views of models. You can specify one, two, or four viewports. Viewports with orthogonal views can be linked, which links orientation and rotation.

Virtual sharp
A sketch point at the intersection of two entities after the intersection itself has been removed by a feature such as a fillet or chamfer. Dimensions and relations to the virtual sharp are retained even though the actual intersection no longer exists.

Weldment
A multibody part with structural members.

Weldment cut list
A table that tabulates the bodies in a weldment along with descriptions and lengths.

Wireframe
A view mode in which all edges of the part or assembly are displayed.

Zebra stripes
Simulate the reflection of long strips of light on a very shiny surface. They allow you to see small changes in a surface that may be hard to see with a standard display.

Zoom
To simulate movement toward or away from a part or an assembly.

Learn Mold-Tooling Designs with SOLIDWORKS Mold Making Textbook

3

A

B

C

D

E

F

G

H

I

T

U

V

W

X

Y

Z

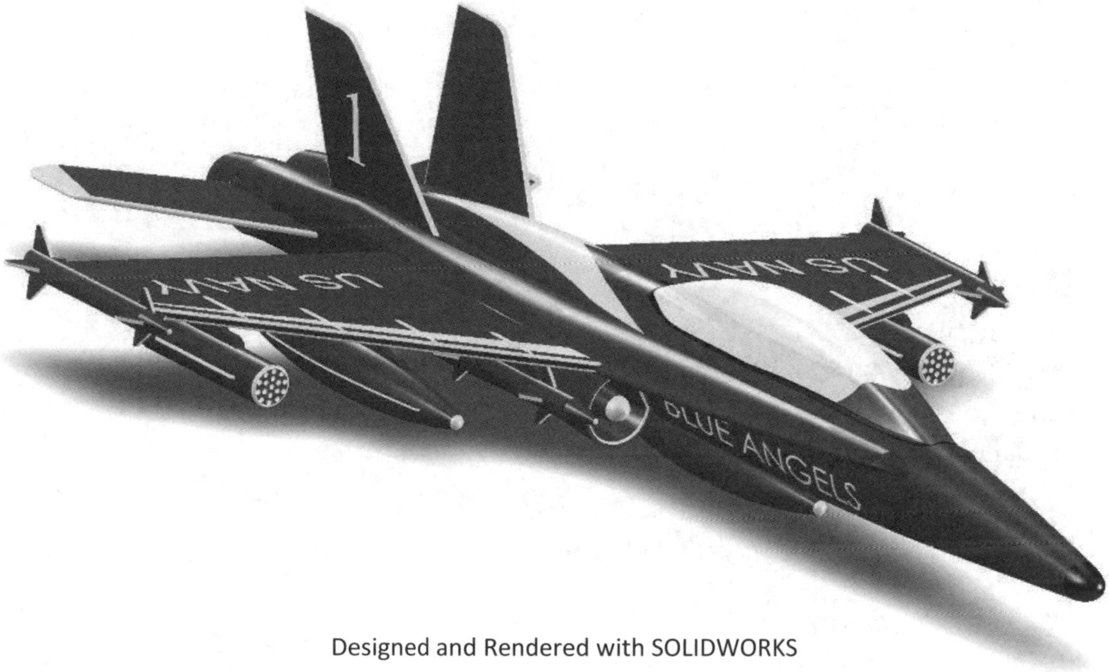

Designed and Rendered with SOLIDWORKS

SOLIDWORKS Quick Guide

STANDARD Toolbar

- Creates a new document.
- Opens an existing document.
- Saves an active document.
- Make Drawing from Part/Assembly.
- Make Assembly from Part/Assembly.
- Prints the active document.
- Print preview.
- Cuts the selection & puts it on the clipboard.
- Copies the selection & puts it on the clipboard.
- Inserts the clipboard contents.
- Deletes the selection.
- Reverses the last action.
- Rebuilds the part / assembly / drawing.
- Redo the last action that was undone.
- Saves all documents.
- Edits material.
- Closes an existing document.
- Shows or hides the Selection Filter toolbar.
- Shows or hides the Web toolbar.
- Properties.
- File properties.
- Loads or unloads the 3D instant website add-in.
- Select tool.
- Select the entire document.
- Checks read-only files.
- Options.
- Help.
- Full screen view.
- OK.
- Cancel.
- Magnified selection.

SKETCH TOOLS Toolbar

- Select.
- Sketch.
- 3D Sketch.
- Sketches a rectangle from the center.
- Sketches a centerpoint arc slot.
- Sketches a 3-point arc slot.
- Sketches a straight slot.
- Sketches a centerpoint straight slot.
- Sketches a 3-point arc.
- Creates sketched ellipses.

Quick Reference Guide to SOLIDWORKS Command Icons & Toolbars

SKETCH TOOLS Toolbar

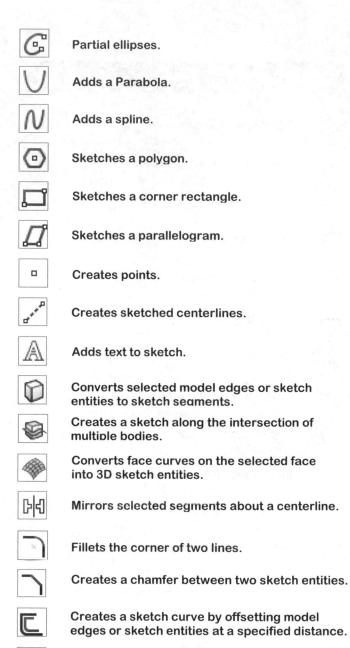

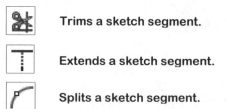

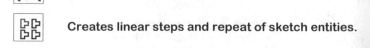

- 3D sketch on plane.
- Sets up Grid parameters.
- Creates a sketch on a selected plane or face.
- Equation driven curve.
- Modifies a sketch.
- Copies sketch entities.
- Scales sketch entities.
- Rotates sketch entities.
- Sketches 3 point rectangle from the center.
- Sketches 3 point corner rectangle.
- Sketches a line.
- Creates a center point arc: center, start, end.
- Creates an arc tangent to a line.
- Sketches splines on a surface or face.
- Sketches a circle.
- Sketches a circle by its perimeter.
- Makes a path of sketch entities.
- Mirrors entities dynamically about a centerline.
- Insert a plane into the 3D sketch.
- Instant 2D.
- Sketch numeric input.
- Detaches segment on drag.
- Sketch picture.
- Partial ellipses.
- Adds a Parabola.
- Adds a spline.
- Sketches a polygon.
- Sketches a corner rectangle.
- Sketches a parallelogram.
- Creates points.
- Creates sketched centerlines.
- Adds text to sketch.
- Converts selected model edges or sketch entities to sketch segments.
- Creates a sketch along the intersection of multiple bodies.
- Converts face curves on the selected face into 3D sketch entities.
- Mirrors selected segments about a centerline.
- Fillets the corner of two lines.
- Creates a chamfer between two sketch entities.
- Creates a sketch curve by offsetting model edges or sketch entities at a specified distance.
- Trims a sketch segment.
- Extends a sketch segment.
- Splits a sketch segment.
- Construction Geometry.
- Creates linear steps and repeat of sketch entities.
- Creates circular steps and repeat of sketch entities

Quick Reference Guide to SOLIDWORKS Command Icons & Toolbars

SHEET METAL Toolbar

- Add a bend from a selected sketch in a Sheet Metal part.
- Shows flat pattern for this sheet metal part.
- Shows part without inserting any bends.
- Inserts a rip feature to a sheet metal part.
- Create a Sheet Metal part or add material to existing Sheet Metal part.
- Inserts a Sheet Metal Miter Flange feature.
- Folds selected bends.
- Unfolds selected bends.
- Inserts bends using a sketch line.
- Inserts a flange by pulling an edge.
- Inserts a sheet metal corner feature.
- Inserts a Hem feature by selecting edges.
- Breaks a corner by filleting/chamfering it.
- Inserts a Jog feature using a sketch line.
- Inserts a lofted bend feature using 2 sketches.
- Creates inverse dent on a sheet metal part.
- Trims out material from a corner, in a sheet metal part.
- Inserts a fillet weld bead.
- Converts a solid/surface into a sheet metal part.
- Adds a Cross Break feature into a selected face.
- Sweeps an open profile along an open/closed path.
- Adds a gusset/rib across a bend.
- Corner relief.
- Welds the selected corner.

SURFACES Toolbar

- Creates mid surfaces between offset face pairs.
- Patches surface holes and external edges.
- Creates an extruded surface.
- Creates a revolved surface.
- Creates a swept surface.
- Creates a lofted surface.
- Creates an offset surface.
- Radiates a surface originating from a curve, parallel to a plane.
- Knits surfaces together.
- Creates a planar surface from a sketch or a set of edges.
- Creates a surface by importing data from a file.
- Extends a surface.
- Trims a surface.
- Surface flatten.
- Deletes Face(s).
- Replaces Face with Surface.
- Patches surface holes and external edges by extending the surfaces.
- Creates parting surfaces between core & cavity surfaces.
- Inserts ruled surfaces from edges.

WELDMENTS Toolbar

- Creates a weldment feature.
- Creates a structure member feature.
- Adds a gusset feature between 2 planar adjoining faces.
- Creates an end cap feature.
- Adds a fillet weld bead feature.
- Trims or extends structure members.
- Weld bead.

Quick Reference Guide to SOLIDWORKS Command Icons & Toolbars

DIMENSIONS/RELATIONS Toolbar

Inserts dimension between two lines.

Creates a horizontal dimension between selected entities.

Creates a vertical dimension between selected entities.

Creates a reference dimension between selected entities.

Creates a set of ordinate dimensions.

Creates a set of Horizontal ordinate

Creates a set of Vertical ordinate dimensions.

Creates a chamfer dimension.

Adds a geometric relation.

Automatically Adds Dimensions to the current sketch.

Displays and deletes geometric relations.

Fully defines a sketch.

Scans a sketch for elements of equal length or radius.

Angular Running dimension.

Display / Delete dimension.

Isolate changed dimension.

Path length dimension.

BLOCK Toolbar

Makes a new block.

Edits the selected block.

Inserts a new block to a sketch or drawing.

Adds/Removes sketch entities to/from blocks.

Updates parent sketches affected by this block.

Saves the block to a file.

Explodes the selected block.

Inserts a belt.

STANDARD VIEWS Toolbar

Front view.

Back view.

Left view.

Right view.

Top view.

Bottom view.

Isometric view.

Trimetric view.

Dimetric view.

Normal to view.

Links all views in the viewport together.

Displays viewport with front & right

Displays a 4 view viewport with 1st or 3rd

Displays viewport with front & top.

Displays viewport with a single view.

View selector.

New view.

FEATURES Toolbar

- Creates a boss feature by extruding a sketched profile.
- Creates a revolved feature based on profile and angle parameter.
- Creates a cut feature by extruding a sketched profile.
- Creates a cut feature by revolving a sketched profile.
- Thread.
- Creates a cut by sweeping a closed profile along an open or closed path.
- Loft cut.
- Creates a cut by thickening one or more adjacent surfaces.
- Adds a deformed surface by push or pull on points.
- Creates a lofted feature between two or more profiles.
- Creates a solid feature by thickening one or more adjacent surfaces.
- Creates a filled feature.
- Chamfers an edge or a chain of tangent edges.
- Inserts a rib feature.
- Combine.
- Creates a shell feature.
- Applies draft to a selected surface.
- Creates a cylindrical hole.
- Inserts a hole with a pre-defined cross section.
- Puts a dome surface on a face.
- Model break view.
- Applies global deformation to solid or surface bodies.
- Wraps closed sketch contour(s) onto a face.
- Curve Driven pattern.
- Suppresses the selected feature or component.
- Un-suppresses the selected feature or component.
- Flexes solid and surface bodies.

- Intersect.
- Variable Patterns.
- Live Section Plane.
- Mirrors.
- Scale.
- Creates a Sketch Driven pattern.
- Creates a Table Driven Pattern.
- Inserts a split Feature.
- Hole series.
- Joins bodies from one or more parts into a single part in the context of an assembly.
- Deletes a solid or a surface.
- Instant 3D.
- Inserts a part from file into the active part document.
- Moves/Copies solid and surface bodies or moves graphics bodies.
- Merges short edges on faces.
- Pushes solid / surface model by another solid / surface model.
- Moves face(s) of a solid.
- FeatureWorks Options.
- Linear Pattern.
- Fill Pattern.
- Cuts a solid model with a
- Boundary Boss/Base.
- Boundary Cut.
- Circular Pattern.
- Recognize Features.
- Grid System.

MOLD TOOLS Toolbar

- Extracts core(s) from existing tooling split.
- Constructs a surface patch.
- Moves face(s) of a solid.
- Creates offset surfaces.
- Inserts cavity into a base part.
- Scales a model by a specified factor.
- Applies draft to a selected surface.
- Inserts a split line feature.
- Creates parting lines to separate core & cavity surfaces.
- Finds & creates mold shut-off surfaces.
- Creates a planar surface from a sketch or a set of edges.
- Knits surfaces together.
- Inserts ruled surfaces from edges.
- Creates parting surfaces between core & cavity surfaces.
- Creates multiple bodies from a single body.
- Inserts a tooling split feature.
- Creates parting surfaces between the core & cavity.
- Inserts surface body folders for mold operation.

SELECTION FILTERS Toolbar

- Turns selection filters on and off.
- Clears all filters.
- Selects all filters.
- Inverts current selection.
- Allows selection of edges only.
- Allows selection filter for vertices only.
- Allows selection of faces only.
- Adds filter for Surface Bodies.
- Adds filter for Solid Bodies.
- Adds filter for Axes.
- Adds filter for Planes.
- Adds filter for Sketch Points.
- Allows selection for sketch only.
- Adds filter for Sketch Segments.
- Adds filter for Midpoints.
- Adds filter for Center Marks.
- Adds filter for Centerline.
- Adds filter for Dimensions and Hole Callouts.
- Adds filter for Surface Finish Symbols.
- Adds filter for Geometric Tolerances.
- Adds filter for Notes / Balloons.
- Adds filter for Weld Symbols.
- Adds filter for Weld beads.
- Adds filter for Datum Targets.
- Adds filter for Datum feature only.
- Adds filter for blocks.
- Adds filter for Cosmetic Threads.

Adds filter for Dowel pin symbols.

Adds filter for connection points.

Adds filter for routing points.

SOLIDWORKS Add-Ins Toolbar

Loads/unloads CircuitWorks add-in.

Loads/unloads the Design Checker add-in.

Loads/unloads the PhotoView 360 add-in.

Loads/unloads the Scan-to-3D add-in.

Loads/unloads the SOLIDWORKS Motions add-in.

Loads/unloads the SOLIDWORKS Routing add-in.

Loads/unloads the SOLIDWORKS Simulation add-in.

Loads/unloads the SOLIDWORKS Toolbox add-in.

Loads/unloads the SOLIDWORKS TolAnalysis add-in.

Loads/unloads the SOLIDWORKS Flow Simulation add-in.

Loads/unloads the SOLIDWORKS Plastics add-in.

Loads/unloads the SOLIDWORKS MBD SNL license.

FASTENING FEATURES Toolbar

Creates a parameterized mounting boss.

Creates a parameterized snap hook.

Creates a groove to mate with a hook feature.

Uses sketch elements to create a vent for air flow.

Creates a lip/groove feature.

SCREEN CAPTURE Toolbar

Copies the current graphics window to the clipboard.

Records the current graphics window to an AVI file.

Stops recording the current graphics window to an AVI file.

EXPLODE LINE SKETCH Toolbar

Adds a route line that connect entities.

Adds a jog to the route lines.

LINE FORMAT Toolbar

Changes layer properties.

Changes the current document layer.

Changes line color.

Changes line thickness.

Changes line style.

Hides / Shows a hidden edge.

Changes line display mode.

Did you know??

* Ctrl+Q will force a rebuild on all features of a part.

* Ctrl+B will rebuild the feature being worked on and its dependents.

2D-To-3D Toolbar

Makes a Front sketch from the selected entities.

Makes a Top sketch from the selected entities.

Makes a Right sketch from the selected entities.

Makes a Left sketch from the selected entities.

Makes a Bottom sketch from the selected entities.

Makes a Back sketch from the selected entities.

Makes an Auxiliary sketch from the selected entities.

Creates a new sketch from the selected entities.

Repairs the selected sketch.

Aligns a sketch to the selected point.

Creates an extrusion from the selected sketch segments, starting at the selected sketch point.

Creates a cut from the selected sketch segments, optionally starting at the selected sketch point.

ALIGN Toolbar

Aligns the left side of the selected annotations with the leftmost annotation.

Aligns the right side of the selected annotations with the rightmost annotation.

Aligns the top side of the selected annotations with the topmost annotation.

Aligns the bottom side of the selected annotations with the lowermost annotation.

Evenly spaces the selected annotations horizontally.

Evenly spaces the selected annotations vertically.

Centrally aligns the selected annotations horizontally.

Centrally aligns the selected annotations vertically.

Compacts the selected annotations horizontally.

Compacts the selected annotations vertically.

Creates a group from the selected items.

Deletes the grouping between these items.

Aligns & groups selected dimensions along a line or an arc.

Aligns & groups dimensions at uniform distances.

Evenly spaces selected dimensions.

Aligns collinear selected dimensions.

Aligns stagger selected dimensions.

SOLIDWORKS MBD Toolbar

Captures 3D view.

Manages 3D PDF templates.

Creates shareable 3D PDF presentations.

Toggles dynamic annotation views.

MACRO Toolbar

Runs a Macro.

Stops Macro recorder.

Records (or pauses recording of) actions to create a Macro.

Launches the Macro Editor and begins editing a new macro.

Opens a Macro file for editing.

Creates a custom macro.

SMARTMATES icons

Concentric & Coincident 2 circular edges.

Concentric 2 cylindrical faces.

Coincident 2 linear edges.

Coincident 2 planar faces.

Coincident 2 vertices.

Coincident 2 origins or coordinate systems.

TABLE Toolbar

Adds a hole table of selected holes from a specified origin datum.

Adds a Bill of Materials.

Adds a revision table.

Displays a Design table in a drawing.

Adds a weldments cuts list table.

Adds an Excel based Bill of Materials

Adds a weldment cut list table.

REFERENCE GEOMETRY Toolbar

Adds a reference plane.

Creates an axis.

Creates a coordinate system.

Adds the center of mass.

Specifies entities to use as references using SmartMates.

SPLINE TOOLS Toolbar

Inserts a point to a spline.

Displays all points where the concavity of selected spline changes.

Displays minimum radius of selected spline.

Displays curvature combs of selected spline.

Reduces numbers of points in a selected spline.

Adds a tangency control.

Adds a curvature control.

Adds a spline based on selected sketch entities & edges.

Displays the spline control polygon.

ANNOTATIONS Toolbar

Inserts a note.

Inserts a surface finish symbol.

Inserts a new geometric tolerancing symbol.

Attaches a balloon to the selected edge or face.

Adds balloons for all components in selected view.

Inserts a stacked balloon.

Attaches a datum feature symbol to a selected edge / detail.

Inserts a weld symbol on the selected edge / face / vertex.

Inserts a datum target symbol and / or point attached to a selected edge / line.

Selects and inserts block.

Inserts annotations & reference geometry from the part / assembly into the selected.

Adds center marks to circles on model.

Inserts a Centerline.

Inserts a hole callout.

Adds a cosmetic thread to the selected cylindrical feature.

Inserts a Multi-Jog leader.

Selects a circular edge or an arc for Dowel pin symbol insertion.

Adds a view location symbol.

Inserts latest version symbol.

Adds a cross hatch patterns or solid fill.

Adds a weld bead caterpillar on an edge.

Adds a weld symbol on a selected entity.

Inserts a revision cloud.

Inserts a magnetic line.

Hides/shows annotation.

DRAWINGS Toolbar

- Updates the selected view to the model's current stage.
- Creates a detail view.
- Creates a section view.
- Inserts an Alternate Position view.
- Unfolds a new view from an existing view.
- Generates a standard 3-view drawing (1st or 3rd angle).
- Inserts an auxiliary view of an inclined surface.
- Adds an Orthogonal or Named view based on an existing part or assembly.
- Adds a Relative view by two orthogonal faces or planes.
- Adds a Predefined orthogonal projected or Named view with a model.
- Adds an empty view.
- Adds vertical break lines to selected view.
- Crops a view.
- Creates a Broken-out section.

QUICK SNAP Toolbar

- Snap to points.
- Snap to center points.
- Snap to midpoints.
- Snap to quadrant points.
- Snap to intersection of 2 curves.
- Snap to nearest curve.
- Snap tangent to curve.
- Snap perpendicular to curve.
- Snap parallel to line.
- Snap horizontally / vertically to points.
- Snap horizontally / vertically.
- Snap to discrete line lengths.
- Snap to angle.

LAYOUT Toolbar

- Creates the assembly layout sketch.
- Sketches a line.
- Sketches a corner rectangle.
- Sketches a circle.
- Sketches a 3 point arc.
- Rounds a corner.
- Trims or extends a sketch.
- Adds sketch entities by offsetting faces, edges curves.
- Mirrors selected entities about a centerline.
- Adds a relation.
- Creates a dimension.
- Displays / Deletes geometric relations.
- Makes a new block.
- Edits the selected block.
- Inserts a new block to the sketch or drawing.
- Adds / Removes sketch entities to / from a block.
- Saves the block to a file.
- Explodes the selected block.
- Creates a new part from a layout sketch block.
- Positions 2 components relative to one another.

CURVES Toolbar

- Projects sketch onto selected surface.
- Inserts a split line feature.
- Creates a composite curve from selected edges, curves and sketches.
- Creates a curve through free points.
- Creates a 3D curve through reference points.
- Helical curve defined by a base sketch and shape parameters.

VIEW Toolbar

- Displays a view in the selected orientation.
- Reverts to previous view.
- Redraws the current window.
- Zooms out to see entire model.
- Zooms in by dragging a bounding box.
- Zooms in or out by dragging up or down.
- Zooms to fit all selected entities.
- Dynamic view rotation.
- Scrolls view by dragging.
- Displays image in wireframe mode.
- Displays hidden edges in gray.
- Displays image with hidden lines removed.
- Controls the visibility of planes.
- Controls the visibility of axis.
- Controls the visibility of parting lines.
- Controls the visibility of temporary axis.
- Controls the visibility of origins.
- Controls the visibility of coordinate systems.
- Controls the visibility of reference curves.
- Controls the visibility of sketches.
- Controls the visibility of 3D sketch planes.
- Controls the visibility of 3D sketch.
- Controls the visibility of all annotations.
- Controls the visibility of reference points.
- Controls the visibility of routing points.
- Controls the visibility of lights.
- Controls the visibility of cameras.
- Controls the visibility of sketch relations.
- Changes the display state for the current configuration.
- Rolls the model view.
- Turns the orientation of the model view.
- Dynamically manipulate the model view in 3D to make selection.
- Changes the display style for the active view.
- Displays a shade view of the model with its edges.
- Displays a shade view of the model.
- Toggles between draft quality & high quality HLV.
- Cycles through or applies a specific scene.
- Views the models through one of the model's cameras.
- Displays a part or assembly w/different colors according to the local radius of curvature.
- Displays zebra stripes.
- Displays a model with hardware accelerated shades.
- Applies a cartoon affect to model edges & faces.
- Views simulations symbols.

TOOLS Toolbar

- Calculates the distance between selected items.
- Adds or edits equation.
- Calculates the mass properties of the model.
- Checks the model for geometry errors.
- Inserts or edits a Design Table.
- Evaluates section properties for faces and sketches that lie in parallel planes.
- Reports Statistics for this Part/Assembly.
- Deviation Analysis.
- Runs the SimulationXpress analysis wizard powered by SOLIDWORKS Simulation.
- Checks the spelling.
- Import diagnostics.
- Runs the DFMXpress analysis wizard.
- Runs the SOLIDWORKSFloXpress analysis wizard.

ASSEMBLY Toolbar

- Creates a new part & inserts it into the assembly.
- Adds an existing part or sub-assembly to the assembly.
- Creates a new assembly & inserts it into the assembly.
- Turns on/off large assembly mode for this document.
- Hides / shows model(s) associated with the selected model(s).
- Toggles the transparency of components.
- Changes the selected components to suppressed or resolved.
- Inserts a belt.
- Toggles between editing part and assembly.
- Smart Fasteners.
- Positions two components relative to one another.
- External references will not be created.
- Moves a component.
- Rotates an un-mated component around its center point.
- Replaces selected components.
- Replaces mate entities of mates of the selected components on the selected Mates group.
- Creates a New Exploded view.
- Creates or edits explode line sketch.
- Interference detection.
- Shows or Hides the Simulation toolbar.
- Patterns components in one or two linear directions.
- Patterns components around an axis.
- Sets the transparency of the components other than the one being edited.
- Sketch driven component pattern.
- Pattern driven component pattern.
- Curve driven component pattern.
- Chain driven component pattern.
- SmartMates by dragging & dropping components.
- Checks assembly hole alignments.
- Mirrors subassemblies and parts.

To add or remove an icon to or from the toolbar, first select:

Tools/Customize/Commands

Next, select a **Category**, click a button to see its description and then drag/drop the command icon into any toolbar.

SOLIDWORKS Quick-Guide©
STANDARD Keyboard Shortcuts

Rotate the model

* Horizontally or Vertically: ______ Arrow keys
* Horizontally or Vertically 90º: ______ Shift + Arrow keys
* Clockwise or Counterclockwise: ______ Alt + left or right Arrow
* Pan the model: ______ Ctrl + Arrow keys
* Zoom in: ______ Z (shift + Z or capital Z)
* Zoom out: ______ z (lower case z)
* Zoom to fit: ______ F
* Previous view: ______ Ctrl+Shift+Z

View Orientation

* View Orientation Menu: ______ Space bar
* Front: ______ Ctrl+1
* Back: ______ Ctrl+2
* Left: ______ Ctrl+3
* Right: ______ Ctrl+4
* Top: ______ Ctrl+5
* Bottom: ______ Ctrl+6
* Isometric: ______ Ctrl+7

Selection Filter & Misc.

* Filter Edges: ______ e
* Filter Vertices: ______ v
* Filter Faces: ______ x
* Toggle Selection filter toolbar: ______ F5
* Toggle Selection Filter toolbar (on/off): ______ F6
* New SOLIDWORKS document: ______ F1
* Open Document: ______ Ctrl+O
* Open from Web folder: ______ Ctrl+W
* Save: ______ Ctrl+S
* Print: ______ Ctrl+P
* Magnifying Glass Zoom ______ g
* Switch between the SOLIDWORKS documents ______ Ctrl + Tab

SOLIDWORKS Quick-Guide©
Sample Customized Keyboard Shortcuts

SOLIDWORKS Sample Customized Hot Keys

Function Keys

F1 ______ SW-Help
F2 ______ 2D Sketch
F3 ______ 3D Sketch
F4 ______ Modify
F5 ______ Selection Filters
F6 ______ Move (2D Sketch)
F7 ______ Rotate (2D Sketch)
F8 ______ Measure
F9 ______ Extrude
F10 ______ Revolve
F11 ______ Sweep
F12 ______ Loft

Sketch

C ______ Circle
P ______ Polygon
E ______ Ellipse
O ______ Offset Entities
Alt + C ______ Convert Entities
M ______ Mirror
Alt + M ______ Dynamic Mirror
Alt + F ______ Sketch Fillet
T ______ Trim
Alt + X ______ Extend
D ______ Smart Dimension
Alt + R ______ Add Relation
Alt + P ______ Plane
Control + F ______ Fully Define Sketch
Control + Q ______ Exit Sketch

SOLIDWORKS Quick-Guide by Paul Tran – Sr. Certified SOLIDWORKS Instructor